SCIENCE & TECHNOLOGY ENCYCLOPEDIA

SCIENCE &
TECHNOLOGY
ENCYCLOPEDIA

The University of Chicago Press

The University of Chicago Press *Science and Technology Encyclopedia* has been created for both students of science and for those who are simply enthusiasts – as a study aid for college and university students, a first stop for general enquiries, and a fascinating mine of knowledge for browsers. The 6,500 alphabetically organized entries provide clear, essential information on a vast variety of scientific and technological subjects, from earth science to life science, and from astronomy to particle physics. The 250 black and white illustrations, including technical "cutaways" and artworks, convey visual information far beyond the descriptive power of many thousands of words.

When choosing a single-volume encyclopedia, the most important consideration for the user is the criteria by which the articles have been selected. Key areas of science and technology, including math, physics, chemistry, biology and information technology, have been given the greatest attention. The articles, compatible with and complementary to what is learned in class, are up-to-date and clearly written.

An encyclopedia, however, must contain more than just a comprehensive coverage of key areas if it is to capture and maintain an interest. Equally full in their treatment are the articles that cover cutting-edge technology, such as linear motors and virtual reality, many of which are illustrated to help clarify the accompanying article, making this encyclopedia ideal for home reference as well as an important resource at school or college. For this reason also, unlike most encyclopedias of this type, the University of Chicago Press *Science and Technology Encyclopedia* features biographical articles on more than 850 famous scientists, providing information on their careers and achievements.

Cross-references

The University of Chicago Press *Science and Technology Encyclopedia* has more than 20,000 individual cross-references, indicated by SMALL CAPITAL letters, which take the reader from one article to other articles that provide useful, related information. For example, contained within the "satellite television" article are cross-references to TELEVISION, SATELLITES, GEOSTATIONARY ORBIT, TRANSMITTER and ANTENNA.

Alphabetical order

The order of articles is strictly alphabetical, except that Mc is treated as if it were spelt Mac, and abbreviations as if spelt out in full. Articles that have more than one word in the heading, such as "saturated solution", are ordered as if there were no space between the words.

Units

Metric units have been used throughout with corresponding imperial measurements following in brackets.

aa Type of LAVA with a blocklike structure. *See also* BLOCKY LAVA

abacus Counting device used in the Middle and Far East for addition and subtraction. The most common form of abacus consists of beads strung on wires and arranged in columns that represent units, tens, hundreds, and so on. Various types have been used for thousands of years.

abdomen In VERTEBRATES, that portion of the body between the chest and the pelvis containing the abdominal cavity and the abdominal viscera, including most of the digestive organs. In ARTHROPODS it is the posterior part of the body, containing the reproductive organs and part of the digestive system.

Abel, John Jacob (1857–1938) US biochemist, best known for the first identification (1898) of a HORMONE, EPINEPHRINE. Abel was professor of pharmacology (1893–1932) at Johns Hopkins. He made his key discovery after many years of studying the chemical composition of body tissues. Later, Abel isolated AMINO ACIDS from blood by the process of dialysis, and discovered INSLUIN in crystalline form.

aberration In astronomy, the apparent slight change of position of a star due to the effect of the Earth's orbital motion and the finite velocity of light. A telescope must be inclined by an angle of up to about 20° to compensate for it. Aberration of light was discovered in 1729 and used to prove that the Earth orbits the Sun.

aberration In physics, a defect in lens or mirror images. **Spherical** aberration occurs when rays falling on the periphery of a lens or curved mirror are not brought to the same focus as light at the center; the image is blurred. **Chromatic** aberration occurs because the varying wavelengths of colored light are not brought to the same focus; the image is falsely colored.

abiotic factors Various nonliving factors – favorable or harmful – that contribute to the environment of living things. They include, for example, air, water, heat, light, and soil. *See also* ECOSYSTEM; ENVIRONMENT

ablation In geology, a measure of glacial loss through melting, evaporation, wind erosion, or calving (formation of icebergs). Sometimes used in geomorphology to mean the loss of surface soil or rock by weathering. *See also* ACCUMULATION; EROSION

abortion Loss of a fetus before it is sufficiently advanced to survive outside the UTERUS; commonly called a miscarriage. A medical abortion is the termination of pregnancy by drugs or surgery. The ethics of termination remains an extremely contentious issue.

abrasion In geology, mechanical wearing down of rock surface by the dragging of rock particles across it. Common agents of abrasion are the bed load of streams, rock debris at the base of glaciers, and sand transported by wind or waves. *See also* EROSION

abrasive Hard and rough substance used to grind and polish surfaces. Some abrasives are used as fine powders, others in larger fragments with sharp cutting edges. Most natural abrasives are minerals, such as diamond, garnet, emery, corundum, pumice, flint, and quartz. Synthetic abrasives include silicon carbide, aluminum oxide, synthetic diamond, and boron carbide.

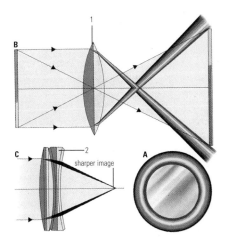

▲ **aberration** The effects of chromatic aberration result in colored fringes around the edges of the lens (running from violet on the outer edge to red at the inner edge), and parts of the image may not be sharp (A). This aberration occurs with a single convex lens (1) because it behaves like a prism and bends blue light more than red light (B). Combining the convex lens with a weaker concave lens (2) made of a different glass cancels out this effect, and both red and blue rays are brought into the same focus to produce a sharper, more distinct image (C).

abscissa In mathematics, the distance of a point from the *y* axis in a CARTESIAN COORDINATE SYSTEM. That is the *x* coordinate of the pair (*x*,*y*), which locates a point in the plane.

abscission Deliberate shedding of a part of a plant. It commonly occurs with fruit and leaves, and is controlled by abscisic acid and other growth substances. An abscission layer of cells forms in an abscission zone at the base of the part. As these cells break down, the part falls off the plant, usually dislodged by the wind or rain.

absolute alcohol Pure ETHANOL (ethyl alcohol). It is made from distilled ethanol (a constant-boiling mixture containing 95.6% ethanol) by the addition of sodium metal or another drying agent.

absolute zero Temperature at which all parts of a system are at the lowest energy permitted by the laws of QUANTUM MECHANICS; zero on the Kelvin temperature scale, which is $-273.15°C$ ($-459.67°F$). At this temperature the system's ENTROPY, its energy available for useful work, is also zero, although the total energy of the system may not be zero.

absorbent Substance able to absorb large quantities of other substances. Absorbents are usually porous materials: examples are activated charcoal and zeolites. They have many uses, including separating, purifying, decolorizing, and deodorizing.

absorption Taking up chemically or physically of one substance into another. The absorbed matter permeates all of the absorber. This includes a gas taken in by a liquid, such as carbon dioxide by sodium hydroxide, and a liquid or gas absorbed by a solid, such as water by a GEL. The process is often utilized commercially, an example being the purification of natural gas by the absorption of hydrogen sulphide in aqueous ethanolamine. Absorption occurs in all scientific disciplines; in nuclear physics, neutrons produced by fission are absorbed by elements such as boron. *See also* ADSORPTION

absorption spectrum *See* SPECTRUM

abyssal Term to describe oceanic features occurring at great depths, usually more than 3,000m (10,000ft) below sea level. Abyssal **plains** cover about 30% of the Atlantic and nearly 75% of the Pacific ocean floors. They are covered by deposits of biogenic oozes formed by the remains of microscopic plankton, and nonbiogenic sediments (red clays). The gradient is less than 1:1,000, except for the occasional low, oval-shaped abyssal **hills**. The plains are characterized by stable temperatures

from $-1°C$ to $5°C$ (30–41°F) and the relative absence of water currents. The abyssal **zone** is the deepest area of the ocean. It receives no sunlight, so there are no seasons and no plants, but there are many forms of life, such as glass sponges, crinoids (sea lilies), and brachiopods (lamp shells).

Acanthocephala Phylum of spiny-headed parasitic worms once thought to be NEMATODES. They are identified by a retractable spiny proboscis and an elongated, cylindrical body. The young are parasitic in arthropods and the adults are parasitic in vertebrates, attaching themselves to the intestinal lining. Average length: 1–2cm (0.4–0.8in). *Echinorhyncus* is the main genus.

acceleration Amount by which the VELOCITY of an object increases in a certain time. For example, a stone dropped over a cliff accelerates from zero velocity at a rate of 9.81m (32.2ft) per second per second, this acceleration being due to the pull of the Earth's gravity. The rate of acceleration can be found by applying the equation acceleration = (change in velocity)/(time taken for change). For example, if an object increases in velocity from 5 meters per second (ms^{-1} or m/s) to $7ms^{-1}$ in one second, its velocity has increased by $2ms^{-1}$ in one second; the acceleration is thus 2 meters per second per second, written as $2ms^{-2}$ or $2m/s^2$. When an object rotates around an axis, its angular acceleration is given as the change in angular velocity divided by elapsed time, expressed in radians per second per second (written rad s^{-2} or rad/s^2). A RADIAN is a measure of angle: 2π radians = $360°$.

acceleration of free fall (acceleration due to gravity, symbol *g*) Acceleration experienced by bodies falling freely in the Earth's gravitational field. It varies from place to place around the globe, but is assigned a standard value of $9.80665ms^{-2}$ ($32.1740fts^{-2}$). Ignoring air resistance, the acceleration does not vary with the size or shape of the falling body. Pilots can lose consciousness at vertical accelerations in excess of $3g$.

accelerator (particle accelerator) In PARTICLE PHYSICS, machine for increasing the energy of charged particles by increasing their speed by use of alternating electric fields in an evacuated chamber. The particles must enter the high-frequency field as it begins increasing (negative particles) or decreasing (positive particles) for maximum energy increase. Magnetic fields are used to focus the particles into a narrow stable beam and to maintain the required curvature of the beam. As the particle velocity rises, a relativistic increase in mass occurs. Accelerators are used mostly in particle physics experiments, in which high-energy particles are forced to collide with other particles. The way the fragments of particles produced behave following the collision provides physicists with information on the forces found within atoms. In a LINEAR ACCELERATOR, the particles travel in a straight line, usually accelerated by an electric field. In a CYCLOTRON, particles are accelerated in a spiral path between pairs of D-shaped magnets with an alternating voltage between them. In a SYNCHROCYCLOTRON, the accelerating voltage is synchronized with the time it takes the particles to make one revolution. A SYNCHROTRON consists of an (often very large) circular tube with magnets to deflect the particles in a curve and radio-frequency fields to accelerate them. The most advanced modern accelerators are **colliders**, in which beams of particles moving in opposite directions are allowed to collide with each other, thus achieving higher energy of interaction. See also BUBBLE CHAMBER; CERN

accelerator, electrostatic Early type of particle ACCELERATOR in which a constant high voltage is applied between a pair of electrodes in an evacuated tube. A large static charge builds up, producing an electric field along the tube. Charged particles are accelerated in the electric field between the electrodes, gaining an energy eV, where e is the electron charge and V the applied voltage. The VAN DE GRAAF GENERATOR (developed 1931) and the COCKCROFT-WALTON GENERATOR (1932) can be used as high-voltage sources.

accelerometer Device used to measure acceleration. A simple example is a plumb bob suspended from the accelerating object, whose angle with the vertical is proportional to the acceleration. A more sophisticated version, used in ballistic missiles, is an electromechanical device that translates the acceleration into electric current.

acclimatization Temporary adjustment of an organism to a new environment, climate or circumstance. It involves a gradual, natural change in physiology allowing an organism to exploit new regions, but unlike EVOLUTION, does not involve any genetic change.

accommodation Process by which the EYE focuses on objects at various distances. In the human eye, focusing is achieved when the muscles of the CILIARY BODY contract or relax to change the shape (curvature) of the lens and bring light rays into focus on the light-sensitive RETINA found at the back of the eye. Presbyopia, an eye condition that occurs after middle age in which people have difficulty in focusing on distant or very near objects, is caused by impaired accommodation.

accretion A continual growing or building up; a term frequently used to describe certain modes of geological DEPOSITION. The term is also used in astronomy to describe the gradual building up of larger celestial bodies from smaller ones by gravitational attraction; also the accumulation of matter by a star or other celestial object. Accretion is an important factor in the evolution of starts, planets, and comets. See also STELLAR EVOLUTION

accumulation In geology, the addition of snow and ice to a GLACIER. It usually occurs near the head of the glacier where the air temperature is well below freezing. Snow from avalanches and precipitation falls onto the glacier and becomes compressed under its own weight, forming ice. See also ABLATION

accumulator (secondary CELL or storage BATTERY) Voltaic cell (battery) that can be recharged. The commonly used car battery is a lead-acid accumulator.

acetaldehyde (ethanal, CH_2CHO) Colorless, volatile, flammable liquid manufactured by catalytic oxidation of ethene or ethanol, or catalytic hydration of acetylene. It is used mainly to make other organic compounds. Properties: r.d. 0.788: m.p. $-123.5°C$ ($-190.3°F$): b.p. 20.8°C (69.4°F).

acetate (ethanoate) Salt or ester of ACETIC ACID. A compound containing the ion CH_3COO or the group CH_3COO-. It is used in synthetic acetate fibers, in lacquers, and in acetate film.

acetic acid (ethanoic acid, CH_3COOH) Colorless, corrosive liquid made by the oxidation of ethanol, either by catalysis or by the action of bacteria. It is the active ingredient in vinegar, and has many uses in the organic chemicals industry. Properties: r.d. 1.049: m.p. 16.6°C (61.9°F): b.p. 117.9°C (244.4°F)

acetone (propanone, CH_3COCH_3) Colorless, flammable liquid made by oxidizing propan-2-ol (isopropyl acholcol). It is a raw material for the manufacture of many organic chemicals and is a widely used solvent. Properties: r.d. 0.79: m.p. $-94.8°C$ ($-138.6°F$): b.p. 56.2°C (133.2°F).

acetylcholine (ACh) Chemical compound released by certain nerve cells that serves as a transmitter in the NERVOUS SYSTEM. It is involved in the transmission of impulses across the junction (SYNAPSE) between nerve and muscle cells that trigger the contraction of muscles, and between nerve cells.

acetyl coenzyme A (acetyl CoA) Derivative formed when PYRUVIC ACID is broken down during the formation of fat from carbohydrates. CoA acts as the carrier of the acetyl group, which is the basic building block for the fatty acids.

acetylene (ethyne CH≡CH) Colorless, flammable gas manufactured by the CRACKING of petroleum fractions. The simplest ALKYNE, it is explosive if mixed with air. When burned with oxygen, it produces extremely high temperatures up to 3,480°C (6,300°F), as is used in oxyacetylene torches for cutting and welding metals. It is polymerized to manufacture plastics, synthetic fibers, resins, and the synthetic rubber neoprene. It is also used to produce acetaldehyde and acetic acid. Properties: r.d. 0.625: m.p. −80.8°C (−113.4°F): b.p. −84°C (−119.2°F).

acetylsalicylic acid *See* ASPIRIN

achene Type of fruit that contains only one seed. It is a dry, indehiscent fruit (that is, it does not open spontaneously) formed from a single carpel, with the seed separated from the wall of the fruit. Examples include members of the buttercup family and clematis. Winged achenes are called **samaras**.

Achilles tendon Strong band of elastic CONNECTIVE TISSUE at the back of the ankle. One of the largest tendons in the body, it connects the calf muscles to the heel bone.

achromatic lens Lens designed to reduce or eliminate chromatic ABERRATION. Chromatic aberration occurs in a simple lens because the different component wavelengths (colors) of white light are brought to a different focus, resulting in a distorted image surrounded by colored fringes. An achromatic lens brings all colors approximately to the same focus. It is usually constructed by combining two lenses made from different kinds of glass (and therefore having different refractive indexes), so that the errors introduced by one lens are canceled out by the other.

acid Chemical compound containing hydrogen that can be replaced by a metal or other positive ION to form a SALT. Acids dissociate in water to yield aqueous hydrogen ions (H^+), thus acting as proton donors; the solutions are corrosive, have a sour taste, give a red color to indicators, and have a pH below 7. **Strong** acids, such as sulfuric acid, are fully dissociated into ions whereas **weak** acids, such as ethanoic acid, are only partly dissociated. *See also* BASE

acid number (acid value) Number of milligrams of potassium hydroxide necessary to neutralize the free fatty acids in one gram of a specified substance. The acid number is used to measure the fatty-acid content of fats, oils, etc.

acid rain Rain that is highly acidic because of sulfur oxides, nitrogen oxides, and other air pollutants dissolved in it. Normal rain is slightly acidic, with a pH of 6. (The pH scale ranges from 1 for extremely acidic to 14 for extremely basic or alkaline, with 7 being neutral.) Acid rain may have a pH value as low as 2.8. Acid rain can severely damage both plant and animal life. Certain lakes, for example, have lost all fish and plant life because of acid rain.

acid rock Any IGNEOUS ROCK that contains a high proportion of SILICA (more than 66%). The main silica mineral is QUARTZ, but some FELDSPARS are also quite rich in silica. Acid igneous rocks are usually light in color. The most common is GRANITE. When they are weathered to form soil, they are generally not very fertile.

acid salt Salt of an acid that has two or more replaceable hydrogen atoms in which not all of the hydrogens are replaced by a metal or its equivalent. For example, the dibasic acid carbonic acid (H_2CO_3) forms acid salts such as sodium hydrogencarbonate (sodium bicarbonate, $NaHCO_3$). The tribasic acid phosphoric(V) acid (orthophosphoric acid, H_3PO_4) forms two series of acid salts, such as sodium dihydrogen phosphate (NaH_2PO_4) and di-sodium hydrogen phosphate (Na_2HPO_4).

acoustics Study of sound, especially the behavior of SOUND WAVES. Architects apply acoustics in the design of public rooms such as concert and lecture halls to acquire a clearer sound. Engineers use it in their designs of sound detectors such as microphones and of sound producers such as loudspeakers and devices using ULTRASONICS. Audiologists use acoustics to assess degrees of abnormality in the hearing of their patients. Crafters of musical instruments use it to improve their designs. *See also* ANECHOIC CHAMBER

acoustic shielding Deadening of unwanted sounds, usually by introducing sound-absorbing materials in the path of the sound waves. Sound enters a room in three ways: through the air, through the structure, or by the diaphragm action of floors, walls, and ceilings. Airborne sounds are absorbed by thick and absorbent walls in which there are no cracks or ducts; doors and windows should fit tightly and be seated in rubber or felt liners. Windows should be double-glazed. Structure-born vibrations and diaphragm effects are reduced by using double walls with insulating material but as few ties as possible between them.

acquired characteristic Feature that develops during the lifetime of an organism. The enlarged muscles of a manual worker are an example of an acquired characteristic. Because they are not genetically controlled, acquired characteristics cannot be passed on to offspring – sthat is, they cannot be inherited (although the French naturalist Jean Baptiste LAMARCK incorrectly thought that they could).

acquired immune deficiency syndrome (AIDS) Disease caused by a RETROVIRUS, called Human Immunodeficiency Virus (HIV), that mainly attacks T-4 cells (which help the production of ANTIBODIES) and renders the body's IMMUNE SYSTEM incapable of resisting infection. The first diagnosis was in New York in 1979. In 1983–84 scientists at the Pasteur Institute in France and the National Cancer Institute in the US isolated HIV as the cause of the disease. The virus can remain dormant in infected T-cells for up to 10 years. Early symptoms include severe weight loss and fatigue. The condition progresses to full-blown AIDS, characterized by secondary infections, neurological damage, and cancers (such as Kaposi's sarcoma and B-cell lymphomas). AIDS is transmitted only by a direct exchange of body fluids. Methods of transmission include sexual intercourse, sharing of contaminated needles by intravenous drug users, and by infected mothers to their babies in the UTERUS. Before effective screening procedures were introduced many hemophiliacs were infected through transfusions of contaminated blood. Development of a vaccine has been hampered by the virus's constantly shifting genetic composition. Research has centered on treatment, and on delaying the onset of the full-blown syndrome. In 1986–87 the first chemical was produced that was partially successful in controlling symptoms, zidovudine or azidothymidine (AZT). More recently the emphasis has been on reducing viral load—the amount of virus within the body—using a combination of zidovudine and newer drugs known as protease inhibitors. By late 1997 an estimated 11.7 million people had died from AIDS and more than 30 million were infected. AIDS raises many legal and moral issues of disclosure and discrimination.

acre Unit of area measurement in English-speaking countries, equal to 4,840sq yd (0.405ha).

acridine Organic chemical compound, the molecule of which is a PYRIDINE ring held between two BENZENE rings. It is used in making drugs and dyes.

acromegaly Condition in which overproduction of PITUITARY growth hormone during adulthood causes enlarged hands, feet, and facial features.

acrylic Type of PLASTIC, one of a group of synthetic short-chain unsaturated CARBOXYLIC ACID derivatives. Variation in the reagents and the method of formation yields either hard and transparent, soft and resilient, or liquid products. Their transparency, toughness and dimensional stability make acrylics useful for molded structural parts, lenses, adhesives, and paints.

ACTH Abbreviation of ADRENOCORTICOROPHIC HORMONE.

actin Protein involved in cellular contractile processes. Found mainly in muscle cells, it reacts with MYOSIN to form ACTOMYOSIN.

actinide series Group of radioactive elements with atomic numbers from 89 to 103; ACTINIUM (89), THORIUM, PROTACTINIUM, URANIUM, NEPTUNIUM, PLUTONIUM, AMERICIUM, CURIUM, BERKELIUM, CALIFORNIUM, EINSTEINIUM, FERMIUM, MENDELEVIUM, NOBELIUM, and LAWRENCIUM (103). Each element is analogous to the corresponding LANTHANIDE SERIES (rare-earth) group. The most important of the group is uranium. Those having atomic numbers greater than 92 are called **transuranium elements**.

actinium (symbol Ac) Radioactive metallic element, the first of the ACTINIDE SERIES, first discovered in 1899. It is found associated with uranium ores. Actinium-227, a decay product of Uranium-235, emits beta particles (electrons) during disintegration. Properties: at.no.89; r.d.10.07 (calc.); m.p.1,100°C (1,900°F); b.p. 3,200°C (5,800°F); most stable isotope ^{227}Ac (half-life 21.8 yr).

action potential Change that occurs in the electrical potential between the outside and the inside of a nerve fiber or muscle fiber when stimulated by the transmission of a nerve impulse. At rest the fiber is electrically negative inside and positive outside; this is called **resting potential**. When the nerve or muscle is stimulated, the charges are momentarily reversed.

activation energy Smallest amount of energy necessary to make a chemical reaction take place. As chemical bonds are broken and formed during a reaction, the energy of the system increases from that of the reactants, reaches a maximum, and then decreases to that of the products. The difference between the energy of the reactants and the maximum is the activation energy. Often this energy has to be supplied to the reaction mixture in the form of heat, although some chemical reactions take place spontaneously by bringing reactants together.

activator Agent that accelerates or augments chemical activity, for example, an impurity that increases LUMINESCENCE.

active transport Energy-requiring process by which molecules or ions are transported across the membranes of living CELLS against a concentration gradient. It is particularly important in the uptake of food across the gut lining, in the reabsorption of water and salts from the urine in the kidney before excretion, in secretion of substances from gland cells, in the transmission of nerve impulses, and in the uptake of minerals by the plant root. The energy is supplied by the chemical ADENOSINE TRIPHOSPHATE (ATP), which is broken down by an enzyme at the site of active transport to form ADENOSINE DIPHOSPHATE (ADP), with the release of energy. Active transport enables cells to maintain an internal chemical environment that is of a different composition from that of their surroundings.

actomyosin Complex of the proteins ACTIN and MYOSIN. This forms the basic unit of contraction in muscle cells.

acupuncture System of medical treatment in which long needles are inserted into the body to assist healing, relieve pain or for anesthetic purposes. Of ancient Chinese origin, it enables surgery to be done with the patient conscious and free of pain. It has so far defied scientific explanation.

ACV Abbreviation of AIR-CUSHION VEHICLE

acyclic In organic chemistry, term used to describe any compound that has no rings in its molecular makeup. *See also* CYCLIC

▲ **adaptation** The various honeycreepers of Hawaii in the Pacific evolved from one species of bird now long extinct (center). Over millions of years the honeycreepers evolved different methods of feeding. This ensured the island's various habitat niches could be exploited, resulting in less competition among the birds and therefore allowing more to survive. The main adaptation was the dramatic change in the shape of the beaks. A few species evolved beaks best suited to feed on nectar (1), others feed purely on insects (2), while some feed on fruit (3) or seeds (4).

acyl group In organic chemistry, group consisting of an ALKYL GROUP joined to CO–. An example is the ethanoyl (acetyl) group, $CH_3CO–$.

Adams, John Couch (1819–92) English astronomer. While analyzing the motion of URANUS, he observed its path was not in agreement with its calculated orbit, and Adams believed that the discrepancies were caused by the gravitational influence of an undiscovered planet. It was Johann Galle who located NEPTUNE, as it was subsequently called, near a position predicted by the French astronomer Urbain LEVERRIER.

adaptation Adjustment by a living organism to its surroundings. Animals and plants adapt to changes in their environment through variations in structure, reproduction or organization within communities. Some such changes are temporary (ACCLIMATIZATION), while others may involve changes in the genetic material (DNA) and be inherited by offspring (EVOLUTION). The word is also used to describe a particular characteristic, of body size, shape, color, physiology or behavior, that helps an organism survive in its environment.

adaptive radiation In biology, the EVOLUTION of different forms of living organisms from a common ancestral stock, as different populations adapt to different environmental conditions or modes of life. Eventually the populations may become so different that they constitute separate SPECIES. If, for instance, a species becomes distributed over several types of surroundings, the populations of each area may develop specialized features suited to the new environment. Examples are the many different kinds of finches in the Galápagos Islands, which diversified to specialize in different kinds of food, feeding methods, and HABITATS. *See also* ADAPTATION

addiction Inability to control the ingestion or inhalation of a drug or other substance, resulting in physiological or psychological dependence. Symptoms of addiction include cravings and habitual or compulsive behavior. It is most frequently associated with drug addiction. In a medical context, addiction requires a **physical** dependence. When the dose of a drug is reduced or withdrawn from the user, the addict experiences withdrawal symptoms. **Psychological** dependence on activities such as gambling or exercise are difficult to distinguish from a personality disorder or manic state. The wide variety of theories on addiction, such as the psychological theory of addictive personality or the physiological concentration on the brain, produce a range of therapeutic treatments.

Addison, Thomas (1793–1860) British physician who identified the ADRENAL HORMONE deficiency known as Addison's disease (1855) and Addison's anemia. He was the co-author of *Elements of the Practice of Medicine* (1839). *See also* ADRENAL GLANDS

addition Arithmetical operation signified by + (plus sign). This is said to be a BINARY operation, since at least two numbers (or elements) are required for the operation to make sense.

addition reaction Chemical reaction in which two substances combine to form a third substance, with no other substance being produced. Addition reactions are most commonly used in organic CHEMISTRY, particularly by adding a simple molecule across a carbon-carbon double bond in an UNSATURATED COMPOUND. For example, the addition reaction between chlorine (Cl_2) and ethene ($CH_2=CH_2$) produces 1,2-dichloroethane ($ClCH_2(CH_2)Cl$). Some plastics (such as polystyrene) are made by addition polymerization, usually initiated by the action of a free radical. *See also* SUBSTITUTION

adenine Nitrogen-containing organic BASE of the PURINE group found combined in the NUCLEIC ACIDS.

adenoids Masses of lymphatic tissue in the upper part of the PHARYNX (throat) behind the NOSE; part of a child's defenses against disease, they normally disappear by the age of ten. When infected, the adenoids become swollen, and may need to be removed surgically.

adenosine diphosphate (ADP) Nucleotide chemical involved in energy-generating reactions during cell METABOLISM. ADP consists of the PURINE base ADENINE linked to the sugar D-ribose, which in turn carries two phosphate groups. It is formed by the HYDROLYSIS of ADENOSINE TRIPHOS-PHATE (ATP) – with the release of energy – or the phosphorylation of ADENOSINE MONOPHOSPHATE (AMP) – which requires the input of energy; both reactions are catalysed by ENZYMES.

adenosine monophosphate (AMP) Nucleotide chemical involved in energy-generating reactions during cell METABOLISM. AMP consists of the PURINE base ADENINE linked to the sugar D-ribose, which in turn carries a phosphate group. It is formed by the HYDROLYSIS of ADENOSINE TRIPHOSPHATE (ATP) – with the release of energy – catalyzed by ENZYMES.

adenosine triphosphate (ATP) Nucleotide chemical consisting of ADENINE, D-ribose, and three phosphate groups. It is found in all plant and animal cells, and is fundamental in the biochemical reactions required to support life. In animals, during RESPIRATION ATP can be broken down through the process of HYDROLYSIS, coupled with PHOSPHORYLATION, to form ADENOSINE DIPHOS-PHATE (ADP) and phosphate, or ADENOSINE MONOPHOSPHATE (AMP) and pyrophosphate. In either case, the reactions, which are catalyzed by ENZYMES, yield large amounts of energy. The energy is used either to turn simple molecules into more complex ones required by individual cells, or to control an activity such as muscle contraction. Conversely, energy is used up when ATP is indirectly synthesized from ADP and AMP by complex biochemical reactions known as the ELECTRON CHAIN SYSTEM during the KREBS CYCLE. In plants, ATP is synthesized during PHOTOSYNTHESIS.

adenovirus Type of VIRUS that contains DNA. Adenoviruses infect cattle, chickens, monkeys, rodents, and human beings, in whom they cause infections of the respiratory tract, similar to the common cold, and gastrointestinal infections. They may also be the cause of a type of cancerous tumor.

Ader, Clément (1841–1926) French pioneer of flight. A self-taught engineer and inventor, he was the first to demonstrate that a heavier-than-air machine could become airborne.

ADH Abbreviation of ANTIDIURETIC HORMONE

adhesion Attraction of molecules of one substance to the molecules of another. GUM, GLUE and paste use the property of adhesion to join various substances together. *See also* COHESION

adhesion In medicine, fibrous band of CONNEC-TIVE TISSUE developing at a site of inflammation or damage; it may bind together adjacent tissues, such as loops of intestine, occasionally causing obstruction. Most adhesions result from inflammation or surgery.

adhesive Substance, or two substances, that have the ability to stick items together. Natural adhesives, usually called GLUES, are made by boiling down animal skin, bones, horns, and hooves. It consists of a jelly of hydrolyzed COL-LAGEN (fibrous protein) – mainly gelatin – mixed with many other animal substances. It dries to form a hard, tough skin or film. There are also vegetable glues made from starch (flour and water), rubber, soybeans, and other sources; synthetic adhesives include EPOXY RESIN and a hardener that react together, and THERMOSETTING and THERMOPLASTIC RESINS.

adiabatic process Thermodynamic change without any gain or loss of heat or mass into or out of a system during expansion or compression of the gas or fluid composing the system. Truly adiabatic changes must take place in short time intervals so that the heat content of the system remains unchanged, or else the system must be perfectly insulated (a practical impossibility). *See also* ISOTHERM

adipose tissue (fatty tissue) CONNECTIVE TISSUE made up of body cells that store globules of fat.

adolescence Period in human development from PUBERTY to adulthood.

ADP Abbreviation of ADENOSINE DIPHOSPHATE

adrenal gland One of a pair of small endocrine glands situated on top of the KIDNEYS. They produce many STEROIDS that regulate the blood's salt and water balance and are concerned with the METABOLISM of carbohydrates, proteins, and fats, and the HORMONES ADRENALINE and noradrenaline, which are closely involved in preparing the body to meet conditions of stress, such as pain, shock, intense cold or physical danger. *See also* ENDOCRINE SYSTEM

adrenaline Name once used for EPINEPHRINE; still used in informal or nontechnical contexts.

adrenocorticotrophic hormone (ACTH) Protein hormone secreted by the anterior lobe of the PITUITARY GLAND at the base of the brain. It stimulates the cortex of the ADRENAL GLANDS to release steroid hormones.

adsorption Attraction of a gas or liquid to the surface of a solid or liquid, unlike ABSORPTION which implies incorporation (the absorbed substance spreading into the absorbing one – for example, a sponge soaking up water). The amounts adsorbed and the rate of adsorption depend on the structure exposed, the chemical identities and concentrations of the substances involved, and the temperature.

adventitious root Root that grows in an unusual place on a plant. Normally roots grow underground at the base of a plant. But some plants produce roots from their stems and branches. With ivy, for example, adventitious roots provide nourishment and enable the plant to cling onto other plants or buildings for support.

Aeolian formation In geology, structure created by wind-transported material. It may be a DUNE on a riverbank or ripple marks in sand on a beach or desert, or the growth phase of dune building. It also can be used to describe shapes carved in rock by the wearing away of softer materials. *See also* BUTTE

aeration Bringing air into contact with a fluid by bubbling through or by agitation. Compressed air, providing oxygen to promote bacterial action, is blown into a reagent tank in the treatment of sewage. Aeration is also used in the fermentation and soft drinks industries, as well as in the manufacture of PENICILLIN and other ANTIBIOTICS.

aerial *See* ANTENNA

aerobe Minute organism that usually grows only in the presence of free atmospheric OXYGEN. Some aerobes can, however, remain alive even in the absence of oxygen and are called **facultative** ANAEROBES.

aerobic Connected with or dependent on the presence of free oxygen or air. An aerobic organism can only function normally in the presence of oxygen and depends on it for breaking down GLUCOSE and other foods to release energy. This process is called **aerobic respiration**. **Aerobic exercise**, which includes running, swimming, skiing, bicycling, and playing such games as tennis and soccer, provides cordiovascular health because it produces an elevated heart rate that causes the heart to pump more richly oxygenated blood to the muscles. *See also* ANAEROBIC

aerodynamics Science of gases in motion and the forces acting on objects, such as aircraft, in motion through the air. An aircraft designer must consider four main factors and their interrelationships: weight of the aircraft and the load it will carry; LIFT to overcome the pull of gravity; DRAG, or the forces that retard motion; and THRUST, the driving force. An airplane must have sufficient thrust to propel itself at speeds high enough for its wings to generate enough lift to overcome the pull of gravity. Air resistance increases as the square of an object's speed and must be minimized by streamlining to limit turbulence (which increases drag). Engineers use a WIND TUNNEL and computer systems to predict aerodynamic performance.

aerolite STONY METEORITE, as opposed to an iron (SIDERITE) one. Aerolites are composed of silicate materials, with less iron and nickel than siderites.

aeronautics Study of flight and the control of AIRCRAFT involving AERODYNAMICS, aircraft structures, and methods of propulsion. Aeronautics started with the study of the BALLOON, which mainly concerned the raising of a load by means of BUOYANCY. It later included the heavier-than-air flight of gliders, planes, HELICOPTERS, and ROCKETS. A helicopter utilizes LIFT provided by a rotor (rotating wing). Gliders and planes use wings to provide lift, but a minimum forward speed is essential to maintain altitude. A plane is pulled forward by a propeller, or is pushed by the reaction forces of hot expanding gases from one or more JET or rocket engines. The increased speeds of modern aircraft to supersonic (speeds in excess of that of sound; 1,225km/h or 760mph) and the accompanying shock waves this produces have brought changes in wing and fuselage designs to improve streamlining. While with hypersonic speeds (in excess of five times the speed of sound or Mach 5), the forces involved again change fundamentally, requiring further design adjustments.

aeronomy Study of the Earth's upper ATMOSPHERE, including its composition, density, temperature, and chemical changes as recorded by satellites.

aerosol Suspension of liquid or solid particles in a gas. Fog – millions of tiny water droplets suspended in air – is a liquid-based example; airborne dust or smoke is a solid-based equivalent. Manufactured aerosols are used in products such as deodorants, cosmetics, paints, and household sprays to name just a few. They use propellants to force the product through a nozzle. Chlorofluorocarbons (CFCs), once commonly used as propellants, are being phased out because they damage the ozone layer.

afforestation Planting of trees in non-wooded areas. Trees may be planted for several reasons: to reduce EROSION and hold the soil together, to act as windbreaks to shelter arable crops, and to produce timber, particularly softwoods for wood pulp. Usually trees are planted where they have not grown before (at least in the recent past), such as moorland and steppes.

afterbirth PLACENTA, UMBILICAL CORD and fetal membranes expelled from the UTERUS after childbirth. The expulsion is brought about naturally by contraction of the uterus.

afterimage Image retained by the brain and seen even after looking away from an object. The color of the afterimage is the complement of that of the object viewed. Television and motion pictures make use of this phenomenon.

agar Complex substance extracted from seaweed; its powder forms a "solid" gel in solution. It is used as a thickening agent or emulsifier in foods; as an adhesive; and as a medium for growing bacteria, mold, YEAST, and other microorganisms; as a medium for TISSUE CULTURE; and as a gel for ELECTROPHORESIS.

Agassiz, Alexander (1835–1910) US marine zoologist, b. Switzerland. He was influential in the development of modern systematic zoology, and

made important studies of the seafloor. In 1873 he succeeded his father, Louis, as curator of the Harvard Museum of Natural History, to which he also made major financial contributions.

agate Microcrystalline form of quartz with parallel bands of color. Extracted from rock cavities, it is regarded as a semiprecious stone and is used for making ornaments and jewellery. Hardness 6.5; s.g. 2.6.

agglomerate In geology, coarse volcanic rock that includes both rounded and angular fragments in a finer matrix, thus combining the characteristics of CONGLOMERATES and BRECCIAS.

agglutination Clumping of BACTERIA or red blood cells by ANTIBODIES that react with ANTIGENS on the cell surface.

aggregation In mathematics, treatment of a number of terms in a mathematical expression as a single unit, denoted by enclosing the terms in parentheses. Thus $7(x^2 = 2)$ is $7 \ x^2 = 14$.

Agnatha Class or subphylum of jawless FISH. Only the lamprey and the hagfish live today, but

Outer shroud with seal and gas entry point

Ram forces cup in position after gas is injected

C

A

4
5
D
1
2

Tool for straining metal of cup to create seal

B

3

Dome base withstands pressure

▲ **aerosol** Dating from 1939, disposable aerosols are now widely used. First the can is filled with the product to be sprayed and the propellant – now usually CFC free – (A), before it is sealed (B). A valve, while in a closed position (C), retains the product. When the pushbutton (1) is depressed (D), the valve is opened (2) and the product is forced up the dip tube (3) and out of the nozzle (4) as a fine spray (5).

many fossil species date from as long ago as 450 million years.

Agricultural Revolution Series of changes in farming practice in the 18th and early 19th centuries that incorporated technological changes to make farming and land use more efficient. The main changes involved crop rotation, new machinery, increased capital investment, scientific breeding, land reclamation, and enclosure of common land.

agronomy Science of soil management and improvement in the interests of agriculture. It includes the studies of particular plants and soils and their interrelationships. Agronomy involves disease-resistant plants, selective breeding, and the development of chemical fertilizers.

AIDS Acronym for ACQUIRED IMMUNE DEFICIENCY SYNDROME

Aiken, Howard (1900–73) US mathematician who did pioneering work in the development of computers. When Aiken was working for IBM 1939 he designed an electromechanical automatic calculating machine. By 1944 he and his associates had built the Mark 1 Automatic Sequence Controlled Calculator, one of the first programmable computers. The input medium was punched cards or punched paper tape.

aileron Hinged control surface on the outer trailing edge of each wing of an AIRCRAFT. By moving a control, the pilot deflects the ailerons down or up in opposite directions to increase or decrease lift, causing the airplane to roll, or bank.

air Gases above the Earth's surface. See ATMOSPHERE

air conditioning Process of controlling the temperature, humidity, flow, and sometimes odor and dust content of AIR in any enclosed space. Air is cooled by refrigeration or heated by steam or hot water. Odors and dust are removed by filters and the moisture content is adjusted by humidifiers before the air is circulated by fans that also remove stale air.

aircraft Any vehicle capable of traveling in the Earth's ATMOSPHERE. By far the most common aircraft is the airplane, or plane. This is a heavier-than-air flying machine that depends upon fixed wings for LIFT in the air, as it moves under the THRUST of its engines. This thrust may be provided by an airscrew (propeller) turned by a piston or turbine engine, or by the exhaust gases of a JET engine or rocket motor. GLIDERS differ from planes only in their complete dependence upon air currents to keep them airborne. The main

body of a plane is the fuselage, to which are attached the wings and tail assembly. Engines may be incorporated into or slung below the wings, but are sometimes mounted on the fuselage towards the tail or, as in some fighter aircraft, built into the fuselage near the wings. The landing gear or undercarriage, with its heavy wheels and strong shock absorbers, is usually completely retractable into the wings or fuselage. Wing design varies with the type of plane; high-speed fighters having slim, often swept-back or adjustable wings that create minimal air resistance (DRAG) at high speeds. At the other extreme heavy air freighters need broader wings in order to achieve the necessary lift at take-off. A delta wing is a broad wing, or fuselage extension, that is aerodynamically suited for both large and small high-speed planes. A plane is steered by the pilot moving flaps and AILERONS on the wings, and rudder and elevators on the tail assembly. This deflects the pressure of air on various AIRFOIL surfaces, causing the plane to rise or descend, to bank (tilt) or swing and turn in the air. A pilot needs to refer to many control instruments, including an altimeter showing altitude, an artificial horizon showing tilt, and indicators for air speed, ground speed, rate of climb, and rate of banking and turning. Engine and airframe systems are monitored by numerous additional gauges, meters, and warning lights. RADAR systems aid navigation, an AUTOPILOT keeps the aircraft steady on a fixed course, and pressurized cabins allow passenger planes to fly at altitudes exceeding 10,000m (32,800ft). See also AIRSHIP; AERODYNAMICS; BALLOON; HELICOPTER

aircraft carrier Military vessel with a wide open deck that serves as a runway for the launching and landing of aircraft. A modern nuclear-powered carrier may have a flight deck about 300m (1,000ft) long, a displacement of about 75,000 tonnes, a 4,000-man crew, and carry 90 aircraft of various types – launched one a minute with each of four steam catapults. Some carriers have large, angled decks to permit launching and landing simultaneously.

air-cushion vehicle (ACV) Vehicle that is lifted from the surface by air forced out from under the craft. The best-known example is a HOVERCRAFT.

airfoil Any shape or surface, such as a wing, tail or propeller blade on an aircraft, the major function of which is the deflection of airflow to produce a pressure differential or LIFT. A typical airfoil has a leading and trailing edge, and an upper and lower camber.

airglow Faint permanent glow of the Earth's IONOS-

PHERE resulting from the recombination of molecules, such as oxygen and nitrogen, that have been ionized by ultraviolet radiation from the Sun and probably by cosmic-ray and solar-wind particles.

airlock Hermetically sealed passageway between two leakproof doors connecting an airless environment, such as the sea depths or outer space, with a pressurized cabin or work area. Entry is made through the outer door, which is then closed; air is pumped into the airlock until its pressure is equal to that inside and then the inner door opens.

air mass Large body of the ATMOSPHERE nearly uniform horizontally in temperature, pressure, and vapor content, and therefore in weather effects. High and low pressure zones on weather maps correspond to air masses, covering large areas horizontally although only a few miles vertically. FRONTS, or frontal zones, occur between air masses, the movement of which brings the weather. Air masses are often classified in terms of origin as polar (cold) or tropical (warm), maritime (wet, over oceans) or continental (dry, over land). *See also* CIRCULATION, ATMOSPHERIC; FRONT

airplane *See* AIRCRAFT

air pollution *See* POLLUTION

air pressure *See* ATMOSPHERIC PRESSURE

air sac In birds, one of the many thin-walled sacs connected to the lungs or to cavities in the bones. Air sacs improve the efficiency of air ventilation in the lungs and contribute to the lightness of the bones. In insects, air sacs connected to the tracheae (breathing tubes) improve gas exchange by increasing the surface area available for respiration.

airship (dirigible) Powered, lighter-than-air craft able to control its direction of motion. A rigid airship, or zeppelin, maintains its form with a framework of girders covered by fabric or aluminum alloy. A gas that is less dense than air, nowadays helium, provides LIFT. Non-rigid airships, or blimps, have no internal structure. They rely on the pressure of the contained gas to maintain the shape. A semi-rigid airship has a rigid or jointed keel to stiffen the envelope.

air speed indicator Instrument that measures the speed of an aircraft relative to the air through which it passes. It measures the difference in pressures of air rushing through a **Pitot tube** (generally mounted on the wing) and static air from a vent on the fuselage; air-speed is usually indicated in miles per hour or knots.

Airy, Sir George Biddell (1801–92) British astronomer who also made discoveries in OPTICS, such as a cylindrical LENS for correcting ASTIGMATISM. He was Astronomer Royal from 1853 to 1881.

alabaster Fine-grained, massive variety of GYPSUM (calcium sulfate), snow-white and translucent in its natural form. It can be dyed or made opaque by heating. For centuries alabaster has been used for making statues and other ornaments.

alanine ($CH_3C(NH_2)COOH$) Colorless, soluble AMINO ACID that occurs widely in PROTEINS, for example those derived from silk.

albedo Fraction of light or other radiation that is reflected from a surface. An ideal reflector has an albedo of 1; those of real reflectors are less. The albedo of snow varies from 0.45 to 0.90; that of the Earth, viewed from satellites, is 0.35. Much has been

◄ **airfoil** (A) As air passes over the top edge of an airfoil (1) it has to travel further than the air flowing beneath it (2), an area of low pressure forms above the wing that generates lift (3). In modern high-lift airfoils (B) the center of pressure is further toward the rear of the wing (4). By moving the point of maximum lift backward the airfoil has a more even distribution of lift allowing a plane to fly slower without stalling.

◀ **algae** Many unicellular (single-celled) algae are said to be motile, that is they move in response to changes in their environment, in particular to light. This is achieved with whiplike flagella (1) which propel them through the water. The illustration shows three types of alga. (A) *Gonyaulax tamarensis*, (B) *Chlamydomonas* and (C) *Prymnesium parvum*. Algae have been found in rocks over 2,700 million years old. They are a vital source of oxygen.

ACIDS. Their systematic names are formed using the suffix –al. *See also* OXIDATION

Alembert, Jean le Rond d' (1717–83) French philosopher and mathematician. He was a religious sceptic and physicist, and formulated d'Alembert's principle, used to solve certain problems in mechanics. He wrote a systematic *Treatise on Dynamics* (1743) as well as several pieces on DIFFERENTIAL EQUATIONS.

Alfvén, Hannes (1908–95) Swedish physicist who shared the 1970 Nobel Prize for physics with Louis Néel of France. He laid the foundation of what has become known as magnetohydrodynamics (mhd); he advanced the theory of frozen-in flux, which states that if a magnetic field passes through a PLASMA, the movement of the plasma is constrained by the magnetic lines of flux.

deduced about the surfaces of planets by comparing their albedos with those of known substances.

albino Person or animal with a rare hereditary absence of pigment from the skin, hair, and eyes. The hair is white and the skin and eyes are pink because, in the absence of pigment, the blood vessels are visible. The eyes are sensitive to light.

albumen (albumin) Type of water-soluble PROTEIN occurring in animal tissues and fluids. It coagulates if heated. Principal forms are EGG albumen (egg white), milk albumen, and blood albumen.

alchemy Primitive form of CHEMISTRY practiced in western Europe from early Christian times until the 17th century, popularly supposed to involve a search for the **Philosopher's Stone** – capable of transmuting base metals into gold – and the elixir of life. In fact, alchemy involved a combination of practical chemistry, astrology, philosophy, and mysticism. Similar beliefs existed during the same period in China, India, and the Islamic World.

alcohol Organic compound having a hydroxyl (–OH) group bound to a carbon atom, for example ETHANOL (C_2H_5OH). Alcohols are used to make dyes and perfumes and as SOLVENTS in lacquers and varnishes.

aldehyde Member of a class of organic compounds, characterized by the group –CHO. They include formaldehyde, acetaldehyde, and benzaldehyde, and are used in the manufacture of synthetic resins, and to make dyes, perfumes, and other chemicals. Aldehydes can be made by oxidizing primary ALCOHOLS. They are generally reducing agents, being oxidized to CARBOXYLIC

algae Large group of essentially aquatic photosynthetic organisms, belonging to the kingdom PROTOCTISTA. Algae are found in salt and fresh water throughout the world, and are a primary source of food for mollusks, fish, and other aquatic animals. Algae are directly important to humans as food (especially in Japan) and as FERTILIZERS. They range in size from unicellular microscopic organisms such as those that form green pond scum to huge brown seaweeds more than 45m (150ft) long. *See also* CYANOBACTERIA; CHLOROPHYTE; RHODOPHYTE

algal bloom Sudden increase in the amount of ALGAE growing in fresh water. The common species involved are CYANOBACTERIA (blue-green algae), although other types of phytoplankton may also rapidly grow in numbers. It is usually caused by an increase in nitrates and other nutrients in the water, often from artificial fertilizers washed off nearby fields or from effluent containing sewage. Algal blooms often occur in lakes, and may be prolific enough to prevent light from reaching the lower depths of the water. They are a factor in EUTROPHICATION.

algebra Branch of MATHEMATICS dealing with the study of equations that are written using numbers and alphabetic symbols, which themselves represent quantities to be determined. An algebraic equation may be thought of as a constraint on the possible values of the alphabetic symbols. For example, $y + x = 8$ is an algebraic equation involving the variables x and y. Given any value of x the value of y may be determined, and vice-versa. The term algebra literally means "to find the unknown" from the Arabic *al-jabr*. *See also* ALGEBRAIC OPERATION; BOOLEAN ALGEBRA; SIMULTANEOUS EQUATIONS

algebraic function (algebraic equation) Function that is expressible using rational powers of the variable. For instance,

$$f(x) = \pi x^3 + x^{1/4} - 2/x$$

is an algebraic function. In contrast, $\log x$ is a transcendental function expressed only by an infinite series. An algebraic equation is any expression that can be written $f(x) = 0$, where f is an algebraic function. All POLYNOMIALS are algebraic. *See also* NUMBER, TRANSCENDENTAL

algebraic operation Operation of ordinary algebra: that is, the arithmetical operations ADDITION, SUBTRACTION, MULTIPLICATION, and DIVISION. Operations that involve infinite series and functions such as $\log x$ are not algebraic – they depend on the use of limits. The term algebraic is also applied to arithmetical operations carried out with due regard to sign. Thus, the algebraic sum of a and $- b$ is $a - b$, the algebraic product of $- a$ and $- b$ is $+ ab$, etc. *See also* ARITHMETIC

algorithm Step-by-step set of instructions or procedures followed to obtain a result from an initial set of input DATA. The term is used in COMPUTER PROGRAMMING to refer to a set of instructions in a computer-readable format that incorporates an established series of steps to obtain the solution to a problem.

alicyclic In organic chemistry, term used to describe an ALIPHATIC COMPOUND that has one or more rings in its molecular make-up. An example is cyclohexane (C_6H_{12}). *See also* CYCLIC

alimentary canal Digestive tract of an animal that begins with the MOUTH, continues through the ESOPHAGUS to the STOMACH and INTESTINES, and ends at the ANUS. It is about 9m (30ft) long in humans. *See also* DIGESTIVE SYSTEM

aliphatic compound Any organic chemical compound the carbon atoms of which are linked in straight chains, not closed rings. They include the ALKANES (paraffins), ALKENES (olefines), and ALKYNES (acetylenes).

alkali Soluble BASE that reacts with an ACID to form a SALT and water. A solution of an alkali has a pH greater than 7 and turns litmus dye blue. Alkali solutions are used as cleaning materials. Strong alkalis include the hydroxides of the ALKALI METALS (lithium, sodium, potassium, rubidium, and cesium) and ammonium hydroxide. The carbonates (salts of carbonic acid) of these metals, and ammonium carbonate, are weak alkalis.

alkali metals Univalent metals forming Group I of the periodic table: LITHIUM, SODIUM, POTASSIUM, RUBIDIUM, CESIUM, and FRANCIUM. They are soft, silvery-white metals that tarnish rapidly in air and react violently with water to form hydroxides, which make strongly basic solutions.

alkaline-earth metals Bivalent metals forming Group II of the periodic table: BERYLLIUM, MAGNESIUM, CALCIUM, STRONTIUM, BARIUM, and RADIUM. They are all light, soft, and highly reactive. All, except beryllium and magnesium, react with cold water to form hydroxides (magnesium reacts with hot water). Radium is important for its radioactivity.

alkaloid Member of a class of complex nitrogen-containing organic compounds found in certain plants. They are sometimes bitter and highly poisonous substances, used as DRUGS. Examples include codeine, morphine, nicotine, and quinine.

alkane (PARAFFIN) HYDROCARBON compound with the general formula C_nH_{2n+2}. Alkanes have a single carbon-carbon bond and form a HOMOLOGOUS SERIES whose first members are METHANE (CH_4), ETHANE (C_2H_6), PROPANE (C_3H_8), and BUTANE (C_4H_{10}). Because alkanes are SATURATED COMPOUNDS they are relatively unreactive. Alkanes are used as fuels.

alkene (olefin) Unsaturated HYDROCARBON compound with the general formula C_nH_{2n}. Alkenes have a carbon-carbon double bond and form a homologous series whose first members are ETHENE (ethylene, $CH_2=CH_2$) and PROPENE (propylene, $CH_3CH=CH_2$). They are reactive, particularly in ADDITION REACTIONS. Alkenes are made by the DEHYDRATION of alcohols, and are used as fuels and to make POLYMERS.

Al-Khwarizmi (active 820) Persian mathematician. He is credited as being the first to solve the QUADRATIC EQUATION a x^2 + bx + c = 0. In his book *Calculation with Hindu Numerals* he described a number notation that was used at the time. Following the translation of this work (in the 13th century) the Indian number system was adopted in Europe. It is the system that we use today. Consisting of ten DIGITS, including zero, this number system is usually (in fact falsely) referred to as the Arabic system and the digits 0,1,...9 are called the Arabic numerals.

alkyd Synthetic resin, generally made from phthalic acid and glycerol. Alkyds can be molded at high speed under low pressure and cured quickly. They are widely used as paint resins and in products where good insulation, high strength and good temperature, voltage, and humidity stability are important, such as in car ignition systems. *See also* POLYESTER

alkyl group In organic chemistry, a group consisting of an ALKANE with one hydrogen atom removed. For example, the alkanes methane (CH_4) and ethane (C_2H_6) give rise to the alkyl groups methyl (CH_3 –) and ethyl (C_2H_5 –).

alkyl halide *See* HALOALKANE

alkyne (acetylene) Unsaturated HYDROCARBON compound with the general formula C_nH_{2n-2}. Alkynes have a carbon-carbon triple bond and form a homologous series whose first members are ETHYNE (acetylene, $CH\equiv CH$) and propyne (methyl acetylene, $CH_3C\equiv CH$). Ethyne is very reactive and prepared by adding water to calcium dicarbide. It is used as a fuel gas, particularly in oxyacetylene welding. It produces a range of POLYMERS when heated to high temperatures.

allantois Sac-like membrane that grows from the hindgut in the embryos of birds, mammals, and reptiles. It is involved in respiration and the absorption of nutrients (in mammals), and in birds and reptiles acts as a bladder for the storage of waste products within the egg. *See also* PLACENTA

allele One of two or more alternative forms of a particular GENE. Different alleles may give rise to different forms of the characteristic for which the gene codes; for example, different flower color in peas is due to the presence of different alleles of a single gene. *See* MENDEL, GREGOR JOHANN

allergy Disorder in which the body mounts a hypersensitive reaction to one or more substances (allergens) not normally considered harmful. Common allergens include pollen, animal hair, fungi, dust, occasionally foods, and some drugs, such as PENICILLIN. Typical allergic reactions are sneezing (hay fever), wheezing, and difficulty in breathing (ASTHMA) and skin eruptions and itching (ECZEMA). A tendency to allergic reactions is often hereditary. Treatment usually requires the identification of the allergen and its avoidance if possible, or a course of desensitization. Drugs can be useful in relieving acute symptoms.

allopathy Treatment of diseases by methods and remedies that produce effects contrasting with those caused by the disease. The term is usually applied to conventional MEDICINE, as opposed to HOMEOPATHY.

allopatric speciation Formation of one or more new SPECIES through geographical isolation. One of the criteria that defines a species is that its members are capable of interbreeding. But if members of a species become separated by a geographical barrier, such as an ocean or a mountain range, they can no longer interbreed (the separated groups are allopatric) and in time may diverge so much from each other that they are no longer even capable of interbreeding. They have therefore become different species.

allotropy Property of some chemical elements that enables them to exist in two or more distinct physical forms. Each form (called an **allotrope**) can have different chemical properties but can be changed into another allotrope – given suitable conditions. Examples of allotropes are molecular oxygen and ozone; white and yellow phosphorous; and graphite and diamond (carbon).

alloy Combination of two or more metals. An alloy's properties are different from those of its constituent elements. Alloys are generally harder and stronger, and have lower melting points. Combinations with the lowest melting points are called **eutectic** mixtures. Most alloys are prepared by mixing when molten. Some mixtures that combine a metal with a nonmetal, such as STEEL, are also referred to as alloys. The proportions in which the component metals are mixed greatly affects the properties of an alloy.

alluvial fan Generally fan-shaped area of ALLUVIUM (water-borne sediment) deposited by a river when the stream reaches a plain on lower ground,

▲ **allotropy** Carbon can exist in two forms, diamond and graphite, known as allotropes of carbon. The top diagram shows the structure of diamond, revealing its strong, interlocked structure. The bottom diagram depicts graphite, showing its layered structure. Although the carbon atoms bond strongly in sheets, the bonds between the layers are weak, enabling us to utilize graphite as a lubricant.

and the water velocity is abruptly reduced. Sand, silt and gravel are the main constituents of the alluvium, with the coarser material deposited near the head of the fan. Organic matter is also transported, making the soil highly fertile. Valuable minerals such as CASSITERITE (tin ore, SnO_2), diamonds, gold, and platinum are often found in alluvial fans.

alluvium Fine mud or silt deposited by rivers. It accumulates on river beds but can be spread over large areas of a river valley at times of flood. All rivers have some alluvium deposits, but large rivers such as the Mississippi or Amazon have alluvial plains that are several kilometers wide. Alluvial deposits at the foot of steep mountains often accumulate in a triangular or cone shape, rather like an inland DELTA.

alp Area of grassland vegetation high up on the side of any mountain range. The alpine pasture is used by grazing animals in the summer but is snow-covered in winter.

Alpha Centauri Brightest star in the constellation Centaurus, and the third-brightest star in the sky. It is a visual BINARY STAR, meaning that both it and its partner star, the fainter Proxima Centauri, are visible through a telescope. It is located at a distance of 4.3 light years.

alpha particle (alpha ray) Stable, positively charged particle emitted spontaneously from the nuclei of certain radioactive ISOTOPES when undergoing RADIOACTIVE DECAY called **alpha decay**. They were discovered and investigated by Ernest RUTHERFORD and later identified as the nuclei of helium atoms consisting of two PROTONS and two NEUTRONS bound together (Helium-4). Their penetrating power is low compared with that of BETA PARTICLES (electrons) but they cause intense IONIZATION along their track. Thus ionization is used to detect them. Alpha particles can be stopped by a sheet of paper.

Alpine folding Most recent large-scale example of folding. It ocurred when the Alps, Himalyas, Rockies, Andes, and other large mountain ranges were formed.

alternating current (AC) *See* ELECTRIC CURRENT

alternation of generations Two-generation cycle by which plants and some algae reproduce. The asexual diploid SPOROPHYTE form produces haploid SPORES that, in turn, grow into the sexual (GAMETOPHYTE) form. The gametophyte produces the egg cell that is fertilized by a male gamete to produce a diploid zygote that grows into another sporophyte.

alternative energy *See* RENEWABLE ENERGY

alternator Electrical GENERATOR that produces an alternating current.

altimeter Instrument for measuring ALTITUDE. Pilots, scientists, surveyors and mountain climbers use altimeters. The simplest type is a form of aneroid BAROMETER. As altitude increases, air pressure decreases, so the barometer scale can be calibrated to show altitude. Some aircraft have a RADAR altimeter. This instrument measures the delay between sending a radio signal to the ground and receiving its reflection. From this delay, the instrument calculates the distance to the ground.

altitude Generally, the vertical height of a point above mean sea level. In astronomy, the ANGULAR DISTANCE of a celestial body above the observer's horizon. It is measured in degrees from 0° (on the horizon) to 90° (at the zenith) along the GREAT CIRCLE passing through the body and the zenith. If the object is below the horizon, the altitude is negative.

altruism Principle of acting for the welfare or interests of someone else. The word was coined by the French philosopher Auguste Comte from the Italian *altrui* (others). It is used in biology to describe the behavior of an animal that acts in a way that enhances the prospects of survival or reproduction of another individual at the expense of its own interests. Most cases of altruism among animals involve close relatives or young, with whom the animal shares many GENES. So it is, in effect, perpetuating its own genes in the process.

alum Double sulfate of ALUMINUM (or another trivalent metal) and another, univalent metal. The most important is potash alum, potassium aluminum sulfate; it is used as a mordant in dyeing.

alumina (aluminum oxide, Al_2O_3) Mineral used as an abrasive, electrical insulator, and furnace lining. The impure hydrated form of alumina, BAUXITE, is the chief ore of ALUMINUM. Other forms of alumina include corundum, two impure varieties of which are the gemstones SAPPHIRE and RUBY.

aluminum (symbol Al) Metallic, silvery white element of group III of the periodic table, first obtained in pure form in 1827. It is the most common metal in the Earth's crust; the chief ore is BAUXITE. Bauxite undergoes two processes to yield aluminum. It is first refined to obtain pure alumina (Al_2O_3), which is then smelted to produce aluminum. The Bayer process, involving digestion, clarification, precipitation, and calcination, is the most common refining process yielding half the original amount of ore. Smelting involves dissolving alumina in melted cryolite and passing a current through the mixture (ELECTROLYSIS) to yield pure aluminum and carbon dioxide (the Hall-Heroult process). Alloyed with other metals, it is used extensively in machined and molded articles, particularly where lightness is important, as in aircraft. It is protected from oxidation (corrosion) by a thin, natural layer of oxide. Properties: at.no. 13; r.a.m. 26.98; r.d. 2.69; m.p. 660.2°C (1,220.38°F); b.p. 1,800°C (3,272°F); most common isotope ^{27}Al. *See also* ANODIZING

Alvarez, Luis Walter (1911–88) US physicist in the field of subatomic particles who won the 1968 Nobel Prize for physics for developing the liquid-hydrogen BUBBLE CHAMBER. He used it to identify many "resonances," very short-lived particles. He also helped construct the first proton linear ACCELERATOR. Alvarez worked on the MANHATTAN PROJECT to develop the atom bomb and invented the radar guidance system for aircraft landings. He and his son, the geologist Walter Alvarez, were in the group who formed the theory that dinosaurs became extinct as a result of changes to the atmosphere after a massive meteorite crashed into the Earth.

alveolus (pl. alveoli) One of a cluster of microscopic air sacs that open out from the alveolar ducts at the far end of each bronchiole in the LUNGS. The alveolus is the site for the exchange of gases between the air and the bloodstream, and its walls contain a network of CAPILLARY blood vessels. *See also* GAS EXCHANGE; RESPIRATORY SYSTEM

Alzheimer's disease Degenerative condition characterized by memory loss and progressive mental impairment; it is the commonest form of DEMENTIA. Sometimes seen in the middle years, Alzheimer's becomes increasingly common with advancing age. Many factors have been implicated, but the precise cause is unknown.

AM Abbreviation of AMPLITUDE MODULATION

amalgam Solid or liquid alloy of mercury with other metals. Dentists once commonly filled teeth with amalgams usually containing copper and zinc. Most metals dissolve in mercury, although iron and platinum are exceptions.

amatol High explosive made from ammonium nitrate and TNT.

amber Hard yellow or brown translucent fossil resin, mainly from pine trees. Amber is most often found in alluvial soils, in lignite beds or around sea-shores, especially near the Baltic Sea. The resin occurs as rods or as irregular nodules, sometimes with embedded fossil insects or plants. Amber can be polished to a high degree and is used to make necklaces and other items of jewelry.

americium (symbol Am) Radioactive metallic element of the ACTINIDE SERIES, first made in 1944 by neutron bombardment of plutonium. It is used in home smoke detectors, and ^{241}Am is a source of gamma rays. Properties: at.no. 95; r.a.m. 243.13; r.d. 13.67; m.p. 995°C (1,821°F); b.p. unknown; most stable isotope ^{243}Am (half-life 7,650 yr).

amethyst Transparent, violet variety of crystallized QUARTZ, containing more iron oxide than other varieties. It is found mainly in Brazil, Uruguay, Ontario, Canada, and North Carolina, USA. Amethyst is valued as a semiprecious gem and may be made synthetically.

amide Organic chemical compound that contains the amide group $-CONH_2$. Amides are generally made by heating the ammonium salt of a CARBOXYLIC ACID. For example, heating ammonium ethanoate (acetate) produces ethanamide (acetamide, CH_3CONH_2), which is a volatile solid.

amine Any of a group of organic compounds derived from AMMONIA by replacing HYDROGEN atoms with alkyl groups. Methylamine (CH_3NH_2), has one hydrogen replaced. Replacement of two hydrogens gives a secondary amine and of three hydrogens, a tertiary amine. Amines are produced in the putrefaction of organic matter and are weakly basic. ANILINE is an aromatic amine compound used in dyeing.

amino acid Organic ACID containing at least one carboxyl group (COOH) and at least one amino group (NH_2). Amino acids are of great biological importance because they combine together to form PROTEINS. Amino acids form PEPTIDES by the reaction of adjacent amino (NH_2) and carboxyl (COOH) groups. Proteins are polypeptide chains consisting of hundreds of amino acids. About 20 amino acids occur in proteins; not all organisms are able to synthesize all of them. **Essential amino acids** are those that an organism has to obtain ready-made from its environment. There are ten essential amino acids for humans: ARGININE, HISTIDINE, ISOLEUCINE, LEUCINE, LYSINE, METHIONINE, PHENYLALANINE, THREONINE, TRYPTOPHAN, and VALINE.

ammeter Instrument for measuring electric current in AMPERES (amps). An ammeter is connected in series in a circuit. In the moving-coil type for direct current (DC), the current to be measured passes through a coil suspended in a magnetic field and deflects a needle attached to the coil. In the moving-iron type for both direct and alternating current (AC), current through a fixed coil magnetizes two pieces of soft IRON that repel each other and deflect the needle. Digital ammeters are now also commonly used. *See also* ELECTRIC CURRENT

ammonia (NH_3) Colorless, nonflammable pungent gas, a compound of hydrogen and nitrogen. It is used to make nitrogenous fertilizers. The gas is extremely soluble in water, forming an alkaline solution of ammonium hydroxide (NH_4OH) that can give rise to ammonium salts containing the ion NH_4^+. Ammonia solutions are used in cleaning and bleaching. Chief properties: r.d. 0.59; m.p. −77.7°C (−107.9°F); b.p. −33.4°C (−28.1 °F).

ammonite Any member of the order Ammonitida, an extinct group of shelled CEPHALOPOD MOLLUSKS. Most ammonites had a spiral shell, and they are believed to be related to the nautiloids – the only surviving form being the pearly nautilus. They are common as fossils in marine rocks.

ammonium chloride (NH_4Cl) Ammonium salt of hydrochloric acid. Formerly called sal ammoniac, ammonium chloride is a volatile, highly soluble white solid, made by reacting ammonia with hydrogen chloride. It is used in dry batteries and as a mordant in the dyeing and printing of cloth.

amnion Membrane or sac that encloses the EMBRYO of a reptile, bird or mammal. The embryo floats in the amniotic fluid within the sac. *See also* UTERUS

amniotic fluid Liquid filling the amniotic cavity that protects the fetus from injury. This fluid is released at delivery when the membranes rupture.

amoeba (genus *Amoeba*) Microscopic, almost transparent, single-celled PROTOZOA of the phylum Rhizopoda. Amoebas have a constantly changing, irregular shape. Found in ponds, damp soil, and animal intestines, they consist of a thin outer cell membrane, a large nucleus, food and contractile VACUOLES and fat globules. They reproduce by binary FISSION. Length: up to 3mm (0.1in). Species include the common *Amoeba proteus* and *Entamoeba histolytica*, which causes amoebic DYSENTERY.

amorphous substance Non-crystalline solid; its atoms or molecules have no regular order. Supercooled liquids such as glass, rubber, and some plastics are amorphous. Many powders appear amorphous but are microcrystalline in structure. *See also* CRYSTAL

AMP Abbreviation of ADENOSINE MONOPHOSPHATE

ampere (Symbol A) SI unit of ELECTRIC CURRENT. It is defined as the current in a pair of straight, parallel conductors of infinite length and 1m (39in) apart in a vacuum that produces a force of 2 × 10^{-7} newton per meter (N/m) in their length. This force may be measured on a current balance instrument, the standard against which current meters, such as an AMMETER, are calibrated.

Ampère, André Marie (1775–1836) French physicist and mathematician. He was professor of chemistry and physics at Bourg and later of mathematics at the École Polytechnique in Paris. He founded electrodynamics (now called ELECTROMAGNETISM) and performed experiments to investigate the magnetic effects of ELECTRIC CURRENTS. He was the first to devise techniques for detecting and measuring currents, and constructed an early type of galvanometer. Ampère's law – proposed by him – is a mathematical description of the magnetic force between two electric currents. His name is also commemorated in the fundamental unit of current, the AMPERE (A), used in the SI system of units.

amphetamine Name for a number of DRUGS (1-phenyl-2-aminopropane and its derivatives) that stimulate the CENTRAL NERVOUS SYSTEM, inhibit sleep, and suppress appetite. These drugs (known as "pep pills" or "speed"), if used long-term, can lead to drug abuse and ADDICTION. After the immediate sense of well-being wears off, the user can experience fatigue and depression.

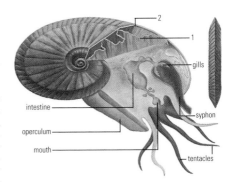

▲ **ammonite** Similar to that of the modern nautilus, which lives in the open end of its shell, the now-extinct ammonites had a soft anatomy. As the animal grew it secreted more and more shell and moved forward into the new part, walling off the old section with a septum (1). The walled-off chambers were used for buoyancy, being supplied with air from a tissue filament or iphuncle (2) that connected them all. The septa met the shell wall in suture lines that had identifiable patterns for each species and became increasingly complex as the group advanced.

amphibian Member of a class (Amphibia) of egg-laying VERTEBRATES, whose larval stages (tadpoles) are usually spent in water but whose adult life is normally spent on land. Amphibians have smooth, moist skin and are cold-blooded. Larvae breathe through GILLS; adults usually have LUNGS. All adults are carnivorous but larvae are frequently herbivorous. There are three living orders: Urodela (NEWTS and SALAMANDERS); Anura (FROGS and TOADS); and Apoda or CAECILIANS.

amphibole Any of a large group of complex rock-forming minerals characterized by a double-chain silicate structure (Si_4O_{11}). They all contain water as OH^- ions and usually calcium, magnesium and iron. Found in IGNEOUS and METAMORPHIC ROCKS, they form wedge-shaped fragments on cleavage. Their orthorhombic or monoclinic crystals are often needle-like or fibrous. Common varieties are hornblende, tremolite, actinolite, and anthophyllite. Some varieties are used in commercial ASBESTOS.

Amphineura Class of MOLLUSKS that includes the CHITONS, small dwellers on rock surfaces. It also includes *Neopilina*, a small marine mollusk recently dredged from great depths, which is so primitive as to show some similarities to the ANNELID worms.

amphoteric In chemistry, term used to describe a substance that has both acidic and alkaline properties. Aluminum hydroxide is a typical amphoteric compound. It behaves like the base $Al(OH)_3$ when it reacts with acids (to form aluminum salts); and it behaves like the acid H_3AlO_3 when it reacts with alkalis (to form aluminates). Zinc hydroxide is also amphoteric. In organic chemistry, AMINO ACIDS can also be regarded as amphoteric because they contain both an acid $-COOH$ group and a basic $-NH_2$ group.

amplifier Device for changing the magnitude of a signal such as voltage, current or mechanical motion but not the way it varies. Amplifiers are used in RADIO and TELEVISION transmitters and receivers, and in audio equipment. The first electronic amplifiers used thermionic valves activated by heat, such as the triode (having three electrodes) invented in 1906 by the American physicist Lee De Forest. Valve amplifiers were once widely used but are now almost completely replaced by amplifiers that employ TRANSISTORS. *See also* THERMIONICS

amplitude modulation (AM) Form of RADIO transmission. Broadcasts on the short-, medium-, and long-wave bands are transmitted by amplitude modulation. The sound signals to be transmitted are superimposed on a constant-amplitude radio signal called the **carrier**. The resulting modulated radio signal varies in amplitude according to the strength of the sound signal. *See also* FREQUENCY MODULATION (FM)

ampulla Dilated end of a body duct or canal.

amu Abbreviation of ATOMIC MASS UNIT

amygdala Almond-shaped mass of nerve cells deep in the cerebral hemispheres of the BRAIN. It is concerned with emotion.

amyl alcohol ($C_5H_{11}OH$) ALCOHOL that has eight isomers. Commercial amyl alcohol, a mixture of these, has a sharp odor and is used as a solvent.

amylase Digestive enzyme secreted by the SALIVARY GLANDS (salivary amylase) and the PANCREAS (pancreatic amylase). It aids digestion by breaking down starch into MALTOSE (a disaccharide) and then GLUCOSE (a monosaccharide).

anabolic steroid Any of a group of HORMONES that stimulate the growth of TISSUE. Synthetic versions are used in medicine to treat OSTEOPOROSIS in older women and some types of ANEMIA; they may also be prescribed to aid weight gain in elderly, sick or severely ill patients. These drugs are associated with a number of side-effects, including acne, fluid retention, liver damage, and masculinization in women. Some athletes, such as sprinters and weight-lifters, have been known to abuse anabolic steroids in order to increase muscle bulk.

anabolism *See* METABOLISM

anacoustic zone (zone of silence) Region above 160km (100mi) altitude where distance between air molecules becomes greater than the wavelength of sound, so that no sound waves can be propagated. With increasing altitude, high-frequency (short-wavelength) sounds disappear first, and only lower tones can be heard.

anaerobe Minute organism that grows only in the absence of free atmospheric oxygen. Anaerobic BACTERIA can be a hazard in food canning because they can multiply in foods even under vacuum. *See also* AEROBE.

anaerobic Connected with the absence of oxygen or air or not dependent on oxygen or air for survival. An anaerobic organism, or ANAEROBE, is a microorganism that can survive by releasing energy from GLUCOSE and other foods in the absence of oxygen. The process by which it does so is called **anaerobic respiration.** Most anaerobes can survive in oxygen but do not need it for RESPIRATION. *See also* AEROBIC

analgesic DRUG that relieves or prevents pain without causing loss of consciousness. It does not cure the cause of the pain, but helps to deaden the sensation. Some analgesics are also NARCOTICS, and many also have valuable anti-inflammatory properties. Common analgesics include aspirin, codeine, and morphine. *See also* ANESTHESIA

analog computer Machine that processes continuously variable information. Information is usually first converted into proportional electrical quantities. These are manipulated by amplifiers and other circuits that perform various mathematical operations. In other words, the COMPUTER solves problems by dealing with quantities (voltages) that are analogous (similar) to the quantities in the problem. Analog computers are time-consuming to set up and operate. Most work once done on analog computers is now carried out on digital computers, which are simpler and quicker to use.

analog signal In telecommunications and electronics, transmission of information by means of variation in a continuous waveform. An analog signal varies (usually in amplitude or FREQUENCY) in direct proportion to the information content of the signal.

analysis In chemistry, any method of determining the composition of a substance. There are two main types of chemical analysis. **Qualitative** analysis is concerned with finding out what elements or compounds are present in a sample. It makes use of various specific chemical tests. **Quantitative** analysis is used to determine how much of a known substance is present in a sample. There are various quantitative techniques, including volumetric analysis (which depends on measuring volumes of reactants), gravimetric analysis (which relies on weighing), SPECTROSCOPY and spectrometry (which identify elements by their spectra), and CHROMATOGRAPHY.

analysis In mathematics, field that is concerned with the concepts of CONVERGENCE, CONTINUITY, DIFFERENTIATION, and INTEGRATION. **Real** analysis is the branch of mathematics where such ideas are applied to real NUMBERS or functions of real numbers. Similarly, **complex** analysis applies these concepts to complex numbers.

analytic geometry *See* COORDINATE GEOMETRY

anaphase Stage or stages in cell division. In MITOSIS, anaphase occurs when chromatids (from the CHROMOSOMES) separate and go to each end of the spindle. In MEIOSIS, there are two anaphases. In the first, paired chromosomes separate and move apart. In the second, as in mitosis the chromatids separate and move to the poles of the spindle.

anaphylaxis (anaphylactic shock) Acute, serious allergic reaction that occurs after a person eats, inhales, or is injected with a substance (ANTIGEN), such as penicillin, insect venom, or certain foods, to which they have been previously sensitized and against which they have developed ANTIBODIES. Second exposure to the offending substance can lead to paleness, faintness, palpitations, hives, swelling, difficulty in breathing, and, if untreated, SHOCK.

anatomy Branch of biological science concerned with the structure of an organism. The study of anatomy can be divided in several ways. On the basis of size, there is **gross** anatomy, which is studying structures with the naked eye; **microscopic** anatomy, studying finer detail with a light MICROSCOPE; **submicroscopic** anatomy, studying even finer structural detail with an electron microscope; and **molecular** anatomy, studying with sophisticated instruments the molecular makeup of an organism. Microscopic and submicroscopic anatomy involve two closely related sciences: HISTOLOGY, the study of tissue that makes up a body organ, and CYTOLOGY, the study of cells that make up tissue. Anatomy can also be classified according to the type of organism studied: plant, invertebrate, vertebrate or human anatomy. *See also* PHYSIOLOGY

andalusite One of many crystalline forms of aluminum silicate, occurring in contact metamorphic rock and in other deposits. It is mined commercially in the USA, Kazakhstan, and South Africa to make temperature-resistant and insulating porcelains.

Anderson, Carl David (1905–91) US physicist who shared the 1936 Nobel Prize for physics with Victor HESS. In 1932 he discovered the first known particle of antimatter, the POSITRON or anti-electron, while studying cosmic rays. He later also helped discover the MUON, an ELEMENTARY PARTICLE.

andesite Fine-grained, intermediate volcanic rock, found most frequently in recent or ancient continental margin areas. It is largely composed of finely crystalline FELDSPARS, with occasional larger crystals called **phenocrysts**. It is chemically and mineralogically similar to DIORITE.

androecium Male part of a flower. It comprises the STAMENS, each consisting of a two-lobed, pollen-carrying ANTHER on a thin stalk called a **filament**. *See also* GYNOECIUM

androgen General name for male sex HORMONES, such as TESTOSTERONE. They are produced mainly by the TESTES following the release of LUTEINIZING HORMONE, although the ADRENAL GLANDS and OVARIES produce much smaller amounts.

Andromeda Large constellation of the N sky, adjoining the Square of Pegasus. The main stars lie in a line leading away from Pegasus, and the star Alpha Andromedae actually forms one corner of the Square. In mythology, Andromeda was the princess rescued by Perseus. The most famous object in the constellation is the ANDROMEDA GALAXY.

Andromeda Galaxy Spiral GALAXY 2.2 million light-years away in the constellation Andromeda, the most distant object visible to the naked eye. The Andromeda Galaxy has a mass of over 300,000 million stars. To the naked eye, it appears as a smudge; through binoculars and small telescopes it is seen to have an elongated structure. Its true diameter is 150,000 light-years, somewhat larger than our own galaxy.

anechoic chamber (dead room) Room designed to be echo-free so that it can be used in acoustic laboratories to measure sound reflection and transmission, and to test audio equipment. The walls, floor and ceiling must be insulated and all surfaces covered with an absorbent material such as asbestos fiber or rubber, often over inward-pointing pyramid shapes to reduce reflection. The room is usually asymmetrical to reduce stationary waves. *See also* ACOUSTICS

anemia Condition in which there is a shortage of HEMOGLOBIN, the oxygen-carrying pigment contained in red blood cells. Symptoms include fatigue, weakness, pallor, breathlessness, faintness, palpitations, and lowered resistance to infection. It may be due to a decrease in the production of hemoglobin or red blood cells or excessive destruction of red blood cells or blood loss. There are several types of anemia. Worldwide, iron deficiency is the commonest cause of anemia. It usually results from poor nutrition, intestinal parasites, heavy periods or chronic bleeding from an ulcer or hemorrhoids. Pernicious anemia is an impairment of the body's ability to absorb vitamin B_{12}, a VITAMIN necessary for the normal development of red blood cells. Deficiencies in other substances, such as folic acid or other B vitamins, may also reduce red blood cell production. In hemolytic anemia there is excessive destruction of these cells. It may be due to hereditary abnormalities (such as SICKLE-CELL ANEMIA), adverse reaction to drugs, or the body's producing antibodies to its own red blood cells. Red blood cells are produced in the bone marrow and damage to the marrow because of radiation, drugs or chemicals, cancer, and so on, may also cause anemia.

anemometer Instrument for measuring wind speed. It consists of three cups attached to an upright length of metal, which in turn drives a mechanism that adjusts a dial. The wind blows the cups around in circles, and the dial shows its speed. More sophisticated instruments keep a record of wind speed and direction. An anemometer should ideally be located in an open space, well away from buildings and trees, and about 10m (30ft) above ground level.

anemophily Type of POLLINATION that relies on the wind to carry the pollen. Most grasses and many trees have anemophilous flowers, which lack scent and nectar, and have small or nonexistent petals. Male and female flowers are usually separate, as with the catkins of many trees, which bloom before the trees come into leaf.

aneroid barometer *See* BAROMETER

anesthesia State of unconsciousness or loss of sensation produced by disease or by various anesthetic drugs used during surgical procedures. During general, or total, anesthesia the patient loses consciousness and feels nothing; in local anesthesia only a specific part of the body is rendered insensate and the patient remains conscious. A general anesthetic may be either an injected drug, such as the barbiturate thiopentone, used to induce unconsciousness, or an inhalation agent such as halothane, which is used to maintain anesthesia for surgery. Local anesthetics such as linocaine numb the relevant part of the body by blocking the transmission of impulses through the sensory nerves that supply it.

anesthetics Branch of medicine involving the use of anesthetic agents to prevent pain during surgical and other procedures and care of the anesthetized patient.

Anfinsen, Christian Boehmer (1916–) US biochemist who in 1972 shared the Nobel Prize for chemistry with Stanford MOORE and William STEIN. Anfinsen carried out research into the relationships between the biological function and molecular structure of PROTEINS.

angiography X-RAY examination of major blood vessels. It is most commonly used to investigate CORONARY ARTERY DISEASE or any disruption of the blood supply to the brain. The technique requires the injection of a radiopaque dye into the bloodstream so that the diseased vessel appears in clear silhouette on the X-ray screen.

angioplasty Surgical repair of diseased or damaged blood vessels. It is a MINIMAL ACCESS SURGERY (keyhole) procedure, in which a balloon-tipped catheter is passed into a main ARTERY, often in the groin, and advanced until it comes to rest in the narrowed vessel; the balloon is then inflated to widen the artery. One of the best-known interventions of this kind, used to treat coronary heart disease, is percutaneous transluminal coronary angioplasty (PTCA), a method of improving blood flow in either of the two arteries supplying the heart muscle.

angiosperm Any of about 250,000 species of plant of the phylum Angiospermophyta, which produce FLOWERS, FRUITS, and SEEDS. Angiosperms include most herbs, shrubs, many trees, fruits, vegetables, and cereals. Their seeds are protected by an outer covering – angiosperm means "plant with enclosed seeds." Angiosperms are further subdivided into two classes, MONOCOTYLEDONS (Monocotyledonae) and DICOTYLEDONS (Dicotyledonae). Many flowering plants are used for food and timber and in medicine.

angiotensin PEPTIDE in the blood that increases blood pressure by inducing contraction of narrow blood vessels. *See also* RENIN.

angle Measure of the inclination of two straight lines or planes to one another; also a means of describing the amount of rotation. One complete revolution is divided into 360° (degrees) or 2π radians. A right angle measures 90° or $\pi/2$ radians. One degree may be subdivided into 60' (minutes), and one minute into 60" (seconds).

angle of incidence In optics, the angle at which a light ray strikes a mirror or other surface. The angle is measured between the ray and the normal – the line at right angles to the surface where the ray meets it. For a mirror, it is equal to the ANGLE OF REFLECTION.

angle of reflection In optics, the angle at which a light ray leaves a reflecting surface. The angle is measured between the ray and the normal – the line at right angles to the surface where the ray leaves it. For a mirror, it is equal to the ANGLE OF INCIDENCE.

angle of refraction In optics, the angle at which a light ray is refracted at the surface between two different transparent media. The angle is measured between the refracted ray and the normal – the line at right angles to the surface where the ray is refracted.

anglesite Sulfate mineral, a form of lead sulfate ($PbSO_4$). It is usually found together with cerussite in hydrothermal veins as an alternation product of galena. Its crystals are usually orthorhombic, and can be tabular or prismatic. It can be colorless, white or gray. Hardness 2.5–3; s.g. 6.4.

angstrom (angstrom unit, symbol Å) Obsolete unit of length, equal to 10^{-10} m or 0.1nm (NANOMETER). It was used to express the wavelength of light and ultraviolet radiation, and interatomic and intermolecular distances. The nanometer is now the preferred unit.

angular acceleration Rate of change of angular velocity. The average angular acceleration of an object whose angular velocity changes from ω_1 to ω_2 over a time t is $(\omega_2 - \omega_1)/t$. Instantaneous angular acceleration is the value approached as t becomes small. The direction of the angular acceleration VECTOR is perpendicular to the plane of motion. The tangential acceleration a_T of a particle at a distance r from a fixed point is $r\alpha$, where α is its angular acceleration.

angular diameter In astronomy, the apparent diameter of a celestial body expressed in angular measure (usually in arc minutes and seconds). It is the angle subtended at the observer by the true diameter of the body under observation. If the distance of the celestial body from the observer is known, its true diameter can be calculated.

angular distance In astronomy, the distance on the CELESTIAL SPHERE between two celestial bodies measured along an arc of a great circle passing through them with the observer at the centre. For example, the angular distance of the Pointers of Ursa Major is 5°.

angular momentum (symbol L) Product of the moment of inertia I and the angular velocity ω of an object or the center of gravity of a system of particles. Angular momentum is a VECTOR quantity that is conserved at all times.

angular velocity Rate of change of an object's angular position relative to a fixed point. Average angular VELOCITY ω of an object moving from angle θ_1 to θ_2 in time t is $(\theta_2 - \theta_1)/t$. Instantaneous angular velocity is the value approached by ω as t tends to zero. The direction associated with the angular velocity of an object is perpendicular to the plane of its motion. The speed v of an object at a distance r from a fixed point is directly proportional to the magnitude ω of its angular velocity: $v=r\omega$.

anhydride Chemical compound derived from another compound by removing water. Thus, sulfur trioxide (SO_3) is the anhydride of sulfuric acid (H_2SO_4).

anhydrite Mineral form of calcium sulfate ($CaSO_4$), usually found in sedimentary rocks in salt beds. Its crystals are orthorhombic, and usually occur in large deposits. It has a glassy or pearly luster and is colorless when pure. Hardness: 3–3.5; s.g. 3. *See also* GYPSUM

aniline (phenylamine, $C_6H_5NH_2$) Highly poisonous, colorless, oily liquid made by the reduction of nitrobenzene. Also called aminobenzene, it is an important starting material for making organic compounds such as drugs, explosives, and particularly dyes. Properties: r.d. 1.02; m.p. $-6.2°C$ (20.8°F); b.p. 184.1°C (363.4°F). *See also* AMINE

animal Living organism of the kingdom Animalia, usually distinguishable from members of the plant kingdom (Plantae) by its power of locomotion (at least during some stage of its existence); a well-defined body shape; limited growth; its feeding exclusively on organic matter; the production of two different kinds of sex cells; and the formation of an embryo or larva during the developmental stage. Higher animals, such as the VERTEBRATES, are easily distinguishable from plants, but the distinction becomes blurred with the lower forms. Some one-celled organisms could easily be assigned to either category. Scientists have classified about a million different kinds of animals in more than twenty PHYLA. The simplest (least highly evolved) animals include the PROTOZOA, SPONGES, jellyfish, and worms. Other invertebrate phyla include ARTHRO-PODS (arachnids, crustaceans, and insects), MOL-LUSKS (shellfish, octopus, and squid), and ECHINO-DERMS (sea urchins and starfish). Vertebrates belong to the CHORDATA phylum, which includes fish, amphibians, reptiles, birds, and mammals.

animal behavior *See* ETHOLOGY

animal classification Systematic grouping of animals into categories based on shared characteristics. The first major classification was drawn up by Aristotle. The method now used was devised by Carolus LINNAEUS, a Swedish botanist, in the 1750s. Each animal is given a two-part Latin name (*see* BINOMIAL NOMENCLATURE), the first part indicating its GENUS, the second its SPECIES. A species is composed of animals capable of interbreeding in nature. A genus includes all similar and related species. The FAMILY takes in all related genera, and an ORDER is made up of all related families. Similar orders are grouped in a CLASS, and related classes make up a PHYLUM. More than twenty separate phyla comprise the animal kingdom. For example, the dog is classified in phylum Chordata; class Mammalia; order Carnivora; family Canidae; genus *Canis* and species *familiaris*.

anion Negative ION that is attracted to the ANODE during electrolysis.

annealing Slow heating and cooling of a metal, alloy or glass to relieve internal stresses and make up crystal dislocations or vacancies that may have been introduced during mechanical shaping, such as rolling or extruding (ejection). Annealing increases the material's workability and durability. Machine tools, wire and sheet metal are annealed during manufacture. *See also* EXTRUSION; TEMPERING

annelid Member of the Annelida phylum of segmented worms. All have encircling grooves usually corresponding to internal partitions of the body. A digestive tube, nerves and blood vessels run through the entire body, but each segment has its own set of internal organs. Annelids form an important part of the diets of many animals. The three main classes are Polychaeta, marine worms; Oligochaeta, freshwater or terrestrial worms; and Hirudinea, leeches.

annihilation In particle physics, the complete conversion of matter into energy. An ELEMENTARY PARTICLE and its ANTIPARTICLE are converted, on collision, into gamma radiation. An electron and positron annihilate to produce two gamma-ray photons, traveling in opposite directions to conserve momentum. Each has an energy of 0.511MeV, which is equivalent to the rest mass of the electron or positron.

annual Plant that completes its life cycle in one growing season, such as sweet pea, sunflower, and wheat. Annual plants overwinter as SEEDS. *See also* BIENNIAL; PERENNIAL

annual parallax Angle subtended at a celestial object by the radius of the Earth's orbit, which is 1 astronomical unit.

annual ring (growth ring) Concentric circles visible in cross-sections of woody stems or trunks. Each year the CAMBIUM layer produces a layer of XYLEM, the vessels of which are large and thin-walled in the spring and smaller and thick-walled in the summer, creating the contrast between the rings. Used to determine the approximate age of trees, the thickness of these rings reveals the environmental conditions during a tree's lifetime.

anode Positive ELECTRODE of an electrolytic cell that attracts ANIONS during ELECTROLYSIS.

anodizing Electrolytic process to coat ALUMINUM or MAGNESIUM with a thin layer of oxide to help prevent corrosion. The process makes the metal the ANODE in an acid solution. The protective coating is insoluble and a good insulator. It can be dyed bright colors, many of which are resistant to sunlight.

anomalistic period Time taken for a celestial body to make one complete revolution around another, starting and finishing at the same point, such as PERIGEE or PERIHELION. It is slightly longer than the SIDEREAL PERIOD–27.55455 days for the Moon and 365.25964 days for the Earth.

anorexia nervosa Abnormal loss of the desire to eat. A pathological condition, it is seen mainly in young women anxious to lose weight. It can result in severe emaciation and may become life-threatening.

anorthosite Coarse-grained, basic IGNEOUS ROCK that forms deep underground and in DIKES and INTRUSIONS. It is composed mainly of FELDSPAR, with some olivine and pyroxine. Usually light in color, it may have bands of darker minerals.

ANS Abbreviation of AUTONOMIC NERVOUS SYSTEM

antagonist In biology, a muscle that acts in opposition to another muscle. For example, the biceps and triceps are an **antagonistic pair**. In medicine, a drug or hormone that acts to oppose the action of another drug or hormone. In each case, the opposing muscle, drug or hormone is called the agonist. The antagonist to a drug may also be termed an ANTIDOTE.

antenna (aerial) Part of a radio system from which the signal is radiated into space (transmitting antenna) or by which it is received from space (receiving antenna). The shape and structure of the antenna depend on the frequency of the radiation and the directional requirements of the system. For example, a beam antenna is required to transmit a radar beam while a simple wire may suffice for an AMPLITUDE-MODULATED (AM) radio receiver.

anthelion Phenomenon in which there appears to be a second Sun in the sky opposite the real Sun. The "ghost" Sun is at the same altitude as the real Sun and is caused by the refraction of sunlight by small crystals of ice in the atmosphere. In polar regions, the faint halo sometimes seen fringing the shadow of an object cast onto fog or a bank of cloud is also called an anthelion.

anther Element of a FLOWER that produces, contains, and distributes POLLEN (which contains the male GAMETES, or sex cells). The pollen is formed within two chambers called **pollen sacs**, and the chambers themselves are located in two lobes. Together with its connecting **filament**, the anther forms a STAMEN.

antheridium Male sex organ of fungi and various primitive plants. As well as in fungi, antheridia are found in algae, bryophytes (liverworts, and mosses), pteridophytes (club mosses, ferns, and horsetails), and cycads. They produce male gametes (sex cells) called **antherozoids**. These move independently in a film of water to the female gametes to achieve fertilization.

anthracene ($C_6H_4(CH)_2C_6H_4$) Organic chemical compound, the molecular structure of which consists of three BENZENE rings linked together. It is produced in the distillation of coal, and is used for the manufacture of dyes.

anthracite Form of COAL consisting of more than 90% CARBON, that is relatively hard, black, and with a metallic luster and subconchoidal fracture. It burns with the hot nonluminous flame of complete combustion. It is the final form in the series of fuels: PEAT, lignite, bituminous coal, and black coal.

anthrax Contagious disease, chiefly of livestock, caused by the microbe *Bacillus anthracis*. The earliest reports of the disease date back to about the first century BC, although it is likely to have existed before then. Human beings can catch anthrax from contact with infected animals or their hides. Symptoms include dark pustules on the skin and severe pneumonia. Treatment is with antibiotics. Since World War II, the disease has been harnessed as a potential biological weapon.

Anthropoidea Suborder of primates including monkeys, apes, and human beings. Anthropoids have flatter, more humanlike faces, larger brains, and are larger in size than prosimian PRIMATES.

antibiotic Substance that is capable of stopping the growth of, or destroying, BACTERIA, and other microorganisms. Many antibiotics are themselves produced by microorganisms (BACTERIA and MOLDS). Antibiotics are GERMICIDES that are safe enough to be swallowed or injected into the body. The introduction of antibiotics, from about the time of World War II, has revolutionized medical science, making possible the virtual elimination of once widespread and often fatal diseases, including typhoid fever, plague, and cholera. Some antibiotics are selective – that is, effective against specific microorganisms; those effective against a large number of microorganisms are known as **broad-spectrum** antibiotics. Some important antibiotics are: PENICILLIN, the first widely used antibiotic, STREPTOMYCIN, and the TETRACYCLINES and CEPHALOSPORINS. Because some bacteria, once sensitive to certain antibiotics, have now become resistant to them, there is a constant search for new antibiotics. *See also* ANTISEPTIC

antibiotic resistance Resistance to ANTIBIOTICS acquired by many BACTERIA and other PATHOGENS. Because they survive while nonresistant strains are killed, they pass their resistance to their progeny and resistance increases in the population. Some bacteria have the ability to pass the genes for antibiotic resistance to other organisms of different species on **plasmids** (small lengths of DNA). Spread of resistance is accelerated by routine prescription of antibiotics to humans and unregulated application to farm animals for the purpose of disease prevention rather than cure. An inadequate dose or failure to complete a course of antibiotics increases the chances of resistant microorganisms surviving to breed. This may lead to the return of epidemics of untreatable infectious diseases.

antibody PROTEIN synthesized in the blood by LYMPHOCYTES in response to the entry of "foreign" substances or organisms (ANTIGENS) into the body. Each episode of bacterial or viral infection prompts the production of a specific antibody to fight the antigens in question. After the infection has cleared, the antibody remains in the blood to fight off any future invasion. This process is a significant part of the body's IMMUNE SYSTEM.

anticline Arch-shaped fold in rock strata, closing upwards. Unless the formation has been overturned, the oldest rocks are found in the center with younger rocks symmetrically on each side of it.

anticoagulant Substance that prevents or counteracts coagulation, or clotting. Anticoagulants are used to treat and prevent diseases caused by blood clots, such as THROMBOSIS. Heparin is a commonly used blood anticoagulant.

anticyclone (high) An area of high pressure where air is subsiding, causing dry weather. Winds are normally light and the weather is settled. Major high-pressure regions are found in the HORSE LATITUDES, about 20° to 30° N and S of the Equator. There are also large anticyclones in the Arctic and Antarctic. The winds of anticyclones move in counterclockwise direction in the Southern Hemisphere, and a clockwise direction in the Northern Hemisphere. *See also* CYCLONE

antidepressant *See* DRUG

antidiuretic hormone (ADH) (vasopressin) HORMONE that controls the concentration of fluid in the body by stimulating the kidneys to absorb water and thereby produce less urine. ADH is secreted by the posterior part of the PITUITARY GLAND, although it is actually produced in the nearby HYPOTHALAMUS of the BRAIN. A deficiency of ADH causes the disorder diabetes insipidus, which is characterized by the production of large quantities of dilute urine.

antidote Term generally used to describe a drug or other agent used to counter the effects of a POISON. Antidotes may be either specific or general in their effect.

antifreeze Substance dissolved in a liquid to lower its freezing point. Ethane diol (ethylene glycol, HOC_2H_4OH) is commonly used in car radiators.

antigen Any substance or organism that is recognized as "foreign" by the IMMUNE SYSTEM. The presence of an antigen triggers the production of an ANTIBODY, part of the body's defense mechanism against disease. The antibody reacts specifically with the antigen and neutralizes it, causes it to destroy itself, or attracts LEUCOCYTES to carry out the destruction.

antihistamine Any one of certain drugs that counteract or otherwise prevent the effects of histamine, a natural substance released by the body in response to injury or more often as part of an allergic reaction. Histamine can produce symptoms such as sneezing, running nose, and burning eyes. *See also* HAY FEVER

antilogarithm *See* LOGARITHM

antimatter Matter made up of antiparticles, which are identical to ordinary particles in every way except that the ELECTRIC CHARGE, SPIN, and MAGNETIC MOMENT are reversed. When an antiparticle, such as a positron (anti-electron), antiproton or antineutron meets its respective particle, both are annihilated. It is possible, though unlikely, that there are stars or galaxies composed entirely of antimatter. *See also* SUBATOMIC PARTICLES

antimony (symbol Sb) Toxic, semimetallic element of group V of the periodic table. Stibnite (a sulfide) is its commonest ore. It is used in some alloys, particularly in hardening lead for batteries and type metal, and in semiconductors. The element has two allotropes: a silvery white metallic form and an amorphous gray form. Properties: at.no. 51; r.a.m. 121.75; r.d. 6.68; m.p. 630.5°C (1,166.9°F); b.p. 1,750 °C (3,182°F); most common isotope ^{121}Sb (57.25%).

antineutron In particle physics, the antiparticle of a NEUTRON. It has no electric charge, and a magnetic moment equal but opposite to that of the neutron. An antineutron is sometimes formed in a bubble chamber when a PROTON collides with an antiproton. *See also* ANTIMATTER

antinode Point of maximum displacement on a standing WAVE. A standing, or stationary, wave is one whose shape remains constant and that does not appear to move through a medium. It occurs when a traveling wave is reflected along its own path. Points of minimum displacement are called nodes. Antinodes and nodes are separated by a quarter of a wavelength.

antioxidant Chemical additive designed to reduce oxidation. Antioxidants are used to prevent fatty foods becoming rancid, deterioration of rubber, formation of gums in gasoline, etc. Most are organic amines and phenols, functioning by terminating the free-radical chain reaction causing oxidation. Sulfur dioxide and ascorbic acid are often used as antioxidants for foodstuffs.

antiparticle *See* ANTIMATTER

antiseptic Chemical that destroys or stops the growth of many microorganisms. Antiseptics are weak germicides that can be used on the skin. The surgeon Joseph LISTER pioneered the use of antiseptics in 1863. One commonly used is ALCOHOL. *See also* ANTIBIOTIC

antitoxin ANTIBODY produced by the body in response to a TOXIN. It is specific in action and neutralizes the toxin. Antitoxin sera are used to treat and prevent bacterial diseases such as TETANUS and DIPHTHERIA.

Antlia (Air Pump) Small constellation of the S sky. It has no star brighter than magnitude 4.3.

anus Opening at the end of the ALIMENTARY CANAL in most animals, through which waste material and undigested food are passed from the body in the form of feces. *See also* DIGESTIVE SYSTEM

aorta Principal ARTERY in the body of higher vertebrates. In humans, carrying freshly oxygenated blood, the aorta leaves the left ventricle of the HEART and descends the length of the trunk, finally dividing to form the two femoral arteries that serve the legs. *See also* CIRCULATORY SYSTEM

apatite Phosphate mineral, usually found as calcium phosphate associated with hydroxyl, chloride or fluoride ions. It occurs in igneous rocks and sedimentary deposits, as prismatic or tabular hexagonal crystals, as granular aggregates or in massive crusts. It is usually too soft for cutting and polishing, but there are two gem varieties. Hardness: 5; s.g. 3.1–3.4.

aperture In photography, a hole that allows light to pass through the LENS onto the film. Modern cameras usually have a diaphragm aperture which works like the iris of a human eye. The photographer can widen or narrow the diaphragm according to a series of points on the lens dial called "f-numbers" or "f-stops." The individual f-numbers represent the focal length of the lens divided by the diameter of the aperture. As the aperture narrows (larger f-number), it gives a longer depth of field.

apex In astronomy, point on the CELESTIAL SPHERE, located in the constellation of Hercules, towards which the Sun appears to be moving. As the Sun slowly orbits the galactic center, nearby stars (as seen from Earth) appear to move away from it because of the Sun's relative velocity.

aphelion Point at which the ORBIT of a celestial body that travels round the Sun is farthest from the Sun. All non-circular solar orbits, such as those of planets, asteroids, comets, and some meteors, have an aphelion, as do the orbits of space probes that orbit the Sun. The point at which they are closest to the Sun is called PERIHELION.

aphotic Describing a region in which there is no light. In the seas and oceans, the aphotic zone exists at depths below which light does not penetrate, generally below 1,500m (5,000ft).

apocrine gland Type of sweat-producing GLAND in human skin, restricted mainly to the armpits and groin. In response to sex and stress stimuli, it secretes a liquid that is quickly decomposed by bacteria and produces an unpleasant smell. *See also* ECCRINE GLAND

apogee Point in the Moon's or an artificial satellite's ORBIT at which it is farthest from Earth.

Apollonius of Perga (262–190 BC) Greek mathematician and astronomer. Known as "The Great Geometer" much of his work built on the foundations laid by EUCLID. In his only surviving book, *On Conic Sections*, he showed that an ELLIPSE, a PARABOLA, and a HYPERBOLA can be obtained by taking plane sections at different angles through a cone. In astronomy, he described the motion of the planets in terms of epicycles. (An epicycle is a path drawn by a point moving on a circle which, in turn, moves so that its center is on the circumference of a larger circle.) This was the basis of the Ptolemaic system used until the time of COPERNICUS.

Apollo program US space program that was set up with the specific aim of landing men on the Moon and returning them safely to Earth. Initiated in May 1961 by President John F. Kennedy, it achieved its objective on July 21, 1969, when Neil ARMSTRONG was the first human being to set foot on the lunar surface during the Apollo 11 mission. There were 17 Apollo missions, six of which included lunar landings. The program terminated with the successful Apollo-SOYUZ link-up in space during July 1975, having placed more than 30 ASTRONAUTS in space and 12 on the surface of the Moon.

appendix In some mammals, finger-shaped organ, about 10cm (4in) long, located near the junction of the small and large intestines, usually in the lower right part of the abdomen. It has no known function in humans but can become inflamed or infected causing appendicitis.

Appleton, Sir Edward Victor (1892–1965) British physicist. He was awarded the 1947 Nobel Prize for physics for his discovery of the Appleton layer of the IONOSPHERE. This layer reflects RADIO waves, and its discovery spurred the development of RADAR.

applied mathematics The use of mathematics to describe or model some real, idealized or hypothetical phenomena. It is a body of knowledge and theory with a mathematical structure that uses the formal rules of pure MATHEMATICS together with physical measurable quantities as input from the natural world. Mathematics has been used to represent, and make predictions about, almost every area of science over the last 3,000 years. Mathematics is practically applied in computing, electronics, engineering, meteorology, aeronautics, space flight, and so on, as well as in more theoretical areas such as MATHEMATICAL PHYSICS and MATHEMATICAL BIOLOGY.

apsis (pl. apsides) Either of two points in an object's ORBIT that are respectively closest to and farthest from the primary body, that which it is orbiting. The closest point is known as the **periapsis** and the farthest is known as the **apapsis**. The apsides of the Earth's orbit are its PERIHELION and APHELION; in the Moon's orbit they are its PERIGEE and APOGEE. The line of apsides is the line connecting these points.

Apus (Bird of Paradise) Far-S constellation. Its brightest star is magnitude 3.8.

aqualung *See* SCUBA DIVING

aquamarine *See* BERYL

aqua regia (Lat. royal water) Mixture of concentrated nitric and hydrochloric acids. It is so called because the acid dissolves the "royal" metal gold.

Aquarius (Water-bearer) Constellation of the S sky; it is situated between PISCES and CAPRICORNUS. Alpha and Beta are jointly the constellation's brightest stars, magnitude 2.9. It also contains two important planetary nebulae, Saturn Nebula and Helix Nebula.

aqueduct Artificial channel for conducting water from its source to its distribution point. While the ancient Romans were not the first to build these conduits, their aqueducts are the most famous because they not only conveyed water over large areas but were also graceful architectural structures in their own right. One of their most extensive water systems, which served Rome itself, consisted of 11 aqueducts and took 500 years to complete. California has the world's largest conduit system: it carries water into its cities over a distance of more than 800km (500mi).

aqueous humor Watery fluid that fills the eyeball between the lens and the CORNEA. It serves a double purpose: to nourish the eyeball and to refract, or bend, light rays, so helping them to focus on the RETINA. *See also* VITREOUS HUMOR

aqueous solution Solution in which water is the solvent. Water dissolves many polar substances, that is, IONIC COMPOUNDS and COVALENT compounds with molecular dipoles, because the water molecules themselves are polar, with a negative charge on the oxygen atoms. They tend to cluster around positive ions (solvation), making the solution energetically favorable.

aquifer Porous, permeable rock, often sandstone or limestone, that stores and transmits water. Much of the world's population obtains its water supply through ARTESIAN WELLS drilled into aquifers.

Aquilia (Eagle) Distinctive constellation of the N hemisphere led by the star Altair of first magnitude. The Milky Way is very rich in this region.

Ara (Altar) Constellation of the S sky, between Scorpius and Triangulum Australe. Its brightest star is Beta, magnitude 2.9.

Arabic numeral Any of the common notations 0 to 9. The system was brought to Europe by the Arabs, but its origin is in India. *See also* AL-KHWARIZMI, NUMBERS

arachnid ARTHROPOD (animal with an exoskeleton and jointed limbs) of the class Arachnida, which includes the SPIDER, tick, mite, scorpion, and daddy longlegs. Arachnids have four pairs of jointed legs, two distinct body segments (cephalothorax and abdomen), and chelicerate jaws (consisting of clawed pincers). They lack antennae and wings.

aragonite Carbonate mineral, calcium carbonate ($CaCO_3$). It is formed under special conditions, generally in caverns and hot springs. Present-day mollusk shells are formed of aragonite. It readily converts to calcite (another mineral of $CaCO_3$). It occurs as orthorhombic system groups of needle-like crystals or massive deposits. Its appearance is glassy white. Hardness 3.5–4; s.g. 2.9.

arc Portion of a curve. For a circle, of radius r, the length (s) of an arc covering θ degrees on that circle is given by
$$s = 2 \times r \times \pi \times \theta/360.$$
If the angle is measured in RADIANS the length of the arc is given by the product of the radius and the angle, that is $s = r\theta$. The arc is said to subtend the angle θ at the center of the circle.

arch Upward-pointing or curving arrangement of masonry blocks or other load-bearing materials, also used in architectural decoration. The ancient Romans invented traditional masonry arches but later cultures extended their repertoire to include many different and sometimes quite elaborate shapes. The basic structure of a masonry arch consists of wedge-shaped blocks ("voussoirs") placed on top of each other and a central **keystone** that holds them together at the top. Modern materials, such as steel, are strong and flexible enough to stand on their own and they can also stretch across much wider spans. The form of an arch is useful for historians who want to date a building.

Archaean Subdivision of Precambrian geological time, from about 4,000 to 2,500 million years ago.

Archaebacteria Subkingdom of the kingdom PROKARYOTAE, comprising BACTERIA that on the basis of both RNA and DNA composition and biochemistry differ significantly from other bacteria. They are thought to resemble the ancient bacteria that first arose in extreme environments such as hot springs, boiling muds, sulfur-rich deep-sea vents, volcanoes, and highly saline seashores, habitats in which they are still found today. Archaebacteria have unique proteinlike cell walls, unique cell membrane chemistry, and RIBOSOMES (microscopic structures that synthesize proteins) of a distinctive shape. They include methane-producing bacteria, which use simple organic compounds such as methanol and ethanoate as food, combining them with carbon dioxide and hydrogen gas from the air, and releasing methane as a by-product. The bacteria of hot springs and saline areas have a variety of ways of obtaining food and energy, including the use of minerals instead of organic compounds. They include both AEROBIC and ANAEROBIC bacteria. Some hot spring bacteria can tolerate temperatures of up to 88°C (190°F) and acidities as low as pH 0.9. One species, *Thermoplasma*, may be related to the ancestor of the nucleus and cytoplasm of the more advanced EUKARYOTE cells. Some taxonomists consider the Archaebacteria to be so different from other living organisms that they belong not even in their own separate kingdom, but in a higher grouping called a DOMAIN. *See also* TAXONOMY

archaeopteryx First known bird. About the size of a crow and fully feathered, its fossilized skeleton is more like that of a reptile than a modern bird, and its beak had pronounced jaws with teeth. It was probably capable only of weak flight.

archegonium Flask-shaped female sex organ of all plants except algae. Bryophytes (liverworts and mosses), pteridophytes (club mosses, ferns, and horsetails), and many GYMNOSPERMS (seed-bearing plants) have archegonia. Female gametes (ova) develop in the base of the flask. At fertilization the cells forming the "neck" of the flask disintegrate to allow male gametes (**antherozoids**) to enter.

Archimedes (*c*.287–212 BC) Greek mathematician and engineer. He developed a method for expressing large numbers and made outstanding discoveries about the determination of areas and volumes, which led to a new accurate method of calculating π (PI). *See also* ARCHIMEDES' PRINCIPLE; ARCHIMEDES' SCREW

▲ **archaeopteryx** The earliest known recognizable bird, archaeopteryx, dates from the upper Jurassic period. The presence of wings and feathers define it as a bird, but the skeleton is quite reptilian. The wings, instead of being the specialized flying limbs of modern birds, were really elongated forelimbs, complete with claws. The tail resembles a lizard's and the skull had teeth. The small breastbone shows it was a poor flyer.

Archimedes' principle Observation by ARCHIMEDES that a body immersed in a FLUID is pushed up by a force equal to the weight of the displaced fluid. He supposedly formulated this principle after stepping into a bath and watching it overflow. According to legend, he became so excited that he ran out naked shouting "Eureka! Eureka!" ("I have found it! I have found it!").

Archimedes' screw Machine used for raising water, thought to have been invented by Archimedes in the 3rd century BC. The most common form of the machine is a cylindrical pipe enclosing a helix, inclined at a 45° angle to the horizontal with its lower end in the water. When the machine rotates, water moves up through the pipe.

arc-lamp Device in which light is produced by an electric arc between two electrodes. Many modern arc-lamps, used as sources of intense light, have metal and oxide electrodes immersed in a gas that becomes luminous during arcing.

area Two-dimensional measurement of a plane, figure or body (such as this page) given in square units, such as cm^2 (sq cm) or m^2 (sq m). The area of a rectangle of sides a and b is ab; the areas of triangles and other polygons can be determined using TRIGONOMETRY. Areas of curved figures and surfaces can be determined by integral CALCULUS.

arenaceous Describing something that is associated with sand; most commonly used to describe types of SEDIMENTARY ROCK. In botany, an arenaceous plant is one that thrives in sandy soil; while in zoology, arenaceous animals live in sand. The shells of some microscopic animals, which consist mainly of sand particles, are also described as being arenaceous.

arête Sharp ridge formed by erosion where the heads of two GLACIERS meet. An example is the Matterhorn on the Swiss-Italian border.

argentite (silver glance) Mineral that is an important ore of silver. Argentite is a dark gray mineral that consists of silver sulfide (Ag$_2$S). It occurs in veins, often together with native silver, in parts of Germany, Mexico, and the USA.

argillaceous Describing something that consists of very fine grains, such as those associated with clay, shale or silt. Argillaceous rocks, such as kaolinite, are sedimentary rocks the particles of which are similar in size to those of clay or silt.

arginine H$_2$NC(NH)NH(CH$_2$)$_3$CH(NH$_2$)COOH) One of the essential AMINO ACIDS. It is a colorless, crystalline compound that occurs in decomposing animal and vegetable proteins, and in the muscles of invertebrates. It can also be made synthetically.

argon (symbol Ar) Monatomic (single-atom), colorless, and odorless gaseous element that is the most abundant NOBLE GAS (inert gas). Argon was discovered in 1894 in air by Lord RAYLEIGH and Sir William RAMSAY. It makes up 0.93% of the ATMOSPHERE by volume, and 99.6% of this is ^{40}Ar, with the remainder being ^{36}Ar (0.34%) and ^{38}Ar (0.06%). Obtained commercially by the fractionation of liquid air, it is used in electric LIGHT BULBS, fluorescent tubes, and argon LASERS, and in arc welding of aluminum and semiconductor production. The element has no known true compounds. Properties: at.no. 18; r.a.m. 39.948; r.d. 0.0017837gcm^{-3}; m.p. −189.4 °C (−308.9°F); b.p. −185.9 °C (−302.6°F).

argument In mathematics, the name given to the dependent variable. For instance, in the function $f(x) = x^2 + 3$, x is the argument.

Aries (Ram) Constellation in the N sky; it is situated between PISCES and TAURUS. Its brightest stars are Alpha, magnitude 2.0, and Beta, magnitude 2.6.

Aristarchus of Samos (310–230 BC) Greek mathematician and astronomer. He tried to calculate the distances of the Sun and Moon from the Earth, as well as their sizes, but although his method wassound the results were inaccurate. He is thought to be the first to propose that the Sun is the center of the Universe (**heliocentric theory**); the idea was not taken up because it did not seem to make the calculation of planetary positions any easier.

▲ **Archimedes' screw**
Thought to have been invented by Archimedes in the 3rd century BC, the Archimedes' screw (which turned inside a close-fitting, sloping pipe) was primarily used for raising water. However, in the 19th century the same principle was applied to propelling ships. During trials, however, this particular design of screw was broken in half, and the result was an immediate increase in the speed of the vessel. In this way, by happy accident, the science of marine propulsion advanced by 50 years.

▶ **Aristarchus of Samos**
The Greek astronomer Aristarchus measured the relative distances of the Sun and Moon. When the Moon is at its first quarter (A), the angle it makes with the Sun (C) is near 90°. By measuring the angle of the Earth (B) Aristarchus could determine from the triangle the relative distances from Earth. He found the angle to be 87°, instead of the true value of 89°52'; but a small error leads to a large discrepancy in the ratio of the distances of the Sun and Moon respectively from Earth. Aristarchus' ratio was 19:1; the true ratio is 370:1.

Aristotle (384–322 BC) Greek philosopher and encyclopedist. He developed a view of the Universe based on the four "elements" (earth, air, fire, and water) and the system of concentric spheres proposed by Eudoxus of Cnidus upon which various celestial bodies were carried. Aristotle added more spheres to account for the motion of all celestial bodies. The celestial bodies were, he said, made of a substance called ether, and are perfect and incorruptible, unlike the Earth. He attempted to estimate the size of the Earth, which he maintained was spherical, but unmoving – the center of the Universe. This "Aristotelian" view of the world remained almost unchallenged until COPERNICUS.

arithmetic Calculations using the operations of addition, subtraction, multiplication, and division. The procedures of arithmetic were put on a formal axiomatic basis by Guiseppe PEANO in the late 19th century. Using certain postulates, including that there is a unique natural number, 1, it is possible to give formal definition of the set of natural numbers and the arithmetical operations. Multiplication can be thought of as repeated additions: subtraction and division are the inverse operations of addition and multiplication. The operations can be extended to negative, rational, and irrational NUMBERS. *See* AXIOMATIC METHOD

arithmetic progression Sequence of NUMBERS in which each term is produced by adding a constant (the common difference d) to the preceding term. It has the general form $a, a + d, a + 2\,d,...$ and so on. The sum of such a progression
$$a + (a + d) + (a + 2\,d) + ...$$
is an arithmetic series. For n terms, the sum has a value $an + dn\,(n-1)/2$. An example of a finite arithmetic progression with 10 terms is the sequence 5, 7, 9, 11, ..., 21, 23. In this case $a = 5$, $d = 2$ and $n = 10$; the sum is 140.

arkose Comparatively young, medium-grained SEDIMENTARY ROCK resembling SANDSTONE, consisting of QUARTZ with up to 30% FELDSPAR. It forms in marine and freshwater sediments and continental deposits, usually from the breakdown of granite. It is pale gray to pinkish in color.

Arkwright, Sir Richard (1732–92) British inventor and industrialist. He introduced powered machinery to the textile industry with his water-driven frame for spinning; he started work on this machine in 1764 and completed it four years later, although it was not patented until the following year.

armature Central part of an ELECTRIC MOTOR or GENERATOR. In most machines the armature consists of several coils of wire on a spindle and rotates in a magnetic field. In an electric motor the armature rotates when an ELECTRIC CURRENT passes through it, thus providing the driving force of the motor.

armor Protection against weaponry. Soldiers' body armor (usually made of metal or thick leather) has been employed since antiquity to prevent injury. Except for helmets (and occasionally bulletproof vests) body armor ceased to be used in European battle in the 17th century. Warships were first armored in the mid-19th century, using metal plating to deflect small-arms fire. Vehicles acquired armor-plating during World War I. Armored vehicles include TANKS, armored personnel carriers, and mechanized ARTILLERY.

Armstrong, Neil Alden (1930–) US astronaut. He was chosen as a NATIONAL AERONAUTICS AND SPACE ADMINISTRATION (NASA) astronaut in 1962 and was the command pilot for the GEMINI 8 orbital flight in 1966. In July 1969, during the APOLLO 11 mission, he became the first man to step onto the Moon, remarking that it was "one small step for man, one giant leap for mankind."

aromatic compound Organic chemical compound that contains atoms of carbon (often six) joined to form a stable ring-shaped molecule, as seen in BENZENE. Some of these compounds have rings also containing nitrogen, oxygen or sulfur.

Arrhenius, Svante August (1859–1927) Swedish chemist and physicist. Arrhenius won the Nobel Prize for chemistry in 1903 for his theory of electrolytic dissociation. His later work was concerned with reaction rates, biochemistry, the structure of the Universe and the revelation that light exerts pressure.

arsenic (symbol As) Toxic, semimetallic element of group V of the periodic table, probably obtained in 1250. Compounds containing arsenic are used as a poison for rodents, insects, and weeds, and to harden lead and make semiconductors. Three allotropes are known: white arsenic, black arsenic, and a yellow, nonmetallic form. Properties: at.no. 33; r.a.m. 74.9216; r.d. 5.7; sublimes 613°C (1,135°F); most common isotope ^{75}As (100%).

arsenopyrite (iron arsenide-sulfide, FeAsS) Sulfide mineral, the major ore of ARSENIC. It is found with precious metal ores in high-temperature veins. It occurs as monoclinic system prismatic crystals or granular masses. It is a metallic white-gray. Hardness 5.5–6; s.g. 6.

arteriography Imaging technique for examining blood vessels. A dye opaque to X rays (radiopaque) is injected into the patient's bloodstream and the arteries are revealed in an X-ray photograph. The technique is particularly useful in detecting ARTERIOSCLEROSIS.

arteriosclerosis Blanket term for degenerative diseases of the arteries, in particular **atherosclero-**

▲ **arthropod** The most numerous invertebrates (animals without a backbone) are the arthropods (joint-legged animals), such as the centipede shown here. They owe their success to the exoskeleton that covers their bodies and allows the development of jointed limbs. Body segments are encased in a rigid protein cuticle (1) and body flexibility is permitted by an overlapping membrane (2). The strength of the exoskeleton ensures that muscles (3) can be anchored to the inside of the cuticle. Groups of muscles (4) are used to move the legs. Centipedes differ from millipedes in that they have one pair of legs attached to each segment rather than the millipede's two.

sis (hardening of the arteries). It is caused by deposits of fatty materials and scar tissue on the ARTERY walls that narrow the channel and restrict blood flow. There is usually an absence of overt symptoms until the disease is well advanced, when there is an increased risk of heart disease, stroke or gangrene. Evidence suggests that predisposition to the disease is hereditary and risks factors include cigarette smoking, inactivity, obesity, and a diet rich in animal fats and refined sugar. Treatment is by drugs and, in some cases, surgery to replace a diseased length of artery.

artery One of the BLOOD VESSELS that carry BLOOD away from the HEART. The pulmonary artery carries deoxygenated blood to the lungs, but all other arteries carry oxygenated blood to the body's tissues. An artery is usually protected and embedded in muscle; its walls are thick, elastic, and muscular and pulsate with the heartbeat. A severed artery causes major HEMORRHAGE.

artesian well Well from which water is forced out naturally under pressure. Artesian wells are bored where water in a layer of porous rock is sandwiched between two inclined layers of impervious rock. The water-filled layer is called an AQUIFER. Water flows up to the surface because parts of the aquifer are higher than the wellhead and the water is under great pressure.

arthritis Inflammation of the joints, with pain and restricted mobility. The most common forms are osteoarthritis and rheumatoid arthritis. **Osteoarthritis**, common among the elderly, occurs with erosion of joint cartilage and degenerative changes in the underlying bone. It is treated with analgesics and anti-inflammatories and, in some cases (especially a diseased hip), by joint replacement surgery. **Rheumatoid arthritis**, more common in women, is generally more disabling. It is an AUTOIMMUNE DISEASE that may disappear by itself but is usually slowly progressive. Treatment includes analgesics to relieve pain. The severest cases may need to be treated with CORTISONE injections, drugs to suppress immune activity or joint replacement surgery. *See also* RHEUMATISM

arthropod Any member of the largest animal phylum, Arthropoda. Living forms include CRUSTACEA, ARACHNIDS, centipedes, millipedes, and INSECTS. Fossil forms include the extinct TRILOBITE. The species (numbering well over 1 million) are thought to have evolved from ANNELIDS. All have a hard outer skin of CHITIN that is attached to the muscular system on the inside, and is periodically shed during growth. The body is divided into segments, modified among different groups, with each segment originally carrying a pair of walking or swimming jointed legs. In some animals some of the legs have evolved into jaws, sucking organs or weapons. Arthropods have well-developed digestive, circulatory, and nervous systems. Land forms use tracheae for respiration.

artificial insemination Method of inducing PREGNANCY without sexual intercourse by injecting SPERM into the female genital tract. Artificial insemination of livestock allows proven sires to impregnate many females at low cost.

artificial intelligence (AI) Science concerned with developing COMPUTERS and COMPUTER PROGRAMS that model human intelligence. The most common form of AI involves a computer being programmed to answer questions on a specialized subject. Such "expert systems" are said to display the human ability to perform expert analytical tasks. A similar system in a WORD PROCESSOR may highlight incorrect spellings, and be "taught" new words. A closely related science, sometimes known as "artificial life," is concerned with more low-level intelligence. For example, a ROBOT may be programmed to find its way around a maze, displaying the basic ability to interact physically with its surroundings.

artificial selection Breeding of plants, animals or other organisms in which the parents are individually selected by humans in order to perpetuate certain desired traits and eliminate others from the captive population. By this means, most of our domestic crops, livestock, and pets have arisen. The many breeds of dog have been developed by artificial selection. Crops such as wheat have been produced by artificial selection over thousands of years. Artificial selection can be accelerated today by techniques such as plant TISSUE CULTURE, in which selected plants can be cloned and grown directly into new plants without having to wait for seed production and germination; and ARTIFICIAL INSEMINATION of livestock. In the near future the cloning of livestock embryos may also be possible. GENETIC ENGINEERING is the ultimate advance in artificial selection, allowing humans to combine specific sequences of DNA to combine desired GENES. See also CLONE

artillery Projectile-firing weapons with a carriage or mount. An artillery piece is generally one of four types: gun, howitzer, mortar or missile launcher. Modern artillery is classified according to caliber; ranging from under 105mm for light artillery to more than 155mm for heavy. The exact origin of artillery using gunpowder is unknown, although such weapons appeared in Europe and the Near East in the 14th century. Used by the Turks to take Constantinople in 1453, artillery

changed the whole strategy and tactics of siege warfare. Advances in the 19th century included smokeless powder, elongated shells, rifling, and rapid-fire breach loading and made artillery indispensable in battle. See also CANNON

Artiodactyla Order of mammals characterized by hooves with an even number (2 or 4) of toes. All herbivores, the order includes giraffes, hippopotamuses, deer, cattle, pigs, sheep, goats, and camels. They are mostly of Old World origin and range in size from the 3.6kg (8lb) mouse deer to the 4.5 ton hippopotamus. See also PERISSODACTYLA

aryl group Organic chemical group consisting of an AROMATIC COMPOUND with one hydrogen atom removed. For example, BENZENE (C_6H_6) gives rise to the phenyl group (C_6H_5 –).

asbestos Group of fibrous, naturally occurring silicate minerals used in insulating, fireproofing, brake linings, and astronaut suits. Asbestos is used in the form of wool, fabric, and various asbestos-cement compounds. Several types exist, the most important being white asbestos. Many countries have banned the use of asbestos because it can cause lung cancer and asbestosis, a lung disease.

Ascomycetes Largest group of the four main classes of FUNGUS, characterized by the formation of asci (enlarged, commonly elongated, cells in which usually eight spores are formed). Examples are yeasts, truffles, and blue and green molds.

ascorbic acid See VITAMIN

asdic See SONAR

asepsis Absence of disease-causing organisms. Aseptic technique is practiced to exclude bacteria during surgery, wound dressing or other medical procedures. In an operating room, clothing, gloves, equipment, and even the air are sterilized.

asexual reproduction Type of reproduction in organisms that does not involve the union of male and female reproductive cells. It occurs in several forms: FISSION – simple division of a single individual, as in bacteria and protozoa; BUDDING – growing out and eventual splitting off of a new individual, as in hydra; spore formation, as in fungi; VEGETATIVE REPRODUCTION – in which a plant sends out runners that take root to form new plants, as in strawberries, or new plants grow from organs such as bulbs. See also CLONE; SEXUAL REPRODUCTION

ash, volcanic Fine particles of LAVA thrown up by a volcanic explosion. The cone of compound volcanoes consists of built-up layers of ash and lava.

asparagine (NH$_2$COCH$_2$CH(NH$_2$)COOH) Crystalline AMINO ACID that occurs in proteins. It was the first amino acid to be discovered (in 1806 in asparagus).

aspartame Artificial sweetener. Consisting chemically of a compound that has two AMINO ACIDS (aspartic acid and phenylalanine) joined by a methylene group, aspartame is nonfattening and much used in diet foods and drinks, and in foods prepared for diabetics. It is 200 times as sweet as sugar and has none of SACCHARIN's disadvantages. It breaks down, however, at high temperatures and is therefore unsuitable for prolonged cooking.

aspartic acid (HOOCCH$_2$CH(NH$_2$)COOH) AMINO ACID, it is formed by the hydrolysis of another amino acid ASPARAGINE.

asphalt *See* BITUMEN

aspirin (acetylsalicylic acid) DRUG widely used to reduce fever, and as an ANALGESIC to relieve minor pain. It is effective against inflammation, as caused by ARTHRITIS, and low-intensity pain such as headache or muscular aches. Recent evidence indicates aspirin can inhibit the formation of blood clots and in low doses can reduce the danger of heart attack and stroke. Aspirin, however, can irritate the stomach and in overdose is toxic and can cause death. It should not be given to children under the age of 12 years.

assay Test to determine the amount of a metal present in a sample of material such as ores and alloys. The term is normally reserved for testing for gold, silver, or platinum.

assembly language COMPUTER LANGUAGE for writing COMPUTER PROGRAMS in a form that is closely related to the form that computers can understand directly. Assembly language is a **low-level language**. Each instruction to be carried out by the computer is represented by a simple code. Programs written using these codes are translated by an "assembler" into a form the computer can understand.

assimilation Process by which an organism uses substances taken in from its surroundings to make new living protoplasm or to provide energy for metabolic processes. It includes the incorporation of the products of food digestion into living tissues in animals, and the synthesis of new organic material by plants during photosynthesis.

associative law Rule of combination in MATHEMATICS in which the result of two or more operations does not depend on the order in which the operations are carried out. Thus, normal ADDITION and MULTIPLICATION of NUMBERS are operations that follow the associative law, since

$$a + (b + c) = (a + b) + c$$

and

$$a \times (b \times c) = (a \times b) \times c.$$

astatine (symbol At) Semimetallic, radioactive element that is one of the HALOGENS (group VII of the periodic table). It is rare in nature, and is found in radioactive decay. Astatine-211 will collect in the THYROID GLAND and is used in medicine as a radioactive tracer. Properties: at.no. 85; r.a.m. 211; m.p. 302°C (575.6°F); b.p. 377°C (710.6°F); most stable isotope ^{210}At (half-life 8.3hr).

asteroid Small body in an independent orbit around the Sun. The majority move between the orbits of Mars and Jupiter, in the main asteroid belt. The largest asteroid (and the first to be discovered) was CERES. The smallest measured are the so-called near-Earth asteroids. There are thought to be a million asteroids with a diameter greater than 1km (0.6mi); below this, they decrease in size to dust particles. Some very small objects find their way to Earth as METEORITES. So far nearly 6,000 asteroids have been catalogued and have had their orbits calculated, and this figure is increasing by several hundred a year. At least 10,000 more have been observed, but not often enough for an orbit to be calculated. Some of the larger asteroids are spherical, but most are irregularly shaped, and a wide variety of compositional types have been identified. Asteroids almost certainly originate from the time of the formation of the SOLAR SYSTEM and are not remnants of a large planet that disintegrated, as was once thought.

asthenosphere Zone within the Earth's upper MANTLE, extending from 100 to 700km (60–450mi) below the surface. It is detected by a slowing of seismic waves passing through the mantle.

asthma Disorder of the respiratory system in which the bronchi (air passages) of the LUNGS go into spasm, making breathing difficult. It can be triggered by infection, air pollution, allergy, certain drugs, exertion or emotional stress. The chest feels tight and breathing is difficult, with coughing and a characteristic wheeze. **Allergic** asthma may be treated by injections aimed at lessening sensitivity to specific allergens. Otherwise treatment is with bronchodilators to relax the bronchial muscles and ease breathing; in severe asthma inhaled steroids may be given. Children often outgrow asthma, while some people suddenly acquire the disease in middle age. In the acute form, the attacks are usually severe and follow longish periods of otherwise good health. The chronic condition is marked less by severe attacks than by a general breathing impairment. *See also* BRONCHITIS; EMPHYSEMA

astigmatism Defect of vision in which the curvature of the LENS differs from one perpendicular plane to another. It can be compensated for by use of corrective lenses.

Aston, Francis William (1877–1945) British physicist awarded the 1922 Nobel Prize for chemistry for his work on ISOTOPES. He developed the MASS SPECTROGRAPH that separates isotopes, and he used it to identify 212 naturally occurring isotopes.

astringent Drug or lotion that contracts the tissues or restricts fluid secretion. Astringents are used in a number of preparations to protect the skin and promote healing.

astrobiology *See* EXOBIOLOGY

astrodynamics Application of CELESTIAL MECHANICS, BALLISTICS, mathematical PERTURBATION theory, and the principle of observation reduction to determine, predict, and correct orbits and trajectories in space.

astrogeology Study of rocks, craters, and other surface features of the Moon, Mars, and other planets. It includes on-site and laboratory studies of Moon rocks, seismological data (Moon only), photographic mappings, magnetic field measurements, and micrometeorite data.

astrolabe Early astronomical instrument for showing the appearance of the CELESTIAL SPHERE at a given moment and for determining the altitude of celestial bodies. The basic form consisted of two concentric disks, one with a star map and one with a scale of angles around its rim, joined and pivoted at their centers (much like a modern planisphere), with a sighting device attached. Astrolabes were used from the time of the ancient Greeks through the 17th century for navigation, measuring time, and terrestrial measurement of height and angles.

astronaut (Rus. *cosmonaut*) Person who navigates or rides in a space vehicle, or is trained to do so. The first man to orbit the Earth was Yuri GAGARIN in 1961. The first man to walk on the Moon was Neil ARMSTRONG in 1969. The first woman in space was Valentina Tereshkova of Russia, in 1963.

astronautics (astronautical engineering) Practical study of the principles of space flight. Astro-

nautics includes ASTRODYNAMICS, propulsion theory, ASTROGEOLOGY, EXOBIOLOGY, communications principles, and the design, materials analysis, guidance, and control of spacecraft.

astronomical telescope *See* TELESCOPE

astronomical unit (symbol AU) Mean distance between the Earth and the Sun, used as a fundamental unit of distance, particularly for distances in the Solar System. 1AU is equal to 149,598,000km (92,956,000mi).

▲ **astronaut** Astronauts on NASA's shuttle use spacesuits (1) that allow the crew to work in space for up to seven hours. The suit is multilayered with eight materials combined. The outside is treated with nylon to stop damage from tiny meteorites. Four layers of aluminum material then provide a heat shield from solar radiation backed by a fire- and tear-resistant layer. The astronaut is protected from the vacuum of space by a pressure suit of nylon coated with polyurethane and is kept comfortable in extremes of heat and cold by water pumped through a network of tubes in a nylon chiffon undergarment. The MMU (manned maneuvering unit) (2) allows an astronaut to move away from the shuttle. Power comes from 24 thrusters arranged at the corners of the MMU. By releasing pressurized nitrogen from two tanks (3) through nozzles, the astronaut can propel himself or herself through the vacuum. The hand controllers regulate rotation (4) and speed (5). A video camera (6) sends pictures to the shuttle and records the work carried out.

astronomy Branch of science studied since ancient times and concerned with the Universe and its components in terms of the relative motions of celestial bodies, their positions on the celestial sphere, physical and chemical structure, evolution and the phenomena occurring on them. It includes CELESTIAL MECHANICS, ASTROPHYSICS, COSMOLOGY, and astrometry (the precise measurement of celestial bodies). Waves in all regions of the ELECTROMAGNETIC SPECTRUM can now be studied either with ground-based instruments or, where no atmospheric window exists, by observations and measurements made from satellites, space probes, and rockets.

History Astronomy was first used practically to develop a calendar, the units of which were determined by observing the heavens. The Chinese had a calendar in the 14th century BC. The Greeks developed astronomy significantly between 600 BC and AD 200. THALES introduced geometrical ideas and PYTHAGORAS saw the Universe as a series of concentric spheres. ARISTOTLE believed the Earth to be stationary but he explained lunar eclipses correctly. ARISTARCHUS put forward a heliocentric theory. HIPPARCHUS used trigonometry to determine astronomical distances. The system devised by PTOLEMY was a geometrical representation of the SOLAR SYSTEM that predicted the motions of the planets with great accuracy. From then on astronomy remained dormant until the scientific revolution of the 16th and 17th centuries, when COPERNICUS stated his theory that the Earth rotates on its axis and, with all the other planets, revolves around the Sun, which had a profound effect upon the religion and philosophy of the day. KEPLER and his laws of planetary motion refined the theory of heliocentric motion, and his contemporary, GALILEO, made use of the TELESCOPE and discovered the moons of Jupiter. Isaac NEWTON combined the sciences of astronomy and physics. His laws of motion and universal theory of GRAVITATION provided a physical basis for Kepler's laws and the work of many astronomers from then on – such as, in the prediction of HALLEY'S COMET and the discovery of the planets Uranus, Neptune, and Pluto. By the early 19th century the science of celestial mechanics, the study of the motions of bodies in space as they move under the influence of their mutual gravitation, had become highly advanced and new mathematical techniques permitted the solution of the remaining problems of classical gravitation theory as applied to the Solar System. In the second half of the 19th century astronomy was revolutionized by the introduction of techniques based on photography and SPECTROSCOPY. These encouraged investigation into the physical composition of stars, rather than their position. Ejnar HERTZSPRUNG and Henry RUSSELL studied the relationship between the color of a star and its luminosity. By this time larger telescopes were being constructed, which extended the limits of the known Universe. Harlow SHAPLEY determined the shape and size of our galaxy and Edwin HUBBLE's study of distant galaxies led to the theory of an expanding Universe. The BIG BANG and STEADY-STATE THEORY of the origins of the Universe were formulated. In recent years space exploration and observation in different parts of the electromagnetic spectrum have contributed to the discovery and postulation of such phenomena as the QUASAR, PULSAR, and BLACK HOLE. There are various branches of modern astronomy:

Gamma-ray astronomy is concerned with the study of gamma rays, which cannot penetrate the Earth's atmosphere and must be studied from satellites. The Sun is the principal source.

Infrared astronomy is concerned with detecting infrared (IR) waves, determining their source and studying their spectra. Most IR radiation is absorbed by the Earth's atmosphere. The Sun and the center of the galaxy are sources of IR radiation.

Optical astronomy is the oldest branch, and studies sources of light in space. Light rays can penetrate the atmosphere but, because of disturbances, many observations are now made from above the atmosphere, as in the case of the HUBBLE SPACE TELESCOPE.

Radar astronomy is used to determine distance, orbital motion, and surface features of objects in the Solar System. Radar pulses are bounced back to Earth from the object.

Radio astronomy uses radio telescopes to detect radio waves from space and determine their source and energy spectrum. The most powerful sources include the Sun, interstellar clouds of hot hydrogen, such as the Orion Nebula, supernova remnants, and pulsars, like the Crab Nebula, quasars, and radio galaxies.

Ultraviolet astronomy studies ultraviolet (UV) waves from space, their source, and spectra. Higher wavelengths can be studied from the ground but lower wavelengths must be studied from satellites and balloons. Sources include the Sun, the Orion Nebula, and Wolf-Rayet stars.

X-ray astronomy is concerned with studying X-ray sources. X rays are absorbed by the Earth's atmosphere so detecting instruments must be carried in satellites. Sources include the Sun, the Crab Nebula, Cygnus A, and the galaxy M87.

astrophysics Branch of ASTRONOMY that studies the physical and chemical nature of celestial bodies and their evolution. Many branches of physics, including nuclear physics, plasma physics, relativity, and spectroscopy, are used to predict properties of stars, planets, and other celestial bodies. Astrophysicists also interpret the information obtained from astronomical studies of the electromagnetic spectrum, including light, X rays, and radio waves.

asynchronous transmission Fastest method of program execution in computers; the beginning of each operation is signalled by the stop bit at the end of the previous one.

atavism Reversion by an organism to a characteristic of its ancestors after an interval of at least one generation in which the trait was absent. The term is no longer in scientific use since the reappearance of ancestral traits is now understood to be the expression of RECESSIVE genes.

atmolysis Process of separating mixed gases by allowing them to pass through a porous material.

atmosphere Envelope of gases surrounding the Earth. It shields the planet from the harsh environment of space, while the gases it contains are vital to life. About 95% by weight of the Earth's atmosphere lies below 25km (15mi) altitude; the mixture of gases in the lower atmosphere is commonly called air. The atmosphere's composition by weight is: nitrogen 78.09%, oxygen 20.9%, argon 0.93%, 0.03% of carbon dioxide, plus 0.05% of hydrogen, other gases, and varying amounts of water vapor. The atmosphere can be seen as concentric shells; the innermost is the **troposphere**, in which dust and water vapor create the clouds and weather. The **stratosphere** extends from 10–55km (8–36mi) and is cooler and clearer and contains ozone. Above, to a height of 70km (43mi), is the **mesosphere** in which chemical reactions occur, powered by sunlight. The temperature climbs steadily in the **thermosphere**, which gives way to the **exosphere** at about 400km (250mi), where helium and hydrogen may be lost into space. The **ionosphere** ranges from about 50km (30mi) out into the VAN ALLEN RADIATION BELTS.

atmospheric pressure Downward force exerted by the atmosphere because of its weight (gravitational attraction to the Earth or other body), measured by barometers and usually expressed in units of millibars (mb). Standard atmospheric pressure at sea level is 1013.25mb. The column of air above 1cm^2 (0.15in^2) of Earth's surface weighs about 1kg (2.2lbs).

atoll Ring-shaped REEF of CORAL enclosing a shallow LAGOON. An atoll begins as a fringing reef surrounding a slowly subsiding island, usually volcanic. As the island sinks the coral continues to grow upwards until eventually the island is below sea level and only a ring of coral is left at the surface.

atom Smallest particle of matter that can take part in a chemical reaction, every element having its own characteristic atoms. The atom, once thought indivisible, consists of a central, positively charged NUCLEUS orbited by negatively charged ELECTRONS. The nucleus (identified in 1911 by Ernest RUTHERFORD) is composed of tightly packed PROTONS and NEUTRONS. It occupies a small fraction of the atomic space but accounts for almost all of the mass of the atom. In 1913 Niels BOHR suggested that electrons moved in fixed orbits. The study of QUANTUM MECHANICS has since modified the concept of orbits: the Heisenberg UNCERTAINTY PRINCIPLE says it is impossible to know the exact position and MOMENTUM of an ELEMENTARY PARTICLE. The number of electrons in an atom and their configuration determine its

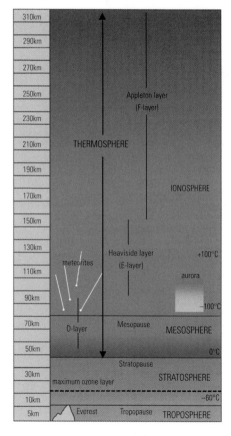

▲ **atmosphere** The Earth's atmosphere is formed of layers. It is believed that the atmosphere has changed three times during the Earth's history. The present atmosphere consists mainly of nitrogen and oxygen.

model of an atom

uranium

hydrogen

nucleus — hydrogen
— proton

electron

◀ **atom** The mass of an atom depends on the size of the nucleus. The nucleus of an atom makes up the great majority of the atom's mass, as electrons weigh hardly anything. A uranium atom, for example, has the greatest mass of all naturally occurring atoms. It has 146 neutrons, 92 protons, and 92 electrons. A hydrogen atom, on the other hand, has the least mass. It has only 1 proton and 1 electron. However, while a uranium atom has a mass almost 240 times greater than that of a hydrogen atom, it is only 3 times larger.

chemical properties. Adding or removing one or more electrons produces an ION.

atomic battery (nuclear battery) Device that converts energy from radioactive particles into electric current. Common sources include tritium (T), Hydrogen–3, and Krypton–85, both emitters of beta particles.

atomic bomb *See* NUCLEAR WEAPON

atomic clock Most accurate type of CLOCK. It is an electric clock regulated by such natural periodic phenomena as emitted radiation or atomic vibration, with the atoms of the metal cesium most commonly used. Clocks that run on the radiation from hydrogen atoms lose no more than one second in about 1,700,000 years.

atomic energy *See* NUCLEAR ENERGY

atomic force microscope (AFM) Type of microscope that depends on the forces between atoms to produce an image of a surface. A spring-loaded arm holds a tiny piece of diamond in contact with the surface while it scans across it. The probe is lowered or raised to keep the tracking force constant, and the necessary movements are analyzed by a computer to produce a "contour map" of the surface displayed on the computer screen. It can scan nonconducting biological specimens, unlike the SCANNING TUNNELLING MICROSCOPE which works only with metallic samples.

atomicity Number of atoms of an element in a particular molecule. For example, neon (Ne) has an atomicity of 1, nitrogen (N_2) has an atomicity

of 2, ozone (O_3) has an atomicity of 3 and certain forms of sulfur (S_8) have an atomicity of 8.

atomic mass number *See* MASS NUMBER

atomic mass unit (symbol u) Unit of mass used to compare RELATIVE ATOMIC MASS (R.A.M.), defined since 1961 as 1/12th the mass of the most abundant isotope of carbon, carbon-12 (which has 6 electrons, 6 protons, and 6 neutrons). It is equal to 1.66033×10^{-27} kg.

atomic number (proton number, symbol Z) Number of protons in the nucleus of an atom of an element, which is equal to the number of electrons moving around that nucleus. It is abbreviated to at.no. It is signified by a subscript number before the symbol, for example, the atomic number for carbon is written $_6C$. The atomic number determines the chemical properties of an element and its position in the periodic table. ISOTOPES of an element all have the same atomic number but a different atomic MASS NUMBER because they have different numbers of neutrons.

atomic spectrum Spectrum of sharp lines characteristic of the element involved and produced by the radiation emitted when electrons jump between energy levels of the ATOM.

atomic volume Quantity arrived at by dividing the RELATIVE ATOMIC MASS (R.A.M.) of an element by its density.

atomic weight Former term for RELATIVE ATOMIC MASS (R.A.M.)

ATP Abbreviation of ADENOSINE TRIPHOSPHATE

atrium (auricle) Either of the two upper chambers of the four-chambered HEART. They are comparatively thin-walled since they only pump blood down into the muscular VENTRICLES of the heart. The term is also used for various other chambers in animals.

atropine ($C_{17}H_{23}NO_3N$) Poisonous ALKALOID drug obtained from certain plants such as *Atropa belladonna* (deadly nightshade). Atropine is used medicinally as an antispasmodic, to dry secretions and regularize the heartbeat during anesthesia, to dilate the pupil of the eye, and to treat motion sickness.

Attenborough, Sir David Frederick (1926–) British naturalist and broadcaster; brother of Sir Richard Attenborough. He was controller of BBC2 television (1965–68). Since 1954, he has travelled on zoological and ethnographical filming expeditions, which have formed the basis of such landmark natural history series as *Life on Earth* (1979), *The Living Planet* (1984), *The Trials of Life* (1990), *The Private Life of Plants* (1995), and *The Life of Birds* (1998). He was knighted in 1985.

attitude Name given to the orientation, usually of ships, aircraft, and spacecraft, with respect to a given set of axes. Three variables specify attitude: **pitch**, the angle with the horizontal made by the long axis of the craft; **yaw**, the angle through which the bow turns left or right of the "forward" direction; and **roll**, the angle of rotation about the long axis.

attrition In geology, a form of EROSION in which the particles produced by the erosive process go on to erode each other. The process may take place in water, where sand particles in the river's load erode each other, or where pebbles on a beach tumble against each other. Eventually particles affected by the process of attrition become completely rounded.

audiometry Technique to evaluate hearing ability. An audiometer is an instrument that measures the sensitivity of the ear to sounds; it is used in the assessment of hearing loss.

audio-visual aid Use of sound and pictures to assist learning. The use of film, television, and computers, together with the ability to record, select, and replay, has made it possible to develop understanding and knowledge more quickly than was possible using only the spoken and written word.

auditory canal Tube leading from the outer EAR to the eardrum. It is about 2.5cm (1in) long.

auditory nerve Bundle of nerves that carry impulses related to hearing and balance from the inner EAR to the brain.

auditory ossicles Three small bones in the middle EAR that transmit vibrations from the eardrum to the cochlea in the inner ear.

Auger, Pierre Victor (1899–1993) French physicist who discovered the **Auger effect**, a change in an atom from an excited to a lower energy state with the emission of an (Auger) electron but without radiation. He became director general of the European Space Research Organization (ESRO) in 1962.

augite *See* PYROXENES

Auriga (Charioteer) Large constellation of the N sky, containing the first-magnitude star Capella. It also contains the eclipsing binary Epsilon Aurigae.

aurora Sporadic display of colored light in the night sky, usually green, caused by charged particles from SOLAR FLARES interacting with atoms and air molecules in the Earth's upper ATMOSPHERE. The charged particles from the Sun are attracted by the Earth's magnetic field into zones called VAN ALLEN RADIATION BELTS. Auroras occur in polar regions and are known as aurora borealis, or "northern lights," in the N, and aurora australis in the S.

auscultation Practice of listening to sounds within the body as an aid to diagnosis. The invention of the stethoscope in 1819 replaced the earlier method of placing the ear against the body.

Australopithecus *See* HUMAN EVOLUTION

autecology Branch of ecology that is concerned with the study of a single species or individual

▲ **aurora** Associated with increased activity from the Sun, aurorae are caused by charged particles emitted from the Sun. When the particles are emitted some will eventually enter two zones of charged particles, called the Van Allen radiation belts, which are created by the magnetic field around the Earth. The Van Allen radiation belts, during periods of increased Sun activity, become disrupted and overloaded and charged particles cascade down into the upper atmosphere striking atoms and molecules and releasing energy in the form of light.

organism and the way in which it interacts with its environment, taking into account both BIOTIC and ABIOTIC FACTORS.

autoimmune disease Any one of a group of disorders caused by the body's production of ANTI-BODIES that attack the body's own tissues. One example of such an autoimmune disease is systemic lupus erythematosus (SLE), an inflammation of the connective tissue occurring most often in young women. The occasional presence of so-called autoantibodies in an individual does not necessarily indicate autoimmune disease. Treatment depends on the specific disease process, but often includes drugs such as CORTISONE, which suppress the immune response.

autolysis In biology, process in which a tissue, cell or part of a cell self-destructs. The process is brought about by ENZYMES that act on cells, produced by LYSOSOMES within the cell itself, normally after the cell has died.

automation Use of self-governing machines to carry out manufacturing, distribution and other processes automatically. By using FEEDBACK, sensors check a system's operations and send signals to a computer that automatically regulates the process. *See also* MASS PRODUCTION; ROBOT

automobile (car) Road vehicle that first appeared in the 19th century. The first cars were propelled by steam, but were not a success. The age of the automobile really dates from the introduction

poor weather. Sensors (9) detect when a wheel is about to lock and release the brake pads for a fraction of a second. An explosive charge inflates the air bag (10), which prevents the driver or passenger from hitting the steering wheel or dashboard. The steering column (11) is designed to collapse so the driver is not impaled. Seat belt tensioners use the impact to pull the belt tight (12), holding the passenger in place. The headrest (13) helps stop whiplash injuries when heads snap back in the aftermath of the impact.

▲ **automobile** A modern car is designed with crumple zones at the front and rear to absorb the energy of a crash and protect the part of the car in which people sit. Side-impact protection bars (1) give strength to the side of the vehicle and spread energy to either side of the passenger cell. Fuel tanks (2) are situated in front of the rear axle to protect the tank if the car is hit from behind. Some manufacturers have replaced the traditional rear brake lights with LEDs (3) which light more quickly. The cover is stepped (4) to prevent the light being obscured by dirt. The suspension, a MacPherson strut system, allows vertical movement through the spring (5) while the wishbone (6) and antiroll bar (7) keep the wheels in position and stop excessive roll respectively. Antilock braking systems (ABS) (8) prevent the wheels locking under heavy braking or in

(1885–86) of the gasoline-driven carriages of Gottlieb DAIMLER and Karl BENZ. The INTERNAL COMBUSTION ENGINE for these cars had been developed earlier by several engineers (most notably Nikolaus Otto in 1876). The main components of a car remain unchanged. A body (**chassis**) to which are attached all other parts, including an **engine** or power plant; a **transmission** system for transferring the drive to the wheels; and steering, braking, and suspension systems for guiding, stopping and supporting the car. Early cars were assembled by a few experts, but modern mass-production was begun in the early 1900s by Henry FORD and R.E. Olds in the USA. In most modern motor factories, component parts are put together on assembly lines. Each worker has a specific task (such as fitting doors or crankshafts). Bodies and engines are made on separate assembly lines that converge when the engine is installed. Overhead rail conveyors move heavy components along the assembly lines, lowering them into position. The final stages of assembly include the fitting of items such as headlights and paint spraying. Electrical, braking and control systems are checked. The assembled car is tested before sale. Recent technology has seen the introduction of robots (properly, robotic arms secured to the workshop floor) on the assembly line. They are usually used for welding and painting. Increasing concern over the environmental impact of the car (such as traffic congestion, pollution and energy consumption) has encouraged governments to examine alternative forms of mass transport, oil companies to produce cleaner fuels and car manufacturers to look at alternative power plants (such as electric- or gas-powered motors).

autonomic nervous system (ANS) In MAMMALS, part of the body's NERVOUS SYSTEM that regulates the body's involuntary functions. In conjunction with the PERIPHERAL NERVOUS SYSTEM it helps prepare the body for action or rest by regulating functions such as heartbeat, digestion and sweating. The ANS is divided into the SYMPATHETIC NERVOUS SYSTEM and the PARASYMPATHETIC NERVOUS SYSTEM. *See also* HOMEOSTASIS; INVOLUNTARY MUSCLE

autopilot (automatic pilot) Electronic and mechanical control system that ensures that an aircraft will follow a preprogrammed flight plan. It monitors the course and speed of the aircraft and corrects any deviations from the flight plan. Systems range from simple wing-levelers in light aircraft to computer-operated units consisting of a GYROSCOPE; an electric SERVOMECHANISM unit; and an ACCELEROMETER, which measures the ACCELERATION of the aircraft.

autotroph Organism that manufactures its own organic food from simple inorganic chemicals in a

process known as **autotrophic nutrition**. Green plants are typical autotrophs; they make sugars by PHOTOSYNTHESIS from carbon dioxide and water using the energy of sunlight. In addition, some organisms, mostly bacteria, obtain energy by oxi-

▲ **autopilot** The diagram shows how an autopilot system works. A preprogrammed flight plan is loaded into the aircraft's computers (1). Two visual display units (2) show the aircraft's position, its intended route and its attitude. The change in movement of small vanes (3) on the outside of the aircraft alert the computers to any change in the aircraft's orientation. The aircraft uses the Global Positioning System (4) to determine its position. The receiver is located on top of the aircraft (5). The computers track the aircraft's route and automatically make any adjustments via servos (6) which control the rudder (7), elevators (8), ailerons (9), flaps (10), and engine throttle settings (11). Pilots can override the system at any time and revert to manual controls (12).

dizing inorganic chemicals (for example sulfur into sulfate). This process is called **chemosynthesis**. In FOOD CHAINS, autotrophs make up the primary producers, providing energy (as food) for organisms higher up the chain. *See also* HETEROTROPH

auxin Plant HORMONE produced mainly in the growing tips of plant stems. Auxins accelerate plant growth by stimulating cell division and enlargement, and by interacting with other hormones. Actions include the elongation of cells (by increasing the elasticity of cell walls, allowing the cells to take up more water) in GEOTROPISM and PHOTOTROPISM, and fruit drop and leaf fall. Synthetic auxins are the basis of rooting powders and selective weedkillers.

average In statistics, the one datum that most typifies an entire set of data. It is the arithmetic MEAN. Other calculations that are also used to express what is typical are the mode (the one that occurs most often), and the median (the middle in a range that thus divides the set of into upper and lower halves).

Aves *See* BIRD

Avogadro, Amedeo, Conte di Quaregna (1776–1856) Italian physicist and chemist. In 1811 Avogadro put forward the hypothesis (now known as **Avogadro's law**) that equal volumes of gases at the same pressure and temperature contain an equal number of molecules. This led later physicists to determine that the number of molecules in one gram molecule (the relative molecular mass expressed in grams) is constant for all gases. This number is called the AVOGADRO CONSTANT. It is both the ratio of the universal gas constant to BOLTZMANN CONSTANT and of Faraday's constant to the charge of the ELECTRON.

Avogadro constant (formerly Avogadro's number, symbol L) Constant equal to 6.022×10^{23}, giving the number of atoms or molecules present in one MOLE of a substance.

Axelrod, Julius (1912–) US biochemist. Axelrod shared the 1970 Nobel Prize for physiology or medicine with Ulf Svante von Euler and Sir Bernard Katz for their work on the chemistry of nerve impulse transmission. He discovered how the NEUROTRANSMITTER NORADRENALINE is inactivated after it has performed its function.

axil Angular space between a growing leaf, bract or branch and the stem it grows from. Buds or shoots that grow in axils are known as axillary buds or shoots.

axiom Assumption used in mathematics or logic as a basis for deductive reasoning. *See* AXIOMATIC METHOD

axiomatic method Method of mathematical reasoning based on logical deduction from assumptions (AXIOMS). The method is fundamental to the philosophy of mathematics: it was used by the Greeks and formalized early in the 20th century by David HILBERT. In an axiomatic system, certain undefined entities (terms) are taken and described by a set of axioms. Other, often surprising, relationships (theorems) are then deduced by the process of logical reasoning. *See also* GÖDEL, KURT

axis Imaginary straight line around which a body rotates, or about which rotation is conceived. In mechanics an axis runs longitudinally through the center of an axle or rotating shaft. In geography and astronomy, it is a line through the center of a planet or star, around which the planet or star rotates. The Earth's axis between the north and south geographic poles is 12,700km (7,900mi) long and is inclined at an angle of 66.5° to the plane in which the Earth orbits the Sun. A mathematical axis is a fixed line, such as the x, y or z axis, chosen for reference.

axon Part of a nerve cell, or NEURON, that carries a nerve impulse beyond and away from the cell body, such as an impulse for movement to a muscle. There is typically only one axon per neuron, and it is generally long and unbranched. It is encased by a fatty, pearly MYELIN SHEATH in all peripheral nerves, and in all central nerves except those of the brain and spinal cord. Axons in peripheral nerves are covered by an additional delicate sheath, a neurilemma, which helps to regenerate damaged nerves.

azimuth ANGLE between the vertical plane through a celestial body and the N-S direction. Astronomers measure the angle E from the N point of the observer's horizon. Navigators and surveyors measure it W from the S point. ALTITUDE and azimuth form an astronomical coordinate system for defining position.

azo and diazo compounds Nitrogenous organic chemical compounds used widely for organic synthesis, particularly in the manufacture of dyes. Diazo compounds include reactive salts having the general formula $Ar\text{-}N_2^+X$, where X is the anion.

azurite Basic copper carbonate mineral found in the oxidized parts of copper ore veins, often as earthy material with MALACHITE. The crystals are blue, brilliant and transparent. In the past the gemstones were used as pigments in wall paintings. Hardness 3.5–4; s.g. 3.77–3.89

Baade, Walter (1893–1960) US astronomer, b. Germany. From Mount Wilson Observatory in the 1943 wartime blackout, he was able to observe individual stars in the Andromeda Galaxy and distinguish the younger, bluer Population I stars from the older, redder Population II stars. He went on to improve the use of CEPHEID VARIABLE stars as distance indicators, and showed that the Universe was older and larger than had been thought.

Babbage, Charles (1791–1871) British mathematician. A professor of mathematics, working on the theory of FUNCTIONS and ALGEBRA, Babbage's main interest was the development of a mechanical calculating machine that could evaluate mathematical tables without error. A small prototype of the "difference engine," which could calculate tables of LOGARITHMS, was made, but a full working model was never completed due to insufficient financial support. He then planned another machine (the "analytical engine") that could carry out any set of arithmetic operations using punched cards (a forerunner of the modern computer). But again lack of money, and the fact that it was too ambitious and intricate for the mechanical devices of the day, ensured it was never completed.

Babbitt metal Any alloy with a high tin content as well as copper and antimony – specifically an alloy invented by Isaac Babbitt (1799–1862) in 1839 as bearing material for steam engines.

Babcock, Harold Delos (1882–1968) US physicist and spectroscopist who developed a modern theory to account for the formation of SUNSPOTS.

bacillus Genus of rod-like BACTERIA present everywhere in the air and soil. One example of a species that is pathogenic in people is *Bacillus anthracis*, which causes ANTHRAX.

backcross Offspring of a first-generation HYBRID and either of its parents. Backcrosses are used by biologists as a means of testing the GENOTYPE of the hybrid.

background radiation RADIATION that is normally present in an environment. Such radiation

must be taken into account when measuring radiation from a particular source. On Earth, background radiation is caused by the decay of naturally occurring radioactive substances in surface rocks. In space, the so-called "microwave background" is attributed to the BIG BANG.

backwash Movement of seawater down a beach after a wave has broken on it. An incoming wave is taken up the beach by the **swash**; backwash takes it back again. On a hard beach the backwash is on the surface; or it may be within the shingle or sand on a softer beach.

bacteria Simple unicellular microscopic organisms belonging to the kingdom Prokaryotae. They lack a clearly defined nucleus and most are without CHLOROPHYLL. Many are motile, swimming about by means of whip-like FLAGELLUM. Most multiply by FISSION. Under adverse conditions many can remain dormant inside highly resistant SPORES with thick, protective coverings. Bacteria may be AEROBIC or ANAEROBIC. Although pathogenic bacteria are a major cause of human disease, many bacteria are harmless or even beneficial to humans by providing an important link in FOOD

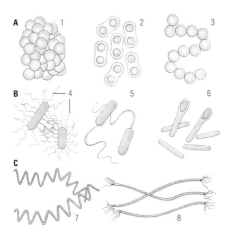

▲ **bacteria** Bacteria occur in three shapes (magnification x 5,000): spherical forms called cocci (A), rod-like bacilli (B), and spiral spirilla (C). **Cocci** can occur in clumps known as staphylococci (1), groups of two called diplococci (2), or chains called streptococci (3). Unlike cocci, which do not move, **bacilli** are freely mobile; some are termed peritrichous and use many flagellae (4) to swim about, while other monotrichous forms use a single flagellum (5). Bacilli can also form spores (6) to survive unfavorable conditions. **Spirilla** may be either cork-screw-shaped spirochaetes, like Leptospira (7), or less coiled and flagellated, such as Spirillum (8).

CHAINS, such as in decomposing plant and animal tissue, and in converting free nitrogen and sulfur into AMINO ACIDS and other compounds that plants and animals can use. Some contain a form of chlorophyll and carry out PHOTOSYNTHESIS. *See also* ARCHAEBACTERIA; EUBACTERIA; PROKARYOTE;

bacteriology Scientific study of BACTERIA. These single-celled organisms were first observed in the 17th century by the amateur microscopist Anton van LEEUWENHOEK, but it was not until the researches of Louis PASTEUR and later Robert KOCH that bacteriology was established as a scientific discipline.

bacteriophage VIRUS that lives on and infects BACTERIA. It has a protein head containing a core of DNA and a protein tail. Since its discovery in 1915 it has been important in the study of GENETICS.

badlands Eroded, barren plateau in an arid or semi-arid area characterized by steep gullies and ravines. Because of the lack of adequate vegetation (due to climate or human intervention), the rainwater runs off very quickly and erodes soft and exposed rock. The best-known examples are the badlands of SW South Dakota and NW Nebraska.

Baekeland, Leo Hendrik (1863–1944) US chemist, b. Belgium. Baekeland invented a type of photographic paper, Velox, capable of being exposed under artificial light. He also invented (1909) the first thermosetting PLASTIC, BAKELITE, a substance that led to the development of the plastics industry.

Baer, Karl Ernst von (1792–1876) Estonian biologist. He described his discovery of the mammalian ovum in De Ovi Mammalium et Hominis Genesi (1827). He also wrote On the Development of Animals (2 vols. 18238, 1837).

Baeyer, Johann Friedrich Wilhelm Adolph von (1835–1917) German chemist. Baeyer specialized in organic chemistry. He was awarded the 1905 Nobel Prize for chemistry for his synthesis of INDIGO. He also investigated uric acid derivatives, discovered the phthalein dyes and devised a "strain" theory to account for the stability and conformation of carbon compounds with five-membered and six-membered rings.

Bailey, Sir Donald Coleman (1901–85) British civil engineer who designed the Bailey bridge for military use. The steel lattice girders can be assembled by groups of six workers from easily manipulable component parts and cantilevered across a gap to provide a bridge that can carry 70 tons over spans of up to 45m (150ft).

Bailey, Liberty Hyde (1858–1954) US botanist. He helped to establish HORTICULTURE as an applied science through the systematic study of cultivated plants. His work had an important influence on the development of GENETICS and plant pathology. He founded and directed the Bailey Hortorium at Cornell University.

Baird, John Logie (1888–1946) Scottish electrical engineer and inventor of TELEVISION. In 1926 he demonstrated the first working television to members of the Royal Institution, London. In 1928 he transmitted television to a ship at sea, and in 1929 was granted experimental broadcasting facilities by the British Broadcasting Corporation (BBC). His 240-line, part-mechanical, television system was used for the world's first public television service by the BBC in 1936. In 1937 it was superseded by a fully electronic model.

Bakelite Trade name (coined by Leo BAEKELAND) for a thermosetting PLASTIC used for insulating purposes and in making paint. It was the first plastic made (1909) by CONDENSATION REACTION, in which many molecules of two chemicals, in this case PHENOL and FORMALDEHYDE, are joined together to form large polymer molecules, by splitting off water molecules.

Baker, Sir Benjamin (1840–1907) British civil engineer. Baker was the principal designer of the Forth railway bridge in Scotland, completed in 1890, which at 518m (1,700ft) had the longest span in the world until 1918. He was also responsible for the construction of much of London's subway system and the first Aswan dam.

Bakewell, Robert (1725–95) British livestock breeder. Bakewell revolutionized animal husbandry through scientific breeding on his estate in England. His methodical inbreeding and selection of sheep and cattle produced better meat animals. (Previously sheep had been raised largely for wool, and cattle for pulling machinery).

baking powder Mixture used in cooking as a substitute for yeast. It contains SODIUM BICARBONATE (sodium carbonate) mixed with an acid component, such as tartaric acid or cream of tartar. During cooking, the acid reacts with the sodium bicarbonate to generate carbon dioxide gas, causing the food to rise without the fermentation effects of yeast.

baking soda Common name for SODIUM BICARBONATE, so-called because it is a constituent of BAKING POWDER.

balance Piece of apparatus for making accurate weighings. In a **beam** balance, two pans are suspended at the ends of a beam pivoted at its center. The mass of the object to be weighed, located in one pan, is counterbalanced by known weights added to the other pan. In a **substitution** balance, there is only one pan suspended from the end of an arm with a fixed counterweight at the other end. The object to be weighed is placed in the pan, and calibrated weights are removed from the arm until it is counterbalanced. An **electronic** balance uses an electromagnetic force to restore the balance of a lever arm, the current flowing in the magnetic coil being a measure of the mass involved. Rough weighings may be made with the aid of a **spring** balance (or newton meter), in which the amount of extension of a vertical spring supporting a pan is a measure of mass.

ball-and-socket joint Type of joint that allows one part to rotate at almost any angle with respect to another. Such joints are used in machinery and occur naturally, as in the human hip joint.

ball bearing Type of bearing for rotating machinery. It consists of two concentric rings of steel, which together form a "race" for a number of steel balls (also referred to as ball bearings) positioned between them. The inner ring is fixed to a shaft and the outer one to a support.

ballistics Science of projectiles, including bullets, shells, bombs, rockets, and guided MISSILES. **Interior** ballistics examines the propulsion and motion of the projectile within the firing device. **Exterior** ballistics investigates the trajectory of the projectile in flight. **Terminal** ballistics is concerned with the impact and effect of the projectile at the target. At each stage, scientists try to maximize the performance of the gun and projectile by improving their design. Ballistic technology has developed alongside ARTILLERY, and with the invention of instruments to monitor variables, such as the ignition and burning of the propellant explosive, the stress on a gun barrel, or the effect of air resistance and gravity on the trajectory.

balloon Unsteerable, lighter-than-air craft. Balloons are used for recreation and for scientific and military purposes. Either hot air or a gas that is lighter than air lifts the balloon from the ground. The first balloons to fly were of the open-necked hot-air type. Unmanned military, meteorological or other scientific balloons are usually filled with hydrogen, least dense of all gases but dangerously flammable. Manned balloons are now generally filled with the safer gas helium, or with hot air. *See also* AIRSHIP

ballpoint pen Pen that employs as its writing point a small ball bearing that rolls against and

picks up semi-solid ink from a reservoir. It was invented by John Loud in 1888 but Lazlo Biro was the first to make a reliable model in 1938, which was patented in 1943.

Baltimore, David (1938–) US microbiologist. He shared the 1975 Nobel Prize for physiology or medicine for his work on the "interaction between tumor viruses and the genetic material of the cell." Working independently of co-laureate Howard TEMIN, he proved the existence of reverse transcriptase, the enzyme necessary for viral genetic information to be incorporated into an animal cell. *See also* REVERSE TRANSCRIPTION

bandwidth Range of frequencies spanned by a radio signal of a particular nominal frequency. If a transmitted radio signal is modulated, the bandwidth is the range of frequencies employed on either side of the CARRIER WAVE signal. It is therefore also the range of frequencies used for a particular radio transmission within a particular waveband. The term can also be used to describe the frequency range over which a device, such as an amplifier or radio receiver, should not significantly differ from its maximum value. In communications and computing, bandwidth can also describe that rate at which data is transmitted (for example, by a modem), usually measured in bits per second.

Banks, Sir Joseph (1743–1820) English botanist. He was the senior scientist of the group that sailed to Tahiti with Captain James Cook aboard HMS *Endeavour* in 1768. At Botany Bay, Australia, in 1770, Banks collected examples of plants hitherto unknown in Europe, including the shrub banksia, named in his honor. Upon his return, he helped set up the Royal Botanic Gardens at Kew, London, and financed international plant-collecting expeditions.

Banting, Sir Frederick Grant (1891–1941) Canadian physician. He shared, with J.J.R. MACLEOD, the 1923 Nobel prize for physiology or medicine for his work in extracting the hormone INSULIN from the PANCREAS. This made possible the effective treatment of DIABETES.

bar Unit of pressure; the pressure created by a column of mercury 75.006cm high at 0°C (32°F). It is equal to 10^5 PASCALS. Standard atmospheric pressure (at sea level) is 1.01325 bars, or 1,013.25 millibars.

Bárány, Robert (1876–1936) Swedish physician, b. Austria. He was awarded the 1914 Nobel Prize for physiology or medicine for his work on the physiology and pathology of the vestibular

▶ **bar code** Bar codes represent information concerning a product and its manufacturer in a series of thick and thin black and white lines (1). Two longer, thinner bars mark the beginning and end of the manufacturer and product codes. A laser (2) is reflected through a glass screen (3) onto the bar code by a rotating, multi-faceted mirror (4). The laser light is scattered by the white lines and absorbed by the black lines. A sensor (5) detects the reflected laser light and compares the relative width of the lines. Because the relative widths are compared, the bar code does not have to be on a flat surface. The sensor passes the information to the till (6) for billing the customer, and a central computer (7) monitors stock levels and orders supplies of new freight.

apparatus of the EAR. His studies of the normal and abnormal functioning of this important balancing mechanism greatly advanced understanding of the ear.

barbiturate DRUG used as a sedative or to induce sleep or sedation. Highly addictive and dangerous in high doses or when combined with other drugs such as alcohol or tranquillizers, most barbiturates are no longer prescribed. Short-acting barbiturates are used in surgery to induce general anesthesia; long-acting formulations are sometimes prescribed for epilepsy.

barchan Crescent-shaped DUNE found in sandy DESERTS throughout the world where the wind is constant in speed and direction. Barchans also occur fairly frequently in coastal regions; they are quickly shifted by the wind, particularly when small.

bar code Coded information consisting of thick and thin black lines, and designed for computer recognition. A laser beam scans the bar code and a light-sensitive detector picks up the reflected signal, which consists of a pattern of pulses. Bar codes are used on many products for sale in shops and supermarkets. The store's computer translates the bar code into information, including the product's name, weight or size. The computer refers to a price list data file to see how much to charge the customer and records the sale so that the shop can monitor its inventory levels.

Bardeen, John (1908–91) US physicist known for his research into SEMICONDUCTORS. He worked with the Bell Telephone Laboratories (1945–51) and was then professor of physics at Illinois University (1951–75). He was the first person to win the Nobel prize twice in the same field, physics: in 1956 he shared it with William SHOCKLEY and Walter BRATTAIN, for their joint invention of the TRANSISTOR, and in 1972 with Leon COOPER and John SCHRIEFFER, for their theory of SUPERCONDUCTIVITY.

barite (baryte or heavy spar) Barium-containing mineral, the chief ore of the metal. Barite consists of barium sulfate ($BaSO_4$). It usually occurs as white crystals, although it may be brown, gray or yellow, and is often found in association with lead or zinc ores. Because of its high density, it is used to make muds for lubricating and cooling oil-well drilling equipment. It is also used in the chemical industry for paper-making, rubber manufacture, and high-quality paints. Hardness 3–3.5; r.d. 4.5

barium (symbol Ba) Silver-white, metallic element of the ALKALINE-EARTH METALS, discovered in 1808 by Sir Humphry DAVY. It is a soft metal the chief sources of which are barite (barium sulfate) and witherite (barium carbonate). Barium compounds are used as rodent poison, pigments for paints, and as drying agents. Barium sulfate ($BaSO_4$) is swallowed to allow X-ray examination of the stomach and intestines because barium atoms are opaque to X-rays; this is called a "barium meal." Properties: at.no. 56; r.a.m. 137.34; r.d. 3.51; m.p. 725°C (1,337°F); b.p. 1,640°C (2,984°F); most common isotope ^{138}Ba (71.66%).

bark Outer protective covering of a woody plant stem. It is made up of several layers. The CORK layer, waxy and waterproof, is the thickest and hardens into the tough, fissured outer covering.

Lenticels (pores) in the bark allow GAS EXCHANGE between the stem and the atmosphere. *See also* CAMBIUM; DENDROCHRONOLOGY

Barkla, Charles Glover (1877–1944) British physicist who was awarded the 1917 Nobel Prize for physics for his discovery that elements have a characteristic X-RAY spectrum, which enables them to be identified. He also formulated laws governing the scattering of X rays and their transmission through solids.

barn (symbol *b*) Scientific unit of area used in nuclear physics to measure the cross-sections in interactions of particles. A barn equals 10^{-24} cm². This area is a measure of the probability that FISSION will occur when a neutron passes near a heavy nucleus.

Barnard, Christiaan (1922–) South African surgeon. He was the first to perform a human heart transplant (3 December 1967). In 1974 he was the first to implant a second heart in a patient and to link the circulations of the two hearts so that they worked together as one.

Barnard's Star RED DWARF star 6 light-years away, in the constellation Ophiuchus. It is the clos-

est star to the Sun after the ALPHA CENTAURI system and was discovered in 1916 by the US astronomer Edward E. Barnard (1857–1923).

barograph Recording BAROMETER. It consists of an aneroid barometer that has, instead of a pointer, a pen that continuously records ATMOSPHERIC PRESSURE on a paper chart wrapped round a revolving drum, usually driven by clockwork.

barometer Instrument for measuring ATMOSPHERIC PRESSURE. There are two main types: the **mercury** barometer, and the **aneroid** barometer, which is the less accurate of the two but is the type normally used in BAROGRAPHS and found in the home. The mercury barometer consists of a glass tube containing mercury, which is inverted over a small reservoir of mercury; as the pressure rises, the weight of the atmosphere forces mercury out of the reservoir and up the tube. The tube has a scale on it, so that the pressure can be read off at the top of the column of mercury. As the pressure falls, the column of mercury gradually moves down. A column of mercury 760mm (30in) in length represents the average pressure at sea level.

barred spiral galaxy Type of GALAXY as classified by Edwin HUBBLE.

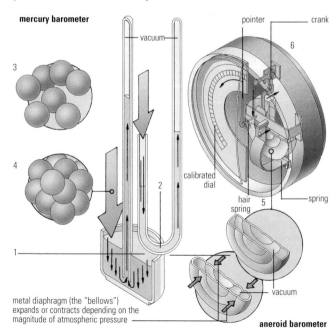

▶ **barometer**
Barometers are instruments used to measure air pressure. A glass tube, open at one end, is filled with mercury (which is used because of its density) and inverted in a bowl of mercury. The column of mercury drops, creating a vacuum above it, until its weight is balanced by the weight of the air pressing on the mercury surface (1). The column of mercury is 760mm tall at sea level and 45° latitude. The bowl of mercury can be replaced by using a curved tube (2). Warmer, less dense air (3) will exert less pressure than cold dense air (4). The aneroid barometer dispenses with the mercury, changes in air pressure being registered by the movement of a flexible metal diaphragm (5) containing a vacuum, which is connected by a series of levers to a sprung indicator pointer and scale (6).

mercury barometer

metal diaphragm (the "bellows") expands or contracts depending on the magnitude of atmospheric pressure

aneroid barometer

barrel Unit of volume used to measure liquids, particularly petroleum (crude oil). For oil and petroleum products, a barrel equals 158.98 liters or 42 gallons. For beer, a barrel equals 163.66 liters or 43.24 gallons. For dry goods (such as grain), a barrel equals $0.1156m^3$ or 3.283 bushels.

barrier island Any long, low island of sand parallel to a shore and permanently separated from it. It may be composed of dunes, swamps, and areas of vegetation.

barrier reef Long, narrow CORAL REEF lying some distance from and roughly parallel to the shore, but separated from it by a deep LAGOON. The Great Barrier Reef off the coast of Queensland, NE Australia, is the most famous.

Barton, Sir Derek Harold Richard (1918–98) British chemist. He did innovative research into **conformational analysis** (study of the geometric structure of complex molecules), discovering that the reactivity of a molecule was related to its preferred shape. For this work he shared the 1969 Nobel Prize for chemistry with Odd HASSEL. One of Barton's many advances in organic chemistry was his use of FREE RADICALS, which enabled him to invent new reactions. Barton was professor (1957–70), Hofman professor (1970–78), and then Emeritus Professor (1978–98) of organic chemistry at Imperial College, London University. He was director (1977–85) of the Institute of Organic Chemistry (CNRS) in France, following which he was professor at Texas A&M University.

baryon Any of several types of ELEMENTARY PARTICLE affected by the STRONG NUCLEAR FORCE. A baryon consists of three QUARKS, which are indivisible elementary particles. Baryons and MESONS (made up of two quarks) are subclasses of HADRONS. The only stable baryons are the PROTON and (provided it is inside a nucleus) the NEUTRON. Heavier baryons are called HYPERONS. *See also* LEPTON

baryte *See* BARITE

basal metabolic rate (BMR) Minimum amount of energy required by the body to sustain basic life processes, including breathing, circulation, and tissue repair. It is calculated by measuring oxygen consumption. Metabolic rate increases well above basal metabolic rate (BMR) during vigorous physical activity, fever or under the influence of some DRUGS (including CAFFEINE). It falls below BMR during sleep, general ANESTHESIA or starvation. BMR is highest in childhood and decreases with age.

basalt Hard, fine-grained, basic IGNEOUS ROCK, which may be an INTRUSIVE or EXTRUSIVE ROCK. Its color can be dark green, brown, dark gray or black. If it originally solidified quickly it can have a glassy appearance. There are many types of basalt with different proportions of elements. It may be compact or vesicular (porous) because of gas bubbles contained in the lava while it was cooling. If the vesicles are subsequently filled with secondary minerals, such as quartz or calcite, it is called **amygdaloidal** basalt. Basalts are the main rocks of ocean floors, and on continental areas form the world's major lava flows, such as the Deccan Trap in India.

base In chemistry, any compound that accepts protons. A base will neutralize an ACID to form a SALT and water. Most are oxides or hydroxides of metals; others, such as ammonia are compounds that yield hydroxide IONS in water. Soluble bases are called ALKALIS. **Strong** bases are fully dissociated into ions; **weak** bases are partly dissociated in solution. *See also* NEUTRALIZATION

base In geometry, the side opposite the vertex from which an altitude is drawn in a TRIANGLE.

base In mathematics, the number of units in a number system that is equivalent to one unit in the next higher counting place in that system. 10 is the base of the decimal system. In base 5 only the digits 0 to 4 can be used in each place. Hence, the counting numbers in base 5 are 1, 2, 3, 4, 10, 11, 12, 13, 14, 20, 21... etc.. The number 45 in base 7 is equal to 33 in base 10 since
$$(4 \times 7) + (5 \times 1) = 33$$
Similarly, the number 2142 in base 5 is equal to 297 in base 10 since
$$(2 \times 125) + (1 \times 25) + (4 \times 5) + (2 \times 1) = 297$$
The BINARY SYSTEM is used in computing since any number can be represented by a string of just two digits 0 and 1, which match states of gates in computer circuitry that can be either open or closed.

base metal Common, nonprecious metal. Typical examples are lead and iron, which like other base metals, tarnish or oxidize on exposure to air or moisture. Such properties distinguish base metals from gold, platinum, silver, and other so-called precious or NOBLE METALS. In chemical terms, base metals are those at the negative end of the ELECTROMOTIVE SERIES. In ancient times, it was an unfulfilled ambition of alchemists to convert base metals into gold. *See also* ALCHEMY

base pair Combination of two nitrogenous BASES in a molecule of DNA. There are four bases in DNA: ADENINE, CYTOSINE, GUANINE, and THYMINE. Using HYDROGEN BONDS, adenine always pairs with thymine, and cytosine always pairs with guanine. The pairs form the "rungs" of the double-helix structure of DNA and form the basis of the GENETIC CODE. RNA also has base pairs, but URACIL replaces thymine.

BASIC (**B**eginners' **A**ll-purpose **S**ymbolic **I**nstruction **C**ode) Computer programming language that is easy to learn and uses many everyday words. It is commonly used by both amateur and professional programmers. To run, BASIC COMPUTER PROGRAMS usually require a separate program called an **interpreter**, which converts the BASIC code into the machine code required by the computer's processor.

basic oxygen process Widely used steelmaking process in which high-pressure oxygen is blown over the surface of the molten charge to help remove impurities. It has superseded the BESSEMER PROCESS and OPEN-HEARTH PROCESS.

basic salt Salt of a BASE that has had one or more, but not all, of its oxide or hydroxide ions replaced by other negative ions. Lead nitrate hydroxide ($Pb(OH)NO_3$) is a basic salt.

Basidiomycota Phylum of FUNGI (formerly the class Basidiomycetes), including the jelly fungus, the bird's nest fungus, the ear fungus, and mushrooms. They reproduce through sexual spores (**basidiospores**) produced by a specialized cell, the BASIDIUM.

basidium Cell that produces spores (**basidiospores**) used in sexual reproduction by certain types of fungi of the phylum BASIDIOMYCOTA. Basidia are cylindrical or club-shaped, often grouped to form the external fruiting bodies (gills) of bracket fungi, mushrooms, and puffballs. Each basidium produces four spores.

basilar membrane Part of the COCHLEA, a structure in the inner EAR.

Basov, Nikolai Gennadiyevich (1922–) Soviet physicist who developed the MASER, which amplifies microwaves, and the LASER, which amplifies light. For these contributions to science, Basov and his co-worker Alexander PROKHOROV shared the 1964 Nobel Prize for physics with Charles TOWNES (who had made similar discoveries independently).

bast Type of vegetable fiber or matting made from the inner bark (PHLOEM) of plants. Examples include jute, hemp and flax.

Bates, Henry Walter (1825–92) British naturalist. Bates' work on NATURAL SELECTION in animal MIMICRY lent support to Charles DARWIN's theory of EVOLUTION. Bates collected over 8,000 previously unrecorded species of insects during 11 years in the Amazon (1848–59) and recounted his experiences in *The Naturalist on the River Amazon* (1863).

Bateson, William (1861–1926) British biologist. He founded and named the science of GENETICS. Bateson translated much of Gregor MENDEL's pioneering work on inheritance in plants, so bringing it recognition. By his own experiments he also extended Mendel's theories to animals, which provided a foundation for the modern understanding of HEREDITY.

batholith Huge mass of IGNEOUS ROCK at the Earth's surface that has an exposed surface of more than 100sq km (40sq mi). It may have originated as an INTRUSIVE igneous structure that was gradually eroded and became surface material. Most batholiths consist of GRANITE rock types, and are associated with the OROGENESIS (mountain-building phases) of PLATE TECTONICS.

bathyal zone Region of the ocean floor from the edge of the CONTINENTAL SHELF, about 133m (436ft) in depth, to about 2,000m (6,560ft), where the ABYSSAL PLAIN begins. Only a feeble light penetrates to the upper layers of this zone.

bathysphere and bathyscaphe Manned vehicles for deep-sea exploration. The **bathysphere**, invented by Otis Barton and William BEEBE in the US, had steel walls and thick, toughened glass windows through which underwater observations could be made. Used mainly during the 1930s, it was released from a surface vessel and was lowered on a steel cable to depths of more than 900m (3,000ft). The **bathyscaphe** was invented by August PICCARD of Switzerland, and was first used in 1948. It consists of a bathysphere slung below a tank called a float. The whole device can be sunk or made buoyant in a controlled manner, while propellers move the craft horizontally. In January 1960, Piccard and Don Walsh of the US Navy made a descent in the bathyscaphe *Trieste* to a depth of 10,916m (35,810ft) (more or less equivalent to the altitude of commercial jets) in the Pacific Ocean's Mariana Trench off the island of Guam. This record still stands.

battery Collection of voltaic CELLS that convert chemical energy into direct current (DC) electricity. The term is also commonly used for a single cell, particularly a dry cell as used in portable electronic equipment. Most primary cell batteries are not rechargeable; some types of primary cell – such as nickel-cadmium (Nicad) batteries – and all accumulators (storage batteries) can be recharged when a current passed through them in the reverse direction restores the original chemical state.

baud Unit for measuring the speed at which a digital communications device or system carries information. One baud is equal to one BIT per second. Although the term baud rate is still widely used, the speed of modern equipment is often expressed in kilobits per second.

Baudot, Jean-Maurice-Emile (1845–1903) French electrical engineer who devised a telegraph code that used the presence or absence of an electrical pulse, rather than the short and long pulses of the MORSE CODE. He also invented a distributor system that permitted several messages to be transmitted along the same wire.

bauxite Rock from which nearly all aluminum is extracted. Bauxite is a mixture of several minerals, such as diaspore, gibbsite, boehmite, and iron. It is formed by prolonged weathering and leaching of rocks containing aluminum silicates. Large deposits occur in the United States, France, Hungary, Guyana, Jamaica, Surinam, Italy, Greece, Russia, Azerbaijan, and Kazakhstan.

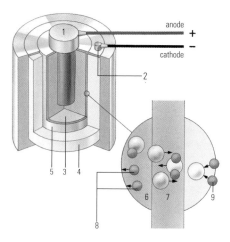

anode +

cathode −

▲ **battery** Sodium sulfur batteries are the newest type of battery and are lighter than nickel cadmium types. They have a carbon anode (1) and a metal cathode (2). The reactants are arranged in rings around the anode. An inner core of sodium (3) is separated from an outer ring of sulfur (4) by a layer of aluminum (5). The sodium (6) reacts with the aluminum (7), giving up electrons (8) which stream to the anode. The cathode gives electrons to the sulfur atoms (9) that bond with sodium ions to form sodium sulfide. The process creates a voltage.

Bawden, Frederick (1908–72) British plant physiologist. In 1936, together with Norman PIRIE, he isolated the tabacco virus in crystalline form (also acheived a year earlier by US biochemist Wendell STANLEY). They showed that the virus consists of 94% protein and 6% RNA, which was much later found to be the infective part of the virus.

bay bar Ridge of sediment that is deposited across the mouth of a bay. It is usually attached to the shore at each end. The deposit may be carried to the bar by LONGSHORE DRIFT, or it may be carried there by a river. When it grows large enough it may completely block the bay from the ocean, so that a DELTA forms on the landward side as further deposits pile up.

Bayliss, Sir William Maddock (1860–1924) British physiologist. He studied blood pressure, circulation, and digestion, and with Professor Ernest STARLING discovered (1902) SECRETIN, a hormone secreted by the DUODENUM.

BCG (Bacille Calmette-Guérin) Vaccine against tuberculosis. It was named after its discoverers, the French bacteriologists Albert Calmette (1863–1933) and Camille Guérin (1872–1964).

BCS theory Theory that explains SUPERCONDUCTIVITY. The theory is named after three American physicists, John BARDEEN, Leon COOPER, AND John SCHRIEFFER. It states that in a superconductor, electrostatic forces between positive sites on the metal lattice and a conducting electron distort the lattice. A second electron is then affected and, as a result, the current is carried not by individual electrons but by bound pairs of electrons called **Cooper pairs**. Such a pair is not affected by the lattice and so carries current indefinitely.

beach Sloping zone of the shore of the sea or a lake, covered by sediment, sand or pebbles, which extends from the low-water line to the limit of the highest storm waves. The sediment is derived from coastal erosion or river ALLUVIUM. Waves breaking on the beach move the sediment back and forth so the heavier pebbles remain on the upper beach and the sand moves downwards towards low water. *See also* LONGSHORE DRIFT

beacon, radio Navigational aid for ships and aircraft. It consists of a transmitter that broadcasts radio signals of a fixed frequency, peculiar to one beacon only. A navigator or pilot, by referring to navigation charts, can find his or her position from the bearings of two or more beacons, using a DIRECTIONAL ANTENNA to receive the signals.

Beadle, George Wells (1903–89) US geneticist. During his study of MUTATIONS in the bread mold *Neurospora crassa*, he and Edward TATUM found that GENES are responsible for the synthesis of ENZYMES and that these enzymes control each step of all biochemical reactions occurring in an organism. For this discovery they shared with Joshua LEDERBERG the 1958 Nobel prize for physiology or medicine.

Beagle, HMS British survey ship that carried Charles DARWIN as the ship's naturalist. The *Beagle* left England in December 1831 and for five years explored parts of South America and the Pacific islands. Darwin's observations formed the basis for his theory of EVOLUTION of species by NATURAL SELECTION.

Bear, Great and Little Popular names for URSA MAJOR and URSA MINOR, respectively.

bearing Supporting component, usually of a resistant metal alloy, used to minimize friction between moving parts of machines. Some bearings are cylindrical shells that separate the surfaces of moving parts by trapping a thin film of lubricant between them. BALL BEARINGS use balls confined to a race to reduce the areas in contact; ROLLER BEARINGS use rollers.

Beaufort wind scale Range of numbers from 0 to 17 representing the force of winds, together with descriptions of the corresponding land or sea effects. The Beaufort number 0 means calm wind less than 1km/h (0.62mph), with smoke rising vertically. Beaufort 3 means light breeze, 12–19km/h (8–12mph), with leaves in constant motion. Beaufort 11 is a violent storm, 103–117km/h (64–72mph) and Beaufort 12 is a hurricane, 118+km/h (73+mph), with devastation. Beaufort 13–17 indicates the force of the hurricane. The scale is named after its inventor, Admiral Sir Francis Beaufort (1774–1857).

Beaumont, William (1785–1853) US surgeon. He gained valuable knowledge of the functioning of the human stomach through a series of experiments he performed on the stomach of Alexis St. Martin, who, as the result of a gunshot wound, had a small opening into his body over his stomach. It provided Beaumont with a unique opportunity to study stomach action.

Becquerel, (Antoine) Henri (1852–1908) French physicist. He was professor of physics at the Paris Museum of Natural History, and later at the École Polytechnique. In 1896 he discovered RADIOACTIVITY in uranium salts, for which he shared the 1903 Nobel prize for physics with Pierre and Marie CURIE. He also studied the rotation of the plane of polarized light in a magnetic field. The BECQUEREL was named after him. *See also* BETA PARTICLE

becquerel (symbol Bq) SI unit of radioactivity. It is defined as the activity of a radioactive material that decays at an average rate of one transition per second. It is named after the French physicist Henri BECQUEREL. It replaces the former unit, the CURIE, which equals 3.7×10^{10} Bq.

bed In geology, a layer of SEDIMENTARY ROCK greater than 1cm (0.4in) in thickness, representing a single episode of sediment deposition. The original sediment of a bed is usually deposited in a horizontal sheet.

bedding plane In geology, the dividing line between two different strata of sedimnetary rock. A bedding plane generally indicates a slight change of rock type; each stratum would have been deposited over a period of years, and then consolidated to form a layer of rock.

bedrock Solid rock that occurs below the soil and any REGOLITH (rock fragments). It is the C or D horizon in a soil profile (*see* SOIL HORIZON). On hillsides the bedrock may be only a few centimeters below the surface, whereas on prairies and plains it may be several meters down.

Beebe, Charles William (1877–1962) US naturalist and explorer, Curator of Ornithology (1899–1919) at the New York Zoological Gardens and Director of Tropical Research (1919–52) at the New York Zoological Society. He led explorations in Central and South America, the West Indies, and the Far East. He also made undersea descents of 1,000m (3,300ft) in his invention, the BATHYSPHERE. His numerous books include *Galápagos* (1923) and *Beneath Tropic Seas* (1928).

behavioral ecology Study of the complex relationship between environment and animal behavior. This involves drawing on the disciplines of natural history, to study the adaptive features of an organism within its habitat. Human behavior is similarly studied. *See also* ADAPTATION; ECOLOGY; ETHOLOGY

behaviorism School of psychology that seeks to explain all animal and human behavior primarily in terms of observable and measurable responses to stimuli. Its method of research often involves laboratory experiments. PAVLOV'S work on conditioned reflexes was a source for the early behaviorists such as J.B. WATSON. Later behaviorists, such as B.F. SKINNER, explain learning and development by "operant conditioning."

behavior therapy (behavior modification) Treatment of psychological disorders by using principles and methods of BEHAVIORISM. It assumes that all behavior, desirable and otherwise, is learned through CONDITIONING and reinforcement. The therapy is designed to change people's behavior by rewarding desirable conduct and punishing or ignoring undesirable behavior.

Behring, Emil Adolph von (1854–1917) German bacteriologist and pioneer immunologist. In 1901 he was awarded the first Nobel Prize for physiology or medicine for his work on serum therapy, especially for developing immunization against DIPHTHERIA (1890) and TETANUS (1892) by injections of ANTITOXINS, a word he introduced. His discoveries led to the treatment of many childhood diseases.

Békésy, Georg von (1899–1972) US physicist, b. Hungary. As director of the Hungarian Telephone System Research Laboratory (1923–46), Békésy worked on problems of communication and the mechanics of human hearing. At Harvard University, he carried on his research (1949–66) on the COCHLEA of the EAR. For this research Békésy received the 1961 Nobel Prize for physiology or medicine. Extending his studies to other senses, he developed a unified picture of the role of sensory and nervous processes in perception.

belemnite Extinct group (order) of CEPHALOPODS that appeared in the Jurassic period (213–144 million years ago) and lived through the Cretaceous period (144–65 million years ago), with a few persisting into the early part of the Tertiary period (65–2 million years ago) up to about 45 million years ago. Entirely marine creatures, belemnites were characterized by a bullet-shaped, cylindrical shell.

Beilstein, Friedrich Konrad (1838–1906) German organic chemist, b. Russia. He carried out research into the isomers of BENEZENE derivatives and devised a test for HALOGENS in organic compounds. But he became known worldwide for his monumental organic chemistry text *Handbook of Organic Chemistry* (1880–82), which listed data about every known organic compound. He produced two further editions, and since 1900 the work has continued to be published regularly.

Bell, Alexander Graham (1847–1922) Scottish-born scientist, inventor of the TELEPHONE. He first worked with his father, the inventor of a system for educating the deaf. The family moved to Canada in 1870, and Bell taught speech at Boston University (1873–77). His work on the transmission of sound by electricity led to the first demonstration of the telephone in 1876 and the founding of the Bell Telephone Company. in 1877.

Bell, Sir Charles (1774–1842) Scottish anatomist and surgeon, the first man to distinguish between sensory and motor nerves in the brain and to describe the facial paralysis known as **Bell's palsy.** He was professor of anatomy and surgery at the Royal College of Surgeons, London, and from 1836 was professor of surgery at Edinburgh University.

Bell, John Stuart (1928–90) British physicist who worked on particle ACCELERATORS at the Atomic Energy Establishment, Harwell, England, and at CERN in Switzerland. He published a series of theoretical papers, culminating in 1964 with one that included equations, now known as **Bell's inequalities**, which predict how measurements of one subatomic particle can affect other distant particles. Such interconnection was found experimentally in the 1980s.

Bell Burnell, (Susan) Jocelyn (1943–) British physicist, b. Northern Ireland. While studying at Cambridge University under Anthony HEWISH she helped construct the radio telescope with which, in 1967, she discovered the first PULSAR. Following this initial discovery, she has held significant teaching posts at various universities, and has received numerous awards for observational astronomy.

Benioff zone Area of deep-focus earthquakes that dips from the Earth's surface to a depth of about 700km (450mi). The zone is thought to indicate areas of active SUBDUCTION of the Earth's tectonic plates. *See also* PLATE TECTONICS

benitoite Glassy, blue to violet mineral, barium titanium silicate ($BaTi(SiO_3)_3$), found in San Benito, Texas. It forms hexagonal system tabular, triangular crystals and is a valuable gem when transparent and without flaws. Hardness 6-6.5; r.d. 3.6.

Bentham, George (1800–84) British botanist. His classification of GYMNOSPERMS (seed plants) provided a foundation for modern systems of classification. With William Hooker (1785–1865) he wrote the multi-volume *Genera Plantarum* (1862–83), a definitive work in its time. Bentham also wrote *Handbook of British Flora* (1858).

benthos Flora and fauna of the seabed. The fauna includes sedentary forms such as sponges, crawling animals such as crabs and snails, and burrowing animals such as worms, and countless bacteria.

Benz, Karl (1844–1929) German pioneer of the INTERNAL COMBUSTION ENGINE. After some success with an earlier TWO-STROKE ENGINE, he built a FOUR-STROKE ENGINE in 1885 that was first applied to a tricycle. Benz achieved great success when he installed the new engine in a four-wheel vehicle in

1893. Benz was the first to make and sell light, self-propelled vehicles built to a standardized pattern. Hundreds had been built by 1900.

benzaldehyde *See* BENZENECARBALDEHYDE

benzene (C_6H_6) Colorless, volatile, sweet-smelling, flammable liquid HYDROCARBON, a product of petroleum refining. A benzene molecule is a hexagonal ring of six unsaturated carbon atoms (benzene ring), and is the simplest AROMATIC COMPOUND. It is a raw material for manufacturing many organic chemicals and plastics, drugs, and dyes. Benzene is carcinogenic and should be handled with caution. Properties: r.d. 0.88; m.p. 5.5°C (41.9°F); b.p. 80.1°C (176.2°F).

benzenecarbaldehyde (benzaldehyde, C_6H_5CHO) Simplest of the AROMATIC compounds in which an aldehyde group, –CHO, is attached directly to a benzene ring. It is a colorless liquid having a smell of bitter almonds; it is used as a chemical reagent, flavoring material, and for making perfumes and dyes. Properties: r.d. 1.04; m.p. −26°C (−14.8°F); b.p. 178.1°C (352.6°F)

benzenecarboxylic acid (benzoic acid, C_6H_5COOH) White, crystalline weak CARBOXYLIC ACID made from toluene. It is used to make dyes and for preserving fruit juices. Properties: r.d. 1.266 (15°C); m.p. 122.4°C (252.3°F); b.p. 249°C (480.2°F).

benzodiazepine Any of a group of mood-altering DRUGS, such as Librium and Valium, that are used primarily to treat severe anxiety or insomnia. These drugs, which intervene in the transmission of nerve signals in the CENTRAL NERVOUS SYSTEM, were originally developed as muscle relaxants. Today they are widely prescribed as tranquilizers. Benzodiazepines may produce a number of side effects, including drowsiness, unsteadiness, and confusion. Also, they are known to cause dependence. For these reasons, they are recommended only for short-term use.

benzoic acid *See* BENZENECARBOXYLIC ACID

benzoin Fragrant, resinous POLYMER once obtained from the balsam resin found in the trees of the genus *Styrax* in tropical SE Asia, but now made synthetically. It is used in perfumes and in decongestant cough syrups.

benzyne (C_6H_4) Unstable organic compound derived from BENZENE. Benzyne molecules consist of a hexagonal ring of carbon atoms joined by two double bonds and one triple bond. It probably exists only briefly in various reactions of AROMATIC COMPOUNDS.

Bergius, Friedrich (1884–1949) German chemist who developed a method of treating coal with hydrogen, under high pressure, to produce oil. It became known as the **Bergius process**, and in 1931 Bergius shared the Nobel Prize for chemistry with Carl BOSCH. He also developed a process for converting wood into sugar by HYDROLYSIS.

bergschrund Deep, wide crevice or a series of parallel narrow crevices in a GLACIER, produced by tension within the ice, often at the point where the moving ice pulls away from the rock slope at the head of the glacier CIRQUE.

berkelium (symbol Bk) Synthetic radioactive, metallic element of the ACTINIDE SERIES. It was first made in 1949 by alpha-particle bombardment of americium-241 at the University of California at Berkeley (after which it is named). Nine isotopes are known. Properties: at.no. 97; r.d. (calcu-

▲ **benzene** The Kekulé structures (named after the German chemist Friedrich Kekulé) for benzene, once generally accepted, usually represent the molecule as a six-membered, hexagonal carbon ring with alternating single and double bonds. However, the carbon-to-carbon bond lengths are equal, intermediate between single and double bonds, and the double bonds do not belong to any particular bond. The six p-electrons are considered to be delocalized and free to move within the overlapping p-orbitals of all six carbon atoms.

Key 1) The Kekulé structures representation for benzene. 2) The double bonds in the Kekulé formulae may be represented by the overlapping of p-orbitals in two different ways. 3) The p-orbitals can overlap equally in both directions, forming a "molecular orbital," which may be represented as a pair of thick rings, or ring-shaped electron clouds.

lated) 14; m.p. 986°C (1,807°F); most stable isotope ^{247}Bk (half-life 1.4×10^3 yr).

berm Narrow ridge of debris above the foreshore. During storms, extra large waves carry debris up the beach so that it forms a narrow horizontal shelf or ridge above the normal high-water mark.

Bernal, John Desmond (1901–71) Irish physicist, remembered for his work in the field of X-RAY CRYSTALLOGRAPHY; he used the technique to study atomic structures. He was professor of physics, and then of crystallography, at London University (1938–68).

Bernard, Claude (1813–78) French physiologist. He defined the role of the PANCREAS in digestion, the role of the LIVER in regulating blood sugar levels and carbohydrate stores, and the regulation of blood supply by vasomotor nerves.

Bernoulli, Daniel (1700–82) Swiss mathematician and physicist, member of a famous family of mathematicians. His work on HYDRODYNAMICS demonstrated that pressure in a FLUID decreases as the velocity of fluid flow increases. This fact, which explains the LIFT of an aircraft wing, has become known as BERNOULLI'S LAW.

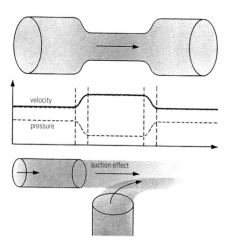

▲ **Bernoulli's law** A moving gas must obey the principle of conservation of momentum. If viscosity is ignored, and if flow is streamlined, this principle predicts the Bernoulli effect (often known as Bernoulli's law): if the flow-rate is speeded up (for example, by making a tube narrower at one place) the pressure falls. This effect is responsible for the familiar suction of gas from an orifice (such as a chimney) by a rapid stream of gas across it.

Bernoulli's law For a steadily flowing FLUID (liquid or gas), the sum of the pressure, kinetic energy per unit volume and potential energy per unit volume is constant at any point in the fluid. Using this relationship, which was formulated by Daniel BERNOULLI, it is possible to measure the velocity of a fluid by measuring its pressure at two points, as with a MANOMETER or PITOT TUBE.

Berthelot, Pierre Eugène Marcelin (1827–1907) French chemist. By making some organic compounds in the laboratory, including acetylene (ethyne), ethanol, methanol, benzene, and methane, he helped to prove that the distinction between organic compounds (as substances formed only in living things) and inorganic compounds was wrong. In the 1860s he did important work in THERMOCHEMISTRY. He created the terms ENDOTHERMIC and EXOTHERMIC reactions.

beryl Mineral, beryllium aluminum silicate. Its crystals are usually hexagonal prisms of the hexagonal system. Gemstone varieties are aquamarine (pale blue-green) from Brazil; emerald (deep green) from Colombia; and morganite (pink) from Madagascar. Cut stones have little brilliance, but are valued for their intense color. Hardness 8; r.d. 2.6–2.8.

beryllium (symbol Be) Strong, light, silver-gray member of the ALKALINE-EARTH METALS, first isolated in 1828. Beryllium occurs in many minerals including aquamarine, emerald, and morganite (all forms of BERYL), and is used in alloys that combine lightness with rigidity. Properties: at.no. 4; r.a.m. 9.012; r.d. 1.85; m.p. 1,285°C (2,345°F); b.p. 2,970°C (5,378°F); most common isotope ^{9}Be (100%).

Berzelius, Jöns Jakob, Baron (1779–1848) Swedish chemist, one of the founders of modern chemistry. Berzelius' accomplishments include the discovery of cerium, selenium, and thorium; the isolation of the elements silicon, zirconium, and titanium; the determination of RELATIVE ATOMIC MASSES; and the devising of a modern system of chemical symbols. He prepared the first table of relative atomic masses and contributed to the founding of the theory of atomic RADICALS. *See also* PERIODIC TABLE

Bessel, Friedrich Wilhelm (1784–1846) German astronomer and mathematician. He was the first astronomer to use a technique called PARALLAX to measure how distant a star is from the Earth using the apparent motion of the star relative to the Earth as the Earth moves in its orbit. The parallax of the binary star Cygni was found to be 0.3 seconds of ARC, which indicated that the star was just over 10 light-years away from Earth. He devised **Bessel functions,** a type of mathematical function used to described the observed perturbations of the planets and stars.

Bessemer process First method for the mass production of steel. The process was patented in 1856 by the British engineer and inventor Sir Henry Bessemer (1813–98) – although a US steelmaker, William KELLY had used a similar process in c.1850. In a **Bessemer converter**, pig iron is converted into steel by blowing air through the molten iron to remove impurities. Precise amounts of carbon and metals are then added to give the desired properties to the steel.

Best, Charles Herbert (1899–1978) Canadian physiologist. He and Frederick BANTING discovered INSULIN in 1921. He was head of the department of physiology at the University of Toronto (1929–65) and chief of the Banting-Best department of medical research there after Banting's death.

beta-blocker Any of a class of DRUGS that block impulses to certain nerve receptors (beta receptors) in various tissues throughout the body, including the heart, airways, and peripheral arteries. These drugs are mainly prescribed to regulate the heartbeat, reduce blood pressure, relieve ANGINA, and improve survival following a heart attack. However, they are also being used in an increasingly wide range of other conditions, including GLAUCOMA, liver disease, thyrotoxicosis, MIGRAINE, and anxiety states. Beta-blockers are not suitable for all heart patients and are not used for patients with asthma or severe lung disease.

beta particle (beta ray) Particle emitted spontaneously by certain radioactive ISOTOPES undergoing RADIOACTIVE DECAY called **beta decay**. Beta particles were discovered in 1876 by Henri BECQUEREL, and are now known to be energetic ELECTRONS. Beta decay results when a NEUTRON (in the nucleus) is converted into a PROTON, with the emission of an electron (beta particle) and an antineutrino or NEUTRINO. They cause IONIZATION along their path and have fairly high penetrating power. Their energy, ranging from 0.003–13 MeV, is characteristic of the emitting isotope.

betatron Type of particle ACCELERATOR consisting of a hollow, evacuated circular ring in which ELECTRONS are confined and accelerated by a rapidly changing magnetic field. When they reach very high energies, on the order of 350MeV, they are focused magnetically to strike a target. The output can be either an electron beam or secondarily produced X rays.

Betelgeuse (Alpha Orionis) Red SUPERGIANT star and the second-brightest in the constellation of Orion. It is a pulsating variable whose diameter

fluctuates between 300–400 times that of the Sun. Characteristics: apparent mag. 0.85 (mean); absolute mag. −5.5 (mean); spectral type M2; distance 500 light-years.

Bethe, Hans Albrecht (1906–) US nuclear physicist, b. Germany. He left Germany when Hitler came to power, going first to Britain and then to the US to become professor of theoretical physics at Cornell University (1935–75) until his retirement. He worked on stellar energy processes and helped to develop the atomic bomb. He is noted for his theories on atomic and nuclear properties and was awarded the 1967 Nobel Prize for physics for his work on the origin of solar and stellar energy. *See also* GAMOW, GEORGE

BeV Abbreviation formerly used for one billion electron volts (eV), a unit of energy equal to a thousand million (10^9) electron volts; now symbolized by GeV.

bicarbonate of soda Popular name for SODIUM HYDROGENCARBONATE

biceps Large muscle on the front of the upper arm, easily felt when the arm is bent. The biceps contracts to raise the forearm towards the upper arm, and also contracts to turn the inturned hand outwards (flexion).

bicycle Two-wheeled vehicle propelled by the rider. The earliest design (a hobbyhorse-type bicycle) dates from c.1790. In 1816 German engineer Karl Drais von Sauerbronn invented a steerable bicycle. It is generally accepted that the first pedal-operated bicycle was invented in 1839 by Kirkpatrick Macmillan, a Scottish blacksmith. In c.1861 French engineers Pierre and Ernest Michaux demonstrated their "boneshaker" velocipede with pedals attached to the front wheel. In 1871 Englishman James Starley produced his "penny-farthing" bicycle. In 1873 J.H. Lawson invented the chain drive that was incorporated into John Starley's modern safety bicycle with tangential-spoked wheels (1885). In 1888 Scottish inventor John Dunlop produced the pneumatic tyre.

biennial Plant that completes its life cycle in two years, producing flowers and seeds during the second year; an example is the onion. This distinguishes it from an ANNUAL, which germinates, flowers, fruits, and dies in one season, and a PERENNIAL, a plant that lives for three years or more.

Big Bang In COSMOLOGY, theory developed from the ideas of Georges LEMAÎTRE and advanced in the 1940s by George GAMOW. According to Big Bang theory, a giant explosion 10 to 20 billion years ago began the expansion of the Universe, which still continues. Everything in the Universe once constituted an exceedingly hot and compressed gas with a temperature exceeding 10 billion degrees. When the Universe was only a few minutes old, its temperature would have been 1,000 million degrees. As it cooled, nuclear reactions took place that led to material emerging from the fireball consisting of c.75% hydrogen and c.25% helium by mass – the composition of the Universe as we observe it today. Slightly denser regions of gas, whose expansion rate lagged behind, collapsed to form GALAXIES when the Universe was perhaps a tenth of its present age. The cosmic microwave BACKGROUND RADIATION detected in 1965 is considered to be the residual radiation of the Big Bang. *See also* OSCILLATING UNIVERSE THEORY; STEADY-STATE-THEORY

Big Dipper *See* URSA MAJOR

bile Bitter yellow, brown or green alkaline fluid, secreted by the LIVER and stored in the GALL BLADDER. Important in digestion, it enters the duode-num via the bile duct. The bile salts it contains emulsify fats (allowing easier digestion and absorption) and neutralize stomach acids. Its color comes from the breakdown products of ERYTHROCYTES (red blood cells).

bile acids Group of steroid acids present in BILE. In humans, the commonest is cholic acid ($C_{24}H_{40}O_5$), which is conjugated by its carboxyl group to the amino groups of the AMINO ACIDS glycine and taurine. The bile acids are emulsifiers for fats and fat-soluble vitamins, thus promoting their absorption by the intestine. *See also* CARBOXYLIC ACIDS

bilirubin Pigment derived from blood. When ERYTHROCYTES (red blood cells) have reached the end of their useful lives, they are broken down, mainly by the LIVER. Useful materials, such as the heme of HEMOGLOBIN, are recycled in the body. Among the waste materials is the green-orange pigment bilirubin, which is converted into stercobilin, the coloring matter of feces.

bimetallic strip Device used in THERMOSTATS and mechanical thermometers. It consists of bonded strips of two metals with dissimilar coefficients of thermal expansion. When heated the bimetallic strip bends because one metal expands more than the other. This reversible distortion is used to move a dial pointer or to open or close a switch in an electrical circuit.

binary fission *See* FISSION

binary star One of two stars in orbit around a common center of mass. **Visual** binaries are those whose components can be seen as separate stars with the naked eye or through a telescope. In an **eclipsing** binary, one star periodically passes in front of the other, so that the total light output appears to fluctuate. Most eclipsing binaries are also **spectroscopic** binaries, systems in which the components are too close for their separation to be measured visually and must be measured spectroscopically. *See also* SPECTROSCOPY

binary system In mathematics, number system having a BASE of 2 (the decimal system has a base of 10). It is most appropriate to computers since it is simple and corresponds to the open (0) and the closed (1) states of a switch, or logic GATE, on which computers are based.

binding energy Energy that must be supplied to an atomic nucleus in order to split it into its constituent nucleons (neutrons and protons). A nucleus must be supplied with its binding energy before it will undergo FISSION (except in the case of

▲ **binary star** Binary stars, such as Sirius A and B, are double stars that revolve around each other. In fact, they revolve about their common center of gravity (G), which is not halfway between them but nearer to the more massive component. Moreover, they do not rotate in circles around G but in co-planar ellipses so that the distance between them varies. The rate of rotation, too, is not uniform. In (2) they are greatly separated and moving slowly. In (6) they are closer and moving more rapidly. (7) shows the proper motion of Sirius A and B over a century. The paths of the two components in space are drawn out to elliptical helices and the diagram also shows that the direction of the proper motion is not perpendicular to the plane of the ellipses.

radioactive decay). The mass of a nucleus is slightly less than the mass of its constituent particles. According to EINSTEIN's law $E = mc^2$, this difference in mass is equivalent to the energy released when the nucleons bind together. This is the energy source of the hydrogen bomb and the FUSION reaction. *See also* NUCLEAR ENERGY

binoculars Optical instrument, used with both eyes simultaneously, that produces a magnified image of a distant object or scene. It consists of a pair of identical telescopes, one for each eye, both containing an objective lens, an eyepiece lens, and an optical system (usually prisms), to form an upright image.

binomial nomenclature System of categorizing organisms by giving them a two-part Latin name. The first part of the name is the GENUS and the second part the SPECIES. For example, *Homo sapiens* is the binomial name for humans. The system was developed by the Swedish botanist Carolus LINNAEUS in the 18th century. *See also* CLASSIFICATION

binomial theorem Mathematical rule for the expansion of the algebraic expression $(x + y)^n$ as a series in powers of the numerical quantities x and y (n is a positive integer). For $n = 2$, its expansion is given by $(x + y)^2 = x^2 + 2xy + y^2$

bioaccumulation Progressive increase in the levels of harmful substances, such as HEAVY METALS and chemicals, notably pesticides, in living organisms. Many such substances are not BIODEGRADABLE. Furthermore these substances cannot be broken down by the organisms themselves, and so tend to accumulate in their bodies, particularly in fatty tissues. For example, in the 1960s mercury that was being discharged by factories, was accummulating at each level of the FOOD CHAIN in organisms around the Japanese town of Minamata. The contamination of the higher sea predators, that is the fish that formed an important element of the diet of the town, resulted in the death of 40 people.

biochemical oxygen demand (BOD) Chemical test for determining the level of pollution of water by organic matter. Two equal samples of water are taken, and the first is treated chemically to "mop up" any dissolved oxygen. Both samples are then incubated in the dark for some days, after which the second sample is given a similar chemical treatment. This allows the amount of mopped-up oxygen to be measured. The difference between the two measurements represents the amount of oxygen that has been used up by living and dead organic matter in the water during the period of storage.

biochemistry Science of the CHEMISTRY of life. It attempts to use the methods and concepts of organic and physical chemistry to investigate living matter and systems. Biochemists study both the structure and properties of all the constituents of living matter (such as FATS, PROTEINS, ENZYMES, HORMONES, VITAMINS, DNA, CELLS, MEMBRANES, and ORGANS) together with the complex reactions and pathways of these in METABOLISM.

biodegradable Property of a substance that enables it to be decomposed by microorganisms. The end result of the decomposition is stable, simple compounds (such as water and carbon dioxide). This property has been designed into products such as plastics to aid trash disposal and reduce pollution.

biodiversity Wide variety of plant and animal SPECIES, genetic strains, and the ECOSYSTEMS that support them. The causes and consequences of biodiversity comprise a major topic of research in ECOLOGY, PALEONTOLOGY, and EVOLUTIONARY BIOLOGY. The maintenance of biodiversity is a major aim of conservationists. It is estimated that less than 10% of the Earth's surface is host to nearly 75% of its species. Many of the world's most species-rich habitats, such as RAINFORESTS and WETLANDS, are also the most threatened by human development. The preservation of the biodiversity of species was a major aim of the EARTH SUMMIT (1992).

bioelectricity Electricity generated in plants and animals. In animals bioelectricity is associated with nerve impulses and muscle contractions. Different electric POTENTIALS are built up within the organism by the process of ionic separation across a membrane. Some fish have electric organs that create more powerful external currents used for sensing or to stun prey.

bioengineering Application of engineering techniques to medical and biological problems. It includes the engineering of devices to aid or replace defective or inadequate body parts, as in the production of artificial limbs and hearing aids.

biofeedback In alternative medicine, the use of monitoring systems providing changing information about body processes to enable them to be controlled voluntarily. By observing data on events that are normally involuntary, such as breathing and the heartbeat, many people learn to gain control over them to some extent in order to improve well-being. The technique has proved helpful in a number of conditions, including headaches and hypertension.

biogas Fuel gas derived from the decomposition of biological material. It is a mixture consisting mainly (up to 60%) of methane, which is combustible, with some carbon dioxide. It is formed by the ANAEROBIC breakdown – termed FERMENTATION – of waste matter such as domestic and agricultural sewage. The waste is contained in a digester where it is acted on by methanogenic bacteria.

biogenesis Biological principle maintaining that all living organisms derive from parent(s) generally similar to themselves. This long-held principle was originally established in opposition to the idea of SPONTANEOUS GENERATION of life. On the whole, it still holds good, despite variations in individuals caused by mutations, hybridization, and other genetic effects. *See also* GENETICS

biogenetic law (recapitulation theory) Principle that the stages that an organism goes through during embryonic development reflect the stages of that organism's evolutionary development.

biogeography Study of the geographical distribution of plants and animals.

biological clock Internal system in organisms that relates behavior to certain rhythms. Functions, such as growth, feeding or reproduction, coincide with certain external events, including day and night, tides, and seasons. This innate sense of timing makes some animals feed during the day when food is available and they can see best. These "clocks" seem to be set by environmental conditions, but if organisms are isolated from these conditions, they still function according to the usual rhythm. If conditions change gradually, the organisms adjust their behavior gradually. *See also* CIRCADIAN RHYTHM

biological control Use of biological methods to control pests. Usually the term describes the deliberate introduction of a parasite or other natural enemy of the pest, and is often preferred to the use of chemical pesticides. Examples include the introduction of ladybirds to kill insects that infested US citrus crops, and the use of cactus moth caterpillars to control prickly pear cactus on Australian farmland. In another method, successfully employed to control screw worm flies that parasitize cattle, male flies sterilized by radiation are released into the natural population to produce a reduction in the fly population (because of the preponderance of sterile matings). Parasitic wasps are also commercially available to gardeners to control white fly and aphids in greenhouses. Care must be taken with biological control so as to avoid causing unwanted ecological "side effects".

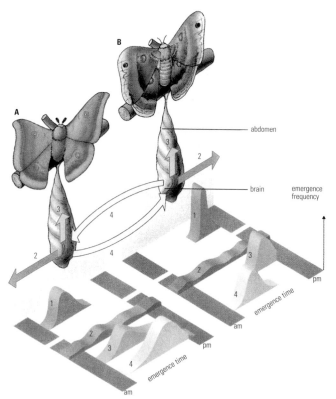

◄ **biological clock** In many insects (A and B) the master clock lies in the brain. It uses hormones to exert its influence on many processes, including the emergence of the adult from the pupa during metamorphosis. The graphs show (1) normal emergence times of the Chinese oak silkmoth (A) and the robin moth (B). When parts of their brains are removed hormonal control is lost and the moths emerge at random times (2). If the brain is replaced in the abdomen of the brainless pupa, normal rhythm is restored (3). If the brains of the two species are exchanged, the moths can be made to swap emergence rhythms (4).

abdomen

brain

emergence frequency

emergence time

pm

am

pm

am

biological shield Thick wall containing such materials as concrete, steel, magnetite, and lead, used to protect workers in nuclear power stations and radiochemical laboratories from the possible harmful effects of radiation. In remote-handling "caves" the biological shield contains a viewing window about 30cm (12in) thick, also with radiation absorbers.

biological warfare Use of disease microbes and their toxins in warfare. The extensive use of mustard gas during World War I prompted the prohibition of biological and CHEMICAL WARFARE by the Geneva Convention (1925). However, many nations have maintained costly research programs for the production of harmful microorganisms and discovery of more effective antidotes to their pathogenic effects. These microbes include plant pathogens for the destruction of food crops and agents of disease that are harmful to human populations. During World War II, the British army's experiments with the predominantly cattle and sheep disease ANTHRAX on a remote Scottish island ensured the island remained uninhabited for some 40 years. Recent developments in GENETIC ENGINEERING have made the production of highly dangerous microorganisms ever more likely.

biology Science of life and living organisms. Its branches include BOTANY, ZOOLOGY, ECOLOGY, PHYSIOLOGY, CYTOLOGY, GENETICS, TAXONOMY, EMBRYOLOGY, and MICROBIOLOGY. These sciences deal with the origin, history, structure, development, and function of living organisms, their relationships to each other and their environment, and the differences between living and nonliving organisms.

bioluminescence Production of light, with very little heat, by some living organisms. Its biological function is varied: in some species, such as fireflies, it is a recognition signal in mating; in others, such as squid, it is a method of diverting predators for protection; and in deep-sea anglerfish it is used as a lure to attract prey. The light-emitting substance (luciferin) in most species is an organic molecule that emits light when it is oxidized by molecular oxygen in the presence of an enzyme (luciferase). Each species has different forms of luciferin and luciferase. *See also* FLUORESCENCE

biomass Total mass (excluding water content) of the plants and/or animals in a particular place. The term is often used to refer to the totality of living things on Earth; or those occupying a part of the Earth, such as the oceans. It may also refer to plant material that can be exploited, either as fuel or as raw material for an industrial or chemical process.

biome Region with a characteristic FAUNA and FLORA determined by climate. Ecologists divide

the Earth (including the seas, lakes, and rivers) into roughly ten biomes. *See also* ECOSYSTEM

biophysics Study of biological phenomena in terms of the laws and techniques of physics. Techniques include those of X-RAY DIFFRACTION and SPECTROSCOPY. Subjects studied include the structure and function of biological molecules, the conduction of electricity by nerves, the visual mechanism, the transport of molecules across cell membranes, muscle contraction (using electron microscopy) and energy transformations in living organisms.

biopsy Removal of a small piece of tissue from a patient so that it can be examined in the laboratory for evidence of disease. A common example is the cervical biopsy (PAP SMEAR), performed in order to screen for the precancerous changes that may lead to cervical cancer.

biorhythm Any regular pattern of changes in METABOLISM or activity in living things. These are usually synchronized with daily, monthly, seasonal or annual changes in the environment. Examples of daily, or circadian, rhythms are the opening and closing of flowers, feeding cycles of animals during the day or the night, the response of marine organisms to the tides and, in humans, changes in body temperature and blood pressure. Monthly changes include the MENSTRUAL CYCLE in women. Annual rhythms include hibernation, migration, and reproductive activity. The internal mechanism of these rhythms (sometimes referred to as the BIOLOGICAL CLOCK) is not yet fully understood.

biosphere (zone of life) Portion of the Earth from its CRUST to the surrounding ATMOSPHERE, encompassing and including all living organisms, animal and vegetable. It is self-sufficient except for energy and extends a few kilometers above and below sea level.

biosynthesis Process in living CELLS by which complex chemical substances, such as PROTEINS, are made from simpler substances. A GENE "orders" a molecule of RNA to be made, which carries the genetic instructions from the DNA. On the RIBOSOMES of the cell, the protein is built up from molecules of AMINO ACIDS, in the order determined by the genetic instructions carried by the RNA.

Biot, Jean Baptiste (1774–1862) French astronomer, mathematician, and physicist who made fundamental discoveries about POLARIZED LIGHT. He developed the technique of estimating the concentration of certain organic substances (particularly sugars) by measuring the angle through which they rotate the plane of polarization of light.

biotic factors Biological factors of an organism's environment or ecosystem that result from the activities of other living organisms. It can be contrasted with ABIOTIC FACTORS.

biotechnology Use of biological processes for medical, industrial or manufacturing purposes. Humans have long used yeast for brewing and bacteria for products such as cheese and yogurt. Biotechnology now enjoys a wider application. By growing microorganisms in the laboratory, new drugs and chemicals are produced. GENETIC ENGINEERING techniques of cloning, splicing, and mixing genes facilitate, for example, the growing of crops outside their normal environment, and the

ripe tomato

new variety

▲ **biotechnology** In a ripe tomato, rotting is caused by an enzyme formed by the copying of a gene in the plant DNA (1) in a messenger molecule mRNA (2). The mRNA is changed into the enzyme (3) which damages the cell wall (4). In a genetically altered tomato, a mirror duplicate of the gene that starts the process is present (5). The result is that two mirror-image mRNA molecules are released (6), and they bind together preventing the

creation of the rotting enzyme. Introducing the necessary DNA through the rigid cell wall is done by using a bacteria (7) that naturally copies its own DNA onto that of a plant. It is easy to introduce the mirror DNA (8) into the bacteria and once the bacteria has infected the cell the DNA is transferred (9). All cells then replicated have the new DNA in their chromosomes and can be grown to create the new variety of plant.

production of vaccines to fight specific diseases. Hormones are also produced, such as INSULIN for treating diabetes.

biotin ($C_{10}H_{16}O_3N_2S$) Member of the VITAMIN B complex. It is vital to the body, but is needed in only minute quantities, which are synthesized by intestinal bacteria.

biotite Common mineral of the MICA group. It is a silicate of aluminum, iron, potassium, and magnesium. Its color ranges from greenish-brown to black. Its lustrous, monoclinic crystals are opaque to translucent, and cleave to form flexible sheets. It is found in IGNEOUS ROCK (such as granite), METAMORPHIC ROCK (such as schist and gneiss) and SEDIMENTARY ROCK.

bird Any one of about 8,600 species of feathered VERTEBRATES of the class Aves. They occupy most natural habitats from deserts and tropics to polar wastes. Birds are warm-blooded and have forelimbs modified as wings, hind-limbs for walking and jaws elongated into a toothless beak. They lay eggs (usually in nests), incubate the eggs and care for young. As a group they feed on seeds, nectar, fruit, and carrion, and hunt live prey ranging from insects to small mammals, although individual species may be very specialized in their diet. Sight is the dominant sense, smell the poorest. Size ranges from the bee hummingbird, 6.4cm (2.5in) to the wandering albatross, whose wingspread reaches 3.5m (11.5ft). The 2.5m (8ft) tall ostrich is the largest of living birds, but several extinct flightless birds were even bigger. Of the 27 orders of birds, the perching birds (Passeriformes) include more species than all others combined. A bird's body is adapted primarily for flight, with all its parts modified accordingly. There are several groups of large flightless birds, including the ostrich, rhea, emu, cassowary, kiwi, and penguin. Birds are descended from Theocodonts (reptiles), and the first fossil bird, ARCHAEOPTERYX, dates from late Jurassic times.

birth (parturition) Bringing forth of live, partly or fully formed offspring. All mammals (except the echidna and the platypus), some reptiles and sharks, and various insects and other invertebrate animals give birth to live young (see VIVIPARITY). All birds, most reptiles, amphibians, and fish, and the majority of insects lay eggs from which the live young later emerge. This process is called hatching (see OVIPARITY). In humans, birth involves the delivery of the FETUS at the end of pregnancy. There are three stages in this process. First, contractions of the UTERUS begin and the CERVIX dilates in readiness; the sac containing the amniotic fluid ruptures. In the second stage the contractions strengthen and the baby is propelled (usually head first) through the birth canal. The third and final stage, once the baby has been born, is the expulsion of the placenta and fetal membranes, together known as the afterbirth.

birth control Alternative term for CONTRACEPTION

bismuth (symbol Bi) Metallic silver-white element of group V of the periodic table, first identified as a separate element in 1753. The chief ores are bismite (Bi_2O_3) and bismuthinite (Bi_2S_3). A poor heat conductor, it is put into low-melting alloys used in automatic sprinkler systems. Bismuth is also used in insoluble compounds to treat gastric ulcers and skin injuries. It expands when it solidifies, a property exploited in several bismuth alloys used in making castings. Properties: at.no. 83; r.a.m. 208.98; r.d. 9.75; m.p. 271.3°C (520.3°F); b.p. 1,560°C (2,840°F); most common isotope ^{209}Bi (100%).

bit In computing, abbreviation for binary digit, a 1 or 0 as used in the BINARY SYSTEM. A bit is the smallest element of information storage. Groups of bits form a BYTE of binary code representing letters and other characters. Binary code is used in computing because it is easy to represent each 1 or 0 using electrical components that can be switched between two states (such as on and off). The code is also easy to store on disk as a magnetic or optical pattern.

bitumen (asphalt) Material used for roadmaking and for proofing wood against rot. It consists of a mixture of hydrocarbons and other organic chemical compounds. Some bitumen occurs naturally in pitch lakes, notably in Trinidad and Venezuela. The material is also made by distilling tar from coal or wood, and a little is obtained during the refining of petroleum.

bivalent In biology, describing a pair of HOMOLOGOUS CHROMOSOMES formed during MEIOSIS. In chemistry, the term is used to describe an atom or group that has a VALENCE of two.

bivalve Any animal of the class Bivalvia, which has a shell with two halves or parts hinged together. The term most usually applies to a class of MOLLUSKS – Pelecypoda or Lamellibranchiata – with left and right shells, such as clams, cockles, mussels, and oysters. It also refers to animals of the phylum Brachiopoda (BRACHIOPOD) with dorsal and ventral shells. Length: 2mm–1.2m (0.17in–4ft).

Bjerknes, Vilhelm Friman Koren (1862–1951) Norwegian physicist who laid the foundation of a revolution in weather forecasting by applying theories of fluid forces and motions to the circulation of the atmosphere with the development of air masses, fronts, and cyclones.

Black, Joseph (1728–99) British chemist and physicist. Rediscovering "fixed air" (carbon dioxide), he found that this gas is produced by RESPIRATION, burning of charcoal, and FERMENTATION; that it behaves as an ACID; and that it is probably found in the atmosphere. He also discovered bicarbonates (hydrogencarbonates) and investigated LATENT HEAT and the concept of specific heat but was unable to reconcile it with the PHLOGISTON THEORY that said burning materials lose a substance to the atmosphere.

blackband iron ore Thin seam of dark-colored iron ore that occurs in layers of coal. Historically it was important to the development of iron and steel industries and often accounted for the location of smelters near coal mines. Once the blackband ore was exhausted, iron ore had to be imported and it became more economical to locate steel works near ports.

black body In physics, a hypothetical body that absorbs all incident radiation and reflects none. Such a body would look "perfectly" black – hence the name. The study of black bodies has been important in the history of physics. WIEN's law, Stefan's law (see BOLTZMANN) and PLANCK's law of black body radiation grew out of this study, as did Planck's discoveries in quantum mechanics.

black earth See CHERNOZEM

Blackett, Patrick Maynard Stuart, Baron (1897–1974) English physicist who was awarded the 1948 Nobel Prize for physics for his research on COSMIC RADIATION. He spent 10 years at the Cavendish Laboratory, Cambridge, developing the Wilson cloud chamber into an instrument to study cosmic radiation. He also used it to prove the existence of the POSITRON, the antiparticle of the ELECTRON. Blackett was professor of physics at Manchester University (1937–53) and then professor of physics at Imperial College, London (1953–65). *See also* ANTIMATTER

black hole Localized region of space from which neither matter nor radiation can escape – in other words, the escape velocity exceeds the velocity of light. The boundary of this region is called the EVENT HORIZON. Its radius, the **Schwarzschild radius**, depends on the amount of matter that has fallen into the region and it increases linearly as the mass increases. A black hole of stellar mass is thought to form when a massive star undergoes total gravitationalcollapse. For stars up to about 1.4 solar masses, gravitational collapse can be physically halted to produce a WHITE DWARF. A slightly more massive object will collapse to form a NEUTRON STAR. If, however, the mass exceeds about 3 solar masses, even after a supernova explosion has blasted away the outer layers of the star, the collapse continues beyond even the neutron star stage. As it contracts below its Schwarzschild radius, the object becomes a black hole and effectively disappears. A star of 3 solar masses has a Schwarzschild radius of about 9km (5.5mi). Inside the event horizon of the black hole, space and time are highly distorted and the stellar matter is increasingly compressed until it forms an infinitely dense singularity. Black holes can have an immense range of sizes. Supermassive black holes with up to a billion solar masses could be the source of energy in quasars and other types of active galaxy. At the other end of the scale, some primordial black holes (see below) could be truly microscopic. Since no light or other radiation can escape from black holes, they are extremely difficult to detect. Any matter encountered by the black hole will most likely go into orbit first rather than being drawn directly into it. A rapidly spinning disk of matter, known as an ACCRETION disk, forms around the object and heats up through friction to such high temperatures that it emits X rays. Black holes may therefore appear as X-ray sources in BINARY STARS. The most famous candidate is Cygnus X-1. Not all black holes result from stellar collapse. During the Big Bang, some regions of space might have become so compressed that they formed so-called **primordial black holes**. Such black holes would not be completely black, because radiation could still "tunnel out" of the event horizon at a steady rate, leading to the evaporation of the hole. Primordial black holes could thus be very hot. *See also* STELLAR EVOLUTION

Black, Sir James Whyte (1924–) British biochemist who in 1964 first isolated BETA-BLOCKERS, now used as drugs to reduce the activity of the heart in the treatment of angina and high blood pressure. He also discovered the drug cimetidine, employed to treat ulcers of the digestive tract by reducing the secretion of acids in the stomach. In 1988 he was awarded the Nobel Prize for physiology or medicine for his work on these drugs.

Blackwell, Elizabeth (1821–1910) US physician, b. England. In 1847 she began to study medicine at the Geneva Medical College in New York. She graduated in 1849, becoming the first woman doctor in the US. She established the New York Infirmary (1857), which combined health services and medical training.

bladder Large, elastic-walled organ in the lower abdomen in which URINE is stored. Urine passes from each KIDNEY by way of two narrow tubes (URETERS) to the bladder, where it is stored until it can be voided. When pressure in the bladder becomes too great, nervous impulses signal the need for emptying. Urine leaves the bladder through a tube called the URETHRA.

◀ **blast furnace** The blast furnace gains its name from the blasts of superheated air that are blown into the furnace. The furnace is charged from the top with coke, iron ore (collectively called sinter) and limestone. Air, heated by hot cowper stoves, is then blown through the blast pipes (tuyères) into the lower part of the furnace. The ore is converted to the liquid pig iron and trickles down into the hearth from where it is discharged every three or six hours. The limestone combines with impurities in the ore, primarily silica, to form a slag. The molten slag floats on the pig iron and is discharged through the slagnotch.

iron ore   coke   limestone

receiving hopper

distribution chute

hot-air stove

iron ore, coke and limestone charge

1,000°C (1,800°F)

exhaust gas recycled to heat stove

tuyère

1,600°C (2,900°F)

stag ladle

hot-metal car

blast furnace Cylindrical smelting furnace. It is used in the extraction of metals, mainly iron and copper, from their ores. The ore is mixed with coke and a FLUX (limestone in the case of iron ore). A blast of hot compressed air is piped in at the bottom of the furnace to force up temperatures to where the reduction of the oxide ore to impure metal occurs. The molten metal sinks to the bottom and is tapped off. Waste "slag" floats to the top of the metal and is piped off. *See also* OXIDATION-REDUCTION

blastula Stage in the development of the EMBRYO in animals. The blastula consists of a hollow cavity (**blastocoel**) surrounded by one or more spherical layers of cells. Commonly called the hollow ball of cells stage, it occurs at or near the end of cleavage (*see* MORULA) and precedes the GASTRULA stage. The blastula stage in mammals is known as a **blastocyst** or **germinal vessel**.

bleach Substance used to remove the color from materials such textiles and paper. Most bleaches act by oxidizing the pigment. Chlorine, or compounds such as calcium and sodium hypochlorites, are commonly used. Other oxidizing bleaches are hydrogen peroxide, used for hair, and sodium perborate. For some applications reducing agents, such as sulfur dioxide, are employed. *See also* OXIDATION

bleaching powder White powder of imprecise composition used for bleaching. It is made by bubbling chlorine gas through calcium hydroxide (slaked lime) solution, and contains calcium chlorate and calcium chloride. It releases chlorine when treated with dilute acid and is used for purifying water and bleaching cloth and wood pulp for paper-making.

bleeding Loss of blood from the circulation. *See* HEMORRHAGE

Blériot, Louis (1872–1936) French aircraft designer and aviator. In 1909 Blériot became the first person to fly an aircraft across the English Channel. The flight from Calais, N France to Dover, S England took 37 minutes. As a designer, Blériot was responsible for a number of innovations, including a system by which the pilot could operate ailerons by remote control.

blindness Severe impairment of, or absence of, vision. It may be due to heredity, accident, disease or old age. Worldwide, the commonest cause of blindness is TRACHOMA. In developed countries it is most often due to severe DIABETES, GLAUCOMA, CATARACT or degenerative changes associated with aging.

blind spot Small area on the RETINA of the EYE where no visual image can be formed because of the absence of light-sensitive cells, the RODS and CONES. It is the area where the OPTIC NERVE leaves the eye.

BL Lacertae object (BL Lac object) Highly luminous object located in the nuclei of some galaxies, the prototype of which, BL Lacertae, was originally classified as a variable star. BL Lac objects are variable at all wavelengths from radio to X rays, sometimes over a time scale of just a few hours, and their optical spectra are unusual in being completely featureless, containing neither absorption nor emission lines. In this respect they differ from QUASARS. BL Lac objects are thought to lie in the centers of gas-free galaxies such as giant ellipticals.

Bloch, Felix (1905–83) US nuclear physicist, b. Switzerland. From 1934 he was professor of theoretical physics at Stanford University. He shared the 1952 Nobel Prize for physics with Edward PURCELL for their separate development of the technique of NUCLEAR MAGNETIC RESONANCE (NMR) used to study the interactions between atomic nuclei. Bloch was the first director (1954–55) of the Conseil Européen pour la Recherche Nucléaire (CERN), the European center in Geneva for research into high-energy PARTICLE PHYSICS.

Bloch, Konrad Emil (1912–) US biochemist, b. Germany. He studied METABOLISM of FATTY ACIDS, and in particular the way CHOLESTEROL is synthesized in the body. For this work he shared the 1964 Nobel Prize for physiology or medicine with Feodor LYNEN.

block mountain Uplift that is the result of block faulting. This is one type of FAULT in which the crustal portions of the Earth are broken into structural blocks of different elevations and positions.

blocky lava Surface of hot molten lava covered with a "skin" of large angular lava blocks formed by the break-up of a solidified surface of a lava flow. Blocky lavas are characteristic of high viscosity lavas with relatively high silica content.

blood Fluid circulating throughout the body that transports oxygen and nutrients to all the cells and removes wastes such as carbon dioxide. In a healthy human, it constitutes about 5% of the body's total weight; by volume, it comprises about 5.5 liters (11.6 pints). It is composed of a colorless, transparent fluid called plasma in which are suspended microscopic ERYTHROCYTES (red blood cells), which make up nearly half the blood volume; LEUCOCYTES (white blood cells); and PLATELETS.

blood-brain barrier Mechanism that prevents harmful substances from reaching the brain via the bloodstream. The mechanism involves various defensive strategies involving the permeability of cells that prevent "foreign" particles from entering the CEREBROSPINAL FLUID. A semi-permeable lipid membrane allows the passage of solutions (containing essential chemicals) but blocks large molecules such as harmful microorganisms and most drugs.

blood clotting Protective mechanism that prevents excessive blood loss after injury. A mesh of tight fibers (of insoluble FIBRIN) coagulates at the site of injury through a complex series of chemical reactions. This mesh traps blood cells to form a clot which dries to form a scab. This prevents further loss of blood, and also prevents bacteria getting into the wound. Normal clotting takes place within five minutes. The clotting mechanism is impaired in some diseases such as HEMOPHILIA.

blood group Any of the types into which blood is classified according to which ANTIGENS are present on the surface of its red cells. There are four

A   B

A   B

1
2
3
4
5

▲ **blood vessel** Arteries (A) and veins (B) conduct blood around the body. They have a common structure consisting of four layers: a protective fibrous coat (1); a layer of smooth muscle and elastic tissue (2), which is thickest in the largest arteries; a thin layer of connective tissue (3); and a smooth layer of cells – endothelium (4). Arteries have thicker walls and a smaller diameter. In veins, inner coat layers are often hard to distinguish. A comparison of the two vessels is shown in half sections of arteries and veins found in the body. Arteries divide into smaller ones, and finally into arterioles (5), where the blood flow can be controlled by autonomic nerves supplying the smooth muscle.

major types: A, B, AB, and O. Each group in the ABO system may also contain the rhesus factor (Rh), in which case it is Rh-positive; otherwise it is Rh-negative. Such typing is essential before BLOOD TRANSFUSION, since using blood of the wrong group may produce a dangerous or even fatal reaction. *See also* LANDSTEINER, KARL

blood plasma *See* PLASMA

blood pressure Force exerted by circulating BLOOD on the walls of blood vessels due to the pumping action of the HEART. Blood pressure is measured, using a gauge known as a SPHYGMO-MANOMETER, in millimeters of mercury. It is greatest when the heart contracts and lowest when it relaxes. High blood pressure (HYPERTENSION) is associated with an increased risk of heart attacks and strokes; abnormally low blood pressure (HYPOTENSION) is mostly seen in people in shock or following excessive loss of fluid or blood.

blood test Analysis of a sample of blood as an aid to diagnosis. There are a great many tests available, from blood typing, in case a transfusion is required, to counts of red or white cells or elaborate assays to trace minute amounts of hormones. Most are performed to evaluate the condition of the blood itself, to detect infection or to probe chemical changes taking place in the body.

blood transfusion Transfer of blood or a component of blood (such as plasma or red cells) from one body to another in order to make up for some deficiency. It is often done to counteract life-threatening SHOCK following excessive blood loss. The BLOOD GROUP of the donor must be compatible with that of the recipient to avoid the donated blood coagulating in the recipient's blood vessels. Donated blood is scrutinized for readily transmissible organisms such as the HEPATITIS B virus and HIV, the virus responsible for AIDS.

blood vessel Closed channel that carries blood throughout the body. An ARTERY carries oxygenated blood away from the heart; these give way to smaller arterioles and finally to tiny capillaries deep in the tissues, where oxygen and nutrients are exchanged for cellular wastes. The deoxygenated blood is returned to the heart by way of the VEINS.

blowhole In geology, hole in a cliff or cave through which seawater spurts at high tide. Blowholes are formed by erosion. Initially, waves acting on a joint or other line of weakness in a cliff face carve out a cave. Continued HYDRAULIC ACTION by pounding waves on the roof of the cave forms a sort of chimney. Eventually this breaks through the surface of the ground near the edge of the cliff. At high tide, incoming waves force water out of the top of the blowhole. In biology, blowhole refers to the nostrils (single or double) of whales, situated towards the back of the skull.

blubber Layer of subcutaneous FAT that surrounds the bodies of many aquatic mammals. The fat, a feature of whales and seals, acts as an energy store and insulating layer to prevent the loss of heat to cold water. During the days of mass international whaling, blubber was rendered down by boiling and used as a source of oil for many applications, all of which can now be supplied by synthetic substances.

blue-green algae See CYANOBACTERIA

blueprint Photographic image on paper with white lines against a blue background, frequently used for engineering drawings. The paper is coated with a solution of ammonium ferric citrate and potassium ferricyanide and exposed to intense light under the sheet of drawings to be copied. The blueprint is "developed" in water.

blue shift Effect in which the lines in the SPECTRUM of a celestial object are displaced towards the blue end of the spectrum. It results from the DOPPLER EFFECT because the object and the observer are moving towards each other. The closing speed can be calculated from the extent of the shift. See also RED SHIFT

bluff Steep slope at the side of a bend in a river. A bluff forms on the outside of a MEANDER by the erosive action of the slow-moving river water. Some bluffs are as much as 100m (330ft) tall.

Blumberg, Baruch Samuel (1925–) US research physician. From 1964 he worked with the American Institute for Cancer Research. He shared the 1975 Nobel Prize for physiology or medicine with D. Carleton GAJDUSEK for discoveries concerning new mechanisms for the origin and dissemination of infectious diseases.

boat Vehicle for passenger and freight transport by water. Today the term "boat" is often reserved for a small craft that can easily be taken out of the water; a larger vessel is called a SHIP. The first boats, made in prehistoric times, included rafts, hollowed-out logs, and vessels made from braided reeds. Among the first maritime peoples were the Phoenicians. They built fleets of galleys for their extensive trading in the Mediterranean and adjoining areas. These galleys, and the later ones of the Greeks and Romans, were propelled by sails, supplemented by oars. The later Viking long-boats, also square-sailed, were slimmer and

▲ **bogie** Modern railroads feature trains, such as the European Eurostar, with bogies that use an articulated air suspension system. This has great weight savings. The bogies (1) link the carriages, which are attached to a metal frame (2) through a ball joint (3). Instead of each carriage having a bogie at either end, there is a single suspension unit between two carriages. Dampers (4) modify articulation to improve passenger comfort. The black rings between the frame and the wheel unit are the air suspension units.

speedier. Lateen (triangular) sails were probably imported from the Persian Gulf and introduced to the West by the empire-building Arabs. Modern boats include sailing vessels, used mainly for pleasure, motorboats, and launches.

BOD Abbreviation of BIOLOGICAL OXYGEN DEMAND

bog Area of permanently waterlogged land that results from poor drainage. The ground is wet and spongy. Vegetation, which contains no trees, is usually partly decomposed, and eventually may form an acid peat lacking nutrients (termed **blanket bog**).

Bode's law Empirical, numerical relationship for the mean distances of the planets from the Sun. It is named after the German astronomer Johann Bode (1747–1826), who popularized it in the late 18th century. If the number 4 is added to 0, 3, 6, 12, 24, 48, 96, and 192, and each sum is divided by 10, the figure arrived at is the mean distance in astronomical units of the planets from the Sun, from Mercury to Uranus, including the asteroid belt. The law does not work for Neptune and Pluto.

bogie Small four- or six-wheeled undercarriage pivoted near the end of a rigid vehicle. Bogies are commonly used in pairs on long railroad cars or buses, so that they can better negotiate curves, bogies are also used at the front of large locomotives to "lead" the vehicle into curves. The term bogie is also used for a short wheel-base railroad car used for carrying minerals.

Bohr, Aage Niels (1922–) Danish physicist. He was educated at the University of Copenhagen and the University of London. From 1956 he was professor of physics at Copenhagen University. He was the director (1963–70) of the Institute of Theoretical Physics, founded by his father, Niels BOHR, and then directed (1975–81) the Nordic Institute for Theoretical Nuclear Physics (Nordita). With Ben MOTTELSON (also of Nordita) and James RAINWATER he shared the 1975 Nobel Prize for physics for devising a "collective model" of the atomic nucleus.

Bohr, Niels Henrik David (1885–1962) Danish physicist who was a major developer of QUANTUM THEORY. He was the first person to apply it with success to the problem of atomic structure. Bohr worked with J.J. THOMSON and Ernest RUTHERFORD in Britain before teaching theoretical physics at the University of Copenhagen. He escaped from German-occupied Denmark during World War II and worked briefly on developing the atom bomb in the US. He later returned to Copenhagen and worked for international cooperation. Bohr used quantum theory to explain the spectrum of hydrogen and in the 1920s helped develop the "standard model" of the quantum theory, known as the Copenhagen Interpretation. He was awarded the 1922 Nobel Prize for physics for his work on atomic structure and in 1957 received the first Atoms for Peace Award. He was the father of Aage BOHR.

boiler Vessel for heating and converting water to steam. The boiler is an essential part of STEAM ENGINES and TURBINES. It consists of a furnace for burning fuel and a container where water is evaporated into steam. The term is also applied to devices used to provide hot, though not boiling, water in central heating systems.

boiling point Temperature at which a substance changes phase (state) from a liquid to a vapor or gas. The boiling point increases as the external pressure increases and falls as it decreases. It is usually measured at a standard pressure of one atmosphere (760mm of mercury, 101,325 pascals). The boiling point of pure water at standard pressure is 100°C (212°F).

bolometer Instrument for detecting and measuring ELECTROMAGNETIC RADIATION such as light and heat. One type uses a thin blackened metal strip,

the electrical resistance of which varies according to the amount of radiation falling on it. Bolometers are often mounted in telescopes to measure the energy given off by stars.

Boltwood, Bertram Borden (1870–1927) US chemist and physicist who did important research in radioactivity which led to the development of the theory of ISOTOPES. Boltwood discovered (1907) ionium – a radioactive isotope of thorium. He was professor of radiochemistry (1910–27) at Yale University.

Boltzmann, Ludwig (1844–1906) Austrian physicist who helped to establish the foundations of statistical mechanics and who was acclaimed for his major contribution to the kinetic theory of gases. His statistical research into the velocity distribution of gas molecules extended the ideas of Scottish physicist James MAXWELL. Boltzmann's general law said a system will approach a state of thermodynamic equilibrium because this is the most probable state. He introduced the "Boltzmann equation" in 1877 relating the kinetic energy of a gas atom or molecule to temperature. The formula involves the gas constant per molecule called the " BOLTZMANN CONSTANT." In 1884 he derived a law, often termed the "Stefan–Boltzmann law," for BLACK BODY radiation discovered by his Viennese teacher, Josef Stefan (1835–93). After being attacked for his belief in the atomic theory of matter, Boltzmann committed suicide.

Boltzmann constant (symbol k) Ratio of the universal GAS CONSTANT to the AVOGADRO CONSTANT, equal to 1.381×10^{-23} joules per kelvin. It shows the relationship between the kinetic energy of a gas particle (atom or molecule) and its absolute temperature. It is named after Ludwig BOLTZMANN.

bomb, volcanic Piece of solid material, usually cooled lava, ejected by a volcano. Volcanic bombs are usually greater than 32mm (1.3in) in diameter, and may be very large.

bond *See* CHEMICAL BOND

bone CONNECTIVE TISSUE that forms the skeleton of most vertebrates, protects their internal organs, serves as a lever during locomotion and when lifting objects, and stores calcium and phosphorus. Bone is composed of a strong, compact layer of COLLAGEN and calcium phosphate and a lighter, porous, inner spongy layer containing MARROW, in which ERYTHROCYTES and some LEUCOCYTES are produced.

bone china Hard-paste PORCELAIN. It consists of kaolin, china stone, and bone ash. Josiah SPODE perfected the manufacture of bone china in the 19th century and was largely responsible for its popularity. Like true porcelain, it is translucent and lends itself well to underglaze decoration.

bony fish *See* FISH

Boole, George (1815–64) English mathematician. Largely self-taught, he was appointed professor of mathematics at Cork University in 1849. He is remembered for his invention of a set of symbols to represent logical operations. *See also* BOOLEAN ALGEBRA

Boolean algebra Branch of mathematics that deals with rules that can be applied to sets and to logical "and" or "or" propositions. For example, in Boolean algebra, xy means x **and** y, while $x + y$ means x **or** y. It is applied widely to computer design, where the BINARY SYSTEM (0 and 1) correspond to the logical propositions a computer uses to function. It was named after George BOOLE.

booster engine ROCKET engine that powers a missile or space vehicle in the early stages of its flight and then drops off, reducing deadweight. Several booster engines may be attached to the first stage of a launch vehicle for extra power.

boot (bootstrap, boot-up) In computing, the loading of a short program that allows the machine to run other programs – most importantly, a DISK OPERATING SYSTEM. With personal computers, the booting program is stored on the hard disk and runs automatically when the computer is switched on. "To boot" also often means "to turn the computer on."

Boötes (herdsman) Prominent constellation of the N sky; it contains the bright orange star Arcturus.

borane HYDRIDE of BORON. The simplest borane is diborane (B_2H_6), made by treating magnesium boride with acid. The higher boranes have up to 10 boron atoms. They are all reactive and readily oxidize, sometimes explosively.

borate Any salt of BORIC ACID or of more complex oxyacids of BORON. Borates are inorganic compounds; many exist as minerals, the most important of which are BORAX, colemanite, and kernite – all used as sources of boron compounds.

borax (hydrated sodium borate, $Na_2B_4O_7.10H_2O$) Most common borate mineral. It is found in large deposits in dried-up alkaline lakes in arid regions as crusts or masses of crystals. It may be color-

Boolean algebra The algebra of logic. Five basic logic statements are illustrated here. In any of them, if A is true the table will show a "1", if A is false, there will be a "0." In the statement AND, C is true (i.e. there is a "1" in the truth table) when A and B are true, but it is false if either of A and B is false. With OR, C is true if either A or B is true and it is only false if both A and B are false. NOT has one input and one output, its function is to reverse "true" and "false;" when applied to AND and OR, their inverses, NAND and NOR, result. The Boolean statements shown here are also illustrated by symbolic circuit elements (inputs on the left and outputs on the right) and by Venn diagrams as in set theory (the outcome is shown by the shaded areas).

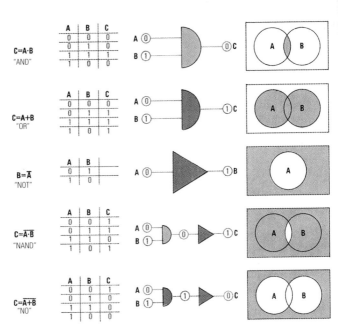

less or white, transparent or opaque. It is used to make heat-resistant glass, pottery glaze, water softeners in washing powders, fertilizers, and pharmaceuticals.

Bordet, Jules Jean Baptiste Vincent (1870–1961) Belgian bacteriologist and immunologist. After working at the Pasteur Institute in Paris (1894–1901), he returned to Belgium where he founded the Pasteur Institute in Brussels (1901). He was awarded the 1919 Nobel Prize for physiology or medicine for his development of the complement-fixation test, an important aid to diagnosis in immunology. He identified the whooping cough bacillus in 1906.

boric acid (boracic acid, H_3BO_3) Soft white crystalline solid that occurs naturally in certain volcanic hot springs. It is used as a metallurgical FLUX, preservative, antiseptic, and an insecticide for ants and cockroaches.

boride Any of a number of chemical compounds formed between a metal and the nonmetallic element BORON. Borides are made by combining the elements at a very high temperature. All are extremely hard and are used as abrasives and refractories.

boring machine Device for making accurate and smoothly finished holes in metal objects. Drilled holes are enlarged with a cutting bore, generally tipped with steel, carbide or diamond, and gripped in an adjustable boring head attached to a rotating spindle. The worktable can usually be moved in two perpendicular directions so that holes can be spaced accurately.

Borlaug, Norman Ernest (1914–) US agronomist. As a director of the Rockefeller Foundation in Mexico, he led a team of scientists experimenting with the improvement of cereal crops. He was awarded the Nobel Peace Prize in 1970 for his accomplishments in the "GREEN REVOLUTION," developing improved wheat seed, a higher-yielding rice, and better ways of using fertilizer and water.

Born, Max (1882–1970) German-British physicist. He was professor of physics at Göttingen University from 1921 but left Germany in 1933 and went to Britain, teaching physics at Cambridge (1933–6) and the University of Edinburgh (1936–53). He returned to Germany in 1954. For his work in QUANTUM MECHANICS, the basis of atomic and nuclear physics, he was awarded the 1954 Nobel Prize for physics, which he shared with Walther BOTHE.

bornite (peacock ore, Cu_5FeS_4) Common copper mineral, a copper iron sulfide. It generally occurs

in masses, and sometimes as crystals (cubic system) in INTRUSIVE IGNEOUS rocks and METAMORPHIC rocks. It is opaque and bronze with an iridescent purple tarnish. Hardness 3; r.d. 5.0.

boron (symbol B) Nonmetallic element of group III of the periodic table, first isolated in 1808 by Sir Humphry DAVY. It occurs in several minerals, notably KERNITE (its chief ore) and BORAX. It has two allotropes (different forms): **amorphous** boron, an impure brown powder, is made by reducing the oxide with magnesium; **metallic** boron, a black to silver-gray hard crystalline material, is obtained by decomposing boron tribromide vapor on a hot metal filament. The element is used in semiconductor devices and the stable isotope ^{10}B is a good neutron absorber, used in NUCLEAR REACTORS and particle counters. Properties: at.no. 5; r.a.m. 10.81; r.d. 2.34 (cryst.), 2.37 (amorph.); m.p. 2,079°C (3,774°F); sublimes 2,550°C (4,622°F); most common isotope ^{11}B (80.22%).

Bosch, Carl (1874–1940) German industrial chemist who adapted Fritz HABER's method of synthesizing ammonia to an industrial scale. Bosch's work involved finding metallic catalysts that could promote high-pressure synthesis. His invention of the **Bosch process** (in which water gas and steam at high temperatures are passed over a catalyst) aided the large-scale preparation of hydrogen. Bosch was awarded the 1931 Nobel Prize for chemistry, with Friedrich BERGIUS, for his high-pressure techniques.

Bose, Sir Jagadis Chandra (1858–1937) Indian plant physiologist and physicist. He invented highly sensitive instruments capable of detecting tissue responses of plants to external stimuli.

Bose, Satyendranath (1894–1974) Indian physicist and mathematician who contributed significantly to the theory of QUANTUM MECHANICS and STATISTICAL MECHANICS. Bose made the initial advances to describe the statistical properties of certain ELEMENTARY PARTICLES (now called BOSONS). These particles, which include all those that mediate force (like the PHOTON that carries the ELECTROMAGNETIC FORCE), all have the property that any number of them can occupy the same quantum state: that is, they do not obey Enrico FERMI's EXCLUSION PRINCIPLE. Bose's work was developed further by EINSTEIN and the statistics that such particles obey is now called **Bose-Einstein statistics**.

boson ELEMENTARY PARTICLE that has an integer SPIN. Named after the physicist Satyendranath BOSE, bosons are those particles that are not covered by the EXCLUSION PRINCIPLE. This means that the number of bosons occupying the same quantum state is not restricted. Bosons are force-transmitting particles, such as PHOTONS and GLUONS (the particles that hold QUARKS together). *See also* FERMION

botanical garden Large garden preserve for display, research, and teaching purposes. Wild and cultivated plants from all climates are maintained outdoors and in greenhouses. Although organized gardens date from ancient Rome, the first botanical gardens were established during the Middle Ages. In the 16th century, gardens existed in Pisa, Bologna, Padua, and Leiden. Aromatic and medicinal herbs were arranged in rows and still exist in the Botanical Garden of Padua. The first US botanical garden was established by John Bartram in Philadelphia in 1728. Famous botanical gardens include the New York Botanical Gardens in the Bronx (established 1891), the Missouri Botanical Gardens in St. Louis (established around 1860) and the Royal Botanical Gardens in Kew, near London (established 1759).

botany Study of plants and algae, including their classification, structure, physiology, reproduction, and evolution. The discipline used to be studied in two halves: **lower** (nonflowering) plants, which included the ALGAE (now in the kingdom PROTOCTISTA), MOSS, and FERNS; and **higher** (seed-bearing) plants, including most flowers, trees, and shrubs. Botany also studies the importance of plants to people.

Bothe, Walther Wilhelm Georg Franz (1891–1957) German physicist. From 1934 he was director of the Max Planck Institute, Heidelberg, and in 1946 was professor of physics at the university there. During World War II he worked on Germany's nuclear energy project and built Germany's first CYCLOTRON. He shared the 1954 Nobel Prize for physics with Max BORN for his development of the **coincidence method**, which can detect two particles emitted simultaneously from the same nucleus during RADIOACTIVE DECAY.

botulism Rare but potentially lethal form of food poisoning caused by a toxin produced by the bacterium *Clostridium botulinum*. The toxin attacks the nervous system, causing paralysis and cessation of breathing. The most likely source of botulism is imperfectly canned food. Botulinum toxin is used medicinally as a treatment for some neuromuscular disorders.

Boulton, Matthew (1728–1809) British engineer and manufacturer. In 1762 he built the Soho iron works near Birmingham, England. In 1775 he went into partnership with James WATT and began manufacturing Watt's steam engines on a commercial basis. In 1790, Boulton patented a steam-powered coining press, and he introduced a new copper coinage in Britain in 1797.

Bouguer anomaly Observation that gravity measured on a great rock mass, such as a mountain range, is higher than average. This is due to the gravitational force exerted by the rock mass itself. It is named after its discoverer, the French mathematician Pierre Bouguer (1698–1758).

boulder clay *See* TILL

boundary-layer flow Behavior of a FLUID in the region immediately next to a solid body immersed in the fluid. In this region, layers of the fluid "slide" over one another, and the measure of this internal friction, viscosity, becomes the determining characteristic of the flow. *See also* AERODYNAMICS; FLUID MECHANICS

Boussingault, Jean-Baptiste (1802–87) French agricultural chemist. After spending 10 years as a mining engineer in Venezuela, he became professor of chemistry at Lyon. He demonstrated that plants obtain nitrogen from the soil, but that legumes can obtain it from the air, and that carbon in plants comes from atmospheric carbon dioxide. He also investigated the nutritional value of various foodstuffs.

Bovet, Daniele (1907–92) Italian pharmacologist. He was awarded the 1957 Nobel Prize for physiology or medicine for his pioneering work on the development of ANTIHISTAMINES and the muscle relaxants used in surgery. He also studied the effects of mental illness on the chemistry of the brain.

bovine spongiform encephalopathy (BSE) ("mad cow disease") In cattle, degenerative disease of the brain caused by infectious particles or PRIONS. The consumption of infected beef or beef products can transmit the disease to other animals and human beings. *See also* CREUTZFELDT-JAKOB DISEASE (CJD); SCRAPIE

bowden cable Device for the flexible transmission of small mechanical movements, generally to the controls of a machine. It consists of a strong steel cable (usually multistranded and lubricated) in a flexible outer tube, often made in the form of a long coiled spring. The outer tube is anchored at the ends so that the inner cable can move independently. Bowden cables are commonly used for throttle (accelerator), clutch, choke, and sometimes brake controls on cars and motorcycles.

Bowen, Norman Levi (1887–1956) Canadian petrologist and mineralogist. His contributions to the study of the origins, structure, and chemistry of rocks were outstanding. His most important book is *The Evolution of Igneous Rocks* (1928).

Bowen's reaction series Explanation of the order of crystallization of minerals within a cooling magma, derived from laboratory experiments on a rock melt carried out by Norman BOWEN. He concluded that two independent lines of mineral crystallization occur; (1) the discontinuous reaction series, consisting of ferromagnesian minerals, where each crystallization "step" produces a different mineral; (2) the continuous reaction series, consisting largely of plagioclase FELDSPARS, where there is a gradual change in feldspar composition as cooling and crystallization take place.

Bowman's capsule (renal capsule) Funnel-shaped end of a NEPHRON in a KIDNEY. It has special epithelial cells (**podocytes**) that allow fluid filtered from the blood to pass into the nephron, from where it eventually passes into the pelvis of the kidney as urine. There are about a million Bowman's capsules in each human kidney.

Boyer, Herbet Wayne (1936–) US biochemist who developed the recombinant DNA technique for synthesizing proteins, the basis of much genetic engineering. He worked in collaboration with Stanley COHEN. Beginning in 1973 they used the

bacterium *Escherichia coli*, into which they inserted DNA made by splicing together segments from two different plasmids. Boyer went on to employ the method for synthesizing useful proteins such as human INSULIN, INTERFERON, AND growth HORMONE. *See also* RECOMBINANT DNA RESEARCH

Boyle, Robert (1627–91) British scientist, b. Ireland. He is often regarded as the father of modern chemistry. At his laboratories in Oxford and London, Boyle conducted research into air, vacuum, metals, combustion, and sound. He was a founding member of the ROYAL SOCIETY. Boyle's *Sceptical Chymist* (1661) proposed an early atomic theory of matter. He made an efficient vacuum pump, which he used to establish (1662) BOYLE'S LAW. Boyle formulated the first chemical definitions of an element and a reaction.

Boyle's law Law that states that the volume of a gas at constant temperature is inversely proportional to the pressure. This means that as pressure increases, the volume of a gas decreases. First stated (1662) by Robert BOYLE, Boyle's law is a special case of the IDEAL GAS LAW (involving a hypothetical gas that perfectly obeys the gas laws).

brachiopod Any of about 260 species of small, bottom-dwelling, marine invertebrates of the phylum Brachiopoda. They are similar in outward appearance to BIVALVE mollusks, having a shell composed of two valves; however, unlike bivalves, there is a line of symmetry running through the valves. They live attached to rocks by a **pedicle** (stalk), or buried in mud or sand. Most modern brachiopods are less than 5cm (2in) across.

bract Modified leaf found on a flower stalk or the flower base. Bracts are usually small and scalelike. In some species, such as dogwood and poinsettia, they are large and brightly colored.

Bragg, Sir (William) Lawrence (1890–1971) English physicist, b. Australia. He was director (1938–53) of the Cavendish Laboratory at Cambridge. With his father, Sir William Henry Bragg, he determined the mathematics involved in X-ray DIFFRACTION, showed how to compute X-ray wavelengths and studied CRYSTAL structure by X-ray diffraction. For these advances, they were jointly awarded the 1915 Nobel Prize for physics.

Brahe, Tycho (1546–1601) Danish astronomer. Under the patronage of King Frederick II of Denmark he became the most skilled observer of the pretelescope era, expert in making accurate naked-eye measurements of the positions of stars and planets. He built an observatory on the island of Hven (1576) and calculated the orbit of the comet

▲ **Boyle's law** When a given quality of gas is compressed, its pressure rises as its volume diminishes. Boyle's law states that at any one temperature, the product pressure × volume remains the same throughout compression or expansion. The curves show this relationship. A gas that obeyed this law exactly, known as an ideal gas, could be imagined as consisting of infinitesimally small, perfectly elastic, colliding particles (rather like extremely tiny steel ball bearings).

Key
P pressure
V volume
T_1 T_2 T_3 etc. different temperatures, higher numbers indicating higher temperatures

seen in 1577. This, together with his study of the supernova, showed that ARISTOTLE was wrong in picturing an unchanging heavens. Brahe could not, however, accept the world system put forward by COPERNICUS. In his own planetary theory (the **Tychonian system**), the planets move around the Sun, and the Sun itself, like the Moon, moves round the stationary Earth. In 1597 he left Denmark and settled in Prague, where Johann KEPLER became his assistant. The accurate series of planetary observations made by Tycho Brahe were used by Kepler in deriving his laws of planetary motion, which ironically demonstrated the validity of the Copernican theory.

braided stream Network of small, shallow interlacing streams. It forms when a river deposits sediment, causing it to divide up into new channels, which then redivide and rejoin, resembling a braided cord.

Braille System of reading and writing for the blind. It was invented by Louis Braille (1809–52), who himself lost his sight at the age of three. Braille was a student, and later a teacher, at the National Institute of Blind Youth in Paris. He developed a system of embossed dots to enable blind people to read by touch. This was first published in 1829, and a more complete form appeared in 1837. There are also Braille codes for music and mathematics.

brain Mass of nerve tissue that regulates all physical and mental activity; it is continuous with the SPINAL CORD. Weighing about 1.5kg (3.3lb) in an adult human (about 2% of body weight), the human brain has three parts – the HINDBRAIN, MIDBRAIN, AND FOREBRAIN. The hindbrain is where basic physiological processes such as breathing and the heartbeat are coordinated. The midbrain links the hindbrain and the forebrain. The forebrain is the seat of all higher functions and attributes (personality, intellect, memory, emotion), as well as being involved in sensation and initiating voluntary movement. *See also* CENTRAL NERVOUS SYSTEM

brain damage Result of any harm done to brain tissue causing the death of nerve cells. It may arise from a number of causes, including oxygen deprivation, brain or other disease or head injury. The nature and extent of damage varies. Sudden failure of the oxygen supply to the brain may result in widespread (global) damage, whereas a blow to the head may affect only one part of the brain (local damage). Common effects of brain damage include weakness of one or more limbs, impaired balance, memory loss, and personality change; epilepsy may develop. Although brain cells do not regenerate, there is some hope of improvement

▲ brachiopod The marine animal known as a brachiopod, or lamp shell, lives in holes in mud flats. It comprises (A) a hinged shell and a stalk with which it grips the rocks. The cross-section (B) shows; lophophore (1) which bears tentacles for feeding; digestive gland (2); mouth (3); and stalks (4). When feeding, *Lingula* (C), rests at the surface of its burrow using feathery cilia to filter water for food particles. When disturbed, its stalk contracts, drawing the animal into the burrow (D), out of sight and reach of its potential predator.

following mild to moderate brain damage, especially with skilled rehabilitation. The survivor of major brain injury is likely to remain severely disabled, possibly even permanently unconscious.

brain disorder Disturbances of physical or mental function due to abnormality or disease of the brain. Some such disorders are hereditary; some are congenital (present at birth); and some are acquired during life. Brain disorders should be distinguished from psychological (psychogenic) mental disturbances in which the functioning of the brain itself is not impaired. Brain disorders are associated with impairment of memory, orientation, comprehension, and judgment, and also by shallowness of emotional expression. Secondary personality changes may occur, depending upon such factors as the strength and type of personality and the amount of psychological and social stress present. Brain disorders are divided into acute and chronic types. **Acute** disorders are temporary, and are generally due to disruption of brain function rather than destruction of brain tissue. They may be caused by such things as infection, drug or alcohol intoxication, and brain trauma. **Chronic** brain disorders are irreversible, and include such things as CONGENITAL DISORDER, hereditary disease, senility, and BRAIN DAMAGE.

brain stem Stalk-like portion of the BRAIN in vertebrates that includes everything except the CEREBELLUM and the CEREBRAL HEMISPHERES. It provides a channel for all signals passing between the spinal cord and the higher parts of the brain. It also controls automatic functions such as breathing and heartbeat.

brake Device for slowing the speed of a vehicle or machine. Braking can be accomplished by a mechanical, hydraulic (liquid) or pneumatic (air) system that presses a nonrotating part into contact with a rotating part, so that friction stops the motion. In a car, the non-rotating part is called a shoe or pad, and the rotating part is a drum or disk attached to a wheel. Some vehicles use electromagnetic effects to oppose the motion and cause braking. A power brake utilizes a vacuum system.

braking, atmospheric Drag exerted on a space vehicle as it re-enters the Earth's atmosphere and encounters increased air density. The friction slows down the vehicle, which must be built to withstand the intense heat created. *See also* ABLATION

Bramah, Joseph (1748–1814) British engineer and inventor. He ran a lock-manufacturing shop which became the basis of the machine-tool industry in Britain, furthering the expansion of manufacturing in Britain in the 19th century. His inventions included an almost pickproof lock, a hydraulic press, and a numbering machine for the printing trade.

branching Growth extension of vascular plants. A branch develops from the stem and consists of the growth from the previous year (branchlet) and new growth (twig). New twigs are produced during the next season of growth, both from the terminal bud at the end of the twig and from lateral buds in leaf axils along it.

Brandt, Georg (1694–1768) Swedish chemist who discovered, isolated, and named COBALT in 1730. Brandt was a research director for the Council of Mines, Stockholm, and later became the warden of the Swedish Mint. He rejected alchemy and discredited many spurious processes for the production of gold.

brass Alloy of mainly copper (55%–95%) and zinc (5%–45%). Brass is yellowish or reddish, malleable, and ductile, and can be hammered, machined or cast. Its properties can be altered by varying the amounts of copper and zinc, or by adding other metals, such as tin, lead, and nickel. Brass is widely used for pipe and electrical fittings, screws, musical instruments, and ornamental metalwork.

Brattain, Walter Houser (1902–87) US physicist. In 1929 he joined the Bell Telephone Laboratories as a research physicist and concentrated on research into SOLID-STATE PHYSICS. In 1956 he shared the Nobel Prize for physics with John BARDEEN and William SHOCKLEY for their development of the TRANSISTOR and research into SEMICONDUCTORS. In 1967 Brattain joined the faculty of Whitman College in Walla Walla, Washington.

Braun, (Karl) Ferdinand (1850–1918) German physicist. From 1895 he was professor of physics and director of the Physical Institute, Strasbourg. In 1897 he invented the oscilloscope – using a CATHODE-RAY TUBE known as the **Braun tube**, the forerunner of the television tube. In 1909, with Guglielmo MARCONI, he shared the Nobel Prize for physics for the development of wireless telegraphy.

Braun, Wernher von (1912–77) US rocket engineer, b. Germany. By 1937 he was director of a rocket research station at Peenemünde, Germany, where he perfected V-2 rocket missiles in the early 1940s. In 1945 he went to the USA, becoming a US citizen in 1955. In 1958 Braun was largely responsible for launching the first US satellite, *Explorer 1*. He later worked on the development of the *Saturn* rocket (for the APOLLO PROJECT) and was deputy associate administrator of NATIONAL AERONAUTICS AND SPACE ADMINISTRATION (NASA) (1970–72).

brazing Process in which metallic parts are joined by the fusion of alloys that have lower melting points than the parts themselves. The bonding alloy is either preplaced or fed into the joint as the parts are heated. Brazed joints are very reliable and are used extensively in the aerospace industry.

breast *See* MAMMARY GLAND

breastbone *See* STERNUM

breathing Process by which air is taken into and expelled from the LUNGS for the purpose of GAS EXCHANGE. When breathing in (inhaling), the intercostal muscles raise the ribs, increasing the volume of the THORAX and drawing air into the lungs. During breathing out (exhalation), the ribs are lowered, and air is forced out through the nose, and sometimes also the mouth.

breccia SEDIMENTARY ROCK formed by the cementation of sharp-angled fragments in a finer matrix of the same or different material. It can also refer to rock formed by movements of the crust (**fault** breccia) and to volcanic rock from a vent (**volcanic** breccia). *See also* CONGLOMERATE

breeder reactor NUCLEAR REACTOR in which more fissile (capable of undergoing FISSION) material is produced than is consumed. In it, surplus high-energy neutrons transmute a "breeding blanket" of nonfissile uranium-238 into fissile plutonium-239.

breeding Process of producing offspring, specifically the science of changing or promoting certain genetic characteristics in animals and plants. This is done through careful selection and combination of the parent stock. Breeding may involve outbreeding (mating of unrelated individuals) or INBREEDING to produce the desired characteristics in the offspring. Scientific breeding has resulted in disease-resistant strains of crops, and in animals that give improved food yields. *See also* GENE; GENETIC ENGINEERING

bremsstrahlung (Ger. braking radiation) ELECTROMAGNETIC RADIATION in the form of X-rays emitted when charged particles slow down or change course rapidly. This happens when high-speed electrons enter the electric field of an atomic nucleus. Such radiation covers a continuous range of wavelengths within the ELECTROMAGNETIC SPECTRUM.

Brenner, Sydney (1927–) British molecular biologist, b. South Africa. He is best known for the work he did in the late 1950s on identifying the CODONS (triplets of bases) in chains of DNA. In 1961 he was also one of the discoverers of MESSENGER RNA (mRNA) and its role in PROTEIN SYNTHESIS in RIBOSOMES.

brewing Preparation of beer and ale by using yeast as a catalyst in the alcoholic FERMENTATION of liquors containing malt and hops. In beer brewing, a malt liquor (wort) is made from crushed germinated barley grains. Hops are added to the boiling wort to impart a bitter flavor, and help to clarify the beer and keep it free from spoilage by microbes. The clear, filtered wort is cooled and inoculated with brewer's yeast, which ferments part of the sugar from the malt into alcohol.

Brewster, Sir David (1781–1868) Scottish physicist who performed significant experiments in OPTICS and POLARIZED LIGHT. He discovered **Brewster's law**, which relates the polarizing angle (the angle at which light strikes a substance) to the REFRACTIVE INDEX of the substance.

brick Hardened block of clay used for building and paving. Usually rectangular, bricks are made in standard sizes either by molding or by machines that cut off extruded sections of stiff clay that are conveyed into a continuously operating kiln where they are baked at temperatures of up to 1,300°C

(2,372°F). The first, sun-dried bricks were used in the Tigris-Euphrates basin about 5,000 years ago. Mechanized processes replaced the old hand-molding methods during the 19th century.

bridge In civil engineering, structure providing a continuous passage over a body of water, roadway or valley. Bridges are built for people, vehicles, pipelines or power transmission lines. Prehistoric in origin, the first bridges were probably merely logs over rivers or chasms. Modern bridges take a great variety of forms including beams, ARCHES, CANTILEVERS, SUSPENSION BRIDGES, and CABLE-stayed bridges. They can also be movable or floating pontoons. They can be made from a variety of materials, including brick or stone (for arches), steel and/or concrete.

bridge In dentistry, a partial denture held in place by anchorage to adjacent teeth. Depending on the situation it may be permanently installed or removable; anchorages are of various sorts. Usually, only a few teeth in a row are artificially replaced in this manner.

Bridges, Calvin Blackman (1889–1938) US geneticist. He helped to prove the chromosomal basis of HEREDITY and sex. His work with Thomas Hunt MORGAN on the fruit fly (*Drosophila*) proved that inheritable variations could be traced to observable changes in the CHROMOSOMES. These experiments resulted in the construction of "gene maps."

Bridgman, Percy Williams (1882–1961) US physicist. From 1919 he was a professor at Harvard University. He was awarded the 1946 Nobel Prize for physics for his work in high-pressure physics. He is known also for his work on crystal properties and for his writings on the philosophy of science. His books include *The Nature of Modern Physics* (1927); his collected papers were published in 1964.

Briggs, Henry (1561–1630) English mathematician who introduced common LOGARITHMS. In 1614, with John NAPIER's agreement, he suggested that Napier's logarithms be altered to a base of 10. In 1624 he published *Arithmetica Logarithmica*.

Bright, Sir Charles Tilston (1832–88) British engineer who helped to develop submarine telegraph cables. In 1856 he became engineer-in-chief of the Atlantic Telegraph Co. and in 1858 supervised the laying of the first transatlantic cable from Ireland to Newfoundland. With the engineer Joseph Clark he also invented an asphalt covering for submarine cables.

Bright, Richard (1789–1858) British physician who was the first to describe the kidney disorder

(NEPHRITIS) that used to be known as **Bright's disease**. He became a physician at Guy's Hospital, in London, in 1820.

Brindley, James (1716–72) British engineer and pioneer canal-builder who constructed the first economically viable major canal in England, from Worsley, Lancashire to Manchester. He was eventually responsible for a network of about 565km (350 mi) of canals, an advancement which helped to hasten the Industrial Revolution in Britain. None of his written records or working drawings has survived.

Britannia metal Tin alloy used for tableware, similar to pewter but containing about 7% antimony, 2% copper but no lead. It is a lustrous, hard, malleable alloy that is easy to cast.

British thermal unit (BTU) Amount of heat energy necessary to raise the temperature of 1lb of water from 59.5°F to 60.5°F. One BTU is the equivalent of 778.3 foot-pounds, 252 CALORIES or 1,055 JOULES. It has been largely superseded by the metric unit of energy, the joule.

broadcasting Transmission of sound or images by radio waves to a widely dispersed audience through RADIO or TELEVISION receivers. The first US commercial radio company, KDKA, began broadcasting in Pittsburgh in 1920. In 1922 the first radio advertisement was aired marking the beginning of the commercial broadcast industry. In 1924 the first experimental coast-to-coast network was connected, and from then on stations and audiences both expanded rapidly. In 1927 there were two major networks. By 1998 there were almost 5,000 commercial AM stations in operation; FM, which was first broadcast in 1941, surpassed AM in number in 1983 and boasted 5,477 by 1998. The development of television was interrupted by World War II, but it took off as soon as the war ended, and by 1998 there were more than 1,100 commercial television stations. In 1962 the *Telstar* communications satellite relayed the first transatlantic television signals, and live international broadcasts are now a regular feature of network schedules. However, the advent of cable (which served 60% of US households by the 1990s), videocassettes, and the internet have cut into the time Americans spend listening to or watching radio or television broadcast networks.

Broca, Paul (1824–80) French pathologist, anthropologist, and neurosurgeon. As a professor at the Faculty of Medicine in Paris and the French Anthropological Institute, he was also a pioneer in neurosurgery. Broca specialized in aphasia, and located the brain center for articulate speech. That part of the brain, the third convolution of the left frontal lobe, is called Broca's Area in his honor. He also devised methods of classifying hair and skin color and of establishing ratios between brain size and skull size.

Broglie, Louis Victor de (1892–87) French physicist who put forward the theory that all ELEMENTARY PARTICLES have an associated wave. He devised the formula that predicts this wavelength, and its existence was proven in 1927 by others. De Broglie developed this form of QUANTUM MECHANICS, called WAVE MECHANICS, for which he was awarded the 1929 Nobel Prize for physics. Erwin SCHRÖDINGER advanced de Broglie's ideas with his equation that describes the wave function of a particle.

bromide Salt of hydrobromic acid or certain organic compounds containing BROMINE. The bromides of ammonium, sodium, potassium, and certain other metals were once extensively used medically as sedatives. Silver bromide is light-sensitive and is used in photography.

bromination Introduction of a bromine atom into a chemical compound, either by reaction with an unsaturated molecule or by replacing hydrogen, as in the preparation of bromobenzene from BENZENE. Organic bromides are widely used as intermediates in dye and drug synthesis, alkali metal bromides as sedatives, and silver bromide in photography.

bromine (symbol Br) Volatile, liquid element of the HALOGEN group (elements in group VII of the periodic table), first isolated in 1826. Bromine is the only nonmetallic element that is liquid at room temperature. It is extracted from soluble bromides, especially magnesium bromide, found in seawater. The bromine is displaced by the more active halogen chlorine and is extracted as a by-product from the saline concentrates left after crystallization of the main salt products. A reddish-brown, fuming liquid having an unpleasant odor, bromine is used in commercially useful compounds, such as those used to manufacture photographic film and antiknock additives for gasoline. Chemically, it resembles chlorine but is less reactive. Properties: at.no. 35; r.a.m 79.904; r.d. 3.12; m.p. $-7.2°C$; $(19.04°F)$; b.p. 58.8°C (137.8°F); most common isotope ^{79}Br (50.54%).

bronchiectasis Enlargement of one or more bronchial tubes usually resulting from bronchitis or tuberculosis. Symptoms are coughing and copious expectoration.

bronchus (pl. bronchi) One of two branches into which the TRACHEA or windpipe divides, with one branch leading to each of the LUNGS. The bronchus divides into smaller and smaller branches, called **bronchioles**, which extend throughout the lung, opening into the air sacs or ALVEOLI. The bronchi are supported and kept open by rings of CARTILAGE.

Bronowski, Jacob (1908–75) British biologist and man of letters. He lectured at many US universities and was made a fellow of Jesus College, Cambridge, in 1967. Among his books are *Science and Human Values* (1958), *Nature and Knowledge* (1969) and three studies of William Blake. His celebrated BBC television series on man's intellectual history was broadcast on US public television and published as *The Ascent of Man* (1973).

bronze Traditionally an ALLOY of copper and no more than 33% tin. It is hard and resistant to corrosion, but easy to work. It has long been used in sculpture and in bell-casting. Other metals are often added to introduce specific properties and uses, such as aluminum in aircraft parts and tubing, silicon in marine hardware and chemical equipment, and phosphorus in springs, gunmetal, and electrical parts.

brown coal *See* LIGNITE

brown dwarf Star with mass less than 0.08 solar masses, in which the core temperature does not rise high enough to initiate thermonuclear reactions. Such a star is, however, luminous, for as it slowly shrinks in size it radiates away its gravitational energy. The surface temperature of a brown dwarf is below the 2,500 K lower limit of a RED DWARF.

Brownian movement Random, zigzag movement of particles suspended in a FLUID (liquid or gas). It is caused by the unequal bombardment of the larger particles, from different sides, by the smaller molecules of the fluid. The movement is named after the Scottish botanist Robert Brown (1773–1858) who in 1827 observed the movement of plant spores floating in water.

Brown, Robert (1773–1858) Scottish botanist. As the naturalist on Flinder's voyage to Australia (1801–05), he returned to England with valuable botanical collections of thousands of new species. He described the flora of Australia in *Prodomus Florae Novae Hollandiae et Insulae Van Diemen* (1810). Brown's studies enabled him to outline the difference between GYMNOSPERMS and ANGIOSPERMS. He is best known for establishing, in *A Brief Account of Microscopical Observations* (1828), that minute particles suspended in a liquid or gas are continuously in motion. This motion has since been called BROWNIAN MOVEMENT.

brucite Mineral form of magnesium hydroxide $(Mg_3(OH)_6)$, derived from periclase. The crystals are either glassy or waxy, can be white, pale green, gray or blue, and occur in fibrous masses or plate aggregates. It is found in SERPENTINE, SCHISTS, and other METAMORPHIC ROCKS. It is used to make magnesia refractory materials.

bruise (contusion) Area of discoloration on the SKIN due to escape of blood from damaged underlying vessels. It is caused by injury, usually a blow.

Brunel, Isambard Kingdom (1806–59) British marine and railway engineer. A man of remarkable foresight, imagination, and daring, he revolutionized engineering in Britain. In 1829 he designed the Clifton Suspension Bridge (completed 1864) and in 1833 he became chief engineer for the Great Western Railway, for whom he designed the Royal Albert Bridge across the River Tamar. He is also famous for his ships: *Great Western* (designed 1837), the first trans-Atlantic wooden steamship, *Great Britain* (1843), the first iron-hulled screw-driven steamship, and *Great Eastern* (1858), a steamship powered by screws and paddles that was the largest vessel of its time. His father was Sir Marc Isambard BRUNEL.

Brunel, Sir Marc Isambard (1769–1849) Architect and engineer. A refugee from the French Revolution, he went to the USA in 1793. He was chief engineer in New York City before moving to England in 1799, where he invented machinery for making ships' pulley blocks. He was responsible for the construction of the first tunnel under the River Thames (1825–43). He was the father of Isambard Kingdom BRUNEL.

bryophyte Any member of the phylum Bryophyta – small, green, rootless, non-VASCULAR PLANTS, including MOSSES and liverworts. Bryophytes grow on damp surfaces exposed to light, including rocks and tree bark, almost everywhere from the Arctic to the Antarctic. There are about 24,000 species. *See also* ALTERNATION OF GENERATIONS

bryozoan (moss animal) Any member of the phylum Ectoprocta – small invertebrate, mainly marine, animals. Growing to about 1mm (0.05in) in length, they live in colonies attached to rocks, seaweed or large shells.

BSE Abbreviation for BOVINE SPONGIFORM ENCEPHALOPATHY

bubble chamber Device for detecting and identifying ELEMENTARY PARTICLES. It consists of a sealed chamber filled with a liquefied gas, usually liquid hydrogen, kept just below its boiling point by high pressure in the chamber. When the pressure is released, the boiling point is lowered and a charged particle passing through the superheated liquid leaves a trail of tiny gas bubbles that can be illuminated and photographed before the pressure is restored. If a magnetic field is applied to the chamber, the tracks are curved according to the charge, mass, and velocity of the particles, which can thus be identified. Donald GLASER received the 1960 Nobel Prize for physics for inventing the bubble chamber, and it was further developed by Luis ALVAREZ. *See also* CLOUD CHAMBER

Büchner, Eduard (1860–1917) German biochemist. In 1896 Büchner discovered that it was not the actual yeast cells, but rather the ENZYMES contained within them, that actively caused the alcoholic FERMENTATION of sugar. In 1903 he isolated zymase, the enzyme mixture responsible for fermentation. He was awarded the 1903 Nobel Prize for chemistry.

Buckminsterfullerene (buckyball) Allotrope of CARBON that consists of many carbon atoms bonded together in the shape of a hollow sphere. The simplest Buckminsterfullerene has 60 carbon atoms arranged as 12 regular pentagons and 20 hexagons (like the panels on a modern soccer ball). It can be made by exposing graphite to a laser beam or electric arc in an inert atmosphere. It is a yellow crystalline solid that dissolves in benzene. It gets its name from the US architect Richard Buckminster FULLER, who invented the GEODESIC DOME structure that the substance's molecules resemble. Derivatives containing metal atoms trapped inside the carbon-atom sphere are being assessed as semiconductors. Stable polymers have also been produced. *See also* ALLOTROPY

buckyball *See* BUCKMINSTERFULLERENE

bud In botany, a small swelling or projection consisting of a short stem with overlapping, immature leaves covered by scales. Leaf buds develop into leafy twigs, and flower buds develop into blossoms. A bud at the tip of a twig is a **terminal** bud and contains the growing point; **lateral** buds develop in leaf AXILS along a twig.

budding Method of ASEXUAL REPRODUCTION that produces a new organism from an outgrowth of the parent. Hydras (small, freshwater polyps), for example, often bud in spring and summer. A small bulge appears on the parent and grows until it breaks away as a new individual.

buffer solution Solution to which a moderate quantity of a strong acid or a strong base can be added without making a significant change to its pH value (acidity or alkalinity). Buffer solutions usually consist of either a mixture of a weak acid and one of its salts, a mixture of an acid salt and its normal salt or a mixture of two acid salts.

bulb In botany, a food storage organ consisting of a short stem and swollen scale leaves. Food is stored in the scales, which are either layered in a series of rings, as in the onion, or loosely attached to the stem, as in some lilies. Small buds between the scale leaves give rise to new shoots each year. New bulbs are produced in the AXILS of the outer scale leaves. *See also* ASEXUAL REPRODUCTION

bulbil Small BULB-like structure that grows on a plant. Bulbils may develop above the ground at the base of a plant or on stems from an axillary BUD, or they form instead of flowers (as with some onions). Bulbils that become detached and fall to the ground can develop into new plants, and so function as a method of ASEXUAL REPRODUCTION.

bulk modulus Physical constant of solids and fluids that indicates their elastic properties when they are under pressure over their entire surfaces. The bulk modulus is also called the **incompressibility:** if a solid or a fluid (liquid or gas) has a high bulk modulus, then it is difficult to compress.

Bullard, Sir Edward Crisp (1907–80) English geophysicist. He is noted for his work in GEOMAGNETISM, especially his theory of the geomagnetic dynamo, based on convective motion in the Earth's core. He was an early supporter of the theory of PLATE TECTONICS, and produced a computer-generated map matching the continental shelves of Africa and South America.

Bunsen, Robert Wilhelm (1811–99) German chemist. Bunsen did important work with organoarsenic compounds, and discovered an arsenic poisoning antidote. He later evolved a method of gas ANALYSIS. With his assistant, Gustav KIRCHHOFF, Bunsen used SPECTROSCOPY to discover two new elements (cesium and rubidium). He invented various kinds of laboratory equipment, such as the Bunsen cell, a carbon-zinc electric cell that was used in arc lamps. Bunsen also improved a gas burner that was later named after him.

Bunsen burner Gas burner widely used in science laboratories. It is named after the German chemist Robert BUNSEN. The burner is a 5-in (13-cm) upright tube, usually of brass, attached to a gas source. It has a variable air inlet at its base to control the intensity of its flame.

buoyancy Upward pressure exerted on an object by the FLUID in which it is immersed. The object is subjected to pressure from all sides, but the pressure on its lower part is greater because of the increasing depth of the fluid. The result of all these pressures is a force acting upwards that is equal to the weight of the fluid displaced. *See also* ARCHIMEDES' PRINCIPLE

Burgess shale Layer of siltstone in a quarry in Yoho National Park in E British Columbia, Canada. Discovered in 1909 by a US scientist, Charles D. Walcott, it contains a large number of fossils of animals that lived in the CAMBRIAN period, more than 500 million years ago. The fossils and the silt that covered them accumulated at the foot of an undersea cliff. The silt has preserved traces of many of the soft bodies of sea creatures, which normally do not survive. The fossils include the oldest known chordate, a forerunner of all animals with backbones. They also include a number of kinds of animals that have completely vanished, and apparently do not belong to any of the 32 or so phyla (major groups) of animals we know today.

burglar alarm Device that gives warning of an actual or attempted illegal entry to premises. In a typical system electrified metal tape is placed at all doors and other entrances to a building. The tape is in circuit with an electromechanical relay, which trips and so energizes an alarm if any tape is disturbed or broken. Other alarm systems employ photoelectric cells (magic eyes), ultrasonic sensors or laser beams.

burnout, rocket Point on a rocket's trajectory at which the fuel supply is exhausted or cut off. In solid fuel rockets, controlled burnout is achieved by lowering the pressure in the combustion chamber or by opening vents to drop the temperature below the combustion point. In liquid fuel rockets the fuel supply is simply turned off.

Bush, Vannevar (1890–1974) US electrical engineer. From 1932–38 he was vice president and dean of engineering at Massachusetts Institute of Technology. He headed the World War II Office of Scientific Research and Development, overseeing the initial stages of the atom bomb project. From 1938–55 he was president of the Carnegie Institute.

bushel Unit of dry measure equal in volume to 9.6 gallons of water, and to four pecks. It is traditionally used to measure grain and fruit.

butadiene (buta-1,3-diene, $CH_2=CHCH=CH_2$) Flammable hydrocarbon made from butenes or by "cracking" NAPHTHA. It is copolymerized with STYRENE to produce synthetic rubbers. Properties: r.d. 0.62; m.p. −108.9°C (−164.0°F); b.p. −4.4°C (24.1°F).

butane (C_4H_{10}) Colorless, flammable gas, the fourth member of the ALKANE series of HYDROCARBONS. It has two ISOMERS: n-butane is obtained from natural gas; isobutane (2-methylpropane) is a by-product of PETROLEUM refining. Butane can be liquefied under pressure at normal temperatures and is used in the manufacture of fuel gas and synthetic rubber. Properties: b.p. (n-butane) −0.3°C (31.5°F) and (isobutane) −10.3°C (13.46°F).

Butenandt, Adolf Friedrich Johann (1903–95) German biochemist who did important research into sex hormones. In 1931 Butenandt isolated the male sex hormone ANDROSTERONE, and investigated (1934) the chemical structure and properties of the female sex hormone PROGESTERONE. For this work he was awarded the 1939 Nobel Prize for chemistry, but had to decline the award because of a Nazi decree prohibiting acceptance.

butene (butylene) Any of three hydrocarbon gases having the general formula C_4H_8: butene-1 ($CH_3CH_2CH=CH_2$), butene-2 ($CH_3CH=CHCH_3$), and isobutene ([$CH_3]_2C=CH$) or 2-methylpropene. Butene is made from petroleum and used as a starting material for other organic compounds.

butte Isolated, flat-topped, steep-sided hill. It is formed when a remnant of hard rock overlies and protects softer rock underneath from being worn down, while the surrounding areas continue to be eroded, and is most often found in arid areas, particularly in the plains of the western United States.

byte Binary number used to represent the letters, numbers, and other characters in a computer system. Each byte consists of the same number of BITS. Byte is a contraction of the words "by eight," and originally meant an eight-bit byte, such as 01101010 (representing j on most systems). Many computers now use 16-, 32-, or 64-bit bytes. *See also* GIGABYTE; MEGABYTE

cable Wire for mechanical support, for conducting electricity or for carrying signals. In civil and mechanical engineering, a cable is a rope made of twisted strands of steel wire. They range in size from small BOWDEN CABLES to massive supporting cables on the decks of SUSPENSION BRIDGES. In electrical engineering, a cable is a CONDUCTOR consisting of one or more insulated wires, which may be either single or multi-stranded. They also range greatly in size, from the small cables used for domestic wiring to the large, armored underwater cables. These are used for telephone, radio, television, and data signals. In a COAXIAL CABLE one conductor is cylindrical and surrounds the other. Fiber optic cables consist of OPTICAL FIBERS and carry signals in the form of coded pulses of light.

A   B

1   2   3   4

▲ **cable** Wire cables are made from preformed wires that are wound into strands and twisted together. The two common ways of winding wire rope are indicated by the lay of the rope. In lang lay (A) the wires in the strand and the strands themselves are laid in the same direction. Regular lay (B) has the wires in the strand laid in one direction and the

completed strands laid perpendicular to the wires. Cross sections of typical cables:
1) Fiber core cable for haulage and driven conveyors
2) Wire rope for bridges and structural engineering
3) Steel core triangular strand cable for earth-moving machinery
4) Locked coil cable for aerial ropeways

cable car Passenger vehicle for ascending and descending mountains, often used as a tourist facility. A car is slung from a trolley pulley running on a steel cable. It is moved along by a steel traction cable, continuously looped around drums that are turned by an electric motor. In San Francisco the public-transit cars that haul people up and down the city's hills are powered by a continuously moving cable that runs below the streets.

cable television Generally refers to community ANTENNA television. CATV is not broadcast, but picks up signals at a central antenna and delivers them to individual subscribers via COAXIAL CABLES. Originally designed for areas with poor reception and no local station, cable television, run by private operators, now serves to increase the variety of local viewing by transmitting channels brought by microwave relay. In the US, there are currently over 11,000 cable operating systems, serving approximately 60% of all US households. Channels are required to carry a proportion of community access programming. Viewers pay a monthly subscription for basic cable services such as HBO (Home Box Office). For an additional monthly charge, they can also receive network services, known as pay TV.

cadmium (symbol Cd) Silver-white, metallic element in group II of the periodic table, first isolated in 1817. Cadmium is found in greenockite (a sulfide) but is mainly obtained as a by-product in the extraction of zinc and lead. Malleable and ductile, its main uses are as a protective electroplated coating, an absorber of neutrons in NUCLEAR REACTORS and in nickel-cadmium BATTERIES. Chemically it resembles zinc. Properties: at.no. 48; r.a.m. 112.4; r.d. 8.65; m.p. 320.9°C (609.6°F); b.p. 765°C (1,409°F); most common isotope ^{114}Cd (28.86%). *See also* ELECTROPLATING

Caelum Constellation of the s sky, adjoining Columba. It has no star above magnitude 4.5. One variable, R Caeli, ranges from magnitude 6.7 to 13.7 with a period of 391 days.

caffeine ($C_8H_{10}N_4O_2$) White, bitter ALKALOID that occurs in coffee, tea, and other substances such as cocoa and ilex (holly) plants. In beverages it acts as a mild, harmless stimulant and DIURETIC, although an excessive dose can cause insomnia and delirium.

Cainozoic Alternative name for the CENOZOIC era of geological time.

caisson Watertight structure or chamber used in underwater excavation or construction. It allows people to work underwater and makes it easier to remove excavated material. Caissons are frequently used in building bridges, where they keep water out during the construction of a foundation. They may be open, with "walls" that project above the surface of the water, or closed at the top and submerged (**pneumatic** caissons), using air pressure to exclude water.

calamine Pinkish, odorless powder of zinc carbonate and some ferric oxide, dissolved in mineral oils and used in skin lotions to alleviate such disorders as chicken pox, hives, and other skin rashes.

calcareous Describing a rock that is rich in calcium carbonate ($CaCO_3$). CHALK is a common calcareous rock, as are other LIMESTONES, although these may also contain many other minerals. Soils that develop on such rocks are also termed calcareous.

calciferol (vitamin D_2) Fat-soluble VITAMIN formed by the action of sunlight on the skin and further converted into compounds necessary for the intestinal uptake of dietary calcium for bone formation.

calcination Process of heating solids to high temperatures (but not to their fusion point) to remove volatile substances, partially to oxidize them, or to render them friable (easily crumbled). Lead, zinc, calcium, copper and iron ores calcine to sintered oxides, which are used as pigments or as intermediates in metal extraction.

calcite Mineral, calcium carbonate ($CaCO_3$). Calcite is a major constituent of CALCAREOUS SEDIMENTARY ROCK, especially LIMESTONE. The crystals are in the hexagonal system and vary in form from tabular (rare) to prismatic or needle-like. Calcite is usually glassy white but may be red, pink or yellow. It reacts with dilute hydrochloric acid. Hardness 3; r.d. 2.7.

calcitonin HORMONE secreted by the THYROID GLAND in mammals. It lowers the concentration of calcium in the blood. It opposes the action of PARATHYROID HORMONE.

calcium (symbol Ca) Common silver-white metallic element of the ALKALINE-EARTH METALS; first isolated in 1808. Calcium occurs in many rocks and minerals, notably limestone and gypsum, and in bones. In the body it helps to regulate the heartbeat and is essential for strong bones and teeth. The metal, which is soft and malleable, has few commercial applications, but its compounds are widely used. Chemically it is a reactive element, combining readily with oxygen, nitrogen, and other nonmetals.

Properties: at.no. 20; r.a.m. 40.08; r.d. 1.55; m.p. 839°C (1,542°F); b.p. 1,484°C (2,703°F); most common isotope ^{40}Ca (96.95%).

calcium carbide (calcium acetylide, CaC_2) Chemical made commercially by heating coke and calcium oxide (lime, CaO) in an ELECTRIC FURNACE. It reacts with water to yield ETHYNE (acetylene, C_2H_2), a gas that burns with a luminous flame; for this reason it was once much used in carriage and locomotive headlights. Calcium carbide is also used to manufacture ACETIC ACID (CH_3COOH) and ACETALDEHYDE (CH_3CHO).

calcium carbonate ($CaCO_3$) White compound, insoluble in water, that occurs naturally in such forms as MARBLE, CHALK, LIMESTONE, and CALCITE. It also forms the shells of mollusks. Crystals are in the hexagonal system and vary in form from tabular (rare) to prismatic or needle-like. Calcium carbonate is used in the manufacture of cement, iron, steel, and lime, to neutralize soil acidity and as a constituent of antacids.

calcium hydrogencarbonate ($Ca(HCO_3)_2$) Salt that is responsible for temporary HARDNESS OF WATER. The hardness is removed when the water is heated and insoluble calcium carbonate is precipitated from the water, forming a "fur" inside pipes and kettles. This also happens when water drips inside caves and forms STALAGMITES and STALACTITES.

calcium hydroxide (slaked lime, $Ca(OH)_2$) White solid obtained by the action of water on calcium oxide. It is used in mortar, plaster, cements, and in water softening and agriculture. *See also* LIME WATER

calcium oxide (quicklime, CaO) White solid made by heating calcium carbonate ($CaCO_3$) at high temperatures. It is used industrially to treat acidic soil and make porcelain and glass, bleaching powder, caustic soda, mortar, and cement, and in the recovery of AMMONIA from the ammonia-soda process. Calcium oxide reacts with water to form CALCIUM HYDROXIDE (slaked lime, $Ca(OH)_2$), which dissolves in water to give lime water, used in laboratories as a test for carbon dioxide gas (CO_2). Calcium oxide was used in the 19th century for producing limelight illumination, an incandescence that occurs when coal gas is burned in air against a block of quicklime.

calcium sulfate ($CaSO_4$) Chemical compound that occurs naturally as the mineral anhydrite. The hydrated form ($CaSO_4.2H_2O$) is the mineral GYPSUM, which loses water when heated to form plaster of Paris (calcium sulfate, $(CaSO_4)_2.H_2O$). This, when mixed with water, quickly sets hard to a crystalline form of the hydrate.

calculus Branch of mathematics involving the techniques of DIFFERENTIATION and INTEGRATION. **Differential calculus** deals with differentiation: the process of finding the instantaneous rate of change of a function at any time. Limiting values of the increment of the function are taken as the increment of one of its variables tends to zero: the ratio of these increments is the DERIVATIVE. Differential calculus is used to find slopes of curves. **Integral calculus** deals with the operation of integration: the finding of a function, one or more derivatives of which are given. To take a simple example, the INTEGRAL of x with respect to x is $^1/_2x^2 + c$, where c is a constant. Integral calculus is used to determine the areas or volumes enclosed by a given boundary. For more than 200 years it was believed that calculus was discovered (independently) by Gottfried LEIBNIZ and Isaac NEWTON. However in 1934 a note written by Newton

▲ **calorimeter**
Calorimeters can measure the amount of heat absorbed or released during a chemical reaction. In a high-pressure flow calorimeter, the apparatus is contained in a vacuum (1) for insulation. A constant flow of liquid or gas enters the calorimeter (2). A platinum resistance thermometer (3) measures the temperature of the substance on entry. A heater (4) puts a known amount of energy into the liquid or gas inside a radiation shield (5) that further lessens any dispersion of energy. The change in temperature is measured by a second thermometer (6) again shielded (7).

was discovered in which he acknowledged that his formulation of calculus was based on a method of drawing tangents first developed by Pierre de FERMAT.

caldera Large CRATER formed when a VOLCANO collapses due to the migration of the MAGMA under the Earth's crust. The caldera of an extinct volcano, if fed by floodwater, rain, or springs, can become a CRATER LAKE.

calibration Testing of scientific instruments for the purpose of affixing measuring scales. For example, thermometers are calibrated in degrees Fahrenheit or Celsius, or in kelvins, and pressure gauges are calibrated in Nm^{-2} or lb/sq ft. All calibrations are based on such fundamental standard units as length and temperature.

californium (symbol Cf) Radioactive, metallic element of the ACTINIDE SERIES, first made in 1950 at the University of California, Berkeley, by alpha-particle bombardment of the curium isotope curium-242. Californium presents biological dangers because one microgram releases 170 million neutrons a minute. Properties: at.no. 98; most stable isotope ^{251}Cf (half-life 800 yr). *See also* TRANSURANIC ELEMENTS

Callisto Second-largest of Jupiter's GALILEAN SATELLITES, with a diameter of 4,800km (2,980mi), and the outermost of them. The surface of the satellite is dark. It is the most heavily cratered object so far discovered. Since the main period of cratering was early in the Solar System's history, this indicates that no subsequent geological activity has disturbed Callisto's surface. As well as the dense craters, there are several large, multiringed impact features, the largest of which is Valhalla, which has an overall diameter of 4,000km (2,500mi).

callus In botany, a protective mass of undifferentiated plant cells formed at the site of a wound in a woody plant. Callus tissue is also formed at the base of cuttings as they start to take root. Callus tissue is important as the starting point for TISSUE CULTURE of plants.

Calmette, Albert Léon (1863–1933) French bacteriologist whose name is remembered, along with his compatriot Charles Guérin , in *bacille Calmette Guérin* (BCG). Since 1923, BCG vaccine has been used to immunize children against tuberculosis. It is a "live" vaccine, containing attenuated TB bacteria, and over the years 250 million people have benefited from it.

caloric theory of heat *See* HEAT

calorie Unit of HEAT. A calorie is the amount of heat required to raise 1g of water 1°C between the temperatures of 14.5 and 15.5°C (58.1 and 59.9°F). The SI system of units uses the JOULE (1 calorie = 4.184 joules) instead of the calorie. A dietitian's "calorie" is the kilocalorie, which is 1,000 times the size of a calorie.

calorimeter Apparatus used for experiments involving heat measurements. It is usually a conducting container, such as a copper vessel, that is thermally insulated. There are many types designed for special purposes, such as measuring calories in a food, specific heat capacities, specific latent heats, heats of reaction, and heats of formation. A **bomb** calorimeter ignites sealed oxygen to measure heats of combustion, like calories in food.

Calvin, Melvin (1911–97) US chemist. Calvin conducted experiments using radioactive carbon-14 as a trace to label carbon dioxide and track the process by which plants turned it into glucose by means of PHOTOSYNTHESIS. The series of reactions that take place during photosynthesis are known as the CALVIN CYCLE. Calvin received the Nobel Prize for chemistry in 1961.

Calvin cycle Sequence of reactions that take place during the dark stage of PHOTOSYNTHESIS. The overall effect of the cycle is the REDUCTION of carbon dioxide to form carbohydrate. First, an ENZYME (ribulose biphosphate carboxylase) causes carbon dioxide to combine with ribulose biphosphate to form a six-carbon compound that quickly decomposes to give two molecules of glycerate 3-phosphate (a three-carbon compound). Then, after being changed to glyceraldehyde 3-phosphate, this re-forms ribulose biphosphate with the release of the sugars fructose and glucose. The whole cycle takes place in the CHLOROPLASTS in leaves or other green parts of the plant. It was worked out, using radioactive tracers, by Melvin CALVIN.

calyx SEPALS of a flower, taken as a group. Usually green in color, the calyx protects the flower when it is a bud. When the flower opens, it surrounds the CARPELS, PETALS, and STAMENS, and forms the outer whorl of the PERIANTH.

cam Mechanical device consisting of an eccentric projection on a rotating shaft, shaped so as to give some desired reciprocating linear motion to another component. Cams are used in many different kinds of machinery.

cambium Layer of cells parallel to the surface of stems and roots of plants that divides to produce new cells to allow for growth in diameter of the stem and roots. Once a plant cell has differenti-

▲ **cam** A disc cam can be regarded as a lever of variable length that changes rotary motion into an up-and-down or side-to-side (reciprocating) motion. Disc cams are often used to operate the valves in a car engine, as shown above.

ed, it is unable to divide again. To allow for growth, certain bands of cells remain undifferentiated, retaining their ability to divide. The cambium is made up of such cells. There are two main types of cambium, the **vascular** cambium and the **cork** cambium. The vascular cambium produces new PHLOEM on the outside and XYLEM on the inside, leaving narrow bands of thin-walled cells that allow nutrients and gases to diffuse to the center of the plant. The cork cambium forms a cylinder just below the epidermis, and produces cork cells to replace the epidermis, which ruptures as the stem and root expand, forming the bark and the corky outer layer of the older root. *See also* MERISTEM

Cambrian Earliest period of the PALEOZOIC era, lasting from about 590 million to 505 million years ago. The rocks of this period are the earliest to preserve the hard parts of animals as fossils. Cambrian rocks contain a large variety of fossils, including all the animal phyla with the exception of the vertebrates. At that time the animals lived in the seas, while the land was barren. The commonest animal forms were trilobites, brachiopods, sponges, and snails. Plant life consisted mainly of seaweeds.

Camelopardalis (Giraffe) Barren constellation of the N sky. Its brightest star, Beta, is of magnitude 4.0. NGC 1502 is an open cluster visible with binoculars.

camera Apparatus for taking photographs, consisting essentially of a lightproof box containing photographic film. When a shutter is opened, usually briefly, light from the scene outside is focused by a lens system onto the film. The amount of light falling on the film is controlled by the shutter speed and by the diameter of the lens APERTURE, which can also be varied using an adjustable iris diaphragm. Most cameras also have a rangefinder, enabling a focused image to be produced for a given object distance, and a built-in exposure meter to determine the correct combination of shutter speed and aperture for the prevailing light conditions. Both of these functions are automated in a camera with a computerized program (to set the correct combination of shutter speed and aperture) and an autofocus lens system. Recently digital cameras have been developed, in which a CHARGE-COUPLED DEVICE (CCD) "records" the image electronically. The image can then be downloaded onto a computer and a print made.

camera lucida Apparatus for drawing and copying in perspective, developed in 1812 by William WOLLASTON. A prism is set between the draftsman's eye and the paper in such a way that light is reflected from the object being copied to form an image on the paper.

camera obscura Optical device consisting of a darkened room into which an inverted image is thrown through a convex lens. A portable version of this was used by 17th- and 18th-century artists to trace scenes from nature. By the 19th century it had become a box, fitted with a lens and mirror and placed on a tripod. It eventually developed into the modern CAMERA.

camouflage In zoology, means by which some animals blend in with their natural environment, usually by resembling their surroundings in form and/or color. Camouflage enables animals to conceal themselves from their prey or to protect themselves from predators. The leaf insect is a notable example. It is flat and green and the female has large leathery forewings with markings that resemble a pattern of leaf veins. Camouflage is a form of MIMICRY.

camphor ($C_{10}H_{16}O$) Organic compound, the molecule of which has a complex ring structure. It has a strong odor, which also occurs in its original source, the wood and leaves of the camphor tree, *Cinammonum camphora*, native to Taiwan. Camphor is used in medicine for liniments, in the manufacture of celluloid, lacquers, and explosives, and as an ingredient of mothballs.

canal Artificial waterway for irrigation, drainage, navigation or in conjunction with hydroelectric DAMS. The first canals were built 4,000 years ago in ancient Mesopotamia. In 1825, the Erie Canal in upstate New York linked the Hudson River with the Great Lakes and thus opened the midwest to commerce and growth.

Cancer (Crab) Constellation of the N sky, between Gemini and Leo. It contains two open clusters: M44, the Praesepe or Beehive Nebula (NGC 2632), and M67 (NGC 2692). The brightest star is Beta Cancri, magnitude 3.5.

cancer Group of diseases featuring the uncontrolled proliferation of cells in a process called tumor formation. Malignant (cancerous) cells spread (metastasize) from their original sites to other parts of the body. There are many different cancers. Known causative agents (carcinogens) include smoking, certain industrial chemicals, asbestos dust, and radioactivity. Viruses are implicated in the causation of some cancers. Some people have a genetic predisposition towards particular types of cancer. Treatments include surgery, CHEMOTHERAPY with cell-destroying drugs and RADIOTHERAPY (or sometimes a combination of all three). Early diagnosis holds out the best chance for successful treatment.

Cancer, Tropic of Line of latitude, 23.5° N of the Equator, that marks the N boundary of the tropics. It indicates the farthest N position at which the Sun appears directly overhead at noon. The Sun is vertical over the Tropic of Cancer on about 21 June, the summer SOLSTICE in the Northern Hemisphere.

candela (symbol cd) SI unit of luminous intensity. It is defined as the luminous intensity of a source of monochromatic light of frequency 540×10^{12} Hz of radiant intensity 1/683 watt per steradian in a given direction.

Canes Venatici (Hunting Dogs) Constellation of the N sky. The only star above magnitude 4.0 is Cor Caroli, magnitude 2.9. It does contain several galaxies, including the famous Whirlpool Galaxy, M51.

canine Any one of four sharp "stabbing" TEETH in the frontal dentition of most mammals. In humans they are also called eyeteeth.

Canis Major (Great Dog) Constellation of the S sky situated S of Monoceros. It contains the bright open cluster M41 (NGC 2287). The brightest star is Alpha Canis Majoris or SIRIUS (Dog Star), the brightest star in the sky, with a magnitude of –1.47

Canis Minor (Little Dog) Constellation of the N sky. It is notable only for containing the eighth-brightest star in the sky, Procyon, which has a magnitude of 0.38.

canning Method of FOOD PRESERVATION by sealing it in cans. Cans are made from tin-plated steel sheet, or from aluminum. Any bacteria present in the food are killed, either by dry heat or by steam heat. A temperature of about 100°C (212°F) is necessary for fruit, and somewhat higher temperatures for meat and vegetables. Once a can is sealed it must remain airtight to prevent infection by bacteria, particularly those that cause FOOD POISONING.

Cannon, Walter B. (Bradford) (1871–1945) US physiologist. He studied the regulation of hunger and thirst in animals, introduced the term HOMEOSTASIS to describe the ability of an organism to maintain its internal environment, and proposed a theory that placed the seat of emotions in the THALAMUS of the brain.

cannon ARTILLERY piece consisting of a metal tube, used to aim and fire missiles propelled by the explosion of gunpowder in the closed end of the cylinder. Cannon, first used in the 14th century, were originally made of bronze or iron. Major improvements during the 19th century included the introduction of steel barrels, more powerful chemical propellants, breech-loading mechanisms, and standardized parts. By World War I recoil devices were employed to absorb the shock of firing.

Canopus (Alpha Carinae) Second-brightest star observed from the Earth; it has a magnitude of –0.72. Its luminosity and distance are not accurately known, but one estimate classifies it as a bright giant, 800 times as luminous as the Sun, and 74 light-years away.

cantilever Either a projecting beam that is rigidly supported at one end with force applied at the free end (such as a diving board), or a BRIDGE supported by two projecting beams, joined in the center by a connecting member and supported on piers and anchored by counterbalancing members.

Cantor, Georg (1845–1918) German mathematician, b. Russia. Cantor was professor of mathematics (1869–1913) at the University of Halle, Germany. His work on the concept of INFINITY challenged the existing deductive processes of mathematics. Cantor developed SET THEORY and provided a new definition of irrational numbers (*see* NUMBER, IRRATIONAL). **Cantor's paradox** arises with the attempt to construct a universal set that includes all other sets because every set has more subsets than members and yet the universal set should contain the subsets as well.

canyon Deep, narrow depression in the Earth's crust. Land canyons are the result of erosion by rivers of comparatively recent origin flowing through arid terrain. Marine canyons may be formed when a riverbed and the surrounding terrain is submerged, or by turbulence produced by deep-water currents.

capacitance (symbol *C*) Ability of an ELECTRIC CIRCUIT or component to store charge. Capacitance is measured in farads (1 farad is a capacitance needing a charge of 1 coulomb to raise its potential by 1 volt), but most capacitances are small enough to be measured in microfarads (one millionth of a farad) or picofarads (one million-millionth of a farad).

capacitor (condenser) ELECTRIC CIRCUIT component that has CAPACITANCE. Having at least two metal plates, it is used principally in alternating-

▶ **cancer** Cancer can spread in two ways. First, by direct growth into adjacent tissues, called "direct extension" (A), when cancer cells penetrate into bone, soft connective tissue, and the walls of veins and lymphatic vessels. Alternatively, a cancer cell separates from its tumor and is transported to another part of the body. This spread of cancer is called "metastasis" (B). In metastasis, after the tumor has grown to some size, cancer cells or small groups of cells enter a blood or lymph vessel through the vessel wall (1). They travel through the vessel until they are stopped by a barrier, such as a lymph node, where additional tumors may develop, before releasing more cells which may develop on other lymph nodes. Such cancers, usually carcinomas, may also invade the blood stream and establish more distant secondary growths. Another type of cancer, sarcomas, tends to spread via venous blood vessels, frequently establishing tumors in the lungs, gastrointestinal tract, or the genito-urinary tract (2). In abdominal cancers (or transcoelomis-spread cancers), metastasis may also arise as a result of travel across certain body cavities, such as the peritoneal, oral, or pleural cavities (3).

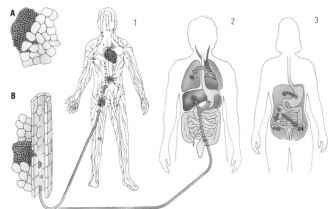

current (AC) circuits. The various types include parallel-plate condensers and electrolytic capacitors. *See also* ELECTRIC CURRENT

capillarity Movement of a liquid in a narrow opening caused by the surface tension between the liquid and the surrounding material. This is most often seen in a vertical, narrow glass capillary tube, but capillarity also occurs in various directions – as when a sponge or blotting paper soaks up water.

capillary Smallest of BLOOD VESSELS, connecting arteries and veins. Capillary walls consist of only a single layer of cells, so that fluid containing dissolved oxygen and other nutrients (as well as carbon dioxide and other wastes) can pass easily between the blood and surrounding tissues.

capillary constant Measure of surface tension defined, for two immiscible fluids in contact, by $2T/g(D-d)$, where T is the surface tension for the interface, g is the acceleration due to gravity and D and d are the densities of the fluids. Usually one of these is air. The capillary constant is equal to the rise of a liquid in a capillary tube multiplied by the radius of the tube.

Capricorn, Tropic of Line of latitude, 23.5°s of the Equator, which marks the s boundary of the tropics. It indicates the farthest s position at which the Sun appears directly overhead at noon. The Sun is vertical over the Tropic of Capricorn on about 22 December, which is the summer SOLSTICE in the Southern Hemisphere.

Capricornus Constellation of the s sky, situated on the ECLIPTIC between SAGITTARIUS and AQUARIUS. Usually referred to as Capricorn (Ram) only for astrological purposes, this constellation contains the faint globular cluster M30 (NGC 7099). Its brightest star is Delta Capricorni (Deneb Algiedi), an eclipsing BINARY STAR with magnitude 2.9.

capsule In botany, a dry type of fruit that releases its seeds when it ripens. The capsule is formed by the fusion of several CARPELS, which may split apart to allow the seeds to scatter. Other capsules have a lid that opens, or holes through which the seeds escape. The spore-containing structures of mosses and liverworts are also called capsules. In zoology, the sticky layer that surrounds the cell walls of some bacteria is called a capsule, as are various other surrounding structures such as the sheath of connective tissue surrounding the bones in a joint and the membranous envelope around the kidneys and spleen.

capsule, space That part of a spacecraft that carries the payload, which may be a crew of astronauts or a variety of scientific instruments. A space capsule is designed to withstand extremes of temperature and pressure, shocks and vibration during acceleration, spinning or tumbling, radiation, and meteoric impacts, while maintaining a stable environment for the payload. Some of the features that help to achieve these ends are an ABLATION heat shield for re-entering the atmosphere; slow, continuous rotation to reduce exposure to the Sun; and highly reflecting exteriors. The capsule of a manned spacecraft contains the flight controls, and television cameras and instruments for monitoring the activities of the crew and controlling their environment.

car *See* AUTOMOBILE

carapace In zoology, the back of the shell of certain animals. The shieldlike part of the skeleton that covers and protects the head and thorax of crabs and other crustaceans is called a carapace, as is the dorsal (upper) part of the shell of a tortoise or turtle. In a tortoise, it is made up of bony plates joined to the backbone and ribs, covered by an outer horny layer.

carat Unit of weight used for gemstones. Traditionally it varied from country to country, perhaps because of its method of origin: the name carat

▲ **capillarity** The cohesion between a liquid and a solid results in the liquid surface curving near the solid, to meet it at a definite angle. Water curves upwards against glass, and the force of cohesion is exerted along the water surface, tending to lift it. The lifting force is proportional to the circumference of the water surface; in a narrow tube this force becomes powerful enough to lift a tall column of water.

derives from that of the seeds of the Mediterranean locust tree, which were once used to weigh gemstones. By 1913 the carat was fixed internationally at 200 milligrams. The carat measure of gold alloys (usually gold-silver or gold-copper) indicates the number of parts of gold in 24 parts. Thus 22-carat gold contains 22 parts of gold in 24 parts of the alloy.

carbide (acetylide) Inorganic compound of carbon with metals or other more ELECTROPOSITIVE elements. BORON and SILICON both form extremely hard carbides which are used as abrasives. Many TRANSITION ELEMENTS also form carbides, in which carbon atoms occupy spaces between adjacent atoms in the metal lattice. Some electropositive metals form ionic carbon compounds; the best known is CALCIUM CARBIDE (CaC_2), which reacts with water to give ACETYLENE (C_2H_2).

carbohydrate Organic compound of carbon, hydrogen, and oxygen that is a constituent of many foodstuffs. The hydrogen and oxygen atoms are in the ratio of 2:1, as they are in water (H_2O). The simplest carbohydrates are the sugars, usually with five or six carbon atoms in each molecule. Glucose and fructose are MONOSACCHARIDES, naturally occurring sugars; they have the same formula ($C_6H_{12}O_6$) but different structures. One molecule of each combines with the loss of water to make sucrose ($C_{12}H_{22}O_{11}$), a DISACCHARIDE which also occurs naturally in sugar cane and sugar beet. Starch and cellulose are POLYSACCHARIDES, carbohydrates consisting of hundreds of glucose molecules linked together. *See also* SACCHARIDE

carbolic acid *See* PHENOL

carbon (symbol C) Common, nonmetallic element of group IV of the periodic table. Carbon forms a vast number of compounds, which, with hydrogen (hydrocarbons) and other nonmetals, form the basis of organic CHEMISTRY. Until recently, it was believed there were two crystalline carbon ALLOTROPES: GRAPHITE and DIAMOND. In 1996 a third type, BUCKMINSTERFULLERENE ("buckyballs"), was discovered, the molecules of which are shaped like geodesic domes. Various amorphous (noncrystalline) forms of carbon also exist, such as coal, coke, charcoal, soot, and lampblack. Amorphous carbon has many uses, for example as pigment for inks and a filler for rubber for tires. A recently made synthetic form of carbon is CARBON FIBER. The radioactive isotope ^{14}C is used for CARBON DATING of archaeological specimens. Properties: at.no. 6; r.a.m. 12.011; r.d. 1.9–2.3 (graphite), 3.15–3.53 (diamond); m.p. 3,550°C (6,422°F); sublimes at 3,367°C (6,093°F); b.p. 4,200°C (7,592°F); most common isotope ^{12}C (98.89%).

carbonate Salt of CARBONIC ACID (H_2CO_3), which is formed when carbon dioxide (CO_2) dissolves in water, as when rain falls through the air. Carbonic acid is an extremely weak acid and both it and many of its salts are unstable, decomposing readily to release carbon dioxide. Nevertheless, large parts of the Earth's crust are made up of carbonates, such as CALCIUM CARBONATE and DOLOMITE.

carbonation In geology, the process of chemical EROSION of rocks by rainwater containing carbon dioxide. When carbon dioxide gas in the atmosphere dissolves in rainwater, it forms CARBONIC ACID. This will dissolve CALCAREOUS rocks such as chalk and limestone. The dissolved chemicals can drip into caves where they may form STALACTITES and STALAGMITES.

carbon black Form of the element carbon made by heating or burning hydrocarbon gases in a restricted supply of air. The product contains a little hydrogen, oxygen, and sulfur as well as carbon. Carbon black is used to reinforce rubber (for vehicle tires) and other materials, and in various dark pigments for paint and inks.

carbon cycle Circulation of carbon in the biosphere. It is a complex chain of events. The most important components are the taking up of carbon dioxide from the atmosphere by green plants during PHOTOSYNTHESIS, and the return of carbon dioxide to the atmosphere by the respiration and eventual decomposition of animals that eat the plants. *See illustration on page 84*

carbon dating (radiocarbon dating) Method of determining the age of organic materials by measuring the amount of radioactive decay of an ISOTOPE of carbon, carbon-14 (^{14}C). This radio-isotope decays to form nitrogen, with a half-life of 5,730 years. When a living organism dies, it ceases to take carbon dioxide into its body, so that the amount of ^{14}C it contains is fixed relative to its total weight. Over the centuries, this quantity steadily diminishes. Refined chemical and physical analysis is used to determine the exact amount remaining, and from this the age of a specimen is deduced.

carbon dioxide (CO_2) Colorless, odorless gas that occurs in the ATMOSPHERE (0.03%) and as a product of the combustion of fossil fuels and respiration in plants and animals. It is taken up by green plants in PHOTOSYNTHESIS. In its solid form (dry ice) it is used in refrigeration; as a gas it is used in carbonated beverages, fire extinguishers, and provides an inert atmosphere for welding. If inhaled in large amounts, carbon dioxide can cause suffocation. Research indicates that its increase in the atmosphere leads to the GREENHOUSE EFFECT and

GLOBAL WARMING. Properties: m.p. $-56.6°C$ ($-69.9°F$); sublimes $-78.5°C$ ($-109.3°F$).

carbon fiber Form of CARBON made by heating textile fibers to very high temperatures. The result is fibers (typically 0.001cm in diameter) that are, weight for weight, some of the strongest of all known fibers. They are too short to be woven into a super-strong yarn. Instead they are incorporated into plastics, ceramics, and glass, which give the materials great strength and resistance to breakage.

carbonic acid (H_2CO_3) Extremely weak acid formed when carbon dioxide (CO_2) dissolves in water. It forms two series of salts, CARBONATES, and BICARBONATES. Carbonates are of great importance in nature both in rock formations and in the shells of various animals.

Carboniferous Fifth geologic period of the PALEO-ZOIC era, lasting from 360 to 286 million years ago. It is divided into two series; the Lower Carboniferous is characterized by marine limestones with a coral-rich fauna; the Upper Carboniferous is dominated by river and deltaic sediments containing coal seams formed from swampy forests of conifers and tree ferns. In North America, the Lower Carboniferous is known as the Mississippian period and the Upper Carboniferous as the Pennsylvanian period.

carbon monoxide (CO) Colorless, odorless, poisonous gas formed during the incomplete combustion of fossil fuels, occurring for example in coal gas and the exhaust fumes of cars. Carbon monoxide poisons by combining with the HEMOGLOBIN in ERYTHROCYTES (red blood cells) thus preventing them from carrying oxygen round the body. (This happens if the inhaled air contains only 0.1% of carbon monoxide by volume.) It is used as a reducing agent in metallurgy. Properties: density 0.968 (air = 1); m.p. $-205°C$ ($-337°F$); b.p. $-191.5°C$ ($-312.7°F$).

carbon tetrachloride (tetrachloromethane, CCl_4) Colorless, nonflammable liquid with a characteristic odor; prepared by the chlorination of methane or the catalytic reaction of carbon disulfide and chlorine. It is used as a refrigerant, insectiside, degreaser, and dry-cleaning fluid. Properties: r.d. 1.59; m.p. $-23°C$ ($-9.4°F$); b.p. 76.8°C (170.2°F)

carbonyl compound Compound containing the carbonyl group (–CO). Organic carbonyls include KETONES, ALDEHYDES, and CARBOXYLIC ACIDS. Inorganic carbonyls contain carbonyl groups bound directly to metal atoms.

carborundum Originally a trade name for silicon carbide (SiC) abrasives and refractories prepared by heating silica (SiO_2) with carbon in an electric furnace. It is used in grinding wheels, grinding papers,

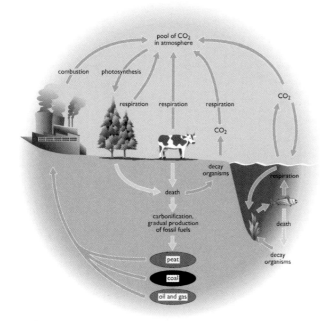

◀ **carbon cycle**
Elemental carbon is in constant flux. Gaseous carbon dioxide (CO_2) is first incorporated into simple sugars by photosynthesis in green plants. These may be broken down (respired) to provide energy, a process that releases CO_2 back into the atmosphere. Alternatively, animals that eat the plants also metabolize the sugars and release CO_2 in the process. Geological processes also affect the Earth's carbon balance, with carbon being removed from the cycle when it is accumulated within fossil fuels, such as coal, oil, and gas. Conversely, large amounts of carbon dioxide are released into the atmosphere when such fuels are burned.

abrasive grains and powders, valve-grinding compounds, and in refractory bricks and blocks. Nearly as hard as diamond, it oxidizes slowly at temperatures above 1,000°C (2,000°F).

carboxylic acid Member of a class of organic chemical compounds containing the group –COOH. The commonest example is ACETIC ACID (ethanoic acid, CH_3COOH), which is present in vinegar. These acids are weakly acidic, forming salts with bases and esters with alcohols. Esters of high-molecular weight, carboxylic acids, such as stearic, lauric, and oleic acids, are present in animal and vegetable fats; for this reason carboxylic acids are often called FATTY ACIDS. The systematic names are formed using the suffix -oic.

carburation Mixing of air and fuel vapor in the correct proportions for an INTERNAL COMBUSTION ENGINE. In a gasoline engine, air sucked into the cylinders passes through the CARBURETOR, creating a partial vacuum that draws gasoline in through a jet. The resulting air/gasoline mixture rapidly burns in the cylinders and the expansion of the gases produced provides the power.

carburetor Component of some gasoline-powered INTERNAL COMBUSTION ENGINES, used to vaporize and mix fuel with air in the correct proportion for proper combustion. Efficient CARBURATION is essential for smooth running and efficient engine performance. No carburetor is needed in engines that use FUEL INJECTION.

carcinogen External substance or agent that causes CANCER, including chemicals, radiation, and some viruses. Many different chemical carcinogens have been found to cause cancer in animals, but more research is needed to confirm their ability to cause cancer in humans. It is thought that tobacco smoke causes lung cancer because of carcinogenic hydrocarbons.

Cardano, Girolamo (1501–76) Italian mathematician and physician. He wrote several influential texts on mathematics. In *De Subtilitate Rerum* (1550) he attempted to summarize the existing scientific beliefs of his time. His most notable contribution is thought to be the solution of the general cubic equation

$$ax^3 + bx^2 + cx + d = 0$$

often called **Cardano's solution**. TARTAGLIA, however, had solved the problem 10 years before Cardano announced his solution.

cardiac cycle Sequence of events in between each heartbeat. Blood enters the heart while it is relaxed, filling the ATRIA and VENTRICLES. Contraction of the ventricles forces blood out of the heart and, at the end of the contraction, the ventricles again relax and the heart starts to fill again, ready for the next cycle.

cardiac muscle *See* MUSCLE

cardinal number *See* NUMBER, MUSCLE

cardiology Branch of medicine that deals with the diagnosis and treatment of the diseases and disorders of the HEART and vascular system.

caries Decay and disintegration of teeth or BONE substance. Tooth decay is caused mainly by acids produced when bacteria present in the mouth break down sugars in food. Regular brushing, a reduced sugar intake and fluoride prevent caries. Eating between meals encourages tooth decay, because it maintains a supply of food for the bacteria and keeps acidity levels high.

Carina (Keel) Part of the dismembered constellation Argo Navis, the ship Argo of the S sky. It is the brightest and richest part of Argo, and contains CANOPUS, the second-brightest star in the sky. Apart from Canopus, its brightest stars are Beta, magnitude 1.7, and Epsilon, magnitude 1.9.

Carlson, Chester (1906–68) US physicist, inventor of XEROGRAPHY (1938). He patented it in 1940, but had great difficulty in convincing anyone of its commercial value. Finally, in 1947, he signed an agreement with the Haloid Company (now the Xerox Corporation). His royalties made him a multimillionaire.

carnivore In the wider, ecological sense of the term, a carnivore is any animal that eats other animals. In the narrower, taxonomic sense, it is any member of the order of flesh-eating mammals (Carnivora). Mustelids – weasels, martens, minks, and the wolverine – make up the largest family. The cats are the most specialized killers among the carnivores; dogs, bears, and raccoons are much less exclusively meat eaters; and civets, mongooses and their relatives also have a mixed diet. Related to the civets, but in a separate family, are the hyenas, large doglike scavengers. More distantly related to living land carnivores are the seals, sea lions, and walruses; they evolved from ancient land carnivores who gave rise to early weasel- or civet-like forms. Extinct carnivores include the saber toothed cats, which died out during the Pliocene epoch 2 million years ago.

carnivorous plant *See* INSECTIVOROUS PLANT

Carnot, (Nicolas Léonard) Sadi (1796–1832) French engineer and physicist whose work laid the foundation for the science of THERMODYNAMICS.

His major work, *Réflexions sur la puissance motrice du feu* (1824), provided the first theoretical background for the STEAM ENGINE and introduced the concept of the second law of thermodynamics (involving ENTROPY), which was formulated later by Rudolf CLAUSIUS. Carnot's work was ignored until quoted and extended in 1834 by the railroad engineer Emile Clapeyron and recognized in 1848 by William KELVIN.

Carnot cycle In thermodynamics, a cycle of events that demonstrates the impossibility of total efficiency in HEAT ENGINES. Named after Sadi CARNOT, it shows how an engine can never convert all the heat energy supplied to it from its burning fuel into mechanical energy. Some heat energy always remains unused in a "cold sink." In an INTERNAL COMBUSTION ENGINE, this can be thought of as the engine itself. More properly, the cold sink is the entire Universe.

carnotite $(K_2(UO_2)_2(VO_4)_2.nH_2O)$ Secondary vanadate mineral, potassium uranium vanadate, an ore of uranium and radium, important for nuclear energy. It occurs in the Colorado Plateau, Australia, and Congo as yellow-green crusts or cavity fillings in sandstone and in fossilized wood. It is finely crystalline (monoclinic), dull or earthy; r.d. 3–5.

carotene Plant pigment that is converted to vitamin A by the liver. It occurs in various fruits and vegetables (such as carrots).

carotenoid One of a group of fat-soluble plant pigments ranging in color from yellow to red. Carotenoids also occur in some animal fats. They include isomers of CAROTENE, a pigment that is converted in the liver into vitamin A, necessary for normal vision and healthy skin.

Carothers, Wallace Hume (1896–1937) US chemist who discovered the synthetic polyamide fiber now called NYLON.

carpal One of the small bones in the wrist.

carpel Female reproductive element of a FLOWER. A carpel consists of a STIGMA (which receives pollen), a STYLE, and an OVARY (containing ovules). A group of carpels make up the GYNOECIUM, the complete female reproductive structure within a flower.

Carrel, Alexis (1873–1944) US surgeon and experimental biologist, b. France. Carrel was a member (1912–39) of the Rockefeller Institute, New York City. He was awarded the 1912 Nobel Prize for physiology or medicine for his development of **anastomosis**, a surgical technique for stitching together blood vessels end to end. Carrel and Charles LINDBERGH invented an artificial heart.

carrier In medicine, apparently healthy person who harbors a disease that, while it does not affect him or her, can be transmitted to other people. Carriers are difficult to detect and are often unaware of the infection they carry.

carrier wave High-frequency, ELECTROMAGNETIC RADIATION that carries information signals from a radio or television transmitting ANTENNA. The fixed wavelength and frequency of the transmitter's wave is the signal, or channel, to which a receiving antenna must be tuned in order to receive its broadcast. To prevent interference, the frequencies of transmitter carrier waves differ by 10kHz.

Carson, Rachel Louise (1907–64) US writer and marine biologist. *The Sea Around Us* (1951) won a National Book Award. *Silent Spring* (1962) directed public attention to the dangers of agricultural pesticides and was a pioneering work in the development of the environmental movement. Other works include *Under the Sea Wind* (1941) and *The Edge of the Sea* (1955).

Cartesian coordinate system System introduced by René DESCARTES in which the position of a point is specified by its distances from intersecting lines (the axes). In the simplest Cartesian system the axes are perpendicular (called y and x axes). The position of a point is then given by a pair of numbers (x, y). The **abscissa**, x, is the point's distance from the y axis, measured in the direction of the x axis, and the **ordinate**, y, is the distance from the x axis. The axes in such a system need not be at right angles but should not be parallel to each other. Three axes are needed to describe the location of points in three dimensions.

cartilage (gristle) Flexible supporting tissue made up of the tough protein COLLAGEN. In the vertebrate embryo, the greater part of the SKELETON consists of cartilage, which is gradually replaced by BONE during development, except in areas of wear such as the ends of bones and the INTERVERTEBRAL DISCS, where caps of cartilage help to protect the bone below. In humans, cartilage is also present in the larynx, nose, and external ear, and rings of cartilage help to support the windpipe (TRACHEA) and BRONCHI.

cartilaginous fish *See* FISH

Cartwright, Edmund (1743–1823) British inventor. Cartwright invented the power LOOM (patented 1785) after visiting Sir Richard ARKWRIGHT's cotton spinning mill, although it was not used commercially until the early 19th-century. He also invented a wool-combing machine (patented 1789) and an alcohol engine (1797).

casein Principal protein in milk, containing 0.71% phosphorus, 0.72% sulfur, and about 15 AMINO ACIDS. It constitutes about 80% of the proteins in cow's milk and about 40% in human milk. Obtained from milk by the addition of either acid or the enzyme rennet, casein is used to make plastics, cosmetics, paper coatings, adhesives, paints, glues, textile sizing, cheeses, and animal feed.

Cassini, Giovanni Domenico (1625–1712) French astronomer. He was the first to measure accurately the dimensions of the SOLAR SYSTEM. He discovered the division in the rings of SATURN that now bears his name, and also four of Saturn's satellites. He measured Jupiter's rotation period.

Cassiopeia Distinctive constellation of the N sky, representing in mythology the mother of Andromeda. The five leading stars make up a "W" or "M" pattern, on the opposite side of the North Star from URSA MAJOR. Alpha (Schedar) is its brightest star, magnitude 2.2.

cassiterite Translucent black or brown mineral, tin oxide (SnO_2); the major ore of tin. It occurs in PLACER DEPOSITS, chiefly in the Malay peninsula, and in pegmatites and other INTRUSIVE IGNEOUS ROCKS. It takes the form of short tetragonal prismatic crystals, or masses, and radiating fibers. Hardness 6–7; r.d. 7.

casting Forming objects by pouring molten metal into molds and allowing it to cool and solidify. Many castings are made by pouring metal into sand or clay molds. Specialized processes, such as plastic molding, composite molding, CIRE PERDUE casting, and die casting, give greater dimensional accuracy, smoother surfaces, and finer detail.

cast iron General term applied to various grades of iron, especially **gray** iron and **pig** iron (directly out of a BLAST FURNACE). It includes a wide range of iron-carbon-silicon alloys containing from 1.7–4.5% carbon with varying amounts of other elements. Gray iron (so-called because its fracture looks grayish) is the most widely used for casting a number of vehicle engines, machinery parts, and many other products.

castration Removal of the sexual glands from an animal or human – the TESTES in a male, the OVARIES in a female. In human beings it has been a form of punishment, a way of sexually incapacitating slaves to produce eunuchs or artificially creating soprano voices (castrati) and as a way of stopping the spread of cancerous growths. It can make livestock animals tamer and improve the quality of their meat; it prevents breeding.

catabolism *See* METABOLISM

catalysis Modification of the rate of a chemical reaction, by the addition of a CATALYST, which is not consumed in the reaction. Catalytic action can reveal the reaction mechanism; many industrial processes rely on catalysis to accelerate reactions and occasionally to inhibit undesirable ones. Catalysis plays a key part in most biochemical reactions. *See also* BIOCHEMISTRY

catalyst Substance that speeds up the rate of a chemical reaction without itself being consumed. Many industrial processes rely on catalysts, for example the HABER PROCESS for manufacturing AMMONIA uses iron as a catalyst. Metals or their compounds catalyze by adsorbing gases to their surface, forming intermediates that then readily react to form the desired product while regenerating the original catalytic surface. The METABOLISM of all living organisms depends on biological catalysts called ENZYMES. Without these catalysts most

▲ **casting** Metal alloys used to make turbine blades must withstand the huge temperatures and forces inside jet engines. The random crystalline structure formed when the alloy cools normally (as seen in the overflow ,1) can be a source of weakness. The strongest structure is achieved by making a blade from a single crystal (2). This can be done by using heating elements (3). After the molten alloy is poured, the elements move up the sides of the mould (4) ensuring the alloy cools from the bottom and forms a single crystal.

reactions would happen so slowly that life would not be possible.

catalytic converter Antipollution device used in internal combustion engines. It consists of a bed of catalytic agents through which flow the gaseous exhaust of fuel combustion. Converters located in silencers reduce harmful unburned hydrocarbons and carbon monoxide. These converters are adversely affected by tetraethyl lead found in some types of gasoline.

catalytic cracking *See* CRACKING

cataract In medicine, opacity in the lens of an eye, causing blurring of vision. Most cases are due to degenerative changes in old age but it can also be congenital, the result of damage to the lens, or some metabolic disorder such as diabetes. Treatment is by surgical removal of the cataract followed by implantation of an artificial lens.

cataract In physical geography, term usually applied to that section of a rapidly flowing river where the running water falls suddenly in a sheer drop. When the drop is less steep, the fall is known as a cascade.

catastrophe theory Mathematical technique published in 1972 by the French mathematician René Thom (1923–). The theory concerns processes that do not always take place through continuously changing conditions but, instead, may at some point undergo a catastrophe – an event in which certain mathematical quantities change by a sudden, discontinuous leap. The technique is particularly useful for describing situations in which small changes in initial conditions or input can cause sudden or drastic changes in a system's behavior or output.

catechol (1,2-dihydroxybenzene, $C_6H_4(OH)_2$) Organic chemical compound. It forms clear aqueous solutions which soon turn brown because, under the influence of light, they react strongly with oxygen. For these reasons catechol is used to protect stored materials against oxidation, and in photographic developers. It also has uses in making dyes.

catecholamine Any of a group of AMINES with important biological activity. Catecholamines are derived from DOPA (dihydroxyphenylalanine) and include nerve transmitters and hormones such as dopamine, adrenaline (epinephrine), and noradrenaline (norepinephrine).

catena Variation on the vertical profile of soils derived from the same parent material, usually resulting from topographic changes, such as a hill

nitric oxide (NO)

carbon monoxide (CO)

hydrocarbons

water (H_2O)

carbon dioxide (CO_2)

nitrogen (N)

3

4

1

2

◀ **catalytic converter**
A catalytic converter is placed in the exhaust system (1) to reduce the pollution produced by internal combustion engines (2). It comprises a ceramic honeycomb structure (3), which maximizes the surface area of the converter, covered in catalysts – normally platinum and rhodium (4). As exhaust gases, primarily carbon monoxide, nitric oxide, and hydrocarbons from the cylinder, pass through the converter, they react with the catalysts. The platinum and rhodium accelerate oxidation and reduction in the hot gases. The pollutants are either oxidized or reduced into water, carbon dioxide, and nitrogen.

slope, which controls other soil-forming factors such as microclimate and drainage. *See also* SOIL PROFILE

catheter Fine tube introduced into the body to deliver or remove fluids. The most common is the **urinary** catheter, fed into the bladder by way of the URETHRA.

cathode Negative ELECTRODE of an electrolytic cell or ELECTRON TUBE. It attracts positive ions (CATIONS) during ELECTROLYSIS, the process that uses electrical energy to bring about chemical changes.

cathode ray Radiation emitted by the CATHODE of a thermionic (heated) ELECTRON valve containing a gas at low pressure. The rays were identified in 1897 by J. J. THOMSON as streams of charged particles having extremely low mass, later called electrons. Some electrons are emitted because the cathode is heated but most because CATIONS (positive IONS) formed in the valve collide with the cathode. *See also* CATHODE-RAY TUBE; THERMIONICS

cathode-ray tube Evacuated ELECTRON tube used for TELEVISION picture tubes, OSCILLOSCOPES, AND display screens in radar sets and computers. An electron gun shoots a beam of electrons, focused by anodes. The electrons strike a fluorescent screen and produce a spot of light. In a television tube, an electrostatic or magnetic field deflects the beam so that it scans a number of lines on the screen, controlled by the incoming picture signals.

cation Positive ION that is attracted to the CATHODE during ELECTROLYSIS.

catkin Drooping, scaly spike of unisexual flowers without petals, such as those of the hazel or poplar. This deciduous flower cluster is also typical of birches and some beeches.

CAT scan Abbreviation of COMPUTERIZED AXIAL TOMOGRAPHY

cat's eye *See* CHRYSOBERYL

Cauchy, Augustin Louis, Baron (1789–1857) French mathematician who created complex analysis and, following EULER, formalized many of the ideas of CALCULUS. Cauchy defined the notion of a limit and of a continuous function and formalized the definitions of derivative and integral that are still used today. He produced more than a thousand papers and it is said that more theorems and concepts are named after Cauchy than any other mathematician.

caustic potash *See* POTASSIUM HYDROXIDE

▲ **cathode-ray tube** Television receivers are a type of cathode-ray tube. Three electron guns (1) receive color signals from a colour decoder that splits the color signal into red, green, and blue. The guns fire three beams of electrons through vertical and horizontal deflection coils (2) onto the screen of a "shadow mask tube" (3). This is made up of about a million dots, a third of which glow red when bombarded, a third blue, and the remaining third, green. The dots compose the color picture received by the television. The beam of electrons scans hundreds of lines on the screen (525 in the US, 625 in Europe) making up the moving pictures. The beam scans from left to right, starting top left and finishing bottom right (4).

caustic soda *See* SODIUM HYDROXIDE

cave Natural underground cavity. There are several kinds of caves, including **coastal** caves, formed by wave erosion, **ice** caves, formed in glaciers, and **lava** caves. By far the largest caves are formed in carbonate rocks such as limestone. Such rocks are impervious, but dissolve in underground streams or ground water formed by rain.

Cavendish, Henry (1731–1810) British chemist and physicist. Cavendish discovered hydrogen and the compositions of water and air; he estimated the Earth's mass and density, and calculated the gravitational constant by a method now known as the "Cavendish experiment." He also measured the specific gravity of carbon dioxide and hydrogen, and stated the inverse square law for the interaction of charged particles. The Cavendish Laboratory at Cambridge University, England, is named after him.

Caventou, Joseph Bienaimé (1795–1877) French chemist who collaborated in 1817 with Pierre Joseph Pelletier (1788–1842) to isolate and name CHLOROPHYLL (choosing the Greek words for "green leaf"). They are considered the founders of alkaloid chemistry, having isolated veratrine, strychnine, brucine, quinine, and cinchonine. Caventou also extracted caffeine from coffee beans in 1822. The son of an apothecary, Caventou studied in Paris where he was made professor of toxicology (1835–60) at the École de Pharmacie. He and Pelletier concentrated on research into plant chemistry in order to find the component that renewed air. Together they discovered that chlorophyll was the compound which turned plants green, and it was later shown that the molecules absorbed the Sun's light energy for PHOTOSYNTHESIS.

Caxton, William (1422–91) First English printer. Following a period in Cologne, Germany (1470–72), where he learned printing, he set up his own printing press in 1476 at Westminster. He published more than 100 items on a wide variety of subjects, many of them his own translations from French, Latin, and Dutch.

Cayley, Arthur (1821–95) British mathematician. He published more than 900 papers in pure mathematics and geometry, but is best know for the invention of matrices. *See also* MATRIX

CD-I (COMPACT DISC interactive) System for storing and reproducing sounds and still and moving pictures on a disc resembling an ordinary CD. The system is described as interactive because, when playing a game, for example, actions by the user control what appears on the screen. CD-i players have either remote or wired control units, and use a television set for reproducing the sound and image.

CD-ROM (COMPACT DISC read-only memory) Optical storage device for computer data and programs. It resembles a COMPACT DISC (CD) used in stereo systems. A CD-ROM can store much more data than a comparably priced portable MAGNETIC DISK. Computer games, encyclopedias and other software are now available in this form. To use a CD-ROM, the disc is placed in a specialized player connected, or built in, to a computer.

cecum Dilated pouch at the junction of the small and large INTESTINES, terminating in the APPENDIX. It has no known function in humans. In rabbits and horses, the cecum contains large numbers of microorganisms that help to break down the cellulose cell walls of the plants they eat, thus making more nutrients available to their hosts.

celestial equator Great circle on the CELESTIAL SPHERE, lying midway between the CELESTIAL POLES in the same plane as the Earth's Equator.

celestial mechanics Branch of ASTRONOMY concerned with the relative motions of stars and planets that are associated in systems, such as the Solar System or a BINARY STAR system, by gravitational fields. Introduced by Isaac NEWTON in the 17th century, celestial mechanics, rather than general RELATIVITY, is usually sufficient to calculate the various factors determining the motion of planets, satellites, comets, stars, and galaxies around a center of gravitational attraction.

celestial meridian Circle on the CELESTIAL SPHERE passing through the CELESTIAL POLES, the ZENITH, and the NADIR and crossing the horizon at the N and S points.

celestial poles Two diametrically opposite points at which the extension of the Earth's axis meets the CELESTIAL SPHERE. The celestial sphere rotates about a line through the celestial poles.

celestial sphere Imaginary sphere of infinite radius used to define the positions of celestial bodies as seen from Earth, the center of the sphere. The sphere rotates, once in 24 hours, around a line that is an extension of the Earth's axis. The position of a celestial body is the point at which a radial line through it meets the surface of the sphere. The position is defined in terms of coordinates, such as declination and right ascension or altitude and azimuth, which refer to great circles on the sphere, such as the CELESTIAL EQUATOR or the ecliptic.

celestine Mineral, strontium sulfate ($SrSO_4$), with distinctive pale blue or white, glassy, orthorhombic crystals, sometimes occurring in fibrous masses. It is found chiefly in SEDIMENTARY ROCK and also as gangue material in ore veins. There are deposits in Britain, Sicily, and the United States. It is an important source of strontium and some of its compounds.

cell In biology, basic unit of which all plant and animal tissues are composed. The cell is the smallest unit of life that can exist independently, with its own self-regulating chemical system. Most cells consist of a MEMBRANE surrounding jelly-like CYTOPLASM with a central NUCLEUS. The nucleus is the main structure in which DNA is stored in CHROMOSOMES. Animal cells vary widely in shape. An ERYTHROCYTE (red blood cell), for instance, is a biconcave disk, whereas a NEURON has a long fiber. The cells of plants and algae are enclosed in a cell wall, which gives them a more rigid shape. Bacterial cells also have a cell wall, but do not have nuclei or chromosomes; instead, they have a loop of DNA floating in the cytoplasm. More advanced cells (those that have nuclei), often have other membrane-bounded structures inside the cell, such as MITOCHONDRIA and CHLOROPLASTS. *See also* EUKARYOTE; PROKARYOTE

cell In physics, device from which electricity is obtained due to a chemical reaction. A cell consists of two ELECTRODES (a positive ANODE and a negative CATHODE) immersed in a solution (ELECTROLYTE). A chemical reaction takes place between the electrolyte and one of the electrodes. In a primary cell, current is produced from an irreversible chemical reaction, and the chemicals must be renewed at intervals. In a secondary cell (BATTERY), the chemical reaction is reversible, and the cell can be recharged by passing a current through it. *See also* ELECTROLYSIS

cell differentiation Way in which an embryo's cells become specialized for specific functions. Initially all embryonic cells are similar but soon begin to differentiate into specific kinds such as bone cells, brain cells, heart cells, muscle cells, and so on. Differentiation also occurs in mature plants and animals during regeneration of lost or damaged tissue.

cell division Process by which living cells reproduce and thereby allow an organism to grow. In EUKARYOTE cells, a single cell splits in two, first by division of the NUCLEUS (occurring by MITOSIS or MEIOSIS), then by fission of the CYTOPLASM. For growth and asexual reproduction, where the daughter cells are required to be genetically identical to their parents, mitosis is used. Meiosis results in daughter cells having half the number of chromosomes (HAPLOID). This type of division results in the production of haploid sex cells or GAMETES, which allows genetic information from two parents to be combined at FERTILIZATION, when the DIPLOID number of chromosomes is restored. *See also* ALTERNATION OF GENERATIONS

cellophane Flexible, transparent film made of regenerated CELLULOSE and used mostly as a wrapping material. It is made by dissolving wood pulp or other plant material in an ALKALI, to which carbon disulfide is added to form viscose. This is forced through a narrow slit into a dilute acid, where it forms a film of cellulose.

cell potential Potential difference between the electrodes of an electric CELL. It depends on the internal resistance (r) of the cell and the external resistance (R) through which the current flows. It is given by the formula $ER/R(r + R)$, where E is the ELECTROMOTIVE FORCE on open circuit.

cellular telephone *See* MOBILE TELEPHONE

celluloid Hard plastic invented by US scientist John Hyatt (1837–1920) in 1869. Hyatt made the plastic by mixing cellulose nitrate with pigments and fillers in a solution of camphor and alcohol. When heated, it can be molded into a variety of shapes, and hardens on cooling. The first major plastic, it was used for early motion picture film. It is highly flammable.

cellulose ($[C_6H_{10}O_5]_n$) POLYSACCHARIDE CARBOHYDRATE that is the structural constituent of the cell walls of plants and algae. Consisting of parallel unbranched chains of GLUCOSE units cross-linked together into a stable structure, it forms the basic material of the paper and textile industries.

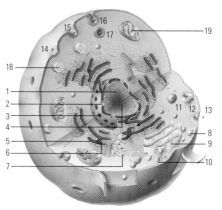

▲ **cell** Animal cells are made up of many different components called organelles. The most prominent is the nucleus (1), which contains all the information of the cell in the form of chromosomes. It is surrounded by the nuclear membrane (2), which contains many pores (3) that allow the nucleus to communicate with the rest of the cell. The center of the nucleus, the nucleolus (4), generates ribosomes (5), which provide the cell with protein. They are found on the rough endoplasmic reticulum (6), a system of flattened sacs and tubes connected to the nuclear membrane. It brings the messenger RNA molecules, which control the creation of protein, to the ibosomes. The smooth endoplasmic reticulum (7) produces small spheres called vesicles (8) that provide the Golgi apparatus (9) with protein. The Golgi apparatus modifies, sorts, and packs large molecules into other vesicles that bud off (10). They are sent to other organelles or secreted from the cell. The fusion of such vesicles with the cell membrane allows particles to be transported out of the cell (exocytosis, 11–13) or brought in (endocytosis, 14–17). Lysosomes (18) break down the molecules entering the cell into enzymes. The mitochondria (19) power the cell, using oxygen and food to generate energy in the form of adenosine triphosphate (ATP). ATP is used in many metabolic processes that are essential for the cell to function.

▲ **centrifuge** A laboratory centrifuge has mounts (1) for test tubes (2). As it turns at several thousand revolutions per minute, driven by a motor, the mixture (3) separates out into layers as the densest material (4) is forced to the bottom of the test tube and the least dense material (5) rises to the top.

cellulose nitrate (nitrocellulose or guncotton) Organic compound made by reacting the natural polymer CELLULOSE with nitric acid. It was chiefly used in the 19th and 20th centuries (together with camphor) to make CELLULOID – one of the first plastics – or as a propellant explosive (also called cordite).

cellulose plastic THERMOPLASTIC derivative of CELLULOSE. Cellulose plastics include: CELLULOSE NITRATE (or nitrocellulose), used in explosives and propellants; various cellulose acetates for making textile fibers and packaging films; and ethyl cellulose, used in the manufacture of shock-resistant materials. *See also* RAYON

Celsius, Anders (1701–44) Swedish astronomer who invented the CELSIUS, or centigrade, scale in 1742. He was a strong supporter of the introduction of the Gregorian calendar. In 1733 he published a collection of 316 observations of the AURORA Borealis made by himself and other astronomers.

Celsius Temperature scale based on the freezing point of water (0°C) and the boiling point of water (100°C). The interval between these points is divided into 100 degrees. The name "Celsius" officially replaced "centigrade" in 1948. Degrees Celsius are converted to degrees FAHRENHEIT by multiplying by 1.8 and then adding 32. The scale was devised by Anders CELSIUS.

cement In geology, a mineral material, such as CALCITE, that fills open pore space in sediments and binds grains to form a SEDIMENTARY ROCK. In construction, the term usually refers to Portland cement. This is made by heating a mixture of limestone and clay, grinding it, and adding GYPSUM.

cementite (Fe_3C) Hard, iron-carbon compound present as a major component of steel. The first of all steel-making processes, practiced in the Hittite civilization of the 2nd millennium BC, is the cementite process; the ancient smiths beat carbon from hot wood ashes into wrought iron, so forming cementite.

Cenozoic Most recent era of GEOLOGICAL TIME, beginning about 65 million years ago and extending up to the present-day. It is subdivided into the TERTIARY and QUATERNARY periods. It is the era during which the modern world with its present geographical features, and plants and animals developed.

Centaurus (Centaur) Brilliant constellation of the S sky. It has 13 stars above magnitude 3.5, the brightest of which is ALPHA CENTAURI. The third-brightest star in the sky, it has a magnitude of –0.27. Centaurus also contains Omega Centauri, the brightest globular cluster galaxy in the sky.

center of gravity Point at which the weight of a body can be considered to be concentrated and around which its weight is evenly balanced. An object in free flight rotates around its center of gravity, which in turn moves along the trajectory that would be followed by a small object of the same mass as the object. In a uniform gravitational field, the center of gravity is the same as the CENTER OF MASS.

center of mass Point at which the whole mass of an object or group of objects is considered to be concentrated. The gravitational forces between spherical celestial bodies can be correctly calculated by regarding their masses as being located at their centers.

centigrade *See* CELSIUS

centimeter Unit of length defined as one hundredth part of a meter. *See* WEIGHTS AND MEASURES.

central nervous system (CNS) In some advanced INVERTEBRATES, a neural pathway along which are located clusters of NEURONS called GANGLIA; these are involved in the movement of limbs, wings, and so on. In VERTEBRATES, part of the NERVOUS SYSTEM that comprises the brain and spinal cord, and is connected to the PERIPHERAL NERVOUS SYSTEM, a branching network of sensory and motor nerves. In humans, the CNS coordinates all neural activity including that producing movement, thought, emotion,and REFLEX ACTIONS. *See also* AUTONOMIC NERVOUS SYSTEM (ANS)

central processing unit (CPU) Part of a digital computer that controls all operations. In most modern computers, the CPU consists of one complex INTEGRATED CIRCUIT (IC), a chip called a MICROPROCESSOR. A CPU contains temporary storage circuits that hold data and instructions; an arithmetic and logic unit (ALU) that performs calculations; and a control unit that organizes operations.

centrifugal force *See* CENTRIPETAL FORCE

centrifuge Rotating device used for separating substances. In laboratories, centrifuges separate particles from suspensions, and erythrocytes (red blood cells) from plasma. In the food industry, centrifuges separate cream from milk and sugar from syrup. In each case, the denser substance is forced to the outside of a rotating container. A spin dryer uses the same principle to remove water from clothes.

centriole Dense body consisting of microtubules near the nucleus of a cell. It occurs in all cells except those of ANGIOSPERMS and the sperm cells of ferns and GYMNOSPERMS. During CELL DIVISION centrioles reproduce before the rest of the cell and move to each pole to form the spindle.

centripetal force In circular or curved motion, the force acting on an object that keeps it moving in a circular path. For example, if an object attached to a rope is swung in a circular motion above a person's head, the centripetal force acting on the object is the tension in the rope. Similarly, the centripetal force acting on the Earth as it orbits the Sun is gravity. In accordance with Newton's laws, the reaction to this can be regarded as a **centrifugal force**, equal in magnitude and opposite in direction.

centromere Part of a CHROMOSOME that appears only during CELL DIVISION. When chromosomes shrink during MEIOSIS or MITOSIS, the centromere appears as a narrowing that contains no genes. It connects the chromosomes to the SPINDLE fibers.

centrosome Region in a cell where microtubules are broken down and assembled during CELL DIVISION. Located alongside the nucleus, the centrosome contains two CENTRIOLES. At METAPHASE, the centrioles separate and the two regions of the centrosome containing them move to opposite sides of the cell as the microtubules form a SPINDLE between them. The spindle eventually divides the chromosomes into the two daughter cells. *See also* HAPLOID

cephalochordates (lancelets or amphioxus) Small marine animals that look and swim rather like fish but have no head or paired fins. They are related to vertebrates, having a NOTOCHORD throughout their lives. Their method of filter-feeding is, however, much more like that of many animals without backbones. Cephalochordates inhabit shallow, temperate and tropical seawater.

cephalopod Any of more than 600 species of predatory marine MOLLUSK of the class Cephalopoda, including squid, nautilus, octopus, and cuttlefish. Each has eight or more arms surrounding the mouth, which typically has a beak. The nervous system is well developed, permitting great speed and alertness; the large eyes have an image-forming ability equal to that of vertebrates. Most squirt an inky fluid to alarm attackers. Cephalopods move by squirting water from their mantle edge. Their heavily yolked eggs develop into larval young that resemble the adults. Members of this class vary dramatically in size from 4cm (1.5in) to the giant squid, which may reach 20m (65ft).

cephalosporin Class of semisynthetic ANTIBIOTIC drugs derived from fungi of the genus *Cephalosporium*. Similar to PENICILLIN, they are effective against a wide spectrum of BACTERIA, including some that have become resistant to penicillin.

▲ **cephalopod** The squid (1), cuttlefish (2), and octopus (3) are all swimming mollusks of the Cephalopoda class. They have advanced, powerful eyes, and tentacles lined with sucker pads that are used to catch fish and small crustaceans. The horny jawed mouth is powerful enough to break up their prey before it is digested in the gut.

Cepheid variable One of an important class of VARIABLE STARS that pulsate in a regular manner, accompanied by changes in LUMINOSITY. Cepheids periodically expand and contract, changing in size by as much as 30% in each cycle. The average luminosity is 10,000 times that of the Sun. Cepheids became important in cosmology in 1912, when Henrietta LEAVITT discovered a simple relationship between the period of light variation and the absolute magnitude of a cepheid. This relationship, the **period-luminosity law**, enables the distances of stars to be ascertained, not only in our galaxy but in other galaxies too.

Cepheus Indistinctive constellation of the N sky, it adjoins URSA MAJOR. It contains the star Delta, important as a prototype CEPHEID VARIABLE.

ceramic In art and technology, article made from inorganic compounds formed in a plastic condition and hardened by heating in a furnace. **Earthenware** is a porous ceramic made from kaolin, ball clay, and crushed flint. **Porcelain** is made from kaolin and feldspar, and heated to a higher temperature. It is nonporous and translucent. Special ceramics are made from pure aluminum oxide, silicon carbide, titranates, and other compounds.

cerebellum Part of the BRAIN, often known as the "little brain," located at the base of the CEREBRUM. It is involved in maintaining muscle tone, balance, and coordinated movement.

cerebral cortex Deeply fissured outer layer of the CEREBRUM. The cortex, or "gray matter," is the most sophisticated part of the BRAIN, responsible for the appreciation of sensation, for initiating voluntary movement, and for all higher functions, such as the emotions and intellect.

cerebral hemispheres Lateral halves of the CEREBRUM, the largest parts of the BRAIN and the sites of higher thought. Because nerve fibers cross over from one cerebral hemisphere to the other, the right side controls most of the movements and sensation on the left side of the body, and vice versa. Damage to the cerebral hemispheres often produces personality changes.

cerebral palsy Disorder mainly of movement and coordination caused by damage to the BRAIN during or soon after birth. It may feature muscular spasm and weakness, lack of coordination, and impaired movement or paralysis and deformities of the limbs; intelligence is not necessarily affected. The condition may result from any one of a number of causes, such as faulty development, oxygen deprivation, birth injury, hemorrhage or infection.

cerebrospinal fluid Clear fluid that cushions the BRAIN and SPINAL CORD, giving some protection against shock. It is found between the two innermost MENINGES (membranes) in the four ventricles of the brain and in the central canal of the spinal cord. A small quantity of the fluid can be withdrawn by lumbar puncture to aid diagnosis of some brain disorders.

cerebrum Largest and most highly developed part of the BRAIN, consisting of the CEREBRAL HEMISPHERES separated by a central fissure. It is covered by the CEREBRAL CORTEX. It coordinates all higher functions and voluntary activity.

Cerenkov, Pavel Alekseevich (1904–90) Russian physicist. Working at the Institute of Physics of the Soviet Academy of Science, he discovered that light (CERENKOV RADIATION) is emitted by charged particles travelling at very high speeds, a phenomenon known as the **Cerenkov effect**. He was awarded the 1958 Nobel Prize for physics with his co-workers, the Russian physicists I.M. Frank and I.Y. TAMM.

Cerenkov radiation Light emitted when energetic particles travel through a transparent medium, such as water, at a speed higher than the velocity of light in that medium. This action is called the **Cerenkov effect**. It is analogous to the SONIC BOOM, and a cone of light is emitted trailing the path of the particle. It is named after Pavel CERENKOV, who discovered the radiation in 1934. Cerenkov radiation is used in a Cerenkov counter or detector of energetic particles.

Ceres Largest ASTEROID and the first to be discovered, by Giuseppe PIAZZI on 1 January 1801. Its diameter is 913km (567mi). It orbits in the main asteroid belt, at an average distance from the Sun of 414 million km (257 million mi), the distance of the "missing" planet predicted by BODE'S LAW.

cerium (symbol Ce) Soft, ductile, iron-gray metallic element, the most abundant of the LANTHANIDE SERIES. It was first isolated in 1803. The chief ore is monazite. It is used in alloys (for use in lighter flints), in catalysts, nuclear fuels, in the making of glass and as the core of carbon electrodes in arc lamps. Properties: at.no. 58; r.a.m. 140.12; r.d. 6.77; m.p. 798°C (1,468°F); b.p. 3,257°C (5,895°F). The most common isotope is ^{140}Ce (88.48%).

cermet One of a group of hard and brittle heat-resistant materials. Cermet is a combination of ceramic material and metal. Applications include the manufacture of drilling tools, heat shields, and turbine blades.

CERN (European Laboratory for Particle Physics) Nuclear research center sited on the Franco-Swiss border w of Geneva. It was founded in 1954 as an intergovernmental organization, when it was called Conseil Européen pour la Recherche Nucléaire. It is the principal European center for research into PARTICLE PHYSICS.

cerussite (white lead) Mineral form of lead carbonate ($PbCO_3$). It forms in oxidized regions of mineral veins, particularly of GALENA. It has prismatic or needle-shaped crystals (orthorhombic) that are usually colorless or white, but may be green or gray with a resinous or vitreous luster. Cerussite is mined as lead ore in Mexico; it is also used as a white pigment. Hardness: 3–3.5; r.d. 6.6.

cervical smear *See* PAP SMEAR

cervix Neck of the UTERUS projecting downwards into the VAGINA. It dilates (expands) widely to allow the passage of a baby during BIRTH.

Cesalpino, Andrea (1519–1603) Italian physiologist and botanist. His *De Plantis* (1583) was one of the first attempts to classify plants according to their flowers and fruits, and was based on a comparative study of his own large collection.

cesium (symbol Cs) Rare, silver-white, metallic element in group I of the periodic table; the most alkaline and electropositive element. Discovered in 1860, cesium is ductile and used commercially in photoelectric cells. The isotope ^{137}Cs is used in cancer treatments. Properties: at.no. 55; r.a.m. 132.9055; r.d. 1.87; m.p.28.4°C (83.1°F); b.p. 678°C (1,252.4°F); most common isotope ^{133}Cs (100%).

cestode Any member of the CLASS Cestoda, which contains the endoparasitic TAPEWORMS. Cestodes usually live in the intestines, holding on to their lining by terminal hooks. They have no mouths and absorb their food through the body wall from their hosts. They are HERMAPHRODITE, one segment fertilizing another. Ripe segments drop off and are voided by the host, after which, to survive, they usually need to be eaten by a second host.

cetacean Any member of the order Cetacea of aquatic mammals found in all oceans and some rivers. The larger cetaceans are WHALES, the smaller ones are PORPOISES or DOLPHINS. Cetaceans are streamlined, with a pair of front flippers and horizontal tail flukes. Their bodies are insulated by a thick layer of BLUBBER. Length: from 1.5m (4.9ft) to more than 30m (100ft); weight: from 36kg (80lb) to 100 tons.

Cetus (Whale) Fourth-largest constellation, but inconspicuous. It lies across the Equator. Its brightest star is Beta, magnitude 2.0

CFC Abbreviation of CHLOROFLUOROCARBON

CGS system System of metric units based on the centimeter, gram, and second. The DYNE is the unit of force, the ERG the unit of energy. It has been largely superseded by the SI UNITS system.

Chadwick, Sir James (1891–1974) British physicist who discovered and named the NEUTRON and worked on the development of the atomic bomb. Chadwick worked on radioactivity with Ernest RUTHERFORD at the Cavendish Laboratory, Cambridge. In 1920 Rutherford had predicted a particle without electric charge in the nucleus of an atom, and in 1932 Chadwick proved the neutron's existence and calculated its mass. For this, he received the 1935 Nobel Prize for physics. He established a school of nuclear physics in Liverpool, constructing Britain's first CYCLOTRON there in 1935. During World War II, he moved to the USA to head British research for the MANHATTAN PROJECT to develop the atom bomb. *See also* NUCLEAR WEAPON

chaetognath Any member of the phylum Chaetognatha, which includes arrow worms – small marine animals that are not at all closely related to any other group. They are between 3mm and 10cm (0.125–4in) long, and have narrow bodies, with bristly jaws for seizing their prey (PLANKTON). Although invertebrate animals, they swim like fish, by means of fins. They are HERMAPHRODITE. Some species are useful indicator organisms for marine biologists because they are sensitive to the temperature, salinity, and depth of the water in which they live.

Chain, Sir Ernst Boris (1906–79) British biochemist, b. Germany. Chain shared the 1945 Nobel Prize for physiology or medicine with Howard FLOREY and Alexander FLEMING for the isolation and development of PENICILLIN as an ANTIBIOTIC. He also studied spreading factor, an enzyme that aids the dispersal of fluids in tissue.

chain reaction Self-sustaining nuclear FISSION reaction in which one reaction is the cause of a second, the second of a third and so on. The initial conditions are often critical, in that the quantity of fissionable material must exceed the CRITICAL MASS. Such a reaction produces energy for all commercial NUCLEAR REACTORS. *See also* NUCLEAR WEAPON

chalcanthite Mineral that consists mainly of hydrated copper sulfate ($CuSO_4.5H_2O$), although it is rarely used as a source of copper. It occurs as greenish-blue triclinic crystals or as fibrous veins or stalactites. It is soluble in water and has a nauseating taste. Hardness 2.5; r.d. 2.25.

chalcedony Microcrystalline form of QUARTZ. When cut and polished, it is used by gem engravers. It is waxy and lustrous, and there are white, gray, blue, and brown varieties. It is often colored by artificial methods. Some varieties contain impurities giving a distinctive appearance, such as AGATE (colored bands), ONYX (striped), and bloodstone (dark green with red flecks).

chalcocite Dark gray, metallic, soft mineral, copper sulfide (Cu_2S); one of the chalcocite group. It is a major ore of copper, found mainly in sulfur deposits. The crystals occur in orthorhombic granular masses, or rarely in prismatic form. Hardness 2.5 to 3.0; r.d. 5.7.

chalcopyrite (copper pyrites) Opaque, brass-colored, copper iron sulfide ($CuFeS_2$); the most important copper ore. It is found in sulfide veins and in IGNEOUS and certain METAMORPHIC ROCKS (*see* CONTACT METAMORPHISM). The crystals are tetragonal but often occur in masses. Hardness 3.5–4; r.d. 4.2.

chalk Porous, fine-grained rock, mainly composed of CALCAREOUS skeletons of marine microorganisms, especially COCCOLITHS and FORAMINIFERA. It varies in properties and appearance; pure forms, such as CALCITE, contain up to 99% calcium carbonate. It is used in making putty, plaster, and cement, and harder forms are occasionally used for building.

Challenger expedition (1872–76) British expedition in oceanographic research. The *Challenger* ship had a staff of six naturalists headed by Charles Wyville THOMSON. It sailed nearly 128,000km (69,000 nautical mi) making studies of the life, water, and seabed in the three main oceans.

Chamberlain, Owen (1920–) US physicist. After working on the development of the atom bomb between 1942 and 1946 he became professor of physics at the University of California, Berkeley.

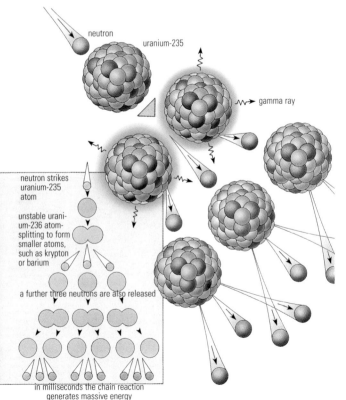

neutron

uranium-235

gamma ray

neutron strikes uranium-235 atom

unstable uranium-236 atom-splitting to form smaller atoms, such as krypton or barium

a further three neutrons are also released

in milliseconds the chain reaction generates massive energy

◄ **chain reaction** To start the chain reaction, a neutron hits an atom of uranium-235 (^{235}U). This becomes an unstable uranium-236 atom, which splits to become krypton and barium atoms with the release of three further free neutrons, which strike uranium-235 atoms releasing nine more neutrons, and so on. In microseconds the chain reaction attains enormous proportions and generates large amounts of energy.

With Emilio SEGRÈ, using the BEVATRON particle ACCELERATOR, he confirmed the existence of the antiproton and was awarded the Nobel Prize for physics in 1959. *See also* ANTIMATTER

Chameleon Small constellation of the S sky, near the South Pole. Its brightest stars are only of magnitude 4.1.

Chandrasekhar, Subrahmanyan (1910–95) US astrophysicist, b. India. While working on the evolution of stars, he calculated the maximum mass of a WHITE DWARF star before it collapses further to form a NEUTRON STAR. Called the **Chandrasekhar limit**, it equals 1.44 times the mass of the Sun. He shared the 1983 Nobel Prize for physics with William FOWLER.

change of state (change of phase) In physics, change that takes place when matter turns from one physical PHASE (that is a gas, liquid or solid) into another. The most typical changes of state include EVAPORATION and CONDENSATION, SUBLIMATION, boiling, and melting and solidification. Change of state always involves the absorption or release of heat energy. However, because of the phenomenon of LATENT HEAT, a change in temperature does not necessarily accompany a change of state.

Channel Tunnel (Chunnel) Railway tunnel under the English Channel, 49km (31mi) long. The first Channel tunnel was proposed in 1802 by a French engineer. A start was made in 1882 but soon abandoned for defense reasons. Another false start was made in the 1970s. In 1985 Eurotunnel, a joint French-English private company, was granted a 55-year concession to finance and operate the tunnel. The French and English sections were linked in 1990 and the tunnel became operational in 1994. It consists of two railway tunnels and one service tunnel, and links Folkestone, S England, with Calais, N France. A tunnel fire in November 1996 raised safety fears.

chaos theory Theory that attempts to describe and explain the highly complex behavior of apparently chaotic or unpredictable systems that show an underlying order. The behavior of some physical systems is impossible to describe using the standard laws of physics. This is because the mathematics needed to describe these systems is too difficult for even the largest supercomputers. Such systems are sometimes known as "nonlinear" or "chaotic" systems, and they include complex machines, electric circuits, and natural phenomena such as the weather. Nonchaotic systems can become chaotic, as when smoothly flowing water hits a rock and becomes turbulent. The lack

of an adequate description means that a standard prediction of their behavior is also impossible. Chaos theory provides mathematical methods needed to describe chaotic systems, and even allows some general prediction of a system's likely behavior. Chaos theory also shows, however, that even the tiniest variation in the starting conditions of a system can lead to enormous differences in the state of the system some time later. Thus because it is impossible to know the precise starting conditions of a system, accurate prediction is also impossible. *See illustration on page 98*

Chappe, Claude (1763–1805) French engineer who, with his brother, Ignace (1760–1829), invented a SEMAPHORE telegraph system in 1791. It consisted of a series of towers fitted with telescopes and semaphore arms with which letters of the alphabet could be signaled. During the French Revolutionary Wars their invention was used to relay information about the state of the war between France and Austria.

charcoal Porous form of CARBON, made traditionally by heating wood in the absence of air, and used in W Europe until late medieval times for smelting iron ore. Today, charcoal is chiefly used for its absorptive properties, to remove pollutants from air, to decolorize food liquids such as syrups, and to separate chemicals. Artists use charcoal sticks for sketching.

Charcot, Jean Martin (1825–93) French physician and founder of NEUROLOGY. He made classic studies of hypnosis and hysteria, and taught Sigmund FREUD and the French psychologist Pierre Janet (1869–1947). Charcot's work centered on discovering how behavioral symptoms of patients relate to neurological disorders.

charge *See* ELECTRIC CHARGE

charge-coupled device (CCD) Type of SILICON CHIP designed to capture images. The CCD is divided into a number of microscopic areas (pixels), arranged in rows. When a PHOTON hits a pixel, it knocks off an electron from a silicon atom, which becomes charged. An opposite charge in a layer on the base of the CCD confines this charged silicon atom, and in this way a charge builds up in each pixel relative to the number of photons hitting it. The contents of each pixel are read off 50 times a second, a row at a time, forming an electrical signal that can be used to create television pictures. CCDs are found in VIDEO cameras, FAX machines, and digital CAMERAS. CCDs are much more sensitive to light than photographic film, and are used in most large telescopes to detect light from distant stars.

Charles, Jacques Alexandre César
(1746–1823) French physicist, inventor, and mathematician who was the first to use a hydrogen BALLOON. He discovered the law relating the expansion of a gas to its temperature rise. GAY-LUSSAC published this work some 15 years after Charles' discovery and it is alternately known as CHARLES' LAW or Gay-Lussac's law. He is credited with inventing a thermometric HYDROMETER and improving Fahrenheit's aerometer and Gravesande's hydrometer.

Charles' law Volume of a gas at constant pressure is directly proportional to its absolute temperature. The relationship was discovered by Jacques CHARLES in 1787. The law is a special case of the IDEAL GAS LAW. It is sometimes called Gay-Lussac's law, because Joseph GAY-LUSSAC established it more accurately in 1802.

Charnley, Sir John (1911–82) British orthopedic surgeon who is best known for developing a low-friction, total hip replacement. He devised the artificial ball-and-socket joint as treatment for osteoarthritis of the hip, a major cause of crippling disability in the elderly. The final Charnley joint of 1963 had a socket made of high molecular weight polyethylene (HMWP) and a "ball" of polished steel.

Charon Satellite of PLUTO, discovered in 1978. Its diameter of 1,270km (790mi) makes it the largest satellite in relation to its primary (Pluto) in the Solar System. Estimates of Charon's mass vary from 8% to 16% of Pluto's. It is unique among planetary satellites in having a synchronous orbit with a period (6.4 days) that matches the rotation period of its primary; Charon therefore keeps the same face turned towards Pluto, and hangs motionless in Pluto's sky. Charon seems to have a grayish surface, probably of water ice. There is almost certainly no permanent atmosphere, but some of the nitrogen outgassed by Pluto when at PERIHELION could be captured and retained by Charon as a temporary atmosphere.

chelation Chemical reaction in which a certain type of organic compound, termed a **chelating**

▶ **chaos theory** is used to explain seemingly complex phenomena, which can be mathematically modeled by applying simple numerical formula a great many times. Some chaotic systems are fractal, that is they exhibit self-similar geometries or structures – in other words a small section of the structure will resemble the whole, and if it is possible to describe the part mathematically the whole can be described also. The Sierpinsky gasket (1) is "constructed" by the repeated division of equilateral triangles (2–3), and is a fractal structure. The apparently complex structure of living things, such as the cauliflower, are also self-similar, a single floret (4) representing a whole head (5). The smoke from an extinguished candle (6) moves in a complex pattern that is difficult to understand, unless the effects of turbulence (7) on the laminar flow of the smoke are modeled. The climate of the Earth is extremely complex, but is governed by simple rules (8). Sunlight falling on the sea causes water to evaporate (9) and form clouds (10) which reflect the sunlight and prevent it reaching the sea or land. Temperature drops and rain may fall (11). If we could measure weather conditions on a large enough scale and make extremely detailed mathematical models, using computers, we could predict the weather perfectly.

agent, combines with a metal ION by forming coordinate bonds with two or more atoms of the organic compound. Tartaric acid (CHOHCOOH)$_2$ and ethylenediamine (CH$_2$NH$_2$)$_2$ are chelating agents. *See also* COORDINATION COMPOUND

chemical bond Any mechanism that holds together atoms to form molecules. There are several types that arise either from the attraction of unlike charges or from the formation of stable configurations through electron-sharing. The total number of bonds that an atom can form is dependent upon its VALENCE. The main types are IONIC BOND, COVALENT BOND, METALLIC BOND, and HYDROGEN BOND.

chemical engineering Application of engineering principles to the making of chemical products on an industrial scale. Unit processes of chemical engineering include OXIDATION-REDUCTION, HYDROGENATION, NITRATION and SULFONATION, ELECTROLYSIS, POLYMERIZATION, ION EXCHANGE, and FERMENTATION.

chemical equation Set of symbols used to represent a CHEMICAL REACTION. Equations show how atoms are rearranged as a result of a reaction, with **reactants** on the left-hand side and **products** on the right-hand side. For example, the formation of magnesium oxide when magnesium burns in oxygen is represented by
$$2Mg + O_2 \rightarrow 2MgO$$
The number of atoms of an element on the left-hand side of an equation must equal the number on the right.

chemical equilibrium Balance in a REVERSIBLE REACTION, when two opposing reactions proceed at constant equal rates with no net change in the system. The initial rate of the reactions falls off as the concentrations of reactants decrease and the build-up of products causes the rate of the reverse reaction to increase.

chemical reaction Change or process in which chemical substances convert into other substances. This usually involves the breaking and formation of CHEMICAL BONDS. Reaction mechanisms include ENDOTHERMIC, EXOTHERMIC, ADDITION, CONDENSATION, combination (formation of a COMPOUND), DECOMPOSITION, and OXIDATION-REDUCTION reactions.

chemical warfare Employment of chemical weapons such as poison and nerve gases, defoliants, and herbicides. Together with BIOLOGICAL WARFARE, chemical warfare was banned by the Geneva Convention (1925) after the use of mustard gas during World War I. Despite this, many

▲ chemical reaction
The application of heat (A) alone to aluminum sulfate (1) and potassium sulfate (2) produces no reaction, but they combine when dissolved in water (3). If heating is continued to evaporation, alum (4) forms. Dry copper sulfate crystals (B) do not conduct current; but dissolve them in water (C) and electrolysis can proceed. Metals may react with a liquid (D): a grain of sodium (5) dropped into a water bath melts, generating hydrogen. Solutions and other liquids react readily (E). Phenolphthalein (6) added to a solution of alkali (7) produces a pink solution (8). When this is added to an acid solution (9), the pink disappears.

nations have developed chemical weapons. Mustard gas was used by Saddam Hussein to kill thousands of Kurds after the Iran–Iraq War. Herbicides, such as the defoliant Agent Orange, were used by US forces during the Vietnam War. During the Gulf War, Allied troops were inoculated and heavily protected against possible chemical warfare. Worries have recently surfaced about its terrorist uses, such as the Sarin nerve gas attack on the Tokyo subway system in 1995.

chemistry Branch of science concerned with the properties, structure, and composition of substances and their reactions with one another. Today, chemistry forms a vast body of knowledge with a number of subdivisions: the major division is between organic and inorganic. **Inorganic** chemistry studies the preparation, properties, and reactions of all chemical elements and their COMPOUNDS, except those of CARBON. However, such simple carbon compounds as CARBONATES, CARBIDES, and carbon oxides come within the province of inorganic chemistry. The historic separation from organic chemistry is, in any case, a false one, because many "inorganic" compounds are found in living organisms, such as common salt (NaCl) in human blood. In education and industry, however, the distinction is frequently still made. **Organic**

chemistry studies the reactions of carbon compounds. Because of carbon's unique ability to bond repeatedly with itself to form chains and rings of atoms (silicon is the only other element that has this property to any degree), carbon is able to form compounds of extreme complexity. For this reason it is predominantly the element of life, forming the "backbone" structures of such compounds as PROTEINS, FATS, AND CARBOHYDRATES. Organic compounds are about 100 times more numerous than nonorganic ones. Organic chemistry also studies an immense variety of molecules, including those of industrial compounds such as plastics, rubbers, dyes, drugs, and solvents. **Analytical** chemistry deals with the composition of substances. PHYSICAL CHEMISTRY deals with the physical properties of substances, such as their boiling and melting points. Its subdivisions include ELECTROCHEMISTRY, THERMOCHEMISTRY, AND chemical KINETICS.

chemoreceptor Tiny region on the outer membrane of some biological cells that is sensitive to chemical stimuli. The chemoreceptor transforms a stimulus from an external molecule into a sensation, such as smell or taste.

chemotherapy Treatment of a disease (usually cancer) by a combination of chemical substances, or DRUGS, that kill or impair tumors or disease-producing organisms in the body. Specific drug treatment was first introduced in the early 1900s by Paul EHRLICH.

chemotropism Growth or movement of a plant or plant part in response to a chemical stimulus. In **positive** chemotropism, the movement is towards the chemical; in **negative** chemotropism movement is away from the chemical. An example occurs during pollination. The ovary releases sugars into the style of the flower, and these act positively to cause pollen to produce a pollen tube that moves down the style.

chemurgy Term used to describe the development of new chemical products for industry from organic raw materials, especially those of agricultural origin.

chernozem (black earth) Humus-rich type of dark soil typical of the grasslands of the steppes of Eurasia and the prairies of North America. It is prized for its agricultural qualities such as good structure and high nutrient content.

chert Impure, brittle type of flint. A cryptocrystalline variety of SILICA, its color can be white, yellow, gray or brown. It occurs mainly in limestone and dolomite but its origin is unknown.

Chevreul, Michel Eugène (1786–1889) French chemist and proponent of theories on color that influenced the painter Delacroix and the neo-impressionists. Chevreul studied the properties of animal fats and helped to develop the commercial manufacture of soap.

chiastolite Variety of andalusite, aluminum silicate (Al_2OSiO_4), found in METAMORPHIC ROCKS. It has elongated prismatic crystals, which in cross-section show a black cross on a gray ground. Hardness 7.5; r.d. 3.1–3.2.

chicken pox (varicella) Infectious disease of childhood caused by a virus of the HERPES group. After an incubation period of two to three weeks, a fever develops and red spots (which later develop into blisters) appear on the trunk, face, and limbs. Recovery is usual within a week, although the possibility of contagion remains until the last scab has been shed.

childbirth *See* BIRTH

chilopod (centipede) Any member of the order Chilopoda, ARTHROPOD animals that have long, flat, segmented bodies, each bearing one pair of legs (the total number of legs being much less than 100, despite the name). The head has a large pair of claws with which the centipede seizes its prey and injects it with venom.

china clay *See* KAOLIN

china stone Partially decomposed granite, frequently used as a FLUX to produce VITRIFICATION and translucency, or mixed with silica and lime to form a glaze in the manufacture of PORCELAIN.

chinook Warm, dry wind experienced on the E side of the Rocky Mountains in Canada and the United States and Canada, and in the Alps. It commonly blows during winter and spring.

chip General term used to describe an INTEGRATED CIRCUIT.

chiropody *See* PODIATY

chiropractic Alternative medical practice based on the theory that the NERVOUS SYSTEM integrates all of the body's functions, including defense against disease. Chiropractors aim to remove nerve interference by manipulations of the affected musculoskeletal parts, particularly in the spinal region, thus improving the nerve pathway.

chiropteran Any member of the order Chiroptera, which contains the 178 genera of bats. Bats are the only MAMMAL to have true flight (although a few others can glide). A bat's wing is formed by a sheet of skin stretched over a frame of greatly elongated bones. Bats are able to navigate in complete darkness by means of a kind of SONAR, which uses echoes of the bat's own supersonic squeaks to locate obstacles and prey. Bats are nocturnal and found in all tropical and temperate regions. Most are small, although they range in wingspan from 25cm to 147cm (10–58in).

chitin Hard, tough substance that occurs widely in nature, particularly in the hard shells (EXOSKELETONS) of ARTHROPODS such as crabs, insects, spiders, and their relatives. The walls of HYPHAE – the microscopic tubes of fungi – are composed of slightly different chitin. Chemically chitin is, like CELLULOSE, a polysaccharide, derived from glucose.

Chladni, Ernst Florens Friedrich (1756–1827) German physicist who made studies of sound, particularly the way in which it vibrates metal plates and diaphragms. Patterns produced in fine powders on sound-vibrated plates are called **Chladni's figures.**

chlamydia Small, virus-like BACTERIA that live as PARASITES in human beings and animals and cause disease. One strain, *C. trachomatis*, is responsible for TRACHOMA, the leading cause of BLINDNESS in the developing world; it is also a major cause of pelvic inflammatory disease (PID) in women. *C. psittaci* causes PSITTACOSIS, a disease of birds that can be transmitted to human beings. Chlamydial infection is the most common SEXUALLY TRANSMITTED DISEASE in many developed countries.

chlorate Salt of chloric acid, containing the ion ClO_3^-. Chlorate salts are good oxidizing agents and are a useful source of oxygen, which they give off when heated. Sodium and potassium chlorates are used in explosives and also as weedkillers. They are manufactured by ELECTROLYSIS.

chloride Salt of HYDROCHLORIC ACID or some organic compounds containing CHLORINE, especially those with the negative ion Cl^-. The best-known example is common table salt, sodium chloride (NaCl). Most chlorides are soluble in water, except mercury(I) (mercurous) and silver chlorides.

chlorine (symbol Cl) Common nonmetallic element, one of the HALOGENS (elements in group VII of the periodic table), first discovered in 1774. It occurs in common salt (NaCl). It is a greenish-yellow poisonous gas extracted by the electrolysis of brine (salt water). It is widely used, in a process called **chlorination**, to disinfect water used for drinking and in

swimming pools. Chlorine is also used to bleach wood pulp, and in the manufacture of plastics, chloroform, pesticides, and other compounds. Chemically, it is a reactive element and combines with most metals. Properties: at.no. 17; r.a.m. 35.453; m.p. −101°C (−149.8°F); b.p. −34.6°C (−30.28°F). The most common isotope is ^{35}Cl (75.53%).

chlorite Complex silicate mineral of iron, magnesium, and aluminum, formed in low-grade METAMORPHIC ROCKS. It is similar to MICA and may be formed from it. Its scaly crystals are usually green, but may also be yellow, white, brown or even black. Hardness: 2–3; r.d. 2.6–3.3, depending on iron content.

chloroethene (vinyl chloride, CH_2CHCl) Gas with an ether-like odor manufactured by the chlorination of ETHYLENE. It polymerizes to form POLYVINYL CHLORIDE (PVC) and is widely used in this form. Properties: m.p. −153.8°C (−244.8°F); b.p. −13.4°C (7.9°F).

chlorofluorocarbon (CFC) Chemical compound in which hydrogen atoms of a HYDROCARBON, such as an ALKANE, are replaced by atoms of fluorine, chlorine, and sometimes bromine. CFCs are inert, stable at high temperatures and are odorless, colorless, nontoxic, noncorrosive, and nonflammable. Under the trade name of Freons, CFCs were widely used in aerosols, fire-extinguishers, refrigerators, and in the manufacture of foam plastics. The two most common are Freon 11 (trichlorofluoromethane, $CFCl_3$) and Freon 12 (dichlorodifluoromethane, CF_2Cl_2). When CFCs are used they slowly drift into the STRATOSPHERE and are broken down by the Sun's ultraviolet radiation into chlorine atoms that destroy the OZONE LAYER. It often takes more than 100 years for CFCs to disappear from the atmosphere. Growing environmental efforts led to a 1990 international agreement by governments to reduce and eventually phase out the use of CFCs and other chemicals harming the ozonosphere, and to develop safe substitutes.

chloroform *See* TRICHLOROMETHANE

chlorophyll Group of green pigments present in the CHLOROPLASTS of plants and ALGAE that absorb light for PHOTOSYNTHESIS. There are five types: chlorophyll *a* is present in all photosynthetic organisms except bacteria; chlorophyll *b* in plants and CHLOROPHYTES; and chlorophylls *c*, *d*, and *e* present in some algae. It is similar in structure to HEMOGLOBIN, with a magnesium atom replacing the iron atom.

chlorophyte (green algae) Any member of the phylum Chlorophyta, a large group of marine and freshwater ALGAE. Chlorophytes have several fea-

▲ **chloroplast** Found mostly in the cells of plant leaves, chloroplasts absorb sunlight and use it to manufacture special types of sugar. They are able to move about in order to receive the maximum amount of light possible. A section through a leaf reveals that during the day (left), chloroplasts have moved to the outer and inner walls in the direct line of light. During the night (right), they move to the inner and side walls only. Chloroplasts are responsible for the green color of most plants.

tures that make them more like typical green plants than other types of algae: they have cup-shaped CHLOROPLASTS that contain CHLOROPHYLL *b*; they have cell walls made of CELLULOSE; and they store food in the form of STARCH. Some produce cells with flagella (*see* FLAGELLUM) at some stage in their lives. Green algae range in size from microscopic, single-cell types (some of which have flagella) to large, complex SEAWEEDS.

chloroplast Microscopic green structure within a plant cell in which PHOTOSYNTHESIS takes place. The chloroplast is enclosed in an envelope formed from two membranes and contains internal membranes to increase the surface area for reactions. Molecules of the light-absorbing pigment CHLORO-PHYLL are embedded in these internal membranes.

chloroquine Synthetic drug used mainly to treat MALARIA and also sometimes HEPATITIS, rheumatoid ARTHRITIS, AND some skin conditions. It can have marked side-effects, such as blurred vision.

chlorosis In plant pathology, a yellowing or blanching of leaves of green plants caused by deficiencies of minerals, especially magnesium or potassium, or by plant parasites.

chlorpromazine Synthetic drug first developed to counter nausea and vomiting but subsequently found to have a powerful calming effect on the CENTRAL NERVOUS SYSTEM. It is used to treat psychotic conditions and also in terminal illness and occasionally in ANESTHESIA. Its long-term use is associated with a number of unpleasant side-effects, including tremor, movement disorders, and HYPOTENSION.

choke In electronics, coil of low-resistance wire that acts as an INDUCTOR. Chokes are used to limit the flow of high-frequency alternating current (AC)

in radio circuits or to smooth the waveform of the direct current (DC) from a RECTIFIER. A familiar choke in domestic use is the "starter" of a fluorescent lighting unit. The term is also used to describe a mechanical valve-like device in many INTERNAL COMBUSTION ENGINES that reduces the amount of air in the air-fuel mixture in order to provide a rich enough mixture for the engine to start.

cholera Infectious disease caused by the bacterium *Vibrio cholerae*, which is transmitted in contaminated water. Cholera, which is prevalent in many tropical regions, produces almost continuous, watery diarrhea often accompanied by vomiting and muscle cramps, and leads to severe dehydration. Untreated it can be fatal, but proper treatment, including fluid replacement and antibiotics, can result in a high recovery rate. There is a vaccine available.

cholesterol White, fatty STEROID. It occurs in large concentrations in the brain, spinal cord, and liver. It is synthesized in the liver, intestines, and skin and is an intermediate in the synthesis of vitamin D and many hormones. GALLSTONES are composed mainly of cholesterol. A diet rich in animal products may produce an excess of cholesterol deposits in blood vessels, and this can lead to ARTERIOSCLEROSIS.

choline Organic compound, sometimes classified as a B vitamin. It is necessary for the synthesis of ACETYLCHOLINE and FATTY ACIDS by the liver. Its dietary sources are egg yolk and some vegetable oils. It can be synthesized in the body from the AMINO ACID serine.

cholinesterase Enzyme that breaks down and inactivates the NEUROTRANSMITTER ACETYLCHOLINE at the nerve SYNAPSE.

Chondrichthyes Class of cartilaginous FISH including Elasmobranchii (shark, ray, **and** skate) and Holocephali (chimaera). These marine fish have cartilaginous (composed of CARTILAGE) skeletons, a well-developed lower jaw, paired fins, separate gill openings, no air bladder, bony teeth, and plate-like (placoid) scales. Chondrichthyes evolved during the late DEVONIAN period.

chordate Any member of the phylum Chordata, a large group of VERTEBRATES and some marine INVERTEBRATES, which, at some stage in their lives, have rod-like, cartilaginous supporting structures (NOTOCHORDS). Invertebrate chordates are divided into three subphyla: Tunicata (seasquirts); Cephalochordata (amphioxus); and Hemichordata (acorn worms).

chorion Outermost of the protective membranes of the embryonic system of birds, reptiles or mammals, or of the insect egg. In birds it is the moist lining between the shell and the ALLANTOIS (the organ for GASEOUS EXCHANGE). In placental mammals the chorion contributes to PLACENTA formation, and through it the EMBRYO receives nourishment, oxygen, and water.

chorionic gonadotrophin (CG) HORMONE produced by the PLACENTA in the UTERUS of most higher mammals in order to maintain the CORPUS LUTEUM during pregnancy. The detection of this hormone, present in urine, is the basis of most methods of pregnancy testing.

chorionic villus sampling Method of checking for fetal abnormalities during early pregnancy. The CHORION is a membrane that surrounds the fetus in the UTERUS. Small pieces of VILLI (outgrowths) of the membrane are removed via the woman's abdomen or cervix. These can be examined to study the chromosomes and, if necessary, DNA of the fetus to detect any abnormalities. The procedure can be carried out much earlier than AMNIOCENTESIS so that if an abnormality is found the parents can be offered a termination at an early stage of pregnancy.

chromatid Either of the two duplicate strands into which each CHROMOSOME in a biological CELL nucleus divides in the first phase of MITOSIS or MEIOSIS (cell division). When the nucleus is about

to split, the pairs of identical chromatids are separated by a long fibrous structure made of proteins, called a **mitotic spindle**. The separated chromatids become identical "daughter" chromosomes of the same kind as those of the parent cell on opposite sides of the nucleus. *See also* HOMOLOGOUS CHROMOSOME

chromatin Substance that makes up CHROMOSOMES in the nucleus of a cell. It consists of DNA and some RNA, as well as HISTONES and other proteins. In a metabolically active cell nucleus, the chromatin expands to create a region in which MESSENGER RNA (mRNA) may be formed.

chromatography Name given to a number of techniques of chemical ANALYSIS by which substances are separated from one another, identified and measured. Chromatography was invented (1906) by the Russian botanist Mikhail Tsvett (1872–1920). There are several types, all involving a moving phase consisting of a liquid or gaseous mixture of the substances to be separated, and a stationary phase consisting of a material that differentially absorbs the substances in the mixture. *See also* GAS CHROMATOGRAPHY; PAPER CHROMATOGRAPHY

chromatophore Pigment-containing cell that occurs in the skin of some lower animals. Accumulation or dispersion of granules of pigment in

▶ **chromatography**
Gas-liquid chromatographs can separate the components of tiny amounts of an unknown mixture. A sample of the mixture (1) is injected (2) into a stream of helium (3), or another inert gas. Heating ensures the vaporized gas mixes fully with the helium. After impurities are removed (4) the gas mixture passes into a tube (5) packed with coated granules of silicon (6). A liquid with a very high boiling point (7) covers the 4mm (0.15in) granules. The components of the vaporized mixture have different solubilities (8) and so pass through the liquid around the silicon, and the whole tube, at different speeds. The whole tube is kept at a high temperature

to prevent the vaporized gas condensing. As the now-separated parts of the mixture exit the tube (9) they enter a detector (10). Hydrogen (11) and oxygen

(12) are added and the gas stream is then burned (13). During burning each compound produces ions that pass a charge between an anode (14)

and a cathode (15). This charge is measured and can be compared to known results to determine the make-up of the initial mixture.

most soluble

least soluble

chromatophores under nervous or hormonal stimulation allows some animals, such as chameleons, to change color to suit their surroundings. Chromatophores that contain black pigment are called **melanophores**.

chromite Black metallic mineral, ferrous chromic oxide ($FeOCr_2O_3$), the only important ore of chromium. It separates from magma when IGNEOUS ROCKS first form, occurring as octahedral crystals and as granular masses. It is weakly magnetic and opaque. Hardness 5.5; r.d. 4.6.

chromium (symbol Cr) Gray-white TRANSITION ELEMENT, first isolated in 1797. Its chief ore is CHROMITE. Chromium is a dull gray metal but takes a high polish and is extensively used as an electroplated coating. It is also an ingredient of many special steels, and chromium compounds are used in tanning and dyeing. It forms two series of salts, termed chromium(I) (chromous), and chromium(III) (chromic). It also forms chromates(VI), containing the ion CrO_4^{2-}. Properties: at.no. 24; r.a.m. 51.996; r.d. 7.19; m.p. 1,890°C (3,434°F); b.p. 2,672°C (4,842°F); most common isotope ^{52}Cr (83.76%).

chromosome Structure carrying the genetic information of an organism, found only in the cell nucleus of EUKARYOTES. Thread-like and composed of DNA, chromosomes carry a specific set of GENES. Each species usually has a characteristic number of chromosomes; these occur in pairs, members of which carry identical genes, so that most cells have a DIPLOID number of chromosomes. GAMETES (sex cells) carry a HAPLOID number of chromosomes. *See also* HEREDITY

chromosphere Layer of the Sun's atmosphere between the PHOTOSPHERE and the CORONA. The chromosphere is about 10,000km (6,000mi) thick and is normally invisible because of the glare of the photosphere shining through it. It is briefly visible near the beginning and end of a total solar

ECLIPSE as a spiky red rim around the Moon's disk and at other times it can be studied by SPECTROSCOPY or with the use of a special filter. At its base the temperature of the chromosphere is about 4,000K, rising to 100,000K at the top, where it gives way to the corona. Powerful magnetic fields are believed to cause this rise in temperature.

chronometer Instrument for accurately measuring time. The chronometer was an important navigational aid for ships before the advent of radio time signals. Knowing the precise time at which a star reaches a certain position enables a navigator to work out a ship's longitude.

chrysalis Intermediate stage in the life cycle of all INSECTS that undergo complete METAMORPHOSIS. The chrysalis is usually covered with a hard case, but some pupae, such as those of the silk moth, spin a silk COCOON around themselves. Within the chrysalis, feeding and locomotion stop and the final stages of the development take place. *See also* PUPA

chrysoberyl Oxide mineral, beryllium aluminum oxide ($BeAl_2O_4$). It is found in beryllium-rich pegmatite dikes. Its hexagonal system crystals are prismatic or tabular. It can be transparent green, yellow or brown, but bright yellow-green is most highly valued. Gem varieties include cat's eye and alexandrite. Hardness 8.5; r.d. 3.69.

chrysotile Fibrous serpentine mineral, from which comes most of the world's supply of ASBESTOS. Serpentines are hydrated magnesium silicates ($3MgO2SiO_2.2H_2O$). The variety called chrysotile has crystalline, tubular fibers that are particularly suitable for spinning and weaving into heat-resistant fabrics.

chyle Milky fluid found in the lymphatic vessels in the intestine and consisting of LYMPH containing minute fat droplets absorbed during DIGESTION. It travels through the LYMPHATIC SYSTEM and drains into the bloodstream from the thoracic duct in the neck.

chyme Stomach contents, a mixture of partly digested food, gastric acid, and digestive enzymes. It passes from the stomach into the small intestine in semiliquid form. *See also* DIGESTION; DIGESTIVE SYSTEM

chymotrypsin Substance produced in the body that aids in the digestion of food. Chymotrypsin is an ENZYME that breaks down proteins. It is made in the SMALL INTESTINE by complex chemical reactions, from chymotrypsinogen. This, together with other digestive substances, is secreted by the PANCREAS through the pancreatic duct into the

▼ **chromosome** The 46 chromosomes in somatic (non-reproductive) cells contain a single sex-determining pair that consist of an X and Y chromosome in males, or an XX pair in females. Ova contain only the X chromosome, while spermatozoa contain X or Y chromosomes in equal proportions. At fertilization therefore there is a 50% chance of an XX or XY pair being formed.

DUODENUM, which lies between the stomach and the small intestine. The secretion is controlled partly by the vagus nerve (the AUTONOMIC NERVOUS SYSTEM) and partly by HORMONES, which are released automatically when food enters the duodenum. *See also* TRYPSIN

cilia (sing. cilium) Small, hairlike structures on cell walls used for propulsion and feeding. Cilia are present in great quantities on some lining cells of the body, such as those along the respiratory tract. The wafting movements of the cilia help to propel foreign particles towards the exterior. They are also found on single-celled PROTOZOA, some of which, such as the trypanosome of sleeping sickness, are known as CILIATES.

ciliary body Band of tissue around the lens of the EYE. The ciliary body supports the lens and contains the ciliary muscles, which contract to change the shape of the lens and focus the eye (*see* ACCOMMODATION). The AQUEOUS HUMOR, the liquid within the eyeball in front of the lens, is produced by the ciliary body.

ciliate Any one of the approximately 8,000 species of the phylum Ciliophora, characterized by hairlike CILIA used for locomotion and food collecting. Ciliates are the largest and the most complex of the PROTOZOA. They are found in both aquatic and terrestrial habitats and many are carnivorous. Ciliates have two nuclei (**macronucleus** and **micronucleus**) and a variety of organelles, such as a cystome (mouth). Subclasses include the Holotrichs (*Paramecium*), Spirotrichs (*Stentor*), and Peritrichs (*Vorticella*).

cinder cone Conical hill or mountain composed largely of unconsolidated material, mostly ash ejected by a volcano. It is characteristic of volcanoes that produce large amounts of gas and ash rather than lava. The coarser the ejected material, the steeper the sides of the cone.

cinnabar Deep red mineral, mercury(II) (mercuric) sulfide (HgS), the major ore of mercury. Its crystal system is trigonal. It occurs as rhombohedral crystals, often twinned, and as granular masses. It is found in hydrothermal veins and volcanic deposits. The ore is reduced to mercury by roasting. Hardness 2–2.5; r.d. 8.1.

cipher *See* CRYPTOGRAPHY

circadian rhythm Internal "clock" mechanism that normally corresponds roughly with the 24-hour day. It relates most obviously to the cycle of waking and sleeping, but is also involved in other cyclic variations, such as body temperature, hormone levels, metabolism, and mental performance.

It can be disrupted by irregular events such as shift-working and jet lag. *See also* BIOLOGICAL CLOCK

Circinus (Compasses) Small constellation of the s sky in the region of Alpha and Beta Centauri. Its brightest star is Alpha, magnitude 3.2.

circle Plane geometric figure that is the locus of points equidistant from a fixed point (the center). This distance is the radius (r). The area of a circle is πr^2 and its perimeter (circumference) is $2\pi r$. *See also* CONIC

circuit *See* ELECTRIC CIRCUIT

circuit breaker Automatic switch that disconnects the electricity supply if dangerous conditions occur. A common type of circuit breaker switches off the current if it exceeds a certain value for more than a specified time. The circuit breaker performs a similar function to a FUSE, but is not destroyed in operation. Instead it can be reset when the fault condition has been cleared.

circular motion In physics, any type of movement that can be related to or analyzed in terms of the motion of an object around a CIRCLE.

circulation *See* CIRCULATORY SYSTEM

circulation, atmospheric Flow of the ATMOSPHERE around the Earth. It is caused by temperature differences in the atmosphere, and the rotation of the Earth, which transfers heat from warm zones (the tropics) to cooler zones (towards the poles). The poleward circulation due to CONVECTION gives rise to large-scale eddies such as a CYCLONE and anticyclone, low-pressure troughs and high-pressure ridges. The eddies also take part in the longitudinal atmospheric circulation around the Earth, with the Earth's rotation maintaining easterly winds towards the Equator and westerly winds towards the poles. Narrow JET STREAMS blow swiftly over middle latitudes in the STRATOSPHERE, and move farther towards the poles during the summer. *See also* CORIOLIS EFFECT

circulatory system Means by which OXYGEN and NUTRIENTS are carried to the body's tissues and CARBON DIOXIDE and other waste products are removed. It consists of BLOOD VESSELS that carry the BLOOD, propelled by the pumping action of the HEART. In humans and other mammals, blood travels to the LUNGS, where, by the process of GAS EXCHANGE, it picks up oxygen and loses carbon dioxide. It then flows to the heart, from where it is pumped out into the AORTA, which branches into smaller ARTERIES, arterioles and CAPILLARIES. Oxygen and other nutrients diffuse out of the blood, and carbon dioxide and other tissue wastes pass into the capillaries, which join to form VEINS leading back to the heart. Blood then returns to the lungs and the entire cycle is repeated. This is known as a double circulatory system, as the blood is pumped first to the lungs, and then to the rest of the body. Because the lungs are delicate, the double cycle allows the heart to pump blood strongly through the body without risk of damaging them. In fish and many other animals, there is a single circulatory system, with blood passing through the GILLS and on to the rest of the body without an extra boost from the heart. Both these circulatory systems are closed; the blood remains confined within the blood vessels. Insects and many other invertebrates have an open circulatory system, where the blood flows freely within the body cavity, but passes through a series of open blood vessels and heart(s), the pumping of which maintains a directional flow.

circumference Distance round the boundary of a plane geometric figure, nearly always applied to a CIRCLE, for which it has the value $2\pi r$, where r is the radius.

cire perdue (Fr. lost wax) Method of CASTING metal objects (usually bronzes) used since classical antiquity. First, the object is molded in clay or plaster in roughly the desired shape but slightly smaller. This core is covered in wax, not more than 1.5cm (0.5in) thick, and further modeling is undertaken. Straws or wax pipes are fitted to admit the molten metal and release hot gases. The whole is then covered in a heat-proof mold, the wax melted away and the metal poured into the space it occupied.

cirque Bowl-shaped, steep-sided hollow in a mountain area that is becoming or is already glaciated. The eroded area is usually cut into the bedrock by repeated freezing and the result of glacial movement. When the ice melts a lake may form in the base of the cirque. Such lakes are known variously as cirque lakes, **corries** (in Scotland) or **cwm** (in Wales). They are often fed by the retreating glacier that created them. *See also* EROSION; GLACIER

cirrocumulus *See* CLOUD

cirrostratus *See* CLOUD

cirrus *See* CLOUD

citric acid ($C_6H_8O_7$) Colorless crystalline solid with a sour taste. It is found in a free form in citrus fruits such as lemons and oranges, and is used for flavoring, in effervescent salts, and as a mordant (color-fixer) in dyeing. Properties: r.d. 1.54; m.p. 153°C (307.4°F).

citric acid cycle *See* KREBS CYCLE

citrine Semiprecious stone, a glassy wine-yellow variety of QUARTZ. It is translucent and has an appearance similar to TOPAZ. It is also known as false topaz or topaz quartz.

civetone Chemical compound smelling strongly of musk, obtained from the perianal glands of civets and used in perfumery. Chemically it is a KETONE, containing the group –CO.

civil engineering Field of engineering dealing with large structures and systems. Civil engineers provide facilities for living, industry, and transportation, such as roads, bridges, canals, airports, water-supply systems, dams, harbors, and tunnels. Civil engineering requires a thorough knowledge of surveying, construction, material properties, soil properties, and hydraulics. Important divisions of the field are architecture, irrigation, transportation, soil and foundation, geodetic, hydraulic, and coastal and ocean engineering.

cladistics System of classifying organisms on the basis of homologies, or features shared due to common descent from the same ancestor. Organisms that share several characteristics are put into taxonomic groups known as **clades**. A **cladogram** is a diagram, resembling a sideways family tree, that shows how pairs or groups of clades diverged from common ancestors. Not all taxonomists agree about the validity of cladistics.

class In biology, part of the CLASSIFICATION of living organisms, ranking above ORDER and below PHYLUM. Class names are printed in roman (not italic) letters, with an initial capital; for example Mammalia is the class in which all mammals are found.

classification Organization of organisms into categories based on appearance, structure, genetic sequence or evolution. The categories, from the most inclusive to the most exclusive, are KINGDOM, PHYLUM, CLASS, ORDER, FAMILY, GENUS, SPECIES, and sometimes variety. Some categories also include subphyla, subfamilies, and so on. For example, the domestic dog is classified as kingdom Animalia, phylum Chordata, class Mammalia, order Carnivora, family Canidae, genus *Canis*, species *Canis familiaris*. *See also* BINOMIAL NOMENCLATURE; TAXONOMY

Clausius, Rudolf Julius Emanuel (1822–88) German physicist, regarded as the founder of

THERMODYNAMICS. Using the work of Nicolas CARNOT, Clausius was the first to formulate the second law of thermodynamics: that heat cannot pass from a colder to a hotter object. He also introduced the term ENTROPY.

clavicle (collarbone) Thin, slightly curved bone attached by ligaments to the top of the STERNUM (breastbone); it is present in most vertebrates. In humans, the clavicle and SCAPULA (shoulder blade) make up the SHOULDER girdle, linking the arms to the axis of the body.

clay Mineral consisting of hydrous silicates of aluminum and magnesium, including kaolinite and halloysite, usually mixed with some quartz, calcite or gypsum. It is formed by the WEATHERING of surface granite or the chemical decomposition of feldspar. Soft when wet, it hardens on firing and is used to make pottery, stoneware, tiles, pipes, bricks and molds, and as a filler for paper, rubber, and paint.

cleavage In embryology, progressive series of CELL DIVISIONS that transforms a fertilized OVUM (egg) into the earliest embryonic stage, called the BLASTULA. The ovum is divided into **blastomeres** (smaller cells), each containing a DIPLOID number of CHROMOSOMES.

cleavage, rock Formation of definite planes through a rock. It is caused by compression associated with folding and metamorphism and results in the rock splitting easily parallel to the cleavage. The line of cleavage follows the alignment of minerals within the rock. *See also* FOLD

cleistogamous flower Small, closed, self-fertilizing flower. Cleistogamous FERTILIZATION takes place within a flower that does not open, and enables seed production in the absence of normal cross-POLLINATION. Sweet violet, oxalis and impatiens are examples of such flowers.

Clerk, Sir Dugald (1854–1932) Scottish engineer. He invented the **Clerk cycle** or **two-stroke cycle** (*see* TWO-STROKE ENGINE), which he patented in 1881. It is still much used on lightweight motorcycles, lawn mowers, and outboard motors on small boats.

client-server Type of relationship between computers in a COMPUTER NETWORK. A **client** computer makes requests of a designated **server** computer. The server performs the requested functions and delivers the results via the network to the client.

cliff Geological formation consisting of a steep slope, sometimes of exposed rock. Cliffs can be nearly vertical or even overhanging. They may result from the exposure of rock strata due to geological FAULTS (*see* ESCARPMENT) or, as in the case of shoreline cliffs, they may be caused by water and weather EROSION.

climate Overall WEATHER conditions of a place or region prevailing over a long time. The daily weather records are added together and then averaged out to give a general pattern of the climate. The major factors influencing climate are temperatures, air movements, incoming and outgoing radiation and moisture movements. Climates are defined on different scales, ranging from **macroclimates**, which cover the broad climatic zones of the globe, down to **microclimates**, which refer to the conditions in a small area, such as a forest or a field. *See also* CONTINENTAL CLIMATE; MARITIME CLIMATE; MEDITERRANEAN CLIMATE

climate modeling Use of a computer to simulate the Earth's climate. Physical data, such as temperature, pressure, wind direction and so on, are manipulated mathematically by a powerful computer to give a model of the Earth's whole climatic system. Researchers can vary various parameters to see what changes occur. In this way they can study the effects of the GREENHOUSE EFFECT and possible GLOBAL WARMING, or what would happen if there was a significant change in the amount of radiation from the Sun. Such modeling can be limited by scientist's understanding of the key factors controlling climates. *See also* CHAOS THEORY

climatology Scientific study of the Earth's climates. **Physical** climatology investigates relationships between temperature, pressure, winds, precipitation, and other weather phenomena. **Regional** climatology considers latitude and other geographical factors, such as the influence of large landmasses in the climatic study of a particular place or region.

climax community In ecology, the stable community that exists at the end of a SUCCESSION. During succession, a series of plants and their associated animal life gradually colonize an area, with newcomers usually displacing many of the species already present. Eventually, at the climax, colonization ceases and a relatively stable community persists in equilibrium with the local conditions.

clint Flat block of limestone, typically 1–2m (3–7ft) across, separated from other clints by vertical cracks, called GRIKES, which together form a limestone pavement.

clitoris *See* VULVA

cloaca Common cavity into which intestinal, urinary, and genital tracts open in fish, reptiles, birds, and some primitive mammals.

clock Instrument for measuring time. The earliest timekeeping instruments had no moving parts, being designed to measure the movements of the Sun, Moon, and stars. Examples include neolithic stone columns and ancient Egyptian sundials. The ancient Egyptians also used water clocks, which were pottery bowls bearing graduated time-marks from which water slowly leaked. Candle clocks and sandglasses were later types of nonmechanical clocks. The central feature of all mechanical clocks is an **escapement** mechanism, which enables a clock to tick off time at discrete intervals. This movement is transmitted through a series of gears to the hands, which are pushed forward a small distance with every escapement movement. Motive power for mechanical clocks has been provided by falling weights and coiled springs. Other modern clocks include those using an electrically oscillated QUARTZ crystal as the basis of time-division. The latter are accurate, but even more accurate are ATOMIC CLOCKS, which rely upon the natural oscillations of atoms (usually those of cesium) and which measure time to an accuracy of thousandths of a second per year.

cloisonné Enameling technique in which the design is constructed out of wires soldered to a plate, and the cells (**cloisons**) thus formed are filled with colored enamel paste and fired.

clone Set of organisms obtained from a single original parent through some form of ASEXUAL REPRODUCTION or by artificial PROPAGATION. Clones are genetically identical and may arise naturally from PARTHENOGENESIS in animals. Cloning is often used in plant propagation (including TISSUE CULTURE) to produce new plants from parents with desirable qualities such as high yield, or from plants which have been genetically engineered. It is now possible to produce animal clones from tissue culture. In 1997 scientists in Scotland announced that they had cloned a sheep. Using nuclear transfer (transfer of a cell NUCLEUS) technology, they produced an embryo from a single udder cell of an adult sheep. The embryo was then implanted into a surrogate mother, and Dolly, an identical twin of the sheep that donated the cell, was born in February 1997. In 1998 Dolly gave birth naturally, thus allaying fears that, as a result of being cloned from an older sheep, she would not be able to reproduce. *See also* ARTIFICIAL SELECTION; GENETIC ENGINEERING

closed-circuit television (CCTV) Television system in which the television camera, receiver, and associated controls are most often directly linked by cables. It has many applications, such as monitoring machinery, watching customers in shops, surveillance of military installations or hospital patients, and so on. CABLE TELEVISION is an extended form of closed-circuit television.

cloud Visible mass of water particles or ice crystals suspended in the lower atmosphere. Clouds are formed when water from the Earth's surface becomes vapor through the process of EVAPORATION. As the water vapor rises into the atmosphere it cools and condenses around microscopic salt and dust particles, forming droplets. Where the atmosphere is cold (below the freezing temperature of water) the droplets turn to ice. There are 10 different classifications of clouds: **cirrus** are high, white, and threadlike. They do not give rain, but their presence can be a sign of an approaching DEPRESSION. Cirrus occurs at heights of above 6,000m (20,000ft) and sometimes as high as 12,000m (39,500ft). **Cirrocumulus** are high clouds consisting of small patches of ice crystals, often stretched out in a linear form. **Cirrostratus** are white and almost transparent sheets of ice crystals. **Altocumulus** are grayish-white globular clouds found between 2,400m (8,000ft) and 6,000m (20,000ft). **Altostratus** are gray/blue and streaky, and often cover the whole sky. They often produce drizzle. **Nimbostratus** are low, thick and dark, and usually shed rain or snow. **Stratocumulus** are masses of white, gray or dark clouds, within which vertical patches of cumulus may cause heavy rain. **Stratus** are low-lying and gray, with thicker examples producing light rain. **Cumulus** are white and fluffy-looking, with a characteristic dome shape. **Cumulonimbus** are vast, towering, dark clouds that generally produce thunderstorms. They have bases almost touching the ground and extend upward to 75,000ft (22,875m). They can also produce tornadoes. By day, clouds reflect the rays of the Sun back into the atmosphere, keeping the ground cool. At night, clouds trap and re-radiate heat rising from the Earth, keeping surface temperatures warm. There is no real difference between low clouds, which may shroud hills or the tops of skyscrapers, and FOG and MIST, which both form at low level.

cloud chamber Instrument used to detect and identify charged particles. It was invented in the 1880s by the British physicist Charles WILSON to study atomic radiation and was refined over several decades. The principle is the same as that of the later BUBBLE CHAMBER, except that the liquefied gas is replaced by air supersaturated with water or alcohol vapor, and the tracks left are droplets that form around the ionizing particle. The tracks are deflected by a magnetic field and photographed for analysis.

club moss Any of about 200 species of small, evergreen, spore-bearing plants of the phylum LYCOPODOPHYTA. Unlike the more primitive true MOSSES, club mosses have specialized tissues for transporting water, food, and minerals. They are related to ferns and horsetails. The leaves are arranged in tight whorls around the aerial stems. Millions of years ago their ancestors formed the trees that dominated the CARBONIFEROUS "coal" forests.

cluster, galaxy *See* GALAXY CLUSTER

cluster, globular *See* GLOBULAR CLUSTER

cluster, open *See* OPEN CLUSTER

cluster, stellar *See* STELLAR CLUSTER

clutch Any device placed between the rotating parts of an ENGINE or motor and the driveshaft to facilitate their quick connection or disconnection. In a car, for example, temporary disengagement of the engine is essential during gear-changes. The clutch usually consists of a pair of friction plates, although there are fluid clutches as well.

clutch, electromagnetic Device that uses MAGNETISM to connect two rotating shafts. Some forms are disc clutches with energized coils and magnetic clutch plates. **Eddy current** clutches induce rotational movement in the shaft to be engaged and rotated. **Hysteresis** clutches also transmit rotation without slip. Other electromagnetic clutches employ magnetic metal particles, either dry-flowing or suspended in a liquid, to induce torque.

cnidarian Any one of the 9,000 species of the phylum Cnidaria (commonly called the **coelenterates**). Marine invertebrates, cnidarians include corals, jellyfish, and sea anemones. Characterized by a digestive cavity that forms the main body cavity, they may have been the first animal group to reach the tissue level of organization. The cnidarians are radially symmetrical, jellylike, and have a nerve net and one body opening. Reproduction is sexual and asexual; REGENERATION also occurs.

CNS Abbreviation of CENTRAL NERVOUS SYSTEM

coagulation Clumping together of colloidal particles (*see* COLLOID), as in the clotting of blood. The particles in a colloidal solution are completely dispersed and will not settle out. But the solution can be destabilized by adding ions, particularly those with a high charge. This process occurs when river water containing colloids of silt reaches the sea, where the ions in sea water coagulate the silt and it precipitates to form a delta. Heat is also used to coagulate colloids, particularly those containing proteins.

coal Blackish, carbon-rich deposit formed from the remains of fossil plants. In the Carboniferous

and Tertiary periods, swamp vegetation subsided to form PEAT bogs. Sedimentary deposits buried the bogs, and the increasing pressure and heat produced LIGNITE (brown coal), then bituminous coal, and finally ANTHRACITE if temperature increased sufficiently. This is termed the **coal rank series**; each rank of coal represents an increase in carbon content and a reduction in the proportion of natural gases and moisture. Lignite, which has a low carbon content, is a poorer fuel than anthracite. Most coal seams are inter-stratified with shale, clay, sandstone, and sometimes limestone.

coal tar By-product from the manufacture of COKE. The tar comes from the bituminous coal used in this destructive DISTILLATION process. Coal tar is a volatile substance, important for its organic chemical constituents. These are extracted by further distillation. Such chemicals, called **coal-tar crudes**, include xylene, toluene, naphthalene, and phenanthrene. They are the basic ingredients from which many products, including explosives, dyes, drugs, and perfumes, are made.

coaxial cable CABLE consisting of a central conductor with a surrounding insulator and tubular shield. Most television sets have an aerial connect-

▲ **coal** The process of making coal begins with plant debris (1). Dead vegetation lies in a swampy environment and forms peat (4), the first stage of coal formation. Underwater bacteria remove some oxygen, nitrogen, and hydrogen from the organic material. Debris carried elsewhere and deposited by water forms a product called cannel coal (2). Algal material collected underwater forms boghead coal (3). If the dead organic material is buried by sediment, the weight on top of the peat and the higher temperature will turn the peat into lignite (5). With more heat and pressure at increasing depths, lignite becomes bituminous coal (6) and then anthracite (7).

ed by coaxial cable. Multiple coaxial cables, used for long-distance communications systems, contain many separate coaxial cables. They are used to transmit high-frequency signals.

cobalt (symbol Co) Gray TRANSITION ELEMENT, first discovered in 1737. It is found in cobaltite and smaltite, but most is obtained as a by-product during the processing of other ores. Cobalt is a constituent of vitamin B_{12}. It is used in high-temperature steels, paints (especially cobalt blue), the manufacture of jet engines, cutting tools and (as an alloy) in magnets. Cobalt-60 (half-life 5.26 yr) is an artificial isotope used as a source of GAMMA RADIATION in radiotherapy and tracer studies. Properties: at.no. 27; r.a.m. 58.9332; r.d. 8.9; m.p. 1,495°C (2,723°F); b.p. 2,870°C (5,198°F); most common isotope ^{59}Co (100%).

COBOL (Common Business-Oriented Language) COMPUTER LANGUAGE developed in 1959 for processing business data. Revised and improved since then, COBOL is now widely used.

cocaine White crystalline ALKALOID extracted from the leaves of the coca plant. Once used as a local anesthetic, it is now primarily an illegal nar-

▲ **coaxial cable** The magnetic fields induced by the current flowing in a normal two-core cable (1) will interfere with each other and may induce unacceptable variations in the level of current flowing in the wire (2). This is particularly a problem when the current represents a communications signal, part of a computer network or the connection between a television and its aerial. Here coaxial cable is used, where one conductor (4) is wound round the other (5), separated by a layer of insulating material (6). The electromagnetic flux no longer causes such a damaging distortion of any transmitted signals (7).

cotic, with stimulant and hallucinatory effects. It is psychologically habit-forming, but increasing doses are not needed, as the body does not develop tolerance. Habitual use of cocaine results in physical and nervous deterioration. A smokable derivative, crack, is a highly volatile, addictive, and lethal substance.

coccolith Any microscopic, single-celled flagellate of the Coccolithophorida, a class of ALGAE of the phylum Chrysophyta. The cell is covered with round, chalky platelets only one or two thousandths of a millimeter in diameter. Many limestone and chalk cliffs are made up entirely of the remains of such platelets.

coccus Spherical BACTERIUM with an average diameter of 0.5–1.25 micrometers. Some, such as *Streptococcus* and *Staphylococcus*, are a common cause of infection.

coccyx Triangular bone at the lower end of the vertebral column. It is formed by the fusion of three to five small VERTEBRAE. *See also* SPINE

cochlea Fluid-filled structure in the inner EAR that is essential to HEARING. It has a shape like a coiled shell, and is lined with hair cells that move in response to incoming sound waves, stimulating nerve cells to transmit impulses to the brain. Different groups of hair cells are stimulated by different pitches of sound, thus helping the brain to analyze the sound.

Cockcroft, Sir John Douglas (1897–1967) British physicist who, with his colleague Ernest WALTON, was the first person to split the atom. He and Walton constructed a particle ACCELERATOR and in 1932 created the first man-made nuclear reaction by bombarding lithium atoms with protons. They shared the 1951 Nobel Prize for physics for their use of particle accelerators to study atomic nuclei. Cockcroft also contributed to the development of NUCLEAR REACTORS and in 1946 became the first director of the UK's Atomic Energy Research Establishment at Harwell.

Cockcroft-Walton generator Device in which a high constant voltage is obtained by multiplying a low alternating voltage using a circuit of rectifiers and capacitors. *See also* ACCELERATOR, ELECTROSTATIC

Cockerell, Sir Christopher Sydney (1910–) British engineer who invented the HOVERCRAFT. In 1948 Cockerell started a boat-building business and, in the early 1950s, began research into the development of AIR-CUSHION VEHICLES. By 1954 he had developed the hovercraft. He filed his first patent in 1955. In 1959 the first SRN1 passenger-carrying hovercraft crossed the English Channel.

cocoon Case or wrapping produced by larval forms of animals (such as some moths, butterflies, and wasps) for the resting or pupal stage in their life cycle. Some spiders spin a cocoon that protects their eggs. Most cocoons are made of silk, and those of the domestic silkworms provide most of the world's commercial silk. *See also* CHRYSALIS; PUPA

codeine White crystalline ALKALOID extracted from OPIUM by the methylation of MORPHINE, and with the properties of weak morphine. It is used in medicine as an ANALGESIC to treat mild to moderate pain, as a cough suppressant and to treat diarrhea.

codon Unit of the genetic code in a molecule of DNA or MESSENGER RNA (mRNA). In messenger RNA it consists of a triplet of bases that usually specifies a particular AMINO ACID as part of a POLYPEPTIDE chain being built into a protein. Some codons carry instructions (such as "start" and "stop") involved in the process of protein building. *See also* GENE

coefficient Term multiplying a specified unknown quantity in an algebraic expression. In the expression $1 + 5x + 2x^2$, 5 and 2 are the coefficients of x and x^2 respectively. In physics, a coefficient is a number that specifies a given property of a substance such as the elasticity coefficient or coefficient of friction.

coelenterate Common name for a CNIDARIAN.

coeliac disease Disorder in which the SMALL INTESTINE fails to absorb food properly. It is caused by intolerance of GLUTEN, a protein in wheat and rye products. Symptoms include depression, diarrhea and malnutrition.

coelom True body cavity, or space between an inner digestive tube and outer body wall, that provides room for additional organs. Indications of a coelom appear in roundworms, but the first true coelom appears among earthworms and other segmented worms.

coenocyte Type of cell that has many nuclei. Coenocytes occur when only the nucleus of a cell divides repeatedly, without division of the whole cell or cell wall. The fibers of striated muscle, for example, are made up of such cells, giving a type of tissue called **syncytium**. Coenocytes are also found in some algae and fungi.

coenzyme Nonprotein organic molecule, usually containing a vitamin and phosphorus. When combined with an **apoenzyme**, a protein molecule, it activates an ENZYME. A coenzyme always regains its original structure even though it may have been altered during a reaction.

coevolution Evolution of complementary features in two different species; a result of interaction between the two. Examples include the evolution predator defenses by a prey species, subsequent evolution by the predator of ways to thwart the prey's defenses, and so on. In other instances, both species gain from the association, and evolve structures or behavior of mutual benefit. The classical example occurs in plants that are pollinated by insects. The plant's flowers have a color or smell that attracts insects, and a shape that makes it easy for insects to reach the nectar and in doing so remove pollen. The insects have evolved to be able to see and smell the flowers, and have mouthparts adapted to reach the nectar.

Cohen, Stanley N. (1935–) US biochemist and physician. In the early 1970s he worked with Herbert Boyer in the development of the recombinant DNA technique employed in GENETIC ENGINEERING. They used the recently discovered restriction enzymes to cut and splice DNA in certain bacteria. *See also* RECOMBINANT DNA RESEARCH

Cohen, Stanley (1922–) US biochemist. He joined Rita LEVI-MONTALCINI in 1952 after receiving a doctorate in biochemistry at the University of Michigan (1948). Together they discovered and isolated a protein that stimulated nerve cell and fiber growth. Cohen also isolated further substances that encourage epidermal growth. Cohen and Levi-Montalcini were awarded the Nobel Prize for physiology or medicine for their work in 1986.

coherence In physics, the property of ELECTROMAGNETIC RADIATION that has a constant phase relationship between two or more sets of waves. If the peaks and troughs of one wave are always the same distance from those of another wave, the waves are coherent. It follows that the waves must also have exactly the same FREQUENCY. A pair of coherent waves produce INTERFERENCE. The light from a LASER is coherent.

cohesion Mutual attraction between the component atoms, ions or molecules of a substance. Weak cohesive forces permit the fluidity of liquids; those of solids are much stronger. Liquids form droplets because of surface tension caused by cohesion. *See also* ADHESION

coke Stiff, porous, gray mass of carbon mixed with small amounts of minerals, sulfur, and residual volatiles. It is left behind as a residue in coking ovens after coal has been heated without air to drive off most of the volatile matter. It is used mainly in steelmaking for fueling BLAST FURNACES. Its high calorific value is due to the high carbon content – more than 90%.

col In geology, a low point between two mountain peaks or on an upland ridge. In meteorology, a calm area between two DEPRESSIONS and two ANTICYCLONES.

colchicine ALKALOID obtained from the corm of the autumn crocus (*Colchicum autumnale*). It is employed in genetic and cellular research, as it inhibits MITOSIS (cell division). This property makes colchicine valuable in cancer research and as an IMMUNOSUPPRESSIVE DRUG.

cold-blooded *See* POIKILOTHERMIC

coleopteran Any member of the order Coleoptera, the largest order of the animal kingdom, generally the BEETLES. They are insects that undergo complete METAMORPHOSIS. Most possess two pairs of wings, the forward pair usually being hard covers that fold over the flying pair. More than 350,000 species are known, many of them pests called weevils. Beetles include some of the largest and many of the smallest of all insects. Their larvae are either maggot-like, or elongated with six legs.

coleoptile Sheath that covers and protects the shoot from an embryo of a grass seed. The sheath protects the shoot as it forces its way upwards through the soil and remains in place until it is split open by the development of the first true leaves.

collagen Protein substance that is the main constituent of bones, tendons, cartilage, connective tissue, and skin. It is made up of inelastic fibers.

collar bone *See* CLAVICLE

collimator Lens system used to produce a parallel or near parallel beam of light, subatomic particles or other radiation. Light passing through a fine slit at the focus of a convex lens is rendered parallel by the lens. A collimated beam is essential for many optical purposes.

colloid Substance composed of fine particles that are dispersed throughout a second substance. A SOL is a solid dispersed in a liquid, an AEROSOL is a solid or a liquid in a gas, an EMULSION is a liquid in a liquid, and a FOAM is a gas in either a liquid or a solid.

colon Part of the large INTESTINE, the digestive tract that extends from the small intestine to the RECTUM. The colon absorbs water from digested food and allows bacterial action for the formation of feces. *See also* DIGESTIVE SYSTEM

colony In biology, group of similar animals or plants living together for mutual benefit. Individuals perform like or varied functions and may be structurally separated or united.

colophony Alternative name for ROSIN

color Sensation experienced when light of sufficient brightness and of a particular wavelength strikes the RETINA of the EYE. Normal daylight (white light) is made up of a spectrum of colors, each of which has a different wavelength. These colors can be arbitrarily assigned to seven bands – red, orange, yellow, green, blue, indigo, and violet – of decreasing wavelength. A pure spectral color is called a **hue**. Any actual color is caused by light of a mixture of wavelengths. It can be matched by (that is, indistinguishable from) a suitably chosen pure wavelength mixed with white light of appropriate brightness. If the color is not pure but contains some white, it is "desaturated." **Saturation** is the degree to which a color departs from white and approaches a pure hue. A color is also defined by its luminosity, or brightness. Any color of light can be matched by a mixture of three primary colors: red, green, and blue-violet; red and green light mix to produce yellow; red and blue-violet mix to form purple, and green and blue-violet mix to form cyan. The production of color by paints and pigments is more complicated. The colors of these result from the absorption of particular wavelengths and reflection of the rest. The "subtractive" primary colors used in printing inks are **magenta** (green-absorbing), **cyan** (red-absorbing), and **yellow** (blue-absorbing).

color film Photographic film available either as positive film for projection or as negative film for making color prints. It contains three layers of light-sensitive chemicals that respond, in turn, to the optical primary colors (roughly red, green, and blue) present in the light passing through the lens of the camera. Polaroid color film contains additionally a number of colored dyes that are released by the light-sensitive chemicals as they react to the optical primaries, so developing a color print directly.

colorimetry Scientific method in which an unknown color is compared with known colors. It is used primarily in chemical ANALYSIS, and for checking the color of products such as paints, dyes, and pharmaceuticals.

color vision Ability of the EYE to detect the different wavelengths (COLORS) of light. This is achieved by three types of cone cell in the RETINA, "red," "green," and "blue," which respond to light of those colors. The cone cells secrete one of three different pigments that break down and generate nerve impulses that the brain then interprets to give the range of colors we perceive.

colostomy Operation to bring a part of the large intestine, called the COLON, out through the wall of the abdomen in order to bypass the lower section of the bowel. An artificial opening (**stoma**) is created so that fecal matter is passed into a collecting appliance worn on the exterior of the body. The site of the colostomy varies according to the location of the disease. A colostomy may be temporary, in which case it can be closed to restore continuity of the bowel; usually, however, it will be permanent, if the lower part of the bowel has been removed. *See also* DIGESTION

colostrum Milk-like fluid produced by the female MAMMARY GLAND soon after (or even before) the birth of young; it precedes true milk, which is produced about three days after birth. Colostrum is rich in protein and the antibodies it contains give the newborn some immunity from infection.

Columba (Dove) Indistinctive constellation of the S sky, adjoining Canis Major and Carina. Its brightest star is Alpha, magnitude 2.6.

columbite Black oxide mineral of IRON, MANGANESE, and NIOBIUM. Niobium is replaced by TANTALUM to form tantalite, with similar properties to columbite but denser. They are found in granite PEGMATITES in W Australia and Madagascar. Columbite crystallizes in the orthorhombic system, forming prismatic crystals with one distinct cleavage. Hardness: 6.0 to 6.5; r.d. 5.2 (columbite) to 8.0 (tantalite).

columbium *See* NIOBIUM

coma Spherical cloud of gas and dust surrounding the nucleus of an active COMET.

Coma Berenices (Berenice's Hair) Constellation of the N sky, adjoining Boötes. Its brightest stars are of 4th and 5th magnitudes.

combe Small hollow on a steep chalk hillside. Generally found on an ESCARPMENT, combes are formed when WEATHERING opens up and enlarges a joint in the rock. Weathered material that slips down to the bottom of the slope may form a mound known as a combe rock.

combustion Burning, usually in oxygen. The combustion of fuels is used to produce heat and light. An example is a fire. In solids and liquids, the reaction speed may be controlled both by the rate of oxygen flow to the surface, and by CATALYSTS. Industrial techniques harness the energy produced using combustion chambers and furnaces.

comet Small, icy Solar System body in an independent orbit around the Sun. F.L. Whipple's (1906–) "dirty snowball" theory of comets is largely accepted. The solid nucleus of a comet is small, that of HALLEY'S COMET measures just 16 × 8km (10 × 5mi), and comprises rock and dust particles

embedded in ice. As the comet approaches the Sun and gets warmer, evaporation begins, and jets of gas and dust form the luminous, spherical coma. Later, radiation pressure from the Sun and the SOLAR WIND may send dust and gas streaming away as a tail, as much as 150 million kilometres in length. There are three main types of comet: **short-period** comets often have their aphelia at approximately the distance of Jupiter's orbit. **Long-period** comets (such as Halley's) have aphelia near or beyond Neptune's orbit. The third type are either called **non-periodic**, or **hyperbolic**. Their periods may be as much as several million years and have such great orbital eccentricities that their paths are almost parabolic. It is now thought that comets orginated with the rest of the Solar System. *See also* METEOR

commensalism Situation in nature in which two species live in close association but only one partner benefits. One of the species (the **commensal**) may gain from increased food supply, or by procuring shelter, support or means of locomotion, but the other (the **host**) neither gains nor loses from the relationship. *See also* MUTUALISM; PARASITE; SYMBIOSIS

▲ **comet** There are three main types of comet. Short-period comets (A) often have their aphelia at approximately the distance of Jupiter's orbit (1). Their periods amount to a few years and all short-period comets are faint. Long-period comets (B) have aphelia near or beyond Neptune's orbit (2) – Halley's is the only conspicuous member of the class. Comets with very long periods (C) have such great orbital eccentricities that the paths are almost parabolic. Because only a short arc can be measured, it is impossible to calculate the periods of these comets accurately. All the really brilliant comets, apart from Halley's, are of this type.

communications Processes for sharing information and ideas. Facial expressions, hand signals, sirens, writing, speech, song, and movies are examples. The 15th-century invention of the printing press revolutionized communications. The 20th century has witnessed its own communications revolution, primarily in terms of increased access. Mass communication, principally through the medium of TELECOMMUNICATIONS, has facilitated rapid, international communications. The invention of the TELEPHONE, RADIO, TELEVISION, and the COMPUTER NETWORK have enabled worldwide exchange of information, and in some cases have shaped 20th-century history. The INTERNET is the latest in a long line of technological innovations. *See also* SATELLITE, COMMUNICATIONS

community In ecology, naturally occurring group of plants and animals living within a restricted area. Communities are usually named after a prominent physical characteristic or a dominant species.

Community Antenna Television (CATV) System in which cables connect a central antenna to numerous television receivers, providing a wide range of commercial programs telecast by distant stations. Subscribers do not need separate antennas, but pay monthly fees for the cable service. CATV began about 1950. *See also* CABLE TELEVISION

commutative law Property of an operation and a set of elements in mathematics. Two elements a and b are said to commute under the operation $*$ if $a * b = b * a$: that is if the result of the operation is the same regardless of the order of the elements. Addition and multiplication of numbers is commutative, since $a + b = b + a$ and $ab = ba$. Matrix multiplication and vector cross products are examples of operations that do not obey the commutative law.

commutator Cylinder made up from copper bars, part of a DC ELECTRIC MOTOR or GENERATOR. It makes contact via brushes between the electric terminals and the rotary ARMATURE winding. It also serves to reverse the current in these windings.

compact disc (CD) Disc used for high-quality DIGITAL sound reproduction. It is a plastic disc with a shiny metal layer and a transparent protective plastic coating. The sound signal consists of millions of tiny pits pressed into one side of the metal. When it is played, a narrow laser beam is reflected from the rotating disc's surface. A sensor detects changes in the beam and forms an electrical signal of pulses. This is processed and decoded to form an ANALOG sound signal that can be amplified for reproduction on loudspeakers. *See also* CD-ROM

comparator Measuring device used to inspect a manufactured component for deviation from a specified value, normally by direct matching or by comparing it with a master part against preset tolerances. The decision to accept or reject a part may be done by an operator or the comparator can make the decision automatically.

compass Direction-finding instrument also used to show direction of a MAGNETIC FIELD. It consists of a horizontal magnetic needle pivoted so that its north-seeking end can turn to point towards magnetic north. Adjustments can be made to give true north. The first use of the magnetic compass was once attributed to the ancient Chinese, but this now seems unlikely. It was not until the 12th century that a compass was used in Europe; the "needle" was a piece of LODESTONE. In navigation today, the magnetic compass is often replaced by the motor-driven GYROCOMPASS.

competition In ecology, the relationship that exists between two or more organisms or groups of organisms when a shared resource is limited. For example, there may be direct confrontation over

▲ **compact disc** A compact disc player reads digital information from a compact disc (1) using a focused laser (2). Music or other information is written on the underside of the disc in a spiral track of pits (3), representing a digital code of zeros and ones. The disc spins and the laser, mounted on a swing arm (4), moves as the disc plays. The laser light passes through a semi-silvered mirror (5) and is focused on the disc (6). When the laser hits a flat area it is reflected back via the mirror to a sensor (7) and the information sent to a chip. When the laser hits a pit, it is scattered.

the amount of living space, food supply or the availability of mates. There are two main types: **intraspecific competition** between members of the same SPECIES, and **interspecific competition** between different species. The usual outcome is the classic "survival of the fittest," with at least a temporary reduction in the numbers of the less fit organism, and for this reason competition is a significant force in the process of EVOLUTION.

compiler COMPUTER PROGRAM that translates the symbols of a programming COMPUTER LANGUAGE into instructions readable directly by a computer. Most programs are written in HIGH-LEVEL LANGUAGES, such as C, Pascal or BASIC, which are made up of words and symbols easily comprehended by humans. A compiler takes these programs and renders them into a form that is readable by the circuits of the computer itself.

complementary angles Two angles whose sum is a right-angle (90°). Supplementary angles have a sum of 180°.

complementary colors Any two COLORS that combine to make white light. Any color can be described in terms of the combination of red, green or blue (the "additive primary" colors) that has the same effect on the eyes. Since red, green and blue in the right proportions add to make white, it follows that the complementary color of red is blue-green (**cyan**), of blue is **yellow** (which is produced by red and green combined) and of green is **magenta** (redblue). In painting, combinations of colors get darker, not lighter. It remains true, however, that complementary colors tend to neutralize each other.

complex compound *See* COORDINATION COMPOUND

complex number *See* NUMBER, COMPLEX

composite Material such as concrete, fiberglass, or plywood, made by combining two or more other materials. A composite usually has qualities, such as strength, superior to those of the materials from which it is made.

compound Substance formed by chemical combination of two or more ELEMENTS that cannot be separated by physical means. Compounds are produced by the rearrangement of VALENCE electrons (outer ELECTRONS of an atom) seeking to attain more stable configurations. They usually have properties quite different from those of their constituent elements. IONIC COMPOUNDS have IONIC BONDS – they are collections of oppositely charged IONS. The ions are packed together in a regular arrangement called a CRYSTAL LATTICE. Ionic com-

pounds, such as sodium chloride, are solids at room temperature and have high melting and boiling points, due to the strong electrostatic forces of attraction holding the lattice together. COVALENT BONDING occurs where nonmetal atoms share electrons. Such compounds can be classified as simple molecular structures (such as carbon dioxide) – with low melting and boiling points; or giant molecular structures (such as graphite and diamond). Their properties depend on the arrangement of the atoms in the macromolecule. *See also* FUNDAMENTAL FORCE; MOLECULE

compound eye Type of animal eye that consists of many individual **facets**, common in adult crustaceans and flying insects. Each facet is an individual visual unit called an **ommatidium**. It has a lens and several light-sensitive cells with nerve fibers that converge to form the optic nerve leading to the animal's brain. The ommatidia in the eyes of day-flying insects each produce a separate image to produce a mosaic of the whole visual field. In night-flying insects, each ommatidium "sees" a large part of the visual field so that individual images overlap to produce a bright view lacking in detail. Compound eyes are made up of from 12 to more than 1,000 facets, depending on the species of animal. *See also* SIGHT

compressed air Air kept at pressures much higher than that of the atmosphere. It is prepared by pumping air into a reservoir from a pump or COMPRESSOR, and is used extensively as a source of power for machines such as PNEUMATIC hammers and other moving tools. It is easily transported in metal bottles, and many factories have their own compressors.

compression-ignition engine *See* DIESEL ENGINE; ENGINE

compression of gas Reduction in the volume of gas achieved by applying external pressure. Some gases, including carbon dioxide, can be liquefied by pressure at room temperatures. Other gases must be cooled before they can be liquefied by pressure. The "critical temperature" of a gas is the highest temperature at which it can be liquefied by an increase of pressure.

compressor Machine that delivers air or gas at pressure. It is used for furnace blast systems, ventilation, and refrigeration systems, pneumatic drills, and for inflating vehicle tires. **Reciprocating** and **rotary** compressors are the two basic types.

Compton, Arthur Holly (1892–1962) US physicist. Working in the field of X RAYS, he discovered what is now known as the COMPTON

▲ **computer** Components of a computer include (1) the central processing unit (CPU), (2) RAM, (3) BIOS and ROM chips, (4) the mother board, (5) expansion cards, (6) video card, (7) expansion slots, (8) optical disk drive, (9) floppy disk drive, (10) hard disk, (11) monitor, (12) keyboard, (13) mouse, (14) power supply, and (15) loudspeaker.

monitor containing a CATHODE-RAY TUBE; a keyboard and a mouse; and a printer. Computer programs are usually stored on disks and transferred to the machine's RAM when required. The keyboard and mouse are called **input devices**, since they allow the user to feed information into the computer. The **keyboard** enables the user to enter letters, numbers, and other symbols. The **mouse** is a small device moved by hand that enables the user to control the computer by positioning a pointer on the monitor screen, to select functions from a list. A magnetic disk drive, such as a HARD DISK, acts as both an input and output device. It can supply programs and data to the computer, and store its output. Many computers have CD-ROM drives. Many other **peripherals** are used, such as a scanner that converts images into an electronic signal so that they can be stored and displayed by the computer, or other HARDWARE for storing and manipulating sounds. The modern computer market is dominated by PCs – the generic term used to refer to machines based on the original IBM personal computer produced in the early 1980s. Most of these machines use an operating system, such as DOS (DISK OPERATING SYSTEM) or Windows, produced by the giant SOFTWARE corporation, Microsoft. Other popular operating systems include Apple Macintosh (MacOS), Linux, and UNIX.

EFFECT: wavelengths of X rays increase when the rays collide with electrons. This helped to prove that X rays could act as particles. He was awarded the 1927 Nobel Prize for physics jointly with the British physicist Charles WILSON. As head of the early phase of the MANHATTAN PROJECT to develop the atom bomb, Compton helped to create the first sustained nuclear CHAIN REACTION.

Compton effect Increase in the wavelength of X ray or gamma ray PHOTONS as they collide with electrons. The effect was discovered (1923) by Arthur COMPTON and is evidence for the dual nature (wave and particle) of energy.

Compton, Karl Taylor (1887–1954) US physicist. An outstanding researcher (particularly in the fields of CRYSTALLOGRAPHY, IONIZATION, and photoelectricity), he was a professor at Princeton University (1915–30) and president of the Massachusetts Institute of Technology (1930–48).

computer Device that processes data (information) by following a set of instructions called a COMPUTER PROGRAM. Read Only Memory (ROM) and Random Access Memory (RAM) chips act as permanent and temporary electronic memories for storing data. A typical desktop computer system consists of a main unit containing a central processor together with memory chips and storage devices (usually MAGNETIC DISKS); a

computer-aided design (CAD) Use of COMPUTER GRAPHICS to assist the design of, for example, fabrics, integrated circuits, buildings and vehicles. With CAD, designers can make alterations and analyze the effect of such changes. In fabric design, colors may be varied and the effect judged on the computer screen. In vehicle design, road conditions may be simulated and tested on the design before a prototype is actually built. In the design of integrated circuit boards and microprocessors, CAD programs can automatically find the most efficient way to route the highly complex interconnections between billions of components.

computer-aided instruction (CAI) Use of computers to assist teaching. It applies to education facilitated by computers, not to education in the use of computers. The area is in rapid growth, as technology advances. Two important developments have been CD-ROMs, enabling instruction to be accompanied by audio and video material, and the Internet, allowing students to access information from worldwide sources.

computer-aided manufacture (CAM) Use of computers to control industrial production. Its main applications involve the control of robots and automated machine tools in factories to achieve rapid and consistent manufacturing without the possibility of human error. Often CAM is linked to COMPUTER-AIDED DESIGN (CAD).

computer-aided software engineering (CASE) Use of computers to automate the writing of COMPUTER PROGRAMS.

computer graphics Illustrations produced on a COMPUTER. Simple diagrams and shapes may be produced by typing on the keyboard. Complex images require a mouse or an equivalent input device such as a graphics pad, painting or drawing SOFTWARE, and sometimes special graphics HARDWARE.

computer interface Way that a COMPUTER PROGRAM or system interacts with its user. The simplest form of computer interface is the simple screen and keyboard, where the user controls the computer by typing in commands. The most common type for personal computers is the GRAPHICAL USER INTERFACE (GUI).

computerized axial tomography (CAT) Method of taking X rays that provides images of "slices" through the body. Inside a CAT scanner is an X-ray source, which produces a narrow beam of radiation. This passes through a patient's body and is detected by an electronic sensor. The X-ray source and detector are rotated around the patient's body so that views are taken from all angles. A computer analyzes the output of the sensor from each of these angles, and uses the information to build up a picture of the slice of the body, showing whether it is damaged or diseased.

computer language System of words and rules used to program a COMPUTER. Most computers work using the BINARY SYSTEM language (using 1s and 0s) called MACHINE CODE. Rather than using a machine code, a language consisting of words and symbols that relate more directly to normal language can be used to instruct a computer. A COMPILER, assembler or other such program then translates this into machine code. Several kinds of programming language have been designed for different purposes. Fortran is for scientific and mathematical use, COBOL for business programs, ALGOL for mathematical applications, and BASIC and Pascal were originally for use by learners. Today, the majority of applications for personal computers are written in a language called C, or derivatives of it. *See also* ASSEMBLY LANGUAGE; COMPUTER PROGRAMMING

computer memory Part of a COMPUTER that stores information in "words" or their "BITS," each of which has an identification number (address)

assigned to it for immediate use by the CENTRAL PROCESSING UNIT (CPU). It may consist of magnetic cores, tapes, drums or disks. Developments in INTEGRATED CIRCUITS have revolutionized computer memory technology, permitting greater retrieval speeds and miniaturization.

computer network Number of computers linked together and able to communicate with each other. A typical local area network (LAN) links computers within the same building, enabling staff to exchange data and share printers. Connections are usually through electric cables that link all the machines. A wide area network (WAN) covers longer distances and may link LANs. The inter-

▲ **computer** Pressing a key (1) or pair of keys changes the current flowing through the key's circuit. A microprocessor (2) scans the circuits and detects when they change. A scan code is transmitted by the microprocessor to the memory buffer in the keyboard (3). The scan code then travels through the cable connecting the keyboard to its controller chip (4) in the body of the computer. The controller chip informs the central processing unit (CPU) (5), which finds the keyboard program in read-only memory (ROM) (6) and cancels the scan code in the keyboard's memory buffer. ROM converts the scan code into the PC's language, ASCII (7), and then instructs the monitor (8) to display the character, in this instance an uppercase E.

connections are made through public TELEPHONE services via electronic units called MODEMS, or through the INTEGRATED SERVICES DIGITAL NETWORK (ISDN), a dedicated high-speed line that carries digital signals. *See also* INTERNET

computer program (SOFTWARE) Set of step-by-step instructions that enables a COMPUTER to carry out a task. A typical computer can carry out many tasks including word processing, calculating, drawing, and communicating, as well as providing games and other forms of entertainment. Programs can be written in a variety of COMPUTER LANGUAGES. A simple program for carrying out a specific calculation might involve only a few lines of code and take five minutes to write. But a team of programmers might take a year, or more, to write and test the vast amounts of code needed for a complex program. It is usually stored on a magnetic DISK. To make a computer perform a particular task, a program is loaded into the computer's RAM.

computer programming Preparation of a COMPUTER so that it can perform a specific task. Before being given data, a computer must be given a set of instructions, or a COMPUTER PROGRAM, telling it how to process the data. Each instruction is a single step and all information must be in the BINARY SYSTEM form. For computer programmers, languages have been developed that make the task of programming increasingly intuitive.

computer virus Sequence of COMPUTER PROGRAM code that is able to copy itself from one computer to another and is usually designed to disrupt the normal operation of a computer. Some viruses attached to widely circulated free programs find their way into computers all over the world. A virus may remain undetected for months and then suddenly go into action. A "Friday the thirteenth" virus, for example, waits until this "unlucky" date before causing chaos. The most destructive viruses cause loss or alteration of data held on the computer. A less destructive virus may simply flash a message on the screen from time to time.

concentration (formerly molarity) Measure of the concentration of a SOLUTION, expressed as the number of MOLES. (It is different from **molality**, which is the number of moles in a liter of solvent.) For example, a solution containing 0.2 moles per liter of solution has a concentration of 0.2 molar (written 0.2M).

concentric circles Two or more circles lying in the same plane and having the same point as their center.

conception Start of pregnancy. The term is used to describe the fertilization of an OVUM (egg cell)

by the SPERM, or the implantation of the fertilized ovum in the wall of the UTERUS.

concrete Hard, strong building material made by mixing Portland CEMENT, sand, gravel, and water. Concrete is an important building material for general, large-scale construction and for making smaller items including slabs and pipes. It can be strengthened by steel rods, which are embedded in it to produce **reinforced** concrete. **Prestressed** concrete contains tensioned wires instead of steel. Its modern use dates from the early 19th century, although the Romans made extensive use of concrete.

condensation Formation of a liquid from a gas or vapor, caused by cooling or an increase in pressure. More particularly, it is the changing of water vapor in the air into water droplets, forming mist, cloud, rain or drops on cold surfaces.

condensation nucleus Small liquid or solid particle, such as dust, on which water vapor in the atmosphere begins to condense in tiny droplets or ice crystals, resulting in CLOUD formation.

condensation reaction Type of CHEMICAL REACTION in which two simple molecules combine to form a third more complex molecule, with the elimination of a simple substance. The molecules of the substance produced are larger than those of the combining molecules, while the eliminated substance (typically ammonia or water) consists of very small molecules. Condensation reactions occur mainly in organic CHEMISTRY. An example is the reaction between a KETONE and hydroxylamine to produce an OXIME and water. A similar process involving MONOMERS can be used to produce a POLYMER in a type of reaction called condensation POLYMERIZATION. *See also* ADDITION REACTION

condenser *See* CAPACITOR

condenser, steam Device for changing steam to water by cooling. Condensers are used in steam power plants where the steam from a TURBINE or steam engine must be condensed to water, which may then be pumped back for reuse in the boiler.

conditioning In experimental psychology, learning in which human or animal subjects learn to respond in a certain way to a stimulus. Most of the procedures and terminology of **classical (or Pavlovian) conditioning** stem from the work of Ivan PAVLOV. A response elicited by a provided stimulus is called a **conditioned reflex**. **Operant (or instrumental) conditioning** was first described by B. F. SKINNER in the 1930s. It involved punishing or rewarding certain actions to positively or negatively reinforce behavior. It can give rise to new behavioral patterns, unlike classical conditioning which can modify only the existing behavior of animals. Operant conditioning occurs naturally, and many emotional responses are learned on this principle.

Condon, Edward Uhler (1902–74) US physicist. A former director of the National Bureau of Standards, he also applied QUANTUM MECHANICS to investigations of the atom and its nucleus. He wrote a controversial report on UFOs.

conductance (Symbol G) Ability of a material to conduct electricity. In a direct current (DC) circuit, it is the reciprocal of electrical RESISTANCE. For example, a conductor of resistance R has a conductance of 1/R. In an alternating current (AC) circuit, it is the resistance divided by the square of IMPEDANCE (the opposition of a circuit to the passage of a current); $G = R/Z^2$. SI units of conductance are SIEMENS (symbol S). *See also* ELECTRIC CURRENT

conduction Thermal conduction is the transfer of heat from a hot region of a body to a cold region. If one end of a metal rod is placed in a flame, the heat energy received causes increased vibratory motion of the molecules in that end. These molecules bump into others farther along the rod, and the increased motion is passed along until finally the end not in the flame becomes hot. **Electrical** conduction is the progress of charged particles through a substance resulting in an electric charge. In metals it is the flow of free ELECTRONS. In gases it is the flow of IONS. *See also* SEMICONDUCTOR; SUPERCONDUCTIVITY

conductivity Measure of the ease with which a material allows electricity (**electrical** conduction) or heat (**thermal** conduction) to pass through it. For a solid substance, the electrical conductivity is the CONDUCTANCE (ability to conduct electricity) between the opposite faces of a unit cube at a specific temperature. It is the reciprocal of RESISTIVITY (a material's ability to oppose the flow of an ELECTRIC CURRENT) and is expressed in SIEMENS per meter. For an ELECTROLYTE, conductivity is the ratio of the current density to the field strength. Thermal conductivity of a metal rod is determined by $H = -k(A/l)(T_2 - T_1)$, where H equals the rate of conductivity, k is constant, A is the cross-section of the rod, l its length, and $T_2 - T_1$, the difference in temperature.

conductor Substance or object that allows easy passage of free ELECTRONS, thereby allowing heat energy or charged particles to flow easily. Conduc-

tors have a low electrical RESISTANCE. Metals, the best conductors, have free electrons that become an ELECTRIC CURRENT when made to move. The resistance of a metallic conductor increases with temperature because the lattice vibrations of atoms increase and scatter the free conduction electrons.

condyle Ball-shaped end of a bone that fits into a socket in another bone to form a movable JOINT. In many mammals the joints between the ends of the mandible (lower jaw) and the skull have condyles, permitting the up-and-down and side-to-side movements in chewing.

cone In botany, conical, spheroidal or cylindrical fruit- or seed-bearing structure borne by CONIFER trees and comprising clusters of stiff, overlapping, woody scales that separate to release seeds from naked ovules developed at their base.

cone In geometry, a figure swept out by a line (the **generator**) that joins a point moving in a closed curve in a plane, to a fixed point (the **vertex**) outside the plane. In a right circular cone the vertex lies above the center of a circle (the **base**), and the generator joins the vertex to points on the circle. Such a

▲ **cone** An important family of mathematical curves results from sectioning a cone at various angles. A horizontal section (1) gives a circle; an inclined one (2) gives an ellipse. A section parallel to one side of the cone (3) gives a parabola and still greater inclination (4) a hyperbola. All these have the same general equation $ax^2 + by^2 +$ $2hxy + 2gx + 2fy = c$. With h^2 greater than ab, it is a hyperbola; with $h^2 = ab$, a parabola; h^2 less than ab gives an ellipse of which the circle (h = 0, a = b) is a special case. The terms a, b, c, h, g, and f are chosen constants; with b = c = h = g = 0, a = 1 and 2f = 1, the equation becomes $y = -x^2$. As $h^2 = ab$ (both = 0) this is the equation of a parabola.

cone has a volume $\frac{1}{3}\pi r^2 h$ and a curved surface area πrs, where h is the vertical height, s the slant height, and r the radius of the base.

configuration, electron Arrangement of ELECTRONS around an atom. It is usually written in a notation using 1, 2, 3, ... for the principal shells and s, p, d, and f for the subshells. The number of electrons is written as a superscript. The electron configuration of the helium atom is $1s^2$; that of the sodium atom is $1s^2 2s^2 2p^6 3s^1$.

congenital disorder Abnormal condition that is present from birth. It may be due to faulty development, infection, or the effects of the mother's exposure to some drugs or other toxic substances during pregnancy. Important examples of congenital disorders include spina bifida, with incomplete development of the spine, and congenital herpes, acquired from the mother during the passage through the birth canal.

conglomerate In geology, a SEDIMENTARY ROCK made up of rounded pebbles embedded in a fine matrix of sand or silt, commonly formed along beaches or on river beds.

congruence Equivalence of shape and size. Two congruent geometric figures will coincide exactly when superimposed. In contrast, two figures are said to be **similar** if the figures cannot be superimposed without one being scaled or reflected.

conic (conic section) Curve formed by the intersection of a plane with a CONE. CIRCLES, ELLIPSES, PARABOLAS and HYPERBOLAS are conic sections. Alternatively a conic is the locus of a point that moves so that the ratio of its distance from a fixed point (the **focus**) to its distance to a fixed line (the **directrix**) is constant. This ratio is called the eccentricity (e): $e = 1$ gives a parabola, $e > 1$ a hyperbola, $e < 1$ an ellipse, and $e = 0$ a circle.

conifer Cone-bearing tree, generally evergreen, such as pines, firs, and redwoods. Some are the Earth's largest plants, reaching heights of up to 99m (325ft). They are major natural resource of the Northern Hemisphere. *See also* GYMNOSPERM

conjugation SEXUAL REPRODUCTION by FUSION of GAMETES. It is characteristic of certain simple animals, lower plants, and bacteria. In some algae, for example, a temporary conjugation tube forms a passageway for the contents of one cell to enter another.

conjunction In astronomy, celestial configuration characterized by a coincidence in the longitudes of two celestial objects, usually two planets

or a planet and the Sun, as viewed from the Earth; this is called **planetary** conjunction. **Inferior** conjunction occurs when either Mercury or Venus lies between the Earth and the Sun; **superior** conjunction occurs when they lie on the opposite side of the Sun to the Earth.

conjunction Logical proposition produced by joining two simple propositions by "and." An example is the proposition "Bill is kind and Bill is honest"; it is true only if both parts are separately true. The conjunction of two simple propositions, P and Q, is written PQ (or, equivalently $P \wedge Q$), read " P and Q ." *See also* DISJUNCTION

connective tissue Type of animal tissue with a supporting and packing function; it helps to maintain a body's shape and hold it together. Connective tissue comprises a small number of cells (such as FIBROBLASTS) and protein fibers within a noncellular substance known as the **extracellular matrix**. BONE, LIGAMENT, CARTILAGE, SKIN and ADIPOSE TISSUE are all types of connective tissue.

conservation Term that has come to mean a number of different, if associated, things in the preservation of nature and natural resources. Conservation requires planning and organization to make the best use of resources or to preserve the natural landscape and wildlife. It may involve a combination of: 1) placing restrictions on the use of materials or areas; 2) using alternative materials; 3) providing areas of protection; and 4) applying methods that limit or control the amount of change or damage to the natural environment. The term is also used to describe the preservation, and sometimes renovation, of ancient and historic structures.

conservation, laws of Physical laws stating that some property of a closed system is unaltered by changes in the system. The most important are the laws of **conservation of matter** and of **energy**. The former states that matter can be neither created nor destroyed; the total mass remains constant when chemical changes occur. The total energy of a system also remains the same; energy is converted from one form into another. Both these laws are only approximately true. Mass and energy are interconvertible according to the equation $E = mc^2$. What is conserved is the total mass and its equivalent in energy. A further law of conservation involves charge, as electric charge can be neither created nor destroyed.

conservation law, nuclear In a nuclear interaction, the total charge, SPIN, or other specific QUANTUM NUMBERS of the interacting particles must equal that of the resulting particles. In

STRONG NUCLEAR FORCE interactions all quantum numbers are conserved. In WEAK NUCLEAR FORCE interactions several of the laws break down, notably that for PARITY.

constant In mathematics and science, a quantity or factor that does not change. It may be universal, such as π: the ratio of the circumference of a circle to its diameter; it may apply only in a particular circumstance, as a symbol that has a fixed value in an algebraic equation; or it may be a defined physical value of a substance, such as the BOLTZMANN CONSTANT for gases or the speed of light (c) in a vacuum.

constantan High-melting point alloy containing 45% nickel and 55% copper. It is used in THERMO-COUPLES and in conjunction with iron and copper to form BIMETALLIC STRIPS – the sensing element in some THERMOSTATS.

constellation Grouping of stars that forms an imaginary figure in the sky viewed from Earth's surface. The stars in a constellation lie at very different distances from the Earth, so the groupings have no physical significance. There are 88 constellations that have been assigned boundaries on the CELESTIAL SPHERE by the International Astronomical Union in 1930. Constellations range from the vast Hydra down to the tiny but brilliant Crux. Many of the names are drawn from ancient mythology, while more recent mapping of the Southern Hemisphere skies introduced some modern-sounding names such as the Octant and the Telescope. *See also individual constellations*

construction engineering Branch of CIVIL ENGINEERING responsible for preparing a site, directing the placement of materials, and organizing personnel and equipment.

constructive margin In PLATE TECTONICS, the boundary between two lithospheric plates, which are moving apart as new crust is being formed. Constructive margins are associated with shallow-focus earthquakes, high heat flow, and eruptions and injections of basalt. Characteristically they produce submarine MIDOCEAN RIDGES, but Iceland is an example where a midocean ridge has surfaced above sea level. *See also* LITHOSPHERE

consumer *See* FOOD CHAIN

contact lens Lens worn on the eye to aid or correct defective vision. The earliest contact lenses were made of glass, but since 1938 specialized plastic has generally been used. In the early 1970s "soft" lenses were introduced, providing the wear-

▶ **The law of conservation of matter**
states that matter is neither created nor destroyed in a chemical reaction. This conservation of mass can be shown by a classic experiment in which a candle is burned inside a weighed bell jar (A). At the end of the experiment, the weight of the jar and its contents (B) are the same as at the beginning, although a part of the candle – made up largely of carbon and hydrogen – has "disappeared" as volatile reaction products (water and carbon dioxide). It was only after scientists accepted the principle of conservation of mass in the late 1700s that a quantitative approach to chemistry became possible.

▶ **The law of conservation of energy**
can be explained by considering a ball weighing 1kg released at the top of a 100m high drop. At first the total energy of the ball is its potential energy (PE). As the ball begins to descend, it progressively loses potential energy and gains kinetic energy (KE) to keep its total energy constant. Thus energy is conserved. A) Kinetic energy increases from zero to a maximum B) Potential energy decreases from the maximum value to zero C) Total energy, which is the sum of the kinetic and potential energies, remains constant.

er with greater comfort. Lenses cover only the CORNEA and float on a layer of tears.

contact metamorphism Recrystallization of rocks surrounding an igneous INTRUSION in response to the heat supplied by the intrusion. There is no significant increase in pressure and the affected rocks do not melt in the process.

contactor Device for the repeated switching on and off of ELECTRICAL CIRCUITS. Often, contactors contain a magnetic coil that diverts and extinguishes the electrical arc formed when a circuit is opened.

continent Any one of seven (or six) large land masses on the Earth's surface. The continents are Europe and Asia (or Eurasia), Africa, North America, South America, Australia, and Antarctica. They are concentrated in the Northern Hemisphere. They cover about 30% of the Earth above sea level and extend below sea level, forming CONTINENTAL SHELVES. All continents have four

180 million years ago

135 million years ago

present-day

▲ **continental drift**
About 200 million years ago, the original Pangaea land mass began to split

into two continental groups, which further separated over time to produce the present-day configuration.

components which make up the continental crust: 1. **Shields** – areas of relatively level land rising no more than a few hundred meters above sea level and consisting of crystalline rocks that are generally very old, up to 3.8 billion years; 2. **Stable platforms** – areas of the continent that have a thin covering of SEDIMENTARY ROCKS, which are generally horizontal; 3. **Sedimentary basins** – broad, deep depressions filled with sedimentary rocks formed in shallow seas that sometimes covered parts of the ancient shields or their margins; 4. **Folded mountain belts** of younger sedimentary rocks, which typically occur along their margins and consist of long, linear zones of intensely folded and faulted rocks that have been metamorphosed and intruded by igneous and volcanic activity. The continental crust is composed of rocks that are less dense than the basaltic rocks in ocean basins and they appear to be riding on the asthenosphere of the underlying MANTLE, moving over the surface of the Earth very slowly by CONTINENTAL DRIFT. Its thickness is mainly between 30–40km (20–25mi) except under large mountain chains where thickness can be 70km (45mi). Continental crust is thought to originate from the SUBDUCTION of early crust, causing partial melting that released lighter material that punched up through the crust. This material was too light to be subducted and the continents grew by the collision and fusion of these microcontinents.

continental climate Type of CLIMATE found in the interior of large continents, that is, hot summers with convectional rainfall and very cold, dry winters, with only light falls of snow. Summer temperatures are about 20°C (68°F), and winter temperatures are −10°C to −20°C (14°F to −4°F) in the coldest months. Total annual precipitation is about 500mm (20in). The natural vegetation consists of grassland. Areas of continental climate are found on the North American prairies, in Poland and Hungary, and on the steppes of Russia. In the Southern Hemisphere there are no areas of real continental climate because the land masses are relatively narrow, so oceanic influences are present throughout. *See also* MARITIME CLIMATE; MEDITERRANEAN CLIMATE

continental drift Theory that the CONTINENTS change position very slowly, moving over the Earth's surface at a rate of a few centimeters per year, adding up to thousands of kilometers over geological time. Until the mid-20th century most scientists believed that the continents were in a fixed position. Early supporters of continental drift claimed that the jigsaw shapes of the present day continents could be pieced together to form an ancient landmass that at some time in the past split and drifted apart. Evidence for the theory included matching the outlines of continents, rock types, geological structures, and fossils. Fossil magnetism in rocks is used to calculate the ancient latitude of continents over time. Continental drift became accepted with the development of PLATE TECTONICS in the 1960s. In recent years continental movement has been confirmed by direct measurements made by global positioning satellites using laser beams. *See also* PANGAEA

continental margin Region of the ocean floor that lies between the shoreline and the deep-ocean floor. It includes the CONTINENTAL RISE, the CONTINENTAL SHELF, and the CONTINENTAL SLOPE.

continental rise Gently sloping region of the CONTINENTAL MARGIN at the foot of the CONTINENTAL SLOPE. It is an area of thick deposits of sediments carried down by currents off the CONTINENTAL SHELF.

continental shelf Gently seaward-sloping part of the CONTINENTAL MARGIN between the shoreline and the CONTINENTAL SLOPE, at a depth of about 150m (500ft). A continental shelf can be quite narrow, as, for example, off the w coast of South America, but in places can be more than 150km (90mi) wide – for example, in the North Sea and around Britain. The shallow waters contain rich food for fish, and therefore some of the world's major fishing grounds are on continental shelves. Other areas of continental shelf, such as the North Sea and the Gulf of Mexico, have been exploited for oil and natural gas.

continental slope Relatively steep slope in the seabed that lies between the CONTINENTAL SHELF and the CONTINENTAL RISE leading into the areas of much deeper water. In many places the continental slope is cut into by deep submarine canyons.

continuous spectrum Unbroken sequence of colors, merging one into the other, produced when light is decomposed by refraction through a prism. An incandescent solid, liquid, or dense gas emits a continuous spectrum. The spectrum of a star consists of a continuous spectrum crossed by **absorption lines**, which are produced when particular wavelengths of radiation from a hot source are absorbed by a cooler, intervening medium.

continuum *See* SPACE-TIME

contour In cartography, a line on a map joining places of equal elevation. Closely spaced contours indicate a steep slope, few or no contours mean flat or almost flat ground.

contraception (birth control) Use of devices or techniques to prevent PREGNANCY. **Oral con-**

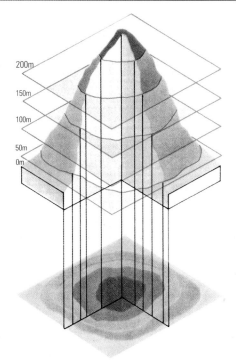

▲ **contour** Recording a three-dimensional shape on a flat surface is a problem that can be solved by the convention of contour scaling. This is shown when the cross-sections of a hill at 50-, 100-,150 and 200-m heights are projected onto a map of the hill. The hill can be envisaged fairly well from such a map, although the "coarseness" of the contour intervals loses some finer detail. The steepness of the sides can be judged by the closeness of the contour lines on the map.

traceptives include the female contraceptive pill, a hormone preparation containing synthesized ESTROGEN and PROGESTERONE that prevents ovulation – the release of an OVUM (egg) – and thickens the cervical mucus, thereby reducing the chances of fertilization should ovulation take place. The convenience and efficacy of the various versions of the "pill" make it the most reliable form of contraception available, although it has been associated with some serious and some minor side effects and should be prescribed on an individual basis. Development of a male pill, containing TESTOSTERONE and progestin, is underway. Although initial tests have shown the male pill to be effective in reducing sperm count, it will not be available for some years. **Barrier contraceptives** include the male and female condom and the diaphragm. The male condom is a latex sheath that covers the penis and collects the ejaculated semen; the female condom lines the inside of the vagina, preventing any sperm entering the UTERUS. The use of condoms is widely advocated because they help protect against some sexually transmitted diseases, including ACQUIRED IMMUNE DEFICIENCY SYNDROME (AIDS). Devices such as diaphragms or caps cover the CERVIX, thus preventing sperm entering the uterus. The intrauterine device (IUD) is a small spring made from plastic or metal inserted into the uterus. It stops the fertilized egg embedding itself in the uterine lining. Emergency contraception, known as the morning-after pill, can be taken up to 72 hours after unprotected sexual intercourse; it prevents the fertilized ovum from embedding itself in the uterus. It is not suitable for regular use. Techniques to prevent pregnancy include the less effective "rhythm method." This involves the avoidance of intercourse on days when conception is most likely (when the woman is ovulating). It is not a reliable method because ovulation cannot always be predicted accurately. Another technique is *coitus interruptus* – the withdrawing of the penis from the vagina before ejaculation. This is also very unreliable as small amounts of semen are often released from the penis before ejaculation. *See also* SEXUAL REPRODUCTION

contractile root Type of ADVENTITIOUS ROOT that develops from the base of a BULB or CORM. New bulbs or corms form higher in the soil at the tops of the contractile roots, which then shorten to pull the bulbs or corms down to a better level.

contractile vacuole *See* VACUOLE

control column Main steering component in an AIRCRAFT that controls roll and pitch. In both airplanes (joystick) and helicopters (collective pitch stick) it combines with foot controls that change direction above the vertical axis. Vertical movement in helicopters is controlled by the cyclic pitch stick.

control systems Means by which a process is made to conform to prescribed instructions, either by maintaining the values of certain parameters at a constant level or by making them change according to a predetermined plan. Control systems may be mechanical, electromechanical, electronic, fluidic (operated by liquid or gas pressure), or a combination of any of these means. All systems depend on either feed-forward (such as a cutting tool that follows the shape of a model) or on FEEDBACK (for example a governor that reduces the input of fuel to an engine when the power exceeds a certain level). Many of the more complex systems used in industry are computer-controlled. *See also* ELECTRONIC CONTROL SYSTEM

convection Transfer of heat by the flow of currents within fluids due to KINETIC THEORY. Warm fluids have a natural tendency to rise (because they are less dense), whereas cooler fluids tend to fall. This movement subsides when all areas of the fluid are at the same temperature. Convection in the ATMOSPHERE is the main method of heat transfer from one part of the Earth to another. Liquid convection is used in a car's cooling system and some domestic central heating systems.

convection cell Organized circular flow of FLUID, such as air or water, based on thermal changes in density and gravitational attraction, rising away from the heat source and sinking in cooler outer areas.

convection current In geology, heat generated from radioactivity deep within the Earth's MANTLE causing rock to flow towards the CRUST. At the top of the mantle the rising rock is deflected laterally below the crust before sinking. This mantle convection is thought to be the process driving PLATE TECTONICS. In meteorology, convection is the process by which air that is warmed close to the ground rises because it is less dense than the surrounding air. This can lead to condensation, cloud formation, and rain, which is called **convectional rainfall**. It is frequently associated with cumulonimbus clouds and thunderstorms.

convergence In mathematics, property of an infinite series (or sequence) having a unique and finite limiting value. Thus, for the series
$$1 + \tfrac{1}{2} + \tfrac{1}{2}^2 + \tfrac{1}{2}^3 + \dots$$
the sum of the first two terms is 1.5, the first three 1.75, and the first four 1.875; as more and more terms are evaluated, the sum approaches the limiting value of 2. Such a series is said to converge. *See also* DIVERGENCE

convergent evolution Tendency of several different species to resemble each other and to develop similar characteristics in their attempt to adapt to similar environments. The most commonly cited example is the wings of bats and birds. *See also* EVOLUTION

convergent margin In PLATE TECTONICS, the boundary between two lithospheric plates that are moving towards each other. It can form either a SUBDUCTION margin or a collision zone. *See also* LITHOSPHERE

convulsion Series of violent, involuntary contractions of the muscles, sometimes accompanied by loss of consciousness. Such a seizure may indicate EPILEPSY. However, isolated seizures can arise from many other causes, including intoxication, brain abscess, or an abnormally low blood-sugar level (**hypoglycemia**). Young children may suffer febrile seizures, brought on by high fever.

Cooke, Sir William Fothergill (1806–79) British electrical engineer. He collaborated with the physicist Charles WHEATSTONE to develop the first commercial electric telegraph in 1839. In 1845 they patented a single-needle apparatus.

coolant Liquid or gas used to transfer heat from a hot to a cool region or to remove excess heat. For example, in NUCLEAR REACTORS heat is transferred from the core to the generator by liquid sodium, pressurized water or a gas.

Coolidge, William David (1873–1975) US physicist. Coolidge developed (1908) a method for drawing tungsten into filaments for making LIGHT BULBS and radio valves. In 1916 he patented an X-ray tube (**Coolidge tube**) capable of producing accurate amounts of radiation. Coolidge also devised portable X-ray units and worked on construction techniques for industrial quality control. During World War II he devised a submarine-detection system with Irving LANGMUIR. Coolidge also worked on atom bomb research. *See also* ELECTRON TUBE

cooling tower Tall construction that uses atmospheric air to cool warm water in refrigeration and steam-power generation systems. There are three types of towers: **atmospheric** cooling towers, which depend on wind currents; **natural draft** towers, which depend on natural convection currents; and **mechanical draft** towers, which use fans to provide forced or induced convection.

Cooper, Leon Neil (1930–) US physicist. From 1974 he was a professor at Brown University, in Providence, Rhode Island. With John BARDEEN and John Robert SCHRIEFFER, he shared the 1972 Nobel Prize for physics for research in SUPERCONDUCTIVITY. They demonstrated that electrons in a superconductor arrange themselves in so-called "Cooper pairs" rather than randomly dispersing energy and momentum.

coorbital satellite Satellite orbiting a planet at the same average distance as another satellite. Although the distances – and hence periods – are the same, the orbital inclinations and eccentricities may differ slightly.

coordinate geometry (analytical geometry) Branch of mathematics combining the methods of pure GEOMETRY with those of ALGEBRA. Any geometrical point can be given an algebraical value by relating it to coordinates, marked off from a frame of reference. Thus, if a point is marked on a square grid so that it is x_1 squares along the x axis and y_1 squares along the y axis, it has the coordinates (x_1, y_1). POLAR COORDINATES can also be used. It was introduced in the 17th century by René DESCARTES. *See also* CARTESIAN COORDINATE SYSTEM; CONIC

coordinate system Reference system used to locate a point in space. A point can be defined by numbers representing distances or angles measured from lines or points of reference. In a CARTESIAN COORDINATE SYSTEM a point is defined by distances from intersecting axes. In a POLAR COORDINATE system, a point is located by a distance from a fixed point (the origin) and an angle from a reference line. Two numbers are required to define a position in a plane: three numbers are required in three-dimensional space, and so on. A coordinate system enables curves or surfaces to be defined algebraically.

coordination compound (complex compound) Type of chemical compound in which one or more groups of molecules each form a **coordinate bond** (two atoms sharing two electrons from one of the atoms) with a central metal atom, usually a TRANSITION ELEMENT. The complex may be either a complex ion or a neutral molecule, as in nickel carbonyl ($Ni(CO)_4$). The coordinating species are known as LIGANDS. Some coordination compounds, such as heme and CHLOROPHYLL, have biochemical importance. *See also* CHELATION; ION

coordination number Number of chemical bonds made to a central atom in a specified complex salt, such as the number of ANIONS surrounding a CATION.

copepod Any marine or freshwater CRUSTACEAN of the subclass Copepoda. Copepods are possibly the most numerous type of animal in the world and are a major component of the marine FOOD CHAIN. Some are parasitic on aquatic animals, especially fish. Their segmented, cylindrical bodies have a single median eye and no carapace. Length: 0.5–2mm (0.02–0.08in); length of parasitic forms may be more than 30.5cm (1ft). There are 7,500 species.

Copernicus, Nicolaus (1473–1543) (Mikołaj Kopernik) Polish astronomer. The accepted view in his day was the geocentric (Earth-centered) Universe with its complicated system of orbits, as explained by PTOLEMY nearly 1,500 years before. Through his study of planetary motions, Copernicus developed a heliocentric (Sun-centered) theory of the Universe. In the **Copernican system** (as it is now called) the planets' motions in the sky were explained by their orbits around the Sun. The motion of the sky was simply a result of the Earth turning on its axis. An account of his work, *De revolutionibus orbium coelestium*, was published in 1543. Most astronomers considered the new system as merely a means of calculating planetary positions, and continued to believe in ARISTOTLE's view of the world. *See also* GALILEO; KEPLER, JOHANNES

copper (symbol Cu) Red-pink TRANSITION ELEMENT. Reddish copper occurs native (free or uncombined) and in several ores including cuprite (an oxide) and chalcopyrite (a sulfide). The ores are separated from the surrounding rock and are then concentrated. They are smelted into a molten mass, called **matte**, and pure copper (**blister copper**) is extracted by oxidizing the impurities with air. Ores are often treated with acids and the copper recovered by ELECTROLYSIS. It is malleable, a good thermal and electrical CONDUCTOR, second only to silver, and is used extensively in boilers, pipes, electrical equipment, and alloys, such as brass and bronze. Copper tarnishes in air, oxidizes at high temperatures and is attacked only by oxidizing acids. It forms two series of salts, termed copper(I) (cuprous) and copper(II) (cupric). Properties: at.no. 29; r.a.m. 63.546; r.d. 8.96; m.p. 1.083°C (1,981°F); b.p. 2,567°C (4,653°F); most common isotope ^{63}Cu (69.09%).

copper oxide Chemical compound of copper and oxygen that exists in two forms: copper(I) (cuprous) oxide (Cu_2O), is a brilliant red powder found in nature as the mineral cuprite; and copper(II) (cupric) oxide (CuO), which is black and decomposes into copper(I) oxide and oxygen when heated. It is added to furnace melts to act as an oxidant.

copper plating Application of a coating of COPPER onto another material. Copper plating by ELECTRODEPOSITION is used to prepare steel for a top coating (also by ELECTROPLATING – a form of electrodeposition) of nickel, chromium or silver. Steel and brass are copper plated in the production of decorative oxidized copper finishes. Steel is sometimes copper plated to prevent surface hardening. The main copper salts used in electroplating solutions are copper sulfate and copper cyanide in the form of COORDINATION COMPOUNDS such as potassium cuprocyanide ($K_2Cu(CN)_3$).

copper sulfate Chemical compound that exists in two forms. Copper(I) (cuprous) sulfate (Cu_2SO_4) is a light gray powder that reacts instantly with atmospheric moisture to produce copper(II) (cupric) sulfate. This is usually encountered as the bright blue crystals of the pentahy-

drate, blue vitriol ($CuSO_4.5H_2O$), used for copper plating, for preserving wood, and for killing algae in ponds. It can also be dehydrated to the colorless anhydrous salt ($CuSO_4$), which, because it absorbs water and turns blue, is used as a desiccant and, more widely, as a moisture indicator.

coral Small, coelenterate marine animal of the class Anthozoa in the phylum Cnidaria, often found in colonies. The calcium carbonate skeletons secreted by each animal POLYP accumulate to form a CORAL REEF. Reef-building corals are found only in waters with temperatures in excess of 20°C (68°F).

coral reef Rock-like formation found in shallow tropical seas. Such reefs are formed from gradual build up of the calcium carbonate secreted by living CORAL organisms as protection against predators and wave action. When the individual coral dies, its skeleton remains, and thus the reef expands. The way in which the coral, and therefore the reef, grows is strongly influenced by the depth, prevailing currents, and the temperature of the surrounding sea-water. The main types are **fringing** reefs and BARRIER REEFS.

cordierite Silicate of aluminum, magnesium, and iron, that occurs mostly in METAMORPHIC ROCKS. It forms crystals in the orthorhombic system. Hardness 7, r.d. 2.6.

cordillera Chain of mountains. The term is used particularly of parallel mountain ranges, such as those of the Rockies, Sierra Nevada, and Coastal Ranges in w North America or of the Andes in South America.

cordite Smokeless, high-explosive shell and bullet propellant made by mixing CELLULOSE NITRATE (guncotton) – a highly nitrated form of CELLULOSE – and NITROGLYCERINE, both of which are themselves explosives. For stability, mineral jelly and acetone are added in small amounts. Cordite is generally manufactured in strands or cords, which give it its name.

core In geology, a cylindrical rock sample that has been gathered by drilling; on land usually by a rotary drill, from the sea often by a metal cylinder, with a cutter on the bottom, which is forced into the rock. The core is used to identify the various layers in the rock or sediment. The mechanism used for collecting the core is called a **corer.**

core Central area of the EARTH from a depth of 2,900km (1,800mi) that accounts for 16% of the Earth's volume and 31% of its mass. Information about the core is obtained from the measurement of SEISMIC WAVES. These indicate that the outer part of the core is liquid, because shear (S) waves will not travel through it, whereas the inner core from 5,150km (3,200mi) to the center of the Earth is interpreted as solid because seismic velocities are lower. It is believed that the change from liquid to solid core occurs because of immense pressure conditions. The core is thought to be composed of iron-nickel alloy (90% iron, 10% nickel). Temperature estimates for the core vary from 4,000 to 7,000°C. CONVECTION in the iron liquid outer core is thought to be responsible for producing the Earth's magnetic field.

Cori, Carl Ferdinand (1896–1984) US biochemist, b. Czechoslovakia. Cori shared the 1947 Nobel Prize for physiology or medicine with his wife **Gerty Theresa Cori** (1896–1957) and Bernardo HOUSSAY for their discovery of how the chemical energy of GLYCOGEN, a carbohydrate stored in the liver and muscle, is broken down into a form that can be used by the body.

Coriolis effect Apparent force on particles or objects due to the rotation of the Earth under them. It causes objects in motion including ocean and atmospheric currents, to be deflected towards the right in the Northern Hemisphere and towards the left in the Southern Hemisphere and therefore impacts greatly on the world's weather systems. The direction of water swirling round in a drain or whirlpool demonstrates this effect. It is named after the French mathematician Gaspard Gustave de Coriolis (1792–1843).

cork Outer dead, waterproof layer of the BARK of woody plants. The bark of the cork oak (*Quercus ruber*) is the chief source of commercial cork.

corm Fleshy underground stem that produces a plant such as the gladiolus or crocus. In most plants, new corms form on top of old ones, which last for one season. They are an important element in forms of ASEXUAL REPRODUCTION.

Cormack, Allan Macleod (1924–) US physicist, b. South Africa. His early work concerned X-ray images of the human body and, beginning in 1963, he worked out a system of using a computer to construct a cross-sectional "picture" of tissues and organs. The scanning technique, called computer-assisted X-ray tomography (CAT), was also developed independently in Britain by Godfrey HOUNSFIELD. Cormack and Hounsfield shared the 1979 Nobel Prize for physiology or medicine.

cornea Transparent membrane at the front of the EYE. It is curved and acts as a fixed LENS, so that light entering the eye is to some extent focused before it reaches the eye lens.

Cornu, Marie Alfred (1841–1902) French physicist who worked in optics and astronomy. He devised a photographic method of measuring light intensities in DIFFRACTION patterns and calculated the velocity of light.

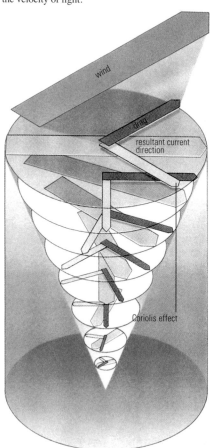

▲ **Coriolis effect** A wind will tend to drag ocean surface water after it. However, as soon as the water starts to move, the Coriolis effect throws it off at an angle. Each successive thin sheet of water beneath the surface layer begins to move because it is linked by friction to the layer above. But each layer is thrown even farther off course by the Coriolis effect. So the surface current moves at an angle of 45° to the wind at about 2–3% of the wind speed. Lower layers move more slowly, at an increasingly eccentric angle. The resulting vertical profile of the motion is the Ekman spiral. At a depth called the Ekman depth, the water flows in the opposite direction to that of the surface current and at 0.043 times the speed of the surface current.

corolla PETALS of a flower. The corolla encloses the STAMENS and CARPELS. The petals, which may be free or fused, together form the inner whorl of the PERIANTH.

corollary In mathematics, a theorem that is incidentally proved in proving some other theorem or proposition. More generally, it is an immediate or natural consequence or a conclusion that is easily arrived at.

corona Outermost layer of the Sun's atmosphere, extending for many millions of kilometers into space. It springs into view as a white halo during a total solar ECLIPSE; at other times it can be observed in visible light only by using a special instrument called a **coronagraph**. The corona emits strongly in the X-ray region, and has been studied by X-ray satellites. Its overall shape changes during the 11-year solar cycle, from regular and symmetrical at solar minimum to uneven with long streamers at solar maximum. The corona has a temperature of 1–2 million K.

Corona Australis (Southern Crown) Small constellation of the s sky. Although none of its stars are above magnitude 4, it is easy to identify by its arc of stars.

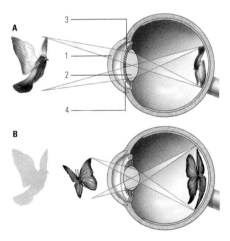

▲ **cornea** Focusing of light rays from distant objects (A) is mainly done by the cornea (1) with a little help from the lens (2). Ciliary muscles (3) encircling the lens relax and stretch ligaments (4), which pull the lens flat.

Rays from a near object (B) are bent by a thick lens produced when the ligaments slacken as the ciliary muscles contract. This process, which is called accommodation, is essential for sharp focusing.

Corona Borealis (Northern Crown) Constellation of the N sky, adjoining Boötes. Its brightest star is Alpha, magnitude 2.2. The "bowl" of the crown contains the famous variable star R Coronae Borealis, the prototype of a group of stars that fade suddenly and unpredictably.

coronary artery disease Disease of the coronary blood vessels, the ARTERIES supplying blood to the HEART muscle. *See also* ARTERIOSCLEROSIS

coronary thrombosis Formation of a blood clot in one or other of the coronary ARTERIES supplying the HEART, preventing blood (and with it oxygen and nutrients) from reaching the heart. It leads to death of part of the heart muscle – a HEART ATTACK. A major attack, marked by severe chest pain, sweating, and sometimes collapse, may be fatal.

corposant *See* ST. ELMO'S FIRE

corpus luteum Mass of yellow tissue formed in the GRAAFIAN FOLLICLE in the OVARY of a mammal after the OVUM (egg cell) is released. If the ovum is fertilized, the corpus luteum secretes PROGESTERONE, a hormone needed to prepare the uterus for pregnancy. If the ovum is not fertilized, the corpus luteum becomes inactive.

corrasion Mechanical EROSION caused by loose material, such as sand or pebbles, during transportation. The material is carried by wind, water or ice, and it scrapes against the bed or sides of the river or glacial valley, or against the cliffs and the shore.

correlation In STATISTICS, a number that summarizes the direction and degree of relationship between two or more dimensions or variables. Correlations range between 0 (no relationship) and 1.00 (a perfect relationship), and may be positive (as one variable increases, so does the other) or negative (as one variable increases, the other decreases).

Correns, Karl Erich (1864–1933) German botanist and geneticist. In 1900 he rediscovered, along with two others, the works of Gregor MENDEL that outlined the principles of HEREDITY. Correns conducted experiments to determine the validity of Mendel's laws and helped to provide evidence that proved his theories.

correspondence In geometry, property of two geometric figures in which angles, lines, and points in one bear a similar relationship to angles, lines, and points in another. In set theory, two sets *A* and *B* are said to be in one-to-one correspondence if every element of *A* can be mapped to an element of *B* by a single function, with no two elements of *A* mapping to the same element in *B*.

corrie *See* CIRQUE

corrosion In chemistry, gradual tarnishing of the surface or major structural decomposition by chemical action on solids, especially metals and alloys. It commonly appears as a greenish deposit on copper and brass, a reddish-brown deposit (RUST) on iron or a gray deposit on aluminum, zinc, and magnesium. Rust is the most important form of corrosion because of the extensive use of iron and steel, and their susceptibility to attack. Some metals, such as aluminum and magnesium, corrode readily, forming an oxide that then protects the undersurface from further corrosion.

corrosion In geology, chemical EROSION leading to the disintegration of rocks, such as by the action of running water or by SOLUTION. It is particularly common in limestone.

cortex In animal and plant anatomy, outer layer of a gland or tissue. Examples are the cortex of the ADRENAL GLANDS; the CEREBRAL CORTEX or outer layer of the BRAIN; and the cortical layers of tissue in plant roots and stems lying between the BARK or EPIDERMIS and the hard wood or conducting tissues.

corticosteroid Any HORMONE produced by the CORTEX (outer layer) of the ADRENAL GLANDS. There are two types of corticosteroids. **Mineralo-corticoids**, such as aldersterone, regulate the balance of fluids and salts in the body. **Glucocorticoids**, such as CORTISONE and hydrocortisone, regulate the use of carbohydrates, fats, and proteins. Some corticosteroids have been synthesized.

cortisone HORMONE produced by the CORTEX of the ADRENAL GLANDS and essential for carbohydrate, protein, and fat metabolism, kidney function, and disease resistance. Synthetic cortisone is used to treat adrenal insufficiency, rheumatoid arthritis, and other inflammatory diseases, rheumatic fever, asthma, severe allergies, and skin complaints. A potent, versatile drug, it can have unwanted side effects such as body swelling.

corundum Translucent to transparent mineral in many hues, aluminum oxide (Al_2O_3). It is found in IGNEOUS and METAMORPHIC ROCKS, occurring as pyramidal or prismatic crystals in the rhombohedral class and as granular masses. It is the hardest natural substance after DIAMOND. Gemstone varieties are sapphire and ruby. It is used in watches and motors and is an important industrial abrasive. Hardness 9; r.d. 4.

Corvus (Crow) Small constellation of the N sky, adjoining Hydra and Crater. Its brightest stars are of the third magnitude.

cosecant In TRIGONOMETRY, ratio of the length of the HYPOTENUSE to the length of the side opposite an acute angle in a right-angled triangle. The cosecant of angle A is usually abbreviated cosec(A) and is equal to the reciprocal of sin(A).

cosine In TRIGONOMETRY, ratio of the length of the side adjacent to an acute angle to the length of the HYPOTENUSE in a right-angled triangle. The cosine of angle A is usually abbreviated cos(A).

cosmic dust Very fine particles of solid matter in any part of the Universe, including meteoric dust and interstellar material that can absorb starlight to form dark NEBULAE in galaxies. Spherical dust particles about 0.05mm (0.002in) in diameter, found in certain marine sediments, are thought to be the remains of some 5,000 metric tons of cosmic dust that fall to Earth each year.

cosmic microwave background Weak ELECTROMAGNETIC RADIATION from the Universe. It was first detected in 1965 by Arno PENZIAS and Robert WILSON. It is considered by many scientists to be the remnant of the early radiation from the hot, early Universe created by the BIG BANG. BACKGROUND RADIATION increased in wavelength as the Universe expanded, and today it peaks in the millimeter region of the ELECTROMAGNETIC SPECTRUM.

cosmic radiation (cosmic rays) Streams of SUBATOMIC PARTICLES from space that constantly bombard the Earth at velocities approaching the speed of light. **Primary** cosmic RADIATION is high-energy radiation that comes from the Sun and other sources in outer space. It consists mainly of atomic nuclei and PROTONS. When primary cosmic rays strike gas molecules in the upper atmosphere they yield showers of **secondary** cosmic radiation, which consists of energetic protons, NEUTRONS, and PIONS. Further collisions yield MUONS, ALPHA PARTICLES, POSITRONS, ELECTRONS, GAMMA RADIATION, and PHOTONS. Cosmic radiation contributes to BACKGROUND RADIATION.

cosmology Branch of scientific study that brings together the disciplines of astronomy, mathematics, and physics in an effort to understand the makeup and evolution of the Universe. Once considered the province of theologians and philosophers, it is now an all-embracing science, which has made great strides in the 20th century. The discovery by Edwin HUBBLE in the 1920s that galaxies are receding from each other promoted the concept of an EXPANDING UNIVERSE, which perhaps started with an explosion – the BIG BANG theory. Associated with this is the OSCILLATING UNIVERSE THEORY. The other main theory is the STEADY-STATE THEORY, which says that matter is continually being created.

Observations of the most remote regions of space, made in the mid-1990s, have shown astronomers what they consider to be a picture of the Universe at only a few hundred thousand years old. Such observations have been generally taken as confirmation of the Big Bang theory.

cosmonaut *See* ASTRONAUT

Cosmos (Gk. order) Universe considered as an ordered whole. PLATO and ARISTOTLE conceived of the Universe as ordered by an intelligent principle. The conviction of an ordered nature became the basis of modern natural science.

cotangent In TRIGONOMETRY, ratio of the length of the side adjacent to an acute angle to the length of the side opposite the angle in a right-angled triangle. The cotangent of angle *A* is usually abbreviated cot(*A*) and is equal to the reciprocal of tangent(*A*).

cotyledon First leaf or pair of leaves produced by the embryo of any ANGIOSPERM (flowering plant). Its function is to store and digest food for the embryo plant, and if it emerges above ground to PHOTOSYNTHESIZE for seedling growth. *See also* DICOTYLEDON; MONOCOTYLEDON

coulomb (symbol C) SI unit of ELECTRIC CHARGE. It is defined as the charge carried by a current of 1 AMPERE in 1 second. It is named after Charles COULOMB.

Coulomb, Charles Augustin de (1736–1806) French physicist. His invention of the TORSION BALANCE led to his experiments in electrostatics and the discovery of the inverse square law that bears his name. *See* COULOMB'S LAW

Coulomb's law One of the basic inverse square laws of physics. It states that the force between two point electric charges is proportional to the product of the charges and inversely proportional to the square of the distance between them. It can be stated mathematically as: F is proportional to Q_1Q_2/d^2, where F is the force, Q_1 and Q_2 are the charges and d is the distance between them. The constant of proportionality is called the PERMITTIVITY of the medium between the charges.

country rock In geology, rock that has been intruded into by a subsequent igneous INTRUSION.

couple Two equal and opposite parallel forces that do not act in the same line. The forces produce a turning effect or TORQUE.

coupling Mechanical fastening connecting shafts together for power transmission. A **flexible** coupling is used to compensate for misalignment of shaft axes and mispositioning of shaft center line; a **rigid** coupling is used for maximum power transfer with minimum misalignment.

Cousteau, Jacques Yves (1910–97) French oceanographer. Best known as the co-inventor (with Emile Gagnan) of the AQUALUNG, Cousteau also invented a process of underwater television and conducted a series of undersea living experiments (Conshelf I–III, 1962–65). Many of the expeditions made by his research ship *Calypso* were filmed by him for television and film. Although immensely popular around the world with the public, academics accused him of showmanship and questioned the validity of his research. In his later life he became a prominent figure in various ecological movements. *See also* SCUBA DIVING

covalent bond Chemical bond in which two atoms share a pair of ELECTRONS, one from each atom. Covalent bonds with one shared pair of electrons are called **single** bonds; **double** and **triple** bonds also exist. The molecules tend to have low melting and boiling points and to be soluble in nonpolar solvents. Covalent bonding is most common in organic compounds.

covellite Mineral form of copper sulfide (CuS). It occurs with other copper minerals and is mined as an ore. It forms hexagonal, platy crystals that have a deep indigo blue color, often tinged with purple. It cleaves into thin, flexible plates. Hardness: 1.5–2.0; r.d. 4.7.

crab nebula NEBULA located about 6,500 light-years away in the constellation Taurus. It is the remnant of a SUPERNOVA noted by Chinese astronomers in July 1054, when it shone as brightly as Venus, visible even in daylight. The nebula was discovered in 1731 by the British astronomer John Bevis and independently by the French astronomer Charles Messier in 1758. It gained its popular name from a sketch by Lord Rosse.

cracking Stage in oil refining during which the products of the first DISTILLATION are treated to break up large hydrocarbons into smaller molecules by the controlled use of heat, CATALYSTS, AND often pressure. The cracking of petroleum yields heavy oils, gasoline, and gases such as ETHANE, ETHYLENE, and PROPYLENE, which are used in the manufacture of plastics, textiles, detergents, and agricultural chemicals. Cracking is therefore a means of yielding greater amounts of the lighter hydrocarbons, which are in much greater demand, from heavier fractions such as lubricating oil.

crane Machine for lifting and placing heavy loads. The large tower cranes seen on building sites have a long jib, or arm, counterbalanced by a weight on a shorter arm. Other heavy cranes, such as those moving on wheels or tracks in foundries, or those in railroad yards, form a bridge over the load. Climbing and crawling cranes are used in the construction of high-rise buildings and large bridges. Small cranes, or jacks, are frequently mounted on towtrucks for roadside repairs.

cranial nerves Twelve pairs of nerves that arise in the BRAIN and serve the muscles, glands, and sensory apparatus of the head and neck as well as organs in the chest and abdomen. Traditionally in anatomy the cranial nerves are identified by Roman numerals in addition to their Latin names.

cranium Dome-shaped part of the SKULL that protects the BRAIN. It is composed of eight bones that are fused together.

Crater (Chalice) Small constellation of the N sky, adjoining Hydra and Corvus. Its brightest stars are of 4th magnitude.

crater Roughly circular depression in the surface of the Earth, the Moon, or some other planets, usually with steep sides. It is formed either by meteoric impact, when shock waves blast out a hole in the ground, or at the vent of a volcano, when lava is expelled explosively. Space probes have revealed such craters on Mars and Mercury, and some of the larger asteroids and satellites of other planets.

crater lake Accumulation of water, usually by precipitation of rain or snow but sometimes ground water, in a volcanic crater (CALDERA). Should an eruption occur, the resulting mud flow (**lahar**) is often more destructive than a lava flow, owing to its greater speed. A noted example is Crater Lake in Crater Lake Park, Oregon. The waters of the lake were formed by precipitation and are maintained solely by rain and snow; there is no outlet.

craton Stable area of the Earth's CRUST that has not been affected by TECTONIC activity during the previous 1,000 million years. Cratons form the SHIELD areas of the Earth's continents.

crazing (crackling) Method of decorative glazing of ceramics, in which fine cracks in the glaze result from different shrinkage rates of the clay and the glaze during firing. It was originated by the Chinese.

creatine AMINO ACID compound found mostly in the muscle tissue of vertebrate animals. Combined with phosphoric acid, it plays an important part in muscle contraction.

creep In geology, slow downward movement of material. This slow movement of soil is not associated with erosion. It may occur on all sorts of slopes including those that are covered with vegetation. The steady downflow of soil and rock carries the vegetation with it. Creep may be caused by the growth of plant roots, the burrowing of small animals or trampling of larger ones, frost movement, or successive drying and wetting that leads to shrinkage and swelling of the soil.

creep In metallurgy, continuing deformation of a metal under stress so that it appears to flow like an extremely viscous liquid. Metal components can be tested for "creep rate" – the percentage elongation per 1,000 hours at a particular temperature. Engineers have to consider both the applied stress and the temperature of metal structures to avoid possible failure through creep.

Cretaceous Last period of the MESOZOIC era, lasting from 144 to c.65 million years ago. Dinosaurs flourished until the end of this period, when they died out in a mass extinction that killed off a vast range of vertebrate and invertebrate life forms. The first true placental and MARSUPIAL mammals and flowering plants appeared. The chalk rocks of NW Europe were deposited during the Upper Cretaceous.

Creutzfeldt-Jakob disease (CJD) Rare, incurable disease of the brain that causes physical and mental deterioration, usually progressing to death within a year of onset. Now known to be caused by an abnormal protein called a PRION, it is one of a group of diseases that also includes SCRAPIE in sheep and BOVINE SPONGIFORM ENCEPHALOPATHY (BSE) – "mad cow disease." Normally CJD attacks older people, but in March 1996 came the first reports of young people in the UK dying of a variant of the disease. Eighteen months later scientists confirmed that the agent giving rise to the new variant (nvCJD) is identical to that of BSE, confirming the link between the human disease and the consumption of infected beef or beef products.

crevasse Deep crack in a GLACIER. It is the result of stress within the glacier or the movement of the glacier over uneven terrain.

Crick, Francis Harry Compton (1916–) British biophysicist. In the 1950s, in association with James WATSON and Maurice WILKINS (1916–), he established the double-helix molecular structure of deoxyribonucleic acid (DNA). The three were jointly awarded the Nobel Prize for physiology or medicine in 1962.

crinoid Any member of the class Crinoidea, a group of primitive ECHINODERMS that includes the sea lily and the feather star. A crinoid's mouth, on a small disc with the other main organs, is surrounded by long feathery arms. It feeds on particles that fall through the water and respires by means of its tube-feet. They first appeared in the Ordovician period (510 million years ago) and fossil crinoids are an important constituent of Palaeozoic limestones.

critical angle Any angle at which a significant transition occurs. In optics, it is the ANGLE OF INCIDENCE with a medium at which total internal REFLECTION occurs. In telecommunications, it is the angle at which a radio wave just misses being reflected by the IONOSPHERE.

critical mass Minimum mass of fissionable (able to undergo FISSION) material required in a fission bomb or NUCLEAR REACTOR to sustain a CHAIN REACTION. The fissionable material of a fission bomb is divided into portions less than the critical mass; when brought together at the moment of detonation they exceed the critical mass and the high-energy chain reaction begins.

critical point Temperature and pressure above which a liquid and its vapor PHASE (a gas) can no longer coexist. If a gas is slowly compressed at temperatures above the critical temperature or cooled at pressures above the critical pressure, it changes gradually from gas to liquid rather than suddenly separating into two phases.

Crookes, Sir William (1832–1919) British chemist and physicist. He invented the RADIOMETER (which measures the intensity of ELECTROMAGNETIC RADIATION) and the CROOKES TUBE, which led to the discovery of the electron by J.J. THOMSON. He was the first to suggest that CATHODE RAYS consist of negatively charged particles. He also discovered the element THALLIUM. *See also* X RAY

Crookes tube Highly evacuated tube developed by Sir William CROOKES for the study of electrical discharges at low pressure. It was used in 1897 by J.J. Thomson to demonstrate the existence of the ELECTRON, and was also used to generate X rays.

cross-bedding In geology, a sedimentary structure comprising a series of inclined parallel or near parallel layers within a bed, found mainly in sandstones but also in some limestones and CONGLOMERATES. Cross-bedding forms where sediment is being moved as it is being deposited, building sand dunes,

sand waves or ripples. It is useful for interpreting ancient environments as the direction and strength of former currents can be determined from the pattern and type of structure.

crossing over During CELL DIVISION, a mechanism by which pairs of HOMOLOGOUS CHROMOSOMES exchange strands of each chromosome, called CHROMATIDS. At the end of the first PROPHASE in MEIOSIS, the diverging chromosomes remain in contact at a number of places, called **chiasma**. Chromatids split and rejoin at each chiasma, with the result that sections of chromatids are exchanged. This alters the distribution of GENES along the chromosomes and so gives rise to genetic variation in the resulting GAMETES, essential in the process of EVOLUTION.

cross-pollination Transfer of POLLEN from the flower of one plant to another. The process may be used by plant breeders to hybridize compatible plant species, thus producing "synthetic" varieties. The result of the process is called a HYBRID.

crucible process Method of making high-quality steel, invented in 1740 by Benjamin Huntsman (1704–76). Steel, charcoal and wrought iron are melted in a fire-clay crucible. Coke fires were once used but were replaced after 1870 by the SIEMENS gas furnace, which could heat 100 crucibles to over 1,700°C (3,100°F). Electric furnaces have now replaced this process.

crude oil *See* PETROLEUM

cruise missile (aerodynamic MISSILE) Self-propelled GUIDED MISSILE that travels, generally at low altitudes, using an advanced guidance system incorporating terrain contour matching (TERCOM). Modern cruise missiles can be extremely accurate and have the advantage of being able to fly low enough to avoid conventional radar defenses. The siting of American cruise missiles on European soil during the 1980s led to large-scale public demonstrations. In the 1991 Gulf War, the US Navy used Tomahawk cruise missiles on ground targets in Iraq.

cruiser Warship smaller, lighter and faster than a BATTLESHIP, ranging in size from 7,500–21,000 metric tons. After World War I, arms limitation treaties restricted its guns to 200mm (8in). Since World War II, cruisers have replaced battleships as the major warships of a modern navy. The USS *Long Beach* (launched 1961) was the world's first nuclear-powered cruiser.

crust In geology, thin outermost solid layer of the EARTH. The crust represents less than 1% of the Earth's volume and varies in thickness from as little as 5km (3mi) beneath the oceans to about 70km (45mi) beneath mountain chains such as the Himalayas. Oceanic crust is generally 7km (4.5mi) thick and basaltic in composition, whereas continental crust is mainly between 30–40km (20–25mi) thick and of granitic composition. The crust is defined by its seismicity (the velocity of EARTHQUAKE waves that pass through it). The lower boundary of the crust is defined by a marked increase in seismic velocity, known as the Mohorovičić discontinuity. *See also* MOHO; SIAL; SIMA

crustacean Any member of the class Crustacea, a class of about 30,000 species of ARTHROPOD. The class includes the DECAPODS (crabs, lobsters, shrimps, and crayfish), isopods (pill millipedes and woodlice) and many varied forms, most of which have no common names. Most crustaceans are aquatic (marine or freshwater) and breathe through gills or the body surface. They are typically covered by a hard EXOSKELETON. They range in size from the Japanese spider crab, which grows to 3m (12ft) across, to the ocean plankton, as small as 1mm (0.04in) in diameter.

Crux (Cross) Smallest constellation in the s sky. Although small, it is a distinctive constellation. Its brightest star is Alpha, magnitude 0.76.

cryogenics Branch of physics that studies materials and effects at low temperatures. Some materials exhibit highly unusual properties such as SUPERCONDUCTIVITY or SUPERFLUIDITY at such temperatures. Low temperatures can be reached in different ways, but among the most common is the JOULE-THOMSON EFFECT, when temperatures of 0.3K can be reached. To cool gases even further requires magnetic processes such as the ADIABATIC PROCESS. The applications of cryogenics vary from computer technology to medicine. Cryogenics has been used in a few cases to freeze human bodies in the hope that future technology may be able to revive the subjects. *See also* CRYOSURGERY

cryolite (kryolite) Brittle, icy-looking, red, brown or black halide mineral, sodium-aluminum fluoride (Na_3AlF_6) found in pegmatite DIKES (Greenland has the only large deposit) and used in ALUMINUM processing. It occurs as crystals in the monoclinic system, occasionally the cubic system, sometimes as granular masses. The crystals are frequently twinned. Cryolite is also a source of aluminum salts and fluorides.

cryosurgery (cryotherapy) Surgery that utilizes CRYOGENICS, and is carried out using a freezing probe, cooled by liquid nitrogen, that destroys diseased tissue on contact. It is particularly effective in treating detached retinas and in the removal of tumors and of some skin blemishes.

crystal Solid with a regular geometrical form and with characteristic angles between its faces, having limited chemical composition. The structure of a crystal, such as common salt, is based upon a regular 3-D arrangement of atoms, ions or molecules (an ionic or CRYSTAL LATTICE). Crystals are produced when a substance passes from a gaseous or liquid PHASE to a solid state, or comes out of solution by evaporation or precipitation. The rate of CRYSTALLIZATION determines the size of crystal formed. Slow cooling produces large crystals, whereas fast cooling produces small crystals.

crystal index Set of three numbers that characterize the type of planar configuration of atoms in a lattice. The three integers give the ratio of the reciprocals of the intercepts of the planes on three axes.

crystal lattice Three-dimensional arrangement of atoms, ions, or molecules in a crystalline substance. The term is sometimes used in a more restricted sense to denote the diagrammatic abstraction of the pattern in which the atoms, ions or molecules are positioned. *See also* CRYSTAL

crystalline substance Solid substance in which the atoms or molecules form a regular ordered lattice. Most solids exist in the crystalline state, which is more stable, but it does not follow that they have crystals in the usual sense; metallic copper, for example, is crystalline only in that the copper atoms are arranged in a regular lattice. The BONDS in crystals can be any of the major types. *See also* AMORPHOUS SUBSTANCE; CHEMICAL BOND

crystallization Process of forming CRYSTALS by a substance passing from a gas or liquid to the solid state (SUBLIMATION or fusion) or coming out of solution (PRECIPITATION or EVAPORATION). In the fusion method a solid is melted by heating, and crystals form as the melt cools and solidifies. Ice crystals and monoclinic sulfur are formed in this way. Crystallization is an important laboratory and industrial technique for purifying and separating compounds.

crystallography Study of the formation and structure of crystalline substances. It includes the study of CRYSTAL formation, chemical bonding in crystals and the physical properties of solids. In particular, crystallography is concerned with the internal structure of crystals including substances that were not previously thought capable of forming crystals, such as DNA. *See also* X-RAY CRYSTALLOGRAPHY

crystal optics Optical properties of CRYSTALS. The transmission of light by crystals differs from that by glass since the REFRACTIVE INDEX may depend on the direction of incidence of the light (crystals are not, in general, ISOTROPIC). **Uniaxial** crystals (belonging to the tetragonal, hexagonal, and trigonal systems) have two principal refractive indices and display double refraction. **Biaxial** crystals (orthorhombic, monoclinic, triclinic) have three principal refractive indices.

crystal packing Manner in which atoms, ions, or molecules are packed together in a crystal. In simple terms, crystal structures can often be considered geometrically, as regularly arranged spheres each of which is equal in diameter. The most efficient ways of packing spheres are the arrangements known as hexagonal and cubic close-packing. In these packing configurations every sphere has 12 nearest neighbors and 73% of the available space is filled.

crystal structure Regular arrangement of atoms linked together by CHEMICAL BONDS in repeating patterns. Because of this regularity, and sometimes a corresponding asymmetry, the physical properties of a crystal, such as strength and conductivity, vary in a regular manner depending on the crystal axis along which they are measured. Single crystals are distinguished from noncrystalline (amorphous) and from polycrystalline materials, the physical properties of which are either invariant (because of complete mixing) or vary in an irregular manner.

crystal systems Six types of crystal structure can be grouped from the 32 classes of crystal symmetry. The elements of symmetry found in crystals are axes, planes, and centers. The six systems are: cubic, tetragonal, hexagonal (collectively called uniaxial), and orthorhombic, monoclinic, and triclinic (called biaxial). A seventh system, the trigonal, is recognized by many mineralogists. The trigonal system has the same set of reference axes as the hexagonal system.

cube In mathematics, the result of multiplying a given number by itself twice. Thus the cube of a is $a \times a \times a$, written a^3. The cube is also described as the third power of a number. Cube is also the name of a regular six-sided solid figure (all its edges are equal in length and all its faces are squares).

cube root (symbol $\sqrt[3]{}$) Number that must be multiplied by itself twice over to give a specified number. For example, the cube root of 64 is 4, since $4 \times 4 \times 4 = 64$. In this case, the cube root is written $\sqrt[3]{64} = 4$. In algebraic terms, the cube root of x^3 is x and the cube root of x is $\sqrt[3]{x}$ or $x^{1/3}$.

cubic equation Algebraic equation of the third degree, that is one in which the highest power of the unknown variable is three. An example is
$$2x^3 + x^2 + 7 = 0.$$
Cubic equations have three ROOTS (solutions), two of which may be complex, or all three may be real. *See* TARTAGLIA, NICCOLÓ FONTANA

Cugnot, Nicolas Joseph (1725–1804) French engineer who, in 1769, invented the first self-propelled road vehicle. This three-wheeled, two-cylinder steam tractor was designed to pull guns, and traveled at nearly 5km/h (3mph).

cumulonimbus *See* CLOUD

cumulus *See* CLOUD

cuprite Reddish-brown, brittle, translucent oxide mineral, cuprous oxide (Cu_2O). It is formed by the OXIDATION of other ores such as copper sulfide and so is commonly found near the surface. Its octahedral, dodecahedral and cubic crystals occur in the cubic system, and it also occurs in grains. Hardness 3.5–4; r.d. 6.1.

cupronickel Alloy containing 75 parts of copper to 25 parts of nickel, widely used for coins.

curare Poisonous, resinous extract obtained from various tropical South American plants of the genera *Chondodendron* and *Strychnos*. Most of its active elements are ALKALOIDS. Causing muscle paralysis, it is used on the poisoned arrows of Native South Americans when hunting. It has also

▲ **crystal structure**
Crystals grow in a random fashion although the internal structures conform to six basic geometrical systems. The external shape of a crystal is dependent upon the internal arrangement of the atoms and molecules within the crystals.

Crystals are built up by repeating the basic unit cell. The unit cell consists of a number of atoms or molecules in a specific spatial arrangement. These arrangements are responsible for the internal structure and external shape of the crystal.

been used as a muscle relaxant in surgery, but has been replaced by safer agents.

Curie, Marie (1867–1934) Polish scientist who specialized in work on RADIOACTIVITY. Marie and her husband Pierre Curie (1859–1906) (who specialized in the electrical and magnetic properties of crystals, and formulated a law relating magnetism to temperature) worked together on a series of radiation experiments in Paris. In 1898 they discovered RADIUM and POLONIUM. In 1903 they shared the Nobel Prize for physics with Antoine BECQUEREL. In 1911 Marie Curie became the first person to be awarded a second Nobel Prize (this time for chemistry), for her work on radium and its compounds. She died of leukemia caused by laboratory radiation. Their daughter, Irène JULIOT-CURIE (1897–1956), and her husband, Frédéric Joliot-Curie (1900–58), were awarded the 1935 Nobel Prize for chemistry for producing artificial radioactive substances by bombarding elements with ALPHA PARTICLES.

curie (symbol Ci) Unit formerly used to measure the activity of a radioactive substance. Named after Marie CURIE, it is defined as that quantity of an ISOTOPE that decays at the rate of 3.7×10^{10} disintegrations per second. The curie has been replaced by an SI unit, the much smaller BECQUEREL.

curium (symbol Cm) Synthetic, radioactive metallic element of the ACTINIDE SERIES. It was first made in 1944 by Glenn SEABORG by the alpha particle bombardment of plutonium-239 in a cyclotron ACCELERATOR. Silvery in color, curium is chemically reactive, intensely radioactive, and is toxic if absorbed by the body. A modern use is to provide power for orbiting satellites. Properties: at.no. 96; r.d. (calculated) 13.51; m.p. 1,340°C (2,444°F); 14 isotopes, most stable ^{247}Cm (half-life 1.6×10^7 yr).

current Broad, slow drift of moving water, distinguishable from the water surrounding it by differences in temperature and/or salinity. A current can be caused by prevailing winds that sweep surface water along, forming drift currents. Currents are affected by the CORIOLIS EFFECT, by the shape of the ocean bed or by nearby land masses. Currents at a deeper level are the result of variation in density of the water, which in turn varies according to temperature and salinity.

Curtiss, Glen Hammond (1878–1930) US aviation pioneer. He began, as did the WRIGHT BROTHERS, by building bicycles. By 1908 he had made the first US flight of more than 1km (0.6 mi). Curtiss built planes during World War I (1914–18), and his JN-4 ("Jenny") was a well-known trainer.

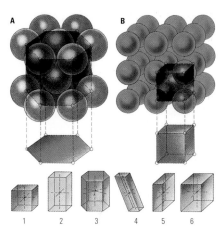

▲ **crystal systems** There are six main crystal structure systems, cubic (1), tetragonal (2), hexagonal (3), monoclinic (4), triclinic (5) and orthorhombic (6). Crystals were first studied by systematically examining their external shape and optical properties. The two examples in the diagram show a complex and simple structure. The molecular structure found in a quartz crystal, SiO_2, is shown in (A) and it can be seen how the hexagonal system is related to the stereochemical arrangement of the atoms. This crystal structure has three equal axes inclined at 120° to each other with a fourth unequal and perpendicular to the other three. The structure of the halite crystal (B), sodium chloride NaCl, was the first to be elucidated, having three axes equal and mutually perpendicular to each other.

His company and that of the Wrights merged to become the Curtiss-Wright Corporation.

curvature of the spine Exaggerated shaping of the SPINE. There are three major types. **Scoliosis**, or lateral curvature, can be due to bad posture or to abnormality. **Lordosis**, an accentuation of the inward curve of the neck region or more commonly of the lower back region, results in a swayback appearance. **Kyphosis**, an accentuation of the outward curve behind the chest, can in severe form result in a hunchback appearance.

curve A continuous one-dimensional set of points. In mathematical terms, a straight line is also a curve, and in analytic geometry, a curve is the locus of points that satisfy a mathematical equation.

Cushing, Harvey (1869–1939) US surgeon. His pioneering techniques for surgery on the brain and spinal cord helped advance NEUROSURGERY. He first described the condition produced by oversecretion of adrenal hormones that is now known as **Cushing's syndrome**. This syndrome is characterized by weight gain in the face and trunk, high blood pressure, excessive growth of facial and body hair, demineralization of bone, and diabetes-like effects.

cuticle Exposed outer layer of an animal. In humans this refers to the EPIDERMIS, especially the dead skin at the edge of fingers. In botany cuticle refers to the waxy layer on the outer surface of the epidermal cells of the leaves and stems of vascular plants. It helps to prevent excessive water loss.

Cuvier, Georges, Baron (1769–1832) French geologist and zoologist, a founder of comparative anatomy and paleontology. His scheme of classification stressed the form of organs and their correlation within the body. He applied this system of classification to fossils, and came to reject the theory of the gradual development of the Earth and animals, favoring instead a theory of catastrophic changes. *See also* PUNCTUATED EQUILIBRIUM

cyanide Salt or ester of hydrocyanic acid (prussic acid, HCN). The most important cyanides are sodium cyanide (NaCN) and potassium cyanide (KCN), both of which are deadly poisonous and have a characteristic smell of almonds. Cyanides have many industrial uses – in ELECTROPLATING, for the heat treatment of metals, in the extraction of silver and gold, in photography, and in insecticides and pigments.

cyanobacteria (formerly blue-green algae) One of the major BACTERIA phyla, distinguished by the presence of the green pigment CHLOROPHYLL and the blue pigment **phycocyanin**. They perform PHOTOSYNTHESIS with the production of oxygen. Analysis of the genetic material of CHLOROPLASTS shows that they evolved from cyanobacteria by ENDOSYMBIOSIS. Many cyanobacteria perform NITROGEN FIXATION. They occur in soil, mud, and deserts; they are most abundant in lakes, rivers, and oceans. Some produce toxic BLOOMS.

cybernetics Study of communication and control systems in animals, organizations, and machines. The term was first used in this sense in 1948 by Norbert WIENER. Cybernetics makes analogies between processes in the brain and nervous system and those in computers and other electronic systems, analyzing, for example, the mechanisms of FEEDBACK and data processing in both. Thus a household thermostat might be compared with the body's mechanisms for temperature control and respiration. Cybernetics combines aspects of mathematics, neurophysiology, computer technology, INFORMATION THEORY, and psychology.

cyberspace Popular term for the perceived "virtual" space within computer memory, especially if rendered graphically. The term is a product of science fiction, where it usually refers to situations involving direct interface between brain and computer. During the mid-1990s the term became widespread in reference to the INTERNET and the WORLD WIDE WEB.

cycad Any member of the phylum Cycadophyta, primitive palm-like shrubs and trees that grow in tropical and subtropical regions. They have feathery palm- or fern-like leaves (poisonous in most species) at the top of thick (usually unbranched) stems. In addition to their main roots, they also have special roots containing CYANOBACTERIA that carry out NITROGEN FIXATION. These plants first flourished about 225 million years ago. Most of the 100 or so surviving species are less than 6.1m (20ft) tall.

cyclamates White, odorless, soluble crystalline salts, calcium cyclamate ($[C_6H_{11}NHSO_3]_2Ca.2H_2O$) and sodium cyclamate ($CH_6H_{11}NHSO_3Na$). They have about 30 times the sweetening power of ordinary sugar (SUCROSE). Both were used as artificial sweeteners until they were shown to cause cancer in laboratory animals.

cycle In physics, series of changes through which any system passes which brings it back to its original state. For example, alternating current starts from zero voltage, rises to a maximum, declines through zero to a minimum and rises again to zero. In the INTERNAL COMBUSTION ENGINE, the TWO-STROKE ENGINE completes one cycle each downward plunge and return; the FOUR-STROKE ENGINE takes two such movements.

cyclic In organic chemistry, describing a substance the molecules of which contain one or more rings of atoms. In a **homocyclic** compound, all the atoms in the ring are the same; examples include benzene (C_6H_6) and cyclohexane (C_6H_{12}). In a **heterocyclic** compound, there are different atoms in the ring; examples include pyridine (C_5H_5N) and pyrrole (C_4H_5N).

cyclohexane (C_6H_{12}) Colorless liquid hydrocarbon that occurs naturally in crude oil, but is made commercially by combining hydrogen with benzene, using a catalyst. It is one of a type of chemical compounds called **conformations**; they can exist in two forms differing only in the arrangement of their bonds, and may "flip" from one form to the other. Properties: m.p. 6°C (43°F); b.p. 81°C (178°F).

cyclone (low) System of atmospheric low pressure that occurs when a cold air mass moving s from the Arctic meets a warm air mass moving N from the tropics to form a circulating air mass. A cyclone is characterized by relatively low pressure at the center, and by counterclockwise wind movement in the Northern Hemisphere and by clockwise motion in the Southern Hemisphere. There are two types of cyclone: the DEPRESSION, associated with temperate latitudes; and the tropical cyclone, which is much more violent but usually affects a smaller area. *See also* ANTICYCLONE

cyclotron Particle ACCELERATOR in which charged particles such as PROTONS or heavier ions are accelerated outwards in a spiral. It has two D-shaped ELECTRODES across which is applied an alternating voltage. The charged particles are attracted by electrostatic forces to the electrode of opposite charge, but since the polarity alternates, the particles are swung outwards at increasing speed. The cyclotron was invented by Ernest LAWRENCE in order to obtain high energy particles without having to use an extremely long LINEAR ACCELERATOR.

Cygnus One of the most distinctive constellations, often nicknamed the NORTHERN CROSS.

cylinder Solid figure or surface formed by rotating a rectangle around one side as axis. If the vertical height is h and the radius of the base r, then the volume is $\pi r^2 h$ and the curved surface area $2\pi rh$.

cylinder, engine One of the cylindrical chambers in an INTERNAL COMBUSTION ENGINE. Inside each, a tightly fitting piston is pushed down by the pressure of the expansive force of burning fuel. The movement of the piston is lubricated by oil from the sump. The cylinders become hot in use and can be cooled either by air or water cooling systems. Reciprocating STEAM ENGINES also have cylinders.

cylindrical polar coordinates System of COORDINATES used to represent the position of a point in space. There are three coordinates, r, θ, and z. The point's projection on a reference plane is represented by the POLAR COORDINATES r and θ (where r is the distance of the projected point from the origin and θ its angle to some reference direction in that plane), and z is the point's height above (or below) the reference plane.

cyst Hollow cavity or sac in the body that contains liquid or semisolid matter. It may occur in a glandular organ, such as the breast or prostate, or in the skin. It may be caused by infection or by a blocked duct. A cyst may be removed for medical or cosmetic reasons.

cysteine ($HSCH_2CH_2COOH$) Crystalline AMINO ACID that occurs in animal proteins, especially those in the hair, hooves, and the keratin of the skin. The $HS-$ group in it makes it the catalytic element in some enzymes.

cystic fibrosis Hereditary glandular disease in which the body produces abnormally thick mucus that obstructs the breathing passages, causing chronic lung disease. There is a deficiency of pancreatic enzymes, an abnormally high salt concentration in the sweat, and a general failure to gain weight. The disease is treated with antibiotics, pancreatic enzymes, and a high-protein diet; sufferers must undergo vigorous physiotherapy to keep the chest as clear as possible. Heart-lung transplantation is sometimes recommended where there is severe lung damage. There have been experiments using GENE THERAPY.

cytochromes Proteins containing heme, an iron group, as in HEMOGLOBIN. They are fundamental to the process of RESPIRATION in all living cells that need atmospheric oxygen.

cytokinin (kinetin or kinin) Any of a group of plant hormones that stimulate cell division. Cytokinins work in conjunction with AUXINS to promote swelling and division in the plant cells producing lateral buds. They can also slow down the aging process in plants, encourage seeds to germinate and plants to flower, and are involved in plant responses to drought and waterlogging. They are used commercially to produce seedless grapes, to stimulate germination of barley in brewing, and to prolong the life of green-leaf vegetables. Cytokinins, such as zeatin, are derived from ADENINE.

cytology Study of living CELLS and their structure, behavior, and function. The discipline began with Robert HOOKE's microscopic studies of cork in 1665, and the various forms of MICROSCOPE are still the main tools of cytology. In the 19th century, a cell theory was developed that suggested that cells are the basic units of organisms. Recently cytological study has focused on the chemistry of cell components (cytochemistry).

cytoplasm Jelly-like matter inside a CELL and surrounding the NUCLEUS. Cytoplasm has a complex constituency and contains various bodies known as organelles, with specific metabolic functions. The proteins needed for cell growth and repair are produced in the cytoplasm. *See also* METABOLISM

cytosine Organic base, first isolated in 1894. Derivatives of cytosine made in the body are important in cellular METABOLISM and in the formation of RNA and DNA, thus being vital for the retention of genetic characteristics. *See also* GENE

Daguerre, Louis Jacques Mandé (1798–1851) French painter and inventor. In 1829 Daguerre and Niepce invented the daguerreotype, an early photographic process in which a unique image is produced on a copper plate without an intervening negative. Their process was announced in 1839, shortly before William Fox Talbot developed the calotype.

Daimler, Gottlieb (1834–1900) German engineer and automobile manufacturer. In 1822, with Wilhelm MAYBACH, Daimler established a research laboratory where a year later they developed a lightweight gasoline INTERNAL COMBUSTION ENGINE. Daimler used this to power a motorcycle and then his first car (1886). In 1890 he founded the Daimler Motor Company, which made Mercedes cars, and in 1926 it became Daimler-BENZ and Company.

Dale, Sir Henry Hallett (1875–1968) British physiologist who shared the 1936 Nobel Prize for physiology or medicine with Otto LOEWI for discoveries relating to the chemical transmission of NERVE IMPULSES. Dale found that the chemical acetylcholine served to transmit nerve impulses across the tiny gap (SYNAPSE) from one NERVE CELL to another. His writings include *Adventures in Physiology* (1953) and *An Autumn Gleaning* (1954).

Dalén, Niels Gustav (1869–1937) Swedish engineer responsible for the development of automatic lighting for lighthouses and railroad signals by means of a "sunvalve," a gas tube responsive to the Sun's radiation. Dalén was awarded the Nobel Prize for physics in 1912, but was blinded in a laboratory explosion a year later. He also invented a substance that absorbs acetylene and prevents the risk of explosion in acetylene tanks.

Dalton, John (1766–1844) British chemist and physicist. His early interest in meteorology yielded important information on the trade winds, the cause of rain and the AURORA borealis. He described color blindness (sometimes called Daltonism) based on his own experiences and those of his brother. His study of gases led to DALTON'S LAW of partial pressures. His atomic theory states that each element is made up of indestructible,

identical, small particles; he also constructed a table of RELATIVE ATOMIC MASSES (r.a.m.).

Dalton's law Pressure exerted by each gas in a mixture of gases does not depend on the pressures of the other gases, provided no chemical reaction occurs. The total pressure of such a mixture is therefore the sum of the partial pressures exerted by each gas (as if it were alone in the same volume as the mixture occupies).

Dam, Carl Peter Henrik (1895–1976) Danish biologist. In 1934 he discovered vitamin K, the fat-soluble vitamin needed for blood clotting. He isolated it from hempseed and the seeds of other plants, and he also discovered it in liver. For this work, he received the 1943 Nobel Prize for physiology or medicine, which he shared with Edward DOISY. In addition, he examined the roles of other vitamins and LIPIDS.

dam Barrier built across a stream, river, estuary, or part of the sea. A dam confines water, or checks its flow, for irrigation, flood control or HYDROELECTRICITY generation. The first dams were probably constructed by the Egyptians at least 4,500 years ago. Common types are **gravity, arch,** and **buttress** dams. Gravity dams are anchored by their own weight. Single-arch dams are curved (convex to the water they retain) and are supported at each end by the river banks. Multiple-arch and buttress dams are supported by buttresses rooted in the bedrock. The highest dam is Russia's Rogun, 335m (1,100ft) high. The Chapeton Dam in Argentina has a record structural volume of 296,000,000m^3 (387,568,000yd^3). The cheapest source of electricity that is used commercially comes from HYDROELECTRICITY made possible by dams such as the Hoover Dam on the Colorado River.

damping Gradual slowing down of the oscillations in a vibrating mechanical system. It occurs as the system loses energy to overcome friction or other resistance; this is why a swinging pendulum gradually slows down. A mechanism may be deliberately damped to facilitate reading an oscillating needle or to slow down an oscillating balance. Damping that stops the oscillation in the shortest possible time is called CRITICAL damping.

D and C operation (abbreviation for dilatation and curettage) Procedure in which the CERVIX (neck of the womb) is dilated so that the lining of the womb can be scraped away (curettage). It is done to treat disease or an incomplete miscarriage, to terminate a pregnancy, or to secure a sample of tissue for testing (BIOPSY).

Darby, Abraham (1677–1717) English iron manufacturer. In 1709 Darby developed the first COKE-fired BLAST FURNACE, at Coalbrookdale, Shropshire. The relatively cheap and abundant supplies of coke fuelled the INDUSTRIAL REVOLUTION. His grandson, **Abraham Darby III** (1750–91), built the world's first cast-iron bridge, at Ironbridge Gorge, Shropshire, England.

dark adaptation Slow change in sensitivity of the human EYE that takes place when someone passes into dim light from bright light. It is caused by a shift in functional dominance from cone cells to rod cells in the RETINA of the eye as the overall illumination is reduced. The complete process takes 35–40 minutes. *See also* LIGHT ADAPTATION

dark matter Unseen matter inferred to exist in galactic halos and in the space between galaxies, which is thought to make up at least 90% of the mass of the Universe. It could exist in various forms, such as BLACK HOLES, BROWN DWARFS, or unknown atomic particles, all of which are termed **cold dark matter**; or it could be in the form of a "sea" of fast-moving NEUTRINOS, known as **hot matter**; or a mixture of both.

dark reaction One of the two distinct phases of PHOTOSYNTHESIS; the other being LIGHT REACTION. During the dark reaction, which can take place in light or in darkness, carbon dioxide is reduced to carbohydrate (sugars) by means of the CALVIN CYCLE.

Darwin, Charles Robert (1809–82) British naturalist, originator of a theory of EVOLUTION based on NATURAL SELECTION. In 1831 he joined a round-the-world expedition on HMS BEAGLE. Observations made of the FLORA and FAUNA of South America (especially the Galápagos Islands) formed the basis of his work on animal variation. The development of a similar theory by Alfred WALLACE led Darwin to present his ideas to the Linnean Society in 1858, and in 1859 he published *The Origin of Species*, one of the world's most influential science books. Darwin argued that organisms reproduce more than is necessary to replenish their population, creating competition for survival. Darwin argued that the variations that prove helpful in the struggle to survive are passed down to the offspring of the survivors. He termed this process NATURAL SELECTION. NEO-DARWINISM supplemented his ideas with modern research into HEREDITY, especially MUTATION.

Darwinism *See* EVOLUTION

DAT Acronym for DIGITAL AUDIO TAPE

data (sing. datum) Information, such as lists of words, quantities or measurements, or codes representing a picture. A COMPUTER PROGRAM works by processing data, which may be entered using a keyboard or other input device, stored as a data file on a COMPUTER DISK, downloaded from the INTERNET, or come from many other sources.

database Collection of DATA produced and retrieved by a computer. The data are usually stored on MAGNETIC DISK or DIGITAL AUDIOTAPE. A database program enables the computer to generate files of data and later search for and retrieve specific items or groups of items. For example, a library database system can list, on screen, all the books by a specified author, or all the books on a particular subject. The computer can then display further details of any selected book. *See also* INFORMATION STORAGE AND RETRIEVAL

data compression Technique in computing to reduce the amount of storage space occupied by data. Methods include representing common characters by fewer BITS than normal and storing frequently used words as shorter words (**tokenization**). Long sequences of repeated characters can be replaced by a single character and a count of how many there are, a technique called **run-length encoding**. By such means, text files can be reduced by up to 50% and digitized images by about 90%. Compression techniques can be divided into two main types: **lossy** and **nonlossy**. With nonlossy compression, there is no loss in the quality of the data. With lossy compression, the compression is greater but the quality of the data is reduced; when the data is uncompressed, it will be slightly different from before it was compressed. Lossy compression is used chiefly for images, video, and music.

data processing Systematic sequence of operations performed on DATA, especially by a computer, in order to calculate new information or revise or update existing information stored on magnetic or optical DISK, magnetic tape, and so on. The data may be in the form of numerical values representing measurements, scientific or technical facts, or lists of names, places or book titles. The main processing operations performed by a computer are arithmetical addition, subtraction, multiplication, and division, and logical operations that involve decision-making based on comparison of data. For the latter, an instruction might read: "If condition a holds, then follow programmed instruction P; if a does not hold, then follow instruction Q."

data protection Measures taken to guard DATA against unauthorized access. Computer technology now makes it easy to store large amounts of data containing, for example, a person's medical or financial records. Many governments have passed data-protection legislation ensuring that such databases are registered and the information that they contain is used only for the purpose for which it was originally given.

data transmission Supply of information to or from a computer via hard wire (telephone lines) or electromagnetic (radio or microwave) broadcast from earth stations or satellites. Computers or other electronic devices at each end of the transmission path regulate the information by encoding, decoding, and synchronizing the transmitted signals.

dating, radioactive (radiometric dating) Any of several methods using the laws of RADIOACTIVE DECAY to assess the very considerable ages of archaeological remains, fossils, rocks, and the Earth itself. The specimens must contain a very long-lived radioisotope of known HALF-LIFE, which, with a measurement of the ratio of radioisotope to a stable isotope (usually the decay product), gives the age. In potassium-argon dating, the ratio of potassium-40 to its stable decay product argon-40 gives ages over ten million years. In rubidium-strontium dating, the ratio of rubidium-87 to its stable product strontium-87 gives ages up to several billion years. In CARBON DATING, the proportion of carbon-14 (half-life 5,730 years) to stable carbon-12 absorbed into once-living matter, such as wood or bone, gives ages up to several thousand years.

◄ **Davy** Perhaps best known for the invention of a miner's safety lamp (Davy lamp), Sir Humphry Davy was a significant chemist who inspired Michael Faraday. The lamp comprised a two-layer, metal-gauze chimney that ensured that any methane gas in the mine would not ignite the flame inside the lamp.

datolite Colorless or white orthosilicate mineral, basic calcium borosilicate (CaBSiO$_4$[OH]), found in cavities in trap-rock. It occurs as monoclinic system crystals or, rarely, as granular masses. Hardness 5–5.5; r.d. 2.9.

Davisson, Clinton Joseph (1881–1958) US physicist who worked on THERMIONICS, ELECTRON DIFFRACTION, and the ELECTRON MICROSCOPE for Bell Telephone Laboratories. In 1927, working with Lester Germer (1896–1971), he was able to confirm (by means of DIFFRACTION by crystals) the wave nature of moving electrons, which had been hypothesized by Louis de BROGLIE. He shared the Nobel Prize for physics with Sir George THOMSON in 1937.

Davy, Sir Humphry (1778–1829) British chemist who discovered that ELECTROLYTIC CELLS produce electricity by chemical means. This led to the use of ELECTROLYSIS to isolate the elements sodium, potassium, barium, strontium, calcium, and magnesium. He also proved that all acids contain hydrogen, which is responsible for their acidic properties. An investigation into the conditions under which firedamp (methane and other gases) and air explode led to his invention of the miner's SAFETY LAMP.

Dawkins, Richard (1941–) English zoologist. He studied at Oxford before accepting a teaching post in California, where he remained until returning to Oxford in 1970. Dawkins' first popular science book, *The Selfish Gene* (1976), examines animal behavior in the context of EVOLUTION, proposing that genes govern behavior in order to survive. *The Blind Watchmaker* (1986) is a good introduction to NEO-DARWINISM.

DDT (dichlorodiphenyltrichloroethane) Colorless, crystalline, organic halogen compound, first used as an insecticide in 1939 against the Colorado potato beetle. It acts as a contact poison, disorganizing the nervous system. It was initially effective against most insect pests and had dramatic effects in reducing malaria. However, many insect species developed resistant populations, and it proved to have long-lasting toxic effects on birds and other animals. As a result, it is now banned in many countries, though still used for malaria control in others.

deafness Partial or total HEARING loss. **Conductive** deafness, where there is faulty transmission of sound to the sensory organs deep in the EAR, is usually due to infection or to inherited abnormalities of the tiny bones (OSSICLES) of the middle ear. **Perceptive** deafness may be inborn or acquired. It is due to injury or disease of the COCHLEA, auditory nerve or hearing centers in the BRAIN. The treatment of deafness depends on its cause and ranges from the removal of impacted wax to cochlear implant or delicate microsurgery to correct some congenital defect. Hearing aids, sign language and lipreading are among techniques that help the deaf to communicate.

deamination Removal of an amino ($-NH_2$) group from a compound, particularly an AMINO ACID. It takes place in the liver of animals, where enzymes cause the amino group to be converted into ammonia, which is either excreted as it is or converted into urea or uric acid and then excreted in the urine.

death Cessation of life. Traditionally in medicine, death has been pronounced on cessation of the heartbeat. However, modern resuscitation and life-support techniques have led to the revival of some people whose hearts have stopped. In a tiny minority of such patients, while for a short time breathing and the heartbeat can be maintained artificially, the potential for life is extinct. In this context death may be pronounced when it is clear that the brain no longer controls vital functions.

DeBakey, Michael Ellis (1908–) US surgeon who pioneered surgical techniques for treatment of CIRCULATORY SYSTEM disorders. In 1966 DeBakey implanted the first mechanical device into a human chest to assist the HEART. He had already devised the roller pump (1932), a part of the machinery used in open-heart surgery, and developed a method of grafting frozen blood vessels to replace those that were diseased as a means of correcting aortic aneurysms; later (1956), he replaced the grafts with plastic tubing.

de Broglie, Louis Victor *See* BROGLIE, LOUIS VICTOR DE

Debye, Peter Joseph Wilhelm (1884–1966) US physical chemist, b. Netherlands. He is best known for his work on molecular structure and dissociation in solution. He also pioneered X-RAY CRYSTALLOGRAPHY. He was awarded the 1936 Nobel Prize for chemistry.

decapod Any of about 8,500 species of the order Decapoda, mainly marine, higher crustaceans, including shrimps, lobsters, and crabs. They have ten legs including the chelae (pincers), which are enlarged modified legs. The order Decapoda has been represented since the Permian or Triassic period.

decay (rot) Partial or complete deterioration of a substance caused by natural changes. Plant rot, caused by soil-borne bacteria and fungi, can affect any plant part, making it spongy, watery, hard or dry. *See also* DECOMPOSITION. The term is also used in physics to describe RADIOACTIVE DECAY.

decibel (symbol dB) Logarithmic unit, one tenth of a *bel*, used for comparing two power levels. It is frequently used for expressing the loudness of a sound in terms of a particular reference level. The faintest audible sound (corresponding to an excess air pressure of 2×10^{-5} pascal) is given an arbitrary value of 0dB. The human pain threshold is about 120dB. Ordinary conversations occur at about 50–60dB.

deciduous Term describing the annual or seasonal loss of all leaves from a TREE or shrub; it is the opposite of EVERGREEN.

decimal fraction Number in the DECIMAL SYSTEM written as a digit to the right of a decimal point. The number 52.437 represents an INTEGER (whole number, 52) added to a decimal fraction (0.437). It is composed of
$$52 + (4 \times 10^{-1}) + (3 \times 10^{-2}) + (7 \times 10^{-3})$$
which may also be written
$$52 + (4/10 + 3/100 + 7/1000), 52 + 437/1000$$
or as the improper fraction 52437/1000. Decimal fractions are added, subtracted, multiplied, and divided like integers, but the decimal point must be correctly positioned after each operation.

decimal place Number of places to the right of a decimal point required to specify a real number (*see* NUMBER, REAL) to certain accuracy. The number 4.893302 to three decimal places is written 4.893 (3d.p.). To five decimal places the same number is 4.89330 (5d.p.). The convention is to round up the last decimal place if the DIGIT after that place is five or greater, and to round down if it is four or less. Thus the numbers 239.705 and 239.706 are both 239.71 (2d.p.) to two decimal places, whereas the number 239.704 to two decimal places is 239.70 (2d.p.). *See also* SIGNIFICANT FIGURES (S.F.)

decimal system Commonly used system of writing numbers using a base ten and the Arabic numerals 0 to 9. It is a positional number system in which each position to the left represents an extra power of ten. Thus 6,741.83 is $(6 \times 10^3) + (7 \times 10^2) + (4 \times 10^1) + (1 \times 10^0) + (8 \times 10^{-1}) + (3 \times 10^{-2})$. Numbers to the right of the decimal point are represented by a digit multiplied by ten raised to a negative power. Note that $10^0 = 1$.

declination (dec. symbol δ) Angular distance of a celestial object N or S of the CELESTIAL EQUATOR. It is calculated positively from 0 to 90 (from the

Equator to the north celestial pole), and negatively from 0 to −90 (from the Equator to the south celestial pole). *See also* CELESTIAL SPHERE

declination, magnetic (deviation) Amount by which the direction indicated by a magnetic compass differs from the direction of true north. It arises because the magnetic north pole is not coincident with the true North Pole. As a result, declination varies from place to place on the Earth's surface. Also, because the north magnetic pole slowly moves over time, declination varies from year to year.

decoder In telecommunications and electronics, a device that converts the information content of a signal into a more usable form. In a satellite-television receiver, it decodes encrypted signals; in a color TELEVISION, it separates the red, blue, and green components; in stereo FM radio it separates the left and right sound channels.

decomposition Natural degradation of organic matter into simpler substances, such as carbon dioxide and water. Organisms of decay are usually bacteria and fungi. Decomposition recycles nutrients by releasing them back into the ecosystem.

decompression Rapid decrease in ATMOSPHERIC PRESSURE, experienced, for example, when the pressurized cabin of an aircraft is ruptured, or when a diver returns to the surface too quickly. When someone works under pressures greater than atmospheric pressure, the respiratory gases are compressed and abnormally large amounts are dissolved in the blood and tissues. Sudden release of this pressure causes the gases to bubble off and interrupt the body's oxygen supply.

decompression chamber Airtight compartment capable of withstanding high pressures, used to house one or more divers or other personnel working in pressurized environments while they are gradually returned to normal atmospheric pressure. Decompression chambers are necessary to avoid **decompression** sickness (THE BENDS), in which nitrogen is released into the blood and tissue causing acute pain in the joints, dizziness, nausea, and paralysis. Deep-sea divers usually work in teams so that work progresses while divers who have been relieved are progressively decompressed in submerged chambers at various depths.

Dee, John (1527–1608) English mathematician and occultist. He was an associate of the alchemist Edward Kelley and a favorite of Queen Elizabeth I. He also did preparatory work on the Gregorian CALENDAR.

deep freezing Method of FOOD PRESERVATION, usually at −5°C (23°F) or below. Some foods can last for years if properly frozen. Three methods are widely used for commercial deep freezing: in **blast** freezing, a flow of cold air is passed over the food; in contact freezing, the food is placed between refrigerated plates or in a refrigerated alcohol bath; and in **vapor** freezing, liquid nitrogen or solid carbon dioxide (dry ice) is made to vaporize in the food compartment. In home freezing, various refrigerated cabinets are used.

Deep Sea Drilling Project (Ocean Drilling Program) US project to investigate the evolution of OCEANIC BASINS by studying the composition of the oceanic CRUST. The *Glomar Challenger*, equipped with well-drilling equipment and a satellite-controlled navigation system, was in operation from 1968 until the late 1970s. Measurements of the cores recovered, some of which were obtained at over 900m (3,000ft) beneath the seabed, have provided evidence for the theory of SEAFLOOR SPREADING and have shown that the present oceanic basins are relatively young. Seismic and magnetic studies are also carried out.

deficiency disease Illness caused by lack of an essential nutrient in the DIET. Scurvy is caused by a lack of vitamin C, rickets by a deficiency of vitamin D and pellagra by insufficient niacin (nicotinic acid) in the diet. Rare in the developed world, such disorders are treated by supplying the missing nutrient.

De Forest, Lee (1873–1961) US inventor of the audion triode ELECTRON TUBE. This three-electrode tube, invented in 1907, was one of the most important inventions in electronics. It could amplify signals and had numerous applications. Tubes became essential in radio, television, radar, computer, and other electronic systems, and remained so until the TRANSISTOR was invented in 1947.

deforestation Clearing away of forests and their ECOSYSTEMS, usually on a large scale. It may be done to create open areas for farming or building, or to make use of the timber from the trees. There is an immediate danger that the vital topsoil will be eroded by wind (as in the dust bowl in the s Great Plains in the 1930s) or, in hilly areas, by rain. Proposals to clear whole regions of the rainforests of Amazonia in Brazil, which play a key role in maintaining the oxygen balance of the Earth, could, if fully implemented, cause an environmental catastrophe. *See also* REFORESTATION

degaussing Demagnetization of a magnetized substance, performed by surrounding the substance with a coil carrying an alternating electric current of decreasing magnitude. The name is derived from the CGS unit of magnetic flux, the **gauss**.

degeneration In biology, term used to describe the effects disease may have on cells or groups of cells, resulting in loss of function or in extreme cases death. In evolution, the term refers to the gradual appearance of a VESTIGIAL STRUCTURE, such as the appendix in humans or the wings of emus, which serve little or no function in the evolved species.

degenerative disease Progressive wasting ailment in which a patient's condition deteriorates at a greater or lesser rate, depending on the treatment used and the response to it; often such a condition can be arrested, although it cannot be reversed.

degree In mathematics, unit of angular measure equal to one three-hundred-and-sixtieth (1/360) of a complete revolution (360°). One degree is written 1°, and can be divided into 60 parts called minutes or "minutes of arc" (one minute is denoted 1'), which may in turn be divided into 60 parts called seconds or "seconds of arc" (one second is denoted 1", so 3,600" = 60' = 1°). 360° are equal to 2π radians. In physics and engineering, a degree is one unit on any of various scales, such as the CELSIUS temperature scale or the **Baumé scale** of SPECIFIC GRAVITY.

dehiscent Describing a FRUIT or seed pod that opens spontaneously to release its seeds. Often violent, dehiscence disperses the seeds over a wide area. The explosive release of spores from a SPORANGIUM or pollen from an ANTHER is also described as dehiscence. *See also* INDEHISCENT

dehydration Removal or loss of water from a substance or tissue. Water molecules can be removed by heat, catalysts or a dehydrating agent such as concentrated sulfuric acid, which removes hydrogen and oxygen in the form of water from another chemical. Dehydration (drying) is used to preserve foodstuffs. In medicine, dehydration is excessive water loss from the body, often a symptom or result of disease or injury.

Delbrück, Max (1906–81) US biologist, b. Germany. He shared the 1969 Nobel Prize for physiology or medicine with Alfred HERSHEY and Salvador LURIA for work on reproduction in VIRUSES.

delocalization In nuclear physics, spreading of VALENCE electrons across more than one CHEMICAL BOND in a compound. Molecules with delocalized electrons tend to be more stable than molecules with electrons that are not delocalized.

Delphinus (Dolphin) Small but distinctive constellation of the N sky. All its leading stars are of fourth magnitude.

delta Fan-shaped body of sediment deposited at the mouth of a river. A delta is formed when a river deposits its sediment load as its speed decreases while entering the sea or a lake, and the waves, tides, and currents are not sufficiently strong to carry the material away. The shape and size of a delta is controlled by a combination of factors, including climate, water discharge, sediment load, rate of subsidence of the sea or lake, and the nature of the river mouth processes. Most deltas are extremely fertile areas, but are subject to frequent flooding.

dementia Deterioration of personality and intellect that can result from disease of or damage to the brain. It is characterized by memory loss, impaired mental processes, personality changes, confusion, lack of inhibition, and a deterioration in personal care and hygiene. Dementia can occur at any age, although the disease process is more common in the elderly. As well as being a degenerative disease in the brain, the causes of dementia include disease of the blood vessels, brain injury, chronic poisoning, and infection. *See also* ALZHEIMER'S DISEASE

De Morgan, Augustus (1806–71) British mathematician. He contributed to the development of set theory and wrote a number of textbooks, including *Elements of Arithmetic* (1830). With his contemporary, George BOOLE, he created a mathematical theory of LOGIC.

dendrite Short branching projection from a nerve cell (NEURON). It carries impulses to the cell body and transmits impulses to other nerve cells over short gaps called SYNAPSES. There may be more than one dendrite per neuron.

dendritic drainage Treelike branching pattern of streams and their tributaries. The pattern is common in regions where the land is essentially flat and the rock is homogeneous.

dendrochronology Means of estimating time by examination of the ANNUAL RINGS (growth rings) in trees. This data can be related to wood used in buildings, for instance. It may also indicate the HYDROLOGY of the region in which the tree grew, thus fixing points in the climatic history of that region. Chronology based on the bristle-cone pine extends back over 7,000 years.

dendron Major projection that protrudes from a nerve cell (NEURON) and is concerned with nerve transmission. Found mostly on MOTOR NEURONS, dendrons usually branch into smaller DENDRITES, which chemically transmit NERVE IMPULSES across a SYNAPSE to other neighboring neurons.

Deneb (Alpha Cygni) Remote and very luminous white SUPERGIANT star in the constellation of Cygnus. It is 60,000 times more luminous than the Sun and located about 1,500 light-years away.

denitrification Process that chemically reduces ammonia, nitrites or nitrates to yield free nitrogen. In biology, under waterlogged conditions denitrifying BACTERIA change the nitrogen of ammonia into free nitrogen that enters the atmosphere or soil. *See also* NITROGEN CYCLE

denominator In mathematics, the part of a FRACTION below the line that serves as the divisor of the NUMERATOR. In the fraction $^7/_8$, the denominator is 8; 7 is the numerator.

densitometer Instrument for measuring the optical transmission or reflection (optical density) of a material such as a photographic film or plate. It is used in SPECTROSCOPY to determine the positions of spectral lines and bands and to measure their RELATIVE DENSITIES, and thus intensities.

density (symbol ρ) Ratio of mass to volume for a given substance, usually expressed in SI UNITS as kilograms per cubic meter (kgm^{-3}). It is an indication of the concentration of particles within a material. The density of a solid or liquid changes little over a wide range of temperatures and pressures. **Relative density** (symbol r.d.) or **specific gravity** (symbol s.g.) is the ratio of the density of one substance to that of a reference substance (usually water) at the same temperature and pressure. The density of a gas depends strongly on both pressure and temperature. In ECOLOGY, the term is used to describe the number of organisms per unit area.

dental formula System for indicating the type and distribution of TEETH in an animal's jaw. Half of each jaw is considered and represented by two sets of four numbers, one set above and one set below a horizontal line. The numbers above the line stand for the upper jaw and the numbers below the line for the lower jaw. The first number in each set stands for the number of incisors, the second number for the canines, the third number for the premolars and the fourth number for the MOLARS. An adult human with a full set of teeth has, in each half of each jaw, two incisors, one canine, two premolars, and three molars. The dental formula is written: 2 1 2 3
2 1 2 3
Adding all the numbers and multiplying by 2 gives the total number of teeth (32). *See also* DENTITION

dentine Hard yellow matter of all TEETH that consists of crystals of calcium and phosphate. Human beings and other higher animals have **tubular** dentine, so called because a line of dentine-producing cells (**odontoblasts**), surrounding the pulp, sends out tubules into the dentine; these transmit sensations to the nerve.

dentistry Profession concerned with the care and treatment of the mouth, the teeth, and their supporting tissues. As well as general practice, dentistry includes such specialities as oral surgery, prosthodontics, periodontics, orthodontics, and public health.

dentition Type, number and arrangement of TEETH (*see* DENTAL FORMULA). An adult human has 32. The incisors are used for cutting; the canines for gripping and tearing; and the molars and premolars for crushing and grinding food. A HERBIVORE has relatively unspecialized teeth that grow throughout life to compensate for wear (plant fibers are very tough) and are adapted for grinding. A CARNIVORE has a range of specialized teeth related to killing, gripping, and crushing bones. In carnivores, unspecialized milk teeth are replaced by more specialized adult teeth, which have to last a lifetime.

denudation In geology, wearing away of land by WEATHERING and EROSION. Denudation is a broad term and includes all the natural agencies, such as Sun, rain, wind, rivers, frost, ice, and sea, as well as heating and cooling, freezing and thawing, SOLUTION, ABRASION, CORRASION, and CORROSION. In addition to weathering and erosion, the removal of material, that is, the transportation, is also part of denudation. Together with DEPOSITION, denudation is the major process which creates the Earth's landscape.

deoxyribonucleic acid *See* DNA

dependent variable *See* VARIABLE

depletion Special form of depreciation referring to the exhaustion of nonrenewable natural resources (such as oil, minerals or natural gas) as they are exploited for human use. Depletion is usually figured as the percentage of the estimated reserves of the resource that has been used.

deposition In geology, laying down of material that has been removed by DENUDATION. Most of the material will be sediment and therefore a possible source of SEDIMENTARY ROCKS when consolidated. After denudation has taken place, the material is transported, sometimes over quite considerable distances, before being deposited. Most of the material will eventually be dumped on the seabed by rivers. However, there will be some deposition on land – for example, silt on flood plains, TILL deposited by ice, LOESS deposited by wind. Together with denudation, deposition is the major process that creates the Earth's landscape.

depression In meteorology, a region of low ATMOSPHERIC PRESSURE with the lowest pressure at the center. It usually brings unsettled or stormy weather. In the Northern Hemisphere, winds circulate counterclockwise in a depression; they circulate clockwise in the Southern Hemisphere. *See also* CYCLONE

depth perception Ability of the eyes to locate the position of objects in three-dimensional space. The RETINA is two-dimensional, so information about depth is created in the brain. The brain uses "depth clues," which include such factors as linear perspective, PARALLAX, relative size, and the slightly different view each eye has of the object.

derivative Rate of change of the value of a mathematical function with respect to a change in the independent variable. The derivative is an expression of the instantaneous rate of change of the function's value. The derivative of a function $f(x)$, at the point x, is defined to be the value of the ratio $[f(x+h) - f(x)]/h$, in the limit where the number h approaches zero, and is denoted df/dx or $f'(x)$. Derivatives provide a means of describing many processes that can be viewed as changing continuously in time. For instance, if the position of an object at a time, t, is given by $p(t)$, then the VELOCITY (which is defined as the *instantaneous* rate of change of position), obtained by differentiation, is given by the derivative dp/dt. Derivatives tell us how fast a quantity changes at a given time. Geometrically, the derivative can be viewed as the gradient, or slope, of a curve at a particular point. *See also* CALCULUS; DIFFERENTIAL

dermapteran Any INSECT of the order Dermaptera, the earwigs. Slender, flattened, brownish-black insects, they are found in crevices and under tree bark. There are some 900 winged and wingless species worldwide. All have a pair of forceps at the hind end of the abdomen.

dermatitis Inflammation of the skin. In acute form it produces redness, itching, and blisters or oozing. In chronic form it causes thickening, scaling, and darkening of the skin. In **contact** dermatitis, a reaction occurs on contact with a particular substance, such as soap or nettles. In **atopic** dermatitis (often associated with HAY FEVER and ASTHMA), excessive dryness occurs, with redness at the

neck, elbows, and knees. In **stasis** dermatitis, heavy pigmentation and sometimes ulcers develop on the inner sides of the lower legs as a result of poor circulation. The term is sometimes used interchangeably with eczema.

dermatology Branch of medicine that deals with the diagnosis and treatment of skin diseases.

dermis Thick inner layer of the SKIN that lies beneath the EPIDERMIS. Also known as the true skin, it consists mainly of loose CONNECTIVE TISSUE and is richly supplied with BLOOD and LYMPH vessels, nerve endings, sensory organs, and sweat glands.

desalination Extraction of pure water (for drinking, industrial and chemical uses, or for irrigation) from water containing dissolved salts, usually seawater. The commonest and oldest method is to evaporate the water from a salt solution by DISTILLATION, and condense the vapor to form water. Another method is to freeze the salt solution; salt is excluded from the ice crystals, which can then be melted. In the method of REVERSE OSMOSIS, only pure water passes through a semipermeable membrane against which salt water is pressurized. Other methods are based on the migration of ions between electrodes.

Descartes, René (1596–1650) French philosopher and mathematician. Descartes is often regarded as the father of modern philosophy. In 1619 he described an all-embracing science of the Universe. He evolved his own philosophical principles during the 1630s and 1640s. His works include *Discourse on Method* (1637), *Meditations on the First Philosophy* (1641) and *Principles of Philosophy* (1644). His methods of deduction and intuition inform modern metaphysics. By doubting all his ideas, he reached one indubitable proposition: "I am thinking," and from this that he existed: *cogito ergo sum* (I think, therefore I am). Descartes also founded analytic geometry, introduced the CARTESIAN COORDINATE SYSTEM and helped establish the science of OPTICS.

desensitization Medical treatment in which people who are allergic to certain substances (**allergens**) are injected with extracts of an allergen in a series of gradually increasing dosages. It is done in order to build up resistance to the allergen, thus decreasing sensitivity. The term desensitization is also used to describe a type of treatment in psychotherapy. *See also* ALLERGY

desert Area with a dry CLIMATE, which is sometimes defined as having a total annual rainfall of 250mm (10in) or less. The effectiveness of 250mm (10in) of rainfall can vary, however, depending on whether it falls within a short time or over a prolonged period. A desert is almost barren, although there are very few places with absolutely no vegetation – this occurs only on some patches of moving sand dunes or on bare rocky areas. Most deserts contain tufts of grass and scattered, usually thorny, bushes that have the ability to withstand long dry spells. In some deserts there are cactus plants and various species of flowering plant that have a very short life cycle. They spring up quickly after a shower of rain and go through their complete life cycle in a few days before dying and leaving seeds to lie dormant until the next rain. Some deserts contain large areas of sand dunes, but most deserts are rocky. There are bare rock areas, but generally there will be a cover of loose stones with patches of moving sand. The stones and sand are gradually broken down by mechanical WEATHERING and wind action (AEOLIAN FORMATION). The world's major deserts, including the Australian, Kalahari, and Sahara, are found in the HORSE LATITUDES, where the permanent high pressure causes drought throughout the year. Deserts occur in various w coast areas where the influence of cool CURRENTS offshore makes the land even drier, with less than 100mm (4in) rainfall in places; for example, the Atacama Desert in Peru (with the Humboldt Current offshore) and the Namib Desert in s Africa (with the Benguela Current offshore). There are deserts in the middle of the largest continents where no onshore winds can reach to bring any rainfall; for example, the Gobi Desert in Asia, and the Mojave Desert, Imperial Valley, and Death Valley in the USA. Some cold regions are also regarded as deserts because they have less than 250mm (10in) annual precipitation. Antarctica and Greenland are ice deserts, and even the TUNDRA areas of n Canada, n Siberia, and elsewhere have some similarities to deserts, although they have cool summers and very cold winters.

desertification Process by which a DESERT gradually spreads into neighboring areas of semidesert, transforming them into true desert. The change may result from a natural event, such as the destruction of vegetation by fire or a slight climatic change, but it occurs most frequently as a result of human activity. In many semiarid areas and dry grasslands, the vegetation becomes overgrazed by domestic animals, so that the land is left bare. Wind and rain then erode the soil, removing any residual fertility, so that no new vegetation can survive. Once the vegetation and soil have gone, the land becomes desert. An additional problem is the use of trees and bushes for firewood by the local people. With an increasing population there is an increasing need

for fuel, and so landscapes become depleted of trees. Once it has been destroyed, restoring the land is a long and slow process. Extensive desertification has been taking place in the Sahel, the s edge of the Sahara Desert, and there are also examples in s Africa, India, and many other places where there are too many people subsisting on an inadequate landscape.

desiccator Chemical apparatus used for drying substances or for keeping them dry. It consists of a glass vessel with a tight-fitting lid. Specimens are supported on a perforated platform above a lower chamber containing a desiccant, such as anhydrous calcium chloride.

desktop publishing (DTP) Use of a computer to prepare text and pictures for publication. The technique uses COMPUTER PROGRAMS that display documents on the computer screen. The operator can control the type font, size, line length, and column organization of text, and incorporate scanned images if necessary. The result can be output in forms ready for conventional printing and publishing.

destructive distillation Chemical process of breaking down complex organic substances by heating them in the absence of air. Substances such as coal and wood yield such useful products as fuel gases (coal gas) and tar (coal tar), leaving carbon in the form of coke. Before the availability of crude oil (petroleum), this was the usual method of obtaining industrial organic compounds.

destructive margin In PLATE TECTONICS, the boundary between two lithospheric plates (*see* LITHOSPHERE) that are moving towards each other, and where oceanic crust is being recycled into the mantle by SUBDUCTION. Destructive margins can be detected by shallow- to deep-focus EARTHQUAKES (the BENIOFF ZONE), ocean TRENCHES, and volcanic ISLAND ARCS. *See also* CONSTRUCTIVE MARGIN

detergent Synthetic chemical cleansing substance that, unlike soap, does not form scum in hard water. There are several types, the commonest being an alkyl sulfonate. Detergents have molecules that possess a long hydrocarbon chain (tail) attached to an ionized group (head). This chain attaches to grease and other nonpolar substances, while the ionized group has an affinity for water (so the grease is washed away with the water). Domestic detergents often contain additives such as perfumes and optical brighteners.

determinant A single numerical value associated with a MATRIX. Matrices can be used to transform any geometric figure in the plane or higher dimensional spaces – causing rotations, reflections, scalings or stretches. In this geometric interpretation of a matrix, the determinant denotes the increase in area or volume of the figure resulting from the transformation. Matrices can also be used to solve sets of SIMULTANEOUS EQUATIONS. The equations can be solved provided the determinant is not zero. The determinant for the matrix $A = \begin{bmatrix} a & b \\ c & d \end{bmatrix}$, is the number $\det(A) = ad - bc$.

detonator Device using sensitive chemicals (initiating explosives) to set off less sensitive main charges (high explosives). The detonating charge is generally housed in a thin-walled metal or plastic waterproof capsule, such as the "cap" at the base of a rifle cartridge. Detonators may be exploded electrically, by mechanical shock or by igniting a fuse.

detritus In geology, sediment deposited by natural forces. It is classed by the size of the particles. *See also* ALLUVIUM

deuterium Isotope (D or ^2H) of hydrogen whose nuclei (DUETERONS) contain a neutron in addition to a proton. For every million hydrogen atoms in nature, there are about 156 deuterium atoms. Deuterium occurs in water as D_2O (HEAVY WATER), from which it is obtained by ELECTROLYSIS. Heavy water is used as a moderator in some FISSION reactors and it could become a fuel in FUSION NUCLEAR REACTORS. Properties: r.a.m. 2.0144.

deuteron Nucleus of a DEUTERIUM (heavy hydrogen) atom, consisting of one proton and one neutron. The nucleus of ordinary hydrogen contains one proton only.

Devonian Fourth-oldest period of the PALEOZOIC era, lasting from 408 to 360 million years ago. It is sometimes called the Age of Fishes. Numerous marine and freshwater remains include jawless fish and forerunners of today's FISH. The first known land vertebrate, the amphibian *Ichthyostega*, appeared at this time. Land animals included scorpions, mites, spiders, and the first insects. Land plants consisted of tall CLUB MOSSES, HORSETAILS, and FERNS.

De Vries, Hugo (1848–1935) Dutch botanist. His experimental methods led to the rediscovery (1900) of Gregor MENDEL'S laws of HEREDITY and the development of a theory of MUTATION. In his work *The Mutation Theory* (1901–03), De Vries argued that genetic mutation was the chief engine of EVOLUTION.

dew Water droplets formed, usually at night, by CONDENSATION on vegetation and other surfaces near the ground. Hoar frost is formed when temperatures are below freezing. Fog also deposits moisture on exposed surfaces.

Dewar, Sir James (1842–1923) Scottish chemist and physicist who carried out research into materials at extremely low temperatures. In 1872 he invented the vacuum **Dewar** flask, the commercial adaptation of which is the Thermos flask. He built a device that could produce liquid oxygen and demonstrated that liquid oxygen and ozone are magnetic. Dewar also liquified hydrogen in 1898 and solidified it in 1899.

dew point Temperature at which a vapor begins to condense, for example when water vapor in the air condenses into clouds as the air becomes saturated with vapor.

dextran Stable, water-soluble POLYSACCHARIDE sugar. It is used as a substitute or extender for blood plasma in transfusions.

diabetes Any metabolic disorder in which there is exaggerated thirst and copious output of urine. The term generally refers to **diabetes mellitus**, a disease characterized by lack of the pancreatic hormone INSULIN needed for sugar METABOLISM. Failure of insulin production leads to a buildup of sugar in the bloodstream (**hyperglycemia**) and subsequently in the urine. Symptoms include abnormal thirst, overproduction of urine, and weight loss; degenerative changes occur in the blood vessels. Untreated, the condition progresses to diabetic coma and death. There are two forms of the disease. Type 1, or insulin-dependent, diabetes is the more severe form. It usually begins in childhood and is an AUTOIMMUNE DISEASE. People with type 1 diabetes owe their survival to insulin injections. In type 2 diabetes, which mostly begins after the age of 40, there is some insulin output but not enough for the body's needs. This milder form can usually be controlled with dietary restrictions, possibly together with oral drugs. Careful treatment of diabetes, including the control of high blood pressure (HYPERTENSION), can keep at bay serious complications such as blindness and kidney failure. Susceptibility to diabetes mellitus is inherited and the disease is slightly more common in males.

diagenesis Physical and chemical processes whereby sediments are transformed into solid rock, usually at low pressure and temperature. Pressure results in compaction, forcing grains together, and eliminating air and water, and cementation binds individual particles together by the precipitation of a secondary mineral, commonly an iron oxide, calcite or silica.

dialysis Process for separating particles from a solution by virtue of differing rates of DIFFUSION through a SEMIPERMEABLE MEMBRANE. In the artificial kidney unwanted (smaller) molecules of waste products (urea and some salt) are separated out to purify the blood. **Electrodialysis** employs a direct electric current to accelerate the process, which is especially useful for isolating proteins.

diamagnetism Type of MAGNETISM possessed by all materials in which an applied **magnetic field** induces a weak magnetization in the opposite direction.

diamond Crystalline form of carbon (C); the hardest natural substance. Diamond is found in kimberlite pipes and in alluvial deposits as cubic system octahedral crystals. It is brilliant, transparent to translucent, colorless or of many hues including yellow, green, blue, and brown, depending on the impurities contained. Bort, a variety of diamond inferior in crystal and color, and carborondo, an opaque gray to black variety called black diamond, as well as all other non-gem varieties are used in industry. Industrial diamonds are used as abrasives, bearings in precision instruments such as watches, and in the cutting heads of drills for mining. Synthetic diamonds, made by subjecting graphite with a catalyst to high pressure and temperatures of about 3,000°C (5,400°F), have been made since 1955, but are fit only for industrial applications. Diamonds are weighed in carats (0.2gm) and in points (1/100 carat). The largest producer of diamonds is Australia. Hardness 10; r.d. 3.5.

diapause Time break in the development of an INSECT. During diapause, growth stops and METABOLISM slows down. It occurs because of interruptions in an innate rhythm brought about by unfavorable changes in the environment, possibly because of seasonal changes in climate. In this way, the developing insect survives until conditions improve. *See also* BIOLOGICAL CLOCK

diaphragm Sheet of muscle that separates the ABDOMEN from the THORAX, the cavity of the chest, in mammals. During exhalation (breathing out) it relaxes and allows the chest to subside; on inhalation it contracts and flattens, causing the chest cavity to enlarge.

diastole Phase of the cardiac cycle in which the HEART muscle is relaxed and is filling with blood.

diathermy Medical treatment that uses heat generated by high-frequency (short-wave) alternating electric current. The heat stimulates local blood circulation and tissue repair. It is used to treat back pain, to aid muscle and tendon repair, and is often used in eye and neurosurgery.

diatom Any of a group of tiny, microscopic, single-celled ALGAE (phylum Bacillariophyta) characterized by a shell-like cell wall made of silica. The shell (**frustule**) consists of two halves that fit together like the two halves of a petri dish. Diatom frustules occur in a wide variety of highly symmetrical shapes. Diatoms live in nearly every environment that has water and is exposed to sunlight, including virtually all bodies of salt and fresh water and even soil, damp rocks, tree bark, and the undersides of icebergs. Light can pass through the transparent cell wall, and is used for PHOTOSYNTHESIS. The microscopic shells of long-dead diatoms accumulate in, for example, diatomite (kieselguhr), a mineral with abrasive, absorbent, and refractory uses.

diazepam Tranquilizer in the benzodiazepine group that also acts as a skeletal muscle relaxant. It is prescribed for acute anxiety, delirium tremens, and epilepsy and is also used in premedication for surgery. Side effects include drowsiness, lethargy, vertigo, and skin rashes.

diazo compound *See* AZO AND DIAZO COMPOUNDS

dicarboxylic acid In organic chemistry, a CARBOXYLIC ACID with two carboxyl groups (–COOH). Examples include ethanedioic acid (oxalic acid, $(COOH)_2$), hexanedioic acid (adipic acid, $(CH_2)_4(COOH)_2$), and phthalic acid $(C_6H_4(COOH)_2)$.

dichroism Property of some CRYSTALS that makes them appear to have a different color according to the direction of the incident light. The crystal allows light vibrations in one plane to pass through but absorbs light vibrations at right angles to this plane. Tourmaline is a natural dichroic crystal; polaroid is a synthetic material with this property. *See also* POLARIZED LIGHT

dicotyledon Any member of the class Dicotyledonae within the phylum Angiospermophyta – ANGIOSPERMS (flowering plants). Dicotyledonous plants are characterized by two seed leaves (COTYLEDONS) in the seed embryo. Other general features of dicotyledons include: broad leaves with branching veins; FLOWER parts (sepals, petals, stamens) arranged in whorls of fours or fives; VASCULAR BUNDLES arranged in a ring in the stem and ROOT; and a main central root (TAPROOT). There are about 250 families of dicotyledons, including the rose, daisy, and magnolia families. *See also* MONOCOTYLEDON

dielectric Nonconducting material such as an electrical insulator that separates two conductors in a CAPACITOR. The PERMITTIVITY of a dielectric is a measure of the extent to which it can resist the flow of charge. The dielectric strength is the maximum field that the dielectric can withstand without breaking down by becoming ionized. For many low-power applications such as tuning capacitors in radios, air is used as the dielectric.

Diels, Otto Paul Hermann (1876–1954) German chemist. Diels discovered (1906) the suboxide of carbon (tricarbon dioxide, C_3O_2), studied the structure and synthesis of cantharidine (a poison contained in the body fluid of blister beetles), and did research on STEROLS (solid, higher alcohols such as cholesterol). He was professor of chemistry (1916–48) at Kiel University. He shared the 1950 Nobel Prize for chemistry with the German chemist Kurt Alder (1902–58), with whom he discovered (1928) the synthesis of dienes (organic compounds containing two double bonds), known as the **Diels-Alder reaction**.

diesel engine INTERNAL COMBUSTION ENGINE in which the heat for igniting the fuel oil is produced by compressing air. The engine, invented by Rudolf DIESEL in the 1890s, is also referred to as a compression-ignition engine. The fuel-air mixture burns rapidly and expands to drive the pistons. Diesel engines usually have electric heaters called **glow plugs** for warming the cylinders during starting. Once the engine has fired, the heaters may be switched off.

Diesel, Rudolf (1858–1913) German inventor of the DIESEL ENGINE. Diesel developed, patented and built the engine in the 1890s. He later established an engine factory in Augsburg, Germany.

diesel fuel (diesel oil) Petroleum product heavier than kerosene but lighter than heating oil, used to power DIESEL ENGINES in industrial and agricultural vehicles such as trucks, buses, tractors, locomotives, and ships. There are many grades of diesel fuels, which, unlike gasoline, burn unevenly. They are graded against standardized mixtures of hexadecane and alpha methylnaphthalene to establish a cetane number. A cetane number of 30 to 45 is desirable.

diet Range of food and drink consumed by a person or animal. The human diet falls into five main groups of nutrients necessary for the maintenance of health: PROTEINS, CARBOHYDRATES, FATS, VITAMINS, and MINERALS. An adult requires daily about 1g (0.035oz) of protein for 1kg (2.2lb) of body weight. Beans, peas, fish, eggs, milk, and meat are important protein sources. The body breaks these down into AMINO ACIDS from which it synthesizes its own protein. Carbohydrates (starches and sugars), stored in

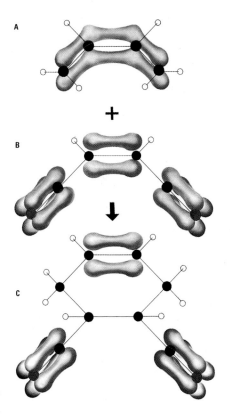

▲ Diels-Alder reaction
The Diels-Alder reaction serves an important role in many syntheses because it smoothly and stereospecifically unites two carbon compounds. A six-membered ring is formed by 1,4-addition of an alkene unit to a conjugated diene. The alkene unit (known as the dienophile) has electron-withdrawing groups attached which activate the diene. The illustration shows the Diels-Alder reaction using molecular orbital models. The addition reaction mechanism is shown by the rearrangement of electron clouds.

Key
A) butadiene. This is a conjugated diene (two double bonds separated by a single bond). It is shown in the s-cis form and undergoes 1,4 addition easily.
B) 1,2-dinitrile ethene. This is the dienophile, and the electron-withdrawing or accelerating groups are nitrile groups. It is these that activate the diene.
C) The diene and dienophile react to form 2,3-dinitrile benzene. The configurations of the diene and dienophile are retained in the product. This means that the reactants come together to give cis addition.

▲ **differential** The differential of a motor car transmits the rotary movement produced by the engine through right-angles to the half-shafts (6) driving the wheels. The pinion (2) of the propeller shaft (1) rotates the crown wheel (3) turning the pinion (4) of the bevel gears (5). The differential gears allow the wheels to turn at different speeds when the car turns a corner and the outer wheel rotates faster than the inner one.

the body as GLYCOGEN and fat, are the chief sources of energy and are found in cereals, root vegetables, and sugars. Carbohydrates make up the bulk of most diets. Fats in the diet, such as butter and vegetable oils, are a concentrated source of energy and aid in the absorption of the fat-soluble vitamins (vitamins A, D, E, and K). Vitamins function as coenzymes in important body processes. Water and minerals such as iron, calcium, potassium, and sodium are also essential. The specialized study of the human diet and nutrition is known as DIETETICS.

dietetics Specialized study of the human DIET and nutrition and its clinical management. A dietician plans what people should eat when they are sick, making sure they get what they need to aid recovery.

differential In mathematics, small change occurring in the value of a mathematical expression due to a small change in a variable. If $f(x)$ is a function of x, the differential of the function in response to a small change dx in x, is written df and is given by $f'(x)$ dx, where $f'(x)$ is the DERIVATIVE of $f(x)$.

differential In a vehicle, set of circular GEARS that transmits power from the engine to the wheels. When a car is turning a corner, the differential allows the outside drive wheel to rotate faster than the inner one.

differential calculus *See* CALCULUS

differential equation Equation containing DERIVATIVES. Differential equations are used in almost all areas of APPLIED MATHEMATICS. *See also* CALCULUS

differential geometry Type of GEOMETRY that uses CALCULUS notation to analyze geometric concepts such as curves and surfaces. For example, a curve, such as the path of a projectile, or the orbit of a spaceship, can be described mathematically in terms of three mutually perpendicular VECTORS, which vary independently along the length of the curve.

differentiation In mathematics, name given to the method of evaluating the DERIVATIVE of some given function. The techniques of differentiation and INTEGRATION together constitute the subject of CALCULUS and are frequently required in almost all branches of APPLIED MATHEMATICS. *See also* DIFFERENTIAL EQUATION

diffraction Spreading of a wave, such as a light beam or a sound wave, on passing through a narrow opening or hitting the edge of an obstacle. Diffraction provides information on the wavelength of light and the internal structure of the diffracting CRYSTALS. All waves are diffracted by obstacles. *See also* ELECTRON DIFFRACTION

diffraction grating Substrate consisting of parallel equidistant lines (as many as 1,500 per mm) for producing spectra by DIFFRACTION of light. Transmission gratings are ruled on glass and are transparent; reflection gratings are opaque – they are ruled on metal films. Such gratings are much used in SPECTROSCOPY. *See also* DIFFRACTION; INTERFERENCE

diffusion Movement of a substance in a mixture from regions of higher concentration to regions of

▲ **diffraction** When waves (1) pass through a narrow gap (2) they are diffracted (3) and spread out in concentric rings (4). An example would be ocean waves passing through the entrance of a harbor.

lower concentration, resulting from the random motion of the individual atoms or molecules. Diffusion ceases when there is no longer a concentration gradient. Its rate increases with temperature, since average molecular speed also increases with temperature. The process occurs quickly in gases and liquids, much more slowly in solids.

digestion Process of breaking down food mechanically and chemically into smaller molecules that can be readily absorbed. Digestion occurs mainly by means of chemical agents called ENZYMES. *See also* DIGESTIVE SYSTEM

digestive system (alimentary system) Group of organs of the body concerned with the DIGESTION of foodstuffs. In humans the digestive system begins with the mouth, where the action of teeth and enzymes in saliva begin the process of breaking down the food. It continues with the ESOPHAGUS, which carries food into the STOMACH. The stomach leads to the small intestine, which then opens into the COLON. After food is swallowed, it is pushed through the digestive tract by PERISTALSIS. On its journey, food is transformed into small molecules that can be absorbed into the bloodstream and carried to the tissues. CARBOHYDRATE is broken down to sugars, PROTEIN to AMINO ACIDS, and FAT to FATTY ACIDS and GLYCEROL. Indigestible matter, mainly cellulose, passes into the rectum, and is eventually eliminated from the body (as feces) through the ANUS.

digit Any numeral from 0 to 9.

digital Describing information expressed in terms of numbers. DATA, such as words, pictures, and music, are represented as a series of numbers (1s and 0s) in a BINARY SYSTEM, such as a COMPUTER. Digital also refers to any way of representing information as numbers as opposed to a continuously varying ANALOGUE fashion. For example, the figures of a digital clock are the counterpart to the ANALOGUE, or continuous, display of a traditional clock face. A COMPACT DISC stores information digitally, in the form of pits in its surface that represent 0s and 1s. A vinyl LP, by contrast, stores information in analogue form, as a "wavy" groove.

digital audio tape (DAT) Technology for recording data in DIGITAL form on MAGNETIC TAPE. DATs are smaller and longer than analog cassettes. They are used primarily for computer backups and studio-recording. In the latter, the original sound signal is encoded to form patterns of equal-strength pulses. These DIGITAL SIGNALS are recorded onto tape. On playback, the patterns are detect-

ed and decoded to produce an identical signal. *See illustration on page 152*

digital signal Group of electrical or other pulses in a COMPUTER or communications system. They may represent data, sounds or pictures. Pulses in a

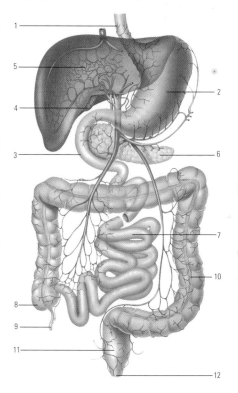

▲ **digestive system**
The digestion and absorption of food in humans takes place within the digestive tract, a coiled tube some 10m (33ft) long that links mouth to anus.

Food is passed down the esophagus (1) to the stomach (2), where it is partially digested. Chyme is released into the duodenum (3), the first part of the 7m (23ft) of small intestine. The duodenum receives bile secreted by the gall bladder (4) in the liver (5), and enzymes secreted by

the pancreas (6). Most absorption takes place in the jejunum and ileum, the remaining parts of the small intestine (7). Any residue passes into the cecum (8), the pouch at the start of the large intestine. At one end of the cecum is the 10-cm (4-in) long vermiform appendix (9), which serves no useful purpose in humans. Water is re-absorbed in the colon (10). Feces then form and collect in the rectum (11) before being expelled as waste matter through the anus (12).

stream of digital signals are represented by 1s and 0s in the BINARY SYSTEM.

dike In engineering, a barrier or embankment designed to confine or regulate the flow of water. Dikes are used in reclaiming land from the sea by sedimentation (as practiced in The Netherlands), and also as controls against river flooding.

dike In geology, an INTRUSION of IGNEOUS ROCK forming a seam that cuts across the structure of the surrounding rock, and is often vertical or near-vertical. The cross-cutting nature indicates that a dike is younger than the rocks it has intruded. At the edge of a dike the rock is finer-grained than at the center, as a result of cooling and solidifying more quickly; this is termed a **chilled margin**. The rock immediately next to the dike may be altered by CONTACT METAMORPHISM, for example in intruded limestones there may be a thin zone of marble alongside each dike.

dimension In mathematics, the spatial dimension is the number specifying the extent of an object in different directions. A figure having length only is said to be one-dimensional; a figure having area but not volume, two-dimensional; and a figure having volume, three-dimensional. More generally the dimension of a figure equals the number of coordinates that are needed to specify all points of the figure. For example, in a plane, which is two-dimensional, two coordinates are required (x, y). *See also* COORDINATE SYSTEM

dinitrogen oxide (nitrous oxide, N_2O) Colorless gas that has a pleasant odor and is sometimes used as an anesthetic or analgesic during surgical or dental operations. It is also known as "laughing gas" since it produces exhilaration, sometimes accompanied by laughter. It is also used in making pressurized foods.

dinosaur (Gr. terrible lizard) Any of a large number of REPTILES that lived during the MESOZOIC era, between 225 and 65 million years ago. They appeared during the TRIASSIC period, survived the JURASSIC and became extinct at the end of the CRETACEOUS. Dinosaurs were mostly egg-laying animals, ranging in size from 30in (91 cm) to the 90ft (27m) diplodocus. There were two orders: **Saurischia** ("lizard hips"), included the bipedal carnivores and the giant herbivores; the **Ornithis-**

◀ **digital audio tape** A digital audio tape (DAT) recorder records sound, an analogue signal, in digital form. The analogue signal enters via a microphone (1) and passes through a converter (2), which transcribes the sound wave into a series of zeros and ones. Two magnetic heads, tiny electromagnets (3) in a rotating drum (4), receive the digital signal as electrical pulses that polarize diagonal strips of magnetic tape (5) (moving right to left, 6) as it is scanned diagonally (7). The heads align the magnetic elements of the tape representing a zero or a one. Each head records one half of a stereo recording. One head records parallel to the tape (8), one perpendicular (9) to avoid interference. The units of an blank tape are jumbled (10). When a tape (11) is played, the head reads the polarization of the tape.

cia ("bird hips") were smaller herbivores. Their posture, with limbs vertically beneath the body, distinguish them from other reptiles. There is evidence that some birds are the living descendants of ornithischians. Many theories are advanced to account for their extinction. It is possible that, as the climate changed, they were incapable of swift adaptation. A more catastrophic theory is that they died because of the devastating atmospheric effects from the impact of a large meteor.

diode Electronic component with two ELECTRODES, used mainly as a RECTIFIER to convert alternating current (AC) to direct current (DC). The SEMICONDUCTOR diode, which has largely replaced the ELECTRON TUBE, usually has a single *p-n* junction. It allows electrons to flow freely in only one direction (*n*-type material to *p*-type), but only a small current flows in the reverse voltage direction. A **Zener** diode blocks current flow until a critical voltage is reached, when it breaks down and conducts. *See also* ELECTRIC CURRENT

dioecious plant Plant bearing either female, **pistillate** flowers (with CARPELS only) or male, **staminate** flowers (with STAMENS only); but not both.

Diophantus (active AD 250) Greek mathematician, believed to have produced one of the earliest works in ALGEBRA. He is best known for his investigation and description of a set of algebraic equations now called **Diophantine equations**. Work in this area is known as **Diophantine analysis**.

diopter Optical unit of magnifying power of a lens system. The power of a lens is equal to one meter divided by the focal length in meters. An algebraic sign (+ or −) indicates whether the system is **convergent** or **divergent**; for example, a diverging lens of focal length 1/3 meter has a power of −3 diopters. The power of a combination of lenses is the sum of their individual powers.

diorite Intermediate coarse-grained IGNEOUS ROCK, similar to GRANITE in its texture but made up mainly of plagioclase feldspar and hornblende, with BIOTITE or augite. It is usually dark gray.

dip In geology, the steepest angle of a tilted plane of a rock (such as bed, joint, cleavage, and fault). It is measured from the horizontal, 90° being vertical, using a CLINOMETER.

dip, magnetic *See also* INCLINATION, MAGNETIC

diphtheria Acute, infectious disease characterized by the formation of a membrane in the throat that can cause asphyxiation; there is also the release of a toxin that can damage the nerves and heart. It is caused by a bacterium, *Corynebacterium diphtheriae*, which often enters through the upper respiratory tract. It is treated with antitoxin and ANTIBIOTICS. Once common throughout the temperate regions, the disease is now rare in countries where there has been an active program of immunization.

diplococcus Genus of BACTERIA characterized by gram-positive, spherical cells that grow in pairs. It includes the pneumococcus, which causes pneumonia.

diploid Term describing a nucleus or CELL that has its CHROMOSOMES in pairs. Almost all animal cells are diploid, except GAMETES (reproductive cells), which are HAPLOID (containing only a single set of chromosomes). The cells of ANGIOSPERMS (flowering plants) and GYMNOSPERMS are also diploid. Algae and lower plants, such as ferns and mosses, have two generations in their life cycle, one diploid and the other haploid. In a diploid cell, the chromosomes of each pair carry the same GENES. *See also* ALTERNATION OF GENERATIONS

diplopod Any member of the class Diplopoda, ARTHROPODS found worldwide. They are known as millipedes. Some attack plants, while others are predatory carnivores. They may have as many as 200 pairs of legs.

dipole Separation of electric charge in a molecule. In a COVALENT BOND between two atoms the electron pair is not necessarily equally shared between them. In hydrogen chloride (HCl), the electrons are attracted towards the more electronegative chlorine atom, giving it a partial negative charge and leaving an equal positive charge on the hydrogen atom. Such dipoles contribute to the chemical properties of molecules. Magnetic substances have dipoles consisting of north and south poles; they may be ordered or disordered to give strong or weak magnetic properties.

dip pole Either of two imaginary points on the Earth's surface where the direction of the Earth's magnetic field is vertical – downwards at the north magnetic pole and upwards at the south pole – and are offset in respect to the geographical poles, each by a different amount so that the S dip pole is not exactly opposite the N dip pole.

dipteran Any member of the order Diptera, the true flies. Adult flies have soft bodies and one pair of wings, the other being reduced to knob-like **halteres**; they have COMPOUND EYES and sucking mouthparts. They have a complete life cycle: the adult lays eggs that hatch into LARVAE (maggots); these pupate and become adults. Within the

Diptera order are many important pests, such as mosquitoes, house flies, and others that attack human beings, animals or crops or carry diseases. *See also* INSECT

Dirac, Paul Adrien Maurice (1902–84) British physicist renowned for his work on QUANTUM MECHANICS. He extended this to combine RELATIVITY and quantum-mechanical descriptions of electron properties (*see* WAVE FUNCTIONS). This work led to his prediction of the existence of the POSITRON, which was later discovered by Carl ANDERSON. Dirac shared the 1933 Nobel Prize for physics with Erwin SCHRÖDINGER for their work in QUANTUM THEORY. He was professor of physics at Florida State University from 1971 until his retirement.

direct current (DC) *See* ELECTRIC CURRENT

directional antenna Radar or radio antenna for receiving or transmitting signals in one prime direction. A radio direction-finder for example, has a rectangular loop antenna that receives maximum signal when it is broadside to the station, and minimum signal when it faces the station end-on. The orientation of the antenna is indicated on a direction scale.

direction finding (df) In NAVIGATION, a method that determines position by the intersection of two projected bearing lines obtained from known landmarks. In communications, radio df instruments can determine a bearing by using a DIRECTIONAL ANTENNA which locates two radio signals from transmitters at known points.

direct motion Orbital motion of a planet, comet or other heavenly body around the Sun or of a satellite around its primary (the planet around which it orbits) in a west-to-east direction. Also the apparent west-to-east movement of a planet against a stellar background, as seen from the North Pole.

disaccharide Type of sugar (including common sugar) formed by the condensation of two MONOSACCHARIDES with the removal of water. Cane sugar (sucrose) is a disaccharide which, on HYDROLYSIS with dilute acid, yields both glucose and fructose (monosaccharides). Lactose (the sugar in milk) and maltose are other important disaccharides. *See also* POLYSACCHARIDE

disc In astronomy, circular appearance of the Sun, Moon or a planet, especially when viewed through a telescope. It contrasts markedly with the sharp point source image of a star.

discharge (symbol Q) Measure of waterflow at a particular point, measured in cubic meters per

second ($m^3 s^{-1}$). Discharge can be calculated with the formula $Q = V \times A$, where V is the velocity of the water and A the cross-sectional area of the river channel.

discharge tube Glass envelope containing gas at low pressure, which gives light due to the collision of electrons with gas molecules. Discharge lamps, filled with one or more of various gases, are used to give colored light: mercury vapor gives purple, sodium vapor gives yellow and neon gives red. Early discharge tubes (called **Geissler tubes**) were used as toys. A FLUORESCENT LAMP is a discharge tube coated internally with a fluorescent substance.

discontinuity *See* MOHO

disease In medicine, any departure from health, with impaired functioning of the body as a whole or one or more of its parts. Disease may be: **acute**, producing severe symptoms for a short time; **chronic**, lasting a long time; or **recurrent**, returning periodically. Treatment depends on the cause and course of the disease but, in general, may be **symptomatic** (relieving symptoms, but not necessarily combating a cause) or **specific** (attempting to cure an underlying cause). Disease prevention involves eradication of harmful organisms, IMMUNIZATION, personal hygiene, public health measures, health education, and routine medical checks.

disinfectant (germicide) Agent that kills, or inhibits the growth of, bacteria and other microorganisms on inanimate objects. ANTISEPTIC is used on contaminated living tissue. Joseph LISTER introduced carbolic acid (PHENOL) as a medical disinfectant in the 1870s. Today, chlorinated phenols are used in pharmaceutical products. Chlorine and chlorine compounds are commonly used to kill bacteria, especially in water. Iodine is used in food preparation. Alcohol is an effective disinfectant. Ammonium is the most common agent in household cleaners. It also acts as a DETERGENT.

disjunction Logical proposition produced by joining two simple propositions by the word "or." An example is the proposition "John is intelligent or John is modest"; it is false if both parts are separately false, otherwise it is true. The disjunction is used in mathematics – in which the proposition is true if *either* or *both* components are true. The disjunction of two simple propositions, P or Q, is written $P \lor Q$, and read "P or Q." *See also* CONJUNCTION

disk (disc) Form of computer DATA storage. Disks come in many different forms; some use magnetic methods to store data (*see* MAGNETIC DISK, while others use optical systems such as the COMPACT DISC (CD-ROM).

disk operating system (DOS) Computer OPER-ATING SYSTEM, developed in the early 1980s by Bill GATES' Microsoft Corporation for use with early IBM personal computers (PCs). DOS is still used by many personal computers worldwide, although it is rapidly being replaced by Windows-based operating systems (also developed by Microsoft).

dislocation In medicine, displacement of bones forming a joint, accompanied by damage to supporting ligaments and the enclosing capsule. It gives rise to swelling and loss of function; it is treated by manipulation, usually under a general anesthetic. In physics, a dislocation is a fault in the CRYSTAL LATTICE.

dispersion, colloidal *See* COLLOID

dispersion, wave *See* WAVE DISPERSION

displacement In geology, relative movements on either side of a FAULT. It incorporates the direction of change and the specific amount of the movement. **Lateral** displacement is described as strike slip and strike separation, whereas **vertical** displacement is known as dip slip and dip separation.

disproportionation Simultaneous OXIDATION-REDUCTION of the same chemical substance. An example is the disproportionation of copper(I) chloride, involving oxidation to copper(II) chloride and reduction to metallic copper:
$$2CuCl \rightarrow CuCl_2 + Cu.$$

dissection Systematic cutting into plant or animal tissue, usually after death, to explore its anatomy or to discover abnormalities that may help in the understanding of the cause and effect of disease. It is the foundation of a proper understanding of how the healthy body works.

dissociation In chemistry, a reaction in which the molecules of a compound split into smaller components. In many dissociation reactions the smaller components are able to recombine under other conditions; such dissociation is called **reversible** dissociation. For example, heating causes the reversible dissociation of hydrogen iodide into hydrogen and iodine. The formation of IONS (electrically charged particles) when an acid dissolves in water may also be called dissociation. Known as **ionic** dissociation, this reaction explains many properties of electrolytic solutions. The extent to which dissociation proceeds is indicated by the **dissociation constant** (which is equal to the EQUILIBRIUM constant of the reaction).

distance ratio (velocity ratio) In a simple MACHINE, distance moved by the point of applica-tion of the effort (input force) divided by the distance moved by the point of application of the load (output force). *See also* FORCE RATIO

distillation Extraction of a liquid by boiling a solution in which it is contained and cooling the vapor so that it condenses and can be collected. The method is used, in the process of DESALINA-TION, to separate liquids that are in solution or liquid solvents from dissolved solids. Distillation is used to yield drinking water from sea water and to produce alcoholic beverages. **Fractional** distilla-tion, which uses a vertical column for condensa-tion, is the method used in oil refineries to separate the various fractions (portions) of crude oil. *See also* FRACTIONATION

distilled water Pure water prepared by boiling tap water and condensing the steam. It is essential for some chemical reactions, in which even the small traces of dissolved salts (such as the carbon-ates of calcium and magnesium) and other impuri-ties present in tap water are intolerable. For many applications it is being replaced by de-ionized water. *See also* WATER SOFTENER

distilling Production of spiritous liquors by DIS-TILLATION, especially of alcohol (ethanol, ethyl alcohol). In such drinks as wines, FERMENTATION using yeast produces a maximum alcohol content of around 15%. Distillation concentrates the alco-hol to a much higher degree in the distilled spirit: further alcohol may be later added as needed for the spirit's final desired proof content. Most spirits (whisky, brandy, gin, rum, vodka, and so on) are around 40% proof.

distributive law Rule of combination in mathe-matics in which an operation applied to a combina-tion of terms is equal to the combination of the operation applied to each individual term. Symbol-ically this is written
$$a * (x + y) = a * x + a * y$$
where * is an operation such as multiplication and a, x, and y are numbers. Thus, in arithmetic $3 \times (2 + 1) = (3 \times 2) + (3 \times 1)$: the multiplication is distributive over the addition.

distributor In a motor vehicle, an electrical device for distributing synchronized pulses of high-voltage secondary current from the induction coil, via a rotating contact (rotor arm), to the vari-ous sparking plugs of an INTERNAL COMBUSTION ENGINE in their proper firing order.

diuretic Drug used to increase the output of urine. It is used to treat raised blood pressure and edema, the accumulation of fluid that occurs in some diseases.

diurnal rhythm *See* CIRCADIAN RHYTHM

divergence Mathematical property of an infinite series (or sequence) of not having a finite limiting value. Such a series is said to diverge. The harmonic series, $1 + 1/2 + 1/3 + 1/4 + ...$, is an example of a divergent series since the sum has no finite value. *See also* CONVERGENCE

diving bell Hollow structure providing a dry environment for underwater workers. Early diving bells were bell-shaped, filled with compressed air, and open at the bottom to give access to the seabed. The BATHYSPHERE, which made its first dive in 1930, was spherical, made of steel, and could withstand considerable pressure, but it was restricted to depths within reach of supply and winch cables. It has been replaced by the BATHYSCAPHE.

division Arithmetical operation signified by $\div$ or $/$, interpreted as the inverse of MULTIPLICATION. The quotient of two numbers, $a \div b$ (or a/b), is the number that must be multiplied by b to give a. a is called the **dividend** and b the **divisor**.

DNA (deoxyribonucleic acid) NUCLEIC ACID that is the major constituent of the CHROMOSOMES of EUKARYOTE cells and some VIRUSES. DNA is often referred to as the "building block" of life since it stores the GENETIC CODE that functions as the basis of HEREDITY. The molecular structure of DNA was first proposed by James WATSON and Francis CRICK in 1953. It consists of a DOUBLE HELIX of two long strands of alternating sugar molecules and phosphate groups linked by nitrogenous bases. The whole molecule is shaped like a twisted rope ladder with the nitrogenous bases forming the rungs. The sugar is deoxyribose, and the four bases are ADENINE (A), CYTOSINE (C), GUANINE (G), and THYMINE (T). The bases are always paired in the same way: adenine always binds with thymine, guanine with cytosine. This regularity ensures accurate self-replication. During replication the two DNA strands separate, each providing a template for the synthesis of a new strand of RNA (MESSENGER RNA). This process of TRANSCRIPTION, mediated by ENZYMES, results in an identical copy of the original helix. The amount of DNA is constant for all cells of a species of animal or plant. In the process of replication the amount of DNA doubles as the chromosomes replicate themselves before MITOSIS; in the gametes, ovum and sperm (HAPLOID cells), the amount is half that of the body cells (*see* MEIOSIS). A base and its associated sugar and phosphate are known as a NUCLEOTIDE; the whole chain is a polynucleotide chain. The genetic code is stored in terms of the sequence of nucleotides: three nucleotides code for one specific AMINO ACID and a series of them constitute a GENE. *See also* BIOTECHNOLOGY; GENETIC ENGINEERING; MUTATION; RECOMBINANT DNA RESEARCH

DNA hybridization Method of comparing DNA from different organisms in order to discover genetic relationships between species. Single strands of DNA from each species are placed together and allowed to react. Some strands will "hybridize" to form double strands (the usual structure of DNA) and the extent to which they do so indicates how many base sequences are complementary. This in turn is a a measure of how similar the genes are.

Dobzhansky, Theodosius (1900–75) US geneticist, b. Russia. An authority on HUMAN EVOLUTION, he was influential in the development of population GENETICS as a separate study. His writings include *Genetics and the Origin of Species* (1937), *Mankind Evolving* (1962), and *Genetics of the Evolutionary Process* (1970).

dock In marine engineering, a wharf, pier, the waterway alongside or an artificial basin where ships are tended. A **wet** dock has watertight gates so that a ship inside does not rise and fall with the tides. A **dry** dock is an enclosure that can be pumped dry. It enables a ship or boat to be repaired below its water line. Floating docks can tend ships at sea and lift them out of the water for repairs.

Doisy, Edward Adelbert (1893–1986) US biochemist. He isolated the female sex hormones estrone (1929) and estradiol (1935). Doisy was awarded, with Henrik DAM, the 1943 Nobel Prize for physiology or medicine for his chemical analysis of vitamin K, which Dam discovered in 1934 and Doisy isolated in 1939.

doldrums Region of the ocean near the Equator, characterized by calms, and light and variable winds. It corresponds approximately to the belt of low pressure around the Equator.

dolerite Medium-grained, basic, dark-colored INTRUSIVE ROCK found in DIKES, SILLS, and VOLCANIC PLUGS.

Doll, William Richard (1912–) British physician who first demonstrated the correlation between cigarette smoking and lung CANCER. Working with Professor Bradford Hill (1897–1991), Doll announced his findings in 1950 based on an analysis of statistics showing the incidence of lung cancer in smokers and non-smokers. They later showed that the risks decrease when a smoker stops smoking.

dolomite Carbonate mineral, calcium-magnesium carbonate ($CaMg(CO_3)_2$), found in altered limestones. It is usually colorless or white. A calcite-like rhombohedral class prismatic crystal, dolomite is often found as a gangue mineral in hydrothermal veins, particularly associated with galena and sphalerite. Dolomite is also the name given to a pearly white or pink SEDIMENTARY rock, probably formed by the alteration of limestone by seawater, where 90% or more of the calcite has been replaced by calcium magnesium carbonate. Hardness 3.5–4; r.d. 2.8.

Domagk, Gerhard (1895–1964) German chemist. In 1927 he was made director of the research institute of the I.G. Farben industrial works. Domagk is known for his discovery (1932) of the antibacterial properties of the dye prontosil. The first SULFONAMIDE DRUG, sulfanilamide, was synthesized from the dye. Domagk was awarded the 1939 Nobel Prize for physiology or medicine, but was prevented by Nazi decree from accepting it until 1947.

domain In mathematics, a set of values that can be assigned to the independent variable in a FUNCTION or relation; the set of values of the dependent variable is called the **range**. For example, let the function be $y = x^2$, with x restricted to 0, 1, 2, 3, −3, and −5. Then y takes the values 0, 1, 4, 9, and 25 respectively. The domain is {0, 1, 2, 3, −3, −5} and the range is {0, 1, 4, 9, 25}.

domain In certain classification schemes, a category that is recognized as higher than KINGDOM. In these schemes, the two subkingdoms of PROKARYOTAE (ARCHAEBACTERIA and the EUBACTERIA) constitute two domains, called **Archaea** and **Bacteria**, while all other living organisms are included in a third domain, the EUKARYA.

domain *See* MAGNETISM

dominant In genetics, term that describes the ALLELE of a heterozygous pair of alleles that manifests itself over the other. For example, if an offspring has one allele each for brown eyes and blue eyes, the brown-eyed allele will dominate and manifest itself over the RECESSIVE, blue-eyed allele, although he or she will retain the ability to pass on recessive genes to his or her children. *See also* HETEROZYGOTE; HOMOZYGOTE

dongle In computing, device that prevents the illegal use of a commercial COMPUTER PROGRAM. It is usually attached between the computer and its printer (in the printer port). Without the dongle, the particular program – or a pirated copy of it – will not run.

dopamine Chemical normally found in the corpus striatum region of the human BRAIN, where insufficient levels are associated with PARKINSON'S DISEASE. Dopamine is a NEUROTRANSMITTER and a precursor in the production of ADRENALINE and NORADRENALINE. The low levels in Parkinson's disease can be treated by administering the drug L-DOPA.

Doppler, Christian Johann (1803–53) Austrian physicist and mathematician famous for his discovery of the DOPPLER EFFECT. He first described it in a paper on DOUBLE STARS published in 1842. Educated in Salzburg and Vienna, Doppler became professor of physics at Vienna in 1850.

Doppler effect Change in frequency of a wave (such as sound, light or radio) when there is relative motion between the wave source and the observer or detector. The amount of change depends on the velocities of the wave, source, and observer. With a sound wave, the effect is demonstrated by the drop in pitch of a vehicle's siren as it passes the observer. With light, the velocity of the source or observer must be large for an appreciable effect to occur, as when light received from a rapidly receding galaxy has its spectral lines shifted towards the red end of the spectrum. *See also* NAVIGATION; RED SHIFT

Doppler radar Essentially a radar navigational aid that relies on the DOPPLER EFFECT – the change in frequency (or wavelength) due to motion of an object relative to an observer. Electronic acces-

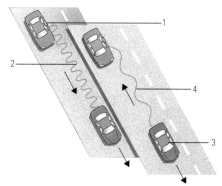

▲ **Doppler radar** The portable radar devices, speed guns (1), used by police to catch speeding motorists use the Doppler effect to measure speed. The "guns" send out a beam of radio waves (2), the frequency of which is known. When the wave strikes a moving car (3) the returning signal (4) has a different frequency. The equipment calculates the difference between the two frequencies and in this way the speed of the motorist is ascertained.

sories are used to measure the difference in frequency between the emitted pulse and the pulse received back. This shows the velocity of the target from which the radar beam has been reflected. This information is displayed on the radar screen.

Dorado (Swordfish) Small constellation of the S sky. It has no stars brighter than the third magnitude but it is notable for containing most of the Large Magellanic Cloud (Nubecula Major).

dormancy Temporary state of inaction or reduced METABOLISM. Animals may become dormant by hibernating during winter months when food resources are scarce; dormant plant seeds for a time cease to grow or develop. An organism can return to a fully active state when conditions, such as temperature, moisture or day length, change.

Dornier, Claude (1884–1969) German aircraft manufacturer who designed and manufactured the first all-metal airplane. While working with Ferdinand von Zeppelin, the AIRSHIP manufacturer, Dornier founded an airplane company and built wooden and metal planes that served in World War I (1914–18). In 1929 he designed the DO-X, a 12-engine passenger plane. From the late 1940s, Dornier's company built US-designed aircraft.

dorsal Upper surface of an organism – the surface that is turned away from the ground or other support. In an animal that walks on all fours, the dorsal surface is its back. By extension, in all VERTEBRATES it is the surface along which the spine runs. In animals that walk on two legs, such as humans, it is therefore the surface that faces backwards, or the posterior surface. On FISH, the dorsal fin is the main fin running along the top of the body.

DOS Acronym for DISK OPERATING SYSTEM

dosimeter Instrument used for measuring RADIATION dose, usually a pocket ELECTROSCOPE. A quartz fiber, after being charged, is viewed against a scale; its deflection across the scale on discharge gives a measure of the dosage of radiation to which it was exposed.

double galaxy Two adjacent galaxies orbiting around a common center of gravity, rather like a BINARY STAR. The components are often linked by tenuous bridges of intergalactic matter.

double helix Structure of DNA (deoxyribonucleic acid), the basic store of genetic information in the cells of each living organism. First demonstrated in 1953 by Francis CRICK and James WATSON, the knowledge of the structure of the DNA molecule has since revolutionized biological science.

double star Two stars that appear close together in the sky. There are two types of double star. If the stars are genuinely close together in space and are connected by gravity, they are known as a BINARY STAR. Two stars that are quite distant from each other, but appear close together as a result of chance alignment, are known as an **optical double**.

Down's syndrome (trisomy 21) Chromosomal abnormality giving rise to degrees of mental retardation ranging from mild to severe; the syndrome also brings a decreased life expectancy and may involve physical problems such as heart and respiratory disorders. The syndrome was first described by the British physician John Down (1828–96). It is caused by the presence of an extra copy of CHROMOSOME 21, so can be detected by counting chromosomes in the cells of the fetus during prenatal testing. The risk of having a child with the syndrome increases with maternal age. The condition was originally termed "mongolism" by Down, a reference to the characteristic flat facial appearance and slanting eyes of affected individuals.

Draco (Dragon) Long, winding constellation of the N sky; it is the eighth largest constellation, and extends between Ursa Major and Ursa Minor.

drag Force opposing the motion of a body through a FLUID (gas or liquid). Aircraft, for example, experience drag (or air resistance) as the friction of air over external surfaces and other parts of the aircraft, such as a fixed undercarriage, not associated with the development of lift. To combat drag and therefore make them more efficient, aircraft and cars are designed to be as streamlined as possible. *See also* AERODYNAMICS

drainage basin (catchment area) Region from which all precipitation, such as rain and melted snow, flows to a single stream or system of streams. By identifying such a basin, scientists can calculate DENUDATION rates and moisture balances from various hydrological measurements such as EVAPORATION rates. The boundary of a drainage basin is called a WATERSHED. The shape of a drainage basin may help in forecasting river flooding.

drainage network Relationship between the sizes, frequency, lengths, and patterns of streams within a DRAINAGE BASIN. The drainage network can by analyzed in any combination of these variables to help describe the characteristics of a drainage basin.

drainage system In geology, the system by which surface water is collected and removed by streams, rivers, and lakes. The pattern of drainage is determined by the type of rock and the slope of the land over which the water flows.

Drebbel, Cornelius Jacobszoon van (1572–1634) Dutch inventor who built the first submarine in 1621. It traveled along the River Thames, London, at a depth of 4–5m (12–15ft). James I is reputed to have ridden in it. Covered with greased leather, the craft was propelled by oars through watertight flaps.

Drew, Charles Richard (1904–50) US physician. His research at the Columbia Medical Center, US in 1940 led to the discovery that BLOOD PLASMA could replace whole blood transfusions. During World War II he supervised the American Red Cross blood-donor project but was not allowed to donate blood himself because he was African American; his continuing protest led to a change in policy.

drift In geology, gradual change in the land mass, related to soil CREEP. In oceanography it indicates slow, oceanic circulation. Sedimentary drift is a general description of surface debris carried by either a river or a glacier. It may also indicate an accumulation such as a snowdrift or sand drift. CONTINENTAL DRIFT is the movement of continents.

drilling rig Platform carrying a derrick and all the associated equipment used for drilling for OIL or natural gas below the seafloor. Most rigs have long legs (usually three or four) located on the seabed, with a working platform up to 30m (100ft) above the water. They are generally sited in comparatively shallow water on the continental shelf up to 200km (125mi) offshore. The gas or oil is transported ashore either in tankers or through undersea pipelines. Drilling rigs are sited in the North Sea and off the coasts of Australia, California, and the N of South America. *See also* OIL EXTRACTION; OIL WELL

drip-feeding Introduction of liquid nourishment directly into the veins of someone who is unable to eat or drink due to unconsciousness, weakness, or illness. Glucose sugar solution is commonly drip-fed; it is held in a sterile container and passes under gravity through a tube which terminates in a vein – usually in the forearm.

drought Condition that occurs when EVAPORATION and TRANSPIRATION exceed PRECIPITATION for long periods. Four kinds are recognized: **permanent**, typical of desert and semiarid regions; **seasonal**, in climates with well-defined dry and rainy seasons; **unpredictable**, an abnormal failure of expected rainfall; and **invisible**, when even frequent showers do not restore sufficient moisture.

drowned coast (submerged shoreline) Coastal strip that has been submerged under the ocean, either because the sea level has risen, or because the land has sunk. Valleys become flooded and hills become islands.

drug In medicine, any substance used to diagnose, prevent or treat DISEASE or aid recovery from injury. Still today many drugs are obtained from natural sources: plants, animals or minerals. Many ANTIBIOTICS, for instance, have been developed from microbes present in the soil. However, since the time of Paul EHRLICH, the German chemist who coined the term "magic bullet," scientists have sought to develop synthetic drugs that will work only on target cells or microorganisms, leaving other tissues unaffected. There is a vast range of drugs in circulation. The most obvious magic bullet effect is that of the antibiotics, which destroy invading bacteria. Some drugs interfere in physiological processes – for instance, anticoagulants render the blood less prone to clotting. Drugs may be given to make good some deficiency, such as hormone preparations that compensate for an underactive gland. There are a number of routes of drug delivery: orally, by injection, as a lotion or ointment, or by inhalation, pessary or transdermal patch. As well as direct effects, many drugs, too, are associated with side effects. If these become too problematic, the dosage may have to be adjusted or a different drug substituted. Also, some people have an ALLERGY to particular drugs. Increasingly, new drugs are being developed and prescribed to combat the effects of certain psychological disorders. The use of some drugs on both psychological and physiological conditions can result in a dependance on that drug, a situation doctors are very careful to avoid where possible. Drugs are subject to extensive clinical trials before being licensed for use. The word is also used to describe illegal or controlled substances, such as COCAINE and ECSTASY (MDMA).

drug addiction Psychological or physical dependence on a drug. Physical addiction is often manifested by symptoms of withdrawal (such as vomiting and convulsions) if the drug dose is decreased or stopped. Long-term drug use often produces tolerance, whereby increasing doses are required to reproduce the psychological effect.

drumlin Smooth, oval-shaped mound of glacial TILL, one end of which is blunt the other tapered. Drumlins usually occur in groups called a "drumlin field" or "drumlin swarm." They are believed to be formed beneath the outer zone of an advancing ice sheet, which deposits and streamlines material. The long axis of a drumlin lies parallel to the direction of the movement of the GLACIER.

drupe (stone fruit) Any FRUIT with a thin skin, fleshy pulp, and hard stone or pip enclosing a single seed. Examples are plums, cherries, peaches, olives, almonds, and coconuts.

dry cell Electric BATTERY, a type of small LECLANCHÉ CELL containing no free fluid. In the most common type the ELECTROLYTE is ammonium chloride paste, the positive ELECTRODE (ANODE) a central carbon rod and the negative electrode (CATHODE) the zinc casing. Dry cells are widely used in flashlights and portable radios.

dry cleaning Method of cleaning fabrics using special solvent fluids and soaps without water. It was discovered by accident in 1849 by a French tailor, Jolly-Belin, when he noticed the cleaning effect of some turpentine he had spilled. Nonflammable solvents in special machines are now used – at first clothing was washed by hand. Clothes are washed with solvent and dried in a warm-air tumbler. Special solvents are used for some stains.

dry ice Popular name for frozen CARBON DIOXIDE.

drying oil Natural oil, such as linseed oil or cottonseed oil, that slowly hardens in air to form a tough film. Drying oils are extensively used in paints, varnishes, lacquers, printing inks, and putty. They are UNSATURATED compounds that harden by OXIDATION or POLYMERIZATION.

dubnium (element 104, symbol Db) Synthetic, radioactive, metallic element, the first of the TRANSACTINIDE ELEMENTS. The longest-lived of its 10 ISOTOPES has a half-life of 70 seconds. Dubnium forms a volatile compound with chlorine. The University of California at Berkeley claimed to have synthesized the element in 1969, but the Joint Institute for Nuclear Research in Dubna, Russia, claimed production five years earlier. A name

compromise of "unnilquodium" (symbol Unq) proved unappealing. "Dubnium" was adopted in 1995 over the US proposal "rutherfordium" (symbol Rf), honoring British physicist Ernest RUTHERFORD, and Russia's proposal of "kurchatovium" (symbol Ku) for Soviet physicist Igor Kurchatov (1903–60). RUTHERFORDIUM has now become element 106. Properties: at.no. 104.

ductility Ability of metals and some other materials to be stretched without being weakened. Copper is said to be a ductile metal as it is easily drawn out to form wire, and silver and gold are even more ductile than copper. The ductility of cast iron varies greatly according to the impurities present.

ductless gland *See* ENDOCRINE SYSTEM

Dulbecco, Renato (1914–) US virologist. He shared the 1975 Nobel Prize for physiology or medicine with David BALTIMORE and Howard TEMIN for his contributions to an understanding of the interaction between tumor viruses and the genetic material of the cell.

Dulong, Pierre Louis (1785–1838) French chemist and physicist. He discovered the explosive nitrogen trichloride. In 1819 he helped to formulate, along with Alexis PETIT, DULONG AND PETIT'S LAW of specific heats, which aided the determination of RELATIVE ATOMIC MASS (r.a.m.).

Dulong and Petit's law Physical rule stating that the product of SPECIFIC HEAT CAPACITY and RELATIVE ATOMIC MASS is approximately equal to 25 (when the specific heat capacity is measured in $J mol-1 K-1$) for the heavier solid elements. The specific heat capacity of an element was one of the easier parameters to measure during the early days of chemistry, so this law was used to obtain an approximation of the RELATIVE ATOMIC MASS

◄ **dune** Sand dunes are sculpted into many different shapes by the prevailing winds (1) and their resultant vortices and eddies (2). Barchans (A) only occur when the prevailing wind has a fairly constant direction. Longitudinal (B) and complex-longitudinal (C) dunes may stretch 50km (30mi). Multi-directional winds create pyramid-shaped dunes (D).

(r.a.m.). The rule was formulated by Pierre DULONG and Alexis PETIT.

dumortierite Silicate mineral, hydrous aluminum borosilicate ($Al_8BSi_3O_{19}[OH]$). It is found scattered in METAMORPHIC ROCKS, and occurs usually in fibrous masses, rarely as orthorhombic crystals. It is glassy to pearly violet or blue. Properties: hardness 7; r.d. 3.3.

dump In computing, information copied from computer memory to an output or storage device. It may be the entire contents of a file copied to another disk, or the contents of the screen being printed out (a **screen** dump).

dune Hill or ridge of wind-blown particles, most often sand. Dunes are found wherever sandy particles carried by the wind are deposited. They occur in a variety of shapes depending on the direction of the wind, whether or not it is constant, and the surrounding landforms. *See also* BARCHAN; SEIF

dunite Coarse-grained IGNEOUS ROCK of color ranging from light yellowish green to an emerald green, composed almost entirely of OLIVINE. It occurs at Dun Mountain, New Zealand, from which it takes its name.

Dunlop, John Boyd (1840–1921) Scottish inventor and pioneer of the pneumatic TIRE. He was a veterinarian in Belfast until, in 1888, he reinvented the pneumatic tire (previously patented in 1846 by Robert Thompson (1822–73) but undeveloped). His invention replaced solid tires.

duodecimal system Number system with 12 as its BASE. It has advantages over the decimal system (base 10) inasmuch as, for example, one-third and two-thirds in duodecimal would be 0.4 and 0.8 respectively – much simpler than the recurring expressions 0.3333... and 0.6666... in the DECIMAL SYSTEM.

duodenum First section of the small INTESTINE, shaped like a horseshoe and part of the DIGESTIVE SYSTEM. The pyloric sphincter, a circular muscle, separates it from the STOMACH. Alkaline BILE and pancreatic juices are released into the duodenum to aid the DIGESTION of food.

Du Pont de Nemours, Eleuthère Irénée (1771–1834) US industrialist, b. France. Founder of a huge chemical company and family empire that continues as E. I. Du Pont de Nemours and Co. He came to the US in 1799 with his father Pierre Samuel Du Pont de Nemours (1739–1817). They started a business producing high-quality gunpowder and prospered from the War of 1812.

Duralumin Alloy of aluminum with copper, magnesium, and manganese composed within the following limits: aluminum – over 90%; copper – about 4%; magnesium – 0.5% to 1.5%; manganese – less than 1%.

Dutton, Clarence Edward (1841–1912) US geologist who developed the principle of isostasy, according to which the level of the Earth's crust is determined by its density. The rise of light materials forms continents and mountains; the sinking of heavy materials forms oceans and basins.

Duve, Christian René de (1917–78) Belgian CELL biologist who shared with Albert Claude and George PALADE the 1974 Nobel Prize for physiology or medicine for a detailed description of the structure and function of the cell and its parts. Analysis of biochemical activity in a cell led him to discover the LYSOSOME, an organelle that acts as the "stomach of the cell."

Du Vigneaud, Vincent *See* VIGNEAUD, VINCENT DU

dwarf star Most common type of star in our galaxy, constituting 90% of its stars, including the Sun. It is also known as a main-sequence star, from its position on the HERTZSPRUNG-RUSSELL DIAGRAM. The term "dwarf" refers to LUMINOSITY rather than size, so dwarfs should be thought of as normal rather than diminutive.

dye Substance, natural or synthetic, used to impart color to various substances, including textiles, leather, hair, wood, and food. Natural dyes have been mostly replaced by synthetic dyes, most of which are derived from COAL TAR. For example alizarin comes from ANTHRACENE and indigo from ANILINE. Dyes are classified according to the way they are applied, which depends on their chemical composition and on the material being colored. **Direct** dyes can be applied directly to a fabric because they bind to the fibers. **Indirect** dyes require a secondary process to fix the dye in the fabric. Substantive dyes, sulfur dyes and vat dyes act directly, ingrain dyes and mordant dyes act indirectly.

dynamics Branch of MECHANICS that deals with objects in motion. Its two main branches are **kinematics**, which studies motion without regard to its cause, and KINETICS, which also takes into account forces that cause motion. *See also* INERTIA; MOMENTUM; MOTION, LAWS OF

dynamic testing Measurement of the forces exerted by moving bodies on resisting members. Dynamic testing is used to study such things as the effects of cars moving over a bridge and the landing shock on an aircraft's undercarriage.

dynamite Solid, blasting explosive. It contains NITROGLYCERINE incorporated in an absorbent base, such as charcoal or wood-pulp. Shock-resistant but easily detonated by heat or percussion, dynamite is used in mining, quarrying, and engineering. Its properties can be varied by adding ammonium nitrate or sodium nitrate. It was invented by Alfred NOBEL in 1866.

dynamo (simple generator) Device that converts mechanical energy into electrical energy by the principle of ELECTROMAGNETIC INDUCTION. A dynamo has a CONDUCTOR, usually an open coil of wire called an armature, that is placed between the poles of a permanent magnet. This armature is rotated within the magnetic field, which causes a current to be induced within the coil. **Alternating-current** (AC) dynamos are usually called ALTERNATORS. *See also* ELECTRIC CURRENT

dynamometer Electrical measuring instrument in which a pointer is deflected as a result of the force exerted between fixed and moving coils, the pointer being attached to the latter. It has applications in AMMETERS, wattmeters, and VOLTMETERS. The term is also used for a torque meter, which measures the power of rotating engines.

dyne (symbol dyn) Unit of force in the CGS (centimeter-gram-second) system of units. One dyne is the force that gives a mass of 1g an acceleration of $1cm\,s^{-2}$. One NEWTON equals 100,000 dynes.

dysentery Infectious disease characterized by diarrhea, with bleeding and abdominal cramps. It is spread in contaminated food and water, especially in the tropics. There are two types: **bacillary** dysentery, caused by BACTERIA of the genus *Shigella*; and **amoebic** dysentery, caused by a type of PROTOZOAN. Both forms are treated with antibacterials and fluid replacement.

dysprosium (symbol Dy) Silver-white, metallic element of the LANTHANIDE SERIES. Its chief ores are monazite and bastnaesite. Its capacity to absorb neutrons makes it important in NUCLEAR REACTOR control systems; dysprosium compounds are also used in lasers. Properties: at.no.

66; r.a.m. 162.5; r.d. 8.54; m.p. 1,409°C (2,568°F); b.p. 2,335°C (4,235°F); most common isotope ^{164}Dy (28.18%).

dystrophy, muscular Any of a group of inherited disorders featuring weakness or wasting of muscles; the muscle fibers degenerate, to be replaced by fatty tissue. The commonest form, **Duchenne** muscular dystrophy, is almost entirely confined to boys, usually developing before the age of four. There is no cure and death is usual before the age of 20.

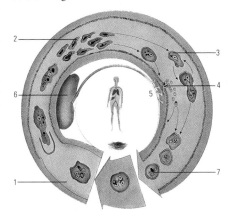

▲ **dysentery** Amoebic dysentery is caused by a microscopic organism (*Entamoeba histolytica*), found in contaminated water or food. *Entamoeba* is a natural inhabitant of the gut; however, under certain conditions, invasion of the gut wall occurs. Ingested *Entamoeba* cysts undergo division and multiplication in the large intestine (1). After division, eight trophozoites (feeding protozoa) are produced (2); non-infective trophozoites remain in the intestine (3) feeding on bacteria and food particles. Infective trophozoites invade the gut wall (4), multiply, and dissolve tissues by protein-digesting enzymes. If organisms enter the bloodstream (5) they can be carried to the lungs (6), liver and brain where abscesses develop. Those released from gut abscesses re-invade tissues or form cysts (7) and are passed in the faeces. The disease is transmitted if flies carry cysts from faeces to food or, more commonly, if contaminated water is drunk (8).

that it remained always over the same place, about 35,400km (22,000mi) above the Atlantic Ocean. It could relay 240 telephone conversations simultaneously between North America and Europe.

e Irrational number that is the base of natural, or Napierian, LOGARITHMS. e is a real decimal number without end equal to 2.718281828459O.... e is the limit of the expression $(1 + 1/n)^n$ as the integer n tends to infinity. In fact the number is transcendental, since there is no ALGEBRAIC EQUATION with e as the solution. e also occurs in EXPONENTIAL functions such as e^x. *See also* NUMBER, IRRATIONAL; NUMBER, TRANSCENDENTAL

ear Organ of HEARING and BALANCE. It converts sound waves to nerve impulses that are carried to the BRAIN. In most mammals it consists of the outer, middle, and inner ear. The **outer** ear carries sound to the eardrum. The **middle** ear is air-filled, and has three tiny bones (OSSICLES) that pass on and amplify sound vibrations to the fluid-filled inner ear. The **inner** ear contains the COCHLEA. Vibrations stimulate tiny hairs in these organs, which causes impulses to be sent via the AUDITORY NERVE to the brain. The inner ear also contains SEMICIRCULAR CANALS, which maintain orientation and balance.

Early Bird Satellite First commercial communications SATELLITE, launched on April 6, 1965. A stationary satellite, it was in a SYNCHRONOUS ORBIT SO

Earth Third major planet from the Sun, and the largest of the four inner, or terrestrial, planets. From space, the Earth is predominantly the blue of the oceans, plus the browns and greens of its land masses, the white polar caps, and a continually changing pattern of white cloud. Some 70% of the surface is covered by water, and it is this and the Earth's average surface temperature of 13°C (55°F) that make it suitable for life. The continental land masses make up the other 30%. Our planet has one natural satellite, the MOON. Monitoring the propagation of SEISMIC WAVES from earthquakes has revealed the Earth's internal structure. Like all the terrestrial planets, there is a dense core rich in iron and nickel, surrounded by a mantle consisting of silicate rocks. The thin, outermost layer of lighter rock is called the CRUST. The inner core rotates at a different rate from the solid outer layers, and this, together with currents in the molten outer core, gives rise to the Earth's MAGNETIC FIELD. The Earth's ATMOSPHERE is divided into a number of layers according to the way in which its temperature varies with altitude. The TROPOSPHERE contains most of the atmosphere, and is where lifeforms are found and weather systems operate. The STRATOSPHERE contains the OZONE LAYER, which absorbs the high-energy ULTRAVIOLET RADIATION from the Sun that is harmful to life. METEORS occur in the MESOSPHERE, and AURORAS in the THERMOSPHERE. The thermosphere is extremely rarefied, and its high temperature indicates the high kinetic energy of its molecules, rather than its

▶ **ear** The ear is divided into three parts – the outer, middle, and inner ear. The outer consists of the pinna and the auditory canal. The pinna funnels sound waves via the canal to the ear drum, tympanic membrane, of the middle ear. The sound waves are amplified and transmitted by tiny bones, the ossicles, which cause the oval window to vibrate. This sets the fluids of the inner ear in motion. Hair cells in the cochlea move and generate nerve impulses, interpreted by the brain as sound. The semicircular canals control balance.

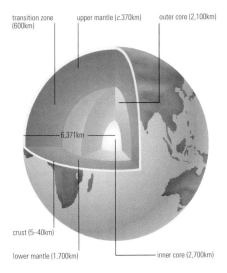

transition zone (600km)

upper mantle (c.370km)

outer core (2,100km)

6,371km

crust (5–40km)

lower mantle (1,700km)

inner core (2,700km)

▲ **Earth** Formed c.4,600 million years ago, the first humans appeared on Earth only 2 million years ago.

The Earth consists of concentric rings, from the uppermost crust to a solid inner core of nickel and iron.

heat content. Ionized atoms and molecules in the mesosphere and thermosphere constitute the IONOSPHERE. Above the thermosphere is the EXOSPHERE, which contains the Earth's MAGNETOSPHERE and VAN ALLEN RADIATION BELTS and merges into interplanetary space. **Data:** Diameter (equatorial) : 12,756km; Mass: 5,378 billion billion tonnes; Volume: 1,803 billion km^3; Density (water=1): 5.44; Orbital period: 365.3 days; Rotation period (equatorial): 23h 56m 04s; Surface temperature: 290K

earthenware Vessels, other utensils, and ornaments made of clay. The material is fired at relatively low temperatures, resulting in porous, opaque pieces. Such items can be made waterproof by glazing the surface.

earthquake Tremor below the surface of the Earth that causes shaking to occur in the CRUST. Shaking lasts for only a few seconds, but widespread devastation can result. According to PLATE TECTONICS, earthquakes are caused by the movement of crustal plates, which produces FAULT lines. The main earthquake regions are found along plate margins, especially on the edges of the Pacific, such as the San Andreas fault. When the shock takes place, three different SEISMIC WAVES are created: primary/push (P), secondary/shake (S), and longitudinal/surface (L). P and S waves originate from the seismic FOCUS (point of origin), up to 690km (430mi) deep. They travel to the surface and cause shaking. On the Earth's surface, they travel as L waves. The surface

point directly above the seismic focus is the EPICENTER, around which most damage is concentrated. A large earthquake is usually followed by smaller "aftershocks." An earthquake beneath the sea can often result in a TSUNAMI. Earthquake prediction is a branch of SEISMOLOGY. Present methods indicate only a probability of earthquake activity, and cannot be used to predict actual events. The world's largest recorded earthquake (1976) at Tangshan, China, killed over 250,000 people and measured 8.2 on the RICHTER SCALE.

earthquake engineering Term used to describe the designing of structures such as buildings, dams, and roads to minimize damage to them during earthquakes. Constructions over FAULT lines are avoided and cushions of concrete or polymeric materials are used as foundations to help buildings float or slide during a earthquake without breaking in areas where strain buildup is evident.

earth sciences General term used to describe all the sciences concerned with the structure, age, composition and atmosphere of the Earth. It includes the basic subject of GEOLOGY, with its subclassifications of GEOCHEMISTRY, GEOMORPHOLOGY, and GEOPHYSICS; MINERALOGY, and PETROLOGY; SEISMOLOGY and VOLCANISM; OCEANOGRAPHY; METEOROLOGY; and PALEONTOLOGY.

Earthshine (Earthlight) Phenomenon most easily observed during the crescent phase of the Moon, when the darkened portion of the lunar disc is illuminated by an ashen light reflected onto it by the Earth.

Earth Summit United Nations Conference on Environment and Development, popularly known as the "Earth Summit," held in Rio de Janeiro, Brazil, in June 1992. Reportedly the largest gathering of world leaders ever held, it was the first serious global acknowledgment of the problems created by climate change and the impact of 20th-century industrial society on the environment. The Rio Declaration laid down principles of environmentally sound development, a "blueprint for action," and imposed limits on the emission of the "greenhouse gases" causing the GREENHOUSE EFFECT.

Eastman, George (1854–1932) US inventor, industrialist, and philanthropist who popularized the art of PHOTOGRAPHY. Eastman invented (1879) a dry-plate process and began to mass produce photographic plates. In 1884 he designed a roll film. The first Kodak camera was produced in 1888. The Eastman Kodak Company was founded in 1892.

East Pacific Rise Part of the midoceanic ridge and a region of seismic activity in the E Pacific

Ocean. It is also known as the Albatross Cordillera. Shallow-focus earthquakes occur along the crest of the rise, which is an area of SEAFLOOR SPREADING.

ebola Virus that causes a hemorrhagic fever. Thought to have been long present in animals, the Ebola virus was first identified in humans during an outbreak in Zaire in 1976. Following an incubation period of between 2 and 21 days, the disease begins with a severe headache, muscle pains, and high fever. A rash develops and there is severe diarrhea and vomiting, with chest pain, and abdominal cramps; hemorrhaging occurs from internal organs. Ebola is acquired through contact with contaminated body fluids. There is no specific treatment and the death rate can be as high as 90%.

Eccles, Sir John Carew (1903–97) Australian physiologist. He made fundamental discoveries concerning the interaction of NEURONS (nerve cells) in the transmission of a nerve impulse. He shared the 1963 Nobel Prize for physiology or medicine with Alan HODGKIN and Andrew HUXLEY.

eccentricity (symbol *e*) One of the elements of an ORBIT. It indicates how much an elliptical orbit departs from a circle. The eccentricity is found by dividing the distance between the two foci of the ellipse by the length of the major axis. A circle has an eccentricity of 0, and a PARABOLA has an eccentricity of 1.

eccrine gland One of two types of SWEAT GLAND in mammals – the other is the APOCRINE GLAND. Both are found within layers of the SKIN. In humans and some apes, eccrine glands are the more numerous and are distributed generally over the body, not associated only with hair, as are apocrine glands.

ecdysone Hormone that stimulates MOLTING in some animals. It is a steroid hormone produced by CRUSTACEANS and INSECTS. By acting at particular sites on genes, it stimulates the production of the proteins that are involved in molting (**ecdysis**) and METAMORPHOSIS.

echinoderm Any member of the phylum Echinodermata, a group of spiny-skinned, marine invertebrate animals, which includes sea urchins, sea cucumbers, and starfish. Radially symmetrical with five axes, they have a skeleton of calcareous plates in their skin. Their hollow body cavity

► **earthquake** The focus of an earthquake – the point of rupture (1) – may be at the Earth's surface or up to 700km (450mi) below. The epicenter of the earthquake (2) is at ground level directly above. Most damage is done by earthquakes that occur at a depth of 10km (6mi) or less. Normally the longer the interval between movements on an active fault line, the greater the eventual shock. The San Andreas fault in California is a 1,200km (750mi) boundary between the Pacific (3) and North American (4) tectonic plates. Although the Pacific plate is heading N-W at an average 6cm (2.5in) a year, most of its movement consists of sudden jumps. Earthquake science (seismology) is not advanced enough to predict such jumps accurately. In the 1989 Loma Prieta quake, the focus was 18km (11mi) below the surface. Although the Pacific plate slipped 2m (7ft) N-W and rode 1m (3.3ft) upwards on the North American plate, most of the energy of the quake was absorbed below ground, so there was only comparatively minor surface cracking (5). Primary, or P-, waves (6), are compressing-dilation (back-and-forth) waves, like sound waves. These travel through typical crustal rocks at approximately 5km/s (3mi/s). Secondary, or S-, waves (7) travel more slowly, at approximately 3km/s (2mi/s), or 7km/s (4mi/s) deeper in the mantle, and are shear (side-to-side) waves. Both of these types of wave are transmitted out in all directions from the focus. Shear waves, however, cannot travel through liquids and so do not penetrate the Earth's molten core. The difference in speed between the waves results in a time lapse between their arrival that grows with distance, and allows scientists to pinpoint the focus of a quake.

includes a complex, internal fluid-pumping system and tube feet. They reproduce sexually, producing bilaterally symmetrical larvae resembling that of CHORDATES; REGENERATION also occurs.

echo Portion of a SOUND or RADAR wave reflected from a surface so that it returns to the source and is heard or observed after a short interval. Sound echoes are often heard as the distorted repetition of the sound in a large empty room. High notes provide a better echo than low notes. Echoes are utilized in ECHOLOCATION, which is used by bats and whales and by navigational equipment. Auditoriums are designed to minimize echoes; they are eliminated by using absorbent material on the walls and by avoiding curved surfaces that can focus the echoes. *See also* ACOUSTICS; ANECHOIC CHAMBER

echocardiography Use of pulses of ultrasound to examine the functioning of the heart. This painless, nonintrusive technique produces a computerized image on a screen. It can detect any structural fault or malfunction that may be linked to heart disease, particularly congenital disorders.

echolocation In animals, system of navigation used principally by whales and bats. The animal emits a series of short, high-frequency sounds, and from the returning ECHO it gauges its environment. Bats and dolphins also use the system for hunting. *See also* SONAR

echo sounder Device that sends ULTRASONIC sound pulses through water and detects their reflection – from the sea bottom, shoals of fish or any submerged object. Continuous chart recordings can provide a profile of the sea bottom. *See also* SONAR

eclampsia Occurrence of convulsions not associated with any other disease, such as epilepsy, in a pregnant woman with abnormally high blood pressure and fluid retention (preeclampsia, toxemia of pregnancy). The condition, which threatens the lives of mother and baby, must be treated immediately or coma and death will follow. Delivery may be hastened or Caesarean section performed. The cause of eclampsia is unknown.

eclipse Astronomical phenomenon that occurs when one celestial body temporarily obscures another, as seen from the Earth. The most familiar kinds are **solar** and **lunar** eclipses. A solar eclipse happens when the Moon passes between the Earth and the Sun so that the Sun's light is blocked from the part of the Earth on which the Moon's shadow, or **umbra**, falls. A lunar eclipse is caused by the Earth when it moves between Sun and Moon so that the Moon passes into the Earth's umbra, and cannot shine by reflected sunlight. Both kinds of eclipses can be **total**, when Earth or Moon pass completely into the other's umbra; or **partial**, when Earth or Moon remain partly in the other's partial shadow, or **penumbra**. If a solar eclipse happens when the Moon is at APOGEE (furthest point from Earth), its apparent size is less than the Sun's, and an **annular**, or ring, eclipse results, with the Sun appearing as a bright ring around the dark Moon. If Earth's and Moon's orbits were in the same plane, solar and lunar eclipses would be monthly, occurring at new and full moons. But because the Moon's orbit is at an angle to Earth's, eclipses are possible only when the Moon crosses Earth's orbital plane and is in line with Earth and Sun. These conditions happen seven times a year at the most. To early peoples eclipses, especially

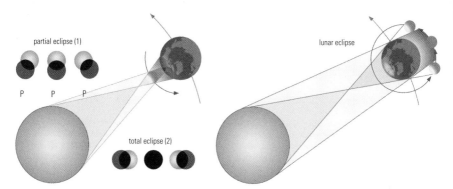

partial eclipse (1)

P   P   P

lunar eclipse

total eclipse (2)

▲ **eclipse** When the Moon passes between the Sun and the Earth it causes a partial eclipse of the Sun (1) if the Earth passes through the Moon's outer shadow (P), or a total eclipse (2) if the inner cone shadow crosses the Earth's surface. In a lunar eclipse, the Earth's shadow crosses the Moon and, again, provides either a partial or total eclipse. Eclipses of the Sun and the Moon do not occur every month because of the 5° difference between the plane of the Moon's orbit and the plane in which the Earth moves.

those of the Sun, were awe-inspiring and portentous events. It is not surprising that the knowledge that eclipses can be predicted added greatly to the authority of its discoverers, the astronomer-priests of Chaldea (Meso- potamia). Today solar eclipses enable astronomers to make detailed studies that are not normally possible, such as checking Einstein's theory of RELATIVITY.

eclipsing binary Type of BINARY STAR in which the orbital plane of the two stars is viewed almost edge-on, leading to mutual eclipses and consequent variations in combined light output. Two eclipses may be expected during an orbital cycle, one usually causing a bigger drop in light than the other since the two stars are rarely equal in size or brightness.

ecliptic GREAT CIRCLE on the CELESTIAL SPHERE, inclined at 23.5° to the CELESTIAL EQUATOR. The ecliptic is the yearly path of the Sun as seen from Earth or the Earth's orbit as seen from the Sun. The plane of the ecliptic thus passes through the center of both the Earth and the Sun. The planets (except PLUTO) are always located near the ecliptic.

ecliptic coordinate system Astronomical COORDINATE SYSTEM that refers to the plane of the ECLIPTIC. The coordinates used are celestial latitude and longitude. Celestial latitude is the angular distance between the object and the ecliptic, measured at right angles to the ecliptic. It is measured from 0° to 90°, positively toward the north ecliptic pole and negatively toward the south ecliptic pole. Celestial longitude is given by the angle between the object and the vernal EQUINOX, measured along the ecliptic from 0° to 360° E from the vernal equinox.

eclogite Coarse- to medium-grained METAMORPHIC ROCK consisting mainly of green pyroxine and red GARNET. A very dense rock, it is formed deep in the Earth's crust under conditions of very high temperature and pressure. It occurs in SCHISTS (Norway) and in diamond-bearing pipes of KIMBERLITE (South Africa).

ecology Study of relationships between organisms and the ABIOTIC FACTORS and BIOTIC FACTORS in their environments. The term was coined in 1866 by Ernst HAECKEL. Ecologists study individuals' ADAPTATIONS to and effects on their environments, as well as phenomena at larger spatial and temporal scales. These include the growth and decline of POPULATIONS, properties of COMMUNITIES such as FOOD WEBS and SUCCESSION, and the flow of energy through and cycling of materials within ecosystems. **Applied ecology** is the practical managament and preservation of natural resources and environments.

ecosphere Similar to BIOSPHERE, the total of all the ECOSYSTEMS of the Earth. The term BIOSPHERE is used solely to indicate the zone of life, but the term ecosphere includes the interaction of living organisms with their abiotic ENVIRONMENT. *See also* ABIOTIC FACTORS

ecosystem Basic unit in ECOLOGY, consisting of a community of organisms in a physical ENVIRONMENT. Study of these systems is based often on energy flow. The Sun provides the energy that **primary producers**, such as green plants, convert into food by PHOTOSYNTHESIS. Plants also need ABIOTIC (nonliving) substances from the water and soil to grow. In a typical FOOD CHAIN, herbivores or PRIMARY CONSUMERS, such as rabbits, eat the plants and in turn serve as food for SECONDARY CONSUMERS, such as foxes. **Decomposers**, such as bacteria, break down dead organic matter into simple nutrients completing the food chain. The chemicals necessary for life are recycled by the processes of the CARBON CYCLE, HYDROLOGICAL CYCLE, NITROGEN CYCLE, and OXYGEN CYCLE. Interference with these finely balanced natural processes, such as POLLUTION, climate change or the loss of a species, can disrupt the entire ecosystem.

ecotone Transitional zone between two neighbouring HABITATS.

ecotype In ecology, variety of a species that has special inherited characteristics that allow it to thrive in a particular habitat. In botany, variety of a plant species the distinguishing characteristics of which are mainly inherited rather than resulting from environmental pressure.

ecstasy (MDMA) (3,4-methylenedioxymethylamphetamine) An AMPHETAMINE-based drug, which raises body temperature and blood pressure by inducing the release of adrenaline and targeting the neurotransmitter, serotonin. Users experience short-term feelings of euphoria, rushes of energy, and increased tactility. Withdrawal can involve bouts of DEPRESSION and INSOMNIA. In the 1980s taking of ecstasy in tablet form was widespread in club culture in Europe and North America. Some deaths have resulted from use of ecstasy.

ECT Abbreviation of ELECTROCONVULSIVE THERAPY

ectoderm One of three so-called GERM LAYERS of tissue formed in the early development of a fertilized OVUM (egg). It is the outer layer of the BLASTULA and later develops, in most animals, into a skin or shell, a nervous system, lining tissue, parts of some sense organs, and various miscellaneous tissues. Other germ layers are the ENDODERM and MESODERM.

ectoparasite *See* PARASITE

ectopic Occurrence of a PREGNANCY outside the UTERUS – in the FALLOPIAN TUBE or elsewhere. The embryo cannot develop normally and spontaneous ABORTION often occurs. If not, urgent surgery is necessary to save the mother from serious, possibly fatal hemorrhage.

ectoplasm In biology, outer layer of the CYTOPLASM of a CELL; it is usually semisolid and transparent.

Eddington, Sir Arthur Stanley (1882–1944) English astronomer and physicist. Eddington pioneered the use of atomic theory to study the internal constitution of stars, and explained the role of radiation pressure in preventing stars from collapsing under gravity. Among his discoveries were the mass-luminosity relationship and the degeneration of matter by WHITE DWARFS. Eddington helped popularize Einstein's theory of RELATIVITY, and in 1919 obtained experimental proof of the general theory that gravity bends light by measuring stars close to the Sun during a solar eclipse.

eddy current Circulatory electric current induced in a CONDUCTOR when subjected to a varying magnetic field. Eddy currents cause a loss of energy in AC GENERATORS and MOTORS: the reaction between the eddy currents in a moving conductor and the field in which it moves retards the motion of the conductor.

Edelman, Gerald Maurice (1929–) US molecular biologist. He shared with Rodney PORTER the 1972 Nobel Prize for physiology or medicine for his work on unravelling the chemical structure of the ANTIBODY GAMMA GLOBULIN. The determination of the structure of gamma globulin was essential to research on how antibodies function in the IMMUNE system.

edentate (Lat. with all the teeth removed) Any of a small order of North and South American mammals found from Kansas to Patagonia. There are *c.*30 species of edentates, including ARMADILLO, SLOTH, and ANTEATER. Only anteaters are truly toothless.

Edison, Thomas Alva (1847–1931) US inventor. With little formal education, he became the most prolific inventor of his generation. In 1876 Edison opened a laboratory in Menlo Park, New Jersey. Here, he invented the carbon transmitter for TELEPHONES (1876), and the phonograph (1877). Using a carbon filament, Edison's invention of the first commercially viable ELECTRIC LIGHT (October 21, 1879) ensured his fame. In New York City, he built (1881–82) the world's first permanent electric power plant for distributing electric light. In 1887 Edison opened a larger research laboratory in West Orange, New Jersey, and in 1892 most of his companies were merged into the General Electric Company. In 1914 he developed an experimental talking motion picture. During World War I Edison worked for the US government on antisubmarine weapons. By the time of his death, he had patented more than 1,300 inventions.

EDTA Ethylenediaminetetraacetic acid (diaminoethane tetracarboxylic acid, $(HOOCCH_2)_2N(CH_2)_2N(CH_2COOH)_2$), a chelating agent (*see* CHELATION). The acid binds with metals such as calcium, magnesium, and iron, and is used in chemical analysis, water softening, and to counteract metallic poisons that have been swallowed. In such applications EDTA acts as a sequestering agent.

effector Term used for any agent or structure that brings about a physiological response. For example, a MOTOR NEURON that causes a muscle to contract or a hormone that stimulates a gland to function are both effectors. The target tissues are effector muscles and effector glands.

efficiency Work a MACHINE does (output) divided by the amount of work put in (input). It is usually expressed as a percentage, and a perfect machine would therefore have an efficiency of 100%. In mechanical systems there are invariably energy losses, such as those caused by FRICTION in the moving parts. The output never equals the input, and the efficiency is always less than 100% (although transformers and other electrical machines with no moving parts do have very high efficiencies). For very simple machines, efficiency can be defined as the FORCE RATIO (mechanical advantage) divided by the DISTANCE RATIO (velocity ratio).

egg For egg-laying animals, the fertilized female GAMETE (OVUM), once it has been laid or spawned by the mother. The developing EMBRYO is surrounded by ALBUMIN, shell, egg case or MEMBRANE, depending on the species. The egg provides a reserve of food for the embryo in the form of YOLK and protects it from a potential harmful external environment. For non-egg-laying animals, including almost all mammals, the term egg is synonymous with ovum.

Ehrenburg, Christian Gottfried (1795–1876) German biologist and founder of the science of micropaleontology. Ehrenburg studied at Berlin University and travelled widely as a naturalist, identifying and classifying many terrestrial and marine plants and microorganisms. He advanced the theory of "complete organisms," stating that all animals – even microscopic ones – have complete organ systems; it was later refuted.

Ehrlich, Paul (1854–1915) German bacteriologist. He shared with Élie METCHNIKOFF the 1908 Nobel Prize for physiology or medicine for his work on immunization, which included the development of basic standards and methods for studying toxins and antitoxins, with special reference to diphtheria antitoxins. His subsequent search for a "magic bullet" against disease and his discovery of salvarsan, a chemical effective against *Treponema pallidum*, the bacterium which causes syphilis, introduced the modern era of CHEMOTHERAPY (a term he coined).

Ehrlich, Paul Ralph (1932–) US biologist and ecologist. After gaining a doctorate from the University of Kansas (1957), he became a professor of biology at Stanford University (1959). In his book *The Population Bomb* (1968), Ehrlich voiced his concern over uncontrolled population growth and its subsequent impact on Earth and its natural resources. In 1976 he was appointed Bing professor of population studies, and in 1990 he shared the Crafoord Prize (given in disciplines of science not covered by the Nobel committee) with Edward Wilson.

Eichler, August Wilhelm (1839–87) German botanist who developed a system of PLANT CLASSIFICATION eventually accepted worldwide. It included division of plants into four major classes. He helped to edit some 15 volumes on the flora of Brazil and published a 2-volume study of the comparative structure of flowers.

Eigen, Manfred (1927–) German chemist who shared the 1967 Nobel Prize for chemistry with Ronald NORRISH and George PORTER for their research into very fast chemical reactions. Eigen devised techniques to disturb the balance of a system near equilibrium in order to make measurements of colors emitted as the system relaxed into a new equilibrium state. These techniques have been used in the study of RADIATION chemistry, and in reactions that use ENZYMES as CATALYSTS.

Eijkman, Christiaan (1858–1930) Dutch medical researcher and physician. In 1929 he shared with Frederick HOPKINS the Nobel Prize for physiology or medicine for his discovery of the antineuritic vitamin. Eijkman was the first to recognize a dietary deficiency disease, demonstrating that beriberi was produced by a lack of a certain dietary substance, later identified as vitamin B_1.

Einstein, Albert (1879–1955) US physicist, b. Germany, best known for his theories of RELATIVITY. In 1905 Einstein published four papers that revolutionized physical science. "The Electrodynamics of Moving Bodies" announced his special theory of relativity. Drawing on the work of Hendrik LORENTZ, Einstein's **special theory of relativity** discarded the notion of absolute motion in favor of the hypothesis that the speed of light is constant for all observers in uniform (unaccelerated) motion. Measurements in one uniformly moving system can be correlated with measurements in another uniform system, if their **relative** velocity is known. It demonstrated the relativity of time and asserted that the speed of light was the maximum velocity attainable in the Universe. A corollary of this special theory – the equivalence of mass and energy ($E = mc^2$) – was put forward in a second paper. A third paper, on BROWNIAN MOVEMENT, confirmed the atomic theory of MATTER. Lastly, Einstein explained the PHOTOELECTRIC EFFECT in terms of quanta or PHOTONS of light. For this insight, which forms the basis of modern QUANTUM THEORY, Einstein received the 1921 Nobel Prize for physics. In 1911 he asserted the equivalence of GRAVITATION and INERTIA. Einstein extended his special theory into a **general theory of relativity** (1916) that incorporated systems in nonuniform (accelerated) motion. This proved highly important in the field of COSMOLOGY. He asserted that matter in space causes

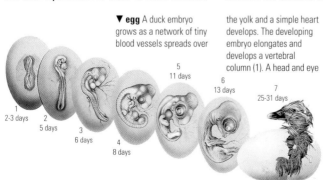

▼ **egg** A duck embryo grows as a network of tiny blood vessels spreads over the yolk and a simple heart develops. The developing embryo elongates and develops a vertebral column (1). A head and eye start to form, and the heart folds into its final position (2). The gut forms, the brain enlarges, and the embryo starts to curl (3–4). The limbs appear, and the tail and mouth form (5). By 13 days (6) it is possible to identify the bird from its bill. Some species of bird hatch shortly after this stage; others, such as the mallard duck (7), continue to develop in the egg.

1
2-3 days
2
5 days
3
6 days
4
8 days
5
11 days
6
13 days
7
25-31 days

curvature in the SPACE-TIME continuum, resulting in gravitational fields. This explained the peculiar motion of the planet Mercury and was confirmed (1919) by Eddington's study of starlight. Fearful of the rise of Nazism, Einstein accepted a post (1933–55) at the Institute of Advanced Study, in Princeton, New Jersey. In 1934 he was deprived of his German citizenship because he was Jewish. In 1940 Einstein became a US citizen. He devoted the rest of his life to a UNIFIED FIELD THEORY that attempted to combine ELECTROMAGNETISM and GRAVITATION into a single theory.

einsteinium (symbol Es) Radioactive, synthetic metallic element of the ACTINIDE SERIES. The isotope ^{253}Es was first identified in 1952 at the University of California at Berkeley; this was after it was found as a decay product of ^{238}U produced by the first large hydrogen bomb explosion. Eleven isotopes have been identified. Properties: at.no. 99; most stable isotope ^{254}Es (half-life 276 days). *See also* TRANSURANIC ELEMENTS

Einthoven, Willem (1860–1927) Dutch physiologist. He became professor at Leiden University in 1886 and was awarded the 1924 Nobel Prize for physiology or medicine for his work on the string GALVANOMETER, the forerunner of the electrocardiograph, the apparatus used in an ELECTROCARDIOGRAM (ECG).

ejector seat Device for escaping from an aircraft in an emergency. In order to parachute successfully from high-speed aircraft it is necessary to blow the seat and the pilot clear of the fuselage and tail section with an explosive charge. In hypersonic

▲ **electric circuit** This simple electric circuit (A) represents a torch. The power source is a cell (1). The switch (2) breaks the circuit when it is in the off position. When closed, in the on position, power flows to the resistor (3), the bulb, which emits light.

craft the pilot can be protected by separating the entire cockpit from the fuselage and parachuting the capsule and occupant together to the ground.

elasticity Capability of a material to recover its size and shape after deformation by STRESS. When an external force (stress) is applied, the material develops STRAIN (a change in dimensions). If a material passes its elastic limit, it will not return to its original shape. *See also* HOOKE'S LAW

elastomer Substance that regains its original size and shape after being deformed. Elastomers include natural RUBBER and various synthetic materials with similar properties. All are POLYMERS in which the molecular chains are folded; stretching the elastomer straightens the chains.

electrical engineering Branch of ENGINEERING that deals with the practical application of ELECTRICITY, especially as related to illumination, communication, COMPUTERS, and automatic control of machinery. Electrical engineers are fully trained in the mathematics of electric circuits and are involved in research, development, design, production, supervision, and improvement of production methods.

electrical switch Any device used to stop and restart the flow of an ELECTRIC CURRENT within an ELECTRIC CIRCUIT. The basic principle always involves an interruption and restarting of the conducting medium, usually the wiring.

electrical unit Standard quantity in which electric energy is measured. The domestic unit of electricity is the kilowatt-hour. An electric heater rated at 1 kilowatt (1,000 watts) consumes 1 unit of electricity when it is switched on for 1 hour.

electric branch circuit Any ELECTRIC CIRCUIT that is an interconnected branch of the main circuit where KIRCHHOFF'S LAWS apply. These state that at any branch junction the sum of the currents flowing to that point is equal to the sum of the currents flowing away.

electric car AUTOMOBILE powered by a rechargeable BATTERY. The electrical energy stored in the battery is converted to mechanical power by ELECTRIC MOTOR. Electric cars are considered to be more environmentally friendly than their gasoline-powered relatives. Firstly, they do not produce exhaust pollutants, thus reducing air pollution. Secondly, they do not consume scarce petroleum resources, and lastly, they are quieter and would reduce noise pollution. The major disadvantage of electric cars is their limited range – they can only travel about 100mi (160km) before the battery needs recharging. Furthermore, the small size of the battery reduces performance. The maximum speed of an electric car is usually less than 60mph (100km/h). The first electric cars were produced in Europe in the late 1800s and many early cars were powered by battery. The INTERNAL COMBUSTION ENGINE, however, proved to be more powerful and cheaper.

electric charge (symbol q or Q) Property shown by some ELEMENTARY PARTICLES. Electric charges (measured in COULOMBS) are described as either **positive** or **negative**. If two particles have a positive charge, or both have a negative charge, they repel one another; if one particle has a positive and one a negative charge, they attract. Electrical charges can be stored on insulated metal spheres (as in a VAN DE GRAAFF GENERATOR), on insulated plates (as in a CAPACITOR) and in chemical solutions (as in an electric BATTERY). An ELECTRIC CURRENT consists of electric charges in motion. *See also* COULOMB'S LAW; ELECTRON; PROTON

electric circuit System of electric CONDUCTORS, appliances or electronic components connected together so that they form a continuously conducting path for an ELECTRIC CURRENT. A circuit must include a source of ELECTROMOTIVE FORCE (EMF) to drive the current. Current flows according to several definite laws, the most important of which is OHM'S LAW. In modern electronic devices, circuits are often printed in copper on a plastic card (PRINTED CIRCUIT). *See also* CAPACITOR; INTEGRATED CIRCUIT (IC); TRANSISTOR

electric constant (symbol ε_0) Physical quantity relating the force between ELECTRIC CHARGES in a vacuum to the size of the charges and the distance between them. It was formerly called the PERMITTIVITY of free space. It appears in COULOMB'S LAW.

electric current (symbol I) Movement of ELECTRIC CHARGES, usually the flow of ELECTRONS along a CONDUCTOR or the movement of IONS through an ELECTROLYTE or gaseous discharge. It flows by convention from a positive (ANODE) to a negative terminal (CATHODE), although electrons actually flow along a wire in the opposite direction. It is measured in AMPERES. Direct current (DC) flows continuously in one direction, whereas alternating current (AC) regularly reverses direction. The frequency of AC current is measured in HERTZ (Hz). The frequency of the domestic supply is different in different parts of the world; in North America it is 60Hz, in Europe 50Hz. *See also* ELECTRICITY

electric displacement (symbol D) Density of ELECTRIC FLUX per unit area.

electric field (electrostatic field) Region around an ELECTRIC CHARGE in which any charged particle experiences a force. An object of opposite charge experiences an attractive force. An object of the same charge as that generating the electric field experiences a force of repulsion. The strength of the field upon unit charge at a distance r from a charge Q is equal to $Q/4\pi r^2\varepsilon$, where ε is the PERMITTIVITY of the medium surrounding the charge. A changing MAGNETIC FIELD can also create an electric field. *See also* ELECTROMAGNETISM

electric flux Quantity of ELECTRIC CHARGE that would be displaced across a CONDUCTOR imagined to be placed in an ELECTRIC FIELD. It is also the amount of charge that would cause the same electric field. It is proportional to the strength of the field and to the PERMITTIVITY of the medium the field passes through. The flux per unit area is called the electric flux density or **electric displacement**. *See also* GAUSS' LAW

electric furnace FURNACE heated to a very high temperature by an electric current. Electric furnaces are used in industry for melting metals and other materials, and for the production of highgrade steels. Three main methods of heating are used: striking an ARC between electrodes in the furnace; producing currents in the material by ELECTROMAGNETIC INDUCTION; and passing a high current through the material, heat being produced because of its RESISTANCE.

electricity Form of energy associated with STATIC or moving ELECTRIC CHARGES. Charge has two forms – positive and negative. Like charges repel each other and unlike attract; these forces are described in COULOMB'S LAW. Electric charges are acted upon by forces when they move in a MAGNETIC FIELD and this movement in turn generates an opposing magnetic field (FARADAY'S LAWS). Electricity and MAGNETISM are different aspects of the same phenomenon, ELECTROMAGNETISM. The flow of charges constitutes an ELECTRIC CURRENT, which in a conductor consists of negatively charged ELECTRONS. For a current to exist in a CONDUCTOR there must be an ELECTROMOTIVE FORCE (EMF) or POTENTIAL DIFFERENCE between the ends of the conductor. If the source of potential difference is a BATTERY, the current flows in one direction only – it is direct current. If the source is an electrical outlet, the current changes direction twice every cycle – it is alternating current (AC). The AMPERE is the unit of current, the COULOMB is the unit of charge, the OHM is the unit of RESISTANCE and the VOLT is the unit of electromotive force. OHM'S LAW and KIRCHHOFF'S LAWS (the summing of voltages and currents in a circuit) are the basic means of calculating circuit values. *See also* ELECTRIC CURRENT; ELECTRONICS

electricity sources Devices that convert other forms of energy into electricity. Most of our electricity is produced in power stations from the chemical energy of fossil fuels. The heat from burning coal, oil or natural gas turns water into steam. The steam drives a TURBINE, linked to an electricity GENERATOR. In a nuclear power station, heat comes from the FISSION of nuclei in a NUCLEAR REACTOR. A BATTERY and FUEL CELL convert chemical energy directly into electricity. SOLAR CELLS convert SOLAR ENERGY into electricity. Wind generators and water turbines produce electricity from the energy of movement in wind and water. *See also* ENERGY SOURCES; HYDROELEC-TRICITY; RENEWABLE ENERGY

electric light Artificial light produced by a flow of electricity in a wire or a gas. In an ordinary LIGHT BULB, light is produced by INCANDESCENCE

▲ **electricity sources** A combined cycle power station burns gas to generate electricity. It is more efficient than traditional fossil fuel power plants. The first turbine (1) sucks in air (2), compressing it before mixing it with the fuel (3) and burning the mixture (4). Exhaust gases spin a second turbine, connected to the first turbine and a generator (5). The energy of the gases is harnessed to power a

second multiple turbine (6) connected to another generator (7). Gases are used to superheat water (8), looping through a special vessel (9). To maximize power generation, superheated steam (10) turns a high-pressure turbine (11) before passing into a lower-pressure turbine (12). Steam is fed into the turbine from the heating loops (13) and is cooled (14) before returning to the circuit.

when a filament, such as tungsten-alloy wire, is heated by an electric current. In a fluorescent tube, the electric current passes through a gas. The gas ATOMS give off invisible ultraviolet rays. These strike a coating on the inside of the tube, causing it to emit light by FLUORESCENCE.

electric motor Machine that converts electrical energy into mechanical energy. In a simple form of electric motor, an ELECTRIC CURRENT powers a set of ELECTROMAGNETS on a rotor in the MAGNETIC FIELD of a permanent MAGNET. Magnetic forces set up between the permanent magnet and the electro-magnet cause the rotor to turn. Electric motors are convenient and relatively quiet, and have replaced many other forms of motive power. They may use alternating current (AC) or direct current (DC), and vary in size from thumb-sized miniatures used in toys, through those in medium-sized domestic appliances (such as vacuum cleaners, food mixers, and hand drills), to large motors such as those used in factories, elevators, and locomotives.

electrocardiogram (ECG) Recording of the electrical activity of the heart traced on a moving strip of paper. It is made with recording apparatus called an **electrocardiograph**. Electrocardiography is used to diagnose heart disease.

electrochemical cell *See* CELL

electrochemical series *See* ELECTROMOTIVE SERIES

electrochemistry Branch of physical chemistry concerned with the relationship between electricity and chemical changes. It includes the properties and reactions of IONS in solution, the CONDUCTIVITY of ELECTROLYTES and the study of the processes occurring in electrochemical CELLS and in ELECTROLYSIS.

electroconvulsive therapy (ECT) Treatment of mental disturbance by means of an electric current passed through the brain to induce convulsions. Given under anesthesia and with the use of a muscle relaxant, it is recommended mainly for severe depression that has failed to respond to other forms of treatment. Electrodes are attached to the scalp and current is applied to one or both sides of the brain in order to alter its electrical activity. The treatment can produce unpleasant side effects, including confusion, memory loss, and headaches, and there is continuing controversy about its use and effectiveness.

electrode Conductor, usually a wire or rod, through which an electric current flows into or leaves a medium. In ELECTROLYSIS, two electrodes

– a positive (ANODE) and a negative (CATHODE) – are immersed in the ELECTROLYTE.

electrodeposition Deposition of an adhering metallic coating onto an ELECTRODE by passing a low-voltage direct current (DC) through an ELECTROLYTE. Its most common application is ELECTROPLATING. *See also* ANODIZING

electrode potential Measure of the tendency of a reaction at an ELECTRODE. An electrode of an element (M) placed in a solution of its ions (M+) constitutes a HALF-CELL. In general a POTENTIAL DIFFERENCE exists between the electrode and the solution, caused by reactions of the type M⇔M+ + e. In practice it cannot be measured absolutely and standard electrode potentials are defined with reference to a hydrogen electrode under specified conditions of concentration, temperature, and pressure. Such reactions are OXIDATIONS, and so the term oxidation potential is often also used. *See also* ELECTROMOTIVE SERIES

electrodialysis *See* DIALYSIS

electrodynamics In physics, study of interactions between electric and magnetic fields and charged bodies. It began in the 19th century with the theoretical work of James Clerk MAXWELL, and later became part of QUANTUM MECHANICS.

electroencephalogram (EEG) Technique for recording the electrical activity of the brain. Electrodes are attached to the scalp to pick up the tiny oscillating currents produced by brain activity. The brainwaves are recorded as a continuous trace on a paper strip. The technique is used mainly in the diagnosis and monitoring of EPILEPSY.

electroforming Making of metal articles by deposition of the metal onto a mould by ELECTROLYSIS. A notable application is the manufacture of metal negative replica molds for vinyl records, where the accuracy of the deposition process provides faithful sound reproduction. *See also* ELECTRODEPOSITION

electroluminescence Production of light by certain substances, particularly phosphors, when they are placed in an alternating electric field. Part of the light from a FLUORESCENT LAMP is due to electroluminescence; the rest is due to PHOTOLUMINESCENCE.

electrolysis CHEMICAL REACTION caused by passing a direct current (DC) through an ELECTROLYTE. This results in positive IONS migrating to the negative ELECTRODE (CATHODE) and negative ions migrating to the positive electrode (ANODE). The

▲ **electric motor** Electric motors work using the interaction of a magnet (1) and a wire with a current passing through it (2). With the current flowing, the magnetic field produced by the loop interacts with the field of the magnet. A downward force acts on the right side, an upward force on the left side. When the loop reaches the vertical the split ring (through which the current reaches the loop) (3) reverses the current and so the magnetic field. Electric motors use multiple coils (4) to ensure constant power. In an electric drill, the turning shaft (5) emerges from the magnetic coils and is then geared (6) through a chuck (7) to the drill bit (8).

type of electron transfer reaction occurring depends on the electrode potentials of the ions present, and the electrode material may also play a part in the reaction. For example, in the electrolysis of copper salts with a copper anode, atoms of the electrode ionize and enter into solution. Electrolysis is an important method of obtaining chemicals, particularly for extracting reactive elements such as sodium, magnesium, aluminium, and chlorine. A commercial use is in ELECTROPLATING.

electrolyte Solution or molten salt that can conduct electricity, as in ELECTROLYSIS (which decomposes the electrolyte). In electrolytes, the current is carried by charged particles called IONS, rather than by electrons. In a lead-acid car battery, for example, the electrolyte is dilute sulphuric acid, which contains negative sulfate ions and positive hydrogen ions.

electromagnet Magnet constructed from a soft iron core around which is wound a coil of insulated wire. A MAGNETIC FIELD is set up when an electric current is passed through the wire, and it disappears when the current is switched off. This creates a magnet that can be turned on and off.

electromagnetic force One of the four FUNDA-MENTAL FORCES in nature. Within an atom, the electromagnetic force binds the negatively charged electrons to the positively charged NUCLEUS. *See also* GRAND UNIFIED THEORY (GUT); UNIFIED FIELD THEORY

electromagnetic induction Use of MAGNETISM to produce ELECTRICITY. If a bar magnet is pushed through a wire coil, an electric current is set up, or "induced," in the coil, as long as the magnet is moving. This is the basic working principle of the DYNAMO, ELECTRIC MOTOR, and TRANSFORMER. *See also* INDUCTANCE; INDUCTION

electromagnetic radiation Energy in the form

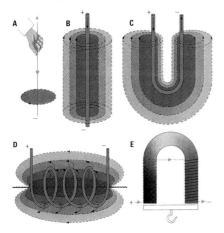

▲ **electromagnetism**
When current flows at uniform velocity along a wire, the magnetic field produced is a set of concentric rings in any plane drawn at right-angles to the wire. The number of lines of force drawn — dotted lines in B, C, and D – by convention, represents the strength of the magnetic field. If the wire is bent, the field follows the change in shape. When the wire is wound (D), in a coil or solenoid, the force lines of the field all point in the same direction through the center of the coil; overall, the field resembles that produced by a bar magnet with the north pole, in this

example, on the right. If an iron bar is inserted into the core of the solenoid, the lines of force prefer to remain within it and this results in a high concentration of lines at the end of the bar. This device, called an electromagnet, can produce a very strong magnetic field which varies with the current in the wire, the number of turns in the solenoid, and its cross-sectional area. In the horseshoe-shaped electromagnet (E), the strong magnetic field induced when current flows through the coil keeps the iron hook firmly in place, and it can be used to lift very heavy objects.

of waves of a wide range of frequencies. Electromagnetic radiation travels through empty space at the speed of light, approximately 300,000km (186,000mi) per second, and through materials at different speeds. Light, radio waves, infrared (heat) and ultraviolet radiation, and X rays are all examples of electromagnetic radiation. In general, electromagnetic waves are set up by electrical and magnetic vibrations that occur in atoms. These waves, which make up the ELECTROMAGNETIC SPECTRUM, can undergo REFLECTION, REFRACTION, INTERFERENCE, DIFFRACTION, and POLARIZATION. Other phenomena, such as the absorption or emission of light, can be explained only by assuming the radiation to be composed of quanta of energy (PHOTONS) rather than waves.

electromagnetic spectrum Range of different types of ELECTROMAGNETIC RADIATION ordered by either frequency or wavelength. It ranges from low-frequency (high-wavelength) radio waves, through microwaves, infrared (heat) waves, light (the visible spectrum), ultraviolet waves, and X rays to very high-frequency gamma rays.

electromagnetism Branch of physics dealing with the laws and phenomena that involve the interaction or interdependence of ELECTRICITY and MAGNETISM. The region in which the effect of an electromagnetic system can be detected is known as an **electromagnetic field**. When a magnetic field changes, an electric field can always be detected. When an electric field varies, a magnetic field can always be detected. Either type of energy field can be regarded as an electromagnetic field. A particle with an electric charge is in a magnetic field if it experiences a force only while moving; it is in an electric field if the force is experienced when stationary.

electrometallurgy Applications of ELECTRO-CHEMISTRY to METALLURGY. It includes the production and refining of metals by ELECTROLYSIS and such techniques as ELECTROPLATING, electrolytic machining and polishing, and ELECTROFORMING. In a wider sense, electrometallurgy includes other uses of electric current, as in alloy-steel production using an ELECTRIC FURNACE.

electrometer Instrument having an electric circuit for measuring differences of ELECTRIC POTENTIAL (voltages) without drawing appreciable current. Modern electrometers are essentially voltage AMPLI-FIERS. *See also* POTENTIAL DIFFERENCE; VOLTMETER

electromotive force (emf) Sum of the POTEN-TIAL DIFFERENCES around an ELECTRIC CIRCUIT. When the circuit is open, so that no current flows, it is equal to the potential difference between the terminals of the power source. When a current does flow, the external potential difference decreases. The emf is equal to the energy expended in moving one unit of charge around the circuit. *See also* ELECTRICITY

electromotive series (electrochemical series) List of METALS and the gas hydrogen, the order of which indicates the relative tendency to be oxidized, or to lose electrons in chemical reactions (*see* OXIDATION-REDUCTION). The series starts with the metal that tends to lose the most electrons in reaction. A standard hydrogen ELECTRODE is arbitrarily assigned an oxidation potential of zero, so that the metals can be compared. Those that lose electrons more readily than hydrogen are termed **electropositive**; those that lose electrons less readily are called **electronegative**. The order of some common metals in the electromotive series is: lithium, potassium, calcium, sodium, magnesium, aluminum, zinc, chrominum, iron, cobalt, nickel, tin, lead, hydrogen, copper, mercury, silver, platinum, and gold.

electromyogram Record of the electrical activity of muscles by means of electromyography (EMG). Muscles produce no currents when at rest, but during activity the currents generated enable a continuous trace of action potentials to be recorded. These traces are valuable for research and in diagnosis of muscle dysfunction.

electron (symbol *e*) Stable ELEMENTARY PARTICLE with a negative charge, having a rest mass of 9.1×10^{-31} kg (1/1836 that of the PROTON). Identified in 1879 by the British physicist Joseph THOMSON, electrons move around the NUCLEUS of an ATOM in complex orbits. In a neutral atom, their total charge balances that of the protons in the nucleus. Removal or addition of an atomic electron produces a charged ION. Chemical bonds are formed by the transfer or sharing of electrons between atoms. When not bound to an atom, electrons are responsible for electrical conduction. Beams of electrons are used in several electronic devices such as CATHODE-RAY TUBES, OSCILLOSCOPES, and ELECTRON MICROSCOPES. In particle ACCELERATORS, electrons and POSITRONS (ANTIPARTICLES of electrons) are made to collide with other particles for the purposes of nuclear research. An electron is classified as a LEPTON. *See also* ELECTROMAGNETIC FORCE

electron diffraction Method by which the structure of gases and the surface of solids are determined. Just as light waves undergo DIFFRACTION when passing through a narrow opening, so also do electrons. The electron wavelength can be adjusted, and the way it behaves after striking the subject material is used to determine the material's CRYSTAL STRUCTURE.

electron emission Liberation of electrons from the surface of a substance. It can occur as a result of heat (THERMIONIC EMISSION), light (PHOTOELECTRIC EFFECT), high electric field (field emission), or bombardment by ions or other electrons (SECONDARY EMISSION). It may also occur from RADIOACTIVE DECAY. In all cases, the electrons must acquire energy from the outside source in excess of the WORK FUNCTION of the substance. Most ELECTRON TUBES and CATHODE RAY TUBES depend on thermionic emission, whereas photoelectric cells rely on photoelectric emission. Field emission is important in the field emission microscope and secondary emission is the principle used by electron multipliers and storage tubes.

electron gun In a CATHODE-RAY TUBE, such as that in a television receiver or an oscilloscope, device that emits electrons. These are focused into a beam that scans a fluorescent screen, forming an image.

electronic circuit Electronic components wired together to form a unit capable of performing a particular function. *See* ELECTRIC CIRCUIT

electronic control system Control system based on ELECTRONIC CIRCUITS. Complex control systems rely on COMPUTERS.

electronic mail (e-mail) Correspondence sent via a COMPUTER NETWORK. In a simple system, messages produced using word processing programs are transmitted over a network (which could be a small company network or the worldwide INTERNET) and stored in a computer called a **mail server**. Users connected to the network can contact the mail server to check whether it is holding mail for them. If it is, they can transfer the messages to their own computer, and print it if they need a permanent record. Electronic mail is transmitted rapidly, and costs are low, even for international communications.

electronic publishing Production of publications by electronic means, generally using a COMPUTER. During production, word processing and graphics programs are used to assemble "pages" on a computer screen. The publication may be printed on paper, either by conventional methods using PRINTING film generated by computer or by various computer-controlled printers (such as a LASER PRINTER). Alternatively the publication may be produced as a CD-ROM. Electronic publications may also be sent to other computer users on-line.

electronics Study and use of circuits based on the conduction of electricity through ELECTRON TUBES and semiconducting devices. John FLEMING invented the DIODE tube, which was modified in 1906 as the triode by Lee DE FOREST. These two devices, with further modifications and improvements, provided the basic components for all the electronics of radio, television, and radar until the end of World War II. A major revolution occurred in 1948 when a team at Bell Telephone Laboratories, led by William SHOCKLEY, produced the first semiconducting TRANSISTOR. SEMICONDUCTOR devices are much lighter, smaller, and more reliable than vacuum tubes. They do not require the high operating voltages of valves and can be miniaturized as an INTEGRATED CIRCUIT (IC). This has led to the production of electronic computers and automatic control devices that have changed the face of both industry and scientific research. *See also* ELECTRIC CIRCUIT; ELECTRICITY

electron microscope MICROSCOPE that "illuminates" the object with a stream of electrons. The "lenses" consist of magnets that focus the electron beam. Smaller objects can be seen, because electrons have shorter wavelengths than light and thus provide greater resolution. In a **transmission** electron microscope, the electrons pass through the target, which is in the form of a very thin slice. In a **scanning** electron microscope, a beam of electrons

is swept across the target, which can be a complete object, but that generally must be metal-coated to reflect the electrons. The image is obtained by converting the pattern into a video display, which may be computer-processed. These microscopes can magnify millions of times.

electron spin resonance (ESR) Method of studying the structure of molecules by finding the positions of electrons within them. ESR is a spectrographic technique and is applicable only to paramagnetic substances. It locates unpaired electrons by their spin and by the way they align themselves in an external magnetic field when the molecule absorbs microwave radiation.

electron transport system (electron transport chain) Sequence of biochemical reactions that transfers electrons, through a series of carriers, in certain metabolic processes. It involves carrier substances that accept electrons and then donate them to the next carrier in the chain, while themselves undergoing a series of OXIDATION-REDUCTION reactions. It forms the last stage of aerobic RESPIRATION in cells, causing hydrogen atoms to

lower (5) electromagnetic lenses. It then passes through an aperture ring (6) and a scan coil (7) before being focused by a projector lens (8) onto the sample (9). The process takes place in a vacuum with air evacuated (10) by a pump. A computer controls the scan coil, which directs the beam across the sample. The sample is placed in an airlock (11) and manipulated into position (12). An image of the sample is created by detecting electrons dislodged (13) from the sample. These electrons correlate to the topography of the sample and are measured by a flash detector (14) when they hit a fluorescent target (15). The image is displayed on a computer monitor (16): here *Lactobacillus bulgaricus* magnified 1,000 times (17).

◄ **electron microscope** In an electron microscope, a beam of electrons (1) streams from the heated tungsten cathode (2) of an electron gun (3) and is focused by upper (4) and combine with oxygen to form water and conserving energy in the form of ATP. PHOTOSYNTHESIS also involves an electron transfer system.

electron tube Electronic device consisting of electrodes arranged within an evacuated glass tube; for special purposes a gas at low pressure may be introduced into the tube. The DIODE, used for rectification, consists of a negative CATHODE, which emits electrons when heated, and a positive ANODE or plate. The TRIODE, used for amplification, has a perforated control grid between the cathode and the anode; a signal fed to the grid provides an amplified signal at the anode. Electron tubes have been largely replaced by TRANSISTORS and other semiconductor devices.

electron volt (symbol eV) Unit of energy equal to the energy acquired by an ELECTRON in falling freely through a POTENTIAL DIFFERENCE of one VOLT. It is equal to 1.602×10^{-19} joules.

electro-osmosis Flow of water caused when an electric current is passed through clay or other porous material in a watery suspension. This phenomenon is caused by the formation of positive and negative IONS, one type of which is bound to the clay while the other (in the water) moves towards the opposite electric pole. *See also* OSMOSIS

electrophoresis Movement of electrically charged colloidal particles through a FLUID from one ELECTRODE to another when a voltage is applied between the electrodes. It is used in the analysis and separation of colloidal suspensions, especially colloidal proteins; clinical medicine uses electrophoresis to measure protein content of body fluids. It is also used as a means of depositing coatings of one material, such as paint, on another. *See also* COLLOID

electrophysiology, cardiac Study of the electrical activity of the HEART muscle. When muscles contract and relax they produce electric currents. The heart produces the principal muscular currents and these are measured for diagnostic purposes by attaching electrodes to the outside of the body and connecting them to an electrocardiograph. *See also* ELECTROCARDIOGRAM (ECG); ELECTROMYOGRAM

electroplating ELECTRODEPOSITION of a coating of one metal on another by making the object to be coated the CATHODE in an electrolytic CELL. Positive IONS in the ELECTROLYTE are discharged at the cathode and deposited as metal. Electroplating is used to produce a decorative or corrosion-resistant layer, as in silverplated tableware and chromium-plated engineering parts.

electroscope Instrument for detecting the pres-

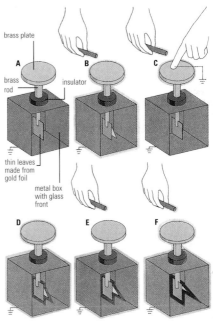

▲ **electroscope** The gold leaf electroscope is a qualitative electrostatic instrument which proved invaluable to early physicists. Among other experiments, it was used to detect the presence, and determine the polarity, of charges.
Key
A) The uncharged electroscope with leaves collapsed
B) A negatively charged rod near the plate attracts positive charges at the plate and repels negative ones to the leaves. The leaves repel each other and diverge to indicate the presence of a charge. The leaves would again diverge if instead a positively charged rod were brought near the plate. This operation is the first in a series which lead to the charging of the electroscope by induction. C) With the charged rod still in place, the plate is earthed, and electrons flow from the leaves through the hand to earth. The leaves collapse. D) When the finger and then the rod are removed the electroscope is positively charged – thus some positive charges at the plate spread the leaves and they repel each other and diverge. E and F) The leaves diverge even further when the rod is brought nearer to the plate indicating that the influence of the charges is greater at close range.

ence of an ELECTRIC CHARGE. The most common type is the **gold-leaf** electroscope, in which two gold leaves hang from a conducting rod held in an insulated container. A charge applied to the rod causes the leaves to separate, and the amount of separation indicates the amount of charge.

▲ elementary particle
The first elementary particle to be known as such was the electron. In the late 19th century, it was known that so-called "cathode rays" (1) could be obtained using a voltage of a few thousand volts in a vacuum. That these rays were in fact electrically charged was established in the 1890s by showing that they were deflected by an electric field (2). The electrons were obviously electrically negative, being attracted by the positive field plate. In the same apparatus, but with an incomplete vacuum, positive particles were also observed, including a second kind of elementary

particle, the proton. The next elementary particle was not discovered until 1932, by which time physicists had developed good techniques for measuring the energies of moving electrons, protons and alpha-particles (helium nuclei). When the alpha-particles (3) from a radioactive element, such as polonium (4), struck a piece of beryllium (5), a light metal, a radiation emerged that was capable of causing a block of wax (6) to emit protons (7) having an energy of 4.5 million electron-volts – surprisingly great. This radiation consisted of neutrons of about the same mass as protons.

electrostatic and electromagnetic systems of units In electricity and magnetism, systems of units based on different definitions of fundamental quantities. These include the centimeter-gram-second (CGS) electrostatic system, applied particularly to electrostatic problems, and the CGS electromagnetic system, for problems involving magnetic fields. These systems are combined in the Gaussian system of units. These systems of units are being superseded by the international system of SI UNITS. *See also* CGS SYSTEM; MKS UNITS

electrostatic generator *See* VAN DE GRAAFF GENERATOR

electrostatics Branch of physics that studies ELECTRIC CHARGE at rest. This is often done on a charged metal sphere (VAN DE GRAAFF GENERATOR) insulated from its surroundings or on the insulated plates of a CAPACITOR.

electroweak theory (Weinberg-Salam theory) Theory that unifies the two FUNDAMENTAL FORCES of ELECTROMAGNETIC and WEAK NUCLEAR FORCES. First postulated in 1967 by Steven WEINBERG and Abdus SALAM, the electroweak interaction involves PHOTONS and vector BOSONS known as W BOSONS and Z BOSONS. The theory suggests that a yet undiscovered heavy particle, given the name HIGGS BOSON, should exist.

element Substance that cannot be split into simpler substances by chemical means. All atoms of a given element have the same ATOMIC NUMBER (at.no.) and thus the same number of ELECTRONS (the factor that determines chemical behavior). The atoms can have different MASS NUMBERS (the number of protons and neutrons in the nucleus) and a natural sample of an element is generally a mixture of ISOTOPES. The known elements range from hydrogen (at.no. 1) to element 112; elements of the first 95 atomic numbers exist in nature, the higher numbers have been synthesized.

element 104 *See* DUBNIUM

element 105 *See* HAHNIUM

element 106 *See* RUTHERFORDIUM

elements 107 to 112 Six radioactive TRANSACTINIDE ELEMENTS that have been synthesized at Darmstadt, Germany, by a team led by Peter Armbruster. In each case, only a few atoms were produced and their half-lives are extremely short.

elementary particle (subatomic particle) Although the term implies a particle that cannot be subdivided, it is used more widely than this. Elementary particles are distinguished from each other by their MASS (usually expressed in equivalent energy units) and a set of QUANTUM NUMBERS, including ELECTRIC CHARGE and SPIN. An elementary particle may be classified by its interaction as either a LEPTON, which does not experience the STRONG NUCLEAR FORCE, or as a HADRON (MESON or BARYON), which is subject to the strong interaction. Mesons consist of two QUARKS and baryons of three (baryons include the PROTON and the NEUTRON). Another method of classification divides elementary particles into BOSONS (which have an integral spin) and FERMIONS (which have nonintegral spin). Many particles also undergo electromagnetic interactions. All elementary particles have an associated antiparticle that has the same mass but an opposite set of quantum numbers. Of the large number of particles known, the only stable ones are the LEPTON, the PROTON, the NEUTRON (when in a nucleus), their antiparticles (*see* ANTIMATTER) and the PHOTON. The others DECAY after a characteristic lifetime to form stabler particles. When an elementary particle collides with its own antiparticle, mutual annihilation (destruction) occurs, with the production of RADIATION energy. Particle-antiparticle pairs can also be created in high-energy interactions *See also* FUNDAMENTAL FORCES

elevator Machine for raising and lowering passengers and freight from one level to another inside or outside buildings. Early elevators were driven by steam or hydraulic power. The first elevator was installed (1856) in a New York store by Elisha OTIS, who had invented a safety system that incorporated ratchets along the elevator shaft to ensure that the body of the elevator could not

drop accidentally. In 1889 the first electric elevator was installed in a New York building.

ellipse CONIC section formed by cutting a right circular cone with a plane inclined at such an angle that the plane does not intersect the base of the cone. When the intersecting plane is parallel to the base, the conic section is a circle. In rectangular Cartesian coordinates (x,y), its standard equation is $x^2/a^2 + y^2/b^2 = 1$ for any nonzero values of a and b. Most planetary orbits are approximately elliptical.

elliptical galaxy Type of regular GALAXY as classified by Edwin HUBBLE.

El Niño (Sp. (Christ) child) Periodic easing or reversing of the TRADE WINDS over the S Pacific Ocean, causing the warm surface waters that have "piled-up" in the W Pacific to flow back and warm the coastal waters of South America by up to 2 to 3°C. It occurs around Christmas time, hence its

▶ **El Niño** The combined influence of land, sea and air on weather conditions can create a global climate rhythm. In the Pacific Ocean, for example (A), trade winds normally blow from E to W (1) along the equator, "dragging" sun-warmed surface waters into a pool off N Australia and thereby depressing the thermocline - the boundary between warm surface waters and the cooler layers beneath (2). High cumulus clouds form above these warm waters, bringing rain in the summer wet season (3). Cooler, nutrient-rich waters rise to the surface off South America (4), supporting extensive shoals of anchovies on which a vast fishing industry has developed. The weather over this cold water region is dry. Every 3–5 years a change occurs in the ocean-atmosphere interaction. The climatic pattern is reversed (B) – an event known as El Niño. The trade winds ease, or even reverse direction (5), during El Niño and the

warm surface waters which have "piled up" in the W Pacific flow back to warm the waters off South America by 2–3°C (3.6–5.4°F) (6). This depresses

the E thermocline (7) and dramatically affects climate. In an El Niño year, drought and bush fires occur over Australia, while floods affect Bolivia and

Peru. The warm waters off South America suppress upwelling of the cold nutrient-rich waters, bringing disaster to the fishing industry.

name. El Niño has a dramatic effect on climate patterns in Australia and Southeast Asia and may be implicated in changed rainfall patterns as far away as Africa. In normal years, trade winds blow E to W along the Equator, dragging sun-warmed surface waters into a pool off N Australia and monsoon rains to Indonesia. In the W Pacific, the Humboldt Current pushes the surface waters away from the coast of Peru, bringing cold, nutrient-rich water to the surface. This upwelled, nutrient-rich water stimulates the production of phytoplankton and swells the population of anchovies, which themselves are prey to seabirds (such as boobies and brown pelicans) and the Peruvian fishing industry thrives. In an El Niño year, the upwelling ceases and the biological productivity of the area collapses. In 1982–83, for instance, the catch of anchovies was reduced by 60%. In addition, mean sea-level along the coast of Latin America may increase by as much as 50cm (20in), causing widespread flooding. Some scientists believe that the frequency and effects of El Niño are increasing.

Elton, Charles Sutherland (1900–91) British ecologist. After work on Arctic animals and their

▲ **embryo** After 3 weeks a human embryo bears a primitive heart and head (A). By the 4th week, the heart is pumping blood around the body and into the placenta, and 25 pairs of tissue blocks (somites) appear, which later give rise to bone and muscle tissue (B). After 5 weeks, limb buds and rudimentary eyes are visible (C). The limbs become well developed, and the tail region recedes by the 6th week (D). The head grows rapidly; eyes, ears, and teeth buds appear by the 7th week (E). The tail portion vanishes and almost all the organs and tissues have developed by the 8th week (F). It is now known as a fetus.

environments, Elton helped to found the Oxford Bureau of Animal Population, England. He also championed the concept of FOOD CHAINS and their effect on species population, summarized in his book *Animal Ecology* (1927). During World War II he did important work on controlling the populations of rats and mice in Britain, to help conserve food stocks. After the war, he did important foundational work on invasive species.

embolism Blocking of a blood vessel by an obstruction called an **embolus** – usually a blood clot, air bubble or particle of fat. The effects depend on where the embolus lodges. For example, a **cerebral** embolism (in the brain) causes a STROKE; a **limb** embolism may cause gangrene. Treatment is with "clot-busting" drugs, anticoagulants or surgery. *See also* ARTERIOSCLEROSIS

embryo Early developing stage of an animal or plant. In animals, the embryo stage starts at FERTILIZATION, and ends when the organism emerges from the egg or from its mother's UTERUS. In mammals, an embryo is sustained through blood supplied by the mother via the PLACENTA. In humans, the embryo is called a FETUS after the first eight weeks of pregnancy. In invertebrate animals the embryo is usually called a LARVA. In plants, the embryo is found in the seed, and the embryo stage ends on GERMINATION. An embryo results when the nuclei of an EGG and a SPERM or male sex cell fuse to form a single cell, called a ZYGOTE. The zygote then divides into a ball of cells called a BLASTULA. The blastula and then the embryo undergoes rapid changes in which the cells differentiate themselves to form features, such as limbs and organs.

embryology Biological study of the origin, development, and activities of an EMBRYO. This science traces the sequence of events from OVUM (egg) to BIRTH, hatching or germination.

emerald Variety of BERYL, varying in color from light to deep green and highly valued as a gemstone. The color is due to the presence of small amounts of chromium but the stone may lose its color if heated. Emeralds were mined in Upper Egypt in 1650 BC; now they are found mainly in Colombia.

Empedocles (*c*.490–*c*.430 BCE) Greek scientist and philosopher. He taught the doctrine of the four elements (earth, water, air, and fire) and he explained change as being alterations in the proportions of the four elements.

emery Impure form of the mineral CORUNDUM (aluminum oxide, Al_2O_3) that occurs as dark granules with MAGNETITE and HEMATITE in them. An unusually hard mineral, it is used as an ABRASIVE.

emulator COMPUTER configured in such a way that it acts like another type of computer. Emulators are often used in the development of new microprocessors.

emulsion SUSPENSION of one liquid in another. For example, when a water-based paint is diluted, the paint forms an emulsion with the water. It does not dissolve in the water, but is dispersed into very fine droplets or particles that take an extremely long time to settle out. Fats can form emulsions with water. During DIGESTION, salts from the bile pass into the gut and act on the fats there, reducing their surface tension so that they can form smaller droplets – an emulsion.

enamel Decorative or protective glazed coating produced on metal surfaces, or a type of paint. Ceramic enamels are made from powdered glass and **calx** (a mixture of tin and lead). Various metal oxides are added to give color. The mixture is melted, quenched, and ground, and then applied to the surface. It is fired to give a fused vitreous coating. Enamel paints consist of zinc oxide, lithopone (brown linseed oil) and high-grade varnish. The finish is hard, glossy, and highly durable. The term "enamel paint" is derived from its resemblance when dry to the finish found on ceramic-enamel products.

enamel In mammals, hard covering of the crown of the TOOTH; it is the hardest body tissue. Strongest at the biting edges, enamel varies in thickness and density over the tooth.

encephalitis Inflammation of the brain, almost always associated with a viral infection; often there is an associated MENINGITIS. The most common cause is the *Herpes simplex* virus. Symptoms include fever, headache, lassitude, and intolerance of light; in severe cases there may be sensory and behavioral disturbances, paralysis, convulsions, and coma. The disease is diagnosed by tests on CEREBROSPINAL FLUID, which is obtained by lumbar puncture and possibly brain scans. The antiviral drug acyclovir is effective against *H. simplex* encephalitis; otherwise treatment is mainly supportive.

endangered species Animals or plants threatened with extinction as a result of such activities as habitat destruction and overhunting. In 1948 the International Union for the Conservation of Nature and Natural Resources (IUCN) was founded in Geneva, Switzerland to protect endangered species. The IUCN publishes the *Red Data Book*, which currently lists more than 1,000 animals and 20,000 plants considered endangered. The Convention on Trade in Endangered Species of Wild Flora and Fauna (CITES) was signed in 1973. It seeks to prevent international trade in *c.*30,000 species. In the US, the Endangered Species Act (1973) gives legal protection to a wide range of wild animals and plants. *See also* BIODIVERSITY; CONSERVATION; ECOLOGY; HABITAT; WETLAND

endemic Term to describe a disease that remains prevalent in a particular region or country. It is most often heard in connection with tropical diseases, which are difficult to eradicate. In ECOLOGY and BIOGEOGRAPHY, the term is used to describe a species native and still restricted to a particular region.

Enders, John Franklin (1897–1985) US microbiologist. He shared the 1954 Nobel Prize for physiology or medicine with Frederick C. ROBBINS and Thomas H. WELLER for the discovery that poliomyelitis viruses can be grown in cultures of various types of tissues. This work was fundamental to the later development of the polio VACCINE.

endocrine system Body system made up of all the **endocrine glands** (ductless glands), which secrete HORMONES directly into the bloodstream to influence body processes. The endocrine system (together with the AUTONOMIC NERVOUS SYSTEM (ANS), and collectively called the **neuroendocrine system**) controls and regulates all body functions. It differs from other body systems in that its ductless glands are not structurally connected to one another. The chief endocrine glands are the PITUITARY GLAND, located at the base of the brain; the THYROID GLAND, located in the throat; the PARATHYROID GLANDS, usually embedded in the thyroid; the ADRENAL GLAND, situated on top of the kidneys; specialized cells called the **islets of Langerhans** in the PANCREAS; and the sex gland or GONAD (TESTIS in males and OVARY in females) and, in pregnancy, the placenta. *See also* EXOCRINE GLAND

endocytosis In biology, process by which a CELL takes in substances. When a cell's MEMBRANE comes into contact with food, a portion of the cytoplasm surrounds the substance and a depression forms within the cell wall. The food is eventually engulfed into a vessicle, which then moves further into the cell where it meets a LYSOSOME. Within the lysosome are digestive enzymes that break down the food. There are two types of endocytosis: **pinocytosis** is the incorporation and digestion of dissolved substances, and **phagocytosis** is the engulfing and digestion of microscopic particles. Phagocytosis is the process by which many PROTOZOA obtain sustinence. In higher animals, PHAGOCYTE cells are an important part of the IMMUNE SYSTEM, helping to expel infectious microorganisms.

endoderm (entoderm) One of the three so-called GERM LAYERS of tissue formed in the early development of a fertilized OVUM (egg). It is the innermost layer of the three and forms the EPITHELIUM of the liver, pancreas, digestive tract, and respiratory system. It is also the inner cell layer of a simple animal body. Other germ layers are the ECTODERM and MESODERM.

endometrium MUCOUS MEMBRANE, well supplied with blood vessels, that lines the UTERUS. It becomes thickened in the latter part of the MENSTRUAL CYCLE, but if no pregnancy takes place it is shed during menstruation.

endoparasite *See* PARASITE

endoplasmic reticulum Network of membranes and channels in the CYTOPLASM of EUKARYOTE cells (those which contain a nucleus, such as the cells of plants, animals, and fungi). These help to transport material inside the cell. Parts of the endoplasmic reticulum are covered with minute granules called RIBOSOMES. These consist of protein and a form of ribonucleic acid (RNA) and are sites of protein synthesis.

endorphin (endogenous MORPHINE) Naturally occurring PEPTIDE neurotransmitter found in the PITUITARY GLAND that has similar pain-relieving effects as morphine and other derivatives of OPIUM. Endorphins block the sensation of pain by binding to pain receptor sites. Two endorphins (**enkephalins**) occur in the brain, spinal cord, and gut. ACUPUNCTURE is thought to provide pain relief by inducing the release of endorphins. *See also* ANALGESIC; NEURON; NEUROTRANSMITTER

endoscope Instrument used to examine internal organs or tissues. Flexible or rigid, it is equipped with an eyepiece, lenses, and its own light source. Examples include a **gastroscope** (for examining the stomach) and a **cystoscope** (for examining the urinary bladder). Although most endoscopy is used for diagnostic purposes, MINIMAL ACCESS SURGERY can be performed using fine instruments passed through the endoscope.

endoskeleton Supporting system of bones and cartilage inside the body of an animal. All VERTEBRATES have an endoskeleton. It supports the body, and surrounds and protects internal organs. It also provides attachment points for muscles in order to facilitate movement. SPONGES have a type of endoskeleton consisting of a network of rigid or semirigid spicules. *See also* EXOSKELETON

endosperm Tissue that surrounds the developing embryo of a seed and provides food for growth. It is **triploid** (each cell has three sets of chromosomes), being derived from the FUSION of one of the male GAMETEs from the germinated pollen grain and two of the HAPLOID nuclei in the embryo sac. *See also* ALTERNATION OF GENERATIONS

endosymbiosis Mutually beneficial relationship in which one organism lives inside another. For example, bacteria were engulfed by EUKARYOTE cells and formed symbiotic relationships with them, eventually becoming so interdependent that the cells behaved as a single organism; the bacteria became MITOCHONDRIA and CHLOROPLASTS. Mitochondria and chloroplasts have their own small RIBOSOMES, which resemble those of bacteria, and also their own DNA distinct from that of the host cell. *See also* SYMBIOSIS

endothelium Tissue consisting of a thin layer of cells that lines various body structures. Like EPITHELIUM, endothelium is made up of platelike cells. It lines the inner surfaces of blood vessels and the heart; lymphatic vessels are also lined with endothelium.

endothermic reaction Chemical reaction in which heat is absorbed from the surroundings, causing a fall in temperature – as in the manufacture of water-gas from coke and steam.

energy In physics, capacity for doing WORK. It is measured in JOULES (J). POWER, the rate at which energy is produced or consumed, is measured in WATTS (W). POTENTIAL ENERGY is an object's ability to do work because of a change in the object's position or shape. KINETIC ENERGY is the energy an object has because it is moving. The many forms of energy include electrical, NUCLEAR, thermal, LIGHT, AND chemical. Energy can be transferred from one body to another through work processes, heating, ELECTROMAGNETIC RADIATION, and ELECTRICITY. Energy can also be converted from one form to another. The law of CONSERVATION of energy states that energy cannot be created or destroyed. The concept of energy began with GALILEO and Sir Isaac NEWTON. The idea that MASS is a form of energy was established by Albert EINSTEIN, who recognized that energy (E) and mass (m) could be transformed into each other according to the relation $E = mc^2$, where c is the velocity of light.

energy level Fixed amount of energy possessed by a nucleus, ELECTRON, atom or molecule. For example, the energy of electrons in an atom is not continuously variable. It can have only a discrete set of values, termed energy levels. The lowest possible energy level is called the **ground state**. If the atom absorbs energy, possibly in the form of a PHOTON (a quantum of light or other ELECTROMAGNETIC

RADIATION), an electron can be raised to an ORBITAL of a higher energy level, when the atom is said to be in an EXCITED STATE. When the electron in an excited atom returns to its previous energy level, the excess energy is emitted, often as a quantum of light. In a similar way, molecules have energy levels involved with rotational and vibrational motion.

energy sources Naturally occurring substances and processes from which we obtain the energy that we need. The energy in all of these sources is derived almost entirely from the Sun. The fossil fuels coal, oil, and natural gas are the remains of life that was dependent for growth on SOLAR ENERGY. HYDROELECTRICITY also derives from solar energy, which maintains the Earth's HYDROLOGICAL CYCLE. Wind power also depends on the Sun, and is caused by uneven heating of the atmosphere. The movements of tides and waves depend upon solar thermal energy and on the gravitational pulls of the Sun and Moon. These movements of the oceans can be harnessed to generate electricity. We use solar energy directly for heating some domestic water supplies, and for providing electricity from PHOTOELECTRIC CELLS. GEOTHERMAL ENERGY is energy obtained from underground hot rocks. Other major energy sources are radioactive metals, such as uranium, plutonium, and thorium, which provide NUCLEAR ENERGY. *See also* ELECTRICITY SOURCES

Engelberger, Joseph (*c.*1920–) US physicist, known as the "father of robotics". His first computer-controlled industrial robots were installed in 1967 at a General Motors factory in the USA, where they were used in the electronics industry, eventually with robots helping to build robots.

engine Machine that produces useful energy of motion from some other form of energy. The term is usually restricted to combustion engines, which burn fuel. These machines include the STEAM ENGINE, DIESEL ENGINE, GASOLINE ENGINE, JET ENGINE, and ROCKET engine. Such engines are distinguished from ELECTRIC MOTORS which, in providing their power, do not directly alter the chemical or physical composition of a substance. Combustion engines are of two main kinds. An external combustion engine burns its fuel outside the chamber in which motion is produced. In the steam LOCOMOTIVE, for example, the fire box is separate from the cylinders. An INTERNAL COMBUSTION ENGINE burns its fuel and develops motion in the same place. In the gasoline engine, for example, this all takes place in the cylinders.

engineering Application of scientific principles for practical purposes, such as the design and construction of buildings and machines. MECHANICAL ENGINEERING includes the design and testing of machinery. CIVIL ENGINEERING includes the preparation of sites and the design and building of structures, such as bridges, tunnels, and harbors. Electrical engineers work on electrical systems. In ELECTRONICS, engineers are concerned with devices such as scientific instruments, radio and television equipment, radar systems, and computers. Nuclear engineers install and operate nuclear reactors. CHEMICAL ENGINEERING includes the design and operation of large-scale chemical processes. There is some overlap of expertise in these many fields of engineering. For this reason, the academic training of all engineers starts with a thorough grounding in the fundamentals of science and an education in general engineering subjects. This is followed by specialized training in the student's chosen field.

engineering, genetic *See* GENETIC ENGINEERING

enstatite ($MgSiO_3$) Orthorhombic mineral of the pyroxene group, commonly found in ultrabasic IGNEOUS rocks such as norites, pyroxenites, gabbros, and peridotites. It varies from colorless to yellowish gray, shading to green if iron is present. Hardness, 5.5; r.d. 3.2–3.5.

enthalpy (symbol *H*) Amount of thermodynamic heat energy possessed by a substance. In any system, enthalpy equals the sum of its internal energy and the product of the pressure and volume. It can be measured in terms of the change (usually uptake) of heat that accompanies a CHEMICAL REACTION that takes place at a constant pressure, and as such is equivalent to the heat of reaction (usually expressed in kilojoules per mole). For a reaction that produces heat (EXOTHERMIC REACTION), the enthalpy is taken to be negative. *See also* THERMODYNAMICS

entomophily POLLINATION of a flower by a pollen-carrying insect. Most entomophilous flowers are scented and brightly colored, often with nectar, to attract insects. The flowers may have different structures to ensure CROSS-POLLINATION.

entropy Quantity that specifies the disorder or randomness of a physical system. In THERMODYNAMICS, it expresses the degree to which thermal energy is available for work – the less available it is, the greater the entropy. The entropy of the Universe is increasing. Energy can be extracted from a system only as it changes to a less ordered state. According to the second law of thermodynamics, an isolated system's change in entropy is either zero or positive in any process. In INFORMATION THEORY, the entropy of an information system increases with the number of random errors in transmitting the message.

E number Names of certain food additives used in the European Union. Food additives approved for use within the European Union must by law be identified in the list of ingredients of any product in which they are used. Such additives might be preservatives, flavor enhancers, stabilizers, or coloring agents.

environment Physical and biological surroundings of an organism. The environment covers both nonliving (ABIOTIC) factors such as temperature, soil, atmosphere, and radiation, and living organisms such as plants, microorganisms, and animals that make up the biotic (living) environment.

enzyme (Gk. *zymosis*, in yeast) Protein that functions as a CATALYST in biochemical reactions. The FERMENTATION properties of yeast cells have

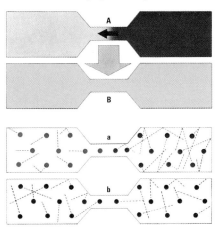

long been utilized in BREWING alcohol. In 1897 German chemist Edward BÜCHNER discovered that cell-free extracts of yeasts could ferment sugars to alcohol. In 1926 the US biochemist James B. SUMNER was the first person to isolate an enzyme in pure crystal form. He extracted urease from the jack bean and proved that enzymes are protein molecules. In the next decade, PEPSIN, TRYPSIN, and CHYMOTRYPSIN were crystallized. Today, more than 1,500 catalysts have been identified, and their AMINO ACID structures revealed through X-RAY CRYSTALLOGRAPHY. In 1969 scientists first synthesized an enzyme, ribonuclease. Chemical reactions can occur several thousand or million times faster with enzymes than without them. The efficiency of an enzyme is measured in terms of its turnover rate. They operate within a narrow temperature range (usually 30°C to 40°C) and have optimal pH ranges. Many enzymes have to be bound to nonprotein molecules. These include **trace elements** (such as metals) and **coenzymes** (such as vitamins). If a coenzyme is tightly attached to the protein enzyme, the unit is termed a **prosthetic group**. The lack or malfunction of enzymes can cause a variety of metabolic diseases. Enzymes are widely used in the manufacture of detergents, antibiotics, and food.

Eocene Second of the five epochs of the TERTIARY period, from 55–38 million years ago. The name implies the "dawn" of life in which the modern families of both plant and animal species appeared. The fossil record shows members of modern plant genera, including beeches, walnuts, and elms, and indicates the apparent dominance of mammals, including the ancestors of camels, horses (notably *Hyracotherium*), rodents, bats, and monkeys. The world climate was warmer than it is now.

ephedrine Drug, chemically similar to ADRENALINE. It stimulates the AUTONOMIC NERVOUS SYSTEM. Once used to treat asthma, it is still found in some cold remedies as a nasal decongestant.

ephemeris (pl. ephemerides) Astronomical table giving the predicted positions of a celestial object, such as a planet or comet, at given intervals. The term also signifies an annual publication supplying such tables along with information concerning the Sun, eclipse and occultation data, data for certain stars, astronomical constants, and so forth.

ephemeris time (ET) System of time reckoning normally employed in the compilation of astronomical data for almanacs and EPHEMERIDES. It takes no account of irregularities in the Earth's rotation and is calculated on the basis of the tropical year. It is slightly in advance of UNIVERSAL TIME.

▲ **entropy** A tendency for heat to pass from a hotter to a cooler body (A), and never in the opposite direction unless externally assisted, can be stated as a quite general principle: the Second Law of Thermodynamics. The change that occurs due to the redistribution of heat – the uniform condition (B) resulting ultimately from the asymmetrical condition (A) – can be expressed as an increase in entropy: the increase in this quantity is defined as the amount of heat transferred divided by the temperature at which the heat-flow occurs. Entropy can be thought of as relative disorder: B is a less organized situation than A. It can also be thought of as a measure of probability. If a given amount of heat energy – which is energy or particle-motion – is to be distributed at random throughout the object in the diagram, it is far more probable that the distribution will be a uniform mixture of speeds (b) than that the high and low speeds will be concentrated in different places (a).

epicenter Spot on the Earth's surface directly above the focus of an EARTHQUAKE. Depending on the character of the focus, the epicenter may be a small circle or a line.

epidemic Outbreak of an infectious disease with rapid spread to a great many people. The study of epidemics, which includes the causes, patterns of contagion, and methods of containment of disease, is known as EPIDEMIOLOGY. An epidemic that sweeps across many countries, as the Black Death did in the Middle Ages, is termed a pandemic.

epidemiology Study of the incidence and patterns of disease with a view to finding means of prevention or control. Traditionally, epidemiological methods were applied to communicable diseases – those, such as CHOLERA or MEASLES, that can be passed from one person to another. However, modern epidemiology is equally concerned with the role of environmental and lifestyle factors in disease causation: for example, the link between smoking and lung CANCER or between ASBESTOS and the cancer known as mesothelioma.

epidermis In animals, outer layer of skin that contains no blood vessels; in plants, the outermost layer of a leaf or of an unthickened stem or root. In plants, the outer surface of the epidermis is usually coated in a waxy layer, the CUTICLE, which reduces water loss. The epidermis of leaves and stems may be perforated by pores, called STOMATA, which allow GAS EXCHANGE.

epididymis Network of ducts in the TESTES where sperm cells mature and are stored. *See also* VAS DEFERENS

epidote Orthosilicate mineral, hydrated calcium iron-aluminum silicate ($Ca_2Fe_3(Al_2O)(OH)(Si_2O_7)$ $[Si_2O_4]$). It is found in METAMORPHIC and IGNEOUS rocks as monoclinic system prismatic crystals and fibrous or granular masses; it is typically pistachio green, glassy, and brittle. Certain large crystals, 7.5–25cm (3–10in), are collector's items. Hardness 6–7; r.d. 3.4.

epidural anesthesia Injection of a local anesthetic into the epidural space surrounding the spinal cord. It is done to anesthetize the spinal nerve roots and so prevent pain during surgery or childbirth.

epiglottis Small flap of CARTILAGE projecting upward behind the root of the tongue. It closes off the LARYNX during swallowing to prevent food entering the airway.

epilepsy Disorder characterized by abnormal electrical discharges in the brain, provoking seizures of varying intensity. Affecting slightly more males than females, epilepsy occurs in up to 2% of the population. It arises either spontaneously or as a result of some brain disorder. It is seen both in generalized forms, involving the whole of the CEREBRAL CORTEX, or in partial (**focal**) attacks arising in one small part of the brain. Attacks are often presaged by warning symptoms, the "aura." Seizure types vary from the momentary loss of awareness seen in *petit mal* attacks (called **absences**) to the major convulsions of *grand mal* epilepsy. They may be triggered by a number of factors, including sleep deprivation, flashing lights or excessive noise. All forms of epilepsy are controlled with anticonvulsant drugs; in some cases surgery may be recommended.

epilimnion Upper layers of water in a lake. It is warmer than the deeper water, and because light penetrates the epilimnion it can support PHOTOSYNTHESIS and plant life. The water may be disturbed by wind and water currents. *See also* HYPOLIMNION

epinephrine *See* ADRENALINE

epiphyte (air plant) Plant that grows on another plant but is not a PARASITE. Epiphytes usually have aerial roots and produce their own food by PHOTOSYNTHESIS. They are common in tropical forests. Examples are some ferns, some orchids, Spanish moss, and many bromeliads.

epithelium Layer of cells, closely packed to form a surface or a lining for a body tube or cavity. Epithelium covers not only the SKIN, but also various internal organs and surfaces such as the intestines, nasal passages, and mouth. Epithelial cells may also produce protective modifications such as hair and nails, or secrete substances such as ENZYMES and MUCUS.

epoch *See* GEOLOGICAL TIME

epoxide In organic chemistry, a compound that has a three-membered ring including oxygen in its molecular makeup. An example is the gas epoxyethane (ethylene oxide, C_2H_4O), which may be regarded as a cyclic ETHER. Polymerization of epoxides with PHENOLS produces EPOXY RESINS, which are used as adhesives and electrical insulators.

epoxy resin Group of thermosetting polymers with outstandingly good mechanical and electrical properties, stability, heat and chemical resistance, and adhesion. Epoxy resins are used as adhesives, in casting and protective coatings. Adhesive epoxy resins are sold in two separate components, a viscous resin and a hardener, which are mixed just before use. *See also* THERMOSETTING RESIN

EPROM (erasable programmable read-only memory) Type of read-only computer memory (ROM) in which data stored on an INTEGRATED CIRCUIT can be erased, usually by means of ultraviolet light. It can then be reprogrammed using pulses of electricity. *See also* PROM; RAM

Epstein-Barr virus Organism responsible for GLANDULAR FEVER, an illness that occurs worldwide in teenagers and young adults. The symptoms include fatigue, headaches, muscular aches, sore throats, and enlarged lymph nodes. Epstein-Barr virus is also implicated in the causation of a number of other diseases, in particular BURKITT'S LYMPHOMA, a common childhood cancer in Central Africa. The virus is spread by droplet infection.

equation Mathematical statement that holds for some subset of all possible values of the variables. An equation of the form $x^2 = 8 - 2x$ is true only for certain values of x ($x = 2$ and $x = -4$). These values are the SOLUTIONS of the equation. This type of equation is contrasted with an **identity**, such as

$$(x + 2)^2 = x^2 + 4x + 4$$

which is true for all values of x. Equations are classified in various ways: for example the highest power of the variable (2 in the case above) is the degree of the equation. Equations are said to be linear, QUADRATIC, cubic, quartic, etc., according to whether their degree is 1, 2, 3, 4, etc. *See also* SIMULTANEOUS EQUATIONS

equation of state Formula that connects the pressure, volume, and temperature of a system containing a given amount of substance. The most simple equation of state is the IDEAL GAS LAW.

equator Name given to two imaginary circles. The **terrestrial Equator** lies midway between the North Pole and South Pole and is the zero line from which latitude is measured. It divides the Earth into the Southern and Northern Hemispheres. The **celestial equator** lies directly above the terrestrial Equator and is used as a reference to determine the position of a star using an astronomical coordinate such as the EQUATORIAL COORDINATE SYSTEM.

equatorial coordinate system Astronomical coordinate system that refers to the plane of the CELESTIAL EQUATOR. Two sets of coordinates are used: right ascension and declination, and hour angle and declination.

equatorial telescope Common TELESCOPE mounting in which the telescope rotates about two perpendicular axes, one of which, the polar axis, is parallel to the Earth's axis, and so points towards the CELESTIAL POLE. The steady movement of celestial bodies, due to the Earth's rotation, can be followed by a daily rotation of the telescope about this axis, usually by means of a clockwork drive.

equilateral triangle Planar figure with three sides all of equal length; the three interior angles are also equal and each of magnitude 60°. *See also* TRIANGLE

equilibrium In physics, a stable state in which forces acting on a particle or object negate each other, resulting in no net force. While thought of as a state of balance or rest, an object with constant velocity is also said to be in equilibrium. The term can also be ascribed to a body with a constant temperature; this is known as **thermal equilibrium**. In ECOLOGY, the term is used to describe a state in which the forces acting to increase a given quantity, such as the number of individuals in a POPULATION or the number of species in a COMMUNITY, roughly or exactly balance the forces acting to reduce the same quantity.

equilibrium constant Ratio of concentrations of products to those of reactants, characterizing the CHEMICAL EQUILIBRIUM of a particular REVERSIBLE REACTION at a specified temperature.

equilibrium sense (vestibular sense) Mammalian ability to remain upright in relation to gravity and to detect changes in position and momentum. The principal organs of equilibrium are contained in the inner EAR – the UTRICLE, which transmits orientation information, and the SEMICIRCULAR CANALS, which are concerned with acceleration and deceleration. These systems help humans to locate their bodies in space.

equinox Either of the two points at which the Sun crosses the celestial EQUATOR. The **vernal** equinox (or **spring** equinox) occurs when the Sun crosses the equator from S to N on or near 21 March each year in the Northern Hemisphere. This point is also called the First Point of Aries. The other equinox is the **autumnal** equinox, which occurs on or near 23 September in the Northern Hemisphere when the Sun crosses the equator from N to S. This point is also called the First Point of Libra. At the equinoxes, the Sun rises due E and sets due W.

equivalence principle Principle that observations made in a closed laboratory (from which the external Universe cannot be observed) cannot distinguish between the effects of GRAVITATION and of an ACCELERATION of the laboratory. It is a cornerstone of Einstein's general theory of RELATIVITY.

equivalent weight Weight of an element, in grams, that combines with 8g of oxygen or with an equivalent weight of some other substances. For example, the equivalent weight of hydrogen is 1 in forming water (H_2O). The equivalent weight of an acid is the number of grams that contains 1g of acidic hydrogen. A similar definition is used for bases.

Equuleus (foal) Small constellation of the N SKY. Its brightest star, Alpha, has a magnitude of 3.9.

Eratosthenes (276–194 BC) Greek scholar who first measured the Earth's circumference by using geometry. Eratosthenes administered the library of Alexandria and was renowned for his work in mathematics, geography, philosophy, and literature. He invented a systematic method of identifying prime numbers in which those numbers with divisors are identified and removed from a list; the remaining numbers are prime numbers. The method is known as **Eratosthenes' sieve**, but is, unfortunately, too slow to decide whether or not a very large number is prime. *See also* NUMBER PRIME

erbium (symbol Er) Silvery, metallic element of the LANTHANIDE SERIES, first isolated in 1843. There are six naturally occurring isotopes, and the chief ores are monazite and bastnaesite. Twelve radioactive isotopes have been identified. Soft and malleable, erbium is used in some specialized alloys, and erbium oxide is used as a pink colorant for glass. Properties: at.no. 68; r.a.m. 167.26; m.p. 1,522°C (2,772°F); r.d. 9.045; b.p. 2,863°C (5,185°F); most common isotope ^{166}Er (33.41%).

erg Unit of energy in the metric CGS (centimeter-gram-second) system of units. One erg is the work done by a force of 1 DYNE acting through a distance of 1 cm. One JOULE equals 10 million ergs.

ergosterol Organic chemical compound, a STEROID, found particularly in yeast and other fungi. When irradiated with ultraviolet light, it is converted to vitamin D.

ergonomics (human-factors engineering) Study of the relationship between human beings and their work environment. In particular, it examines the physical and psychological interaction of man and machine. The science of ergonomics dates from the beginnings of MASS PRODUCTION. In recent years, it has focused on human-computer interaction (HCI). Applications include the design of aircraft and automobile interiors.

Eridanus (River) Large but inconspicuous con-

stellation of the s sky. Achernar is its brightest star, with a magnitude of 0.46.

▲ **erosion** Erosion is the breakdown and transportation of rock due to the action of an outside agent. There are three main forms: river, glacial, and wind. Rivers (1) erode their channels through the flow of water and the abrasion of the load they are carrying against the banks and riverbed. Erosion is most forceful at the outside of bends (2), where the banks are undercut (3) often creating cliffs or bluffs (4) down which material moves. Flood surges dramatically increase the power of the river and correspondingly magnify the erosive force. On a smaller scale, rainwater will move material down a hillside (5). Particles of soil are carried by rivulets and the impact of raindrops throws soil down the slope. Vegetation reduces such erosion by binding the soil together. Where vegetation is removed, as on tracks (6), erosion is accentuated. In arid conditions wind erosion can carve distinctive features. Sand and stones blown by the wind (7) have the same effect as shot-blasting. Mushroom-shaped formations, pedestals (8), are often the result. This is because the rock is untouched above the maximum height at which the erosive sand bounces across the surface (9) as it is carried by the wind.

erosion Alteration of landforms by the wearing away of rock and soil, and the removal of any debris (as opposed to WEATHERING). Erosion is carried out by the agents of wind, water, GLACIERS, and living organisms. In **chemical** erosion, minerals in the rock react to other substances, such as weak acids found in rainwater, and are broken down. In **physical** erosion, powerful forces such as rivers and glaciers physically wear rock down and transport it. Erosion can have disastrous economic results, such as the removal of topsoil, the gradual destruction of buildings, and the alteration of water systems. Inland, erosion occurs most drastically by the action of rivers, and in coastal regions by the action of waves. *See also* GEOMORPHOLOGY

erratic In geology, a rock that has been transported some distance from its source by glacial action, and is therefore unlike the surrounding rocks. The tracing of erratics back to their source can give important information about the movement of ice.

eruption Release of lava and gas from the Earth's interior to the Earth's surface, either on land or under the sea, and into the atmosphere. *See also* VOLCANO

erythrocyte (red blood cell) BLOOD cell that carries oxygen around the body. It contains HEMO-GLOBIN, which combines with oxygen to form OXYHEMOGLOBIN and gives blood its red color. In mammals erythrocytes are usually disk-shaped and have no nucleus. In other vertebrates they are more oval in shape and contain nuclei. Normal human blood contains an average of 5 million such cells per mm^3 of blood, making them the most numerous blood cell. They are manufactured in bone MARROW.

Esaki, Leo (1925–) Japanese physicist who developed the tunnel diode, a SEMICONDUCTOR that allows electrons to cross normally impassable electronic barriers. Ivar GIAEVER extended Esako's research to the field of SUPERCONDUCTIVITY, which led to the prediction of the **Josephson effect** (electrical effects of a current flowing between superconductors). For this work, they shared the 1973 Nobel Prize for physics with Brian JOSEPHSON. The tunnel diode, also called **Esaki diode**, has found extensive applications, which include its use in computer information storage.

escalator Electrically powered moving stairs, driven by chain and sprocket and held in the correct plane by two tracks. It is usually inclined at 30° and limited to a total rise of up to 18m (60ft). As the steps approach the landing, they pass through a protective comblike device. Escalators are used to carry pedestrians between floors in busy areas such as large stores, office buildings, and subway stations. The Otis Elevator Company exhibited the first escalator in 1900 at the Paris Exposition.

escape velocity Minimum velocity required to free a body from the gravitational field of a celestial body or stellar system. Escape velocities are 11.2km s^{-1} (7mi s^{-1}) for the Earth and 2.4km s^{-1} (1.5mi s^{-1}) for the Moon. They can be calculated from the formula: $v = (2GM/R)^{1/2}$ where G is the gravitational constant, M the mass of the planet or system and R the distance of the rocket from the center of mass of the system.

escarpment (scarp) Steep slope or cliff arising from a flat or gently sloping area, produced by faulting and/or differential EROSION.

esker Long, narrow ridge of rock formed by a stream flowing under a GLACIER. Subglacial streams and rivers carry gravel and sand that they deposit on their beds as they tunnel below the ice. When the ice melts, the rivers may move to other channels, leaving the eskers as long ridges 10–25m (30–80ft) high and 5–25m (15–80ft) across; they may be many kilometers long.

esophagus (gullet) Muscular tube, part of the ALIMENTARY CANAL (or gut), that carries swallowed food to the STOMACH. Food is moved down the lubricated channel by a wavelike movement known as PERISTALSIS.

essential amino acid *See* AMINO ACID

essential element Element needed by a living organism so that it can grow, develop, and maintain its body structure normally. All organisms require carbon, hydrogen, nitrogen, and oxygen, which they take in the form of food, air, and water, and which make up most of their body tissues. In addition, most organisms also need calcium, chlorine, magnesium, phosphorus, potassium, sodium, and sulfur, which are termed **major elements**. Required in much smaller quantities are the so-called **trace elements**, which include cobalt, copper, chromium, IODINE, IRON, manganese, selenium, and zinc. All of these additional essential elements – major and trace – have to be supplied by the organism's food.

essential oil Oil found in flowers, fruits or plants. It is the source of their characteristic odor and is widely used in aromatherapy, potpourri, and perfumed toiletries. Many are obtained from plants that originate in dry Mediterranean climates. These oils are usually produced by special glands.

ester Any of a class of organic compounds formed by reaction between an ALCOHOL and an ACID.

estivation In biology, a state of prolonged inactivity in some animals that occurs during a period of hot weather or drought. Bodily activities such as feeding, movement, and respiration slow down, and the animal expends very little energy; in this respect, it is similar to DORMANCY. It occurs in some insects, snails, and lungfish, for example. In botany, the term describes the way petals and sepals are arranged in a flower bud. *See also* HIBERNATION

estrogen Female SEX HORMONE. First produced by a girl at PUBERTY, estrogen leads to the development of the SECONDARY SEXUAL CHARACTERISTICS that turn her body into a woman's: breasts, body hair, and redistributed fat. It regulates the MENTRUAL CYCLE and prepares the UTERUS for pregnancy. Estrogen is also a constituent of the contraceptive pill (*see* CONTRACEPTION).

estrous cycle Physiological changes occurring during the female reproductive cycle of most placental MAMMALS. Controlled by HORMONES, it is evident among animals other than humans. Cycles of different animals vary in frequency and length. Typically, ovulation is associated with the estrous (heat) period. *See also* MENSTRUAL CYCLE

estuary Coastal region where a river mouth opens into the ocean and fresh water from the land is mixed with salt water from the sea. Many estuaries are drowned river valleys, perhaps formed after a rise in sea level at the end of an ice age. They usually provide good harbors and are often breeding grounds for many kinds of marine life. *See also* DROWNED COAST

Eta meson (symbol η) Uncharged ELEMENTARY PARTICLE with zero SPIN. It is grouped under the HADRONS (particles that experience the STRONG NUCLEAR FORCE).

ethane (C_2H_6) Colorless, odorless gas, the second member of the ALKANE series of HYDROCARBONS. It is a minor constituent of natural gas. *See also* SATURATED COMPOUND

ethanol (ethyl alcohol, C_2H_5OH) Colorless, flammable, and volatile liquid, the best-known ALCOHOL, produced by the FERMENTATION of sugars, molasses, and grains, or by the catalytic HYDRATION of ethylene. Also known as grain alcohol, its many uses include alcoholic beverages (such as wine, beer, cider, and spirits), cleaning solutions, antifreeze, rocket fuels, cosmetics, and pharmaceuticals. Properties: r.d. 0.789; b.p. 78.5°C (173.3°F).

ethene *See* ETHYLENE

ether In chemistry, colorless, volatile, flammable liquid that has the $-O-$ group in its structure. The most common is ETHOXYETHANE (diethyl ether). It is prepared by the action of sulfuric acid on ethanol followed by DISTILLATION.

ether In physics, hypothetical medium that was supposed to fill all space and offer no resistance to motion. It was postulated as a medium to support the propagation of electromagnetic radiations, but its existence was disproved by the MICHELSON-MORLEY EXPERIMENT.

ethology Study of animal behavior especially in the natural environment, first outlined in the 1920s by Konrad LORENZ and Nikolaas TINBERGEN. Ethologists study natural processes that range across all animal groups, such as courtship, mating, and self-defense. Field observations and laboratory experiments are both used.

ethoxyethane (ether, diethyl ether, $C_2H_5OC_2H_5$) Volatile, flammable liquid with a sickly smell, once used as a general anesthetic. It is made from ethanol (ethyl alcohol) by dehydration using concentrated sulfuric acid, or by reacting it with a chloroethane (ethyl chloride) and sodium metal (**Williamson's synthesis**). It is used mainly as a solvent. Properties: m.p. $-116.2°C$ ($-177.2°F$); b.p. 34.5°C (94.1°F).

ethyl alcohol *See* ETHANOL

ethylene (ethene, $CH_2=CH_2$) Colorless gas derived from the CRACKING of propane and other compounds. Vast quantities of the gas are used annually for the manufacture of POLYETHENE. Ethylene is also used widely for many other chemical syntheses.

ethyne *See* ACETYLENE

etiology In medicine, the science of the origins and causes of diseases.

Eubacteria Subkingdom of the kingdom Prokaryotae, or perhaps even a separate DOMAIN. Eubacteria include all multicellular BACTERIA, as well as all those that use the energy of sunlight to make their own food (PHOTOSYNTHESIS), deriving their carbon from the air. They do not have the unique types of cell walls, RIBOSOMES, and RNA of the other subkingdom, ARCHAEBACTERIA. *See also* PROKARYOTE

Euclid (*c*.300–*c*.95 BC) Ancient Greek mathematician whose ideas dominated mathematics, particularly GEOMETRY, for almost 2,000 years. He is remembered for his textbooks on geometry,

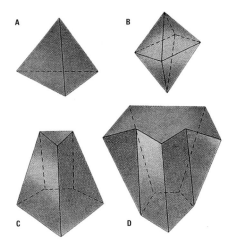

A

B

C

D

▲ **Euler's rule** All solids that do not have holes through them, and have flat faces, obey Euler's rule: $V + F - E = 2$ where V is the number of vertices (corners), F the number of faces and E the number of edges. For the tetrahedron (A) $4 + 4 - 6 = 2$; for the octahedron (B), $6 + 8 - 12 = 2$. The shapes C and D also obey the rules, in fact the size and shape does not matter at all.

ARCHIMEDES, for calculating the areas of plane figures and volumes of solids. He also gave a precise definition of the real numbers in the framework of a general theory of proportion. In astronomy, he put forward the first system for describing the motions of the heavenly bodies. This system introduced the idea of the crystal spheres that carried the celestial bodies round the Earth, a notion subsequently taken up by ARISTOTLE. *See also* NUMBER, REAL

eugenics Study of human improvement by SELECTIVE BREEDING, founded in the 19th century by Sir Francis GALTON. It proposed the genetic "improvement" of the human species through the application of social controls on parenthood, encouraging parents who are above average in certain traits to have more children, while ensuring those who are below average have fewer. As a social movement, eugenics was discredited in the early 20th century owing to its ethical implications and its racist and class-based assumptions. Advances in GENETICS have given rise to the modern field of genetic counselling, through which people known to have defective genes that could cause physical or mental disorders in offspring are warned of the risks.

Euglenophyta Phylum of single-celled algae that includes the genus *Euglena*. Members of this group have both animal and plant characteristics. They swim by means of FLAGELLA. Many species contain CHLOROPLASTS and employ PHOTOSYNTHESIS, but some are colorless and feed on BACTERIA and DIATOMS. Photosynthetic euglenoids can temporarily "close" their chloroplasts and feed like colorless forms.

eukaryote Organism whose CELLS have a membrane-bound nucleus, with DNA contained in CHROMOSOMES. Eukaryotes include all animals, plants, fungi, and PROTOCTISTA. They have a complex CYTOPLASM with an ENDOPLASMIC RETICULUM, and most of them possess MITOCHONDRIA. Most plants and algae also possess CHLOROPLASTS. Other structures specific to eukaryotic cells include microtubules, GOLGI BODIES, and membrane-bound flagella. In some classification systems eukaryotes make up the DOMAIN Eukarya. *See also* PROKARYOTE

Euler, Leonhard (1707–83) Swiss mathematician. An outstanding scholar, Euler worked in many areas of pure and applied mathematics, including analysis, differential equations, geometry, trigonometry, acoustics, fluid dynamics, and celestial mechanics. His analysis of the Moon's motion laid the foundations of lunar theory, and he made important advances in studies of the tides, planetary perturbations, and orbits of comets. His work on optical systems strongly influenced the technical development

especially *The Elements*, which was first printed in 1482 in a Latin translation from the Arabic. His other works include *Data* (on geometry) and *Phaenomena* (on astronomy). Several books have been lost. *See also* EUCLIDEAN GEOMETRY

Euclidean geometry Type of geometry that obeys the AXIOMS outlined in EUCLID's *The Elements*. Starting with a set of self-evident axioms, Euclid discovered many results using rigorous logic. The results were treated as absolute truths applicable to the physical world for almost 2,000 years. In the 19th century it was shown that his axioms are not universal and do not hold in all circumstances. Nikolai LOBACHEVSKY and Georg RIEMANN were among the mathematicians to discover types of geometry where the Euclid axioms did not hold. Their description of geometry is said to be NON-EUCLIDEAN. One striking feature of non-Euclidean geometry is that two straight lines that are parallel in one region of space can meet. In his work on general RELATIVITY Albert EINSTEIN discovered that the geometry of the Universe in which we live is non-Euclidean. Euclidean geometry, however, provides an adequate description of everything in our everyday lives.

Eudoxus of Cnidus (*c*.408–*c*.355 BC) Greek mathematician and astronomer, one of the greatest mathematicians of antiquity. Eudoxus devised the method of exhaustion, later extended by of telescopes and microscopes. Euler is said to have been the most prolific mathematician ever.

Euler diagram Simple diagram used in logic to illustrate syllogisms. Classes of objects are represented by circles so that, for example, a premise of the type "some *a* is *b* " can be represented by an overlap of two circles, one which represents *a* and the other representing *b*. *See also* VENN DIAGRAM

Euler's rule In solid geometry, states that for any polyhedron (many-sided solid figure) $V + F - E = 2$, where V is the number of vertices (corners), F the number of faces, and E the number of edges of the polyhedron. The rule is named after Leonhard EULER.

euphotic zone Uppermost layer of oceans. Much light penetrates this comparatively shallow zone, and many green plants and herbivores are found there, and often sea-dwelling mammals.

Europa Smallest of Jupiter's GALILEAN SATELLITES, with a diameter of 3,138km (1,950mi). It is a predominantly rocky body. Europa's smooth crust of water-ice is crisscrossed by a network of light and dark linear markings. There are very few craters, so old ones must have been removed by some form of geological activity.

European Space Agency (ESA) An organization of European nations to promote space research

and technology for peaceful purposes, founded in 1975. ESA has 14 member states: Austria, Belgium, Denmark, Finland, France, Germany, Ireland, Italy, the Netherlands, Norway, Spain, Sweden, Switzerland, and the United Kingdom. Its headquarters are in Paris. ESA has designed and built the Ariane series of launch rockets, and operates a launch site at Kourou in French Guiana, on the Atlantic coast of South America.

europium (symbol Eu) Silver-white, metallic element of the LANTHANIDE SERIES, being the softest and most volatile member. It was first isolated as the oxide in 1896. Its chief ores are monazite and bastnaesite. The metal is used in the manufacture of color television screens, lasers, and in control rods in NUCLEAR REACTORS. Properties: at.no. 63; r.a.m. 151.96; r.d. 5.25; m.p. 822°C (1,512°F); b.p. 1,597°C (2,907°F); most common isotope ^{153}Eu (52.18%).

Eustachian tube Small channel that connects the middle EAR to the back of the throat. It opens when swallowing, to allow the pressure in the middle ear to remain the same as the pressure of air outside the body. It is named after the Italian anatomist Bartolomeo Eustachio (1520–74).

eutectic *See* ALLOY

eutrophication Process by which a stream or lake becomes rich in nutrients by runoff from the

► **eutrophication** Lakes may be classified according to their fertility or productivity. Young oligotrophic lakes (A) become more fertile, or eutrophic, with age as they are efficient traps of nutrients delivered in run-off water from surrounding land (1). This natural process is hastened when the water contains agricultural fertilizers and industrial organic effluent. Oligotrophic lakes are characterized by clear water, a sandy or pebbly bottom, and a low density of algae and plankton (mainly diatoms (2) and desmids (3)). Typical fish are trout (4) and char (5), which thrive in deep, cold water. In eutrophic lakes (B), nitrate fertilizers and industrial organic effluent (6), which is rapidly broken down to release nitrates, phosphates and potassium, cause a great increase in algal growth, especially green algae (7) (chlorophytes) and cyanobacteria (8). The bottom of the lake becomes muddy, the water becomes turbid, and stagnation eliminates the bottom fish, which are replaced by perch (9) and bream (10). Eventually the algal bloom becomes so important that most of the dissolved oxygen is removed from the water, killing large numbers of fish. The decomposition of large amounts of organic matter will further reduce the oxygen content and a reducing condition exists in which hydrogen sulfide and methane are produced.

land or by artificial means. Compounds of nitrogen, phosphorus, iron, sulfur, and potassium are vital for plant growth in water; in excess, however, they overstimulate the growth of surface CYANOBACTERIA and microorganisms, which consume all the available dissolved oxygen.

Evans, Oliver (1755–1819) US engineer, pioneer of the use of steam power in the USA. In 1802 he constructed the first high-pressure steam engine in the country, and in 1804 he patented an amphibious steam dredger. His earlier inventions included a machine for making carding combs (for textile manufacture), and an "automatic" flour mill which employed bucket elevators and screw conveyors, all of Evan's design. His high-pressure engine was later used for making millstones for the flour mills.

evaporation Process by which a liquid or a solid becomes a vapor. The reverse process is CONDENSATION. Solids and liquids cool when they evaporate because they give up energy (LATENT HEAT) to the escaping molecules.

evaporite Mineral deposit of precipitated salts, formed by the evaporation of saline lakes or confined volumes of salt water, usually in previous geological eras. They are important sources of GYPSUM, anhydrite, rock-salt, sylvite, and small amounts of nitrates and borates.

evapotranspiration Combined processes of EVAPORATION and TRANSPIRATION in which water is transferred from the EARTH's surface to the atmosphere. It is a key factor in the HYDROLOGICAL CYCLE. Water or ice is evaporated by the Sun's rays and by the wind, and plants are cooled by transpiration.

event horizon Boundary of a BLACK HOLE, from which nothing can escape. Observers outside the event horizon can therefore obtain no information about the black hole's interior. The radius of the event horizon is called the **Schwarzschild radius**.

evergreen Plant that retains its green foliage for a year or more (DECIDUOUS plants lose their leaves every autumn or dry season). Evergreens are divided into two groups: narrow-leaved (CONIFERS) and broad-leaved. Conifers include fir, spruce, pine, and juniper. Among the broad-leaved evergreens are holly and rhododendron. Not all conifers are evergreens; exceptions are the deciduous larch (*Larix*) and dawn redwood (*Metasequoia*).

evolution Changes that occur across successive generations of organisms. The causes of evolution include natural selection and genetic drift. Early work on evolutionary theory was initiated by Jean LAMARCK during the early 1800s, but it was not until Charles DARWIN wrote *The Origin of Species* during the mid-1800s that the theory was considered worthy of argument. Present-day evolutionary theory is largely derived from the work of Darwin and MENDEL. *See also* NEO-DARWINISM; PUNCTUATED EQUILIBRIUM

evolutionary biology Study of biology based on the way organisms have evolved rather than on such observable characteristics as their anatomy or physiology. It therefore places more emphasis on the relationships between members of a group.

Ewing, James Alfred (1855–1935) Scottish physicist and engineer. He observed and identified hysteresis, the lag in effect when forces acting on an object are changed, and invented a number of instruments to test magnetic properties.

Ewing, William Maurice (1906–74) US geophysicist. The first person to take seismic measurements in open seas (1935), he aided understanding of marine sediments and ocean basins. He proposed that earthquakes are associated with central oceanic rifts and took the first deep-sea photographs (1939).

excited state Condition of an atom, ion or molecule, when its energy level is higher than that of the ground (lowest) state. For example, an atom can be in an excited state having absorbed a PHOTON. The increased energy causes one of the electrons to occupy an ORBITAL of higher energy; the atom may restore its former state by various emissions.

exclusion principle Basic law of QUANTUM MECHANICS, proposed by Wolfgang PAULI in 1925, stating that no two electrons in an atom can possess the same ENERGY LEVEL and SPIN. More precisely, the set of four quantum numbers characterizing certain ELEMENTARY PARTICLES called FERMIONS must be unique. In atoms, these numbers specify an electron's spin direction, ORBITAL shape, and the energy level at which it resides, or would reside, in a magnetic field. For this work, Pauli was awarded the 1945 Nobel Prize for physics.

excretion Elimination of materials from the body that have been involved in METABOLISM. Such waste materials, particularly nitrogenous wastes, would be toxic if allowed to accumulate. In mammals these wastes are excreted mainly as URINE, and to some extent also by sweating. Carbon dioxide, a waste product of metabolism, is excreted mainly through the lungs during breathing. Defecation, strictly speaking, is not excretion: feces consist mostly of material that has never been part of the body. They are eliminated through the ANUS.

exfoliation Process of weathering in which rock flakes off in small pieces. It is caused by large differences in day and night temperatures, as occur in hot deserts. During the day the rock heats up and expands; during the night it cools and contracts. Eventually the rock is weakened by the repeated expansion and contraction, and pieces flake off. The presence of dew or other water hastens the process. Because the rock flakes off like the skin of an onion, it is also known as onion WEATHERING.

exocrine gland Gland that releases its secretion onto the surface or into a body cavity. Examples of exocrine glands in the skin include SEBACEOUS GLANDS (which secrete waxy **sebum**) and SWEAT GLANDS (which secrete perspiration). MAMMARY GLANDS are temporary exocrine glands that secrete milk. Exocrine glands in the pancreas secrete digestive juices along ducts to the duodenum and intestines. *See also* ENDOCRINE SYSTEM

exoskeleton Protective skeleton or hard supporting structure forming the outside of the soft bodies of certain animals, notably ARTHROPODS and MOLLUSKS. In arthropods, it consists of a thick horny covering attached to the outside of the body and may be jointed and flexible. The exoskeleton does not grow as the animal grows; instead it is shed periodically and the animal generates a new one. *See also* ENDOSKELETON

exosphere Outer shell in the Earth's ATMOSPHERE from which light gases, including hydrogen and helium, can escape. It lies about 400km (250mi) above the surface of the Earth.

exothermic reaction CHEMICAL REACTION in which heat is evolved, causing a rise in temperature. A common example is COMBUSTION. *See also* ENDOTHERMIC REACTION

expanding Universe Theory of the origin and direction in time of the Universe. Physicists have attempted to explain the RED SHIFT phenomena of some stars as resulting from a single, huge explosion that causes these stars to move away from our section of the Universe. The RED SHIFT occurs when, due to the DOPPLER EFFECT, the perceived wavelengths of light from some stars are lengthened because of their outwards movement. There is also the opposite effect, a BLUE SHIFT, but this does not occur as often. Today the balance of opinion is in favor of the theory of the Universe expanding following the BIG BANG. *See also* HUBBLE CONSTANT

expansion In mathematics, a process of replacing an expression by a sum of a finite number of terms, or by an infinite series. For example, the expression $(x + 1)(x + 3)$ has the expansion $x^2 + 4x + 3$; the function $\sin x$ can be expanded into the converging series $x - x^3/3! + x^5/5! - ...$ and so on. *See also* BINOMIAL THEOREM

expansion (expansivity) In physics, a change in the size of an object with a change in temperature. Most substances expand on heating, although there are exceptions – water expands when it cools from 4°C (39°F) to its freezing point at 0°C (32°F). A solid has three coefficients of expansion: **linear**, **superficial**, and **volume**, equal to the fractional increase in length, area or volume (respectively) per unit temperature rise. For a gas, the coefficient of expansion is the rate of change of volume with temperature (at constant pressure), or of volume with pressure (at constant temperature). For a solid, the coefficient of expansion is usually small; for a gas, it is much larger. *See also* CHARLES' LAW

expiration (exhalation) Process by which air (or water in the case of fish) is expelled from the lungs (or gills). In mammals it involves a reduction in the volume of the chest cavity by relaxation of the DIAPHRAGM muscles and contraction of the muscles between the ribs. This "squeezes" the lungs and forces gases out until the pressure in the lungs is the same as atmospheric pressure. Expiration and its opposite process, **inspiration**, are fundamental in aiding GAS EXCHANGE, part of the complete process of RESPIRATION.

explosive Substances that react rapidly and violently, emitting heat, light, sound, and shock waves. Chemical explosives are highly nitrated compounds or mixtures that are unstable and decompose violently with the evolution of much gas. Nuclear explosives are radioactive metals, the atoms of which can undergo a CHAIN REACTION of nuclear FISSION, or light particles such as hydrogen that undergo nuclear FUSION, to release vast amounts of radiant energy and devastating shock waves. *See also* NUCLEAR WEAPON

explosive decompression Sudden drop in pressure of an aircraft or spacecraft cabin due to meteorite puncture or system failure. A pressure drop to below 47mm Hg (pressure of water vapor at 38°C) results in the boiling of body liquids. Lungs collapse completely within a few seconds, blood pressure drops, and the large bowel relaxes. Death is inevitable.

exponent Superscript number placed to the right of a symbol indicating its power; for example, in

plastic pellets
mould
piping
Archimedes' screw
ram
metal slug
mould
lubricant

▲ **extrusion** An example of hot extrusion (top) shows how pellets of plastic are fed into an Archimedes' screw where they are heated and then pushed through a mould while molten to create piping. Hot extrusion is a constant process. Cold extrusion (below) uses the force of a ram to push a metal slug into the shape of the mould. To aid extraction from the mould the metal slug is coated with a lubricant which has to be removed after the process.

$a^4 (= a \times a \times a \times a)$, 4 is the exponent. Certain laws of exponents apply in mathematical operations. For example, $3^2 \times 3^5 = 3^{(2+5)} = 3^7$; $3^4/3^3 = 3^{(4-3)} = 3^1$; $(3^2)^5 = 3^{(2 \times 5)} = 3^{10}$; $3^{-5} = 1/3^5$.

exponential In general a function of x of the form a^x, where a is a constant. The exponential function e^x (where e is the base of natural logarithms, 2.7182818...) can be represented by a power series $1 + x + x^2/2! + x^3/3! + \ldots$.

extensor MUSCLE that makes a limb extend or straighten. See also FLEXOR

extinction Dying out of a species or population. Extinction is part of the evolutionary process in which species of plants and animals die out, often to be replaced by others. The rate at which extinctions have occurred is very variable. Periods of the Earth's history when extinction rates have been high are called **mass extinctions**. Extinctions brought about by human impact on the environment do not necessarily involve the replacement of extinct species by others. Extinctions can occur locally or globally. **Local extinction** refers to the loss of a species in a particular place, with other populations of the same species possibly, but not necessarily, occuring elsewhere. **Global extinc-**

tion is the permanent loss of a species througout its entire range. See also EVOLUTION; PUNCTUATED EQUILIBRIUM

extracellular fluid Any of various body fluids that exist outside CELLS in body spaces lined with moisture-exuding membranes. Such fluids are also found in blood, in lymph, in various body tissues including muscle, and in the channels and cavities of the brain and spinal cord.

extraction In chemistry, physical separation of a liquid or solid mixture by selectively dissolving some components with a solvent. Specific ingredients for perfumes and flavorings may be extracted in this way from plant oils. Proteins are separated from carbohydrates in soy beans by controlled solvent extraction.

extrapolation In mathematics, an approximate method of finding values beyond those measured. Most of economics is extrapolation, in that statistics are compiled up to the present year, and using them as a guide, decisions are made concerning future years. See also INTERPOLATION

extrusion Operation of forcing copper, aluminum, magnesium, their alloys, or plastics at the optimum temperature through a die to manufacture specific shapes such as rods, tubes, and various hollow or solid sections. Plastic extrusion can produce composite sheets, and coat film and wire.

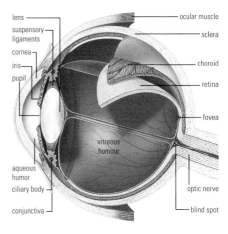

lens
suspensory ligaments
cornea
iris
pupil
ocular muscle
sclera
choroid
retina
fovea
vitreous humour
aqueous humor
ciliary body
conjunctiva
optic nerve
blind spot

▲ **eye** Light entering the human eye passes through a lens, which is controlled by muscles and ligaments. The lens focuses the image of the object onto the retina. The retina comprises light-sensitive nerve cells, called rods and cones, which convert the image into nerve impulses. These impulses are relayed to the brain via the optic nerve.

extrusive rock (volcanic rock) Type of IGNEOUS ROCK formed by MAGMA that has reached the surface of the ground. Extrusive rock generally comes from the vents of VOLCANOES or fissures producing LAVA flows. It cools quickly and is therefore fine-grained or even glassy. The term does not generally apply to PYROCLASTIC rocks.

eye Organ of SIGHT. It converts light energy to nerve impulses that are transmitted to the visual center of the brain. Most of the mass of a human eye lies in a bony protective socket, called the **orbital cavity**, which also contains muscles and other tissues to hold and move the eye. The eyeball is spherical and composed of three layers: the **sclera** (white of the eye), which contains the transparent CORNEA; the **choroid**, which connects with the IRIS, PUPIL, and LENS and contains blood vessels to provide nutrients and oxygen; and the RETINA, which contains RODS AND CONES for converting the image into nerve impulses. The AQUEOUS HUMOR (a watery liquid between the cornea and the iris) and the VITREOUS HUMOR (a jellylike substance behind the lens) both help to maintain the shape of the eye.

Fabre, Jean Henri (1823–1915) French ento-mologist. He made accurate and important studies of the anatomy and behavior of INSECTS and ARACHNIDS, especially bees, ants, beetles, grasshoppers, and spiders. He emphasized the importance of the theory of inherited instincts in insects. His most important work was embodied in the ten volumes of his *Souvenirs Ento-mologiques* (1879–1907).

Fabricius, Hieronymus (1537–1619) (Giro-lamo Fabrici) Italian anatomist. He was a pupil of Gabriello FALLOPIUS, whom he succeeded as pro-fessor of anatomy at Padua in 1562. His book, *On the Formed Foetus* (1600), describes his pioneer-ing research in EMBRYOLOGY. Fabricius also gave the first complete description of the valves in veins, although he misunderstood their function.

facies In geology, all the features of a rock or a stratum of rock that show the history of the rock's formation. Geologists often distinguish the age by the facies. The term is also applied to gradations of IGNEOUS ROCK.

factor In mathematics, any number that divides exactly into another given number without remain-der. For example, the factors of 72 are 1, 2, 3, 4, 6, 8, 9, 12, 18, 24, 36, and 72 itself.

factor analysis In statistics and psychometrics, mathematical method for reducing a large number of measurements or tests to a smaller number of "factors" that can completely account for the results obtained on all the tests, as well as for the correlations between them.

factorial (symbol !) Number formed by the prod-uct of a given number and all the whole numbers less than it. For example, "6 factorial" is $6! = 6 \times 5 \times 4 \times 3 \times 2 \times 1 = 720$. By definition, 0 factor-ial is $0! = 1$.

Factor VIII (anti-hemophilic factor) Blood-clot-ting factor that is deficient or absent in people with HEMOPHILIA. Factor VIII is a protein that brings about the conversion of prothrombin to thrombin, which in turn creates FIBRIN fibers around which a blood clot forms. It is extracted from blood plasma

or genetically engineered and used for treating hemophilia.

faculae *See* SUNSPOT

Fahrenheit, Gabriel Daniel (1686–1736) Ger-man physicist and instrument-maker. He invented the alcohol THERMOMETER (1709) and the first mer-cury thermometer (1714), for which he devised the FAHRENHEIT temperature scale. He also showed that the boiling points of liquids vary with changes in atmospheric pressure and that water can remain liquid below its freezing point.

Fahrenheit Temperature scale based on the freezing point of water (32°F) and the boiling point of water (212°F). The interval between these points is divided into 180 equal parts. Although replaced by the CELSIUS scale, the Fahrenheit scale is still sometimes used for nonscientific measure-ments. Degrees Fahrenheit are converted to degrees Celsius by substracting 32 and then divid-ing by 1.8. The system was devised by the German physicist Gabriel FAHRENHEIT.

fallopian tube (oviduct) In mammals, either of two narrow ducts leading from the upper part of the UTERUS into the pelvic cavity and ending in finger-like projections called **fimbriae**, which almost encircle each OVARY. After ovulation, an OVUM enters and travels through the fallopian tube where FERTILIZATION can occur. The fertilized ovum continues into the uterus where it becomes implanted. *See also* SEXUAL REPRODUCTION

Fallopius, Gabriello (1523–62) Italian anatomist who became professor of anatomy and botany at Padua in 1551. His book *Anatomical Observations* (1561) contains the first description of the FALLOPIAN TUBES and the SEMICIRCULAR CANALS of the inner EAR.

fallout (radioactive fallout) Radioactive contami-nation in the atmosphere following a leakage or accident at a NUCLEAR REACTOR, or explosion of a NUCLEAR WEAPON. Large, windborne particles fall to Earth after a few hours, sometimes up to several hundred kilometers from the source (**local** fallout); lighter particles entering the TROPOSPHERE are detected after a longer period at about the same lat-itude as the source (**tropospheric** fallout). Any particles entering the STRATOSPHERE eventually fall over the Earth's surface, often many years later (**stratospheric** fallout).

false color image Image to which colors are added to emphasize or give a value for certain fea-tures. The colors are usually added by a computer, typically using data transmitted to Earth from satellites or space probes or from medical imaging equipment. Colors may be added to indicate such things as temperature (with, for example, red for hot, and greens and blues for cool temperatures) or altitude (to generate a CONTOUR map).

family In biology, part of the CLASSIFICATION of living organisms, ranking above GENUS and below ORDER. Family names are printed in Roman (ordinary) letters with an initial capital; for example Felidae is the family of all cats.

farad (symbol F) Unit of CAPACITANCE. It is equal to the capacitance of a capacitor that acquires a charge of 1 coulomb when a POTENTIAL DIFFERENCE of 1 volt is applied across the plates. It is a very large unit, and the microfarad (symbol μF), or one millionth of a farad, and the picofarad (symbol pF), or one million-millionth of a farad are commonly used.

Faraday, Michael (1791–1867) British physicist and chemist. Faraday worked as Sir Humphry DAVY's assistant at the Royal Institution in London, where, in 1825, he became director of the laboratories. He liquefied chlorine, discovered benzene and two compounds of carbon with chlorine, and enunciated the laws of electrolysis (FARADAY'S LAWS). Moving from chemistry to electricity, he discovered ELECTROMAGNETIC INDUCTION, made the first DYNAMO, built a primitive electric motor, and studied nonconducting materials (DIELECTRICS). The unit of CAPACITANCE (the FARAD) is named after him. Faraday's work was developed by James Clerk MAXWELL into a single mathematical theory of ELECTROMAGNETISM.

Faraday's laws Two laws of ELECTROLYSIS and three of ELECTROMAGNETIC INDUCTION, formulated by Michael FARADAY. In modern form, the electrolysis laws state that (1) the amount of chemical change during electrolysis is proportional to the charge passed, and (2) the amount of chemical change produced in a substance by a certain amount of electricity is proportional to the electrochemical equivalent of that substance. Faraday's laws of induction state that (1) an electromagnetic force is induced in a conductor if the magnetic field surrounding it changes, (2) the electromagnetic force is proportional to the rate of change of the field, and (3) the direction of the induced electromagnetic force opposes the change in the external field.

Farman, Henri (1874–1958) French aviation pioneer and manufacturer. In 1908 he won a prize for the first circular flight of 1km (0.6mi) and in 1909 he established a flight record of 179km (112mi). In 1912 he founded an aircraft factory that later became one of the largest in Europe and with his brother Maurice established the first air passenger service between London and Paris.

farming *See* AGRICULTURE

farsightedness *See* HYPERMETROPIA

fast neutron High-energy NEUTRON that occurs in a nuclear FISSION reaction called **fast fission**. Their high energy is caused by undergoing fewer collisions during an initial fission reaction, and thereby unable to sustain a CHAIN REACTION. Fast fission is used in certain NUCLEAR REACTORS called **fast reactors**.

fat Semisolid organic substance made and used by plants and animals to store energy. In animals, fats also serve to insulate the body and protect internal organs. Fats are soluble in organic solvents such as ether, carbon tetrachloride, chloroform, and benzene. They are triglycerides: ESTERS of one molecule of glycerol connected to three molecules of FATTY ACIDS (carboxylic acids), such as palmitic,

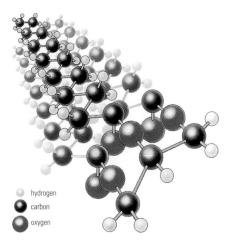

hydrogen
carbon
oxygen

▲ **fat** Fats, found widely distributed throughout the animal and plant kingdoms, are generally esters of fatty acids and the trihydric alcohol, glycerol. Three molecules of the fatty acid combine with one molecule of glycerol, with the liberation of three molecules of water, to form the fat. The "glyceride structure" is an important common feature of fats. Those fats having one or more double bonds in the chains are known as unsaturated fats, while those only with single bonds are known as saturated fats. In both animals and plants, fats are found in the food storage areas of the organism. They are easily hydrolysed in the cells to give energy.

lauric, and stearic acid, each having 12 to 18 carbon atoms. Research indicates that the consumption of high levels of animal fats can increase the risk of heart disease. Vegetable oils are similar to fats, but they are viscous liquids rather than semisolids and have a higher proportion of molecules with double carbon–carbon (C=C) bonds in the chain – that is, they are unsaturated. *See also* LIPID; SOAP

fathom Unit used in measuring depth, especially of water. One fathom equals 1.83m (6ft). It is also a quantity of material that has a cross-section 6 feet square but varies in length. Originally, a fathom was the distance spanned by a person's arms.

fatigue In general, mental or physical tiredness after activity. In physiology, an inability to function at normal levels of physical and mental activity. Muscle fatigue results from the accumulation of LACTIC ACID in the muscle tissue and the depletion of GLYCOGEN (stored carbohydrates).

fatigue, metal Weakening of the crystalline structure of a metal due to repeated rhythmic straining, bending or vibration. It can cause superficial cracks that may spread inward, causing the metal to crumble. Components have to be designed to reduce and delay fracture due to metal fatigue.

fatty acids Organic compounds, so called because they are present widely in nature as constituents of FAT. They are CARBOXYLIC ACIDS containing a single carboxyl group (–COOH). Examples of saturated fatty acids (lacking double bonds

in their hydrocarbon chain) are acetic acid and palmitic acid, the latter being a common fat constituent; unsaturated fatty acids (having one or more double carbon–carbon bonds – C=C) include oleic acid. Both saturated and unsaturated types have molecules shaped like a long, straight chain. *See also* LIPID

fault In geology, a crack or fracture in the EARTH's CRUST along which movement has occurred. The movement will be slow and quite small, only a few centimeters, though fault movements often continue for thousands of years. In such cases, uplift or downthrow of hundreds of meters is possible. Faulting is caused by PLATE TECTONICS, when movements in the Earth's crust create stress and tension in the rocks, causing them to stretch and crack. Vertical movements cause **normal** and **reverse** faults, and horizontal movements cause **tear** faults. The vertical change of height on opposite sides of a fault is called the **throw**, and the horizontal movement is called the **heave**. Faults often occur in groups and if two or more roughly parallel faults cause a block of land to rise, it is called a HORST, or block mountain. If the land sinks between two or more parallel faults, it creates a GRABEN, or RIFT VALLEY.

fault plane Surface along which a FAULT occurs. In a normal fault, caused by tension pulling two masses of rock apart before one mass slips downward, the fault plane is either vertical or inclined so that the mass moves downward on the dip side of the fault plane.

▲**fault** The Earth's crust is subjected to enormous forces, and the stress creates faults. In a tear fault (1) the stresses cause horizontal movement. The forces build up until they are released in a sudden movement (2) often causing earthquakes. In a normal fault (3) the rocks are pulled apart, causing one side to slip down along the plane of the fault. In a reverse fault (4) the rocks on either side of the fault are forced together. One side rises above the other along the fault plane. In a horst fault (5) the central section is left protruding due to compression from both sides or the sinking of the bracketing rock. A rift valley (6) has a sunken central section, formed either by compression or the outward movement of the two valley sides.

▶ **fax** A fax machine converts text or images fed into the machine (1) into a digital code (2). The code is created by shining light on tiny strips of the document in turn (3). Sensors (4) detect the amount of light that bounces back. Where ink is present little light is reflected creating an electrical pulse of low voltage. A high voltage results when light is reflected from white paper. The digital code is converted by a modem in the fax into an analogue signal (5) and transmitted to the receiving fax machine (6) via the telephone network. A modem in the second machine converts the analogue code back into a digital code (7), a printer (8) interprets the digital code and produces the hard

facsimile copy (9). Each machine has its own

number which is dialled in via a numerical keyboard

(10) on the sending machine.

fauna Collective name for all the animals that are found in a particular area, or that lived at a particular time.

fax (facsimile transmission) Equipment by which text, photographs, and drawings can be transmitted and received through a TELEPHONE system. The image, on paper, is scanned to translate it into a series of electrical pulses. Inside the fax machine, a MODEM converts the pulses into a digital form that can be transmitted through the telephone system. At the receiving end, the fax machine's modem converts the signals back into pulses, and prints these as dots to build up a copy of the original document. *See also* SCANNING

feather One of the skin appendages that makes up the plummage of birds. They are composed of the fibrous protein KERATIN, and provide insulation and enable flight. They are usually replaced at least once a year.

feces *See* EXCRETION

feedback In technology, process by which an electronic or mechanical control system regulates itself. Feedback works by returning part of the "output" of the system to its "input." In other words, the process compares how the system is performing with how it is set up to perform. If there is any difference between the two, the feedback system makes adjustments accordingly. A governor on an engine, for example, uses feedback to keep the speed within a specified limit, regardless of the load carried. The high-pitched feedback heard in electronic music systems occurs when sound from the loudspeaker is picked up by a microphone. *See also* BIOFEEDBACK

Feigenbaum, Mitchell (1945–) US mathematician who introduced chaos theory into calculations involving systems that behave in a seemingly random and unpredictable way. It was first applied to meterology and gave rise to the famous "butterfly effect" (a butterfly flapping its wings in one part of the world may cause a tornado in another part), which merely emphasizes the sensitivity of the calculations to the completeness of the original data. *See also* CHAOS THEORY

feldspar Important group of common rock-forming minerals that all contain aluminum, silicon, and oxygen, but with varying proportions of potassium, sodium, and calcium. They are essential constituents of IGNEOUS ROCK. ORTHOCLASE and MICROCLINE are potassium feldspars of monoclinic and triclinic system crystals, respectively. Members of the PLAGIOCLASE series (sodium and calcium feldspars) have physical properties similar to microcline, but with crystals frequently twinned. Hardness 6–6.5; r.d. 2.5–2.8.

femur Thigh bone, extending from the hip to the knee of four- and two-legged VERTEBRATES, including humans. It is the longest and strongest bone of the human SKELETON. Its rounded, smooth head articulates with the pelvis at the hip socket.

fen Tract of low-lying marshy land where peat is formed below the surface. The soil is only slightly acid and much drainage is needed before a fen can become arable.

Fermat, Pierre de (1601–65) French mathematician. With Blaise PASCAL he helped to formulate the theory of probability and, by showing that light travels along the shortest optical path (**Fermat's principle**), he laid the foundation for geometric optics. In mathematics, Fermat is best known for his work in number theory. It was also recently discovered that differential CALCULUS, thought to be an invention of NEWTON, was communicated to him by Fermat in a series of letters. Leibnitz is believed to have developed calculus independently of both men. *See also* FERMAT'S LAST THEOREM

Fermat's last theorem Hypothesis, first stated by FERMAT, that for all integers $n > 2$, there are no natural NUMBERS x, y, and z that satisfy the equation $x^n + y^n = z^n$. Fermat wrote in one of his books (Diophantus' text *Arithmetica*) that he had found a "remarkable" proof of this theorem, which the margin of the book was too small to contain. He died without revealing his proof to anyone. Subsequent attempts to find a valid proof of the result baffled mathematicians for more than 350 years. In 1993 Andrew WILES of Princeton University, drawing together strands from diverse areas of mathematics, announced a proof of the theorem. This alleged proof was soon found to contain a gap, but further work has repaired this and the complete proof was published and widely accepted in 1995. See also NUMBER, NATURAL

fermentation Energy-yielding metabolic process by which sugar and starch molecules are broken down to carbon dioxide and ethanol in the absence of air (ANEROBIC respiration), catalysed by ENZYMES usually in microorganisms such as yeast. Uses include bread-making, wine-making, beer-brewing, and cheese maturation. The intoxicating effect of crushed fruits stored in a warm place (where they would ferment) may have been known as early as 4000 BC.

Fermi, Enrico (1901–54) US physicist, b. Italy. He worked mainly in the fields of atomic behavior and structure, and in QUANTUM THEORY. He discovered the element NEPTUNIUM and proved that neutron bombardment of most elements produces RADIOISOTOPES of those elements. Fermi produced the first self-sustaining CHAIN REACTION in uranium and in 1942 built the world's first nuclear reactor. He helped to develop the atom bomb for the MANHATTAN PROJECT and later worked on developing the hydrogen bomb. With Paul DIRAC, he studied quantum statistics. For his work with RADIOACTIVITY, Fermi was awarded the 1938 Nobel Prize for physics; the element FERMIUM was named after him. The US Atomic Energy Commission named their special award for outstanding work in NUCLEAR PHYSICS after Fermi, who received the first one in 1954. *See also* NUCLEAR WEAPON

fermion ELEMENTARY PARTICLE that has a half-integer SPIN. Named after the physicist Enrico FERMI, fermions are those particles that obey the EXCLUSION PRINCIPLE. This means that only one fermion can occupy a certain quantum state. Nuclear structures tend to be made up of fermions; LEPTONS (such as ELECTRONS and NEUTRINOS) and QUARKS are all fermions. *See also* BOSON

fermium (symbol Fm) Radioactive metallic TRANSURANIC ELEMENT of the ACTINIDE SERIES. It was first identified in 1952 as a decay product of uranium ^{255} after it was produced in the first large hydrogen bomb explosion. Ten isotopes have been identified, but fermium has only been produced in trace amounts. Properties: at.no. 100; most stable isotope ^{257}Fm (half-life 80 days).

fern Any member of the estimated 9,000–15,000 species of the nonflowering plant phylum Filicinophyta. Ferns are extremely diverse in their habitats, in their shapes, and in the ways they reproduce. Ferns are characterized by their ALTERNATION OF GENERATIONS: the conspicuous SPOROPHYTE, which possesses leafy fronds, stems, RHIZOMES, and roots and reproduces by minute SPORES usually clustered on the leaves; and the inconspicuous GAMETOPHYTE, which resembles moss and produces GAMETES (sperm and ova). Fronds unroll from curled fiddle-heads and may be divided into leaflets. Ferns were growing in the Devonian period, some 400 million years ago.

ferric Description of compounds in which the element iron has a greater VALENCE (usually three) than in FERROUS compounds.

ferricyanide *See* HEXACYANOFERRATE

ferrite One of the crystalline forms (ALLOTROPES) of the metal iron, present together with other forms in steel, wrought and cast iron. It is also called α-iron, and has highly magnetic properties. The name ferrite is also given to chemical compounds such as nickel ferrite ($NiFe_2O_4$) and zinc ferrite ($ZnFe_2O_4$), which are found naturally in rocks and also synthesized as magnetic materials. *See also* FERROMAGNETISM

ferro-alloys Combinations of silicon, manganese, chromium, molybdenum, vanadium, titanium, and several other elements, added to molten steel to confer such properties as greater strength and corrosion resistance. Individually they are named after their major constituents; for example, ferrochromium contains about 70% chromium, about 6% carbon, and about 2% silicon as well as iron. *See also* ALLOY

ferrocene ($Fe(C_5H_5)_2$) Red, crystalline organo-iron compound, systematic name dicyclopentadienyl iron(II). Its molecules consist of an iron(II) (ferrous) ion sandwiched between two five-membered carbon rings. It is, therefore, an example of a **sandwich compound**.

ferrocyanide *See* HEXACYANOFERRATE

ferroelectrics Crystalline materials that are naturally electrically polarized. Within such a material lie regions or domains that are spontaneously polarized in specific directions, although these directions can often be changed by the application of an electric field. All ferroelectrics exhibit the PIEZOELECTRIC EFFECT; this leads to many applications of the materials in research and industry.

ferromagnetism Form of MAGNETISM exhibited by substances (such as iron, cobalt, and nickel) with high magnetic PERMEABILITIES. Below a certain temperature, called the Curie temperature, a ferromagnetic material in an applied magnetic field becomes magnetized in the direction of the field. The magnetism increases with the applied field, up to a limit called the **saturation magnetization**. Some of the magnetization persists when the applied field is removed.

ferrous Description of compounds in which the element iron has a lesser VALENCE (usually two) than in FERRIC compounds.

fertility drugs Drugs taken to increase a woman's chances of conception and pregnancy. One of the many causes of female sterility results from insufficient secretion of pituitary hormones, and this malfunction is often treated with either human chorionic gonadotropin or clomiphene citrate, although use of the latter has resulted in multiple births. In cases where fertilization of the ovum does occur, but where the uterine lining is unable to support the developing fetus, the hormone progesterone may be used. There are many cases of infertility that cannot be corrected with drugs.

fertilization Key process in SEXUAL REPRODUCTION during which the nuclei of female and male

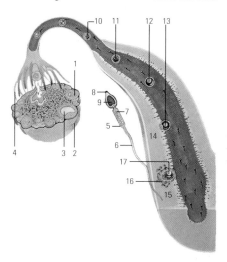

▲ **fertilization**
Mammalian fertilization begins with ovulation, in which an ovum (1) develops in an ovary (2) into a follicle (3). The follicle consists of the ovum, a sac of liquid, and follicle cells. The pressure in the follicle increases until it bursts, releasing the ovum into the fallopian tube (4). During ovulation, the hormone estrogen is produced by the collapsed follicle. Estrogen causes the lining of the uterus to thicken and extend its network of blood vessels, from which the fertilized ovum will be nourished. Fertilization occurs when millions of sperm (5) are ejaculated from the male's penis during copulation. The sperm use their long tails called flagella (6), powered by mitochondria (7), to swim up the uterus. The first sperm to reach the ovum penetrates the ovum membrane with enzymes secreted by the acrosomal vesicle (8). This triggers the formation of a membrane (9), making the ovum impenetrable to other sperm. The sperm nucleus (10) fuses with the ovum nucleus bringing together the hereditary traits of the parents. The fertilized ovum then goes through a stage of division (11–13) to form the embryo (14). The embryo moves down to the uterus, where it releases enzymes that break down the lining (15), creating a hole in which the embryo sits (16). The embryo indirectly obtains its oxygen and food from the mother's blood as it flows through the uterus. The embryo (17) develops the placenta which is comprised of millions of tiny appendages called villi. The oxygen and food are absorbed from the mother's blood, via capillaries in the villi, into the embryo's blood.

GAMETES (sex cells) fuse to form a ZYGOTE. The zygote contains the genetic material (CHROMOSOMES) from both parents (*see* HEREDITY). In most animals, the female sex cell is called the OVUM and the male cell SPERM. After fertilization, the zygote begins to divide through a number of stages to form an EMBRYO. Fertilization of the female ovum by the male sperm can be external (outside the body, as in most fish, amphibians, and aquatic invertebrates) or internal (inside the body, as in reptiles, birds, mammals, and insects). In plants, the male gamete is found in POLLEN, and for most higher plants POLLINATION occurs before fertilization. While most animals and plants undergo **cross-fertilization**, in which the male gamete fuses with the female gamete of another animal or plant, some organisms undergo **self-fertilization**, whereby the male gamete fuses with the female gamete of the same flower or plant, or where an HERMAPHRODITE animal fertilizes itself. *See also* CELL DIVISION; DIPLOID; HAPLOID

fertilizer Organic or inorganic substance containing nutrients – mainly nitrogen, phosphorus, and potassium – that is added to soil to aid plant growth. **Organic** fertilizers include manure and compost, fish and bone meal, and guano. **Nitrogen** fertilizers are the most widely used inorganic fertilizers. Phosphorus fertilizers (**phosphates**) are made from the mineral apatite. **Potassium** fertiliz-

ers are extracted from deposits of potassium chloride. Nitrogen fertilizer in surface water promotes the growth of algae that degrades water quality and can cause EUTROPHICATION. Inorganic fertilizers suppress nitrogen-fixing bacteria, making agriculture increasingly dependent on artificial fertilizer.

Fessenden, Reginald Aubrey (1866–1932) US engineer, physicist, and inventor who was a pioneer in radio and echo-sounding. He is thought to have broadcast the first radio program (using speech signals, not Morse code) in 1906 from a transmitter he constructed in Brant Rock, Massachusetts. Fessenden developed a new type of wireless system using continuous waves. Among his 500 or so patents were AMPLITUDE MODULATION (AM), the high-frequency ALTERNATOR, the electrolytic detector, the heterodyne system of radio reception and the fathometer.

fetus EMBRYO in a mammal after the main adult features are recognizable. In humans it dates from about 8 weeks after CONCEPTION. *See also* BLASTULA

fever Elevation of the body temperature above the normal 37°C (98.6°F). It is mostly caused by bacterial or viral infection and can accompany virtually any infectious disease.

Feynman, Richard Phillips (1918–88) US physicist, one of the most important theoretical physicists of modern times. He helped to develop the atom bomb during the MANHATTAN PROJECT, before going to Cornell University with Hans BETHE where he worked on QUANTUM ELECTRODYNAMICS (QED). His invention of **Feynman diagrams** greatly facilitated the solution of electromagnetic interactions between ELEMENTARY PARTICLES. He was professor of physics at the California Institute of Technology from 1950 until he died. He shared the 1965 Nobel Prize for physics with Julian SCHWINGER and Shin'ichirō Tomonaga for their independent work on quantum electrodynamics. With Murray GELL-MANN, Feynman developed a theory of weak INTERACTIONS, such as those that occur in the emission of electrons from radioactive nuclei. Feyman also did research on the structure of protons and the properties of liquid helium. In 1986 he was a key member of the committee that investigated the *Challenger* space shuttle disaster.

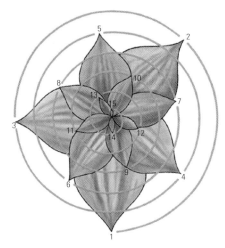

▲ Fibonacci sequence
The leaves of a plant when seen from above grow in a spiral pattern. The angles between one leaf and the next follow a strict mathematical series known as the Fibonacci

sequence. This ensures that each leaf on the plant stem receives the maximum amount of sunlight available. The numbers on the diagram indicate the order in which the leaves have appeared.

fiber Fibrous material of animal, vegetable, mineral or synthetic origin. Natural fibers can be made into yarn, textiles, and other products, including carpets, rope, and felt. The fibers consist of long narrow cells. **Animal** fibers are based on protein molecules and include wool, silk, mohair, angora, and horsehair. **Vegetable** fibers are based mainly on CELLU-

LOSE and include cotton, linen, flax, jute, sisal, and kapok. The mineral ASBESTOS is a natural, inorganic fiber. Regenerated fibers are manufactured from natural products, modified chemically. For example, RAYON is made from cellulose fiber obtained from cotton or wood. **Synthetic fibers** are made from a molten or dissolved plastic RESIN by forcing it through fine nozzles (spinnerets). The result is a group of filaments that are wound onto bobbins. These fibers can be used as single-strand yarn, or spun to form multistrand yarn. The yarn may be woven into textiles. Many textiles, particularly for clothing, use both synthetic and natural fibers woven together. Some synthetic fibers are made into rope, carpets, and other products. Synthetic fibers include NYLON and other polyamides, POLYESTERS, and ACRYLICS. Nylon, the first synthetic fiber, was introduced in 1938. Other synthetic fibers, such as glass, carbon or metals, can be used to reinforce resins to produce extremely strong materials to be used in sports equipment, for example.

fiberglass Glass in the form of fine filaments. It is used widely for heat insulation (as glass wool), for fabrics, and with a plastic resin to make a construction and repair material called GRP (glass-reinforced plastic). GRP, a rigid composite material, is commonly referred to as fiberglass and is a popular medium for car bodies, boats, aircraft parts, and containers. It resists heat, corrosion, rot, and most chemicals and can be, weight for weight, stronger than steel. Short lengths of fiberglass are made by blasting air or steam into molten glass. Most forms of fiberglass are made by forcing molten glass through fine metal nozzles or spinnerets. The resulting continuous filaments are usually bundled together to form strands. These may then be chopped, twisted or woven according to the product required.

fiber optics Branch of OPTICS concerned with the transmission of data and images by reflecting light through very fine glass OPTICAL FIBERS.

Fibonacci, Leonardo (c.1170–c.1240) Italian mathematician. He wrote *Liber abaci* (c.1200), the first Western work to propose the adoption of the Arabic (originally Indian) numerical system. He produced the mathematical sequence known as the FIBONACCI SEQUENCE.

Fibonacci sequence Mathematical SEQUENCE in which each term of the sequence is formed by the addition of the two terms preceding it. Thus, if the nth term of the sequence is denoted x_n, the sequence is defined by the equation $x_{n+2} = x_n + x_{n+1}$, with the first two terms being $x_1 = 1$ and $x_2 = 1$. The sequence begins 1, 1, 2, 3, 5, 8, 13, 21,..., the next number being 34, the sum of 13 and 21,

and so on. As n becomes very large the ratio of successive terms becomes $1:(\sqrt{5} + 1)/2$. This ratio is called the golden ratio. The Fibonacci sequence can be seen in nature in successive spiral shell segments and sunflower petals radiating from the center of the flower. *See also* GOLDEN MEAN

fibrillation Rapid, uncontrolled beating of part of the heart muscle. **Atrial** fibrillation is seen in a number of conditions, including rheumatic heart disease, ATHEROSCLEROSIS, and hyperthyroidism. **Ventricular** fibrillation, which occurs mostly after a heart attack, can be fatal.

fibrin Insoluble, fibrous protein that is essential to BLOOD CLOTTING. Developed in the blood from a soluble protein, FIBRINOGEN, fibrin forms at the site of a wound in the form of a mesh, which then dries and hardens so that the bleeding stops.

fibrinogen Soluble protein synthesized by the liver and released into the bloodstream. It is converted to FIBRIN (an insoluble protein) by THROMBIN (an enzyme) during the BLOOD CLOTTING process.

fibroblast Type of living cell found in the CONNECTIVE TISSUES of the body. Fibroblasts are mobilized to restore tissue damaged as a result of injuries. The elongated or irregularly shaped cells lay down the fibrous protein COLLAGEN and other structural materials of the connective tissues.

fibro-cartilage Form of CARTILAGE found mainly in parts of the skeleton that are subject to extreme physical tension or compression. These include INTERVERTEBRAL DISCS and the parts of TENDONS that are attached to the ends of the long bones.

fibroid Benign (noncancerous) tumor, consisting of fibrous tissue and muscle, in the wall of the uterus. It may grow to be enormous, causing pain and very heavy menstrual periods. If so, it may be removed in a procedure known as **myomectomy**; or HYSTERECTOMY may be recommended.

fibula Long thin outer bone of the lower leg of four- and two-legged VERTEBRATES, including humans. It articulates with the other lower leg bone, the TIBIA, just below the knee; the lower end forms the projection on the outer side of the ankle.

field In computing, a particular item of DATA, usually part of a record, which in turn is part of a DATABASE.

field In physics, region in which an object is affected by a FORCE as a result of the presence nearby of another object or objects. There are various kinds of fields, including ELECTRIC FIELDS, MAGNETIC FIELDS, and GRAVITATIONAL FIELDS.

file In computing, a block of stored data. A file may contain information (such as a group of addresses), a document or a complete program. It is usually stored on a MAGNETIC DISK, but tape or other similar media may be used. With random-access files, any item of data can be accessed immediately. With serial files, the data must be read through from the beginning until the required item is reached. *See also* DATABASE

film, photographic Sensitized strips of CELLU-LOSE acetate or other plastic, coated on one side with a light-sensitive emulsion, used to record photographic images. The emulsion of a black-and-white film consists of a suspension of finely divided grains of silver bromide in gelatin. After exposure the film is kept in darkness until the latent image is made visible by developing and fixing. Film is rated according to its "speed" or sensitivity to light. This is a measure of the size of the silver bromide grains contained in the emulsion. A "fast" black-and-white film (high ASA or DIN number) gives rise to a grainy, high-contrast image, whereas "slower" film captures a greater range of tones.

filter Porous device for separating solid particles from a liquid or a gas. The process is known as filtration. Many complex forms of filter have been devised for various uses. Most cars have a number of filters, for air, gasoline, and oil. These operate either by trapping solid particles in porous materials such as paper or meshes, or by circulating the material to be filtered through a maze, the pockets of which trap particles, as in the air filter.

filtration Process of removing soilids from liquids, or solids from gases, by passage through a suitable medium such as filter paper, glasswool, or sand. *See also* FILTER

fire Combustion of flammable materials, usually accompanied by flames or smoke. Chemically it is an example of rapid OXIDATION, and a supply of air or oxygen is generally necessary for a fire to continue to burn. Early humans learned to control fire for warmth, cooking, making pottery, and metal-working. Although fundamental to much technology, uncontrolled it can cause great devastation.

firearm Weapon from which a projectile is fired, generally by the expansion of gases after the rapid combustion of GUNPOWDER or other explosive. Although occasionally used to describe an ARTILLERY piece, the term generally means a small arm – a weapon carried and fired by one person or a small group of people. Firearms were used in Europe in the 14th century but were ineffective in close combat until about 1425, when a primitive trigger to bring a lighted match into contact with the gunpowder charge was invented. Such firearms, called **matchlocks**, were heavy and cumbersome, and needed a constantly lit match. The lighter flintlock (which used the spark produced by flint striking steel to ignite the powder) superseded the matchlock in the mid-17th century and became the main infantry weapon for more than 100 years. During the 19th century there were great changes. In 1805 the Rev. Alexander Forsyth (1769–1843) discovered the explosive properties of mercury fulminate; together with the percussion cap invented in 1815, it provided a surer, more efficient means of detonation, permitting the development by 1865 of both the center-fire cartridge (which has been the type of ammunition used in firearms ever since) and breech loading, not previously practicable. Another major 19th-century advance was rifling, the cutting of spiral grooves along the inside of a barrel to make the bullet spin in flight, thus vastly increasing accuracy. During the 1830s Samuel Colt perfected the revolver, a pistol that could fire several shots without the need to reload. By the 1880s magazine rifles were also in use and were made more effective when a bolt action was incorporated after 1889. The next step was toward a weapon that could fire a continuous stream of bullets. Manually operated systems had been tried (such as the gatling gun), but the first modern machine gun was

▲ **filtration** Filtration removes solid particles from solution. A mixture (A) placed in a filter funnel (1) drains through the filter paper (2) by gravity, leaving sediment (3) trapped while the filtrate (4) flows through. In the Buchner funnel (B), the same principle is enhanced by suction (5). In the industrial rotary filter (C) the spinning action of the drum drives the mixture onto the fine mesh filter (6).

the Maxim gun, invented in the 1880s, which used the recoil energy of the fired bullet to push the next round into the breech and recock the weapon. Guns of this type dominated the trench warfare of World War I, and by World War II more portable automatic weapons, light machine guns such as the Bren gun and submachine guns, were in use. Most 20th-century firearms operate on principles established in the 19th century. Newer developments, under the stimulus of the emergence of aerial warfare and the TANK, include gas-operated rifles, recoilless rifles, firearms with several rotating barrels and extremely high rates of fire, and small firearms that use explosive bullets.

fireball Exceptionally bright METEOR. Fireballs have been loosely defined as meteors brighter than the planets; with the modern estimate of the maximum brightness of Venus, this would mean that all meteors brighter than magnitude –4.7 should be classified as fireballs.

firebrick Brick formed in a variety of shapes for use in structures, such as FURNACE linings, that are exposed to high temperatures. Firebrick is composed of FIRECLAY and other nonmetallic, high melting-point minerals, particularly those rich in silica, aluminum oxide, and carbon.

fireclay Clay that can withstand high temperatures without becoming deformed. It is used for making FIREBRICK, crucibles, and many refractory shapes. Fireclay approaches KAOLIN in composition; better grades contain at least 35% alumina (Al_2O_3) when fired.

firedamp In coal mining, flammable and explosive gas emitted by coal seams. It is composed mostly of methane (natural gas, CH_4) but also contains some hydrogen, oxygen, and carbon dioxide, and occasionally a little ethane. Firedamp has caused many disastrous pit explosions; it is detectable with a Davy lamp (after Sir Humphry DAVY), the flame of which elongates when the gas is present.

fire retardant Chemical solution, consisting mainly of boric acid and borax, impregnated into flammable materials to slow down the process of combustion. In construction, fire-resistant buildings are made of reinforced concrete or protected steel. ASBESTOS is no longer considered to be a safe building material.

fireworks (pyrotechnics) Controlled EXPLOSIVES (and their spectacular display) that are intended for visual rather than destructive effects. They range from light and sparkle-emitting capsules and coils, to rockets that burst into miniature "galaxies of stars" after a short near-vertical flight. Fireworks

began with the ancient Chinese, who first used them in the 8th or 9th centuries. They spread to Europe in the 12th and 13th centuries, along with the military uses of GUNPOWDER. Besides gunpowder, potassium chlorate mixtures came to be used as explosives and propellants; iron filings and magnesium and aluminum powders were incorporated to provide sparkle and brilliance; and fireworks were colored brightly by the inclusion of salts of the metals sodium (yellow), barium (green), strontium (red), and copper (blue). Fireworks find serious applications in distress rockets and flares, and Very lights.

firing Heating of ceramics until they are chemically changed; the process prevents a return to the plastic state. Many ceramic clays contain matter that must vitrify (fuse) to render the ware nonporous. This may require temperatures of up to 1,450°C (2,650°F). Firing is done in KILNS, or ovens.

firn (névé) Hard-packed mountain snow that has been converted into granular ice in a mountain GLACIER and, with an accumulation of broken rock materials at its base, digs out round basins called CIRQUES.

Fischer, Emil Hermann (1852–1919) German chemist. Fischer's important organic chemistry research contributed substantially to the knowledge of PURINE substances. He was the first to recognize the importance of STEREOCHEMISTRY and was fundamental in establishing ENZYME chemistry. Fischer synthesized glucose and fructose in addition to several other sugars, and for this work, and his work on purines, he was awarded the 1902 Nobel Prize for chemistry.

Fischer, Ernst Otto (1918–) German chemist. He became professor of chemistry in Munich in 1957. He shared the 1973 Nobel Prize for chemistry with Geoffrey WILKINSON for their work on organometallic chemistry.

Fischer, Hans (1881–1945) German biochemist who received the 1930 Nobel Prize for chemistry for his structural studies of CHLOROPHYLL and of the red BLOOD pigment hemin. His research indicated the close relationship between these two substances, and he was able to synthesize hemin and almost completely to synthesize one of the chlorophylls. He also studied carotene.

fish The classification of fish varies. They are usually divided into three classes: Agnatha, which are **jawless** fish, including the hagfish and lamprey; Chondrichthyes (**cartilaginous** fish), which includes shark, skate, ray, and chimera; and the much more numerous Osteichthyes (**bony** fish), including subclasses of soft-rayed fish (lungfish

▼ **fish** There are more than 22,000 species of bony fish. Although they vary in shape and the way they swim, they share many common features.

All have a tail, with equal upper and lower lobe sizes, which provides neither up nor down thrust. Such fish achieve natural buoyancy by adjusting their density using the swim bladder. The fish can expand or contract the swim bladder by secreting gas into or absorbing gas out of it, so adjusting the volume and external pressure, and counteracting the tendency to sink or float to the surface.

8and lobefin), and the very successful teleost fish, such as salmon and cod. There are more than 22,000 species of bony fish, and they represent about 40% of all living vertebrates. They are divided into 34 orders and 48 families. All fish are cold-blooded (POIKILOTHERMIC), aquatic, vertebrate animals characterized by fins, gills for breathing, a streamlined body almost always covered by scales or bony plates on to which a layer of mucus is secreted, and a two-chambered heart. Fish are the most ancient form of vertebrate life, with a history of about 450 million years. They reproduce sexually, and FERTILIZATION may be external or internal. The eggs develop in water or inside the female, according to species. Fish have lateral line organs, which are fluid-filled pits and channels that run under the skin of the body. Sensitive fibers link these channels to the central nervous system and detect changes of pressure in the water and changes of strength and direction in currents. About 75% of all fish live in the sea; the remainder are freshwater species that live in lakes, rivers, and streams. A few fish, such as the SALMON and EEL, divide their lives between salt and freshwater habitats.

Fisher, Alva John (1862–1947) US engineer who invented the domestic washing machine. His original machine, called the "Thor", was marketed in 1910. It used a small electric motor driving a horizontally mounted cyclindrical drum, which held water and washing within a metal tank. A gear box reversed the direction of rotation periodically, to prevent the washing clumping together in the drum. There was no provision for heating the water.

fission In zoology, form of ASEXUAL REPRODUCTION in some single-celled organisms. In **binary** fission, the parent cell, such as a bacterium, diatom or protazoan, simply divides in two to produce two identical daughter cells. **Multiple** fission produces 4, 8, or, in

the case of some protozoa, more than 1,000 daughter cells, each developing into a new organism.

fission, nuclear In physics, form of NUCLEAR ENERGY in which a heavy atomic NUCLEUS splits into two, with the release of two or three NEUTRONS and large amounts of heat and nuclear RADIATION. Fission may occur spontaneously or be made to occur by bombarding certain nuclei with low-energy (slow) neutrons or PROTONS. The only naturally occurring fissionable material is URANIUM-235. Natural uranium is not capable of sustaining a CHAIN REACTION because its major isotope, U^{238}, tends to absorb neutrons before they can split another atom. Currently, there are two main methods of creating a sustained chain reaction in uranium fuel. In the first, the neutrons released by fission of U^{235} are slowed down by a **moderator** composed of light atoms, such as DEUTERIUM, which does not absorb neutrons. This method is used in most NUCLEAR REACTORS. Alternatively, the uranium may be **enriched** (increasing the amount of fissionable material). This is the method employed in fast breeder reactors. Nuclear fission and nuclear FUSION are used in NUCLEAR WEAPONS.

Fitch, John (1743–98) US inventor. He designed the first steamboat in the USA. In 1787 Fitch launched his boat on the Delaware River. It reached a speed of 5km/h (3mph). He made several improvements to the design, but failed to attract public recognition.

Fitzgerald, George Francis (1851–1901) Irish physicist who studied ELECTROLYSIS and ELECTROMAGNETISM and who is noted for his theory of ELECTROMAGNETIC RADIATION. As an explanation of the MICHELSON-MORLEY EXPERIMENT to determine the Earth's movement through the ETHER, Fitzgerald suggested the theory that objects change length (the **Lorentz-Fitzgerald contraction**) due to this type of movement. This was important to Albert EIN-STEIN when he devised his theory of relativity. *See also* LORENTZ TRANSFORMATION

fixer, photographic Solution in which photographic film or paper is immersed after development to remove unexposed and unreduced silver halide and render the image stable. Sodium or ammonium thiosulphate (hypo) is usually used, often acidified to avoid staining.

Fizeau, Armand Hippolyte Louis (1819–96) French physicist, the first to determine with accuracy the speed of light in both air and water. Fizeau also took, with the French physicist Jean Léon FOUCAULT, the first clear photograph of the Sun. He researched the POLARIZATION of light, the expansion of crystals, and the DOPPLER EFFECT.

fjord (fiord) Narrow, steep-sided inlet on a sea coast. These deeply cut valleys were formed by GLACIERS as they moved toward the sea, and then flooded when the ice melted and sea levels rose.

flagellate Any member of the class Mastigophora. Flagellates are single-celled organisms that possess, at some stage of their development, one or several whiplike structures (FLAGELLUM) for locomotion and sensation. They are divided into two major groups; the **phytoflagellates** resemble plants (in that they obtain their energy through photosynthesis), the **zooflagellates** resemble animals (in that they obtain their energy through feeding). Most have a single nucleus. Reproduction may be asexual (FISSION) or sexual. See also ASEXUAL REPRODUCTION; SEXUAL REPRODUCTION

flagellum (pl. flagella) Long, whip-like extension of a cell. There may be a single flagellum or a group of them. Many cells, such as SPERM cells and some BACTERIA, PROTOZOA, and single-celled algae, beat their flagella as a means of locomotion – they "swim" through fluids in this way. In sperm and protozoa, the structure and movement of the flagella resembles that of CILIA, whereas in bacteria, the flagellum has a different structure and a rotary action. *See also* FLAGELLATE

flame test In chemical analysis, test for the presence of metallic elements whose main spectral emission lines give characteristic colors in a Bunsen flame. A salt of the metal is introduced into the flame by means of platinum wire first dipped in hydrochloric acid. Some metals and their colors are: lithium, red; sodium, yellow; potassium, lilac; strontium, crimson; barium, apple-green; copper, blue-green; lead and arsenic, blue.

Flamsteed, John (1646–1719) British astronomer, the first Astronomer Royal (1675). His book *British Catalogue of Stars* contained a record of his observations and listed nearly 3,000 stars; it was the most accurate of the time and became the standard work for many years.

flash point Lowest temperature at which a flammable liquid heated under test conditions gives off sufficient vapor to be ignited (as a flash) when a flame is applied.

F-layer (Appleton layer) Region within the IONOS-PHERE of the Earth's atmosphere.

Fleming, Sir Alexander (1881–1955) Scottish

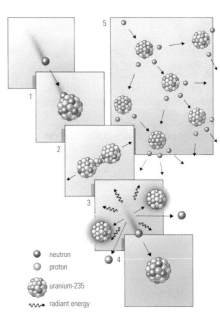

neutron
proton
uranium-235
radiant energy

▲ **fission, nuclear** Most nuclear power stations use uranium-235 as fuel. When a uranium-235 nucleus is struck by a slow-moving neutron (1), it absorbs the neutron to form uranium-236. This is unstable and splits violently (2) forming two smaller nuclei, generating radiant energy (some in the form of heat), and releasing several neutrons (3). These neutrons can then start the process again (4), splitting further nuclei, which in turn release yet more neutrons (5). Such a process is known as a chain reaction and can spread at lightning speed. In a nuclear reactor, many of the neutrons are absorbed to prevent the chain reaction from running out of control and causing an excessive release of energy. Atom bombs are designed to encourage the chain reaction to spread extremely rapidly.

bacteriologist, discoverer of PENICILLIN (the first ANTIBIOTIC). In 1922, while at St. Mary's Hospital in London, Fleming had discovered lysozyme, a natural antibacterial substance found in saliva and tears. During research on staphylococci in 1928, Fleming noticed that a mold, identified as *Penicillium notatum*, liberated a substance that inhibited the growth of some bacteria. He named it penicillin, the most powerful antibiotic known. In 1938 he became professor of bacteriology at the University of London. Howard FLOREY and Ernst CHAIN refined the drug's production, and in 1941 it was produced commercially. In 1944 Fleming was knighted, and in 1945 he shared the Nobel Prize for physiology or medicine with Florey and Chain.

Fleming, Sir John Ambrose (1849–1945) English electrical engineer who invented the thermionic ELECTRON TUBE. Fleming's tube was a rectifier, or DIODE. It consisted of two electrodes in an evacuated glass envelope resembling an electric light bulb. The diode permitted current to flow in one direction only. It could detect radio signals but could not amplify them. He is also remembered for FLEMING'S RULES for motors and generators.

Fleming's rules In physics, aids for remembering the relationships between the directions of the current, field, and motion in ELECTRIC MOTORS and GENERATORS. In the left-hand rule (for motors), the forefinger represents field, the second finger current, and the thumb, motion; when the digits are extended at right-angles to each other, the appropriate directions are indicated. The right-hand rule applies the same principles to generators. Sir John FLEMING devised the rules.

flexor MUSCLE that makes a limb bend. A flexor usually has two points of attachment on either side of a joint. For example, the biceps muscle attaches to the forearm and the upper arm. Flexing the biceps causes the arm to bend. Like all flexors, the biceps has an antagonistic ("opposite") muscle, called an EXTENSOR, that causes the arm to straighten in this case the triceps.

flight Movement through the air. To maintain level flight through the air, the two main requirements are thrust and lift. Thrust overcomes air resistance (DRAG) and gives forward motion. LIFT overcomes the downward pull of gravity. In an AIRCRAFT, thrust is provided by an engine-driven propeller or by the rapid emission of hot expanding gases from JET or ROCKET engines. Lift is provided by AIRFOILS, which cause the air to move faster over their more highly curved upper surfaces than under their lower surfaces. The slower-moving air exerts a higher pressure, resulting in an overall upward force. The wings of birds have a similar shape (in cross-sec-

tion) to an airfoil. Instead of a propeller, birds, insects, and bats beat their wings, angling them to push air behind them. *See also* HELICOPTER

flight recorder Device for automatically recording data during the operation of an aircraft. Investigators analyze the data after a crash or malfunction. A small aircraft may have just a simple cockpit voice recorder (CVR). This records all cockpit sounds, and also all radio contact with air traffic control. Larger aircraft carry a separate flight data recorder (FDR). Control settings, instrument readings, and other data are recorded on magnetic wire. The FDR, often called a "black box," is usually colored orange for ease of location at a crash site.

flight simulator Device that duplicates the instrument behavior and physical attitude of an aircraft in flight, used for training pilots and aircrew. A more sophisticated simulator, in addition to having an exact replica of the controls of a particular aircraft, may also have a visual display of the terrain being flown over to contribute to realism.

flint Granular variety of QUARTZ (SiO_2) of a fine crystalline structure. It is usually smoky brown or dark gray, although the variety known as CHERT is a paler gray. It commonly occurs in rounded nodules and is found in chalk or other SEDIMENTARY ROCKS containing calcium carbonate. Of great importance to early humans during the Stone Age, when struck a glancing blow, flint can be flaked, leaving sharp edges appropriate for tools and weapons; two flints struck together produce a spark that can be used to make fire.

flint clay Hard, flinty FIRECLAY. It is a KAOLINITE, usually found at greater depths than most clays. It is used almost exclusively in the production of FIREBRICK and crucibles.

flip-flop (bistable circuit) Electronic circuit that can be in either of two stable states. Much used in computers, a flip-flop (or bistable circuit) changes state when it receives an electric pulse. It can be used as a one-bit storage device, and is employed in delay elements (which store a signal briefly before transferring it to an output line).

floodplain Level land alongside a river consisting of ALLUVIUM deposited by the river when in flood. Such plains usually have extremely fertile soil and are often used, as near the Nile River, for intensive cultivation.

floppy disk Portable MAGNETIC DISK for use in personal COMPUTERS. Protected by a plastic coating, floppy disks are traditionally still measured in inches, and are commonly either 3.5in (7.9cm) or, now less common, 5.25in (13.3cm) in diameter. To access the data on a disk it is inserted into a computer's disk drive. Data is stored magnetically on both sides of the disk, and is read by magnetic heads in the computer as the disk rotates at some 300rpm. Modern disks can store up to 1.4 MEGABYTES of data, which can be overwritten when it is no longer required.

flora Collective term used to describe all the plants that are found in a particular area, or that occurred at a particular time.

Florey, Sir Howard Walter (Baron Florey of Adelaide) (1898–1968) British pathologist, b. Australia. He shared, with Alexander FLEMING and Ernst CHAIN, the 1945 Nobel Prize for physiology or medicine for his part in the development of PENICILLIN. Florey isolated the antibacterial agent from the mold discovered by Alexander Fleming. This made possible the large-scale preparation of PENICILLIN.

Flory, Paul John (1910–85) US chemist whose discovery of a method of analyzing POLYMERS accelerated the development of plastics and earned him the 1974 Nobel Prize for chemistry. With Wallace CAROTHERS he developed the first NYLON and a synthetic rubber, NEOPRENE. Flory also discovered similarities between the elastic properties of organic tissues and plastics.

flotation process Method of separating mineral ores from unwanted rock and other material. The ore as mined is finely ground and then mixed with water, to which chemicals called **wetting agents** are added. The mixture is aerated to make a froth. The required ore particles stick to the bubbles and float, and are skimmed off the surface; waste rock sinks to the bottom. The method works because of differences in wetting properties (SURFACE TENSION) between the various components.

flow, solid In geology, the slide, flow or creep of solid, unsuspended material downward on a slope. The movement within the solid is achieved by rearrangement between or within its particles.

flower Reproductive structure of all ANGIOSPERMS (flowering plants). It has four sets of organs set in whorls on a short apex (RECEPTACLE). Typically the SEPALS are leaflike structures that protect the bud; they form the CALYX. The petals, often brightly colored, form the COROLLA; the STAMENS are stalks (filaments) tipped by ANTHERS (pollen sacs); the CARPELS form the PISTIL with an OVARY, STYLE, and STIGMA. Flowers are bisexual if they contain stamens and carpels, and unisexual if only one of these is present. Reproduction occurs, following POLLINA-

TION, when POLLEN is transferred from the anthers of one flower to the stigma of another flower of another plant (**cross-pollination**), or to the same flower or flower of the same plant (**self-pollination**). A pollen tube grows down into the ovary where FERTILIZATION occurs and a seed is produced. The ovary bearing the seed ripens into a FRUIT containing the seed, and the other parts of the flower wilt and fall.

flowering plant *See* ANGIOSPERM

fluid Any substance that is able to flow. Of the three common states of matter, GAS, and LIQUID are considered fluid, while any SOLID is not. *See also* PHASE

fluid flow Behavior of a moving FLUID, determined by its velocity, pressure, and density. These three quantities are related by three basic equations: the **equation of continuity**, which relates the amount of fluid flowing into a given space with the amount flowing out of that space; EULER's **equation of motion**, which shows how the velocity of the fluid changes with time at a given point in space; and the **adiabatic equation**, which describes the exchange of heat between different parts of the fluid. In incompressible fluid flow, which applies to most liquids, these equations take on a particularly simple form. Compressible flow equations are necessary for high-speed aerodynamic calculations. Often a fluid is treated as "ideal," meaning that no internal friction or viscosity is assumed. The equations for realistic fluids are so complicated that complete solutions to most problems do not exist; numerical solutions must be attempted by computer techniques.

fluidics Use of devices operated by a FLUID (gas or liquid) for controlling processes and instruments. Fluidic systems simulate electronic circuits, but use a fluid instead of electrons. Fluidic circuits were developed in the USA in the 1960s for use in rocket and aircraft guidance.

▶ **flower** A typical flower has four main parts: sepals, petals, stamens, and carpels. The sepals (1) protect the developing bud and lie outside the showy petals (2). Each male stamen is made up of an anther (3), which contains the pollen grains, borne on a filament (4). The female carpels are found at the center of the flower, each containing ovaries (5) and a style (6), which supports the stigma (7).

fluidization Powdering of a solid so that it can be processed as if it were a fluid. The technique not only allows solids to be conveniently transported, but also hastens gas-solid reactions – the gas is injected from below and the powder is suspended to form a fluidized bed, an ideal condition for industrial drying or roasting.

fluid mechanics Study of the behavior of FLUIDS (liquids and gases). **Fluid statics** includes the study of pressure, density, and the principles of PASCAL and ARCHIMEDES. **Fluid dynamics** includes the study of streamlines, BERNOULLI'S LAW, and the propagation of waves. Engineers use fluid mechanics in the design of bridges, dams, and ships. Physicists use it in studying the structure of the atomic nucleus, and astronomers use it to explain the spiral structure of some galaxies.

fluid mining Method of obtaining minerals that involves dredging or HYDRAULIC MINING. Gravel (PLACER) deposits containing ore can be broken down with high-pressure streams of water or flooded so that dredgers with digging and processing equipment can operate. Many offshore deposits, including gold, tin, iron-bearing sands, shell, and gravel, are mined by dredging.

fluorescence Emission of radiation, usually light, from a substance when its atoms have acquired excess energy from a bombarding source of radiation, usually ultraviolet light or electrons. Unlike PHOSPHORESCENCE, fluorescence ceases when the source of energy is removed. Mercury vapor is a fluorescent substance used in the fluorescent lamps of highway lights; television tubes use fluorescent screens.

fluorescent lamp Glass tube coated on the inside with a chemical phosphor and containing, at low pressure, small quantities of mercury. When an electric current is passed through the tube, the atoms of mercury emit ultraviolet radiation, which impinges on the phosphor coating, causing it to glow brightly. *See also* DISCHARGE TUBE; FLUORESCENCE

fluoridation Addition of inorganic fluorides to the water supply to reduce tooth decay. The additive is usually sodium fluoride, at a concentration of about one part per million. Since its inception in the 1930s fluoridation has been adopted in many countries.

fluoride Any salt of hydrogen fluoride (HF); more particularly, fluoride compounds added to drinking water or toothpaste in order to build up resistance to tooth decay. Fluoride protection appears to result from the formation of a fluorophosphate complex in the outer tooth layers, which become resistant to penetration by acids made by mouth bacteria.

fluorine (symbol F) Gaseous, toxic element of the HALOGEN group (elements in group VII of the periodic table), first isolated in 1886. Its chief sources are fluorspar and cryolite. The pale yellow element, obtained by ELECTROLYSIS, is the most electronegative of all elements and the most reactive of nonmetallic elements, attacking many compounds. It is in FLUORIDE in drinking water and used in making FLUOROCARBONS and in extracting uranium. Properties: at.no. 9; r.a.m. 19; m.p. $-219.6°C$ ($-363.3°F$); b.p. $-188.1°C$ ($-306.6°F$); single isotope ^{19}F.

fluorite (fluorspar) Mineral, calcium fluoride (CaF_2). It has cubic system crystals with granular and fibrous masses. It is brittle and glassy and its color varies; it can be yellow, purple or green, and is frequently banded. It is found in mineral veins as a gangue mineral with metallic ores. It is used as a FLUX in STEEL production and in the ceramics and chemical industries. The deep-purple banded variety is known as Blue John. Hardness 4; r.d. 3.1.

fluorocarbon Organic compound that is produced by replacing the hydrogen atoms of HYDROCARBONS with fluorine atoms. They are inert, have low toxicity, and an ability to withstand high temperature, making them ideal for use in plastics such as PTFE (POLYTETRAFLUOROETHYLENE) or Teflon. Many of these chemicals also contain chlorine and are called CHLOROFLUOROCARBONS (CFCs), the breakdown of which by sunlight damages the OZONE LAYER.

fluorocarbon plastic Plastic made from a class of chemically inert compounds composed entirely of carbon and FLUORINE. The best known of these plastics is POLYTETRAFLUOROETHENE (PTFE), also known as Teflon. Fluorocarbons are valued for their nonflammability, low chemical activity, and low resistance to friction.

fluorspar *See* FLUORITE

flux In ceramics, any substance that promotes vitrification when mixed with clay. When the ware is fired, the flux melts, filling the porous clay form. As the piece cools, it hardens, becoming glossy and nonporous. Fluxes include felspathic rock, silica, and borax. In METALLURGY, a flux is added to the charge of a smelting furnace to purge impurities from the ore and to lower the melting point of the slag.

flux In physics, rate of flow of energy or matter past a surface at right angles to the surface. *See also* MAGNETIC FLUX

flywheel Massive solid or heavy-rimmed wheel attached to a machine's drive shaft to minimize fluctuations in rotational velocity and to store energy. The inertia of the large wheel tends to absorb energy in the effective portion of an engine's cycle and to release energy in the least effective portion of the cycle. The flywheel was initially developed by James WATT.

FM Abbreviation of FREQUENCY MODULATION

f-number In photography, measurement used to describe the relative opening or APERTURE of a lens, often known as the f-stop or focal ratio. It is calculated by dividing the FOCAL LENGTH of a lens by the diameter of the lens opening. For example, a lens with a 50mm focal length with the aperture set at f/2 indicates an aperture diameter of 25mm, whereas f/4 indicates an aperture of 12.5mm. The higher the f-number (f/32 is usually the highest), the less light reaches the film. The f-numbers allow a photographer to gauge the amount of light to which a film is exposed.

foam Suspension of gas bubbles separated from each other by thin films (0.1–1mm thick) of liquid or solid. Foaming is not possible in a pure liquid and a promoter must be added – such as soap or proteins, which are used in such edible foams as marshmallow.

focal length Distance from the midpoint of a curved mirror or the center of a thin LENS to the focal point of the system. For converging systems, it is given a positive value; for diverging systems, a negative value.

focal point (focus) Point on the axis of a LENS or curved mirror to which incident light rays, parallel and close to the axis, are converged (**real focus**) or from which they appear to be diverged (**virtual focus**) after REFLECTION or REFRACTION.

focus Either of two points on the major axis of an ELLIPSE such that the distance from one focus to any point on the ellipse and back to the other focus is constant. In geology, the term is also used to describe the point of maximum displacement caused by an EARTHQUAKE. In optics it is the FOCAL POINT.

fog Mass of water droplets immediately above the Earth's surface that reduces visibility to less than 1km (0.6mi). A light fog is called mist or haze. Fog is caused by water vapor condensing as a result of the air becoming cooler. This condensation takes place around particles of dust. There are four main types of fog: **advection** fog develops from air flowing over a surface of a different temperature, such as steam fog that results from cold air passing over warm water; **frontal** fog forms when warm rain falls through cold air near the ground; **radiation** fog occurs when the ground cools on a still, clear night, and is most common in valleys; **upslope** fog develops when air cools as it ascends a slope. See also DEW POINT

fold In geology, a bend in a layer of rock. An upfold is an ANTICLINE; a downfold, a SYNCLINE. A monocline (flexure) slopes in one direction only and usually passes into a FAULT. Folds occur as part of the process of PLATE TECTONICS, where rock strata buckle and bend under pressure. If the compression is fairly gentle and even, the resulting fold is "symmetrical." If the pressure is uneven, then asymmetrical folds will form. In many cases, the folds are pushed over to form **recumbent folds**. Eventually the rock strata may break under the pressure, to form an overthrust or a NAPPE.

foliation Crude layering of rocks produced under compression. The layering is approximately parallel to the bisecting planes of folds, and often results in the rock splitting because of the parallel orientation of the mineral layers. Mica and slate are good examples of foliation.

folic acid Yellow crystalline derivative of glutamic acid; it forms part of the VITAMIN B complex.

▲ **fold** Tectonic forces warp the Earth's surface and are powerful enough to bend strata of rock. An anticline (1) is created when the rock is pushed upward. When the rock is forced down, a syncline (2) is formed. If the force continues an overfold (3) forms. If the strata are warped too much, they can in effect snap (4) making an overthrust fold or nappe, which is both a fold and a fault. At the top of an anticline, the rocks are "stretched" (5), while at the base of a syncline they are compressed. The extra joints and small cracks at the top of the anticline make it more prone to weathering and erosion. If this process continues over many thousands of years, an anticline can be eroded to form what appears to be a syncline (7).

Found in liver and green vegetables, it is crucial for growth and is used in the treatment of ANEMIA.

follicle In botany, fruit that splits along one side to release its seeds when it is ripe (unlike a pod, which splits along both sides). Such follicles occur in clusters called **etaerios**. They are found in plants such as delphiniums and larkspur.

follicle In zoology, group of cells forming a sac that envelops and protects a structure within it. A hair follicle surrounds and nourishes the root of a hair. A GRAAFIAN FOLLICLE encloses a developing OVUM.

follicle-stimulating hormone (FSH) HORMONE produced by the anterior PITUITARY GLAND located at the base of the brain of mammals. In females it regulates OVULATION by stimulating the GRAAFIAN FOLLICLES found in the OVARY to produce eggs (*see* OVUM). In males, FSH promotes SPERMATOGENESIS, the process by which SPERM are produced. FSH is an ingredient in most FERTILITY DRUGS.

fontanelle Space between the bony plates of an infant's SKULL, often called a "soft spot." There are in fact two fontanelles in a baby's skull, one in the center front, one toward the rear, which close over as the skull plates grow toward each other. By the age of two they are joined, but the seams between the bones do not completely fuse until old age.

food Material taken into an organism to maintain life and growth. Important substances in food include PROTEINS, FATS, CARBOHYDRATES, MINERALS, and VITAMINS. *See also* FOOD CHAIN

food chain Series of organisms, each organism consuming the previous member. In its simplest form, the main sequence of a food chain is from green plants (PRIMARY PRODUCERS) to HERBIVORES (PRIMARY CONSUMERS) and then to CARNIVORES (SECONDARY CONSUMERS). DECOMPOSITION, brought about by organisms of decay such as bacteria and fungi at each stage and at the end of the chain, breaks down waste and dead matter into forms that can be absorbed by plants, so perpetuating the chain. *See also* PHOTOSYNTHESIS

food poisoning Acute illness caused by consumption of food that is itself poisonous or that has become poisoned or contaminated with BACTERIA. Frequently implicated are SALMONELLA bacteria, found in many different strains in cattle, pigs, poultry, and eggs, and LISTERIA, sometimes found in certain types of cheese. Symptoms include abdominal pain, DIARRHEA, nausea, and vomiting. Treatment includes rest, fluids to prevent dehydration and, possibly, medication to curb vomiting. *See also* BOTULISM; GASTROENTERITIS

food preservation Treatment of foodstuffs to prolong the time for which they can be kept before spoiling. Smoke-curing and heat-drying preserve meat by removing water and forming protective outer layers. Salting, pickling, and FERMENTATION preserve food chemically. Chemical preservatives, such as sodium benzoate, can also be added to foods. In CANNING, meats and vegetables are sterilized by heat after being sealed into airtight cans. Cold storage at 5°C (41°F) prolongs the life of foods temporarily, while deep-freezing at −5°C (23°F) or below greatly extends the acceptable storage period. The packing of foods in sealed plastic containers is a method of temporary preservation, and in the modern technique of **freeze-drying**, the frozen foods are placed in a vacuum chamber and the water in them is removed as vapor, the foods can then be fully reconstituted at a later date. Since the early 1990s, **irradiation** (the preservation of food, especially soft fruit, by subjecting them to low levels of radiation in order to kill microorganisms) has been increasingly used.

food technology Application of scientific techniques to the generation, mass production, packaging, preparation, and preservation of all types of food. Generating new and better forms of food is an entire science in itself. Vegetables are grown to achieve optimum size and shape through the scientific provision of controlled environments, pest control measures, and constantly metered nutrition through roots, stems, and leaves. Animal stock may be afforded similar idealized conditions of growth and maturity, but can be further scientifically bioengineered through the careful introduction of hormones and other chemical substances intended to cause effects beneficial to the eventual consumer. In the world of edible plants the genetic modification of strains of grains, legumes, and ordinary table vegetables has been undertaken in the hope that greater yield and resistance to disease and aging is achieved. There are fears, however, that **genetically modified** foods can alter the delicate balance of existing environments and their food chains. For example, it has been shown that weeds can also benefit from a close proximity to genetically modified crops by becoming more resistant to certain herbicides. Additionally there are fears that animals fed on genetically modified crops may develop unforeseen side effects. Improving genetic strains is also important in the mass production of all forms of meat farming (including fish farming). Mechanical and electronic machinery for the mass production of both plant and animal foodstuffs evolves year by year, sometimes radically. The presentation of food to the consumer is an aspect into which considerable technological effort has been expended in terms of packaging and preparation, both to look inviting on the store shelf and to be appetizing on the table. The modern kitchen is full of mechanical and electric gadgets, many useful in food preparation but some aimed at FOOD PRESERVATION. *See also* BIOTECHNOLOGY

food web The often complex pattern of interrelating FOOD CHAINS. When created schematically on paper the lines that link the various animals and plants, and that make the individual food chains, form a crisscross pattern resembling a web.

foot (symbol ft or ') Unit of measurement equal to 0.3048m. There are 12 INCHES to 1ft.

foramen Cavity or aperture in a body part or organ. The foramen magnum, for example, is an aperture in the skull that allows the spinal column to pass through. In gorillas it is positioned towards the back, whereas in humans it is set well forward.

foraminifer Any member of the order Foraminiferida – amoeboid protozoan animals that live among plankton in the sea. They have multi-chambered chalky shells (**tests**), which may be spiral, straight or clustered, and vary in size from microscopic to 5cm (2in) across, according to species. Many remain as fossils and are useful in geological dating. When foraminifera die, their shells sink to the ocean floor to form large deposits, a source of chalk and limestone.

force (symbol F) Loosely speaking, a push, a pull or a turn. A force acting on an object may (1) balance an equal but opposite force or a combination of forces to maintain the object in equilibrium (so that it does not move), (2) change the state of motion of the object (in magnitude or direction), or (3) change the shape or state of the object. There are four FUNDAMENTAL FORCES in nature.

force ratio (mechanical advantage) Factor by which a simple MACHINE multiplies an applied FORCE. It is the ratio of the load (output force) to the effort (input force). *See also* DISTANCE RATIO

Ford, Henry (1863–1947) US industrialist. He developed a gasoline-engined car in 1892 and founded Ford Motor Company in 1903. It achieved huge success with the economical and inexpensive Model T (1908), which from 1913 was produced on an assembly line, lowering production costs enormously.

forebrain One of the three primary parts of the developing BRAIN, clearly distinguishable, together with the MIDBRAIN and HINDBRAIN, in the early embryo. It becomes overlaid in the adult human by the cerebral hemispheres, which are massive developments from the forebrain.

forensic medicine Branch of medicine concerned with the detailed investigation of any assault, injury or sudden death that may be subject to legal action. Its practitioners include forensic pathologists.

forest Large area of land covered with a dense growth of trees and plants. Earth's first forests developed about 365 million years ago. In the early 1800s forests covered 60% of the Earth; today they account for about 30%. DEFORESTATION is a major environmental concern, since forests make such a vital contribution to Earth's atmosphere and also regulate water flow. Forests have been an important source of timber, food, and other resources since prehistoric times. The forest ECOSYSTEM has five basic strata; forest floor, herb layer, shrub layer, understory, and canopy. Forests are classified into three general formations: **tropical hardwood** forests, including RAIN FORESTS, are predominantly EVERGREEN. They account for 7% of the Earth's landmass, but about 50% of the Earth's species. **Temperate** hardwood forests are mostly DECIDUOUS and are found in temperate climates. **Boreal forests**, consisting mainly of evergreen CONIFERS, lie in the far N.

forestry (silviculture) Management of FOREST resources for human benefit, in particular the production of timber through REFORESTATION. It also includes the conservation of soil, water, and wildlife. In 1907 the first US national forests were created, under the administration of the National Forest System. Today, more than 180 million acres of forest is owned and managed by the US Forest Service. In 1960 the service was directed to manage the national forests according to the principles of multiple use and sustained yield; to produce a continuous supply of timber while preserving the natural environment. In 1964 the Forest Service controversially adopted the commercial practice of clearcutting, which came under severe criticism from conservationists. In 1974 the Forest and Rangeland Renewable Resources Planning Act was passed to establish procedures for continually reviewing the management of US forests.

forging Shaping of metal by hammering or by applying pressure against a shaped die. Blacksmiths forge horseshoes and other iron items by hammering the red-hot metal on an anvil. In mass manufacturing processes, pressure from a hydraulic forging press or blows from a forging hammer shape metal parts by forcing them against hard-metal dies, which are carved in the shape of the required piece. Most metals are forged hot, but cold forging is also practiced.

formaldehyde (methanal, HCHO) Colorless, flammable, poisonous gas with a penetrating odor. It is the simplest ALDEHYDE and is produced by the oxidation of METHANOL by air. It was discovered (1867) by August von HOFMANN. Most formaldehyde is in the form of **formalin**, an aqueous solution of 35–40% formaldehyde used as a preservative for biological specimens. Formaldehyde is used in the manufacture of dyes and plastics. Synthetic resins made from formaldehyde helped to create the first plastic objects. Properties: r.d. 0.82; m.p. −92°C (−133.6°F); b.p. −19°C (−2.2°F).

formic acid (methanoic acid, HCOOH) Colorless, corrosive, pungent liquid CARBOXYLIC ACID used to produce insecticides and for dyeing, tanning, and electroplating. It occurs naturally in a variety of sources – stinging ants, nettles, pine needles, and sweat. The simplest of the carboxylic acids, it can be produced by the action of sulfuric acid on sodium formate (methanoate). Properties: r.d. 1.22; m.p. 8.3°C (46.9°F); b.p. 100.8°C (213.4°F).

formula In chemistry, a **molecular** formula represents the elements and the number of their atoms in a compound. For example, the formula H_2O indicates that water is made up of two atoms of hydrogen to one atom of oxygen. A **structural** formula not only represents the elements and the number of atoms, but also indicates the structure of the compound and sometimes the number of bonds between molecules. For example, the structural formula of ethene (CH_2CH_2) indicates the two carbohydrate molecules are attached by a double bond.

formula In mathematics, rule or relationship expressed in mathematical symbols. An example is the formula $V = 4/3\pi r^3$ for the volume of a sphere.

Forssmann, Werner (1904–79) German physician and physiologist. He shared the 1956 Nobel Prize for physiology or medicine with A. F. COURNAND and D. W. RICHARDS. This was belated recognition for his development of the cardiac catheter, a tube that is inserted into a vein in the arm and advanced until it reaches the heart. It is used in the diagnosis of heart and lung diseases.

FORTRAN (Acronym for **for**mula **tran**slation) High-level COMPUTER LANGUAGE generally used for scientific applications. FORTRAN has been in general use for more than 40 years, but has developed greatly from its original form.

fossil Any direct evidence of the existence of an organism more than 10,000 years old. Fossils mostly consist of original structures, such as bones or shells or wood, often altered through MINERALIZATION or preserved as molds and casts. Imprints such as tracks and footprints are also fossils. Leaves are often preserved as a carbonized film outlining their form. Occasionally organisms are totally preserved in frozen soil (such as mammoths), peat bogs, and asphalt lakes, or trapped in hardened resin (such as insects in AMBER). Fossil excrement, called COPROLITE, frequently contains undigested and recognizable hard parts. Very few animals and plants that die become fossilized. Since fossils reveal evolutionary changes through time, they are essential clues for geological dating.

fossil fuels Term to describe COAL, OIL, and NATURAL GAS – FUELS that were formed millions of years ago from the fossilized remains of plants or

▲ **fossil** Animal fossils are not only made from their bones or shells, often their tracks can also be fossilized. Typically, a footprint (1) is left behind in soft mud, which partially hardens to form a cast. If the mud becomes flooded (2), sediment is laid over the mud especially quickly (3) helping to preserve the shape of the footprint. Over the course of time, the mud and sediment become compressed and turn to rock (4). The original mud-based rock forms a mold of the footprint (5) and the sediment-based rock forms a cast (6).

animals. By their very nature fossil fuels are a non-renewable energy source

Foucault, Jean Bernard Léon (1819–68) French physician and physicist who invented the GYROSCOPE. He used a pendulum (FOUCAULT'S PENDULUM) to prove that the Earth spins on its axis. He also devised a method to measure the absolute velocity of light, which he did, and in 1850 he showed it to be slower in water than in air. Foucault, with Armand FIZEAU, took the first clear photograph of the Sun. He also noted the occurrence of eddy currents, which became known as **Foucault currents**. They are electrical currents induced in a conducting medium when the conductor and a magnetic field passing through it are in relative movement.

Foucault's pendulum Heavy metal bob hung from a long, fine wire, first used by Jean FOUCAULT to demonstrate the Earth's rotation. As the Earth turns on its axis, the plane of swing of the pendulum, as observed from the Earth's surface, rotates because of the heavy bob's inertial resistance to change of absolute direction, until it has returned to its original orientation. Its period of rotation depends on its latitude.

foundry Workshop in which metals are processed by melting and casting in molds. Up to the 18th century, foundries were usually small, mainly casting gun barrels and bells. From 1760, however, they grew in size as the use of coke provided an efficient fuel to heat large furnaces. Foundries became increasingly sophisticated, especially where iron and steel were concerned, after the innovations of men such as Henry Cort (1740–1800) and Henry BESSEMER.

four-color process Method of printing that reproduces COLOR images by using only four color printing plates. When printed one after the other the four colors recreate a complete range of colors. The color image is "scanned" and separated into red, green, and blue primary colors. Films are made representing each color and a fourth is made representing the shades of gray and black in order to strengthen these tones in the final image. The film is then used to create the printing plates.

Fourier, Jean Baptiste Joseph (1768–1830) French mathematician and physicist, scientific adviser to Napoleon in Egypt from 1798 to 1801. His work on the mathematical analysis of heat flow led to his devising a technique, now known as **Fourier analysis**. In this technique, complex periodically varying functions are decomposed into sums of waves and a problem is analyzed using these waves, which are easier to manipulate than the original function. *See also* FOURIER SERIES

Fourier series Series of sine and cosine functions used to represent other PERIODIC FUNCTIONS. Most bounded, periodic functions (on the domain $-\pi \leqslant \pi$) can be analyzed as a sum of simple harmonic components. Thus if $f(x)$ is such a function, with x taking values between $-\pi$ and π, so that
$$f(x + 2\pi) = f(x)$$
it may be expressed as
$$f(x) = a_0/2 + (a_1\cos x + b_1\sin x) + (a_2\cos 2x + b_2\sin 2x) + ...$$
In this series the nth coefficients a_n and b_n are given by
$$a_n = 1/\pi \int_{-\pi}^{+\pi} f(x)\cos(nx)dx$$
and
$$b_n = 1/\pi \int_{-\pi}^{+\pi} f(x)\sin(nx)dx.$$

four-stroke engine INTERNAL COMBUSTION engine in which the operation of each piston is in four stages. This kind of operation is called the **four-stroke cycle**, or the **Otto cycle**, after its inventor, Nickolaus OTTO. Most internal combustion engines of the sort found in cars, including DIESEL ENGINES, use the four-stroke principle. Each of the four stages corresponds to one movement of a reciprocating piston along a cylinder. First, the piston descends to admit the fuel-air mixture (intake stroke). Second, the piston rises, compressing the mixture (compression stroke). Third, the mixture is ignited by spark or compression, and expanding gases from the explosion force the piston downward (ignition or power stroke). Fourth, the piston again rises to expel the spent gases from the cylinder (exhaust stroke). *See also* TWO-STROKE ENGINE

fourth dimension Time considered as an additional dimension, together with the three dimensions of space, in a full description of the motion of a particle. In RELATIVITY theory this four-dimensional framework is called SPACE-TIME.

fovea *See* BLIND SPOT; OPTIC NERVE

Fowler, William A. (Alfred) (1911–95) US physicist and astrophysicist. Fowler is best known for his contributions to the understanding of the way in which chemical elements are built up (from lighter to heavier) within stars as they evolve. He developed these ideas (published 1957) while working with three British astronomers: Fred HOYLE, and Geoffrey and Margaret Burbidge. In 1983 he shared the Nobel Prize for physics with Subrahmanyan CHANDRASEKHAR.

fps system Abbreviation of the "foot-pound-second" system of units. In the fps system, formerly used in much of the English-speaking world, the FOOT is the unit of length, the POUND the unit of mass and the SECOND the unit of time. It has been replaced by the METRIC SYSTEM or SI UNITS for all scientific and many commercial purposes.

fractal Geometrical figure in which an identical motif repeats itself on smaller and smaller scales. Such figures are said to be "self-similar." The term "fractal" was coined by Benoit MANDELBROT, and fractal geometry is closely associated with CHAOS THEORY. There are many examples of fractal-like objects in nature, from shells and cauliflowers to mountains and leaves, in addition to those produced mathematically in computer graphics.

fraction Quotient written in the form of one number divided by another, such as 3/4. In general, a fraction is a/b, where a is the numerator and b the denominator. If a and b are whole numbers, the quotient is a simple fraction. If a is smaller than b, it is a **proper** fraction; if b is smaller than a, it is an **improper** fraction. In an **algebraic** fraction, the denominator, or the numerator and denominator, are algebraic expressions, for example $x/(x^2 + 2)$. In a **composite** fraction, both the numerator and denominator are themselves fractions. Two fractions a/b and c/d can be added, subtracted, multiplied, and divided according to the rules: $a/b + c/d = (ad + bc)/bd$; $a/b - c/d = (ad - bc)/bd$; $a/b \times c/d = ac/bd$; and $a/b \div c/d = ad/bc$.

fractionation Separation of a mixture using CRYSTALLIZATION (**fractional crystallization**) or DISTILLATION (**fractional distillation**). In fractional crystallization, the mixture is dissolved in a hot solvent. Then, as the solution gradually cools, the components of the mixture crystallize out one at a time – the least soluble one first. In fractional distillation, a liquid mixture is boiled and the vapor allowed to pass up a long vertical condenser, called a **fractionating column**. The lowest-boiling (most volatile) components in the mixture pass up to the top of the column, while the least volatile ones remain near the bottom. The various components – called **fractions** – can be drawn off at the appropriate levels. The technique is the key process in OIL refining.

frame of reference Mathematical coordinate system for describing events in space and time with respect to a given observer. In the theory of RELATIVITY, this frame of reference is four-dimensional, and the description of events in other frames of reference depends on the relative velocities and accelerations of those frames with respect to the frame of reference of the observer.

francium (symbol Fr) Radioactive, metallic element of group I of the periodic table; first discovered in 1939. The heaviest member of the ALKALI METALS, it occurs naturally in uranium ores and is a decay product of ACTINIUM. It is quite rare and its chemical properties are unknown, but it is thought to resemble CESIUM. There are 21 isotopes identi-fied. Properties: at.no. 87; most stable isotope ^{223}Fr (half-life 22 minutes).

Franck, James (1882–1964) US physicist, b. Germany. With Gustav HERTZ, he experimented with electron bombardment of gases, providing support for the theory of atomic structure proposed by Niels BOHR and information for the quantum theory of Max PLANCK. Franck and Hertz shared the 1925 Nobel Prize for physics for these studies on the changes of energy occurring when atoms collide with electrons. In 1935 he became a US citizen. He worked on the development of the atom bomb with the MANHATTAN PROJECT but in 1945 drew up the "Franck petition" saying the bomb should not be used against Japanese civilians.

Frankland, Sir Edward (1825–99) British chemist who formulated the theory of VALENCE. This states that each element may be assigned a whole number (1, 2, 3 or more) or valence that represents the number of hydrogen atoms with which an atom of the element will combine. He also discovered, with Sir Joseph LOCKYER, the gas helium in the Sun.

Franklin, Rosalind (1920–58) British crystallographer whose studies of DNA were instrumental in enabling Francis CRICK and James WATSON to determine its structure. She worked at King's College, London and designed a special low-angle camera to take X-ray diffraction photographs of strands of DNA. She deduced that DNA has a long, chain-like molecule, possibly helical. She went on to determine the hollow, cylindrical structure of the tobacco mosaic virus.

Franklin, Benjamin (1706–90) US statesman and inventor, one of the founders of the republic. He was a successful printer in Philadelphia and published *Poor Richard's Almanac* (1732–57). He conducted scientific research, notably in electricity, which he identified as lightning. He also described differences between positive and negative electricity and invented the lightning rod.

Frasch process Method of mining SULFUR by pumping superheated water and air into the sulfur deposits, melting the mineral, and forcing it to the surface. Named after the German-born chemist Herman Frasch, the process was first put to practical use in Louisiana and made the US independent of imported sulfur.

Fraunhofer, Joseph von (1787–1826) German physicist, the founder of astronomical SPECTROSCOPY. He studied the DIFFRACTION of light through narrow slits and developed the earliest form of DIFFRACTION GRATING. In 1814 he observed and began to map the dark lines in the Sun's spectrum, now called **Fraunhofer lines**, and was the first to appreciate their significance.

free fall State of motion of an unsupported body in a gravitational field. *See* GRAVITY

free radical Short-lived (less than 1ms) molecule that has an unpaired ELECTRON and, therefore, rapidly binds with other molecules. Occurring as by-products of normal CELL chemistry, free radicals can cause extensive damage in the body, even though cells have some protective enzymes. They are thought to play a role in aging and in a number of disease processes, including cancer and cardiovascular problems. Free radicals are used commercially to initiate POLYMERIZATION.

freezing point Temperature at which a substance changes PHASE (state) from liquid to solid. The freezing point for most substances increases as pressure increases. The reverse process, from solid to liquid, is melting; melting point is the same as freezing point.

Freon Trade name for a type of CHLOROFLUORO-CARBON (CFC). Freons are used as refrigerants, AEROSOL propellants, cleaning fluids and solvents. They are all clear, stable, and inert liquids. Implicated in the destruction of OZONE in the IONOS-PHERE, the use of Freons is being phased out.

frequency Rate of occurrence. In statistics, the number of times a numerical value, event, or special property occurs. In physics, the frequency is the number of oscillations (or waves) passing a given point per second (measured in HERTZ), including those of sound, light and radio waves, a swinging PENDULUM, and vibrating springs. The product of the frequency and the period of a wave (or WAVELENGTH) is constant and equals the wave velocity.

frequency modulation (FM) Form of RADIO transmission. It is the variation of the frequency of a transmitted radio carrier wave by the signal being broadcast. The technique makes radio reception fairly free from static interference and, although restricted in range to receivers in line of sight of the transmitter, has become the most favored transmission method as a result. *See also* AMPLITUDE MODULATION (AM)

Fresnel, Augustin Jean (1788–1827) French physicist and engineer. His pioneer work in optics was instrumental in establishing the wave theory of light. He researched into the conditions governing INTERFERENCE phenomena in POLARIZED LIGHT, studied double refraction, and devised a way of producing circularly polarized light.

Fresnel lens Type of lens, used in spotlights and lighthouses, that consists of a piece of heat-resistant glass cast with concentric portions of lenses of different diameters and approximately the same focal length. The Fresnel lens is two or three times more efficient than a plano-convex lens.

friction Resistance encountered when surfaces in contact slide or roll against each other, or when a fluid (liquid or gas) flows along a surface. Friction

▶ **Fresnel lens** Once commonly used in lighthouses and spot lights, the concentric circles of the Fresnel lens (A) help to focus light more effeciently than a standard convex lens (B).

A   **B**

is directly proportional to the force pressing the surfaces together and the surface roughness. Before the movement begins, it is opposed by static friction up to a maximum "limiting friction" and then slipping occurs. The streamlined design of aircraft and cars reduces air (fluid) friction. *See also* AERODYNAMICS

friction, coefficient of Number characterizing the force necessary to slide or roll one material along the surface of another. If an object has a weight N and the coefficient of FRICTION is μ, then the force (F) necessary to move it without acceleration along a level surface is $F = \mu N$. The coefficient of static friction determines the force necessary to initiate movement; the coefficient of kinetic friction determines the (lesser) force necessary to maintain movement.

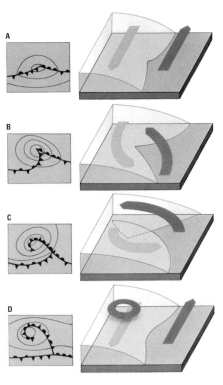

A

B

C

D

▲ **front** A front forms in temperate latitudes where a cold air mass meets a warm air mass (A). The air masses spiral round a bulge causing cold and warm fronts to develop (B). The warm air rises above the cold front and the cold air slides underneath the warm (C). Eventually, the cold areas merge, and the warm air is lifted up or occluded (D).

Frisch, Karl von (1886–1982) Austrian zoologist. He shared with Konrad LORENZ and Nikolaas TINBERGEN the 1973 Nobel Prize for physiology or medicine for his pioneering work in ETHOLOGY (the study of animal behavior patterns). He deciphered the "language of bees" by studying their dance patterns (WAGGLE DANCE) in which one bee tells others in the hive the direction and distance of a food source. In his earlier work he showed that fish and bees see colors, fish can hear, and bees can distinguish various flower scents.

front In meteorology, the boundary between two air masses of different temperatures or of different densities. **Cold** fronts occur as a relatively cold and dense air mass undercuts warmer air. The upward movement of the warm air usually results in condensation, cloud formation, and rainfall. With a **warm** front, the warmer air gradually rises over colder air. As the air rises it cools and forms clouds. Eventually it may give rise to some warm rain. An **occluded** front is a combination of a warm and a cold front in a DEPRESSION. It occurs when the cold front catches up with the warm front and undercuts it. The effects of the warm front and the cold front together are to cause heavier and more prolonged rain than either front would yield alone.

frost Ice formed by water vapor on a surface when the temperature of the surface is below 0°C (32°F). Frost often looks light and feathery, and it often forms in delicate, lacy loops. Frost can do great damage to crops; the US Weather Service issues advisories when it is forecast, and farmers and fruitgrowers try to protect their crops against its effects. Frost is also a cause of EROSION, of the mechanical WEATHERING of rock masses, and of the upthrusts of ground called **frost heaves** that damage highways.

frostbite Freezing of living body tissue in sub-zero temperatures. Frostbite is an effect of the body's defensive response to intense cold, which is to shut down blood vessels at the extremities in order to preserve warmth at the core of the body. Consequently, it mostly occurs in the face, ears, hands, and feet. In superficial frostbite the affected part turns white and cold; it can be treated by gentle thawing. If freezing continues, ice crystals form in the tissues, the flesh hardens, and there is no sensation. Deep frostbite, which causes tissue death, requires urgent medical treatment. No attempt should be made at rewarming if there is a risk of refreezing because this increases the likelihood of GANGRENE.

fructose (fruit sugar) Simple white, crystalline sugar ($C_6H_{12}O_6$) found in honey, sweet fruits, and the nectar of flowers. It is a MONOSACCHARIDE that is sweeter than the DISACCARIDE sucrose (cane sugar), of which it is a component along with glucose. It is made commercially by the HYDROLYSIS of beet or cane sugar and is used in foods as a sweetener. Its derivatives play a crucial role in providing energy to living organisms.

fruit Seed-containing, mature OVARY of an ANGIOSPERM (flowering plant), in which the seeds are surrounded by the PERICARP (fruit wall). Fruits serve to disperse plants and are an important food source (they provide vitamins, acids, salts, calcium, iron, and phosphates). They can be classified as simple, aggregate or multiple. **Simple** fruits, dry or fleshy, are produced by one ripened ovary of a single PISTIL (unit comprising a STIGMA, STYLE, and OVARY) and include legumes (peas and beans) and nuts. **Aggregate** fruits develop from several simple pistils; examples are raspberry and blackberry. **Multiple** fruits develop from a flower cluster; each flower produces a fruit which merges into a single mass at maturity; examples are pineapples and figs. Although considered fruits in culinary terms, apples and pears are regarded botanically as "false" fruits (PSEUDOCARPS), as the edible parts are created by the RECEPTACLE and not the carpel walls. Some fruits spontaneously disperse their seeds (*see* DEHISCENT), others keep their seeds and the fruit is dispersed as a whole (*see* INDEHISCENT).

FSH Abbreviation of FOLLICLE-STIMULATING HORMONE

Fuchs, Klaus (1912–88) German physicist and Soviet spy. Interned in Britain at the start of World War II, he was released and naturalized. He began to pass secrets to the Soviet Union in 1941. In 1943 he went to the US to work on the atom bomb, returning to the UK in 1946 to head the theoretical physics division of the atomic research center at Harwell. He was imprisoned in 1950 for his espionage activities, and his British citizenship was revoked. Released in 1959, he went to East Germany to work at a nuclear research center.

fuel Substance that is burned or otherwise modified to produce energy, usually in the form of heat. Apart from FOSSIL FUELS (COAL, OIL, and GAS) and firewood and charcoal, the term also applies to radioactive materials used in NUCLEAR REACTORS.

fuel cell ELECTROCHEMICAL CELL for direct conversion of the energy of oxidation of a fuel to electrical energy. Suitably designed electrodes are immersed in an ELECTROLYTE, and the fuel (such as hydrogen) is supplied to one and the oxidizer (such as oxygen) to the other. Electrode reactions occur leading to oxidation of the fuel, with production of electric current. Fuel cells are used in space vehicles.

fuel injection Method of introducing fuel into the cylinders of an INTERNAL COMBUSTION ENGINE,

▲ **fruit** Fruits are the ripened ovary of a flowering plant, which is enclosed by the fruit wall (the pericarp). Most fruits, such as blackcurrants (1), cherries (2), strawberries (3), oranges (4), and peppers (5) contain more than one seed. The avocado (6), however, contains only one seed, about the size of a golf ball. The largest fruit, the double coconut (7) is also single-seeded.

utilizing a pump rather than piston-created suction. It distributes fuel more evenly for greater power and less likelihood of engine knock or vapor lock.

fuel rod (fuel element) Cylindrical case containing a fissionable (able to undergo FISSION) material such as uranium-235 placed in the core of a NUCLEAR REACTOR. Fuel rods have to be able to withstand the heat generated by the nuclear reaction and be sufficiently strong so that there is no danger of them splitting and releasing radioactive material. Spent fuel rods are generally made safe in a reprocessing plant.

fulcrum Point about which a LEVER pivots. Its use in the PRINCIPLE OF MOMENTS was recognized by ARCHIMEDES, who is reputed to have said: "Give me a firm place on which to stand and I will move the Earth."

fuller's earth Clay-like substance that contains more than 50% SILICA. Originally used to remove oil and grease from wool (the process of **fulling**), it is now used to bleach petroleum and in refining vegetable oils.

fulminate Explosive, a salt of fulminic acid (HONC), especially fulminate of mercury, $(Hg(ONC)_2)$, which is formed by the action of nitric acid on mercury metal in the presence of alcohol. It is a highly unstable compound used in explosive DETONATORS and blasting caps.

fumarole Vent in the ground that emits gases and vapors, usually found in volcanic areas. The term also refers to a hot spring or GEYSER that emits steam. A fumarole can be defined in terms of the composition of its gases, such as a chlorine fumarole.

function In mathematics, a rule that assigns to each element of a given set a uniquely specified value. The given set is the **domain** of the function, and the set of values is the **range**. Two or more elements of the domain may be assigned the same value, but a function must assign only one value to each element of the domain. We also say that a function f maps each element x of the domain to a corresponding element (or value) y in the range. Here x and y are variables, with y dependent on x through the functional relationship f. The dependent variable y is said to be a function of the independent variable x. Examples of functions are the natural logarithm, whose domain is the set of positive real numbers and whose range is the set of real numbers, and the square-root function, with domain and range the nonnegative real numbers. *See also* NUMBER, REAL; TRIGONOMETRIC FUNCTION

fundamental forces Four basic forces that exist in physics. The most familiar, and the weakest, is GRAVITATION. The gravitational force between the Earth and an object accounts for the object's WEIGHT. Much stronger is the ELECTROMAGNETIC FORCE, which operates between electrically charged particles. It holds atoms together and binds them to each other chemically. The two other forces recognized both operate only on the subatomic level: the WEAK NUCLEAR FORCE, associated with the decay of particles, is intermediate in strength between the gravitational and electromagnetic force; the STRONG NUCLEAR FORCE associated with the "glue" that holds nuclei together is the strongest force known in nature.

fundamental unit One unit out of the three that are usually required to make up a system of units. For example, the CGS SYSTEM (now adapted to provide the basis of the SI UNITS system) comprises the centimeter, gram, and second fundamental units.

fungicide Chemical that kills fungi. For example, creosote is used on wood to prevent dry rot.

fungus Any of *c*.100,000 species of the kingdom Fungi, which are unable to photosynthesize and which reproduce by means of spores and never produce cells with FLAGELLA. They include mushrooms, molds, and yeasts. Fungi have relatively simple structures, with no roots, stems, or leaves. Their cell walls contain the POLYSACCHARIDE CHITIN. The main body of a typical multicellular fungus consists of an inconspicuous network (MYCELIUM) of fine filaments (HYPHAE), which contain many nuclei, and which may or may not be divided into segments by cross-walls. The hypha nuclei are HAPLOID. The mycelia may develop spore-producing, often conspicuous, fruiting bodies, mushrooms and toadstools. Fungal PARASITES depend on living animals or plants;

SAPROPHYTES utilize the materials of dead plants and animals, and symbionts obtain food in a symbiotic relationship with plants (*see* SYMBIOSIS). Fungi feed by secreting digestive ENZYMES onto their food, then absorbing the soluble products of digestion. Many fungi cause diseases in crops, livestock, and, humans (athlete's foot). Molds and yeasts are used in the production of beer and cheese; some fungi, such as *Penicillium*, are sources of ANTIBIOTICS.

Funk, Casimir (1884–1967) US biochemist who discovered VITAMINS. In 1912 he found vitamins B_1, B_2, C, and D, coined the term "vitamine" (the AMINES of life) and presented a paper formulating the idea of vitamin deficiency disorders. Funk also contributed to the knowledge of cancer and of the hormones of the sex glands.

fur Soft, dense hair covering the skin of certain mammals. Such mammals include mink, fox, ermine, muskrat, wolf, bear, squirrel, and rabbit. Some of the animals are hunted and killed for their pelts, which, when manufactured into clothing, may command high prices. Some fur-bearing animals are now protected by law because overhunting has made extinction likely.

furlong Unit of measurement, equal to 220yds (201m). There are 8 furlongs to 1 mile. The word derives from a "furrowlong"; it represented the standard length of a furrow on a square field of 10 acres in medieval times.

furnace Enclosed space that is raised to high temperatures usually by the combustion of fuels or by electric heating. Most furnaces are used in the extraction of metals or the making of alloys. Some furnaces heat ovens, boilers, and kilns. A domestic boiler, especially if it burns solid fuel, is sometimes called a furnace. There are three main types of electric furnace. An arc furnace relies on the heat generated by an electric arc (spark), often between two large carbon electrodes, which are slowly consumed. A resistance furnace is heated by passing an electric current through a heating element or directly through metallic material. An induction furnace uses ELECTROMAGNETIC INDUCTION to cause a current to flow in a metallic charge. The resulting heat is sufficient to melt the metal.

fuse In electrical engineering, a safety device to protect against overloading. Fuses are commonly strips of easily melting metal placed in series in an electric circuit such that, when overloaded, the fuse melts, breaking the circuit and preventing damage to the rest of the system.

fuselage Body of an AIRCRAFT, to which the wings, tail assembly, and landing gear are attached; it houses the crew, cargo, passengers, and controls. Its structural strength lies in a stress-bearing skin, internal bulkheads, lengthwise stringers and formers.

fusel oil Poisonous, clear, colorless liquid with a disagreeable smell. It consists of a mixture of amyl alcohols, obtained as a by-product of the fermentation of plant materials containing sugar and starch. Fusel oil also occurs as a dangerous impurity in badly made spirit drinks. It is used as a solvent for waxes, resins, fats, and oils, and in the manufacture of explosives and pure amyl alcohols.

fusion, nuclear Type of nuclear reaction in which nuclei of light atoms (such as hydrogen) combine to form one or more heavier nuclei with the release of large amounts of energy. The process takes place in the Sun and other stars, and has been reproduced on Earth in the HYDROGEN BOMB. In a self-sustaining fusion reaction, the combining nuclei are in the form of a PLASMA. It is the maintenance of this state of matter that has proved difficult in harnessing fusion reaction as a controlled source of NUCLEAR ENERGY, though attempts have been made in TOKA-MAK reactors. *See also* FISSION, NUCLEAR

fuzzy logic System of logic able to represent statements that are true or false depending on context. For example, the statement "this is warm" applied to the inside of a freezer that is not working is true only in the context that the freezer is not literally "freezing." Fuzzy logic is based on a notion of degrees of truth rather than precise truth, which is the concern of other systems of logic. Computer-controlled devices programmed on the principles of fuzzy logic are able to put into context information they receive from external sources, thereby responding flexibly to the environment. Such ability is more associated with human reasoning than classic machine reasoning. For example, dishwashers have been fitted with microprocessors that program the release of detergent according to how dirty the water is following an initial washing cycle.

▲ **fusion, nuclear** In experiments to generate power by nuclear fusion, the aim is usually to produce energy by fusing tritium and deuterium. These are isotopes of hydrogen and the process can only occur at a temperature above 100 million°C and enormous pressure. Deuterium has one proton, one neutron, and one electron while tritium has one proton, two neutrons, and an electron.

In a fusion reactor the mixture of the two isotopes is heated by intense radio emissions, ion bombardment, and electrical pulses (1). The plasma which results is suspended in a magnetic field (2). The tritium and deuterium nuclei fuse (3) creating a helium nucleus (4), a loose neutron (5), radiation (6), and massive amounts of energy when the products hit the edge of the plasma (7).

g Symbol for the ACCELERATION OF FREE FALL

G Symbol for the GRAVITATIONAL CONSTANT

gabbro Coarse-grained basic IGNEOUS ROCK. It can be regarded as the plutonic equivalent of BASALT, being much coarser because of its slow crystallization. The constituents are sodium and calcium FELDSPAR and the dark minerals OLIVINE and PYROXENE.

Gabor, Dennis (1900–79) British physicist, b. Hungary. He was awarded the 1971 Nobel Prize for physics for his invention of HOLOGRAPHY. He developed the basic technique in 1947, but it was not until the invention of the laser in 1960 by Charles TOWNES that holography became commercially feasible.

gadolinium (symbol Gd) Silvery white, metallic element of the LANTHANIDE SERIES, first isolated as the oxide in 1880. Chief ores are gadolinite, monazite, and bastnaesite. A malleable and ductile metal, its specialized uses include neutron absorption (important in many NUCLEAR REACTORS) and the manufacture of certain alloys. Properties: at.no. 64; r.a.m. 157.25; r.d. 7.898; m.p. 1,311°C (2,392°F); b.p. 3,233°C (5,851°F); most common isotope ^{158}Gd (24.87%).

Gagarin, Yuri Alekseyevich (1934–68) Russian cosmonaut, the first man to orbit the EARTH (April 12, 1961). Gagarin made the single orbit in 1 hour 29 minutes. He died in a plane crash.

Gaia hypothesis Scientific theory that relates the Earth's many and varied processes – chemical, physical, and biological. Popular in the 1970s, when it was proposed by UK scientist James Lovelock (1919–), the Gaia hypotheses showed the Earth as one single living organism, with its components made up of all living things. It was popular with ecological groups because it explained the delicate interrelationship between people and the environment.

Gajdusek, Daniel Carleton (1923–) US physiologist. He shared the 1976 Nobel Prize for physiology or medicine with Baruch Blumberg (1925–) for work on the neurological disease kuru. Working among the Fore tribe in Papua New Guinea, which was being decimated by this disease, he postulated its transmission by the ritual eating of human brains. Further work showing that kuru could be transmitted to animals led him to conclude that it was caused by a slow-acting agent that could remain dormant for years.

galactic cluster *See* OPEN CLUSTER

galactic coordinate system Astronomical COORDINATE SYSTEM that refers to the plane of the galaxy. The coordinates are galactic latitude and longitude. Galactic latitude is the angular distance of a celestial object above or below the galactic plane. Galactic longitude is the angular distance around the galactic equator from a defined zero point. This point is at right ascension 17hr 42.4min, declination −28°55'.

galaxy Huge assembly of stars, dust, and gas, an example of which is our own galaxy. There are three main types, as originally classified by Edwin HUBBLE in 1925. **Elliptical** galaxies (E) are round or elliptical systems, showing a gradual decrease in brightness from the center outward. Graded E0 to E7

▲ **galaxy** Hubble's classification of galaxies according to their forms has three main groups: **Elliptical**, which vary from nearly spherical (E0) to flattened ellipsoid (E7), contain little gas and dust but many red stars; **Spiral**, which vary from galaxies with large, bright nuclei and tightly coiled spiral arms (Sa) to those with smaller nuclei and more prominent arms (Sc); and **Barred spiral**, which have a conspicuous bar intersecting the nucleus. The sequence SBa to SBc shows large bright nuclei giving way to less noticeable ones with the arms progressively less tightly coiled. SO type galaxies resemble the spirals in most features but lack the arms.

according to increasing ellipticity, elliptical galaxies consist of old stars free of gas and dust. **Spiral** galaxies (S) are flattened, disk-shaped systems in which young stars, dust, and gas are concentrated in spiral arms coiling out from a central bulge, the **nucleus**. They are subdivided Sa to Sc to designate tightly wound spirals (a) to loose spirals (c). **Barred spiral** galaxies (SB) are distinguished by a bright central bar from which the spiral arms emerge. They too are subdivided SBa to SBc in the same way as spirals. In addition there are **lenticular** galaxies, systems intermediate between ellipticals and spirals, having a disk and nucleus, similar to spiral galaxies, but with no apparent spiral arms. IRREGULAR GALAXIES (Ir) are systems with no symmetry. Current theories suggest that all galaxies were formed from immense clouds of gas at roughly the same time, soon after the BIG BANG. Galaxies can exist singly or in clusters (GALAXY CLUSTER), which contain anywhere from just a few to thousands of members. Our galaxy is a star system containing the SOLAR SYSTEM. It is spiral in shape and about 100,000 light-years in diameter. Our Sun and Solar System are located at the edge of one of the spiral arms, about 30,000 light-years from the center. The spiral arms form a disk-shaped system, with a bulging core (or nucleus) in the direction of Sagittarius. The stars of the spiral arms form the MILKY WAY, running around the sky. However, it is uncertain how many spiral arms the galaxy has, or whether it is a normal or a barred spiral. The whole galaxy is rotating, but the rotational rate varies with the distance from the cener. Our Sun circles the center at about 250km (150mi) s^{-1}, taking 220 million years per orbit. The galaxy's age is thought to be about 10 billion years, during which time the Sun would have completed about 50 revolutions. The galaxy is a member of a cluster known as the Local Group, which includes the ANDROMEDA GALAXY and the MAGELLANIC CLOUDS. *See also* DOUBLE GALAXY; RADIO GALAXY

galaxy, double *See* DOUBLE GALAXY

galaxy, radio *See* RADIO GALAXY

galaxy cluster Group of associated GALAXIES, consisting of several separate systems moving together through space. A concentration of galaxy clusters is termed a **galaxy supercluster**.

Galen (*c.*129–*c.*199) Greek physician whose work and writings provided much of the foundation for the development of medical practice. He tried to systematize all that was known of medical practice and to develop a theoretical framework for an explanation of the body and its disorders. He made numerous anatomical and physiological discoveries, including ones concerning HEART valves, secretions of the KIDNEY, RESPIRATION, and nervous function. He was among the first to study physiology by means of detailed and ingenious animal experimentation. Galen's often wildly inaccurate theories influenced medical practice for centuries.

galena Gray, brittle, metallic mineral, lead sulfide (PbS). It is the major ore of lead. It occurs as granular masses, and commonly as cubic or octahedral crystals with a lead-gray streak. It is widely found in hydrothermal veins and as a replacement in limestone and dolomite rocks. Hardness 2.5–2.7; r.d. 7.5.

Galilean satellites Four chief SATELLITES of JUPITER: GANYMEDE, CALLISTO, IO, and EUROPA, named collectively after GALILEO, who observed them in 1610. Apart from the Earth's Moon, they are the brightest satellites in the SOLAR SYSTEM, and all would be naked-eye objects were it not for the glare from Jupiter itself.

Galileo (1564–1642) (Galileo Galilei) Italian scientist who helped to found the modern experimental method. He was a lecturer at Pisa, later moving to Florence and then to Padua. According to legend, he observed that a hanging lamp in Pisa Cathedral took the same time to complete one oscillation however long the swing, and he suggested the pendulum could be used for timekeeping. Galileo later studied falling bodies and disproved ARISTOTLE'S view that they fall at different rates according to weight. He also discovered the parabolic flight path of projectiles. In 1609 he used one of the first astronomical TELESCOPES to discover SUNSPOTS, lunar craters, Jupiter's major satellites, and the phases of Venus. In *Sidereus Nuncius* (1610) he announced his support for the Copernican view of the Universe, with the Earth moving around the Sun. He was forbidden by the Roman Catholic church to teach that this system represented physical reality, as opposed to a calculational device, but in *Dialogue on the Two Great World Systems* (1632), Galileo defied the pope by making his views even more explicit. As a result, he was brought before the Inquisition and forced to recant.

Galileo SPACE PROBE to Jupiter, launched in October 1989. Galileo flew past Venus once and the Earth twice. The probe passed the asteroids Gaspra and Ida in October 1991 and August 1993 respectively, photographing both. Galileo went into orbit around Jupiter in 1995, after dropping off a sub-probe to enter the planet's atmosphere. *See also* SPACE EXPLORATION

gall Abnormal swelling or protuberance of plant tissue stimulated by an invasion of any of a wide variety of parasitic or symbiotic organisms, including bacteria, fungi, insects, and nematodes. Most gall organisms stunt but do not kill the affected plants. The gall wasp larva makes a spongy "apple" gall on oak trees.

gall bladder Muscular sac, found in most VERTE-BRATES, which stores BILE. In humans it lies beneath the right lobe of the LIVER and releases bile into the DUODENUM by way of the bile duct. Its release is signaled by a hormone that is secreted when food is present in the duodenum.

Galle, Johann Gottfried (1812–1910) German astronomer, the first to observe and identify Neptune as a planet (1846), although its presence had been postulated by other scientists. Galle also suggested a system for determining the scale of the SOLAR SYSTEM based on the observation of asteroids.

gallium (symbol Ga) Gray, metallic element of group III of the periodic table. It was predicted by Dmitri MENDELEYEV and discovered in 1875. Chief sources are bauxite and some zinc ores. The metal, which melts at low temperature, is used in LASERS, TRANSISTOR SEMICONDUCTORS, and high-temperature thermometers. Properties: at.no. 31; r.a.m. 69.72; r.d. 5.9; m.p. 29.78°C (85.60°F); b.p. 2,403°C (4,357°F); most common isotope ^{69}Ga (60.4%).

Gallo, Robert Charles (1937–) US scientist who identified the VIRUS that causes AIDS. Now called the HUMAN IMMUNODEFICIENCY VIRUS (HIV), the organism was discovered by Gallo in 1984, a year later than its independent discovery at the Pasteur Institute.

▲ **galvanometer** The galvanometer is an instrument for measuring extremely small currents. A coil is wound around an iron core (1) suspended between the poles of a horseshoe magnet (2).

When a current flows in the coil, it induces a magnetic field in the core which causes it to move (3) in the field of the horseshoe magnet registering as a deflection of the attached pointer (4).

gallstone (cholelithiasis) Hard mass, usually composed of bile pigments, cholesterol, and calcium salts, that forms in the GALL BLADDER. Gallstones may be present for years and give no trouble. However, they may cause severe pain, or they may become lodged in the bile duct, causing obstructive JAUNDICE or cholecystitis (inflammation). Treatment is by removal of the stones or of the gall bladder.

Galois, Évariste (1811–32) French mathematician who laid the foundation of GROUP THEORY. In 1830 he submitted several papers, some of which were lost, and one, on algebraic equations, that was dismissed by the mathematical establishment as incomprehensible and thus ignored. During the turmoil that followed the French Revolution, Galois was arrested twice. After being released from jail the second time, he was challenged to a duel by a political opponent. Reputedly, the night before the fatal duel, Galois tried to note down as many of his thoughts as possible. These notes and a few other papers were discovered 14 years later by Joseph Liouville, who recognized them as works of genius. Galois set out the theory of groups and laid down conditions for the solvability of various algebraic equations.

Galton, Sir Francis (1822–1911) British scientist and scholar. He pioneered the study of human differences and statistical analysis of data. In *Hereditary Genius* (1869) he described his theories of HEREDITY and advocated EUGENICS.

Galvani, Luigi (1737–98) Italian physician and physicist who became a lecturer in anatomy at Bologna and did pioneer studies in electrophysiology, or "animal electricity." His experiments with frogs' legs indicated a connection between muscular contraction and electricity. He believed a new type of electricity was created in the muscle and nerve, but the correct explanation was given by Alessandro VOLTA. Galvani's name was given to the galvanometer and to several processes involving electricity.

galvanizing Coating of iron or steel articles with zinc, either applied directly in a bath of molten zinc, electroplated from cold zinc sulfate solutions, or dusted on and baked. Corrugated sheets, nails, pails and wire netting may be protected from the corrosive effects of rusting by galvanizing. The metal to be protected is first derusted but if, after galvanizing, the zinc coating is pierced, protection is still provided because of the preferential attack by corrosion on zinc rather than on iron. *See also* ELECTROPLATING

galvanometer Instrument for detecting, comparing or measuring ELECTRIC CURRENTS. It is based on the principle that a current through a CONDUCTOR (usually a coil) creates a MAGNETIC FIELD that reacts with a magnet giving a deflection. The main types are the **moving-coil** galvanometer (the most commonly used) and the **moving-magnet instrument**.

gamete Reproductive (sex) CELL that, during FERTILIZATION, fuses with another reproductive cell of the opposite sex to form a ZYGOTE, the start of a new individual. Female gametes (*see* OVUM) are usually motionless; in many organisms, the male gametes (in this example SPERM) have a tail (FLAGELLUM) that enables them to swim to the ova. All gametes are HAPLOID – they contain only a single set of CHROMOSOMES due to MEIOSIS. On fusion, the resulting zygote is DIPLOID. *See also* SEXUAL REPRODUCTION

game theory In mathematics, analysis of problems involving conflict. Its application includes problems in business management, sociology, economics, and military strategy as well as "true" games such as poker and chess. The theory was first introduced by Émile Borel and developed by John VON NEUMANN in 1928.

gametophyte Generation of PLANTS and ALGAE that bears the female and male GAMETES (reproductive cells). In flowering plants (ANGIOSPERMS) these are the germinated pollen grains (male) and the embryo sac (female) inside the OVULE. *See also* ALTERNATION OF GENERATIONS; FERN; MOSS

gamma globulin One of the protein components of the serum of mammalian blood. It contains approximately 85% of the circulating ANTIBODIES of the BLOOD. It gives temporary IMMUNITY in patients who have been exposed to certain diseases, such as measles.

gamma radiation (gamma rays) Form of very short wavelength ELECTROMAGNETIC RADIATION emitted from the nuclei of some RADIOACTIVE atoms. High-energy gamma rays have even greater powers of penetration than X RAYS, so that sources of this radiation must be stored inside thick lead containers. These sources are used in medicine to treat cancer and in the food industry to kill microorganisms. Gamma rays can be detected with a GEIGER COUNTER. *See also* FOOD PRESERVATION; RADIOACTIVITY; RADIOTHERAPY

gamma-ray astronomy *See* ASTRONOMY

gamma-ray burst Intense short burst of GAMMA RADIATION, first announced in 1971 although satellites of the US Air Force had discovered them six years earlier. Lasting from a few milliseconds to a few tenths of a second, and with intensities of up to 1 million ELECTRON VOLTS, they occur uniformly across the sky at a frequency of about one per day. In 1997 the Bepo SAX satellite's X-ray camera discovered the position of a gamma-ray burster which was later identified optically by the Keck Observa-

tory. Their origin is unknown, although the merging of two NEUTRON STARS is a possible source.

Gamow, George (1904–68) US nuclear physicist, cosmologist, and author, b. Russia. His most famous work, co-authored with Ralph Alpher (1921–) and Hans BETHE, pictured a hot and dense early UNIVERSE in which the nuclei of elements were built up from protons; this was a cornerstone of the BIG BANG theory. His greatest prediction, borne out by the detection of the COSMIC MICROWAVE BACKGROUND RADIATION, was that the Universe should by now have cooled to a few degrees above ABSOLUTE ZERO. He also contributed to the deciphering of the genetic code, developed the QUANTUM THEORY of radioactivity, and proposed the liquid-drop model of atomic nuclei. With Edward TELLER, he put forward the **Gamow-Teller theory** of beta decay and a theory of the internal structure of RED GIANT stars.

ganglion Cluster of nervous tissue containing numerous cell bodies and SYNAPSES, usually enclosed in a fibrous sheath of MYELIN. In a VERTEBRATE, most ganglia occur outside the CENTRAL NERVOUS SYSTEM. *See also* NEURON

gangue Waste material in an ORE or deposit, as opposed to metal-containing or other useful minerals.

Ganymede Largest of Jupiter's GALILEAN SATELLITES, with a diameter of 5,262km (3,270mi), and the largest satellite in the SOLAR SYSTEM. Ganymede contains a high proportion of water-ice. Its surface is a mixture of dark, heavily cratered terrain and a brighter, less cratered terrain covered with meandering, intersecting grooves that suggest recent geological activity. The most prominent feature is the dark Galileo Regio, some 4,000km (2,500mi) across.

garnet Two series of common orthosilicate minerals: the **pyralspite series** and the **ugradite series**. Garnet is found in METAMORPHIC ROCKS and PEGMATITES as cubic system crystals, rounded grains, and granular masses. It is brittle, glassy and occurs in many hues; some varieties (particularly the red form) are important as gemstones. Hardness 6.5–7.5; r.d. 4.

garnierite Ore of nickel. It is a lustrous green SERPENTINE, also containing magnesium and silicon. It occurs in ultrabasic rocks as a decomposition product of OLIVINE.

gas State (PHASE) of MATTER in which molecules are free to move in any direction; a gas always spreads (by DIFFUSION) to fill a container of any size. Because of their low densities, most gases are poor CONDUCTORS of heat and electricity (although at high voltages a gas may be ionized and become electrically conducting). When cooled, gases become liquids. Some, such as carbon dioxide, can be liquefied by pressure alone. Others require cooling below their critical temperature before they can be liquefied (*see* CRITICAL POINT). All gases follow certain laws, such as AVOGADRO'S law, BOYLE'S LAW, CHARLES' LAW, GRAHAM'S LAW, and the IDEAL GAS LAW. *See also* LIQUID; PLASMA; SOLID

gas, natural *See* NATURAL GAS

gas analysis Determination of the nature and quantity of constituents of gas mixtures. Three methods of analysis are available: 1) the gases can be selectively absorbed in chemical solutions and the decrease in volume measured; 2) in GAS CHROMATOGRAPHY the rates of diffusion of gases through such media as charcoal can be utilized; 3) in the instrumental method such physical properties of the gases as density can be used as the basis of analysis.

gas chromatography Type of CHROMATOGRAPHY in which the carrier medium is a gas. The technique is used for analyzing or separating a mixture

of gases, often derived from a heated liquid mixture. A "carrier" gas (the **mobile phase**) – often hydrogen – sweeps the mixture around a very long coiled tube containing the **stationary** phase, which differentially absorbs the components of the mixture. In **gas-liquid** chromatography (GLC), the stationary phase is a liquid such as a nonvolatile oil coating a solid support; in **gas-solid** chromatography (GSC), it is a solid such as diatomite (kieselguhr). The components of the mixture are detected as they leave the apparatus in turn or, if the chromatograph has been calibrated with known substances, they are identified by the time they take to pass through. This method of chromatography is used to analyze tiny amounts of a sample.

gas constant (symbol R) Constant in the universal gas equation (*see* IDEAL GAS LAW). R, also called the universal molar gas constant, has a value of $8.314510 \, \text{JK}^{-1} \, \text{mol}^{-1}$.

gas exchange In biology, the uptake and output of gases, especially oxygen and carbon dioxide, by living organisms. In animals and other organisms that

▶ **gas** The kinetic theory of gases explains the behavior of a gas by considering it as consisting of very small, perfectly elastic, particles (molecules) which collide with one another (A) and with the walls of the container (B). All will have different velocities but useful statistical statements can be made about the gas as a whole. The pressure – the force exerted on a unit area of the walls due to the particles hitting it – clearly depends on the number of particles per unit volume and on their velocities and mass. In the simplest case pressure is proportional to the average kinetic energy and thus to the average of the squares of the speeds. The temperature is also proportional to the average kinetic energy and thus also to the pressure. The graph (C) shows the statistics of velocity for two different temperatures (C_1 and C_2 are average speeds). When the gas is a mixture, both constituents adopt the same average kinetic energy; the lighter of the two kinds of particles moving faster. One result (D_1) is that, if there is a small hole in the container, the lighter gas escapes faster than the heavier one (in proportion to the square root of the particle mass – Graham's law of diffusion) and the mixture is partially separated (D_2).

obtain their energy by the breakdown of food in the chemical reactions of AEROBIC RESPIRATION, gas exchange involves the uptake of oxygen and the output of carbon dioxide. In plants, algae and photosynthetic bacteria, which carry out PHOTOSYNTHESIS, the opposite may occur when photosynthesis is active, with a net carbon dioxide uptake and net oxygen output. At the cellular level, gas exchange takes place by DIFFUSION across cell MEMBRANES in solution. In multicellular animals there may be a specialized transport system for transporting these gases between the cells and the outside environment. By using up one or other of these gases and generating the other, respiration and photosynthesis maintain their own concentration gradients for diffusion. Where transport systems are involved, the gradient may be enhanced by other means. For example, in vertebrate ERYTHROCYTES (red blood cells), the affinity of the pigment HEMOGLOBIN for oxygen changes, as the local oxygen concentration varies, enabling it to take up oxygen in the lungs and release it to respiring cells. *See also* BREATHING; CIRCULATORY SYSTEM; RESPIRATORY SYSTEM; VENTILATION

gas law *See* IDEAL GAS LAW

gasohol Gasoline to which alcohol has been added. A typical gasohol composition is 90% unleaded gasoline and 10% ethanol (grain alcohol). This gasohol can be used in cars with CATALYTIC CONVERTERS and has about the same OCTANE NUMBER as unleaded gasoline. The chief advantage of gasohol is that the ethanol can be produced from renewable biomass (organic matter), a relatively cheap resource. Gasoline, on the other hand, is produced from expensive petroleum.

gasoline (gas) Major HYDROCARBON fuel, a mixture consisting mainly of hexane, octane, and HEPTANE. One of the products of OIL refining, gasoline is extracted from crude oil (PETROLEUM). Frequently other fuels and substances are added to gasoline to alter its properties. Most modern cars have high-compression INTERNAL COMBUSTION ENGINES, and the mixture of air and gasoline vapor supplied to it can explode too quickly, pushing against a rising instead of a descending piston. This effect is known as engine KNOCK, and to eliminate it many manufacturers add LEAD(IV) tetraethyl (an antiknock agent) to slow the rate of combustion. However, this additive has been found to be implicated in atmospheric pollution and causes brain damage, and so many manufacturers omit it and instead improve gasoline's OCTANE NUMBER – a measure of its antiknock properties – by altering the mixture of hydrocarbons in it. *See also* OIL REFINERY

gasoline engine Most common type of INTERNAL COMBUSTION ENGINE.

gastric juice Fluid comprising a mixture of substances, including PEPSIN and HYDROCHLORIC ACID, secreted by GLANDS of the stomach. Its principal function is to break down proteins into POLYPEPTIDES during DIGESTION.

gastropod Member of the class Gastropoda, a class of mollusks, which includes the snail, slug, whelk, limpet, abalone, and sea slug. Many possess a single spiral shell, produced by chemical precipitation from the MANTLE. Many types of gastropod live immersed in seawater, breathing through gills. Some freshwater snails, however, breathe through lungs and need to surface periodically for air. Like slugs, sea slugs are entirely without shells and are often brilliantly colored.

gastrula Early stage in the development of the EMBRYO in animals; it follows the BLASTULA stage. The gastrula is a hollow sac with two layers (*see* GERM LAYER) of cells, the inner ENDODERM and the outer ECTODERM. The sac cavity is called the **archenteron** and its mouth the **blastopore**.

gas turbine *See* TURBINE

gate In ELECTRONICS and FLUIDICS, circuit having two or more inputs but a single output energized only under specific input conditions. Gates are used extensively in LOGIC CIRCUITS in which the sequence of operations, or their combinations, are determined by the rules of symbolic logic.

Gates, Bill (William Henry) (1955–) US businessman. In 1975 he co-founded Microsoft Corporation, which in the 1980s became the predominant computer software producer. He is noted for his innovative thinking and for his aggressive marketing and business tactics. In 1999 Microsoft was found guilty of using monopolistic practices regarding software it has produced for the Internet.

gauge Term for many kinds of measuring device. For example, calipers are width gauges, dipsticks are depth gauges and thermometers are temperature gauges, although here the term is restricted usually to mechanical thermometers. Pressure gauges are of various kinds, the Bourdon gauge being the type used to measure fluctuations of air pressure in aneroid BAROMETERS and variations in gas and liquid pressure in industry. The term is also used of the perpendicular distance between railway lines, from the largest actual gauge (the "5-foot gauge") to miniature and model.

gauge boson Type of ELEMENTARY PARTICLE that mediates interactions between other particles. The four gauge bosons are: PHOTONS, for electromagnetic interactions in QUANTUM ELECTRODYNAMICS;

GLUONS, for strong interactions in QUANTUM CHROMODYNAMICS; intermediate vector BOSONS designated W and Z, for WEAK NUCLEAR FORCE interactions; and GRAVITONS, for gravitational interactions.

Gauss, Karl Friedrich (1777–1855) German mathematician. While still a teenager, Gauss made many discoveries, including the idea of non-EUCLIDEAN GEOMETRY (thirty or so years before Nikolai LOBACHEVSKY). He discovered the fundamental theorem in complex analysis a decade before Baron CAUCHY and discovered QUATERNIONS before William HAMILTON. Gauss chose not to publish these ideas, refusing to publish his work unless he was satisfied that it was perfect. He made contributions in many areas of mathematics, including algebra, number theory, probability theory, analysis, and the theory of series. He also studied electricity and magnetism, gravitation, optics, and astronomy. The unit of magnetic flux density is named after him.

Gauss' law Total ELECTRIC FLUX over a closed surface equals the ELECTRIC CHARGE within that surface divided by the PERMITTIVITY of the medium. The law, stated by Karl GAUSS, also applies to surfaces in a magnetic field and similar statements can be made for a gravitational field.

Gay-Lussac, Joseph Louis (1778–1850) French chemist and physicist who did pioneer research into the behavior of gases. He discovered the law of combining gas volumes (GAY-LUSSAC'S LAW) and the law of gas expansion, often also attributed to Jacques CHARLES (who discovered it earlier but did not publish his results – CHARLES' LAW). Gay-Lussac prepared (with Louis Jacques THÉNARD) the elements POTASSIUM and BORON, investigated fermentation and hydrocyanic (prussic) acid, and invented a HYDROMETER.

Gay-Lussac's law When gases react, their volumes are in a simple ratio to each other and to the volume of products, at the same temperature and pressure. It was named after Joseph GAY-LUSSAC.

gear Wheel, usually toothed, attached to a rotating shaft. The teeth of one gear engage those of another to transmit and modify rotary motion and TORQUE. The smaller member of a pair of gears is called a pinion. If the pinion is on the driving shaft, speed is reduced and turning force increased. If the larger gear is on the driving shaft, speed is increased and turning force reduced. A screw-type driving gear, called a **worm**, gives the driven gear a greatly reduced speed. Smooth wheels can be used as gears, but may slip against each other.

Geber (c.720–815) □(Jabir or Jabir ibn-Hayyan) Arab alchemist whose works were studied by the English philosopher Roger Bacon (c.1214–92). The identification of Geber with Jabir is uncertain, but Geber is known to have performed scientific experiments as well as attempting to convert mercury into gold. He felt that a TRANSMUTATION of baser metals into nobler ones was possible. He prepared nitric acid and distilled vinegar (to obtain acetic acid).

gegenschein (counterglow) Faint luminous patch, visible at a point along the ecliptic, that is diametrically opposite to the Sun. Best seen in tropical regions, it is a phenomenon allied to the zodiacal light, and often appears as an extension of it.

Geiger, Hans Wilhelm (1882–1945) German physicist who, with Ernest RUTHERFORD, devised in 1908 a method of detecting and counting alpha particles. This device, with modifications, became the GEIGER COUNTER. Geiger and Ernest Marsden measured the deflection of alpha particles by thin metal foil in 1909, and this phenomenon was the basis of Rutherford's discovery of the atomic NUCLEUS.

Geiger counter (Geiger-Müller counter) Instrument used to detect and measure the strength of radiation by counting the numbers of ionized particles produced. It is a type of IONIZATION CHAMBER in which a high voltage is applied across a pair of electrodes. Radiation and particles entering the chamber ionize gas atoms and the resulting ions, gaining energy from the electric field between the electrodes, magnify the effect by producing many more ions. These make a current that quickly subsides, so a pulse of current is produced for each particle. Each pulse activates a counting circuit enabling up to about 10,000 particles per second to be counted.

Geissler tube Early form of DISCHARGE TUBE.

gel Homogeneous COLLOID consisting of minute particles dispersed or arranged in a liquid to form a fine network throughout the mass. Their appearance can be elastic or jellylike, as in GELATIN or fruit jelly, or quite rigid and solid, as in silica gel, a material resembling coarse white sand used as a dehumidifier and in the manufacture of silicone and rubber.

gelatin Colorless or yellowish protein obtained from COLLAGEN (a fibrous protein) in animal cartilages and bones by boiling in water. It is used in photographic film emulsions, capsules for medicines, as a culture medium for bacteria, and in foods such as jellies.

Gell-Mann, Murray (1929–) US theoretical physicist who was awarded the 1969 Nobel Prize for physics for his application of group theory to ELEMENTARY PARTICLES, which led to the prediction of the QUARK as the basic constituent of the BARYON and MESON. His theory also predicted the existence of a baryon called the **omega-minus particle**, subsequently discovered in 1964. Gell-Mann had also introduced in 1953 the concept of "strangeness" to explain the longevity of certain quarks.

gem Any of about 100 minerals, either opaque, transparent or translucent, valued for their beauty, rarity, and durability. Transparent stones, such as DIAMOND, RUBY, EMERALD, and SAPPHIRE, are the most highly valued. PEARL, AMBER and CORAL are gems of organic origin. During the Middle Ages, many gems were thought to have magical powers. The art of cutting and polishing gemstones to bring out their color and brilliance was developed in the 15th century, particularly in Italy. Stones with a design cut into them are **intaglios**; with a design in relief, **cameos**. The study of gems is known as gemmology. *See also* SEMIPRECIOUS STONES

Gemini (Twins) Large constellation of the N sky, lying between TAURUS and CANCER. It contains the star cluster M 35 (NGC 2168) and the planetary nebula NGC 2392. Its brightest stars are Castor (Alpha Geminorum, magnitude 1.57) and Pollux (Beta Geminorum, magnitude 1.14).

Gemini project Series of US manned experimental space flights that followed the Mercury and preceded the APOLLO PROJECT. Gemini 3 was a two-person craft first crewed by Virgil Grissom and John Young on March 23, 1965; they made three Earth orbits. Subsequent flights and experiments, ending with Gemini 12 (November 11–15, 1966), demonstrated space walks, space docking maneuvers, and tests of longevity including a 14 day flight by Frank Borman and James LOVELL, Jr, in December 1965.

gemma In botany and zoology, a bud that will give rise to a new individual. The term also refers to a multicellular reproductive structure found in algae, liverworts, and mosses.

gene Unit by which hereditary characteristics or traits are passed on from one generation to another in plants and animals. A gene is a length of DNA that codes (*see* GENETIC CODE) for a particular protein or peptide. Genes are found along the CHROMOSOMES of the cells of plants, animals, fungi, and protoctists (in bacteria the DNA is not contained in chromosomes). In most cell nuclei genes occur in pairs, one located on each of a chromosome pair. In cases where different forms of a gene (ALLELES) are present in a population, some forms may be RECESSIVE to others (DOMINANT genes) and will not be expressed unless present on both members of a chromosome pair. *See also* GENETICS; GENETIC ENGINEERING; HEREDITY

gene bank Collection of genetic material kept for possible future use. Material stored includes cultures of bacteria and molds; seeds, spores, and tubers; frozen sperm, eggs, and embryos; and even live plants and animals. The material can be used in plant and animal BREEDING, GENETIC ENGINEERING, and in medicine. Live species are used for restocking natural habitats in which species are extinct or in danger of EXTINCTION.

gene pool All of the GENES, and their various ALLELES, that occur in all the members of a particular species or population of species at any given moment. For a species to be vigorous and survive, the gene pool must be large enough for natural variation to take place through CROSS-FERTILIZATION. A large population does not necessarily imply a large gene pool, however. For example, there is concern about the viability of some sizeable populations of cheetahs that, despite their numbers, have insufficient variation in the gene pool.

general theory of relativity *See* RELATIVITY

generator Device for producing electrical energy. The most common is a machine that converts the mechanical energy of a TURBINE or INTERNAL COMBUSTION ENGINE into electricity by employing ELECTROMAGNETIC INDUCTION. There are two types of generators: alternating current (AC) and direct current (DC), often called an ALTERNATOR and a DYNAMO. Each has an ARMATURE (or ring) that rotates within a magnetic field, creating an induced current. The direct current generator has a COMMUTATOR on the armature that divides the induced current as it changes direction, giving rise to a pulsating direct current. *See also* ELECTRIC CURRENT

gene therapy (gene replacement therapy, GRT) Treatment of disease by giving patients, healthy GENES in place of missing or defective ones. The first human being to undergo gene therapy, in the USA in 1990, was a four-year-old suffering from a rare enzyme deficiency that destroys a person's natural immunity. Capitalizing on techniques of GENETIC ENGINEERING, the gene therapist repairs or replaces a faulty gene. The healthy gene is packaged into some kind of vector (usually a suitably doctored virus) so that it can be targeted at affected cells. Originally gene therapy was intended to fight hereditary disorders, such as cystic fibrosis or sickle cell disease, but researchers are also seeking to direct it against other disorders, such as cancer, where a gene has become faulty over a period of time. Although thousands of patients have been treated, mostly in the USA, gene therapy has yet to live up to its early promise. Few of the inserted genes reach their target cells and those that do may not always work efficiently. Also, there are problems with the use of viruses as gene shuttles.

The body treats the virus as "foreign" and some patients have suffered severe immune reactions. There is, too, a theoretical risk that the virus itself could spread and cause cancer.

▲ genetic fingerprinting Using a technique known as genetic fingerprinting, a person can be accurately identified. The process allows a person's DNA to be represented in visual form (1). Each DNA pattern is unique (like a fingerprint) – with the exception of identical twins. In a case of disputed paternity, DNA fingerprinting allows the relationship to be settled beyond doubt. DNA is present in all cells, so a sample can be taken from blood (2), skin, or even sweat. DNA is separated out (3), and an enzyme that divides DNA is added. The enzyme attacks the mini-satellite region between the genes (4). The genes are then sorted by size by an electric field (5). Gel electrophoresis exploits the fact that small sections of DNA carry a charge to force them through a gel. The size of the sections controls how far they travel, giving a pattern unique to each individual. A child combines DNA from both parents, so will have a partially similar pattern. Paternity is confirmed by the matching marks (6).

genetic code Arrangement of information stored in GENES. It is the ultimate basis of HEREDITY and forms a blueprint for the entire organism. The genetic code is based on the genes that are present, which in molecular terms depends on the arrangement of nucleotides in the long molecules of DNA in the cell CHROMOSOMES. Each group of three nucleotides specifies, or codes, for an AMINO ACID or for an action such as "start" or "stop". By specifying which PROTEINS to make and in what quantities, the genetic code not only **directly** controls production of structural materials, but also, by coding for ENZYMES, **indirectly** codes for the production of other cell materials.

genetic drift Chance alteration in the GENE POOL of a small, usually isolated, population – the smaller and more isolated the population, the greater the chance of genetic drift. Genetic drift may greatly affect the genetic characteristics of the population. For example a specific gene may either randomly disappear or become widespread depending on the specific genetic traits of individuals and the reproduction habits of the population.

genetic engineering Construction of a DNA (deoxyribonucleic acid) molecule containing a desired GENE, which is then introduced into a bacterial, fungal (yeast), plant or mammalian cell, so that this cell then produces the desired protein. The technique has been used successfully to alter bacterial genetic material to produce substances such as human growth hormone, insulin and enzymes for biological washing powder. Many people are concerned about the ethics of genetic engineering, given the possibility that scientists may one day be able to alter the genetic structure of humans, or accidentally release genetically engineered, disease-causing bacteria into the environment.

genetic fingerprinting Forensic technique pioneered by Alec JEFFREYS using genetic material, specifically the DNA within sample body cells, to identify individuals. It is used mainly in paternity suits to detect the true father of a child, and sometimes in criminal cases to establish identity from traces of the assailant's semen, hair, or skin. The chances of two individuals who are not blood relations having genetic "fingerprints" that could be confused are estimated to be at least a billion to one. The technique's first use in a court of law was in the late 1980s.

genetics Study of HEREDITY. Geneticists study how the characteristics of an individual organism depend on its GENES, how the characteristics are passed down to the next generation, and how changes may occur through MUTATION. A person's behavior, learning ability, and physiology may be explained partly by genetics, although the environment in which an individual grows up may have a considerable influence, too. The deliberate modification of GENETIC CODES by scientists is called genetic manipulation, or GENETIC ENGINEERING. Scientists can modify the genes of livestock and crops in order to improve the quality or quantity of the produce. Genetics was founded by Gregor MENDEL, following experiments with successive generations of peas.

genome Entire complement of genetic material carried within the CHROMOSOMES of a single cell. In effect, a genome carries all the genetic information about an individual coded in sequence by the DNA that makes up the chromosomes. It has also been applied to the whole range of GENES in a particular species. *See also* GENETICS; HUMAN GENOME PROJECT

genotype Genetic makeup of an individual. The particular set of GENES present in each cell of an organism is distinct from the PHENOTYPE, the observable characteristics of the organism.

genus Part of the CLASSIFICATION of living organisms, ranking below FAMILY and above SPECIES. A genus is a group of closely related biological species with common characteristics. Genus names are printed in *italic* with a capital initial letter; for example the domestic cat belongs to the genus *Felis*.

geochemistry Study of the chemical composition of the Earth, particularly the abundance and distribution of the chemical elements and their isotopes, and the changes that have resulted in it from chemical and physical processes.

geochronology Dating of rocks or of Earth processes. Absolute dating techniques (RADIOMETRIC DATING) involve the measurement of radioactive decay to determine the actual date in years for a given rock. Relative dating involves the use of fossils, sediments or relationships between structures to place rock sequences and geological events in order. *See also* CARBON DATING

geode Hollow rock nodule with inner walls lined with crystals, generally quartz or calcite. It is formed by gelatinous silica and mineral-bearing water within a cavity.

geodesic dome Architectural structure of plastic, metal or even cardboard, based in shape upon triangular or polygonal facets, that evenly distribute the tension. They were originated (1947) and perfected by R. Buckminster Fuller. A well-known example was the US Pavilion at Montreal's Expo 67. *See also* BUCKMINSTERFULLERENE

geodesic surveying Method of surveying that covers areas large enough to involve consideration of the Earth's curvature. Geodesic surveying is carried out in order to establish geographical features, such as national boundaries, and as a basis for the mapping of whole states or countries. A MAP is made using various systems for representing the Earth's curved surface as a flat drawing.

geodesy In geophysics, the determination of the size and shape of the Earth, its gravitational field, and the location of fixed points. *See also* GEODESIC SURVEYING

geography Science that studies the spatial relationships between the surface of the Earth and humankind. It includes the size and distribution of land masses, seas, resources, climatic zones, and plant and animal life. Because it seeks to relate all the Earth's features to human existence, geography differs from other Earth sciences such as GEOLOGY, METEOROLOGY, and OCEANOGRAPHY, which study this planet's features as specific phenomena.

geological time Timescale of the history of the Earth, divided into periods of time millions of

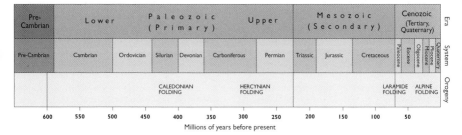

▲ **geological time** The 4.6 billion years since the formation of the Earth are divided into four eras that are further split into periods and, in the case of the most recent era, epochs. The present era is the Cenozoic ("new life"), extending backwards through "middle life" and "ancient life" to the Pre-Cambrian. Although traces of ancient life have since been found, it was largely the proliferation of fossils from the beginning of the Paleozoic era onwards that first allowed precise subdivisions to be made.

231

years long. Until recently, only methods of relative dating were possible. These include the worldwide study, comparison and correlation of sequences of rock formations and the fossils they contain. The information thus gathered is used to distinguish earlier from later deposits, to estimate periods of passed time, and to reconstruct geologic and climatic events by assuming that geological processes in the past were the same as today. Geological time is divided into four eras: PRE-CAMBRIAN, PALEOZOIC, MESOZOIC, and CENOZOIC, which are further divided into periods. Periods, in turn, are subdivided into series or epochs. Epochs are subdivided into stages and then zones.

geology Study of the materials of the Earth, their origin, arrangement, classification, change, and history. Geology is divided into several categories, the major ones being MINERALOGY (arrangement of minerals), PETROLOGY (rocks and their combination of minerals), STRATIGRAPHY (arrangement and succession of rocks in layers), PALEONTOLOGY (study of fossilized plant and animal remains), GEOMORPHOLOGY (study of landforms), STRUCTURAL GEOLOGY (classification of rocks and the forces that produced them), and **environmental geology** (geological study applied to the best use of the environment by humans).

geomagnetism Physical properties of the Earth's MAGNETIC FIELD. Geomagnetism is thought to be caused by the metallic composition of the Earth's CORE. The gradual movements of magnetic north result from currents within the MANTLE.

geometric isomer *See* ISOMER

geometric mean Geometric mean of n numbers is the nth root of their product. An example is the square root of the product of two numbers: the geometric mean of 8 and 2 is $\sqrt{(8 \times 2)} = 4$. The geometric mean of 5, 8, and 25 is $\sqrt[3]{(5 \times 8 \times 25)} = 10$.

geometric progression Sequence of numbers in which the ratio of each term to the preceding one is a constant (called the common ratio). It has the form $a, ar, ar^2, ar^3, ...$, where r is the common ratio. The sum of these terms, $a + ar + ar^2+...$, is a **geometric series**. If there are n terms, the sum equals $a(1 - r^n)/(1 - r)$. Infinite geometric series converge to $a/(1 - r)$ if r lies between -1 and $+1$.

geometry Branch of mathematics concerned broadly with studies of shape and size. To most people, geometry is the EUCLIDEAN GEOMETRY of simple plane and solid figures. It is, however, a much wider and more abstract field with many subdivisions. COORDINATE GEOMETRY (also called **analytical geometry**), introduced in 1637 by René

DESCARTES, applies algebra to geometry and allows the study of more complex curves than those of Euclidean geometry. PROJECTIVE GEOMETRY is, in its simplest form, concerned with projection of shapes – operations of the type involved in drawing maps and in perspective painting – and with properties that are independent of such changes. It was introduced by Jean Victor Poncelet (1788–1867) in 1822. Even more abstraction occurred in the early 19th century with formulations of NON-EUCLIDEAN GEOMETRY by Janos Bolyai (1802–60) and N.I. LOBACHEVSKY. DIFFERENTIAL GEOMETRY – based on the application of CALCULUS and ideas such as the curvature and lengths of curves in space – was also developed during this period. Geometric reasoning was also applied to spaces with more than three dimensions. TOPOLOGY is a general form of geometry, concerned with properties that are independent of any continuous deformation of shape.

geomorphology Scientific study of features of the Earth's surface and the processes that have formed them.

geophone Seismic detection device placed on or in the ground to measure the echoes produced by volcanic or seismic (earthquake) activity.

geophysics Study of the characteristic physical properties of the Earth as a whole system. It uses parts of CHEMISTRY, GEOLOGY, ASTRONOMY, SEISMOLOGY, METEOROLOGY, and many other disciplines. From the study of seismic waves, geophysicists have worked out the structure of the Earth's interior.

geostationary orbit Location of an artificial SATELLITE so that it remains above the same point on a planet's surface because it completes one orbit in the same time as it takes the planet to rotate once on its axis. Communications and REMOTE-SENSING satellites are often placed in geostationary orbits above the Earth. *See also* SYNCHRONOUS ORBIT

geostrophic current General term for an ocean or wind current whose direction is affected by the meeting of its pressure gradient with the CORIOLIS EFFECT.

geosyncline Great basin or trough in which deposits of sediments and volcanic rock, thousands of meters thick, have accumulated during slow subsidence over long geological periods. *See also* SYNCLINE

geothermal Referring to heat from the interior of the Earth. Geothermal activity is responsible for many geological phenomena, including hot springs, GEYSERS, and VOLCANOES. Some of this natural heat is being harnessed and used in the form of GEOTHERMAL ENERGY.

geothermal energy Heat contained in the Earth's rocks. It is produced by RADIOACTIVITY and by the movement of tectonic plates (*see* PLATE TECTONICS). It is released naturally in GEYSERS and VOLCANOES, and is used as a power source for generating electricity in several places, including the United States, Iceland, Italy, and New Zealand.

geotropism In plant growth, the response to the stimulus of gravity. Plant stems are generally negatively geotropic and grow upward; roots are positively geotropic and grow downward. For example, growth curvature is caused by the accumulation of the plant hormone AUXIN in the tissue on the lower side of the stem; growth increases on that side and the stem bends upward.

Gerhardt, Charles Frédéric (1816–1856) French chemist best known for his attempt to systematize organic chemistry under the theory of types. This based all compounds on four main types and was influential in the development of structural chemistry. He published this and other ideas in *Traité de Chimie Organique* (4 vols, 1853–56).

germ Popular term for any infectious agent. Germs can be pathogenic bacteria, fungi or viruses. The word "germ" is used in biology to denote a rudimentary stage in plant growth, such as an embryo in a seed, or a bud.

germanium (symbol Ge) Gray-white, metalloid element of group IV of the periodic table, predicted by the Russian chemist Dmitri MENDELEYEV and discovered in 1886. Germanium is a by-product of smelting zinc ores or may be obtained from the combustion of certain coals. It is important in transistors, rectifiers, and similar semiconductor devices. Properties: at.no. 32; r.a.m. 72.59; r.d. 5.35; m.p. 937°C; (1,719°F); b.p. 2,830°C (5,126°F); most common isotope ^{74}Ge (36.54%).

germicide Any substance for destroying disease-causing microorganisms. Germicides include ANTISEPTICS, disinfectants and ANTIBIOTICS.

germination Growth of the embryo in a SEED to produce a new plant following FERTILIZATION. It may occur immediately or after a period of DORMANCY. In order to germinate, a seed or spore needs favorable conditions of temperature, light, moisture, and oxygen. The process begins after the seed has taken up water. Germination is complete when a root appear outside the seed coat.

germ layer Any one of three layers of cells found in a developing EMBRYO. The three layers are the ECTODERM, MESODERM, and ENDODERM, each one giving rise to particular tissues and organs.

gerontology Study of aging. The capacity of an organism to repair tissues effectively, by the replacement of worn-out cells, diminishes with age, so that "growing old" is inevitable (even when, as in many large trees, it is delayed for some thousands of years). The effect of certain types of medicine has been to relieve and treat the

electricity generating station

cold water pumped down to cavity

steam is forced up second bore-hole to the surface

water is turned to steam by natural heat of the rocks

bore-hole

natural cavity 300°C (570°F)

▲ **geothermal energy** An artificial hot spring can be used as a source of heat energy. A bore-hole is drilled several hundred meters into a natural cavity in the Earth, in which the temperature may be as high as 300°C (570°F). Water pumped down the bore is heated by the natural heat of the rocks, turns to steam and is forced up a second bore-hole. At the surface the steam drives turbines to produce electricity.

disease burdens of age. In itself, it has not increased life expectancy.

gestation Carrying a developing EMBRYO in a UTERUS between conception and birth. Gestation periods are specific to each mammalian species and range from 12 days (Virginia opossum) to 22 months (Indian elephant). In humans and many apes the period is approximately 38 weeks.

geyser Hot spring that erupts intermittently, throwing up jets of superheated water and steam to a height of about 60m (200ft), followed by a thunderous roar. Geysers occur in the United States – best known is Yellowstone Park's Old Faithful – Iceland, and New Zealand.

Giaever, Ivar (1929–) US physicist, b. Norway. He shared the 1973 Nobel prize for physics with Brian JOSEPHSON and Leo ESAKI for their research into tunneling effects in SEMICONDUCTORS and superconductors. Esaki developed the tunnel diode, which enables electrons to cross normally impassable electronic barriers, and Giaever applied this to superconductors, demonstrating new effects that led to a better understanding of SUPERCONDUCTIVITY.

▲ **geyser** A plume of hot water and steam, a geyser is the result of the boiling of water at depth in a series of interconnecting chambers by volcanic heat (A). The expansion of steam produced drives the water and steam above it out at the surface (B), and this is followed by a period of refilling and heating, making it a periodic phenomenon (C).

giant star Star belonging to a luminosity class midway between main-sequence stars and SUPERGIANTS. Giant stars lie directly above the MAIN SEQUENCE on the HERTZSPRUNG-RUSSELL DIAGRAM and are characterized by large dimensions, high luminosity, and low density. They are found throughout the entire surface temperature range. Those of spectral type G–M (the RED GIANTS) contrast with corresponding DWARF STARS of the main sequence.

Giauque, William Francis (1895–1982) US chemist. Giauque is noted for discovering the isotopes of oxygen and for his studies of phenomena at extremely low temperatures. In 1926 he suggested a method for producing temperatures approaching ABSOLUTE ZERO, until then believed impossible in practice. He received the 1949 Nobel Prize for chemistry.

gibberellin Any of a group of organic compounds (HORMONES) found in plants that stimulate cell division, stem elongation, the breaking of DORMANCY (by triggering the production of enzymes essential for germination) and response to light and temperature. Gibberellins interact with some AUXINS to promote cell enlargement and have been used to increase crop yields substantially.

Gibbs, Josiah Willard (1839–1903) US theoretical physicist and chemist. While a professor at Yale, he devoted himself to establishing the basics of physical chemistry. His application of THERMODYNAMICS to physical processes led to STATISTICAL MECHANICS. He evolved the concepts of free energy and chemical equilibrium, devised the phase rule, which specifies the number of distinct PHASES OF MATTER that can co-exist in various circumstances, and developed VECTOR ANALYSIS.

gigabyte (symbol Gb) In computing, 1,024 MEGABYTES. Often used for expressing the capacity of a COMPUTER memory and disk drives, the gigabyte is usually rounded to 1,000 megabytes.

Gilbert, Walter (1932–) US molecular biologist who determined the sequence of NUCLEOTIDES in DNA. In 1966 Gilbert identified a repressor substance, a protein that regulates gene activity. He then experimented with enzymes that split the DNA molecule at known places, and was gradually able to discover the sequence of nucleotides along the molecule. For devising this technique he shared the 1980 Nobel Prize for chemistry.

Gilbert, William (1544–1603) English physicist and physician to Queen Elizabeth I. His *De Magnete, Magneticisque Corporibus, et de Magno Magnete Tellure, Physiologia Nova* (1600) laid the foundation for the scientific study of MAGNET-ISM. He was the first to recognize terrestrial magnetism and mistakenly concluded that a type of magnetism keeps the planets in their orbits. He coined the terms magnetic pole, electric attraction, and electric force.

Gilchrist, Percy Carlyle (1851–1935) British chemist who developed, with his cousin Sidney Gilchrist THOMAS, a process for smelting iron ore rich in phosphorous in a BESSEMER converter. The process was patented and extended to the open-hearth furnace.

gilding Process by which a thin layer of gold is used to cover another surface. Silver is the metal most commonly gilded, but base metals are also used extensively. The original method of gilding was by a chemical process, but in the 19th century ELECTROPLATING, a cheaper if less effective method, was introduced.

gills Organs through which most FISH, some larval amphibians, such as tadpoles, and many aquatic invertebrates obtain oxygen from water. When a fish breathes it opens its mouth, draws in water, and shuts its mouth again. Water is forced over the gills, through the gill slits, and out into the surrounding water. Oxygen is absorbed into small capillary blood vessels, and at the same time, waste carbon dioxide carried by the blood diffuses into the water through the gills. The gills of young tadpoles, axolotls (a kind of salamander) and many invertebrates are on the outside of their bodies.

ginkgo (maidenhair tree) Member of the order Ginkgoales, the oldest living species of GYMNOSPERM. Gingkos flourished throughout the world during the Mesozoic era (245–74 million years ago). The only surviving species is *Ginkgo biloba*. It has fan-shaped leaves, small, foul-smelling fruits and edible, nutlike seeds. Height: to 30m (100ft).

girder In engineering, a main load-supporting beam. Steel or iron girders may be made of a single piece or of laminated strips built up of plates, latticework or bars. When the cross-section is formed in the shape of an I by riveting and welding plates or by rolling the girder, greater stiffness is achieved and larger spans are possible. Girders may also be made of reinforced or prestressed concrete.

glacial groove Deep, wide, usually straight furrow cut in bedrock. It is caused by the abrasive action of large rock fragments dragged along the base of a moving GLACIER. The grooves are larger and deeper than glacial striations, ranging in size from a deep scratch to a glacial valley.

glaciation Covering of a large region or a landscape by ice. The term is also used to refer to an ICE AGE.

glacier Large mass of ice, mainly recrystallized snow, that moves slowly by CREEP downslope or outward in all directions due to the stress of its own weight. The flow terminates where the rate of melting is equal to the advance of the glacier. There are three main types: the **mountain**, or **valley**, glacier, originating above the snow line in mountain regions; the PIEDMONT, which develops when valley glaciers spread out over lowland; and the ICE SHEET and ICE CAP.

glaciology Study of ice. No longer restricted to the study of glaciers, this science deals with ice and the action of ice in all its natural forms. Glaciology, therefore, draws upon the knowledge of many other related subjects, notably physics, chemistry, geology, and meteorology.

gland Cell or tissue that manufactures and secretes special substances. The glands of animals are of two basic types. EXOCRINE GLANDS make such substances as hydrochloric acid, mucus, sweat, sebaceous fluids, and ENZYMES, and secrete these usually through ducts to an external (such as the SKIN) or internal body surface. Endocrine glands (*see* ENDOCRINE SYSTEM) contain cells that secrete HORMONES directly into the bloodstream.

glandular fever Acute disease, usually of young people, caused by the EPSTEIN-BARR VIRUS. Symptoms include fever, painful enlargement of the LYMPH nodes and pronounced lassitude (which may persist for months or even years). There may also be a sore throat, skin rash, and digestive disorder. It is also known as **infectious mononucleosis** due to the presence in the blood of an increased number of white blood cells called MONOCYTES.

Glaser, Donald Arthur (1926–) US physicist who in 1953 invented the BUBBLE CHAMBER and used it to devise a new method for studying ELEMENTARY PARTICLES; for this work he was awarded the 1960 Nobel Prize for physics. In 1959 Glaser joined the physics faculty at the University of California at Berkeley where he did research in applying physics to molecular biology.

Glashow, Sheldon Lee (1932–) US physicist and educator. He attended Cornell and Harvard, joining the Harvard faculty in 1966. He shared the 1979 Nobel Prize for physics with Steven WEINBERG and Abdus SALAM for their independent formulation of a unified "electroweak" theory that describes both electromagnetic interaction and the WEAK NUCLEAR FORCE of ELEMENTARY

PARTICLES. He also proposed that QUARKS make up particles that interact strongly. *See also* FUNDAMENTAL FORCES

glass Brittle, transparent material. Although glass behaves in a similar way to a solid, it is not crystalline. It is really a liquid that has been cooled so quickly that the particles have not had time to organize themselves into a regular pattern (*see* SUPERCOOLING). It is made by melting together silica (sand), sodium carbonate (soda), and calcium carbonate (limestone). Glass melts slowly, can be worked only while hot and pliable, and must be cooled gradually to prevent strains or breakage. Glass appears to have been invented in the E Mediterranean, probably about 2500 BC in Egypt, for jewelry and small containers. There are many types of glass. Soda-lime glass is used in the manufacture of bottles and drinking vessels. Flint glass, which refracts light well, is used in lenses and prisms. Toughened glass – made by laminating with plastic – is used in car windshields. Glass is also used in OPTICAL FIBERS, for the transmission

of information such as telephone calls and in investigative medicine for the ENDOSCOPE.

glass fiber *See* FIBERGLASS

Glauber, Johann Rudolf (1604–68) German chemist and physician. He made valuable contributions to chemistry, mainly concerning the preparation of salts. He prepared hydrochloric acid, hydrated sodium sulfate (known as Glauber's salt) and tartar emetic.

glaucoma Condition in which the pressure within the eye is increased due to an excess of AQUEOUS HUMOR. Most frequently found in people over 40, the disease poses a threat to vision and cannot be cured, but is managed with drugs and surgery.

glauconite Complex silicate mineral of iron, magnesium and aluminum, a member of the MICA family. It forms very small, thin, narrow crystals (monoclinic), which occur as rounded grains in SEDIMENTARY ROCKS. It is bluish or dark green with a usually dull luster; in sufficient quantity, it can color the entire rock green. This is known as greensand. Hardness 2.0–2.5; r.d. 2.7–2.9.

Glenn, John Herschel, Jr. (1921–) First US astronaut to orbit the Earth. On February 20, 1962, Glenn made three orbits of the Earth in the spacecraft *Friendship 7*. Nearly five hours after launch from Cape Canaveral, Florida, the craft splashed down in the Pacific Ocean. Glenn later moved into politics, becoming a US Senator from Ohio.

glider Aircraft, usually with no power source of its own. Gliders usually have long narrow wings to provide maximum lift with minimum drag. Gliders may be launched in a variety of ways. They may be pulled by a winch or an elastic shock cord, or towed by a plane, car or ground crew. Once airborne, gliders descend relative to the surrounding air. If there is a rising "updraft" of warm air (thermal), a glider may gain altitude for a while, thus prolonging its flight.

Global Positioning System (GPS) Navigation system based on a network of Earth-orbiting SATELLITES. There are 24 satellites, each with a 24-hour orbit and its own identification signal. The satellites transmit continuous time signals. A user needs to be within range of four satellites to fix a position. One satellite sends out a reference signal and the other three give bearings indicating direction. A computer uses the tiny time differences in receiving the signals from each satellite to calculate the position. Ordinary users can find their position to within 100m (330ft), although the military can pinpoint position to within 10m (33ft).

global warming Trend toward higher average temperatures on Earth's surface. During the past few million years, there have been several periods when surface temperatures have been significantly higher or lower than at present. During cold periods (ICE AGES) much of the land area has been covered by glaciers. The Earth is currently in the middle of a warm period (**interglacial**) that began about 10,000 years ago. Since the 1960s, some scientists have called attention to signs that the Earth is becoming unnaturally warmer as the result of an increased GREENHOUSE EFFECT caused by human activity.

Globigerina Genus of one-celled marine PROTOZOA whose empty shells are an important component of ocean floor ooze. The shell is spiraled into a lumpy sphere with needlelike extensions. One species is an indicator of sea temperature, and therefore of ocean paleoclimatology. Shells coiling to the left indicate that the animals lived in cold water; shells coiling to the right were of animals that lived in warm water.

globular cluster Near-spherical cluster of very old stars in the halo of our GALAXY. Globular clusters contain anything from 100,000 to several million stars, concentrated so tightly near the center that they cannot be fully separated by ground-based telescopes. Globulars are at least 10 billion years old, as deduced from the highly evolved state of their stars. This extreme age indicates that globulars formed while our galaxy was condensing from a huge cloud of gas. Globulars can also be seen around other galaxies. They are more plentiful around elliptical galaxies than spirals: our galaxy has about 140 known globular clusters, whereas the largest elliptical galaxies may have thousands.

globulins Family of globular proteins that are soluble in dilute salt solutions, found in blood serum. Some globulins serve as ANTIBODIES; others are carriers of lipids, hormones, and minerals.

glomerulus Mass of capillary blood vessels within a BOWMAN'S CAPSULE, the funnel-shaped end of a NEPHRON in the KIDNEY. Fluid passes from the blood in the capillaries into the capsule and then down the tubule of the nephron. It eventually passes out of the kidney along the ureter as urine as part of the process of EXCRETION.

glottis Tube less than 25mm (1in) long between the VOCAL CORDS at the lower end of the PHARYNX. It opens into the TRACHEA and its dimensions are changed by the vibration of the vocal cords.

glove-box Chamber with a viewing window into

which are let, and firmly sealed, rubber gloves. It is used in laboratories for handling toxic and other harmful materials, including radioactive chemicals and metals, and cultures and toxins of pathogenic microbes.

glucagon Hormone secreted by cells in the ISLETS OF LANGERHANS in the PANCREAS. It helps to regulate blood sugar levels. *See also* INSULIN

glucose (dextrose, $C_6H_{12}O_6$) Colorless, crystalline SUGAR that occurs in fruit and honey. It requires no digestion before being absorbed into the bloodstream. The storage carbohydrate GLYCOGEN in the bodies of animals is converted to glucose before being utilized as an energy source in cellular RESPIRATION. A MONOSACCHARIDE sugar (having a single sugar unit), glucose is prepared commercially by the HYDROLYSIS of starch using hydrochloric acid. It is used in foods, especially confectionery, as a sweetener, and in tanning and pharmaceuticals.

glucoside Carbohydrate-containing compound that yields a GLUCOSE and a nonsugar component (either an alcohol or phenol) when decomposed by the process of HYDROLYSIS. The natural glucosides are important in plant metabolism, and many POLYSACCHARIDES, such as cellulose, starch, and glycogen, are regarded chemically as glucosides. They may be used as drugs, coloring agents, and aromatics.

glue Name for any natural ADHESIVE.

gluon Particle that can be thought of as the "glue" that holds QUARKS together. Quarks are the fundamental particles that combine to form ELEMENTARY PARTICLES such as PROTONS and NEUTRONS. The gluon is the carrier of STRONG NUCLEAR FORCES. It is analogous to the PHOTON, which is the carrier of ELECTROMAGNETIC INTERACTIONS. *See also* FUNDAMENTAL FORCES

glutamic acid ($COOH(CH_2)_2CHNH_2COOH$) Colorless AMINO ACID commonly found in proteins. One of its functions is to increase the solubility of its associated proteins by providing them with one negative charge. It also helps in the removal of poisonous ammonia from the body. Its sodium salt (sodium glutamate) is used as a food flavoring. *See also* GLUTAMINE

glutamine ($COOHCH(NH_2)(CH_2)_2CONH_2$) Colorless soluble AMINO ACID. It is formed when one of the acidic carboxyl groups of GLUTAMIC ACID couples with ammonia, a poison to the body, which is carried safely to the liver where it is converted to urea, and freely expelled as urine.

gluten Main protein substance present in wheat flour, gray in color and elastic in texture. Not present in barley, oats or maize, gluten contributes the elasticity to dough. Insoluble in water, it can be washed out of flour; it is then used as an additive to chocolate and coffee.

glycerol (glycerine) Syrupy, sweet liquid (1,2,3–trihydroxypropane, $HOCH_2CH(OH)CH_2OH$) obtained from animal and vegetable fats and oils, as when making soap, or from propene (propylene). It is used in the manufacture of various products including plastics, explosives, cosmetics, foods, antifreeze, and paper coatings.

glyceryl trinitrate In medicine, a VASODILATOR mainly used in treating the symptoms of angina pectoris. It reduces the amount of blood returning to the heart, thereby reducing the amount of work the heart has to do. It is also used as an explosive, when it is known as NITROGLYCERINE.

glycine (NH_2CH_2COOH) Colorless, soluble crystalline AMINO ACID; the principal amino acid in sugar cane and the simplest, structurally, of the alpha-amino acids.

glycogen CARBOHYDRATE stored in the animal body, principally by the liver and muscles. Glycogen is sometimes known as animal starch, and, like starch and cellulose, it is a POLYMER of GLUCOSE. When the body needs energy, glycogen is broken down to glucose, which is further metabolized to carbon dioxide and water, providing ADENOSINE TRIPHOSPHATE (ATP), a source of chemical energy, in the process. *See also* METABOLISM; RESPIRATION **glycol** (ethylene glycol, ethane-1,2-diol, $(CH_2OH)_2$) Colorless, odorless liquid containing two HYDROXYL groups. It is a viscous liquid used in plastics, solvents, antifreeze, and artificial fibers.

glycolysis Series of biochemical reactions in which glucose is converted to PYRUVATE. It takes place during RESPIRATION in CELLS. The nine stages of glycolysis are accompanied by the net release of two molecules of the energy-containing substance ADENOSINE TRIPHOSPHATE (ATP) per glucose molecule. During AEROBIC RESPIRATION, the pyruvate enters the KREBS CYCLE, with the ultimate yield of 12 more molecules of ATP. During ANAEROBIC RESPIRATION, the pyruvate is converted to lactic acid. *See also* ELECTRON TRANSPORT SYSTEM; PHOSPHORYLATION

GMT Abbreviation of GREENWICH MEAN TIME

gneiss METAMORPHIC ROCK with a distinctive layering or banding formed by high pressure or high temperature conditions of regional metamorphism. The darker minerals are likely to be hornblende, augite, mica or dark feldspar.

Goddard, Robert Hutchings (1882–1945) US physicist and pioneer in ROCKET development. He developed and launched (1926) the first liquid-fueled rocket, and developed the first smokeless powder rocket and the first automatic steering for rockets.

Gödel, Kurt (1906–78) US mathematician, b. Czechoslovakia. Gödel is best known for his "undecidability" or "incompleteness" theorem, the proof of which first appeared in a German technical journal in 1931. The theorem states that an axiom based mathematical system can contain statements that can be neither proved nor disproved using results from within the system. Put simply, Gödel's theorem means that the truth of some statements in mathematics can never be decided. His work ended nearly a century of attempts to establish axioms that would provide a rigorous basis for all mathematics. In 1940 he emigrated to the USA; he was a professor, and a close friend of EINSTEIN, at the Institute for Advanced Study at Princeton University from 1953 to his death.

goethite Hydroxide mineral, iron oxyhydroxide, ($FeO[OH]$). It is found in secondary oxidized deposits with orthorhombic system slender plates and velvety needles. It also occurs in a massive fibrous form with uniform surfaces. It is black, brilliant and earthy. Hardness 5–5.5; r.d. 3.3–4.3.

gold (symbol Au) Naturally occurring TRANSITION ELEMENT. It is also obtained as a by-product in the electrolytic refining of copper. Gold is used in jewelry, in connectors for electronic equipment, and as a form of money for international transactions. Gold leaf can be made as thin as 0.00001mm (0.0000004in). Gold in the form of a COLLOID is sometimes used in coloring glass. It is often alloyed with copper or silver to make it harder. The gold content of an alloy is expressed as fine (parts per 1,000) or in carats (parts per 24). Gold forms two series of salts termed gold(I), or aurous, and gold(III), or auric. The isotope ^{198}Au (half-life 2.7 days) is used in RADIOTHERAPY. The metal is unreactive, being unaffected by oxygen and common acids. It dissolves in aqua regia, a mixture of nitric and hydrochloric acids. Properties: at.no. 79; r.a.m. 196.9665; r.d. 19.30; m.p. 1,063°C (1,945°F); b.p. 2,800°C (5,072°F); most common isotope ^{197}Au (100%).

golden mean (golden section or golden ratio) Classical ratio created when a line is divided into two parts in such a way that the ratio of the shorter to the longer is as the longer to the whole, that is, $a/b = b/(a + b)$, where $a + b$ is the line's length. The ratio is approximately 3:5, or exactly $1:(\sqrt{5} + 1)/2$ (*see* FIBONACCI SEQUENCE). The golden mean was cited by the Roman historian Vitruvius as the basis of proportion in Classical Greek architecture. The ratio, considered pleasing to the eye, is observed in many Renaissance paintings and buildings.

Goldmark, Peter Carl (1906–77) US scientist, b. Hungary. In 1940 he demonstrated a color television system that he had devised, and he developed the first system to find commercial acceptance. In the late 1940s he invented the $33^1/_3$ rpm long playing phonograph record (the LP). Goldmark later developed an electronic VIDEO recorder used as an educational aid, and a system that enabled photographs to be transmitted from space to Earth.

Golgi body Collection of microscopic vesicles or packets observed near the nucleus of many living CELLS. It is a part of a cell's inner membrane structure, or ENDOPLASMIC RETICULUM, specialized for the purpose of packaging and dispatching proteins made by the cell.

Golgi, Camillo (1843–1926) Italian histologist (specialist in the structure of CELLS). In 1873 he developed a method of staining tissue with silver nitrate for microscopic study. With this he discovered the GOLGI BODY within the CELL. Golgi shared the 1906 Nobel Prize for physiology or medicine with Santiago Ramón y Cajal (1852–1934) for his work on the structure of the NERVOUS SYSTEM.

gonad Primary reproductive organ of male and female animals, in which develop the GAMETES or sex cells. Thus, the gonad in the male is a TESTIS and in the female an OVARY. Hermaphrodite animals possess both types.

gonadotrophins General term for two HORMONES secreted by the PITUITARY GLAND, which stimulate the development and function of the GONADS (ovary and testis) and similar hormones made by the PLACENTA. Some gonadotrophins are used to treat INFERTILITY.

Gondwanaland Name given to the S supercontinent that began to break away from the single land mass PANGAEA about 200 million years ago. It later became South America, Africa, India, Australia, and Antarctica. The name comes from a historical region in central India. The N supercontinent, which eventually became North America and Eurasia without India, was LAURASIA. The occur-

rence of tillites (glacial deposits) in separate parts of India, South America, and Africa provides evidence that these were once a single continent.

goniatite Any CEPHALOPOD mollusk belonging to the order Goniatitida. Goniatites have coiled shells divided into chambers by zigzag partitions. They first appeared in the upper Devonian period, about 350 million years ago. The group became extinct at the end of the Permian period. They are useful as zone fossils. Goniatites and AMMONITES belonged to the same subclass, Ammonoidea.

goniometer Instrument used mainly by MINERAl collectors to help in the identification of crystal forms by measuring the angles of related sets of crystal faces. These angles are characteristic for certain minerals.

Goodall, Jane (1934–) British zoologist, famous for her studies of chimpanzees in the wild. She began observing the primates in the Africn bush in 1960, and her work became a model of how to study animal behavior. Among her many discoveries was that chimpanzees use primitive tools, such as grass stems to extract termites from their mounds. She also noted that the animals are carnivorous, and engaged in cannibalistic warfare among themselves.

Gorgas, William Crawford (1854–1920) US surgeon whose successful mosquito-control program in Panama eradicated MALARIA and YELLOW FEVER there and made possible the building of the Panama Canal.

gorge Narrow, deep-sided valley. Gorges form only in hard rock, such as carboniferous limestone, otherwise erosion would soon break down the steep sides. A very large gorge is known as a CANYON.

Gould, Stephen Jay (1941–) US paleontologist, evolutionary biologist, and science writer, who proposed that EVOLUTION could occur in sudden spurts rather than gradually. Called PUNCTUATED EQUILIBRIUM, Gould's theory suggested that sudden accelerations in the evolutionary process could produce rapid changes in species over the comparatively short time of a few hundred thousand years. Gould has also written many popular science books, including *Wonderful Life* (1990) and *Bully for Brontosaurus* (1991), which often use examples taken from the modern world to explain his theories.

Graaf, Regnier de (1641–73) Dutch anatomist and doctor after whom the GRAAFIAN FOLLICLES were named. Graaf investigated the functions of the PANCREAS and the nature of the sexual organs. He was the first user of the term "OVARY."

Graafian follicle (ovarian follicle) Fluid-filled, cystlike cavity found in the OVARIES of MAMMALS. It surrounds and protects the developing OVUM. Once the ovum is released into a FALLOPIAN TUBE, the follicle develops a yellowish mass of tissue called the CORPUS LUTEUM.

graben Elongated, trenchlike, down-dropped block of the Earth's crust bordered by two or more similarly trending normal FAULTS. The Basin and Range province in Utah and Nevada consists of grabens and HORSTS that form sedimentary basins and mountain ranges. A long graben found at the Earth's surface can also be called a RIFT VALLEY, such as the Great Rift Valley in East Africa

graft, tissue Organ or tissue transplanted to replace a part of the body that is damaged or non-functioning. The replacement tissue may be taken from elsewhere on the patient's body (an **autograft**) or from a donor (**allograft**). Except in the case of an identical twin, the use of donor organs may set up rejection, the phenomenon where the IMMUNE SYSTEM attacks "foreign" tissues.

grafting In horticulture, method of plant PROPAGATION. A twig of one variety, called the **scion**, is established on the roots of a related variety, called the **stock**. Most fruit trees are propagated by a similar process called **budding**, in which the scion is a single bud. Grafting is a way of producing new plants that are genetically identical to the parent from which the scions are cut.

Graham, Thomas (1805–69) British chemist who is best remembered for the rule he formulated (GRAHAM'S LAW). He also discovered DIALYSIS.

Graham's law The DIFFUSION rate of a GAS is inversely proportional to the square root of its density. This principle, formulated by Thomas GRAHAM, is used in separating isotopes by the diffusion method; it has important industrial applications.

gram (symbol g) SI UNIT of mass defined as one thousandth of a kilogram. *See also* PHYSICAL UNITS

gram-atom Quantity of an element the mass of which, in grams, is equal to its RELATIVE ATOMIC MASS (R.A.M.). It has been replaced by the SI unit, the MOLE. For example, one gram-atom of hydrogen (H, r.a.m. = 1) is 1g.

Gramme, Zénobe Théophile (1826–1901) Belgian engineer who developed the first practical industrial DYNAMO in 1869. His knowledge of electrical theory was limited, but even so he developed an ELECTRIC MOTOR from his dynamo. He was one of a group of scientists who, in 1872, succeeded in transmitting direct current electricity over long distances.

gram molecular volume *See* MOLAR VOLUME

Gram's stain In microbiology, a differential staining method named after the Danish physician Hans Christian Gram (1853–1938). It aids in the categorizing and identification of BACTERIA, which are said to be Gram-positive or Gram-negative, depending on whether or not they retain the original violet stain at the end of the process. If Gram-negative, the stain washes out to a red counterstain.

grand unified theory (GUT) Theory that would demonstrate that three of the four FUNDAMENTAL FORCES – the WEAK NUCLEAR FORCE, STRONG NUCLEAR FORCE, and ELECTROMAGNETIC FORCE – are really slightly different aspects of the same fundamental force. The weak nuclear force and electromagnetic force have been incorporated as the **electroweak force** (*see* ELECTROWEAK THEORY), as demonstrated by particle ACCELERATOR experiments. In order to achieve a GUT, the electroweak force must be unified with the strong nuclear force. If GRAVITATION could be incorporated, then a UNIFIED FIELD THEORY would be produced. *See also* SUPERSTRING THEORY; SUPERSYMMETRY

granite Coarse-grained, acid, IGNEOUS ROCK from deep within the Earth, composed chiefly of feldspar and QUARTZ, with some mica or hornblende. Its color is usually light gray, although feldspar may redden it. Its durability makes it a valuable construction material. Granite is thought to have solidified from magma (molten rock), but the occurrence of some granite with features normally associated with rocks of metamorphic origin suggests that not all granites are igneous. It crystallizes at great depths where the pressure is high; it becomes exposed at the Earth's surface only by erosion of surface rocks or by movements in the Earth's crust.

granulite (leptites) Granular METAMORPHIC ROCK that derives largely from quartz and feldspar. Granulites often have a banded appearance.

granuloma Growth or nodule of connective tissue and capillaries usually associated with an infection such as tuberculosis or syphilis.

graph In mathematics, a diagram representing a relationship between numbers or quantities. Many graphs use the CARTESIAN COORDINATE SYSTEM. Usually axes are drawn at right angles and marked with scales representing the variable quantities. On a temperature conversion graph, for example, the vertical axis might represent degrees Fahrenheit, with degrees Celsius along the horizontal axis. The graph itself would be a line drawn on the diagram. To convert from one reading to the other, you would have to trace from the appropriate scale to the graph, and from there to the other scale. Other forms of graph include bar charts, in which a series of figures is represented by lines of various lengths. In pie charts, the quantities are represented by sectors of a circle, which resemble the slices of a round pie.

graphical user interface (GUI) COMPUTER PROGRAM that enables a user to operate a COMPUTER using simple symbols. Early personal computers used operating systems (the programs that organize how the computer stores data and runs all other programs) that were text based. Commands were often obscure combinations of letters and numbers that made using the systems difficult for the uninitiated. A GUI replaces these commands with a screen containing symbols called icons. The user manipulates these using an onscreen pointer controlled by a MOUSE. Other important features of GUIs are menus, from which commands are selected, and windows displaying program events. The GUI makes it possible for files to be transferred and programs to be opened with a simple movement of the hand or the finger. The first GUI was designed by Xerox in the 1970s. Microsoft's Windows operating system is now the most commonly used GUI, although it was predated by the Macintosh operating system used on Apple computers.

graphics pad (graphics tablet or bit pad) Computer input device consisting of a sensitive pad and a pen linked to the computer. When the pen is moved over the pad, digital signals indicating the pen's position pass to the computer. The device is generally used to input drawings and diagrams. *See also* COMPUTER GRAPHICS

graphite (plumbago) Dark gray, soft crystalline form of CARBON that occurs naturally in deposits of varying purity and is made synthetically by heating petroleum coke. It is used in pencils (a mixture of graphite and clay is the "lead"), lubricants, electrodes, brushes of electrical machines, rocket nozzles, and as a MODERATOR to slow down neutrons in NUCLEAR REACTORS. Graphite is a good CONDUCTOR of heat and electricity. It owes its lubricating properties to overlapping scale-like crystals, which tend to slide, giving it a smooth slippery feel. Properties: r.d. 2–2.25.

graptolite Any of an extinct group of colonial, marine, drifting organisms; sometimes consid-

ered a separate phylum but also thought to be related to the CHORDATES. They are found most frequently as flattened fibers of CARBON, resembling pencil marks, in black shales of the ORDOVICIAN and SILURIAN ages. Their uncompressed skeletons etched out of limestone show that graptolites were composed of many small tubes regularly arranged along branches, which are presumed to have been attached to a common bladder-like float. Graptolites first appeared in the Middle Cambrian and the last are known from the Lower Carboniferous. They are important zone fossils, used for correlating and dating rocks of the Lower Paleozoic.

grass Any one of about 8,000 species of the class Monocotyledonae (*see* MONOCOTYLEDON), being nonwoody plants with fibrous roots and long, narrow leaves enclosing hollow, jointed stems; the stems may be upright or bent, lie on the ground, or grow underground. the flowers are small, without PETALS and SEPALS. The leaves grow from the base, and so removal of the tips does not inhibit growth, making grass suitable for lawns and pastures. Cereal grasses, such as rice, millet, maize, and wheat, are cultivated for their edible seeds. Others are grown as food for animals and for erosion control and ornament. Family Poaceae/Gramineae.

gravel Mixed pebbles and rock fragments, 2mm to 60mm (0.1–3in) in diameter. Gravel beds are generally the remains of ancient seashores or river beds. Usually gravel is used as an aggregate in CONCRETE.

gravimeter (gravity meter) Instrument used in geophysical surveys to measure the strength of the local GRAVITATIONAL FIELD. Essentially, it is a heavy mass suspended from a spring. In the vicinity of a dense underground feature such as an ignesus deposit, the weight of the gravimeter mass increases slightly by gravitational attraction, so causing the spring to extend slightly.

gravimetric analysis In geology, study of the magnitude of the Earth's GRAVITATIONAL FIELD in a particular area. Small variations in this field can be caused by the density of the rocks beneath the surface and can provide information about structures that are otherwise inaccessible. The instrument used for such a survey is known as a GRAVIMETER. In chemistry, gravimetric analysis is a method of determining the composition of a substance by making weighings.

gravitation One of the four FUNDAMENTAL FORCES in nature. Gravitation is extremely weak compared with the others. The gravitational force F between two masses m_1 and m_2 a distance d apart was found by Isaac NEWTON to be $F = Gm_1m_2/d^2$, where G is a constant of proportionality called the universal GRAVITATIONAL CONSTANT. A more complete treatment of gravitation was developed by Albert EINSTEIN, who showed in his general theory of RELATIVITY that gravitation is a manifestation of the curvature of SPACE-TIME.

gravitational constant (symbol G) Constant in Newton's law of GRAVITATION. Its value is 6.67259×10^{-11} Nm^2 kg^{-2}.

gravitational field Region around an object that has mass in which there is an attractive force on any other object. The force divided by the mass of the second object is the gravitational field strength. A massive object, such as the Earth, has a powerful gravitational field, which we call the force of gravity. Weak gravitational forces exist between even very small particles.

gravitational red shift *See* RED SHIFT

gravitational waves Gravitational equivalent of electromagnetic waves, postulated by Einstein's general theory of RELATIVITY as being emitted from a massive accelerating body, such as an exploding or collapsing star, and traveling at the speed of light. Experimental results put forward as evidence for such waves have not yet been generally accepted.

graviton Hypothetical ELEMENTARY PARTICLE thought to be continuously exchanged between bodies of mass as the carrier of the gravitational force. The graviton is analogous to the PHOTON – the quantum of electromagnetic waves.

gravity Gravitational force of attraction at the surface of a planet or other celestial body. The Earth's gravity produces an acceleration of 9.8 ms^{-2} (32fts^{-2}) for any unsupported body. If the mass M and radius R of a planet are known, the acceleration due to its gravity (g) at its surface can be determined from $g = GM/R^2$, where G is the universal constant of gravitation. The WEIGHT of a body is a measure of the force with which the Earth's gravity attracts it. Unlike MASS, which remains constant at normal speeds, weight (or the force of gravity) varies with altitude.

gravity anomaly Deviation in gravity from the expected value. Gravity measurements over deep ocean trenches are lower than average; those in mountainous regions are higher than average. Higher values are also found over deposits of dense minerals.

gravity-assisted flight Use of a close approach to a planet to obtain extra velocity from its gravitational pull. A rocket "grazing" a planet is accelerated and whipped off in a new direction with greatly increased velocity, perhaps unattainable by conventional means. All probes to the giant outer planets use this "whiplash" technique, often flying by inner planets, including the Earth, several times in order to do so.

Gray, Asa (1810–88) US botanist. He made many contributions to PLANT CLASSIFICATION, and his donation of a valuable collection of books and plants to Harvard University in 1865 led to the establishment of that school's department of botany. His *Manual of Botany for the Northern United States* (1848) is considered a classic.

Gray, Elisha (1835–1901) US inventor. He received his first patent for a self-adjusting telegraph relay in 1867. He also invented and patented the telegraphic switch, telegraphic repeater, type-printing telegraph, and teolautograph. He claimed priority as the inventor of the speaking telephone, but Alexander Graham BELL's patent rights were upheld by the US Supreme Court.

Gray, George William (1926–) British chemist who, with physicist Cyril Hilsum, developed LIQUID CRYSTALS for digital displays at the UK government's Malvern Radar Establishment, but success did not come until the 1980s when Gray was able to produce the necessary stable liquid crystals.

gray (symbol Gy) SI unit of absorbed radiation dose. One gray is equivalent to supplying 1 joule of energy per kilogram of irradiated material. It superseded the rad (1 gray = 100 rad). It was named after the British biologist L. H. Gray.

gray matter Darker-colored nerve tissue that forms the CEREBRAL CORTEX of the BRAIN and is also present in the SPINAL CORD. It is distinct from the so-called white matter, which contains more nerve fibers (NEURONS) and larger quantities of the whitish insulating material called MYELIN.

graywacke Any of a variety of SANDSTONES that consist of a mixture of rock fragments, feldspar, and quartz, strongly bonded together in a mud matrix. They are characterized by poor sorting of angular or subangular particles. They are usually interpreted as indicating short-distance, rapid-sediment depostition.

grease Semisolid lubricant consisting of a mixture of a liquid lubricant with either a saponified FAT or a dispersion of metallic soaps. There are five types: water soluble; water-resistant; synthetic;

special-purpose; and multipurpose. Although liquid lubricants are usually preferred, greases are used in situations where a piece of machinery cannot be relubricated or where there is a need to avoid leakage, which may occur if oil were used.

great circle Circle on a sphere the center of which is coincident with the center of the sphere. On the Earth, the equator is a great circle, as are all meridians. On the CELESTIAL SPHERE, the CELESTIAL EQUATOR is a great circle, as are the meridians (circles passing though both celestial poles).

Great Red Spot (GRS) Large oval area in the s hemisphere of JUPITER. Since it was first observed by Robert HOOKE in 1664, it has varied in size and color, from about 11,000 to 14,000km wide by 24,000 to 40,000km long (7,000–9,000mi by 15,000–25,000mi), and from a deep red to a light pink, occasionally fading away altogether. It is an atmospheric phenomenon, a huge anticyclone, projecting above the surrounding cloud-tops. The red color may be produced by phosphorus.

green belt Area of open land maintained as a barrier between adjoining built-up areas. Green belts provide insulation from factories and intensive commercial areas. They are a source of recreational space and aid in the replenishment of atmospheric oxygen.

green flash Greenish or bluish hue of the upper rim of the Sun seen just as it is about to disappear at setting, or appear at rising, due to refraction, scattering, and absorption of light in the atmosphere.

greenhouse effect Raised temperature at a planet's surface as a result of heat energy being trapped by gases in the atmosphere. Certain gases cause the atmosphere to act like the glass in a greenhouse. As a result, the temperature of a planet's surface may be higher than it otherwise would be – on Earth about 33°C (59°F) higher. The main gases that produce the greenhouse effect on Earth are water vapor and CARBON DIOXIDE. Scientists suspect that increased discharge of carbon dioxide from human activity (notably gasoline-powered car engines and industry) is contributing to GLOBAL WARMING.

green revolution Intensive plan of the 1960s to increase crop yields in developing countries by introducing higher-yielding strains of plants and new fertilizers. The effort began in Mexico in the 1940s and was introduced to parts of India, SE Asia, the Middle East, and Latin America.

Greenwich Mean Time (GMT) Local time at Greenwich, London, located on the prime MERIDI-AN. It has been used as the basis for calculating standard time throughout the world since 1884.

Gregg, Sir Norman McAlister (1892–1966) Australian ophthalmic surgeon. In 1941 he discovered that RUBELLA (German measles) in a pregnant woman can cause physical defects in her child.

grid (control grid) Perforated electrode placed between the cathode and the anode of an ELECTRON TUBE (valve) for controlling the flow of electrons in the valve. This effect is the basis of electronic amplification. "Grid" also refers to the supply network for electricity or gas, as in the national grid.

Griess, Johann Peter (1829–88) German chemist who discovered the diazo reaction that enabled azo dyes to be prepared from a wide range of intermediate chemical compounds. *See also* AZO AND DIAZO COMPOUNDS

Grignard, François Auguste Victor (1871–1935) French chemist who shared the 1912 Nobel Prize for chemistry with Paul Sabatier (1854–1941) for his discovery and investigation of GRIGNARD REAGENTS. He also synthesized organic compounds of aluminum and mercury.

Grignard reagent Organic chemical compound containing the metal magnesium. The Grignard reagents are named after François GRIGNARD who discovered them in 1900. Examples are methylmagnesium iodide (CH_3MgI) and phenylmagnesium bromide (C_6H_5MgBr). Grignard reagents undergo many reactions to form carbon-carbon bonds and for this reason they are invaluable in organic synthesis.

grike (gryke) Enlarged vertical crack in an area of bare limestone formed by solution along JOINTS. The resulting limestone blocks are called CLINTS and the two features together form a limestone pavement.

grit In geology, a coarse sandstone. It is usually formed in a river delta where material eroded from a landmass has been deposited rapidly before the fragments have been rounded or sorted.

gritstone Medium- to coarse-grained SEDIMENTARY ROCK with angular grains that may rub off easily. Most gritstones are formed in water. The chief mineral components are QUARTZ, FELDSPAR, and MICA. Those with more than 75% quartz are called **quartz** gritstone, whereas more than 25% feldspar gives rise to **feldspathic** gritstone; the former has an orange color and the latter is brown, sometimes tinged with pink.

groin Artificial dam of rocks or wooden pilings that juts out from a beach face, built in an attempt to combat LONGSHORE DRIFT. Groins are now considered a less effective and more expensive way of maintaining beaches than a beach-nourishment program, in which material swept along the beach is replaced by new material.

grooming Behavior pattern of self-cleaning of the body surface, usually stereotyped, practised by many animals. **Mutual** grooming, such as fur grooming among monkeys and apes, serves to cement pair-bonding and social bonds in groups.

ground state Most stable energy state of an atom, ion, or molecule. It is the state of an atom when the orbiting electrons move in orbits such that the atom's energy is at a minimum.

groundwater Water that lies beneath the surface of the Earth. The water comes chiefly from rain, although some is of volcanic or sedimentary origin. Groundwater moves through porous rocks and soil and can be collected in wells. Groundwater can dissolve minerals from the rocks and leave deposits in other places, creating such structures as CAVES, STALAGMITES, STALACTITES, and cavities called SINK-HOLES.

group theory Branch of mathematics applicable to sets with symmetric properties. A group is a set with elements (together with an operation) that must obey four rules: closure; associativity; every element must have an inverse; and there must be an identity element. The theory was developed by the French mathematician Évariste GALOIS. Group theory is particularly useful in QUANTUM MECHANICS, SPECTROSCOPY, and theories of elementary particles, but is used widely in mathematics to describe many other natural phenomena that have symmetry.

growth Process by which an organism increases in size and mass. It involves CELL DIVISION or enlargement, or both.

growth hormone (somatotrophin) Hormone produced by the anterior lobe of the PITUITARY GLAND; it is involved in general growth of the body. Oversecretion results in ACROMEGALY in the adult, and gigantism in the young; undersecretion results in dwarfism in children.

growth ring *See* ANNUAL RING

Grus (Crane) Insignificant constellation of the S skies. Its brightest star is Alpha, magnitude 1.7.

gryke *See* GRIKE

guanine One of the nitrogenous bases in DNA and RNA. Guanine is a derivative of PURINE and in DNA is always paired with the pyrimidine derivative CYTOSINE. *See* BASE PAIR

guano Dried excrement of seabirds and bats. It contains phosphorous, nitrogen and potassium, and is a natural fertilizer. It is found mainly on certain coastal islands off South America and Africa, and on some Paficic islands.

guided missile MISSILE controlled throughout its flight to target either by exterior or interior control systems. There are four types: (1) surface-to-surface, (2) surface-to-air, (3) air-to-air, (4) air-to-surface. The first guided missiles were built in Germany during World War II. The first of these was the V1, which was powered by a pulse-jet engine and flew at low speeds, making it easy to destroy in the air. The V2 was a far more sophisticated weapon – rocket-powered, with an automatic pilot and electronic guidance; it could reach a height of 100km (60mi) and speeds of over 5,800km/h (3,600mph) and deliver a ton of high explosive. Postwar development improved upon the V2, with missiles ranging from the huge intercontinental ballistic missiles (ICBMs) with ranges of 10,000km (6,000mi) and nuclear warheads, to small hand-launched anti-tank missiles. The main strategic capability of nuclear missiles resides in the submarine systems developed by the superpowers, the first of which was Polaris. The multiple independently targeted re-entry vehicles (MIRVs) – ICBMs with many independently targetable sub-missiles – were developed in the late 1960s. The CRUISE MISSILE can evade radar by flying at low heights and its computer guidance compares satellite photographs with the terrain over which it flies, bringing the missile potentially to within 10m (33ft) of its target.

Guillaume, Charles Édouard (1861–1938) Swiss physicist who was awarded the 1920 Nobel Prize for physics for his discovery of invar, an alloy of iron and nickel with an extremely small coefficient of expansion. He also developed another alloy, platinite, which expands at about the same rate as glass.

Gulf Stream Relatively fast-moving current of the N Atlantic Ocean. It flows from the straits of Florida, along the E coast of North America, then E across the Atlantic Ocean (at which point it is known as the **North Atlantic Drift**) to the NW European coast. It was long considered to be one wide mass of water, but research now indicates that it is made up of many thin streams that cause local variations in the water temperature. The current has a warming effect on the coastal climates along its course.

Gullstrand, Allvar (1862–1930) Swedish ophthalmologist. He received the Nobel Prize for physiology or medicine in 1911 for his contributions to dioptrics, the study of refraction of light by the eye. He invented many important ophthalmological tools, including the slit lamp, which aided the study of the structure and function of the CORNEA, and improved corrective lenses for use after surgery.

gum In botany, the secretions of some plants. Some gums are soluble in water; others absorb water and swell. Gums are chemically complex, consisting mainly of various SACCHARIDES bound to organic acids. Common examples are gum arabic, AGAR, AND tragacanth. Many substances of similar appearance, such as some RESINS, are classed as gums.

gums Soft tissue surrounding the base of the teeth. The gums are liable to inflammation, called GINGIVITIS, and to infection from food debris.

guncotton (cellulose nitrate, nitrocellulose) Highly nitrated and highly explosive form of CELLULOSE. It is made by soaking cotton cellulose in mixed nitric and sulfuric acids. When dry, guncotton is dangerously unstable. It is used mainly as an ingredient of explosive propellants such as CORDITE.

gunmetal Types of wear-resistant BRONZE once used to make heavy field and naval guns, but now used in gears, bearings, and pump spindles. A typical gunmetal alloy contains about 88% copper, 10% zinc, and 2% tin.

gunpowder Mixture of saltpeter (potassium nitrate), charcoal, and sulfur. When ignited, it expands violently due to the almost instantaneous conversion of the solid ingredients into gases such as carbon dioxide, carbon monoxide, nitrogen, nitrogen oxides, sulfur oxides, and steam. The sudden release of enormous volumes of these gases gives the reaction its explosive force. It was used extensively in FIREARMS until about 1900, after which it was replaced by smokeless powders such as CORDITE.

Gutenberg, Beno (1889–1960) US seismologist, noted for his analyses of EARTHQUAKE waves. He showed that 75% of earthquakes occur in the Circum-Pacific belt. He worked with Charles Richter (1900-85) to develop the RICHTER SCALE.

Gutenberg, Johann (1400–68) German goldsmith and printer credited with the invention of printing from movable metallic type. He experimented with printing in the 1430s, and his innovations included a new type of press and a typemetal alloy. He made the first printed Bible, known as the Gutenberg Bible (c.1455).

Guth, Allan Harvey (1947–) US physicist and cosmologist. He is best known for modifying the conventional BIG BANG theory of the origin of the Universe by postulating a rapid expansion phase (called an inflation) 10^{-30} seconds after the initial bang. One consequence of the theory, announced in 1980, is that the Universe could have evolved out of an infinitesimal region – virtually out of nothing. As Guth said, "The Universe may be the ultimate free lunch!"

guyot (tablemount) Flat-topped, submarine mountains. They rise 1,000m (3,300ft), with tops about 2.5km (1.5mi) below sea level. Before becoming submerged, they are believed to have been volcanic islands with peaks flattened by wave erosion. If so, this is evidence for the subsiding of the ocean floor.

gymnosperm Seed-bearing plant with naked seeds borne on scales, usually in cones. Most trees commonly referred to as EVERGREENS are gymnosperms. However, larch and some other CONIFERS are DECIDUOUS. All living seed-bearing plants are divided into two main groups: gymnosperms and ANGIOSPERMS (with seeds enclosed in an ovary). In the modern "Five KINGDOMS" classification system, gymnosperms no longer form a

▲ **guyot** Formed originally from seamounts (submarine mountains rising at least 1,000m (3,300ft) above their surroundings), guyots have flat tops at depths down to 2,500m (8,200ft). These tops are often much too large to be explained as ancient craters filled to the rim by sediment and other material. It has been proposed that guyots were originally volcanoes above sea level (A) which after extinction were, over many years, worn flat by the action of waves (B) and which then sank as the sea level rose or as the sea bed subsided (C). This theory has subsequently been confirmed by the presence of beach pebbles.

continuously driven GYROSCOPE. The spinning axis of the gyroscope is horizontal and its direction indicates true N, irrespective of the course or attitude of the craft. *See also* COMPASS

gyroscope Symmetrical spinning disk that can adapt to any orientation, being mounted in gimbals (a pair of rings with one swinging freely in the other). When a gyroscope is spinning, a change in the orientation of the gimbals does not change the orientation of the spinning wheel. This means that changes in direction of an aircraft or ship can be determined on board by a gyroscope without external references. A gyrostabilizer is used to stablize the roll of a ship or aircraft, and a gyrocompass is a gyroscope modified to act as a compass. When torque (twisting force) is applied to a fast-spinning gyroscope, such as leaning it out of the vertical, this produces a phenomenon known as PRECESSION: the gyroscope rotates about a fixed point with the axis of spin describing a cone around the vertical. This tendency to resist changes in the spin axis accounts partly for the stability of bicycles and of the orbits of stars and planets.

▲ **gyrocompass** A laser gyrocompass measures rotation by comparing the wavelength of lasers (1). A current is passed from an anode (2) to two cathodes (3) creating two lasers in a gas-filled triangular chamber (4) drilled in a solid glass block (5). Part of the lasers are bled out at one end of the gyrocompass (6) and the wavelength measured. If the gyrocompass rotates to the left the path of the laser traveling to the left is reduced fractionally reducing its wavelength. The opposite occurs to the other laser. A sensor (7) compares the two lasers in order to measure the rotation.

taxonomic group. Instead they comprise three distinct phyla, the Coniferophyta (such as pine, spruce, and cedar); the Ginkgophyta (a single species, the GINKGO); and the Gnetophyta (strange plants such as *Welwitschia*, *Ephedra*, and *Gnetum*, a genus of woody shrubs, vines, and large-leaf tropical trees). The CYCADS are no longer considered to be gymnosperms.

gynoecium Collective name for the female elements of a FLOWER. The gynoecium is composed of a flower's CARPELS, consisting of the STIGMA, STYLE, and OVARY. *See also* ANDROECIUM

gypsum Most common sulfate mineral, hydrated calcium sulfate ($CaSO_4.2H_2O$). It is formed by precipitation from evaporating seawater. Huge beds of gypsum occur in SEDIMENTARY ROCKS, where it is associated with halite. It can be clear, white or tinted and it crystallizes in the monoclinic system as prismatic or bladed CRYSTALS. Varieties are ALABASTER (massive); selenite (transparent and foliated); and satinspar (silky and fibrous). It is a source of plaster of Paris. Hardness 2; r.d. 2.3.

gyrocompass Navigational aid incorporating a

gimbal rings / revolving wheel

▲ **gyroscope** Stability is the vital contribution to the science of measurement made by the gyro. This instrument depends on a rapidly spinning wheel with a heavy rim which is suspended, with a minimum of friction, in a system of gimbal rings that allow it to rotate on its axis in any plane. If momentum is maintained at a given speed (by an electric motor for instance), the axis of the wheel maintains the position it took up when first spun. As the Earth revolves, the axis continues to point to a particular position in space, although the gimbals change their angle relative to it. Early applications were gun-sighting at sea and torpedo steering.

LEPTONS, they have a measurable size and a sub-structure of QUARKS. *See also* ANTIMATTER

Haeckel, Ernst Heinrich (1834–1919) German scientist and philosopher. Haeckel made the famous statement that "ontogeny recapitulates phylogeny" – that the stages of an embryo's development represent past stages of the organism's evolution. The theory has been superseded, but it provided a stimulus for work in the field of embryology.

Haber, Fritz (1868–1934) German physical chemist whose early work involved ELECTROCHEM-ISTRY and THERMODYNAMIC gas reactions. With Carl BOSCH he invented (1908–09) the HABER PROCESS for converting atmospheric nitrogen into ammonia. He was awarded the 1918 Nobel Prize for chemistry.

Haber process Industrial process (invented by Fritz HABER and Carl BOSCH) in which nitrogen from the atmosphere is "fixed" by synthesizing ammonia. A mixture of nitrogen and hydrogen is passed over a heated catalyst at a pressure of about 1,000 atmospheres. The chemical reaction, $N_2 + 3H_2 \rightarrow 2NH_3$, is EXOTHERMIC, which in turn speeds up the reaction. *See also* NITRO-GEN FIXATION

habitat Place in which an organism normally lives. A habitat is defined by characteristic physical conditions and the presence of other organisms.

hacker Person who obtains unauthorized access to a computer DATABASE through a communications system. A hacker may read or even alter the information in the database. Government departments, companies and other organizations provide employees with access to their computer databases through the public telephone system. Anyone with suitable equipment and SOFTWARE can gain access from a computer if they know the right telephone numbers and are able to discover the right passwords to enter after contact is made. *See also* MODEM

Hadley cell Atmospheric circulation cell, named after the British scientist George Hadley. Hadley proposed it (1735) to explain the trade winds, in which winds rise and flow poleward from the equator and then descend and flow equatorward, transferring heat by CONVECTION.

hadron Class of ELEMENTARY PARTICLE with STRONG NUCLEAR FORCE interaction. The group can be divided into BARYONS, such as the NEUTRON and PROTON, and MESONS, such as the PION and KAON. Hundreds of hadrons have been discovered, mostly since about 1950. With the exception of the proton and antiproton, they are all unstable. Unlike

hagfish (slime eel) Eel-like, primitive, jawless fish found in temperate-to-cold marine waters. It has underdeveloped eyes, and four to six fleshy whiskers around its sucking mouth; it also secretes a slimy mucus from pores along its sides. It is a scavenger and feeds on dead or dying fish. Length: 80cm (30in). Family Myxinidae

hafnium (symbol Hf) Silver TRANSITION ELE-MENT, first discovered in 1923. Hafnium's chief source is as a by-product in obtaining the element ZIRCONIUM. It is used as a MODERATOR in NUCLEAR REACTOR control rods. Properties: at.no. 72; r.a.m. 178.49; r.d. 13.31; m.p. 2,227°C (4,041°F); b.p. 4,602°C (8,316°F); most common isotope ^{180}Hf (35.24%).

Hahn, Otto (1879–1968) German chemist, who co-discovered nuclear FISSION in 1939 with Fritz STRASSMANN (for which Hahn won the 1944 Nobel Prize for chemistry). With Lise MEITNER, he discovered protactinium and several isomers. During World War II he remained in Germany and, after the war, was appointed president of the Max Planck Institute in Berlin.

Hahnemann, Christian Friedrich Samuel (1755–1843) German physician. He popularized HOMEOPATHY, a system of medical treatment based on the idea that "like cures like," or that a disease should be treated with minute doses of agents that produce the symptoms of the disease.

hahnium (symbol Ha) Synthetic, radioactive, TRANSACTINIDE ELEMENT. It has atomic number 105; six isotopes have been synthesized. It was first reported by a Soviet team at the Joint Institute for Nuclear Research at Dubna. They claimed the isotopes of mass numbers 260 and 261, as a result of bombarding AMERICIUM with neon ions. In 1970 a team at the University of California claimed the isotope 260 (half-life 1.6 seconds) obtained by bombarding CALIFORNIUM with nitrogen nuclei. The element is named after Otto HAHN.

hail PRECIPITATION from clouds in the form of balls of ice. Hailstorms are associated with atmospheric turbulence extending to great heights together with warm, moist air nearer the ground. Hailstones are usually less than 1cm (0.4in) across but some have exceeded 13cm (5in).

hair Outgrowth of mammalian SKIN; it has insulating, protective, and sensory functions. In MAMMALS, a thick coating of hair is usually called FUR. Hair grows in a FOLLICLE, a tubular structure extending down through the EPIDERMIS to the upper DERMIS. New cells are continually added to the base of the hair; older hair cells become impregnated with KERATIN and die. A muscle attached to the base of the hair allows it to become erect in response to nerve signals sent to the follicle. Erecting the hairs traps a thicker layer of air close to the skin. There are about one million hairs on the head of an average person. Each hair grows about 5-10mm a month, and grows for about three years before falling out and being replaced. *See also* CILIA

Haldane, John Burdon Sanderson (1892–1964) British scientist whose work formed the basis of the mathematical study of population genetics. His book *The Causes of Evolution* (1933) examined the theory of NATURAL SELECTION in the light of modern genetic research. In 1932 he was the first to estimate the mutation rate of a human gene. His book *Daedalus, or Science and the Future* (1924) was an early attempt to popularize science.

Hale, George Ellery (1868–1938) US astronomer who organized a number of observatories. He planned and secured finance for the 1-m (40-in) refracting telescope at the Yerkes Observatory in Williams Bay, Wisconsin, completed in 1897, the 2.5-m (100-in) Hooker reflector at Mount Wilson Observatory in Pasadena, California, completed in 1917, and the 5-m (200-in) reflector at Mount Palomar Observatory, near San Diego, California, which entered regular service in 1949. Each was in turn the largest telescope of its day. The 5-m (200-in) was named the Hale telescope in his honor. Hale was an accomplished solar observer, and he also invented the spectroheliograph.

half-cell ELECTRODE immersed in an ionic solution. A full ELECTROLYTIC CELL is formed of two connected half-cells – different solutions being separated by a membrane or conducting bridge that allows electricity to flow but prevents mixing. *See also* ELECTRODE POTENTIAL

half-life Time taken for one-half of the nuclei in a given amount of a radioactive ISOTOPE to decay (change into another element or isotope). Only

the half-life is measured because the decay is never total. Half-lives remain constant under any temperature or pressure, but there is a great variety among different isotopes. Oxygen-20 has a half-life of 14 seconds and uranium-234 of 250,000 years. A radioactive isotope disintegrates by giving off alpha or beta particles, and a measurement of this rate of emission is a way of recording decay. The term "half-life" also refers to particles that spontaneously decay into new particles, such as a free neutron transforming into a proton and an electron. *See also* CARBON DATING; RADIOACTIVE DECAY

halide Salt containing one of the HALOGENS (elements in group VII of the periodic table); examples are sodium fluoride, hydrogen bromide, and potassium chloride and iodide. The haloalkanes (alkyl halides) are organic compounds, such as chloromethane (methyl chloride, CH_3Cl), containing an alkane group bound to a halogen atom.

halite Mineral form of sodium chloride (NaCl), or common (rock) salt. It is found in evaporite SEDIMENTARY ROCKS, in association with gypsum and in salt domes and dried lakes. It is colorless, white, orange or grey, with a glassy luster. It has a cubic system of interlocking cubic crystals, granules, and masses. It is important as table salt and as a source of CHLORINE. Hardness 2.5; r.d. 2.2.

Hall effect Generation of an ELECTROMOTIVE FORCE across a conductor or semiconductor when it is carrying a current perpendicular to a magnetic field around it. The voltage is at 90° to both the direction of the current and the direction of the magentic field. The effect was discovered in 1879 by the US physicist Edwin H. Hall (1855–1938) and is used to test the electrical properties of materials and in ascertaining the strength of magnetic fields.

Halley, Edmond (1656–1742) British astronomer and mathematician. His most famous achievement in astronomy was to realize that comets could be periodic, following his observation of HALLEY'S COMET in 1682. It was Halley who persuaded Isaac NEWTON to write the *Principia*, and he also financed its publication. Halley's other achievements are numerous. He founded modern GEOPHYSICS, charting variations in the Earth's magnetic field, and establishing the magnetic origin of the AURORA borealis. In meteorology he showed that ATMOSPHERIC PRESSURE decreases with altitude, and studied monsoons and trade winds. In mathematics, he showed how to use mortality statistics to cost life insurance policies.

Halley's comet Bright periodic COMET. It takes 76 years to complete an orbit that takes it from within Venus's orbit to outside Neptune's. It was observed by Edmond HALLEY in 1682; later he deduced that it was the same comet that had been seen in 1531 and 1607, and predicted it would return in 1758. There are historical records of every return since 240 BC. Several space probes were sent to Halley on its 1986 return. The closest approach was by the probe Giotto, whose cameras showed the nucleus to be an irregular object measuring 15 × 8km (9 × 5mi). The main constituent was found to be water-ice, but the surface was covered with a dark deposit. Some craters were visible.

hallucinogen Drug that causes hallucinations, unusual perceptions without external cause. Hallucinogenic drugs, such as mescaline and LSD (lysergic acid diethylamide), have been used in primitive religious ceremonies and are illicitly taken today.

haloalkane (alkyl halide) Organic chemical compound consisting of an ALKANE in which one or more hydrogen atoms have been substituted by a HALOGEN element. Haloalkanes are used as solvents and in organic synthesis. Examples include iodomethane (methyl iodide, CH_3I), dibromomethane (methylene bromide, CH_2Br_2), trichloromethane (chloroform, $CHCl_3$), and tetrachloromethane (carbon tetrachloride, CCl_4). They are made by halogenating an alcohol or by the direct reaction between a halogen and an alkane in ultraviolet light. Chlorofluorocarbons (CFCs) are haloalkanes. *See also* HALOGENATION

halogen Element (FLUORINE, CHLORINE, BROMINE, IODINE, and ASTATINE) belonging to group VII of the periodic table. They react with most other elements and with organic compounds; reactivity decreases down the group. The halogens are highly electronegative and react strongly because they require only one electron to achieve the "stable 8" inert gas configuration; they produce crystalline salts (HALIDES) containing negative ions of the type F^- and Cl^-. The name halogen means salt-producer.

halogenation Introduction of one of the HALOGEN elements – fluorine, chlorine, bromine, iodine or astatine – into an organic compound either by addition or by substitution of an atom or group of atoms. This reaction is widely used in making pharmaceuticals and dyes.

halon Any of several organic gases used in fire extinguishers. Chemically halons can be considered as simple hydrocarbons (such as METHANE or ETHANE) that have had some or all of their hydrogen atoms replaced by a HALOGEN. They are similar to CHLOROFLUOROCARBONS (CFCs) but up to ten times more destructive of the OZONE LAYER. The most commonly used halon is bromotrifluoromethane ($CBrF_3$, halon 1301).

halophyte Any plant, usually a GYMNOSPERM (seed-bearing plant), that is able to live in a salty environment. Typical examples include mangrove trees, sea lavender, and rice grass.

Hamilton, William D. (1936–) New Zealand biologist. He developed a theory in GENETICS that sought to explain the evolution of ALTRUISM in animals in terms of Darwinian NATURAL SELECTION. Hamilton theorized that if an organism sacrificed itself to save its relatives, it was doing so to ensure that at least some of its own GENE variants, or ALLELES, were passed on to the following generation. The success with which an individual's alleles are passed on is called **inclusive fitness**. His theories were taken up by Richard DAWKINS.

Hamilton, Sir William Rowan (1805–65) Irish mathematician. In 1827 Hamilton was appointed professor of astronomy at Trinity College, Dublin, even before he had formally graduated with his first degree. He worked in the field of complex numbers, but his greatest achievement was the creation of QUATERNIONS.

hamstring *See* ACHILLES TENDON

hanging valley Tributary valley that ends high up the face of a larger main valley, possibly with a stream running through it and ending in a waterfall. Most hanging valleys result from glacial deepening of the main valley. *See also* GLACIER

haploid Term describing a CELL that has only one member of each CHROMOSOME pair. All human cells except GAMETES are DIPLOID, having 46 chromosomes. A gamete (OVUM and SPERM) is haploid, having 23 chromosomes. The body cells of many lower organisms, including many algae and single-celled organisms, and also moss plants, are also haploid. *See also* ALTERNATION OF GENERATIONS; MEIOSIS

hard disk Rigid MAGNETIC DISK for storing COMPUTER PROGRAMS and DATA. The built-in hard disk drive in a typical personal COMPUTER consists of a number of hard platters coated with a magnetic material set on a common spindle. They are housed inside a sealed container, with a motor to spin the stack of platters, a head to write (record) and read (replay) each side of each platter, and associated electronic circuits. Hard disk capacity is continually being increased: most computers are now sold with a disk that ranges from 1 to 5 gigabyte capacity.

Harden, Sir Arthur (1865–1940) British biochemist. He showed that each stage in the FERMENTATION of sugar is catalyzed by a specific ENZYME and that the essential first step is the attachment of a phosphorus-containing group to the sugar. For this research he shared the 1929 Nobel Prize for chemistry with Hans von Euler-Chelpin (1873–1964). His work laid the foundation for the elucidation of the KREBS CYCLE, which describes the process of glucose METABOLISM.

hardness Resistance of a material to abrasion, cutting or indentation. The MOHS SCALE is a means of expressing the comparative hardness of materials.

▲ **hard disk** A computer hard disk is made up of multiple rotating platters (1) each one of which has circular magnetic tracks (2) that are read and written on by a magnetic head (3) held by an arm (4). The disks spin at 100 times per second. The magnetic heads, tiny electromagnets, align magnetic particles on the surface of the platters to represent a digital code of zeros and ones (5). The magnetic tracks on the platters are divided into sectors (6) and when information is written on the hard disk files are split into different sectors on the platters (7). A file allocation table tells the chip (8) controlling the hard disk where information is held on the platters.

hardness of water Reluctance of water to produce a lather with soap, due to various dissolved salts, mainly those of calcium and magnesium. These salts give rise to an insoluble precipitate, which causes "fur" or "scale" in boilers, pipes, and kettles. Lather is inhibited until all the dissolved salts are precipitated as scum, which floats on the surface. Hardness may be temporary (removed by boiling), caused by calcium bicarbonate; or permanent (not affected by boiling), caused by calcium sulfate.

hardware Equipment as opposed to the programs, or software, with which a COMPUTER carries out various functions. The computer, keyboard, printer and electronic circuit boards are examples of hardware.

Hargreaves, James (1722–78) British inventor and industrialist. In 1764, while working as a weaver at Stanhill, Lancashire, England, Hargreaves invented the SPINNING JENNY. This machine greatly speeded the spinning process of cotton by producing eight threads at the same time. In 1768 local spinners destroyed his house and equipment, fearing that Hargreaves and his machines threatened their jobs. Hargreaves moved to Nottingham and, with Thomas James, built a spinning mill and became one of the first great factory owners.

harmonic progression Sequence of the form $1/a$, $1/b$, $1/c$, ..., where a, b, c, and so on, form an ARITHMETICAL PROGRESSION. The simplest is formed by the reciprocals of the positive integers: 1, 1/2, 1/3, 1/4, ... PYTHAGORAS discovered that taut strings with lengths proportional to these terms (and with identical diameter and tension) vibrate with harmonic musical tones.

harmonics (overtones) In acoustics, additional notes the frequencies of which are multiples of a basic (fundamental) note. When a violin string is plucked or a drumskin tapped, the sounds emitted correspond to vibrations of the string or skin. The loudest sound (note) corresponds to the fundamental mode of vibration. But other, weaker notes, corresponding to subsidiary vibrations, also sound at the same time. Together these notes make up a harmonic series.

Harriot, Thomas (1560–1621) English mathematician who is best known as one of the founders of the British school of algebra. Harriot (or Horiot) introduced simplified notation and contributed to the theory of equations. He also made astronomical telescopes, observed Halley's comet (1607) and discovered the moons of Jupiter independently of Galileo.

Harvey, William (1578–1657) English physician and anatomist who discovered the circulation of the blood. This landmark in medical history marked the beginning of modern physiology. His findings, published (1628) in *De Motu Cordis et Sanguinis* (*On the Motions of the Heart and Blood*), were ridiculed at first and only later generally accepted. Harvey also carried out important studies in EMBRYOLOGY. *See also* CIRCULATORY SYSTEM

Hassel, Odd (1897–1981) Norwegian chemist who shared the 1969 Nobel Prize for chemistry with Derek BARTON for his work on conformational analysis (the study of the three-dimensional structures of molecules). In 1930 he began research on aromatic organic compounds and discovered the two stereoisomers of cyclohexane.

haustorium Specialized invasive sucker-like or tubelike structure in a parasitic plant or fungus. It is used to penetrate the outer tissues of a host plant in order to absorb nourishment.

Hawking, Stephen William (1942–) British theoretical physicist. He studied at Oxford and later at Cambridge, where he became a professor of mathematics. During the course of his research into RELATIVITY, Hawking studied the BIG BANG origin of the Universe and did much pioneering work on the theory of BLACK HOLES. He put forward the existence of **mini black holes** (objects the size of a subatomic particle that have a mass of a thousand million tons). He found that black holes, by abiding to quantum theory, should lose energy – by emitting subatomic particles called **Hawking radiation** – and eventually "evaporate." This work also helped to link the laws of classical thermodynamics with quantum mechanics. He published a best-selling, popular account of his work in *A Brief History of Time* (1988). Other works include *Superpsace and Supergravity* (1981) and *The Very Early Universe* (1983).

Hawksbee, Francis (died *c.*1713) English experimental physicist and student of Robert BOYLE. Hawksbee was responsible for the double-cylindered air pump, with which he showed that air glows at low pressure when excited by an electric discharge. He also demonstrated that the transmission of sound through air is dependent on pressure.

Haworth, Sir Walter Norman (1883–1950) British organic chemist and Fellow of the Royal Society. Educated at Manchester and Göttingen, he held academic posts in London, St Andrews, and Newcastle and was Director of Chemistry at

▲ **hardness of water**
Furring in kettles and water pipes is caused by the presence in water (1) of dissolved calcium carbonate (2), usually from the chemical weathering of limestone. In hot or boiling water, calcium carbonate precipitates, forming solid limescale deposits on surfaces, such as the inside of kettles. Calcium carbonate also prevents soap from lathering. In an ion exchange tank (3), the tank is filled with grains of sodium-coated material with which the water has to come into contact (4). Sodium ions (5), which are more reactive than calcium, are exchanged for calcium ions (6). Because of the different properties of the sodium ions, the sodium salts formed remain in solution even when boiled.

Birmingham from 1925–48. He determined the constitution of vitamin C and various carbohydrates, sharing the 1937 Nobel Prize for chemistry with Paul KARRER for this work.

hearing Process by which SOUND waves are experienced. Humans and many other mammals hear when sound waves enter the EAR canal and vibrate the eardrum. The vibrations are transmitted by three small bones (ossicles) to the COCHLEA. In the cochlea, receptors generate nerve impulses that pass via the auditory nerve to the brain to be interpreted as sound.

hearing aid Electronic sound-reproducing device to increase the sound intensity at the ear. Modern aids use a small crystal microphone, a microminiaturized battery-powered amplifier and an earpiece (often shaped to fit into the auditory canal). The device consists either of two parts connected by a thin wire (the battery and amplifier being hidden in the clothing) or of one tiny unit placed in or behind the ear or housed in eyeglass frames.

heart Muscular organ that pumps BLOOD throughout the body by means of the CIRCULATORY SYSTEM. In humans, the heart is located behind the breastbone between the lower parts of the lungs. Divided longitudinally by a muscular wall, the right side contains only deoxygenated blood, the left side only oxygenated blood. Each side is divided into two chambers, an ATRIUM and a VENTRICLE, both separated by valves. The average heartbeat rate for an adult at rest is 70–80 beats per minute.

heart attack (myocardial infarction) Death of part of the heart muscle due to coronary occlusion, the blockage of a coronary ARTERY by a blood clot (thrombosis). It is accompanied by gripping chest pain, sweating, and vomiting. The greatest danger is within the first few hours. Modern drug treatment treats abnormal heart rhythms and dissolves the clots in the coronary arteries. **Heart failure** occurs when the heart is unable to pump blood at the rate necessary to supply body tissues. It may be due to high BLOOD PRESSURE (HYPERTENSION) or heart disease. Symp-

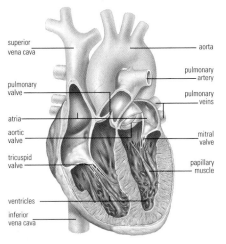

superior
vena cava
aorta
pulmonary
artery
pulmonary
valve
pulmonary
veins
atria
aortic
valve
mitral
valve
tricuspid
valve
papillary
muscle
ventricles
inferior
vena cava

▲ **heart** The human heart consists of four chambers – two atria and two ventricles – and four sets of valves. Blood from around the body passes into the right atrium, via the vena cavae. Flow of blood into the right ventricle is controlled by the tricuspid valve. Pulmonary arteries carry blood from the right ventricle to the lungs, while the pulmonary veins carry oxygenated blood back from the lungs to the left atrium. In a similar way, the mitral valve controls the flow of blood between the left atrium and the left ventricle. The aorta conducts the oxygenated blood from the left ventricle to all parts of the body.

toms include shortness of BREATH, EDEMA, AND fatigue. Treatment is with a DIURETIC and heart drugs. *See also* ANGINA

heart-lung machine Apparatus used during surgery to take over the function of the heart and lungs. Its use allows the patient's own circulation to be interrupted briefly – for example, to enable open-heart procedures. It consists of a pump to circulate blood around the body and special equipment to add oxygen to the blood and remove carbon dioxide from it.

heart transplant *See* ORGAN TRANSPLANT

heat Form of energy associated with the constant vibration of atoms and molecules. It was once thought to be a material substance called caloric that was contained in all objects and could flow from one to another. The KINETIC THEORY, presently accepted, holds that the degree of hotness of a body depends on the extent of vibration of its atoms. Heat is distributed in three forms: CONVECTION through FLUIDS (gases and liquids); CONDUCTION through solids; and RADIATION mainly through space. Heat and TEMPERATURE are sometimes confused; temperature is only one factor on which the heat content of an object depends. Heat and mechanical energy are interchangeable – for example hands can be warmed by rubbing them because friction produces heat.

heat balance (heat budget) Balance between the Sun's heat received by the Earth's atmosphere and that returned back into space. About two-thirds of solar radiation is absorbed by clouds, the atmosphere, and the Earth, and about one-third is reflected back, mainly by clouds. The heat absorbed powers the circulation of the atmosphere, oceans, and water cycle. Eventually the heat is reradiated into space, maintaining the Earth's heat equilibrium. *See also* CIRCULATION, ATMOSPHERIC; GREENHOUSE EFFECT

heat capacity (thermal capacity, symbol C) Ratio of the heat supplied to an object to its resultant rise in temperature. It is measured in joules/kelvin. For most purposes, the alternative SPECIFIC HEAT CAPACITY is used.

heat engine Any engine that converts heat energy, usually through the burning of a fuel, into useful mechanical energy. Thus, all INTERNAL COMBUSTION ENGINES are heat engines.

heat exchange (heat transfer) Flow of heat energy from one object to another. This flow of energy occurs at all times when two or more bodies at different temperatures are in thermal contact. Three methods of heat transfer may be distinguished: CONDUCTION, CONVECTION, and RADIATION. In conduction, heat is transferred from molecule to molecule within a body, as in an iron rod stuck in a fire; in convection, heat is transferred by circulation of FLUID, as in boiling; in radiation, heat is transferred in the form of electromagnetic waves, as in sunlight. **Heat exchangers** are essential in many industrial processes where heat may be extracted from one source for use elsewhere without the two sources combining. The most simple heat exchangers use conduction to achieve this. For example, a hot fluid is run through tubes running parallel to one another (and with a high surface area). This arrangement of tubes is immersed or built around a container into, and through, which another cold fluid flows. By conduction the heat is transferred from the hot fluid to the cold.

▲ **heat exchange** The three ways in which heat moves all take place when a pan is heated (A) – conduction through the metal walls of the pan (1), convection by fluid motion (2) and radiation from the heat source to the pan (3). In theory an insulated good conductor with ice at one end and boiling water at the other varies in temperature linearly with distance along the bar (B), as in the straight-line graph. With poor insulation a curve represented by the dotted line results. A vacuum flask (C) has a vacuum (4) to prevent conduction and convection and silvered walls (5) to minimize heat loss by radiation.

heat pump Device employed mainly in refrigerators and central cooling systems. In a refrigeration cycle, a liquid boiling at about $-30°C$, usually a FREON, is evaporated at low pressure in metal coils. This absorbs heat from the surrounding space, so producing the cooling effect. The freon vapor then passes to a pump driven by an electric motor, which delivers it at raised temperature and pressure to a condenser outside the refrigerated area. As the vapor condenses to a liquid it gives up heat to the surrounding space. The liquid, still under pressure, passes to a receiver, from which the suction intake of the pump withdraws it at lowered pressure back to the evaporator in the refrigerated area, so completing the cycle.

heat shield In aerospace technology, plate or apron attached to the leading edge of a spacecraft to absorb or dissipate heat upon re-entry into the Earth's atmosphere. It is constructed of ablative materials, which chip off or vaporize to dissipate heat and provide an insulating layer of gases to protect the spacecraft. Materials used include quartz and various plastics.

Heaviside-Kennelly layer (E-layer) Region within the IONOSPHERE of the Earth's ATMOSPHERE.

heavy metal Any metal of high density (high ATOMIC NUMBER). They are used in electron microscopy to "stain" biological specimens in order to reveal details of structure. The term may also refer to metallic pollutants in soil that restrict plant growth and to toxic metals such as cadmium and lead.

heavy water (deuterium oxide, symbol D_2O)

Water in which the role of hydrogen atoms is played by DEUTERIUM, the isotope of HYDROGEN of RELATIVE ATOMIC MASS (R.A.M.) approximately 2 (normal hydrogen has a relative atomic mass of approximately 1). It occurs in small proportions in ordinary water, from which it is obtained by ELECTROLYSIS. Heavy water is used as a MODERATOR in some NUCLEAR REACTORS.

hectare Unit of area, equal to 2.47 acres. One hectare equals $10,000m^2$ ($107,640ft^2$) while 100 hectares are $1km^2$ ($0.4mi^2$)

Heisenberg, Werner Karl (1901–76) German physicist and philosopher, professor at Leipzig and later in Berlin. He is best known for his discovery in 1927 of the UNCERTAINTY PRINCIPLE (1927). He was awarded the 1932 Nobel Prize for physics for his theory of matrix mechanics, a form of QUANTUM MECHANICS.

helicopter Aircraft that gains lift from power-driven rotor(s). The helicopter is capable of vertical take-off and landing (VTOL), hovering, and

▲ **helicopter** A helicopter rotor head transfers the power of the engines to the rotor blades via a number of gears (1) and the rotor shaft (2). The swash plate controls the tilt of the rotor (3) and also the pitch of the blades (4). The upper (5) and lower (6) swash plates are controlled by hydraulic cylinders (7) attached to the lower plate. The upper plate is connected to the rotor blades by control rods (8). The pitch of the blades controls the amount of lift generated, while the attitude of the whole rotor controls which direction the helicopter moves horizontally. If the rear of the swash plate is raised (9) the rotor dips toward the nose of the helicopter, causing the aircraft to travel forward.

forward, backward, and lateral flight. To maneuver, helicopters have a complex system of hydraulic cylinders, control rods, and plates that alter the orientation of the rotor blades. *See also* ROTOR, HELICOPTER

heliosphere Region of space in which the SUN's magnetic field and the SOLAR WIND dominate the interstellar medium. It is similar to a planet's MAGNETOSPHERE, with an outer bow shock, and a boundary called the **heliopause** where the energy of the solar wind particles has fallen to the level of galactic cosmic ray particles. Measurements returned by the Voyager probes suggest that the heliosphere extends for 100–150AU.

helium (symbol He) Gaseous, nonmetallic element, a NOBLE GAS, discovered in 1868. The element was first obtained in 1895 from the mineral clevite; the chief source today is from natural gas. It is also found in some radioactive minerals and in the Earth's ATMOSPHERE (0.0005% by volume). Helium is the second-most common element in the Universe and the second lightest element

(after HYDROGEN). It has the lowest melting and boiling points of any element. A colorless, odorless nonflammable gas, it is used in meteorological balloons, to make artificial "air" (mixed with oxygen) for deep-sea divers, and in welding, semiconductors, LASERS, metallurgy, and other applications requiring an INERT atmosphere. Liquid helium is used in CRYOGENICS (physics dealing with low temperatures). The element forms no chemical compounds. Properties: at.no. 2; r.a.m. 4.0026; r.d. 0.178; m.p. $-272.2°C$ ($-458°F$); b.p. $-268.9°C$ ($-452.02°F$); single isotope ^4He.

helix Curve generated when a point moves over the surface of a CYLINDER in such a way that the path it traces is inclined at a constant angle to the axis of the cylinder, as in a coiled spring or the thread of a bolt.

helix, double *See* DOUBLE HELIX

Helmholtz, Hermann Ludwig Ferdinand von (1821–94) German anatomist, physicist, and physiologist. He made contributions in ACOUSTICS and OPTICS, expanding the three-colour theory of vision formulated by Thomas YOUNG and inventing an OPHTHALMOSCOPE and an ophthalmometer. His experiments on the speed of nerve impulses led to a study of animal heat, which in turn led to work on the principle of conservation of energy.

Helmont, Jan Baptista van (1580–1644) Flemish chemist, physician, and physiologist. He believed in the TRANSMUTATION of metals and that water was the overall basic substance of the Earth. He carried out quantitative experiments and coined the word "gas" (Greek *chaos*). Helmont also discovered the gas that we now call carbon dioxide.

hematite One of the most important iron ores, containing mainly ferric oxide (Fe_2O_3). Containing 70% iron by weight, it occurs in several forms, frequently in small dome-shaped masses called **kidney-ore**. Hematite varies in color from steel-gray to black, but can ocassionally be found in a red form. Deposits are found on all the world's continents. Hardness 5–6; r.d. 4.9–5.3.

hemoglobin Protein present in ERYTHROCYTES (red blood cells) of vertebrates. Hemoglobin carries oxygen to all cells in the body. It is scarlet when combined with oxygen to form OXYHEMOGLOBIN and bluish-red when deoxygenated. Oxygen attaches to the hem– part of the protein, which contains iron; the –globin part is a globular PROTEIN. Worn-out erythrocytes are destroyed by the liver, and the iron is used again to make more hemoglobin. *See also* BLOOD

hemophilia Hereditary blood-clotting disorder in which there is prolonged external or internal bleeding, often without apparent cause. The most common form, **hemophilia A**, is due to an inability to synthesize FACTOR VIII, a substance essential to blood clotting. Bleeding into the joints may cause joint deformities. The gene for the disorder is passed on almost exclusively from mother to son. About one in 10,000 boys are born with hemophilia. It is treated by injections of Factor VIII. A rarer form, **hemophilia B** or Christmas disease, is caused by a deficiency of blood Factor IX. It too is inherited and follows the same course as hemophilia A.

hemiplegia Paralysis of one side of the body. It can be caused by brain injury, and may be accompanied by spasticity.

Hemiptera Order of the class Insecta (*see* INSECT), having two pairs of wings and symmetrical mouthparts that are adapted for piercing and sucking.

Hench, Philip Showalter (1896–1965) US physician. He shared the 1950 Nobel Prize for physiology or medicine with Edward KENDALL and Tadeus REICHSTEIN for their pioneering work on hormones of the adrenal cortex. Their work led to discovery of the treatment of rheumatoid ARTHRITIS with the hormone CORTISONE. *See also* ADRENAL GLAND

Henry, Joseph (1797–1878) US physicist whose contribution to ELECTROMAGNETISM was essential for the development of the commercial telegraph. His work on INDUCTION led to the development of the TRANSFORMER. In 1850 he introduced a system of using the telegraph for sending weather reports, making possible the establishment of the US Weather Bureau. The unit of inductance is named after him.

Henry, William (1774–1836) British chemist and physician whose most important work was his study of the solubility of gases in water under varying temperatures and pressures. This led to his formulation in 1803 of HENRY'S LAW.

Henry's law Principle in physical chemistry stating that the mass of a gas dissolved in a fixed quantity of a liquid is directly proportional to the pressure of the gas on the liquid, at constant temperature. Thus the more a gas is compressed, the more it is absorbed in a liquid. The principle was named after William HENRY.

henry (symbol H) Unit of INDUCTANCE equal to the inductance of a closed loop that gives rise to a MAGNETIC FLUX of one weber for each ampere of current that flows.

heparin Anticoagulant substance produced in the body that inhibits the production of THROMBIN, an ENZYME involved in blood clotting. A purified extract is used after surgery to minimize the risk of thrombosis and to prevent the formation of further clots in anyone who has had a pulmonary EMBOLISM.

hepatitis Inflammation of the liver. It is usually due to a generalized infection. Early symptoms include lethargy, nausea, fever, jaundice, and muscle and joint pains. Five different hepatitis viruses have come to light: A, B, C, D, and E. The most common single cause, accounting for up to 40% of hepatitis cases worldwide, is the hepatitis A virus (HAV), which colonizes the digestive tract, causing an acute illness. Much more serious, however, is infection with the hepatitis B virus (HBV). Formerly known as serum hepatitis, it can lead to chronic inflammation or complete failure of the liver and, in some cases, to liver cancer. Hepatitis C can also become chronic. HDV, common in the Mediterranean region, can only replicate in the presence of HBV. HEV is endemic in some tropical countries. Vaccines are available against HAV and HBV.

heptane Any chemical compound having the formula C_7H_{16}. Such compounds are saturated hydrocarbons (ALKANES). Normal heptane, or *n*-heptane, is used in car fuels to determine OCTANE ratings.

herb Seed-bearing plant (SPERMATOPHYTE), usually with a soft stem that withers away after one growing season. Most herbs are ANGIOSPERMS (flowering plants).

herbaceous Term for any nonwoody plant. Herbaceous plants can be either ANNUAL (dying completely after one season), BIENNIAL (taking two seasons to complete its lifecycle) or PERENNIAL (living for three or more seasons).

herbicide Chemical substance used to kill weeds and other unwanted plants. There are two kinds: selective herbicides kill the weeds growing with crops, leaving the crops unharmed; nonselective herbicides kill all the vegetation.

herbivore Animal that feeds solely on plants. Although strictly most herbivores are insects, the term is closely associated with mammals, especially UNGULATES (hoofed mammals). Mammalian herbivores are characterized by broad molars and blunt-edged teeth, which they use to pull, cut, and grind their food. Their digestive systems are adapted to the assimilation of CELLULOSE. In a standard FOOD CHAIN, herbivores are PRIMARY CONSUMERS.

hereditary disorder (genetic disorder) Any disorder passed on from parents to their young in defective genes or chromosomes. Some 3,000 such disorders are known, including DOWN'S SYNDROME, CYSTIC FIBROSIS, and SICKLE-CELL DISEASE.

heredity Transmission of physical and other characteristics from one generation of plants or animals to another. Characteristics such as blue eyes and red hair may be specific to the individuals involved; other features, such as erect posture and the possession of external ears, may be typical of the type of organism. The combination of characteristics that makes up an organism and makes it different from others is set out in the organism's GENETIC CODE, passed on from parent(s) to offspring. The first studies of heredity were carried out during the 19th century by Gregor MENDEL. *See also* ALLELE; CHROMOSOMES; GENE; SEXUAL REPRODUCTION

hermaphrodite Organism that has both male and female sexual organs. Most hermaphrodite animals are INVERTEBRATES, such as the earthworm and the snail. They reproduce by the mating of two individuals, each of which receives SPERM from the other. Sometimes, in both plant and animal hermaphrodites, self-FERTILIZATION takes place.

hernia (rupture) Protrusion of an organ, or part of an organ, through its enclosing wall or through connective tissue.

heroin Drug derived from MORPHINE. Compared to morphine, heroin requires a smaller dose to produce similar effects more quickly, and is less likely to cause nausea and vomiting. It is prescribed for the relief of pain in terminal illness and severe injuries. Illicit heroin users soon become addicted.

Hero of Alexandria (active *c*.1st–2nd century AD) Greek engineer who built an early steam turbine, called an **aeolipile.** He also devised mechanisms to control doors automatically and to power moving statues.

Herschbach, Dudley Robert (1932–) US physical chemist who used intersecting beams of atoms or molecules to study atomic and molecular motions in chemical changes and to record the energy of the reactant and product molecules. He also invented devices to identify the products. Herschbach shared the 1986 Nobel Prize for chemistry with Yuan Tseh Lee and John Polanyi for their work in chemical reaction dynamics. They originally used their new technique for the reaction between potassium atoms and iodomethane (methyl iodide) molecules and later employed it for numerous other systems.

Herschel, Sir John Frederick William (1792–1871) English scientist and astronomer, son of Sir William HERSCHEL. At first assisting his father with observations and telescope-making, he went on to extend his father's work on DOUBLE STARS and NEBULAE, discovering more than 500 more nebulae and clusters. In 1834 he took one of William's telescopes to the Cape of Good Hope

◀ **herbivore** Mammalian herbivores may conveniently share a habitat without competing for resources. On the African plains, giraffes (1) browse in branches up to 20ft (6m) above the ground. Elephants (2) too can browse tree canopies, using their trunks to pluck off vegetation. Eland (3) attack the middle branches with their horns, twisting twigs to break them off, while gerenuk (4) stand on their hind legs to reach higher branches. The black rhino (5) uses its hook-like upper lip to feed on bark, twigs, and leaves (white rhinos have lengthened skulls and broad lips for grazing the short grasses that they favor). The wart hog (6) and dik-dik (7) eat buds and flowers, and will also dig up roots and tubers. Such sharing of a single resource also occurs among grazers. Migrating zebra (8) crop the taller, coarse grasses; wildebeest (9) feed on the leafy center layer, allowing small gazelles (10) to reach the tender new shoots.

and undertook a systematic survey of the s sky, discovering more than 1,200 doubles and 1,700 nebulae and clusters. He combined these and his father's observations into a *General Catalogue of Nebulae and Clusters*. He made the first good direct measurement of solar radiation, and inferred the connection between the Sun and aurorae. His *Outlines of Astronomy* (1849) was a standard textbook for many decades. He was also a pioneer of photography and its application in astronomy.

Herschel, Sir William (1738–1822) (originally Friedrich Wilhelm Herschel) English astronomer, b. Germany. He became famous in 1781 by discovering the planet Uranus, which he wanted to name after King George III; the next year he was appointed the king's private astronomer. He discovered two SATELLITES of Uranus (1787), and two of Saturn (1789). He observed and catalogued many double stars, hoping to measure their distances by detecting the parallactic movement of the nearer against the further if they were optical doubles. He observed more than 2,000 nebulae and clusters and published catalogues of them. Herschel realized that the Milky Way is the plane of a disk-shaped stellar Universe, whose form he calculated by counting the numbers of stars visible in different directions, and also noted the motion of the Sun toward a point in the constellation Hercules. In 1800 he discovered and investigated the properties of INFRARED RADIATION.

Hershey, Alfred Day (1908–97) US biologist who shared the 1969 Nobel Prize for physiology or medicine for his part in discovering the way VIRUSES reproduce. Through his study of bacteriophages, Hershey confirmed that the NUCLEIC ACID RNA is solely responsible for infection and subsequent viral development. He was director of the Cold Spring Harbor Laboratory in New York from 1962–74.

Hertz, Gustav (1887–1975) German physicist, who shared the 1925 Nobel Prize for physics with his colleague James FRANCK for their research on the electron bombardment of gases, work that supported the QUANTUM THEORY. Hertz was Professor of Physics at Berlin and Director of the Siemens Research Laboratory until 1945. He worked as head of a research laboratory in the Soviet Union from 1945 to 1954, returning then to East Germany to become Director of the Physics Institute at Leipzig. He was the nephew of Heinrich HERTZ

Hertz, Heinrich Rudolf (1857–94) German physicist who discovered radio waves. He was assistant to Hermann HELMHOLTZ and held professorships at Karlsruhe and later at Bonn. He discovered, broadcast and received the radio waves predicted by

James Clerk MAXWELL. He demonstrated that heat and light are also part of the ELECTROMAGNETIC SPECTRUM. The unit of frequency, the HERTZ (Hz), is named after him. He was the uncle of Gustav HERTZ.

hertz (symbol Hz) SI unit of wave FREQUENCY named after Heinrich HERTZ. One oscillation per second is equivalent to 1Hz. This unit replaces cycle per second. Commonly used multiples are the kilohertz (kHz) equal to 1,000Hz and the megahertz (MHz) equal to 1,000,000Hz.

Hertzsprung, Ejnar (1873–1967) Danish astronomer who devised (1908) a diagram that illustrated the relationship between the LUMINOSITIES, surface temperatures, and SPECTRAL CLASSIFICATION of stars. He anticipated the work of Henry RUSSELL in the US a few years later. The diagram is now known as the HERTZSPRUNG–RUSSELL DIAGRAM.

Hertzsprung–Russell diagram (HR diagram) Plot of the absolute MAGNITUDE of stars against their SPECTRAL CLASSIFICATION; this is equivalent to plotting their LUMINOSITY against their surface temperature or color index. Brightness increases from bottom to top, and temperature increases from right to left. The diagram was devised by Henry RUSSELL in 1913, independently of Ejnar HERTZSPRUNG, who had had the same idea some years before. The HR diagram reveals a pattern in which most stars lie on a diagonal band, the MAIN SEQUENCE, which includes our Sun. There is another region of GIANT STARS, which lies above the main sequence. There are further groupings above the main sequence called SUPERGIANTS and subgiants. Below the main sequence are subdwarfs and WHITE DWARFS. The HR diagram is an important aid for interpreting astrophysical data.

Herzberg, Gerhard (1904–) Canadian scientist, b. Germany. Educated at Darmstadt, Göttingen and Bristol, he settled in Canada in 1935. He joined the National Research Council in Ottawa in 1949, and became their first Distinguished Research Scientist in 1969. He was awarded the 1971 Nobel Prize for chemistry for his work on the electronic structure and geometry of molecules, particularly FREE RADICALS.

Hess, Victor Francis (1883–1964) US physicist, b. Austria. As a result of his investigations into the ionization of air, he suggested that radiation similar to X-rays, later named COSMIC RADIATION, comes from space. He shared the 1936 Nobel Prize for physics with US physicist Carl ANDERSON, who had worked indpendently on cosmic rays. Hess then emigrated to the US in 1938 to become Professor of Physics at Fordham University in New York City.

Hess, Walter Rudolf (1881–1973) Swiss physiologist. He shared with António de Egas MONIZ the 1949 Nobel Prize for physiology or medicine for his work on how specific, highly circumscribed areas of the BRAIN control body functions, such as blood circulation, muscle relaxation, and breathing rate.

Hess's law In physical chemistry, law referring to the heat produced or absorbed during CHEMICAL REACTIONS leading to the preparation of any chemical compound. It states that the net value of this heat is the same by any chemical route.

heterocyclic compound *See* CYCLIC

heterodyne In radio, the production of one WAVE FREQUENCY as the difference of two others. It is used in radio-frequency OSCILLATORS and in receivers. The heterodyne system was invented by Reginald A. FESSENDEN in 1905.

heterostyly Condition found in some ANGIOSPERMS (flowering plants) of having STYLES of different lengths. The classic example of heterostyly is the primrose *Primula vulgaris*, in which there are two distinct flower forms, each occurring on different plants. Each form has a different-sized style and STAMEN from the other so that pollen collected by insects from the ANTHER of one type is dropped on the STIGMA of the other. Heterostyly increases the likelihood of CROSS-POLLINATION because pollen cannot be passed between plants with sex organs of the same size.

heterotroph Organism that obtains its energy from the digestion of organic matter (usually plant or animal tissue) through a process known as **heterotrophic nutrition**. All animals and fungi are heterotrophs; the digestive process breaks down the tissue to provide material with which the organism can synthesize essential nutrients, such as carbohydrates, protein, fat, vitamins, and minerals, in the case of humans.

heterozygote Organism possessing two contrasting forms (ALLELES) of a GENE in a CHROMOSOME pair. In cases where one of the forms is DOMINANT and one RECESSIVE only the dominant form will be expressed in the PHENOTYPE. *See also* HOMOZYGOTE

Hevesy, George Karl von (1855–1966) Hungarian-Swedish chemist, who was awarded the 1943 Nobel Prize for chemistry. Co-discoverer of the element HAFNIUM (Hf) in 1923, he was also an early researcher into the uses of radioactive ISOTOPES, including their use as "tracers" in living tissue. In the early 1940s he fled from Germany, where he had been teaching, to live in Sweden. He

worked with Ernest RUTHERFORD in England and later with Niels BOHR in Copenhagen.

Hewish, Antony (1924–) British radio astronomer. He and Martin RYLE developed the technique of **aperture synthesis,** which allows an array of radio dishes to synthesize the resolution of a much larger telescope, in 1960. In 1967 his student Jocelyn Bell Burnell (1943–) obtained the first signal from a PULSAR. For subsequent work on pulsars, Hewish shared (with Ryle) the 1974 Nobel Prize for physics.

hexacyanoferrate Complex ION of iron. There are two ions: hexacyanoferrate(II) ($[Fe(CN)_6]^{4-}$), formerly called ferrocyanide, and hexacyanoferrate(III) ($[Fe(CN)_6]^{3-}$), or ferricyanide. The potassium salts are used in chemical ANALYSIS for the presence of iron. Iron(III) hexacyanoferrate(II) (ferric ferrocyanide) is the dark blue pigment Prussian blue. The pigment was once used a source of HYDROCYANIC ACID, hence its former name prussic acid.

hexagon Six-sided plane figure. Its interior angles add up to 720°. A regular hexagon has all sides and interior angles equal; each interior angle is 120°.

Heyrovský, Jaroslav (1890–1967) Czechoslovak chemist. He was awarded the 1959 Nobel Prize for chemistry for the discovery of **polarography**, an electrochemical method of analysis that can be used with substances that undergo electrolytic OXIDATION (loss of electrons) or REDUCTION (gain of electrons) in solution.

hibernation Dormant (sleeplike) condition adopted by some animals to survive the harsh conditions of winter. Adaptive mechanisms to avoid starvation and extreme temperatures include reduced body temperature, a slower heartbeat, slower breathing rate, and slower rate of METABOLISM.

Higgs boson ELEMENTARY PARTICLE the existence of which is necessitated by the ELECTROWEAK THEORY. Predicted by particle physicist Peter Higgs (1929–), this boson has not yet been detected. It should have a finite mass but no SPIN.

high *See* ANTICYCLONE

high-definition television (HDTV) Form of TELEVISION in which the picture is made up of 1,250 or 1,125 scanning lines instead of the previous 625 or 525. The increased number of scanning lines makes the television image much clearer and sharper (comparable with the visual clarity of movie film), and relies on digital transmission along OPTICAL FIBERS, rather than the transmission of electronic signals by RADIO waves. The use of such cables greatly increases the total number of channels available, and also allows for an interactive facility between receiver and transmitter. The screens of many HDTV sets are proportionately wider and flatter than former models.

high-fidelity (hi-fi) Sound recording and reproduction with the minimum of distortion, usually involving the use of high-quality components and circuits. The original sound source may be a COMPACT DISC (CD), phonograph record, MAGNETIC TAPE, a radio signal or a live performance. An audio amplifier is required to "magnify" the signals, and one or more loudspeakers are needed to make the source audible. Domestic high-fidelity sound systems have been produced since the 1920s, but did not become common until the advent of long-playing phonograph records in the 1950s and 1960s.

high-level language COMPUTER LANGUAGE used for programming that is reasonably close to spoken English. The high-level language is converted to a form a computer can use by a COMPILER. Most languages in use today (such as Basic, Pascal or C) are high-level languages.

high-temperature physics Production and analysis of the effects of temperatures above 50,000°C (90,000°F). At such temperatures atoms begin to be ionized (stripped of their ELECTRONS) and a fourth state of matter, called a PLASMA, is achieved. Normal stars are plasmas; so is the matter at the heart of an exploding hydrogen bomb. To make controlled thermonuclear FUSION (joining of light nuclei and release of energy) a reality, physicists must find a way to confine a plasma at temperatures of over a million degrees. *See also* MAGNETIC BOTTLE; NUCLEAR REACTOR; NUCLEAR WEAPON

Hilbert, David (1862–1943) German mathematician, b. Russia, one of the most important figures in 20th-century mathematics. While professor of mathematics at Göttingen, Hilbert wrote *Foundations of Geometry*, which put geometry on a rigorous axiomatic basis. His *Numbers Information* systematized all the known results of algebraic number theory. He developed the concept of an infinite dimensional space, now called a **Hilbert space**, which was a tool necessary for the development of QUANTUM THEORY. In 1900 Hilbert put forward a list of 23 problems for the mathematicians of the new century to tackle. Most of these have led to profound new understandings in mathematics. Hilbert's dream was that it would be possible either to prove or to disprove all mathematical questions starting with only one set of well-defined rules and assumptions. In 1931 Kurt GÖDEL showed that this could not be achieved.

hindbrain One of the three primary parts of the

BRAIN. In the embryo, the hindbrain develops into the CEREBELLUM, the part of the brain concerned with unconscious muscle coordination, and also into the PONS and the MEDULLA OBLONGATA. *See also* FOREBRAIN; MIDBRAIN

Hinshelwood, Sir Cyril Norman (1897–1967) British chemist, president of the Royal Society (1955–60). From 1937–64 he was Professor of Chemistry at Oxford, and Research Fellow at Imperial College, London, from 1964. He shared the 1956 Nobel Prize for chemistry with Nikolai SEMENOV for their simultaneous work on chemical reaction kinetics.

Hinton, Christopher, Baron Hinton of Bankside (1901–83) British nuclear engineer. From 1946, as deputy controller of NUCLEAR REACTOR atomic energy production, he created a new industry in Britain with the world's first full-scale atomic power station at Calder Hall, which was opened in 1956.

hip Joint on each side of the lower trunk, into which the head of each FEMUR (thigh bone) fits; the hip bones form part of the PELVIS.

Hipparchus (*c*.190–127 BC) Greek astronomer. Hipparchus estimated the distance of the Moon from the Earth and drew the first accurate star map with more than 800 stars. He divided stars into orders of magnitude based on their brightness – a system fundamentally in use today. He also developed an organization of the Universe, which, although it still had the Earth at the center, provided for accurate prediction of the positions of the planets.

Hippocrates (460–377 BC) Greek physician, often called "the father of medicine." Although little is known about him, and the writings known as the *Hippocratic Collection* probably represent the works of several people, he exerted a tremendous influence. He freed medicine from superstition, emphasizing clinical observation and providing guidelines for surgery and for the treatment of fevers. He is credited with providing the **Hippocratic oath**, a code of professional conduct still followed by doctors today.

histamine Substance derived from the AMINO ACID HISTIDINE, occurring naturally in many plants and in animal tissues, and released on tissue injury. Histamine's several functions in the body include dilation of the capillaries. It is implicated in allergic reactions (such as hay fever), which can be treated with ANTIHISTAMINE drugs.

histidine ($C_3H_3N_2CH_2CHNH_2COOH$) Colorless, soluble, crystalline AMINO ACID, which is the precursor of HISTAMINE.

histochemistry Study of the distribution of biochemical substances and reactions in living tissue, usually through visual examination. *See also* BIOCHEMISTRY

histogram (bar chart) Chart in which bars represent the frequency (in absolute terms or as a percentage of the total) with which certain values (or ranges of value) occur within a given set of data. *See also* GRAPH

histology Biological, especially microscopic, study of TISSUES and structures in living organisms.

histone Basic protein found associated with DNA in CHROMOSOMES. Histones function during CELL DIVISION to condense CHROMATIN and coil the chromosomes. There are five kinds, each containing large amounts of the AMINO ACIDS arginine, histidine and lysine; they are soluble in water.

HIV *See* HUMAN IMMUNODEFICIENCY VIRUS

Hodgkin, Sir Alan Lloyd (1914–) British physiologist. He shared the 1963 Nobel Prize for physiology or medicine with John Eccles (1903–) and Andrew Huxley for his part in discoveries concerning the ionic mechanisms involved in the excitation and inhibition of the membranes of nerve cells (NEURONS). This work is fundamental to an understanding of the function of neurons themselves.

Hodgkin, Dorothy Mary Crowfoot (1910–94) British chemist. After teaching at Oxford and Cambridge universities, she spent some time at the University of Ghana, returning to Oxford in 1960. In 1964 she was awarded the Nobel Prize for chemistry for her determination of the structure of VITAMIN B_{12} by X-RAY CRYSTALLOGRAPHY. She also determined the structure of PENICILLIN and INSULIN.

Hodgkin's disease Rare type of cancer characterized by painless enlargement of the LYMPH NODES, lymphatic tissue, and spleen, with subsequent spread to other areas. Fever is a common symptom; and weight loss, anemia, loss of appetite, and night sweats may occur. It is named after the pathologist Thomas Hodgkin (1798–1866), who first described it. The condition is more common in men. Treatment varies with the stage reached by the disease but in general consists of RADIOTHERAPY, surgery, drug therapy or a combination of these. Caught in the early stages, it may be curable.

Hoff, Jacobus Hendricus van't (1852–1911) Dutch physical chemist, professor in Amsterdam, Leipzig, and then Berlin. His research involved advanced investigations of the carbon atom, the theory that gas laws are also applicable to dissolved substances, and the chemical application of THERMODYNAMICS. In 1901 he was awarded the first Nobel Prize for chemistry for his studies on the mechanism of CHEMICAL EQUILIBRIUM, reaction rates, and osmotic pressure. *See also* IDEAL GAS LAW; OSMOSIS

Hofmann, August Wilhelm von (1818–92) German chemist and teacher. After serving as an assistant to Justus von LIEBIG he was appointed director of the new Royal College of Chemistry in London. Most of his work was based on the compounds of COAL TAR and their derivatives. Hofmann discovered a method, known as the **Hofmann reaction**, for converting an AMIDE organic compound into an AMINE. He also discovered FORMALDEHYDE. One of his students, William Perkin (1838–1907), made the first ANILINE dye, the color mauve.

Hofstadter, Robert (1915–) US physicist who shared the 1961 Nobel Prize for physics with Rudolf MÖSSBAUER for their investigations of nuclear structure. Hofstadter proposed that the PROTON and the NEUTRON (both of which make up the NUCLEUS) have a positively charged central core surrounded by a cloud of ELEMENTARY PARTICLES (PIONS).

Holland, John Philip (1840–1914) US engineer, b. Ireland. He invented the first modern, gasoline-driven SUBMARINE. He offered a design to the US Navy Department in 1875, but it was rejected as impracticable. After further development, the successful *Fenian Ram* was launched in 1881. This was not practical for extended operations, but its design was the basis for the *Holland*, which had an INTERNAL COMBUSTION ENGINE for surface cruising and an electric motor for submerged operations. The US government bought the *Holland* in 1900.

Hollerith, Herman (1860–1929) US computer pioneer. In 1890 Hollerith invented a mechanical tabulating machine for use in analyzing the results of the US census of that year. The machine used punched cards to record and process data. These gave rise to the HOLLERITH CODE (later employed by early electronic computers), which uses 12 bits per alphanumeric character. Hollerith's firm, the Tabulating Machine Company, later became a founder of IBM.

Hollerith code Computer code consisting of 12 levels, or bits per character, which defines the relation between an alphanumeric character and the punched holes in an 80-column computer data card.

Holley, Robert William (1922–93) US biochemist. He shared with Har KHORANA and Marshall NIRENBERG the 1968 Nobel Prize for physiology or medicine for his work on the way GENES determine the function of CELLS. Holley described the first full sequence of subunits in nucleic acid, the genetic material of a cell. This was an important step toward the understanding of gene action.

Holmes, Arthur (1890–1965) British geologist and geophysicist who studied the age of rocks by measuring their radioactivity. He devised the first quantitative GEOLOGICAL TIMESCALE in 1913 and estimated the Earth's age by means of temperature measurement.

holmium (symbol Ho) Metallic element of the LANTHANIDE SERIES, first identified spectroscopically in 1878. Its chief ore is monazite (a phosphate). The element has few commercial uses. Properties: at.no. 67; r.a.m. 164.930; r.d. 8.803 (25°C); m.p. 1,470°C (2,680°F); b.p. 2,720°C (4,930°F); most common isotope ^{165}Ho (100%).

Holocene (Recent or Post-Glacial Epoch) Division of the GEOLOGICAL TIME extending from roughly 10,000 years ago to the present. It includes the emergence of human beings as settled members of communities, with the first known villages dating from about 8,000 years ago.

holography Process of making a **hologram**. One or more photographs are formed on a single film or plate by means of interference between two parts of a split LASER beam. An apparently meaningless pattern is formed, but when light falls on it (ideally laser light, but possibly ordinary light), the reflected or transmitted light reconstructs a 3-D image or images. The concept of holography has been appreciated since Dennis GABOR's inventions in the 1940s, but progress was not possible until the development of a sufficiently coherent light source – the laser.

homeopathy Unorthodox medical treatment involving the administration of minute doses of a drug or remedy that causes effects or symptoms similar to those that are being treated. It is the opposite of allopathy, in which curatives are administered to create the opposite effect. Homeopathy was popularized in the 18th century by the German physician Christian HAHNEMANN.

homeostasis In biology, processes that maintain constant conditions within a cell or an organism in response to either internal or external changes.

homeothermic (endothermic) Describes an animal whose body temperature does not fluctuate as

the temperature of its surroundings fluctuates, often referred to as "warm-blooded." Mammals and birds are warm-blooded. They maintain their

▲ **homeothermic**
Mammals are homeothermic – they maintain a constant body temperature, usually at around 37°C (98.6°F) by generating metabolic heat. In the cold, heat loss is reduced by a layer of insulating fur or blubber (fatty tissues), but temperature is further regulated by the circulatory system. A simple heat exchanger (A) dissipates internal heat (1) to the outside (2), but most mammals have more complex systems (B). The warm blood in an artery (1) branches out into capillaries, where heat is lost through the skin (2). Cooled blood returns through the veins (3). Heat loss can be minimized by diverting warm blood through transverse blood vessels (4). Some mammals, like whales, that live in very cold climates have counter-current heat exchange systems (C) that further reduce heat loss. A vein returning cold blood (1) divides and surrounds an artery (2). The arterial blood warms up the venous blood – lessening the shock it might cause to the major organs – and is in its turn cooled down so that less body heat is lost to the surface.

body temperature through METABOLISM. *See also* POIKILOTHERMIC

hominid Any member of the family Homonidae. Members include modern humans and our earliest ancestors. *See also* HOMO

Homo Genus to which humans belong. Modern humans are classified *Homo sapiens sapiens*. *See also* HOMO ERECTUS; HOMO HABILIS; HUMAN EVOLUTION; NEANDERTHAL

Homo erectus Species of early human dating from about 1.5 million to 0.2 million years ago. The "Ape Man of Java" was the first early human fossil to be found, late in the 19th century. Both it and Peking man, another early discovery, represent more advanced forms of *Homo erectus* than older fossils found more recently in Africa. *See also* HUMAN EVOLUTION

homogenization Process that reduces a substance contained in a FLUID (liquid or gas) to small particles and redistributes them evenly through it. For example, the fat in milk can be broken down so thoroughly by homogenization that particles do not recombine and cream does not rise.

Homo habilis Species of early human, discovered by Louis LEAKEY in 1964 in the Olduvai Gorge, East Africa. Its fossil remains have been estimated to be between 1.8 and 1.2 million years old, being contemporary with those of AUSTRALOPITHEUS. The development of hand and skull is much more like that of modern human. *See also* HUMAN EVOLUTION

homologous Term that describes similarity in essential structure of organisms based on a common genetic heritage. It often refers to organs that now have a different superficial appearance and function in different organisms. For example, despite appearances, a human arm and a seal's flipper are homologous, having evolved from a common origin. *See also* EVOLUTION

homologous chromosome One of a pair of CHROMOSOMES that share the same genetic structure, but not always the same ALLELES for a given characteristic. Found in DIPLOID cells, pairs of homologous chromosomes are composed of one chromosome from the female parent, the other from the male. During MEIOSIS, when GAMETES (sex cells) are formed, homologous chromosomes undergo a complex process known as CROSSING OVER, when strands of the chromosomes (CHROMATIDS) break from the original chromosome and exchange places with the other member of the pair. This "reshuffling" gives rise to genetic variation,

which together with random MUTATION is an essential part of EVOLUTION.

homologous series Series of related organic chemical compounds that share similar properties and the same general formula, differing only by a single, unchanging collection of atoms. A notable example is the ALKANE hydrocarbon group. They all have the general formula C_nH_{2n+2}. Therefore each member of the group changes by a CH_2 group – methane (CH_4), ethane (C_2H_6), propane (C_3H_8), and butane (C_4H_{10}).

homozygote Organism possessing identical forms of a GENE on a CHROMOSOME pair. It is a purebred organism and always produces the same kind of GAMETE (sex cell). *See also* HETEROZYGOTE

Hooke, Robert (1635–1703) English philosopher, experimental physicist, and inventor, considered the best mechanic of his generation. Interested in astronomy, he deduced that Jupiter rotated and claimed to have stated the laws of planetary motion before Isaac NEWTON. Hooke made many improvements to astronomical instruments, watches, and clocks. He also studied the elastic properties of solids, which led to HOOKE'S LAW. Among his inventions were a practical optical system and the Gregorian (reflecting) MICROSCOPE, with which he discovered plant CELLS. He also coined the biological term "cell."

Hooke's law Relationship between STRESS and STRAIN in an elastic material when it is stretched. The law states that the stress (force per unit area) is proportional to the strain (a change in dimensions). The law, which holds only approximately and over a limited range, was discovered in 1676 by Robert HOOKE. *See also* ELASTICITY

Hopkins, Sir Frederick Gowland (1861–1947) British biochemist. He shared with the Dutch physician Christiaan EIJKMAN the 1929 Nobel Prize for physiology or medicine for his work on VITAMINS. Hopkins pointed out that some diseases, such as scurvy and rickets, might be caused by the deficiency in the diet of a substance necessary for proper health, the so-called **vitamin concept**. He also investigated the function of LACTIC ACID in muscular activity.

horizon In geology, a term used in three ways: 1) a continuous horizontal surface or time-plane between two STRATA that has no thickness; 2) a horizontal layer generally less than 1m (3ft) in thickness that is characterized either by a distinct fossilized FLORA or FAUNA or by a distinctive rock type; 3) a distinct layer in soil is also called a horizon. *See also* SOIL HORIZON

horizon, celestial GREAT CIRCLE on the CELESTIAL SPHERE, the plane of which contains the line through an observer's position at right angles to the vertical. It lies midway between the observer's ZENITH and NADIR and cuts the observer's meridian at the N and S points.

horizon coordinate system Astronomical COORDINATE SYSTEM referred to the plane of the observer's horizon. The coordinates are AZIMUTH and ALTITUDE.

hormone Chemical substance secreted by living CELLS that affects the METABOLISM of cells in other parts of the body. In MAMMALS, hormones are secreted by glands of the ENDOCRINE SYSTEM and released directly into the bloodstream. They exercise chemical control of physiological functions, regulating growth, development, sexual maturity, and functioning, metabolism, and (in part) emotional balance. Hormones circulate in the body in minute quantities and usually exert their effects at considerable distances from their sites of origin. Most are slow to take effect, exert widespread action, and are also slow to disappear from the system. The secretion and activity of the various hormones are closely interdependent; one can stimulate or inhibit the secretion of another, or two or more can act together to produce a certain effect. In general they maintain a delicate equilibrium that is vital to health. The HYPOTHALAMUS, which lies at the base of the brain and is adjacent to the PITUITARY GLAND, is responsible for overall coordination of the secretion of many hormones. Familiar hormones include THYROXINE, ADRENALINE, INSULIN, ESTROGEN, PROGESTERONE, TESTOSTERONE, and ANTIDIURETIC HORMONE (ADH). In plants, hormones control many aspects of metabolism, including cell elongation, direction of growth, cell division, initiation of flowering, development of fruits, leaf fall, and responses to environmental factors such as light, water, and gravity. The most important plant hormones include AUXIN, GIBBERELLIN, CYTOKININ, and ABSCISIC ACID, a growth inhibitor that prevents buds from opening too soon in spring and keeps seeds DORMANT through adverse seasons. *See also* HOMEOSTASIS

hormone replacement therapy (HRT) Use of the female HORMONES ESTROGEN and progestogen to relieve symptoms in women who are either menopausal or who have had both ovaries removed. HRT is effective against hot flushes, vaginal dryness, and the mood changes associated with the MENOPAUSE; it also gives some protection against disease of the heart and blood vessels, and OSTEOPOROSIS. However, except in people who have undergone HYSTERECTOMY, the estrogen component causes a thickening of the lining of the uterus, which may lead to an increased risk of cancer of the ENDOMETRIUM. The progestogen component of the treatment causes a regular shedding of the lining, similar to MENSTRUATION, which may lessen this risk.

hornblende Black or green mineral found in IGNEOUS and METAMORPHIC ROCK. The most common form of AMPHIBOLE, it contains iron and silicates of calcium, aluminum, and magnesium. Hardness 5.5; r.d. 3.2.

hornfels Group of hard, fine-grained METAMORPHIC ROCKS formed originally in a high-temperature area in near contact with an IGNEOUS INTRUSION. They all contain QUARTZ and MICA, with a third component that gives them their individual names: **chiastolite** hornfels, **cordierite** hornfels, **garnet** hornfels, and **pyroxine** hornfels. All are dark gray or black, although garnet hornfels has reddish patches.

horn Defensive or offensive structure that grows from the head of some mammals. A typical horn is made up of a central bony core covered by a layer of the skin protein KERATIN; in the rhinoceros, the entire horn is made of keratin. Horns are generally retained for life, although deer shed their antlers annually.

horology Science of measuring TIME, or the art of constructing CLOCKS and watches.

horse latitudes Either of two areas of high pressure found near 25°N and 25°S of the equator. The air rises at or near the equator because of low pressure and then circles in the atmosphere, but much falls near the tropics of Cancer and Capricorn. The falling air gives rise to fairly permanent regions of high pressure, known as the horse latitudes. Both the TRADE WINDS, which blow from the horse latitudes toward the equator, and the westerly winds, which blow from the horse latitudes toward the poles, are caused by the high pressure.

horsepower (symbol hp) Unit indicating the rate at which work is done, adopted by James WATT in the 18th century. He defined it as the weight, 550lb (250kg), a horse could raise 1ft (0.3m) in one second – a rate of 550ft-lb/s. At the output shaft of an engine or motor, it is termed "brake horsepower" or "shaft horsepower." In large reciprocating engines it is termed "indicating horsepower" and is determined from the pressure in the cylinders. The electrical equivalent of one horsepower is 746 watts.

horsetail Any of about 30 species of flowerless, rushlike plants of the phylum Sphenophyta. They are allied to FERNS and grow in all continents except Australasia. The hollow jointed stems have a whorl of tiny leaves at each joint. Spores are produced in a conelike structure at the top of a stem. Horsetails date from the Carboniferous period.

Horsley, Sir Victor Alexander Haden (1857–1916) British physiologist and surgeon who introduced refinements in NEUROSURGERY. He performed the first successful removal of a tumor of the spinal cord and carried out research on the functions of the thyroid gland, on rabies vaccine, and on functions of various regions of the brain.

horst In geology, an elongated upthrust block bounded by parallel normal FAULTS on its long sides. *See also* GRABEN

horticulture Growing of vegetables, fruits, seeds, herbs, ornamental shrubs, and flowers, on a commercial scale, in nurseries, orchards, and gardens. The origins of horticulture lie in the small intensive kitchen gardens of the medieval farming system. The techniques employed include asexual, or vegetative, PROPAGATION by leaf, stem, and root cuttings, and by stem and bud grafting. Fruit trees, shrubs and vines, including those bearing apples, pears, plums, cherries, citrus fruits, and grapes, are usually propagated by GRAFTING the fruiting stock on to a hardier rootstock. In the case of apple trees, the rootstock is usually French crab apple. Plants propagated sexually by seed include some fruits and vegetables, including corn and many other types that produce seed abundantly. Often the seed needs to be overwintered, or stored at low temperatures and high humidity, before it will begin GERMINATION. Seed itself is a major horticultural crop. Close scientific control of POLLINATION is essential for producing crops of specific quality.

host Organism infected by a PARASITE, either **definitive**, in which the parasite reaches sexual maturity, or **intermediate**, in which it does not.

hot spot Area of high volcanic activity. Some are located on CONSTRUCTIVE MARGINS, such as Iceland; others occur within plates, for example Hawaii. Hot spots are thought to be caused by upwelling of MAGMA in the MANTLE beneath (mantle plumes). *See also* PLATE TECTONICS

Hounsfield, Sir Godfrey Newbold (1919–) British engineer who, beginning in 1967 and independently of Allan CORMACK in the USA, developed a CAT scanner (first used in 1972). CAT stands for COMPUTERIZED AXIAL TOMOGRAPHY, in which an X-ray machine takes a sequence of photographs of organs and tissues which a computer then assembles into cross-sectional images.

Hounsfield and Cormack shared the 1979 Nobel Prize for physiology and medicine.

Houssay, Bernardo Alberto (1887–1971) Argentinian physiologist who greatly advanced knowledge of the ENDOCRINE SYSTEM. He shared with Carl CORI the 1947 Nobel Prize for physiology or medicine for discovering the role played by the HORMONE of the anterior lobe of the PITUITARY GLAND in regulating the METABOLISM of sugar. He demonstrated the complex interlocking action of various hormones of the body – in this case, a pituitary hormone with the sugar-metabolizing hormone INSULIN.

hovercraft (air-cushion vehicle) Fast, usually amphibious craft invented by Sir Christopher COCKERELL. A horizontal fan produces a cushion of air that supports the hovercraft just above the ground or water. Vertical fans propel the craft. Most hovercraft are powered by gas turbine engines, similar to those used in aircraft, or by diesel engines. Hovercraft travel at speeds of up to 160km/h (100mph). They are invaluable for negotiating swamps, and are used as marine ferries and as military vehicles. Smaller, less speedy craft have been developed as runabouts. A **hovertrain** is a high-speed train that uses an air cushion to support it just above a special track.

Hoyle, Sir Fred (Frederick) (1915–) English astrophysicist and cosmologist. With Hermann BONDI and Thomas GOLD he developed the STEADY STATE THEORY, in which the continuous creation of matter drives the expansion of the Universe. Although it was subsequently displaced by the BIG BANG theory (which owes its name to a disparaging remark by Hoyle), it sparked research by Hoyle and others into nucleosynthesis (the production of chemical elements from other chemical elements via nuclear reactions) in stars, which has been of lasting value. Hoyle has often attracted controversy with unorthodox ideas, such as his suggestion that life was brought to Earth by comets.

HR diagram Abbreviation of HERTZSPRUNG-RUSSELL DIAGRAM

Hubble, Edwin Powell (1889–1953) US astronomer. At Mount Wilson Observatory near Pasedena, California, Hubble used the 2.5-m (100-in) telescope for NEBULA investigations. He found that although some nebulae were clouds of gas lying within our galaxy, many others (the so-called spiral nebulae) were resolvable as independent star systems. These results were published in 1925, along with the HUBBLE CLASSIFICATION. Also in 1925, Hubble measured the distance to the ANDROMEDA GALAXY. He attributed the RED SHIFT of the spectral lines of galaxies to the recession of galaxies, and hence to the expansion of the Universe, upon which modern cosmology is based. In 1929 he announced HUBBLE'S LAW, which claims a linear relation between the distance of galaxies from us and their velocity of recession, deduced from the red shift in their spectra. *See also* EXPANDING UNIVERSE

Hubble classification Classification of GALAX-IES according to their shape or structure. The system was introduced by Edwin HUBBLE in 1925. The three major categories are elliptical (E), spiral (S), and barred spiral (SB); each category has several subdivisions depending on the observable shape or structure of the galaxy. Irregular galaxies (Ir) were not included in Hubble's original classification.

Hubble constant (symbol H_0) Rate at which the velocity of recession of galaxies (as shown by the RED SHIFT) increases with distance from us, according to HUBBLE'S LAW. The zero subscript specifies that it is the expansion rate at the present time that is meant, for the rate changes with time. The inverse of the Hubble constant is called the **Hubble time**, which gives a maximum age for the Universe on the assumption that there has been no slowing of the expansion. The constant is named after Edwin HUBBLE. *See also* EXPANDING UNIVERSE

Hubble's law Law proposed by HUBBLE claiming a linear relation between the distance of galaxies from us and their velocity of recession, deduced from the RED SHIFT in their spectra. The figure linking velocity with distance is the HUBBLE CONSTANT.

Hubble Space Telescope (HST) Optical TELE-SCOPE that was placed in Earth orbit by the SPACE SHUTTLE in 1990. It was hoped that bypassing the Earth's atmosphere, the telescope would provide scientists with the clearest images of distant objects ever seen. However, images transmitted back to Earth revealed that the telescope's main mirror was incorrectly shaped. A shuttle repair team corrected the fault in 1993, and it was again repaired in 1997 and 1999. Hubble now produces accurate images of celestial bodies.

Huggins, Charles Brenton (1901–) US physician. He shared the 1966 Nobel Prize for physiology or medicine with Francis ROUS for his pioneering work in CANCER chemotherapy and his development of ways of investigating and treating cancers through the use of HORMONES.

Huggins, Sir William (1824–1910) British astronomer, the first to use the SPECTROSCOPE to determine the chemical composition of stars, comets, and novas. He developed spectroscopic photography and applied the DOPPLER shift *(see* RED SHIFT) of the spectral lines of stars to the measurement of stellar motion. He distinguished between nebulae that are uniformly gaseous and those that contain stellar clusters.

Hulst, Hendrik van de (1918–) Dutch astronomer. He carried out research on interstellar matter and the Solar CORONA. In 1944 he predicted that hydrogen should emit radio waves with a wavelength of 21cm, and in 1951 he and Jan OORT used DOPPLER shifts at this wavelength to map the galaxy.

human body Physical structure of a human. It is composed of water, PROTEIN, and other organic compounds, and some inorganic material (minerals). It has a bony framework, the SKELETON, consisting of more than 200 bones, sheathed in SKELETAL MUSCLE to enable movement. A protective bony case, the SKULL, surrounds the large BRAIN. The body is fueled by nutrients absorbed from the DIGESTIVE SYSTEM and oxygen from the LUNGS, which are pumped around the body by the CIRCULATORY SYSTEM. Metabolic wastes are eliminated mainly by the process of EXCRETION. Continuation of the species is enabled by the REPRODUCTIVE SYSTEM. Overall control is exerted by the NERVOUS SYSTEM, working closely with the ENDOCRINE SYSTEM. The body surface is covered by a protective layer of SKIN. *See also* ANATOMY

human evolution Process by which humans developed from prehuman ancestors. The FOSSIL record of human ancestors is patchy and unclear. Some scientists believe that our ancestry can be traced back to one or more species of Australopithecines *(see* AUSTRALOPITHECUS) that flourished in S and E Africa 4–1 million years ago. Other scientists believe that we are descended from some as yet undiscovered ancestor. The earliest fossils that can be identified as human are those of *Homo habilis* (handy people), which date from 2 million years ago. The next evolutionary stage was *Homo erectus* (upright people), who first appeared 1.5 million years ago. The earliest fossils of our own species, *Homo sapiens* (wise people), date from 250,000 years ago. An apparent side-branch, the NEANDERTHALS (*Homo sapiens neanderthalensis*), existed in Europe and w Asia 130,000–30,000 years ago. Fully modern humans, *Homo sapiens sapiens*, first appeared 100,000 years ago. All human species apart from *Homo sapiens sapiens* are now extinct.

Human Genome Project International scientific program initiated in 1988 to map the human GENOME by analyzing the 100,000 or so GENES making it up and charting their positions within the human chromosomes and their functions and biochemical nature. The study is expected to take 15 years and cost $3 billion. Researchers are located in more than 20 centers around the world, and their efforts are coordinated primarily in the US and France by the Human Genome Organization (HUGO). The study is expected to increase the understanding of human evolution as well as of genetic disorders such as Alzheimer's disease and muscular dystrophy. It will also identify genetic similarities between humans and other species.

human immunodeficiency virus (HIV)

Organism that causes ACQUIRED IMMUNE DEFICIENCY SYNDROME (AIDS). A RETROVIRUS, only identified in 1983, HIV attacks the IMMUNE SYSTEM, leaving the person unable to fight off infection. There are two distinct viruses: HIV-1, which has now spread worldwide; and HIV-2, which is concentrated almost entirely in West Africa. Both cause AIDS, with people infected with HIV-1 generally developing symptoms sooner than those with HIV-2. There are three main routes of transmission: from person to person by sexual contact; from mother to baby during birth; and by contact with contaminated blood or blood products (for instance, during transfusions or when drug users share needles). People can carry the virus for many years before developing symptoms.

Humboldt, Baron Friedrich Heinrich Alexander von

(1769–1859) German scientist and explorer. His speciality was mineralogy, however on his scientific trips in Europe and Central and South America he studied volcanoes, the origins of tropical storms, and the increase in magnetic intensity from the equator toward the poles. His principal, five-volume work, *Kosmos* (1845–62), is a comprehensive description of the physical Universe.

Humboldt current

(Peru current) Current in the SE Pacific Ocean. An extension of the West Wind Drift, the Humboldt current flows N along the W coast of South America. A cold current, it lowers temperature along its course until it joins the warm Pacific South Equatorial current.

humerus

Bone in the human upper arm, extending from the SCAPULA (or shoulder blade) to the elbow. A notch, or depression, called the **olecranon cavity**, on the posterior roughened lower end of the humerus, provides the point of articulation for the ULNA, one of the forearm bones. *See also* HUMAN BODY

humidity

(relative humidity) Measure of the amount of water vapor in air. It is the ratio of the actual vapor pressure to the saturation vapor pressure at which water normally condenses, and is usually expressed as a percentage. Humidity is measured by a HYGROMETER.

humus

Dark brown, organic substance, resulting from partial decay of plant and animal matter. It improves soil by retaining moisture, aerating, and increasing mineral nutrient content and bacterial activity. Often made in a compost heap or obtained naturally, types of humus include PEAT MOSS, leaf mold, and soil from woods.

Hunter, John

(1728–93) Scottish anatomist and surgeon who worked with his brother **William** (1718–83) on the LYMPHATIC SYSTEM, the course of the OLFACTORY NERVES and the coagulation of blood. In the 1760s, John Hunter was the first to ligate (tie off) an artery.

Huntington's disease

(formerly Huntington's chorea) Rare hereditary disease of the brain characterized by jerky involuntary movements (chorea), disturbed behavior, and disintegration of personality and intellect, culminating in dementia. Fifty percent of the children of an affected parent are at risk. The disease usually occurs in early middle age.

hurricane

Wind of Force 12 or greater on the BEAUFORT WIND SCALE; intense and devastating tropical cyclone with winds ranging upward from 118km/h (73mph), known also as a **typhoon** in the Pacific. Originating over oceans 10 to 20° from the equator, hurricanes have a calm central hole called the **eye**, surrounded by inward spiraling winds and cumulonimbus CLOUDS. Weather satellites often give adequate warning of their approach, but the wind and rain of hurricanes take many lives, and cause extensive damage, particularly in the Caribbean and the SE and E US coasts.

Hutton, James

(1726–97) Scottish geologist. Hutton sought to formulate theories of the origin of the Earth and of atmospheric changes. He concluded that the Earth's history could be explained only by observing current forces at work within it, thereby laying the foundations of modern geological science.

Huxley, Sir Andrew Fielding

(1917–) British physiologist. He analyzed the electrical and chemical events in NEURON (nerve cell) discharge and for this work he shared the 1963 Nobel Prize for physiology or medicine with John Eccles (1903–) and Alan HODGKIN.

Huxley, Sir Julian Sorell

(1887–1975) British biologist, grandson of Thomas HUXLEY. His researches were chiefly on the behavior of birds and other animals in relation to EVOLUTION. His writings greatly promoted public interest in the subject. He served as secretary of the Zoological Society of London (1935–48) and as the director general of the United Nations Educational, Scientific and Cultural Organization (UNESCO) (1946–48). His books include *The Individual in the Animal Kingdom* (1911) and *Evolutionary Ethics* (1943).

Huxley, Thomas Henry (1825–95) British biologist. Huxley was a champion of DARWIN's theory of EVOLUTION and of the scientific method in research. His works include *Zoological Evidences as to Man's Place in Nature* (1863), *Manual of Comparative Anatomy of Vertebrate Animals* (1871) and *Evolution and Ethics* (1893).

Huygens, Christiaan (1629–95) Dutch physicist and astronomer. In 1655 he discovered Saturn's largest satellite, Titan, and the next year explained that the planet's appearance was due to a broad ring surrounding it. He was concerned with the development of the TELESCOPE and introduced the convergent eyepiece. Huygens' many contributions to physics include the idea that light is a wave motion and the theory of the pendulum; he built the first pendulum clock.

hybrid Offspring of two parents of different GENE composition. It often refers to the offspring of different varieties of a species or of the cross between two separate species. Most interspecies hybrids, plant or animal, are unable to produce fertile offspring.

hybridization CROSSBREEDING of plants or animals between different species to produce offspring that differ in genetically determined traits; it is commonly the result of human intervention. Changes in climate or in the environment of an organism may give rise to natural hybridization, but usually humans use specialized techniques to produce hybrids that may be hardier or more resistant to disease than the original forms, or more economical.

hybrid vigor Inordinate strength demonstrated in some HYBRID offspring, not characteristic of the parents.

Hydra (Water snake) Largest constellation in the S sky. Its brightest star is Alpha (Alphard), magnitude 2.0. There is one bright open cluster, M48.

hydration Assimilation of water by an ION or compound. A hydrate is a compound, such as copper(II) sulfate (cupric sulfate) that has a fixed amount of water weakly bonded to each crystal ($CuSO_4.5H_2O$). ETHYLENE can be hydrated to produce ETHANOL.

hydraulic action Erosive effect of water. The force of moving water can be an effective agent of EROSION. It is important both in rivers and on coastlines. In rivers the movement of currents causes hydraulic action, though if a load of silt or sand were added the process would be called CORRASION. Along the coast, breaking waves batter cliffs and may open out JOINTS in the rock, forming caves or undermining the cliff and leading to collapse.

hydraulics Physical science and technology of the behavior of FLUIDS in both static and dynamic states. It deals with practical applications of fluid in motion and devices for its utilization and control. *See also* FLUID MECHANICS; HYDRODYNAMICS

hydrazine (N_2H_4) Fuming corrosive liquid obtained by the reaction between chloramine (NH_2Cl) and ammonia (NH_3), or by the oxidation of UREA. It is used as a jet and rocket fuel, as a reducing agent and as a corrosion preventer. Properties: r.d. 1.011, m.p. 1.4°C (34.5°F); b.p. 113.5°C (236.3°F).

hydrazone In organic chemistry, a compound formed between hydrazine (N_2H_4) and an ALDEHYDE or KETONE. Hydrazones are condensation products (*see* CONDENSATION REACTION) containing the group $=C=NNH_2$. They are used in QUALITATIVE ANALYSIS to identify aldehydes and ketones. In practice, the substance phenylhydrazine is generally employed as the analytical reagent.

hydride Chemical compound of HYDROGEN with another ELEMENT, especially a more electropositive element (one that loses ELECTRONS and forms positive IONS). The hydrides of the electropositive metals, such as lithium hydride (LiH) and sodium hydride (NaH), are saltlike ionic compounds containing the negative ion H^-. They are useful reagents for HYDROGENATION reactions.

hydrocarbon Organic compound containing only CARBON and HYDROGEN. Many thousands of different hydrocarbons exist, including open-chain compounds, such as the ALKANES (paraffins), ALKENES (olefins), and ALKYNES (acetylenes). Aromatic hydrocarbons have properties similar to BENZENE; they contain at least one benzene ring, as in naphthalene and anthracene. Petroleum, natural gas and coal tar are sources of hydrocarbons.

hydrocephalus Increase in volume of CEREBROSPINAL FLUID (CSF) in the ventricles of the brain. A condition that exerts dangerous pressure on brain tissue, it can be due either to an obstruction that prevents the CSF from draining away, or to a failure of natural reabsorption. In babies it is congenital, often associated with SPINA BIFIDA; in adults it may arise from brain injury or disease. It is treated by insertion of a shunting system to drain the CSF harmlessly into the abdominal cavity.

hydrochloric acid Solution of the pungent, colorless gas hydrogen chloride (HCl) in water. It is obtained by the action of sulfuric acid on common salt, as a by-product of the chlorination of hydrocarbons, or by combination of HYDROGEN and CHLORINE. Hydrochloric acid is used to produce other chemicals and has uses in the food, oil, and metallurgical industries. It is also produced in humans by cells in the stomach lining as it is needed for the enzyme PEPSIN to digest proteins. The concentrated commercial acid contains up to 35% hydrogen chloride.

hydrocyanic acid (prussic acid) Solution of hydrogen cyanide (HCN) in water; a deadly poisonous, weak acid. It is a liquid with an odor of bitter almonds, sometimes used as a fumigant to destroy rats and other pests of buildings. It also has uses in the chemical industry but is so dangerous to handle that its solid salts, potassium and sodium cyanide, are preferred, even though these, too, are highly poisonous.

hydrodynamics In physics, branch of MECHANICS that deals with the motion of FLUIDS (liquids and gases). It is of great importance in industry, particularly chemical, petroleum, and water engineering. The properties of fluids studied include their adhesion, cohesion, viscosity, and type of flow (viscous and turbulent). *See also* HYDRAULICS; FLUID MECHANICS

hydroelectricity Electricity generated from the motion of water, whether from dams, tides or waves. In nearly all installations this energy of movement, or kinetic energy, drives the blades of a water TURBINE, which in turn drives the rotor of an electric GENERATOR.

hydrofluoric acid Colorless, corrosive fuming liquid consisting of a solution of the gas hydrogen fluoride (HF) in water, prepared by distilling a mixture of sulfuric acid (H_2SO_4) and fluorspar (FLUORITE). It is used in frosting and etching glass and in FLUOROCARBONS.

hydrofoil Boat or ship the hull of which is lifted clear of the water, when moving at speed, by submerged wings. Hydrofoils usually have gas-turbine or diesel engines that power propellers or water jets. Speeds range from 30 to 60 knots.

hydrogen (symbol H) Gaseous, nonmetallic element, first identified as a separate element in 1766 by Henry CAVENDISH who called it "inflammable air." Colorless and odorless, it is usually classified with the ALKALI METALS in group I of the periodic table. Isotopes include DEUTERIUM and the radioactive TRITIUM. Hydrogen is the lightest and most abundant element in the Universe (76% by mass), mostly found on Earth combined with oxygen in water. It is used to manufacture AMMONIA, to harden fats and oils, and in ROCKET fuels. The hydrogen bomb (*see* NUCLEAR WEAPON) is produced by the nuclear FUSION of hydrogen isotopes. Properties: at.no. 1; r.a.m. 1.00797; r.d. 0.0899; m.p. $-259.1°C$ $(-434.4°F)$; b.p. $-252.9°C$ $(-423.2°F)$; most common isotope ^1H (99.985%).

◄ **hydrofoil** Hydrofoils use the lift of underwater "wings" (1) to push the boat out of the water. They work in water the same way airfoils work in air. A low-pressure zone is created above the top surface of the hydrofoil, pushing the wing upward (2).This lessens the drag allowing the boat to travel faster. The Boeing Jetfoil (shown) scoops up water (3) and uses gas turbines (4) to drive high-pressure pumps (5) that throw the water from the rear, creating thrust. The jetfoil can reach speeds of up to 75km/h (45 mph).

▲ **hydrological cycle** The world's water balance is regulated by the constant recycling of water between the oceans, the atmosphere, and the land. The movement of water between these three "reservoirs" is called the hydrological cycle. The oceans play a vital role in this cycle: 74% of total precipitation falls over the oceans and 84% of total evaporation comes from the oceans. Water vapor in the atmosphere circulates around the planet, transporting energy as well as water itself. When the vapor cools it falls as rain or other precipitation.

hydrogenation Chemical reaction between hydrogen and an element or compound, sometimes under pressure, usually in the presence of a metal CATALYST such as nickel, platinum or palladium. An unsaturated compound, such as benzene, may accommodate the extra hydrogen, but a saturated species will break up (**destructive** hydrogenation). On an industrial scale, hydrogenation of coal produces oil, and hydrogenation of liquid vegetable oils yields solid fats (such as margarine).

hydrogen bomb *See* NUCLEAR WEAPON

hydrogen bond CHEMICAL BOND formed between certain hydrogen-containing molecules. The hydrogen atom must be bound to an electronegative (electron-withdrawing) atom; the bond is formed between the positive charge on hydrogen and the negative charge on an atom in an adjacent molecule. Hydrogen bonding occurs in water, and in many biological systems.

hydrogencarbonate (bicarbonate) Salt of CARBONIC ACID, containing the hydrogencarbonate ion (HCO_3^-). An example is SODIUM HYDROGENCARBONATE.

hydrogen chloride *See* HYDROCHLORIC ACID

hydrogen cyanide *See* HYDROCYANIC ACID

hydrogen peroxide (H_2O_2) Liquid compound of hydrogen and oxygen, usually sold in aqueous solution. It is prepared by electrolytic OXIDATION of sulfuric acid and by methods involving REDUCTION

of oxygen. Hydrogen peroxide solution is used as a bleach and a disinfectant. Concentrated hydrogen peroxide is dangerously explosive and is used as an oxidizer for ROCKET fuel. Properties: r.d. 1.44; m.p. −0.9°C (30.4°F); b.p. 150°C (302°F).

hydrogen sulfide (H_2S) Colorless, poisonous gas with the smell of rotten eggs. It is produced by decaying matter, found in crude oil and prepared by the action of sulfuric acid on metal sulfides. It is used in traditional QUALITATIVE ANALYSIS. Properties: m.p. −85.5°C (−121.9°F), b.p. −60.7°C (−77.3°F).

hydrography Science of charting the water-covered areas of the Earth. Navigational charts have been made since the 13th century, but these were accurate only for coastlines. Interest in the charting of oceanic areas away from coasts only developed in the 19th century. Now, detailed bathymetric (deep-sea) surveys are available for geological studies. Hydrographic offices are run by governments of maritime nations to furnish their mariners with nautical charts; the work of these offices is coordinated by the International Hydrographic Bureau, founded in 1921.

hydrological cycle Circulation of water around the Earth. Water is evaporated from the sea; most falls back into the oceans, but some is carried over land. There it falls as PRECIPITATION (rain, snow, and so on) and, either by EVAPORATION, EVAPOTRANSPIRATION, or TRANSPIRATION by plants, is returned to the ATMOSPHERE. Alternatively, it gradually finds its way back to the sea, by surface runoff, infiltration, or seepage. Less than 1% of the world's water is involved in this cycle; 97% of all water is in the oceans, and much of the remainder is held as snow or ice.

hydrology Study of the Earth's waters, their sources, circulation, distribution, uses, and chemical and physical composition. The HYDROLOGICAL CYCLE is the Earth's natural water circulation system. Hydrologists are scientists and engineers involved in various aspects of hydrology, including the provision of fresh water, building dams and irrigation systems, and controlling floods and water pollution.

hydrolysis Chemical reaction in which molecules of a substance are split into smaller molecules by reaction with water, often assisted by a CATALYST. For example, in digestion, ENZYMES catalyze the hydrolysis of CARBOHYDRATES, PROTEINS, and FATS into smaller, soluble molecules that the body can assimilate.

hydrometallurgy Treatment of ores by wet processes to extract their metals. It usually includes LEACHING by water and additives such as sulfuric acid, separation of the waste, purification of leach solution and precipitation of metal from solution by chemicals or electrolytic means. Most metals are extracted by this method.

hydrometer Instrument for measuring density or RELATIVE DENSITY (R.D.) of a liquid. It consists of a long-necked, sealed glass bulb that is weighted. The neck is calibrated to indicate relative density, which is read off against the surface of the liquid in which it floats. Hydrometers are used to check concentrations of liquids in storage batteries, freezing points of radiator solutions and the "proof" of alcohol.

hydrophily POLLINATION of a FLOWER through the agency of water. A rare type of pollination employed by aquatic plants, it can occur in one of two ways. Either the pollen grains themselves are carried on or under the surface of the water, or the male flowers break off and float along until they come into contact with a female flower.

hydrophone Specialized MICROPHONE designed to detect sounds and ultrasonic waves underwater. Initially used to detect submarines, hydrophones have, more recently, been used in oceanography, for example, for listening to whale song.

hydrophyte (aquatic plant) Plant that grows only in water or in damp places. Examples include water lilies, water ferns, water milfoil, water hyacinth, duckweed, and various pondweeds.

hydroplane Speedboat able to skim over water by rising out of it at speed until only a small area at the stern remains immersed. This is achieved by means of a flat-bottomed hull, which is powered by a high-performance outboard motor (*see also* HYDROFOIL). The term is also used to describe a horizontal, fin-like planning surface on a submarine that controls the vessel's vertical position underwater.

hydroponics (soil-less culture or tank farming) Method of growing plants with their roots in a mineral solution or a moist inert medium (such as gravel) containing the necessary nutrients, instead of soil. Added nutrients include nitrogen, phosphorus, and potassium, with trace elements such as calcium, sulfur, and magnesium. The technique has been in use for more than 100 years and is employed on a commercial scale to grow such crops as tomatoes.

hydrosphere Water on the surface of the Earth, including oceans, lakes, streams, and groundwater.

hydrostatics In physics, the branch of mechanics that deals with liquids at rest. Its practical applications are mainly in water engineering and in the design of such equipment as hydraulic presses, rams, elevators, and vehicle braking and control systems. *See also* STATICS

hydrothermal process Geological activity produced by the action of very hot subterranean water, at temperatures between 300 and 500°C (570–930°F). Superheated water can dissolve many minerals, such as those of copper, lead, and zinc, depositing them as veins in rock crevices. It can also trigger chemical changes, for example turning the mineral olivine into serpentine. Much of this hydrothermal activity takes place in the deep ocean seabeds. Hot water welling from the floor of oceanic trenches is responsible for colonies of life 2.5km (1.5mi) below the surface, where tube worms, crustaceans, and fish thrive without the benefit of the Sun's warmth.

hydrothermal vent (smoker) Fissure in the seafloor through which mineral-rich water escapes. Hydrothermal vents form in areas of volcanic activity. Seawater seeps down through the bottom sediments into the hot rocks beneath and dissolves various minerals. When the water erupts through a vent, it cools rapidly and the dissolved minerals are precipitated out of solution as a "smoke." Unique species of bacteria and animals have been found living in the mineral-rich water around such vents.

hydrotropism Plant growth in response to water stimulus. It is not a strong TROPISM.

hydroxide Inorganic chemical compound containing the ion OH⁻; which acts as a BASE (a substance that accepts protons and reacts with an acid to yield a salt and water). The strong inorganic bases such as potassium hydroxide (KOH) dissociate in water almost completely to provide many hydroxide ions.

hydroxyl group Monovalent chemical grouping consisting of a hydrogen atom joined to an oxygen atom. Represented by the formula –OH, the hydroxyl group is an important functional group in organic compounds such as alcohols and phenols. It should not be confused with the hydroxide ion OH⁻.

hygrometer Instrument used to measure the water-vapor content, or HUMIDITY, of the atmosphere. One type, the psychrometer, compares the wet- and dry-bulb temperatures of the air. The wet bulb is covered by a cloth, which is kept damp by water from a small reservoir. Evaporation from the cloth reduces the temperature of the wet bulb and the difference in temperature between the wet bulb and the dry bulb shows the relative humidity. Other types of hygrometer measure absorption or condensation of moisture from the air, or chemical or electrical changes caused by that moisture.

hygroscopic Term describing a substance that, on standing, reacts with or absorbs water vapor from the air. Magnesium chloride and concentrated sulfuric acid are typical examples. Some hygroscopic solids such as magnesium chloride, said to be deliquescent, absorb sufficient water to dissolve, yielding a concentrated solution.

hymen In ANATOMY, MEMBRANE that covers or partly covers the entrance to the VAGINA. Intact at birth, it normally opens spontaneously before PUBERTY or at first penetration during sexual intercourse.

hymenopteran Member of the order of insects Hymenoptera, containing sawflies, ants, wasps, hornets, and bees. All have a complete life cycle – egg, larva, pupa, and adult – and two pairs of membranous wings, which move together in flight. They are found worldwide, living solitarily or in social groups. The larvae (grubs or caterpillars) feed on plants or are parasitic or predaceous.

hyperbaric chamber Sealed chamber in which oxygen is used at pressures higher than normal in the treatment of certain conditions. It is most often used in the treatment of decompression sickness, carbon monoxide poisoning or gas gangrene.

hyperbola Plane curve defined relative to a fixed point (the **focus**) and a fixed line (the **directrix**) so that the distance from any point on the curve to the focus is a constant multiple of the distance to the

directrix. The curve has two branches and is a CONIC section. Its standard equation in CARTESIAN COORDINATES x and y is $x^2/a^2 - y^2/b^2 = 1$, where a and b are real, nonzero numbers.

hypermetropia (hyperopia, farsightedness) Common disorder of vision in which near objects appear blurred. It occurs when parallel light rays entering the EYE are brought to a focus behind the RETINA, often because the eyeball is too short front to back. It can be rectified by wearing glasses or contact lenses with convex lenses. *See also* MYOPIA

hyperon Class of ELEMENTARY PARTICLES with an anomalously long lifetime. The group includes the lambda particle (Λ), sigma particles (Σ^+, Σ^0, Σ^-), xi particles (Ξ^0, Ξ^-), and the omega-minus particle (Ω^-). A lambda can replace a NEUTRON in a nucleus to form a **hypernucleus**. *See also* BARYON

hyperplasia Increased production of cells in an organ or tissue, causing it to enlarge in size. *See also* HYPERTROPHY

hypersensitivity Condition in which a person reacts excessively to a stimulus. Most hypersensitive reactions are synonymous with ALLERGIES, the most common being HAY FEVER, a reaction to pollen. Some people become hypersensitive to drugs, such as penicillin, or to chemicals after initial contact.

hypertension Persistent high BLOOD PRESSURE. Hypertension may be of unknown cause (**essential** or **primary** hypertension) or may be secondary to a variety of conditions, including kidney disease. Treatment is by salt restriction and antihypertensive drugs; weight reduction is recommended where appropriate. Untreated, hypertension predisposes to heart disease, stroke, and kidney failure. *See also* HYPOTENSION

hypertext Computing term that describes a method by which one piece of DATA is linked to another piece of data. Hypertext is most commonly seen on the WORLD WIDE WEB (WWW) and such things as interactive CD-ROMs. It usually manifests itself in the form of highlighted words (usually either in a different color or underlined), which, when selected, take the viewer to associated material. This material may be many things, such as more text, a visual image or a sound clip. Hypertext can be seen as performing a similar function in MULTIMEDIA environments as cross-references do in printed encyclopedias and dictionaries.

hyperthermia Abnormally high body temperature, usually defined as being 41°C (106°F) or more. It is usually due to overheating (as in HEATSTROKE) or FEVER.

hypertrophy Enlargement of a tissue or an organ due to an increase in size of its cells. Muscles enlarge in this way in response to increased use. *See also* HYPERPLASIA

hyperventilation Rapid breathing that is not brought about by physical exertion. It reduces the carbon dioxide level in the blood to undesirably low levels, producing dizziness, tingling in the limbs and a feeling of tightness in the chest; if prolonged, it may cause loss of consciousness.

hypha (pl. hyphae) Very fine, theadlike projections found in FUNGI. Collectively a number will form the MYCELIUM, which is responsible for the reproduction and feeding of the fungus.

hypnotic drug (soporific) Drug that depresses the CENTRAL NERVOUS SYSTEM, inducing sleep. They include tranquilizers, BARBITURATES, chloral hydrate, and some ANTIHISTAMINES.

hypodermic syringe Surgical instrument for injecting fluids beneath the skin into a muscle or blood vessel. It comprises a graduated tube containing a piston plunger, connected to a hollow needle. The whole is provided sterile – free from microorganisms – and must be resterilized after use. Many hypodermic syringes are disposable: designed to be used once and thrown away. *See also* INJECTION

hypolimnion Lower, colder layers of water in a lake, beneath the EPILIMNION. There is not enough light for PHOTOSYNTHESIS to occur and levels of dissolved oxygen are low.

hypotension Condition in which the BLOOD PRESSURE is abnormally low. It is commonly seen after heavy blood loss from any cause or excessive fluid loss due to prolonged vomiting or diarrhea. It also occurs in many kinds of serious illness. Some people experience a drop in blood pressure when standing up from a lying down position. Temporary hypotension may cause sweating, dizziness, and fainting. *See also* HYPERTENSION

hypotenuse Side opposite the right angle in a right-angled TRIANGLE. It is the longest side of the triangle.

hypothalamus Region at the base of the BRAIN, adjoining the PITUITARY GLAND, containing centers that regulate body temperature, fluid balance, hunger, thirst, and sexual activity. It is also involved in emotional activity, sleep, and the integration of HORMONE and nervous activity.

hypothermia Fall in body temperature to below 35°C (95°F). Most at risk are newborns (particularly if premature) and the aged living in conditions of inadequate heating. Insidious in onset, it can progress to coma and death. Hypothermia is sometimes induced during surgery to lower the body's oxygen demand. It occurs naturally in animals during HIBERNATION.

hypothyroidism Deficient functioning of the THYROID GLAND. Congenital hypothyroidism can lead to cretinism in children. In adults the condition is called **myxedema**. More common in women, it causes physical and mental slowness, weight gain, sensitivity to cold, and susceptibility to infection. It can be due to a defect of the gland itself or to a lack of iodine in the diet, both of which can lead to goiter or to inadequate stimulation of the thyroid gland by the pituitary. It is treated with the HORMONE thyroxine.

hypsometer Instrument for testing a THERMOMETER'S accuracy by using the boiling point of water. The name hypsometer is also given to an instrument that measures height above sea level as a function of the boiling point of water (which decreases with increasing height).

hysterectomy Removal of the UTERUS, possibly with surrounding structures. It is performed to treat fibroids or cancer or to put an end to heavy menstrual bleeding. A woman who has undergone hysterectomy does not menstruate and is no longer fertile. *See also* MENSTRUATION

hysteresis Phenomenon occurring in the elastic behavior of substances in which the STRAIN is greater when the STRESS is decreasing than when it is increasing because of a lag in the effect. When the stress is removed, a residual strain remains. The phenomenon also occurs in the magnetic behavior of materials: the magnetization is greater when the magnetizing field is decreasing. This phenomenon is particularly important in FERRO-MAGNETISM; the magnetization of ferromagnetic materials lags behind the magnetizing force.

iatrogenic disease Illness or disorder caused by medical treatment. It can be applied to any untoward condition that is caused by medical staff or equipment, therapeutics, or especially drugs.

ice Water frozen to 0°C (32°F) or below, when it forms complex six-sided CRYSTALS. Ice is less dense than water and floats. When water vapor condenses below freezing point, ice crystals are formed. This occurs mainly in high, cirrus CLOUDS, but also in the gray portions of other clouds. Clusters of these ice crystals form snowflakes. It also occurs near the ground and gives rise to FROST.

ice ages Prolonged period of colder climatic conditions, during which snow and ice covered large areas of the Earth. There is evidence of about 20 ice ages having occurred throughout Earth's history, the earliest dating back to 2.3 billion years ago. The most recent began about 2 million years ago and is often referred to as the "Ice Age." During this period, ice advanced for several thousand years, and then retreated for about the same period. An advance of ice is called a **glacial phase**, and the retreat is an **interglacial**. The Earth is experiencing an interglacial at present, although there are still ice-age conditions over Greenland and Antarctica. During the glacial phases, ice covered the US as far s as New York City and St. Louis, MO. Nearly half of Europe was also covered by glaciers. Areas farther s experienced PERMAFROST and TUNDRA conditions. When the ice melted, finally disappearing about 12,000 years ago, the climate warmed up, enabling different types of vegetation grow. The last ice age produced many of the landforms seen in the N continents and affected sea level on a global scale. *See also* GLACIER

iceberg Large, drifting piece of ice, broken off from a GLACIER, or polar ICE CAP. As the ice moves slowly outward from a snow-cap, some reaches the sea. It floats and is gradually broken off by the movement of tides and waves. Some icebergs may be several hundred meters across, but they gradually melt and shrink as they float across the oceans. Icebergs from Greenland are quite tall, whereas those from the Antarctic are flat and tabular in shape. Up to 80% or 90% of an iceberg is below the surface of the water. As they melt, icebergs sometimes turn over; this readjusts the balance. One of the major areas for icebergs is on the route of the cold Labrador Current, which brings them s along the coast of Greenland, towards Newfoundland, where they can be a hazard to shipping.

icebreaker Ship with a heavy bow and armored sides, designed with powerful engines that enable it to make a passage through ice. For breaking thick ice, the bow is so designed that the forward motion of the ship drives it up onto the ice, and the weight of the ship breaks it. Propellers fore and aft allow great maneuverability. Icebreakers are used regularly to clear channels in the Great Lakes and in the Baltic Sea, and they have also been used in Arctic and Antarctic explorations. The Soviet icebreaker *Lenin*, launched in 1959, was the first nuclear-powered nonnaval vessel.

ice cap Small ICE SHEET, often in the shape of a flattened dome, which spreads over the mountains and valleys of polar islands. The floating ice fields surrounding the North Pole are sometimes incorrectly called an ice cap.

Iceland spar (calcium carbonate, $CaCO_3$) Transparent form of CALCITE that has the property of **birefringerence** (or double refraction) – that is, it bends light two ways so that an image seen through it appears double. It is used for polarizing PRISMS and in polarizing MICROSCOPES and other optical instruments. *See also* POLARIZED LIGHT

ice sheet (continental GLACIER) Large expanse of snow and ice covering a land mass; for example, Antarctica. Ice sheets can cover entire mountain ranges and can spread across lowlands and oceans. The ice may become very thick: in Greenland and Antarctica there are places where the ice measures 3,000m (9,800ft) in depth. An ICEBERG formed when ice breaks from the margins of such a sheet may be many square kilometers in area.

ichthyology Zoological study of FISH, including their classification, structure, distribution, and ecology. The ancient Greeks, particularly Aristotle, are regarded as the pioneers of this science in the West.

iconoscope CATHODE-RAY TUBE used in early TELEVISION cameras, invented by Vladimir ZWORYKIN. Light entering the tube strikes a PHOTOELECTRIC PLATE, causing the plate to become positively charged. The other side of the plate is scanned by a beam of ELECTRONS from an ELECTRON GUN. Those electrons not held in the plate bounce back to be collected and transmitted as the television signal.

ideal gas law (gas laws) Law relating pressure, temperature, and volume of an **ideal (perfect) gas**: $pV = nRT$, where n is the number of molecules of the gas and R is the GAS CONSTANT of proportionality. This law implies that at constant temperature (T), the product of pressure and volume (pV) is constant (BOYLE'S LAW); and at constant pressure, the volume is proportional to the temperature (CHARLES' LAW).

idiopathic In medicine, term used to describe a condition that arises spontaneously, or from unknown causes.

idocrase (versuvianite) Orthosilicate mineral, consisting of hydrous calcium, iron, magnesium, aluminum silicate. Found in impure limestone, it forms tetragonal system crystals and is glassy green, brown, yellow or blue. Hardness 6.5; r.d. 3.4. Its transparent green or brown crystals are cut as gems.

Idrisi (*c*.1099–1165) Arab geographer. He worked at the court of Roger of Sicily after 1145 and wrote the *Kitab Rujjar* (Book of Roger, 1154), a geographical study of the world. He also made a map of the world engraved on silver.

igneous rock Term denoting rocks of volcanic origin that were formed from MAGMA. If the magma solidified beneath the Earth's surface, cooling would have been slow; if it solidified at the surface, cooling would have been more rapid. The faster the cooling process, the smaller are the crystals that make up the rock. All igneous rocks are crystalline. In addition to grain size, the other important differences found in igneous rocks are related to the chemical content of the minerals in the rock. Acidic minerals, including QUARTZ, some FELDSPARS, and some MICAS, are generally light in color. Conversely, basic minerals are generally dark in color; they include some feldspars, biotite mica, HORNBLENDE, and AUGITE. Igneous rocks formed at depth are called **plutonic**, those formed at the surface are called **volcanic**, and those formed between are **hypabyssal**. Volcanic rocks are EXTRUSIVE, and the commonest variety is BASALT. Hypabyssal rocks are found in SILLS and DIKES, and they are INTRUSIVE ROCKS. Plutonic rocks form in large bodies of magma called BATHOLITHS, stocks, and "bosses" (circular intrusions, lying at a steep angle to the surface). The most common plutonic rock is GRANITE, which is acidic in mineral content; the basic equivalent is called GABBRO. Igneous rocks do not have BEDDING PLANES, and they do not contain fossils. They are mostly quite resistant to erosion, and often form high ground.

ignition, engine Electrical components for igniting fuel in a gasoline engine. Air and gasoline vapor are usually mixed in a CARBURETOR and delivered to the CYLINDER, where the mixture is compressed and ignited by a spark produced by the SPARK PLUG.

ileum Major part of the small intestine, *c*.4m (13ft) long. Its inner wall is lined with fingerlike projections (villi), which increase the surface area available for the absorption of nutrients.

ilium Broad flat bone on either side of the PELVIS. Joined by the SACRUM, the fused bones of the base of the spine, it is fused with the ISCHIUM and PUBIS in a triangular suture at the hip socket. The ilia form the back of the "basin" of the pelvic girdle and anchor the muscles that form the frontal wall of the abdomen. *See also* HUMAN BODY

illusion Lack of correspondence between the perception of an object and the physical reality. Illusions include such phenomena as hallucinations and can occur in any of the five senses, although optical and visual illusions are the most common. **Optical** illusions are caused by physical properties such as REFRACTION, which explains why a rod lowered into water appears to bend. A **visual** illusion is the misinterpretation by the brain of something perceived by the eye. **Geometric** illusions occur when patterns of lines or shapes present conflicts of interpretation, such as the illusion in which two lines of

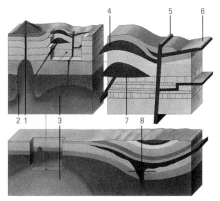

▲ **igneous rock** Many different shapes of igneous rocks can be found. Among the most common are the neck (1), which is a circular vertical feed channel of a volcano. A stock (2) is a large mass of rock which solidified at great depth. A batholith (3) is a large body of granite with no detectable bottom. A laccolith (4, 7) is a dome-shaped mass which has forced the rock above it to arch. A dike (or dyke) (5) is a vertical, sheet-like mass of rock and a sill (6) is a horizontal, sheet-shaped body of rock. A lopolith (8) is a saucer-shaped mass of rock.

equal length appear unequal because of the diagonal lines that are attached to their ends in different ways. Other visual illusions concern the perception of color, brightness, motion, and depth.

ilmenite (iron titanate, FeTiO$_3$) Black, oxide mineral, a major ore of the metal TITANIUM. It is found in basic IGNEOUS ROCKS and beach sands as tabular to fine, scaly crystals in the rhombohedral system, and as compact masses or granules. It has a metallic luster and is magnetic. Hardness 5–6; r.d. 4.7.

image, optical Representation of an object produced by an optical instrument. A **real** image has rays of light passing through it and is formed by points to which the rays from the object converge. Rays of light do not pass through the points composing a **virtual** image: the image lies at the place from which the light rays, if traced backward, appear to diverge. Images formed in a MIRROR are examples of virtual images. A real image can be projected onto a screen and recorded in a photographic emulsion; a virtual image, such as that produced by a plane mirror, cannot. *See also* LENS

image intensifier Electronic device in which gamma rays, X rays, and ultraviolet waves from a source can be converted into a visible image or in which the intensity of a visible image can be increased. The radiation falls onto a special surface and releases electrons that are then accelerated and focused onto a fluorescent screen, producing an enhanced visible image. *See also* ELECTROMAGNETIC SPECTRUM; PHOTOMULTIPLIER

imaginary number *See* NUMBER, IMAGINARY

imago Adult, reproductive stage of an INSECT that has undergone full METAMORPHOSIS. Imagos are the winged insects, such as butterflies and dragonflies, that emerge from PUPAS or develop from NYMPHS.

immune system System by which the body defends itself against disease. It involves many kinds of LEUCOCYTES (white blood cells) in the blood, lymph, and bone marrow. Some of the cells (B-cells) make ANTIBODIES against invading microbes and other foreign substances (ANTIGENS), or neutralize TOXINS produced by PATHOGENS, whereas other antibodies encourage PHAGOCYTES and MACROPHAGES to attack and digest invaders. Another type of cell, T-CELL, provides a variety of functions in the immune system. *See also* ACQUIRED IMMUNE DEFICIENCY SYNDROME (AIDS)

immunity Resistance to attack by disease-causing microorganisms. It can be acquired naturally, as from an infection that stimulates the body to produce protective ANTIBODIES (a newborn baby

carries some ANTIBODIES from its mother). Alternatively, it can be conferred by IMMUNIZATION.

immunization Practice of conferring IMMUNITY against disease by artificial means. **Passive** immunization may be conferred by the injection of an antiserum containing ANTIBODIES. **Active** immunization involves vaccination with dead or attenuated (weakened) microorganisms to stimulate production of specific antibodies and so provide lasting immunity.

immunoglobulin PROTEIN found in the bloodstream that plays a role in the body's IMMUNE SYSTEM. Immunoglobulins act as ANTIBODIES for specific ANTIGENS. They can be obtained from donor plasma and injected into patients most at risk of particular diseases.

immunology Study of IMMUNITY and all the defence mechanisms of the IMMUNE SYSTEMS mounted by the body for its protection. Immunologists are also concerned with related phenomena such as allergic responses, the rejection of transplanted organs and AUTOIMMUNE DISEASE.

immunosuppressive drug Any drug that suppresses the body's IMMUNE SYSTEM's responses to infection or "foreign" tissue. Such drugs are used to prevent rejection of transplanted organs and to treat AUTOIMMUNE DISEASE and some cancers.

impedance (symbol Z) Property of a component in an ELECTRIC CIRCUIT that opposes the passage of current. In a direct current (DC) circuit, the impedance corresponds to the RESISTANCE (R). In an alternating current (AC) circuit with CAPACITANCE or INDUCTANCE, the additional property of reactance (X) has to be allowed for, as expressed in the equation $Z^2 = R^2 + X^2$. All of these quantities are measured in OHMS.

imperial system Units of measurement developed in the UK. *See* FPS SYSTEM

impermeable Term that describes rock that does not allow water to pass through it. Slate is one example. *See also* IMPERVIOUS; PERMEABILITY

impervious Term that describes rocks that contain no cracks or fissures for water to flow through, although the rock itself may absorb water. Clay is one example. *See also* IMPERMEABLE

implication Logical proposition of the type "if P then Q," connecting two simple propositions P (the **antecedent**) and Q (the **consequent**). In the form used in mathematical LOGIC the two propositions need not be connected. This is called **materi-**

al implication – an example is "if the Earth is flat then gold is a metal." A material implication is false only when the antecedent is true and the consequent false, otherwise it is true. In normal discourse **formal** implication is used, in which the simple propositions are related in meaning; for example "if it does not rain, we will go for a walk." An implication is written as $P \rightarrow Q$ (or $P \Rightarrow Q$), read "P implies Q."

imprinting Form of learning that occurs within a critical period in very young animals. A complex relationship develops between the newborn infant and the first animate object it encounters; this is usually a parent. The future emotional development of the infant depends upon this relationship. Imprinting in birds has been studied by Konrad LORENZ, who believed that it is an irreversible process.

inbreeding Mating of two closely blood-related organisms. It is the opposite of outbreeding and over successive generations causes much less variation in GENOTYPE and PHENOTYPE than is normal in a wild population. A form of GENETIC ENGINEERING, it can be used to improve breeds in domestic plants and animals. In humans, it can have harmful results, such as the persistence of HEMOPHILIA in some European royal families.

incandescence Emission of light by a substance at a high temperature. Whether or not a hot object is seen as incandescent depends to some extent on the dark adaptation of the human EYE, but an incandescent object is never at a temperature below 400°C (750°F). An object like a FLUORESCENT LAMP can emit bright light without being incandescent.

inch (symbol in or ") Unit of measurement equal to 2.54cm. There are 12in to 1 ft.

incisor Any of the chisel-shaped cutting TEETH between the canines in the front of the mouths of mammals. There are eight in humans, four in the center front of each jaw. *See also* DENTITION

inclination, magnetic (magnetic dip) Angle between the direction of the Earth's magnetic field and the horizontal, measured by a free-floating magnet. At the north magnetic pole the inclination is zero; at the magnetic equator it is 90°. *See also* DECLINATION, MAGNETIC

inclined plane Simple machine that consists of a ramp. It is easier to push a load (mass *m*) up a slope (at an angle θ to the horizontal) than it is to lift it vertically – there is a MECHANICAL ADVANTAGE. The minimum force needed to push a load up an inclined plane is *mg* sinθ (where *g* is the acceleration of free fall).

incomplete dominance In GENETICS, a situation in which neither GENE for a particular characteristic is DOMINANT. As a result, the organism shows the influence of both genes. For example, a plant with genes for both red flowers and white flowers may have pink flowers.

incubation In biology, process of maintaining stable, warm conditions to ensure that eggs develop and hatch. Incubation is carried out naturally by birds and by some reptiles. It is accomplished by sitting on the eggs, by making use of volcanic or solar heat or the warmth of decaying vegetation, or by covering the eggs with an insulating layer of soil or sand.

incubation period In medicine, time lag between becoming infected with a disease and the appearance of the first symptoms. In many infectious diseases, the incubation period is quite short – anything from a few hours to a few days – although it may also be more variable. In leprosy, for example, incubation periods range from a few months to as much as 20 years.

indehiscent Describing a FRUIT or seed pod that does not open spontaneously to release its seeds when ripe. Release of the seeds takes place when the fruit rots or is eaten by an animal. Fungi that do not release their spores are also termed indehiscent. *See also* DEHISCENT

independent assortment Separation of the ALLELES of a GENE into the GAMETES (sex cells) with no regard to the way in which the alleles of other genes have separated. It takes place during MEIOSIS – the type of CELL DIVISION that gives rise to gametes. Theoretically, as a result, all possible combinations of alleles should occur with equal frequency, giving rise to maximum genetic variation. It does not always happen in practice because alleles on the same CHROMOSOME are usually inherited together.

independent variable *See* VARIABLE

index fossil Any FOSSIL of limited time distribution that clearly marks certain beds or strata of rocks. Its occurrence in rocks located kilometers apart proves that these deposits were formed at the same time. These fossils are important in mapping rock formations and in locating valuable resources.

indicator In chemistry, substance used to indicate an ACID or an ALKALI. It does this usually by a change of color. Indicators, such as the dye LITMUS, can detect a change of pH that measures a solution's acidity (litmus turns red) or alkalinity (turns blue). Universal indicator (liquid or paper) undergoes a spectral range of color changes from pH 1 to 13.

indigo (indigotin, $C_{16}H_{10}O_2N_2$) Violet-blue dye traditionally obtained from plants of the genus *Indigofera*. It has been manufactured synthetically since the early 1900s.

indium (symbol In) Silver-white, metallic element of group III of the periodic table. Its chief source is as a by-product of zinc refining. Malleable and ductile, indium is used in certain alloys with low melting points, in semiconductors, to wet glass and coat metals, and as a mirror surface. Properties: at.no. 49; r.a.m. 114.82; r.d. 7.31; m.p. 156.6°C (313.9°F); b.p. 2,080°C (3,776°F); most common isotope ^{115}In (95.77%).

inductance Property of an electric circuit or component that produces an ELECTROMOTIVE FORCE (EMF) following a change in the current. The SI unit of inductance is the HENRY. **Self-inductance** (symbol L) occurs when the current flows through the circuit or component, and **mutual inductance** (symbol M) when current flows through two circuits or components that are linked magnetically. *See also* ELECTROMAGNETIC INDUCTION; INDUCTION

induction In medicine, initiation of LABOR before it starts of its own accord. It involves perforating the fetal membranes and administering the hormone oxytocin to stimulate contractions of the UTERUS. It is undertaken where there is medical risk to the mother or baby in waiting for labor to commence naturally.

induction In physics, process by which magnification or electrification is produced in an object. In ELECTROMAGNETIC INDUCTION, an electric current is produced in a CONDUCTOR when placed within a varying MAGNETIC FIELD. The magnitude of the current is proportional to the rate of change of MAGNETIC FLUX. In a TRANSFORMER, the alternating current in the primary coil creates a changing magnetic field that induces a current in the secondary coil. A DYNAMO, or generator, contains a constant magnetic field, within which a conducting coil (armature) is rotated. In MAGNETIC INDUCTION, a magnetic field is created in an unmagnetized metal by another nearby magnetic field. *See also* INDUCTANCE

induction coil Type of TRANSFORMER that produces high-voltage alternating current from lower-voltage current. A common example is the induction coils used to ignite SPARK PLUGS in most gasoline INTERNAL COMBUSTION ENGINES. All induction coils have two windings of wire, called the primary and secondary winding, the latter having more turns of wire. When the current in the **primary wire** is interrupted, a magnetic field, due to INDUCTION, is created around both windings,

with the second windings creating greater voltage because of the greater number of turns.

inductor Device that introduces INDUCTANCE into an electric circuit. A typical inductor is a TORUS – a doughnut-shaped ring of metallic or ceramic material, around which is wound a coil carrying an electric current.

Indus (Amerindian) Small constellation of the s sky; the only star above magnitude 3.5 is Alpha.

industrial chemistry Chemical reactions and processes involved in the large-scale manufacture of chemicals in industry. While the engineering aspects of the subject belong to the study of CHEMICAL ENGINEERING, the industrial chemist is concerned with the chemical problems associated with the transfer of small-scale laboratory reactions to industrial plants.

industrial engineering Branch of ENGINEERING concerned with the application of engineering principles to industrial processes. Industrial engineers try to discover the most efficient ways for a firm to use the available workforce, equipment, and materials. The engineers' work includes industrial planning, management organization, arranging technical training programs and devising quality control methods.

inequality Mathematical statement that one expression is less than or greater than another. The symbol $>$ represents "is greater than," and $<$ stands for "is less than." An example is $2x + 4 > 12$, which is equivalent to $12 < 2x + 4$. Inequalities of this type may be handled in a somewhat similar way to equations: thus, in the case above, $x > 4$. The symbols $\geq$ and $\leq$ indicate "greater than or equal to" and "less than or equal to," respectively.

inert gas *See* NOBLE GAS

inertia Property possessed by all MATTER that is a measure of the way an object resists changes to its state of motion. Isaac NEWTON formulated the first law of motion, sometimes called the law of inertia, stating that a body will remain at rest or in uniform motion unless acted upon by external forces.

inertial guidance In aeronautics, system using components that respond to changes in INERTIA to control missiles or spacecraft the orbits of which are largely above the Earth's atmosphere. The main components include GYROSCOPES for reference and for detecting changes in orientation, engines or jets for correcting differences between planned and actual flight paths, and ACCELEROMETERS for determining velocity and position. The system can be supplemented by radar observations and control to correct gyro drift. *See also* NAVIGATION

infection Invasion of the body by disease-causing organisms that become established, multiply, and give rise to symptoms. *See also* IMMUNE SYSTEM

infertility Inability to reproduce. In a woman it may be due to a failure to ovulate (release an OVUM for FERTILIZATION), obstruction of the FALLOPIAN TUBE or disease of the ENDOMETRIUM; in a man it is usually due to inadequate SPERM production. In plants, the term refers to inability to reproduce sexually. Infertility occurs in a HYBRID between different species, which are unable to produce viable GAMETES (eggs and male sex cells). Such plants cannot be grown from seed. However, many can reproduce by ASEXUAL REPRODUCTION. *See also* IN VITRO FERTILIZATION

infinity (symbol ∞) Abstract quantity that represents a number without bound. In analysis, $1/x$ approaches (positive) infinity as x approaches zero through positive values. In SET THEORY, the set of all integers is an example of an infinite set.

inflammation Reaction of body tissue to infection or injury, with resulting pain, heat, swelling, and redness. It occurs when damaged cells release a substance called HISTAMINE, which causes blood vessels at the damaged site to dilate. LEUCOCYTES (white blood cells) invade the area to engulf bacteria and remove dead tissue, sometimes with the formation of pus. Inflammation can become chronic.

inflorescence FLOWER or flower cluster. Inflorescences are classified into two main types according to branching characteristics. A **racemose** inflorescence has a main axis and lateral flowering branches, with flowers opening from the bottom up or from the outer edge in; types include panicle, raceme, spike, and umbel. A **cymose** inflorescence has a composite axis with the main stem ending in a flower and lateral branches bearing additional, later-flowering branches.

information storage and retrieval Branch of COMPUTER science that is concerned with the function of data sets. **Database** systems retrieval involves searching on large computer files for specific DATA that may be organized into a variety of fields. Once a perfect match is found the information is made available. This type of system allows specific sets of data to be retrieved independently or with other sets. **Document-retrieval** systems store and retrieve entire documents, which can be searched for and retrieved using the document name or key words within the document. **Reference-retrieval** systems do not store documents themselves but references to documents. A search will make available the location of the relevant documents. This system is most commonly used in libraries where printed matter, such as books and periodicals, can be quickly located. *See also* DATABASE

information technology (IT) COMPUTER and TELECOMMUNICATIONS technologies used in processing information of any kind. Word processing, the use of a DATABASE and the sending of messages over a COMPUTER NETWORK all involve the use of information technology. Television stations use information technology to provide viewers with TELETEXT services. *See also* WORD PROCESSOR

information theory Mathematical study of the laws governing communication channels. It is primarily concerned with the measurement of information and the methods of coding, transmitting, storing, and processing this information. Some of the concepts of information theory have been used in psychology to elucidate the processes involved in sensory perception and memory.

infrared astronomy *See* ASTRONOMY

infrared radiation Radiations occupying the ELECTROMAGNETIC SPECTRUM between the red end of the visible spectrum and microwaves. Often called heat radiation, it was first discovered by Sir William HERSCHEL in 1800. Its wavelength range is about 750nm to 1mm. Infrared ASTRONOMY uses infrared radiation to study celestial objects, the military uses it for missile guidance systems and night sights, and in medicine it is used for thermal imaging.

infrasound (infrasonics) SOUND waves having a frequency of 15 hertz (cycles per second) or less, below the range of human HEARING. This range is also known as the **subsonic range**, because such low frequency waves are felt as vibrations rather than heard as sounds. *See also* ULTRASONICS

inhibitor Compound that stops or substantially reduces the rate of a CHEMICAL REACTION. Inhibitors are as specific in their action as CATALYSTS and are widely used to prevent CORROSION, OXIDATION or POLYMERIZATION.

injection In medicine, use of a syringe and needle to introduce drugs or other fluids into the body to diagnose, treat or prevent disease. Most injections are either **intravenous** (into a vein), **intramuscular** (into a muscle), or **intradermal** (into the skin).

ink Colored liquid used for writing, drawing or printing. It may be colored by a suspended pigment or a soluble dye. The latter, often based on the compound ANILINE, is used in inks for ballpoint pens. Printing ink usually contains finely divided carbon black suspended in a drying oil, often with added synthetic RESINS. Some inks dry by evaporation of a volatile solvent.

inlier Area of old rocks surrounded by younger ones. The old rocks may have resisted EROSION, and stand as hills above the surrounding eroded area. Alternatively, the upper and outer rocks of an ANTICLINE may have been eroded, leaving the older, once lower rocks surrounded by younger ones. *See also* OUTLIER

inner ear *See* EAR

inoculation *See* IMMUNIZATION; VACCINATION

inorganic chemistry *See* CHEMISTRY

insanity (madness) Imprecise term for severe mental illness. The condition has many manifestations, and is known by various other names that change from culture to culture and from age to age. Mental illnesses fall into one of three categories: PSYCHOSIS, which includes schizophrenia, manic depression, paranoia, and functional psychoses; NEUROSIS, including states of extreme anxiety or phobia, overt obsessional or compulsive states and the hysterical or depressive neuroses; and **personality disorders**, such as the behavioral disorders of childhood, alcoholism, and drug addiction.

insect Any of more than a million species of small, INVERTEBRATE animals of the order Insecta; including the beetle, bug, butterfly, ant, and bee. There are more species of insects than all other species combined. Adult insects have three pairs of jointed legs, usually two pairs of wings, and a three-segmented body (**head**, **thorax**, and **abdomen**) with a horny outer covering or **exoskeleton.** The head has three pairs of mouthparts, a pair of compound eyes, three pairs of simple eyes and a pair of antennae. Most insects can detect a wide range of sounds through ultrasensitive hairs on various parts of their bodies. Some can "sing" or make sounds by rubbing together parts of their bodies. Most insects are HERBIVORES; many are serious farm and garden pests. Some prey on small animals, especially other insects, and a few are scavengers. There are two main kinds of mouthparts – chewing and sucking. Reproduction is usually sexual. Most insects go through four distinct life stages, in which complete METAMORPHOSIS is said to take place. The four stages are OVUM (egg), LARVA (caterpillar or grub), PUPA (chrysalis), and IMAGO (adult). Young grasshoppers and some other insects, called NYMPHS, resemble wingless miniatures of their parents. The nymphs develop during a series of molts (incomplete metamorphosis). Silverfish and a few other primitive, wingless insects do not undergo metamorphosis. *See feature on page 194*

insecticide Substance used to destroy or control insect pests. Insecticides may be **stomach** poisons, such as lead arsenate and sodium fluoride; **contact** poisons, such as DDT and ORGANOPHOSPHATES; or **systematic** poisons, such as octamethylpyrophosphoramide, which are toxic to insects that eat plants into which the poisons have been absorbed. Organophosphates are preferred to chlorinated hydrocarbons (such as DDT) because they break down into nontoxic substances. They are, however, hazardous to humans and must be handled with care.

▶ **insect** The internal anatomy of all insects, such as the honeybee, is contained and protected within the confines of the tough, flexible exoskeleton. The typical insect body contains organs of digestion, respiration, circulation, excretion, and reproduction. There are muscles through which movement is effected and a nervous system which co-ordinates and controls insect actions on the basis of information received by the sense organs, most important of which are the large compound eyes and the feelers and antennae. All insect bodies comprise three parts: the head, thorax, and abdomen.

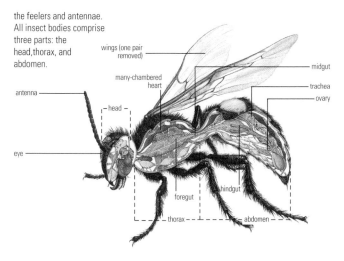

wings (one pair removed)
many-chambered heart
antenna
head
eye
midgut
trachea
ovary
foregut
hindgut
thorax
abdomen

insectivore In the wider, ecological sense of the term, an insectivore is any animal that eats insects. In the narrower, taxonomic sense, it is a small order of carnivorous MAMMALS (Insectivora), many of which eat insects. Almost worldwide in distribution, some species live underground, some on the ground and some in streams and ponds. Most insectivores have narrow snouts, long skulls, and five-clawed feet. Three families are always placed in the order: Erinaceidae, comprising moon rats, gymures, and hedgehogs; Talpidae, comprising moles, shrew moles, and desmans; and Soricidae, the shrews. Six other families – including tree shrews, tenrecs, and solenodons – are also often included in the order.

insectivorous plant (carnivorous plant) Any of several plants that have poorly developed root systems and are often found in nitrogen-deficient sandy or boggy soils. They obtain the missing nutrients by trapping, "digesting", and absorbing insects. Some, such as the Venus flytrap (*Dionaea muscipula*), are active insect trappers; when triggered, its hinged leaves close on its prey. The sundews (*Drosera*) snare insects with a sticky substance, and then enclose them in their leaves. Bladderworts (*Utricularia*) suck insects into underwater bladders. Other plants have vase-shaped leaves, such as the pitcher plant (*Sarracenia flava*) into which insects fall.

inselberg Steep-sided, round-topped hill rising abruptly from a plain, found in semi-arid, tropical, and subtropical regions. Inselbergs are probably formed by exfoliation EROSION of old mountains.

insemination, artificial Use of instruments to introduce donor semen into a female's reproductive tract in order to bring about FERTILIZATION. First developed for livestock breeding purposes, it is now routinely used to help INFERTILE couples. When SPERM from the woman's husband is used, it is known as AIH (artificial insemination by the husband); if it is acquired from an unknown donor, it is termed AID (artificial insemination by a donor).

insolation Quantity of solar radiation reaching the Earth, as measured in the SOLAR CONSTANT. It consists of a broad range of ELECTROMAGNETIC RADIATION, from infrared to X and gamma rays.

instinct Behavior that is innately determined, as opposed to learned. In the 19th century instincts were often cited to explain behavior, but the term fell into disrepute with the advent of behaviorism. The term was revived in the work of such ethologists as Konrad LORENZ.

instrument landing system (ILS) Combination of three radio systems that guide an aircraft pilot to a landing when visibility is poor. A **glide slope** beam sent from the runway indicates the proper angle of descent, and the **localizer** beam indicates its direction. The **outer marker** beam, set about 8km (5mi) from the runway, and the **middle marker**, at *c*.0.8km (0.5mi), show distance.

insulation Technique for reducing or preventing the transfer of heat, electricity, sound or other vibrations. Wool, fiberglass and foam plastic are good **heat**-insulating materials because they contain air. This trapped air reduces the transfer of heat by CONDUCTION. Free-moving air is not an effective insulator because it circulates and transfers heat by CONVECTION. Water is also a good heat insulator. A diver's wet suit keeps the wearer warm by trapping a layer of water around the body. **Electrical** insulators are substances that provide a high resistance to an ELECTRIC CURRENT. They are used to prevent contact between CONDUCTORS in a circuit. Common insulators are rubber, polyvinyl chloride (PVC), mica, teflon, glass, asbestos, thermoplastics, and porcelain. **Sound** insulating materials absorb sound. The physical structure of sound insulators determines their effectiveness at various frequencies. For example, a blanket over the head will block out mainly high frequencies, giving a muffled sound quality. **Noise** insulation, from engine vibrations in a car for example, is achieved by mounting the engine on springs or rubber fittings.

insulin HORMONE secreted by the ISLETS OF LANGHERHANS in the PANCREAS, important in the maintenance of blood-glucose levels. Insulin has the effect of lowering the blood-glucose level by helping the uptake of glucose into cells, where it is used up, and by causing the liver to convert glucose to GLYCOGEN, which is then stored in the liver. Glucose, unmetabolized because of lack of insulin, accumulates in excess amounts in the blood and urine, resulting in DIABETES.

integer Negative or positive whole NUMBER and zero, for example ..., −3, −2, −1, 0, 1, 2, 3, ... There is an infinite number of such integers. The positive integers are also called the natural numbers. The existence of negative integers and zero allow any integer to be subtracted from any other integer giving an integer result.

integral (symbol ∫) Mathematical symbol used in CALCULUS representing a summing operation. The integral of a function $f(x)$, written $\int f(x)dx$ can be interpreted as the area enclosed between the curve $y = f(x)$ and the x axis. INTEGRATION (finding the integral of a function) is formally carried out by dividing the area into a number of rectangular strips parallel to the y axis, and taking the limit of the sum of their areas as the number of strips increases and their width decreases. In fact, the symbol for an integration, ∫, stands for "sum." A **definite** integral has a numerical value that represents the area bounded by the *x* axis, *y* = *f*(*x*) and the two lines *x* = *a* and *x* = *b*. If *a* and *b* are unspecified the integral is said to be indefinite. In many cases the DERIVATIVE of the **indefinite** integral of a function gives the original function itself, so integration can be viewed as the inverse of DIFFERENTIATION.

integral calculus *See* CALCULUS

integrated circuit (IC) Complete miniature ELECTRONIC CIRCUIT incorporating SEMICONDUCTOR devices such as the TRANSISTOR and RESISTOR. The components of a **monolithic** integrated circuit occupy the surface layers of a single crystal of silicon (commonly called a SILICON CHIP). **Hybrid** integrated circuits have separate components attached to a ceramic base. Components in both types are joined by conducting film. Among their many uses, integrated circuits are found in pocket calculators and microcomputers. *See also* PRINTED CIRCUIT

Integrated Services Digital Network (ISDN) High-speed TELEPHONE lines designed to carry digital information. There are various grades of ISDN that can carry information more than a thousand times faster than conventional analog voice lines. ISDN lines connect directly to a computer and do not need a MODEM.

integration In mathematics, name given to a set of techniques used to evaluate a given INTEGRAL. Integration together with DIFFERENTIATION forms the branch of mathematics known as CALCULUS.

intensive care unit (ICU) Specialized hospital unit treating patients with acute, life-threatening conditions. It is characterized by the use of continuous (electronic) monitoring and life-support technology, together with intensive nursing and the judicious use of drugs. The intention is to offer short-term support to critically ill or injured patients whose condition is deemed to be treatable.

interaction, nuclear Interaction in which ELEMENTARY PARTICLES can take part and by which they may be classified. HADRONS are subject to the STRONG NUCLEAR FORCE, which acts over a tiny range (10^{-13} cm, approximately the diameter of a PROTON). Two hadrons inside this range interact, in about 10^{-23} second, by producing other particles or being deflected. LEPTONS are subject to the WEAK NUCLEAR FORCE and a much lower probability of interaction. *See also* FUNDAMENTAL FORCE

intercom (intercommunication system) System that permits selective loudspeaker voice communication via wires between any pair of several stations,

usually in the same building. The stations may be either "master stations," which may initiate calls to any in a group of stations, or "slave stations," which may initiate calls only to a master station.

interface *See* COMPUTER INTERFACE

interference In OPTICS, the interaction of two or more wave motions, such as those of light and sound, creating a disturbance pattern. The beams totally or partly reinforce or cancel each other out, producing an interference pattern. **Constructive** interference is the reinforcement of the wave motion because the component motions are in phase. **Destructive** interference occurs when two waves are out of phase and cancel each other. Interference of light radiation from a LASER is used in HOLOGRAPHY to create a hologram image.

interferometer Instrument in which a wave, especially a light wave, is split into component waves, that are made to travel unequal distances to recombine as INTERFERENCE patterns. The patterns have such uses as quality control of lenses and prisms, and the measurement of wavelengths.

▲ **interferometer** The Michelson (1890) interferometer consists of a diagonal mirror (1) which divides the beam from the lamp in half; one half is transmitted to the mirror (C) and the other reflected to the mirror (D). A glass plate (2) is inserted to equalize the path lengths. The beams then return and recombine towards the left where 'fringes' form.

an observer sees both the image of C and D. If the planes of these two are parallel, circular fringes appear. Twyman (1918) modified Michelson's interferometer for looking for flaws in glass prisms and lenses (B). The prism to be tested (3) is placed in the light path. If it has imperfections the resulting image will be distorted.

interferon Protein produced by body CELLS when infected with a VIRUS. Interferons can help uninfected cells to resist infection by the virus, and may also impede virus replication and PROTEIN SYNTHESIS. In some circumstances they can inhibit cell growth; human interferon is now produced by GENETIC ENGINEERING for therapeutic use, to treat some cancers, hepatitis, and multiple sclerosis.

interglacial Interval between ICE AGES; the period of glacial retreat. *See also* GLACIER

internal combustion engine Engine, widely used in cars and motorcycles, in which fuel is burned inside, so that the gases formed can produce motion. An internal combustion engine may be a TWO-STROKE ENGINE or a FOUR-STROKE ENGINE. In the most common type of engine, a mixture of gasoline vapor and air is ignited by a spark from a SPARK PLUG. The gases produced in the resulting explosion drive a PISTON down along a CYLINDER. A crankshaft changes the reciprocating (to-and-fro) movement of the pistons into rotary motion. In the WANKEL ROTARY ENGINE, the gases produced in the explosions drive a triangular rotor. The DIESEL ENGINE is another form of internal combustion engine.

Internet (Net) Worldwide communications system consisting of hundreds of small COMPUTER NETWORKS, interconnected by telephone systems. It is a network of networks, in which messages and data are sent and received using short local links from place to place around the world. This enables users to send a message to the other side of the world by ELECTRONIC MAIL (E-MAIL) for the cost of a local phone call. *See also* WORLD WIDE WEB (WWW)

interphase Stage following cell division (MEIOSIS OR MITOSIS) in which the nucleus "rests." The nucleus is not dividing and adopts its final form in each of the daughter cells.

interplanetary matter Material in the space between the planets. It is made up of atomic particles (mainly PROTONS and ELECTRONS ejected from the Sun via the SOLAR WIND) and dust particles (mainly from COMETS but some are believed to be possibly of cosmic origin) in the plane of the ecliptic. The particles in the inner Solar System constitute the zodiacal dust cloud, scattered sunlight from which is responsible for the zodiacal light. An extremely tenuous ring of dust, over 50 million km (30 million mi) wide, orbits the Sun just outside the Earth's orbit. There is also dust in METEOR streams.

interpolation Mathematical procedure for finding intermediate unknown values of a FUNCTION lying between two or more known values. One simple method is to assume that the unknown val-

◀ **internal combustion engine** An in-line, four cylinder gasoline engine is the most common internal combustion engine used in cars. Air is sucked through a filter (1) into the carburettor (2), where it mixes with the gasoline. This mixture then enters the cylinders through the dual inlet valves (3) on each cylinder (4). The spark plug (5) then ignites the mixture forcing the piston rapidly down. The burnt gases are expelled through the outlet valves (6). The reciprocating motion of the pistons is converted into rotation by the crankshaft (7). The crankshaft also turns the timing belt (8) that controls the opening of the valves, through the cams (9) located on the camshaft (10), and also the firing of the spark plugs.

ues lie on a straight line between the known values; there are, however, many other interpolation techniques available. *See also* EXTRAPOLATION

intersection Point, or LOCUS of points, common to two or more geometrical figures. Two nonparallel lines in the same plane meet in a point; two nonparallel planes meet in a line. The intersection of two SETS is a new set consisting only of elements that are in both sets.

interstellar cloud Relatively dense concentration of INTERSTELLAR MATTER, especially dust, that usually appears in the form of a dark NEBULA. Such clouds are responsible for absorption leading to reddening or dimming of starlight, and they also may be regions in which stars are born.

interstellar matter Matter within galaxies occupying the space between stars. It includes bright NEBULAE (clouds of gas excited to luminescence by nearby stars) and dark nebulae (not excited). Farther away from stars, and filling interstellar space, is gas – mostly hydrogen – and particulate matter at exceedingly low densities, about $10–21 kg/m^3$. The interstellar matter in our galaxy accounts for 5% of its total mass.

interstitial cells Cells that form the connective tissue between other tissues or organized groups of cells. In cnidarians (such as jellyfish), for example, interstitial cells take the form of embryonic cells in the spaces between the columnar cells that form the body structure. In the TESTES of vertebrates, interstitial cells between the seminiferous tubules secrete androgen male sex HORMONES.

intervertebral disk Ring of CARTILAGE that separates the VERTEBRAE – the bones of the SPINE. Intervertebral disks allow, with the exception of the axis and atlas bones (*see* HUMAN BODY), small amounts of movement, thus providing the spine with a degree of flexibility. A second function is to absorb shock, particularly for the skull and brain. The disks consist of a fibrous ring containing a soft, pulpy substance. Strain or injury to the spine may rupture or herniate a disk so that the pulpy interior protrudes and presses on a spinal nerve. This condition, commonly called a slipped disk, causes pain and immobility.

intestine Lower part of the ALIMENTARY CANAL, beyond the STOMACH, part of the digestive system. Food is moved through the intestine by the wavelike action known as PERISTALSIS. Food undergoes the final stages of digestion and is absorbed into the bloodstream in the **small** intestine, a coiled tube that extends from the stomach to the large intestine. In the **large** intestine (cecum, colon, and rectum)

water is absorbed from undigested material, which is then passed out of the body through the anus.

intravenous drip (IV) Apparatus for delivering drugs, blood and blood products, nutrients, and other fluids directly into the bloodstream. A hollow needle is inserted into an appropriate vein and then attached to a length of tubing leading from a bag containing the solution. The infusion is run into the vein at a controlled rate that may be regulated by a simple clip or by a computer-operated infusion pump.

intrusion In geology, emplacement of rock material that was either forced or flowed into spaces among other rocks. An igneous intrusion, sometimes called a **pluton**, consists of MAGMA that never reached the Earth's surface but filled cracks and faults, before cooling and hardening.

intrusive rock Any IGNEOUS ROCK that forms by slow cooling under the Earth's surface (INTRUSION). In general, they are coarser grained than volcanic rocks that cooled on the surface.

invar Alloy containing about 64% iron, 36% nickel, and small quantities of carbon. It has a very low thermal expansion, which leads to its use in instruments for measuring, such as surveying rods.

inverse square law Any of several laws in physics in which a quantity is inversely proportional to the square of the distance to the source of the quantity. A typical example is COULOMB'S LAW, which states that the force between two electric charges is proportional to the product of the charges and inversely proportional to the square of the distance between them. Newton's law of GRAVITATION is another example.

inversion Atmospheric condition in which a property of the air, such as moisture content or temperature, increases with altitude. In a **temperature** inversion, the air temperature rises with altitude and a cap of hot air encloses the cooler air below. With little wind or turbulence to break up the condition, POLLUTION can build up often to a dangerous extent.

invertebrate In zoology, term for an animal without a backbone. There are more than a million species of invertebrates, divided into 30 major groups. One of these is Arthropoda (ARTHROPODS), the largest animal PHYLUM in terms of number of species. Most invertebrates are INSECTS, but the term also covers CRUSTACEANS and ARACHNIDS. MOLLUSKS make up the second-largest group of invertebrates.

in vitro fertilization (IVF) Use of artificial techniques that join an OVUM with SPERM outside the woman's body to help some infertile women to carry and give birth to children. The basic technique of IVF involves removing ova from a woman's OVARIES, fertilizing them in the laboratory and then inserting them into the woman's UTERUS. In ZYGOTE intrafallopian transfer (ZIFT), a fertilized ovum (zygote) is returned to the FALLOPIAN TUBE, from where it makes its own way to the uterus. In GAMETE intrafallopian transfer (GIFT), the ova are removed, mixed with sperm, then both ova and sperm are inserted into a Fallopian tube to be fertilized in the natural setting. These IVF procedures result in live births in up to 20% of cases.

involuntary muscle One of three types of MUSCLE in VERTEBRATES, so called because, unlike SKELETAL MUSCLE, it is not under the conscious control of the brain but is stimulated by the AUTONOMIC NERVOUS SYSTEM and by HORMONES in the bloodstream. It is of two kinds: **smooth** muscle is the muscle of the ALIMENTARY CANAL, blood vessels, and bladder; **cardiac** muscle powers the HEART. *See also* HUMAN BODY

iodine (symbol I) Nonmetallic element that is the least reactive of the HALOGEN group. The black, volatile solid gives a violet vapor and has an unpleasant odor that resembles CHLORINE. Iodine, only found in combination with other elements, was discovered in 1811. Existing as iodide in seawater, seaweeds, and other plants, it is also extracted from Chile saltpeter and oil-well brine. Iodine is essential for the functioning of the THYROID GLAND. It is used as a medical antiseptic and in photography (light-sensitive silver iodide). The radioactive isotope ^{131}I (half-life 8 days) is used in the treatment of thyroid gland disorders. Properties: at.no. 53; r.a.m. 126.9; r.d. 4.93; m.p. 113.5°C (236.3°F); b.p. 184.4°C (363.9°F); most stable isotope ^{127}I (100%).

iodoform (triiodomethane, CHI_3) Yellow, crystalline compound with a characteristic odor. Is is used as an antiseptic.

Io Innermost of Jupiter's four Galilean satellites, and the second smallest, with a diameter of 3,630km (2,980mi). It has the highest density (3.53 g/cm³) of the four, indicating a lack of water-ice in its makeup. Io is by far the most volcanically active world in the Solar System. The active regions contain volcanic vents and CALDERAS, lava flows, and volcanic plumes that send material to heights of up to 300km (200mi). A thin atmosphere of sulfur dioxide surrounds Io, and a doughnut-shaped plasma TORUS of ionized sulfur and oxygen surrounds its whole orbit and interacts with Jupiter's MAGNETOSPHERE. In 1996 the Galileo SPACE PROBE detected an iron core and magnetic field, and found the surface features to have changed significantly since the satellite was first imaged during the 1979 VOYAGER PROJECT missions.

ion ATOM or group of atoms with an electric (positive or negative) charge resulting from the loss or gain of one or more ELECTRONS. Positive ions are

◄ in vitro fertilization
In vitro fertilization is a technique developed to help couples who are unable to conceive a baby normally. The mother, or egg donor, is given hormone treatment to cause many ovarian follicles to develop (1). While the ovary is observed through a laparoscope, a long needle (2) is inserted into each follicle, which is then sucked up (3) and transferred to a petri dish where sperm are added (4). The cells are monitored under a microscope. As the zygote divides (5) to a stage where is consists of 16 cells, it is introduced into the mother's uterus through a long plastic tube so that it can implant in the wall (6) and progress to full term in the usual way.

ovary   uterus   fallopian tube

called **cations** and move toward the CATHODE in ELECTROLYSIS; negative ions are called **anions** and move toward the ANODE. Many crystalline inorganic solids are composed of arrays of ions of opposite charge. The process of forming ions is called IONIZATION.

ion engine Type of ROCKET engine that uses IONS, rather than hot gases, as the means of propulsion (ion propulsion). Ions are produced by projecting metal atoms (such as cesium or mercury) through an electric field; the ions are then accelerated to high speeds by an even more powerful field. A separate beam of electrons is projected behind the engine to neutralize the positively charged ions (or else the engine would acquire a high electric charge). The engine, which has been successfully tested in space, would be used to power long-distance spacecraft.

ion exchange Method of water softening employed widely in domestic and industrial units. Permanent HARDNESS OF WATER is caused mainly by chlorides and sulfates of calcium and magnesium. In ion exchange softeners, the calcium and magnesium ions are exchanged by chemical reactions with those of sodium. Since sodium salts are much more soluble, this softens the water.

ionic bond (electrovalent bond) Type of chemical bond in which IONS of opposite charge are held together by electrostatic attraction.

ionic compound Substance formed by IONIC BONDS, chemical bonds of positively and negatively charged IONS. Salts, bases and some acids are ionic compounds. Examples include magnesium oxide (MgO) and sodium fluoride (NaF). As crystalline solids, with negative and positive ions arranged alternately in the lattice, such compounds have high melting points above 250°C (480°F) and boiling points above 500°C (930°F). They are usually soluble in water but insoluble in organic solvents. As solids, ionic compounds are nonconductors of electricity, but in solution and in their molten state, they are good CONDUCTORS.

ionic equilibrium Equilibrium state existing in solutions of ELECTROLYTES. Strong acids and bases are completely dissociated into IONS when in AQUEOUS SOLUTION, but weak acids or bases are only partly dissociated. Electrolyte, AB, when placed in solution, may partly dissociate into A^+ and B^-. $[A^+]$ $[B^-]/[AB]$, where [] indicates activities, is a constant, termed the **equilibrium constant**.

ionic radius Effective radius of a particular type of ION. An ionic solid (CRYSTAL) is composed of two or more species of ions, which may be regarded as spheres packed together in a regular pattern. By considering the CRYSTAL STRUCTURES and parameters of numbers of compounds, it is possible to obtain a consistent set of ionic radii.

ionic transport *See* TRANSPORT

ionization Process in which neutral atoms or molecules are converted into IONS. Positive ions can be formed by supplying energy to detach ELECTRONS from the atom, as by the action of X rays, ULTRAVIOLET RADIATION or high-energy particles. The minimum energy to form an ion is the ionization energy (or potential). The opposite process – electron capture by a neutral species to yield a negative ion – is much less probable. Both types of ions can also be produced by breaking bonds, which can be induced by PHOTONS, or can occur spontaneously, as in the dissociation of acids and salts when dissolved in water.

ionization chamber Instrument for measuring the intensity of ionizing particles or radiation, such as X rays. The gas-filled chamber contains two electrodes across which a voltage is applied. The passage of radiation through the chamber ionizes the gas and the ions formed move toward the charged electrodes. The current thus produced, which is amplified in an associated circuit, is proportional to the radiation intensity. A similar process is used in a GEIGER COUNTER, which detects and measures radioactivity.

ionizing radiation ELECTROMAGNETIC RADIATION having sufficient energy to react with atoms to produce IONS. There are two natural sources: radioactive elements in the Earth's crust and atmosphere, which emit beta and gamma radiation; and COSMIC RADIATION from outer space, which give rise to the IONOSPHERE. *See also* IRRADIATION

ionosphere Region around a planet in which there are free ELECTRONS and IONS produced by ULTRAVIOLET RADIATION and X rays from the Sun. The degree of IONIZATION is greatly affected by solar activity. The Earth's ionosphere extends from 80km (50mi) to 1,000km (625mi) above the surface (that is, to the limits of the ATMOSPHERE in the VAN ALLEN RADIATION BELTS). It is divided into distinct layers distinguished by the concentration of electrons. The lowest layers, D and E, or Heaviside-Kennelly layers, undergo molecular ionization, while the upper layer, the F, or Appleton layer, undergoes atomic ionization. These layers reflect radio waves of long wavelengths while shorter wavelengths can pass through undisturbed. The reflecting power of these layers makes long-range radio broadcasting possible up to frequencies of about 30MHz.

ion propulsion *See* ION ENGINE

iridium (symbol Ir) Silver-white TRANSITION ELEMENT, first discovered in 1804. A platinum-type metal, iridium is hard and brittle and the most corrosion-resistant metal. It is used in making surgical instruments, scientific instruments, pen tips, and electrical contacts. Properties: at.no. 77; r.a.m. 192.22; r.d. 22.42; m.p. 2,410°C (4,370°F); b.p. 4,130°C (7,466°F); most common isotope ^{193}Ir (62.6%).

iris Colored part of the EYE. It controls the amount of light that enters the PUPIL in the center of the eye by increasing or decreasing the size of the pupil. These changes are brought about by muscles in the iris contracting or relaxing.

iron (symbol Fe) Common TRANSITION ELEMENT, known from the earliest times. Its chief ores are HEMATITE (Fe_2O_3), MAGNETITE (Fe_3O_4), and PYRITE (FeS_2). It is obtained in a BLAST FURNACE by reducing the oxide with carbon monoxide from coke (carbon), using limestone to form a SLAG. The pure metal – a reactive soft element – is rarely used; most iron is alloyed with carbon and other elements in the various forms of STEEL. The element has four allotropic forms, one of which is ferromagnetic. It forms two series of salts, termed iron(II), or ferrous, and iron(III), or ferric. Properties: at.no. 26; r.a.m. 55.847; r.d. 7.86; m.p. 1,535°C (2,795°F); b.p. 2,750°C (4,982°F); most common isotope ^{56}Fe (91.66%).

iron meteorite (siderite) METEORITE consisting mainly of iron (90–95%) and some nickel, with traces of other metals. Iron meteorites are classified into a number of groups according to the proportions of these other metals that they contain, and each group is thought to correspond to a different parent ASTEROID.

iron oxide One of three compounds that exist in three OXIDATION states: iron(II) oxide (ferrous oxide, FeO); iron(III) oxide (ferric oxide, Fe_2O_3), which occurs naturally as HEMATITE; and ferrosoferric oxide (Fe_3O_4), which has iron in both oxidation states.

iron pyrites *See* PYRITE

ironstone Ancient medium- to fine-grained SEDIMENTARY ROCK that contains iron-rich minerals such as HEMATITE and SIDERITE; MAGNETITE and PYRITE may also be present. Ironstones were formed some 2–3 billion years ago in the Precambrian era. In **banded** ironstone, the basic rock is CHERT interspersed with layers of ferruginous minerals. **Oolitic** ironstone, formed in sediments at the bottom of ancient seas, consists of small rounded grains (OOLITHS) in which ferruginous minerals have replaced lime-rich ones. In both types, the iron-containing minerals result in a deep red color.

irradiation Exposure to nuclear or ELECTROMAGNETIC RADIATION. Materials are often irradiated with high-energy neutrons in NUCLEAR REACTORS, to make them temporarily RADIOACTIVE. More portable sources of such radiation are RADIOISOTOPES such as cobalt-60 and cesium-137, which are used in the irradiation treatment for various forms of cancer. Treatment also involves the use of particle ACCELERATORS, including proton and neutron beam machines. Radioactive sources are used industrially, such as in FOOD PRESERVATION. IONIZING RADIATION is used to destroy bacteria and microorganisms in some foodstuffs, while in other foods, such as soft fruits, it helps to increase shelf life. The process of irradiation in the food industry is still being closely monitored. Some people fear it not only destroys harmful bacteria but also beneficial nutrients.

irrational number *See* NUMBER, IRRATIONAL

irregular galaxy Type of GALAXY characterized by an apparent absence of symmetrical or definite structure. Only about 3% of observed galaxies fall into this category, and it is possible that some represent intermediate stages in development, while others may be distorted by internal or external disturbances.

irrigation Artificial watering of land for growing crops. Irrigation enables crops to grow in regions with inadequate precipitation. The first irrigation systems date from before 3000 BC, when canals and ditches were built to carry water from rivers to nearby fields in Egypt, Asia, and the Middle East. Later, simple mechanical devices, including the ARCHIMEDES' SCREW, were used to transfer water from rivers and streams into irrigation channels. Today, most water for irrigation is surface water or GROUNDWATER. Surface water comes from streams, rivers, and lakes. Building a DAM across a river causes large quantities of water to accumulate in an artificial lake or reservoir. This water can be used for irrigation when rainfall is scarce. Groundwater is obtained from WELLS. In some regions, fresh water for irrigation is obtained by DESALINATION. Canals, ditches, pumps and pipes convey water to fields.

ischium U-shaped bone at the base and on either side of the PELVIS. It is fused with the ILIUM and the PUBIS in the hip socket; its lower edge meets the pubis. In primates the ischium takes most of the body's weight when sitting. *See also* HUMAN BODY

ISDN Abbreviation of INTEGRATED SERVICES DIGITAL NETWORK

island arc Chain of volcanoes that occurs along one side of a deep ocean TRENCH. The volcanoes rest on the plate of LITHOSPHERE that is not moving down into the Earth. Their andesitic (*see* ANDESITE) lavas may be formed from the partially melted material of the descending plate. Northern Japan and the Aleutian Islands are examples.

islets of Langerhans Clusters of cells in the PANCREAS that produce two hormones: INSULIN to raise the level of GLUCOSE in the blood and GLUCAGON to counteract the effect of insulin when necessary.

isobar Line on a weather map connecting points of equal pressure, either at the Earth's surface or at a constant height above it. The patterns of isobars depict the variation in ATMOSPHERIC PRESSURE, showing areas of high and low pressure on the map.

isogamy In biology, state of having male and female reproductive cells that act like GAMETES (sex cells), but are similar in size and structure. Isogamy is found in algae, some protozoans, and primitive plants. It is distinct from **anisogamy,** where male and female sex cells differ in appearance.

isohyet Line on a weather map connecting points with equal rainfall. Isohyets are usually drawn for a particular period of time, to show the distribution of rainfall over a few months in the summer or winter, or perhaps for a whole year.

isoleucine $(CH_3CH_2CH(CH_3)CH(NH_2)COOH)$ One of the essential AMINO ACIDS found in protein.

isomers Chemical compounds having the same molecular formula but different properties due to the different arrangement of atoms within the molecules. **Structural** isomers have atoms connected in different ways. **Geometric** isomers, also called cis-transisomers, differ in their symmetry about a double bond. **Optical** isomers are mirror images of each other.

isomorphism In CRYSTALLOGRAPHY, resemblance between crystals of chemical compounds having similar crystal structures. In biology, isomorphism refers to the similarity of shape observed in unrelated groups due to CONVERGENT EVOLUTION.

isoprene (2-methylbuta-1,3-diene, $CH_2=C(CH_3)$ $CH=CH_2$) Organic chemical compound. It is a synthetic product that closely resembles natural RUBBER.

isosceles triangle TRIANGLE having any two of its sides equal in length; the angles opposite these sides are also equal.

isostasy Theory describing the maintenance of an equilibrium in the total mass of the Earth's CRUST despite its crustal movement. There exists a balance between the land masses and the MANTLE of the Earth on which the continental plates float, so that the plates rise and sink on the semimolten surface of the mantle in such a fashion that the relative mass weighing upon the Earth's crust is constant. The spread of the continental plates by the upwelling of material from deep within the Earth's crust is balanced by the submergence of the opposite edges of the plates. *See also* CONTINENTAL DRIFT; PLATE TECTONICS

isotherm Line of equal temperature on a map (usually a weather map). The patterns of isotherms depict how the temperature changes across the area of the map.

isotope Variety of a chemical element differing in the numbers of particles in the nuclei of their ATOMS. Atoms of different isotopes have the same number of electrons or protons (same ATOMIC NUMBER) but a different number of neutrons, so that both the MASS NUMBER and the mass of the nucleus are different for different isotopes. The RELATIVE ATOMIC MASS (R.A.M.) of an element is an average of the relative atomic masses of the element's naturally occurring isotopes. Isotopes are distinguished by writing the mass number by the name or symbol of the element, for example, ^{14}C or carbon-14. The isotopes of an element all have similar chemical properties, but physical properties vary slightly. Most elements have two or more naturally occurring isotopes, some of which are radioactive (radioisotopes). Many radioisotopes can be produced artificially by bombarding elements with high-energy particles such as alpha particles. Radioisotopes are used in medicine, research, and industry. Isotopes are also used in radioactive DATING. *See also* RADIOACTIVITY

isotropic Term meaning the possession of physical properties that are the same regardless of the direction in which they are measured. For example, heat conductivity and viscosity are isotropic in water, but they may not be so in a very thick oil.

IVF Abbreviation of IN VITRO FERTILIZATION

Jabir *See* GEBER

Jackson, John Hughlings (1835–1911) British neurologist. He studied speech defects as a symptom of brain disease, which he located in the left cerebral hemisphere. He also traced motor spasms to a brain disorder. Such spasms are now called Jacksonian EPILEPSY.

Jacob, François (1920–) French biologist. With Jacques MONOD he discovered that a substance that they named MESSENGER RNA (mRNA) carried hereditary information from the CELL nucleus to the sites of PROTEIN SYNTHESIS and that certain GENES, called **operator genes,** control the activity of others. They shared the 1965 Nobel Prize for physiology or medicine with another French biologist, André LWOFF. *See also* HEREDITY

Jacquard, Joseph Marie (1752–1834) French silk weaver who perfected (1801–06) a LOOM that could weave patterns automatically. By 1812 there were 11,000 Jacquard looms in use in France. The loom, which was the first automated machine, was controlled by PUNCHED CARDS.

jade Semiprecious silicate mineral of two major types: **jadeite,** which is often translucent, and **nephrite,** which has a waxy quality. Both types are extremely hard and durable. Jade is found mainly in Burma and comes in many colors, most commonly green and white. Hardness 5–6; r.d. 3–3.4.

Jansky, Karl (1905–50) US engineer who in 1931, while investigating interference to telephone communications, discovered unidentifiable radio signals from outer space. Jansky concluded that they were stellar in origin and that the source lay in the direction of SAGITTARIUS, the direction of the center of our galaxy. His discovery is considered to be the beginning of RADIO ASTRONOMY. A unit measuring radio emission is named after him.

Janssen, Pierre Jules César (1824–1907) French astronomer who specialized in solar SPECTROSCOPY. He developed special instruments, including a spectrohelioscope for observing solar PROMINENCES without waiting for an ECLIPSE. He

discovered the helium line in the solar spectrum, and produced a photographic atlas of the Sun. In 1893 he established an observatory on Mont Blanc.

jasper Opaque semiprecious stone, usually red to brown, sometimes yellow or gray to green. Jasper is found in Greece, Siberia, Libya, and the Nile River valley. It was once believed that stomach complaints could be cured by wearing these stones. A dark form of jasper was traditionally used as a touchstone to estimate the gold content of precious alloys.

jaw Either of two bony structures that hold the TEETH and frame the mouth in humans and most vertebrates. The upper jawbone (MAXILLA) is fused to the SKULL. The lower jawbone (MANDIBLE) is movable, having two ligament hinges, one at each side of the temple near the entrance to the ear.

Jeans, Sir James Hopwood (1877–1946) British mathematician and physicist. In astronomy, he worked on stellar dynamics, the formation and evolution of stars, DOUBLE-STAR systems, and spiral GALAXIES. In 1917 he proposed that our Solar System was formed from matter pulled out of the Sun by the gravitational attraction of a passing star, a theory later developed by Harold Jeffreys (1891–1989), but since abandoned. Jeans wrote many popular books on astronomy, including *The Universe Around Us* (1929).

Jeffreys, Alec John (1950–) British geneticist. He was responsible for discovering particular elements of DNA that enabled GENETIC FINGERPRINTING to be used as an accurate method of identification.

jejunum Portion of the small INTESTINE, about 25–30cm (10–12in) long, into which food passes from the DUODENUM. The jejunum precedes the ILEUM, which at the lower end opens into the large intestine. *See also* DIGESTIVE SYSTEM

jellyfish *See* CNIDARIAN

Jenner, Edward (1749–1823) British physician, pioneer of the science of IMMUNOLOGY. Aware that cowpox, a minor disease, seemed to protect people from smallpox, in 1796 Jenner inoculated a healthy boy with cowpox from the sores of an infected dairymaid. The boy developed the disease but six weeks later, when inoculated with smallpox, was found to be immune to infection. This finding established the principle of VACCINATION as a lifesaving medical technique.

jet Dense variety of LIGNITE coal formed from wood buried on the seafloor. It is often polished and used in jewelry.

JET (Joint European Torus Research Project) Nuclear FUSION power research center set up in 1977 at Culham, England. Sponsored by the European Union, Sweden, and Switzerland, scientists hope to develop a machine that will heat PLASMA to immense temperatures within a MAGNETIC BOTTLE and thereby fuse atoms of hydrogen, with the subsequent release of nearly limitless energy. The machine uses the TOKAMAK design.

jet engine Engine that derives forward motion by the rapid discharge of a jet of fluid (gas or liquid) in the opposite direction. In a jet engine, fuel burns in oxygen from the air to produce a fast-moving stream of exhaust gases. These are ejected from the back of the jet engine and produce a forward thrust (propulsion) in accordance with NEWTON'S LAWS OF MOTION. ROCKETS work on similar principles, but have their own oxygen supply so they can work outside the Earth's atmosphere.

Jet Propulsion Laboratory (JPL) Space center in Pasadena, California, for the development and control of unmanned spacecraft. The California Institute of Technology runs JPL for the NATIONAL AERONAUTICS AND SPACE ADMINISTRATION (NASA). JPL scientists sent the Surveyor probes to the Moon in the 1960s. Other notable achievements include the GALILEO and MARINER PROJECTS, the VIKING SPACE MISSION and the VOYAGER PROJECT.

jet stream Strong current of air blowing through the atmosphere at a high altitude. The main jet streams are in the middle and subtropical latitudes. The temperate jet stream flows from W to E, though along a wavelike route, going N and S as it travels all around the world. It can blow at speeds of up to 370km/h (230mph) and so is an important factor for aircraft. If planes can fly with the jet stream, which often blows at a height of 10,000m to 13,000m (33,000–42,600ft) above the Earth's surface, it can increase the speed of the journey and save fuel costs. Jet streams influence the weather systems in the lower layers of the atmosphere, as depressions tend to follow their route.

Jex-Blake, Sophia Louisa (1840–1912) British physician who fought vigorously and successfully to get legislation through Parliament enabling women to become doctors and to practice medicine and surgery. She founded a medical school for women in London in 1874 and another in Edinburgh in 1886.

Jobs, Steven (1955–) US computer pioneer. Jobs and his partner Stephan WOZNIAK are remembered for developing the Apple series of microcomputers that, beginning in 1976, were among the first personal computers. The Apple II of 1977 employed an "open system", which allowed the addition of add-on hardware. VisiCalc of 1979 was the first user-friendly spreadsheet program.

joint In anatomy, place where one BONE meets another. In movable joints, such as those of the knee, elbow, spine, fingers, and toes, the bones are separated and cushioned from one another by pads of CARTILAGE. In fixed joints cartilage may be present in infancy but disappears later as the bones fuse together, as in the SKULL. In the movable joints of bony VERTEBRATES, the bones are held together by LIGAMENTS, which form a fibrous capsule. Lining the capsule is the synovial membrane, which secretes SYNOVIAL FLUID into the joint to lubricate it.

joint In geology, an upright or near-upright crack in a rock. Joints occur in SEDIMENTARY ROCKS, such as limestone, along lines of weakness caused by shrinking or weathering. Their position influences the structure of cliffs and escarpments. Joints form in IGNEOUS ROCKS, such as basalt, as magma solidifies, and cooling and contraction takes place.

▶ **jet engine** Turbofan engines are used on many civil jet aircraft. Fuel entering the engine (1) mixes with compressed air and burns in the combustion chamber (2). The expanding gases rotate high-speed (3) and low-speed (4) turbines. These in turn drive a compressor (5), which forces air into the combustion chamber, and fans (6), which push air round the combustion chamber and into the tail pipe, providing thrust by means of displacement.

Joliot-Curie Name of two French physicists. In 1926 **Irène Curie** (1897–1956) married **Frédéric Joliot** (1900–58), a fellow assistant to her parents, Pierre and Marie CURIE. Irène worked on the physical aspects of RADIOACTIVITY, while Frédéric concentrated on the chemical. Their discovery of artificial RADIOISOTOPES earned them the 1935 Nobel Prize for chemistry. They were active in the French resistance and became communists. Frédéric was responsible for the construction of France's first NUCLEAR REACTOR, but later was dismissed for political reasons. *See also* ISOTOPE

Josephson, Brian David (1940–95) British physicist who shared the 1973 Nobel Prize for physics with Leo ESAKI and Ivar GIAEVER for his discovery of the **Josephson effect.** In 1962 he predicted that an electric current would flow between two superconductors separated by a thin layer of insulator (the Josephson current), in addition to the current already expected on the basis of QUANTUM THEORY. He also predicted several other related phenomena. This hypothesis was later verified experimentally, and has helped in the understanding of SUPERCONDUCTIVITY.

Joule, James Prescott (1818–89) British physicist. **Joule's law** (1841) relates the current flowing through a wire to its heat loss and laid the foundation for the law of CONSERVATION of energy. The JOULE is named after him.

joule (symbol J) SI unit of energy. One joule is the work done by a force of one NEWTON acting over a distance of one meter. It was named after James JOULE and replaced the erg.

Joule-Thomson effect (Joule-Kelvin effect) Fall in temperature that occurs when a gas expands through a small opening into an area of lower pressure. It occurs because the gas has to do work to overcome intermolecular forces so that it can expand. The effect is employed in refrigerators. It is named after its discoverers, James JOULE and William Thompson KELVIN.

jugular In anatomy, term that applies to any structure in the neck and especially to any of several VEINS. The external jugular veins receive blood from the outside of the cranium, the neck, and the deep tissues of the face. Others receive blood from the back of the neck, the larynx, tissues below the lower jaw, the brain, and the face.

Juno Asteroid discovered (1804) by Karl Harding. It is the tenth-largest asteroid, with a diameter of 244km (152mi).

Jupiter Fifth major planet from the Sun and the largest of the giant planets. It is one of the brightest objects in the sky. Through a telescope, Jupiter's yellowish elliptical disk is seen to be crossed by brownish-red bands, known as belts and zones. The most distinctive feature is the GREAT RED SPOT (GRS). Spots, streaks and bands are caused by Jupiter's rapid rotation and turbulent atmosphere. Eddies give rise to the spots, which are cyclones or (like the GRS) anticyclones. Hydrogen accounts for nearly 90% of Jupiter's atmosphere and helium for most of the rest. The pressure at the cloud tops is 0.5 bar. At 1,000km (600mi) below the cloud tops there is an ocean of liquid molecular hydrogen. At a depth of 20,000–25,000km (12,500–15,000mi), under a pressure of 3 million bars, the hydrogen becomes so compressed that it behaves as a metal. At the center of Jupiter there is probably a massive iron-silicate core surrounded by an ice mantle. The core temperature is estimated to be 30,000K. The deep metallic hydrogen "mantle" gives Jupiter a powerful magnetic field. It traps a large quantity of PLASMA (charged particles); high-energy plasma is funneled into radiation belts. Its MAGNETOSPHERE is huge, several times the size of the Sun, and is the source of the planet's powerful radio emissions. Jupiter has 16 known SATELLITES, the four major ones being the GALILEAN SATELLITES. Knowledge of the planet owes much to visits by space probes: Pioneers 10 and 11, Voyagers 1 and 2, Ulysses, and GALILEO. **Data:** Diameter (equatorial) : 142,800km; Mass: (Earth=1) 317.9; Volume: (Earth=1) 1,319; Density (water=1): 1.33; Orbital period: 11.86 years; Rotation period (equatorial): 9h 50m 30s; Temperature at cloud tops: 125K

Jurassic Central period of the MESOZOIC era; it began 213 million years ago and lasted 69 million years. In this period there were large saurischian DINOSAURS such as *Atlantosaurus* and *Allosaurus*; and ornithischian dinosaurs such as *Camptosaurus* and *Stegosaurus*. Plesiosaurs, pterosaurs, and ARCHAEOPTERYX (believed to be the precursor to modern birds) also date from this period. Primitive mammals had begun to evolve; they were the ancestors of later MARSUPIAL and placental species.

juvenile hormone (neotenin) Hormone found in insects that prevents the development of adult characteristics when larval insects molt. It is produced in the corpora allata glands behind the brain. These organs become inactive at the last molt, and METAMORPHOSIS then begins.

kame Deposit formed by a subglacial stream near the terminal margin of a melting GLACIER. Streams flowing down mountainsides onto glaciers deposit sand and gravel at the point where the stream first reaches the ice, or where the water goes down into a crevasse. The accumulations of sand and gravel gradually fall down to the valley floor as the ice melts. Kames are generally small features, 20–30m (65–100ft) in length or breadth.

Kamerlingh-Onnes, Heike (1853–1926) Dutch physicist who was the first to liquefy HELIUM. In the late 1880s, he began studying gases at low temperatures and in 1908, using a liquid hydrogen cooling system, he liquefied helium and found its temperature to be 4°C (7°F) above ABSOLUTE ZERO. He discovered that at this temperature some metals such as MERCURY and LEAD lose all electrical RESISTANCE and become superconductors. He was awarded the 1913 Nobel Prize for physics. *See also* SUPERCONDUCTIVITY

kaolin (china clay) Especially fine and pure clay composed chiefly of the mineral KAOLINITE, a hydrous silicate of aluminum. It is important in the manufacture of pharmaceuticals, ceramics, and coated paper.

kaolinite Sheet silicate mineral of the kaolinite group, hydrous aluminum silicate $(Al_2Si_2O_5(OH)_4)$. It is a product of the weathering of feldspar and has triclinic system tabular crystals, claylike masses and particles. It is white with a dull luster and may be tinted by impurities. Hardness 2–2.5; r.d. 2.6.

kaon (K meson) ELEMENTARY PARTICLE that is either a charged or a neutral MESON with zero SPIN.

Kapitza, Peter Leonidovich (1894–1984) Russian physicist. He did extensive research on MAGNETISM (in England with Ernest RUTHERFORD) and discovered that helium-2, the stable form below −271°C (−456°F), has almost no resistance to flow – a phenomenon called SUPERFLUIDITY. He then returned to the Soviet Union and worked on satellite research and nuclear FISSION. In 1978 he was awarded the Nobel prize for physics for his work on low-temperature physics and magnetism.

Kármán, Theodore von (1881–1963) US research engineer, b. Hungary. He is best known for his work in the application of mathematics to aeronautics and space rocketry. He became a US citizen in 1936 and helped to found the NASA JET PROPULSION LABORATORY (JPL), the International Council of the Aeronautical Sciences, and the International Academy of Astronautics.

Karrer, Paul (1889–1971) Swiss chemist, b. Russia. He shared the 1937 Nobel Prize for chemistry with Sir Walter HAWORTH for research on COENZYMES, CAROTENOIDS, and VITAMINS. In 1930 Karrer determined the formula of beta CAROTENE, the precursor of vitamin A, and in 1931 elucidated the structure of the vitamin itself. He also studied vitamin E and demonstrated that lactoflavin was part of the vitamin B_2 complex.

karst Limestone plateau characterized by irregular protuberant rocks, sinkholes, caves, disappearing streams, and underground drainage. This type of landscape forms where solution WEATHERING wears away much of the limestone, and where rivers erode the rock. Most rivers are underground, and there are caves and large caverns. The largest caverns may collapse to form gorges, and gradually the entire area of limestone may be worn away. Such topography is named after its most typical site in the Karst region of former Yugoslavia. The barren, forbidding plateau usually deters people from settling on it, and it usually holds little arable land.

karyotype Arrangement, structure and number of CHROMOSOMES in the nucleus of a CELL. All the body cells of an organism have the same karyotype. The number of chromosomes varies widely with species, for example 8 in fruit flies, 12 in kangaroos, 46 in humans, and 200 in crayfish.

Kastler, Alfred (1902–84) French physicist, b. Germany. He was awarded the 1966 Nobel Prize for physics for the discovery and development of optical methods for studying RESONANCE in atoms. Kastler developed the technique of optical pumping, in which gas atoms absorb energy from bombardment with light and radio waves and release it a few milliseconds later, the manner of release varying according to the atomic species. This technique led to a new knowledge of atomic structure and the development of the MASER and LASER.

Katz, Sir Bernard (1911–) British biophysicist. He shared the 1970 Nobel Prize for physiology or medicine with Julius AXELROD for their work on the chemistry of NERVE transmission. Katz discovered how the NEUROTRANSMITTER acetylcholine is released by neural impulses, causing muscles to contract.

Kay, John (1704–64) British engineer. In 1733 he patented his famous flying shuttle which, by enabling a weaver to throw the shuttle automatically from side to side across the warp of a loom, doubled output. His invention, which was a major step in the development of automatic weaving, was seen as a threat to handloom weavers and he was forced to emigrate to France.

Keeler, James Edward (1857–1900) US astronomer. As head of the Allegheny Observatory (1891–98) he confirmed that SATURN's rings are not solid units but are composed of meteoric particles. While director of the Lick Observatory (1898–1900), he used the new Crossley reflecting telescope to establish that a spiral GALAXY is the most common type of observable galaxy.

Keith, Sir Arthur (1866–1955) British anatomist. He carried out work on the anatomy of the heart and applied his knowledge of general anatomy to an important study of human origins, reconstructing prehistoric humans from fossil remains. His books include *The Antiquity of Man* (1915) and *A New Theory of Human Evolution* (1948).

Kekulé von Stradonitz, Friedrich August (1829–96) German chemist who discovered the ring structure of BENZENE. He became professor first at Ghent, and later at Bonn. He worked on the structure of organic molecules and on the concept of VALENCE, particularly with respect to benzene and other aromatic compounds.

Kelly, William (1811–88) US steelmaker. In about 1850 he invented a method of converting molten iron into steel by OXIDIZATION. The process, later known as the BESSEMER PROCESS, was taken up and patented by the British engineer Sir Henry Bessemer in 1856.

kelp Any of several brown SEAWEEDS commonly found on Atlantic and Pacific coasts, a type of brown ALGAE. A source of iodine and potassium compounds, kelps are now used in a number of industrial processes. Giant kelp (*Macrocystis* sp.) exceeds 45m (150ft) in length. Phylum Phaeophyta.

kelvin (symbol K) SI unit of TEMPERATURE. The Kelvin temperature scale has a zero point at ABSOLUTE ZERO and degree intervals (kelvins), the same size as degrees CELSIUS. The freezing point of water occurs at 273K (0°C or 32°F) and the boiling point at 373K (100°C or 212°F).

Kelvin, William Thomson, 1st Baron (1824–1907) British physicist and mathematician after whom the absolute scale of TEMPERATURE is named. He is known particularly for his work in heat and electricity. The success of the Atlantic submarine telephone cable was due to Kelvin's research into the transmission of electric currents for which, in 1866, he was knighted. In THERMO-DYNAMICS, Kelvin was able to resolve conflicting interpretations of the first and second laws. He published more than 600 scientific papers.

Kendall, Edward Calvin (1886–1972) US chemist who worked on the biological effects of the HORMONES of the cortex of the ADRENAL GLANDS, in particular the STEROID hormone CORTI-SONE, which he isolated. He shared the 1950 Nobel Prize for physiology or medicine with Philip Showalter HENCH and Tadeus REICHSTEIN.

Kendrew, Sir John Cowdery (1917–97) British molecular biologist who determined the structure of MYOGLOBIN, the protein that serves as an oxygen store in muscle. Kendrew worked (1948–70) at the Cambridge Medical Research Council (MRC) Laboratory. Using X-RAY CRYS-TALLOGRAPHY he was able to elucidate the arrangement of AMINO ACIDS in the myoglobin HELIX. For this work Kendrew shared the 1962 Nobel Prize for chemistry with Max PERUTZ, who worked on the structure of HEMOGLOBIN. In 1974 he became the first director of the European Molecular Biology Organization laboratory (Embl) at Heidelberg.

Kennelly, Arthur Edwin (1861–1939) US electrical engineer. In 1902, after Guglielmo MARCONI had experimented with radio waves, Kennelly noticed that the waves could reach beyond the Earth's horizon. He proposed that they did this by bouncing off a layer of ions high in the ATMOS-PHERE. The British physicist Oliver Heaviside (1850–1925) made a similar proposition, and the layer was first called the HEAVISIDE-KENNELLY LAYER (now known as the E-layer). It is found in the IONOSPHERE.

Kepler, Johannes (1571–1630) German mathematician and astronomer. He supported the Sun-centered Solar System put forward by COPERNICUS. In 1600 he went to Prague, becoming assistant to Tycho BRAHE, whom he succeeded as Imperial Mathematician to the Holy Roman Emperor, Rudolf II, with the task of completing tables of planetary motion begun by Tycho. From Tycho's accurate observations, he concluded in *Astronomia nova* (1609) that Mars moved in an elliptical orbit, and went on to establish the first of his laws (*see* KEPLER'S LAWS) of planetary motion. Erroneous reasoning led him nevertheless to the second of these laws, and his desire to match celestial and musical harmony led to the third law, published in *Harmonices mundi* (1619). The *Rudolphine Tables*, based on Tycho's observations and Kepler's laws, appeared in 1627 and remained the most accurate until the 18th century. Other works include *De stella nova*, on the supernova of 1604 , and *Dioptrice* (1611), on optics and the theory of the telescope.

Kepler's laws Three fundamental laws governing the motions of the planets around the Sun (and other celestial bodies in closed orbits), first worked out by Johannes KEPLER between 1609 and 1619, and based on observations made by Tycho BRAHE. **Law 1** deals with the shape of a planetary orbit: the orbit of each planet is an ellipse with the Sun at one of the foci. **Law 2** explains the varying speed of planetary motion such that the planet moves faster the nearer it is to the Sun: the line joining the planet to the Sun (the radius vector) sweeps out equal areas in equal times. This is known as the law of areas. **Law 3** relates the size of the orbit to the period of revolution: the square of the period of revolution (P) is directly proportional to the cube of the mean distance of the planet from the Sun (a). In practice this means that P^2/a^3 is approximately constant for all bodies orbiting the Sun. *See diagram*

keratin Fibrous PROTEIN present in large amounts in the superficial cells of the SKIN, where it serves as a protective layer. Hair and fingernails are made up of modified epidermal cells filled with keratin, which is also the basis of claws, feathers, and the horns of some animals. *See also* EPIDERMIS

kerosene (paraffin) Distilled PETROLEUM product heavier than GASOLINE but lighter than DIESEL FUEL. Kerosene, known historically as an illuminant (paraffin oil), is now used in camping stoves, trac-

tor fuels, and gas-turbine fuels for JET ENGINES and turboprop aircraft.

ketone Organic chemical compound containing a carbonyl (C=O) group joined to two hydrocarbon groupings. Ketones are liquids or low-melting solids. Examples include acetone (propanone, CH_3COCH_3) and butanone (methyl ethyl ketone, $CH_3COC_2H_5$). Ketones can be made by oxidizing secondary alcohols. They are reactive, forming addition compounds and undergoing various condensation reactions, but they do not oxidize easily. They are used as solvents.

ketone bodies Three chemical compounds: acetoacetic acid, hydroxybutyric acid, and acetone. When present in the blood in high concentrations, they increase the blood's acidity. Known as **ketosis**, this occurs in starvation and DIABETES mellitus. The ketone bodies are products of poor carbohydrate METABOLISM, and they accumulate in the body tissue and fluids, especially the urine.

kettle hole Steep-sided basin formed when a chunk of ice left behind by a receding glacier is covered by rocks and debris previously pushed forward by the glacier. The ice melts and the rocks fall through, creating a kettle pot-shaped depression.

khamsin Hot s wind that blows across N Africa from the Sahara during spring and summer. It is very dry, with relative humidity sometimes as low as 10%. The name means "fifty" in Arabic, the number of days during which the wind is supposed to blow.

Khorana, Har Gobind (1922–) US biochemist who shared the 1968 Nobel Prize for physiology or medicine with Robert HOLLEY and Marshall NIRENBERG for discoveries about the way in which GENES determine cell function. They established

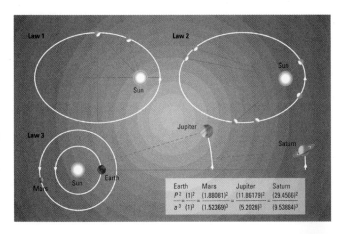

▶ **Kepler's laws** The three laws proposed by Johannes Kepler involve the planets of the Solar System. The first law states that the orbit of any of the planets forms an ellipse, with the Sun at one of the foci. The second law shows how planetary motion increases the closer a planet gets to the Sun. The third law relates the size of the planetary orbit to the period of revolution.

that most CODONS, combinations of three of the four different bases found in DNA and RNA, eventually cause the inclusion of a specific AMINO ACID into the cell proteins.

kidney In vertebrates, one of a pair of organs responsible for regulating blood composition and the EXCRETION of waste products. The kidneys are at the back of the abdomen, one on each side of the backbone. The human kidney consists of an outer cortex and an inner medulla with about one million tubules (NEPHRONS). Nephrons contain numerous CAPILLARIES, which filter the blood entering from the renal ARTERY. Some substances, including water, are reabsorbed into the blood. URINE remains, which is passed to the URETER and on to the BLADDER. *See also* HOMEOSTASIS

kidney machine (artificial kidney) Equipment designed to perform **hemodialysis** (removing toxic wastes from the blood) in kidney failure. Plastic tubing is used to pipe blood from the body into the machine, where waste products are filtered out by the technique of DIALYSIS. Usually two or three sessions are required a week, each lasting about four hours. Today many more patients are kept alive by the newer technique of **peritoneal dialysis**, in which fluids are passed through the peritoneal cavity to reduce blood urea.

kidney stones Small, hard, pebblelike masses formed in the KIDNEY from an accumulation of mineral substances; known medically as calculi. Passage of a stone through the URETER (tube from the kidney to the bladder) causes the excruciating pain of **renal colic**. They can be removed by surgery or, if the stones are small, they can be passed out in the urine after treatment, which includes drinking large amounts of water.

kidney transplant *See* ORGAN TRANSPLANT

kiln In ceramics, an oven for firing ware. Early kilns were holes in the ground into which the ware was placed and covered by a large fire. Later, special wood- or coalfired, oven-type kilns were built; today most kilns use gas or electricity.

kilocalorie *See* CALORIE

kilogram (symbol kg) SI unit of mass defined as the mass of the international prototype cylinder of platinum-iridium kept at the International Bureau of Weights and Measures near Paris; 1 kilogram is equal to 1,000g (2.2lb).

kilowatt hour (symbol kWh) Unit of electrical energy equal to the energy used when a power of 1 kilowatt (1,000 watts) is expended for 1 hour. It is

the unit in which domestic and industrial users of electricity are charged.

kimberlite Ultrabasic rock found in diamond-bearing pipes, mainly in South Africa. It is a mica PERIDOTITE consisting chiefly of olivine and phlogopite; it weathers to the yellow and blue grounds of diamond mines.

kinesthetic sense (proprioception) Internal sense that conveys information from the muscles and tendons of the body. Specialized receptors connect to a tract of the NERVOUS SYSTEM that provides information about the contraction and expansion of muscles. The kinesthetic sense is the position sense, because it allows human beings to know the positions of their limbs without visual confirmation. *See also* CEREBELLUM

kinetic energy Energy that an object possesses because it is in motion. It is the energy given to an object to set it in motion; it depends on the mass (m) of the object and its velocity (v), according to the equation K.E. = $\frac{1}{2}mv^2$. On impact, it is converted into other forms of energy such as heat, sound, and light. *See also* POTENTIAL ENERGY

kinetics In physics, one of the branches of DYNAMICS. In chemistry, a branch of physical chemistry that deals with the rates of chemical reactions. By studying rates at different pressures and temperatures, chemists can determine how reactions take place.

kinetic theory Theory in physics dealing with matter in terms of the forces between particles and the energies they possess. The following principles are basic to the simplest form of the kinetic theory: matter is composed of tiny particles; these are in constant motion; they do not lose energy in collision with each other or the walls of their container; there are no attractive forces between the particles or their container; and at any time the particles in a sample may not all have the same energy. More advanced forms of the theory take into account VAN DER WAALS FORCES between the particles.

kingdom In biology, the most widely adopted CLASSIFICATION for living organisms is the Five Kingdoms system, in which the kingdom is the topmost level, or **taxon**. The five kingdoms are Animalia (ANIMAL), Plantae (PLANT), Fungi (FUNGUS), PROKARYOTAE, and PROTOCTISTA. Two subkingdoms are often recognized within Prokaryotae, ARCHAEBACTERIA and EUBACTERIA, but the bacteria are so diverse that many taxonomists think they comprise more than one kingdom. Some believe that they merit the status of a new, even higher category, DOMAIN. *See also* EUKARYOTE

Kinsey, Alfred Charles (1894–1956) US zoologist, noted for his studies on human sexual behavior. He was a professor of zoology and later director of the Institute for Sex Research, Indiana University. He is best known for *Sexual Behavior in the Human Male* (1948) and *Sexual Behavior in the Human Female* (1953).

Kipp's apparatus Chemical apparatus for generating a stream of any gas that can be made by reacting a liquid with a solid, for example hydrogen sulfide gas from hydrochloric acid and iron(II) (ferrous) sulfide or carbon dioxide from hydrochloric acid and calcium carbonate (marble chips). When the outlet tap is turned off, internal gas pressure separates the liquid and solid so that the reaction, and gas generation, cease.

Kirchhoff, Gustav Robert (1824–87) German physicist. He worked with Robert BUNSEN and developed the SPECTROSCOPE, with which they discovered the elements CESIUM and RUBIDIUM. He also examined the solar spectrum and worked on BLACK BODY radiation. He is famous for two laws (KIRCHHOFF'S LAWS) that apply to multiple-loop electric circuits.

Kirchhoff's laws Two rules, based on the laws of the conservation of charge and energy, that apply to multiple-loop electric circuits. Essentially they state that (1) charge does not accumulate at one point and thin out at another, and (2) around each loop the sum of the ELECTROMOTIVE FORCES equals the sum of the POTENTIAL DIFFERENCES (voltages) across each of the resistances.

Kitasato, Shibasaburo (1852–1931) Japanese bacteriologist who isolated the bacilli that cause the diseases tetanus, anthrax (1889), and dysentery (1898). In 1890 he prepared a diphtheria antitoxin. He also discovered, independently of Alexandre YERSIN, the infectious organism that causes bubonic plague. He trained under Robert KOCH in Germany.

Kjeldahl method Rapid method for measuring the proportion of nitrogen in an organic compound, named after the Danish chemist Johan Kjeldahl (1849–1900). Nitrogen in the sample is converted into ammonium sulfate by heating it and a catalyst in concentrated sulfuric acid. Sodium hydroxide is added and the whole is boiled, liberating ammonia. This is dissolved in acid, and the quantity of ammonia, and therefore the nitrogen, is determined by TITRATION.

Klaproth, Martin Heinrich (1743–1817) German chemist who began as an apothecary's apprentice. He became the first professor of chemistry at Berlin University. He was a pioneer in analytical techniques, recognizing and naming a number of elements, including uranium (1789). He also recognized and named titanium (1795), tellurium (1798), and cerium (1803), some of which had already been isolated, although left unnamed, by other scientists.

Klebs, Edwin (1834–1913) German physician and pathologist known for his work in bacteriology. In 1884 he and Friedrich LÖFFLER discovered the diphtheria bacillus. He held professorships in Europe and in the US. His research included original observations on tuberculosis, syphilis and malaria.

klystron ELECTRON TUBE that makes use of a stream of electrons moving at controlled speeds. Klystrons are used in ultra-high-frequency (UHF) radio and microwave circuits, such as RADAR transmitters, where they operate at frequencies up to 400,000MHz.

K-meson *See* KAON

knock, engine (pre-ignition or pinking) Metallic noise produced in an INTERNAL COMBUSTION ENGINE, caused by detonation of part of the compressed fuel-air mixture before the spark from the SPARK PLUG ignites the remainder. It lowers the efficiency of the engine but can be overcome by adding an anti-knocking agent to the fuel. The traditional additive is LEAD(IV) TETRAETHYL, but because this compound releases lead into the atmosphere and thereby contributes to pollution, its use has declined with the introduction of unleaded gasoline.

knot Unit of measurement equal to one NAUTICAL MILE per hour – 1 knot equals 1.852km/h (1.15mph). The speeds of ships and aircraft are generally expressed in knots, as are those of winds and currents.

Kocher, Emil Theodor (1841–1917) Swiss surgeon who was awarded the 1909 Nobel Prize for physiology or medicine for his work on the physiology, pathology, and surgery of the THYROID GLAND. He was the first to remove the thyroid gland in goiter cases.

Koch, Robert (1843–1910) German bacteriologist. He was awarded the 1905 Nobel Prize for physiology or medicine for his discovery of the bacillus that causes TUBERCULOSIS. This work laid the foundation for methods of determining the causative agent of a disease.

Kolbe, Adolf Wilhelm Hermann (1818–84) German chemist. In 1843 he converted carbon disulfide to acetic acid, one of the first syntheses of an organic compound from inorganic chemicals. He also developed a theory of RADICALS and predicted the existence of secondary and tertiary alcohols.

Kolff, Willem Johan (1911–) US physician,b. Netherlands. In the German-occupied Netherlands during World War II he invented the KIDNEY MACHINE. A member of the Dutch Resistance, Kolff established blood banks for treating the wounded. He developed his machine to treat total kidney failure, and by 1943 had perfected his ideas, using artifical sausage skins as the dialysis membrane.

Köppen, Wladimir Peter (1846–1940) German meteorologist and climatologist, b. Russia, who introduced a CLIMATE classification system still used today. In 1884 he produced a world map of temperature belts. In 1900 he introduced his system of climate classification, which divided climate into five major categories according to temperature and rainfall.

Kornberg, Arthur (1918–) US biochemist. He was medical director of the US Public Health Service (1951) and chairman of the department of biochemistry at Stanford University. In 1959 he shared the Nobel Prize for physiology or medicine with Severo OCHOA for work on the synthesis of RNA and DNA, an important contribution to the study of GENETICS.

Korolev, Sergei Pavlovich (1906–66) Soviet engineer. He was chief designer at the Scientific Research Institute, Moscow and directed the design and manufacture of the VOSTOK and SOYUZ manned spacecraft, including Vostok I in which, in 1961, Yuri GAGARIN made the first manned space flight.

Krebs cycle (citric acid or tricarboxylic acid cycle) Biochemical pathway by which most EUKARYOTE organisms obtain much of their energy by oxidizing foodstuffs. Occuring in the MITOCHONDRIA of CELLS, the Krebs cycle comprises a number of complex chemical reactions, many of which release energy, in association with a process called the ELECTRON TRANSPORT SYSTEM, as ADENOSINE TRIPHOSPHATE (ATP) becomes ADENOSINE DIPHOSPHATE (ADP). ATP provides chemical energy for metabolic reactions. The Krebs cycle is an essential part of the process of cell RESPIRATION and METABOLISM. The cycle is named after Sir Hans KREBS. *See also* GLYCOLYSIS

Krebs, Sir Hans Adolf (1900–81) British biochemist, b. Germany. He shared with Fritz LIPMANN the 1953 Nobel Prize for physiology or medicine for his discovery of the KREBS CYCLE, the process that results in the production of energy in living organisms (RESPIRATION).

kryolite *See* CRYOLITE

krypton (symbol Kr) Gaseous, nonmetallic element that is a NOBLE GAS (inert gas). Discovered in 1898, krypton makes up about 0.0001% of the Earth's atmosphere by volume and is obtained by the fractional DISTILLATION of liquid air. It is used in FLUORESCENT LAMPS, LASERS and electronic heart valves. The isotope ^{86}Kr was used from 1960–83 to define the standard linear meter. Properties: at.no. 36; r.a.m. 83.80; density 3.73g dm^{-3}; m.p. $-156.6°C$; ($-249.9°F$); b.p. $-152.3°C$; ($-242.1°F$); most common isotope ^{84}Kr (56.9%).

Kuhn, Richard (1900–67) German chemist who shared the 1938 Nobel Prize for chemistry with Paul KARRER. They researched CAROTENOIDS and the isolation of vitamin B$_2$. He was forced by the Nazis to refuse the award and did not receive it until after World War II.

Kuiper, Gerard Peter (1905–73) US astronomer, b. The Netherlands. After World War II Kuiper embarked upon an observational program to revitalize lunar and planetary astronomy. He discovered the satellites Miranda (of Uranus) in 1948 and Nereid (of Neptune) in 1949. He found methane in the atmospheres of Uranus, Neptune and Titan, and carbon dioxide in the atmosphere of Mars. He was involved with many US missions to the Moon and planets, in particular with the Ranger and Mariner probes. The Kuiper Airborne Observatory is named in his honor.

kurchatovium *See* DUBNIUM

Kuroshio current (Japan current) Western boundary CURRENT of the North Pacific gyre that flows along the E coast of Japan. It is comparable to the GULF STREAM along the E coast of North America.

Kusch, Polykarp (1911–93) US physicist, b. Germany. He shared the 1955 Nobel Prize for physics with Willis LAMB for his precise measurement of the MAGNETIC MOMENT of the ELECTRON, which he discovered had a higher value than was predicted by the theory of Paul DIRAC. This was an important step in the theory of QUANTUM ELECTRODYNAMICS (QED).

kyanite (disthene) Mineral form of aluminum silicate (Al_2SiO_5). It occurs in SCHISTS and other METAMORPHIC ROCKS, generally as elongated, blue-bladed crystals (triclinic) or aggregates, although it may also be white, green or gray. Color may vary within a single crystal, which is usually transparent with a vitreous luster. Kyanite is used in making high-grade PORCELAIN for electrical insulators; good clear crystals are used as gemstones. Hardness: 4–5 along the crystal, 6–7 across; r.d. 3.5–3.7.

labeling (radioactive tracing) Technique for following the path of a biochemical or chemical reaction by replacing one of the key elements involved by one of its RADIOSOTOPES, called a RADIOSOTOPE TRACER. The method relies on the fact that the substitution of a different isotope for the normal form of an element does not affect its chemical identity or properties. But if the new form is radioactive, its presence can be traced by the radioactivity it emits (using a GEIGER COUNTER). For example, any compound containing hydrogen can be labeled by substituting tritium (a radioisotope of hydrogen of mass 3) for the normal hydrogen, or an oxygen compound can be labeled by substituting the oxygen-18 isotope for normal oxygen-16.

labium In anatomy, term meaning a lip or lip-shaped organ. In arthropods it refers to the lower lip, which may be modified in insects to form sucking mouthparts. In human anatomy, it denotes the labia majora and labia minora, skin folds of the female genitals. In botany, it refers to the lower lip of plants of the Labiatae family.

labor *See* BIRTH

Lacaille, Nicolas Louis de (1713–62) French astronomer. From 1751 to 1753 he surveyed the Southern Hemisphere skies from the Cape of Good Hope, discovering 14 new s constellations. In 1761 Lacaille made an accurate measurement of the Moon's distance from Earth, taking into account the Earth's OBLATENESS; this made possible a more accurate method of determining terrestrial longitude.

laccolith Dome of INTRUSIVE ROCK formed over the STRATA it has penetrated. The base is typically horizontal while the upper surface is convex. They are generally less than 16km (10mi) in diameter with thicknesses of 30–900m (100–3,000ft). Laccoliths are formed of MAGMA; they are smaller than BATHOLITHS.

lachrymal gland (lacrimal gland) Organ near the EYE that produces TEARS. It is located in the orbital cavity in a slight depression, and is controlled by the AUTONOMIC NERVOUS SYSTEM; it pro-

duces slightly germicidal tears that flow through ducts to the surface of the eye to lubricate it.

lacquer Varnish used for ornamental or protective coatings, forming a film by loss of solvent through evaporation. Lacquer is usually composed of a cellulose derivative, such as CELLULOSE NITRATE, in combination with a RESIN. The solvent may be an alcohol.

lactation Secretion of milk by female MAMMALS to feed their young. In pregnant women, HORMONES induce the BREASTS to enlarge, and PROLACTIN (a hormone of the PITUITARY GLAND) stimulates breast cells to begin secreting milk. The milk "comes in" the breast immediately after the birth of the baby. Its flow is stimulated by suckling, which, in turn, triggers the release of the hormone OXYTOCIN that controls the propulsion of milk from the breast.

lactic acid (2-hydroxypropanoic acid, CH_3 $CHOHCOOH$) Colorless, organic acid formed from LACTOSE in milk by the action of bacteria. It is also produced in muscles when ANAEROBIC respiration occurs, due to insufficient oxygen, and causes muscle fatigue. Lactic acid is used in foods and beverages, in tanning, dyeing, and adhesive manufacture. Properties: r.d. 1.206; m.p. 18°C (64.4°F); b.p. 122°C (251.6°F).

lactose (milk sugar) DISACCHARIDE present in milk, made up of a molecule of GLUCOSE linked to a molecule of galactose. It is important in cheese making, when lactic bacteria turn it into LACTIC ACID, so souring the milk in the production of cheese curd.

Laënnec, René Théophile Hyacinthe (1781–1826) French physician who invented the STETHOSCOPE. In 1816, when trying to examine fat patients, he discovered that he could hear their chest sounds better by listening through a roll of paper. Within a year he had developed the instrument into a funnel-shaped tube of wood. The binaural stethoscope, with two ear pieces, was invented in 1828 by Laënnec's compatriot Pierre Poirry.

lagoon Stretch of shallow water partially or completely cut off from the sea. If it is completely cut off, there will be some percolation of water through the ridge or spit separating the lagoon from the sea. Lagoons are often found behind spits, bars, and CORAL REEFS.

Lagrange, Joseph Louis (1736–1813) French mathematician. He became professor of mathematics in Turin at the age of 19 and later succeeded Leonhard EULER as head of the Berlin Academy of Sciences. In *Analytical Mechanics* (begun in 1753 but not finished until 1788), he introduced a new general approach to mechanics, using the idea that the motions of bodies are found by minimizing a function known as the **Lagrangian function**. **Lagrangian dynamics** is used in many areas of applied mathematics and physics.

Lagrangian points One of the points at which a celestial body can remain in a position of equilibrium with respect to two much more massive bodies orbiting each other. At these points the forces acting on the smaller body cancel out. There are five Lagrangian points in the orbital plane of two large bodies.

lake Inland body of water, generally of considerable size and too deep to have rooted vegetation completely covering the surface, occupying a hollow in the Earth's surface. Usually freshwater, lakes may have an in- or outflowing river and are not always permanent features on the landscape. The expanded part of a river and a reservoir behind a dam are also termed "lakes." A particularly large lake that is natural and saline may be called a "sea" (for example, the Dead Sea).

Lalande, Joseph Jérôme Le Français de (1732–1807) French astronomer whose main achievement was a catalogue of more than 47,000 stars. In 1762 Lalande became professor of astronomy at the Collège de France, a post he held for 46 years. Made director of the Paris Observatory in 1768, Lalande observed Neptune in 1795, 51 years before it was located by Johann GALLE, without realizing it was a planet. In 1802 he established the Lalande Prize for achievement in astronomy.

Lamarck, Jean Baptiste Pierre Antoine de Monet, Chevalier de (1744–1829) French biologist. His theories of EVOLUTION (**Lamarckism**), according to which ACQUIRED CHARACTERISTICS are inheritable, influenced evolutionary thought throughout most of the 19th century, but were subsequently disproved by Charles DARWIN. In *Philosophie zoologique* (1809) he proposed that new biological needs of an organism promote a change in habits from which develop new physical structures. These are then transmitted to offspring as permanent characteristics. For example, he erroneously believed giraffes' long necks evolved because they reached for leaves.

Lamb, Willis Eugene, Jr. (1913–) US physicist who applied new techniques to measure the lines of the hydrogen spectrum. He found that the actual positions (wavelengths) varied from the positions predicted by the theory of Paul DIRAC. For this research, which was an important step toward the theory of QUANTUM ELECTRODYNAMICS (QED), Lamb shared the 1955 Nobel Prize for physics with Polykarp KUSCH. He also developed microwave techniques for examining the helium spectrum.

laminar flow (streamline flow) Orderly flow of FLUID (liquid or gas) without turbulence. The fluid flows in layers that slide past each other. As the velocity of the layers increases, or as the VISCOSITY of the fluid decreases, a point is reached when laminar flow breaks up into turbulent flow. For a given fluid this point occurs at a certain value of the REYNOLDS NUMBER. *See also* AERODYNAMICS; FLUID MECHANICS

lamprophyre Member of a group of medium-grained IGNEOUS ROCKS that occur in DIKES and SILLS. Their composition is rich in augite, biotite, and other colored silicates, which form relatively large crystals, with little or no feldspar.

Land, Edwin Herbert (1909–91) US physicist, inventor of the "instant" camera. He worked with light POLARIZATION and by 1936 had developed Polaroid, a light-polarizing material later used in sunglasses. In 1947 Land developed a camera that produced a finished print within one minute.

Landau, Lev Davidovich (1908–68) Soviet physicist. His many contributions included the basic theories describing FERROMAGNETISM and liquid HELIUM. In 1927 he proposed a concept for energy called the **density matrix** that was later used extensively in QUANTUM MECHANICS. In the 1930s he originated the theory that underlies the superfluid behavior of liquid helium. He received the 1962 Nobel Prize for physics for his research into condensed matter, especially helium. *See also* SUPERFLUIDITY

land bridge Temporary, natural connection between continents that can allow animal migration. One existed across the Bering Strait until comparatively recently, and a modern example is the isthmus of Panama. Before the theory of CONTINENTAL DRIFT, such bridges were proposed as a possible major influence on fossil distribution.

land reclamation Transformation of land that, for one reason or another, is unusable for building or cultivation. Although the term generally refers to reclamation of land that has been beneath the sea (that is, with artificial islands or merely by extending the shoreline), it also applies to the beautifying of derelict industrial sites (with canals and recreation areas) and techniques of irrigation and fertilization.

Landsat (formerly Earth Resources Technology Satellite, ERTS) Series of scientific SATELLITES. The satellites observe the Earth's surface in different segments of the ELECTROMAGNETIC SPECTRUM, yielding valuable information concerning crops, soils, forests, population patterns, and other such matters. It is in a nearly circular polar orbit.

landslide Relatively rapid displacement of rock or soil slipping over a definite surface. It may be caused by an earthquake, but is generally the result of rain soaking the ground; the effects are exaggerated if there is an impervious layer beneath the surface. In colder climatic areas, the saturated rock may be affected by freezing and thawing of the water, and this increases the likelihood of landslides.

Landsteiner, Karl (1868–1943) US pathologist, b. Austria. He discovered the four different BLOOD GROUPS (A, B, AB, and O) and demonstrated that certain blood groups are incompatible with others; clots form if blood of such groups are mixed. His findings explained why some blood transfusions in the past had been beneficial, whereas others had been fatal. He was awarded the 1930 Nobel Prize for physiology or medicine. In 1940 he helped to identify the RHESUS FACTOR.

Langevin, Paul (1872–1946) French physicist. In 1905 he was the first to interpret PARAMAGNETISM and DIAMAGNETISM in terms of the behavior of electrons in atoms. During World War I, he built the first submarine detector based on ULTRASONIC waves, and it was later developed into SONAR.

Langley, Samuel Pierpont (1834–1906) US astronomer who showed that mechanical flight was possible. He did this by building a large steam-powered model aircraft in 1896, which achieved the most successful flights up to that time. In 1880 he developed a BOLOMETER to measure INFRARED RADIATION.

Langmuir, Irving (1881–1957) US physical chemist who invented a gas-filled, tungsten lamp. He showed that the shortness of a lamp's life was due to the evaporation of the filament, and that this could be reduced by introducing nitrogen into the bulb instead of a high vacuum. He also devised the atomic hydrogen welding process and techniques to produce rain by cloud seeding. His researches in surface chemistry led to the general theory that CHEMICAL REACTIONS occurred between adjacent substances that were absorbed. For this work Langmuir received the 1932 Nobel Prize for chemistry.

lanolin Purified, fatlike substance derived from sheep's wool and used with water as a base for ointments and cosmetics.

lanthanide series (lanthanide elements, rare-earth metals) Series of 15 rare, metallic elements with atomic numbers from 57 to 71. They are placed in group III of the periodic table. They are, in order of increasing atomic numbers: LANTHANUM (sometimes not considered a member), CERIUM, PRASEODYMIUM, NEODYMIUM, PROMETHI-UM, SAMARIUM, EUROPIUM, GADOLINIUM, TERBIUM, DYSPROSIUM, HOLMIUM, ERBIUM, THULIUM, YTTERBIUM, and LUTETIUM. Their properties are similar and resemble those of lanthanum, from which the series takes its name. These shiny metals occur in MONAZITE and other rare minerals.

lanthanum (symbol La) Silver-white, metallic element of the LANTHANIDE SERIES, first identified in 1839. Its chief ores are monazite and bastnaesite. Soft, malleable and ductile, lanthanum is used as a CATALYST in CRACKING crude oil, in alloys, and to manufacture optical glasses. Properties: at.no. 57; r.a.m. 138.9055; r.d. 6.17; m.p. 920°C (1,688°F); b.p. 3,454°C (6,249°F); most common isotope ^{139}La (99.91%).

lanugo Soft, downy hair that covers the human fetus and those of other mammals during development in the womb. It is shed, and virtually disappears, at birth.

lapis lazuli (lazurite) Glassy, deep blue, semiprecious gemstone, a silicate mineral found in metamorphosed limestones (*see* METAMORPHIC ROCK). It occurs rarely as crystals in the cubic system, but more often as granular masses. Hardness 5–5.5; r.d. 2.4.

Laplace, Pierre Simon, Marquis de (1749–1827) French astronomer and mathematician. Laplace made significant advances in PROBABILITY theory: defining the probability of an event to be the number of ways that event can happen divided by the total number of possible outcomes. His study and application of Newton's law of GRAVITATION to the Solar System was summarized in his book *Celestial Mechanics* (1798–1827). He used Newton's law to interpret the PERTURBATIONS of planets and satellites, the shape and rotation of Saturn's rings, and the stability of the Solar System. He also did fundamental work in the study of heat, magnetism, and electricity.

laptop Type of MICROCOMPUTER that is portable and carries its own power supply. Modern laptops utilize the latest developments in miniaturization of INTEGRATED CIRCUITS, giving them the ability to process information as quickly as most desktop COMPUTERS. Their independent power supply comprises rechargeable batteries, and they all have a built-in LIQUID-CRYSTAL display. The miniaturization of their components and a quality of manufacture that can withstand being carried about ensures that laptops are comparatively expensive.

larva Developmental stage in the life cycle of many INVERTEBRATES and some other animals. A common LIFE CYCLE, typified by the butterfly, is egg, larva, PUPA, IMAGO (adult) (*see* INSECT). The larva fends for itself and is mobile, but is distinctly different in form from the sexually mature adult. It has a well-developed alimentary system and often stores food until it metamorphoses (or pupates) to become an adult. Names for the larval stage in different organisms include MAGGOT, caterpillar, and tadpole. *See also* METAMORPHOSIS

larynx (voice box) Triangular cavity located between the TRACHEA and the root of the tongue. Inside it are the vocal cords, thin bands of elastic tissue that vibrate when outgoing air passes over them, setting up resonant waves that are changed into sound by the action of throat muscles, the shape of the mouth and the tongue.

laser (acronym for **l**ight **a**mplification by **s**timulated **e**mission of **r**adiation) Optical MASER, a source of a narrow beam of intense, coherent (same wavelength) light or ULTRAVIOLET or INFRARED RADIATION. It was first developed by

▲ **laser** Normal emission of light occurs when an electron in a high-energy orbit falls to a low energy orbit (A). Stimulated emission (B) is triggered by light emitted from another atom. In a laser (C), most atoms are brought to a high-energy state by pumping in energy. Some begin to produce light by normal emission, and mirrors at each end reflect the light to and fro, producing stimulated emission until all

the atoms are in a low-energy state. The light leaves the laser through one of the mirrors. Ordinary light (D) is a mixture of frequencies moving in various directions, whereas laser light (E) has a single frequency and moves in the same direction with all waves in phase. The first laser (F) contained a synthetic ruby crystal surrounded by a flash tube (to pump in light energy) and a pair of reflecting mirrors.

Theodore MAIMAN in 1960. The source can be a solid, liquid or gas. A large number of its atoms are excited to a higher energy state. One PHOTON of radiation emitted from an excited atom then stimulates the emission of another photon of the same frequency and direction of travel, which in turn stimulates the emission of more photons. The photon number multiplies rapidly to produce a narrow, coherent, monochromatic laser beam of very high energy content. It has applications in medicine, scientific research, engineering, telecommunications, HOLOGRAPHY, and other fields.

laser printer Computer printer with a LASER diode to control image formation. The laser scans lines across an electrically charged drum. The beam flashes on and off according to whether each point is to be light or dark. Exposed areas become discharged. The charged areas attract toner powder, thus forming an image. Charged plain paper picks up the powder image from the drum. A heated roller fuses the powder onto the paper to make the image permanent. Many so-called laser printers use an array of LIGHT-EMITTING DIODES instead of a laser diode.

laser surgery Surgical treatment carried out using a LASER beam. The high energy in an extremely narrow laser beam can burn through body tissues to make a fine cut. The heat also seals blood vessels, so there is much less bleeding than when a knife is used. A less powerful laser beam can remove colored marks, such as tattoos, from the skin. Heat destroys the pigment because it absorbs much energy from the beam, whereas the surrounding skin is virtually unaffected. Surgeons also use lasers to treat some forms of skin cancer and eye problems.

latent heat (symbol L) Heat absorbed or given out by a substance as it changes its PHASE (of matter) at constant temperature – from a solid to a liquid state or from a liquid to a gas. When ice melts, its temperature remains the same until it has been completely transformed into water; the heat necessary to do this is called the **latent heat of fusion.** Similarly the heat necessary to transform water into steam at constant temperature is called the **latent heat of vaporization.**

laterite Reddish, hard-baked soil found in tropical parts of the world. It results when hot, wet conditions wash away most of the nutrient content of the soil, leaving behind hydrated oxides of iron and aluminum. The top is baked hard by the Sun, and this factor, together with the leached nature of the soil, means that it is poor for agriculture. Some laterites, however, contain sufficient iron to be of commercial value.

latex Milky fluid produced by certain species of plant, the most important being that produced by the RUBBER tree. Rubber latex is a combination of gum resins and fats in a watery medium. It is used in paints, special papers and adhesives, and to make sponge rubbers. Synthetic rubber latexes are also produced.

lathe Machine tool that turns a workpiece (either wood or metal) against a cutting tool in order to shape it.

latitude Distance N or S of the EQUATOR, measured at an angle from the Earth's center. All lines of latitude are parallel to the equator, which is the zero line of latitude. Each degree of latitude is about 111km (69mi). The tropics of CANCER and CAPRICORN are 23.5° away from the equator, and the Arctic and Antarctic circles are at 66.5°, which is 23.5° away from the poles. In astronomy, celestial latitude is the angular distance on the CELESTIAL SPHERE of a celestial body from the ECLIPTIC; it is measured at right angles to the ecliptic, from 0° at the ecliptic to 90° at the ecliptic poles, positive to the N and negative to the S.

Laue, Max Theodor Felix von (1879–1960) German physicist. He became professor at Zurich and later director of the Institute for Theoretical Physics in Berlin. Using IONS in a crystal as a grating, Lane produced X-ray INTERFERENCE patterns, showing that X RAYS are waves and providing a method of investigating CRYSTAL STRUCTURE. For this discovery of X-ray diffraction in crystals, he received the 1914 Nobel Prize for physics. *See also* X-RAY CRYSTALLOGRAPHY; X-RAY DIFFRACTOMETER

laughing gas *See* DINITROGEN OXIDE

launch window Interval of time during which atmospheric and other conditions allow a space vehicle to be launched to achieve its required TRAJECTORY and orbit. Several launch windows may be available during a single launch opportunity.

Laurasia Ancient N continent, containing North America and Eurasia, that formed when a rift, the Tethyan trench, split PANGAEA (the Earth's single original land mass) from E to W along a line slightly N of the equator. The name is a combination of Laurentian, a geological period in North America, and Eurasia. CONTINENTAL DRIFT and PLATE TECTONICS caused Laurasia to split up into the separate continents that exist today.

lava Molten rock or MAGMA that reaches the Earth's surface and flows out through a volcanic vent in streams or sheets. There are three main types of lava: **vesicular**, like pumice; **glassy**, like obsidian; and **even-grained**. Chemically, lavas range from acidic to ultrabasic. If it is acidic it is viscous and slow-flowing, but if it is basic it is more fluid. Basic lavas form lava flows and often erupt out of fissures, as well as volcanic craters. Basic lavas are most common at points where plate margins are moving apart, but acidic lavas occur at colliding plate margins and are often associated with very explosive VOLCANOES. Once they reach the surface, lavas cool and solidify. If the lava is very gaseous, a rough, jagged surface will result, which is described as "scoriaceous."

Laveran, Charles Louis Alphonse (1845–1922) French physician. He contributed greatly to tropical medicine and was awarded the 1907 Nobel Prize for physiology or medicine for his work on the role played by PROTOZOA in causing diseases. Laveran discovered the parasite (*Plasmodium*) responsible for MALARIA.

Lavoisier, Antoine Laurent (1743–94) French chemist who founded modern chemistry. His careful experiments enabled him to demolish the PHLOGISTON THEORY (which said the vital essence phlogiston was lost during combusion) by demonstrating the function of OXYGEN in combustion. Lavoisier named both oxygen and hydrogen and showed how they combined to form water. In collaboration with Claude Louis Berthollet (1748–1822) and others, he published *Methods of Chemical Nomenclature* (1787), which laid down the modern method of naming substances. His *Elementary Treatise on Chemistry* (1789) established the basis of modern chemistry. He also helped to establish the METRIC SYSTEM.

Lawrence, Ernest Orlando (1901–58) US physicist. In 1930 he became professor at the University of California at Berkeley, where he built the first CYCLOTRON (a particle ACCELERATOR). Lawrence developed larger cyclotrons with higher energies, and his work in this area earned him the 1939 Nobel Prize for physics. The element LAWRENCIUM was named after him, as was the Lawrence Berkeley Radiation Laboratory in California.

lawrencium (symbol Lr) Radioactive, metallic element, one of the ACTINIDE SERIES. It was first made (1961) at the University of California at Berkeley by bombarding CALIFORNIUM with boron nuclei. The element has been made in only trace amounts but has been identified chemically. Properties: at.no. 103; r.a.m. 262; most stable isotope ^{256}Lr (half-life 27 seconds).

laxative Any agent used to counteract constipation. There are various kinds available, including bulk-forming drugs (mainly prescribed for people with bowel disease), stimulant laxatives, fecal softeners, and saline purgatives. However, modern medical thinking is that laxatives are rarely necessary. Constipation should be kept at bay by regular exercise and a DIET high in fiber.

layering Method of plant PROPAGATION that induces or encourages root formation on a stem or branch while it is still attached to the parent plant; blackberries and loganberries multiply naturally in this way.

lazurite *See* LAPIS LAZULI

L-dopa (levodopa) Naturally occurring ISOMER of an AMINO ACID used to relieve some symptoms of PARKINSON'S DISEASE. It sometimes suppresses the trembling, unsteadiness and slowness of movement that characterize the condition. Side effects include involuntary facial movements, mood swings, loss of appetite, nausea, vomiting, and dizziness.

leaching In geology, process by which chemicals and nutrients are removed from a soil. Rainwater, especially in warm climatic regions, will dissolve anything soluble and wash it away. Once removed, these solubles can only be replaced very slowly. As a result, leached soils become coarse and are infertile. Saline soils can, however, be reclaimed for agriculture by deliberately leaching out salts. The term is also used to describe the industrial process of extracting a soluble material from a solid by washing the solid with solvents.

lead (symbol Pb) Metallic element of group IV of the periodic table. Its chief ore is GALENA (lead sulfide), from which lead is obtained by roasting. Exposure to lead from paints, pipes, gasoline, and other sources can lead to lead poisoning. Soft and malleable, it is used as a shield for X rays and nuclear radiation, and in plumbing, batteries, cable sheaths, and alloys such as pewter and solder. Chemically, lead is unreactive, a poor CONDUCTOR of electricity, and resists corrosion. It forms two series of salts, termed lead(II), or plumbous, and lead(IV), or plumbic. Properties: at.no. 82; r.a.m. 207.19; r.d. 11.35; m.p. 327.5°C (621.5°F); b.p. 1,740°C (3,164°F); most common isotope ^{208}Pb (52.3%).

lead(II) acetate (Pb(CH$_3$COO)$_2$, formerly sugar of lead) Sweet-tasting, but highly poisonous, white crystalline solid that is one of the very few lead salts soluble in water.

lead(II) oxide (lead monoxide, PbO) Yellow, crystalline compound that does not dissolve in water. It is made by heating molten lead in air, which at low temperatures produces the form called **massicot** and at high temperatures produces

litharge. Lead(II) oxide is AMPHOTERIC; it dissolves in acids to produce lead(II) salts and in alkalis to give plumbates. It is used to make glazes for pottery, in paints and varnishes, and in the manufacture of some rubber products.

lead(II) sulfate (PbSO$_4$) White, crystalline solid that is virtually insoluble in water. It occurs in nature as the mineral anglesite. With lead(II) hydroxide and water it forms the basic sulfate, which has been employed as a white paint pigment, but its toxicity has resulted in a decline in this use.

lead(IV) oxide (lead monoxide, PbO$_2$) Brown or black amorphous solid that does not dissolve in water, but reacts slowly with concentrated acids. It is used in some kinds of safety matches and as the anode (positive plate) material in lead-acid accumulators.

lead(IV) tetraethyl *See* TETRAETHYL LEAD

lead sulfide (PbS) Toxic, black powder; a compound of LEAD. It occurs naturally as the mineral GALENA. It is insoluble in water, and is used as a SEMICONDUCTOR, and in ceramics.

leaf Part of a plant, an organ that contains the green pigment CHLOROPHYLL, and is involved in PHOTOSYNTHESIS and TRANSPIRATION. It usually consists of a blade and a stalk (**petiole**), which attaches it to a stem or twig. Most leaves are simple; some are compound, and divided into leaflets. Modifications include succulent types with fleshy tissue for water storage, tendrils that coil around supports, and needles, common in conifers.

Leakey, Louis Seymour Bazett (1903–72) Kenyan paleoanthropologist and archaeologist. Leakey and his wife, **Mary** (1913–96), conducted important research on HUMAN EVOLUTION. Many of their most important discoveries were made in Olduvai Gorge, Tanzania. Here, they unearthed (1959) a HOMINID fossil *Zinjanthropus* or *Australopithecus boisei* about 1.75 million years old, and found (1964) the remains of *Homo habilis* – thought to be a direct ancestor of *Homo sapiens sapiens*. Leakey proposed the "Out of Africa" model of human development. After his death, Mary and their son, Richard LEAKEY, continued to make important finds.

Leakey, Richard Erskine Frere (1944–) Kenyan paleoanthropologist and archaeologist, son of Mary and Louis LEAKEY. In 1972, at Lake Turkana, Kenya, Leakey discovered a 1.9 million year-old skull of *Homo habilis*. Other discoveries include a *Homo erectus* skeleton about 1.6 million years old. After serving as head of the Kenyan

National Museums, Leakey became director of the Kenyan Wildlife Service (1988–94). *See also* HUMAN EVOLUTION

leather Animal hide, treated to make it hardwearing and resistant to decay. Most leather is made from cattle hide, but many other kinds of skin are used too. The skin is first cured, via a drying process or the application of salt. It is then washed and prepared for tanning, a process that usually consists of treating the skins with a solution of chromium salts or plant extract (TANNIN).

Leavitt, Henrietta Swan (1868–1921) US astronomer noted for her work on CEPHEID VARIABLE stars and stellar magnitudes. She discovered that the periods of Cepheids are related to their true brightness. The **Cepheid period-luminosity law** has since been used to determine the distances of stars.

Leblanc, Nicolas (1742–1806) French chemist who devised (1790) a process for producing SODA ash (sodium carbonate, Na_2CO_3) from salt (sodium chloride, NaCl) by treating it with sulfuric acid. Leblanc was physician to Louis Philippe, Duc d'Orléans, who helped him to establish a factory to produce soda ash. Leblanc's process was used extensively in the 19th century.

Le Châtelier's principle Principle announced (1888) by the French chemist Henry Louis Le Châtelier (1850–1936). It states that if a system in a state of EQUILIBRIUM is disturbed, the system tends to neutralize the disturbance and restore the equilibrium.

lecithin Substance containing FATTY ACIDS and CHOLINE and found in many animal tissues, especially in nerves, semen, and the liver.

Leclanché cell Primary electric CELL invented (1867) by French engineer Georges Leclanché (1839–82). Its ANODE was a zinc rod and its CATHODE a carbon plate surrounded by packed manganese dioxide. These electrodes were dipped into a solution of ammonium and zinc chlorides. It is the basis of the dry cell (BATTERY).

LED Abbreviation of LIGHT-EMITTING DIODE

Lederberg, Joshua (1925–) US geneticist. In 1958 he shared, with George BEADLE and Edward TATUM, the Nobel Prize for physiology or medicine for work that initiated the study of bacterial genetics. Lederberg discovered that sexual recombination of genetic materials occurs in bacteria and that genetic materials are linked in groups in bacteria as well as in other organisms.

Lee, Tsung-Dao (1926–) US physicist, b. China. He and his colleague Chen Ning YANG showed that among the WEAK NUCLEAR FORCES of ELEMENTARY PARTICLES, the law of conservation of PARITY (that nature, in effect, makes no distinction between right- and lefthandedness) does not always hold. This was subsequently verified by experiment. For their prediction, Lee and Yang were awarded the 1957 Nobel Prize for physics.

Leeuwenhoek, Anton van (1632–1723) Dutch scientist. Leeuwenhoek was a scientific amateur who built simple MICROSCOPES with a single LENS, made so accurately that they had better magnifying powers than the compound microscopes of his day. He investigated many microorganisms and their life histories, and described various microscopic structures such as spermatozoa.

left-hand rule *See* FLEMING'S RULES

Legendre, Adrien Marie (1752–1833) French mathematician. Legendre studied the theory of numbers and through his work on quadratic residues discovered the law of reciprocity. His most influential work was *Elements of Geometry* (1794).

legionnaire's disease Lung disease caused by infection with the bacterium *Legionella pneumophila*. It takes its name from the serious outbreak that occurred during a convention of the American Legion held in Philadelphia in 1976. Identified the following year, the bacterium thrives in water and may be found in defective heating, ventilation, and air-conditioning systems. It is inhaled in fine water droplets present in the air. The disease is pneumonialike in its effects and is treated with antibiotics.

legume Member of the pea family (Leguminosae) of ANGIOSPERMS the roots of which bear nodules that contain nitrogen-fixing bacteria. The family includes many trees, shrubs, vines, and herbs. The fruit is typically a pod (legume) containing a row of seeds. Important food species include the pea, runner bean, soybean, lentil, broad bean, kidney bean, and haricot bean. *See also* NITROGEN CYCLE; NITROGEN FIXATION; ROOT NODULE

Leibniz, Gottfried Wilhelm (1646–1716) German philosopher and mathematician. Leibniz became interested in mathematics in the 1670s, when he met Christiaan HUYGENS and Robert BOYLE who described current problems in mathematics to him. While working as a lawyer, he developed CALCULUS (independently of Isaac NEWTON), combinatorial analysis, and a theory of limits. He also made many practical inventions, including a calculating machine (1671). Leibniz is recognized as one of the greatest mathematicians in history. In philosophy, he created a rationalist form of metaphysics.

Leloir, Luis Federico (1906–97) Argentinian chemist, b. France. Leloir was awarded the 1970 Nobel Prize for chemistry for his research into the biochemical processes that break down complex sugars into simpler CARBOHYDRATES. In 1947 he helped to establish the Institute of Biochemical Research in Buenos Aires, where he began work on the production of LACTOSE. This research eventually led to the discovery of sugar NUCLEOTIDES, fundamental factors in the natural process of carbohydrate synthesis. Leloir also synthesized GLYCOGEN and demonstrated the necessity of certain liver enzymes in the manufacture of glycogen in the body.

Lemaître, Abbé Georges Édouard (1894–1966) Belgian astrophysicist and mathematician who first formulated the BIG BANG theory of the origin of the Universe. He saw the Universe as originally analogous to a radioactive atom, with all the energy and matter concentrated into a kernel that Lemaître called the "primeval atom." He argued that an EXPANDING UNIVERSE would have originated in the explosion of that primeval atom. His book *The Primeval Atom: An Essay on Cosmogony* (1950) describes the theory.

Lenard, Philipp Eduard Anton (1862–1947) German physicist, b. Hungary. Lenard was awarded the 1905 Nobel Prize for physics for his studies of the properties of CATHODE RAYS. His work was important in the development of ELECTRONICS and NUCLEAR PHYSICS. He also did research on ULTRAVIOLET LIGHT, the electrical conductivity of flames and PHOSPHORESCENCE.

lens Piece of transparent glass, plastic, quartz or organic matter, bounded by two surfaces, usually both spherical, that changes the direction of a light beam by REFRACTION and can produce an IMAGE. A **converging** lens is convex in form (bulging at the center) and bends light rays toward the lens axis. A **diverging** lens is concave (thinnest at the center) and bends rays away from the axis. The optical image may be rightside-up or inverted, real or virtual, and magnified or reduced in size. Lens images suffer from various ABERRATIONS such as blurring and false colors. These characteristics also apply to the lens of the EYE.

lenticular galaxy Type of GALAXY intermediate in form between spiral and elliptical types. They resemble a convex lens in cross-section, hence their name.

Lenz's law Electromagnetic law deduced in 1834 by the Russian physicist Heinrich Lenz (1804–65). It states that an induced electric current flows in a direction that tends to oppose the charge producing it. *See also* INDUCTION

Leo (Lion) Large constellation of the N sky, lying between CANCER and VIRGO. The brightest stars are Alpha Leonis or Regulus, of the first magnitude, and Beta Leonis or Denebola, of the second magnitude.

Leonids METEOR SHOWER that occurs in November, associated with Comet Tempel–Tuttle. The shower is periodic, providing modest displays in most years, but producing magnificent meteor storms at intervals of 33 years – rates of 60,000 meteors per hour were recorded in 1966 on the night of November 16–17.

leopard (panther) Solitary spotted big cat found throughout Africa and S Asia. It has a round head, a short nose, and a long, thin tail. The coat may be yellow and white with dark spots, or almost completely black. A good climber and swimmer, it feeds on birds, monkeys, antelopes, and cattle. Length: to 2.45m (8ft); weight: to 90kg (200lb). Family Felidae; species *Panthera pardus*.

▲ **lens** A convex lens (1) focuses light on a single point (2) by diffracting the beams of light toward each other. A concave lens (3) diffracts parallel rays of light, making them diverge. A magnifying glass (4) is a convex lens. The glass makes the rays of light diverge, making it appear that they come from a larger image (5) than is the actual case. Cameras use a combination of convex and concave lenses (6) to focus light on the film without separation of the colors of the spectrum.

leprosy (Hansen's disease) Chronic, progressive condition affecting the skin and nerves, caused by infection with the microorganism *Mycobacterium leprae*. There are two main forms. **Lepromatous** leprosy is a contagious form in which raised nodules appear on the skin and there is thickening of the skin and peripheral nerves. The disease is characterized by numbness, weakness, loss of function, and, finally, deformity of the affected parts. In **tuberculoid** leprosy, there is loss of sensation in parts of the skin, sometimes with loss of pigmentation and hair. The visible signs of leprosy, including deformity, paralysis, and loss of digits, are the effects of nerve damage, possibly arising from injuries of which the person has been unaware (due to loss of sensation). Leprosy can be treated with a combination of drugs, but nerve damage is irreversible. Leprosy is now confined almost entirely to the tropics and subtropics. Of the 12 to 20 million cases worldwide, about 2,000 are in the United States.

lepton One of a class of ELEMENTARY PARTICLES that comprises three families: the ELECTRON and the electron-NEUTRINO; the MUON and the muon-neutrino; and the TAU PARTICLE and the tau-neutrino, together with all their antiparticles (antileptons). Leptons are governed by the WEAK NUCLEAR FORCE, which is the force involved in RADIOACTIVE DECAY. *See also* ANTIMATTER

lesion Any abnormality in a body tissue caused by injury or disease. Examples are wounds of any kind, sores, blisters, ulcers, and tumors.

letterpress printing Method of RELIEF PRINTING in which raised type is assembled and laid into a large frame (**chase**) by a compositor, then placed either on a flat-bed press (which prints one sheet at a time) or processed into a curved plate and placed on a rotary press, which is capable of higher speeds. Ink is transferred from the raised type onto the paper. Today, this method is used only for more specialist PRINTING. *See also* LINOTYPE; LITHOGRAPHY

leucine ([CH$_3$]$_2$ CHCH$_2$ CHNH$_2$ COOH) White, soluble, essential AMINO ACID found in proteins.

leucite Gray or white FELDSPAR mineral, a potassium aluminum silicate (KAl(SiO$_3$)$_2$). Unstable at high pressures, it has a restricted occurrence. It can be found in potassium-rich lava flows and VOLCANIC PLUGS. Hardness 5.5–6; r.d. 2.5.

leucocyte White blood cell, a colorless structure containing a NUCLEUS and CYTOPLASM. There are two types of leucocytes, LYMPHOCYTES and PHAGOCYTES. Lymphocytes produce ANTIBODIES, and phagocytes destroy harmful organisms (by engulfing them) when the body is infected. Normal blood contains 5,000–10,000 leucocytes per mm^3 of blood. Excessive numbers of leucocytes are seen in such diseases as LEUKEMIA. *See also* IMMUNE SYSTEM

leukemia Any of a group of CANCERS in which the bone marrow and other blood-forming tissues produce abnormal numbers of immature or defective LEUCOCYTES (white blood cells). This overproduction suppresses output of normal blood cells and PLATELETS, leaving the person vulnerable to infection, anemia, and bleeding. Leukemias are classified into **acute** and **chronic** forms and also according to the type of leucocyte involved. **Acute lymphoblastic** leukemia (ALL) is predominantly a disease of childhood; **acute myelogenous** leukemia (AML) is mainly seen in older adults. Both forms are potentially curable. The outlook is less hopeful with the chronic forms, which are seen in older people. However, chronic leukemia can often be kept under control for some years. *See also* IMMUNE SYSTEM

levee Natural embankment formed alongside a river by the deposition of silt when the river is in flood. Silt is deposited all over the FLOODPLAIN, but most is deposited near to the river banks, and so a slightly higher area is created alongside the river. Levees can help to prevent flooding, and are sometimes built up and strengthened artificially. Some levees grow so large that flood water cannot return to the main channel. When this happens, the river may change course. Levees may be only a few centimeters in height, but on some larger rivers, such as the Mississippi, they may reach heights of 15m (50ft).

lever Simple MACHINE used to multiply the force applied to an object, usually to raise a heavy load. A lever consists of a rod and a point (FULCRUM) about which the rod pivots. In a crowbar, for example, the applied force (effort) and the object to be moved (load) are on opposite sides of the fulcrum, with the point of application of the effort farther from it. The lever multiplies the force applied by the ratio of the two distances. *See also* DISTANCE RATIO; FORCE RATIO

Leverrier, Urbain Jean Joseph (1811–77) French astronomer. Originally a chemist, he turned to celestial mechanics and made a study of planetary motions and PERTURBATIONS. In 1845 Leverrier predicted that a planet (which he prematurely named Vulcan) lay within the orbit of Mercury, but despite some claimed sightings it was never found. More successful was his prediction the next year, independently of John Couch Adams, that an unknown planet (NEPTUNE) was responsible for discrepancies between the calculated and observed orbital motion of Uranus.

Levi-Montalcini, Rita (1909–) Italian neurologist. She moved to Washington University in 1947 to study nerve tissue growth, where, with the help of Stanley COHEN, they isolated a chemical in a mouse tumour that encouraged such growth. In 1986 she and Cohen received the Nobel Prize for physiology or medicine for their work.

Lewis, Gilbert Newton (1875–1946) US chemist. Working first in THERMODYNAMICS, he wrote a standard textbook on the subject. Later, he worked out a theory of VALENCE for non-IONIC COMPOUNDS based on shared electrons (covalence), and extended the concept of acids and bases. Lewis was the first to prepare HEAVY WATER (deuterium oxide), since used as a moderator in NUCLEAR REACTORS. *See also* COVALENT BOND

Leyden jar Earliest and simplest device for storing STATIC ELECTRICITY. It was the original electrical condenser (CAPACITOR), developed in about 1745 in Leiden (then Leyden), the Netherlands. It consists of a foil-lined, glass jar, partly filled with water and closed with a cork through which protrudes a knobbed brass rod or chain wired to the foil. Electricity (originally produced by friction) is passed into the jar along the rod or chain.

liana Any ground-rooting, woody vine that twines and creeps extensively over other plants for support; it is common in tropical forests. Some species may reach a diameter of 60cm (24in) and a length of 100m (330ft).

Libby, Willard Frank (1908–80) US chemist who developed the CARBON DATING technique. Libby worked (1941–45) in the war research division at Columbia University, New York, on the separation of ISOTOPES for the atom bomb. This work led to his interest in NUCLEAR PHYSICS and to his development of radioactive carbon-14 DATING, for which he was awarded the 1960 Nobel Prize for chemistry. He is the author of *Radiocarbon Dating* (1955).

Libra (Scales) Obscure constellation of the s sky, situated on the ECLIPTIC between Virgo and Scorpius. It formerly contained the First Point of Libra, the intersection of the ecliptic and the equator marking the N autumnal EQUINOX. Because of PRECESSION, this point has shifted W into Virgo. The brightest star, Beta Librae, is of magnitude 2.7.

libration Small oscillation of a celestial body about its mean position. The term is used most frequently in connection with the Moon. As a result of libration it is possible to see, at different times, 59% of the Moon's surface.

lichen Plant consisting of a FUNGUS in which microscopic (usually single-celled) ALGAE are embedded. The fungus and its algae form a symbiotic association in which the fungus contributes support, water, and minerals, while the algae contribute food produced by PHOTOSYNTHESIS. *See also* SYMBIOSIS

Liebig, Baron Justus von (1803–73) German chemist. He was the first to realize that animals get their energy from the combustion of food (respiration), although he incorrectly thought that muscular power is a result of protein oxidation. Liebig also showed that plants derive their minerals from the soil, and introduced synthetic FERTILIZERS into agriculture.

lie detector Alternative name for a POLYGRAPH

life Widely accepted as the ability of organisms to obtain energy from the Sun or from food and to use this for growth and reproduction. All living organisms interact with their environment: they take in energy and materials from their environment and respond to changes in that environment. When they reproduce, they pass on instructions for their characteristics to the next generation in the form of DNA or RNA. Debate continues over the status of VIRUSES, which do not grow, and, although they do contain DNA or RNA, use the host cell's machinery to reproduce themselves. The current theory on life's origin is that giant molecules, similar to proteins and nucleic acids, reacted together in the watery surface environment of the young Earth that is now commonly called the **primordial soup**. In this environment the giant molecules multiplied by consuming smaller molecules. Despite much research, the evolution and development of cellular life out of this molecular prelife remains a controversial subject and has yet to be explained fully.

life cycle Sequence of stages in the development of an organism from its formation to the creation of its offspring, followed by death. In organisms that reproduce sexually, formation corresponds to the fusion of sex cells (GAMETES) from the parents. The resulting ZYGOTE develops into an organism similar to the parents, which goes on to mature and produce new gametes that create the next generation. In organisms that reproduce asexually, formation occurs through FISSION (simple division), BUDDING, or VEGETATIVE REPRODUCTION. In plants, there is an ALTERNATION OF GENERATIONS in which the plants (GAMETOPHYTE generation) differ from the spore-forming (SPOROPHYTE) generation.

lift In AERODYNAMICS, force that acts upwards on the undersurface of an AIRFOIL, or wing. As it travels forwards, the leading edge of an airfoil splits the airstream. Because the upper part of the airstream is forced to travel farther over the airfoil's more highly curved upper surface before rejoining the lower part of the airstream, it travels faster and consequently has a lower pressure, in accordance with BERNOULLI'S LAW. The lift force is a result of the upward pressure underneath the airfoil being greater than the downward pressure on the top.

lifting magnet Powerful ELECTROMAGNET used for lifting and transferring heavy metal objects. Such a magnet is suspended from the jib of a crane.

ligament Bands of tough, fibrous CONNECTIVE TISSUE that join bone to bone at the JOINTS. In the wrist and ankle joints, for example, they surround the bones like firm inelastic bandages. Ligaments, which contain a tough inorganic substance known as COLLAGEN, form part of the supporting tissues of the body.

ligand Ion, molecule or group of atoms linked to a metal atom or ion by coordinate chemical bonds in a metal complex. Thus chloride ions act as ligands in the ion $(CuCl_4)^{2-}$, and carbon monoxide molecules are ligands in such compounds as nickel carbonyl $(Ni(CO)_4)$. Some molecules, such as 1,2-diaminoethane (ethylene diamine, H_2NCH_2 CH_2NH_2), can coordinate at two points in the molecule. *See also* COORDINATION COMPOUND

light Part of the total ELECTROMAGNETIC SPECTRUM that can be detected by the human eye. Visible light is in the wavelength range from about 400nm (violet) to 770nm (red). The speed (symbol c) at which ELECTROMAGNETIC RADIATION (including light) travels in a vacuum is equal to 299,792,458m s^{-1}. Light exhibits typical phenomena of wave motion such as REFLECTION, REFRACTION, DIFFRACTION, POLARIZATION, and INTERFERENCE. The properties of light were investigated in the 17th century by Isaac NEWTON, who was the first to split white light into its component colors (SPECTRUM) with a PRISM. Newton believed in a corpuscular or particle theory of light, but the wave theory was well established by the second decade of the 19th century after the work of Thomas YOUNG and Augustin Jean FRESNEL on the diffraction and interference of light. At the beginning of the 20th century, experiments on the PHOTOELECTRIC EFFECT and the work of Max PLANCK revived the idea that light can behave like a stream of particles. This dilemma was resolved by the QUANTUM THEORY, according to which light consists of elementary particles of zero rest mass called PHOTONS. When light interacts with matter, as in the photoelectric effect, energy is exchanged in the form of photons and so light seems to consist of particles. Otherwise, it behaves as a wave. The quantum theory also explains how all electromagnetic radiations result from the acceleration of electric charges. *See also* ETHER; HOLOGRAPHY; LASER

light adaptation Shift in functional dominance from rod cells to cone cells within the RETINA of the EYE as overall illumination increases. Unlike DARK ADAPTATION, adaptation to light is swift but usually uncomfortable. Humans can discern color and form soon after emerging from prolonged darkness.

lightbulb Device consisting of a metal filament that emits light when heated to INCANDESCENCE by electricity. An early form of lightbulb dates to 1860, when Joseph SWAN made a workable model. In 1879 Thomas EDISON patented an improved design in the US. Modern lightbulbs comprise a filament enclosed by a glass envelope or a vacuum containing a NOBLE GAS gas such as argon. The filament is usually made of TUNGSTEN because of its high tensile strength, low evaporation, and high efficiency in converting electrical energy into light as it glows white hot. *See also* ELECTRIC LIGHT

light-emitting diode (LED) SEMICONDUCTOR device that produces light. An LED consists of a *p-n* junction DIODE that emits light when electrons and holes in the semiconductor material (typically gallium arsenide phosphide) recombine at the junction. In this way it converts electrical energy applied across the junction into light. LEDs are used for display purposes in various kinds of electronic apparatus.

lightning Visible flash of light accompanying an electrical discharge between clouds or between clouds and the ground, most commonly produced in a THUNDERSTORM. A typical discharge consists of several lightning strokes, initiated by leaders that follow an irregular path of least resistance, the **lightning channel**. Intense heating by the discharge expands the channel rapidly up to a diameter of 13–25cm (5–10in), creating the sound waves of thunder.

light reaction (light-dependent reaction) One of the two distinct phases of PHOTOSYNTHESIS; the other being DARK REACTION. During this stage, the green pigment CHLOROPHYLL absorbs sunlight and uses its energy to break down water into oxygen gas, electrons, and hydrogen ions. The energy of the electrons is used to convert ADENOSINE DIPHOSPHATE (ADP) to ADENOSINE TRIPHOSPHATE (ATP) and NADPH (the hydrogenated form of nicotinamide adenine dinucleotide phosphate). Together ATP and NADPH ensure the dark reaction can take place.

light-year (symbol l.y.) Unit of astronomical distance equal to the distance traveled in free space or a VACUUM by light in one tropical year. One light-year is equal to 9.4607×10^{12} km (5.88×10^{12} mi).

lignin Complex, noncarbohydrate substance that occurs in the woody tissues (especially XYLEM) of plants, often in combination with CELLULOSE. It makes the CELL walls stronger, helping them to resist compression and tension. It is lignin that gives WOOD its strength. To obtain pure cellulose for the paper and rayon industries, the lignin has to be removed from wood. Lignin is also a major source of vanillin, a white, crystalline ALDEHYDE used as a flavoring and in perfumes and pharmaceuticals.

lignite Soft COAL, brown-black in color, with a carbon content higher than that of PEAT but lower than those of other coals. Sometimes the shape of its original woody structures is visible.

Lilienthal, Otto (1849–96) German engineer and pioneer of glider design. A glider designed by Sir George CAYLEY had carried a passenger in 1853, but this aircraft had no controls. In 1891 Lilienthal became the first person to control a glider in flight. He used an aircraft of his own design with cotton fabric wings. He made about 2,500 more flights before his death in a crash.

Lillie, Frank Rattray (1870–1947) US zoologist and embryologist, b. Canada. Lillie was noted for his discoveries in the field of fertilization and the role of HORMONES in SEX DETERMINATION. He spent much time helping to develop the Marine Biological Laboratory and the Oceanographic Institution at Woods Hole, Massachusetts. His book *The Development of the Chick* (1908) is a leading work in EMBRYOLOGY.

limb Jointed extension of the VERTEBRATE body, used for locomotion and manipulation of the physical environment. The human forelimb (arm) is particularly specialized in the latter function. The form of the limbs is regarded as a criterion of evolutionary development. *See also* PENTADACTYL LIMB; HUMAN BODY

limbic system Collection of structures in the middle of the BRAIN. Looped round the HYPOTHALAMUS, the limbic system is thought to be involved in emotional responses, such as fear and aggression, and the production of mood changes, as well as the laying down of memories.

lime *See* CALCIUM HYDROXIDE; CALCIUM OXIDE

limestone SEDIMENTARY ROCK formed on the bed of a warm sea by an accumulation of dead sea creatures. At least 50% of a limestone consists of CALCITE ($CaCO_3$). There are many different types, including Carboniferous limestone, CHALK, dolomitic limestone and oolitic limestone. They all contain fossils or fragments of fossils. Because of the calcium carbonate content, limestones are soluble and permeable, and generally give rise to dry landscapes. Chalk is pure limestone and forms rounded hills and vertical cliffs. Carboniferous limestone is hard and forms high ground; oolitic hills are generally more gentle. Carboniferous limestone, like chalk and oolitic limestone, has numerous dry valleys, but also contains CAVES with STALACTITES and STALAGMITES. Limestone is used to make cement, as a source of commercial lime and as a building material.

lime water Saturated solution of CALCIUM HYDROXIDE (slaked lime, $Ca(OH)_2$). It is used to detect carbon dioxide (CO_2), which forms a milky white precipitate of calcium carbonate ($CaCO_3$) when bubbled through lime water. Excess carbon dioxide changes the calcium carbonate into calcium hydrogencarbonate (calcium bicarbonate, $Ca(HCO_3)_2$), which is soluble, and so the white precipitate disappears.

limit In mathematics, value or values approached by the terms of a sequence. The limit of a SERIES is the value approached by the sum as more and more terms are included. The concept of limits is fundamental to differential and integral CALCULUS. *See also* CONVERGENCE

limnology Science of freshwater LAKES, ponds, and streams. These bodies of water are explored in terms of chemistry, physics, and biology. The plants, animals and environment are quantitatively examined in light of food cycles, food chains, habitats, and ZONATION of organisms. Freshwater bodies are subject to greater extremes of temperature and are therefore more fragile ecosystems and more specialized than those in marine environments. *See also* HYDROLOGY

limonite (brown iron or bog ore, $FeO(OH).nH_2O$) One of the most important sources of iron. Limonite may appear lustrous black, brown to yellow and is streaky, porous, and often mixed with sand or clay.

Lindbergh, Charles Augustus (1902–74) US aviator. Lindbergh became an international hero when, in *The Spirit of St Louis*, he made the first nonstop transatlantic solo flight, from New York to Paris (1927) in 33 hours 30 minutes. In 1932 his baby son was kidnapped and murdered. In 1936 he and Alexis CARREL invented an artificial heart.

Lindemann, Frederick Alexander, 1st Viscount Cherwell (1886–1957) British physicist, b. Germany. Involved in aircraft research (1914–18), he discovered how to regain control of an aircraft in an uncontrolled spin. During World War II he was Winston Churchill's scientific adviser. Lindemann developed the Clarendon Laboratory of Oxford University into a major research facility.

Lindley, John (1799–1865) British botanist, known for his CLASSIFICATION of plants in a natural system in which all characteristics of a plant are considered. Among Lindley's best-known books are *Theory and Practice of Horticulture* (1842) and *The Vegetable Kingdom* (1846), which included his new classification of plants.

line Continuous set of points. The term can be synonymous with "curve," or can be taken to mean a straight line – the shortest distance between two points in EUCLIDEAN GEOMETRY. In a rectangular CARTESIAN COORDINATE SYSTEM, a straight line has the equation $y = mx + b$, where m is the gradient and b the intercept on the y axis.

linear accelerator Type of particle ACCELERATOR in which charged particles travel in straight lines through a vacuum chamber. Early linear accelerators were electrostatic ACCELERATORS. In more modern types, a high-frequency, alternating electric field is used to accelerate the particles. Final energies depend on the length of the chamber, and can reach several GeV (million ELECTRON VOLTS). The linear accelerator at Stanford University is 3.2km (2mi) long and produces electrons with energies of 20GeV.

linear function Mathematical function given by a POLYNOMIAL involving no powers of its variables higher than one; for example, $f(x) = 7x + 3$ is linear. The graph of a linear function, in two dimensions, is a straight line.

linear motor Type of ELECTRIC MOTOR that is being developed to power high-speed trains. Its principle is similar to that of rotary ELECTRIC MOTORS, but instead of a number of coils rotating (a rotor) within a fixed ELECTROMAGNET (a stator), the linear motor is "unfolded" so that the "rotor" windings are fixed to the locomotive's path, and the "stator" windings are incorporated into the locomotive, which is both levitated and driven forward.

linear programming Mathematical procedure in which a multivariable linear function is analyzed to find maximum and minimum values. It is useful in business planning and industrial engineering to produce optimal control conditions, as in inventory control, where costs must be minimized in terms of storage time, warehouse space, customer delivery schedules, reorder times and transportation expenses.

▶ **linear motor** A linear motor is, in effect, a conventional motor unwound, so that its electromagnetic coils will move a flat stator (usually an aluminum plate in which magnetic fields are involved) or another set of electromagnets. Although most commonly used to open sliding doors or in factory applications, linear motors are being considered for use in the field of rapid public transport. A passenger carriage (1) carries a set of powerful electromagnets (2) to lift and propel the vehicle, and a second set to guide it and provide additional braking force (3). The track contains a series of electromagnets (4) that are energized to a polarity opposing the ones in the vehicle slightly ahead of the carriage (5). The magnetic forces lift the carriage and pull it forward. As it does so, the magnets ahead of the

vehicle are switched on to continue to attract it forward (6), while the

magnets behind the vehicle have their polarity reversed (7) to help repel

the magnets in the carriage and push it forward.

lines of force Lines in an ELECTRIC FIELD or MAGNETIC FIELD whose direction at every point is that of the field.

line spectrum *See* SPECTRUM

linkage group Group of inherited characteristics or GENES that occur on the same CHROMOSOME and that remain connected in such a way that they are usually inherited together through successive generations.

Linnaeus, Carolus (1707–78) (Carl von Linné) Swedish botanist and taxonomist. His *Systema Naturae* (1735) laid the foundation of the modern science of TAXONOMY by including all known organisms in a single CLASSIFICATION system. Linnaeus was one of the first scientists to define clearly the differences between species, and he devised the system of BINOMIAL NOMENCLATURE, that gave standardized Latin names to every organism.

Linotype Method of setting type for printing, by the Linotype machine, invented (1884) by Ottmar MERGENTHALER in Baltimore. It was first used by the New York *Herald Tribune* in 1886, and was subsequently used by newspapers all over the world. It is operated by a keyboard like that of a typewriter, which sets letters as one complete line of hot-metal type, called a **slug**. The Linotype system has now been replaced by PHOTOSETTING.

lipid One of a large group of organic compounds in living organisms that are insoluble in water but soluble in alcohol. They include animal FATS, vegetable OILS, and natural WAXES. Lipids form an important food store and energy source in plant and animal cells. Storage fat is composed chiefly of **triglycerides**, which consist of three molecules of FATTY ACID linked to GLYCEROL.

Lipmann, Fritz Albert (1899–1986) US biochemist b. Germany, who isolated and partially explained the molecular structure of COENZYME A, derived from the B vitamin pantothenic acid. For this and other work on METABOLISM, he shared the 1953 Nobel Prize for physiology or medicine with Hans KREBS (1900–81), who had discovered the KREBS CYCLE. In 1950 Lipmann demonstrated the formation of citric acid from oxalo-acetate and acetate, and found that this process required coenzyme A. During his career, Lipmann also investigated the way in which a cell acquires energy and the key role played in that process by ADENOSINE TRIPHOSPHATE (ATP).

Lippershey, Hans (1570–1619) German spectacle maker who discovered that a convex LENS and a concave lens used in conjuction can make distant objects appear closer, thereby inventing the TELESCOPE. Its potential for astronomy was recognized by Jacques Bovedere of Paris, who reported it to GALILEO who then built his own.

Lippmann, Gabriel (1845–1921) French physicist who was awarded the Nobel Prize for physics in 1908. The first person to produce a color photograph of the visible SPECTRUM, Lippmann also invented the direct color process of PHOTOGRAPHY. The process used a plate consisting of a grainless emulsion in contact with a layer of liquid mercury. His work was important in the development of holograms. *See also* HOLOGRAPHY

Lipscomb, William Nunn, Jr. (1919–) US chemist who made fundamental discoveries about the bonding of molecules, for which he was awarded the 1976 Nobel Prize for chemistry.

liquefaction of air Process achieved by cooling air to its critical temperature of $-147°C$ $(-232.6°F)$, at or below which it can be liquefied by pressure. The method involves repeated compression, followed by rapid ADIABATIC expansion, which results in progressive drops in temperature due to the JOULE-THOMSON EFFECT.

liquefied petroleum gas (LPG) Liquefied gas of light HYDROCARBONS, principally propane and butane, produced in the DISTILLATION of crude oil and the refining of NATURAL GAS. It is used as a fuel and raw material in chemical industries and as a bottled gas for domestic heating and cooking.

liquid State of MATTER intermediate between a GAS and a SOLID. A liquid substance has a relatively fixed volume but flows to take the shape of its container. The state that a substance assumes depends on the temperature and the pressure at which it is kept; a substance that is liquid, such as water, at room temperature can be changed into a vapor (its gaseous state, steam) by heating or into a solid (ice) by cooling. Unlike the molecules in gases, those in a liquid are weakly bonded, and liquids, like solids, are compressed only very slightly by pressure.

liquid crystal Substance that can exist halfway between both the liquid and the solid states with its molecules partly ordered. By applying a carefully controlled electric current, liquid crystals turn dark. They are used in liquid crystal displays (LCDs) to show numbers and letters, as in pocket calculators.

Lister, Joseph, 1st Baron (1827–1912) English surgeon who introduced the principle of antisepsis, which complemented the theory of PASTEUR that bacteria cause infection. Using carbolic acid (phenol) as the ANTISEPTIC agent, and employing it in conjunction with the heat sterilization of instruments, Lister brought about a dramatic decrease in postoperative fatalities.

liter (symbol l or L) Metric unit equal to a cubic decimeter, one thousandth of a cubic meter. Another definition, used from 1901–68, was that 1 liter equaled the volume of 1kg of pure water at 4°C (39°F). A liter is equivalent to 0.2642 gallons.

lithium (symbol Li) Rare, silver-colored metallic element, one of the ALKALI METALS, first isolated in 1817. Ores include lepidolite and spodumene. Chemically it is similar to sodium. The element, which is the lightest of all metals, is used in alloys and in glasses and glazes; its salts are used in medicine. Properties: at.no. 3; r.a.m. 6.941; r.d. 0.534; m.p.180.5°C (356.9°F); b.p. 1,347°C (2,456.6°F); most stable isotope ^{7}Li (92.58%).

lithography Method of PRINTING from a flat inked surface. This technique, invented in the 1790s, is based on the fact that grease and water do not mix. In traditional lithography, the design is made on a prepared plate or stone with a greasy pencil, crayon or liquid. Water applied to the surface is absorbed where there is no design. Oil-based printing ink, rolled over the surface, sticks to the design, but not to the moist areas. Pressing paper onto the surface produces a print. Multicolor prints are made using a separate plate or stone for each color ink used. Today, many magazines, books and newspapers are produced by a process called OFFSET PRINTING, which uses a modern form of lithography, involving photographically produced metal plates.

lithosphere Upper layer of the solid EARTH that includes the CRUST and the uppermost MANTLE. It is of variable thickness, from 60km (40mi) extending down to approximately 200km (125mi). Rigid, solid and brittle, the lithosphere is made up of a number of tectonic plates that move independently and give rise to PLATE TECTONICS.

litmus Purplish dye in neutral aqueous solutions, used to indicate acidity (by turning red) or alkalinity (by turning blue). It is most familiar in the form of litmus paper used as an acid-base indicator. The dye is extracted from lichens. *See also* pH

littoral zone Beach area between high and low tides. It also refers to the benthic division – between high tide and a depth of 200m (656ft). The larger zone is divided into the **eulittoral**, from high tide to a 50m (164ft) depth, and the **sublittoral**, from 50–200m (164–656ft). The lower edge of the eulittoral is the lowest limit at which abundant attached plants can grow.

liver Large organ located in the upper right abdomen of VERTEBRATES. Weighing up to 2kg (4.5lbs) in an adult human, it is divided into four lobes and has many functions. It is extremely important in the control of the body's internal environment (HOMEOSTASIS). It receives nutrients from the intestine and is a site of METABOLISM of proteins, carbohydrates, and fats. It synthesizes BILE and some vitamins, regulates the blood-glucose level, produces blood-clotting factors, breaks down worn-out ERYTHROCYTES (red blood cells) and removes toxins from the blood. The many metabolic reactions that go on in the liver are the body's main source of heat, which is distributed around the body by the blood. HOMEOTHERMIC (warm-blooded) animals maintain a constant body temperature by producing too much heat, then controlling its dissipation. The liver has phenomenal powers of regeneration: if part of it is removed, it regrows within weeks. *See also* INSULIN

load In geology, sediment carried along by ice, a river or the sea. Ice sheets and large GLACIERS can

▲ **liquid crystal**
Liquid-crystal displays use the property of liquid crystals to twist the polarization of light to produce numbers or symbols. Incoming light (1) is first regimented in one place by a polarizer (2) before it passes through the first of two plates of glass (3) on which are fixed electrodes (4). Seven electrodes are needed to represent Arabic numerals.

The liquid crystal is between the two plates of glass and twists light (5) passing through uncharged electrodes (6). This light can pass through the second polarizer (7) and be seen. The light passing through the charged electrodes (8) is not twisted (9), is blocked by the polarizer, and cannot be seen, so forming the components of the number (10).

carry a greater load than a thin glacier, so that they dump larger amounts of TILL when the ice melts. The load carried by a river depends on the type and amount of material available and the speed of the current. It may be in solution or in suspension, or be rolled along the river bed. A stream or slow-moving river can normally transport only mud and sand, whereas a fast-moving torrent can shift large boulders.

Lobachevsky, Nikolai Ivanovich (1792–1856) Russian mathematician. He was educated at Kazan University and appointed professor there in 1816. His outstanding achievement (announced in 1826) was the creation of one of the first comprehensive systems of NON-EUCLIDEAN GEOMETRY.

lobotomy, prefrontal Surgical operation on the BRAIN. It was first performed in the 1930s to treat psychiatric disorders by severing tracts of nerve fibers leading to the frontal lobes. It may cause irreversible deterioration of personality. Drugs have almost entirely replaced this operation.

Local Group of galaxies Small group of about 30 GALAXIES that includes our galaxy, the two MAGELLANIC CLOUDS, the ANDROMEDA GALAXY, and the Triangulum spiral. They are distributed over a roughly ellipsoidal space, about 5 million light-years across. The members are gravitationally bound so that, unlike more distant galaxies, they are not receding from us or from each other.

lock Structure built into a stretch of inland waterway to raise or lower water levels to correspond with the surrounding countryside. Each lock consists of two sets of lock gates. A vessel enters the lock, the gates are closed, and sluices are opened to admit or release enough water to bring the vessel to the same level as the water beyond the second pair of gates.

lock Mechanism fitted to a door or case to prevent entry to people without keys. Locks date back more than 4,000 years and were used by the Egyptians. The lever tumbler lock, an early version of the modern mortise, or dead lock, dates back to the 18th century. This system employs a tumbler and bolt. When the right key is inserted, the tumbler, a simple lever, is raised to a height whereby the key is able to engage the bolt, turn, and move the bolt back in the lock. Turning the door handle makes a cam move the latch across. The lock most commonly used today is the Yale, named after its inventor Linus YALE Jr. It is a type of cylinder lock that has many pins of differing lengths within the barrel. When the correct key is inserted in the lock the pins align allowing the key to turn and the latch to contract. Other locks include those that require the insertion of a punched card to release the

mechanism, or those that are activated by cards with magnetic strips. It is expected that in the future locks will be activated only by recognizing a particular person's unique fingerprints or eye characteristics.

Lockyer, Sir Joseph Norman (1836–1920) British astronomer who discovered the element HELIUM, independently of Jules JANSSEN, in the Sun's spectrum. In 1868 he developed a technique for examining PROMINENCES at the edge of the Sun and attributed lines in the solar spectrum to a new element, which he named helium – 40 years before it was discovered on Earth.

locomotive Engine that moves under its own power, usually used on rails for pulling cars. In 1804 British mining engineer Richard TREVITHICK built the first STEAM ENGINE locomotive for transporting heavy loads at an ironworks. His design was improved upon by George STEPHENSON, who built the *Rocket* in 1829. The first steam locomotive providing railroad service for passengers was George Stephenson's *Locomotion*, built in 1825. Electric-powered locomotives arrived in the late 1800s, and today obtain electricity from overhead cables or a rail running alongside the tracks. Diesel, diesel-electric and gas-turbine locomotives were introduced in the 1900s. Diesel locomotives burn oil to power the locomotive wheels directly. Diesel-electric locomotives use a diesel engine to generate the electricity with which the locomotive is powered. Gas-turbine locomotives use a TURBINE to create power.

locus In geometry, the path traced by a specified point when it moves according to a given formula or condition. For example, a circle is the locus of a point in a plane moving in such a way that its distance from a fixed point (the center) is constant.

lode ORE formation consisting of a closely spaced series of veins, usually in stratified layers. Often several different minerals occur along the same lode. Lodes were probably formed as a result of liquids or gases, heated by volcanic activity, forcing their way through existing rocks, and then cooling and solidifying.

lodestone *See* MAGNETITE

Lodge, Sir Oliver Joseph (1851–1940) British physicist and author, the first principal of Birmingham University. Lodge investigated ELECTROMAGNETS and LIGHTNING, and perfected and named the **coherer**, the radio-wave detector used in the early days of wireless telegraphy. He supported the belief that the Sun might be a source of RADIO WAVES. He also wrote several books on psychical research and spiritualism.

Loeb, Jacques (1859–1924) US biologist, b. Germany, who studied the chemical processes of living organisms. He stressed the importance of physical and chemical laws in understanding life phenomena in both animals and plants. He achieved popular notoriety for experiments on artificial PARTHENOGENESIS in animals.

loess Fine-grained clay, a sedimentary material made up of rock fragments. It is earthy, porous and crumbly, usually yellowish or brown in color. It consists of very small angular particles of mainly quartz and calcite from glaciated areas, blown by the wind and often built up into thick layers. The largest expanse is in N China in the valley of the River Huang He, where dust from the Gobi Desert has accumulated. The Huang is sometimes called the Yellow River because of all the loess it transports.

Loewi, Otto (1873–1961) German pharmacologist and medical researcher. He shared the 1936 Nobel Prize for physiology or medicine with Henry DALE for discovering that a chemical substance (later shown by Dale to be acetylcholine) mediated the transmission of NERVE IMPULSES from one NEURON to another.

Löffler, Friedrich August Johannes (1852–1915) German bacteriologist, best known for his discovery (in association with Edwin KLEBS) of the organism that causes DIPHTHERIA, *Corynebacterium diphtheriae* (Klebs-Löffler bacillus). He also made several discoveries in veterinary medicine, such as the cause of swine fever and plague, the causative organism of glanders, and the fact that foot-and-mouth disease is caused by a virus.

logarithm Aid to calculation devised by John NAPIER in 1614 and developed by the English mathematician Henry Briggs (1561–1631). The logarithm of a number (n) is the power (x) to which a base (b) must be raised to equal n, i.e., if $b^x = n$. The logarithm "of n in base b" is written $\log_b n = x$. The most widely used bases are base 10 and the so-called natural logarithm base e (2.71828...). Logarithms to the base 2 are used in COMPUTER science and information theory. A related notion is that of the antilogarithm: n may be called the antilogarithm of x in base b. Thus 100 is the antilogarithm of 2 in base 10.

logarithmic function Mathematical function the value of which depends on the LOGARITHM, to a particular base, b, of the independent variable, x, and is thus given by $f(x) = \log_b x$. For natural logarithms, where the base is exponential e, the logarithmic function is the inverse of the EXPONENTIAL function.

logic, mathematical (symbolic logic) Analytical branch of logic that attempts to give the inferences of classical logic a more precise mathematical basis. It is concerned with statements, which may be true or false, and with the relationships between statements and the operations that can be made on them, including CONJUNCTION, DISJUNCTION, and negation. The statements and their relationships and operations are reduced to a symbolic notation and are manipulated according to prescribed mathematical rules. Thus inferences can be analyzed and their validity checked.

logic circuit Network of conductors linking modules (such as amplifiers) and components (such as transistors, diodes, resistors, and capacitors), used to route and process electronic signals (such as voltages) according to the rules of symbolic logic and BOOLEAN ALGEBRA. Also, in computing and formal mathematics, a logic circuit is a network of connected logical decisions or operations called GATES.

Lomonosov, Mikhail Vasilievich (1711–65) Russian scientist and poet who helped to found the University of Moscow (1755). Lomonosov was one of the first Russian scientists to anticipate the law of CONSERVATION of mass, atomic theory, and a kinetic theory of heat. He formulated Russian classical literary theory in his *Letter on the Rules of Russian Versification* (1739).

lone pair In the molecule of a compound, two unbonded electrons associated with one of the component atoms. For example, the nitrogen atom in ammonia (NH_3) and the carbon atom in carbon monoxide (CO) each have a lone pair of electrons. These electrons can form bonds in COORDINATION COMPOUNDS, in which their molecules act as LIGANDS.

Long, Crawford Williamson (1815–78) US physician. In 1842 he began using ether as an ANESTHETIC but did not publish or publicize his work until 1849, after William Morton and others had been credited with the discovery.

long bones In human anatomy, LIMB bones characterized by a long shaft terminating at both ends in swellings (**epiphyses**) that articulate with other bones. Long bones include the RADIUS and ULNA in the forearm, the HUMERUS in the upper arm, the FEMUR in the thigh and the TIBIA and FIBULA in the lower leg. *See also* HUMAN BODY

longitude Angular measurement around the Earth, usually in degrees E or W of an imaginary N–S line through Greenwich, England. This line through Greenwich is called the prime MERIDIAN. All N–S lines are called meridians, or lines of longitude. Map makers and navigators refer to a position on the Earth by quoting its LATITUDE and longitude. In astronomy, celestial longitude is a coordinate on the CELESTIAL SPHERE or on the surface of a celestial body. It is measured E from the vernal EQUINOX to the GREAT CIRCLE passing through the pole of the ECLIPTIC and the object to be measured. It is reckoned in degrees from 0 to 360.

longitudinal wave Type of WAVE, such as a sound wave, in which the particles of the transmitting medium are displaced along the direction of energy propagation, that is, in the direction of wave motion.

longshore drift Movement of sand and pebbles along a seacoast. The material is carried along the beach by the waves hitting the coast obliquely (swash) but is swept back at right angles to the beach by the backwash. If the carrying power of the waves decreases, the material may be deposited to form a SPIT. To combat the effect of longshore drift GROINES are constructed. Usually wooden wall-like structures, they run perpendicular to the coast at regular intervals and so stop the EROSION of the coast. Another increasingly used method to combat the effect is to replace the material in a process known as "beach nourishment."

loom Frame or set of frames on which threads are woven into cloth. The loom enables a set of threads, called the **weft**, to be passed over and under a set of lengthwise threads, called the **warp**. The simplest kind of loom is a single frame on which weaving is done by hand. Such looms have been used for more than 7,000 years. By the 1700s, weavers in Europe were using complex, hand-operated looms to produce many kinds of cloth. One of the most important developments came in 1785, when English clergyman Edmund CARTWRIGHT invented a loom powered by a steam engine to speed cloth production. However, it took another 30 years of development before power-operated looms became widespread. Meanwhile, French weaver Joseph JACQUARD had invented a hand loom that could weave intricate designs. These were determined by patterns of holes in PUNCHED CARDS. Changing the cards changed the pattern produced. Today's most advanced commercial looms are computer controlled and have mechanisms that thread the weft through the warp at speeds of 100km/h (60mph).

loran (**lo**ng **ra**nge **n**avigation) Radio navigational system for guiding ships and aircraft. Pairs of TRANSMITTERS emit signal pulses that are picked up by an aircraft or ship's receiver. By measuring the difference in time between the signals reaching the receiver, the vessel's position can be plotted. The most widely used system is called loran C. The transmitters are about 970km (600mi) apart. The range of the system is 1,800–4,500km (1,100–2,800mi). The system has, in many cases, been superseded by the GLOBAL POSITIONING SYSTEM.

lordosis *See* CURVATURE OF THE SPINE

Lorentz, Hendrik Antoon (1853–1928) Dutch physicist and professor at Leiden. His early work was concerned with the theory of ELECTROMAGNETIC RADIATION devised by James Clerk MAXWELL. Lorentz tried to explain the fact that the speed of LIGHT appears the same to all observers by suggesting that moving objects contract in the direction of their movement. He derived equations that culminated in the LORENTZ-FITZGERALD CONTRACTION, a theory that was later relevant to Albert EINSTEIN's theory of RELATIVITY. Lorentz also worked on the ZEEMAN EFFECT, for which he and Pieter ZEEMAN were awarded the 1902 Nobel Prize for physics. *See also* FITZGERALD, GEORGE FRANCIS; LORENTZ TRANSFORMATION

Lorentz-Fitzgerald contraction (Lorentz contraction, Fitzgerald contraction) Contraction of a moving body in the direction of its motion. The theory was proposed independently by both Hendrik LORENTZ and George FITZGERALD. The theory held that any moving body is contracted in the direction of its motion by a factor $\sqrt{(1 - v^2/c^2)}$, where v is the velocity of the body and c is the velocity of light. It was proposed as an explanation for the negative result of the MICHELSON-MORLEY EXPERIMENT. Albert EINSTEIN used the theory as the mathematical basis of his special theory of RELATIVITY, but showed that it was an effect produced by differences in the position of an observer rather than a real deformation of matter.

Lorentz transformation Relation developed by Hendrik LORENTZ connecting the space and time coordinates of an event as observed from two frames of reference, especially at relativistic velocities. It was shown by Albert EINSTEIN in 1905 to be a consequence of the special theory of RELATIVITY.

Lorenz, Konrad (1903–89) Austrian pioneer ethologist. Unlike psychologists, who studied animal behavior in laboratories, Lorenz studied animals in their natural habitats. He observed that instinct played a major role in animal behavior – for example in IMPRINTING, by which an animal may learn to identify its parents. Some of his views are expressed in his book *On Aggression* (1966). In 1973 he shared, with Nikolaas TINBERGEN and Karl von FRISCH, the Nobel Prize for physiology or medicine. *See also* ETHOLOGY

loudness Magnitude of the sensation produced when the human EAR responds to a SOUND. There is no simple relationship between loudness and the intensity of the sound; the response of the ear also depends to a certain extent on the frequency.

loudspeaker Device for converting changing electric currents into SOUND. The most common type has a moving coil attached to a stiff paper cone and suspended in a strong magnetic field. By ELECTROMAGNETIC INDUCTION, the changing currents in the coil cause the cone to vibrate at the frequency of the currents, thus creating sound waves. Many modern loudspeakers, which are used in radio, stereo, and television sets, give faithful sound reproduction between 80 and 20,000 HERTZ.

Lovell, James Arthur, Jr. (1928–) US astronaut. His first space flight was in Gemini 7 (1965, part of the GEMINI PROJECT), which made the first successful space rendezvous (with Gemini 6). He was a member of the crew of APOLLO 8 (1968), the first manned mission around the Moon, and commander of the unsuccessful Apollo 13 mission (1970).

Lovell, Sir Alfred Charles Bernard (1913–) British astronomer. From 1951–81 he was director of the Jodrell Bank experimental station for radio ASTRONOMY near Manchester, England, and oversaw the construction there of the world's first large steerable RADIO TELESCOPE. He used the telescope to pioneer the exploration of radio emission from space.

low *See* CYCLONE

Lowell, Percival (1855–1916) US astronomer. In 1894 he built an observatory (now the Lowell Observatory) in Flagstaff, Arizona. From there he observed Mars, producing intricate maps of the so-called canals, which he ascribed to the activities of intelligent beings. He went on to study the orbits of Uranus and Neptune, and calculated that their orbital irregularities were caused by an undiscovered Planet X. His predictions led to the discovery of Pluto in 1930. It was at Lowell's urging that the US astronomer Vesto Slipher (1875–1969) carried out the observations that led to the discovery of galactic RED SHIFTS in 1916.

LSD (lysergic acid diethylamide) Hallucinogenic drug, causing changes in mental state, sensory confusion, and behavioral changes when the drug blocks the action of SEROTONIN in the brain. First synthesized in the 1940s, LSD was made illegal in the US (and in the UK as well) in the mid-1960s.

lubricant Oil, grease or other substance placed between moving parts to reduce friction and dissi-

pate heat. Most lubricants are now derived from petroleum. Solid lubricants are usually graphite or molybdenum disulfide; synthetic silicones are used where high temperatures are involved.

luciferin Proteins found in animal tissue that produce light when oxidized in the presence of the ENZYME luciferase. It is found in the light-generating organs of glowworms and fireflies. *See also* BIOLUMINESCENCE

Ludwig, Karl Friedrich Wilhelm (1816–95) German physiologist. He studied the cardiovascular system and invented the kymograph (1847), which records changes in arterial BLOOD PRESSURE. He also invented a device to measure arterial and venous blood flow, and investigated the functions of oxygen in the blood.

lumen (symbol lm) SI unit measuring the amount of visible light emitted in one second in a unit solid angle (one steradian) from a source of unit intensity (one CANDELA).

luminescence *See* FLUORESCENCE; PHOSPHORESCENCE

luminosity Absolute brightness of a STAR, given by the amount of energy radiated from its entire surface per second. It is expressed in watts (joules per second), or in terms of the Sun's luminosity. **Bolometric** luminosity is a measure of the star's total energy output, at all wavelengths. Absolute **magnitude** is an indication of luminosity at visual wavelengths. *See also* SPECTRAL CLASSIFICATION

lunar eclipse *See* ECLIPSE

lunar module Part of the Saturn rocket-launched spacecraft, devised by NASA and used in the APOLLO PROJECT, which detached itself from the command/service module when in lunar orbit and landed, with its passengers, on the Moon's surface. This happened for the first time on July 21, 1969. The upper part (ascent stage) of the lunar module later took off from the Moon using its thrust jet (with the descent stage acting as launch pad) and rejoined the command/service module.

lunar probe Rocket-launched space vehicle that carried no passengers and was aimed at either reaching the Moon's surface or going into orbit around it. The first were launched by the Soviet Union (1959) and the US (1964), and in 1966 both nations landed probes to explore the Moon's surface. An early Russian probe was the first to send photographs back to the Earth of the dark side of the Moon.

lunar roving vehicle (Moon buggy) Electrically powered, four-wheeled passenger vehicle first used in August 1971 by astronauts of the APOLLO PROJECT. It was designed to cope with the reduced gravity and rocky, uneven surface of the Moon, enabling the astronauts to explore local regions of the Moon and carry lunar rock samples back to the LUNAR MODULE for passage back to Earth.

lungs Organs of the RESPIRATORY SYSTEM of vertebrates, in which the exchange of gases between air and blood takes place. They are located in a cavity (the **pleural cavity**) within the rib cage. This cavity is lined by two sheets of TISSUE (the PLEURA), one coating the lungs and the other lining the walls of the thorax. Between the pleura is a fluid that cushions the lungs and prevents friction. Light and spongy, lung tissue is composed of tiny air sacs, called ALVE-OLI, which are served by networks of fine CAPILLAR-IES. *See also* GAS EXCHANGE; VENTILATION

Lunokhod Name of two unmanned Russian scientific lunar surface craft. Luna 17, launched in October 1970, carried Lunokhod 1, which crawled over the surface of the Moon collecting scientific data. Lunokhod 2 was carried by Luna 21, launched in January 1973.

Luría, Salvador Edward (1912–91) US biologist, b. Italy. He shared the 1969 Nobel Prize for physiology or medicine with Max DELBRÜCK and Alfred HERSHEY for contributing to the knowledge of the growth, reproduction, and mutation of VIRUSES.

luteinizing hormone (LH) Hormone produced by the anterior (front) lobe of the PITUITARY GLAND, just below the brain. In women it stimulates OVU-LATION by the development of the CORPUS LUTEUM and secretion of PROGESTERONE. In men it promotes the secretion of TESTOSTERONE from the TESTES.

lutetium (symbol Lu) Metallic element of the LAN-THANIDE SERIES, first isolated in 1906, together with YTTERBIUM. Its chief ore is monazite. The element is used as a CATALYST, but has no other commercial uses. Properties: at.no. 71; r.a.m. 174.97; r.d. 9.842 (25°C); m.p. 1,656°C (3,013°F); b.p. 3,315°C (5,999°F); most common isotope ^{175}Lu (97.41%).

lux (symbol lx) SI unit of illumination, equal to one LUMEN per square meter.

Lwoff, André Michel (1902–94) French microbiologist. He shared the 1965 Nobel Prize for physiology or medicine with François JACOB and Jacques MONOD for work on the genetics of VIRUS-

◄ **lungs** Breathing introduces air into the lungs for the exchange of oxygen and carbon dioxide. Air is funneled into the trachea, the flexible windpipe ringed with cartilage. The trachea forks into the left and right bronchi. Each bronchus branches into several small segments, or bronchi, terminating in 250,000 respiratory bronchioles, each about 0.5mm (0.02in) in diameter. Beyond lie the alveolar ducts leading into hollow alveoli. Here in the alveoli, networked with capillaries only one cell in width, diffusion occurs across a fine membrane. Stale blood is reoxygenated and makes its way back to the heart to be pumped to each living cell in the body. Carbon dioxide is eliminated from the lungs in expired air. The inset diagram shows the position of the diaphragm during breathing.

superior vena cava
larynx
trachea
aorta
pulmonary artery
left main bronchus
bronchi
right lung
heart
bronchiole
inferior vena cava
pulmonary venule
alveolar sac
inhalation
exhalation
lungs
diaphragm
alveoli
capillaries

ES. Lwoff studied the means by which viruses destroy the bacteria they infect.

Lycopodophyta Phylum of about 1,000 species of VASCULAR PLANTS, which includes the CLUB MOSSES, spike mosses, and quillworts. Related to FERNS, lycopods have branching underground stems (RHIZOMES) and upright shoots supported by roots. Some species are EPIPHYTES.

lye Strongly alkaline solution. Originally it applied solely to POTASSIUM CARBONATE (potash), but was extended to include SODIUM HYDROXIDE and POTASSIUM HYDROXIDE. Household lye is usually sodium hydroxide.

Lyell, Sir Charles (1797–1875) British geologist. He was influential in shaping 19th-century ideas about science and wrote the popular books *Principles of Geology* (1830–33), *Elements of Geology* (1838) and *The Geological Evidence of the Antiquity of Man* (1863).

lymph Clear, slightly yellowish fluid derived from BLOOD and similar in composition to PLASMA. Circulating in the LYMPHATIC SYSTEM, it conveys LEUCOCYTES (white blood cells) and some nutrients to the tissues.

lymphatic system System of connecting vessels and organs in vertebrates that transports LYMPH through the body. Lymph flows into tiny and delicate lymph capillaries and from these into lymph vessels, or lymphatics. These extend throughout the body, joining to form larger vessels leading to LYMPH NODES that collect lymph, storing some of the LEUCOCYTES (white blood cells). Lymph nodes empty into large vessels, linking up into lymph ducts that empty back into the CIRCULATORY SYSTEM. The lymphatic system plays a major role in the body's immune system. At the lymph nodes, MACROPHAGES (large leucocytes) trapped in a network of fibers remove foreign particles, including bacteria, from the lymph. The lymph nodes also supply LYMPHOCYTES to the blood. The lymph vessels provide a route for certain nutrients to enter the bloodstream. Fat droplets that are too large to enter the capillaries of the VILLI in the small intestine pass into the lymph vessels for transport to the CIRCULATORY SYSTEM.

lymph node (lymph gland) One of many masses of tissue occurring along the major vessels of the LYMPHATIC SYSTEM. They are filters and reservoirs that collect harmful material, notably bacteria and other disease organisms, and often become swollen when the body is infected. Clusters of lymph nodes are found in the throat, neck, armpits, and groin.

lymphocyte Type of LEUCOCYTE (white blood cell) found in vertebrates. Produced in the bone MARROW, it is mostly found in the LYMPH and blood and around infected sites. In human beings lymphocytes form about 25% of leucocytes and play an important role in combating disease. There are two main kinds: **B-lymphocytes**, responsible for producing antibodies, and **T-lymphocytes**, which have various roles in maintaining the IMMUNE SYSTEM.

lymphoma Any malignant growth of the LYMPH NODES. HODGKIN'S DISEASE is a lymphoma.

Lynen, Feodor Felix Konrad (1911–79) German biochemist who shared the 1964 Nobel Prize for physiology or medicine with Konrad BLOCH for research into the metabolism of FATTY ACIDS and CHOLESTEROL. He helped to ascertain the mechanism of fatty-acid breakdown and explain the pathway of the biosynthesis of several compounds, including natural rubber and cholesterol. He was also the first to isolate acetyl COENZYME A, important for the process of RESPIRATION.

Lysenko, Trofim Denisovich (1898–1976) Russian agronomist and geneticist. He expanded the theory of LAMARCK to include his own ideas of plant genetics (**Lysenkoism**). Lysenko promised the Soviet government vast increases in crop yields through the application of his theories. He enjoyed official sanction under Stalin, but in the late 1950s his influence waned.

lysergic acid diethylamide *See* LSD

lysine ($H_2N(CH_2)_4CH(NH_2)COOH$) Soluble, essential AMINO ACID found in proteins.

lysis Destruction of a CELL. A living cell may be destroyed when its outer cell wall is dissolved by LYSOSOMES or when it is digested in phagocytosis by PHAGOCYTES – a normal reaction against invading bacterial cells.

lysosome Small membranous sac that occurs within the cytoplasm of an animal CELL. It contains enzymes that control the breakdown of foreign substances entering the cell, the digestion of the contents of food VACUOLES and the breakdown of the cell itself after it dies.

maar Coneless volcanic crater that is formed by a single explosive eruption, and not accompanied by a flow of LAVA. A crater ring surrounds the hole. *See also* VOLCANO

McAdam, John Loudon (1756–1836) Scottish engineer who is attributed with inventing the macadam road surface. He proposed that roads should be raised above the surrounding ground, with a base of large stones covered with smaller stones and bound together with fine gravel. McAdam applied these ideas when he was appointed surveyor general of roads in Bristol, sw England, in 1815. They proved successful, and his road-making techniques were generally adopted after a parliamentary inquiry in 1823. McAdam was made surveyor general of roads in Britain in 1827.

McClintock, Barbara (1902–92) US botanist and geneticist. She received a doctorate in botany from Cornell University (1927). While undertaking research during the 1940s and 1950s on the coloration of kermals of maize, she discovered that genes can transfer position along chromosomes. As the so-called TRANSPOSONS change site they affect the behaviour of neighboring genes. This work was to advance knowledge of the mechanisms of MUTATION. She won the Nobel Prize for physiology or medicine in 1983.

McClung, Clarence Erwin (1870–1946) US zoologist who discovered the mechanism of chromosomal SEX DETERMINATION. He postulated that CHROMOSOMES determine sex and that the X chromosome is the crucial element. This stimulated further research, and by 1905 Edmund WILSON had developed the theory, now generally accepted, that both X and Y chromosomes form the basis of sex determination.

McCormick, Cyrus Hall (1809–84) US inventor of a mechanical reaper in 1831. It was introduced into England in 1851. Large-scale manufacture of the reaper, together with McCormick's invention of the twine binder and side-rake, revolutionized the harvesting of cereal crops.

Mach, Ernst (1838–1916) Austrian physicist and philosopher. Interested in the physiology and psychology of the senses, Mach believed that physical phenomena should be explained only by data perceived by the senses. He did research on rapid air flow, and the MACH NUMBER was named after him.

machine Device that modifies or transmits a force in order to do useful work. In a basic, or simple, machine, a force (**effort**) opposes a larger force (**load**). The ratio of the load (output force) to the effort (input force) is the machine's FORCE RATIO, formerly called mechanical advantage. The ratio of the distance moved by the load to the distance moved by the effort is the DISTANCE RATIO, formerly known as the velocity ratio. The ratio of the work done by the machine to that put into it is the EFFICIENCY, usually expressed as a percentage. The three primary machines are the INCLINED PLANE (which includes the screw and the wedge), the LEVER, and the wheel (which includes the PULLEY and the WHEEL AND AXLE).

machine code In computing, instructions that the CENTRAL PROCESSING UNIT (CPU) of a computer can execute directly, without the need for translation. Machine-code statements are written in a binary-coded (low-level) COMPUTER LANGUAGE. Programmers usually write COMPUTER PROGRAMS in a HIGH-LEVEL LANGUAGE (such as FORTRAN or C), which a COMPILER program then translates into machine code for execution. *See also* BINARY SYSTEM

machine tools Power-driven machines for cutting and shaping metal and other materials. Shaping may be accomplished in several ways. These include shearing, pressing, rolling, and cutting away excess material using lathes, shapers, planers, drills, milling machines, grinders, and saws. Other techniques include the use of machines that use electrical or chemical processes to shape the material. Advanced machine-tool processes include cutting by means of LASER beams, high-pressure water jets, streams of PLASMA (ionized gas), and ULTRASONICS. Today, computers control many cutting and shaping processes carried out by machine tools and ROBOTS.

Mach number Ratio of the speed of a body or FLUID (gas or liquid) to the local speed of sound. Mach 1 therefore refers to the local speed of SOUND. An aircraft flying at below Mach 1 is said to be subsonic, or flying at less than the speed of sound. SUPERSONIC FLIGHT means flying at speeds above Mach 1. Mach numbers are named after Ernst MACH, who investigated supersonic speeds and shock waves.

McIndoe, Sir Archibald Hector (1900–60) New Zealand plastic surgeon. One of the pioneers in this field, he founded a hospital for plastic surgery in East Grinstead, Surrey, England, which later became internationally famous. As one of the foremost plastic surgeons of his generation, McIndoe was much in demand during World War II and was appointed consultant in plastic surgery to the Royal Air Force.

Macintosh, Charles (1766–1843) Scottish chemist and inventor after whom the mackintosh coat is named. In 1823 he invented a method of making waterproof cloth by joining two pieces of fabric using a solution of rubber in coal-tar naphtha. However, it was not until 1839, with the advent of vulcanized rubber, which withstood temperature changes, that the mackintosh became a practical garment.

Macleod, John James Rickard (1876–1935) Scottish physiologist who shared the 1923 Nobel Prize for physiology or medicine with Sir Frederick Grant BANTING for the discovery of INSULIN and studies of its use in treating DIABETES. The actual discovery was made by Banting and Charles BEST working in Macleod's laboratory at the University of Toronto. Macleod's works include *Diabetes* (1913) and *Carbohydrate Metabolism and Insulin* (1926).

McLuhan, (Herbert) Marshall (1911–80) Canadian academic and expert on communications. His view that the forms in which people receive information (such as, television, radio, and computers) are more important than the messages themselves was presented in his books: *The Mechanical Bride: Folklore of Industrial Man* (1951), *Understanding Media* (1964), and *The Medium Is the Message* (1967).

McMillan, Edwin Mattison (1907–91) US physicist who shared the 1951 Nobel Prize for chemistry with Glenn SEABORG for the discovery of NEPTUNIUM and other TRANSURANIC ELEMENTS. McMillan worked on the MANHATTAN PROJECT developing the atom bomb at Los Alamos, New Mexico, then on the CYCLOTRON (a type of particle ACCELERATOR) with Ernest LAWRENCE at the University of California at Berkeley. He developed the SYNCHROCYCLOTRON, an advanced type of accelerator, for which he shared the 1973 Atoms for Peace Prize with the Soviet physicist Vladimir Veksler (1907–66). *See also* NUCLEAR WEAPON

Macmillan, Kirkpatrick (1813–78) Scottish blacksmith who invented the first pedal-operated bicycle in 1839. Cranks attached to the rear axle of the machine were connected to pedals near the front. Kirkpatrick's invention became popular in 1846, when it was manufactured by a Scottish cooper called Gavin Dalzell.

macrocephaly Excessive size of the head. It is often caused by HYDROCEPHALUS and is associated with mental subnormality.

macromolecule MOLECULE up to 1,000 times greater in diameter than the molecules of most

substances. Many proteins, nucleic acids, plastics, resins, rubbers, and natural, and synthetic fibers are made up of such giant units, each of which contains thousands of atoms.

macronutrient Element needed in relatively large amounts by growing plants. Macronutrients include carbon, hydrogen, oxygen, and nitrogen, as well as calcium, magnesium, phosphorus, potassium, and sulfur. *See also* MICRONUTRIENT

macrophage Large LEUCOCYTE (white blood cell) found mainly in the liver, spleen, and lymph nodes. It engulfs foreign particles and microorganisms by phagocytosis. Working together with other LYMPHOCYTES, it forms part of the body's IMMUNE SYSTEM.

mad cow disease *See* BOVINE SPONGIFORM ENCEPHALOPATHY (BSE)

Magellanic clouds Two nearest GALAXIES to our own, visible to the naked eye like isolated patches of the Milky Way in the S sky. The Large Magellanic Cloud (LMC) lies in the constellations Dorado and Mensa; the Small Magellanic Cloud (SMC) lies in the constellation Tucana. Their distances from Earth are about 169,000 light-years and 190,000 light-years, respectively. Although usually classified as irregular galaxies, both clouds are somewhat bar-shaped with faint outer structures that can be interpreted as rudimentary spiral arms.

maggot Name commonly given to the legless LARVA of a fly. It is primarily used to describe those larvae that infest food and waste material; others are generally called grubs or caterpillars.

magic number Number of PROTONS or NEUTRONS in a very stable atomic NUCLEUS. For both particles, the magic numbers are 2, 8, 20, 28, 50, and 82 (corresponding, for protons, to the elements helium, oxygen, calcium, nickel, tin, and lead). Proton number 114 is also a magic number, as are 126 and 184 for neutrons. A shell model of the nucleus (analogous to electron shells in the ATOM) has been proposed to account for the extra stability conferred by the magic numbers.

magic square Any square MATRIX divided into cells and filled with numbers or letters in ways once thought to have special magical significance. The most familiar lettered square is the SATOR square, composed of the words SATOR, AREPO, TENET, OPERA, and ROTAS. Arranged both vertically and horizontally the words read the same and form a cross through the middle that reads TENET. In arithmetical magic squares, the numbers are generally arranged so that each column, every row, and the two main diagonals add to the same constant sum.

magma Molten rock beneath the surface of the Earth that forms IGNEOUS ROCKS when it solidifies. Below the surface, cooling is slow and large crystals form as the rock solidifies. Large masses of magma are in BATHOLITHS, and thin seams of magma create SILLS and DIKES. If magma reaches the surface, it flows out as LAVA. The different igneous rocks formed by magma vary according to their chemical content as well as the depth of formation. *See also* EXTRUSIVE ROCK

magnesia *See* MAGNESIUM OXIDE

magnesite Carbonate mineral, magnesium carbonate ($MgCO_3$). It is an alteration product of CALCITE or DOLOMITE in SEDIMENTARY ROCKS and is found in microcrystalline masses and crystals like ICELAND SPAR. Magnesite is white, colorless or of light tint, with a glassy to dull luster. Hardness 3.5–5; r.d. 3.1.

magnesium (symbol Mg) Silver-white, metallic element, one of the ALKALINE-EARTH METALS. The eighth most abundant element in the Earth's crust, it was first isolated (1808) by Sir Humphry DAVY. It is necessary in human and animal diets. Magnesium is always found combined, and its chief sources are MAGNESITE and DOLOMITE. Magnesium burns in air with an intense white flame, and is used in optical instruments, flashbulbs, fireworks, flares, and incendiaries. Magnesium alloys are light and used in aircraft fuselages, jet engines, missiles, and rockets. Chemically, the element is similar to CALCIUM. Hydrated magnesium sulfate is called Epsom salts. Properties: at.no. 12; r.a.m. 24.312; r.d. 1.738; m.p. 648.8°C (1,200°F); b.p. 1,090°C (1,994°F); most common isotope ^{24}Mg (78.7%).

magnesium oxide (magnesia, MgO) White, neutral, stable powder formed when magnesium is burned in oxygen. It is used industrially in firebrick, and medicinally in antacids. Magnesium carbonate ($MgCO_3$), found as the mineral MAGNESITE and also used as an antacid, is often also called magnesia.

magnet Object that produces a MAGNETIC FIELD, an area around the magnet in which other magnetizable objects experience a force. Lodestones, which are naturally magnetic, were used as early magnets, and strong magnetic materials were later recognized as containing iron, cobalt, or nickel. A typical **permanent** magnet is a straight or horseshoe-shaped magnetized iron bar with the ends called the north and south magnetic poles. The Earth is a giant magnet, its magnetic lines of force being detectable at all latitudes. An ELECTROMAGNET is much stronger than a permanent one and is used for raising heavy steel weights and scrap. A **superconducting** magnet, the strongest of all, is made from special alloys cooled to very low temperatures.

magnetic anomaly Small variations in the Earth's MAGNETIC FIELD caused by large iron objects or deposits.

magnetic bottle Configuration of MAGNETIC FIELDS used to contain the PLASMA in a fusion reactor or experimental device. It applies particularly to a linear configuration in which the ends are stoppered with magnetic mirrors, where a very high magnetic field causes the plasma to be reflected.

magnetic declination (magnetic deviation, symbol D) Angle between the direction of magnetic north (as indicated by a compass) and the direction of true north. In any one place, magnetic declination varies slowly with time as the position of magnetic north gradually changes. There are also small annual variations. Angles to the east of true north are taken as positive; angles to the west are negative.

magnetic disk Plastic disk coated with magnetic material and used for storing COMPUTER PROGRAMS and DATA (information) as a series of magnetic spots. Most computers contain a HARD DISK unit for general storage. There is also a unit for inserting floppy disks. These lower-capacity, flimsy magnetic disks are protected by a plastic cover. Hard magnetic disks can

▲ **magnetic disk** A magnetic floppy disk (1) uses a magnetic head, a tiny electromagnet (2), to polarize magnetic particles in the surface of the disk (3). The polarization represents zeros and ones. A screw (4) moves the head across the disk which is spun by a motor (5).

When the disk is inserted, a lever moves aside the protective window over the disk (6). An LED (7) checks whether the disk is write-protected. If the light can pass through a window on the disk (8) it is protected, and no new data can be stored on the disk or old data removed.

store larger amounts of data and come in cartridges that slot into a special drive unit. *See also* CD-ROM

magnetic disturbance Change in the shape of the MAGNETOSPHERE. The most commonly observed phenomena that result from this are the polar AURORAS and interference in radio and television communications.

magnetic equator That part of the Earth's MAGNETIC FIELD where a finely balanced, magnetic needle will remain absolutely horizontal. Because of the Earth's magnetic asymmetry, the magnetic equator does not coincide with the geographical EQUATOR.

magnetic field Region surrounding a MAGNET, or a conductor through which a current is flowing, in which magnetic effects, such as the deflection of a compass needle, can be detected. A magnetic field can be represented by a set of lines of force (flux lines) spreading out from the poles of a magnet or running around a current-carrying conductor. These lines of force can be seen if iron filings are sprinkled on to a sheet of paper below which a magnet is placed. The filings line up along the lines of force, the density of the lines being greatest where the field is strongest. The direction of a magnetic field is the direction a tiny magnet takes when placed in the field. **Magnetic poles** are the field regions in which MAGNETISM appears to be concentrated. If a bar magnet is suspended to swing freely in the horizontal plane, one pole will point north; this is called the north-seeking or **north pole**. The other pole, the south-seeking or **south pole**, will point south. Unlike poles attract each other; like poles repel each other. **Earth's magnetic poles** are the ends of the huge "magnet" that is Earth.

magnetic field reversal Reversal of polarity whereby the Earth's N magnetic pole becomes the S and vice versa. Analyses of the magnetic direction of land and ocean basaltic lavas and seafloor sediments have shown that the Earth's main MAGNETIC FIELD has undergone frequent and rapid reversals. The field has changed many times in the past four million years. *See also* PALAEOMAGNETISM

magnetic flux (symbol Φ) Measure of the strength and extent of a MAGNETIC FIELD. The flux through an area A at right angles to a uniform magnetic field is $\Phi = \mu H A$, where μ is the magnetic PERMEABILITY of the medium and H is the magnetic field intensity. Magnetic flux density is the flux per unit area (symbol B), which equals μH. Change of magnetic flux through an electric circuit induces an ELECTROMOTIVE FORCE.

magnetic moment Measure of the strength of a permanent magnet or current-carrying coil. It is

the maximum turning force (TORQUE) exerted on a magnet, coil, or moving electric charge in a MAGNETIC FIELD divided by the strength of the field. Charged particles and atomic nuclei also have magnetic moments.

magnetic permeability *See* PERMEABILITY

magnetic pole *See* MAGNETIC FIELD

magnetic recording Formation of a record of sounds on a wire or tape by means of a pattern of magnetization. In a tape recorder, ferromagnetic tape (*see* MAGNETIC TAPE) is fed past an electromagnet that is energized by the amplified currents produced by a MICROPHONE. By ELECTROMAGNETIC INDUCTION, variations in magnetization (from the oscillating current produced by the sound) are induced in the particles of ferromagnetic material on the tape. When played back, the tape is fed past a similar electromagnet, which converts the patterns into sound that is in turn fed to an AMPLIFIER and LOUDSPEAKER.

magnetic resonance imaging (MRI) Medical scanning system that utilizes the phenomenon of NUCLEAR MAGNETIC RESONANCE (NMR) to produce three-dimensional images of the brain, spinal cord, and other soft tissues in the body. The benefit of this scanning system is that it does not expose the patient to potentially harmful X-ray radiation. The scanner's magnet causes the nuclei within the atoms of the patient's body to line themselves up in one direction. A brief radio pulse is then beamed at the nuclei, causing them to change their orientation. As they realign themselves to the magnet, they give off weak radio signals that can be recorded and converted electronically into images.

magnetic storm (geomagnetic storm) Disturbance in the Earth's MAGNETIC FIELD. Since the field encompasses all the Earth, the effects of such a storm are global: AURORAS are seen, both in areas where such displays are normal and in others as well; and radio signals are disturbed.

magnetic tape Medium for recording electrical signals, particularly audio signals in TAPE RECORDERS, video signals in video recorders, and data in COMPUTERS. It consists of a thin plastic tape coated on one side with fine particles of a ferromagnetic material, such as iron, iron oxide, or chromium oxide. *See* MAGNETIC RECORDING; VIDEOTAPE

magnetism Properties of matter and of electric currents associated with a field of force (MAGNETIC FIELD) and with a north–south polarity (magnetic poles). All substances possess these properties to some degree because orbiting ELECTRONS in their atoms produce a magnetic field in the same way as

▲ **magnetism** A simple way to magnetize materials, such as iron and its alloys, is to stroke them with a bar magnet (A), the nearness of which, coupled with its movement, tends to align the magnetic domains within the material. In such a way they reinforce each other rather than keeping their normally random arrangements. The north-seeking ends of the domains try to follow the movement of the original magnet's south pole so that the right-hand side of the new magnet becomes a north pole. The domains lie with their north poles to the right, and so their south poles are to the left. Another way to magnetize a bar of suitable material is simply to hit it (B). The domains receive a mechanical shock and the Earth's field tends to align them with itself. Adapting the technique shown in (A) it is possible to make a bar magnet of suitable material using two magnets (C). In this case, the right-hand side acquires a south pole, and the left-hand side a north pole.

an ELECTRIC CURRENT produces a magnetic field; similarly, an external magnetic field will affect the electron orbits. All substances possess weak magnetic properties and will tend to align themselves with the field, but in some cases this DIAMAGNETISM is masked by the stronger forms of magnetism: PARAMAGNETISM and FERROMAGNETISM. **Paramagnetism** is caused by electron SPIN and occurs in substances having unpaired electrons in their atoms or molecules. The most important form of magnetism, **ferromagnetism**, is shown by substances such as iron and nickel, which can be magnetized by even a weak field due to the formation of tiny regions, called domains, that behave like miniature magnets and align themselves with an external field. These domains are formed as a result of strong, interatomic forces caused by the spin of electrons in unfilled, inner electron shells of the atoms. **Permanent magnets**, which retain their magnetization after the magnetizing field has been removed, are ferromagnetic. ELECTROMAGNETS have a ferromagnetic core around which a conducting coil is wound. The passage of a current through the coil magnetizes the core.

magnetite (ferrosoferric oxide, Fe_3O_4) Oxide mineral, iron(II)-iron(III) oxide. The most magnetic mineral discovered to far, it is a valuable iron ore, found in IGNEOUS and METAMORPHIC ROCKS. It displays cubic system octahedral and dodecahedral crystals, and granular masses are also common. It is black, metallic, and brittle. Permanently magnetized deposits are called **lodestone**. Hardness 6; r.d. 5.2.

magneto Machine that converts mechanical energy into a high-voltage alternating electric current. Magnetos are used when batteries are impractical, such as in the ignition systems of gasoline engines in small motorcycles, lawn mowers, outboard motors, or for electrically detonating explosives. A magneto has a primary winding in which a rotating magnet produces a low voltage, which in turn induces a high voltage in a surrounding secondary winding. A make-and-break device in the primary winding creates the repeated changes of magnetic flux that induce the large voltage in the secondary winding. The magneto is connected to the rotating shaft of the engine. Initially, the inner magnet is rotated by hand (often by pulling on a spring-return cord); once running, the magnet is rotated by the engine itself. *See also* INDUCTION

magnetochemistry Branch of chemistry concerned with investigating the magnetic properties of compounds. In particular, magnetic measurements made on TRANSITION-ELEMENT complexes, which are often paramagnetic due to unpaired electrons, give information on their structure and ELECTRON configuration.

magnetohydrodynamics (MHD) Study of the motion of FLUIDS, usually PLASMAS, under the influence of electric and magnetic fields. An MHD generator consists of a plasma fluid flowing in a magnetic field, thus generating an electric current. The fluid is seeded with easily ionized elements (such as potassium) to increase the power yield.

magnetometer Instrument for measuring MAGNETIC FIELD strengths. It usually consists of a short, bar magnet with a nonmagnetic pointer attached to its center so that it is at right angles to the axis of the magnet. The magnet is pivoted, like a compass needle, at its center, and the pointer travels over a calibrated scale. Field strengths of magnets are compared by measuring the deflections of the pointer.

magnetosphere Region of space surrounding a planet in which the planet's MAGNETIC FIELD predominates over the SOLAR WIND and controls the behavior of PLASMA (charged particles) trapped within it. The boundary of the magnetosphere is called the **magnetopause**, outside which is a turbulent magnetic region called the **magnetosheath**. Downwind from the planet, the solar wind draws the magnetosphere out into a long, tapering **magnetotail**. Mercury, Earth, and the giant planets have magnetospheres. The Earth's contains the VAN ALLEN RADIATION BELTS of charged particles.

magnetron Vacuum tube containing an ANODE and a heated CATHODE. The flow of electrons from cathode to anode is controlled by an externally applied magnetic field. When incorporated in a resonant system, a magnetron can act as an OSCILLATOR. It can generate high frequencies and high power in short bursts, and is used in radar sets and microwave ovens.

magnification Measure of the enlarging power of a MICROSCOPE or TELESCOPE. It is the size of an object's image produced by the instrument compared with the size of the object viewed with the unaided eye. In an astronomical telescope, magnification is equal to the ratio of the FOCAL LENGTH of the objective (the lens or lenses nearest the object) to the focal length of the eyepiece. With a very short eyepiece focal length increases the magnification but reduces the field of view and can produce a distorted image.

magnifying glass Convex lens that increases the apparent size of objects examined through it, usually between two and ten times. Maximum magnification is obtained when the object is situated just within the focal length of the lens. A magnifying glass generally has a handle attached to a circular clip holding the lens.

magnitude In astronomy, numerical value expressing the brightness of a celestial object on a logarithmic scale. **Apparent** magnitude is the magnitude as seen from Earth, determined by the eye, or photographically or photometrically, and it ranges from positive through zero to negative values, the brightness increasing rapidly as the magnitude decreases. First magnitude is exactly 100 times brighter than sixth magnitude (just visible to the naked eye). **Absolute** magnitude indicates intrinsic luminosity and is defined as the apparent magnitude of an object at a distance of 10 parsecs (32.6 light-years) from the object.

Maiman, Theodore Harold (1927–) US physicist who successfully built the first optical MASER, which he called the LASER. This was done in 1960 after he had first made improvements to the maser devised by Charles TOWNES. In 1962 Maiman founded the Korad Corporation to construct high-powered lasers.

mainframe Large, fast and powerful central COMPUTER and its associated storage devices. Users may be provided with small terminal units resembling personal computers; these have SOFTWARE that enables the user to access the data on the mainframe. The terminal units are linked to each other and to the mainframe by a COMPUTER NETWORK. Today's personal computers are much more powerful than early mainframe machines, so many tasks once carried out by mainframes are now done on the desktop computer, or even on a portable machine (LAPTOP).

main sequence In astronomy, region on the HERTZSPRUNG-RUSSELL DIAGRAM where most stars lie, including the Sun. It runs diagonally from hot, bright stars at the upper left down to cool, faint stars at the bottom right. The position of a star on the main sequence depends on its mass, the most massive stars being the brightest. Stars spend most of their lives on the main sequence, producing energy from the fusion of hydrogen to helium in their cores. The zero-age main sequence is where stars lie when they first start to burn hydrogen. As stars use up their internal hydrogen they move away from the main sequence. The more massive a star, the sooner it evolves off the main sequence. All stars on the main sequence are termed DWARFS, irrespective of whether they are larger or smaller than the Sun. *See also* STELLAR EVOLUTION

▲ **magnetosphere** The magnetosphere is the region in which the Earth's magnetic field can be detected. It would be symmetrical were it not for electrically charged particles streaming from the Sun (A), which distort it to a teardrop shape. The particles meet the Earth's magnetic field at the shock front (1). Behind this is a region of turbulence, and inside the turbulent region is the magnetopause (2), the boundary of the magnetic field. The Van Allen belts (3) are two zones of high radiation in the magnetopause. The inner belt consists of high-energy particles produced by cosmic rays and the outer belt of solar electrons.

malachite ($Cu_2CO_3(OH)_2$) Carbonate mineral, basic copper carbonate, found in weathered copper ore deposits. Malachite has a monoclinic system, and is green in color; it is sometimes used as a gemstone. Hardness 3.5–4; r.d. 4.

malaria Parasitic disease resulting from infection with one of four species of *Plasmodium* PROTOZOA. Transmitted by the *Anopheles* mosquito, malaria is characterized by fever and enlargement of the spleen. The parasite multiplies in erythrocytes (red blood cells), causing their destruction. Attacks of fever, chills and sweating typify the disease and occur as new generations of parasites develop in the blood, with the frequency of attacks related to the species involved. The original antimalarial drug, QUININE, has given way to synthetics such as chloroquine, although resistant strains of the main malaria parasite, *P. falciparum*, are spreading rapidly. With 270 million people infected, malaria is one of the world's most widespread diseases, claiming two million lives a year, mostly in the tropics.

malic acid ($HOOCCH(OH)CH_2COOH$) Organic chemical compound. It has several important roles in the body's METABOLISM, notably in the KREBS CYCLE, which is concerned with the production of energy during RESPIRATION.

malleability Property of metals (or other substances) that can be permanently shaped by hammering or rolling without breaking. In some cases, it is increased by raising temperature.

mallee Scrubby Australian vegetation that consists mainly of shrubs of the genus *Eucalyptus*. They have leathery leaves that conserve moisture during dry periods.

malnutrition Condition resulting from a diet that is deficient in necessary components such as PROTEINS, FATS or CARBOHYDRATES. It can lead to DEFICIENCY DISEASES, increased vulnerability to infection, and death. Worldwide, high death rates from malnutrition are linked to poverty, famine or war.

Malpighi, Marcello (1628–94) Italian physiologist. He was a founder of microscopic anatomy, demonstrating how blood reaches the tissues through tiny vessels (CAPILLARIES), as William HARVEY had conjectured. Malpighi explained the network of capillaries on the surface of the lung. The Malpighian tubules of the kidney and the MALPIGHIAN LAYER are among the structures named after him.

Malpighian layer Innermost layer of the EPIDERMIS of the skin. A fibrous basement membrane separates the Malpighian layer from the DERMIS beneath. The layer contains polygonal cells, which continually divide by MITOSIS. As the cells grow older, they move slowly through the epidermis to replace dead surface skin cells that have been worn away. The layer is named after Marcello MALPIGHI.

maltose (malt sugar, $C_{12}H_{22}O_{11}$) Disaccharide that contains two molecules of the simple sugar GLUCOSE. It is produced by the HYDROLYSIS of STARCH by the enzyme AMYLASE and by the breakdown of starches and GLYCOGEN during digestion.

Malus, Étienne Louis (1775–1812) French physicist who, in 1809, announced his discovery of the POLARIZATION of light by reflection. In 1810 Malus also produced a theory of double REFRACTION of light in crystals. Earlier in his career, in 1798, he had taken part in the invasion of Egypt as a member of the engineer corps in the army of Napoleon I.

mammal Class (Mammalia) of VERTEBRATE animals characterized by mammary glands in the female and full, partial or vestigial hair covering. Mammals are HOMEOTHERMIC (warm-blooded). They have a four-chambered heart with circulation to the lungs separate from the rest of the body. As a group, mammals are active, alert, and intelligent. They usually bear fewer young than other animals and give them longer and better parental care. Most mammals before birth grow inside the mother's body and are nourished from her by means of a PLACENTA. When born, they continue to feed on milk from the mother's mammary glands. There is a wide range of features, shapes, and sizes among mammals. Mammals include 17 orders of PLACENTALS, one MARSUPIAL order – all live-bearing – and an order of egg-laying MONOTREMES. They probably evolved about 180 million years ago from a group of warm-blooded reptiles. Today, mammals range in size from shrews weighing a few grams, to the blue whale, which can weigh up to 150 tons.

mammary gland Organ of a female MAMMAL that secretes milk to nourish newborn young. In males the glands are rudimentary and nonfunctional. The human female breast, which develops during puberty, is made up of about 15–20 irregularly shaped lobes separated by connective and fat tissues. Lactiferous ducts lead from each lobe to the nipple.

mammography X-ray procedure used to screen breast tissue for CANCER. It can detect a malignant tumor at an early stage of development.

Mandelbrot, Benoit B. (1924–) US mathematician, b. Poland. Mandelbrot has made major contributions to CHAOS THEORY. He is best known for coining the term FRACTAL to describe irregular geometrical figures that look the same when magnified (regardless of how large the magnification). This property is called self-similarity. His book *The Fractal Geometry of Nature* (1982) contains examples of fractal-like objects seen in the world around us, such as ferns, trees, and rivers. The Mandelbrot set, a well-known fractal object, is named after him.

mandible In vertebrates, the lower JAW. In insects and CRUSTACEA, one of a pair of mouthparts used for crushing and eating.

mandrake Plant of the potato family, native to the Mediterranean region and used since ancient times as a medicine. It contains the ALKALOIDS hyoscyamine, scopolamine, and mandragorine. Leaves are borne at the base of the stem, and the large greenish-yellow or purple flowers produce a many-seeded berry. The branched root resembles a distorted human figure, and for centuries the plant was believed to utter an almost human shriek when picked. Height: 40cm (16in); family Solanaceae; species *Mandragora officinarum.*

manganese (symbol Mn) Gray-white TRANSITION ELEMENT, first isolated in 1774. Its chief ores are pyrolusite, manganite, and hausmannite. It is also found in MANGANESE NODULES. The metal is used in alloy steels, ferromagnetic alloys, fertilizers and paints, and as a gasoline additive. It forms two series of salts, termed manganese(II), or manganous, and manganese(III), or manganic. It also forms manganate(VI), containing the ion MnO_4^{2-}, and manganate(VII) (permanganate), MnO_4^-. Properties: at.no. 25; r.a.m. 54.938; r.d. 7.20; m.p. 1,244°C (2,271°F); b.p. 1,962°C (3,564°F); most common isotope ^{55}Mn (100%).

manganese nodules Concretions with diameters averaging 4cm (2in) that occur in red clay and ooze on the floor of the Pacific Ocean, with the major concentrations around the Samoan Islands. They contain high concentrations of manganese. They are thought to have formed as agglomerates from colloidal solution (*see* COLLOID) in the ocean's waters. The commercial mining of the nodules is being delayed by legal controversy over international rights to this valuable mineral resource.

manganite (MnO[OH]) Hydroxide mineral of the diaspore group, manganese oxyhydroxide. It occurs as low-temperature veins and secondary deposits and has monoclinic system striated prisms, which are often crusts of small crystals. Its color is black or gray with a submetallic luster. Hardness 4; r.d. 4.3.

Manhattan Project Codename given to the development of the US atom bomb (*see* NUCLEAR WEAPON) during World War II. Work on the bomb, advocated by Albert EINSTEIN and other scientists, was carried out in great secrecy by a team headed by Enrico FERMI and including J. Robert OPPENHEIMER. This took place at several locations, notably Oak Ridge, Tennessee, and Los Alamos, New Mexico. Research was also carried out at the University of Chicago, the University of California at Berkeley and Columbia University. The first test took place on July 16, 1945, near Alamogordo, New Mexico, and in August 1945 the first bombs were dropped on the Japanese cities of Hiroshima and Nagasaki.

man-made fiber *See* SYNTHETIC FIBER

manometer Device for measuring pressure. It consists of a U-shaped tube containing a liquid, one end open to the atmosphere and the other end attached to the vessel whose pressure is to be measured. If the gas pressure in the vessel is greater than atmospheric, it will force the liquid down on

▲ **mammary gland** In humans, there is one pair of mammary glands, which are composed of a mass of epithelial ducts (1, shown magnified) surrounded by fibrous tissue. In women, these ducts enlarge and spread, differentiating into milk-producing tissue. This process occurs under complex hormonal control from the anterior and posterior pituitary glands (A), the placenta during pregnancy (B), and from the ovaries (C). Full development of the glands involves extensive growth of mammary ducts from which specialized lobules proliferate (2). Each lobule, lined by milk-producing cells (3), opens into the ducts leading to the gland nipple. Three to four days after childbirth, the hormone prolactin enables milk containing fat droplets (4) and protein to become available to the child.

A

B

10
9
8
7
6
5
4
3
2
1
0

10
9
8
7
6
5
4
3
2
1
0

1   2   3

▲ **manometer** A manometer is essentially a U-shaped liquid column gauge used to measure the difference between two fluid pressures. The well type manometer is illustrated here, one column of which has a relatively small diameter; the second serves as a reservoir. The difference in cross-sectional area (sometimes as much as 1500 times) ensures that while the level of the reservoir does not change appreciably with change of pressure, the level of the small diameter column does; this makes for adequate accuracy. Small adjustments to the position of the vertical scale, with the aid of the level indicator, compensate for the little reservoir change that does occur. Readings of differential pressure may then be made directly on the vertical scale.
(A) Both columns are at atmospheric pressure, so the levels are the same.
(B) The reservoir is pressurized. Its level is indicated so the scale may be adjusted to the new level.

Key
1) reservoir
2) level indicator
3) pressure indicating column

the side nearest to the vessel and up on the side open to the atmosphere. The difference in heights of the liquid in the two arms of the tube shows the difference in pressure.

mantissa Decimal part of a LOGARITHM

mantle Layer of the EARTH between the CRUST and the CORE, down to 2,900km (1,800mi). Its boundaries are determined by major changes in seismic velocity. The mantle forms the greatest bulk of the Earth, 82% of its volume and 68% of its mass. Several distinct parts of the mantle have been recognised. The uppermost part is rigid, solid, and brittle and together with the Earth's crust forms the LITHOSPHERE. The upper mantle also has a soft zone, which is called the ASTHENOSPHERE. Temperature and pressure are in delicate balance, so that much of the mantle material is near melting point, or partially melted and capable of flowing. Recent studies have suggested this layer is responsible for the production of basaltic MAGMAS. The remainder of the mantle is thought to be more solid, but still capable of creeping flow. In the lower mantle, several changes in seismic velocity can be detected, interpreted as mineral phase changes, in which the atomic packing is rearranged into denser and more compact units. The chemical constitution of the mantle is uncertain, but it is thought to be made up of iron-magnesium silicates.

map Graphic representation of part or all of the Earth's surface. The science of mapmaking is called cartography. Maps are usually printed on a flat surface using various kinds of projections. To make a map, a cartographer uses information from land surveys, aerial photographs, and other sources. A map may be formed by drawing on paper, engraving lines into plastic, or constructing an image on a computer screen.

map projection Any systematic method of drawing the Earth's meridians and parallels on a flat surface. Only on a globe can areas and shapes be represented with any fidelity. On flat maps of large areas distortions are inevitable. Projections are either geometrical derivations (cylindrical, conical or azimuthal) or networks and grids derived mathematically in the transposition from globe to flat surface.

marble Metamorphic rock composed largely of recrystallized LIMESTONES and DOLOMITES. The term is more loosely used to refer to any crystalline calcium carbonate ($CaCO_3$) rock that has a good pattern and color when cut and polished. The color is normally white, but when tinted by serpentine, iron oxide, or carbon can vary to shades of yellow, green, red, brown or black. It has been a favorite building and sculpting material of many ancient and modern civilizations.

marcasite Brass-yellow mineral, iron sulfide (FeS_2), found commonly in near-surface deposits. It is found as tabular or prismatic crystals in the orthorhombic system, as cockscomb aggregates, and as stalactitic, globular, and radiating forms. It has a metallic luster and is brittle. Hardness 6–6.5; r.d. 4.9.

Marconi, Guglielmo (1874–1937) Italian physicist who developed RADIO. By 1897 Marconi was able to demonstrate radio telegraphy over a distance of 19km (12mi) and, after forming a wireless telegraph company, established radio communication between France and England in 1899. By 1901 radio transmissions were being received across the Atlantic Ocean, and in 1909 Marconi was awarded the Nobel Prize for physics (which he shared with C. F. Braun). His later work on short-wave radio transmission formed the basis of nearly all modern long-distance communication.

Marcus, Rudolph (1923–) US physical chemist, b. Canada. Marcus developed what is now known as the "Marcus theory" for electron transfer (OXIDATION-REDUCTION) reactions of solvent molecules. His formula calculates the size of the energy barrier the electrons must surmount to move from one molecule to another. The theory is applicable to such processes as photosynthesis and the electrical conductivity of polymers. In 1992 Marcus received the Nobel Prize for chemistry.

marigold Any of several mostly golden-flowered plants, mainly of the genera *Chrysanthemum*, *Tagetes*, and *Calendula*, all of the daisy family (Asteraceae/COMPOSITAE). Those most commonly cultivated are the French marigold (*Tagetes patula*) amd the African marigold (*T. erecta*).

marijuana NARCOTIC drug prepared from the dried leaves of the Indian hemp plant (*Cannabis sativa*); it is different from hashish, which is prepared from resin obtained from the flowering tops of the plant. Some people smoke marijuana like tobacco. In moderate doses it can act as a relaxant and euphoriant. It has proved beneficial to some sufferers of MULTIPLE SCLEROSIS (MS). Possession of the drug is illegal in the United States and many other countries. *See also* ADDICTION

marine biology Science and study of life in the sea. It covers organisms that live in the ocean waters, on the seabed, and along shores. Marine biologists explore the ways in which these organisms fit into their environments, as well as their interactions with the human environment.

marine engine Power plant used to propel seagoing SHIPS, and as auxiliaries in smaller sailing vessels. In the 19th and early 20th centuries, marine engines were coal-fired, piston-driven STEAM ENGINES. Today, large steam engines are still used, but these are usually oil-fired steam TURBINE engines; steam, raised in a boiler, turns a turbine, which is geared down to turn propeller shafts. Small, medium and large diesel engines, and small gasoline engines (including outboard motors) are also in general use. A few ships, notably icebreakers and submarines, are fitted with nuclear engines in which steam for turbines is raised by heat from a NUCLEAR REACTOR.

Mariner project Series of US SPACE PROBES to the planets. **Mariner 2** flew past Venus in 1962, discovering that the planet was hot. **Mariner 4** flew past MARS in July 1965, photographing craters on its surface. **Mariner 5** passed Venus in October 1967, making measurements of the planet's atmosphere. **Mariners 6 and 7** obtained further photographs of Mars in 1969. **Mariner 9** became the first probe to enter planetary orbit in November 1971. It made a year-long photographic reconnaissance of the planet's surface, and obtained close views of the two moons, Phobos and Deimos. **Mariner 10**, the last of the series, was the first two-planet mission, passing Venus in February 1974 and then encountering Mercury three times, in March and September 1974 and March 1975. *See also* SPACE EXPLORATION

Mariotte, Edme (1620–84) French physicist, plant physiologist, and priest, noted for his independent discovery of BOYLE'S LAW. This law, which states that at constant temperature the volume of a gas is inversely proportional to its pressure, is known as **Mariotte's law** in France. Mariotte also concluded that plants synthesize materials by chemical processes, a theory verified after his death. He was a founder of the Academy of Sciences in Paris.

maritime climate CLIMATE that is strongly influenced by proximity to the sea. Oceans warm up more slowly than land, because the heat of the Sun is spread out through a great depth of water and because ocean currents allow the heat to move vertically as well as horizontally. However, oceans also retain heat for much longer than land. For this reason, climatic conditions near an ocean tend to be much warmer in winter and slightly cooler in summer than in inland areas at the same latitudes. In addition to the effects on temperature, the sea also influences precipitation, which is greater on the coast than in inland locations.

marl (mudrock) Extremely fine-grained, SEDIMENTARY ROCK that is an intermediate between a clay and a limestone. It contains up to 60% of calcareous minerals such as calcite, with some quartz and particles of silt. Red marl owes its color to the presence of iron oxide, whereas green marl contains chlorite or glauconite. They formed from deposits at the bottoms of lakes or the sea. Marls of marine origin may occur in beds with gypsum and halite.

marmot Stocky, terrestrial rodent of the squirrel family, native to N.America, Europe, and Asia. Most marmots have brown to gray fur, short, powerful legs, and furry tails. Length, excluding tail: 30-60cm (12-24in); weight: 3-8kg (6.6-16.5lb). Family Sciuridae.

marrow Soft tissue containing blood vessels, found in the hollow cavities of BONE. The marrow found in many adult bones, including the shafts of LONG BONES, is somewhat yellowish and functions as a store of fat. The marrow in the flattish bones, including the ribs, breastbone, skull, spinal column, and the ends of the long bones, is reddish and contains cells that give rise eventually to ERYTHROCYTES (red blood cells) as well as to most of the LEUCOCYTES (white blood cells), but not LYMPHOCYTES or PLATELETS.

Mars Fourth major planet from the Sun, and the second smallest of the inner planets. Mars appears red to the naked eye because of the high iron content of its surface crust. When viewed through a telescope it appears as a small orange-red disk with lighter and darker markings, and white patches often visible at one pole or the other. The atmosphere consists mainly of 95% carbon dioxide, 2.5% nitrogen and 1.5% argon, with smaller quantities of oxygen, carbon monoxide, and water vapor. Its axial tilt is similar to the Earth's, so it passes through a similar cycle of SEASONS, and seasonal variations in its appearance may be observed through a telescope. The surface temperature on Mars varies between extremes of 130K and 290K. Much information has been returned by various Mars probes – spacecraft that have orbited or landed on the planet. The surface of Mars reveals a long and complex history of geological activity. The major difference in terrain is between the largely smooth, lowland volcanic plains of the northern hemisphere and the heavily cratered uplands of the south. The biggest volcanic structure on Mars, and in the whole Solar System, is Olympus Mons, which is hundreds of kilometers across and 27km (17mi) high. Other geographical features, such as some giant canyons, are channels in which rivers once flowed. Perhaps all the water is now frozen into the subsurface permafrost layer; more likely, the planet's low gravity has simply not been enough to hang on to all of its air and water, which have slowly been escaping. The variable polar ice caps appear to be composed of solid carbon dioxide with underlying caps of water-ice. Mars has two tiny

SATELLITES in very close orbits, Phobos and Deimos. In 1996 scientists investigating a METEORITE, thought to have originated on Mars, found fossilized microorganisms, which some believe indicates the presence of primitive life on the planet. In 1997 the US Pathfinder probe reached Mars and launched a small vehicle onto the planet's surface. It analyzed the geological formations it encountered, sending information back to Earth. **Data:** Diameter (equatorial): 6,787km; Mass: (Earth=1) 0.11; Volume: (Earth=1) 0.15; Density (water=1): 3.94; Orbital period: 686.9 days; Rotation period (equatorial): 24h 37m 23s; Surface temperature (average): 220K *See also* VIKING SPACE MISSION

marsh Flat, grassland area, devoid of peat, and saturated by moisture during one or more seasons. Typical vegetation includes grasses, sedges, reeds, and rushes. Marshes are sometimes the breeding sites of mosquitoes and other disease carriers, but they are valuable WETLANDS and maintain water tables in adjacent ECOSYSTEMS.

marsh gas *See* METHANE

Marsigli, Count Luigi Ferdinando (1658–1730) Italian soldier, naturalist, and oceanographer. After long service in the Austrian army, during which he was captured, he was demoted in 1704. Thereafter, Marsigli engaged in scientific study, founding the Accademia delle Scienze in Bologna (1712) and becoming a member of the Royal Society (1722), to which he was presented by Isaac NEWTON.

marsupial Mammal of which the female usually has a pouch (marsupium), within which the young are suckled and protected. At birth, the young are in a very early stage of development. Most marsupials are Australasian, and include such varied types as the kangaroo, koala, wombat, tasmanian devil, bandicoot, and marsupial mole. The only marsupials to live outside Australasia are the opossums and similar species found in the Americas. *See also* MONOTREME

Martin, Archer John Porter (1910–) British biochemist who developed partition CHROMATOGRAPHY, particularly for use in amino-acid analysis. For this research, Martin shared the 1952 Nobel Prize for chemistry with Richard SYNGE. Later, he developed gas-liquid chromatography.

Martin, Pierre Émile (1824–1915) French engineer who pioneered the OPEN-HEARTH PROCESS of steel manufacture. The **Siemens-Martin** process preheats the air-blast by a heat-regeneration method and the excess carbon is removed by adding a calculated amount of iron ore. By applying this process to an open-hearth furnace charged with a mixture of pig iron and wrought iron, Martin was able to obtain the desired carbon content simply by dilution.

mascon Any of several high-density regions of the MOON's surface that produce stronger gravitational effects than the surrounding areas, causing noticeable perturbations in the orbits of spacecraft. The word is derived from the term "mass concentration" and the phenomenon seems to indicate that the lunar interior is not uniform.

maser (acronym for **m**icrowave **a**mplification by **s**timulated **e**mission of **r**adiation) Device using atoms artificially kept in states of higher energy than normal to provide amplification of radio signals. The principle of the maser was discovered by Charles TOWNES of Columbia University, for which he shared the 1964 Nobel Prize for physics with Nikolai BASOV and Alexander PROKHOROV, who also worked in this field. The first maser used electrostatic (charged) plates to separate high-energy ammonia atoms from low-energy ones. Radiation of a certain frequency stimulated the high-energy ammonium atoms to emit similar radiation and strengthen the signal. The very narrow frequency emitted made the ammonia maser one of the most accurate "atomic clocks" known. *See also* LASER

Maskelyne, Nevil (1732–1811) British astronomer. In 1765 he was appointed the fifth Astronomer Royal, a post he held for 46 years. Maskelyne introduced a method of determining longitude at sea by the measurement of lunar position, and in 1767 he published the first Nautical Almanac, an annual book for the same purpose. In 1774 he measured the mass of the Scottish mountain of Schiehallion by seeing how far it caused a plumbline to deviate, from which he calculated the average density of the Earth to be 4.7g cm^{-3} (the true value is 5.52g cm^{-3}).

mass (symbol m or M) Measure of the quantity of matter in an object. Scientists distinguish between two types of mass: **gravitational** mass is a measure of the mutual attraction between bodies such as the Earth, as expressed in Newton's law of GRAVITATION. **Inertial** mass is a measure of a body's resistance to change in its state of MOTION, as expressed in Newton's first law of motion. Newton meters (spring balances) and platform scales measure gravitational mass; INERTIA balances measure inertial mass. EINSTEIN's general theory of RELATIVITY is based on the principle of EQUIVALENCE: that the inertial mass and the gravitational mass of an object are equivalent. *See also* WEIGHT

mass action, law of Principle that a chemical reaction rate is proportional to the product of the concentrations of each reactant.

mass defect Difference in MASS between the total rest mass of the PROTONS and NEUTRONS, when free, from which a particular NUCLEUS is formed and the slightly lower mass of the nucleus itself. The mass defect is equivalent to the energy needed to separate the particles, and to indicate the stability of the nucleus. *See also* BINDING ENERGY; RELATIVITY

mass number (nucleon number, symbol A) Total number of NUCLEONS (NEUTRONS and PROTONS) in the NUCLEUS of an ATOM. It is usually shown as a superscript number before the chemical symbol of an element. Thus, the lightest element, hydrogen, has only one proton in its nucleus, and its nuclear notation is written ^1H. Heavier elements have both protons and neutrons in their nuclei: one ISOTOPE of uranium has a mass number of 238, meaning each of its nuclei contains 238 nucleons (protons plus neutrons), and is written ^{238}U. An atom also has an ATOMIC NUMBER representing the number of protons only. This is written as a subscript; for example $_6\text{C}$.

mass production Manufacture of goods in large quantities by standardizing parts, techniques, and machinery. In 1798 US inventor Eli WHITNEY introduced mass production when he was given the contract to make 10,000 guns for the army. Whitney built machinery to make identical parts for all the guns, so that all the components could be produced quickly and were guaranteed to fit. It also meant that damaged parts could be replaced with standard spares. The assembly line, with a conveyor belt carrying the work through a series of assembly areas, is another feature of some mass-production processes. In 1912 Henry FORD installed this system in Detroit, Michigan, for the mass production of his Model T car. Many modern mass-production processes depend on computer-controlled machines.

mass spectrograph (mass spectrometer) Instrument for separating ions according to their masses (or more precisely, according to their charge-to-mass ratio). In the simplest types, the ions are first accelerated by an electric field and then deflected by a strong magnetic field; the lighter the ions, the greater the deflection. The mass spectrograph detects the lines of the resulting mass spectrum on a photographic plate; the mass spectrometer does this electrically. By varying the field, ions of different masses can be focused in sequence on to the photographic plate or detector and a record of charge-to-mass ratios is obtained.

mastectomy In surgery, removal of all or part of the breast. It is performed as a treatment for breast cancer. Simple mastectomy involves the breast alone; if the cancer has spread, radical mastectomy may be undertaken, removing also the lymphatic tissue from the armpit.

Masters and Johnson US physician William Howell Masters (1915–) and his partner and ex-wife, the psychologist Virginia Johnson (1925–), became noted for laboratory studies of the physiology and anatomy of human sexual activity. They also conducted clinical marriage counseling for people with sexual problems. Their works include *Human Sexual Response* (1966) and *Human Sexual Inadequacy* (1970).

mastoid (mastoid bone) Raised region behind the ear, containing air cells that communicate with the middle ear. In mastoiditis, the region becomes infected. This once-common disorder was frequently treated by surgery, but today is usually cured quickly with antibiotics.

mathematical biology Modern branch of applied MATHEMATICS and theoretical BIOLOGY dealing with the growth of organisms, the development of patterns and the spread of populations and disease. Also, recent advances have been made in medicine where mathematics has been used to describe wound healing, the events leading to heart attacks and the growth of tumors within the body.

mathematical induction Method of proving that a mathematical statement is true for any positive integer n by proving (1) that it is true for a base value, for example, 1, and (2) that if it is true for a value k then it is also true for $k + 1$. If (1) and (2) hold, then it follows in a finite number of steps that the statement is true for any positive integer n. For instance, the formula for the sum of the first n natural numbers is a result that can be proved using mathematical induction. The technique is a powerful way of proving that a result holds for an infinite number of values.

mathematical model Set of equations and concepts used to describe and predict a given phenomenon or behavior. Mathematical models can be of practical or theoretical use (or both). Practically, mathematical models can be used to design new materials, to predict the weather, and to test aircraft, bridges, and other structures before they are built. Theoretically, mathematics is the language used to describe the properties of matter in PHYSICS and CHEMISTRY and, more recently, the formation of patterns and the behavior of animals in MATHEMATICAL BIOLOGY.

mathematics Study concerned with the properties of numbers, space, and shape and with deductions made from assumptions about abstract entities. Mathematics is often divided into **pure** mathematics, which involves purely abstract reasoning based on axioms, and **applied** mathematics, which involves the use of mathematics to model processes in engineering, physics, chemistry, biology, and economics. The distinction between the two branches comes from the fact that the former does not require any observation of, or data from, the natural world. However, the two fields cannot be separated because results from pure mathematics are often useful in practical applications, and studying real problems often stimulates pure mathematics research. The main divisions of pure mathematics are GEOMETRY, ALGEBRA, and ANALYSIS. *See also* ARITHMETIC; TRIGONOMETRY

matrix (pl. matrices) Rectangular array of numbers in rows and columns. The number of rows need not equal the number of columns. Matrices can be added, subtracted, and multiplied according to certain rules. They are useful in the study of transformations of COORDINATE SYSTEMS and in solving sets of SIMULTANEOUS EQUATIONS. *See also* DETERMINANT

matter What things are made of. Ordinary matter is made up of ATOMS, which are combinations of ELECTRONS, PROTONS, and NEUTRONS. Atoms, in turn, make up ELEMENTS, an ordered series of substances having from one proton in their nuclei (hydrogen) to a hundred or more. Many other subatomic or ELEMENTARY PARTICLES can be produced at high energies and live for short periods of time. Elements other than hydrogen and helium (such as carbon, oxygen, and iron) were originally built up by nuclear FUSION in stars, including the Sun. All matter exerts an attractive force on other matter, called GRAVITATION. Charged particles exert an attractive or repulsive ELECTROMAGNETIC FORCE that accounts for nearly all everyday phenomena – the sense of touch, for instance, is dependent on the repulsion of MOLECULES at close range. The STRONG NUCLEAR FORCE is responsible for binding the protons and neutrons in an atomic NUCLEUS, and the WEAK NUCLEAR FORCE is responsible for beta decay. *See also* ANTIMATTER; MATTER, STATES OF

matter, states of Classification of matter according to its structural characteristics. Four states of matter are generally recognized: SOLID, LIQUID, GAS, and PLASMA. Any one ELEMENT or compound may exist sequentially or simultaneously in two or more of these states: water, ice, and water vapor can all exist at one temperature and pressure. **Solids** may be crystalline (have a regularly repeated molecular structure), as in salt and metals; or amorphous, as in tar or glass. **Liquids** have molecules that can flow past one another, but which remain almost as close as in a solid. In a **gas**, molecules are so far from one another that they travel in relatively straight lines until they collide with the walls of their container. In a **plasma**, atoms are torn apart into electrons and nuclei by high temperatures, such as those in stars.

Mauchly, John William (1907–80) US physicist and engineer who, with John Presper Eckert, Jr., (1919–), invented and built (1946) the first commercial general-purpose COMPUTER – the Electronic Numerical Integrator and Computer (ENIAC). Taking up 15,000sq ft, ENIAC was bought as a military calculator by the US War Department. The two men, who met in 1943 on the faculty of the University of Pennsylvania, next built the Electronic Discrete Variable Computer (EDVAC), whose ability to store PROGRAMS helped to launch the computer revolution, and then in 1951 the Universal Automatic Computer (UNIVAC) for the US Census Bureau. In 1948 they formed the Eckert-Mauchly Computer Corporation that was sold two years later to Remington Rand.

Maudslay, Henry (1771–1831) British engineer who became a leading builder of MARINE ENGINES. Maudslay made many innovations in engineering practice, including the production of accurate plane surfaces, the use of a LATHE slide rest, an accurate bench MICROMETER and a practical screw-cutting lathe.

Maury, Matthew Fontaine (1806–73) US naval officer, a proponent of OCEANOGRAPHY. As superintendent of the US Naval Observatory and Hydrographic Office (1842–61), he produced charts showing winds and currents for the Atlantic, Pacific, and Indian oceans. Maury proved the feasibility of laying a transatlantic cable, and wrote *The Physical Geography of the Sea* (1855). *See also* HYDROGRAPHY

maxilla Upper jawbone of a vertebrate, or one of a pair of mouthparts in various arthropods. Centipedes, crustaceans, insects and millipedes have maxillae that move sideways to manipulate food. They lie close to the mouth, behind the MANDIBLES. In insects, a second pair of maxillae are united to form the LABIUM; in crustaceans, the two pairs of maxillae are separate. *See also* JAW

Maxim, Sir Hiram Stevens (1840–1916) US inventor of the Maxim gun. A mechanical genius, he took out his first patent at the age of 26 and invented numerous devices before settling in London in 1881. In 1883 Maxim built the first fully automatic machine gun, the Maxim gun. It was the first machine gun to use the recoil energy of the fired bullet to eject successfully the shell and to place another cartridge in the firing chamber. *See also* FIREARM

Maxwell, James Clerk (1831–79) Scottish mathematician and physicist who did outstanding theoretical work revealing the existence of ELECTROMAGNETIC RADIATION. In 1871 Maxwell became the first Cavendish Professor of Experimental Physics at Cambridge University, England. He used the theory of the electromagnetic field to derive what are now called **Maxwell's equations**. These predict the existence of electromagnetic waves. Maxwell theoretically established (correctly) the nature of Saturn's rings, and did further work in thermodynamics and statistical mechanics. A unit of magnetic flux, the maxwell, was named after him – it has been replaced by the SI unit, the WEBER.

Maybach, Wilhelm (1846–1929) German engineer. Maybach worked with Gottlieb DAIMLER in the design and manufacture of INTERNAL COMBUSTION ENGINES and automobiles, for which he designed many improvements in fuel injection, gears, and so on. He opened his own factory in 1909, building engines for airships and, later, Maybach Cars.

Mayer, Julius Robert von (1814–78) German physicist and physician who determined the mechanical equivalent of heat (credited to James JOULE), and propounded a statement of the law of conservation of energy (credited to Hermann HELMHOLTZ). Lack of recognition led Mayer to attempt suicide and he was committed to an asylum.

Mayer, Maria Goeppert (1906–72) US physicist, b. Germany. In 1949 she proposed that protons and neutrons in an atomic NUCLEUS are arranged in orbits or shells, much as electrons are arranged around the nucleus. This provided an explanation of the apparently anomalous properties of some nuclei. Mayer shared the 1963 Nobel Prize for physics with Johannes H.D. Jensen (1907–73) and Eugene P. WIGNER.

ME (abbreviation of myalgic encephalomyelitis) Condition defined as extreme fatigue that persists for six months or more and is not relieved by rest. Also known as chronic fatigue syndrome and postviral fatigue syndrome, it may be accompanied by many other nonspecific symptoms, such as fever, headache, muscle and joint pains, gastric upset, dizziness, and mood changes. Often occurring after a VIRUS infection, it ranges in severity from chronic weariness to total physical collapse. The cause of the condition is unknown. There is no specific treatment, and recovery may take months or even years.

mean In statistics, mathematical AVERAGE of a set of values representing some data or observations. The mean of *n* such quantities is calculated as their sum divided by *n*. For instance, the mean of the numbers 4, 8, 9, 12, and 17 is 10.

meander Naturally occurring looplike bend in the course of a river. A river will wind round any obstacle, such as hard rock. Once a meander has been created, it will become accentuated by the erosive action of the river. On the outside of a bend there will be lateral CORRASION, which will gradually work out sideways. On the inside of the bend there is likely to be some deposition, which will build up a flat flood plain. Meanders gradually move downstream. Sometimes meanders make complete loops, which, when cut off, form OXBOW LAKES. The name comes from the winding River Maeander in Turkey.

mean free path Average distance a molecule moves between successive collisions. The concept is most widely used in the KINETIC THEORY of gases, for which the mean free path is usually about 10^{-7}m at normal temperature and pressure.

mean sea level Mean height of the ocean calculated from measurements at regular intervals at points on the open coast. It is used as a standard or fixed geodetic point. *See also* GEODESY

measles (rubeola) Extremely infectious viral disease of children. The symptoms (fever, nasal discharge, skin rash, and spots inside the mouth) appear about two weeks after exposure. Hypersensitivity to light is characteristic. Complications such as pneumonia and encephalitis occasionally occur, and middle-ear infection is a hazard of the disease, which was once a major cause of infant mortality in Western countries. Vaccination produces life-long IMMUNITY to measles.

measurement *See* WEIGHTS AND MEASURES

mechanical advantage *See* FORCE RATIO

mechanical engineering Field of ENGINEERING concerned with the design, construction, and operation of machinery. Mechanical engineers work in many branches of industry, including transportation, power generation, and tool manufacture. Achievements in mechanical engineering include the development of wind and water TURBINES; STEAM ENGINES and INTERNAL COMBUSTION ENGINES; assembly line production; and the control of machines using HYDRAULICS, FLUIDICS, and COMPUTERS.

mechanics Branch of physics concerned with the behavior of MATTER under the influence of FORCES. It may be divided into solid mechanics and fluid mechanics. Another classification is as statics – the study of matter at rest – and DYNAMICS – the study of matter in motion. In **statics**, the forces on an object are balanced and the object is said to be in equilibrium; static equilibrium may be stable, unstable or neutral. **Dynamics** may be further divided into kinematics – the description of motion without regard to cause – and KINETICS – the study of motion and force. Classical dynamics rests primarily on the three laws of MOTION formulated by

Isaac NEWTON. Modern physics has shown these laws to be special cases approximating to more general laws. Relativistic mechanics deals with the behavior of matter at high speeds, approaching that of light, whereas QUANTUM MECHANICS deals with the behavior of matter at the level of atoms and molecules. *See also* QUANTUM THEORY; RELATIVITY

Medawar, Sir Peter Brian (1915–87) British biologist. He shared the 1960 Nobel Prize for physiology or medicine with Sir Frank Macfarlane Burnet (1899–1985) for the discovery of acquired immune tolerance. Medawar confirmed Burnet's theory that an organism can acquire the ability to recognize foreign tissue during embryonic development, and if that tissue is introduced in the embryonic stage it may be reintroduced later without inducing an immune reaction. *See also* IMMUNE SYSTEM

median In statistics, the middle item in a group found by ranking the items from smallest to largest. In the series 2, 3, 7, 9, 10, for example, the median is 7. With an even number of items, the MEAN of the two middle items is taken as the median. Thus, in the series 2, 3, 7, 9, the median is 5.

medicine Practice of the prevention, diagnosis, and treatment of disease or injury; the term is also applied to any agent used in the treatment of disease. Pictographs dating from the earliest times show medical procedures, and we can deduce from ancient skeletons that the practice of medicine goes back to prehistoric humans. The Egyptians included magico-religious elements along with empirical therapies for many ailments, and the Hebrews placed a marked emphasis on hygiene. The Greeks had become thoroughly secular in their approach to medicine by the 6th century BC, and several medical schools had been built by that time. The Romans and then the Arabs carried on and developed further the Greek medical tradition. During the Renaissance this knowledge was brought back to Europe. The dawn of modern medicine coincided with accurate anatomical and physiological observations first made in the 17th century. By the 19th century practical diagnostic criteria had been developed for many diseases; BACTERIA had been discovered and research undertaken for the production of immunizing sera in attempts to eradicate disease. The great developments of the 20th century include CHEMOTHERAPY, new surgical procedures, including organ transplants, and sophisticated diagnostic devices such as radioactive TRACERS and various scanners. *See also* ANATOMY; PHYSIOLOGY

medulla oblongata Principal structure of the hindbrain, joining the BRAIN to the SPINAL CORD. Medulla functions are necessary to maintain life: it controls respiration and heartbeat. The medulla oblongata also contains part of the reticular formation that plays a role in arousal states, such as consciousness, wakefulness, and attention. General anesthetics, such as chloroform, probably work by depressing medulla activity.

medusa In zoology, free-swimming stage in the lifecycle of CNIDARIA. There are many types of medusa. Some are small creatures, transient stages in the life cycle of the Cnidaria, which soon change into the dominant, POLYP stage. Other medusas are larger and live longer, the polyp stage being the briefer of the two. Yet other medusas, including many jellyfish, have no polyp stage, and when they reproduce give rise directly to medusa offspring.

meerschaum (sepiolite) Clay silicate mineral, hydrated magnesium silicate, found mainly in Asia Minor. It is opaque white in color, and has a fibrous texture. Hardness 2–2.5; r.d. 2.

megabyte (symbol Mb) In computing, 1,024 kilobytes. Often used for expressing the capacity of computer memory or storage devices, the megabyte is sometimes rounded to stand for 1 million BYTES.

meiosis In biology, process of cell division that reduces the CHROMOSOME number from DIPLOID to HAPLOID. Meiosis involves two nuclear divisions. The first meiotic division halves the chromosome number in the two resulting cells; the second division then forms four haploid cells. In most organisms, the resulting haploid cells are the GAMETES, or sex cells, the OVA and SPERM. However, in others, which show ALTERNATION OF GENERATIONS (with two distinct generations in the life cycle, one haploid and the other diploid), these haploid cells may give rise to a new generation of plants or algae, the haploid "gametophyte" generation, that will later produce gametes by MITOSIS (normal cell division that does not involve halving of the chromosome number). *See also* ANAPHASE; CROSSING OVER; METAPHASE

Meitner, Lise (1878–1968) Austrian physicist. Working with Otto HAHN, Meitner discovered PROTACTINIUM and the fission of URANIUM, and she investigated beta decay and nuclear isomerism. She did research with Fritz STRASSMANN on the products resulting from neutron bombardment of uranium. Although her work contributed to the development of the atom bomb, Meitner refused to assist the project. She also researched the radioactive disintegration products of ACTINIUM, THORIUM, and RADIUM. In 1938 Meitner fled the Nazi regime, working in Sweden until 1960, when she retired to England.

melamine ($C_3N_6H_6$) Organic chemical compound that reacts with formaldehyde (HCHO) to form a resin POLYMER, one of the thermosetting PLASTICS. Melamine resins are used as adhesives, for impregnating paper, for treating wool against shrinkage, and as constructional plastics.

melanin Dark pigment found in the skin, hair, and parts of the eye. The amount of melanin determines skin color. Absence of melanin results in an ALBINO.

melanocyte-stimulating hormone (MSH) Hormone, produced by the pituitary gland, that stimulates certain cells in the skin to synthesize MELANIN.

melanoma Type of skin CANCER. Melanomas are highly malignant tumors formed by melanocytes, cells in the skin that make the dark pigment MELANIN. They may also occur in mucous membranes and in the eye. Untreated, they may spread to the liver and lymph nodes. Excessive exposure to sunlight has been cited as a causative factor. Treatment is usually by surgery.

Melloni, Macedonio (1798–1854) Italian physicist who studied INFRARED radiation. Melloni improved the thermopile and used it to detect infrared radiation. He also showed that rock salt is transparent to infrared, and by 1850 had demonstrated that infrared undergoes reflection, refraction, polarization, and interference in the same way as does visible light. Melloni laid the foundation for James Clerk MAXWELL's hypothesis of a spectrum of ELECTROMAGNETIC RADIATION beyond the visible region.

meltdown Out-of-control heating of a NUCLEAR REACTOR resulting in melting of the reactor core. The temperature of a reactor core is constantly monitored, and any dangerous rise in temperature automatically initiates a shutdown procedure. Other precautions against the possible disastrous results of a meltdown include surrounding the reactor core with a containment vessel of thick concrete, and a system of pipes that flood the core in such an emergency. Two known partial meltdowns have occurred: in 1979 at Three Mile Island, Pennsylvania, and in 1986 at Chernobyl, Ukraine.

melting point Temperature at which a substance changes from SOLID to LIQUID. The melting point of the solid has the same value as the freezing point of the liquid; for example, the melting point of ice, 0°C (32°F), is the same as the freezing point of water.

meltwater Water originating from melted ice or snow. When temperatures rise, snow produces meltwater, which swells rivers and may contribute to flooding. Meltwater from glaciers may add to the water in an existing river or supply streams that form a new river.

membrane In biology, boundary layer inside or around a living CELL or TISSUE. Cell membranes include the plasma membrane surrounding the cell, the network of membranes inside the cell (ENDO-PLASMIC RETICULUM), and the double membrane surrounding the NUCLEUS. The multicellular membranes of the body comprise mucous membranes of the respiratory, digestive, and urogenital passages, synovial membranes of the joints, and the membranes that coat the inner walls of the abdomen, thorax, and the surfaces of organs. *See also* EPITHELIUM

Mendel, Gregor Johann (1822–84) Austrian naturalist who discovered the laws of HEREDITY and in so doing laid the foundation for the modern science of GENETICS. Mendel grew pea plants and observed how various characteristics, such as tallness, flower color, and seed shape, appeared in their offspring. He proposed that the characteristics were inherited, carried by "factors" (now known to be genes) contributed by each parent. His study was published in *Experiments with Plant Hybrids* (1866), which was rediscovered in 1900.

mendelevium (symbol Md) Radioactive, metallic TRANSURANIC ELEMENT. It was first synthesized in 1955 at the University of California at Berkeley (one atom at a time) by the alpha-particle bombardment of einsteinium-253. Properties: at.no. 101; r.a.m. 258; most stable isotope ^{258}Md (half-life 2 months).

Mendeleyev, Dmitri Ivanovich (1834–1907) Russian chemist who devised the PERIODIC TABLE. Mendeleyev demonstrated that chemically similar elements appear at regular intervals if the elements are arranged in order by RELATIVE ATOMIC MASS (R.A.M.). He classified the 60 known elements and left gaps in the table, predicting the existence and properties of several unknown elements later discovered, including GALLIUM and SCANDIUM. His work was published in Russian in 1869, and was soon translated. His textbook, *The Principles of Chemistry*, was written between 1868 and 1870 and translated in 1905, rapidly becoming a standard work. The element MENDELEVIUM is named after him.

Mendelism Theory of inheritance originally proposed by Gregor MENDEL. It established the science of classical GENETICS.

meninges Three membranes that cover the BRAIN and SPINAL CORD. The outermost membrane, the **dura mater**, is a protective covering. Within it is the second, the **arachnoid**; the innermost membrane, the **pia mater**, is a delicate layer on the surface of the brain and spinal cord containing blood vessels that supply the nervous system with nourishment. Between the arachnoid and the pia mater is the subarachnoid space, in which flows the cerebrospinal fluid.

meningitis Inflammation of the MENINGES (membranes) covering the brain and spinal cord, resulting from infection. Bacterial meningitis is more serious than the viral form. Symptoms include headache, fever, nausea, and stiffness of the neck. The disease can be fatal.

meniscus Curved upper surface of a liquid in a container, due to surface tension. The surface of water in an air-water-glass system is concave, whereas in an air-mercury-glass system the mercury meniscus is convex.

menisectomy Surgical removal of damaged cartilage from the knee joint.

menopause Stage in a woman's life marking the end of the reproductive years, when the MENSTRU-AL CYCLE becomes irregular and finally ceases, generally around the age of 50. Popularly known as the "change of life," it may be accompanied by unpleasant effects such as hot flushes, excessive bleeding, and emotional upset. HORMONE REPLACE-MENT THERAPY (HRT) is designed to relieve menopausal symptoms.

Mensa (table) Constellation of the far S sky. Although it has no star above magnitude 5, part of the Large MEGALLANIC CLOUD extends into it.

menstrual cycle Cycle in female humans and some higher primates of reproductive age, during which the body is prepared for pregnancy. In women, the average cycle is 28 days. At the beginning of the cycle, HORMONES from the PITUITARY GLAND stimulate the growth of an ovum (egg cell) contained in a follicle in one of the two OVARIES. At approximately mid-cycle, the follicle bursts, the egg is released (ovulation) and travels down the FALLOPIAN TUBE to the UTERUS. The follicle (now called the CORPUS LUTEUM) secretes two hormones, PROGESTERONE, and ESTROGEN, during this secretory phase of the cycle, and the ENDOMETRIUM thickens, ready to receive the fertilized egg. Should fertilization (conception) not occur, the corpus luteum degenerates, hormone secretion ceases, the endometrium breaks down and menstruation occurs; the unfertilized egg is discharged in the blood flow from the VAGINA. In the event of conception, the corpus luteum remains and maintains the endometrium with hormones until the PLACENTA is formed. Menstruation, therefore, marks the end of the cycle. It is customary, for the purposes of CONTRACEPTION and calculating the date of an expected birth, to count the first day of bleeding as the first day of the next menstrual cycle. The onset of the menstrual cycle (menarche) is called PUBERTY (age 10–15 years); it ceases with the MENOPAUSE (around 50 years). *See also* FOLLICLE STIMULATING HORMONE (FSH); LUTEINIZING HORMONE (LH)

menthol ($C_{10}H_{19}OH$) White, waxy crystalline compound having a strong odor of peppermint. Its main source is oil of peppermint, *Mentha arvensis*. It IS an ingredient of decongestant ointments and nasal sprays and is used to flavor toothpaste and cigarettes.

Mercalli scale, modified Scale of 12 points used for measuring earthquake intensity. Named after the Italian seismologist Giuseppe Mercalli (1850–1914), it is based on damage done at any point and so varies from place to place. Earthquake magnitude, on the other hand, is a function of the total energy released. *See* RICHTER SCALE

mercaptan *See* THIOL

Mercator, Gerardus (1512–94) Flemish cartographer. His huge world map of 1569 employed the system of projection now named after him, in which lines of longitude, as well as latitude, appear as straight, parallel lines. His scholarship was matched by his practical abilities, and he was respected as the greatest geographer of his time, patronized by Emperor Charles V (1519–56). Mercator also produced a globe (1541) and a huge, unfinished, atlas of the world (1595).

Mercury Innermost planet of the Solar System, and the smallest of the four inner planets. Mercury is difficult to observe because it is very close to the Sun, but through a telescope light and dark areas can be made out. The planet has no known satellite. Very little was known about Mercury's surface until the Mariner 10 probe made three close approaches to the planet in 1974 and 1975 and returned pictures of nearly half the surface. These showed a heavily cratered, lunarlike world, marked by many craters, valleys, and ridges. The surface temperature at noon on the equator can reach over 800K at the perihelion, while on the night side it can fall to 90K. In 1991–92 radar mapping of Mercury's polar regions revealed what may be water-ice on the floors of craters permanently in shadow, where the temperature is estimated never to rise above 112K. There is a very tenuous atmosphere, mainly of helium and sodium. The presence of a weak magnetic field and the planet's high density for its size suggest that Mercury has a very large iron-rich core, proportionally much bigger than for any other Solar System body. **Data:** Diameter: 4,878km; Mass: (Earth=1) 0.0553; Volume: (Earth=1) 0.0562; Density (water=1): 5.43; Orbital period: 89.97 days; Rotation period: 58d 15h 30m; Surface temperature: 100–800K

mercury (quicksilver, symbol Hg) Liquid, metallic element, known from earliest times. The chief ore is cinnabar (a sulfide), from which it is extract-ed by roasting. The silvery element is the only metal that is liquid at normal temperatures. It is a dangerous cumulative poison. Mercury is used in barometers, thermometers, laboratory apparatus, mercury-vapor lamps, and mercury cells. It forms two series of salts, termed mecury(I), or mercurous, and mercury(II), or mercuric. Mercury compounds are used in pharmaceuticals and some dentists still fill teeth with a mercury alloy (AMALGAM), although increasingly composite resins are used. Mercury's toxic effects have always plagued workers in industries using its compounds. The industrial dumping of mercury wastes into the sea allows it to enter the human food cycle, because fish and birds ingest the mercury and these are eaten, passing on the poison to the next level in the food chain. Symptoms of mercury poisoning include loss of coordination, balance, and peripheral vision, sensory disturbance, and deformed offspring. Properties: at.no. 80; r.a.m. 200.59: r.d. 13.6; m.p. $-38.87°C$; ($-37.97°F$); b.p. 356.58°C ($673.84°F$); most common isotope ^{202}Hg (29.8%).

mercury switch In electrical engineering, switch in which mercury is contained in a tube that when tilted causes the mercury to bridge two contacts. An alternative design consists of mercury cups into which movable contacts dip.

Mergenthaler, Ottmar (1854–99) US inventor, b. Germany. In 1884 Mergenthaler patented his LINOTYPE machine for setting solid lines of type, once used by most newspapers. In 1885 the machine was improved by a device that automatically justified the type. The system has been superseded by PHOTOSETTING.

meridian Circle that runs through the N and S poles, at right angles to the equator. *See also* CELESTIAL MERIDIAN; LONGITUDE

meristem In plants, a layer of cells that divides repeatedly to generate new tissues. It is present at the growing tips of shoots and roots, and at certain sites in leaves. In MONOCOTYLEDONS, the leaf meristem is at the base, explaining why grasses continue to grow when the leaf tips are removed by grazing or mowing. *See also* CAMBIUM

mesa Flat area of upland rather like a table, so called after the Spanish mesa, meaning "table." They often have horizontal strata of SEDIMENTARY ROCK. They are found primarily in the American Southwest; well-known examples are the Mesa Verde in Colorado and the Enchanted Mesa in New Mexico.

mesentery Sheet of tissue, or double layer of the PERITONEUM, which attaches the abdominal organs to the rear wall of the abdomen.

mesoderm One of the three so-called GERM LAY-ERS of tissue formed in the early development of a fertilized OVUM (egg) of higher animals. It is the middle layer, and in later development of the embryo, the mesoderm gives rise to muscles, blood, and connective tissues. Other germ layers are the ECTODERM and ENDODERM.

meson ELEMENTARY PARTICLE, member of a sub-group of HADRONS, all of which have either zero or integral SPIN. They include the pions, kaons, and eta mesons. There is no restriction on the number of mesons produced or destroyed in a nuclear reaction or present in a particular energy state.

mesophyll Soft tissue located between the two layers of EPIDERMIS in a plant leaf. In most plants, mesophyll cells contain chlorophyll-producing structures called CHLOROPLASTS, which are essential to PHOTOSYNTHESIS.

mesophyte Plant that grows under average moisture conditions, thriving where there is a good balance of water and evaporation. Such plants have well-developed root and leaf systems. *See also* HYDROPHYTE; XEROPHYTE

mesosphere Middle shell of gases in the Earth's ATMOSPHERE between the STRATOSPHERE and the THERMOSPHERE.

Mesozoic Third era of geologic time, extending from about 248 million to 65 million years ago. It is divided into three periods: TRIASSIC, JURASSIC, and CRETACEOUS. For most of the era, the continents are believed to have been conjoined into one huge land mass called PANGAEA. There was much volcanic activity and mountain building throughout the period, which was also characterized by the variety and size of its reptiles. For this reason it is sometimes called the "Age of Reptiles."

messenger RNA (mRNA) Type of RNA (ribonucleic acid) that carries the GENETIC CODE for PROTEIN SYNTHESIS. The mRNA transcribes the code from DNA and carries it to RIBOSOMES within the cell, where AMINO ACIDS are assembled to make polypeptides and proteins.

Messier catalogue List of 109 STAR CLUSTERS, NEBULAE, and GALAXIES compiled by the French astronomer Charles Messier (1730–1817). Messier drew up the list so that he and other comet hunters would not confuse these permanent, fuzzy-looking objects with COMETS. The first edition of the catalogue was published in 1774, with supplements in 1780 and 1781. Objects in the catalogue are given the prefix M, and are still widely known by their Messier numbers.

metabolism Chemical and physical processes and changes continuously occurring in a living organism. They include the breakdown of organic matter (CATABOLISM), resulting in energy release, and the synthesis of organic components (ANABOLISM) to store energy and build and repair TISSUES. *See also* BASAL METABOLIC RATE (BMR)

metabolite Any chemical substance that is involved in the metabolic processes of cells in organisms. These substances function in the various biological energy exchanges necessary for growth, maintenance, and reproduction. *See also* METABOLISM

metacarpals In humans and other primates, bones of the palm of the hand. They articulate with the CARPALS (wrist bones) and PHALANGES (finger bones). In four-footed animals, they are the bones of the forefoot.

metal Any element that is a good conductor of heat and electricity – the atoms of which are bonded together within crystals in a unique way. Mixtures of such elements (ALLOYS) are also metals. About three-quarters of the elements are metals. Most are hard, shiny materials that form oxides (combined with oxygen). Many corrode if exposed to moist air. MERCURY is exceptional in being a liquid at room temperature. SODIUM and POTASSIUM are examples of metals that are soft and chemically very reactive: they tarnish quickly in air and are most familiar as their salts. The lightest metal is LITHIUM, which is also very reactive. The heaviest is OSMIUM, which is 22.6 times denser than water and is one of the PLATINUM group, relatively unreactive metals that include RUTHENIUM, RHODIUM, PALLADIUM, and IRIDIUM. Malleability and ductility are further metallic characteristics. GOLD is the most malleable of all metals; it can be beaten so thin as to be virtually translucent. Some metals have very high melting points and various high-temperature applications: TUNGSTEN, with the highest melting point of all at 3,410°C (6,170°F), is employed as LIGHTBULB filaments. ALUMINUM, followed by IRON, are the two most abundant and useful of metals. TITANIUM, although rarely seen as the metal, is more commonly distributed than the more familiar COPPER, ZINC, and LEAD. Other metals of great economic importance, because of their radioactivity, are URANIUM and PLUTONIUM. *See also* ALKALI METALS; ALKALINE-EARTH METALS; METALLIC BOND; TRANSITION ELEMENTS

metal fatigue Progressive fracture of metals subjected to repeated cycles of STRESS. Fatigue can occur early in the metal's life, with stresses building up as it cools after the manufacturing process. The phenomenon is of particular relevance to aeronautical engineers because metals can break, tear, or otherwise deform permanently under repeated or reversed loads at stress levels much lower than for a single loading.

metallic bond In chemistry, a bond that holds atoms together in a METAL. Inside the crystals of metals, positive IONS are held by the electrostatic attraction of a cloud or "sea" of surrounding electrons that can, under various influences, move. Such electron movement under the influence of an applied voltage constitutes an electric current, so accounting for the electrical CONDUCTIVITY of metals.

metallography Study of the structure of metals and alloys, using optical and electronic microscopes and X-ray diffraction techniques. Examination by microscope reveals the size and shape of crystals and the distribution of nonmetallic inclusions. An X-RAY DIFFRACTOMETER is used to study the arrangement of the atoms in a metallic specimen.

metalloid (semimetal) Element having some properties typical of METALS, and some normally associated with NONMETALS. In moving from left to right across the PERIODIC TABLE and moving down the groups, there is a transition from metallic to nonmetallic elements. Metalloids occur as borderline cases in groups III–VI. Examples are SILICON, GERMANIUM and ARSENIC. Some metalloids are SEMICONDUCTORS.

metallurgy Science and technology concerned with METALS. Metallurgy includes the study of:

methods of the extraction of metals from their ORES; physical and chemical properties of metals; ALLOY production; and the hardening, strengthening, corrosion-proofing, and ELECTROPLATING of metals. Analyses by chemical methods and SPECTROSCOPY are made to determine the composition of alloys during manufacture. Among the physical tests for metals is the measurement of resistance to impact and tension. Works processes include melting, SINTERING, STAMPING, FORGING, dipping and ELECTROPLATING, GALVANIZING, ANODIZING, and SHERARDIZING.

metamorphic rock Broad class of rocks that have been changed by heat or pressure, or both heat and pressure, from their original nature – SEDIMENTARY, IGNEOUS, or older metamorphic. Heat is the result of volcanic activity, and pressure is the result of earth movements. Metamorphic rocks are generally hard and resistant to erosion, and are likely to form high ground. In metamorphic rocks, minerals may recrystallize or be compressed, and the new rock may look completely different from the original sedimentary or igneous rock. Thus, the metamorphic rock SLATE is made from sedimentary SHALE, the metamorphic GNEISS from igneous GRANITE.

metamorphosis Change of form or structure during the development of various organisms, such as the changing of a caterpillar into a moth, or a tadpole into a frog. Sometimes the change is gradual, as with a grasshopper, and is known as incomplete metamorphosis. Complete metamorphosis involves a change in habit or environment

▲ **metamorphosis** When common frogs mate, fertilization and egg laying occur in water (1). Within an hour, the jelly around the egg swells to produce frogspawn (2). The eggs develop (3) and produce embryos (4) that hatch as long-tailed tadpoles with external feathery gills six days after fertilization (5). Mouths and eyes develop later and the tails become powerful means of propulsion. Hind legs are well formed by week eight (6); meanwhile, the tadpole has changed from a herbivore to a carnivore. Via an intermediary gill and lung stage, the tadpole changes from gill- to lung-breathing, its internal lungs growing as its external gills are absorbed; the process is complete when the gills fully disappear at month three, by which time the forelegs are well developed (7). Metamorphosis is complete when the young frog loses its tail.

and usually involves the more distinct stages of LARVA, PUPA, and IMAGO (adult stages).

metaphase Stage in cell division that occurs during MITOSIS and MEIOSIS. During metaphase, the nuclear membrane breaks down, the spindle forms, and CHROMOSOMES become attached to the center of the spindle. In the first metaphase of meiosis, paired chromosomes attach to the spindle. In the second metaphase of meiosis and in mitosis, individual chromosomes attach to the spindle. The division then proceeds to ANAPHASE.

metasomatism Production of mineral deposits by the movement of hot fluids from an igneous body through cracks or pores in the surrounding rock. Veins of lead and tin ores surrounding the Cornish granites were formed in this way.

metastable state Uneasy stability of a system that can easily be disturbed, making it descend to a lower energy state (a coin balanced on its edge is an example). A supersaturated solution or supercooled liquid, which remains liquid below its freezing point, is also metastable because the addition of a speck of dust or a tiny seed crystal will make it rapidly crystallize.

metastasis Spread of malignant TUMORS beyond their original sites.

metatarsals Bones of the feet (hind feet in quadrupeds). They articulate at one end with the TARSALS (ankle bones) and at the other with the PHALANGES (toe bones).

Metchnikoff, Élie (1845–1916) Russian microbiologist. Metchnikoff shared the 1908 Nobel Prize for physiology or medicine with Paul EHRLICH for work on the mechanism of IMMUNITY, which included the discovery that LEUCOCYTES (white blood cells) are important in the body's resistance to infection and disease. He was also noted for his theories concerning longevity. He wrote *Immunity in Infectious Diseases* (1905).

meteor (shooting star) Brief streak of light in the night sky caused by a **meteoroid** (a solid particle, usually the size of a grain of dust) entering the Earth's upper atmosphere at high speed from space. They occur at altitudes of about 100km (60mi). The typical meteor lasts for a few tenths of a second to a second or two, depending on the meteoroid's impact speed, which can vary from about $11-70 \text{km s}^{-1}$ ($7-45 \text{mi s}^{-1}$). A few meteors per hour may be seen on any clear, moonless night at any time of year. But at certain times of the year there are METEOR SHOWERS, which occur when the Earth passes through a meteor stream – dust parti-

cles spread around the orbit of a COMET. Most of the meteors appearing during the year are sporadic meteors, not associated with cometary orbits. *See also* METEORITE

meteorite That part of a large **meteoroid** (a solid particle moving in interplanetary space) that survives passage through the Earth's atmosphere and reaches the ground. Most of a meteoroid burns up in the atmosphere to produce a METEOR, but about 10% reaches the surface in the form of meteorites or micrometeorites. Meteorites generally have a pitted surface and a fused charred crust. There are three main types: IRON METEORITES (siderites); STONY METEORITES (aerolites); and STONY-IRON METEORITES. Some are tiny particles; others weigh up to 200 metric tons. Meteoroids weighing more than 100 metric tons that do not break up are not decelerated as much as lighter bodies, and produce impact craters, known as meteoric craters.

meteorology Study of weather conditions and a branch of the study of CLIMATOLOGY. Meteorologists study and analyse data from a network of weather ships, aircraft, and satellites in order to compile maps showing the state of the high- and low-pressure regions in the Earth's atmosphere. They also anticipate changes in the distribution of the regions and forecast the future weather. Wind strength and direction can be predicted accurately by measuring the differences in air pressure over the surface of the Earth.

meteor shower Appearance of METEORS from the same point in the sky, the **radiant**, at around the same time each year. Nearly all showers are named after the constellation in which their radiant lies. There are a dozen or so major showers and many minor ones. During major showers there is a build-up of activity to a maximum, when 10–100 shower meteors may be visible each hour from any one location. A shower occurs when the Earth passes through a meteor stream. In an **annual** shower, the meteor rates vary little from one year to the next because the meteoroids in the stream are spread evenly around the orbit. With streams in which the meteoroids are bunched together in a swarm, meteor numbers are low except when the Earth intersects the swarm. Such **periodic** showers may produce a meteor storm (*see* LEONIDS).

meter Instrument that measures a particular quantity. For example, a gas meter measures the amount of gas that has flowed in a certain time, and a voltmeter measures the voltage between two points in an electric circuit.

meter (symbol m) SI unit of length. Conceived as being one ten-millionth of the surface distance between the North Pole and the equator, it used to be defined by two marks on a platinum bar kept in Paris. It is now defined as the length of the path traveled by light in a vacuum during 1/299,792,458 of a second. One meter equals 39.3701 inches.

methadone Synthetic opiate drug, the use of which is widespread as an institutionalized solution to HEROIN addiction. Similar to MORPHINE and heroin in chemical structure, it is an addictive drug, although the effects of withdrawal are believed to be milder than with morphine or heroin. *See also* OPIUM

methanal *See* FORMALDEHYDE

methane (CH_4) Colorless, odorless HYDROCARBON, the simplest ALKANE (paraffin). It is the chief constituent of NATURAL GAS, from which it is obtained, and of FIREDAMP. Methane explodes when mixed with oxygen and ignited. It is produced naturally by decomposing organic matter, such as in marshes, which led to its original name, marsh gas. In the air, it contributes to the GREENHOUSE EFFECT and to increases in global temperature. Methane is used in the form of natural gas as a fuel and, in its pure form, as a starting material for the manufacture of many chemicals. Properties: m.p. $-182.5°C$ ($-296.5°F$); b.p. $-164°C$ ($-263.2°F$).

methanoic acid *See* FORMIC ACID

methanol (methyl alcohol, CH_3OH) Colorless, poisonous, flammable liquid that is the simplest of the ALCOHOLS. It is obtained synthetically from carbon monoxide and hydrogen, by the oxidation of natural gas, or by the destructive distillation of wood, giving it the informal name of wood alcohol. It is used as a solvent and to produce rocket fuel, and as a gasoline additive. Properties: m.p. $-93.9°C$ ($-137°F$); b.p. $64.9°C$ ($148.8°F$).

methionine ($CH_3S(CH_2)_2CH(NH_2)COOH$) Soluble, essential AMINO ACID found in proteins. It is unusual in that it contains sulfur. It is needed for growth in babies and for nitrogen equilibrium in adults.

methotrexate Chemical substance (an antimetabolite) that interferes with the synthesis of the vitamin FOLIC ACID, an essential component of cells. It is used to treat malignant TUMORS, LEUKEMIA, rheumatoid ARTHRITIS, and psoriasis.

methyl alcohol *See* METHANOL

methylated spirit Industrial form of ETHANOL (ethyl alcohol). It contains 5% METHANOL (methyl alcohol), which is extremely poisonous, and enough pyridine to give it a foul taste. It is dyed purple and used as a solvent and fuel.

metric system Decimal system of WEIGHTS AND MEASURES based on a unit of length called the METER (m) and a unit of mass called the KILOGRAM (kg). Larger and smaller metric units are related by powers of 10. Devised by the French in 1791, the metric system is used internationally by scientists – SI UNITS are metric measurements. The system is not used much by nonscientists in the US, where the "imperial" system is the norm, but it is used in most other Western countries.

MeV (abbreviation of million ELECTRON VOLTS) Unit of energy used in particle physics to express the energy of elementary particles.

Meyer, Julius Lothar (1830–95) German chemist who worked on the relationship between ATOMIC VOLUME and atomic weight (RELATIVE ATOMIC MASS). He devised the periodic table at the same time as Dmitri MENDELEYEV, with whom he received the Davy Medal in 1882.

mica Group of common rock-forming minerals of the sheet silicate (SiO_4) type, characterized by a platy or flaky appearance. All contain aluminum, potassium and water in the form of OH^- ions; other metals such as iron and magnesium may be present. Micas have perfect basal cleavage; common members are MUSCOVITE and the BIOTITE group, which includes a number of varieties. **Sheet** mica is used for insulation and as a resonant diaphragm in some accoustical devices, and **ground** mica is sometimes pressed into sheets that can also be used as sheet mica. Scrap and ground mica is used in lubricating oil, and in wallpaper, fancy paints, ornamental tiles, roofing, and Christmas-tree snow.

mica schist One of a number of METAMORPHIC ROCKS that have QUARTZ and MICA as their main constituents. Like all SCHISTS, mica schists have the bulk of their mineral components arranged in parallel, giving them a characteristic striped appearance. They often contain other metamorphic minerals such as garnet. Mica shists are found worldwide.

micelle Roughly spherical group of large molecules that come together in a COLLOID. For example, a molecule of detergent or soap has a hydrophilic ("water-loving") polar head and a long hydrophobic ("water-hating") nonpolar tail. When detergents or soaps dissolve in water, the molecules clump together to form micelles with the nonpolar tails at the center, surrounded by a sphere of polar heads linked to water molecules. Bile salt in the intestine make the product of fat digestion form micelles, which are thereby more easily absorbed.

Michel, Hartmut (1947–) German biochemist. He was the first person to reduce a photosynthetic action center (a 4-protein complex) to pure crystalline form. Next Michel, along with Robert Huber (1937–) and Johan Deisenhoffer (1943–), developed a process that used X-ray technology to determine the exact size of these very large molecules. For this work all three were awarded the 1988 Nobel Prize in chemistry.

Michelson, Albert Abraham (1852–1931) US physicist, b. Germany. In 1887 he conducted an experiment with Edward MORLEY to determine the velocity of the Earth through the ETHER, using an INTERFEROMETER of his own design. The result of the MICHELSON-MORLEY EXPERIMENT is now seen as a key piece of evidence for the theory of RELA-TIVITY. Michelson also determined the speed of light and became the first US scientist to win a Nobel Prize, receiving the 1907 award for physics.

Michelson-Morley experiment Notable experiment performed in 1887 by Albert MICHEL-SON and Edward MORLEY to determine the motion of the Earth through the ETHER. That such a motion was not detected discredited the ether theory and led to a crisis in physics that was resolved by EIN-STEIN's theory of RELATIVITY.

Michurin, Ivan Vladimirovich (1855–1935) Russian plant breeder. His unorthodox theories of HEREDITY (Michurinism), which included belief in the theory of ACQUIRED CHARACTERISTICS, were for a time accepted as the official science of GENETICS by the Soviet Union. Through GRAFTING he managed to produce new strains of fruit, but his belief that grafting produces heritable changes has since been discredited.

microbiology Study of microorganisms, their structure, function, and significance. Mainly concerned with single-cell forms, such as VIRUSES, BACTERIA, PROTOZOA, and FUNGI, it has immense applications in medicine and the food industry. Microbiology began in the 17th century with the invention of the microscope, which enabled scholars to view microorganisms for the first time. Pioneers in the field include Robert HOOKE, Anton van LEEUWENHOEK, and Louis PASTEUR.

microchip General term for either an INTEGRATED CIRCUIT or a SILICON CHIP

microclimate *See* CLIMATE

microcline ($KAlSi_3O_8$) FELDSPAR mineral with a triclinic CRYSTAL STRUCTURE. It has a characteristic checkered twinning, seen through a microscope. It is common in granites. Hardness 6–6.5; r.d. 2.56.

microcomputer Small COMPUTER with its CEN-TRAL PROCESSING UNIT (CPU) on an INTEGRATED CIRCUIT (SILICON CHIP) called a MICROPROCESSOR. The term "microcomputer" is often shortened to **micro**, and it generally has the same meaning as "personal computer" (PC). It usually refers to a desktop computer or to a portable (LAPTOP) computer that is used by only one person at a time. *See also* COMPUTER NETWORK

microelectronics Electronic systems designed and produced without wiring or other bulky components. They allow a high packing density that greatly reduces the size of component assemblies. In the years following World War II, the application of such newly developed devices as the semiconductor TRANSISTOR and DIODE saw the beginnings of the microelectronics industry. This accelerated with the development of the PRINTED CIRCUIT, in which metal connections between miniaturized components are etched out in a single piece on a circuit board. Even further reduction in size, or **micro-miniaturization**, was later achieved with INTE-GRATED CIRCUITS, in which complex circuits can be made smaller than a fingernail. Molecular electronics is a new development that promises to be the ultimate in size reduction, in which complex circuits of microscopic size are grown inside single crystals. The applications of microelectronics have been felt most in the COMPUTER industry, but pocket calculators, miniaturized radios, and miniaturized television sets are evidence of the importance of microelectronics to consumers. *See also* ELEC-TRONICS; NANOTECHNOLOGY

micrometer In astronomy, an instrument that provides an accurate measure of the apparent distance between stars. A **filar** micrometer can be fitted to the eyepiece of a TELESCOPE.

micrometer (symbol μm) Unit of length equal to one millionth of a meter, formerly called the **micron**. In modern scientific usage, the micrometer has been largely replaced for small measurements by the nanometer (symbol nm), equal to one billionth of a meter.

micrometer gauge Hand-held caliper instrument used in engineering machine shops to check the thickness of metal parts. The caliper of the micrometer is placed over the part to be measured and a screw is turned by the thumb, advancing a spindle until it touches the metal part (which is then just held by the caliper). The thickness is read off the uncovered part of a scale, to an accuracy of about 0.002mm (0.0001in).

microminiaturization *See* MICROELECTRONICS

micronutrient Element needed in relatively small amounts by growing plants. Micronutrients include iron, boron, zinc, copper, molybdenum, and chlorine. *See also* MACRONUTRIENT

microorganism (microbe) Organism that can be seen only with the help of a MICROSCOPE. Microorganisms include BACTERIA, PROTOZOA, RICKETTSI-AE, VIRUSES, and some microscopic ALGAE and FUNGI. Disease-causing micro-organisms are called PATHOGENS.

microphone Device for converting sound into varying electric currents with the same pattern of FREQUENCY and amplitude. Many telephones use a carbon microphone, in which the sound pressure causes a variation in the electrical resistance of carbon granules held between a diaphragm and a carbon block. Live music performers often use a moving coil microphone, in which a coil attached to a diaphragm vibrates in a stationary magnetic field. The recording industry prefers the condenser microphone, which employs a CAPACITOR. Crystal microphones use the PIEZOELECTRIC EFFECT.

microprocessor Entire CENTRAL PROCESSING UNIT (CPU) contained on an INTEGRATED CIRCUIT (SILICON CHIP) used to control the operation of a computer or other equipment, such as a washing machine.

microscope Optical device for producing an enlarged image of a minute object. The first practical microscope was made in 1668 by Anton van LEEUWENHOEK. The modern, **compound** microscope has two converging lens systems, the objective and the eyepiece, both of short focal length. The object to be examined is placed close to the objective lens and illuminated by a strong source of light. The objective produces a magnified image that is further magnified by the eyepiece to give the image seen by the observer. Most microscopes have three objective lenses on a turret, and they may be interchanged to give a choice of low, medium or high magnification. Because of the nature of visible light, an optical microscope can magnify objects only up to 2,000 times, and this is possible only by using lenses immersed in oil. Another type of instrument, the **polarizing** microscope, illuminates mineral specimens with POLAR-IZED LIGHT. For extremely small objects, an **optical** microscope cannot be used, because it is incapable of resolving them. The ELECTRON MICROSCOPE was invented to solve this problem. Smaller objects can be seen, because electrons have shorter wavelengths than light and thus provide greater resolution. These microscopes can magnify up to a million times.

Microscopium (microscope) Small constellation of s sky. Its brightest star is of magnitude 4.7.

microsurgery Delicate surgery performed under a binocular microscope using specialized instruments, microneedles as small as 2mm long, and sutures 20 micrometers in diameter. It is used in a number of specialized areas, including the repair of nerves and blood vessels, eye, ear, and brain surgery, and the reattachment of severed parts. Microsurgery was first used by the Swedish surgeon G. Holmgren in the 1920s for ear operations, and by the 1940s it was used for corneal transplants. During the 1950s, techniques were developed to suture tiny blood vessels and nerves in such a way as to retain their function, making possible the first successful limb re-implantation by microsurgery in 1962.

microswitch Electrical switch that requires only a small effort and movement to open or close it. The

▲ **microscope** An optical microscope magnifies a sample (1) held on a slide (2). Light from below (3) illuminates the sample, which is magnified by a lens (4) that can be changed. A prism arrangement (5) direct the image onto an eyepiece (6). The position of the sample can be altered physically by turning knobs (7) that bring the image into position and focus.

prefix "micro" refers to the smallness of the movement, rather than the switch. It is used for automatic, or semi-automatic, applications in electromechanical machines in which a moving actuator opens or closes the switch to stop or start an operation.

microtome Precision device for slicing thin sections of specimens for examination under a MICROSCOPE.

microtubule Thin, protein filament in the CYTOPLASM of a CELL, which may form pairs or bundles, that helps to maintain the shape of the cell. During cell division (MEIOSIS and MITOSIS), the microtubules from the spindle are responsible for the movement of the chromosomes. Microtubules also occur in CENTRIOLES, CILIA, and FLAGELLA.

microwave Form of ELECTROMAGNETIC RADIATION having a wavelength between 1mm (0.04in) and 1m (3.3ft) and a frequency range to about 300,000MHz. This puts microwave radiation between infrared and shortwave radio waves in the ELECTROMAGNETIC SPECTRUM. Microwaves are being increasingly used today for such purposes as RADAR, RADIO and TELEVISION broadcasting, high-speed microwave heating, and cellular telephones. *See also* BACKGROUND RADIATION

microwave heating Use of MICROWAVES to heat materials. Microwave ovens produce heat by causing the water molecules in food to vibrate.

Mid-Atlantic Ridge Underwater mountain range along the margin between the American crustal plate on one side and the European and African plates on the other. It runs from Iceland to near the Antarctic Circle along the middle of the Atlantic Ocean. It is formed at a CONSTRUCTIVE MARGIN, where two plates are diverging and new magma is flowing upward and outward to form new ocean crust. Iceland is located on the ridge itself and was formed by the outpourings of volcanic lava. *See also* PLATE TECTONICS

midbrain (mesencephalon) One of the three divisions of the embryonic BRAIN of backboned animals. In a fully formed human being, the midbrain becomes overlaid by the greatly developed cerebral lobes. It is thick-walled and is concerned particularly with sight and hearing. *See also* FOREBRAIN; HINDBRAIN

middle ear *See* EAR

midocean ridge Great median ridge of the SEAFLOOR where new LITHOSPHERE is being formed. The ridges are the spreading edges of the tectonic plates that cover the Earth. They form a world-encircling system that extends, with several side branches, along the MID-ATLANTIC RIDGE, up, around, and down through the Indian Ocean (the Mid-Indian Ridge) and across the Pacific (the Pacific-Antarctic Ridge). *See also* OCEANIC BASIN; PLATE TECTONICS; SEAFLOOR SPREADING

migmatite Coarse-grained, METAMORPHIC ROCK, often found in the shield of an ancient continent. It is formed at high temperatures deep within the Earth's crust, and has a granular texture consisting of dark schist or gneiss, with lighter folded bands of granitic rock.

migration Periodic movement of animals or humans, usually in groups, from one area to another, in order to find food, breeding areas or better conditions. Animal migration involves the eventual return of the migrant to its place of departure. Fish migrate between freshwater and saltwater or from one part of an ocean to another. Birds usually migrate along established routes. Mammals usually migrate in search of food. For thousands of years, the deserts of central Asia widened inexorably, and this phenomenon resulted in the human migration of prehistoric tribes to China, the Middle East, and Europe.

milk Liquid food secreted from MAMMARY GLANDS by the females of nearly all mammals to feed their young. The milk of domesticated cattle, sheep, goats, horses, camels, and reindeer has been used as food by humans since prehistoric times, both directly and to make butter, cheese, and fermented milks such as yogurt. Milk is a suspension of fat and protein in water, sweetened with lactose sugar. The proportions of these constituents vary with each mammal, those of cows' milk being water 87.1%, protein 3.4%, fat 5.9%. In modern dairying, cows' milk is pasteurized at about 72°C (160°F) for 16 seconds to kill all harmful microbes, then bottled under aseptic conditions. Nevertheless, unrefrigerated milk sours after a day or so because of the action of lactic acid bacteria. *See also* PASTEURIZATION

Milky Way Faint band of light visible on clear dark nights encircling the sky along the line of the galactic equator. It is the combined light of an enormous number of stars, in places obscured by clouds of interstellar gas and dust. It is, in fact, the disk of our GALAXY, viewed from our vantage point within it.

Miller, Stanley Lloyd (1930–) US biochemist. In 1953 he carried out the first deliberate attempt to create the chemical components of life, by passing a mixture of AMMONIA, METHANE, water vapor, and HYDROGEN (to simulate the Earth's early atmosphere) through water that was continuously sparked by a corona discharge. Within days, AMINO ACIDS were formed.

Millikan, Robert Andrews (1868–1953) US physicist who determined the value of the electronic charge. He also verified the photoelectric equation of Albert EINSTEIN and found a precise value for PLANCK'S CONSTANT. Millikan was awarded the 1923 Nobel Prize for physics for his measurement of the electronic charge and his work on the PHOTOELECTRIC EFFECT.

milling, ore Stage in the treatment of crude ORE during which valuable MINERALS are extracted mechanically. Primary and secondary crushers break down large fragments. Fine grinding (comminution) is done with rotating cylindrical mills that contain steel balls, rods or pebbles. The fine powder is then sorted by a variety of methods.

Milne, John (1850–1913) British geologist whose invention (1880) of the SEISMOGRAPH helped to found the science of SEISMOLOGY. He helped to establish many seismological stations, particularly those for recording earthquakes in Japan. His books include *Earthquakes* (1883) and *Seismology* (1898).

Milstein, César (1927–) British molecular biologist and immunologist, b. Argentina, who shared the 1984 Nobel Prize for physiology or medicine for helping to develop ANTIBODIES that can be commercially produced for drugs and diagnostic tests. Milstein studied ENZYMES at the University of Cambridge. In 1975 he and his Cambridge colleague, the German immunochemist Georges Köhler, developed a technique for cloning MONOCLONAL ANTIBODIES (MABs) that combat diseases by targeting their sites. Milstein and Köhler shared the Nobel Prize with the immunologist Niels Jerne (1911–).

mimicry Form of animal protection through deception. The mimic, generally a harmless, edible species, imitates the warning shape or coloration of a "model," a poisonous or dangerous species. When coloration increases an animal's chances of survival, it is commonly referred to as protective coloration. **Batesian** mimicry, named after the British naturalist Henry BATES, is exemplified by hoverflies, which mimic inedible wasps. Similarly, the viceroy butterfly imitates the inedible monarch butterfly. The less common Müllerian mimicry, named after the German naturalist Fritz Müller, involves two or more unpalatable species that share a similar pattern, thus reinforcing it as one of the warnings to predators. A third form is aggressive mimicry, in which a predatory or parasitic species resembles a harmless one, thus allowing the former to remain undetected by its prey or host.

mine In civil engineering, excavation from which MINERALS (mainly coal and metal ores) are extracted. Underground mines are of two main types: shaft mines and drift mines. **Shaft** mines are sunk vertically in the Earth's crust until they reach the depth of the seams to be exploited, which are reached by tunnels or galleries. The gold mines of South Africa, the world's deepest mines, are shaft mines. **Drift** mines are generally shallower, the seams being reached by a drift, or gradually sloping shaft, which leads on to a gallery system. In OPEN-PIT MINING, the seams are near or on the surface and are exposed by giant dragline machines. *See also* MINING

mine In military technology, concealed or buried explosive device detonated through contact with individuals or vehicles. Underwater mines are used either to protect or to blockade coastal areas. There is now much opposition to the use of land mines as they are very difficult to remove from an area after the conflict has ceased. Each year, many thousands of civilians are killed by mines left behind in war zones.

mineral Natural, homogeneous and, with a few exceptions, solid and crystalline materials that form the Earth and make up its ROCKS. Most are formed through inorganic processes, and more than 3,000 minerals have been identified. They are classified on the basis of chemical makeup, crystal structure, and physical properties such as hardness, specific gravity, cleavage, color, and luster. Some minerals are economically important as ORES from which metals are extracted.

mineral acid Strong, inorganic acid such as HYDROCHLORIC ACID (HCl), NITRIC ACID (HNO_3), and SULFURIC ACID (H_2SO_4).

mineralization FOSSIL-forming process whereby preserved parts of plants and animals are altered by the minerals in the soil. Circulating water dissolves certain constituents of bones and shells, which are replaced by silica, iron or other compounds. This replacement can be so exact as to preserve the finest structures. It is a type of PETRIFICATION.

mineralogy Investigation of MINERALS, naturally occurring inorganic substances found on Earth and elsewhere in the Solar System. Major subdivisions are: CRYSTALLOGRAPHY, which studies the composition and atomic arrangement of minerals; **paramagnetic** mineralogy, which deals with the associations and order of crystallization of minerals; **descriptive** mineralogy, concerned with the physical properties used in identification of minerals; and **taxonomic** mineralogy, the classification of minerals by chemical and crystal type. *See also* GEOCHEMISTRY; PETROLOGY

mineral oil (liquid paraffin, liquid petrolatum) Colorless liquid without taste or smell, distilled from PETROLEUM. It is used as a lubricant, and as a solvent in the manufacture of plastics. It is used medicinally as a laxative.

mineral water Originally natural waters valued for their mineral content, now also used to describe carbonated drinks containing salts or various flavorings. "Medicinal waters" from many famous spas are bottled and sold. They are classified according to location, use, and chemical content. Synthetic mineral waters were developed to imitate natural ones.

minimal access surgery Term used to encompass operations that do not involve cutting open part of the body in the traditional way. Minimal access, or "keyhole," procedures are performed either by means of an ENDOSCOPE, an instrument for viewing the interior of the body, or by passing miniature instruments through a fine catheter, or tube, that is fed into a large blood vessel. Certain keyhole operations are said to be safer and less traumatic than conventional surgery. There is a growing list of such procedures, many of which require only local ANESTHESIA, and can be done on an outpatient basis, or at least with a much shorter hospital stay and faster recovery. The surgical LASER – the so-called bloodless scalpel – is used in many minimal access procedures.

mining Process of obtaining metallic and nonmetallic materials from the Earth's crust. Included are underground, surface and underwater methods. Mining mostly involves the physical removal of rock and earth. Petroleum, gas, and some sulfur are extracted by techniques that are different from those that would normally be put under the heading "mining." Any mining operation comprises four stages: prospecting, exploration, development, and exploitation. Once a valuable deposit has been found and delineated, decisions are made on modes of entry, subsidiary developments, and removal techniques. In underground mining, various methods are employed to obtain minerals buried deep in the ground. Access may be through oblique or vertical shafts or horizontal tunnels. Cross cuts are made at various levels and the ore body is divided into blocks by vertical "rises" that connect different levels. The ore is broken up by the use of hand tools, blasting or machinery. Coal, potash and rock salt are generally mined by the "room and pillar" method, in which passageways open into rooms supported by tenant rock. Underwater mining is done by the use of large conveyor belts that are capable of bringing to the surface many tons of material each minute. The parent ships sift the material and deposit the unwanted material overboard. *See also* MINE

Minkowski, Hermann (1864–1909) Russian mathematician who worked in Germany. He contributed to the mathematical foundations of EINSTEIN's work on special RELATIVITY and formulated the concept of SPACE-TIME.

Miocene Geological epoch beginning about 25 million and ending about 5 million years ago. It falls in the middle of the TERTIARY period and is marked by an increase in grasslands over the globe at the expense of forests, and the development of most of the modern mammal groups. No British Miocene deposits are known, although some folding took place caused by the rising of the Alps to the S.

Mir Russian (formerly Soviet) SPACE STATION orbiting the Earth. The main part of Mir was launched in 1986. It weighs nearly 21 tons and has six docking ports for the attachment of scientific modules or cargo vessels. Over the years various modules were added to enlarge the space station, which was permanently staffed by three-person crews who occupied Mir for many months at a time. A series of mishaps beginning in 1997 plagued the project, and it was abandoned in 1999.

Mira Ceti (Omicron Ceti) RED GIANT star in the constellation of Cetus. It is a long-period VARIABLE STAR, with a mean period of 332 days, but this is subject to irregularities. Its mean range is magnitude 3.5 to 9.1, but maxima as bright as 2.0 and minima as faint as 10.1 have been observed. Mira Ceti is the most celebrated long-period variable.

mirage Type of optical illusion sometimes seen near the Earth's surface when light is refracted as it passes between cool, dense air to warmer, less dense air. Mirages are most commonly seen shimmering on desert sands or hot, dry roads; the shimmer is a refracted image of the sky. *See also* REFRACTION

mirror Highly polished surface that produces an image of objects in front of it because of the laws of REFLECTION. Most mirrors are made of glass "silvered" on one side with SILVER, MERCURY or ALUMINUM. They can be flat (plane) or curved (spherical or parabolic). **Plane** mirrors produce a virtual image that is the same size but inverted laterally (left to right). **Spherical** mirrors may be concave (curving inwards) or convex (bulging outwards). The image can be rightside up or inverted, real or virtual, depending on the position of the object in relation to the focal point of the mirror; it may also be either magnified or reduced in size. A spherical mirror suffers spherical ABERRATION that is absent in a concave parabolic mirror, as used in reflecting TELESCOPES. *See also* IMAGE, OPTICAL

miscarriage Popular term for a spontaneous ABORTION, the loss of a fetus from the UTERUS before it is sufficiently developed to survive.

missile Unmanned and self-propelled flying weapons, powered by ROCKET, ram jet or turbo-jet. Ballistic missiles travel in the outer atmosphere and can be powered only by rockets. GUIDED MISSILES carry self-contained guidance systems or can be controlled by radio from the ground. CRUISE MISSILES, a type of guided missile, travel in the lower atmosphere and can be powered by jet engines. Unguided missiles are free flying, with no control other than initial aim and amount of fuel.

Mississippian In North America, name given to the earlier part of the CARBONIFEROUS period.

mist Water droplets in the atmosphere at ground level that decrease visibility. By standard definition, visibility in a mist is between 2,000–1,000m (6,560–3,280ft). When visibility drops below 1,000m (3,280ft), the mist is referred to as FOG. *See also* SMOG

mistral Cold, winter, N wind that sweeps down the Rhône valley from the Massif Central of France. It is caused by cold air from Europe blowing southwards into a low-pressure area over the Mediterranean. When the mistral blows, temperatures sometimes fall as low as freezing point and strong gusts of up to 60 km/h (37mph) make conditions very unpleasant. Most farmhouses have rows of trees on their N side as protection from the wind, and many fields of fruits and vegetables shelter behind rows of cypress trees.

mitochondrion Structure (organelle) inside a CELL containing ENZYMES necessary for energy production. Mitochondria are found in the cytoplasm of most types of cell (but not in bacteria). *See also* RESPIRATION

mitosis Nuclear division of a CELL resulting in two genetically identical daughter cells with the same number of chromosomes as the parent cell. Mitosis is the normal process of TISSUE growth, and is also involved in ASEXUAL REPRODUCTION. *See also* ANAPHASE; METAPHASE; MEIOSIS

Mitscherlich, Eilhardt (1794–1863) German chemist who discovered (1819) that compounds of similar composition tend to have the same crystal structure – the phenomenon known as ISOMORPHISM. This helped in the understanding of the relationship between the molecular composition and crystalline structure of compounds. In 1823 Mitscherlich discovered the monoclinic form of sulfur and in 1827 he discovered selenic acid.

◄ **mirror** A plane mirror reflects all light rays at the same angle (A), whereas a matt surface (B) scatters light. The brain imagines that light rays reaching the eye from a mirror come to it in straight lines, and it therefore sees an image at that point where the rays would original if their paths were not bent by the mirror (C). The image is seen laterally inverted (D), because the reflected rays reach opposite sides of the eye. A noninverted image can be seen in two plane mirrors at right-angles (E). Although an image reflected once appears laterally inverted (1, 2), one reflected twice (3) is seen correctly (F).

Mitscherlich named BENZENE and was the first to synthesize nitrobenzene (1832); he was also among the first to recognize CATALYSIS.

mixture In chemistry, a material made up of two or more substances that retain their specific identities (such as air containing oxygen, nitrogen, and other gases). The identities remain separate no matter in what proportion or how closely the components are mixed. *See also* COMPOUND; SOLUTION

mks system METRIC SYSTEM of units based on the meter, kilogram, and second. *See also* WEIGHTS AND MEASURES; SI UNITS

mobile telephone Portable radio device that connects users to the public TELEPHONE system. They are also called cellular phones, because they operate within a network of radio cells. The first generation operated with analog signals; the second generation has been designed to operate with DIGITAL SIGNALS.

True mobile telephones (that operate from anywhere on the planet's surface by communicating directly with SATELLITES) became available in the 1990s.

Möbius strip Shape or figure that can be made by giving a long strip of paper a half-twist, then joining the ends together. It is of great interest in TOPOLOGY, being a one-sided surface (a line drawn along the strip of paper will appear both sides, returning to the starting point to meet itself). It was invented by August Ferdinand Möbius (1790–1868).

mode In a statistical sample, the value of a VARIABLE that occurs most often. For example, in a survey on beer-drinking, a sample of 100 men were asked how many beers they drank in a week. The results were as follows (with numbers of men in parentheses): 0 beers (1); 10 beers (14); 12 beers (19); 15 beers (29); 20 beers (18); 25 beers (12); 30 beers (7). The mode is 15. If there are two modes, the distribution is said to be **bimodal**.

▲ **mobile telephone**
Mobile telephone networks use a system of cells (1). By having a transmitter (2) in each cell, the same frequencies can be used in each cell, allowing an enormous capacity for calls. Where there are many users, such as in the heart of a city (3), the cells are much smaller – further multiplying the number of frequencies available. A

digital cellular telephone (4) sends digital information (5) to a transmitter tower (6). Digital cellular phones are better than analogue versions because they reduce background noise and interference and are more difficult to bug. The transmitter passes the message to the systems' central exchange (7). If the call is for another mobile phone, the exchange sends

a message (8) to the other transmitters, which in turn send out a message to locate the receiving phone. The transmitter locating the required phone (9) sends a confirmation message to the exchange (10), which then connects the conversation (11 - dotted line). When a mobile phone moves out of range of a transmitter (12), a complicated procedure

ensures the conversation can continue seamlessly. When the cellular phone's signal to the transmitter becomes weaker the exchange sends a message (13) to the transmitters in the surrounding cells to see which is receiving the strongest signal. It then transfers the conversation to that transmitter (14 - dotted line).

modem (**mo**dulator-**dem**odulator) Electronic device for sending and receiving COMPUTER signals through a TELEPHONE system. The streams of electrical pulses produced by computers cannot pass through a telephone system. So they are fed into a modem, which uses the pulses to modulate a continuous tone (carrier) by a process called FREQUENCY MODULATION (FM). The modulated carrier can then be sent through a telephone system. At the other end, another modem extracts the pulses (demodulation), so that they can be fed into a receiving computer. *See also* COMPUTER NETWORK; INTERNET

moderator In nuclear physics, substance consisting of light elements, such as HEAVY WATER or graphite, used in NUCLEAR REACTORS to slow down the FAST NEUTRONS produced by the FISSION reaction so that they can maintain a CHAIN REACTION.

modulation In physics, process of varying the characteristics of one wave system in accordance with those of another. It is basic to RADIO broadcasting. In AMPLITUDE MODULATION (AM), the amplitude of a high-frequency radio CARRIER WAVE is varied in accordance with the amplitude of a current generated by a sound wave. For static-free, short-range broadcasting FREQUENCY MODULATION (FM) is used, in which the frequency of the carrier wave is modulated.

Mohl, Hugo von (1805–72) German botanist who did pioneering research into the anatomy and physiology of plant CELLS. He formulated the idea that the cell nucleus is surrounded by a granular, colloidal substance, which in 1846 he called PROTOPLASM, although the term had been invented by Jan Purkinje (1787–1869) in reference to the embryonic material in eggs. Von Mohl was the first to propose that new cells arise from CELL DIVISION. In 1851 he claimed that the secondary wall of a plant cell is fibrous, and this was subsequently confirmed. He also gave the first distinct account of the function of OSMOSIS.

Moho (Mohorovičić discontinuity) Boundary between the Earth's LITHOSPHERE and MANTLE. It is identified by a change in the velocity of seismic waves passing through the Earth, and is named after the Croatian geophysicist Andrija Mohorovičić (1857–1936), who first recognized it in 1909. The velocity change is explained by a change to more dense rocks in the mantle; above the Moho the waves travel at about 6.4km s^{-1} (4mi s^{-1}), whereas below they travel at 8.2km s^{-1} (5.1mi s^{-1}). The depth of the Moho varies from about 5km (3mi) to 60km (37mi) below the Earth's surface; it is usually deeper beneath landmasses than beneath the sea. Initially interpreted as a

▲ **modem** A modem allows computers to communicate over telephone systems. It converts the digital signals (1) used by computers into an analog signal (2) that travels over phone lines (3). The analog signals (4) are converted back to a digital signal (5) using binary code which is read by the receiving computer.

sharp boundary, it is now known to be a wider zone in some locations, and its exact nature remains a matter of scientific debate. The **Gutenberg** discontinuity (discovered in 1912, by Beno GUTENBERG) separates the mantle and the core, and is *c*.2,900km (1,800mi) deep. EARTHQUAKE and other shock waves are deflected at these boundaries. *See also* MOHOLE PROJECT

Mohole Project Attempt by US geologists and engineers, begun in the 1950s, to penetrate the Earth's crust to the deeper layers of the mantle by drilling a borehole at least 10km (6mi) deep. It was named after the MOHO, the irregular boundary separating the crust and the mantle. Because of rising costs the Mohole was discontinued in 1966.

Mohr, Karl Friedrich (1806–79) German chemist. The foremost pharmacist of his time in Germany, Mohr improved analytical techniques and invented many items of laboratory equipment. He was also among the first to advance the idea of CONSERVATION of energy (in 1837). In 1867 he became professor of chemistry at the University of Bonn.

Mohs scale Range of HARDNESS used by geologists to express the comparative hardness of minerals by testing them against ten standard materials. It was devised in 1812 by the German mineralogist Friedrich Mohs (1773–1839). The hardest mineral, diamond, has a hardness of 10 on the Mohs scale. It can scratch, or mark, any mineral with a lower

Mohs number, including corundum (9), topaz (8), quartz (7), orthoclase (6), apatite (5), fluorite (4), calcite (3), gypsum (2), and talc (1). A mineral hard enough to scratch material 3, but soft enough to be scratched by material 5, would be rated as having hardness 4 on the Mohs scale. Useful tools for determining hardness are the fingernail (about 2.5) and a penknife (about 5.5).

Moissan, Ferdinand Frédéric Henri (1852–1907) French chemist. He trained as an apothecary and won international recognition as the first to isolate FLUORINE (1886). He also developed an electric FURNACE that enabled him to prepare samples of several of the less common metals, such as URANIUM, long before modern metallurgy made them commercially available. He was awarded the Nobel Prize for chemistry in 1906.

molar One of the large back TEETH of MAMMALS, adapted for grinding and chewing food. In most HERBIVORES the cusps (points or ridges on the top surface) are fused to form ridges for grinding plants. In adult humans there are 12 permanent molars, three on each side of each jaw, top and bottom.

molarity See CONCENTRATION

molar volume Volume occupied by one MOLE of a substance. It is approximately the same for all gases at STANDARD TEMPERATURE AND PRESSURE, 22.414 liters.

mold Mass composed of the spore-bearing MYCELIA (vegetative filaments) and fruiting bodies produced by numerous fungi. Many molds live off fruits, vegetables, cheese, butter, jelly, silage, and almost any dead organic material. Roquefort, camembert, and stilton cheeses involve the use of mold. Although many species are pathogenic (disease-causing), penicillin and a few other antibiotics are obtained from molds. See also FUNGICIDE; FUNGUS; SLIME MOLD

mold, metal Mold for CASTING metal, usually made of sand or clay. This is packed over the face of the pattern that forms the cavity for the casting. The mold must be strong, resisting pressure of the hot, liquid metal, and permeable to allow gases to escape the cavity. It must also resist fusion with the metal, which is poured in through special channels.

mole (symbol mol) SI UNIT of amount of substance. It is the amount of substance that contains as many elementary units, such as atoms and molecules, as there are atoms in 0.012kg of carbon-12. The mass of one mole of a compound is its RELATIVE MOLECULAR MASS (molecular weight) in grams.

molecular biology Biological study of the make-up and function of MOLECULES found in living organisms. Major areas of study include the chemical and physical properties of proteins and of NUCLEIC ACIDS such as DNA. See also BIOCHEMISTRY

molecular formula See FORMULA

molecular model Geometrical structure of a molecule made by joining small colored balls, representing atoms, using stiff metal springs or plastic rods, representing single bonds. They are invaluable aids not only in visualizing shapes of molecules but also in indicating "strain" in bonds and possible conformations that the molecule may adapt in reactions.

molecular orbital theory Explanation of how ELECTRONS are distributed in stable MOLECULES. In the simpler VALENCE theory of the CHEMICAL BOND, each ATOM of a molecule is assumed to retain its own electrons. Molecular orbital theory, however, treats each electron as associated with the molecule as a whole and it describes the configuration of electrons for each molecule. These patterns are unique for each molecule and can be considered as regions surrounding the molecule in which there is a probability of finding an electron. The calculations are very complex, and only the simplest molecules can be treated exactly.

molecular weight Former term for RELATIVE MOLECULAR MASS

molecule Smallest particle of a substance (such as a compound) that exhibits the chemical properties of that substance. A molecule may consist of one ATOM, but generally consists of two or more atoms held together by CHEMICAL BONDS. For example, water molecules consist of two atoms of hydrogen bonded to one atom of oxygen (H_2O). A molecule (unlike an ION) has no electrical charge. See also MACROMOLECULE

mollusk Any of more than 80,000 species of INVERTEBRATE animals in the phylum Mollusca. They include the familiar snails, clams, and squids, and a host of less well-known forms. Originally marine, members of the group are now found in the oceans, in fresh water, and on land. Classes of mollusk include: the primitive GASTROPODS, univalves (slugs and snails), BIVALVES (clams, etc.), tusk shells and CEPHALOPODS (squids, etc.). The mollusk body is divided into three: the head, the foot, and the visceral mass. Associated with the body is a fold of skin called the **mantle**, which secretes the limy shell typical of most mollusks. The head is well developed only in snails and in the cephalopods, which have eyes, tentacles, and a well-formed mouth. The visceral mass contains the internal organs of circulation (blood vessels and heart), respiration (gills), excretion (kidney), digestion (stomach and intestine), and reproduction (gonads). The sexes are usually separate but there are many hermaphroditic species. Cephalopods, bivalves and gastropods feature as important fossils in the geological past. *See also* HERMAPHRODITE

molting Process involving the shedding of the outermost layers of an organism and their replacement. Mammals molt by shedding outer skin layers and hair, often at seasonal intervals – human beings do not molt but lose dead, dry skin continuously as it is replaced from below. Birds molt their feathers, and amphibians and reptiles their skin. In all cases the process is controlled by HORMONES. It often serves, in growing animals, to replace worn out tissues of skin that have become too small. The molting of INSECTS and other ARTHROPODS is a more elaborate affair that is fundamental to growth. The process, also called **ecdysis**, involves the resorption into the body of materials from the hard outer cuticle of the EXOSKELETON, so making the cuticle more fragile. The arthropod then swells its body and bursts free from the old cuticle, and slowly reforms a new one around its swollen body, thus increasing in size. *See also* METAMORPHOSIS

molybdenite Sulfide mineral, molybdenum sulfide (MoS_2). A major ore of molybdenum, molybdenite is found in PEGMATITES, IGNEOUS and METAMORPHIC ROCKS. It has hexagonal system tabular prisms, flakes, and fine granules and is lead-gray in color with a metallic luster. Hardness 1–1.5; r.d. 4.7.

molybdenum (symbol Mo) Silver-white TRANSITION ELEMENT, first discovered in 1778. It is obtained from ores containing MOLYBDENITE (MoS_2). The concentrated mineral is roasted to yield molybdenum trioxide, which is mixed with iron in electric FURNACES to make ferromolybdenum. The pure metal is produced as a powder and converted to massive metal. Hard but malleable and ductile, molybdenum is used in alloy steels (as a hardener), X-ray tubes, and missile parts; molybdenum compounds are used as CATALYSTS, and as LUBRICANTS for bearings. It is one of the essential TRACE ELEMENTS for plant growth. Properties: at.no. 42; r.a.m. 95.94; r.d. 10.22; m.p. 2,610°C (4,730°F); b.p. 5,560°C (10,040°F); most stable isotope ^{98}Mo (23.78%).

moment of force *See* TORQUE

moment of inertia (symbol *I*) For a rotating object, the sum of the products formed by multiplying the elements of mass of the rotating object by the squares of their distances from the axis of the rotation. Finding this distribution of mass is important when determining the force needed to make the object rotate.

momentum Product of the mass and linear velocity of an object. One of the fundamental laws of physics is the principle that the total momentum of any system of objects is conserved (remains constant) at all times, even during and after collisions.

monadnock Residual hill rising above a PENEPLAIN, representing the core of the upland area that was eroded to give the plain. The plains of N Australia have many examples.

monazite Mineral containing rare-earth metals (LANTHANIDE SERIES) such as cerium, lanthanum, and thorium. It occurs widely in granitic and other rocks, and is the major source of those metals.

Mond, Ludwig (1839–1909) British chemist and industrialist, b. Germany. Mond experimented with alkalis and developed a PRODUCER GAS (a mixture of carbon monoxide and nitrogen). He also discovered nickel carbonyl, a gas formed from carbon monoxide and metallic nickel. Using nickel carbonyl, he developed a useful industrial method (the **Mond process**) for extracting pure nickel from its ore.

Monera *See* PROKARYOTAE

Moniz, António de Egas (1874–1955) Portuguese physician and medical researcher. He shared the 1949 Nobel Prize for physiology or medicine with Walter HESS for developing the surgical technique known as prefrontal LOBOTOMY. This treatment of severely mentally ill patients, involving cutting selected brain fibers, was made popular by Moniz in the late 1940s and was the foundation of the controversial field of psychosurgery.

monoamine oxidase (MAO) ENZYME widely distributed in animals; other names are adrenaline oxidase and tryaminase. Its function in the body is the breakdown of certain biologically active AMINES, three of the most important of which are the tryptamine derivatives, the catechol amines (such as ADRENALINE and DOPAMINE) and histamine. **Monoamine oxidase inhibitors** are used as antidepressants.

Monoceros (Unicorn) Faint equatorial constellation situated S of Gemini and E of Orion. The Milky Way passes through this group, which also contains several bright and dark nebulae, including the Rosette Nebula (NGC 2237), and some star clusters, such as NGC 2244.

monochromatic Describing ELECTROMAGNETIC RADIATION, such as light, which has a single wavelength or frequency (a single color). Pure monochromatic radiation is not possible, although the light from a LASER occupies a very narrow band of wavelengths and is virtually monochromatic, as are some spectral lines.

monocline In geology, sudden downward turn in the direction of a rock STRATUM within a fold, surrounded by horizontal rock.

monoclonal antibody ANTIBODY produced by cells that are derived by cloning a single original parent cell. Monoclonal antibodies are all identical and have unique sequences of AMINO ACIDS in their protein makeup. They are made by fusing a normal LYMPHOCYTE (an antibody-producing cell) with a cancerous cell derived from lymphatic tissue (LYMPHOMA) or bone MARROW (myeloma). The resulting "hybridoma" multiplies rapidly and produces large quantities of a single antibody. Monoclonal antibodies are used in highly specific VACCINES and in identifying ANTIGENS such as those involved with BLOOD GROUPS.

monocotyledon Any member of the Monocotyledonae subclass of flowering plants (ANGIOSPERMS) characterized by one seed leaf (COTYLEDON) in the seed embryo; the leaves are usually parallel-veined. Lilies, onions, orchids, palms, and grasses are examples of monocotyledons. The larger subclass of plants is the Dicotyledonae (*see* DICOTYLEDON).

monoculture Agricultural practice of growing one crop. Monoculture has been common for a long time in many places, such as plantations or vineyards, but it has become far more widespread during the last 20 or 30 years, especially in the production of cereals. This is a result of the availability of machinery for harvesting large quantities very quickly. There are also many fertilizers that restore fertility to the soil, so that crop rotation (whereby different crops are grown in the same fields every year or so) is unnecessary. Monoculture has created problems, however, as dependence on one crop can be serious if the price suddenly falls because of a glut, or if new pests invade the crop. Moreover, it is being realized that monoculture harms and possibly ruins the soil, in spite of the addition of artificial fertilizers. Monoculture in such areas as the American and Canadian prairies, the steppes in Russia, and East Anglia in the UK is now declining slightly, and there are signs of a return to crop rotation methods.

monocyte Large LEUCOCYTE (white blood cell). A monocyte is characterized by a kidney-shaped nucleus and protoplasm that stains a blue-gray colour. It is a PHAGOCYTE, actively digesting foreign particles such as bacteria and debris from dead cells. A milliliter of normal blood contains up to a million monocytes. *See also* IMMUNE SYSTEM

Monod, Jacques (1910–76) French biochemist. With François JACOB, Monod developed the idea that MESSENGER RNA carries hereditary information from the nucleus of a CELL to the cellular sites during the PROTEIN SYNTHESIS, and the concept of the operator GENE controlling the activity of other genes. Together with André LWOFF, they were awarded the 1965 Nobel Prize for physiology or medicine.

monoecious In botany, having both male and female FLOWERS on the same plant. The male flowers bear STAMENS, and the females bear one or more CARPELS.

monomer Chemical compound composed of single molecules, as opposed to a POLYMER (which is built up from repeated monomer units). For example, an AMINO ACID is a monomer of a protein, and propylene is the monomer from which polypropylene is made.

monorail train Passenger train that runs on a single rail; modern versions are gyroscopically stabilized and propelled by electric motors. The passenger cars are unsupported from the bottom or sides; they generally hang from wheeled axles that run along the rail. Bottom-supported cars, such as those used in zoos and amusement parks, are also referred to as monorails, as are certain types of MAGLEV (**mag**netic **lev**itation) train.

◄ **monotreme** An example of a primitive mammal, the echidna (spiny anteater) is classified as a monotreme. Instead of giving birth to live young like other mammals, it is an egg-layer. The tiny egg (1) is soft-shelled and resembles a reptile's egg. Once the egg is laid, the echidna uses its hind limbs to roll it to a special incubation groove (2). The minute hatchling is c.1.25cm (0.5in) long.

monosaccharide Sweet-tasting CARBOHYDRATE that cannot be broken down by HYDROLYSIS; a simple sugar. GLUCOSE is a monosaccharide. *See also* DISACCHARIDE; POLYSACCHARIDE

monosodium glutamate (MSG) Food additive with a meatlike taste; a white crystalline powder that is the sodium salt of glutamic acid (an amino acid). It is obtained mainly from cereal glutens, and used as a flavor enhancer for meat products.

monotreme Member of the Monotremata order of primitive MAMMALS that lay eggs. The only monotremes are the platypus and two species of echidna, all native to Australasia. The eggs are temporarily transferred to a pouch beneath the female's abdomen where they eventually hatch and are nourished by rudimentary mammary glands. *See also* MARSUPIAL; PROTOTHERIAN

Monotype In printing, TYPESETTING machine that casts letters one at a time. *See also* LINOTYPE; PHOTOSETTING

monsoon Seasonal wind, especially in S Asia. In summer, which is the monsoon season, the winds normally blow from the sea to the land and bring RAIN, but in winter there is a complete change of direction and the winds blow out from the land, giving dry weather. Some monsoon regions are very wet; Cherrapunji in India, for example, receives over 11,000mm (433in) of rain each year. But others can be dry, such as the Thar Desert between India and Pakistan, where the rainfall is usually less than 250mm (10in). The major monsoon areas are in Asia, where the seasonal reversal of wind is greatest. This is because Asia, the largest continent, is adjacent to the largest ocean, the Pacific. In the smaller continents of South America, Africa, Australia, and North America, the monsoonal effects are less marked. These smaller continents do not have such wet summers or such dry winters, and are sometimes referred to as "eastern marginal" rather than true monsoon.

Montgolfier, Joseph Michel (1740–1810) and **Jacques Étienne** Montgolfier (1745–99) French inventors of the hot-air balloon. In 1782 the brothers experimented with paper and linen balloons filled with hot gases collected over a fire. In September 1783, one of their balloons carried some animals about one mile. In November 1783, the brothers launched a balloon that carried Jean-François PILÂTRE DE ROZIER and the Marquis d'Arlandes over Paris. This was the first free flight.

month Time taken for the Moon to travel completely around the Earth. The sidereal month is the time of one revolution with respect to the stars and is equal to 27.32 days. As the Earth is in motion around the Sun, the synodic month – from full moon to full moon – is longer than the sidereal month and is equal to 29.53 days.

montmorillonite General name for a group of clay minerals, hydrous calcium-sodium aluminum-magnesium-iron silicate. The minerals are moisture-sensitive clays weathered out of IGNEOUS ROCKS, and are a common constituent of soil. Montmorillonite is used in the paper industry for carbonless copying paper. It is white, yellowish or gray. Hardness 1–2; r.d. 2.5.

moon Natural SATELLITE of a planet. The Earth's Moon, apart from the Sun, is the brightest object in the sky because of its proximity, being at a mean distance of only 384,000km (239,000mi). With an actual diameter of 3,476km (2,160mi), the Moon has 0.0123 of the Earth's mass and 0.0203 of its volume. The Earth and the Moon revolve around a common center of gravity. The gravitational attraction between the Moon and Earth produces TIDES. Although the Moon looks bright by contrast with the night sky, its surface rocks are dark. It is a cratered world, in many ways a typical SOLAR SYSTEM satellite. As the Moon orbits the Earth, it is seen to go through a sequence of PHASES as the proportion of the illuminated hemisphere visible to us changes. One complete sequence, from one new Moon, say, to the next, is called a **lunation**. The Moon's surface features may be broadly divided into the darker maria, which are low-lying volcanic plains, and the brighter highland regions (sometimes called terrae), which are found predominantly in the S part of the Moon's nearside and over the entire farside. There are impact features of all sizes. Other features include mountain peaks and ranges, valleys, and elongated depressions. The origin of the Moon is uncertain. A current theory, known as the **giant impactor** or **big splash theory**, is that a Mars-sized body collided with the newly formed Earth, and debris from the impact formed the Moon. Subsequent large impacts produced the Moon's basins, and smaller impacts the craters. The chemical composition of material brought back from the Moon has been found to consist mainly of silica, iron oxide, aluminum oxide, calcium oxide, titanium dioxide, and magnesium oxide. Lunar rocks are IGNEOUS ROCKS. The Moon has only the most tenuous of atmospheres. Apollo instruments detected traces of gases such as helium, neon, and argon. The surface temperature variation is extreme, from 100 to 400K. In 1998 an American spacecraft, the *Lunar Prospector*, found frozen water in craters of the Moon. It is believed that there could be between 11 million and 330 million tons of water on the Moon's surface at the poles.

moor Tract of bleak open land, generally on a plateau or mountain, found mainly in Great Britain. The vegetation is likely to be heather or bracken, with some coarse grasses.

Moore, Stanford (1913–82) US biochemist. He shared the 1972 Nobel Prize for chemistry with Christian ANFINSEN and William STEIN for research into the functioning and composition of the ENZYME ribonuclease. Anfinsen had discovered the composition of ribonuclease in the 1950s; subsequently Moore and Stein explained how it catalyzes the digestion of food. They used CHROMATOGRAPHY to analyze AMINO ACIDS and PEPTIDES and to determine the structure of ribonuclease. By 1973, using essentially the same technique, Moore and Stein had analyzed the structure of deoxyribonuclease.

moraine Accumulation of boulders and rock fragments that have been deposited by a GLACIER. Some of the rocks may have been eroded by the ice; some may be the result of freeze-thaw activity; and others are weathered blocks that have simply been transported by the ice. Morainic debris is carried by the ice and dumped at the sides or at the end, or at the bottom of the ice if it falls through crevasses. Accumulations at the side of a glacier are called **lateral moraines**; at the end of the glacier they are called **terminal moraines**. When two glaciers merge, two lateral moraines unite in the middle of the enlarged glacier to form a **medial moraine**. Beneath the ice there will be gradual accumulations of **ground moraine**, but much of it is reduced to small fragments and powder by the effect of the ice grinding it on the bedrock. Ground moraine can be an effective abrasive tool, if scraped along the valley floor by the movement of the ice.

mordant In dyeing processes, chemical that reacts with the dye or the fiber being dyed, or both, to "fix" the dye to the fiber, making it less likely to be washed out. Many mordants are metallic HYDROXIDES or SALTS (such as ALUM).

Morgan, Thomas Hunt (1866–1945) US biologist. He was awarded the 1933 Nobel Prize for physiology or medicine for the establishment of the CHROMOSOME theory of HEREDITY. His discovery of the function of chromosomes through experiments with the fruitfly (*Drosophila*) is related in his book *The Theory of the Gene* (1926, rev. ed. 1928).

Morley, Edward Williams (1838–1923) US chemist who worked with Albert MICHELSON on the MICHELSON-MORLEY EXPERIMENT in 1887. Their experiment demonstrated the absence of a hypothetical substance called "ether," which was supposed to pervade the Universe and carry light waves.

morphine White crystalline ALKALOID derived from OPIUM; first isolated in 1806. It depresses the CENTRAL NERVOUS SYSTEM and is used as an ANALGESIC for severe pain. An addictive drug, its use is associated with a number of side effects, including nausea. *See also* HEROIN

morphology Biological study of the form and structure of living things, especially the external form. It ranges from visible characteristics to microscopic structures, and often focuses on the relation between similar features in different organisms.

Morse code Series of signals used for sending TELEGRAPH messages, either along wires or by radio telegraphy. It consists of dots and dashes created by the interruption of a continuous electric current or radio signal. The dashes are three times the length of the dots. The codes for letters of the alphabet, or numbers, combine dots and dashes. It was named after its creator, US inventor and artist Samuel Morse (1791–1872).

mortar In building construction, a material that is used to bind brick, stone, tile or concrete blocks into a structure. Modern mortar generally consists of a mixture of cement, sand, and water; lime may be added to improve its spreading properties.

mortar In chemistry, a bowl of smooth, hard material in which softer substances are ground or beaten with a pestle, a club-shaped tool.

morula Early stage in the development of the EMBRYO in animals when the cells are in the process of splitting before the BLASTULA stage. The morula consists of a number of blastomeres, the cells formed from the fertilized OVUM as a result of CELL DIVISION.

Moseley, Henry Gwyn Jeffreys (1887–1915) British physicist. His initial studies involved RADIOACTIVITY; later he discovered a relationship between the X-ray spectra of the elements and their atomic numbers. He also discovered that the ATOMIC NUMBER of an element, and not its RELATIVE ATOMIC MASS, determines its major properties.

moss Any of about 14,000 species of small, simple, nonflowering green plants of the Bryophyta division, which typically grow in colonies, often forming dense carpets. They reproduce by means of SPORES produced in a capsule on a long stalk. The spores germinate into branching filaments, from which buds arise that grow into moss plants. Mosses grow on soil, rocks, and tree trunks in a wide variety of land habitats, especially in shady, damp places. *See also* ALTERNATION OF GENERATIONS; BRYOPHYTE; SPHAGNUM

Mössbauer, Rudolf Ludwig (1929–) German physicist. His doctoral thesis, published in 1958, dealt with the emission of GAMMA RADIATION by radioactive nuclei within crystals (*see* MÖSSBAUER EFFECT). For this work, he shared the 1961 Nobel Prize for physics with Robert HOFSTADTER. Mössbauer's discovery has been used to test Albert EINSTEIN's theory of general RELATIVITY.

Mössbauer effect RESONANCE effect observed in certain radioactive substances emitting GAMMA RADIATION. When such a radioactive atom is part of a CRYSTAL LATTICE, the recoil momentum is taken up by the atomic nuclei, so that the gamma rays have a sharply defined energy instead of a range of energies. The rays can then be reabsorbed by a similar nucleus. This effect was first studied by Rudolf MÖSSBAUER. Motion of the source alters the frequency of the rays, by the DOPPLER EFFECT, and the Mössbauer effect may be used to study nuclear energy levels and molecular structure with great precision.

motion, laws of Three laws proposed by Isaac NEWTON in his *Principia* (1687). They form the basis of the classical study of motion and FORCE. According to the **first law**, a body resists changes in its state of motion – a body at rest tends to remain at rest unless acted on by an external force, and a body in motion tends to remain in motion at the same velocity unless acted on by an external force. This property is known as INERTIA. The **second law** states that the change in velocity of a body as a result of a force is directly proportional to the force and inversely proportional to the mass of the body; if the change in velocity, or ACCELERATION, is a, the force is F, and the mass is m, then $a = F/m$. According to the **third law**, to every action there is an equal and opposite reaction.

motor Mechanism that converts energy (such as heat or electricity) into useful work. The term is sometimes applied to the internal combustion ENGINE (which converts heat produced by burning gases into reciprocating or rotary motion), but is more often applied to the ELECTRIC MOTOR (which converts electrical energy into rotary motion). ROCKETS are motors that can leave the Earth's atmosphere because they carry both fuel and oxidizer. Other types of motor presently being developed include the LINEAR MOTOR, a new motor proposed for high-speed train propulsion, and ION ENGINES, which employ ION PROPULSION to drive spacecraft.

motorcycle Powered, two-wheeled vehicle. Gottlieb DAIMLER is usually credited with building the first practical motorcycle. He mounted a

four-stroke INTERNAL COMBUSTION ENGINE on the frame of a wooden bicycle. Since then the motorcycle industry has developed increasingly powerful and efficient engines, and sophisticated frames that have improved the handling of the machines. Motorcycles are classified in terms of engine capacity, generally from 50cc to 1200cc. Smaller capacity engines tend to be air-cooled and use a TWO-STROKE ENGINE, while larger engines (125cc and above) are mostly FOUR-STROKE and are usually water-cooled. Transmission of power to the rear wheel is most commonly by chain, although some motorcycles have either belt or shaft transmission. The clutch, accelerator and front brake controls are usually on the handlebars. Foot pedals control the gear change and rear brake.

motor neuron (motor neurone) NERVE that carries messages to an EFFECTOR organ (usually a muscle) from the CENTRAL NERVOUS SYSTEM (CNS), thereby causing an appropriate response. The cell bodies of some motor neurons form part of the SPINAL CORD. The AXONS (electrochemically conductive extensions) of these nerves, which are sheathed with MYELIN (an insulating substance), pass from the spinal cord to connect with the muscles. These nerves are involved in spinal REFLEX ACTIONS. Motor neurons are, however, also linked to the BRAIN by the descending spinal tracts. The cell bodies of these tracts lie in the CEREBRAL CORTEX, and their axons, which pass down the spinal cord, connect the nerves to the brain, so allowing voluntary muscular control. Some motor neurons act on glands of the ENDOCRINE SYSTEM to make them release their hormone secretions. *See also* NEURON; SENSORY NEURON

Mott, Sir Nevill Francis (1905–96) British scientist. He made important contributions in the field of QUANTUM MECHANICS, especially in the

▲ **motion, laws of** Newton's first law of motion describes inertial effects. An object resists being moved from rest by toppling backwards (1). Once moving steadily forward, the object appears undisturbed, as if at rest (2). When stopped, it resists slowing and tends to continue moving (3).

theory of atomic SCATTERING. In 1977 Mott was awarded the Nobel Prize for physics, jointly with Philip Anderson (1923–) and John VAN VLECK, for their work on the electronic structures of magnetic and disordered systems.

Mottelson, Ben Roy (1926–) Danish physicist, b. US, who shared the 1975 Nobel Prize for physics with Aage Niels BOHR and James RAINWATER for their cooperative research into the structure of atomic nuclei. To explain why atomic nuclei were not spherical, Rainwater had suggested that nuclear particles form an inner nucleus and an outer nucleus, and these two interacting sets produce energy that deforms the symmetrical nucleus. From 1950 to 1953 Mottelson and Bohr, combining the shell and liquid drop nuclear models, provided experimental evidence to support this theory, which made nuclear FUSION a practical possibility.

Moulton, Forest Ray (1872–1952) US astronomer who put forward the PLANETESIMAL THEORY to account for the formation of the SOLAR SYSTEM. Together with Thomas Chamberlain (1843–1928), he proposed that relatively small aggregates (planetesimals) condensed out of the original solar nebula, and that these were attracted to each other by gravity and accreted to form the planets. He also suggested that the small moons of Saturn are captured asteroids.

mountain Part of the Earth's surface that rises conspicuously higher, at least 380m (1,250ft) higher than the surrounding area. Mountains have a restricted summit area, comparatively steep sides, and considerable bare rock surface. They are formed in three main ways. First, FOLD mountains, such as the Himalayas, are formed by a squeezing in of rock layers, caused by movements of the tectonic plates (*see* PLATE TECTONICS) of the Earth's CRUST. Second, BLOCK MOUNTAINS, such as the Sierra Nevada of North America, are formed by vertical movements between geological FAULTS, leading to the tilting of large blocks of STRATA. Third, a typical VOLCANO forms from the molten rock and ash that piles up around its original vent hole. The extent and speed of this process is illustrated by Parácutín, a Mexican volcano that rose out of a cornfield to a height of 450m (1,475ft) from its base between 1943 and 1952. Mountains may occur as single isolated masses, as ranges, and in systems or chains. *See also* OROGENISIS

mouse In computing, input device that can be operated with one hand. It is designed to fit the palm of the hand, with one or more buttons that can be pressed by the fingers of the same hand. When the mouse is moved around a flat surface, it controls the movement of a cursor or pointer on the COMPUTER screen. The buttons can be "clicked" onto icons or other responsive areas on the computer display. *See also* GRAPHICAL USER INTERFACE (GUI)

mouth In animals, the anterior (front) end of the ALIMENTARY CANAL, where it opens to the outside. In humans and other higher animals it is the cavity within the JAWS, containing the TEETH and TONGUE.

MS-DOS Abbreviation of Microsoft Disk Operating System. It was the original operating system produced in 1981 for IBM-compatible personal computers.

mucopolysaccharide Class of POLYSACCHARIDE molecules composed of aminosugars that are linked into repeating units to give a linear, unbranched, polymeric compound. They are structurally similar to GLYCOGEN and STARCH. CHITIN is a mucopolysaccharide that acts like CELLULOSE as a structural polysaccharide for many phyla of lower plants and insects.

mucous membrane Sheet of TISSUE (or EPITHELIUM) lining all body channels that communicate with the air, such as the mouth and respiratory tract, the digestive and urogenital tracts and the various glands that secrete MUCUS. The membrane contains gland cells that secrete mucus, which serves for lubrication and protection.

mucus Slippery, viscous fluid containing mucin, produced by MUCOUS MEMBRANES of the body. It serves for lubrication and protection. Nasal mucus traps airborne particles; mucus of the stomach protects the lining from irritation by hydrochloric acid secreted during DIGESTION.

mudstone Rock made of consolidated mud that, although firmer than clay lacks the laminated structure and tendency to cleave to SHALE and sometimes decomposes into mud when exposed to the atmosphere.

Muir, John (1838–1914) US naturalist, b. Scotland. An advocate of forest preservation, he was influential in the establishment of many national parks. Muir studied the glaciers and forests of the Sierra Nevada and explored Alaska, where he discovered Glacier Bay and Muir Glacier. By 1867 he was urging the US government to endorse a policy of forest conservation, and his writings and lobbying influenced conservation programs of both Presidents Grover Cleveland and Theodore Roosevelt.

Muller, Hermann Joseph (1890–1967) US geneticist. He found that he could artifically increase the rate of MUTATIONS in the fruitfly (*Drosophila*) by the use of X rays. He thus highlighted the human risk in exposure to RADIOACTIVITY. For this work Muller was awarded the 1946 Nobel Prize for physiology or medicine.

Müller, Johannes Peter (1801–58) German physiologist and anatomist. Regarded as one of the founders of modern physiology, he conducted pioneering research into the NERVOUS SYSTEM, LYMPH and BLOOD systems, hearing, and embryology.

Müller, Paul Hermann (1899–1965) Swiss chemist who was awarded the 1948 Nobel Prize for physiology or medicine for his discovery of the use of DDT as an INSECTICIDE. In 1935 he began the search for a perfect insecticide, one toxic for a wide variety of insects with few poisonous effects on other life forms. He concentrated on chlorine-containing compounds and in 1939 tested DDT. In 1944 DDT was successfully employed against a typhus epidemic in Naples, and for more than 20 years was the world's most widely used insecticide. In the 1970s, however, it was implicated as a hazard to animal life because it persists in FOOD CHAINS; its use has been banned in many countries.

Mulliken, Robert Sanderson (1896–1986) US chemist and physicist who was awarded the 1966 Nobel Prize for chemistry for his fundamental work on CHEMICAL BONDS and the MOLECULAR ORBITAL THEORY. He studied at the Massachusetts Institute of Technology and the University of Chicago, becoming a professor at the latter in 1937. He later worked on the development of the atom bomb and devised a scale of electronegativity of the elements.

Mullis, Kary (1944–) US biochemist who discovered how to copy minute quantities of DNA millions of times for research purposes. This technique, POLYMERASE CHAIN REACTION (PCR), has been used in a variety of ways, including tests for HIV and for DNA forensic-science investigations. For his research, Mullis shared the 1993 Nobel Prize for chemistry with Michael SMITH, who worked independently on altering the genetic code.

multimedia COMPUTER system that includes text, audio (sound), and video (picture) components. Often the user can interact with the system to interrogate it or even to control or contribute to what is happening on screen. A multimedia product is usually produced on a CD-ROM, which has the necessary high-storage capacity for the audio and graphics elements. A computer to exploit it therefore has to have a CD-ROM drive, together with a high-resolution color monitor and a sound card. The computer must have a minimum of 8 megabytes of RAM and at least an 80486 processor.

multiple birth Bearing of several offspring. Multiple BIRTHS are common among small MAMMALS but relatively rare in human beings, in whom the most usual form is TWINS. Identical twins develop from a single OVUM that has divided shortly after FERTILIZATION by a single SPERM; fraternal twins arise from two individually fertilized OVA. The use of FERTILITY DRUGS is associated with an increased incidence of multiple births.

multiple myeloma Malignant TUMOR of the bone MARROW. Occurring most often in older people, it leads to anemia, collapse of the vertebrae, blood clotting problems, and damage to the eyes and internal organs. It is treated with CHEMOTHERAPY and RADIOTHERAPY.

multiple sclerosis (MS) Incurable disease in which there is degeneration of the MYELIN SHEATH that surrounds nerves in the brain and SPINAL CORD. Striking mostly young adults (more women than men), it is mainly a disease of the world's temperate zones. Symptoms vary according to which nerves are affected, but may include unsteadiness, loss of coordination and speech, and visual disturbances. Affected people typically have relapses and remissions over many years. Its cause is unknown, but previous infection may play a part in genetically susceptible people.

multiple star Gravitationally connected group of more than two stars. An example is Castor, which has six components. In such systems there is a recognizable hierarchy or order. In Castor, the two main pairs rotate about a common center of gravity, while a pair of cool RED DWARFS rotates about the same center but at a larger distance.

multiplet In SPECTROSCOPY, a set of closely spaced, spectral lines arising from the splitting of a single line by, for example, electron SPIN. In NUCLEAR PHYSICS, a group of elementary particles, all HADRONS with about the same mass, identical in all other properties except electric charge, usually having two or three members. The NUCLEONS and PIONS form multiplets. In STRONG NUCLEAR FORCE interactions, members of a multiplet are all equivalent. A supermultiplet is a larger, more sophisticated and symmetrical grouping of hadrons involving eight QUANTUM NUMBERS, all of whose members have identical spin.

multiplication Arithmetical operation signified by × (interpreted as repeated ADDITION). $a \times b$ is $a + a + \ldots + a$, in which terms are added b times. In $a \times b$ ("a multiplied by b"), a is the **multiplicand**, b the **multiplier**, and the result is the **product**. *See also* ARITHMETIC

mumps Viral disease, most common in children, characterized by fever, pain, and swelling of one or both parotid salivary glands (located just in front of the ears). It has an incubation period of about 18 to 22 days and in children usually takes a relatively mild course. MENINGITIS is an occasional complication, and the infection may involve other organs such as the pancreas. The symptoms are more serious in adults and in men inflammation of the testes (**orchitis**) may occur, with the risk of sterility. Children over 18 months of age can be vaccinated against mumps, MEASLES, and rubella (German measles) (MMR vaccine). One attack of mumps generally confers lifelong immunity.

muon (symbol μ^-) Negatively charged ELEMENTARY PARTICLE, originally thought to be a MESON but now classified as a LEPTON. It has SPIN $^1/_2$, a mass about 207 times that of the ELECTRON, and decays weakly into an electron, a NEUTRINO, and an antineutrino. *See also* ANTIMATTER

Murchison, Sir Roderick Impey (1792–1871) Scottish geologist. His work on the rocks underlying the Old Red Sandstone in S Wales led to the definition of the strata known as the SILURIAN system (corresponding to that period of geological time). He worked with Sir Charles LYELL on the Auvergne volcanics and with Adam SEDGWICK on the structure of the Alps. He later collaborated with Sedgwick on a study of the rocks that were to become known as the DEVONIAN system. He is also responsible for founding the PERMIAN system.

Murdock, William (1754–1839) Scottish engineer and pioneer of gas lighting. In 1792 he began experiments to produce COAL GAS for lighting. He also devised ways of purifying and washing the gas. In 1802 he used coal gas to illuminate the exterior of a factory in the Soho area of Birmingham, England, where he worked for Matthew BOULTON and James WATT. Murdock also invented GEARS and valves for use on STEAM ENGINES.

Musca (fly) Small distinctive, constellation of the S sky. Its brightest star, Alpha, has magnitude 2.7.

muscle Tissue that has the ability to contract, enabling movement. There are three basic types: VOLUNTARY MUSCLE (or skeletal muscle), INVOLUNTARY MUSCLE (or smooth muscle) and cardiac muscle. **Voluntary** muscle is the largest tissue component of the human body, comprising about 40% by weight. It is attached by TENDONS to the BONES of the SKELETON and is characterized by cross-markings known as striations; it typically contains many nuclei per cell. Most voluntary muscles require conscious effort for contraction (when they move the limbs and body). A muscle whose contraction causes a limb or a part of the body to straighten (extend) is called an EXTENSOR. A muscle whose contraction causes a limb or part of the body to bend (flex) is called a FLEXOR. **Involuntary** muscle lines the digestive tract, blood vessels, and many other organs. It is not striated and typically has only one nucleus per cell; it is not under conscious control, and is therefore known as involuntary muscle. **Cardiac** muscle is found only in the HEART and differs from the other types of muscle in that it beats rhythmically and does not need stimulation by a nerve impulse to contract. Cardiac muscle has some striations (but not as many as in voluntary muscle) and has only one nucleus per cell.

muscovite Sheet silicate mineral; a form of hydrous potassium aluminum silicate ($KAl_2(Si_3Al)O_{10}(OH)_2$), and the most common MICA. Found in many kinds of rocks, muscovite crystallizes in hexagonal, tabular forms in the monoclinic system. Its name derives from its use as a glass in Muscovy, Russia. It is clear or tinted with varied lusters. Hardness 2–2.5; r.d. 2.9.

muscular dystrophy Any of a group of hereditary disorders in which the characteristic feature is progressive weakening and atrophy of the muscles. The commonest type, **Duchenne** muscular dystrophy, affects boys, usually before the age of four. Muscle fibers degenerate, to be replaced by fatty tissue. The child develops a characteristic waddling gait and an inward curvature of the lower spine; later there is involvement of the respiratory muscles. There is no cure, and Duchenne victims (one in 3,000 baby boys) usually die before the age of 20.

muskeg Area of boggy ground found in the N TUNDRA regions. Most of the ground in such regions, such as the tundra of N Canada and Russia, is PERMAFROST. But in summer the top few centimeters of soil thaw, and the MELTWATER, unable to run away, forms muskegs. The bogs support lichens and mosses, which are important food for herds of reindeer.

mutagen Anything that causes an increase in the number of MUTATIONS in a population of organisms. Mutagens work by affecting the GENES, thus producing inherited defects. They may damage chromosomes, or change the DNA and affect the DNA code. Mutagens include chemicals (such as colchicine) and ionizing radiation (such as radioactivity and X rays). Some mutagens have been shown to cause CANCER.

mutation Sudden change in an inherited characteristic of an organism. This change occurs in the

DNA of the GENES. Natural mutations during reproduction are rare, occur randomly, and usually produce an organism unable to survive in its environment. Occasionally the change results in the organism being better adapted to its environment, and through NATURAL SELECTION the altered gene may pass on to the next generation. Natural mutation is therefore one of the key means by which organisms evolve. The mutation rate can be increased by exposing genetic material to IONIZING RADIATION, such as X RAYS or ULTRAVIOLET RADIATION, or mutagenic chemicals. *See also* EVOLUTION

mutualism Relationship with mutual benefits for the two or more organisms involved. *See* SYMBIOSIS

mycelium Vegetative body of a FUNGUS, found underground. It is made up of a web of filaments (HYPHAE), sometimes massed like felt. The hyphae are of two types, those that feed and those responsible for reproduction.

mycology Science and study of FUNGUS.

mycoplasma Any of about 15 species of the smallest known form of cellular life; they are usually considered to be BACTERIA. Most are PARASITES on birds and mammals, some live in stagnant water, and some play a part in such diseases as pleuropneumonia in cattle.

mycorrhiza (fungus root) Association between certain fungi and the root cells of some VASCULAR PLANTS. The FUNGUS may penetrate the root cells or form a mesh around them. Water and minerals enter the roots via these threads. Sometimes the fungus digests organic material for the plant. *See also* SYMBIOSIS

myelin sheath Protective layer around the AXONS of peripheral and some central NERVE fibers. It insulates the fiber to prevent loss of electrical impulses and aids conduction. The myelin sheath is composed of specialized cells (SCHWANN CELLS) made up of proteins and fats. The disease MULTIPLE SCLEROSIS features degeneration of the myelin sheath.

mylonite Any of several laminated, fine-grained rocks formed when layers of parent rock fault, granulate or flow. It is chemically stable but partly melted and is reduced to a powder by the movement of rock along a fault line. It generally contains fragments of the parent rock.

myoglobin Protein found in animals. In vertebrates it is the pigment producing the red color of MUSCLE tissue. Like HEMOGLOBIN, myoglobin combines readily with oxygen for use in rapidly contracting muscles. It has been used extensively in research into the structure of PROTEINS. In 1962 John KENDREW shared the Nobel Prize for chemistry for his construction of a three-dimensional crystalline model of sperm whale myoglobin. The prize was shared with Max Perutz, who used a similar process to discover the structure of hemoglobin.

myopia (nearsightedness) Common disorder of vision in which near objects are seen sharply (in focus) but distant objects are hazy (out of focus). It is caused either by the eyeball being too long or by the EYE's lens being too powerful, so that light rays entering the eye focus in front of the RETINA. It is corrected with concave LENSES in glasses or contact lenses. *See also* HYPERMETROPIA

myosin Thick filamentous PROTEIN present in MUSCLE cells, associated with ACTIN in the contractile process.

myriapod Member of the Myriapoda class of ARTHROPODS with bodies made up of many similar segments. Each segment has one or more pairs of legs. Centipedes and millipedes are myriapods.

myxomatosis VIRUS disease of rabbits that is sometimes spread deliberately by farmers to exterminate them, notably in Australia where early campaigns killed many millions of animals. The disease is generally fatal, although there is evidence that some rabbit populations are becoming resistant. Among rabbit populations, the disease is spread by physical contact and by insect VECTORS such as mosquitoes.

nadir Point on the CELESTIAL SPHERE vertically below the observer. It is diametrically opposite the ZENITH.

nail In anatomy, tough outgrowth from the fingers and toes of primates. Nails, like FEATHERS, are made of the same substance as HAIR, the fibrous protein KERATIN.

nanometer (symbol nm) Unit of length, equal to 10^{-9} meter (one-billionth of a meter). It is used in the measurement of intermolecular distances and wavelengths. It has superseded the ANGSTROM as the accepted unit for such measurements.

nanotechnology Micromechanics used to develop working devices the size of a few NANOMETERS, one-billionth of a meter. US scientists have already etched from silicon an electric motor that is smaller than 0.1mm wide and have made workable gears with a diameter less than a human hair. Further developments have resulted in the manipulation of electrons and individual atoms.

napalm Gelatinous PETROLEUM used to make bombs and as fuel for flamethrowers. It was developed during World War II and was widely used by US forces in the Vietnam War. When napalm hits its target, it spreads out, clinging to and burning everything it touches.

naphtha Any of several volatile liquid-hydrocarbon mixtures. In the 1st century AD, "naphtha" was mentioned by Pliny the Elder. Alchemists used the word for various liquids of low boiling point. Several types of products are now called naphtha, including coal-tar naphtha and petroleum naphtha. Petroleum naphtha contains ALIPHATIC COMPOUNDS and boils at higher temperatures than gasoline and at lower temperatures than kerosene.

naphthalene Important hydrocarbon ($C_{10}H_8$) composed of two BENZENE rings joined together. A white, waxy solid, naphthalene is soluble in ether and hot alcohol and is highly volatile. It is used in mothballs, dyes, and synthetic resins, and is obtained from coal tar and the high-temperature CRACKING of

petroleum. It crystallizes in white plates. Properties: m.p. 80°C (176°F); b.p. 218°C (424°F).

Napier, John (1550–1617) Scottish mathematician. He developed "Napier bones," a calculating apparatus that he used to invent LOGARITHMS (1614) and the present form of decimal notation.

nappe Large-scale FOLD in rocks, thrown up by mountain-building processes and transported over large distances. The Alps are a system of several such nappes.

narcotic Any DRUG that induces sleep and/or relieves pain. The term is used especially in relation to OPIUM and its derivatives, such as HEROIN and MORPHINE. These drugs have largely been replaced as sedatives because of their addictive properties, but they are still used for severe pain, notably in terminal illness. Other narcotics include ALCOHOLS (such as ethanol) and BARBITURATES.

Narlikar, Jayant Vishnu (1938–) Indian physicist who has conducted research in such areas as general relativity and gravitation, quantum theory, cosmology, and astrophysics. His many publications include *Action at a Distance in Physics and Cosmology* (1974), written in collaboration with Sir Fred HOYLE.

NASA Acronym for NATIONAL AERONAUTICS AND SPACE ADMINISTRATION

nascent hydrogen Highly reactive form of hydrogen gas. It is generated within a chemical reaction mixture, usually by adding magnesium or zinc to dilute hydrochloric acid, where it acts as a powerful REDUCING AGENT.

Nasmyth, James (1808–90) Scottish inventor and engineer. He developed a foundry for making machine tools and steam-powered machines. In 1842 he patented a steam hammer designed to forge the drive shafts originally intended for the SS *Great Britain*.

National Academy of Sciences US organization made up of elected members (based on original research in several fields of science). The academy acts as an official advisor to the federal government on science and technology matters. It was founded in 1863.

National Aeronautics and Space Administration (NASA) US government agency that organizes civilian aeronautical and space research programs. NASA was established in 1958. Its headquarters are in Washington, D.C., and it operates various field stations. These include the Goddard Space Flight Center in Greenbelt, Maryland, which handles space science research and Earth satellite tracking; the JET PROPULSION LABORATORY in Pasadena, California, which is mission control for NASA space probes and manages NASA's deep-space tracking stations around the world; the Ames Research Center, San Francisco, for aeronautics and planetary science; the John F. Kennedy Space Center in Cape Canaveral, Florida, the main US launch site; and the Lyndon B. Johnson Space Center in Houston, Texas, mission control for manned space missions. NASA employs more than 20,000 scientific and technical staff.

national parks Protected areas where restraint on the killing of wildlife is enforced, and forests, waters, and other natural environments are preserved from commercial use. The US was the first country to set aside large reserves for preservation and recreation, and Yellowstone National Park is the first of these to be established (1872). The degree of protection varies from country to country. The main purpose of the famous African national parks is game preservation.

National Physical Laboratory (NPL) Establishment responsible in the UK for national systems of measurement and for technical aspects of physical standards. Located at Teddington, England, it carries out research into many fields of physics and engineering. It was set up in 1900 under the Department of Industry and sponsored by the Royal Society.

natrolite Hydrated silicate mineral, hydrous sodium aluminum silicate ($Na_2Al_2Si_3O_{10}.2H_2O$). It has orthorhombic system, needlelike crystals, with radiating nodules or compact fibrous masses. It is colorless or white, glassy, and brittle. Hardness 5–5.5; r.d. 2.2. *See also* ZEOLITE

Natta, Giulio (1903–79) Italian chemist who shared the 1963 Nobel Prize for chemistry with Karl ZIEGLER for their work on POLYMERS. In 1953 Natta began studying very large molecules and succeeded in polymerizing POLYPROPYLENE, a substance of great commercial importance.

natural gas Naturally occurring, combustible, gaseous mixture of HYDROCARBONS trapped in pore spaces in sedimentary rocks. It is used as a fuel and in the production of plastics, drugs, antifreeze, and dyes. Natural gas is the gaseous component of PETROLEUM and is extracted from OIL WELLS. Certain wells, however, yield only natural gas. Before natural gas can be used as a fuel, the heavier hydrocarbons of BUTANE and PROPANE are extracted; in liquid form these hydrocarbons are forced into containers as bottled gas. The remaining gas,

called "dry gas," is piped to consumers for use as fuel. Dry gas is composed of the light hydrocarbons METHANE and ETHANE.

natural number *See* NUMBER, NATURAL

natural selection In EVOLUTION, theory that advantageous changes in an organism will tend to be passed on to successive generations. Changes arise out of natural genetic VARIATION, especially MUTATION. Those that give an individual organism a greater capacity for survival and reproduction in a particular environment will help it to produce more offspring bearing the same beneficial characteristic or trait. Thus the proportion of individuals in the population bearing that trait will increase through successive generations. This theory was put forward by Charles DARWIN in his book *The Origin of Species* (1859). It is still regarded as the key mechanism of evolution.

nature reserve Area of land, sometimes including inland waters and estuaries, set aside for the study and CONSERVATION of wildlife, HABITAT or geological features. Reserves are usually looked after by a warden or a conservation officer. The style of management on a reserve varies according to its main interest and purpose; some may allow permission to enter only for a serious study whereas others may encourage people to visit through a nature trail. *See* NATIONAL PARKS

naturopathy System of medical therapy that relies exclusively on the use of natural treatments, such as exposure to sunlight, fresh air, and a healthy diet of organically grown foods. Herbal remedies are preferred to manufactured drugs.

nautical mile Unit used to measure distances at sea. It is defined as the length of one minute of arc of the Earth's circumference. The international nautical mile is equal to 1,852m (6,076.04ft), although in the UK it is defined as 6,080ft (1,853.18m). A speed of one nautical mile per hour is called a KNOT, a term used both at sea and in the air.

navigation Process for determining the position of a vehicle (usually a ship or plane) and its course. Five main techniques are used: dead reckoning (the distance and the direction traveled are used to plot the postion); piloting (generally the navigation of ships using buoys, landmarks, and so on); celestial navigation; INERTIAL GUIDANCE (used in the navigation of missiles); and radio navigation. The latter includes the use of radio beacons, LORAN (long range navigation), RADAR navigation, and satellite navigation systems, such as GPS (GLOBAL POSITIONING SYSTEM). Instruments and charts enable the navigator to determine position, expressed in terms of LATITUDE and LONGITUDE, direction in degrees of arc from true north, speed, and distance traveled. *See also* COMPASS; GYRO-COMPASS; SEXTANT

Neanderthal Middle PALAEOLITHIC variety of human, known from fossils in Europe and Asia. Neanderthals were discovered when a skeleton was unearthed in the Neander Valley in w Germany in 1856. The bones were thick and powerfully built and the skull had a pronounced brow ridge. Neanderthals are now considered to be a separate species of human, possibly a local adaptation during the ice ages, and are not thought to be ancestral to modern humans. Neanderthals predated modern humans in Europe, but were superseded by them about 35,000 years ago. *See also* HUMAN EVOLUTION

neap tide *See* TIDE

near-sightedness *See* MYOPIA

nebula (Lat. cloud) Region of interstellar gas and dust. There are three main types of nebula. **Emission** nebulae are bright diffuse nebulae that emit light and other radiation as a result of ionization (removal of electrons) and excitation of the gas atoms by ultraviolet radiation. The source of the ultraviolet is usually one or more hot stars. In contrast, the brightness of **reflection** nebulae results from the scattering by dust particles of light from nearby stars. **Dark** nebulae are not luminous: interstellar gas and dust absorb light from background stars, producing apparently dark patches in the sky.

neck In vertebrates, the part of the body that connects the head and the trunk. In human beings, it contains the seven cervical VERTEBRAE of the spinal column and, to the front, the PHARYNX, leading to the TRACHEA and OESOPHAGUS.

necrosis Death of plant or animal tissue. It can be caused by disease, injury, or interference with the blood supply.

nectar Sweet liquid secreted by most flowering plants (ANGIOSPERMS). It consists mainly of a solution of GLUCOSE, FRUCTOSE, and SUCROSE in water. The glands (nectaries) that produce it usually lie at the base of the flower petals, but may be found also in parts of the stem or at the leaf bases. Nectar attracts insects, which help with cross-POLLINATION. Bees turn nectar into honey.

Néel, Louis Eugène Félix (1904–) French scientist who shared the 1970 Nobel Prize for physics with Hannes ALFVÉN for their work in SOLID-STATE PHYSICS. Néel studied the magnetic properties of solids and discovered antiferromagnetism, in which the magnetic fields of small groups of atoms are aligned in opposite directions. His discoveries have been widely used in computer memories.

nekton Active swimming organisms at the sea's surface, as opposed to the floating PLANKTON. Nekton includes the large migrating marine animals such as adult squids, fish, and whales. *See also* BENTHOS; PELAGIC

nematocyst In CNIDARIANS such as jellyfish, a tiny fluid-filled sac containing a coiled hollow thread that can be shot out at prey. It forms part of a thread cell (**nematoblast**) in the ECTODERM of the animal. When a "trigger" (**cnidocil**) on the thread cell is touched, the thread is rapidly projected to coil around the prey or to inject poison. The poisonous stings on the tentacles of jellyfish derive from thread cells and their nematocycts.

nematode Wormlike INVERTEBRATES of the phylum Aschelminthes. They are identified by their unsegmented, elongated, cylindrical bodies, pointed at both ends. Covered in thick cuticle and comprising longitudinal muscles, the nematodes are found in a variety of habitats including soil, and marine and freshwater environments. Although some species are free-living, others, such as *Acaris* (roundworms), are parasitic. Reproduction is sexual. Class Nematoda.

neo-Darwinism Development of DARWINISM that incorporates the modern ideas of genetic HEREDITY (notably MENDEL's) with DARWIN's ideas of EVOLUTION through NATURAL SELECTION. It uses research into GENETICS to explain the origin of VARIATION within a species.

neodymium (symbol Nd) Silver-yellow metallic element of the LANTHANIDE SERIES. It was first isolated in the form of its oxide in 1885 and the pure metal was first obtained in 1925. Chief ores are monazite and bastnaesite. It is used to manufacture lasers, and neodymium salts are used to color glass. Properties: at.no. 60; r.a.m. 144.24; r.d. 7.004; m.p. 1,010°C (1,850°F); b.p. 3,068°C (5,554°F); most common isotope ^{142}Nd (27.11%).

Neolithic (New Stone Age) Stage in human cultural development following the PALEOLITHIC. The Neolithic began around 8000 BC in w Asia, and around 4000 BC in Britain. It was during this period that people first lived in settled villages, domesticated and bred animals, cultivated cereal crops, and practised stone-grinding and flint mining.

neon (symbol Ne) Gaseous, nonmetallic element, a NOBLE GAS, discovered in 1898. Colorless and odorless, it is present in the atmosphere (0.0018% by volume) and is obtained by the fractional distillation of liquid air. Its main use is in discharge tubes for advertising signs (emitting a bright red glow while conducting electricity) and in gas lasers, Geiger counters, and particle detectors. The element forms no compounds, except briefly and under extreme conditions. Properties: at.no. 10; r.a.m. 20.179; m.p. $-248.67°C$ $(-415.6°F)$; b.p.$-246.05°C$ $(-410.89°F)$; most common isotope ^{20}Ne (90.92%).

neoplasm Any new or abnormal growth on the body. It may be malignant (cancerous) or benign (noncancerous).

neoprene $(CH_2=CCl.CH=CH_2,$ 2-chlorobuta-1,3diene) Synthetic RUBBER that is very resistant to corrosive chemicals and is widely used in the chemical industry. Neoprene is a POLYMER of chloroprene.

neoteny Persistence in an adult animal of larval characteristics. It includes the retention of GILLS, as in some salamanders, the best example of which is the axolotl of w USA and Mexico. An entire order of tunicates (sea squirts), the Larvacea, is permanently larval, never reaching a typical adult form.

nephritis (Bright's disease) Inflammation of the KIDNEY. It is a general term, used to describe a condition rather than any specific disease. It may be acute (due perhaps to a streptococcal infection) or chronic, often progressing to kidney failure.

nephron Basic functional unit of the mammalian KIDNEY, involved in the formation of URINE. There are more than a million nephrons in a human kidney. Each consists of a GLOMERULUS (a cluster of tiny blood capillaries) cupped in a structure called a BOWMAN'S CAPSULE, with an attached long, narrow tubule. Blood enters the kidney under pressure, and water and wastes are forced into the tubule by a process known as ultrafiltration. Some water and essential molecules are reabsorbed into the bloodstream; the remaining filtrate, urine, is passed to the BLADDER for voiding.

Neptune Eighth planet from the Sun. The mass, orbit and position of an unseen planet had been calculated by Urbain LEVERRIER and, independently, John Couch Adams to account for perturbations in the orbital motion of URANUS. Neptune was actually observed in 1846 by Johann GALLE and Heinrich d'Arrest. Neptune is invisible to the naked eye. Through a telescope, it appears as a small, greenish-blue disk on which very few details can be distinguished. In its size, mass, atmosphere, and color, Neptune resembles Uranus. In 1989 the flyby of the VOYAGER 2 probe in 1989 provided most of our current knowledge of the planet. The upper atmosphere is about 85% molecular hydrogen and 15%

▶ **neon** Neon is an unreactive element, yet it can be used to produce colored lights (A). Positively charged ions (1) strike a negative electrode, causing emission of electrons (2), which "energize" electrons in neon atoms (3). When the energized electrons return to their stable, "ground" state (4), red light is emitted. Similarly, a mercury vapor lamp (B) emits ultraviolet light (5). This may be absorbed by a fluorescent substance (6) that can then release the energy in stages (7).

A

negative electrode

positive electrode

switch

helium. Its predominant blue color is due to a trace of methane, which strongly absorbs red light. Several different atmospheric features were visible at the time of the Voyager encounter. There were faint bands parallel to the equator, and spots, the most prominent of which was the oval Great Dark Spot (GDS), about 12,500km (8,000mi) long and 7,500km (4,500mi) wide. It is about the same size relative to Neptune as the Great Red Spot (GRS) is to Jupiter, and like the GRS it is a giant anticyclone. White, cirrus-type clouds of methane crystals cast shadows on the main cloud deck some 50km (30mi) below. There are also the highest wind speeds recorded on any planet, over 2,000km/h (1,250mph) in places. Neptune's dynamic and changing atmosphere probably has to do with its internal heat source: it radiates more than twice as much heat as it receives from the Sun. Below the visible atmosphere, the composition and structure are less certain. Neptune has eight known SATELLITES and a ring system. **Data:** Diameter (equatorial): 49,528km; Mass: (Earth=1) 17.14; Volume: (Earth=1) 57.67; Density (water=1) 1.77; Orbital period: 164.8 years; Rotation period (equatorial): 16h 7m 0s; Temperature at cloud tops: 55K

neptunium (symbol Np) Radioactive metallic element, the first of the TRANSURANIC ELEMENTS to be synthesized (1940). The silvery element is found in small amounts in uranium ores and is obtained as a by-product in nuclear reactors. Properties: at.no. 93; r.a.m. 237.0482; r.d. 20.25; m.p. 640°C (1,184°F); b.p. 3,902°C (7,056°F); most stable isotope ^{237}Np (half-life 2.2 million years).

Nereid NEPTUNE's outermost satellite, discovered by Gerard KUIPER in 1949. It has the most eccentric orbit of all known satellites, taking it from 1,4 million km (850,000mi) from the planet out to 9,7 million km (6 million mi). It is almost certainly a captured body.

Nernst, Walther Hermann (1864–1941) German chemist who was a professor at Göttingen and then at Berlin. Nernst was awarded the 1920 Nobel Prize for chemistry for his discovery of the third law of THERMODYNAMICS. His other important work was concerned with chain reactions in photochemistry. He wrote *Thermodynamics* (1893) and *The New Heat Theorem* (1918).

nerve Collection of NEURONS providing a communications link between the vertebrate NERVOUS SYSTEM and other parts of the body. Afferent or sensory nerves (SENSORY NEURONS) transmit nervous impulses to the CENTRAL NERVOUS SYSTEM (CNS); efferent or motor nerves (MOTOR NEURONS) carry impulses away from the central nervous system to muscles. Some nerves contain both afferent and efferent fibers.

nerve cell *See* NEURON

nerve gas Extremely poisonous form of gas first developed as a weapon by the Germans in World War II but never used. Nerve gases are often organic chemical compounds containing phosphorus. Their effect is to inactivate the enzyme CHOLINESTERASE, and so prevent NERVE messages reaching the muscles. The result is death within minutes by failure of respiratory and other systems. Inhalation of even small amounts of the gas, or any physical contact with the liquid chemical, is enough to cause death.

nerve impulse Electrical signal that travels along the AXON of a nerve cell. Nerve impulses carry information throughout the NERVOUS SYSTEM. As an impulse passes, sodium and potassium ions flow into and out of the axon's membrane, creating a voltage reduction called the ACTION POTENTIAL. Nerve impulses can travel at up to 150m (500ft) per second.

nervous system Communications system consisting of interconnecting nerve cells or NEURONS that coordinate all life, growth, and physical and mental activity. In simple animals, such as jellyfish and sea anemones, it consists of a **nerve net** without a center, or brain. The vertebrate nervous system consists of the CENTRAL NERVOUS SYSTEM (CNS), comprising the BRAIN and SPINAL CORD, and, serving all parts of the body outside the CNS, the PERIPHERAL NERVOUS SYSTEM. *See also* AUTONOMIC NERVOUS SYSTEM (ANS)

network *See* COMPUTER NETWORK

Neumann, Franz Ernst (1798–1895) German scientist who, in two papers (1845, 1847), formulated the first mathematical theory of ELECTROMAGNETIC INDUCTION, the production of ELECTROMOTIVE FORCE (EMF) by changes in a magnetic field. Neumann also conducted research into optics and crystallography.

Neumann, John von *See* VON NEUMANN, JOHN

neuralgia Intense pain from a damaged nerve, possibly tracking along its course. It may be temporary or chronic. There are various forms, including trigeminal neuralgia, which features attacks of stabbing pain in the mouth area, and post-herpetic neuralgia, which follows an attack of Trigeminal neuralgia (once known as *tic douloureux*) and usually affects people over 60, especially women. When it occurs in younger people, it is often a symptom of serious neurological disease. Most cases of neuralgia are treated with anticonvulsant drugs.

neurology Branch of medicine dealing with the diagnosis and treatment of disorders of the NERVOUS SYSTEM.

neuron (neurone, nerve cell) Basic structural unit of the NERVOUS SYSTEM that enables rapid transmission of NERVE IMPULSES between different parts of the body. It is composed of a cell body, containing a nucleus, and a number of trailing, fingerlike processes. The largest of these is the AXON, which carries outgoing impulses, causing the release of NEUROTRANSMITTERS across a junction (SYNAPSE) resulting in the stimulation of another neuron; the rest are DENDRITES, which receive incoming impulses. *See also* MOTOR NEURON; SENSORY NEURON

neurosis Emotional disorder such as anxiety, depression, or various phobias. It is a form of mental illness in which the main disorder is of mood, but the person does not lose contact with reality as happens in PSYCHOSIS.

neurosurgery Surgical treatment of disorders of the NERVOUS SYSTEM.

neurotransmitter Any one of several dozen chemicals involved in communication between NEURONS (nerve cells) or between a nerve and muscle cells. When an electrical impulse arrives at a nerve ending, a neurotransmitter is released to carry the signal across the specialized junction (SYNAPSE) between the nerve cell and its neighbor. Fast-acting neurotransmitters either prompt nerves to fire or inhibit them from firing; slow acting ones seem to be involved in modifying the activity of whole groups of nerve cells. ACETYLCHOLINE, the neurotransmitter that causes muscle to contract, is the target of NERVE GASES. Some drugs work by dis-

rupting neurotransmission. *See also* ADRENALINE; NERVOUS SYSTEM; NORADRENALINE; SEROTONIN

neutralization In chemistry, mixing of equivalent amounts of an acid and a base in an aqueous medium until the mixture is neither acidic nor basic. It then has a pH of 7, which indicates a neutral solution. Neutralization is often accompanied by the generation of heat. *See also* TITRATION

neutrino (symbol v) Uncharged ELEMENTARY PARTICLE with no, or very low, mass, SPIN $\frac{1}{2}$, and traveling at, or near, the speed of light. Classified as a LEPTON, it has little reaction with matter and is difficult to detect (postulated in 1930 by Wolfgang PAULI, but not directly detected until 1956). Neutrinos are created and destroyed by particle decays that involve the WEAK NUCLEAR FORCE. There are three known species. The **electron** neutrino is closely associated with the ELECTRON and is produced when, for example, PROTONS and electrons react to form NEUTRONS, as in the Sun. The electron neutrino's antiparticle is more common and occurs when a neutron decays. The **muon** neutrino is associated with the MUON, and occurs in high-energy reactions. The **tau** neutrino is associated with the TAU PARTICLE. All have their own antiparticles, called **antineutrinos**.

neutron (symbol n) Uncharged ELEMENTARY PARTICLE that occurs in the atomic nuclei of all chemical elements, except the lightest isotope of HYDROGEN. It was first identified (1932) by James CHADWICK. Outside the nucleus, it is unstable,

A

B

cerebrum

cranial nerves

spinal cord

cervical nerves (8 pairs)

thoracic nerves (12 pairs)

lumbar nerves (5 pairs)

sacral nerves (5 pairs)

sensory root

spinal nerve

vertebra

motor root

sympathetic ganglia of central nervous system

► **nervous system** The nervous system is divided into two parts: central and peripheral. The central nervous system (CNS) (A) includes the brain and spinal cord. It receives information, makes decisions, and transmits instructions. The peripheral nervous system (B) consists chiefly of nerve fibers leading to and from the CNS. It cannot make "decisions" and acts only as a message transmitter.

decaying with a half-life of 11.6 minutes into a PROTON, ELECTRON, and antineutrino. Its neutrality allows it to penetrate and be absorbed in nuclei and thus to induce nuclear TRANSMUTATION and FISSION. It is a BARYON with SPIN $\frac{1}{2}$, and a mass slightly greater than that of the proton.

neutron activation analysis Highly sensitive method of identifying the chemical composition of a substance. The sample is bombarded with high-energy NEUTRONS that are then absorbed by the atoms present. The resulting radioactive nuclei emit radiation of an energy and decay rate characteristic of the original atoms, which can thus be identified. The quantity present can also be found with extreme precision.

neutron bomb See NUCLEAR WEAPON

neutron flux Rate at which NEUTRONS are released during the nuclear FISSION process in a controlled nuclear CHAIN REACTION. In a NUCLEAR REACTOR, MODERATORS of several kinds (especially DEUTERIUM and GRAPHITE) are used to reduce neutron speeds. See also FAST NEUTRON

neutron star Extremely small, dense star that consists mostly of NEUTRONS. It is the last stage of many stars. Neutron stars are formed when a massive star explodes as a SUPERNOVA, blasting off its outer layers and compressing the core so that its component PROTONS and ELECTRONS merge into neutrons. They are observed as PULSARS. They have masses comparable to that of the Sun, but diameters of only about 20km (12mi) and average densities of about 10^{15} g/cm^3. The maximum mass for a neutron star is about 3 solar masses, beyond which it would collapse further, into a BLACK HOLE. See also STELLAR EVOLUTION

névé See FIRN

nevus Discolored patch on the skin that has been present since birth, also termed "mole" or "birthmark." There are many different types, including those caused by clusters of tiny blood vessels just below the skin surface, as in a "strawberry mark."

Newcomb, Simon (1835–1909) US mathematical astronomer, b. Canada. He calculated highly accurate values for astronomical constants, such as the solar PARALLAX, and prepared extremely accurate tables of the motions of the Moon and the planets. It was Newcomb who calculated the discrepancy in the advance of Mercury's PERIHELION that was to provide a proof of Einstein's special theory of RELATIVITY.

Newcomen, Thomas (1663–1729) British engineer and inventor of the atmospheric STEAM ENGINE. Assisted by John Calley, a plumber, he spent more than 10 years perfecting the invention. In 1705 Newcomen built a model of the engine, which consisted of a piston moved by atmospheric pressure within a CYLINDER in which a partial vacuum had been created by condensing steam. A 1698 patent by Thomas SAVERY covered the principle, and Newcomen went into partnership with him. For many years, this type of engine was used to power pumps, particularly to drain mines.

New General Catalogue (NGC) In astronomy, list of stellar clusters, nebulae, and galaxies compiled by Danish astronomer Johan Dreyer in 1888, based on the observations of William HERSCHEL and his son, John. Dreyer later published two Index Catalogues (IC) bringing the number of stellar objects listed in the catalogue to approximately 13,000.

Newlands, John Alexander Reina (1837–98) British chemist. In 1864 Newlands announced his law of octaves, which arranged the chemical elements in a table of eight columns according to increasing relative atomic mass. The law was ridiculed until MENDELEYEV incorporated it in his own PERIODIC TABLE five years later.

Newton, Alfred (1829–1907) British zoologist, b. Switzerland. Newton promoted the first laws for the protection of birds, and was the editor of the ornithological journals *Ibis* and *The Zoological Record*. He wrote *A Dictionary of Birds* (1893–96).

Newton, Sir Isaac (1642–1727) English scientist who proposed a theory of MECHANICS and of GRAVITATION that survived unchallenged until the 20th century. He studied at Cambridge and became professor of mathematics there (1669–1701). Many of his discoveries were made between 1664 and 1666, but not published until later. His main works were *Philosophiae Naturalis Principia Mathematica* (1687), in which he proposed the three laws of MOTION, and *Opticks* (1704). In the former, he also proposed the principle of universal GRAVITATION; in the latter, he showed that white light is made up of colors of the SPECTRUM and put forward his particle theory of light. In the 1660s Newton also created the first system of CALCULUS, but did not publish it until Gottfried LEIBNIZ had published his own system in 1684. He built a reflecting telescope around 1671, and studied alchemy and biblical chronology. Newton was president of the Royal Society (1703–27) and master of the mint after 1701. Newton engaged in acrimonious scientific disputes with Leibniz, Robert HOOKE, and John FLAMSTEED.

newton (symbol N) SI unit of FORCE. One newton is the force that gives a mass of one kilogram an acceleration of one meter per second per second. One kilogram weighs 9.807N; one pound weight equals 4.45N.

Newton's laws of gravitation *See* GRAVITATION

Newton's laws of motion *See* MOTION, LAWS OF

niccolite (kupfernickel) One of the chief ores of NICKEL, consisting of nickel arsenide (NiAs), often also with some cobalt, iron, and sulfur. It has hexagonal system of crystals generally in columnar masses, rarely as tabular crystals. It is found in vein deposits and is copper colored, with an easily tarnished metallic luster. Hardness 5–5.5; r.d. 7.8.

niche In ecology, the status of an organism in a particular ECOSYSTEM or COMMUNITY. From the point of view of the organism, a niche can be defined in terms of its food supply and the presence or absence of competition. Ecologically it can be defined by its temperature and moisture content. It is a fundamental principle of ecology that no two species can occupy exactly the same niche.

nickel (symbol Ni) Silver-white, TRANSITION ELEMENT, first discovered in 1751. Its chief ores are nickel-iron sulfide (pentlandite) and nickel arsenide (niccolite). Refining is complex, and involves the differential separation of the various compounds to obtain concentrated nickel sulfide (or arsenide), which is reduced to the pure nickel. Hard, malleable and ductile, nickel is used in stainless steels, other special alloys, coinage, cutlery, storage batteries, and as a hydrogenation catalyst. The metal is ferromagnetic. Properties: at.no. 28; r.a.m. 58.71; r.d. 8.90 (25°C); m.p. 1,453°C (2,647°F); b.p. 2,732°C (4,950°F); most common isotope ^{58}Ni (67.84%).

nickel carbonyl (Ni(CO)$_4$) Colorless, flammable, toxic liquid. It is used in the Mond process (*see* MOND, LUDWIG) for the production of pure nickel. Properties: r.d. 1.32g cm^{-3}; m.p. -25°C (-13°F); b.p. 43°C (109.4°F).

nickel silver (German silver) Alloy of copper and nickel. It may also contain small amounts of zinc. Nickel silver is hard and does not tarnish easily. It is used for making coins and cutlery, when it is often plated with silver to give E.P.N.S. (electroplated nickel silver).

Nicolle, Charles Jules Henri (1866–1936) French bacteriologist. Nicolle was awarded the 1928 Nobel Prize for physiology or medicine for his discovery (1909) that typhus is transmitted by the body louse. He later distinguished between classical typhus and murine typhus, which is passed on to humans by the rat flea. Nicolle was head of the Pasteur Institute in Tunis (1902–32), which under his direction became a distinguished center for bacteriological research.

Nicol prism Optical device used for the production of POLARIZED LIGHT. It was invented (1828) by the British physicist William Nicol (1768–1851). A Nicol PRISM consists of a rectangular crystal of ICELAND SPAR (calcite) split along a diagonal and rejoined with Canada balsam (a type of resin). Light passing through it is separated into a reflected (ordinary) ray and a transmitted (extraordinary) ray, both polarized in different planes. It is used in the polarizing MICROSCOPE, and also in GEOLOGY to examine thin sections of rock microscopically.

nicotine Poisonous ALKALOID obtained from the leaves of tobacco, used in agriculture as a pesticide and in veterinary medicine to kill external parasites. Nicotine is the principal addictive agent in smoking tobacco.

nicotinic acid Vitamin of the B complex, lack of which causes the disease PELLAGRA.

nimbus CLOUD from which rain is discharged. In meteorological cloud classification it is added as a suffix to the names of clouds that typically produce rain or snow, such as cumulonimbus.

niobium (symbol Nb) Shiny, gray-white TRANSITION ELEMENT, first discovered in 1801. Its chief ore is pyrochlore. Soft and ductile, niobium is used in special stainless steels and in alloys for rockets and jet engines. Properties: at.no. 41; r.a.m. 92.9064; r.d. 8.57; m.p. 2,468°C (4,474°F); b.p. 4,742°C (8,568°F); most common isotope ^{93}Nb (100%).

Nirenberg, Marshall Warren (1927–) US biochemist who found the key to the amino acid GENETIC CODE by deciphering different combinations of three nucleotide bases (called "codons") within long nucleotide chains in DNA and RNA. Each combination is coded to convert a different AMINO ACID to PROTEIN, a key process in transferring inherited characteristics. Nirenberg found he could decipher the unknown configurations by synthesizing a nucleic acid with a known base combination, and then recording the amino acid that it changed to protein. He was awarded the 1968 Nobel Prize for physiology or medicine, together with Robert W. HOLLEY and Har Gobind KHORANA, for his part in discovering how GENES determine cell function.

niter (saltpeter) Commercial form of POTASSIUM NITRATE (KNO$_3$), known and used since medieval times as an ingredient of GUNPOWDER.

nitrate Salt of NITRIC ACID (HNO_3). Nitrate salts contain the nitrate ion (NO_3^-) and some are important naturally occurring compounds, such as saltpeter (potassium nitrate, KNO_3) and Chile saltpeter (sodium nitrate, $NaNO_3$). Nitrates are used as food preservers, fertilizers, explosives, and as a source of nitric acid.

nitration Introduction of a nitro group ($-NO_2$) to the carbon atom of an organic compound, usually replacing a hydrogen atom. Strong nitric acid or a NITRATE salt with sulfuric acid is a nitrating agent. The nitro compounds formed are useful chemical intermediates, as in the production of ANILINE from BENZENE, and several are explosives (such as trinitrotoluene, TNT).

nitre (saltpetre) Commercial form of potassium nitrate (KNO_3), known and used since medieval times as an ingredient of GUNPOWDER.

nitric acid (HNO_3) Colorless liquid, one of the strongest mineral acids. It fumes in air and soon turns yellow as the result of decomposition into NITROGEN DIOXIDE (NO_2). Nitric acid attacks most metals, resulting in the formation of NITRATES, and is a strong oxidizing agent. It is mostly made by oxidizing AMMONIA over a catalyst, and is used for many purposes, including the manufacture of agricultural chemicals, explosives, plastics, dyes, and rocket propellants.

nitric oxide *See* NITROGEN MONOXIDE

nitrile Organic chemical compound containing the cyanide group ($-C\equiv N$) connected to a HYDROCARBON group. Nitriles are used in organic synthesis, particularly for the manufacture of acrylic fabrics and synthetic RUBBER.

nitrobenzene ($C_6H_5NO_2$) Aromatic nitro compound, a pale yellow, highly toxic, oily liquid that smells of almonds. It freezes at 5.7°C (42.3°F) and boils at 211°C (411.6°F). Nitrobenzene is used in the manufacture of ANILINE and other organic chemicals.

nitrocellulose *See* CELLULOSE NITRATE

nitrogen (symbol N) Common gaseous, nonmetallic element of group V of the periodic table, discovered in 1772. Colorless and odorless, it is the major component of Earth's atmosphere (78% by volume), from which it is extracted by fractional distillation of liquid air. The NITROGEN CYCLE is an essential process for the existence of life on Earth. The main industrial use is in the HABER PROCESS, which produces ammonia for fertilizers and making nitric acid. Nitrogen compounds are used in fertilizers, explosives, dyes, foods, and drugs. The element is chemically inert. Properties: at.no. 7; r.a.m. 14.0067; r.d. 1.2506; m.p. −209.86°C (−345.75°F); b.p. −195.8°C (−320.4°F); most common isotope ^{14}N (99.76%).

nitrogen cycle Circulation of nitrogen through plants and animals in the BIOSPHERE. Nitrogen-fixing BACTERIA in the soil or in plant root nodules take free nitrogen from the soil and air to form the nitrogen compounds (nitrates) used by plants in ASSIMILATION (*see* NITROGEN FIXATION). HERBIVORES obtain their nitrogen from the plants, and in turn CARNIVORES obtain nitrogen by eating herbivores. SAPROPHYTES decompose the tissue of all the organisms concerned and the nitrogen is released back into the cycle.

nitrogen dioxide (NO_2) Oxide of nitrogen. It is a pungent-smelling, brown gas that readily forms the dimer (two identical molecules linked together) dinitrogen tetroxide (N_2O_4). It is made by the action of concentrated nitric acid on copper, and dissolves in water to give a mixture of nitrous and nitric acids. It is also formed by the reaction of oxygen with NITROGEN MONOXIDE. Its presence in the atmosphere from the exhaust gases of gasoline engines contributes to the formation of ACID RAIN and the depletion of the OZONE LAYER.

nitrogen fixation Incorporation of atmospheric NITROGEN into chemicals for use by organisms. Nitrogen-fixing microorganisms (mainly BACTERIA and CYANOBACTERIA) absorb nitrogen gas from the air, from air spaces in the soil, or from water, and build it up into compounds of AMMONIA. Other bacteria then change these compounds into NITRATES, which can be taken up by plants. Some nitrogen-fixing bacteria live in symbiotic association with other organisms. Bacteria in the root nodules of plants, especially members of the pea family, exchange nitrogen compounds for carbohydrates. If the remains of these plants are plowed back into the soil, they enrich it with nitrates. Cyanobacteria are important nitrogen fixers in the sea, in fresh water, and in soil, and are particularly important in rice fields. *See also* NITROGEN CYCLE; SYMBIOSIS

nitrogen monoxide (nitric oxide, NO) Oxide of nitrogen, a colorless soluble gas. It is made by reducing nitric acid or by reacting sodium nitrite, sulfuric acid, and iron(II) (ferrous) sulfate. It reacts with oxygen to form NITROGEN DIOXIDE.

nitroglycerine Oily liquid used in the manufacture of explosives. It is made by reacting nitrating mixture (nitric and sulfuric acids) with glycerol (glycerine). Too dangerous to handle in any quantity on its own, it is made into DYNAMITE. It is also used in medicine, known as GLYCERYL TRINITRATE, for treating angina.

nitrous oxide *See* DINITROGEN OXIDE

nivation Type of mechanical WEATHERING of rocks that results from alternate freezing and thawing. At night, when temperatures fall, water lying in cracks in the rock freezes and expands. The following morning the ice melts and contracts, only to freeze again after nightfall. The continual expansion and contraction causes stresses that eventually break the rock along the crack or other line of weakness. Accumulated fragments of rock produced in this way form SCREE on hillslopes.

NMR *See* NUCLEAR MAGNETIC RESONANCE

Nobel, Alfred Bernhard (1833–96) Swedish chemist, engineer, and industrialist. In 1866 Nobel invented DYNAMITE (a mixture of NITROGLYCERINE and kieselguhr). In 1876 he patented a more powerful form of blasting gelatine, gelignite. Nobel was a pacifist and, with the immense fortune he made from the manufacture of explosives, he founded the NOBEL PRIZES, first awarded in 1901.

nobelium (symbol No) Radioactive, metallic element, one of the ACTINIDE SERIES. It was first made in 1958 by bombarding CURIUM with CARBON at the University of California at Berkeley. Seven isotopes are known. Properties: at.no. 102, most stable isotope ^{255}No (half-life 3 minutes). *See also* TRANSURANIC ELEMENT

Nobel Prize Awards given each year for outstanding contributions in the fields of physics, chemistry, physiology or medicine, literature, and economics, and to world peace. Established in 1901 by the will of Alfred NOBEL, the prizes are awarded annually on December 10. The winners are selected by committees based in Sweden and Norway.

Nobili, Leopoldo (1784–1835) Italian physicist, a pioneer of ELECTROCHEMISTRY. He generated electricity using platinum ELECTRODES in an alkaline nitrate ELECTROLYTE and devised the astatic GALVANOMETER to measure the current. Nobili also invented the thermocouple and the thermopile (two types of thermometers).

noble gas (inert gas) Any of a group of colorless, odorless, and unreactive gases. It includes (in order of increasing atomic number) HELIUM, NEON, ARGON, KRYPTON, XENON, and RADON and forms group 0 of the periodic table. The noble gases' low reactivity is due to complete outer electron shells (two electrons for helium and eight each for the rest), thus offering no VALENCE "hooks." The noble gases were once thought to be completely inert (nonreactive), hence their alternative name, but compounds have been identified for krypton, xenon, radon, and argon. They are also called the "rare gases," although argon and helium are relatively common.

noble metal Any of a group of unreactive metals found naturally in the native, metallic state and not usually as chemical compounds or ores. The group includes SILVER, GOLD, PLATINUM, OSMIUM, IRIDIUM, PALLADIUM, RHODIUM, and RUTHENIUM. However, COPPER, also sometimes found as the native metal, is not included, whereas MERCURY, found both as the metal and as ores such as CINNABAR, often is, so that the noble metals are at best a rather loose assemblage. Gold and platinum are precious metals used in jewelry for their rarity, luster, resistance to tarnishing, ductility, and malleability. Osmium, the heaviest of all metals, is alloyed with iridium to make osmiridium, used to tip fountain pen nibs. Platinum and palladium are used in industry as CATALYSTS. *See also* BASE METAL

node In anatomy, a node is a thickening or enlargement of an organ or tissue, such as a lymph node (gland) or the sinoatrial node of nervous tissue in the heart, which controls the heartbeat. In botany, a node is the position on the stem of a plant from which a leaf or leaves grow. In physics, nodes are points of minimum displacement (zero amplitude) on a STANDING WAVE such as the stationary points on a vibrating string.

node In astronomy, point at which one ORBIT cuts another. Specifically, a node is one of the two points in the orbit of a planet or comet at which it cuts the ECLIPTIC, or at which the orbit of a satellite cuts that of its primary. The ascending node is the node at which a celestial body passes from south to north; the descending node is the node at which a celestial body passes from north to south.

Noether, Amalie Emmy (1882–1935) German mathematician. Noether was invited by David HILBERT to lecture and conduct research at the University of Göttingen. She remained there until the Nazi purges of Jewish academics in 1933 forced her to seek refuge in the US. Her main work was in the field of abstract ALGEBRA, where she contributed significantly to the general theory of ideals and to noncommutative algebras.

Noguchi, Hideyo (1876–1928) Japanese bacteriologist known for isolating the causative agent of syphilis in the CENTRAL NERVOUS SYSTEM. Noguchi improved the technique of the WASSERMANN reaction and devised the Wassermann skin test.

noise Any unwanted SOUND, which may come from any source, including traffic and factory machinery. Experiments carried out to determine the effects of noise on workers in manufacturing industries show that noise seriously affects only that output requiring mental concentration. Nevertheless, noise is often regarded as a form of acoustic pollution, and noise-abatement pressure groups lobby for legislation. Noise is usually measured on a DECIBEL scale: conversation is generally about 60dB and pain starts at about 120dB.

non-Euclidean geometry Self-consistent GEOMETRY that uses a different set of AXIOMS from those of EUCLIDEAN GEOMETRY, in particular a set that does not include the parallel postulate. Euclid's fifth postulate is equivalent to the statement that if a point lies outside a line, then only one line can be drawn through that point that does not cut the first line. In the early 19th century Farkas Bolyai and, independently, Nikolai Ivanovich LOBACHEVSKY developed systems of geometry in which an infinite number of parallels could exist. This system, called hyperbolic non-Euclidean geometry, was self-consistent; that is, it had no inherent contradiction in the results obtained from the postulates. Later, Georg RIEMANN introduced a form (elliptical geometry) in which no parallel can exist. Development of these systems cast light on the fundamental nature of geometry. Non-Euclidean geometry is also used in RELATIVITY theory.

nonmetal Element that has none of the properties of a metal. Nonmetals are usually poor conductors of heat and electricity (they are thermal and electrical insulators). Examples include carbon, the halogens, oxygen, nitrogen, phosphorus, and sulfur, all of which are also electronegative. Nonmetal oxides are acidic or neutral - not basic - and they usually form part of the negative ion (anion) in any compound. A few nonmetals have some metallic properties and are termed metalloids; they include elements that are semiconductors, such as silicon and germanium.

nonstoichiometric compound Chemical compound in which the atoms are not combined in simple, whole-number ratios. The composition of such compounds varies, depending on their source. For example, titanium(IV) oxide (expected formula TiO_2) obtained from the mineral rutile generally has the formula $TiO_{1.8}$.

norepinephrine Hormone secreted by nerves in the autonomic NERVOUS SYSTEM and by the ADRENAL GLANDS. Chemically, it is closely related to ADRENALINE and has similar effects on the cardiovascular system. It slows the heart rate and constricts small arteries, thus raising the blood pressure. It is used therapeutically to combat the fall in blood pressure that accompanies shock. *See also* NEUROTRANSMITTER

Norrish, Ronald George Wreyford (1897–1976) British chemist who shared the 1967 Nobel Prize for chemistry with Manfred EIGEN and George PORTER for research into very fast chemical reactions. Although primarily of theoretical importance, their work has potential applications in such areas as the hybridization of large molecules and ULTRASONIC communications.

North Atlantic Drift *See* GULF STREAM

Northern Cross (Cygnus or The Swan) Northern constellation between Pegasus and Lyra. It contains several bright stars, including DENEB (Alpha), one of the three components of the Summer Triangle, and Albireo (Beta), which is a DOUBLE STAR. Chi is a long-period VARIABLE STAR, with a period of 408 days. The MILKY WAY is particularly bright in this region and there are several STELLAR CLUSTERS.

northern lights *See* AURORA

nose In human beings, other primates and some vertebrates, the prominent structure between the eyes; it contains RECEPTORS sensitive to various chemicals (sense of SMELL) and serves as the opening to the respiratory tract, warming and moistening incoming air and trapping dust particles on the MUCOUS MEMBRANES. It has two cavities separated by a wall of CARTILAGE; the external openings are the nostrils.

notochord In CHORDATES and the early embryonic stages of vertebrates, the flexible, primitive backbone; in mature vertebrates, it is replaced by the SPINE.

nova (plural novas or novae) Faint star that undergoes unpredictable increases in brightness by several magnitudes, and then slowly fades back to normal. A typical nova rises to maximum in a few days, and declines thereafter in brightness by a factor of about ten in 40 days. Novas occur in close binary systems, where one component is a WHITE DWARF and the other is a GIANT STAR. Matter drains from the surface of the giant, where gravity is weak, to the surface of the white dwarf, where gravity is strong. Nuclear reactions involving the infalling material cause a violent explosion on the surface of the white dwarf. The energy of a typical explosion is about the same as the Sun emits in 10,000 years. Clouds of gas are ejected during the outburst. The total amount of material lost is estimated to be about one ten-thousandth of the mass of the white dwarf.

nucellus Tissue that forms most of the OVULE in a FLOWER. It consists of nutrient tissue and an embryo sac, enclosed by a skin that has a small hole in it called the **micropyle**.

nuclear energy ENERGY released during a nuclear reaction as a result of the conversion of MASS into energy. The conversion involves the BINDING ENERGY of the NUCLEUS of an ATOM, and occurs as described by the equation $E = mc^2$ (where E is energy, m is mass, and c is the speed of light), which was derived by EINSTEIN in his special theory of RELATIVITY. Nuclear energy is released in two ways: by FISSION and by FUSION. Fission is the process responsible for the atom bomb, and for the fission NUCLEAR REACTORS now contributing to energy requirements throughout the world. The fission reaction was first discovered in 1938 by Otto HAHN and Fritz STRASSMANN; and in 1942 the first sustained nuclear CHAIN REACTION was acheived by Enrico FERMI. Fusion provides the energy for the Sun and the stars and for the hydrogen bomb. It also offers the prospect of cheap energy, once a method has been perfected for controlling fusion reactions. *See also* CRITICAL MASS; MANHATTAN PROJECT; MATTER OF STATES; NUCLEAR WEAPON

nuclear engineering Branch of engineering concerned with the design, construction, and operation of NUCLEAR REACTORS and particle ACCELERATORS as well as coolant, shielding, and moderator devices. *See also* NUCLEAR ENERGY

nuclear fission *See* FISSION, NUCLEAR

nuclear fuel Various chemical and physical forms of URANIUM and PLUTONIUM used in NUCLEAR REACTORS. Fluid fuels are required in homogeneous reactors; heterogeneous reactors use various forms of fuels – pure metals or alloys, as well as oxides or carbides. The fuel must have a high thermal conductivity, be resistant to radiation damage and be easy to fabricate.

nuclear fuel enrichment Separation of the fissionable ISOTOPE of URANIUM, uranium-235, from the more abundant uranium-238 isotope. Gaseous uranium(VI) fluoride undergoes diffusion separation using cascades of barriers with microscopically small pores. The difference in mass between the two isotopes is minimal, but sufficient so that the heavier, slower-moving uranium-238 molecules are concentrated on one side. High-speed centrifugal-force separation methods are also used.

nuclear fusion *See* FUSION, NUCLEAR

nuclear magnetic resonance (NMR) Absorption of ELECTROMAGNETIC RADIATION by certain nuclei when placed in a magnetic field. In this field the NUCLEUS, as a result of its SPIN, can have slightly different energy values. It can make transitions between these energy values, acquiring the energy by absorbing radio-frequency radiation of

the appropriate wavelength. This pheneomenon is used in medicine in the form of MAGNETIC RESONANCE IMAGING (MRI), and in chemical analysis and research in nuclear physics in a method called NMR SPECTROSCOPY.

nuclear physics Branch of physics concerned with the structure and properties of the atomic NUCLEUS. The principal means of investigating the nucleus is the particle SCATTERING experiment, carried out in particle ACCELERATORS, in which a nucleus is bombarded with a beam of high-energy ELEMENTARY PARTICLES and the resultant particles analyzed. Other information is obtained by studying their magnetic moments and SPIN and any FISSION and FUSION products. Study of the nucleus has led to an understanding of the processes occurring inside stars and has enabled the building of NUCLEAR REACTORS and NUCLEAR WEAPONS.

nuclear radiation Refers to all the types of emissions of waves and particles from the nuclei of atoms as the result of natural or induced RADIOACTIVITY. Natural radioactivity is the spontaneous breakdown of unstable nuclei, such as those of radium or uranium-235, with the release of ALPHA and BETA PARTICLES and GAMMA RADIATION. **Alpha** (α) particles are identical to nuclei of atoms of helium gas: they travel very fast, but have little penetrating power and so are relatively harmless. **Beta** (β) rays or particles are electrically charged, and are identical to high-energy ELECTRONS or POSITRONS. They have greater penetrating power. **Gamma** (γ) rays are very high frequency X RAYS, and are dangerously penetrating. Other emissions from nuclei include PROTONS, NEUTRONS, and NEUTRINOS. These and yet other particles, called MESONS, result from the interactions of cosmic rays with the upper atmosphere, and from the bombardment of materials with high-speed particles in particle ACCELERATORS.

nuclear reactor Device in which nuclear FISSION (and sometimes nuclear FUSION – *see* TOKAMAK) reactions are used for power generation or for the production of radioactive materials. In the reactor, the fuel is a radioactive heavy metal: URANIUM-235, uranium-233 or PLUTONIUM-239. In these metals, ATOMS break down spontaneously, undergoing a process called RADIOACTIVE DECAY. Some NEUTRONS released in this process strike the nuclei of fuel atoms, causing them to undergo fission and emit more neutrons. These in turn strike more nuclei, and in this way a CHAIN REACTION is set up. Usually, a material called a MODERATOR is used to slow down the neutrons to a speed at which the chain reaction is self-sustaining. This process occurs within the reactor **core**. The chain reaction is regulated by inserting control rods, which contain neutron-absorbing material such as cadmium or boron, into the core. The heat generated by the nuclear reaction is absorbed by a circulating **coolant** and transferred to a boiler to raise steam. The steam drives a TUR-BINE that turns a GENERATOR that in turn produces electricity. There are a variety of nuclear reactors, named after the type of coolant they use. For example, a **boiling-water** reactor (BWR) and a **pressurized-water** reactor (PWR), presently the most common type of reactor, both use water as the coolant and the moderator. In **advanced gas-cooled** reactors (AGR), the coolant is a gas – most commonly carbon dioxide. **Fast** reactors do not use a moderator, and fission is caused by FAST

NEUTRONS. This type of reactor generates greater temperature and the coolant used is a liquid metal, usually liquid sodium. Sometimes called "fast-breeder" reactors, fast reactors produce ("breed") more fissionable material than they consume. Excess neutrons from the fission of a fuel such as uranium-235, instead of being absorbed in control rods, are used to bombard atoms of relatively inactive uranium-238, which transmutes to (or "breeds") the active ISOTOPE plutonium-239. When the original fuel is spent, the plutonium can be used as a nuclear fuel in other reactors or NUCLEAR WEAPONS. *See* ELECTRICITY SOURCES

nuclear submarine Underwater vessel powered by a NUCLEAR REACTOR. The great advantage of nuclear SUBMARINES is that they use little fuel and may remain submerged for many weeks, even months. Heat produced by the reactor turns water into steam. The steam turns TURBINES, which are linked to electricity GENERATORS that charge the submarine's batteries. The vessel's propeller may be driven by an electric motor or by linking its shaft to a turbine. The first nuclear submarine, the USS *Nautilus*, was launched by the US Navy in 1954. *Nautilus* traveled 100,000km (62,000mi) in its first two years of service, and in 1958 was the first vessel to sail under the ice cap at the North Pole. In 1960 the USS *George Washington* became the first submarine to fire a missile while it was submerged. Submarines are now used to launch guided missiles, as they did during the Persian Gulf War.

nuclear waste Residues, generally from NUCLEAR REACTORS, containing radioactive substances. After URANIUM, PLUTONIUM and other useful fission products have been removed, some long-lived, radioactive elements remain, such as CESIUM-137 and STRONTIUM-90. International regulations exist for disposing of such wastes, with separate regulations applying to liquids, gases, and solids.

nuclear weapon Device that derives a huge explosive force from nuclear FISSION or FUSION reactions. The first nuclear weapons were dropped by the USA on the Japanese cities of Hiroshima and Nagasaki in August 1945. These **atom bombs** consisted of two stable, subcritical masses of URANIUM or PLUTONIUM that when brought forcefully together, caused the CRITICAL MASS to be exceeded, thus initiating an uncontrolled nuclear fission CHAIN REACTION. In such detonations, huge amounts of energy and harmful radiation are released: the explosive force can be equivalent to 200,000 tons of TNT. The much more powerful **hydrogen bomb** (H-bomb or thermonuclear bomb), first tested in 1952, consists of an atom

▲ **nuclear reactor** A pressurized water reactor (PWR) is so named because the primary coolant (1) that passes through the reactor core (2) is pressurized to prevent it from boiling. The uranium-235 fuel is loaded into the reactor in pellets (3) contained by the fuel rods (4). To prevent an uncontrolled chain reaction the fuel rods are separated by control rods of graphite (5). All the rods are loaded into the reactor from above (6). The primary coolant is heated by the fission reaction in the fuel rods

and circulates into a steam generator (7) where it superheats the secondary coolant (8). The secondary coolant leaves the protective containment vessel (9) and drives turbines (10) which produce electricity through a generator (11). A third coolant loop (12) cools the secondary coolant, transferring the heat to a sea, river or lake. Reducing the temperature of the secondary coolant increases the efficiency of the transfer from the primary to the secondary coolant.

bomb that on explosion provides a temperature high enough to cause nuclear fusion in a surrounding solid layer, usually lithium deuteride. The explosive power can be that of several million tons (megatons) of TNT. Devastation from such bombs covers a wide area: a 15-megaton bomb will cause all flammable material within 20km (12mi) to burst into flame. A third type of weapon, the **neutron bomb**, is a small hydrogen bomb, also called an enhanced radiation weapon, that produces a small blast but a very intense burst of high-speed NEUTRONS. The lack of blast means that buildings are not heavily damaged. The neutrons, however, produce intense radiation sickness in people located within a certain range of the explosion, killing those affected within a week. *See also* NUCLEAR ENERGY

nucleic acid Giant, chemical molecules present in all living cells and in viruses. There are two types: DNA (deoxyribonucleic acid) stores the GENETIC CODE that functions as the basis of heredity; and RNA (ribonucleic acid) delivers these coded instructions to the cell's PROTEIN manufacturing sites. Chemically, nucleic acids are polymers of NUCLEOTIDES. *See also* CHROMOSOME; ENZYME; GENE; MESSENGER RNA (MRNA)

nucleon Either of the two kinds of particles found within the NUCLEUS of an atom: a NEUTRON or a PROTON.

nucleoside Subunit of NUCLEIC ACID, consisting of a sugar linked to a nitrogenous base (purine or pyrimidine).

nucleosynthesis Production of all the various chemical elements that exist in the Universe from one or two simple atomic nuclei. It is believed to have occurred by way of a huge release of NUCLEAR ENERGY during cosmogenesis and is still in progress in the Sun and other stars. Starting with hydrogen and helium, repeated nuclear FUSION reactions can account for most of the elements up to iron. Elements heavier than iron can be explained in terms of repeated reactions involving the capture of NEUTRONS by nuclei.

nucleotide Complex, naturally occurring chemical group that contains a nitrogen base linked to a sugar and an acid phosphate. Nucleotides are the building blocks of NUCLEIC ACIDS (such as DNA and RNA). In the molecules of nucleic acids, nucleotides are linked together by bonds between the sugar and the phosphate groups. Nucleotides also occur freely in a CELL as various COENZYMES, such as ADENOSINE TRIPHOSPHATE (ATP), the principal carrier of chemical energy in the metabolic pathways of the body.

nucleus In biology, membrane-bound structure that, in most CELLS, contains the CHROMOSOMES. As well as holding the genetic material, the nucleus is essential for the maintenance of cell processes. It manufactures the RNA used to build RIBOSOMES. Other RNA molecules carry the genetic code from the DNA through pores in the nuclear membrane into the CYTOPLASM, where it is used as a template for PROTEIN SYNTHESIS by the ribosomes. CELL DIVISION involves the splitting of the nucleus and CYTOPLASM. Cells without nuclei include BACTERIA and mammalian ERYTHROCYTES (red blood cells). Instead of chromosomes, bacteria have a naked, circular molecule of DNA in the cytoplasm.

nucleus In physics, central core of an ATOM. Made up of NUCLEONS (PROTONS and NEUTRONS), the nucleus accounts for almost all of an atom's mass. Because protons are positively charged and neutrons have no charge, a nucleus has an overall positive charge; this is cancelled out, however, by the negatively charged ELECTRONS that orbit the nucleus. Following the discovery (1836) of RADIOACTIVITY by Henri BECQUEREL, in 1911 Ernest RUTHERFORD proposed the existence of the nucleus after identifying ALPHA and BETA PARTICLES. The number of protons in a nucleus is called its ATOMIC NUMBER, while the number of nucleons is the MASS NUMBER. *See also* FUNDAMENTAL FORCE; QUANTUM THEORY; NUCLEAR ENERGY; RELATIVE ATOMIC MASS (R.A.M.); SPIN

nuée ardente Cloud of burning hot gas that travels quickly down the side of a VOLCANO during an eruption. Small particles of dust and ash may be carried in the cloud, which kills anyone and anything it passes over. In 1902 a nuée ardente accompanying the eruption of Mount Pelée accounted for the deaths of all but one of the inhabitants of St Pierre on the island of Martinique.

number Symbol representing a quantity used in counting or calculation. All ancient cultures devised their own NUMBER SYSTEMS for the practical purposes of counting and measuring. The development of the idea of place-value notation and the introduction of zero into the base-10 number system culminated in the familiar DECIMAL SYSTEM that we use today. This system seems to have reached the West from India via 12th-century translations of Arabic mathematical writings. From the basic process of counting we get the natural numbers. This concept can be extended to define the INTEGERS, the rational numbers, the real numbers, and the complex numbers. Other classes of number, like PRIME NUMBERS and perfect numbers, arise in a branch of mathematics called NUMBER THEORY.

number, cardinal Number expressing the content of a SET, but not the order of its members. For example, six is a cardinal number in "six books." Two sets have the same cardinal number if their members can be put in one-to-one correspondence – a concept that allows the idea of cardinal numbers of infinite sets. The infinite set of INTEGERS is said to have cardinal number χ_0 (aleph-zero). The set of all real numbers (*see* NUMBERS, REAL) cannot be put into one-to-one correspondence with aleph-zero and is a "larger" infinite set. *See also* NUMBER, ORDINAL

number, complex Number of the form $z = a + bi$, where $i = \sqrt{-1}$, and a and b are real numbers (*see* NUMBER, REAL). a is called the **real part** and b the **imaginary part**. All complex numbers can be represented on an **Argand diagram** using rectangular Cartesian coordinates, the complex number $a + bi$ being identified with the point (a, b) in the plane. Thus, real numbers are positioned on the x axis and purely imaginary numbers are positioned on the y axis. An equivalent description uses POLAR COORDINATES (r,θ) with $z = r\cos(\theta) + r\sin(\theta)i$, where the two descriptions are related through the equations $r = \sqrt{(a^2 + b^2)}$, $\tan(\theta) = b/a$. Complex numbers have applications in many areas of pure and applied mathematics.

number, imaginary Square root ($\sqrt{\ }$) of a negative quantity. The simplest, $\sqrt{-1}$, is usually represented by i. The numbers are so called because when first discovered they were widely regarded as meaningless. However, they are necessary for the solution of many QUADRATIC EQUATIONS, the roots of which can be expressed only as complex numbers (*see* NUMBER, COMPLEX), which are composed of a real part and an imaginary part. They find applications in ALTERNATING CURRENT (AC) theory.

number, irrational Number that cannot be expressed as a fraction. Examples include $\sqrt{2}$ and π (PI). Irrational numbers are, therefore, numbers with infinitely many (nonrepeating) DECIMAL PLACES. (However, the reverse is not true: some rational fractions, such as $\frac{1}{3}$, also have infinite decimal representations.) Irrational numbers, together with the rational numbers (*see* NUMBER, RATIONAL), make up the set of real numbers (*see* NUMBER, REAL).

number, natural Any of the numbers 1, 2, 3, 4, ... as used in counting; they are the simplest numbers, with no fractional, decimal or imaginary part, and of which there is an infinite number. They are all positive numbers, that is, greater than zero. Negative numbers are those less than zero. *See also* DECIMAL FRACTION; INTEGER; NUMBER, IMAGINARY

number, ordinal Number that indicates the position of a member in an ordered sequence or in a group. Thus "first," "second," and "third" are ordinal numbers.

number, perfect INTEGER equal to the sum of all its FACTORS including 1. For instance, the number 28 is a perfect number since its factors are 1, 2, 4, 7, and 14 (excluding 28 itself), and the sum of these equals 28. It is not known whether or not there exist any odd perfect numbers.

number, prime Positive INTEGER that has no FACTORS other than itself and 1. The first few primes are 2, 3, 5, 7, 11, 13, and 17. The INTEGERS 4, 6, 8, ... are not prime numbers, since they can all be divided by 2 (that is, 2 is a factor). In the 4th century BC, EUCLID proved that there are an infinite number of prime numbers. But it remains an unsolved problem to find formulas that will generate the sequence of primes. In 1640 Pierre de FERMAT showed that the numbers $F_n = 2^x + 1$, where $x = 2^n$, for $n = 0, 1, 2, 3,$ and 4, were prime. He conjectured that F_n would be prime for all n, but in the 18th century Leonhard EULER showed that F_5 (= 4,294,967,297) was a composite number (that is, equal to the product of two numbers) and therefore not a prime.

number, rational Number that can be written in the form a/b, where a and b are INTEGERS (and b is not zero). Rational numbers include all integers. A number that cannot be so expressed is called an irrational number (*see* NUMBER, IRRATIONAL). The rational numbers are also called FRACTIONS.

number, real Number that is either a rational or an irrational number. Real numbers are expressible as decimals. They are distinguished from complex numbers, which are of the form $a + bi$ (where $i = \sqrt{-1}$). *See also* NUMBER, COMPLEX; NUMBER, IRRATIONAL; NUMBER, RATIONAL

number, transcendental Irrational number (*see* NUMBER, IRRATIONAL) that is not the solution to any algebraic equation with rational (fractional) coefficients. For example, the numbers π (PI) and e are both transcendental, but the irrational number $x = \sqrt{2}$ is not transcendental because it satisfies the algebraic equation $x^2 - 2 = 0$. *See also* ALGEBRAIC FUNCTION

number, whole *See* INTEGER

number system Counting methods that use various number BASES. The DECIMAL SYSTEM, for example, has a base of 10; the DUODECIMAL SYSTEM a base of 12. Computers do their calculations within the BINARY SYSTEM, base 2.

number theory Branch of mathematics concerned with the properties of natural numbers (*see* NUMBER, NATURAL) or special classes of natural numbers such as prime numbers (*see* NUMBER, PRIME). Typical results from number theory are that there are infinitely many prime numbers (proved by EUCLID), and that any number may be written as the sum of four square numbers (proved by Joseph LAGRANGE). Many of the most important problems in number theory are very easy to pose but have been found to be incredibly difficult to solve. FERMAT'S LAST THEOREM remained unproven for more than 350 years, until Andrew WILES found a proof in 1995. Many problems relating to the prime numbers remain completely elusive: for instance, no one has yet found a formula that will generate all the prime numbers.

numeral Symbol used alone or in a group to denote a NUMBER. **Arabic numerals** are the 10 digits from 0 to 9. **Roman numerals**, as normally used today, consist of seven letters or marks (I =1, V = 5, X = 10, L = 50, C = 100, D = 500, and M = 1,000). The formation of numbers from numerals depends on the NUMBER SYSTEM used.

numerator In mathematics, upper number in a FRACTION. The lower number is known as the DENOMINATOR. For example, in the fraction 2/3, 2 is the numerator, 3 the denominator.

nut In botany, dry, one-seeded FRUIT with a hard woody or stony wall. It develops from a flower that has petals attached above the ovary (inferior ovary). Examples include acorns and hazelnuts.

nutation Oscillating movement (period 18.6 years) superimposed on the steady precessional movement of the Earth's axis, so that the precessional path of each celestial pole on the CELESTIAL SPHERE follows an irregular rather than a true circle. It results from the varying gravitational attraction of the Sun and Moon on the Earth, due to variations in their distances from Earth and in their relative directions. *See also* PRECESSION

nutrition All the processes by which plants and animals take in and make use of food. The science of nutrition involves identifying the kinds and amounts of nutrients necessary for growth and health. Nutrients are generally divided into PROTEINS, CARBOHYDRATES, FATS, MINERALS, and VITAMINS.

nylon Any of numerous synthetic materials consisting of polyamides (with proteinlike structures). Developed in the USA in the 1930s, nylon can be formed into fibers, filaments, bristles or sheets. It is characterized by elasticity and strength and is used chiefly in yarn, cordage, and molded products. Hard and tough or soft and rubbery nylon products can be made by varying their composition. *See also* SYNTHETIC FIBER

nymph Young insect of primitive orders that do not undergo complete METAMORPHOSIS. The term is used to designate all immature stages after the OVUM. The nymph resembles the adult (IMAGO) and does so more closely with each successive molting. Some examples are the aquatic nymphs of dragonflies, mayflies, and damsel flies.

oasis Fertile location that has water in an arid landscape. Usually GROUNDWATER is brought to the surface in a well, but an oasis may occur at the point where a river flowing from a wetter region crosses the desert on its way to the sea, such as the Nile in N Africa or the Indus in S Asia. Most Saharan oases contain large numbers of date palms, and there are also many fields of grains, vegetables, and fruits. There is often a settlement at the oasis, sometimes with as many as 30,000 inhabitants.

Oberon Second-largest of the five main satellites of URANUS, discovered in 1787 by William HERSCHEL. It is 1,524km (947mi) in diameter and consists largely of a mixture of rock and water-ice. Oberon has many craters, some with rays and some with dark floors, and mountains, one of which appears to be 20km (12mi) high. The high density of craters is evidence of a world unaltered by geological activity.

Oberth, Hermann Julius (1894–1990) German physicist, b. Austria-Hungary, who made a fundamental contribution to the development of astronautics. Oberth conducted experiments to simulate and study the effects of weightlessness and designed a long-range, liquid-propellant ROCKET. During World War II he worked on German rocket development under Wernher VON BRAUN. Oberth later worked on the development of solid-propellant, anti-aircraft rockets.

observatory Location of TELESCOPES and other equipment for astronomical observations. Large optical telescopes are housed in domed buildings, usually situated in mountain areas, well away from the lights and smog of cities. These observatories rotate to face any area of the sky; the telescope is exposed by the retraction of a panel in the dome. Radio observatories are open sites containing one or more large RADIO TELESCOPES, usually having the shape of a parabolic dish. The largest radio telescopes have been built in natural mountain hollows: the Arecibo Observatory in Puerto Rico, for example, is 300m (975ft) across. The latest observatories include telescopes carried by orbiting satellites, such as the HUBBLE SPACE TELESCOPE, entirely above the absorbing and distorting effects of Earth's atmosphere. In addition to optical and radio studies, these craft carry out measurements in the infrared, ultraviolet, X-ray, and gamma-ray regions of the spectrum, which do not penetrate Earth's atmosphere and so cannot be studied from ground-based observatories.

obsidian Rare gray to black glassy volcanic rock. High in silica, it is the uncrystallized equivalent of RHYOLITE and GRANITE. It polishes well and makes an attractive semiprecious stone. Hardness 5.5; r.d. 2.4.

occultation In astronomy, temporary concealment of one celestial body by another, especially of a star by the Moon or a planet. Lunar occultation is of particular importance because it has served accurately to pinpoint a number of notable radio sources.

ocean Continuous body of salt water that surrounds the continents and fills the Earth's great depressions. Oceans cover about 71% of the Earth's surface (more than 80% of the Southern Hemisphere), and represent about 98% of all the water on the face of the Earth. There are five main oceans: Atlantic, Pacific, Indian, Arctic, and Antarctic. They may be described by distinct region (LITTORAL, BENTHOS, PELAGIC, and ABYSSAL), or by depth (CONTINENTAL MARGIN, deep sea plain, and deep TRENCHES). The SEAFLOOR has a varied topography. SEAWATER contains salt and other mineral deposits; the salt content, between 3.3% and 3.7%, is the result of washout from the land and interchange with the ATMOSPHERE over the ages. Light penetrates seawater to a maximum depth of about 300m (1,000ft), below which plant life cannot grow. The oceans are constantly moving in CURRENTS, TIDES, and WAVES. They form an integral part of the Earth's HYDROLOGICAL CYCLE and CLIMATE. They are a rich source of FOSSILS, providing invaluable evidence of the evolution of life on Earth. They also provide minerals such as MANGANESE NODULES, OIL, and GAS. The CONTINENTAL SHELF yields sand and gravel. Marine fauna, such as FISH and PLANKTON, are a vital part of the food chain.

Through volcanic activity the oceans have been forming over the last 200 million years. The theory of CONTINENTAL DRIFT (and associated SEAFLOOR SPREADING) has revealed that PANGAEA was surrounded by one vast ocean, Panthallasia. As Pangaea began to split, a smaller and shallower ocean, the Tethys Sea, formed between the continents. By about 65 million years ago, the Atlantic and Indian oceans appeared. The Pacific was separated from the Atlantic and Indian oceans when the North and South American continents joined. The separation of Greenland from North America, and the widening of the North Atlantic, completed the encirclement of the Arctic Ocean.

Total area: 360 million sq km (138 million sq mi). **Total volume**: c.1.4 billion cu km (322 million cu mi). **Average depth**: 3,500m (12,000ft). **Average temperature**: 3.9°C (39°F). *See also* DESALINATION; TIDE; WATER POLLUTION

ocean floor *See* SEAFLOOR

oceanic basin One of two major provinces of the deep ocean floor, lying at over 2km (1.2mi) in depth. The MIDOCEAN RIDGES form the other province. Together they constitute 56% of the Earth's surface. The deep ocean basin is underlain by a basaltic crust, about 7km (4.3mi) thick, and is covered in sediment and dotted by low ABYSSAL hills. *See also* TRENCH

oceanic current *See* CURRENT

oceanography Study of the OCEANS. The major subdisciplines of OCEANOGRAPHY include marine geology (see PLATE TECTONICS), MARINE BIOLOGY, marine METEOROLOGY, and physical and chemical oceanography. The science of oceanography dates from the CHALLENGER EXPEDITION (1872–76). Jacques COUSTEAU'S invention of the SCUBA aided human exploration of the seas. In 1948 the BATHYSCAPHE, invented by August PICCARD, was first used. The DEEP SEA DRILLING PROJECT is a long-term investigation of the evolution of OCEANIC BASINS.

ocelot Member of the flesh-eating mammal order Carnivora, the ocelot (*Felis pardalis*) is found in Central and South America, and sometimes as far north as Texas. It measures from 80–147cm (31–58in) long, and is grouped with all other cats in the family Felidae.

Ochoa, Severo (1905–93) US biochemist, b. Spain. He shared the 1959 Nobel Prize for physiology or medicine with Arthur KORNBERG for work on the synthesis of RNA, which helped greatly in the study of GENETICS.

ochre Yellowish or red pigment made by mixing a type of iron ore (LIMONITE or HEMATITE) with fine clay and grinding the dried product to a powder.

octane number Indication of the resistance of GASOLINE to engine KNOCK (preignition firing). It represents the percentage by volume of iso-octane in a reference fuel, consisting of a mixture of iso-octane and normal heptane, that matches the knocking properties of the fuel being tested. The higher the number, the less likely the possibility of the fuel preigniting. Regular gasoline has an octane number of about 90. Premium grade gasolines have octane numbers between 95 and 96.

odonatan Any member of Odonata, an order of primitive winged insects found worldwide. Damselflies, in the suborder Zygoptera, have thin bodies with wings held vertically along the body when at rest. The long, slender, aquatic nymphs have three, leaflike gills on the abdomen. Dragonflies, in the suborder Anisoptera, have heavy bodies with wings held horizontally when at rest. The stout nymphs have gills at the anal end. All prey on insects; none attack humans. Length: 18–193mm (0.7–8in).

Oersted, Hans Christian (1777–1851) Danish physicist and professor at Copenhagen University. He took the first steps in explaining the relationship between electricity and magnetism, thus founding the science of ELECTROMAGNETISM. The OERSTED unit of magnetic field strength is named after him. In 1825 he became the first scientist to isolate pure metallic ALUMINUM.

oersted Unit of magnetic field strength in the CGS SYSTEM, equal to the magnetic field that would cause a unit magnetic pole to experience a force of one dyne in a vacuum. It is named after Hans OERSTED.

offset printing Method of PRINTING widely used for high-volume publications. In the printing machine, a roller applies ink to the printing plate, which is mounted on a rotating cylinder. The image is then transferred (offset) to a cylinder with a rubber covering, called the blanket. This transfers the image to the paper. Usually, the plates are made by LITHOGRAPHY, and the process is called offset lithography. Separate plates are used for each color of ink. This book was printed using a four-color offset-litho process.

ohm (symbol Ω) SI unit of electrical RESISTANCE, equal to the resistance between two points on a conductor when a constant potential difference of one VOLT produces a current of one AMPERE.

Ohm, Georg Simon (1787–1854) German physicist. He was appointed professor at Munich following his discovery (OHM'S LAW) of the relationship linking current, electromotive force, and resistance in an ELECTRIC CIRCUIT. His name is also given to the SI unit of electrical RESISTANCE.

Ohm's law Statement that the amount of steady current through a material is proportional to the voltage across the material. Proposed in 1827 by Georg OHM, Ohm's law is expressed mathematically as $V = IR$, where V is the voltage in volts, I is the current in amperes and R is the resistance in ohms. In alternating current (AC) circuits where deductances and capacitances are present, the formula is amended to $V = IZ$, where Z is IMPEDANCE.

It holds only approximately and under limited conditions and not for all materials.

oil General term to describe a variety of substances, whose chief shared properties are viscosity at ordinary temperatures, a density less than that of water, flammability, insolubility in water, and solubility in ether and alcohol. There are three main types: **mineral** oils are hydrocarbon mixtures that occur, most notably, in crude or PETROLEUM oil, and that are extracted by refining to be used primarily as fuels; **animal** oils are glycerides of FATTY ACIDS and are used as food, lubricants, and a major ingredient of soap, paints, and varnishes. There are two kinds of **vegetable** (or fixed) oil: drying, such as linseed or poppyseed oil, and nondrying, such as olive and castor oil. All vegetable oils are ESTERS or mixtures of substances called TERPENES. There are also ESSENTIAL OILS that are obtained from plants and that, unlike other vegetable oils, are volatile. Such oils are found in perfumes and food flavorings.

oil-drop experiment Experimental method used by Robert MILLIKAN to measure an ELECTRON'S charge by subjecting charged oil drops to a variable electric field.

oil extraction Process by which PETROLEUM (crude oil) is brought to the Earth's surface and transported to an OIL REFINERY. A borehole is made into the Earth's crust and the oil reservoir emptied with the help of NATURAL GAS pressure or pumps. The discovery of undersea deposits, as in the North Sea, has meant that many new techniques for extracting and transporting the oil have had to be developed in order to avoid WATER POLLUTION and blowouts. *See also* DRILLING RIG; OIL WELL

oil refinery Large, industrial chemical complex in which PETROLEUM (crude oil) is refined. This involves many processes whereby the oil is split into its major fractions, such as gasoline, kerosene, diesel oil, lubricating oil, and bitumen. These are obtained by fractional DISTILLATION, but many refineries carry out further operations on some of these fractions. These include catalytic CRACKING, alkylation – increasing the OCTANE NUMBER of gasoline – HYDROGENATION and POLYMERIZATION.

oil rig *See* DRILLING RIG

oil shale Dark-colored, soft rock containing hydrocarbon compounds that can be distilled off as SHALE oil. Oil shales yielding more than a small percentage of oil can be valuable as fuels and as sources of organic chemicals. They are found in many countries, sometimes in vast deposits; that of the Green River shales in the

American west alone has been estimated to contain 960,000 million barrels of oil.

oil well Shaft through which PETROLEUM (crude oil) is brought to the surface from underground deposits. Wells are usually drilled with a rotating shaft bearing a drill bit, which, for cutting through hard rock, may be diamond-tipped. Rock particles are pumped from the hole as a fluid slurry, which also lubricates and cools the drill. Proven wells are cased with steel piping and topped above ground with an oil take-off arrangement, often of the type known as a "Christmas tree."

Olbers, Heinrich Wilhelm Matthäus (1758–1840) German astronomer and physician. He discovered five comets and the asteroids Pallas (1802) and Vesta (1807). He also devised a new method of calculating a comet's orbit. In 1826 he discussed a longstanding problem, now called **Olbers' paradox**: Why is the sky dark at night? If there are an infinite number of stars, evenly distributed in space, every line of sight should end on a star, and the sky should be roughly as bright as the Sun, all over. Olbers' explanation, that interstellar dust obscures the light, was discarded in the 1970s. In fact, space is not infinite. According to the BIG BANG theory, the Universe is believed to have been born about 15 billion years ago, and so light reaches us only from a region 15 billion light-years in radius. In addition, the light from distant galaxies is moved out of the visible spectrum by the RED SHIFT caused by the expansion of the Universe.

Oldowan Type of stone tool dating from the early PLEISTOCENE epoch (about 2 million years ago). The name comes from the OLDUVAI GORGE in Tanzania, where archaeologists found the first tools of this kind. Made from quartz or basalt stones, the edges were chipped away to form tools capable of chopping, scraping or cutting. Oldowan tools were made for about 1.5 million years.

Olduvai Gorge Site in N Tanzania where remains of primitive humans have been found. Louis LEAKEY uncovered four layers of remains dating from about 2 million years ago to about 15,000 years ago. In 1964 he announced the discovery of *Homo habilis*, whom he believed to have been a direct ancestor of modern humans. The gorge, which is 40km (25mi) long and 100m (320ft) deep, runs through the Serengeti Plain.

olefin *See* ALKANE

oleic acid ($CH_3(CH_2)_7CH=CH(CH_2)_7COOH$) Organic chemical compound that has a long-chain molecule that is unsaturated (containing a double bond). It is the unsaturated acid found most commonly in fatty oils in the form of a glycerol ESTER.

olfaction Alternative name for the sense of SMELL

olfactory nerve Nerve of SMELL, one of the 12 pairs of cranial nerves present in all vertebrate animals. Its nerve cells (NEURONS) lie in the brain, and their long shafts, or AXONS, extend to their sense RECEPTORS in the upper part of the nasal cavity.

Oligocene Extent of geological time from about 38 to 25 million years ago. It is the third of five epochs of the TERTIARY period in the CENOZOIC era. During the Oligocene epoch, the climate cooled. Many modern mammals evolved, including early elephants and *Mesohippus*, an ancestor of the modern horse. Only a few archaic mammals, such as titanotheres, survived into the epoch, and they became extinct before it ended.

oligochaete Any member of the class Oligochaeta. Oligochaetes are ANNELID (segmented) worms, characterized by long, naked bodies bearing a few bristles (chaetae) on each segment. Earthworms are the best-known oligochaetes, but many are freshwater worms as small as 1mm (0.025in) long. Giant Australian earthworms grow up to several meters long.

Oliphant, Sir Mark (1901–) Australian nuclear physicist who conducted much work on the nuclear disintegration of LITHIUM. He was a member of the atom bomb team at Los Alamos (1943–45) and became Australian representative on the United Nations Atomic Energy Commission in 1946.

olive Tree, shrub, or vine and its fruit, especially the common olive tree, *Olea europaea*, native to the Mediterranean region. It has leathery, lance-shaped leaves, a gnarled and twisted trunk, and may live for more than 1,000 years. Height: to 9m (30ft). Family Oleacaea.

olivine Ferromagnesian mineral, $(Mg,Fe)_2SiO_4$, found in basic and ultrabasic IGNEOUS ROCKS. Olivine has orthorhombic system crystals, of usually granular masses. Its color is commonly olive-green, but can be brown or gray. It is glassy and brittle with no cleavage. Hardness 6.5–7; r.d. 3.3.

omnivore Any creature that eats both animal and vegetable foods – examples are human beings and pigs. Omnivorous animals are characterized by having teeth adapted for cutting, tearing, and pulping food. *See also* CARNIVORE; HERBIVORE

oncogene GENE that, by inducing a cell to divide abnormally, contributes to the development of CANCER. Proto-oncogenes are present in all normal cells and in some viruses. Oncogenes arise from the mutation of proto-oncogenes. Some of these mutations are associated with various cancers. The oncogene concept led to the development of the onco-mouse, a laboratory animal that has one of these rogue genes implanted into its cells; it is used in testing anticancer treatments. *See also* GENETICS; RETROVIRUS

oncology In medicine, specialty concerned with the diagnosis and treatment of CANCER.

on-line publishing Distribution of information for public access using COMPUTER NETWORKS instead of physical media, such as paper or CD-ROMs. Access to on-line publications may be free or allowed in return for payment. The content of on-line publications is usually similar to that of traditional publications, but may be enhanced by the use of such features as HYPERTEXT, search facilities, and MULTIMEDIA. The majority of on-line publishing takes place over the INTERNET, particularly by means of the WORLD WIDE WEB.

Onnes, Heike Kamerlingh *See* KAMERLINGH-ONNES, HEIKE

Onsager, Lars (1903–76) US chemist, b. Norway. Onsager was awarded the 1968 Nobel Prize for chemistry for his development of a theory of irreversible chemical reactions. In 1931 he presented an explanation of the motion of IONS in solution that, although it received little attention at the time, has since been called the fourth law of THERMODYNAMICS. From this theory, Onsager developed a technique for producing uranium-235 (the basic fuel of NUCLEAR REACTORS) from natural uranium (which consists mainly of uranium-238). This technique was put into large-scale operation in 1943 with the first gaseous diffusion plant at Oak Ridge, Tennessee, and remained the primary method for obtaining uranium-235 until the 1960s.

ontogeny Total biological development of an organism. It includes the embryonic stage, birth, growth, body changes, and death. *See also* BIOGENETIC LAW; PHYLOGENETICS

onyx Variety of the mineral CHALCEDONY with straight parallel bands. Black and white onyx is often used for making cameos; white and red forms are called **carnelian** onyx; white and brown, **sardonyx**. It is widely distributed, but found mostly in India and South America. It is a form of AGATE.

oocyte Female gametocyte, an animal OVUM before the first polar body is formed. *See* OOGENESIS

oogenesis Process by which a female reproductive cell develops, ready for fertilization. DIPLOID primary egg cells (**oogonia**) divide by MITOSIS to produce many prospective eggs cells (**oocytes**). Each of these oocytes divides by MEIOSIS to produce a HAPLOID secondary oocyte and a **polar body**. The secondary oocyte then divides again to produce an OVUM and a second polar body.

oolite Sedimentary rock composed of OOLITHS.

oolith Small, spherical concretion that occurs in SEDIMENTARY ROCKS. Oolithic grains, up to 2mm (0.1in) in diameter, usually consist of calcium carbonate, dolomite (magnesium calcium carbonate) or chamosite (an iron silicate mineral) and have a radiating, fibrous structure, or one consisting of concentric layers. The center of an oolith may be a grain of shell or quartz, which suggests to geologists that they grow in oceans or streams by accreting layers of calcium carbonate as they roll about in the turbulent water. Ooliths give the rounded grainy structure to oolitic limestone.

Oort, Jan Hendrik (1900–92) Dutch astronomer. He carried out research on the structure and dynamics of stellar systems, especially our galaxy, whose rotation he confirmed in 1927. Oort was also a pioneer of radio astronomy, in particular with respect to the 21cm radiation of interstellar hydrogen, collaborating with Hendrik van de HULST to map the galaxy at this wavelength. In 1950 he proposed the existence of what has come to be called the **Oort cloud** – a spherical region of space surrounding the solar system in which comets are thought to reside.

ooze Fine-grained, deep-ocean deposit containing materials of more than 30% organic origin. Oozes are divided into two main types: **Calcareous** ooze at depths of 2,000–3,900m (6,600–12,800ft) contains the skeletons of animals such as foraminifera and pteropods. **Siliceous** ooze at depths of more than 3,900m (12,800ft) contains skeletons of radiolarians and DIATOMS.

opal Noncrystalline variety of QUARTZ, found in recent volcanoes, deposits from hot springs and sediments. Usually colorless or white with a rainbow play of color in gem forms, it is the most valuable of quartz gems. Hardness 5.5–6.5; r.d. 2.0.

open-chain compound *See* ALIPHATIC COMPOUND

open cluster (galactic cluster) Group of young stars in the spiral arms of our GALAXY, containing from a few tens of stars to a few thousand. They are usually several light-years across. One example is the Pleiades. Related to open clusters are **associa-tions** – loose groupings of stars of common origin. Often one or more open clusters are found in the central parts of associations. *See also* STAR CLUSTER

open-hearth process Method of producing STEEL in a FURNACE heated by overhead flames. The flames come from gas or oil burners, and oxygen may be blown through the furnace to increase its temperature. Pig iron, scrap steel and limestone are heated together. Various impurities form SLAG, which is removed from the surface of the molten metal. Most of the carbon present in the iron forms carbon monoxide gas, which is removed. Various materials are added to the metal to produce steel of the required type. Use of the open-hearth process is declining because faster and cheaper processes are available, notably the BASIC OXYGEN PROCESS.

open-pit mining Stripping surface layers from the Earth's crust to obtain coal, ores, or other valuable minerals. Open-pit MINING is relatively cheap as it does not require the sinking of shafts, tunneling of galleries or underground working. Large, dragline excavators strip away surface layers. Bulldozers and other mechanical shovels distribute minerals and spoil. Trucks, railroad cars, or bucketwheel excavators run at ground level or overhead to carry minerals away for grading and processing. This form of mining is restricted in many countries because of the damage it causes to valuable land. Often mine owners are required to restore the environmental quality of the land by landscaping and replanting after mining has ceased.

operator Mathematical object representing a specific action on a function or variable. The operator itself is meaningless unless a rule is given to explain how the operator acts on the function. For example, the DERIVATIVE operator, written $D = d/dx$, has meaning only when applied to the function $f(x)$, that is $Df(x) = df(x)/dx = f(x)$.

operon Any of a group of GENES on a CHROMOSOME comprising structural genes and an operator gene. **Structural** genes direct the synthesis of ENZYMES involved in the formation of a cell constituent or the utilization of a nutrient. The **operator** gene responds to a **repressor** molecule, and can exist open or closed. When the operator gene is open, the genes it controls are functional, producing PROTEINS. When interacting with the repressor, the operator gene is closed. REGULATORY GENES control the operons.

ophthalmology Branch of medicine that specializes in the structure, function, and diseases of the eyes and surrounding structures. It includes surgical and other treatment of eye disorders and the correction of defective vision. *See also* OPTOMETRY

ophthalmoscope Instrument for examining the interior of the eye, invented by Hermann von HELMHOLTZ in 1851.

opium Drug derived from the unripe seeds of the opium poppy. Its components and derivatives have been used as NARCOTICS and ANALGESICS for many centuries. It produces drowsiness and euphoria and reduces pain, although it is addictive. MORPHINE and CODEINE are opium derivatives.

Oppenheimer, J. (Julius) Robert (1904–67) US theoretical physicist. After early work in nuclear physics at the University of California, Oppenheimer was appointed director (1943–45) of the Los Alamos laboratory in New Mexico, where he headed the MANHATTAN PROJECT to develop the atom bomb. In 1949, however, he strongly opposed the construction of the hydrogen bomb and in 1953 was suspended by the Atomic Energy Commission and declared a security risk. He was subsequently reinstated and presented with the Commission's Enrico Fermi Award (1963) for his contribution to science.

opposition In astronomy, position on the ORBIT of a superior planet at which the Sun, the Earth, and the planet are in alignment, with the Earth located between the planet and the Sun. Only superior planets, with larger orbits than that of the Earth, can be in opposition to the Sun, and these may be observed most clearly at this time.

optical activity Ability of some chemical compounds in solution and some crystals and transparent substances placed in an intense magnetic field to rotate the plane of POLARIZED LIGHT. Optically active crystals are used in polarimeters, instruments for measuring the degree of optical activity of chemical solutions. Optically active compounds are characterized by having molecules that are asymmetrical. Magnetic rotation of the plane of polarized light was discovered by Michael FARADAY in 1845 and is known as the Faraday effect.

optical character recognition (OCR) Technique for reading letters or numbers so that they can be inputted into a COMPUTER. The text to be read is scanned and the text "image" converted into a digital form. A special program then identifies the characters by their shapes. Early optical characters, such as those used for account numbers printed on check forms, had special shapes to prevent ambiguity (for example, the ordinary characters for 1 and I or 0 and O would be easy to confuse). However, modern scanners can deal with most ordinary typefaces and some can even read handwriting, although special optical characters are still retained in applications where errors could be serious.

optical disk In computing, high-density storage device consisting of a disk on which data is recorded and read by a laser. The most common type is a CD-ROM, although an audio compact disk is also a read-only device of this kind. A recording facility is provided by a WORM disk (write once, read many), on which a computer can save data once, and then read it repeatedly but not be able to change it.

optical fiber Fine strand of glass less than 1mm (0.04in) thick. It is able to transmit digital information in the form of pulses of light. Such transmission is possible because light entering a glass optical fiber is conducted, by reflection, from one end of the fiber to the other without loss of energy. Images may be magnified, distorted or scrambled, depending on how bundles of the fibers are configured. In medicine, optical fibers have facilitated the development of the ENDOSCOPE for the internal

trunk cable

▲ **optical fiber** Optical fibres carry information as signals of light. The fibers comprise two layers of the highest quality glass. The inner, glass core (1) of a fiber optic cable is surrounded by an outer cladding of glass (2) of a different refractive index. The cladding confines the light pulses within the core. The light signal cannot leave the core because it always hits the edge of the core at too shallow an angle to escape – an effect known as "total internal reflection". A sheath (3) provides physical protection, and bundles of the sheathed cores (trunk cables) are given strength by a central steel wire (4). Narrow-core fibers (A) are increasingly used because they allow signals to be sent over greater distances without blurring. In a wide-core cable (B) more reflections can occur, causing the pulse to spread out and merge with adjacent pulses. To prevent this, more space must be left between pulses in wider cables and that limits the volume of data that can be transmitted.

observation of organs. They also have an increasingly widespread application for telecommunications as they are able to carry more information than conventional metal cable systems, providing high-speed data links.

optical illusion Effect in which visual information is misleading or misinterpreted. Natural optical illusions are often created by light REFRACTION, such as a MIRAGE of water on a hot, dry road.

optical rotation Angle through which a molar solution of an optically active compound rotates the plane of POLARIZATION of a beam of POLARIZED LIGHT at a given temperature. Many natural products show an ability to rotate polarized light counterclockwise; few rotate clockwise. *See also* OPTICAL ACTIVITY

optic nerve Second cranial nerve, which carries the visual stimuli from the RETINA of the EYE to the visual center in the CORTEX of the brain. That part of the retina where the optic nerve enters the eye is known as the blind spot. About 1 million optic nerve fibers comprise the optic nerve, and these fibers are arranged in such a way that impulses from the left side of the visual field travel to the right side of the brain, and vice versa.

optics Branch of physics concerned with the study of LIGHT and its behavior. Fundamental aspects are the physical nature of light, both as a wave phenomenon and as particles (PHOTONS), and the REFLECTION, REFRACTION, and POLARIZATION of light and its transmission through various media. Optics also involves the study of MIRRORS, LENSES, and lens systems (including that of the eye and those of optical instruments) and of optically active chemicals and crystals that polarize light. More generally, optics deals with the wider subject of the part of the ELECTROMAGNETIC SPECTRUM lying between short RADIO waves and soft X RAYS, called the "optical spectrum."

optometry Science that deals with the examination of eyes and the prescribing of eyewear, such as glasses or contact lenses, in order to correct vision defects. It is distinct from OPHTHALMOLOGY, the medical and surgical treatment of the eyes.

orbit Path of a celestial body in a gravitational field. The path is usually a closed one about the focus of the system to which it belongs, as with those of the planets about the Sun, or the components of a binary system about their common center of mass. Most celestial orbits are elliptical, although the ECCENTRICITY can vary greatly. It is rare for an orbit to be parabolic or hyperbolic. *See also* APSIS; DECLINATION, MAGNETIC; INCLINATION, MAGNETIC; KEPLER'S LAWS; PERTURBATION

orbital In PARTICLE PHYSICS, region in space around an atomic NUCLEUS in which ELECTRONS can move. There is a high probability of finding an electron in such an orbital, which can accommodate one or two electrons and has a shape and energy characterized by the atom's QUANTUM NUMBERS. In molecules, the bonding electrons move in the combined electric field of all the nuclei. The atomic orbitals then become molecular orbitals – regions encompassing two nuclei, having a characteristic energy and containing two electrons. These molecular orbitals, which can be thought of as formed by the overlap of atomic orbitals, constitute CHEMICAL BONDS.

orbiter Spacecraft designed to orbit a moon or planet. The very first spacecraft, the Soviet *SPUTNIK I* (launched in 1957), was an orbiter that circled the Earth. More Earth satellites followed before the Soviet Union and the USA sent orbiters to the Moon in 1966. Unmanned orbiters have also been sent to Mars and Venus. The term is also used to describe the main part of the SPACE SHUTTLE.

order In biology, the CLASSIFICATION of living organisms ranking below CLASS and above FAMILY. Order names are printed in Roman (ordinary) letters, and begin with a capital letter. Among animals, the names of most orders end in -a, such as Anura (frogs and toads) and Chiroptera (bats), although bird orders end in -iformes, such as Columbiformes (pigeons) and Stringiformes (owls). In plants, order names generally end in -ales, such as Rosales (roses).

ordinal number *See* NUMBER, ORDINAL

Ordovician Second-oldest period of the PALEOZOIC era, 505 to 438 million years ago. All animal life was restricted to the sea. Numerous invertebrates flourished, including trilobites, brachiopods, corals, graptolites, mollusks, and echinoderms. Remains of jawless fish in coastal deposits mark the first record of the vertebrates.

ore Mineral or combination of minerals from which metals and nonmetals can be extracted. Ores occur in veins, beds or seams parallel to the enclosing rock or in irregular masses. Industrial rock deposits in beds, such as gypsum and limestone, are not called ores. *See also* METALLURGY; MINING

organelle Part of a CELL, such as a MITOCHONDRION or a flagellum, with a persistent structure and a specific function. It is to a cell what an organ is to an organism.

organ In biology, group of TISSUES that form a functional and structural unit in a living organism. The major organs of the body include the brain, heart, lungs, skin, liver, and kidneys. Leaves, flowers and roots are examples of plant organs.

organic chemistry *See* CHEMISTRY

organic compounds COMPOUNDS that contain the element CARBON; they are about a hundred times more numerous than inorganic compounds. Examples include HYDROCARBONS, basic structures that, when combined with atoms of other elements (such as oxygen and nitrogen), form a vast range of compounds, including those essential to life.

organic agriculture Food or fiber grown without the use of chemical FERTILIZERS or PESTICIDES. Only natural fertilizers such as manures, bone meal, seaweed, and fishmeal are used. These organic fertilizers are thought to be better for the structure of the soil than chemical or inorganic fertilizers, and plants cultivated in this manner do not contain possibly harmful chemicals. Examples include organic cotton and organic produce.

organ of Corti Complex structure within the inner EAR of mammals, birds, and reptiles, concerned with the final reception of inner ear movements resulting from sound waves striking the eardrum. It rests on a platform of membrane and bone that extends along the COCHLEA, and contains sensory hair cells that detect movements caused by sound waves and connect with nerve fibers that carry messages to the brain.

organometallic compound Chemical compound in which one or more organic groups or RADICALS are bonded to an atom of a metal. Metallic carbonates (such as sodium carbonate) and salts of common fatty acids (such as sodium ethanoate) are usually excluded from this classification. Typical examples are metallic alkyl compounds (such as LEAD(IV) TETRAETHYL and triethyl aluminium), GRIGNARD REAGENTS (such as ethylmagnesium iodide) and compounds of transition metals.

organophosphate Chemical in which the molecules contain at least one atom of carbon and one of phosphorus. The best-known organophosphates are some INSECTICIDES, certain nerve gases, essential nucleic acids, and nucleotide enzymes. Organic derivatives of the phosphorus acids are important in the manufacture of FERTILIZERS. These are highly toxic.

organ transplant Surgical implantation of an organ (such as a kidney or a heart) from another individual to substitute for a malfunctioning or diseased organ in the patient. Careful pre-operative preparation and matching followed by post-operative drug treatment are necessary to ensure that the transplanted organ is not rejected by the recipient's immune system. *See also* IMMUNOLOGY

Orion Prominent CONSTELLATION, named after a mythical hunter, situated s of Taurus and Gemini. Its brightest stars are BETELGEUSE and RIGEL. A row of three other stars – Zeta, Epsilon, and Delta – make up Orion's belt. From the belt hangs Orion's sword, marked by the ORION NEBULA.

Orion Nebula Gaseous emission NEBULA in the ORION constellation. It is a mass of gas, lit up by a group of new-born stars called the Trapezium. Because they are very hot, about 50,000K, these stars emit most of their energy in the ultraviolet, which makes the nebula glow. The Trapezium stars have destroyed most of the volatile dust in the nebula, etching out a roughly spherical hollow that is still growing. The visible Orion Nebula is, therefore, a hole in a much larger dark nebula. Within the densest portions of the dark cloud, star formation is still under way.

ornithology Study of BIRDS. Included in general ornithological studies are classification, structure, function, evolution, distribution, migration, reproduction, ecology, and behavior.

orogenesis MOUNTAIN building, especially where the Earth's crust is compressed to produce large-scale FOLDS and FAULTS.

orrery Mechanical model of all or parts of the solar system. Orreries vary from simple ones showing just the Earth, Moon, and Sun, to highly complicated representations of the whole solar system, where the planets with their satellites not only revolve in their orbits but also rotate on their axes. The name is from Charles Boyle, fourth Earl of Orrery (1676–1731), who commissioned one in 1712, although this was not the first to be made.

orthoclase Essential mineral found in igneous and metamorphic rocks containing potassium FELDSPAR. It is a potassium aluminum silicate, $KAlSi_3O_8$, with monoclinic system crystals. Its color is generally white, but there are pink varieties. Hardness 6–6.5; r.d. 2.5–2.6.

orthopteran Any member of the order Orthoptera, an order of insects represented by 20,000 species found worldwide, especially in tropical regions. Orthopterans include katydids, crickets, grasshoppers, and locusts in the suborders Ensifera and Caelifera. The term was formerly used to group together the orders Dictyoptera (mantids and cockroaches), Phasmida (leaf and stick insects) and Grylloblattidae (grylloblattids). Most of the species feed on plants, have chewing mouthparts, and undergo incomplete METAMORPHOSIS from egg to nymph and adult. Earwigs, termites and webspinners are also considered Orthopteroid insects. Length: to 10cm (4in).

oscillating Universe theory Variant of the BIG BANG theory in which it is suggested that the Universe passes through successive cycles of expansion and contraction (or collapse). At the end of the collapse phase, with the Universe packed into a small volume of great density, it is possible that a "bounce" would occur. The Universe would thus oscillate between Big Bang and "Big Crunch" episodes and so be infinite in age. However, for this to happen the density of the Universe would have to be above a certain value (the critical density), which is not thought to be the case.

oscillator In the physics of SOUND, a device for producing sound waves, as in SONAR or in an ultra-sonic generator. In electronics, an oscillator circuit converts direct current (DC) electricity into high-frequency alternating current (AC). Harmonic oscillators generate sinusoidal wave forms. *See also* ELECTRIC CURRENT

oscilloscope (cathode-ray oscilloscope) Electronic instrument in which a CATHODE-RAY TUBE (CRT) system displays how quantities, such as voltage or current, vary over a period of time. The input signal is usually fed to the vertical plates of the CRT. The electron beam that traces the pattern on the screen is moved by a time-base generator within the oscilloscope. The result is generally a curve or graph on the screen. The oscilloscope can be calibrated and used to measure the input signal.

Osler, Sir William (1849–1919) Canadian physician. Osler was the first to identify (1876) platelets in the blood. In 1888 he was appointed professor of medicine at Johns Hopkins University in Baltimore, which he developed into one of the foremost medical schools. Osler wrote *The Principles and Practice of Medicine* (1892), the leading medical text of the time and still in print. Probably the most famous physician of his era, he gave his name to several medical terms, such as Osler's nodes and Osler-Vaquez disease.

osmium (symbol Os) Blue-white, metallic TRANSITION ELEMENT, discovered in 1803. The densest of the elements, osmium occurs associated with platinum; the chief source is as a by-product from smelting nickel. Like IRIDIUM, osmium is used in producing hard alloys. It is also used to make electrical contacts and pen points. Properties: at.no. 76; r.a.m. 190.2; r.d. 22.57; m.p. 3,045°C (5,513°F); b.p. 5,027°C (9,081°F); most common isotope ^{192}Os (41.0%).

osmoregulation Process by which the amount of water and concentration of salts are regulated in the bodies of animals and protozoans. In saltwater environments, water tends to pass from an animal's body through the action of OSMOSIS. This effect is counteracted by the kidneys. Similarly, in freshwater environments the kidneys of higher animals, and the contractile VACUOLES of protozoans, prevent water from passing into the body.

osmosis Diffusion of a SOLVENT (such as water) through a natural or artificial semipermeable membrane (one that allows only the passage of certain dissolved substances) into a more concentrated solution. Because the more concentrated solution contains a lower concentration of solvent molecules, the solvent flows by DIFFUSION to dilute it until concentrations of solvent are equal on both sides of the membrane. Osmosis is a vital cellular

▲ **oscilloscope** An oscilloscope displays an electronic signal on a monitor in analog form. Usually time is represented on the x axis (horizontal), with the y axis (vertical) recording the incoming voltage – here a heartbeat (1). The signal from the object being monitored is converted into an electrical voltage (2). That voltage is shown in visible form on the screen of a cathode ray tube (3) similar to a black and white television. Deflector magnets (4) direct the stream of electrons from the electron gun (5). The magnets sweep the electron beam from left to right over a set period, while variations in the voltage of the external signal cause the wave pattern. The control box (6) allows the period of the x axis and the strength of the signal displayed to be changed.

process. Plant roots absorb water by osmosis, and plant cells are kept firm as a result of the uptake of water by osmosis (turgor); membranes of all living cells use it to control the passage of required substances. *See also* DIALYSIS; PERMEABLE MEMBRANE; REVERSE OSMOSIS; TURGOR PRESSURE

osmotic pressure (symbol Π) Pressure exerted by a dissolved substance by virtue of the motion of its molecules. In dilute solutions, it varies with the concentration and temperature as if the solute were a gas occupying the same volume. It can be measured by the pressure that must be applied to counterbalance the process of OSMOSIS into the solution. Osmotic pressure is given by $\Pi V = nRT$, where n = moles in solution, V = volume, T = temperature, and R = gas constant.

ossicle Any one of the small bones found in the middle EAR of most mammals. The ossicles are the **malleus** (hammer), **incus** (anvil), and **stapes** (stirrup). Vibrations of the eardrum (tympanum) are picked up by the malleus, amplified by the movement of the incus and passed on to the stapes, which connects with the oval window in the COCHLEA of the inner ear.

ossification (osteogenesis) Process of BONE formation in vertebrates. It occurs through the action of special cells called osteoblasts, which secrete bone-forming minerals that combine with a network of COLLAGEN fibers to form the hard bone matrix.

Osteichthyes Class of fish found in almost every environment. Their characteristics include a bony skeleton, a single flap (operculum) covering the gill openings, and erythrocytes (red blood cells) with nuclei. Most members of this class have scales. Osteichthyes first appeared during the DEVONIAN period, when they were heavily armored and lived in fresh water.

osteomyelitis Infection of the BONE, sometimes spreading along the marrow cavity. Today, quite rare except in diabetics, it can arise from a compound fracture, where the bone breaks through the skin, or from infection elsewhere in the body. It is accompanied by fever, swelling, and pain, possibly with abscess formation. The condition may be treated with immobilization, ANTIBIOTICS, and possibly surgical drainage.

osteopathy System of alternative medical treatment based on the use of physical manipulation to rectify damage caused by mechanical stresses. The concept was formulated by Andrew STILL in 1874, and the first school of osteopathy was opened in Kirksville, Missouri, in 1892. Still's emphasis on treating the whole person remains an ideal of osteopathy.

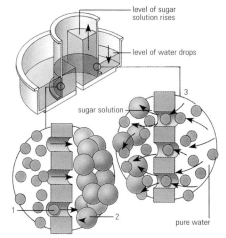

▲ **osmosis** Semi-permeable membranes have small pores in them. While the pores allow tiny molecules through (1), they also block out larger molecules (2). If a semi permeable membrane is used to separate a sugar solution from a container of water (3), initially the concentration of water molecules in the solution will be lower than that in the pure water. Over time, water molecules will pass through the membrane, diffusing into the container, but as the pores are too small to allow the sugar solution to flow the other way, the level of the sugar solution rises. Osmosis, the net flow of a solvent across a membrane, drives many cellular processes in organisms, like maintaining hydrostatic pressure in plant cells to keep them turgid. Artificial membranes are used in kidney dialysis machines.

Ostwald, Wilhelm (1853–1932) German chemist, b. Russia. He became professor of chemistry at Leipzig University (1887–1906). His work on SOLUTIONS led him to formulate Ostwald's dilution law, which deals with the dissociation of weak ELECTROLYTES. His process for preparing nitric acid and his research on catalysis became important in industry. He was awarded the 1909 Nobel Prize for chemistry.

Otis, Elisha Graves (1811–61) US inventor who designed and manufactured elevators and escalators. His first invention (1852) was an automatic safety hoist for an elevator in a factory. Otis's fame dates from 1856, when he installed a passenger elevator in a New York store. In 1861 he patented a steam-powered elevator.

Otto, Nikolaus August (1832–91) German engineer. In 1861 he built a gas-fired engine that won a gold medal at the 1867 Paris Exhibition. Otto later built an INTERNAL COMBUSTION ENGINE based on a four-stroke cycle (also called the **Otto cycle**). This FOUR-STROKE ENGINE was the forerunner of most of today's engines.

outcrop Exposure at the surface of the Earth of an edge of rock stratum. This phenomenon may be caused by the EROSION of soil by water, wind, ice (especially glaciers), or gravity.

outlier Mass of newer rocks surrounded by older rocks. The newer rocks may have become detached from a larger formation by EROSION. *See also* INLIER

output device, computer Apparatus that transmits or records data from a COMPUTER. It may be a MODEM for remote receipt of data, a hard-copy **printer** or recorder (such as a teletypewriter), a **line-printer,** or an alphanumeric display or **monitor** (a TV-type screen on which letters or numbers are presented). Other forms include digital to analog systems, which directly control various mechanisms and processes.

ovary In biology, part of a multicellular animal or a seed-bearing plant that produces egg cells (ova), the female reproductive cells. In vertebrates, it also produces female SEX HORMONES. In human females, there is an ovary on each side of the UTERUS. Controlled by the PITUITARY GLAND, each ovary, as well as the ova, produces two major female sex hormones, ESTROGEN and PROGESTERONE, which in turn control development and functioning of the female reproductive system. In seed-bearing plants, the female sex cells are contained within structures called OVULES inside the ovary. After FERTILIZATION, the ovules develop into seeds and the ovary develops into one or more FRUITS. *See also* HORMONE; MENSTRUAL CYCLE; OVUM

overtone Usually a HARMONIC, a constituent of a musical note with a frequency that is a whole-number multiple of that of the fundamental note. Some instruments produce nonharmonic overtones.

oviducts Tubes that connect the ovaries with the UTERUS and through which egg cells (ova) are released from the OVARY. In mammals, they are known as FALLOPIAN TUBES. These are lined with a tissue called the EPITHELIUM.

oviparity (ovipary) Type of REPRODUCTION in which fertilization of the female ova (eggs) occurs inside the body, but once achieved the eggs are laid and subsequently hatch outside her body. Unlike VIVIPARITY, the developing EMBRYO receives all its nutrition from the egg. It is the most common form of reproduction and occurs in all

animals except MARSUPIALS and PLACENTAL MAMMALS. *See also* OVOVIPARITY

ovoviparity (ovoviviparity) Type of REPRODUCTION in which a female's fertilized ova (eggs) develop and then hatch inside her OVIDUCTS. Unlike VIVIPARITY, during ovoviparity development the ZYGOTES receive no nutrition from the mother as no PLACENTA is present. Ovoviparity is common in many invertebrate animals and some fish and snakes. When the young emerge, the mother appears to be giving birth to live young, although they have only just hatched. *See also* OVIPARITY

ovulation Release of a mature OVUM (egg) from the OVARY ready for FERTILIZATION. In women, one egg (ova) is released into the oviduct midway through the MENSTRUAL CYCLE, stimulated by LUTEINIZING HORMONE from the PITUITARY GLAND.

ovule In seed-bearing plants (SPERMATOPHYTES), the part of the female reproductive organ that contains an egg cell or OVUM and develops into a seed after FERTILIZATION. In ANGIOSPERMS (flowering plants), ovules develop inside an OVARY. In GYMNOSPERMS (conifers), ovules are borne on the inner surface of the female cone without any covering.

ovum Female GAMETE (reproductive cell) produced in an OVARY (in animals) or an OVULE (in plants). After FERTILIZATION by the male gamete (SPERM in animals, POLLEN in plants), it becomes a ZYGOTE that is capable of developing into a new individual.

oxalic acid (ethanedioic acid) Poisonous, colorless, crystalline organic acid ($C_2H_2O_4$) whose salts occur naturally in some plants, such as sorrel and rhubarb. It is used for metal and textile cleaning, and in tanning. Properties: r.d. 1,653; m.p. 101.5°C (214.7°F).

oxbow lake Crescent-shaped section of a river channel that no longer carries the main discharge of water. An oxbow lake forms from a MEANDER. As sediment is deposited, the meander becomes cut off from the river to create a lake. Once formed, the lake gradually shrinks as sediment fills it in; vegetation grows on the new muddy area, and the land can be reclaimed. The name derives from the shape of the lake, said to resemble an ox's collar.

oxidase In biochemistry, any enzyme that catalyses an OXIDATION REDUCTION reaction.

oxidation Chemical reaction that involves a loss of one or more ELECTRONS by an atom or molecule (always part of an OXIDATION-REDUCTION reaction in which those electrons are gained by another atom or molecule). Previously the term was more strictly applied to a reaction in which oxygen combines with another element or compound to form an oxide. Oxidation is brought about by OXIDIZING AGENTS.

oxidation number *See* OXIDATION STATE

oxidation-reduction (redox) Chemical reaction involving simultaneous OXIDATION (a loss of one or more electrons by an atom or molecule) and REDUCTION (a gain of those electrons by another atom or molecule). In general, oxidation and reduction reactions occur together and in the same quantities. For example, in the reaction between iron oxide and carbon, iron (the electron acceptor) is **reduced** by carbon (the reducing agent), and the carbon (the electron donor) is **oxidized** by iron oxide (the oxidizing agent). Oxidation-reduction reactions are important in many biochemical systems.

oxidation State in which an atom is potentially able to form a compound, depending on the number of electrons available to be transferred. It is quantified on a scale of **oxidation numbers** which indicate the degree of IONIZATION. Thus, in the compound sodium oxide, Na_2O, SODIUM (Na^+) has an oxidation number of +1, and OXYGEN (O^{2-}) an oxidation number of -2. In COVALENT and COORDINATION (complex) COMPOUNDS, the oxidation number is the electric charge that the atoms would have had if the compound was ionic; for example, in the ION $(CuCl_4)^{2-}$, regarded as formed from Cu^{2+} and $4Cl^-$, copper has an oxidation number of +2. Oxidation numbers are often used in the names of chemical compounds, as in iron(II) chloride ($FeCl_2$) and iron(III) chloride ($FeCl_3$), formerly called ferrous and ferric chloride. *See also* VALENCE

oxide Any inorganic chemical compound in which OXYGEN is combined with another element. Oxides are often formed by burning the element in air or oxygen; thus magnesium (Mg) burns to form magnesium oxide (MgO). Oxides are acidic, basic or AMPHOTERIC, the latter capable of being acidic or basic. Other forms include PEROXIDES, such as barium peroxide, and the neutral oxides, such as carbon monoxide, which do not react with acids or bases.

oxidizing agent Substance that causes OXIDATION reactions. Thus in the oxidation of carbon to form carbon monoxide, $2C + O_2 \rightarrow 2CO$, oxygen is the oxidizing agent. Other common oxidizing agents include NITRIC ACID, HYDROGEN PEROXIDE, OZONE, potassium dichromate, iron(III) (ferric) compounds, and POTASSIUM MANGANATE(VII). *See also* OXIDATION-REDUCTION

oxime In organic chemistry, a compound formed between hydroxylamine (H_2NOH) and an ALDEHYDE or KETONE. Oximes are condensation products (*see* CONDENSATION REACTION) containing the group $-C=NOH$, and are used in qualitative analysis to identify aldehydes and ketones. For example, acetaldehyde forms the oxime $CH_3CH=NOH$, and acetone forms the oxime $(CH_3)_2C=NOH$.

oxyacetylene welding *See* WELDING

oxygen (symbol O) Common gaseous element that is necessary for the RESPIRATION of plants and animals and for combustion. It was discovered in 1774 by Joseph PRIESTLEY and independently (*c*.1772) by Karl SCHEELE. Oxygen, which is colorless and odorless, is the most abundant element in the Earth's crust (49.2% by weight) and is a constituent of water and many rocks. It is also present in the atmosphere (28% by volume), being extracted by fractional DISTILLATION of liquid air. It can also be obtained by the ELECTROLYSIS of water. Oxygen is used in steelmaking (BESSEMER PROCESS), welding, the manufacture of industrial chemicals, and in apparatus for breathing (oxygen masks) and resuscitation (oxygen tents);

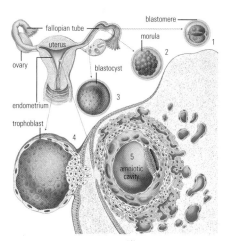

▲ **ovum** It takes about a week for the fertilized ovum to pass down the fallopian tube and implant itself in the uterine lining, the endometrium. Within hours of conception, mitosis begins with the development of a sphere of an increasing number of cells; the sphere starts as the blastomere (1), develops into the morula (2) of about 64 cells. At this stage, it changes into a hollow, fluid-containing ball – blastocyst (3) – with the inner cell mass at one end. It can now begin implantation (4). By the ninth day after conception, the blastocyst has sunk deep into the endometrium (5) and is already receiving nutrition from the mother.

liquid oxygen is used in rocket fuels. It is chemically reactive, and forms compounds with nearly all other elements (especially by OXIDATION). Properties: at.no. 8; r.a.m. 15.9994; r.d. 1.429; m.p. −218.4°C (−361.1°F); b.p. −182.96°C (−297.3°F); most common isotope ^{16}O (99.759%). *See also* OXIDATION-REDUCTION; OZONE

oxygen cycle Interchange of OXYGEN among agencies such as the atmosphere, the oceans, animal and plant processes and chemical combustion. The main renewable source of the Earth's oxygen is the plant process of PHOTOSYNTHESIS, wherein oxygen is liberated. Oxygen is utilized by lifeforms through RESPIRATION, a process essential to most living forms except ANAEROBIC bacteria. *See also* CARBON CYCLE

oxygen debt Insufficient supply of oxygen in the muscles following vigorous exercise. This reduces the breakdown of food molecules that generate energy, causing the muscles to overproduce LACTIC ACID, which creates a sensation of FATIGUE. The automatic rapid breathing (hyperpnea) experienced after exercise creates extra oxygen to reduce the oxygen debt. As exercise ceases, the body uses up this extra supply of oxygen to break down the lactic acid.

oxyhemoglobin Combination of HEMOGLOBIN in ERYTHROCYTES (red blood cells) with OXYGEN from the lungs, in which form oxygen is transported in the blood to all cells of the body. When oxyhemoglobin gives up its oxygen to cells, a chemical reaction is promoted that makes CARBON DIOXIDE (CO_2) from the tissues more soluble in the blood for transport back to the lungs and elimination from the body.

oxytocin Hormone produced by the posterior PITUITARY GLAND in women during the final stage of pregnancy. It stimulates the muscles of the uterus, initiating the onset of LABOR and maintaining contractions during the BIRTH of the child. It also stimulates LACTATION.

ozone (O_3) Unstable, pale-blue, gaseous allotrope of OXYGEN. It has a characteristic pungent odor and decomposes into molecular oxygen. It is present in the atmosphere, mainly in the OZONE LAYER, where it is formed by the action of ultraviolet radiation on oxygen. It acts as a filter and prevents much harmful ultraviolet radiation from reaching the Earth's surface. Ozone, prepared commercially by passing a high-voltage discharge through oxygen, is used as an OXIDIZING AGENT in bleaching, air-conditioning, and purifying water. *See also* ALLOTROPY

ozone layer Region of Earth's atmosphere in which OZONE (O_3) is concentrated. It is densest at altitudes of 21–26km (13–16mi). Produced by incoming sunlight, the ozone layer absorbs much of the solar ultraviolet radiation, thereby shielding the Earth's surface. Aircraft, nuclear weapons, and some aerosol sprays and refrigerants all yield chemical agents that can break down high-altitude ozone, which could lead to an increase in the amount of harmful ultraviolet radiation reaching the Earth's surface. *See also* CHLOROFLUOROCARBON (CFC).

▶ **ozone** A naturally occurring substance, ozone acts as a sunscreen for the Earth because its molecules absorb the Sun's ultraviolet radiation (1). The presence of chlorofluorocarbon pollution (2) causes the layer to break down, allowing ultraviolet rays through. Ozone is created when ultraviolet rays split oxygen molecules. The lone oxygen atoms (3) bond with oxygen molecules (4) to make ozone (5). When, however, chlorofluorocarbons are present, they are split by the ultraviolet rays. The released chlorine atom (6) in turn splits ozone molecules to form a chlorine monoxide molecule (7) and oxygen. The process is continued as the chlorine monoxide absorbs the lone oxygen atoms that previously formed ozone. This frees the chlorine atom, which splits another oxygen molecule (8) creating another chlorine monoxide molecule and oxygen.

○ oxygen atom
⊘ oxygen molecule
⊗ ozone molecule

chlorofluorocarbon

chlorine monoxide

pacemaker (sinoatrial node) Specialized group of cells in the vertebrate HEART that contract spontaneously, setting the pace for the heartbeat itself. If it fails its function can be taken over by an artificial pacemaker, an electronic unit that stimulates the heart by means of tiny electrical impulses to maintain a normal beat.

Page, Sir Frederick Handley (1885–1962) British aeronautical engineer who founded (1909) one of the first aircraft manufacturing companies (Handley Page Ltd.). The company produced civil and military aircraft, such as the Halifax bomber.

Paget, Sir James (1814–99) British surgeon, a founder of the science of PATHOLOGY. Paget discovered *Trichina spiralis*, the parasitic worm that causes trichinosis. He described the bone disease osteitis deformans (Paget's disease), and was among the first experts to recommend surgical removal of bone marrow tumors.

paint Coating applied to a surface for protective, decorative or artistic purposes. Paint is composed of PIGMENT (color) and a liquid **vehicle** (binder or medium). The vehicle suspends the pigment, adheres to a surface and hardens when dry. Pigments are made of metallic compounds, usually OXIDES, or synthetic materials. Vehicles may be oils, water mixed with a binding agent, organic compounds or synthetic RESINS, which may be soluble in water or oil. Water-based paint (emulsion) mixes with water; oil-based paint (gloss) mixes with TURPENTINE or various organic solvents.

pair production Creation of an ELECTRON and a POSITRON from a PHOTON in a strong electric field of a NUCLEUS. Electrons and positrons are only formed when a photon has high quantum energy since, on the basis of the special theory of RELATIVITY ($E = mc^2$), they have a total mass of 16 x 10^{-14} J. *See also* PARTICLE PHYSICS

Palade, George Emil (1912–) US cell biologist, b. Romania. He shared the 1974 Nobel Prize for physiology or medicine with Albert Claude (1899–1983) and Christian de DUVE. Using sophisticated electron microscopic techniques and other methods, Palade discovered RIBOSOMES, which are the sites of PROTEIN synthesis in the cell.

paleobotany Study of ancient plants that have been preserved by carbonization, waterlogging or freezing. Some plants have been preserved almost intact in frozen soils and in AMBER. Pollen is also very resistant and has characteristic patterns for identification. The earliest land plants grew in the CAMBRIAN period, but some water plants flourished earlier. Such plants provide important evidence of prehistoric climates and ECOSYSTEMS.

Paleocene Geological epoch that extended from about 65 to 55 million years ago. It is the first epoch of the TERTIARY period in the CENOZOIC era, when the majority of the DINOSAURS had died out and small herbivores and ungulates were flourishing. Primates and rodents evolved towards the end of the Paleocene epoch. *See also* GEOLOGICAL TIME

paleoclimatology Study of prehistoric CLIMATES. Records of past climates are found in SEDIMENTARY ROCKS, in cores taken through deep layers of ice, and in fossil-bearing cores from the beds of seas and lakes. From such evidence, climatologists have discovered that the Earth is subject to alternate periods of cold, called glacials or ICE AGES, and warmth, called INTERGLACIALS. During the last two million years there have been 17 ice ages.

paleogeography Science that studies ancient GEOGRAPHY. It seeks to determine the physical and biological conditions of the Earth throughout GEOLOGICAL TIME. Paleogeography also seeks to determine how conditions in one period led to conditions in the next and succeeding periods. Much of the data needed for good analysis is lacking, and most studies in this area are general.

Paleolithic (Old Stone Age) Earliest stage of human history, from *c.*2 million years ago until between 40,000 and 8,000 years ago. It was marked by the use of stone tools and covers the period of HUMAN EVOLUTION from *Homo habilis* to *Homo sapiens*. It was followed by the NEOLITHIC.

paleomagnetism Study of changes in the direction and intensity of Earth's MAGNETIC FIELD in GEOLOGICAL TIME. It is important to the theory of CONTINENTAL DRIFT. Since the "magnetic memory" of rocks is measurable, this determines their orientation in relation to magnetic north at the time of their solidification. The Earth's polarity has reversed at least 20 times in the past 4 to 5 million years; earlier changes cannot at present be determined. The gross displacement of large rock formations as measured by their magnetic qualities can be explained by SEAFLOOR SPREADING. *See also* MAGNETIC FIELD REVERSAL

paleontology Study of the FOSSIL remains of plants and animals. Evidence from fossils is used in the reconstruction of ancient environments and in tracing the EVOLUTION of life on Earth.

Paleozoic Second era of GEOLOGICAL TIME, after the PRE-CAMBRIAN era, lasting from 590 million to 248 million years ago. It is subdivided into six periods: CAMBRIAN, ORDOVICIAN, SILURIAN, DEVONIAN, CARBONIFEROUS, and PERMIAN. Invertebrate animals evolved hard skeletons capable of being preserved as fossils in the Cambrian, fishlike vertebrates appeared in the Ordovician, amphibians emerged in the Devonian, and reptiles in the Carboniferous.

palladium (symbol Pd) Silver-white TRANSITION ELEMENT, first discovered in 1803. Malleable and ductile, palladium is found in nickel ores associated with PLATINUM, to which it is chemically similar. It does not tarnish or corrode and is used in ALLOYS for ELECTROPLATING, surgical instruments, dentistry, jewelry, and for CATALYTIC CONVERTERS. Properties: at.no. 46; r.a.m. 106.4; r.d. 12.02; m.p. 1,552°C (2,826°F); b.p. 3,140°C (5,684°F); most common isotope ^{106}Pd (27.3%).

palmitic acid (hexadecanoic acid) FATTY ACID that is a major constituent of vegetable oils (particularly palm oil) and animal fats. Together with STEARIC ACID, it is the fatty acid most often present in soaps. Palmitic acid is a SATURATED COMPOUND, containing only single bonds, and has the formula $CH_3 (CH_2)_{14} COOH$.

palynology Study of SPORES, SEEDS, AND POLLENS; it is a part of such disciplines as archaeology, PALEOGEOGRAPHY, and PALEONTOLOGY. Studies of pollen in lake sediments have provided much information about the vegetation, CLIMATE, AND environmental conditions in past ages. Palynology has also helped to deduce the patterns of land use and settlement of primitive humans.

pancreas Elongated gland lying behind the STOMACH to the left of the midline. It secretes pancreatic juice into the small INTESTINE to aid digestion. Pancreatic juice contains the enzymes AMYLASE, TRYPSIN, and lipase. The pancreas also contains a group of cells known as the ISLETS OF LANGERHANS, which secrete the hormones INSULIN and GLUCAGON concerned in the regulation of blood-sugar level. *See also* DIABETES

Pangaea Name for the single supercontinent that formed about 240 million years ago and that began to break up at the end of the TRIASSIC period. It was surrounded by Pantalassa, the ancestral Pacific. Using calculations based on computer data, present land masses plus their continental shelves can be fitted together into this one continent. Pangaea was first written about by Alfred WEGENER in his theory of CONTINENTAL DRIFT. *See also* GONDWANALAND; LAURASIA

pantothenic acid VITAMIN of the B complex involved in the synthesis of ACETYL COENZYME A, itself concerned in the oxidation of fats, carbohydrates, and several amino acids. A common vitamin, pantothenic acid is found in cereals, yeast, liver, and egg yolks.

paper Sheet or roll of compacted CELLULOSE fibers with a wide range of uses including packaging, writing, wall covering, and clothing. The word "paper" derives from papyrus, the plant that the Egyptians used at least 5,500 years ago to make sheets of writing material. The papyrus reed was soaked, slit into strips, laid at right angles and pounded and pressed into a sheet. The modern process of manufacture originated about 2,000 years ago in China and consists of reducing wood fiber, straw, rags, and grasses to a pulp by the action of an ALKALI, such as SODIUM HYDROXIDE (CAUSTIC SODA). The LIGNIN and other noncellulose material is then extracted and the residue is bleached. After washing and the addition of a filler to provide a smooth and flat surface, the pulp is allowed to settle on a wire mesh to form a thin sheet and dried.

paper chromatography Type of CHROMATOGRAPHY in which absorbent paper is used. The mixture to be analyzed is dissolved and a drop of the solution placed near one edge of sheet of paper. The sheet is suspended vertically with that edge dipping into a trough of SOLVENT. As the solvent rises up the paper by capillary action, the components in the spot move with the solvent front at different rates, depending on their composition. When the paper is dried, the components appear as a line of spots, which may need spraying with a reacting substance to color them and reveal their presence. The distance a substance moves in a given time is characteristic of its identity.

papilla Small, conical protuberance found in mammals and plants. In mammals papillae project into the EPIDERMIS from the DERMIS; they are found in many parts of the body, including the mammalian INNER EAR and on the surface of the TONGUE.

Papin, Denis (1647–1712) French physicist and inventor who helped to develop the STEAM ENGINE.

Pap test (cervical smear) Test for CANCER of the CERVIX, established by US physiologist George Papanicolaou (1883–1962). In this diagnostic procedure, a small sample of tissue is removed from the cervix and examined under a microscope for the presence of abnormal, precancerous cells. Treatment in the early stages of cervical cancer can prevent the disease from developing.

parabola Mathematical curve, a CONIC section traced by a point that moves so that its distance from a fixed point, the focus, is equal to its distance from a fixed straight line, the directrix. It may be formed by cutting a cone parallel to one side. The general equation of a parabola is $y = a \, x^2 + b \, x + c$ where a, b, and c are constants. Any projectile fired from Earth, falling under the influence of GRAVITATION, follows a parabolic trajectory.

paraboloid Mathematical solid figure in which all sections parallel to the AXIS of symmetry are PARABOLAS, and sections at an angle to this axis are other CONIC sections, such as ELLIPSES, HYPERBOLAS or CIRCLES.

Paracelsus (1493–1541) Swiss physician and alchemist, b. Philippus Aureolus Theophrastus Bombastus von Hohenheim. He became professor of medicine at Basel, where his lectures discredited past and contemporary medicine and were preceded by the burning of the works of GALEN and Avicenna (980–1037). According to Paracelsus, the human body is primarily composed of salt, sulfur, and mercury, and it is the separation of these elements that causes illness. He introduced mineral baths and made opium, mercury, lead, and various minerals part of the PHARMACOPOEIA. According to his enemies, Paracelsus died in Salzburg after a drunken bout. Others say he was thrown down a hill by hirelings of jealous apothecaries.

paracetamol (acetaminophen) ANALGESIC drug that lessens pain and is also effective in reducing FEVER. It is used to treat mild to moderate pain, such as headaches, toothache, and rheumatic conditions, and is particularly effective against musculoskeletal pain. Paracetamol is often used as an alternative to ASPIRIN because it is less irritating to the lining of the STOMACH. However, it lacks the anti-inflammatory action of aspirin. Paracetamol overdosage can cause LIVER damage and death.

paraffin (kerosene) Common, domestic fuel that is mainly a mixture of ALKANE hydrocarbons. It is a product of the distillation of PETROLEUM. Less volatile than gasoline, paraffin is used also as a fuel for jet aircraft. Paraffin wax is a white, translucent, waxy substance consisting of a mixture of solid alkanes obtained by solvent extraction. It is used to make candles, waxed paper, polishes, and cosmetics.

paraldehyde (ethanal trimer, $C_6H_{12}O_3$) Colorless liquid formed by the POLYMERIZATION of ETHANAL (acetaldehyde) by the action of SULFURIC ACID. It regenerates ethanal on heating with dilute acids. It has a pleasant odor but disagreeable taste, and is used mainly as a SEDATIVE.

parallax Angular distance by which a celestial object appears to be displaced with respect to more distant objects, when viewed from opposite ends of a baseline. It is used as a measure of an object's distance. The parallax of a star (**annual** parallax) is the angle subtended at the star by the mean radius of the Earth's ORBIT (one ASTRONOMICAL UNIT); the smaller the angle, the more distant the star. *See also* PARSEC

parallel circuit ELECTRIC CIRCUIT in which the components are interconnected so that the CURRENT divides between them and passes through all of them at once. The total RESISTANCE, R, of a number of RESISTORS connected in parallel is given by the expression $1/R = 1/R_1 + 1/R_2 + 1/R_3 + ...$, where R_1, R_2 and R_3 are the resistances of the component resistors. The total CAPACITANCE, C, of capacitors in parallel is given by $C = C_1 + C_2 + C_3 + ...$, where C_1, C_2 and C_3 are the capacitances of the individual capacitors. *See also* SERIES CIRCUIT

parallelogram Quadrilateral (four-sided plane figure) having each pair of opposite sides parallel. Both the opposite sides and the opposite angles of a parallelogram are equal. Its area is the product of one side and its perpendicular distance from the opposite side. A parallelogram with all four sides equal is called a RHOMBUS.

parallelogram of vectors Method of calculating the sum of two VECTOR quantities. The direction and size of the sum of the vectors is ascertained by trigonometry or scale drawing. The vectors are represented by two adjacent sides of a PARALLELOGRAM and the sum is the diagonal through their point of intersection.

parallel processing In computing, a technique that allows the simultaneous execution of two or more tasks. Multitasking is performed by dedicated MICROPROCESSORS in a COMPUTER.

paramagnetism Type of MAGNETISM displayed by such metals as platinum and magnesium, which, when placed in a MAGNETIC FIELD, are magnetized parallel to the field by an extent proportional to the strength of that field. This effect is much weaker than that in ferromagnetic materials such as iron, cobalt, and nickel. *See also* DIAMAGNETISM; FERROMAGNETISM

parasite Organism that lives on or in another organism (the HOST) upon which it depends for survival; this arrangement may be harmful to the host.

Parasites occur in many groups of plants and in virtually all major animal groups. A parasite that lives in the host is called an **endoparasite**; a parasite that survives on the host's exterior is an **ectoparasite**. Many parasites, such as PROTOZOA, fleas, and worms, carry disease or cause sores or lesions that may become infected. The European cuckoo and cowbird rely on other birds to rear their young, and are therefore considered "brood parasites." In **parasitoidism**, the relationship results in the death of the host. For example, various flying insects, such as the ichneumon flies, lay their eggs on or in a host that becomes food for the insect LARVAE. A **hyperparasite** is one that parasitizes another parasite. *See also* COMMENSALISM; MUTUALISM; SYMBIOSIS

parasympathetic nervous system One of two parts of the AUTONOMIC NERVOUS SYSTEM, the other is the SYMPATHETIC NERVOUS SYSTEM. Both are involved with the action of INVOLUNTARY MUSCLES. The parasympathetic nervous system controls muscles that prepare the body for a relaxed state, for example by decreasing heart rate and aiding digestion by encouraging PERISTALSIS. The sympathetic nervous system does the reverse.

parathyroid glands Four small endocrine glands, usually embedded in the back of the THYROID GLAND, that secrete a HORMONE to control the level of calcium and phosphorus in the BLOOD. Overproduction of parathyroid hormone causes loss of calcium from the bones to the blood; a deficiency causes tetany (involuntary muscle spasm). *See also* ENDOCRINE SYSTEM

parathyroid hormone (PTH) HORMONE secreted by the PARATHYROID GLAND. It helps to regulate calcium levels in the BLOOD by promoting the release of calcium from bones into the blood, inhibiting secretion in the kidneys and increasing reabsorption of calcium and magnesium from the intestines.

Paré, Ambroise (*c*.1517–90) French physician regarded as the founder of modern SURGERY. In 1537 Paré was employed as an army surgeon and in 1552 became surgeon to Henry II, one of four French kings he served during his lifetime. As an army surgeon, he introduced new methods of treating wounds, described in his book *The Method of Treating Wounds Made by Harquebuses and Other Guns* (1545), and revived the practice of tying ARTERIES during surgery instead of cauterizing them.

parenchyma Soft tissue made up of nonspecialized, thin-walled CELLS, either spherical or blunt-edged in shape, often with spaces between the cells. It is one of the chief tissues of plant stems, leaves, and fruit pulp; it stores nutrients and water, and helps to support the plant. In animals, parenchyma is similar loose, connective, indeterminate tissue. It packs the spaces between the organs of simple animals such as worms.

parity (symbol *P*) In physics, term used to denote space-reflection symmetry. The principle of conservation of parity states that physical laws are the same in both a left- and right-handed COORDINATE SYSTEM. This was regarded as inviolable until 1956, when Chen Ning YANG and Tsung-Dao LEE showed that it was transgressed by WEAK NUCEAR FORCES between ELEMENTARY PARTICLES (*See also* SPIN). Parity is also used in INFORMATION THEORY to denote a property of sequences of digits that can be employed in message transmission to detect errors.

A

B

C

D

◀ **parasite** Some external parasites live either on, or nearby, humans. The crab louse (A) lives on areas of the body with widely spaced coarse hair. The acarus mite (B) is just about visible to the human eye, and is responsible for causing scabies. The body louse (C) is one of the more dangerous parasites. It is a carrier of epidemic typhus. The common bedbug (D) is found all over the world. During the day it is inactive, but at night it finds humans and sucks their blood.

Parkinson's disease Degenerative brain disease characterized by tremor, muscular rigidity, and poverty of movement and facial expression. It arises from a lack of DOPAMINE, a NEUROTRANSMITTER that is essential for the control of voluntary movement. Slightly more common in men, it is rare before the age of 50. Foremost among the drugs used to control the disease is L-DOPA.

parotid gland One of two SALIVARY GLANDS located one on either side of the mouth just in front of the ear. With ducts opening on the inside of the cheek, it is the gland that becomes swollen during an attack of MUMPS.

parsec (*pc*) Distance at which a star would have a PARALLAX of one second of arc; equivalent to 3.2616 light-years, 206,265 astronomical units, or 30.857×10^{12} km.

parthenogenesis In biological reproduction, development of a female sex cell or GAMETE without FERTILIZATION. Since there is no involvement of a male gamete, it leads to the production of offspring that are genetically identical to the mother. This process produces little VARIATION and occurs naturally among some invertebrate animals, such as aphids and bees. The same process occurs in a few plants, such as dandelions, and is termed **parthenocarpy**.

partial differential equation Type of DIFFERENTIAL EQUATION used when a FUNCTION depends on more than one INDEPENDENT VARIABLE. For example, a wave in two dimensions has an amplitude (height) U that depends on time t and also on the two distance measurements x and y along mutually perpendicular axes. The differential equation representing the wave is
$$\partial^2 U/\partial x^2 + \partial^2 U/\partial y^2 = 1/c^2 \, \partial^2 U/\partial t^2$$
where c is the wave's velocity. When solved, the solution U will give the amplitude of the wave at any point (x, y) and at any time. Symbols such as $\partial^2 U/\partial x^2$ are called **partial derivatives**. These equations are used extensively in physical science.

partial eclipse See ECLIPSE

particle accelerator See ACCELERATOR

particle physics Branch of physics that studies ELEMENTARY PARTICLES and interactions between them. Particle physics and NUCLEAR PHYSICS developed simultaneously, each contributing to the other. Physicists now recognize more than 300 different elementary particles, which can be grouped in various ways, although they are all composed of a few basic building blocks. The particles that are currently regarded as fundamental can be grouped into the LEPTONS, the QUARKS, and the carriers of the FUNDAMENTAL FORCES, such as GLUONS and PHOTONS. Particles termed HADRONS are made up of two or more quarks. For example, PROTONS and NEUTRONS are made up of three quarks, and MESONS are made up of two quarks. Most elementary particles are unstable and decay into other particles.

Pascal, Blaise (1623–62) French mathematician and philosopher. A prodigy, he had written a book on CONIC sections by the time he was 17. Later, with Pierre de FERMAT, Pascal laid the foundations of the mathematical theory of PROBABILITY. He also contributed to CALCULUS and HYDRODYNAMICS, devising **Pascal's law** (1647). This stated that the PRESSURE applied to an enclosed FLUID is transmitted equally in all directions and to all parts of the enclosing vessel, if pressure changes due to the weight of the fluid can be neglected. In 1655 he retired from science to devote himself to religious and philosophical writing, of which *Pensées* is his best-known work. The SI unit of pressure is named after him.

pascal (symbol *Pa*) SI unit of pressure. It is equal to a pressure of 1 NEWTON per sq m, equivalent to 10 DYNES per sq cm, or 1.45×10^{-4} lb per sq in. It is named after Blaise PASCAL.

Paschen-Back effect Phenomenon of the splitting of a line of the SPECTRUM that is observed when atoms are subject to a STRONG NUCLEAR FORCE. It was discovered (1912) by Louis Paschen (1865–1947) and Ernest Back (1881–1959). *See also* ELECTRON; PRECESSION; SPIN; ZEEMAN EFFECT

Pasteur, Louis (1822–95) French chemist, one of the founders of MICROBIOLOGY. He made important contributions to chemistry, bacteriology, and medicine. He discovered that microorganisms can be destroyed by heat, a technique now known as PASTEURIZATION and used to destroy harmful microorganisms in food. Pasteur also discovered that he could weaken certain disease-causing microorganisms – specifically those causing ANTHRAX in animals and rabies in man – and then use the weakened culture to vaccinate individuals against the disease.

pasteurization Controlled heat treatment of food to kill bacteria and other microorganisms, discovered by Louis PASTEUR in the 1860s. Milk is pasteurized by heating it to 72°C (161.6°F) and holding it at that temperature for 16 seconds. Ultrapasteurization is now used to produce UHT (ultra-heat-treated) milk; the milk is heated to 132°C (270°F) for one second to provide a shelf-life of several months.

patella (knee-cap) Large, flattened, roughly triangular bone just in front of the joint where the FEMUR and TIBIA are linked. It is surrounded by bursae (sacs of fluid), which cushion the joint.

pathogen Microorganism that causes disease in plants or animals. Animal pathogens are most commonly BACTERIA and VIRUSES, while common plant pathogens also include FUNGI.

pathology Study of diseases, their causes, and the changes they produce in the CELLS, tissues, and organs of the body. Although pathologists are not involved directly in treating patients, their work is of immense importance to the understanding and treatment of disease.

Pauli, Wolfgang (1900–58) US physicist, b. Austria. He studied QUANTUM THEORY under Niels BOHR. In 1925 Pauli formulated his EXCLUSION PRINCIPLE, which states that each FERMION (such as an ELECTRON or QUARK) has a unique quantum state. IN 1931 he correctly predicted the existence of the neutrino. Pauli received the 1945 Nobel Prize for physics for his achievements in quantum physics.

Pauling, Linus Carl (1901–94) US chemist who won both the Nobel Prize for chemistry and the Nobel Peace Prize. He applied QUANTUM THEORY to the field of chemistry. His early work on the application of WAVE MECHANICS to molecular structure, detailed in his book *The Nature of the Chemical Bond* (1939), led to the Nobel Prize for chemistry in 1954. In the 1950s Pauling worked on the structure of PROTEINS, and his suggestion that the MOLECULES of DNA were arranged in a helical structure anticipated the findings of Francis CRICK and James WATSON. He was known also for advocating massive doses of vitamin C to treat the common cold and for general health. A leading figure in the campaign for nuclear disarmament, Pauling was awarded the 1962 Nobel Peace Prize.

Pavlov, Ivan Petrovich (1849–1936) Russian neurophysiologist. His early work centered on the physiology and neurology of digestion. In 1904 Pavlov received the Nobel Prize for physiology or medicine. He is best known, however, for his classical (Pavlovian) CONDITIONING of behavior in dogs. His major works are *Conditioned Reflexes* (1927) and *Lectures on Conditioned Reflexes* (1928).

Pavo (Peacock) Constellation of the s sky. Its brightest star, Alpha, has magnitude 1.9.

PCB Abbreviation of POLYCHLORINATED BIPHENYL

Peano, Giuseppe (1858–1932) Italian mathematician. He studied at the University of Turin, later becoming a professor there. Peano's main achievement was his AXIOMATIC METHOD for defining the natural NUMBERS, a major development in formal ARITHMETIC. Regarded as one of the founders of mathematical LOGIC, he developed a complete system of notation for logic, which was later adopted as standard. The Peano curve, an early example of a space-filling curve, is named after him.

pearl Hard, smooth, iridescent concretion of CALCIUM CARBONATE produced by certain marine and freshwater bivalve MOLLUSKS. It is composed almost entirely of nacre, or mother-of-pearl, which forms the inner layer of mollusk shells. A pearl, the only GEM of animal origin, results from an abnormal growth of nacre around minute particles of foreign matter, such as a grain of sand.

peat Soil composed of dead and decaying remains of vegetation. It is generally dark brown or black in color, and is thought to be similar to the first stage in the formation of COAL. Peat forms in areas of high rainfall or very poor drainage, and contains a high proportion of water. In very wet conditions, plants do not decay when they die because the conditions are nearly ANAEROBIC, which means that the vegetation is not broken down by BACTERIA. Waterlogged and airless conditions create a very acidic and infertile landscape, which can become very boggy. Only specialized plants, notably SPHAGNUM MOSS, can survive in such areas. Peat is used sometimes as a fuel.

peat moss Decomposed organic matter (HUMUS) obtained from disintegrated SPHAGNUM MOSS (bog moss). The most widely obtainable source of humus, it is dug into soil and added to compost to retain moisture. Its use in horticulture is being reduced as its continuing extraction poses an ecological threat to peat bogs.

pectin Water-soluble POLYSACCHARIDE found in the cell walls and intercellular tissue of certain ripe fruits or vegetables. When fruit is cooked, pectin yields a gel that is the thickening agent of jellies and jams.

pectoral Relating to the chest or breast, as in the pectoral muscles, which cover the front of the chest, or the pectoral girdle, the bones of which make up the shoulder.

pedicel In botany, term for a flower stalk. The pedicel attaches an individual flower to the **peduncle** (the main floral axis), often growing out of the AXIL of a BRACT. In zoology, a pedicel is the second joint of an insect's ANTENNA or any other stalk-like appendage.

pedology Scientific study of SOILS. Pedologists divide soils according to their physical and chemical composition. These SOIL HORIZONS are determined by factors such as the presence of organic matter and the extent of drainage. Soil types are classified by such names as PODZOL and CHERNOZEM, reflecting the pioneering work of Russian pedologists.

Pegasus (Flying horse) Conspicuous constellation of the N sky. Its most famous feature, the Square of Pegasus, is made up of four stars one of which is actually part of ANDROMEDA. Beta (Scheast) is a RED GIANT irregular variable. Pegasus contains the GLOBULAR CLUSTER M15.

pegmatite In geology, any very coarse-grained IGNEOUS ROCK, generally light in color and often of

GRANITIC composition. Pegmatites are the chief sources of GEMS, MICA, and FELDSPAR.

pelagic Zone of an OCEAN and the marine organisms that inhabit it. Pelagic organisms live anywhere in oceans, seas or lakes, except on the bottom (*see* BENTHOS). They are divided into NEKTON (large fish and whales) and PLANKTON (small plants and animals) on which the nekton feed.

Peltier effect Phenomenon of the temperature changes at a junction where an ELECTRIC CURRENT passes from one kind of metal to another. When a current passes through a THERMOCOUPLE (THERMOMETER), the temperature at one junction increases while that at the other decreases, so heat is transferred from one to the other. The effect, used in refrigeration, was discovered in 1834 by the French physicist Jean Charles Peltier (1785–1845). *See also* SEEBECK EFFECT; THERMOELECTRICITY

pelvis (pelvic girdle) Dish-shaped bony structure that supports the internal organs of the lower abdomen in vertebrates, and serves as a point of attachment for muscles that move the limbs or fins. In women, the pelvis is broader than in men, to facilitate childbirth. *See also* ILIUM; ISCHIUM; PUBIS

pendulum Any body suspended at a point so that it swings in an arc. A **simple** pendulum consists of a small heavy mass attached to a string or light rigid rod. A **compound** pendulum has a supporting rod whose mass is not negligible. A simple pendulum was first used to regulate clocks in 1673 by Christiaan HUYGENS. Huygens determined the formula for the period of such a device: $T = 2\pi\sqrt{(l/g)}$, where T is the time for one complete swing, l = length of the wire and g = acceleration of gravity. FOUCAULT'S PENDULUM, devised by Léon FOUCAULT, swings in all directions and was used to demonstrate the Earth's rotation.

peneplain Fairly flat area of land formed over millions of years due to the wearing down of ancient mountains by EROSION. Such long-term erosion, called **denudation**, often results from the action of rivers and may leave areas of more resistant rock, which stand up as monadnocks or residual hills.

penicillin ANTIBIOTIC agent derived from a mold of the genus *Penicillium*. The first antibiotic to be discovered (by Sir Alexander FLEMING in 1928), it was later developed in a soluble form and can now be produced synthetically. It is effective in combating a wide range of bacterial infections. It can produce allergic reactions, including itching and (rarely) potentially fatal shock. Because some microorganisms have become resistant to its action, other antibiotics are often used instead.

penis Male reproductive organ. It contains the URETHRA, the channel through which URINE and SEMEN pass to the exterior, and erectile tissue that, when engorged with blood, causes the penis to become erect during arousal. *See also* SEXUAL REPRODUCTION

Penney, William George, Baron (1909–91) British atomic physicist who helped to develop the first British atomic bomb (*see* NUCLEAR WEAPON). During World War II he was engaged in nuclear research for the British government. In 1944 Penney went to the Los Alamos Scientific Laboratory and in 1945 was official observer of the bombing of Nagasaki. In 1964 he became chairman of the UK Atomic Energy Authority.

Pennsylvanian In North America, name given to the later part of the CARBONIFEROUS period.

pentadactyl limb Limb with five digits (fingers or toes), characteristic of four-legged vertebrates. It is generally used for locomotion and is found in amphibians, reptiles, birds, and mammals, but may be greatly modified (as in the flippers of seals and whales or the wings of bats). It probably evolved from the fins of primitive fish, which are the only modern vertebrates not to retain it.

▲ **pentadactyl limb** The feet of mammals have evolved in many different ways from the basic mammalian foot (A), possessed by the earliest shrew-like mammal. Seals (B) have developed evenly graduated toes for a webbed paddle. Moles (C) have truncated toes for leverage when digging. The camel's two toes (D) are padded for walking on sand. Horses have a hoof (E) instead of claws, and elongated feet for speed, as has the cheetah (F). Bats (G) have enormously elongated digits to support wings. Kangaroos' toes (H) are designed for hopping. Lemurs (I) and sloths (J) have forelimbs for grasping trees.

pentagon Five-sided plane figure. Its interior ANGLES add up to 540°. For a regular pentagon, one whose sides and interior angles are all equal, each interior angle is 108°.

Penzias, Arno Allan (1933–) US astrophysicist, b. Germany. He and Robert WILSON discovered COSMIC MICROWAVE BACKGROUND, the residual BACKGROUND RADIATION emanating from outer space, which scientists agree supports the BIG BANG theory. In 1964, while working at Bell Laboratories, Penzias and Wilson began monitoring radio waves in the Milky Way. They detected a non-varying signal that corresponds to radiation at a temperature of $c.3°K$, or $-270°C$ ($-454°F$), corresponding to thermal energy left over from the Big Bang. For this pioneering discovery, Penzias and Wilson shared the 1978 Nobel Prize for physics with Peter KAPITZA.

pentode ELECTRON TUBE with five electrodes: cathode, anode, and three grids. The grids are made of fine wire mesh and are placed between the anode and the cathode in order to suppress the loss of ELECTRONS. This results in an output signal 1,500 times greater than the input. *See* CATHODE-RAY TUBE

pepsin Digestive ENZYME secreted by GLANDS of the STOMACH wall as part of the GASTRIC JUICE. In the presence of hydrochloric acid, pepsin catalyzes the splitting of PROTEINS in food into POLYPEPTIDES.

peptide Molecule consisting of two or more AMINO ACID molecules linked by bonds between the amino group of one acid and the carboxyl group of another. This type of linkage is called a peptide bond, and peptides containing several amino acids are called POLYPEPTIDES. PROTEINS consist of polypeptide chains having up to several hundred amino acids cross-linked to each other in various ways.

percentage Quantity or amount reckoned as part of a whole, expressed in hundredths; the rate or proportion per hundred. Thus, 75% of a quantity means 75 parts in every 100, or equivalently 3/4 of the total.

Peregrinus, Petrus (active 13th century) French scientist who described the properties of MAGNETISM. A student of Roger Bacon, he distinguished between and named the MAGNETIC POLES, and developed the existing rudimentary magnetic COMPASS into a reliable instrument for NAVIGATION.

perennial Plant with a life cycle of more than two years. It is a common term for flowering HERBACEOUS and woody plants. They include the lily, daisy, and iris, and all trees. *See also* ANNUAL; BIENNIAL

perfect number *See* NUMBER, PERFECT

perfume Manufactured substance that produces a pleasing fragrance. The scents of plants such as rose, citrus, lavender, and sandalwood are obtained from their ESSENTIAL OILS. These are blended and combined with a fixative of animal origin, such as musk, ambergris or civet. Fixatives add pungency and prevent the more volatile oils from evaporating too quickly. Liquid perfumes are usually alcoholic solutions containing 10–25% of the perfume concentrate; colognes and toilet waters contain about 2–6% of the concentrate.

perianth Outer region of a FLOWER. The perianth includes all the structures surrounding the reproductive organs and usually consists of an outer whorl of SEPALS (CALYX) and an inner whorl of PETALS (COROLLA).

pericardium Double membrane that surrounds the HEART, separating it from the rest of the chest cavity and protecting it from mechanical injury. The outer tough fibrous pericardial layer is separated from the inner membrane by lubricating pericardial fluid.

pericarp In SPERMATOPHYTES (seed-bearing plants), the wall of a ripened FRUIT that is derived from the ovary wall. The tissues of the pericarp may be fibrous, stony, or fleshy.

perichondrium In a developing EMBRYO, membrane surrounding the CARTILAGE that eventually becomes BONE. It is well supplied with BLOOD VESSELS and is the source of the osteoblasts (bone-building cells) that invade the embryonic cartilage and lay down the hard bone matrix.

peridot Gem variety of transparent green OLIVINE, a silicate mineral. Large crystals are found on St John's Island in the Red Sea and in Burma.

peridotite Coarse-grained, heavy IGNEOUS ROCK composed of OLIVINE and PYROXENE with small flecks of MICA or HORNBLENDE. It alters readily into SERPENTINE. Rocks that consist mainly of olivine are called DUNITES.

perigee Point in the orbit about the Earth of the Moon or an artificial satellite at which the body is nearest to the Earth. *See also* APOGEE; APSIS

perihelion Point in the orbit of a planet, asteroid, comet or other body (such as a spacecraft) moving around the Sun at which the body is nearest the Sun. *See also* APHELION; APSIS

perimeter Distance around the boundary of a plane closed figure. The perimeter of a circle is its CIRCUMFERENCE; that of a polygon is the sum of the lengths of its sides.

period In astronomy, time taken by one celestial body (such as PLANET or SATELLITE) to complete one revolution around another. It can be measured in several ways, such as the SIDEREAL PERIOD and the SYNODIC PERIOD.

periodic law In chemistry, law first stated (1869) by Dmitri MENDELEYEV asserting that the properties of the ELEMENTS are a periodic function of their RELATIVE ATOMIC MASS (R.A.M.). The groupings of the elements based on this law formed the forerunner of the PERIODIC TABLE. From gaps in these groupings, Mendeleyev was able to predict the existence and properties of undiscovered elements. But his table contained anomalies, which were not resolved until Henry MOSELEY's work on X-RAYS discovered that periodicity was related to ATOMIC NUMBER (rather than relative atomic mass) and the later discovery of ISOTOPES by QUANTUM THEORY.

periodic table Arrangement of the chemical ELEMENTS in order of their ATOMIC NUMBERS in accordance with the PERIODIC LAW stated (1869) by Dmitri MENDELEYEV and later modified by Henry MOSELEY. In the modern form of the table, the elements are arranged into 18 vertical columns and seven horizontal periods. The vertical columns, containing groups, are numbered I to VII (sometimes called IA to VIIA) with a final column numbered 0. The metallic TRANSITION ELEMENTS are arranged in the middle of the table between groups II and III. ALKALI METALS are in group I and ALKALINE-EARTH METALS in group II. METALLOIDS and nonmetals are found from groups III to VII, with the HALOGENS in group VII and the NOBLE GASES (inert gases) collected into group 0. The elements in each group have the same number of VALENCE electrons and accordingly have similar chemical properties. Elements in the same horizontal period have the same number of ELECTRON shells. The elements are arranged in the periods in order of increasing atomic number from left to right. *See complete table on page 396*

peripheral nervous system All parts of the NERVOUS SYSTEM that lie outside the CENTRAL NERVOUS SYSTEM (CNS). It comprises the 12 pairs of cranial nerves, which principally serve the head and neck region, and 31 pairs of spinal nerves with their fibers extending to the furthermost parts of the body. It carries impulses from RECEPTORS to the CNS, and carries back responses from the CNS to EFFECTORS. *See also* AUTONOMIC NERVOUS SYSTEM (ANS)

periscope Optical instrument consisting of a series of MIRRORS or PRISMS that allows a person to view the surroundings from a concealed position by changing the direction of the observer's line of sight. Since World War I, the periscope has been most commonly associated with submarines.

Perissodactyla Order of mammals characterized by hoofs with an odd number of toes. All HERBIVORES, the only living members of the order are horses, tapirs, and rhinoceroses; there are more than 200 extinct forms known from fossils. *See also* ARTIODACTYLA

peristalsis Series of wavelike movements that propel food through the digestive tract. It is caused by contractions of the smooth INVOLUNTARY MUSCLE of the gut wall. The reverse process, antiperistalsis, produces vomiting.

peritoneum Strong membrane of CONNECTIVE TISSUE that lines the body's abdominal wall and covers the abdominal organs. The greater omentum, a fold in the peritoneum containing layers of FAT cells, forms an apron over the INTESTINES. Inflammation of the peritoneum is known as **peritonitis**.

permafrost Land that is permanently frozen, often to a considerable depth. The top few centimeters generally thaw in summer, but the meltwater is not able to sink into the ground because of the frozen subsoil. If the landscape is fairly flat, surface water lies on the ground throughout the summer. Construction work is very difficult, and many methods have been employed in Russia, Canada, and Alaska to overcome the problems. *See also* MUSKEG; TUNDRA

permeability In geology, ability of rock to transmit water; it is not directly related to POROSITY. Limestone, for example, is a permeable but non-porous rock; water percolates only through the joints and fissures.

permeability (symbol μ) In physics, ratio of the MAGNETIC FLUX density in a body to the external MAGNETIC FIELD inducing it. The permeability of free space is called the magnetic constant (symbol μ_0) and has the value $4\pi \times 10^{-7}$ henry per meter. The relative permeability of a substance (symbol μ_r) is the ratio of its permeability to the magnetic constant.

Permian Last geological period of the PALEOZOIC era, lasting from 286 to 248 million years ago. There was widespread geologic uplift, resulting in the formation of PANGAEA. The major climatic characteristics of the period are aridity and glaciation. Many groups of marine invertebrate animals became extinct in the Permian period, but reptiles flourished.

permittivity (symbol ε) In physics, DIELECTRIC property of a material. It is the ratio of the ELECTRIC FLUX density (displacement) in a medium to the intensity of the ELECTRIC FIELD causing it. The permittivity of free space is called the ELECTRIC CONSTANT (symbol ε_0). The relative permittivity of

KEY

atomic number	43
atomic symbol	Tc
name of element	Technetium
atomic weight	[97]

(most stable isotope in brackets)

GROUP I	II											III	IV	V	VI	VII	0
1 **H** Hydrogen 1.00794																	2 **He** Helium 4.0026
3 **Li** Lithium 6.941	4 **Be** Beryllium 9.0122											5 **B** Boron 10.81	6 **C** Carbon 12.011	7 **N** Nitrogen 14.0067	8 **O** Oxygen 15.9994	9 **F** Fluorine 18.998	10 **Ne** Neon 20.179
11 **Na** Sodium 22.9898	12 **Mg** Magnesium 24.305											13 **Al** Aluminum 26.9815	14 **Si** Silicon 28.086	15 **P** Phosphorus 30.9738	16 **S** Sulfur 32.06	17 **Cl** Chlorine 35.453	18 **Ar** Argon 39.948
19 **K** Potassium 39.098	20 **Ca** Calcium 40.06	21 **Sc** Scandium 44.956	22 **Ti** Titanium 47.90	23 **V** Vanadium 50.941	24 **Cr** Chromium 51.996	25 **Mn** Manganese 54.9380	26 **Fe** Iron 55.847	27 **Co** Cobalt 58.9332	28 **Ni** Nickel 58.70	29 **Cu** Copper 63.546	30 **Zn** Zinc 65.38	31 **Ga** Gallium 69.72	32 **Ge** Germanium 72.59	33 **As** Arsenic 74.9216	34 **Se** Selenium 78.96	35 **Br** Bromine 79.904	36 **Kr** Krypton 83.80
37 **Rb** Rubidium 85.4678	38 **Sr** Strontium 87.62	39 **Y** Yttrium 88.906	40 **Zr** Zirconium 91.22	41 **Nb** Niobium 92.906	42 **Mo** Molybdenum 95.94	43 **Tc** Technetium [97]	44 **Ru** Ruthenium 101.07	45 **Rh** Rhodium 102.905	46 **Pd** Palladium 106.4	47 **Ag** Silver 107.868	48 **Cd** Cadmium 112.40	49 **In** Indium 114.82	50 **Sn** Tin 118.69	51 **Sb** Antimony 121.75	52 **Te** Tellurium 127.75	53 **I** Iodine 126.9045	54 **Xe** Xenon 131.30
55 **Cs** Caesium 132.905	56 **Ba** Barium 137.34	57–71 Lanthanide Series	72 **Hf** Hafnium 178.49	73 **Ta** Tantalum 180.948	74 **W** Tungsten 183.85	75 **Re** Rhenium 186.207	76 **Os** Osmium 190.2	77 **Ir** Iridium 192.22	78 **Pt** Platinum 195.09	79 **Au** Gold 196.9665	80 **Hg** Mercury 200.59	81 **Tl** Thallium 204.37	82 **Pb** Lead 207.2	83 **Bi** Bismuth 208.98	84 **Po** Polonium [209]	85 **At** Astatine [210]	86 **Rn** Radon [222]
87 **Fr** Francium [223]	88 **Ra** Radium [226]	89–103 Actinide Series	104 **Db** Dubnium [261]	105 **Ha** Hahnium [262]	106 **Sg** Seaborgium [263]	107 **Uns** Unnilseptium [262]	108 **Uno** Unniloctium [265]	109 **Une** Unnilennium [266]									

LANTHANIDE SERIES (rare earth elements)

57 **La** Lanthanum 138.9055	58 **Ce** Cerium 140.12	59 **Pr** Praseodymium 140.9077	60 **Nd** Neodymium 144.24	61 **Pm** Promethium [145]	62 **Sm** Samarium 150.36	63 **Eu** Europium 151.96	64 **Gd** Gadolinium 157.25	65 **Tb** Terbium 158.9254	66 **Dy** Dysprosium 162.50	67 **Ho** Holmium 164.9308	68 **Er** Erbium 167.26	69 **Tm** Thulium 168.9342	70 **Yb** Ytterbium 173.04	71 **Lu** Lutetium 174.97

ACTINIDE SERIES (radioactive rare earth elements)

89 **Ac** Actinium [227]	90 **Th** Thorium 232.0381	91 **Pa** Protactinium 231.0359	92 **U** Uranium 238.029	93 **Np** Neptunium 237.0482	94 **Pu** Plutonium [244]	95 **Am** Americium [243]	96 **Cm** Curium [247]	97 **Bk** Berkelium [247]	98 **Cf** Californium [251]	99 **Es** Einsteinium [254]	100 **Fm** Fermium [257]	101 **Md** Mendelevium [256]	102 **No** Nobelium [254]	103 **Lr** Lawrencium [256]

a medium is the ratio $\varepsilon/\varepsilon_0$. It is the ratio of the CAPACITANCE of a capacitor (condenser) in which the material is the dielectric to its capacitance with a vacuum between the capacitor's plates.

peroxide Chemical COMPOUND containing two oxygen atoms bonded to each other and another element. An example is barium peroxide (BaO_2), which yields a solution of HYDROGEN PEROXIDE (H_2O_2), a popular bleach, when treated with an acid. Peroxides are powerful OXIDIZING AGENTS.

perpetual motion There are two theoretical forms of perpetual motion. In the first, a mechanism works indefinitely without an exterior supply of ENERGY. This type of machine violates the first law of THERMODYNAMICS, concerning the CONSERVATION of energy. In the second theory, a machine would take HEAT from a reservoir, such as the ocean, and convert all of it into WORK. This contravenes the second law of thermodynamics, which states that such a process must consume more energy than it can gain from the reservoir. *See also* ENTROPY

Perrin, Jean Baptiste (1870–1942) French physicist and chemist. He was awarded the 1926 Nobel Prize for physics for his work on the structure of MATTER and for his investigation of CATHODE RAYS and BROWNIAN MOVEMENT of particles.

personal computer (PC) *See* COMPUTER

Perseus Prominent N constellation. Alpha Persei is magnitude 1.8. Beta is the prototype eclipsing BINARY, **Algol**. Perseus is a rich constellation, crossed by the Milky Way.

Perspex (tradename for polymethyl methacrylate) Colorless, transparent, solid PLASTIC. It is made by POLYMERIZATION of methyl methacrylate, and is widely used as a substitute for glass.

perturbation Irregularity in the orbital motion of a celestial body, brought about by the gravitational attraction of other bodies. Attractions by PLANETS and SATELLITES cause significant displacements of bodies from where they would be in a purely elliptical ORBIT. Perturbations are responsible for dramatic changes in the orbits of COMETS, changing long-period orbits into short-period ones. In ecology, a disturbance that forces an ecological system away from equilibrium. *See also* KEPLER'S LAWS

Perutz, Max Ferdinand (1914–) British biochemist, b. Austria, who studied the X-ray DIFFRACTION of PROTEINS. In 1953 he discovered that adding a heavy ATOM, such as gold or mercury, to each molecule of HEMOGLOBIN produces a slightly different diffraction pattern. By this means he demonstrated the structure of hemoglobin, for which he shared the 1962 Nobel Prize for chemistry with John C. KENDREW who, by a similar method, discovered the structure of MYOGLOBIN.

pervious Term describing rocks through which water can pass by means of cracks and fissures. CARBONIFEROUS limestone and GRANITE are good examples of pervious rocks

pesticide Chemical substance that is used to kill insects, rodents, weeds, and other pests. Among pesticides, a HERBICIDE is used for weeds, an INSECTICIDE for insects and a FUNGICIDE for fungal pests. Pesticides are usually harmful chemicals and an important factor in their manufacture is that they should decompose after they have performed their function. Some previously common pesticides, such as DDT, have been shown to be too toxic and long-lasting, so their use is now restricted. *See also* ORGANOPHOSPHATE

▲ **pesticide** Applied to the land, pesticides, even in small doses, are poisonous to many animals. The concentration of some pesticide poisons increases in a process known as bioaccumulation along the food chain, finally becoming lethal to animals at the end of the chain. A pesticide (1), such as DDT, is applied to water at 0.015 parts per million (ppm) to control midge larvae, but the plankton (2) accumulates 5ppm. The fish population (3,4) builds up still higher concentrations, and finally a grebe (5), which feeds on the fish, accumulates as much as 1,600ppm of the pesticide in its body fat – enough to kill the bird.

petal Part of a FLOWER. The PETALS of a flower are known collectively as the COROLLA. Surrounded by SEPALS, flower petals are usually brightly colored and often secrete NECTAR and scent to attract the insects and birds necessary for CROSSPOLLINATION. Once fertilization occurs, the petals usually drop off. The petals of wind-pollinated flowers are often small and inconspicuous, allowing the STAMENS and STIGMA to be exposed for POLLINATION.

Petit, Alexis Thérèse (1791–1820) French physicist. He and his brother-in-law, François Arago (1786–1853), worked on the REFRACTION of light in gases and eventually accepted Augustin FRESNEL's new wave theory of light. In 1818 Petit and Pierre DULONG won a prize for determining the SPECIFIC HEAT CAPACITY of solids and for devising measurements of temperature, research that led the following year to DULONG AND PETIT's LAW.

Petri dish Shallow, flat-bottomed circular dish, often with a tight-fitting cover, used in laboratories mainly for producing cultures of microorganisms. It is named after the German bacteriologist J.R. Petri (1852–1921).

petrification Fossilizing process in which organic material (such as wood) changes into stone. Petrification is caused by mineral-rich water seeping into the empty spaces of dead, buried trees or animals, which eventually become stone. Although petrified remains can be up to 300 million years old, the stone often reproduces the original living material so clearly that the CELL structure is identifiable. The Petrified Forest National Park in Arizona, USA, has many examples of petrified wood. *See also* FOSSIL; MINERALIZATION

petrochemical Chemical substances derived from PETROLEUM or NATURAL GAS. The refining of petroleum is undertaken on a large scale not only for the fuels obtained (GASOLINE, PARAFFIN, fuel oil, and natural gas) but also for the wide range of chemicals that can be derived. These chemicals include the common ALKANES (paraffins) and ALKENES (olefins), BENZENE, TOLUENE, and NAPHTHALENE.

petroleum (crude oil) FOSSIL FUEL that is chemically a complex mixture of HYDROCARBONS. It accumulates in underground deposits and the chemical composition of petroleum strongly suggests that it originated from the bodies of long-dead organisms, particularly marine PLANKTON. After death these organisms sank to the ocean floor, to be broken down by bacteria in oxygen-poor conditions into simpler organic materials, including hydrocarbons. Petroleum is rarely found

at the original site of formation, but migrates laterally and vertically until it is trapped. Most petroleum is extracted via OIL WELLS from reservoirs in the Earth's crust sealed by upfolds of impermeable rock or by SALT DOMES that form traps.

During petroleum refining, the heavier hydrocarbons, which usually have higher boiling points than lighter ones, are distilled (separated from the original crude substance) as the first stage of refining. The next stage is CRACKING, which breaks the heavy hydrocarbons down into more economically useful products, such as GASOLINE and PARAFFIN. Purification of the various products to remove impurities, such as SULFUR and NITROGEN compounds, completes the refining process. The most versatile end products are ETHENE and PROPENE, which are widely used in the plastics and chemical industries. *See also* NATURAL GAS; OIL REFINERY

petrology Study of ROCKS, including their origin, chemical composition, and where they are found. Formation of the three classes of rocks – IGNEOUS (of volcanic origin); SEDIMENTARY (deposited by water); and METAMORPHIC (either of the other two changed by temperature and pressure) – are studied.

pewter Any of several silver-colored, soft alloys that consist mainly of tin and lead. The most common form has about four parts of tin to one of lead, combined with small amounts of antimony and copper. Some food and drink can become contaminated with poisonous lead from pewter utensils, so other alloys have replaced pewter for such items.

pH (abbreviation of potential of hydrogen) Indication of the acidity or alkalinity of a solution. The **pH scale** expresses a range of pH values based on a logarithmic measure of the concentration or, more properly, activity of hydrogen ions. The scale runs from 0 to 14, and a neutral solution, such as pure water, has a pH of 7. A solution is acidic if the pH is less than 7 and alkaline if greater than 7. The pH may be measured with an INDICATOR or meter. The scale was introduced (1909) by S.P. Sørensen (1868–1939).

phaeophyte Any member of the phylum Phaeophyta, which consists of the brown ALGAE. Phaeophytes are part of the kingdom PROTOCTISTA (which until recently was known as Protista). Classified by some biologists as plants, the simple organisms belonging to this group are mostly marine, found mainly in the intertidal zone of rocky shores. They include familiar seaweeds such as *Fucus* (wracks) and *Ascophyllum* (bladder wrack). The largest, *Macrocystis*, grows to more than 100m (320ft) at up to 0.5m (18in) per day. *Sargassum* forms vast floating masses in the Sargasso Sea in the mid-Atlantic, with their own distinctive communities of animals and microorganisms.

phagocyte Type of LEUCOCYTE (white blood cell) able to engulf other CELLS, such as bacteria. It digests what it engulfs (in a process known as **phagocytosis**) in the defence of the body against INFECTION. Phagocytes also act as scavengers by clearing the bloodstream of the remains of the cells that die as part of the body's natural processes. *See also* MACROPHAGE; MONOCYTE

phalanges Bones in the toes and fingers. In humans there are 14 phalanges in each hand and foot, 2 in each thumb and big toe and 3 in the remaining digits. They are connected to the METACARPALS in the hand and to the METATARSALS in the foot.

phanerogam Early botanical term for seed-bearing, flowering plants, now called SPERMATOPHYTES. *See also* CRYPTOGAM

Phanerozoic Geological timescale that describes the whole of time that has elapsed since the end of the PRECAMBRIAN period. It has lasted for about 570 million years since life first evolved and thus includes the PALEOZOIC, MESOZOIC, and CENOZOIC eras. Because some living organisms had by this time developed hard shells and internal skeletons, most fossils come from this period. *See* GEOLOGICAL TIME

pharmacology Study of the properties of DRUGS and their effects on the body.

pharmacopoeia Reference book listing drugs and other preparations for medical use. Included are details of their formulae, dosages, routes of administration, known side-effects, contraindications, and precautions.

pharynx Cavity at the back of the nose and mouth that extends down towards the ESOPHAGUS and TRACHEA. Inflammation of the pharynx, usually caused by viral or bacterial infection, is known as pharyngitis.

phase In astronomy, proportion of the illuminated hemisphere of a body in the solar system (in particular the Moon or an inferior planet) as seen from Earth. The phase changes as the Sun and the Earth change their relative positions. All the phases of the Moon (new, crescent, half, gibbous, and full) are observable with the naked eye. The inferior planets, Mercury and Venus, show phases from a slender crescent to a fully illuminated disk when observed through a telescope. Of the superior planets, Mars can show quite a marked gib-

bous phase, and Jupiter a very slight gibbous phase, but with Saturn and the planets beyond, no phase is ever discernible. Phase is sometimes expressed as the percentage of the visible disk that is illuminated.

phase In physics, one of the physical states of MATTER. Each homogeneous form is a distinct part within a heterogeneous system and is separated from other parts by boundary surfaces: ice is thus a phase of H_2O, water. A mixture of ice and water is called a two-phase system. Temperature and pressure determine the phase of a substance, and a PHASE DIAGRAM indicates the relationships between phases over a range of temperatures and pressures. A change in the internal ENERGY of a substance causes a phase transition (change in state), as seen in melting or boiling. *See also* BOILING POINT; MELTING POINT

phase In physics, stage in the cycle of an oscillation, such as the wave motion of light or sound waves. This is usually measured from an arbitrary starting point or compared with another motion of the same frequency. Two waves are said to be "in phase" when their maximum and minimum values happen simultaneously. If not, there is a "phase difference" resulting in INTERFERENCE.

phase-contrast microscope Type of MICROSCOPE used for examining transparent specimens. It is widely used in biology for studying CELLS and thin slices of tissue. Light passing through the specimen slows down and becomes out of PHASE with the original light. Differences in thickness or density in the specimen cause DIFFRACTION patterns. A phase-contrast plate converts these to INTERFERENCE patterns, which manifest themselves as large contrasts in brightness of different parts of the microscope image.

phase diagram Graph showing the conditions under which different equilibrium PHASES of substances exist. For example, a curve of MELTING POINT against PRESSURE of a pure solid divides the graph into two regions. Points in one represent temperatures and pressures at which the substance is solid; points in the other represent the liquid conditions. Graphs of composition against temperature are used to indicate such features as SOLUBILITIES and the ranges of stability of alloy phases. *See illustration on page 400*

phenacite (beryllium silicate, Be_2SiO_4) Orthosilicate mineral found in PEGMATITES and high-temperature veins. It has hexagonal system rhombohedral crystals, and is either colorless or glassy white, yellow, red or brown in color. It is sometimes used as a gem. Hardness 7.5–8; r.d. 3.

phenetics Differences or similarities between organisms that can be observed. These characteristics can be used to construct methods of CLASSIFICATION (**phenetic classification**), in contrast with the evolutionary relationships that are more usually employed.

phenol Group of AROMATIC COMPOUNDS that have one or more HYDROXYL (–OH) groups attached to a BENZENE ring. The simplest of the family is also called phenol or CARBOLIC ACID (C_6H_5OH). Phenols are colorless liquids or white solids at room temperature, with higher melting and boiling points than the parent HYDROCARBONS from which they are derived. Phenol is a major constituent of coal tar. It is used by the chemical and pharmaceutical industries for making such products as aspirin, dyes, fungicides, explosives, and as a starting material for nylon and epoxy resins.

phenolphthalein ($C_{20}H_{14}O_4$) Organic compound prepared by a reaction between PHENOL and phthalic anhydride in the presence of SULFURIC ACID. It is used as a chemical indicator (colorless in acid solutions, red-pink in alkalis), in laxatives, and in dyes.

phenotype Physical characteristics of an organism resulting from HEREDITY. Phenotype is distinct from GENOTYPE, since not all aspects of GENETIC makeup manifest themselves.

phenylalanine ($C_6H_5CH_2CH(NH_2)COOH$) Crystalline, soluble AMINO ACID found in PROTEINS. It is essential to human nutrition.

phenylamine *See* ANILINE

phenylketonuria (PKU) Hereditary condition in which the amino acid PHENYLALANINE cannot be metabolized normally because the ENZYME phenylalanine hydroxylase is absent or inactive. Infants with PKU excrete phenylalanine in the urine. Special diets are used to treat the condition.

pheromone Substance secreted externally by certain animals that influences the behavior of members of the same species. Common in mammals and insects, these substances are often sexual attractants and may be a component of body products such as URINE, or they may be secreted by specific GLANDS.

Phillips, Robert (1906–76) US physiologist. An officer in the US navy, Philips was instrumental in introducing methods of combating cholera in the Middle East, Indian sub-Continent, and Southeast Asia. The key to treatment, introduced in the 1940s, was rehydration. At first he recommended intravenous fluids and, in later years, rehydration by mouth. The treatment was also successfully used for children with other causes of diarrhoea.

▶ **phase diagram**
The solid, liquid, and gaseous forms of a particular substance (in the small inset drawing, carbon dioxide is shown) can all exist in mutual contact at a particular temperature and pressure, called the triple point. The three lines show how the boiling and melting and sublimation change with pressure (in sublimation a solid turns to gas without melting). Above the critical point the liquid cannot exist at all; only up to that temperature can the gas be liquefied by simply compressing it. In the larger drawing, the changes of volume are included: here the solid, liquid, and gas states are symbolized as surfaces, as are the conditions for melting (1), boiling (3), and sublimation (6). The pressure/volume behavior (4) of a gas is simple at high temperatures and more complex near the critical point (2). The triple point is shown as a line (5).

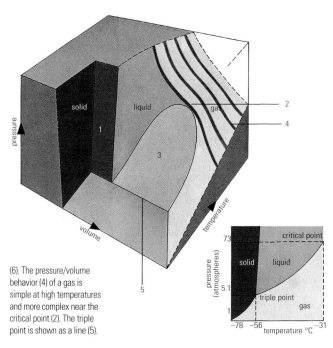

phillipsite (KCa(Al$_2$Si$_6$O$_{16}$).6H$_2$O) White or reddish ZEOLITE mineral. It occurs in vein formations as monoclinic CRYSTALS. Hardness 4.5–5; r.d. 2.2.

phloem Vascular tissue for distributing dissolved food materials in plants. Phloem tissue contains several types of CELLS. The most important are long, hollow cells called **sieve-tube cells.** Columns of sieve tubes are joined end to end, allowing passage of materials from cell to cell. The sieve tubes are closely associated with "**companion cells,**" which have dense CYTOPLASM and many MITOCHONDRIA and are thought to produce the energy needed to transport the food substances (*see* ACTIVE TRANSPORT). Phloem may also contain FIBERS, which help to support the tissue. *See also* XYLEM

phlogiston theory Notion advanced by J.J. Becher (1635–82) and popularized in the 18th century by Georg Stahl (1660–1734). It postulated that combustible materials contained an odorless, colorless, weightless material called **phlogiston**; COMBUSTION involved loss of this phlogiston and it thereby became dephlogisticated in turning to calx (ash). The theory was disproved by Antoine LAVOISIER's discovery of the true nature of combustion.

phlogopite (KMg$_3$Fe$_3$AlSi$_3$O$_{10}$(OH)$_2$) White or brown MICA mineral. It is found in limestones as monoclinic crystals. Hardness 2.5–3; r.d. 2.8.

Phoenix Constellation of the S sky. Its brightest star, Alpha, is magnitude 2.4.

phosgene (carbonyl chloride, COCl$_2$) Colorless, toxic GAS. It was used as a poison gas in World War I, but is now used in the manufacture of various dyestuffs and resins. Properties: b.p 8.2°C (46.8°F), m.p.−118°C (−180.5°F).

phosphate Chemical compounds derived from PHOSPHORIC(V) ACID (H$_3$PO$_4$) and other phosphorus(V) oxyacids, for example salts with the phosphate ion (PO$_4$)$^{3-}$.

phosphine Phosphorus hydride (PH$_3$), a colorless, highly poisonous GAS made by the reaction between calcium phosphide and water or dilute acid. Traces of diphosphine impurity make the gas spontaneously flammable. Pure phosphine finds application in doping SEMICONDUCTOR materials.

phospholipid Type of LIPID that contains a PHOSPHORIC ACID group or groups and an alcohol base. Phospholipids are found in egg yolk and brain tissue.

phosphor Substance capable of LUMINESCENCE (storing energy and later releasing it as light). There are two main types: the zinc phosphors, as used on CATHODE-RAY TUBES; and the oxygen type, as used on FLUORESCENT LAMPS. Zinc sulfide is often mixed with cadmium sulfide and a small quantity of metallic phosphates, silicates or fluorides to produce phosphors. *See also* PHOSPHORESCENCE

phosphorescence Form of luminescence in which a substance emits light of one wavelength without associated heat. Unlike FLUORESCENCE, it may persist for some time after the initial excitation. It is a natural as well as artificial phenomenon; in warm climates, for example, the sea often appears phosphorescent at night, as a result of the activities of millions of microscopic ALGAE, which phosphoresce when disturbed by the movements of the water.

phosphoric acid Group of ACIDS, the chief forms of which are phosphoric(V) (orthophosphoric) acid (H$_3$PO$_4$), phosphoric(VI) (metaphosphoric) acid (HPO$_3$) and heptaoxodiphosphoric(V) (pyrophosphoric) acid (H$_4$P$_2$O$_7$). Phosphoric acid was discovered in 1770 by Karl SCHEELE and J.G. Gahn. **Phosphoric(V)** acid is a colorless liquid obtained by the action of SULFURIC ACID on phosphate rock (calcium phosphate); it is used in fertilizers, soaps, and detergents. **Phosphoric(VI)** acid is obtained by heating phosphoric(V) acid; it is used as a dehydrating agent. **Heptaoxodiphosphoric(V)** acid is formed by moderately heating phosphoric(V) acid or by reacting phosphorus(V) oxide (phosphorus pentoxide, P$_2$O$_5$) with water; it is used as a CATALYST.

phosphorus (symbol P) Common, nonmetallic element of group V of the periodic table, first discovered in 1669. It occurs as PHOSPHATES in minerals; APATITE is the chief source. The element is used in making PHOSPHORIC ACID for detergents and fertilizers (phosphorus is essential for plant growth). Small amounts are used in insecticides and in matches. Phosphorus exhibits ALLOTROPY; allotropes include the highly reactive white phosphorus, which ignites spontaneously on exposure to air, and the more stable form, red phosphorus. Properties: at.no. 15; r.a.m. 30.9738; r.d. 1.82 (white), 2.34 (red); m.p. 44.1°C (111.38°F) (white); b.p. 280°C (536°F) (white); most common isotope ^{31}P (100%).

phosophorus cycle Circulation of phosphorus in the environment, primarily between ABIOTIC and BIOTIC elements. Phosphate salts and phosphorus acids exist in bodies of water and the soil, from which they are absorbed by plants. Animals eat the plants and in this way the phosphates enter FOOD CHAINS, where organisms incorporate them in organic compounds such as DNA. On the death of the organisms, the phosphorus compounds return to the soil and are broken down by bacteria. Over millions of years, some phosphates are incorporated into rocks, which return them to the soil and bodies of water through EROSION and the activities of humans, by employing phosphates as fertilizers.

phosphorylation Biochemical reaction in which a PHOSPHATE group is introduced into a MOLECULE. Controlled by the ENZYME phosphorylase, it is the initial stage in many natural biochemical processes, such as the conversion of ADENOSINE DIPHOSPHATE (ADP) to energy-rich ADENOSINE TRIPHOSPHATE (ATP). Phosphorylation is also an important reaction in the activation or deactivation of enzymes, often under hormone control.

photocell *See* PHOTOELECTRIC CELL

photochemistry Branch of chemistry concerning photochemical reactions caused by LIGHT or ULTRAVIOLET RADIATION and those reactions that produce light. Examples include PHOTOGRAPHY, PHOTOSYNTHESIS, and bleaching by sunlight. Chemical effects can be produced only if light is absorbed by a system. PHOTONS of light are absorbed by reactant MOLECULES, raising them to an electronically excited state that can lead to further reactions. The quantum yield, or efficiency, of a reaction is the number of reactant molecules used, or product formed, per photon absorbed. Photolysis is the decomposition or

dissociation of molecules caused by this exposure to light or ULTRAVIOLET RADIATION.

photochromic glass Form of GLASS, used in spectacle lenses and electronic devices, which darkens on exposure to ultraviolet or bright visible light. It fades to its original clear state when the light is removed. SILVER CHLORIDE or silver bromide crystals throughout the glass interact with light to cause this change.

photocopying Reproduction of the written word, drawings or photographs by machine. In a photocopier, a light shines on the item to be copied, and an optical system forms an image of it. Various techniques may be used to reproduce this image on paper. In a modern, plain-paper copier using XEROGRAPHY, the image is projected onto an electrically charged **drum**, coated with a light-sensitive material such as the element SELENIUM. Light makes such a material conduct electricity, so bright areas of the drum lose their charge. The dark areas, which usually correspond to image detail, retain their charge, and this attracts particles of a fine powder called **toner**. Electrically charged paper in contact with the drum picks up the pattern of toner powder. A heated **roller** fuses the powder so that it sticks to the paper and forms a permanent image.

▶ **photocopying** The document to be photocopied (1) is placed face down on the copier (2) and illuminated by a bright, halogen lamp (3). Light is reflected from a moving mirror (4), via a system of lenses (5) and a secondary mirror (6), onto a revolving drum (7), coated in a photosensitive polymer containing selenium. White parts of the document reflect light. Wherever light strikes the drum the overall negative charge of the drum is dissipated (8), resulting in a pattern of charged areas that will correspond to the darker parts of the original document. The revolving drum picks up the positively charged toner particles (9) only where it has a negative charge. A sheet of paper (10) is given a negative charge by a long electrode and pressed against the drum (11), where the toner is transferred to the paper. The copy passes through a set of hot rollers (12), which fuse the toner to the paper (13). Excess heat is expelled by a fan (14), blowing cold air over the rollers.

Most machines print in one color, normally black, but some machines can make full-color copies.

photodiode SEMICONDUCTOR device that produces an ELECTRIC CURRENT when light shines on it. It consists of a *p-n* junction DIODE with a lens that focuses light onto the junction. It is used to measure light intensity or to detect the presence of light.

photoelectric cell (photocell) Type of electric cell whose operation depends on the amount of light to which it is exposed. It was formerly an ELECTRON TUBE with a photosensitive CATHODE, but now nearly all photoelectric cells are made using two ELECTRODES separated by light-sensitive SEMI-CONDUCTOR material. Photoelectric cells are used as switches (electric eyes), light detectors (burglar alarms), devices to measure light intensity (PHO-TOMETERS, as for cameras) or as power sources (solar cells), most recently in space vehicles.

photoelectric effect Liberation of ELECTRONS from the surface of a material (such as silicon) when ELECTROMAGNETIC RADIATION, such as light, ultraviolet radiation, X-rays or gamma rays, falls on it. The effect can be explained only by the QUANTUM THEORY: PHOTONS in the radiation are absorbed by ATOMS in the substance and enable electrons to escape by transferring energy to them. A PHOTOELECTRIC CELL uses the photoelectric effect, producing electricity when exposed to light.

photoengraving (process-engraving) Process of preparing illustrations for letterpress printing in which the image is transferred by PHOTOGRAPHY onto metal or plastic. It involves two steps: the preparation of a photographic negative copy of the material to be reproduced; and the making of a positive printing plate. Plates are made of zinc, copper or magnesium alloy, coated with a photosensitive solution. The light passing through clear sections of the negative affects the coating, making it insoluble in water. After exposure, the plates are washed, leaving an image formed by the insoluble portions. The non-printing surface is then etched away with an acid.

photogrammetry Use of photographs for measuring distances and areas in SURVEYING. Photographs taken from aircraft or orbiting SATELLITES enable precise measurements to be taken for making maps and charts. The use of infrared film reveals details that are invisible to the naked eye or to ordinary photographic emulsions.

photography Process of obtaining a permanent image of an object, either in black and white or in color, on treated paper or FILM. A CAMERA is used to expose a film for a set time to an image of the object to be photographed. In black and white photography, the film is covered on one side with an emulsion containing a SILVER halide (silver bromide or silver chloride). Exposure makes the silver compound easily reduced to metallic silver when treated with a **developer**. The action of the developer is to produce a black deposit of metallic silver particles on those parts of the film that were exposed to light, thus providing a "**negative**" image. After fixing in "**hypo**" (thiosulfate) and washing, the negative can be printed by placing it over a piece of sensitized paper and exposing it to light so that the silver salts in the paper are affected in the same way as those in the original film. The dark portions of the negative let through the least light, and the image on the paper is reversed back to a positive. Color photography works using a similar, but more complex, process. In 1816 Nicéphore NIEPCE produced the first negative on paper and in 1827, working with Jacques Mandé Daguerre, he produced the first known photograph, on metal. In 1819 Sir John HERSCHEL discovered the hypo. In 1839 Daguerre announced his invention of a direct positive image on a silver plate – the daguerreotype. At the same time, William Henry Fox TALBOT invented the calotype, the first process capable of producing an infinite number of positive prints from a single negative.

photoluminescence Type of luminescence produced by the action of PHOTONS, such as those in infrared, visible or ultraviolet light or in X-rays or gamma rays. Luminescence produced by the action of ELECTRONS is known as ELECTROLUMINESCENCE.

photometer Instrument used to measure the luminous intensity of a light source by comparing it with a light source of known intensity. This is achieved by comparing the positions of the sources when they produce equal illumination on a reference surface. Photometry is the branch of physics concerned with such measurements. Modern photometers usually use PHOTOELECTRIC CELLS. In astronomy, photoelectric photometers measure the brightness of a celestial object by generating an electrical signal that is directly proportional to the intensity of the light from that object falling on it. The unit of luminous intensity is the CANDELA (cd).

photomultiplier Electronic device that converts a light signal or light beam into an equivalent amplified electrical signal. ELECTRONS produced by the PHOTOELECTRIC EFFECT from an illuminated CATHODE bombard a series of other electrodes and are increased in number at each bombardment. An amplified signal is obtained from the final stage.

photon Quantum of ELECTROMAGNETIC RADIA-TION, such as LIGHT, which can be considered to consist of streams of photons. The energy of a photon equals the frequency of the radiation multiplied by PLANCK'S CONSTANT. Absorption of photons by atoms and molecules can cause excitation or ionization. A photon is a stable ELEMENTARY PARTICLE of zero charge and spin 1, travelling at the velocity of light. Virtual photons are continuously exchanged between charged particles and are the carriers of the ELECTROMAGNETIC FORCE. *See also* ANTIMATTER; BLACK BODY; COMPTON EFFECT; PHOTOELECTRIC EFFECT; VIRTUAL PARTICLE

photoperiodism Biological mechanism that governs the timing of certain activities in an organism by reacting to the duration of its daily exposure to light and dark. For example, the start of flowering in plants and the beginning of the breeding season in animals are determined by day length. *See also* BIOLOGICAL CLOCK

photoreceptor In animals, a sensory cell (RECEPTOR) or collection of CELLS that react to the presence of light. Most photoreceptors contain a PIGMENT that undergoes a chemical change when it absorbs light, producing a NERVE IMPULSE. The RODS AND CONES in the retina of the EYE are examples of photoreceptors. In plants, photoreceptors are light detectors containing the blue-green pigment phytochrome. They are involved in many plant processes, such as seed germination, flowering, and leaf fall.

photosetting Composition of text to be printed by photographic means, developed in the 1950s as a speedier and more flexible alternative to setting type in metal. The size of type to be set is obtained from a master negative font by photographic reduction or enlargement. Computer units allied to most photosetters enable the text to be set as fast as the operator can key it in. Once programmed, these units automatically control line width, column length, spacing, justification, and hyphenation. *See also* LINOTYPE; MONOTYPE; PRINTING; TYPE METAL; TYPESETTING

photosphere Visible surface of the SUN. It is a layer of highly luminous gas c.500km (300mi) thick and with a temperature of c.6,000K, falling to 4,000K at its upper level, where it meets the CHROMOSPHERE. Because this is the region of the Sun from which all its output of visible light is emitted, the photosphere is the source of the Sun's visible spectrum. The lower, hotter gases produce the continuous emission spectrum, while the higher, cooler gases absorb certain wavelengths. SUNSPOTS and other visible features of the Sun are situated in the photosphere.

photosynthesis Chemical process occurring in green plants, algae, and many bacteria, by which water and CARBON DIOXIDE are converted into food and OXYGEN using energy absorbed from sunlight. There are two distinct stages of photosynthesis: LIGHT REACTION and DARK REACTION. The reactions take place in the CHLOROPLASTS. During the first part of the process, light is absorbed by CHLOROPHYLL and splits water into HYDROGEN and OXYGEN. The hydrogen attaches to a carrier molecule and the oxygen is set free. The hydrogen and light energy build a supply of cellular chemical energy, ADENOSINE TRIPHOSPHATE (ATP). In the dark reaction, hydrogen and ATP, by a process called the CALVIN CYCLE, convert the carbon dioxide into SUGARS, including glucose, and starch. *See also* AUTOTROPH

phototropism (heliotropism) Growth of a plant in response to the stimulus of light, which increases CELL growth on the shaded side of the plant, resulting in curvature towards the source of light. AUXIN hormones are involved in this process. They are produced in the tip of the shoot, and when light shines on the shoot, auxin travels preferentially down the shaded side of the shoot, where it promotes cell elongation, causing the shoot to bend towards the light. Leaves and stems respond positively to light and roots respond negatively or not at all. Indoor plants lean toward windows; leaves usually grow at right angles to light and are positioned to ensure that overlapping occurs as little as possible.

phyllite Medium-grained METAMORPHIC ROCK formed from sediments at low temperatures and only moderate pressure. It is composed of quartz, feldspar, and mica, with chlorite, which imparts a green or gray color to the shiny foliated structure. It may also contain small crystals of garnet.

phyllotaxis (phyllotaxy) Arrangement of leaves on the STEM of a plant. In nearly all LEAF-bearing plants, each leaf grows out from the stem according to a regular pattern that is characteristic of the species. Most leaves grow either in a spiral arrangement up the stem, or an alternate arrangement with one leaf on one side of the stem and the next leaf on the other side. In an opposite arrangement, two leaves originate from the same leaf NODE. In a whorled arrangement, three or more leaves grow from the same node.

phylogenetics Study of the evolutionary relationships between organisms. In molecular phylogeny, the evolutionary distances between organisms are analyzed by comparing the DNA sequences of specific GENES. These analyses confirm, for example, that seals are more closely related to dogs than to whales. At the most fundamental level, molecular phylogeny has revealed that all known organisms evolved from a common ancestor and can be grouped into five KINGDOMS, whose members are more closely related to each other than to members of other kingdoms.

phylum Part of the CLASSIFICATION of living organisms, ranking above CLASS and below KINGDOM. Phylum names are written in Roman (ordinary) letters and begin with an initial capital letter; for example all animals with backbones (VERTEBRATES) are members of the phylum Chordata. Some phyla are further subdivided into subphyla (**subphylum**). In PLANT CLASSIFICATION, the analogous category is sometimes called division. *See also* TAXONOMY

physical chemistry Study of the physical changes associated with CHEMICAL REACTIONS and the relationship between physical properties and chemical composition. The main branches are THERMODYNAMICS, concerned with the changes of energy in physical systems; chemical KINETICS, concerned with rates of reaction; and molecular and atomic structure. Other topics include ELECTROCHEMISTRY, SPECTROSCOPY, and some aspects of NUCLEAR PHYSICS. *See also* CHEMISTRY

physical units Units used in measuring physical quantities. In specifying a unit, it is necessary to define an instance of that physical quantity and a way in which it can be compared in making a measurement. For example, the KILOGRAM unit of mass is defined as the mass of a specified block of platinum-iridium. Other masses are measured by weighing them and comparing them, directly or indirectly, with this. Units are of two types: **base** units that, like the kilogram, have fundamental definitions, and **derived** units that are defined in terms of these base units. Various systems of units exist, founded on certain base units. They include the FPS SYSTEM (the imperial units, foot, pound, second), CGS SYSTEM (centimeter, gram, second), and MKS SYSTEM (meter, kilogram, second). For all scientific purposes, SI UNITS have been adopted, which include seven base units: METER (length), kilogram (mass), SECOND (time), AMPERE (electric current), KELVIN (temperature), CANDELA (luminous intensity), and MOLE (amount of substance).

physics Branch of science concerned with the study of MATTER and ENERGY. Physics seeks to identify and explain their many forms and relationships. Modern physics recognizes four basic FORCES in nature: GRAVITATION, which was first adequately described by Isaac NEWTON; the ELECTROMAGNETIC FORCE, codified in the 19th century by MAXWELL's equations; the WEAK NUCLEAR FORCE, which is responsible for the decay of some SUBATOMIC PARTICLES; and the STRONG NUCLEAR FORCE, which binds together atomic nuclei. Physics rests on six fundamental theories: Newtonian MECHANICS, THERMODYNAMICS, ELECTROMAGNETISM, STATISTICAL MECHANICS, RELATIVITY, and QUANTUM MECHANICS. Branches of physics include PARTICLE PHYSICS, GEOPHYSICS, BIOPHYSICS, ASTROPHYSICS, and NUCLEAR PHYSICS.

physiology Branch of biology concerned with the functions of living organisms, as opposed to ANATOMY, which is concerned with their structure. Vast in scope, physiology includes the study of single CELLS as well as multicellular organisms.

phytoplankton Free-floating, oceanic plant life as opposed to ZOOPLANKTON, the animal life. Most of the organisms are microscopic, such as DIATOMS.

pi (π) Symbol used for the ratio of the circumference of a CIRCLE to its diameter. It is an IRRATIONAL NUMBER and as such it cannot be expressed as a FRACTION. To seven decimal places it is 3.1415926 (7d.p.). The ratio 22/7 is often used as a rougher approximation.

Piazzi, Giuseppe (1746–1826) Italian astronomer and monk. He discovered the first ASTEROID, CERES, on January 1, 1801, during a lengthy program of taking accurate measurements of star positions (published as a catalogue in two parts, in 1803 and 1814). Piazzi proposed the name "planetoid," but "asteroid" has prevailed.

Piccard, Auguste (1884–1962) Swiss physicist who explored the STRATOSPHERE and deep seas. In 1931 a hydrogen balloon carried a pressurized aluminum sphere, containing Piccard and an assistant, to an altitude of almost 15,800m (51,800ft). This was the first ascent into the stratosphere. In 1932 Piccard set a new altitude record of 16,940m (55,577ft). From 1948 Piccard experimented with designs for a diving vessel called a bathyscaphe. In 1953 Piccard and his son, Jacques, descended in the bathyscaphe *Trieste* to a depth of c.3,100m (10,000ft) in the Mediterranean Sea. In 1960 Piccard and US naval officer Don Walsh used the same craft in the Pacific Ocean to reach a new record of 10,916m (35,810ft) below the surface. *See also* BATHYSPHERE AND BATHYSCAPHE

picric acid (2,4,6-trinitrophenol, $C_6H_2OH(NO_2)_3$) Yellow, crystalline solid, widely used as an EXPLOSIVE before and during World War I. It is now used in the manufacture of dyes and pigments.

Pictor (Painter's easel) Unremarkable constellation of the S sky. Its leading star is Alpha, magnitude 3.3. *Beta Pictoris*, magnitude 3.8, is believed to be a planetary system in the process of formation.

piedmont Describing the foot of a mountain or a place where GLACIERS merge. The term is used of areas near mountains, such as the Piedmont (It. *Piemonte*) region in N Italy or the Piedmont Plateau to the E of the Appalachian Mountains in the USA. A piedmont glacier is formed when two or more glaciers flow from their valleys onto a plain and merge with their neighbors.

Pierce, John Robinson (1910–) US electrical engineer who, beginning independently in 1943 and then in collaboration with Australian-born British engineer Rudi Kompfner, developed the travelling wave tube amplifier. Unlike other types of high-grain amplifier, this device creates no electronic noise and proved to be of great value in handling the complex high-frequency signals of satellite communications. He also invented an electron multiplier and a high-intensity electron gun.

piezoelectric effect Appearance of positive ELECTRIC CHARGE on one side of a nonconducting CRYSTAL and a negative charge on the other when the crystal is squeezed. The pressure polarizes the crystal, slightly separating the positive and negative charge distributions in the crystal. This charge separation results in an ELECTRIC FIELD that can be detected as VOLTAGE between the opposite crystal faces. The effect has been used in phonograph and crystal microphones and cigarette lighters.

pigment Colored, insoluble substance used to impart color to an object and added for this purpose to PAINT and INK. Unlike DYES, pigments are insoluble in the coating vehicle or medium. They generally function by absorbing and reflecting light, although some luminescent pigments emit colored light.

pigmentation In biology, a natural chemical that gives color to animal TISSUE. In humans, the skin, hair, and the IRIS of the eye are colored by the pigment MELANIN. HEMOGLOBIN in erythrocytes (red blood cells) also acts as a pigment.

Pilâtre de Rozier, Jean-François (1756–85) French pioneer of ballooning. In November 1783 he and the Marquis d'Arlandes were the first to fly in an untethered, hot-air BALLOON. During a flight lasting about 25 minutes, they covered a distance of 8km (5mi). In 1785 Pilâtre de Rozier was killed while attempting to cross the English Channel by balloon.

pillow lava LAVA that has occurred when MAGMA is expelled under water, or flows into water before solidifying. It commonly takes the form of a distorted globular mass, resembling the shape of a pillow. It apparently results from the rapid chilling of the outer skin, thus making a more or less spherical "balloon" that grows and flattens under its own weight.

pinchbeck Alloy of COPPER and ZINC, but with much less zinc than is contained in BRASS. Invented by the London watchmaker Christopher Pinchbeck (c.1670–1732), it is used to simulate gold.

Pincus, Gregory Goodwin (1903–67) US biologist who helped to develop the contraceptive pill (*see* CONTRACEPTION) using synthetic HORMONES.

These hormones simulate the physiological state of women when pregnant, thus preventing FERTILIZATION. Pincus wrote several articles on hormones and a book, *The Eggs of Mammals* (1936).

pineal body Small GLAND attached to the undersurface of the vertebrate BRAIN. In human beings, it has an endocrine function, secreting the hormone melatonin, which is involved in daily rhythms. *See also* ENDOCRINE SYSTEM

pingo Cone-shaped mound that stands alone and has a core of ice. It is formed when freezing water expands and is forced upwards. Pingos are generally found in areas of PERMAFROST, where water cannot reach the surface and therefore pushes up the ground as it freezes. Pingos may be up to 90m (330ft) high and 800m (2,600ft) across.

pinna (auricle) Flap of skin and cartilage that comprises the visible, external part of the EAR. It helps to collect sound waves and direct them into the ear canal.

pinocytosis Intake and transport of fluid by living CELLS. Rather than entering and passing through the cell membrane as individual molecules, a droplet becomes bound to the membrane. A pocket then forms, which pinches off to form a vesicle in the CYTOPLASM. The vesicle may then pass across to the far side of the cell, where reverse pinocytosis occurs. It is a type of ENDOCYTOSIS.

pint Liquid or dry measure. One liquid pint is equal to 16 fluid ounces or half a liquid quart; one dry pint is equal to half a dry quart.

pion (pi-meson) ELEMENTARY PARTICLE classified as a MESON. There are three types, forming a nuclear MULTIPLET (triplet). The **charged** pions, π^+ and π^-, have equal mass, about 280 times the electron mass, and are antiparticles of each other; the **neutral** π^0, of slightly lower mass, is its own antiparticle. All have zero SPIN. Charged pions decay into MUONS and muon NEUTRINOS; the π^0 decays into PHOTONS. Virtual pions are thought to be exchanged between NUCLEONS bound together by the STRONG NUCLEAR FORCE.

Pioneer program Series of interplanetary SPACE PROBES launched by the US between 1958 and 1978. The first four were failed Moon probes. **Pioneers 5 to 9** were put into orbit around the Sun to study the SOLAR WIND and SOLAR FLARE activity. **Pioneer 10**, launched in March 1972, was the first probe to travel through the ASTEROID belt; it passed Jupiter in December 1973, sending back photographs. **Pioneer 11** followed it, flying past Jupiter in December 1974 and moving on to Saturn, which it reached in September 1979. Both probes are now on their way out of the solar system. The last two in the series were the Pioneer Venus probes, launched in May and August 1978. **Pioneer Venus 1** went into orbit around Venus in December 1978, making a RADAR map of the surface and photographing the planet's clouds; it burned up in the planet's atmosphere in October 1992. **Pioneer Venus 2** dropped five probes into Venus' atmosphere in December 1978, which sent back data on the planet's atmospheric conditions. *See also* SPACE EXPLORATION

Pirie, Norman Wingate (1907–) British biochemist who in 1936, together with compatriot Frederick Bawden, became the first scientist to crystallize a virus. This achievement had important repercussions on later research into the nucleic acids DNA and RNA. Pirie went on to study the subtle differences between living and nonliving macromolecules, and wrote about the problems of population and food resources.

Pisces (the Fish) Equatorial constellation situated on the ECLIPTIC between Aquarius and Aries. It is hard to see, and consists of a chain of rather faint stars south of Pegasus. Its brightest star, Eta, is magnitude 3.6.

Piscis Austrinus (Southern Fish) Constellation in the S sky. First-magnitude Fomalhaut is the only star brighter than magnitude 4.

pistil Female organ located in the center of a FLOWER. It consists of an OVARY, a slender STYLE, and a STIGMA, which receives POLLEN.

pitch In AERODYNAMICS, angular displacement of the longitudinal relative to the axis of an aircraft relative to the horizon. The elevators control pitch by lowering (or raising) the tail relative to the nose to increase (or decrease) the angle of attack (angle at which air strikes the wing).

pitch In physics, quality of sound that determines its position in a musical scale. It is measured in terms of the FREQUENCY of sound waves (measured in hertz) – the higher the frequency, the higher the pitch. It also depends to some extent on loudness and timbre: increasing the intensity decreases the pitch of a low note and increases the pitch of a high one.

pitchblende *See* URANINITE

pith Central strand of parenchymatous tissue that occurs in the STEMS of most VASCULAR PLANTS. It is usually surrounded by vascular tissue and is believed to function chiefly for storage. The term is also used for the soft core at the center of the heartwood of logs, consisting of the dried remains of the pith. *See also* PARENCHYMA

pitot tube Device for measuring the rate of flow of a fluid, either liquid or gas. It was invented by Henri Pitot (1695–1771). For **liquids**, the device used is generally a MANOMETER, with one open end facing upstream and the other open end out of the stream. The different pressures at the two ends cause a liquid to shift position within the two arms of the tube. For **gases**, a Pitot tube is generally L-shaped, with one end open and pointing towards the flow of gas and the other end connected to a pressure-measuring device. This type of pitot tube is commonly used as an airspeed indicator in aircraft.

pituitary gland Major gland of the ENDOCRINE SYSTEM located at the base of the BRAIN. In human beings it is about the size of a pea and is connected to the HYPOTHALAMUS by a stalk. It produces many HORMONES, some of which regulate the activity of other endocrine glands, while others control growth.

placenta Organ in PLACENTAL MAMMALS that connects the FETUS to the wall of its mother's UTERUS, providing a means of nutrition, GAS EXCHANGE, and EXCRETION for the fetus. Part of the placenta contains tiny blood vessel branches through which oxygen and food are carried from the mother to the embryo, via the UMBILICAL CORD, and wastes are carried from the embryo to the mother's bloodstream to be excreted. The placenta secretes HORMONES that maintain pregnancy. It is discharged from the mother's body as the AFTERBIRTH, immediately after delivery.

placental mammals Mammals whose young develop to an advanced stage attached to the PLACENTA – a life-support organ inside the mother's UTERUS. All mammals except the MONOTREMES and most marsupials are placentals. *See also* VIVIPARITY

placer deposits Concentrations of heavy minerals formed by the action of GRAVITY, usually found in streams. Minerals that occur as placer deposits include gold, copper, rutile, cassiterite, and magnetite.

placoderm Group of primitive jawed fish that existed in the DEVONIAN period, characterized by armored plates on the front part of the body. Most placoderms were quite small, but some grew to 9m (30ft).

plagioclase Type of FELDSPAR (the most abundant group of minerals on Earth). Plagioclase minerals are widely distributed and occur in many IGNEOUS and METAMORPHIC rocks. Off-white, or sometimes pink, green or brown in color, they are composed of varying proportions of the silicates of sodium and calcium with aluminum. They show an oblique cleavage, and have triclinic system crystals. Hardness 6–6.5; r.d. 2.6.

plain Large area of flat or slightly undulating, low-lying land. Some plains, such as a PENEPLAIN, result from the wearing away of higher terrain. Most plains result from deposition of SEDIMENTS, as by rivers and lakes, which leave flat plains when the water dries up. FLOODPLAINS are formed in river valleys (see ALLUVIAL FAN). Glacial plains are large, level areas of TILL, which is left by retreating GLACIERS.

planarian Flatworm (see PLATYHELMINTH) found in marine and fresh water, identified by its triangular head with two light-sensitive eyespots. Planaria have flat, tail-like bodies, with extendable PHARYNXES for sucking in food. Reproduction is hermaphroditic or by asexual splitting, or regeneration. Length: 1in (2.54cm). Phylum Platyhelminthes; class Turbellaria; genera include *Dugesia* and *Polycelis*.

Planck, Max Karl Ernst Ludwig (1858–1947) German theoretical physicist whose revolutionary QUANTUM THEORY helped to establish modern physics. He was professor (1889–1928) at the University of Berlin, where he studied the characteristics of BLACK BODY radiation. In 1900 Planck came to the conclusion that the frequency distribution (SPECTRUM) of the radiation could be accounted for only if the radiation was emitted in separate "packets" called quanta, rather than continuously. PLANCK'S CONSTANT (1900) links wave and particle behavior on the atomic scale. His equation, relating the ENERGY of a quantum to its FREQUENCY, is the basis of quantum theory. Planck was awarded the 1918 Nobel Prize for physics for his work. He remained in Germany during the Nazi era, but spoke out against the persecution of Jews. In 1944 his son, Erwin, was executed for taking part in the unsuccessful plot to assassinate Hitler. He served as president (1930–35) of the Kaiser Wilhelm Society for the Advancement of Science, Berlin, which after World War II became part of the Max Planck Institutes.

Planck's constant (symbol h) Universal constant of value 6.626×10^{-34} joule seconds, equal to the energy of a quantum of ELECTROMAGNETIC RADIATION (a PHOTON) divided by the radiation frequency. It appears in formulae describing physical quantities that can assume only certain discrete values. It is named after Max PLANCK. See also PARTICLE PHYSICS; QUANTUM THEORY

plane In mathematics, a flat surface such that a straight line joining any two points on it lies entirely within the surface. Its general equation in the three-dimensional CARTESIAN COORDINATE SYSTEM is: $ax + by + cz = d$, where a, b, c, and d are constants. Any flat shape or surface such as a wing or craft can also be called a plane, hence airplane and hydroplane. In woodworking, a plane is a tool used to smooth wood by cutting away thin strips as it passes over the surface.

plane geometry Form of GEOMETRY in which all lines, angles, and figures are represented in two-dimensional (PLANE) form. In plane geometry, EUCLID's axioms are valid.

planet Large, nonstellar body in orbit around a STAR, shining only by reflecting the star's light. In our SOLAR SYSTEM there are nine major planets, as opposed to the thousands of small bodies known as ASTEROIDS or minor planets. It is now known that other stars, apart from our SUN, have planets orbiting them. See MERCURY; VENUS; EARTH; MARS; JUPITER; SATURN; URANUS; NEPTUNE; PLUTO

planetarium Structure in which a representation of the stars and planets as visible in the night sky is **projected** onto the inside of a dome for an audience seated below. It is used to demonstrate the positions and motions of celestial bodies. In 1913 the PROJECTOR was invented by Walter Bauersfeld of the Zeiss Optical Company. In 1923 the first planetarium, built by the Zeiss company, began operating in Munich. From then until the early 1980s, planetarium projectors consisted of many individual optical projectors mounted on a complicated rotating frame. In the new generation of projectors, a computer-generated image is projected through a single, stationary fish-eye LENS.

planetesimal In theories of the origins of our SOLAR SYSTEM (and other planetary systems), a body, between a few millimeters and a few kilometers in size, that condensed out of a cloud of gas and dust (the solar NEBULA), away from the center where the Sun was forming. Once they had reached a few kilometers in size, planetesimals' gravitational attraction would allow them to combine with one another, by a process of ACCRETION, to form protoplanets. The term was first used for the very small planetary bodies supposed to have condensed from matter torn out of the Sun by a passing star in the "planetesimal theory" proposed by Forest MOULTON and Thomas Chamberlain in the early 1900s.

plane trigonometry Branch of TRIGONOMETRY that deals with the sides and angles of PLANE triangles and their measurements and relations.

plankton All the floating or drifting life of the ocean, especially that near the surface. The organisms are very small or microscopic and move with the currents. There are two main kinds: PHYTOPLANKTON, which comprises the floating plants such as DIATOMS and dinoflagellates; and ZOOPLANKTON, floating animals such as radiolarians, plus the larvae and eggs of larger marine animals.

plant Any member of the kingdom Plantae, a large kingdom of multicellular organisms whose cells have CELLULOSE cell walls and contain CHLOROPLASTS or similar structures (PLASTIDS). Plants develop from DIPLOID embryos and have a regular alternation of HAPLOID and diploid generations in their life cycles. Most plants are green, contain chloroplasts, and make their own food by PHOTOSYNTHESIS. A few are colorless PARASITES or SAPROPHYTES. Simple plants reproduce by means of SPORES, whereas more advanced plants produce seeds and fruits. Plants show a wide range of biochemistry; some produce chemicals such as ALKALOIDS, NARCOTICS, and even poisons such as CYANIDE; others secrete substances into the soil to prevent other plants growing near them. Many of these chemicals form the bases for the development of DRUGS. Plants are classified on the basis of their MORPHOLOGY (shape and structure). The most important phyla are the Bryophyta (BRYOPHYTES), which include the MOSSES and liverworts; LYCOPODOPHYTA, or CLUB MOSSES; Sphenophyta (HORSETAILS); Filicinophyta (FERNS); Cycadophyta (CYCADS); Ginkgophyta (GINKGO); Coniferophyta (CONIFERS); and Angiospermophyta (ANGIOSPERMS). *See also* ALTERNATION OF GENERATIONS

plant classification System devised to group PLANTS according to relationships among them. The currently accepted system of plant classification includes the following phyla: **Bryophyta**, **Lycopodophyta**, **Sphenophyta**, **Filicinophyta**, **Cycadophyta**, **Ginkgophyta**, **Coniferophyta**, and **Angiospermophyta**. An older system split plants into four divisions: Thallophyta (ALGAE and LICHENS), Pteridophyta (CLUB MOSSES and FERNS), Spermotophyta (which included the GYMNOSPERMS and ANGIOSPERMS), and Bryophyta (LIVERWORTS and MOSSES). *See also* EICHLER, AUGUST WILHELM; TAXONOMY

plant genetics Science of HEREDITY and VARIATION in plants. Research in GENETICS since 1900 has supplied the principles of plant breeding, especially HYBRIDIZATION – the controlled crossing of plant varieties in order to reproduce new and improved plants. Breeders can select for specific characteristics, such as resistance to disease, or size and quality of yield. The development of consistently reliable and healthy first-generation crosses (**F1 hybrids**) has revolutionized the growing of food crops, ornamental annuals, and bedding plants. Genetic engineers grow CELL and TISSUE CULTURES by the replication or cloning of sterile plant types. They also concentrate on isolating individual GENES with the aim of producing new color varieties for traditional flowers, improving the flavor of food crops, breeding resistance to pests and HERBICIDES, and lengthening the shelflife of harvested crops. *See also* CLONE; GENETIC ENGINEERING

plant hormone (phytohormone) Organic chemical produced in plant cells and functioning at various sites to affect plant growth, leaf and fruit drop, healing, cambial growth, and possibly flowering. HORMONES are transported away from the STEM tip. They include abscisic acid (leaf fall), AUXINS (growth), CYTOKININS (leaf and bud growth and development) and GIBBERELLINS (growth). *See also* ABSCISSION

plant pigment Organic compound present in plant cells and tissues that colors the plant. The most common plant pigment is green CHLOROPHYLL, which exists in all higher plants. CAROTENOIDS color plants yellow to tomato red. Located in CHLOROPLASTS and chromoplasts are more than 150 varieties of these durable pigments. Many are essential to PHOTOSYNTHESIS and are a source of vitamin A. Anthocyanins, responsible for pink, red, blue, and purple, are found in the cell SAP. The shorter days and lower temperatures of autumn cause these pigments to combine with other substances and produce the brilliant foliage colors of deciduous trees. *See also* PIGMENTATION

plaque Abnormal deposit building up on a body surface, especially the film of saliva and bacteria that accumulates on TEETH. Dental plaque leads to tooth decay and gum disease.

plasma In biology, liquid portion of the BLOOD in which the CELLS are suspended. It contains an immense number of ions, inorganic and organic molecules such as IMMUNOGLOBULINS, and HORMONES and their carriers. It clots on standing.

plasma In physics, an ionized GAS. Plasma, often described as the fourth state of MATTER, occurs at enormous temperatures, as in the interiors of the Sun and other stars and in FUSION reactors. *See also* ION

plasmid Strand or loop of NUCLEIC ACID, containing GENETIC information. It can be introduced to a host CELL where it will replicate independently of the host's CHROMOSOMES. Plasmids are used in RECOMBINANT DNA RESEARCH.

plasmodium Genus of parasitic PROTOZOA that causes MALARIA. It infects the ERYTHROCYTES (red blood cells) of mammals, birds, and reptiles worldwide, being transmitted by the bite of a female *Anopheles* mosquito.

plaster of Paris (gypsum cement) Powdered form of calcium sulfate hemihydrate, $2CaSO_4.H_2O$, obtained by heating GYPSUM to around 150°C (300°F). After the addition of water, it sets and hardens and is used as a PLASTER for a wide range of purposes, including the setting of broken limbs and the making of molds.

plastic Synthetic material composed of organic molecules, often in long chains called POLYMERS, that can be shaped and then hardened. The weight and structure of the molecules determine the physical and chemical properties of a given compound. Plastics are synthesized from common materials, mostly from PETROLEUM. CELLULOSE comes from cotton or wood pulp, CASEIN from skimmed milk, others from chemicals derived from potatoes, peanuts, and soy beans. There are two main types of plastics: **thermosetting** plastics, such as BAKELITE, stay hard once set; **thermoplastics**, such as POLYETHENE, can be resoftened by heat. New biodegradable plastics, more expensive to produce, are environmentally friendly because they eventually decompose (mainly due to the action of microorganisms) in landfills. The plastics industry began in 1872 with the manufacture of CELLULOID by the US inventor John Wesley Hyatt (1837–1920) and his brothers. There are three basic plastic-forming processes: calendering, casting, and extrusion. **Calendering** is used to make a layered composite material from a thermosetting resin, such as phenol-formaldehyde (Bakelite), and materials such as paper and cloth. CASTING uses thermosetting plastics, such as EPOXY, POLYESTER, and phenolic or UREA RESINS, and enables plastic to be cast around metal parts, as in making screwdrivers. EXTRUSION forces the softened plastic through a die that shapes it into the desired form, such as a tube, rod or sheet. This is employed with THERMOPLASTIC RESINS such as polyethene, POLYSTYRENE, and NYLON.

plastid Type of ORGANELLE found in the CELLS of plants and green algae. Plastids have a double membrane and contain DNA. CHLOROPLASTS and leucoplasts are two examples of plastids.

plateau Fairly flat, raised area of land. Mountains may stand up above the general level of a plateau, or it may be carved by deep river valleys or canyons to form a **dissected** plateau. An **intermontane** plateau is completely surrounded by mountains.

platelet Colorless, usually spherical structures found in mammalian BLOOD. Chemical compounds in platelets, known as factors and cofactors, are essential to the mechanism of BLOOD CLOTTING. The normal platelet count is about 300,000 per cu mm of blood.

plate tectonics Theory or model to explain the distribution, evolution, and causes of the EARTH's crustal features. It proposes that the Earth's CRUST and part of the upper MANTLE (the LITHOSPHERE) are made up of several separate rigid slabs, termed **plates**, which move independently and form part of a cycle in the creation and destruction of crust. The plates collide or move apart at the CONTINENTAL MARGINS, and these produce zones of EARTHQUAKE and volcanic activity. Three types of plate boundary can be identified. At a CONSTRUCTIVE MARGIN (or divergent margin), new basaltic MAGMA originating in the mantle is injected into the plate, which forces the crust to separate, and an oceanic ridge is formed. At a DESTRUCTIVE MARGIN (or convergent margin), plates collide and one plate moves into the mantle at the site of an oceanic TRENCH. The recycling of crust by SUBDUCTION results in the melting of some crustal material and volcanic island arcs (such as the islands of Japan) are produced. Material that cannot

magma pushes up and forces the plates apart

oceanic trench

subduction zone

formation of mountain range or island chain

constructive margin

destructive margin

line of mid-ocean ridge

areas of volcanic activity

◀ **plate tectonics** The discovery that the continents are carried along on the top of slowly moving crustal plates provided the mechanism for continental drift theories to work. The plates converge and diverge along margins marked by seismic and volcanic activity. Plates diverge from mid-ocean ridges, where molten lava pushes up and forces the plates apart at a rate of up to 3.75cm (1.5in) a year; converging plates form either a trench (where the oceanic plates sink below the lighter continental rock) or mountain ranges (where two continents collide).

be subducted is scraped up and fused onto the edge of plates. This can form a new CONTINENT or add to existing continents. Mountain chains are explained as the sites of former subduction or continental collision. At a **conservative margin**, plates move past each other along a transform fault. Plate movement is thought to be driven by CONVECTION CURRENTS in the mantle. *See also* SEAFLOOR SPREADING

platinum (symbol Pt) Lustrous, silver-white TRANSITION ELEMENT, first discovered in 1735. It is chiefly found in certain ores of nickel. Malleable and ductile, platinum is used in jewelry, dentistry, electrical-resistance wire, magnets, thermocouples, surgical instruments, electrodes, and as a catalyst in catalytic converters for car exhausts. It is chemically unreactive and resists tarnishing and CORROSION. These traits are shared with five other elements that, with platinum, are known as the platinum metals and are often found together as alloys: IRIDIUM, OSMIUM, PALLADIUM, RHODIUM, and RUTHENIUM. Properties: at.no. 78; r.a.m. 195.09; r.d. 21.45; m.p. 1,772°C (3.222°F); b.p. 3,800°C (6,872°F); most common isotope ^{195}Pt (33.8%).

platyhelminth Any member of the order Platyhelminthes, comprising the flatworms. Platyhelminths are simple, carnivorous, ribbon-like creatures that, having no circulatory system and sometimes no mouth or gut, feed by ABSORPTION through the body wall. Almost all flatworms are hermaphroditic. The fluke and TAPEWORM are both PARASITES of animals.

Pleistocene Sixth epoch of the CENOZOIC era of GEOLOGICAL TIME. The Pleistocene epoch began about 2 million years ago, during which humans and most forms of familiar mammalian life evolved. Episodes of climatic cooling in this epoch led to widespread GLACIATION in the Northern Hemisphere; the Pleistocene is the best-known glacial period or ICE AGE in the Earth's history. The present HOLOCENE epoch succeeded the Pleistocene around 8000 BC.

pleura Double membrane that lines the space between the LUNGS and the walls of the THORAX. The fluid between the pleura lubricates the two surfaces to prevent friction during breathing movements.

Pliocene Last epoch of the TERTIARY period. It lasted from 5 to 2 million years ago and preceded the PLEISTOCENE. Animal and plant life was similar to that of today. *See* GEOLOGICAL TIME

plumbago *See* GRAPHITE

plumule In plants, embryonic shoot that develops during GERMINATION of a SEED.

Pluto Smallest and outermost planet of the SOLAR SYSTEM. Independently, William H. Pickering (1858–1938) and Percival LOWELL calculated the possible existence of Pluto, its size, ORBIT, and position in the sky. The planet was eventually located by Clyde TOMBAUGH within 5° of Lowell's predicted position, and the discovery was publicly announced in March 1930. Pluto has yet to be visited by a SPACE PROBE and most of our knowledge of it dates from the discovery (1978) of its satellite, CHARON, which allowed, among other things, Pluto's mass and unusual axial tilt to be determined. Pluto is in a much more inclined and eccentric orbit than the other planets in our solar system, and for a 20-year period around PERIHELION (as, for example, from 1979 to 1999) it comes within the orbit of Neptune. Pluto seems to have a mottled surface with light and dark regions, and signs of polar caps. The surface is covered with icy deposits consisting of 98% nitrogen, with traces of methane, and also probably water, carbon dioxide, and carbon monoxide. For a 60-year period in each orbit, when Pluto's high orbital eccentricity brings it closer to the Sun, the surface warms to above 50K, sufficient to release a thin, temporary atmosphere of nitrogen. Pluto is certainly not the massive planet whose supposed perturbations of Uranus and Neptune led to its discovery. The realization of just how small Pluto is has led some astronomers to question its status as a major planet. **Data:** Diameter: 2,300km; Mass: (Earth=1) 0.0021; Volume: (Earth=1) 0.0058; Density (water=1): 2.02; Orbital period: 247.7 years; Rotation period: 6d 9h 17m; Surface temperature (average): 45K

plutonic Describing a rock that has formed beneath the surface of the Earth. Plutonic rocks are IGNEOUS ROCKS consisting of solidified MAGMA, which has cooled slowly, allowing large CRYSTALS to form in the rock. They are coarse-grained. The most common plutonic rock is GRANITE, which is light in color because its mineral content is acidic. The most common of the basic plutonic rocks is GABBRO. Most plutonic rocks form in BATHOLITHS beneath the surface, but because of the EROSION of overlying rocks, they are now commonly found on the surface.

plutonium (symbol Pu) Silver-white radioactive metallic element of the ACTINIDE SERIES. It was first made in 1940 at the University of California at Berkeley by the deuteron (heavy hydrogen) bombardment of URANIUM. It is found in small amounts in uranium ores but mostly produced synthetically: ^{239}Pu (half-life 24,360 years) is made in large quantities in breeder reactors. It is a fissile element (able to split into other elements) used in nuclear reactors and nuclear weapons, such as the atom bomb. The element is very toxic, a strong emitter of ALPHA PARTICLES, and absorbed by bone, making it a dangerous radiological hazard. Properties: at.no. 94; r.d. 19.84;

m.p. 641°C (1,186°F); b.p. 3,232°C (5,850°F); most stable isotope ^{244}Pu (half-life 76 million years). *See also* NUCLEAR REACTOR; TRANSURANIC ELEMENTS

pneumatic Describing a device powered by compressed air normally used to produce rotary or a reciprocating (back and forth) motion to speed up operations such as sawing, grinding, digging, hammering and riveting. The pneumatic drill used in roadworks has a reciprocating, pounding motion at speeds of between 80 and 500 revs per minute.

pneumothorax Presence of air in the pleural space between the LUNGS and the chest wall. It may arise spontaneously or be caused by injury or disease. The lung is liable to collapse since it is prevented from expanding normally.

pod Fruit of any leguminous plant, such as a pea or bean. A pod is an elongated caselike structure filled with SEEDS. It develops from a single CARPEL, and when ripe splits down both sides to release the seeds. In some plants, the pods burst explosively, scattering the seeds widely. *See also* LEGUME

podiatry (chiropody) Treatment and care of the foot. Podiatrists treat such conditions as corns and bunions and devise ways to accommodate foot deformities. They can also perform certain kinds of foot surgery.

podzol Light-colored, infertile SOIL, poor in lime and iron. The name comes from the Russian word for "ash soil." It forms under cool humid conditions and is typical of the taiga forests of North America and Eurasia. The upper layer is light gray; the lower layer is darker and contains HUMUS and minerals, such as ALUMINA, iron, and lime, that have been leached out of the upper layer by rainfall. The minerals form a hard layer that is impervious to water and prevents good drainage. *See also* LEACHING

poikilothermic (ectothermic) Animal whose body temperature fluctuates with the temperature of its surroundings, often referred to as **cold-blooded**. Reptiles, amphibians, fish and invertebrates are cold-blooded. They can control their body temperature only by their behavior – by moving in and out of the shade, or orienting themselves to absorb more or less sunlight. *See also* HOMEOTHERMIC

Poincaré, (Jules) Henri (1854–1912) French mathematician. He contributed more than 500 works in many diverse areas, including ANALYSIS, DIFFERENTIAL EQUATIONS, TOPOLOGY, PROBABILITY, and NUMBER THEORY. Poincaré developed several important techniques for the study of dynamic and chaotic systems. He used the techniques to analyze the motion of planets and the nature of rotating flu-

ids; systems that are now known to exhibit chaotic behavior. He also made contributions to the special theory of RELATIVITY and electromagnetic theory. *See also* CHAOS THEORY

Poiseuille's equation In FLUID MECHANICS, the formula for the relation between the rate of flow, R, of an incompressible fluid of viscosity η, in a tube of radius a and length l is $R = \pi pa^4 /8l\eta$, where p is the difference in pressure between the two ends of the tube. It is named after the French physiologist Jean Louis Marie Poiseuille (1799–1869).

Poisson, Siméon Denis (1781–1840) French scientist who applied MATHEMATICS to many areas of PHYSICS, including sound and heat propagation, ELECTROMAGNETISM, AND MECHANICS. Poisson extended the work of Joseph Louis LAGRANGE on CELESTIAL MECHANICS and worked on INTEGRATION and the FOURIER SERIES. In *Researches on the Probability of Opinions* (1837), he presented the **Poisson distribution**, which gives the PROBABILITY of the occurrence of a specific event among a large number of events. **Poisson's ratio**, which relates lateral contraction to longitudinal extension in stressed materials, is also named after him.

polar coordinate System of coordinates for representing the position of a point P on a PLANE. There are two coordinates, r and θ, where r is the point's distance from the origin and θ is an ANGLE measured counterclockwise from some reference line emanating from the origin. **Spherical** polar coordinates incorporate a second angle, ϕ, which represents the angle of inclination of a line drawn to the point from the plane surface.

polarization Suppression of certain directions of vibration of the ELECTRIC and MAGNETIC FIELDS of ELECTROMAGNETIC RADIATION, so that the wave form is no longer symmetrical about the direction of propagation. *See also* POLARIZED LIGHT

polarized light LIGHT waves that have electromagnetic vibrations in only one direction. **Ordinary** light vibrates in all directions perpendicular to the direction of propagation. Scientists distinguish between three types: plane-polarized, **circularly polarized**, and **elliptically polarized** light, each depending on the net direction of the vibrations. In **plane-polarized** light, the electric vibrations are confined to one direction only, the magnetic vibrations occurring at right angles. Plane-polarized light can be produced by REFLECTION, as from a sheet of glass or a water surface, or by passing light through certain crystals, such as quartz, tourmaline or calcite. Polarizing sunglasses use a Polaroid material to reduce glare by cutting out light polarized by reflection. *See also* LAND, EDWIN HERBERT; PRISM

polarography Method of chemical analysis that can measure minute concentrations of IONS in solution. Two ELECTRODES are inserted into the solution: one, the reference electrode, is kept at a constant VOLTAGE; the other has a variable voltage and consists of a glass capillary tube containing a head of mercury, with tiny drops of mercury on its end becoming the face of the electrode. As these drops form, a current grows to a peak between the two electrodes and then suddenly falls as the drop leaves the capillary tip. By gradually increasing the voltage, a polarogram – a graph of the current versus the voltage, which looks like a series of steps – is produced. From these, the REDUCTION potentials of the ions in the solution may be deduced and, as these are known for each metal ion, the ions can be identified.

polar solution SOLUTION in which the solvent molecules have substantial DIPOLE moments, such as water. Hydrogen bonding and solvent-solute complexes form, encouraging the association of solvent and solute molecules in preferred orientations in solution.

polar winds Prevailing WINDS that blow from the North and South Poles. They are deflected from east to west by the effects of the Earth's rotation, and are often known as the polar easterlies. They carry very cold air and their effect reaches as far as latitude 60°.

pole Generally either of the two points of intersection of the surface of a sphere and its axis of rotation. The Earth has four poles: the North and South geographic poles, where the Earth's imaginary axis meets its surface; and the north and south magnetic poles, where the Earth's MAGNETIC FIELD is most concentrated. A bar MAGNET has a north pole, where the magnetic flux leaves the magnet, and a south pole, where it enters. A pole is also one of the terminals (positive or negative) of a battery, electric machine, generator or circuit. *See also* GEOMAGNETISM; MAGNETISM

pollen Powder-like SPORES, usually yellow in color, that give rise to the male GAMETES (sex cells) in plants. In ANGIOSPERMS (flowering plants), pollen grains are produced in the ANTHERS on the STAMEN, whereas in GYMNOSPERMS (cone-bearing plants), they are produced by the male cones. Pollen has thick resistant walls with a characteristic pattern of spines, plates or ridges, according to species. During POLLINATION, the pollen lands on the STIGMA of a flower, or a female cone, of a compatible plant. It germinates, sending a long pollen tube down through the STYLE to the OVARY. During this process, one of its nuclei divides, giving rise to two male nuclei, one of which will fuse with a female sex cell (actually a HAPLOID nucleus in the embryo sac inside the OVULE) in FERTILIZATION. The other male sex cell fuses with two

more of the female nuclei to form a special tissue, the ENDOSPERM, whose cells contain three sets of CHROMOSOMES (triploid). In many species, this tissue develops into a food store for the EMBRYO in the seed.

pollination Transfer of POLLEN (containing male GAMETES) from the ANTHER to the STIGMA in ANGIOSPERMS (flowering plants), or from the male cone to the female cone in GYMNOSPERMS (cone-bearing plants), resulting in FERTILIZATION. Pollination occurs mainly by wind (ANEMOPHILY) in the cases of most trees and gymnosperms, and by flying insects (ENTOMOPHILY) for angiosperms. Other agents of pollination include grazing mammals, birds, bats, and nonflying insects. **Self-pollination** occurs when the pollen of one flower (or male cone) pollinates the same flower or another flower (female cone) of the same plant. However, plants have developed intricate incompatibility mechanisms to ensure self-pollination occurs as infrequently as possible, such as chemicals that make the pollen and eggs of the same plant sterile. CROSSPOLLINATION occurs between two flowers (or male and female cones) on different plants. It results in a great variety of genetic combinations, thereby improving the chances of survival. See also SEX; SEXUAL REPRODUCTION

A

B

C

D

▶ **pollination** Flowers are adapted to different pollination methods. Non-specialized simple flowers, such as the buttercup (A), can be pollinated by a variety of means. Other flowers can only be pollinated by one method. There are bird-pollinated flowers, such as the hummingbird-pollinated hibiscus (B); bee-specialized flowers, including the gorse (C); and wind-pollinated flowers, such as the catkins found on the hazel (D).

pollution Spoiling of the natural environment, generally by industrialized society. Pollution is usually a result of an accumulation of waste products, although excess of NOISE and heat that has adverse effects on the surrounding environment is also included in the term. *See also* GREENHOUSE EFFECT; GLOBAL WARMING; WATER POLLUTION

polonium (symbol Po) Rare, radioactive, metallic element of group VI of the periodic table, first discovered in 1898 by Marie CURIE. It is found in trace amounts in URANIUM ores and may be synthesized. Properties: at.no. 84; r.d. 9.40; m.p. 254°C (489°F); b.p. 962°C (1,764°F); most stable isotope [209] Po.

polyamide Synthetic POLYMER with the generic name of NYLON, formed from the reaction between a dibasic acid (such as adipic acid) and a diamine (such as hexamethylenediamine). Nylon consists of long-chain synthetic polyamides in which the diamine and acid molecules are joined by AMIDE linkages. It was first produced in 1938 by E.I. du Pont de Nemours & Co. *See also* PLASTIC

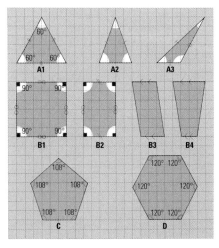

▲ **polygon** Polygons are many-sided plane figures, the sides and angles of which may be equal or unequal, giving symmetrical or asymmetrical shapes. The simplest figure is the equilateral triangle (A1) which has all sides and angles equal. The isosceles triangle (A2) has two sides equal while (A3), which happens to be obtuse, has sides of differing length. The simplest quadrilateral (B1) is the square with equal sides and equal internal angles. The rectangle (B2) has two pairs of equal and opposite sides with equal internal angles. The parallelogram (B3) has two pairs of equal and opposite parallel sides. The trapezium (B4) has only two parallel sides. The regular pentagon (C) has five equal sides and the regular hexagon (D), six.

polycarbonate POLYMER plastic used for making bottles and shatterproof "glass." Polycarbonates are condensation products of carbonyl chloride (PHOSGENE) and a dihydroxy compound such as diphenylol propane. These thermosetting PLASTICS are clear, strong, and tough.

polychaete Any member of the class Polychaeta, marine ANNELID worms characterized by distinctive segmentation of the body. The class includes ragworms and bristleworms. Polychaetes have a distinct head with sensory projections and most segments have outgrowths or bristles.

polychlorinated biphenyl (PCB) Several stable mixtures – liquid, resinous or crystalline – of organic compounds made by the reaction of CHLORINE with biphenyl. They are fire-resistant and are used as lubricants, heat-transfer fluids, and fluids in transformers. They are also used to impregnate capacitors and to give flexibility to protective coatings for wood, concrete, and metal. Since 1973 the use of PCBs has been restricted internationally because they are toxic, and their resistance to decomposition in streams and soils poses a threat to wildlife.

polyester Class of synthetic substances composed of large molecules arranged in either a chain or a network and formed from many smaller molecules through the establishment of ESTER linkages. Polyester fibres are resistant to chemicals and may be washed in alkaline solutions or dry-cleaned. They are made into woven and knitted fabrics. As a clothing material, polyester is valued for its resistance to creasing. *See also* PLASTIC

polyethene (polythene, polyethylene) POLYMER of ETHENE (ethylene), a partially crystalline, lightweight THERMOPLASTIC RESIN, with high resistance to chemicals, low moisture absorption, and good insulating properties. *See also* PLASTIC

polygon PLANE geometric figure having three or more sides intersecting at three or more points (vertices). They are named according to the number of sides or vertices: TRIANGLE (three-sided), quadrilateral (four-sided), PENTAGON (five-sided), and so on. A regular polygon is equilateral (has sides equal in length) and equiangular (has equal angles).

polygraph Electronic device that may be capable of detecting lies when used by a trained examiner. The polygraph monitors such factors as heart rate, breathing rate, and perspiration, all of which may be affected when a person lies.

polyhedron Three-dimensional figure or solid, the surface of which is made up of POLYGONS. The polygon faces are joined together at the edges and the points at which edges meet are called vertices. *See also* EULER'S RULE

polymer Substance formed by the union of from two to several thousand simple molecules (MONOMERS) to form a large molecular structure. Some polymers, such as CELLULOSE, occur in nature; others form the basis of the PLASTICS and synthetic RESIN industry.

polymerase chain reaction (PCR) Chemical reaction that is used to make large numbers of copies of a specific sequence of DNA, or part of a GENE, starting from only one or few DNA molecules. It is speeded up by the ENZYME polymerase. The reaction enables scientists to make large enough quantities of DNA to be able to analyze it or manipulate it. The starting piece of DNA may be extracted from an organism, it may have been "designed" and synthesized in the laboratory, or it may be a forensic sample taken from the scene of a crime. PCR is extremely important in GENETIC ENGINEERING and GENETIC FINGERPRINTING.

polymerization Chemical combination of several molecules to form straight-chain molecules, cross-linked giant molecules or a combination of both, all called POLYMERS. It is the major industrial process in the manufacture of PLASTICS. In nature, large biochemical compounds (such as PROTEINS and CELLULOSE) are also formed by polymerization.

polymorphism In mineralogy, the existence of a substance in two (dimorphic) or more CRYSTAL forms. Polymorphism is caused by different temperatures or pressures or both. The separate forms are called **polymorphs**; for example, DIAMOND and GRAPHITE are polymorphs of CARBON. Polymorphism of the elements is also known as ALLOTROPY. In biology and botany, polymorphism is the existence of two or more types of individual within a single species. For example, some social insects, such as the ant and bee, have several polymorphic types of worker adapted structurally to perform different tasks within their colony. Flowers can also assume more than one form.

polynomial Sum of terms that are powers of a given VARIABLE. In general, a polynomial has the form $P_n(x) = a_n x^n + a_{n-1} x^{n-1} + a_{n-2} x^{n-2} + ... a_2 x^2 + a_1 x + a_0$, where $a_0, ... a_{n-1}, a_n$ are the COEFFICIENTS of the polynomial. The degree of a polynomial is the highest power appearing in the sum with a non-zero coefficient. For instance,

$$P_4(x) = 2x^4 - 3x^3 + x^2 + x + 5$$

is a polynomial of degree four. In this example, the values of the polynomial at $x = 0$, 1, and 2 are $P_4(0) = 5$, $P_4(1) = 6$ and $P_4(2) = 19$, respectively. Polynomials can be represented graphically by plotting values of $y = P_n(x)$ against values of x.

polyp Small, stalked growth occurring on a MUCOUS MEMBRANE, most commonly in the nose, uterus or bowel. Polyps are usually benign, but some, especially those in the intestine, may become cancerous.

polyp Sedentary stage in the life cycle of CNIDARIANS. A polyp has a mouth surrounded by extensible tentacles and a lower end that is adapted for attachment to a surface. It is distinct from the free-swimming MEDUSA. It may be solitary, as in the sea anemone, but is more often an individual of a colonial organism such as CORAL.

polypeptide Chemical consisting of a long chain of AMINO ACIDS linked by PEPTIDE bonds. PROTEINS are polypeptides.

polyploidy State of a cell or nucleus that has more than two sets of CHROMOSOMES. More likely to occur in plants than in animals, polyploidy is found in many cereals. Polyploidy can be introduced artificially using chemicals to produce HYBRID plants with desirable characteristics, such as increased size and hardiness. *See also* DIPLOID; HAPLOID

polypropylene POLYMER of PROPYLENE, resembling polyethene and used chiefly in making fibers, films, and molded products. *See also* PLASTIC

polysaccharide Any of a group of complex CARBOHYDRATES made up of long chains of MONOSACCHARIDE (simple-sugar) molecules, often thousands of them. GLUCOSE is a MONOSACCHARIDE, and the polysaccharides STARCH and CELLULOSE are both POLYMERS of glucose. Higher carbohydrates are all polysaccharides and decompose by HYDROLYSIS into a large number of monosaccharide units. Polysaccharides function as both food stores (starch in plants and GLYCOGEN in animals) and as structural materials (cellulose and PECTIN in the cell walls of plants, and CHITIN in the protective skeleton of insects).

polystyrene Synthetic, organic POLYMER, composed of long chains of the aromatic compound STYRENE. It is a strong THERMOPLASTIC RESIN, acid- and alkali-resistant, nonabsorbent, and an excellent electrical insulator. *See also* PLASTIC

polytetrafluoroethene (PTFE) Chemically inert, solid PLASTIC, known also by the trade names Teflon and Fluon. PTFE is used as a heat-resistant material for heat shields on spacecraft, as a nonstick coating on cooking utensils, and as a LUBRICANT. It is stable up to $c.300°C$ ($572°F$).

polythene *See* POLYETHENE

polyunsaturate Type of FAT or OIL with molecules of long CARBON chains having many double bonds. Nature's main example, considered essential in the human diet, is linoleic acid, found as a glyceryl ESTER in linseed oil and other fats and oils. Polyunsaturated fats exist in fish oils and most vegetable oils, but not those of the palm and coconut. At room temperature, unsaturated oils are liquids and SATURATED FATS, such as lard, are solids. Polyunsaturates, which have low or no CHOLESTEROL content, are widely used in margarines and cooking oils. They are considered healthier than saturated fats, which have been found by medical research to contribute to cardiovascular diseases.

polyurethane POLYMER used chiefly in making flexible foams (in upholstery), rigid foams (in cores of airplane wings), ELASTOMERS, and RESINS (for adhesives and coatings). *See also* PLASTIC

polyvinyl acetate Colorless resin POLYMER used as an ingredient of latex paint and in lacquers, cements, adhesives, and polyvinyl alcohol. It is thermoplastic, flammable, and dissolves in BENZENE. *See also* PLASTIC

polyvinyl chloride (PVC) (polychloroethene) White, tough, solid thermoplastic that is a POLYMER of CHLOROETHENE (vinyl chloride). It can be made by heating chloroethene in water with potassium persulfate or HYDROGEN PEROXIDE. PVC can be softened and made elastic with a plasticizer. Easily colored and resistant to weather and fire, PVC is used to produce a variety of products, including fibers, bottles, shoes, windows, electrical insulation, vinyl flooring, audio discs, food containers, and coatings for raincoats. *See also* PLASTIC

pome Fleshy FRUIT formed from the flower RECEPTACLE or base. It is not developed from the CARPEL, and is therefore a PSEUDOCARP (false fruit). Familiar examples are the apple, pear, and quince.

pons Upper segment of the human BRAIN STEM, which contains the nerve fibers connecting the two halves of the CEREBELLUM. The brain stem is the lower part of the BRAIN and is structurally continuous with the SPINAL CORD. Below the pons is the MEDULLA OBLONGATA, which transmits ascending and descending nerve fibers between the spinal cord and the brain.

population In ecology, a group of organisms of the same species that live in a particular geographical region. The size of a stable population is generally limited by the supply of food and adverse environmental factors (such as disease), that is by the **carrying capacity** of the habitat. The numbers may also be affected when organisms die or move away, or when new individuals are born or move into the region.

porcelain White, glasslike, nonporous, hard, translucent ceramic material. Porcelain is widely used for tableware, decorative objects, laboratory equipment, and electrical insulators. The material is often referred to as "china" because the earliest type was developed by the Chinese in the 7th or 8th century. What is referred to as china is, in fact, true or **hard-paste** porcelain, made of KAOLIN (white china clay) mixed with powdered FELDSPAR. The mixture is fired at *c*.1,400°C (2,550°F). **Soft-paste** porcelain is composed of clay and powdered glass, fired at a comparatively low temperature, lead glazed, and refired. English BONE CHINA is made from a clay paste containing bone ash, which improves the material's translucency.

poriferan Any member of the phylum Porifera – the SPONGES. Poriferans are many-celled, stationary animals that remain permanently fixed to rocks. They feed by filtering food particles from the water.

porosity In geology, degree to which rock is capable of holding GROUNDWATER. This depends on the numbers of AQUIFERS, cracks, fissures, and holes present. Porosity is measured and expressed as a percentage of the rock's total volume. *See also* HYDROLOGY; PERMEABILITY

porphyrin Class of red-pigmented ORGANIC COMPOUNDS consisting of four PYRROLE residues (a ring compound, C_4H_4NH) bonded together. The porphyrins form part of the active nucleus of HEMOGLOBIN, CHLOROPHYLLS *a* and *b*, and the ENZYMES catalase and peroxidase.

porphyry Rock that contains large CRYSTALS (phenocrysts) in a fine-textured igneous MATRIX. They are found in both INTRUSIVE and EXTRUSIVE rocks. Different varieties are named after the phenocrysts.

portal vein Large vein that carries blood from the stomach and surrounding organs to the liver.

Porter, Rodney Robert Gerald (1917–85) British biochemist. Porter and Gerald EDELMAN shared the 1972 Nobel Prize for physiology or medicine for their work on the chemical structure of the GAMMA GLOBULIN molecule.

Porter, Sir George (1920–) British chemist and director (1966–85) of the Royal Institution of Great Britain. Porter shared the 1967 Nobel Prize for chemistry with Ronald NORRISH and Manfred EIGEN for their work on very fast CHEMICAL REACTIONS. He and Norrish disturbed the EQUILIBRIUM between atoms and molecules in CHLORINE gas by irradiating it with very short pulses of energy; they then observed it as it returned to equilibrium.

positron Particle that is identical to the ELECTRON, except it is positively charged, making it the ANTIPARTICLE of the electron. The theory of the electron, stated in 1928 by Paul DIRAC, required the existence of a mirror-image particle. In 1932 this particle was observed in COSMIC RADIATION by Carl ANDERSON, and it is also emitted from certain radioactive nuclei. Electron-positron pairs can be produced when GAMMA RADIATION interacts with matter. *See also* ELEMENTARY PARTICLE

positron emission tomography (PET) Medical imaging technique (used particularly on the brain) that produces three-dimensional images. RADIOISOTOPES, injected into the bloodstream prior to imaging, are taken up by tissues where they emit POSITRONS that produce detectable PHOTONS.

positronium Unstable, bound state of an ELECTRON and a POSITRON (the electron's ANTIPARTICLE) held together by electromagnetic attraction, first observed in 1951. It has a half-life of either 1.39×10^{-7} sec or 1.25×10^{-10} sec, depending on whether the SPIN axes of the two particles are parallel or antiparallel.

potash Any of several POTASSIUM compounds, especially potassium oxide (K_2O), POTASSIUM CARBONATE (K_2CO_3), or POTASSIUM HYDROXIDE (KOH).

Potash is used as FERTILIZER because potassium is an ESSENTIAL ELEMENT for plant growth.

potassium (symbol K) Common metallic element, one of the ALKALI METALS, first isolated in 1807 by Sir Humphry DAVY. Its chief ores are sylvite (a CHLORIDE), carnallite, and polyhalite. The metal is used as a heat-transfer medium in NUCLEAR REACTORS, but has few other commercial uses. Chemically, it resembles SODIUM, being rather more reactive. Potassium in the form of POTASH is used as a fertilizer, because it is an ESSENTIAL ELEMENT for plant growth. The natural element contains a radioisotope ^{40}K (half-life 1.3×10^9 yr), which is used in the radioactive DATING of rocks. Properties: at.no. 19; r.a.m. 39.102; r.d. 0.86; m.p. 63.65°C (146.6°F); b.p. 774°C (1,425°F); most common isotope ^{39}K (93.1%).

potassium-argon dating Method of assessing geological age up to about 10 million years. It is a type of radioactive DATING.

potassium carbonate (potash, K_2CO_3) White solid, usually produced commercially by ELECTROLYSIS of POTASSIUM CHLORIDE, followed by treatment of the resulting POTASSIUM HYDROXIDE with CARBON DIOXIDE. It is used as a fertilizer, to make glass, soap and dyes, and in cleaning and ELECTROPLATING metals.

▶ **positron emission tomography (PET)** used for the detection of certain diseases, particularly those that effect the brain, PET scans are an effective way of avoiding invasive and potentially dangerous surgery. The patient is drip-fed a radioactive chemical (1) that works its way around the body. The scanner (2) emits low-energy gamma radiation (3) to the areas of the body under examination. The gamma radiation causes the radioactive chemical in the blood supply (4) to emit detectable photons (5) which are interpreted by computer software (6) to show how that part of the body is functioning. As the scanner moves along the body cross-sections of the organ, in this example the brain, can be viewed (7).

potassium chloride Colorless or white POTAS-SIUM salt (KCl), extracted from lake brines and from minerals such as sylvite, kainite, and carnallite. It is used as a fertilizer and as a raw material in the production of POTASSIUM HYDROXIDE and POTASSIUM CARBONATE.

potassium hydroxide (caustic potash) White solid (KOH) prepared commercially by ELECTROLYSIS of POTASSIUM CHLORIDE. It is a strongly alkaline substance used for making soaps and detergents. *See also* ALKALI

potassium manganate(VII) (potassium permanganate) Purple, soluble, crystalline salt (KMnO$_4$) made from PYROLUSITE (manganese dioxide) and POTASSIUM HYDROXIDE. It is used as an oxidizing agent, disinfectant, dye, and in the preparation of other chemicals. Properties: r.d. 2.70; decomposes at 240°C (464°F).

potassium nitrate (saltpeter, niter) Colorless solid (KNO$_3$) used as a fertilizer for its NITROGEN content. It is also used in the manufacture of EXPLOSIVES and as a food preservative.

potential, electric *See* ELECTRIC POTENTIAL

potential difference Difference in ELECTRIC POTENTIAL between two points in a circuit or electric field, usually expressed in VOLTS. It is equal to the work done to move a unit electric charge from one of the points to the other. *See also* ELECTROMOTIVE FORCE (EMF)

potential energy Type of ENERGY an object possesses because of its vertical position in the Earth's GRAVITATIONAL FIELD; also, the energy stored in a system like a compressed SPRING or in an oscillating system like a PENDULUM. An object on a shelf, for example, has potential energy given by mgh, where m is its mass, g the acceleration due to gravity, and h the height of the shelf. If the object falls to the ground, its potential energy is converted into KINETIC ENERGY.

potentiometer Instrument used to measure an unknown VOLTAGE or POTENTIAL DIFFERENCE by comparing it with a known one, such as that of a standard electric CELL. It consists of a length of resistance wire fixed taut on a wooden base, and connected across a BATTERY. Both the known and the unknown voltages are in turn used to balance, through a GALVANOMETER and sliding contact, the voltages in a portion of the wire, the ratio of whose lengths equals the ratio of the voltages.

pothole In geology, various formations that are pot-shaped. Most commonly it denotes a circular,

bowl-shaped hollow formed in a rocky stream-bed by CORRASION, the grinding action of sand and stones whirled around by eddies or by the force of the stream. Such potholes are usually found in rapids or at the foot of a waterfall. It is also used to describe the vertical pits worn in LIMESTONE rocks by the action of water, occurring where a JOINT has been enlarged by SOLUTION; they often provide the entrances to CAVE systems.

pound Imperial unit of MASS equal to 0.454kg.

powder metallurgy Manufacture of metal powders and their use in producing metal parts. Powder particles are compressed to the desired shape and then sintered (heated to just below MELTING POINT). The use of powders is more economical than molten metal in making such items as small gears. Melting may also prove impractical, such as when high melting-point metals are used, or when an ALLOY of unfusable materials is involved. Powder METALLURGY is also used to make porous metal parts.

Powell, Cecil Frank (1903–69) British physicist. During the 1930s, he developed a technique to record ELEMENTARY PARTICLES directly onto film. In 1947 Powell used this method at high altitude to investigate COSMIC RADIATION and discovered a new particle, the PION (pi MESON). This discovery supported the theory of nuclear structure proposed by Hideki YUKAWA. He subsequently discovered the antiparticle of the pion and the decay process of KAONS (K mesons). Powell was awarded the 1950 Nobel Prize for physics.

power In physics, rate of doing WORK or of producing or consuming ENERGY. It is a measure of the output of an engine or other power source. James WATT was the first to measure power; he used the unit called HORSEPOWER. The modern unit of power is the WATT. *See also* HORSEPOWER

power In mathematics, the result of multiplying a number or VARIABLE by itself a specified number of times. Thus, $a^2 (= a \times a)$ is the second power of a; a^3 is the third power; a^4 the fourth, and so on. The superscript (2, 3 and 4 above) is called the EXPONENT or power.

praseodymium (symbol Pr) Silver-yellow metallic element of the LANTHANIDE SERIES, first isolated in 1885. Its chief ores are monazite and bastnaesite. Soft, malleable and ductile, praseodymium is used in carbon electrodes for arc lamps. Its green salts are used in colored glasses, ceramics, and enamels. Properties: at.no. 59; r.a.m. 140.9077; r.d. 6.77; m.p. 931°C (1,708°F); b.p. 3,512°C (6,354°F); most common isotope ^{141}Pr (100%).

Pre-Cambrian Oldest and longest era of Earth's history, lasting from the formation of the Earth about 4.6 billion years ago to the beginning of a good FOSSIL record about 590 million years ago. It is often split into the subdivisions ARCHAEAN and PROTEROZOIC. Pre-Cambrian fossils are extremely rare because the earliest life forms are presumed not to have had hard parts suitable for preservation. Also, Pre-Cambrian ROCKS have been greatly changed and deformed by metamorphism. Nonetheless, primitive bacteria have been identified in deposits more than 3 billion years old.

precession Wobble of the axis of a spinning object. This effect occurs as a result of the TORQUE on the spin axis. The Earth precesses about a line through its center and perpendicular to the plane of the ECLIPTIC extremely slowly (a complete revolution taking 25,800 years) at an angle of 23.5°. It makes the EQUINOXES move backwards through the ZODIAC, and the positions of the stars change by a small amount each year. The motion of a GYROSCOPE is another consequence of precession, in which the entire ring containing the spinning wheel and its axle precess around the support pivot. *See also* NUTATION

precipitate Insoluble solid formed in SOLUTION either by direct reaction or by varying the solution composition to diminish the SOLUBILITY of the dissolved compound. This technique (precipitation) is used for the separation and identification of COMPOUNDS in chemical processes.

precipitation In meteorology, all forms of water particles, whether liquid or solid, that fall from the atmosphere to the ground. Distinguished from CLOUD, FOG, DEW, and FROST, in that it must fall and reach the ground, precipitation includes RAIN, drizzle, SNOW (ice pellets and crystals) and HAIL. Measured by rain and snow gauges, the amount of precipitation is expressed in millimeters or inches of liquid water depth. Precipitation occurs with the CONDENSATION of water vapor in clouds into water droplets that coalesce into drops as large as 7mm (0.25in) in diameter. It also forms from melting ice crystals in the clouds. **Drizzle** consists of fine droplets, and snow of masses of many-sided ice crystals. **Sleet** is formed when raindrops freeze into small ice pellets, and hail when concentric layers of ice in cumulonimbus clouds freeze, forming lumps measuring from 0.5–10cm (0.2–4in) in diameter.

predator Animal that gets its food by hunting and killing its PREY. Many, but not all, CARNIVORES are predators (some, such as vultures and hyenas, are scavengers). Predators are SECONDARY CONSUMERS, occupying the top position in a FOOD CHAIN or web. Most predator-prey relationships naturally regulate one another's population. *See also* ECOSYSTEM

prefabrication Assembly of buildings or their components at locations other than their final position. Prefabrication methods are used for kitchen cabinets and appliances, wall and floor panels, roof trusses, window-wall elements, and whole buildings. Crystal Palace was the first building prefabricated in sections and assembled on site (1851).

pregnancy (gestation) Period of time from CONCEPTION until BIRTH; in humans, normally about 40 weeks (280 days). In humans, it is generally divided into three 3-month periods called trimesters. In the first trimester, the EMBRYO grows from a small ball of cells to a FETUS about 7.6cm (3in) in length, during which time the skeleton, brain, and such vital organs as the heart and lungs develop. At the beginning of the second trimester, movements are first felt and the fetus grows to about 36cm (14in). In the third trimester, the fetus attains its full body weight. The gestation period for mammals varies greatly. For instance, a female elephant (cow) gives birth to its calf after 18 to 22 months gestation.

Prelog, Vladimir (1906–) Swiss chemist, b. Yugoslavia. He shared the 1975 Nobel Prize for chemistry with John W. Cornforth (1917–) for their work on STEREOCHEMISTRY, the branch of chemistry that deals with the arrangement of ATOMS in a MOLECULE. Prelog devised a procedure for determining whether a molecule is right- or left-handed.

premolar In the DENTITION of adult human beings and other mammals, the two crushing or cutting TEETH between the CANINES and MOLARS on both sides of the upper and lower jaws; there are usually eight in all.

presbyopia Type of HYPERMETROPIA caused by lack of ACCOMMODATION of the eye, common with aging. As the eye lens becomes older, it loses some of its elasticity and ability to change curvature to focus on near objects. The condition is easily corrected with eyeglasses or contact lenses.

pressure In physics, force on an object's surface divided by the area of the surface. The SI unit is the PASCAL (symbol Pa), which is $1Nm^{-2}$ (NEWTON per sq m). In meteorology, the millibar (symbol mb), which equals 100 pascals, is commonly used.

pressure gauge Instrument used to measure fluid PRESSURE. A liquid-column gauge, such as the mercury BAROMETER and MANOMETER, uses the shifts of liquid in U-shaped tubes to measure pressure; a PITOT TUBE measures gas pressures between two points. A mechanical gauge, like the Bourdon tube or bellows-element gauge, utilizes the elasticity of metals to measure pressure.

pressurization Reducing the volume of a substance by forcing it into an increasingly smaller space (such as compressing a gas), or by restricting expansion of a heated substance (as in pressure-cooking). Pressurization is accompanied by an increase in temperature. Pressure-reduction is accompanied by cooling – this is the principle of the process of REFRIGERATION.

prey Animal hunted and killed for food by a PREDATOR. Prey animals generally occupy the last place but one in a FOOD CHAIN or web. *See also* ECOSYSTEM

Priestley, Joseph (1733–1804) British chemist and clergyman who discovered (1774) OXYGEN (which he called "dephlogisticated air"). He also discovered a number of other gases, including AMMONIA and the oxides of NITROGEN. He studied the properties of CARBON DIOXIDE (then called "fixed air") and invented carbonated drinks. He was an advocate of the later discredited PHLOGISTON THEORY.

primary consumer In ecology, any HETEROTROPH that feeds on PRIMARY PRODUCERS in the FOOD CHAIN but does not occupy the top position. Most HERBI-

VORES are primary consumers, feeding on plants (below them in the food chain) but themselves providing food for SECONDARY CONSUMERS (above them in the food chain). *See also* ECOSYSTEM

primary producer In ecology, an organism at the lowest level of a FOOD CHAIN. Primary producers are AUTOTROPHS. They ultimately feed all the consumers (HETEROTROPHS) above them. Most primary producers are ALGAE or PLANTS. *See also* ECOSYSTEM

Primates Order of MAMMALS that includes monkeys, apes, and human beings. Primates, native to most tropical and subtropical regions, are mostly herbivorous, diurnal (day-active), arboreal (tree-dwelling) animals. Their hands and feet, usually with flat nails instead of claws, are adapted for grasping. Most species have opposable thumbs, and all but humans have opposable big toes. They have a poor sense of smell, good hearing, and acute binocular vision. The outstanding feature of primates is a large complex brain and high intelligence. Primate characteristics are less pronounced in the relatively primitive prosimians (including tree shrews, galagos, lorises, and tarsiers) and are most pronounced in the more numerous and advanced ANTHROPOIDEA (monkeys, apes, and human beings).

prime number *See* NUMBER, PRIME

principle of moments In mechanics, a law that states that the moments of two bodies balanced about a central pivot or FULCRUM are equal (the moment of a body is the product of its mass and its distance from the pivot). This principle may be observed in BALANCES and seesaws.

Pringsheim, Nathanael (1823–94) German botanist who investigated the reproductive systems of lower plants. Pringsheim discovered the occurrence of SEXUAL REPRODUCTION in ALGAE and, after further studies, concluded that NATURAL SELECTION plays only a minor part in EVOLUTION; he believed that VARIATIONS are spontaneous and tend towards greater complexity. He was the first to demonstrate **apospory**, the production of a sexual from an asexual generation without the intervention of SPORES.

printed circuit Network of electric CONDUCTORS chemically etched from a layer of copper foil coating a board of insulating material such as plastic, glass or ceramic. This interconnects components such as CAPACITORS, RESISTORS, AND INTEGRATED CIRCUITS (IC). The printed CIRCUIT board (PCB) represents one stage in the miniaturization of electronic circuits (*see* MICROELECTRONICS). It serves to reduce the maze of connecting wires and to compact more circuitry, such as a complete calculating complex, onto a single board or module.

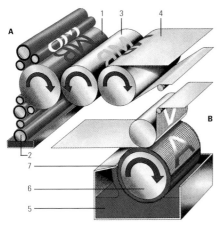

▲ **printing** Image areas of (A) offset lithographic printing plates (1) accept a grease-based ink and non-image areas, damped by rollers (2), reject the ink. The inked image is offset on to a rubber roller (3) and then on to the paper (4). (B) In copper gravure printing, the cylinder (6) has the image etched as "cells" (small depressions beneath the surface), the volume of cell determining the quantity of ink and hence the tonal value. The cylinder is covered with ink (5) and the surface is then cleaned by a "doctor" blade (7), leaving ink only in the etched depressions. When the cylinder contacts the paper the ink in the cells is transferred to it. The gravure process is used principally for high-quality pictorial work.

printing Technique for multiple reproduction of images, such as text and pictures. There are three basic forms of printing – **relief**, **intaglio**, and LITHOGRAPHY (**planographic**). The ancient Chinese and Japanese inked carved wooden blocks to print pictures. The earliest-known printed book is Chinese and dates from AD 868. Each page was printed from a solid wooden block. From the 10th century, the Chinese tried using separate pieces of type, so that each page could be printed from arrangements of standard characters. In *c*.1403 metal type made by casting first appeared in Korea. It enabled large quantities of type to be produced quickly. In the 1400s, Johann GUTENBERG and William CAXTON developed the use of LETTERPRESS PRINTING in Europe. In the 1700s and 1800s printing expanded rapidly. Lithography enabled printers to produce impressive color prints. *The Pencil of Nature* (1844–46) by William Fox TALBOT was the first book to be illustrated with photographs. PHOTOGRAPHY, in turn, aided the development of new techniques of reproduction, such as the halftone process. In 1846 Richard Hoe designed a rotary press that used STEREOTYPE plates. For text, metal printing plates were cast from the pages of type, so that the type could be used for setting other pages. In 1884 Ottmar MERGENTHALER invented the LINOTYPE machine. Other TYPESETTING machines, such as the MONOTYPE, speeded up the process of setting up pages. More recently, production speeds have greatly increased with the application of PHOTOSETTING, in which the type is set photographically on sheet film, and OFFSET PRINTING. Illustrations are often reproduced by PHOTOENGRAVING. Today, many publications are produced using a WORD PROCESSOR to enter the text. DESKTOP PUBLISHING programs allow images of the text and pictures to be arranged on screen. The computer data is used to print sheet film for each page, and the film images are transferred to printing plates.

prion Infective agent that appears to consist simply of a PROTEIN. Prions are thought to cause diseases such as CREUTZFELDT-JAKOB DISEASE (CJD) in humans, BOVINE SPONGIFORM ENCEPHALOPATHY (BSE) in cattle, and SCRAPIE in sheep. It is not understood how prions work, or how they persist in nature: unlike VIRUSES and BACTERIA, they do not contain DNA or RNA. Infection with a prion appears to cause a change in shape of the prion protein to a form that has a harmful effect.

prism In mathematics, solid geometrical figure, the ends of which are CONGRUENT (usually triangular) and perpendicular to the other faces, which are rectangular. The volume of a prism equals the area of either end multiplied by the length of the prism. In physics, a prism is a piece of transparent material, such as glass, plastic, or quartz, in which a light beam is refracted and split into its component colors (SPECTRUM) by dispersion or undergoes internal REFLECTION.

probability Number in the range from zero to one (inclusive) representing the likelihood of a given event. The probability of an event is defined as being equal to the number of ways that event may occur divided by the total number of possible outcomes, assuming that each possibility is equally likely. For instance, if a six-sided die is thrown there are six possible outcomes, three of which give an even number, so the probability of throwing an even number is $3/_6$ (or $1/_2$). A probability of 0 represents an event that is impossible (throwing a seven on a six-sided die), while 1 is the probability of an event that is certain and must occur. When a fair die is thrown, the outcome is independent of the result of previous throws. In this case, the probabilities are said to be independent or unconditional. More generally one can consider conditional probabilities too, where the probability of an event depends on previous outcomes. The theory of probability was first developed (*c*.1654) by Blaise PASCAL and Pierre de FERMAT. *See also* STATISTICS

Proboscidea Order of MAMMALS that has lived on Earth from the EOCENE to the present day, but is now represented only by elephants (genera *Elephas* and *Loxodon*). They have primitive limbs but specialized trunks and teeth. Family Elephantidae.

process-engraving *See* PHOTOENGRAVING

producer gas Mixture of the flammable CARBON MONOXIDE and HYDROGEN and the nonflammable NITROGEN and CARBON DIOXIDE. It is made by the partial combustion of coal or coke in air and steam. It has a lower heating value than other fuels, but can be manufactured with simple equipment. Producer gas is often used as a fuel in large industrial FURNACES.

progesterone Steroid HORMONE secreted mainly by the CORPUS LUTEUM of the mammalian OVARY and by the PLACENTA during pregnancy. Its principal function is to prepare and maintain the inner lining (ENDOMETRIUM) of the UTERUS for pregnancy. Synthetic progesterone is one of the main components of the contraceptive pill (*see* CONTRACEPTION).

program *See* COMPUTER PROGRAM

programming *See* COMPUTER PROGRAMMING

progression *See* ARITHMETIC PROGRESSION; GEOMETRIC PROGRESSION

proinsulin Precursor of the hormone INSULIN. Proinsulin is a POLYPEPTIDE formed in the PANCREAS by the beta cells of the ISLETS OF LANGERHANS. Its long-chain molecule splits to form the insulin molecule.

projector Instrument with a LENS system, used to cast images onto a screen from an illuminated flat object. An **episcope** is a projector for opaque objects, such as a printed page; it uses light that is reflected from the object. A **diascope** is a projector for transparent objects, such as photographic slides and films; it uses light transmitted through the object. An EPIDIASCOPE can project images from either transparent or opaque objects. A motion-picture or **cine** projector produces moving images from many frames (pictures) on a transparent FILM.

prokaryote Any member of the kingdom Prokaryotae (formerly Monera), which includes BACTERIA and CYANOBACTERIA (formerly blue-green ALGAE). They have more simple CELLS than other organisms. DNA is not contained in CHROMOSOMES in the NUCLEUS, but lies in a distinct part of the CYTOPLASM, called the nucleoid. They have no distinct membrane-surrounded structures (ORGANELLES). Cell division is simple and in the rare cases where SEXUAL REPRODUCTION occurs, genetic material is simply transferred from one partner to another; there are no separate sex cells. In photosynthetic prokaryotes, PHOTOSYNTHESIS takes place on the cell membrane. At present, two subkingdoms are recognized, ARCHAEBACTERIA and EUBACTERIA. *See also* ASEXUAL REPRODUCTION; EUKARYOTE

Prokhorov, Alexsandr Mikhailovich (1916–) Russian physicist. He shared the 1964 Nobel Prize for physics with his colleague Nikolai BASOV and the US physicist Charles H. TOWNES. This was for research in quantum electronics that resulted in the development of the MASER and LASER. In 1953 Prokhorov and Basov proposed the maser principle, and Townes constructed the first operational device.

prolactin HORMONE that stimulates the production of milk after childbirth. It is secreted by the anterior PITUITARY GLAND. Prolactin also stimulates the secretion of PROGESTERONE by the CORPUS LUTEUM of the OVARY. *See also* LACTATION

proline ($C_5H_9O_2N$) White, crystalline AMINO ACID found in PROTEINS.

PROM (acronym for **p**rogrammable **r**ead-**o**nly **me**mory) COMPUTER memory device consisting of INTEGRATED CIRCUITS, the contents of which are individually installed after manufacture (and cannot thereafter be reprogrammed). PROMS are used in control circuits for lifts and domestic appliances, such as washing machines. *See also* ROM

promethium (symbol Pm) Radioactive, metallic element of the LANTHANIDE SERIES. It was first made in 1941 by particle bombardment of NEODYMIUM and PRASEODYMIUM. Promethium occurs in URANIUM ores in minute amounts. The isotope ^{147}Pm is used in phosphorescent paints, X-ray tubes, and in nuclear-powered batteries for space vehicles. Properties: at.no. 61; m.p. 1,080°C (1,976°F); b.p. 2,460°C (4,460°F); most stable isotope ^{145}Pm (half-life 17.7 years).

prominence, solar Cloud of matter extending outwards from the Sun's CHROMOSPHERE into the CORONA. Prominences are regions of higher density and lower temperature than the surrounding corona. Near the SUN's limb, they are visible as bright protrusions into the dark corona during total eclipses or with the aid of an instrument such as a coronagraph or a special filter. **Quiescent** prominences, which extend for tens of thousands of kilometers, can last for weeks or months. **Active** prominences are short-lived, high-speed, flamelike eruptions that can reach heights of as much as 700,000km (450,000mi) in just 1 hour. All prominences have spectra showing lines of neutral hydrogen, helium, and ionized calcium. Their forms and behavior are very varied, and their direction of flow is controlled by the Sun's MAGNETIC FIELD. *See also* SOLAR FLARE

promoter In chemistry, substance used in small quantities in conjunction with a CATALYST to increase the catalyst's activity. For example, in the HABER PROCESS, an iron catalyst is used to speed up the combination of hydrogen and nitrogen under pressure to form ammonia; this catalyst is usually promoted with small amounts of potassium oxide.

propagation In botany, method of producing several individual plants from one original. This may be in the form of SEXUAL REPRODUCTION, in which the plant is pollinated and the seed produced in the normal way, or VEGETATIVE REPRODUCTION (vegetative propagation), in which plants reproduce asexually by means of CORMS, BULBS, TUBERS or RUNNERS. Vegetative reproduction can be performed artificially by taking cuttings or by grafting. In zoology, ASEXUAL REPRODUCTION, for example the budding of new individuals in hydra, is also called vegetative propagation. In other areas of science, the term propagation describes a spreading out, for example of waves from a central source or of nerve impulses along the AXON of a NEURON. *See also* POLLINATION

propane (C_3H_8) Colorless, flammable GAS, the third member of the ALKANE series of HYDROCARBONS. It occurs in NATURAL GAS, from which it is obtained; it is also acquired during petroleum

REFINING. Propane is used (as bottled gas) as a fuel, as a SOLVENT, AND in the preparation of many chemicals. Properties: m.p. −190°C (−310°F); b.p. −42°C (−43.6°F).

propanol (propyl alcohol) Colorless ALCOHOL used as a solvent and in the manufacture of various chemicals. It exists as two ISOMERS. Normal propanol ($CH_3CH_2CH_2OH$) is a by-product of the synthesis of METHANOL (methyl alcohol). Propan-2-ol (isopropyl alcohol, $(CH_3)_2CHOH$), is a secondary alcohol which is easily oxidized into PROPANONE.

propanone (acetone) Colorless, flammable liquid (CH_3COCH_3) made by oxidizing propan-2-ol (isopropyl alcohol). It is a raw material for the manufacture of many organic chemicals and is a widely used solvent. Properties: r.d. 0.79; m.p. −94.8°C (−138.6°F); b.p. 56.2°C (133.2°F).

propellant Material that undergoes chemical, nuclear or thermoelectric reactions to propel a ROCKET. **Liquid** propellants consist of a fuel such as PARAFFIN or HYDRAZINE (N_2H_4), which reacts with an oxidizer such as liquid OXYGEN. **Solid** propellants contain both fuel and oxidizer in powder form. **Nuclear** propellants include URANIUM and PLUTONIUM. **Ion** propellants include metallic CAESIUM, which boils off ions into an electric field that accelerates them to high exhaust velocities. The pressurized gas in an AEROSOL is also called a propellant.

propeller Device for producing THRUST, usually mounted on a rotating shaft. The cross-section of an aircraft propeller reveals an AIRFOIL shape, which allows it to function as a rotating wing. It generates forward thrust by producing aerodynamic LIFT. *See also* SCREW PROPELLER

propene *See* PROPYLENE

proper motion Apparent motion of a star on the CELESTIAL SPHERE, as a result of its movement relative to the Sun. Most annual proper motions are smaller than 0.1 arc seconds. BARNARD'S STAR has the largest known proper motion, 10.3 arc seconds per year. Proper motion is determined by comparing the star's position on photographic plates exposed on widely separated occasions, usually many years or decades apart. Much more accurate measurements of proper motions have been obtained by the position-measuring satellite Hipparcos.

prophase Stage in cell division, the first stage of MITOSIS or MEIOSIS. During prophase, CHROMOSOMES become visible, shrink, and split along their length to form CHROMATIDS.

prophylaxis Medical treatment in which the main focus is preventing a disease or other disorder. Prophylactic measures include health screening, IMMUNIZATION, AND, on a more general level, a healthy diet and adequate exercise. The practice of safe sex to avoid the risk of AIDS is a form of prophylaxis.

proportion Relationship between four numbers in which the ratio of the first pair (a and b) equals the ratio of the second pair (c and d), a/b = c/d. A continued proportion is a group of three or more quantities, each bearing the same ratio to its successor, such as the sequence 1:3:9:27:81.

proprioception Perception of internal nervous stimuli through which an animal is aware of its own movement, posture, and internal condition.

prop root (stilt root) Type of ROOT that grows from the STEMS of a plant and gives it extra support. Prop roots continue to be produced as the thin, main stem grows taller, and are often necessary to prevent the plant from falling over. They occur in herbaceous plants such as maize and in some trees, such as mangroves. In other tropical trees, wide, flat-topped prop roots called **buttress roots** grow from the base of the trunk and help to support the shallow-rooted trees.

propyl alcohol *See* PROPANOL

propylene Colorless, aliphatic HYDROCARBON ($CH_3CH:CH_2$), manufactured by the thermal CRACKING of ETHENE. It is used in the manufacture of a wide range of chemicals including vinyl and acrylic RESINS. Properties: b.p. −48°C (−54.4°F); m.p. −185°C (−301°F).

prostaglandin Series of related FATTY ACIDS, with hormone-like action, present in semen, liver, brain, and other tissues. Their biological effects include the lowering of blood pressure and the stimulation of contraction in a variety of smooth muscle tissues, such as the UTERUS.

prostate gland Gland in the male reproductive tract surrounding the URETHRA. It secretes specific chemicals that mix with SPERM cells and other secretions to make up the sperm-containing fluid, SEMEN.

prosthesis Artificial substitute for a missing organ or part of the body. Until the 17th century artificial limbs were either wooden or solid metal, but innovations in metallurgy and engineering design have enabled lighter, jointed limbs to be made. Methods of attachment, too, have greatly improved. More recent prosthetic devices include artificial heart valves made of SILICONE materials.

protactinium (symbol Pa) Rare, radioactive, metallic element of the ACTINIDE SERIES, first identified in 1913. Its chief source is URANINITE. Properties: at.no. 91; r.a.m. 231.0359; r.d. 15.4; m.p. 1,200°C (2,192°F); b.p. 4,000°C (7,232°F); most stable isotope ^{231}Pa (half-life 3.25×10^4 yr).

protective coloration Natural CAMOUFLAGE or warning colors of organisms that serve to blend it in with the surrounding environment or to ward off PREDATORS. Tigers and some moths have permanent protective coloring. Chameleons and some flatfish can change color to match the background. Warning colors of an animal usually mean it is poisonous, aggressive, or distasteful to most predators. Predators learn to recognize and avoid these colorations, which may be mimicked by harmless species. *See also* MIMICRY; PIGMENTATION

protein Organic COMPOUND containing many AMINO ACIDS linked together by covalent, PEPTIDE bonds. Protein molecules consist of POLYPEPTIDE chains. Living CELLS use about 20 different amino acids; because proteins have thousands of amino acids in each MOLECULE, the number of possible proteins is very large. The order of amino acids in proteins is controlled by the GENES in the cell's DNA. The most important proteins are ENZYMES, which determine all the chemical reactions in the cell, and ANTIBODIES, which combat infection. **Structural** proteins include KERATIN and COLLAGEN. **Gas transport** proteins include HEMOGLOBIN. **Nutrient** proteins include CASEIN. METABOLISM is regulated by protein HORMONES. ACTIN and MYOSIN are contractile **muscle** proteins. *See also* NUCLEIC ACID; X-RAY DIFFRACTION

protein synthesis Production within CELLS of PROTEINS from their component AMINO ACIDS. The GENETIC CODE, carried by the DNA of the CHROMOSOMES, determines which amino acids (of the 20 available) are used and in which order they are combined. The code is, in turn, encoded in MESSENGER RNA (mRNA) transcribed from the DNA in the cell's NUCLEUS. The mRNA carries the information to the RIBOSOMES, where protein synthesis actually takes place. *See also* NUCLEIC ACID

Proterozoic Period of GEOLOGICAL TIME, the second of the two subdivisions of the PRECAMBRIAN era. Dating from about 2.5 billion years ago to about 590 million years ago, it saw the emergence of the very earliest life forms such as BACTERIA and ALGAE.

prothallus In some primitive plants, part that carries the sex organs. A flattened green structure, it is found in FERNS, CLUB MOSSES, AND HORSETAILS, and is the independent GAMETOPHYTE generation (*see* ALTERNATION OF GENERATIONS). It is anchored to damp soil by hair-like RHIZOIDS. A single prothallus may bear a male sex organ (ANTHERIDIUM) or a female sex organ (ARCHEGONIUM), or both types. The prothallus initially provides nourishment for

▲ **protein synthesis** occurs in ribosomes, the sites where amino acids become linked together to form protein chains. They attach to the messenger RNA (mRNA) template, which moves relative to them. Amino acids join up in sequence as the bases of the transfer RNA (tRNA) molecules, on which they are carried, must combine with the relevant bases on (mRNA) as it passes the ribosome. Adenine always pairs with thymine, and cytosine pairs with guanine. One mRNA molecule may carry many ribosomes, with protein chains growing on each (this is called a polysome).

the EMBRYO and young SPOROPHYTE after it forms in the archegonium.

prothrombin Precursor of the blood ENZYME THROMBIN. It is converted into thrombin by thromboplastin. Prothrombin plays an essential role in BLOOD CLOTTING.

protoctist Any member of the kingdom Protoctista (formerly Protista), which includes such widely differing groups as ALGAE (including large seaweeds), AMOEBAS, and other PROTOZOA, SLIME MOLDS, and downy mildews.

proton (symbol p) Stable, ELEMENTARY PARTICLE with a positive charge equal in magnitude to the negative charge of the ELECTRON. It forms the NUCLEUS of the lightest isotope of HYDROGEN and, with the NEUTRON, is a constituent of the nuclei of all other elements. Also, like the neutron, it is made up of three QUARKS. The proton is a BARYON with spin $1/2$ and a mass 1836.12 times that of the electron. The number of protons in the nucleus of an element is equal to its ATOMIC NUMBER, also called proton number. Protons also occur in primary COSMIC RADIATION. Beams of high-velocity protons, produced by PARTICLE ACCELERATORS, are used to study nuclear reactions. In 1919, as a result of his work on the disintegration of the nucleus, Ernest Rutherford discovered the proton. Its antiparticle, the antiproton, was discovered in 1955. *See* ANTIMATTER

protonema Early stage in the development of the GAMETOPHYTE of a moss or liverwort. Most protonemata consist of green branched threads that grow on the soil surface. They carry buds that eventually develop into the adult form of the plant. *See also* ALTERNATION OF GENERATIONS

protoplasm Term referring to all the discrete structures within a plant or animal CELL, including the cell membrane. It includes both the NUCLEUS and the CYTOPLASM of cells.

protoplast (energid) In biology, term that describes the living unit of a plant CELL. The protoplast includes the cell NUCLEUS and the CYTOPLASM with its various ORGANELLES, but excludes the cell wall. *See also* PROTOPLASM

prototherian Any member of the Mammalia subclass Prototheria. Prototherians are represented in living fauna only by the MONOTREMES – echidnas and platypuses.

protozoan Any member of the phylum Protozoa, which comprises unicellular organisms found worldwide in oceans or freshwater, free-living and as PARASITES. These microscopic animals have the ability to move (by CILIA or pseudopodia) and have a NUCLEUS, CYTOPLASM, AND cell wall; some contain CHLOROPHYLL. Reproduction is by FISSION or encystment. Length: 0.3mm (0.1in). The 30,000 species are divided into four classes – Flagellata, Cnidospora, Ciliophora, and Sporozoa.

Proust, Joseph Louis (1754–1826) French chemist who specialized in analysis, with little interest in theory. As a result of his research, however, he concluded that COMPOUNDS were of constant and determinate composition. This discovery was very important in the establishment of the atomic theory of chemical composition.

Prozac Trade name for one of a small group of antidepressants, known as selective SEROTONIN reuptake inhibitors (SSRIs), which work by increasing levels of serotonin in the brain. Serotonin, or 5-hydroxytryptamine (5-HT), is a NEUROTRANSMITTER involved in a range of functions. Low levels of serotonin are associated with depression. Prozac is said to restore self-confidence, improve self-image, and boost energy. The DRUG appears to be as effective as older antidepressants and less toxic, but may produce side-effects, such as nausea and diarrhea.

prussic acid *See* HYDROCYANIC ACID

pseudocarp (false fruit) Fruit that is formed from other parts of a FLOWER as well as the OVARY wall. A pseudocarp may incorporate tissue from the CALYX or RECEPTACLE. In a strawberry, for example, the receptacle becomes the fleshy part while the true fruits are the "seeds" embedded in it. Other examples include apples, figs, pears, and rose hips.

pseudomorphism In mineralogy, chemical or structural alteration of a mineral without change in shape. An example is **substitution**, in which the original substance is removed and replaced by another. This is exemplified by petrified wood; the wood has been gradually replaced by SILICA. *See also* PETRIFICATION

psychobiology Field, closely related to physiological psychology, that studies anatomical and biochemical structures and processes as they affect behavior.

psychogenic disorders Mental disturbances believed to have an underlying psychological origin. NEUROSIS and functional PSYCHOSIS are usually thought of as psychogenic.

psychosis Serious mental illness in which, in contrast to NEUROSIS, the patient loses contact with reality. It may feature extreme mood swings between depression and mania, delusions or hallucinations, distorted judgment, and inappropriate emotional responses. The organic psychoses may spring from BRAIN DAMAGE, advanced syphilis, senile DEMENTIA or advanced EPILEPSY. The **functional** psychoses, for which there is no known organic cause, include schizophrenia and manic-depressive psychosis (bipolar disorder).

psychosomatic Descriptive of a physical complaint thought to be rooted, at least in part, in psychological factors. Extending the psychiatric concept of mind and body (psyche and soma) unity, it follows that emotional conflict can manifest itself in physical symptoms. The term has been applied to many complaints, including ASTHMA, migraine, ulcers, and HYPERTENSION.

pteridophyte Commonly used name for any of a group of SPORE-bearing VASCULAR PLANTS. At one time, pteridophytes were taken to include CLUB MOSSES, HORSETAILS, and FERNS. These plants have similar life cycles but in other respects are quite distinct and are now classified as separate phyla. *See also* TRACHEOPHYTE

PTFE Abbreviation of POLYTETRAFLUOROETHENE

PTH Abbreviation of PARATHYROID HORMONE

Ptolemaic system Arrangement of celestial bodies and their motions, proposed by APOLLONIUS OF PERGA and formulated by PTOLEMY *c*.AD 140. In the Ptolemaic system, the Earth is fixed at the center, with the Moon, Mercury, Venus, the Sun, Mars, Jupiter, and Saturn revolving about it. Beyond these planets lies the sphere of fixed stars. Each of the bodies moves around a small circle called the **epicycle**, the center of which in turn revolves around the Earth on a larger circle called the **deferent**. Ptolemy added to each ORBIT

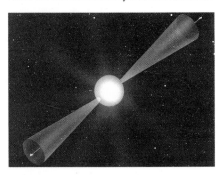

▲ **pulsar** It is believed that pulsars emit radiation in a narrow beam so that the rotation of the star produces regular flashes like a lighthouse.

two more points, the **eccentric** and the **equant**, equally spaced either side of the Earth. He made the epicycle revolve around the eccentric rather than the Earth, and Earth have uniform motion with respect to the equant. The resulting model reproduced the apparent motions of the planets so well that it remained unchallenged until the revival of the heliocentric theory by Nicolaus COPERNICUS in the 16th century. *See also* ECLIPTIC; ECCENTRICITY

Ptolemy (90–168) (Claudius Ptolemaeus) Greek astronomer and geographer. He worked at the library of Alexandria, Egypt, then a great center of Greek learning. His chief astronomical work, the *Almagest*, was largely a compendium of contemporary knowledge, including a star catalogue, drawing heavily on the work of HIPPARCHUS. The PTOLEMAIC SYSTEM is based on the geocentric world system of the ancient Greeks. Ptolemy's *Geography*, which provided the basis for a world map, was a definitive text for many centuries. He was regarded as the most reliable geographical authority by Muslim and Christian scholars until the Renaissance.

ptomaine Any of several basic organic chemical COMPOUNDS, some of which are poisonous to human beings. They are derived from the DECOMPOSITION or putrefaction of animal or vegetable PROTEINS and bear a resemblance to ALKALOIDS. Ptomaine poisoning is a form of food poisoning caused by BACTERIA or TOXINS.

puberty Time in human development when sexual maturity is reached. The reproductive organs take on their adult form and SECONDARY SEXUAL CHARACTERISTICS, such as the growth of pubic hair, start to become evident. Girls develop breasts and begin their MENSTRUAL CYCLE; in boys, there is deepening of the voice and the growth of facial hair. Puberty may begin at any time from about the age of ten, usually occurring earlier in girls than in boys. The process is regulated by HORMONES.

pubis In the PELVIS (pelvic girdle), either of a pair of small bones at the front of the hip bones. They are almost U-shaped and meet at the pubic symphysis, which is closed by a pad of cartilage. Each pubis joins an ILIUM and an ISCHIUM at a triangular suture in the hip socket.

pulley Simple MACHINE used to multiply FORCE or to change the direction of its application. A simple pulley consists of a WHEEL, often with a groove, attached to a fixed structure. Compound pulleys consist of two or more such wheels, some movable, that allow a person to raise objects much heavier than he or she could lift unaided. *See also* FORCE RATIO

Pullman, George Mortimer (1831–97) US industrialist and developer of the railway sleeping carriage. A cabinet maker in his youth, he began converting old railway coaches in 1858 to make long-distance traveling more comfortable. In 1865 the first modern sleeping car went into service. Known as the Pullman car, it was designed by Pullman and Ben Field.

pulsar Celestial object that emits RADIO WAVES in pulses of great regularity. They were first noticed in 1967 by the British astronomer Jocelyn BELL BURNELL, working at the Mullard Radio Astronomy Observatory, Cambridge, England. The first pulsar to be discovered pulsed every 1.337 seconds. Pulsars are believed to be rapidly rotating NEUTRON STARS, the aftermaths of SUPERNOVA explosions. A beam of radio waves emitted by the rotating pulsar sweeps past the Earth and is received in the form of pulses, in the same way as a lighthouse is seen to flash. The **Crab Pulsar** was the first to be seen flashing optically, in 1969. In some pulsars, the pulse amplitude and shape can change with time, and the pulsation can fade out temporarily. Pulsars are gradually slowing down, but some, such as the **Vela Pulsar**, occasionally increase their spin rate abruptly; such an event is called a glitch. More than 500 pulsars are now known, flashing at rates from about 4 seconds to 1 millisecond. The most common period is just under 1 second. Pulsars are concentrated towards the disk of our Galaxy, which is where supernovae most commonly occur. Most pulsars are thought to be more than a million years old.

pulsating star VARIABLE STAR, the LUMINOSITY of which is intrinsically altered as a result of periodic expansions and contractions arising from changes in the star's atmosphere. Most pulsating stars are relatively short-period variables, including the classical CEPHEID VARIABLES and the RR Lyrae variables.

pulse Regular wave of raised pressure in ARTERIES that results from the flow of BLOOD pumped into them at each beat of the HEART. The pulse is usually taken at the wrist, although it may be observed at any point where an artery runs close to the body surface. The average pulse rate is about 70 per minute in a resting adult.

pumice Rhyolitic LAVA blown when it is molten to a low density rock froth by the sudden discharge of GASES during a volcanic action. It tends to be very acidic in chemical content. When ground to a powder and pressed into cakes it is used as a light abrasive.

pump Device for raising, compressing, propelling or transferring fluids (liquids or gases). The lift pump, for raising water from a well, and

▲ **pump** A typical lift pump, such as the ones that were commonly used to raise water from an underground supply, work by employing two one-way valves. As the handle is pumped back and forth, the action is converted into the reciprocal, up-and-down motion of the plunger (1). As the plunger moves up it creates a partial vacuum in a collecting chamber (2), into which water is sucked. As the plunger descends, a valve (3) prevents the water being forced down the bore hole, while a second valve (4) allows the water to flow through the plunger which, when it rises again, forces water from the pump outlet.

the bicycle pump, for inflating tires, are examples of **reciprocating** (to-and-fro) pumps. In many modern pumps, a rotating impeller (set of blades) causes the fluid to flow. **Jet** pumps move fluids by forcing a jet of liquid or gas through them.

punched card In computing, one of the earliest mediums for storing and entering data into COMPUTERS and accounting machines. The information is typed on the keyboard of a machine, which encodes it into a pattern of holes on a paper card, which the machine can interpret. This form of storage is now rarely used. An early form of punched card was developed in the early 19th century by the French weaver Joseph JACQUARD. The cards automatically changed the pattern on the material being woven. *See also* INFORMATION STORAGE AND RETRIEVAL

punctuated equilibrium Theory expounded by Stephen Jay GOULD and Niles Eldridge in 1972, strongly sceptical of the notion of "gradual" change in the EVOLUTION of the natural world, as advocated by such theorists as Charles DARWIN. Fossil records rarely document the gradual development of a new SPECIES, rather its seemingly sudden appearance. Darwin reasons that this is due to gaps in the fossil records. Punctuated equilibrium explains these "missing links" by rejecting a notion of gradual change and by invoking a different model of evolution. Each species is predominantly in a steady state (equilibrium), but this is punctuated by brief but intense periods of sudden change which give rise to new species in relatively short periods of time.

pupa Nonfeeding, developmental stage during which an INSECT undergoes complete METAMORPHOSIS. It generally occurs as part of a four-stage life cycle from the OVUM, through LARVA to pupa, then IMAGO (adult). Most pupae consist of a tough protective outer casing, inside which the tissues of the insect affect a drastic reorganization to form the adult body. Insects that undergo pupation include the many species of butterfly, moth, beetle, and many kinds of fly. The pupa is often called a CHRYSALIS in butterflies and moths. It often acts as a resting or dormant stage, providing a resistant form in which the insect can survive severe weather or drought.

pupil In the structure of the EYE, circular APERTURE through which light falls onto the LENS; it is located in the center of the IRIS. Its diameter changes by reflex action of the iris to control the amount of light entering the eye.

Pupin, Michael Idvorsky (1858–1935) US physicist and inventor, b. Yugoslavia. He devised a means of extending the range of long-distance TELECOMMUNICATIONS by placing loading coils at intervals along the transmitting wire, and invented a method of taking short-exposure X-RAY photographs. His autobiography *From Immigrant to Inventor* (1923) won a Pulitzer Prize.

Puppis (Argo Nevis) Constellation of the S sky. Its brightest star is Zeta (Naos), magnitude 2.2. It has several bright, open clusters.

Purcell, Edward Mills (1912–97) US scientist who shared the 1952 Nobel Prize for physics with Felix BLOCH for their independent work in NUCLEAR MAGNETIC RESONANCE (NMR). Purcell's discovery of NMR in liquids and solids is instrumental in the measurement of the MAGNETIC MOMENTS of nuclei and molecules. He also did research in radio astronomy, astrophysics, and biophysics.

purine White, crystalline, organic COMPOUND ($C_5H_4N_4$) related to URIC ACID. ADENINE and GUANINE are both purines and are two of the four bases that comprise DNA. Another derivative of purine is ADENOSINE TRIPHOSPHATE (ATP), a NUCLEOTIDE essential to the transfer of energy within living cells.

pus Yellowish fluid forming as a result of bacterial infection. It comprises blood SERUM, LEUCOCYTES (white blood cells), dead tissue, and living and dead BACTERIA. An abscess is a pus-filled cavity.

putrefaction DECOMPOSITION of organic matter, especially PROTEINS, by FUNGI, BACTERIA, and OXIDATION. It results in foul-smelling products. Putrefaction of meat, for example, yields HYDROGEN SULFIDE, AMINES, and THIOLS.

PVC Abbreviation of POLYVINYL CHLORIDE

pyramid In geometry, a solid figure having a POLYGON as one of its faces (the base), the other faces being TRIANGLES with a common VERTEX. The volume of a regular pyramid is one-third of the base area multiplied by the vertical height.

pyridine Heterocyclic compound (C_5H_5N) characterized by a six-membered ring structure composed of five CARBON atoms and one NITROGEN atom. It occurs in bone oil and coal tar and is synthetically produced from ACETALDEHYDE (ethanal) and AMMONIA. Pyridine is used as a SOLVENT and converted to sulfapyridine (a drug used to combat bacterial and viral infection), pyribenzamine, and pyrilamine (ANTIHISTAMINE).

pyridoxine VITAMIN B_6, a COENZYME in the metabolism of AMINO ACIDS in the body. Pyridoxine is found in lean meat, egg yolks, milk, and fish. A deficiency can lead to ANEMIA, and it is also taken as a vitamin supplement by those who believe it helps to ease depression.

pyrimidine Colorless liquid ($C_4H_4N_2$) characterized by a heterocyclic ring structure composed of four CARBON atoms and two NITROGEN atoms. It may be prepared from URACIL (a dihydroxy pyrimidine compound) by chemical reactions that remove two OXYGEN atoms. It is used in the manufacture of various SULFONAMIDE DRUGS and BARBITURATES.

pyrite (iron pyrites) Most common and widespread sulfide mineral, iron sulfide (FeS_2), occurring in all types of rocks and veins. It is a brass-yellow color and, because of this, is often called "fool's gold." It was formerly used widely to produce SULFURIC ACID. It crystallizes as cubes and octahedra, and also as granules and globular masses. It is opaque, metallic, and brittle. Hardness 6.5; r.d. 5.0.

pyroclastic Term that describes any rock fragments thrown out by volcanic activity, generally in association with some violent explosive action. Pyroclasts normally include solidified LAVA left behind by a previous eruption of the VOLCANO, as well as rocks from the crust, smaller pieces of cinders, ash, and dust. The largest pyroclastic fragments, known as "volcanic bombs," weigh several metric tons. Smaller pieces are known as "lapilli." When the pyroclastic activity ceases, outpourings of lava often follow. Rocks formed by consolidation of these pyroclastic fragments, such as volcanic BRECCIA and TUFF, are also termed pyroplastic.

pyrolusite Common mineral form of manganese dioxide (MnO_2), an important ore of MANGANESE. It is formed as a PRECIPITATE in lakes and bogs at low temperatures, often in veins of quartz, but also occurs as MANGANESE NODULES on the floor of the Pacific Ocean. It is black or dark gray, usually occurring in masses that have an earthy luster, although it may form prismatic crystals (tetragonal) and fern-like aggregates in the joints of sedimentary rocks. Hardness: massive form 2–6, crystals 6–7; r.d. 4.5–5.

pyrolysis Chemical decomposition of a complex substance into simpler ones by the action of heat alone. Waste materials and coal may be pyrolyzed to produce valuable fuels and chemicals.

pyrometer THERMOMETER for use at extremely high temperatures, well above the ranges of ordinary thermometers. An **optical** pyrometer consists essentially of a small TELESCOPE, a RHEOSTAT (a type of variable resistor) and a filament. When the telescope is aimed at a furnace or other hot object, the filament appears dark against the background. As the electric current flowing through the filament is increased slowly using the rheostat, it grows brighter until it matches the intensity of the furnace. The amount of current is a measure of the temperature.

pyrope ($Mg_3Al_2(SiO_4)_3$) Magnesium aluminum GARNET, the most common form of garnet. It may be purplish or brownish red and is found mainly in South Africa and the USA. Often used in jewelry, it has been given such misleading names as Bohemian garnet and is often mistaken for a RUBY.

pyroxene Any of an important group of rock-forming minerals. They are single-chain SILICATES. Their colors are variable but are usually dark greens, browns, and blacks. Crystals are usually short PRISMS with good cleavages. Hardness 2.3–4; r.d. 5.5–6.

pyrrhotite Magnetic mineral form of iron sulfide, FeS (but variable), containing more sulfur than PYRITE. Its color is a metallic bronze and it forms massive aggregates or hexagonal, flat crystals (monoclinic). It may contain nickel and in such cases is mined as a nickel ore. Hardness: 3.5–4.5; r.d. 4.6.

pyrrole (C_4H_5N) Heterocyclic, organic chemical COMPOUND consisting of a five-membered ring with four CARBON atoms and one NITROGEN atom. A colorless liquid, it smells like TRICHLOROMETHANE (chloroform) and is obtained from COAL TAR and bone oil. It forms part of the molecules of various natural pigments, including CHLOROPHYLL and HEMOGLOBIN.

pyruvic acid ($CH_3COCOOH$) Colorless liquid that can be formed by the distillation of TARTARIC ACID; it has a vinegary odor. Phenylpyruvic acid, a derivative of pyruvic acid, occurs in the urine of patients suffering from PHENYLKETONURIA – a disorder in which the amino acid PHENYLALANINE is not metabolized normally.

Pythagoras (*c*.580–500 BC) Greek philosopher and founder of the Pythagorean school. He was born on the Greek island of Samos and later lived in Croton, Italy, where he founded his school. The Pythagoreans believed that numbers and their interrelationships could be used to represent everything in the Universe. Pythagoras himself is credited with many advances in MATHEMATICS, geometry, medicine, and philosophy. He was probably the first to teach that the Earth is a sphere revolving about a fixed point, and is said to have been the first to discover the mathematical relationship between the length of a string and the PITCH of the sound it makes when vibrating (*see* FREQUENCY). Pythagoras is also credited with the theorem that the square of the HYPOTENUSE of a right-angled TRIANGLE equals the sum of the squares of the other two sides, but this result was actually known to the Egyptians and Babylonians many hundreds of years earlier. Profoundly at odds with contemporary religion, his brotherhood was suppressed at the end of the 6th century BC, but its doctrines were revived (**Neo-Pythagoreanism**) by the Romans *c*.500 years later.

Pyxis (Compass) Constellation of the S sky adjoining VELA and PUPPIS. Its most interesting feature is the recurrent NOVA T Pyxidis.

quadrant In plane geometry, a quarter of a CIRCLE, bounded by radii at right angles to each other and the arc of the circle. In analytic geometry, it is one of the four sections of a plane divided by an x-axis and a y-axis.

quadrat Square frame used to mark out an area of ground for quantifying and studying the plants it encloses. The area is typically 1 or $0.5m^2$. Botanists and ecologists can count the numbers of species within a quadrat and estimate their abundance. Repeated measurements at several locations provide information about overall species distribution.

quadratic equation Polynomial EQUATION of degree two. A quadratic equation has the general form $ax^2 + bx + c = 0$, where a, b, and c are constants, with a nonzero. It has two solutions (also called roots), shown by AL-KHWARIZMI to be given by the formula $x = [-b \pm \sqrt{(b^2 - 4ac)}]/2a$.

quadratic function Mathematical FUNCTION, the value of which depends on the square of the independent VARIABLE, x, and is thus given by a quadratic POLYNOMIAL, as in
$$f(x) = 4x^2 + 17 \text{ or } f(x) = x^2 + 3x + 2.$$

qualitative analysis Identification of the chemical elements or ions in a substance or mixture.

quantitative analysis Identification of the amounts of chemical constituents in a substance or mixture. Chemical methods use reactions such as NEUTRALIZATION and OXIDATION, and determine the concentrations of substances present in terms of volume (VOLUMETRIC ANALYSIS) or weight (GRAVIMETRIC ANALYSIS). Physical methods measure qualities such as density and REFRACTIVE INDEX (how much light is refracted by a medium).

quantum In physics, the smallest quantity of some physical property, such as matter, that a system can possess according to the QUANTUM THEORY.

quantum chromodynamics Study of the properties of QUARKS in which, to explain permissible combinations of quarks to form various ELEMENTARY PARTICLES, each is notionally assigned a color. Quarks are given one of the three primary colors – red, green, and blue. When three quarks combine to form BARYONS, the resulting color is always white. Antiquarks are given one of the three complimentary colors – cyan, magenta, and yellow. When a quark combines with an antiquark to form a MESON, the resulting color is also white. *See also* QUANTUM ELECTRODYNAMICS (QED)

quantum electrodynamics (QED) QUANTUM THEORY of ELECTROMAGNETIC RADIATION and how it interacts with charged particles. For example, the theory predicts that a collision between an ELECTRON and a PROTON should result in the production of a PHOTON of electromagnetic radiation.

quantum field theory Theory that embraces all ELEMENTARY PARTICLES and their interactions. It applies QUANTUM MECHANICS to systems with an infinite number of degrees of freedom. Each type of particle is represented by fields that can oscillate only in quantized modes. Thus the particles become the quanta of the fields – for example, the PHOTON is the quantum of LIGHT (the electromagnetic field).

quantum mechanics Use of QUANTUM THEORY to explain the behavior of ELEMENTARY PARTICLES. In the quantum world, waves and particles are interchangeable concepts. In 1924 the French physicist Louis de BROGLIE suggested that particles have wave properties, the converse having been postulated in 1905 by Albert EINSTEIN. In 1926 the Austrian physicist Erwin SCHRÖDINGER used this hypothesis to predict particle behavior on the basis of wave properties, but a year earlier the German physicist Werner HEISENBERG had produced a mathematical equivalent to Schrödinger's theory without using wave concepts at all. In 1928 the British physicist Paul DIRAC unified these approaches, while incorporating RELATIVITY into quantum mechanics. QUANTUM ELECTRODYNAMICS was derived by the US physicist Richard FEYNMAN in the 1940s. *See also* QUANTUM NUMBERS

quantum numbers Set of four numbers used to classify ELECTRONS and their atomic states. The **principal** quantum number (symbol n) gives the electron's energy level, the ORBITAL quantum number (symbol l) describes its angular momentum, the **magnetic** quantum number (symbol m) describes the energies of electrons in a magnetic field, and the SPIN quantum number (symbol m_s) gives the spin of the individual electrons. *See also* QUANTUM THEORY

quantum theory Together with the theory of RELATIVITY, the foundation of 20th-century physics. It is concerned with the relationship between MATTER and ENERGY at the elementary or subatomic level and with the behavior of ELEMENTARY PARTICLES. According to the theory, all radiant energy is emitted and absorbed in multiples of tiny "packets" or quanta. This breaks down the traditional distinction between energy (as a wave motion) and matter (as occupying a fixed space), because quanta have properties of both. In 1900 Max PLANCK proposed the idea that energy is radiated and absorbed in quanta. He calculated that the energy E of a quantum is related to the frequency of radiation v according to the formula $E = hv$, where h (PLANCK'S CONSTANT) is a constant with the value of 6.626×10^{-34} joule seconds. Using this formula, Albert EINSTEIN quantized light radiation and in 1905 explained the PHOTOELECTRIC EFFECT. In 1913 Niels BOHR used quantum theory to explain atomic structure and the relationship between the energy levels of an atom's electrons and the frequencies of radiation emitted or absorbed by the ATOM.

quark Any one of six particles and their antiparticles (antiquarks) that are the fundamental members of the HADRON group of ELEMENTARY PARTICLES. Quarks always exist in combination; free quarks cannot exist at the temperatures of the modern Universe, though they may have done so in the first moments after the BIG BANG. In 1964 the US physicists Murray GELL-MANN and George Zweig proposed independently that hadrons (particles that experience the STRONG NUCLEAR FORCE) are composed of quarks. A BARYON, such as a PROTON or NEUTRON, consists of three quarks; a MESON is made up of a quark and an antiquark tightly bound together. A quark has a charge of either $-2/3$ or $-1/3$ of the charge on an ELECTRON. It can occur in one of six "flavors": up, down, top, bottom, charmed, and strange. Antiquarks have similar flavors but their charge is opposite to that of their corresponding quark. Each quark and antiquark also has a "color" (*see* QUANTUM CHROMODYNAMICS). *See also* ANTIMATTER

quartz (silica, SiO_2) Rock-forming mineral, the natural form of silicon dioxide. It is widely distributed, occurring in IGNEOUS and METAMORPHIC ROCKS (notably granite and gneiss), and in clastic sediments. It is also a common gangue mineral in mineral veins. Pure quartz is clear and colorless, but the mineral is often opaque or translucent, and may be colored by impurities. It forms six-sided crystals. The most common varieties are colorless quartz (known as rock crystal), rose, yellow, milky, and smoky. The most usual cryptocrystalline varieties, the crystals of which can be seen only under a microscope, are CHALCEDONY and FLINT. Quartz crystals exhibit the PIEZOELECTRIC EFFECT and are used in electronic clocks and watches to keep accurate time. Hardness 7; r.d. 2.65.

quartzite METAMORPHIC ROCK usually produced from sandstone, in which the quartz grains have recrystallized. Fracturing through these grains rather than between them. A hard and massive rock, its color is usually white, light gray, yellow, or buff, but it can be made green, blue, purple, or black by various minerals.

quasar (quasi-stellar object) Compact celestial object that is an extremely luminous source of energy, radiated over a wide range of wavelengths from X-rays through optical to radio wavelengths. First discovered in 196, every quasar shows a very large RED SHIFT which, if it results from the expansion of the Universe, indicates that quasars are the oldest and most distant objects in the Universe. Quasars seem to inhabit the centers of very remote galaxies. Since they vary in brightness with periods of a few days, they can be no larger than a few hundred astronomical units wide. A surrounding galaxy has been seen around the nearest quasars, almost lost in the glare from the brilliant center. Quasar activity is thought to be caused by a BLACK HOLE accreting material at the center of the host galaxy. The most luminous quasars would need to be powered by black holes of 100 million solar masses, swallowing stars at a rate of about one every year. Quasars are bright ultraviolet sources, and this hot radiation seems to come from a swirling ACCRETION disk. Also associated with quasars are emission lines of hydrogen, helium, and iron that indicate motions as rapid as $5,000 \text{km s}^{-1}$, as would be expected near a massive black hole. Because quasars are the most luminous objects in the Universe, they can be seen at greater distances, and hence further back in time, than anything else. Quasars thus provide us with a way of examining the Universe when it was young.

Quaternary Most recent period of the CENOZOIC era, beginning about 2 million years ago and extending to the present. It is divided into the PLEISTOCENE epoch, characterized by a periodic succession of great ice ages, and the HOLOCENE epoch, which started some 10,000 years ago.

quaternions Type of abstract number, discovered by William HAMILTON. An ordinary complex number has the form $a + bi$ (where a and b are real numbers, i is the square root of -1). A quaternion has the form $a + bi + cj + dk$, in which i, j, and k are defined by the equations $i^2 = j^2 = k^2 = -1$, and $-ji = k$).

quinine White, crystalline ALKALOID isolated in 1820 from the bark of the cinchona tree. It was once widely used to treat MALARIA, but has been largely replaced by drugs that are less toxic and more effective.

quinone Aromatic organic compound that contains a BENZENE ring in which two hydrogen atoms have been replaced by oxygen atoms. Paraquinone (1,4-benzoquinone, $O = C_6H_4 = O$) is a yellow crystalline solid obtained by the oxidation of aniline with chromic acid; it is used in the manufacture of hydroquinone for photographic developers.

Rabi, Isidor Isaac (1898–1988) US physicist, b. Austria, who was known for his work on MAGNETISM and QUANTUM MECHANICS. Rabi was awarded the 1944 Nobel Prize for physics for his discovery and measurement of the radio-frequency spectra of atomic nuclei whose magnetic SPIN has been disturbed. This led to the development of new SPECTROSCOPIC methods.

racemization Transformation of an optically active substance into a MIXTURE of equal quantities of two mirror-image forms (**enantiomorphs**), usually by heat or the action of acids or bases. Individually, each enantiomorph rotates the plane of POLARIZED LIGHT through a characteristic angle, but an equal quantity of each will cancel each other's rotatory effect.

rack In engineering, a straight rod on which grooves are cut to mesh with a cylindrical GEAR (pinion). Rotary motion of the pinion is converted into reciprocating (to-and-fro) motion at the rack. It is used extensively for steering systems in AUTOMOBILES.

rack railway Form of mountain transport in which the locomotive powers a cogwheel that engages a cogged rail (rack) set between the track. The first rack railway in Britain came into use in 1812 between Middleton colliery and Leeds in the N of England, a distance of 5.5km (3.5mi).

radar (contraction of **ra**dio **d**etecting **a**nd **r**anging) Electronic system for determining the direction and distance of objects. First developed by the British and Germans during World War II, it works by the transmission of pulses of RADIO WAVES to an object from the radar transmitting equipment. The object reflects the pulses, which are detected by an ANTENNA. By measuring the time it takes for the reflected waves to return, the object's distance may be calculated, and its direction ascertained from the alignment of the receiving radar antenna. Air traffic controllers use radar to indicate aircraft positions. In this case, a rotating antenna transmits the signals and picks up the reflections. The reflected signals from planes are shown as bright moving dots on a circular display called a plan position indicator. In 1946 radar beams from Earth were reflected back from the Moon, marking the beginnings of RADAR ASTRONOMY. *See also* DOPPLER EFFECT; DOPPLER RADAR

radar astronomy Branch of ASTRONOMY in which RADAR pulses, reflected back to Earth from celestial bodies inside the SOLAR SYSTEM, are studied for information concerning the distance from Earth of the bodies, their orbital motion, and large surface features. It differs from RADIO ASTRONOMY, which involves only the reception of radio waves emitted naturally by objects in the Universe. Radar echoes from the Moon were first obtained in 1946. Radar reflections from Venus were received in 1958 and from the Sun in 1959. The resulting measurements provided an accurate new idea of the scale of the Solar System. As the power of radars grew and orbiting PLANETARY PROBES were developed, techniques were developed for radar mapping of planetary surfaces, which proved particularly important for cloud-covered VENUS. Other objects in the Solar System studied by radar include ASTEROIDS, Saturn's rings, planetary SATELLITES, and the nuclei of COMETS.

radial velocity In astronomy, velocity of a celestial body in the line of sight, either towards (positive) or away from (negative) an observer. The radial velocity is determined from the DOPPLER EFFECT on lines in the object's SPECTRUM. *See also* RED SHIFT

radian Angle formed by the intersection of two radii at the center of a CIRCLE, when the length of the arc cut off by the radii is equal to one radius in length. Thus, the radian is a unit of ANGLE equal to $c.57.296°$, and there are 2π radians in $360°$.

radiant heat Form of ENERGY transfer through the atmosphere or space. The Sun's energy reaches the Earth as radiant heat or INFRARED RADIATION, as well as visible light and other ELECTROMAGNETIC RADIATIONS. Domestic electric fires emit radiant HEAT, but they also heat the air by CONVECTION.

radiation Transmission of energy by SUBATOMIC PARTICLES or ELECTROMAGNETIC WAVES. All ELECTROMAGNETIC RADIATION can cross a VACUUM, which distinguishes it from such effects as CONDUCTION and CONVECTION and the transmission of sound. In a vacuum, electromagnetic waves travel at the speed of light, $299{,}792.5$km s^{-1} ($186{,}291$mi s^{-1}), but this speed is slightly decreased as the waves pass through a medium (as when light passes through glass), an effect that explains REFRACTION. This type of radiation also exhibits some of the properties of a stream of PHOTONS. Similarly, streams of particles such as ELECTRONS exhibit properties like those of waves, which allows for electron DIFFRACTION as in an ELECTRON MICROSCOPE.

radiation, heat Energy given off from all solids, liquids, or gases as a result of their temperature, however low. The energy comes from the vibrations of ATOMS in an object and is emitted as ELECTROMAGNETIC RADIATION, often in the form of INFRARED RADIATION.

radiation, nuclear Particles or ELECTROMAGNETIC RADIATION emitted spontaneously and at high energies from atomic nuclei. Possible causes include RADIOACTIVE DECAY, which yields ALPHA PARTICLES (helium nuclei), BETA PARTICLES (electrons), GAMMA RADIATION, and, more rarely, POSITRONS (antielectrons). It can also result from spontaneous FISSION of a NUCLEUS, with the ejection of NEUTRONS or gamma rays. *See also* RADIOACTIVITY

radiator Space heater in a central heating system. Commonly, it is a slim fluted steel or iron container through which hot water or steam is circulated. Electrically powered appliances, such as convector heaters and storage heaters, can also function as radiators. In cars, trucks and other vehicles powered by an INTERNAL COMBUSTION ENGINE, a radiator is a heat exchanger, which dissipates the heat in circulating water (to the surrounding air) to keep the engine cool.

radical (symbol $\sqrt{}$) Expression used to indicate the ROOT of a number. The expression $\sqrt[n]{x}$ represents the nth root (or index) of the number x (the radicand). The CUBE ROOT of ten in radical form is written $\sqrt[3]{10}$. Another way of expressing the nth root of a number x is by $x^{1/n}$.

radical, atomic Single ATOM or group of atoms in which all the atomic VALENCES are not satisfied by chemical bonding. Thus, removal of a hydrogen atom from methane, CH_4, gives the methyl radical CH_3. The term is usually taken to mean a FREE RADICAL, that is, one existing free for a short time in a reaction. Free radicals are highly reactive; they can be produced by ULTRAVIOLET LIGHT, PYROLYSIS, ELECTRON impact, and other means, and play an important part in the mechanisms of many CHEMICAL REACTIONS.

radicle Part of a plant embryo that develops into the ROOT. When the radicle emerges from a germinating seed, its tip is protected by a root cap (**calyptra**), which enables it to break through the seed case (**testa**) and penetrate the soil. In most plants, the radicle originates the whole root system.

radio BROADCASTING or reception of ELECTROMAGNETIC RADIATION in the form of RADIO WAVES. A radio signal of fixed FREQUENCY (the carrier wave) is generated at a TRANSMITTER. The sound to be broadcast is converted by a MICROPHONE into a varying electrical signal that is combined with the carrier by means of MODULATION. FREQUENCY MODULATION (FM) minimizes interference and provides greater fidelity than AMPLITUDE MODULATION (AM). The modulated carrier wave is passed to an ANTENNA from which it is transmitted into the atmosphere. At the receiver, another aerial intercepts the signal, and it undergoes "detection," the reverse of modulation, to retrieve the sound signal. It is amplified to activate a LOUDSPEAKER that reproduces the original sound. A station is selected by tuning to a specified frequency. Radio waves travel at the speed of light and are transmitted not only by line-of-sight (ground waves), but also by reflection from the IONOSPHERE (sky waves). Sky waves enable long-range radio transmission. The ULTRA HIGH FREQUENCY (UHF) and VERY HIGH FREQUENCY (VHF) waves used in TELEVISION broadcasting penetrate the ionosphere with little reflection, and long-range broadcasting is made possible by means of artifical SATELLITES. The most efficient and commonest circuit used in radio is the SUPERHETERODYNE RECEIVER. Walkie-talkies and MOBILE TELEPHONES use radio for communications, television uses radio to send sound and vision signals, and RADAR transmits pulses of radio waves. The development of radio can be traced to the mathematical studies of James Clerk MAXWELL. Heinrich HERTZ devised an apparatus for the transmission and detec-

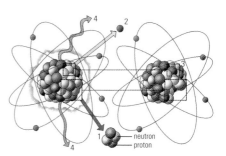

▲ **radiation, nuclear**
Many elements are naturally radioactive, so there is always background radiation even when there is no manmade or accidentally generated radioactivity in the vicinity. Alpha radiation occurs when an unstable atomic nucleus emits an alpha (1) particle, a helium nucleus comprising of two protons and two neutrons. Alpha radiation (α) is stopped by flesh. Beta radiation (β) occurs when an electrically neutral neutron emits a negative electron (which is the beta (2) particle) and becomes in the process a positive proton (3). Beta particles can be stopped by a thin sheet of aluminum. X-rays and gamma (γ) are both electromagnetic radiation (types of light), created when a nucleus loses energy without undergoing structural change (4). X-rays can be stopped by a sheet of lead, while high-energy gamma rays will penetrate thin sheets of lead, but may be stopped by a layer of concrete.

tion of radio waves. In 1895 Guglielmo MARCONI gave a demonstration of the first wireless TELEGRAPH, and in 1901 he sent the first transatlantic message using MORSE CODE. In 1904 Sir John FLEMING invented the thermionic tube, the first vacuum ELECTRON TUBE. In 1906 Lee DE FOREST developed the audion triode tube, which was able to detect and amplify radio waves. It remained at the heart of radio and television manufacture until the invention of the TRANSISTOR in 1947.

radioactive decay Process by which a RADIOISOTOPE loses ELEMENTARY PARTICLES from its NUCLEUS and so becomes a different, more stable element. The starting isotope is called the "parent" element and the resulting isotope is termed the "daughter" element. Each radioactive element has its own constant rate of decay. This rate is measurable, and given as an element's HALF-LIFE. This property is utilized in radioactive DATING. A decaying ISOTOPE releases one, two or three of the natural types of RADIOACTIVITY: ALPHA PARTICLES, BETA PARTICLES, and GAMMA RADIATION.

radioactive series Chain of ISOTOPES, each being a product of the RADIOACTIVE DECAY of its predecessor, starting with a RADIOISOTOPE and ending at one of the stable isotopes of LEAD. The three naturally occurring series are headed by uranium-238, thorium-232, and uranium-235. The neptunium-237 series does not occur in nature, because the parent isotope has a HALF-LIFE (about 2 million years) that is much less than the age of the Earth and has long decayed; this series is artificially produced. Each step in a radioactive series is called a TRANSMUTATION.

radioactive waste *See* NUCLEAR WASTE

radioactivity Disintegration of the nuclei of a RADIOISOTOPE, such as uranium-238, usually with the emission of ALPHA PARTICLES (helium nuclei) or BETA PARTICLES (ELECTRONS), often accompanied by GAMMA RADIATION. The two processes of alpha decay or beta decay cause the radioisotope to be transformed into a chemically different ATOM. **Alpha decay** results in the nucleus losing two PROTONS and two NEUTRONS; **beta decay** occurs when a neutron changes into a proton with an electron (beta particle) being emitted in the process. Thus, the ATOMIC NUMBER (number of protons in the nucleus) changes in both types of decay, and an ISOTOPE of another element is produced, which might also be radioactive. The stability of different isotopes varies widely. It is impossible to predict when a nucleus will disintegrate, but in a large collection of atoms there is a characteristic time (the HALF-LIFE) after which one-half of the total number of nuclei would have decayed. This time varies from millionths of a

second to millions of years, depending on the isotope concerned. The activity of any radioactive sample decreases exponentially with time. *See also* CARBON DATING; NUCLEAR ENERGY

radio astronomy ASTRONOMY of radio emissions, ELECTROMAGNETIC RADIATION with wavelengths from about 1mm to many meters, reaching the Earth from space. Observations down to about 2cm can be made at sea level without serious interference from the atmosphere. At shorter wavelengths it is necessary to observe from high mountains to avoid atmospheric absorption, mainly by WATER VAPOR. In 1931 radio noise from the Milky Way was discovered by Karl JANSKY, and the subject grew rapidly after World War II. The radio emissions first detected from the Milky Way were produced by the process of SYNCHROTRON radiation – the motion of high-speed ELECTRONS in the galaxy's MAGNETIC FIELD. A number of localized radio sources were also found. Some of these turned out to be SUPERNOVA remnants within our galaxy, but optical identifications of sources led to the realization that most were far off in the Universe. These sources are now known as RADIO GALAXIES and QUASARS. The number of radio sources increases with distance, demonstrating that the Universe has been evolving with time. This, combined with the discovery at radio wavelengths of the COSMIC MICROWAVE BACKGROUND, is strong evidence in favor of the BIG BANG theory of the origin of the Universe. In 1967 a new class of individual radio sources was discovered in our galaxy: the fast-flashing PULSARS, which are ultradense NEUTRON STARS formed when massive stars die. Individual spectral lines can also be detected at radio wavelengths, similar to optical spectral lines. In 1951 the emission from HYDROGEN at 21cm was detected, and has proved to be an especially useful means of studying the structure of our galaxy. *See also* RADIO TELESCOPE

radio beacon Strategically located radio TRANSMITTER that emits a regular, nondirectional, and easily recognizable signal by which mobile receivers can obtain a bearing. Such beacons are used particularly by marine and air navigators who collate the information from two or more such beacons to fix their positions. Directional beacons are used at airports to aid landing aircraft.

radiocarbon dating *See* CARBON DATING

radio galaxy GALAXY that is an abnormally high emitter of RADIO WAVES, around a million times stronger than the weak emissions from galaxies such as our own. Radio galaxies are usually giant ellipticals, such as M87 in the constellation Virgo. The emission from a typical radio galaxy is concentrated in two lobes well outside the galaxy, often located symmetrically either side, which appear to have been ejected from the galaxy in explosions. The extended radio lobes are almost invariably connected to the main galaxy by jets that seem to originate from the galactic nucleus. Strong, pointlike radio sources exist in the nuclei of many of these objects. It is not understood why some elliptical galaxies emit such enormous amounts of radio energy. However, it does seem clear that the source of the activity lies in the galaxy's nucleus, and in this respect radio galaxies are similar to radio QUASARS and BL LACERTAE OBJECTS. A radio galaxy's enormous energy output comes from SYNCHROTRON radiation. The central "engine" produces vast quantities of energetic electrons traveling near the speed of light. When these electrons encounter a MAGNETIC FIELD, they spiral around and, in so doing, lose energy. It is this energy that we detect as radio waves. *See also* RADIO ASTRONOMY

radiography Use of X-RAYS to record the interiors of opaque bodies as photographs. Industrial X-ray photographs can show assembly faults and defects in the CRYSTAL structure of metals. In MEDICINE and DENTISTRY, radiography is invaluable for diagnosing bone damage, tooth decay, and internal disease. The images show bones most clearly, as well as some tissue structure and air spaces. Using sensitive modern scanning techniques, cross-sectional outlines of the body can be obtained showing organs, blood vessels, and diseased parts. *See also* COMPUTERIZED AXIAL TOMOGRAPHY (CAT)

radioisotope ISOTOPE that undergoes RADIOACTIVE DECAY. Radioisotopes occur in nature along with the stable isotopes of an ELEMENT, but they can also be made artificially by bombarding other isotopes with high-energy ELEMENTARY PARTICLES in a NUCLEAR REACTOR or a PARTICLE ACCELERATOR. Radioisotopes are formed naturally in the atmosphere as the result of collisions between ATOMS in the air and COSMIC RADIATION. Among the elements with radioactive isotopes of economic importance are COBALT, HYDROGEN, PLUTONIUM, RADIUM, STRONTIUM, and URANIUM.

radioisotope tracer RADIOISOTOPE of an ELEMENT used to trace the course of that element through a biochemical or chemical system – a process called LABELING. As the stable ISOTOPE and radioisotope behave identically from a chemical point of view, the radioisotope will follow exactly the same path as the stable isotope, but its presence in the cells and tissues of the organism can be monitored by the RADIATION it emits. Compounds labeled with radioisotopes are used in research and diagnosis.

radiology Medical specialty concerned with the use of RADIATION and radioactive materials in the diagnosis and treatment of DISEASE. *See also* RADIOGRAPHY; RADIOTHERAPY

radiometer Instrument for detecting and measuring ELECTROMAGNETIC RADIATION, especially INFRARED RADIATION. The **Crookes radiometer** was invented (1875) by the British chemist and physicist William CROOKES. It has pivoted vanes that rotate (in a partial VACUUM) when exposed to light or heat, because the vanes are painted black on one side and the sides absorb different amounts of radiant energy. More sophisticated radiometers can measure radiation in any part of the ELECTROMAGNETIC SPECTRUM. *See also* GEIGER COUNTER

radiosonde Meteorological instrument carried up into the atmosphere by a BALLOON. It monitors weather conditions and transmits the information by RADIO to scientists on the ground. When the balloon reaches a height of 30km (18mi) above the Earth it bursts, and the radiosonde returns to the ground by parachute. It is tracked by radar, in the hope that it can be recovered and reused.

radio source Discrete astronomical source of RADIO WAVES, detected by RADIO TELESCOPE. Radio sources include the Sun, SUPERNOVA remnants, and PULSARS, QUASARS, and certain NEBULAE and GALAXIES, many of which have been identified optically. The strongest sources are the ANDROMEDA GALAXY, Perseus, Taurus A, Puppis A, Virgo A, Centaurus A, Cygnus A, and Cassiopeia A.

radio telescope Instrument used to collect and record radio waves from space. The basic design is the large single **dish** or parabolic reflector, up to 100m (330ft) in diameter. The smoothness and accuracy of the reflecting surface determines the shortest wavelength that can be detected, since surface irregularities must be no more than a small fraction of this wavelength. However, for longer wavelengths, such as around 21cm, open wire mesh can be used for the reflector. Radio waves are reflected by the dish possibly via a secondary reflector, to a **focus**, where they are converted into electrical signals. The signals are amplified and sent to the main control room, where there is further amplification before analysis and recording. A radio telescope needs to have high **sensitivity** (the ability to detect faint signals), and high RESOLUTION (the ability to see fine detail). The largest individual dishes, such as those at Effelsberg Radio Observatory, Jodrell Bank, and Parkes Observatory, are very sensitive. Better resolution can be obtained by linking two or more dishes to form a radio INTERFEROMETER, or by combining many of them to form an APERTURE synthesis array.

radiotherapy In medicine, the use of RADIATION to treat tumors or other pathological conditions.

radio waves Form of very high wavelength ELECTROMAGNETIC RADIATION. Radio waves are identified by their FREQUENCIES, expressed in kilohertz (kHz), megahertz (MHz), or gigahertz (GHz). Audio waves are lower-frequency waves. Signals broadcast in the AMPLITUDE MODULATION (AM) band range from 540kHz to 1,800kHz. FREQUENCY MODULATION (FM) broadcast frequencies range from 88MHz to 108MHz. Video, television, or digital information is broadcast at VERY HIGH FREQUENCY (VHF) or ULTRA HIGH FREQUENCY (UHF).

radium (symbol Ra) White, radioactive, metallic element of the ALKALINE-EARTH METALS, first discovered in pitchblende (1898) by Pierre and Marie CURIE; the metal, which is present in uranium ores, was isolated by Marie CURIE in 1911. Radium has been used in medical RADIOTHERAPY to treat cancer. It has 16 isotopes and they emit alpha, beta, and gamma radiation, as well as heat. The gas RADON is a decay product. Properties: at.no. 88; r.a.m. 226.025; r.d. 5.0; m.p. 700°C (1,292°F); b.p.

1,140°C (2,084°F); most stable isotope ^{226}Ra (half-life 1,622 yr).

radius In anatomy, one of the two forearm bones, extending from the elbow to the wrist. The radius rotates around the ULNA, the other forearm bone, permitting the hand to rotate and be flexible. A projection just above the thumb side of the wrist marks the end of the radius.

radius In geometry, the distance or a line from the center of a circle or sphere to any point on the CIRCUMFERENCE of the circle or surface of the sphere.

radon (symbol Rn) Radioactive, nonmetallic, gaseous element, a NOBLE GAS. It was first discovered (1899) by Ernest RUTHERFORD. The 20 known ISOTOPES, which are ALPHA PARTICLE emitters, are present in the Earth's atmosphere in trace amounts. Radon is mainly used in medical RADIOTHERAPY to treat cancer. Chemically, the element is mostly inert but does form compounds such as FLUORIDES. Properties: at.no. 86; r.d. 9.73; m.p. −71°C (−95.8°F); b.p. −61.8°C (−79.24°F); most stable isotope ^{222}Rn (half-life 3.8 days).

radula Structure found in the mouths of some MOLLUSKS with which they rasp their food. It con-

◄ **radio telescope** Radio telescopes are used to record radio emissions from distant stars, galaxies and other celestial bodies in the Universe that are too distant to be visible with optical telescopes. In these cases astronomers have to rely on longer wavelength radiation that can penetrate the Earth's atmosphere (A). Because of the much longer wavelengths of the radio signals, to resolve distant objects (1) large dishes that can track and follow a particular area of sky are needed (2). The largest dishes are fixed, and observations need to coincide with the rotation of the Earth. To overcome this problem, it is possible to use a series of dishes in an array (called a radio interferometer) many kilometers across (3) and combine the results from each dish electronically to make a synthetic large "dish" (4).

Distant objects in the Universe emit radio waves of different lengths. A false-color image builds a "picture" of the object sists of a tonguelike strip of horny material (modified EPITHELIUM), which bears many rows of chitinous or horny teeth. In slugs and snails, it is used to rasp off pieces of leafy food. In some marine mollusks, it is modified for boring or scraping.

railroad (railway) Form of transport in which carriages (wagons or bogies) run on a fixed track, usually steel rails. Railroads date from the 1500s, when wagons used in mines were drawn by horses along tracks. In 1804 English mining engineer Richard TREVITHICK built the first steam LOCOMOTIVE. In 1825 George STEPHENSON's *Locomotion* became the first steam locomotive to pull a passenger train, on the Stockton and Darlington Railway. In 1830 the first full passenger-carrying railroad, the Liverpool and Manchester Railway, was opened, using Stephenson's *Rocket*. The growth of rail fed the Industrial Revolution. In 1830 *Tom Thumb* became the first steam locomotive to be produced in the US. The first transcontinental railroad was completed in 1869, when the Union Pacific Railroad from Nebraska met the Central Pacific Railroad from California in Utah. The vast distances encouraged the development of sleeping cars, and George PULLMAN's design came into service in 1865. In 1851 India started to build what is now the world's largest rail network. By the 1870s, railroads were a principal means of transport in Europe, North America, India, Russia, and Australia, and were emerging in such countries as Japan, South Africa, and New Zealand. Important technological developments included George WESTINGHOUSE's air brake in 1872. The world's first SUBWAY to carry passengers was the City and South London Railway in 1890. Many cities adopted the idea. STEAM ENGINES are still used in India, but most countries use electric, diesel, diesel-electric, or gas turbine locomotives. Modern developments include high-speed trains, such as the Japanese "Bullet" train or the French TGV (Train à Grande Vitesse), that travel at an average speed of *c*.300km/h (185mph). **Maglev** (magnetic levitation) trains use magnetic forces to hold them above a guide rail. **Air** trains hover above the track by an air cushion. Trains that do not run on wheels are usually propelled by the magnetic forces of a LINEAR MOTOR, or by jet engine.

rain Water drops that fall from the Earth's atmosphere to its surface, as opposed to FOG or DEW, which drift as suspensions, and SNOW or HAIL, which fall in the form of ice particles. CLOUDS are aggregates of minute droplets of moisture, and raindrops are formed when these enlarge by further CONDENSATION and by coalescence with other drops as they fall. Warm air passing over the sea absorbs water vapor and rises in thermal currents, or on reaching a mountain range. The water vapor condenses and forms clouds, which account for the usually heavier annual rainfall on windward, compared to leeward, mountain slopes. Rainfall is measured by a RAIN GAUGE. *See also* PRECIPITATION

rainbow Bright, multicolored band, usually seen as a circular or partial arc formed opposite to the Sun or other light source. The **primary** bow is the one usually seen; in it the colors are arranged from red at the top to violet beneath. A **secondary** bow, in which the order of the colors is reversed, is sometimes seen beyond the primary bow. The colors are caused by reflection of light within spherical drops of falling RAIN, which causes white light to be dispersed into its constituent wavelengths. The colors usually seen are those of the visible SPECTRUM: red, orange, yellow, green, blue, indigo, and violet.

rain forest Dense forest of tall trees that grows in hot, wet regions near the Equator. The main rainforests are in Africa, central and s America, and SE Asia. They comprise 50% of the timber growing on Earth and house 40% of the world's animal and plant species. They also, through PHOTOSYNTHESIS, supply most of the world's oxygen. This is why the present rapid destruction of the rain forests (up to 20 million hectares are destroyed annually to provide timber and land for agriculture) is a cause of great concern. Also, clearing rain forests contributes to the GREENHOUSE EFFECT and may lead to GLOBAL WARMING. There are many species of broad-leaved EVERGREEN trees in rain forests, which grow up to 60m (200ft) tall. The crowns of other trees, up to 45m (150ft) tall, form the upper canopy of the forest. Smaller trees form the lower canopy. Climbing vines interconnect the various levels, providing habitats for many kinds of birds, mammals, and reptiles. Very little light penetrates to the forest floor, which consequently has few plants. Rain forest trees provide many kinds of food and other useful materials, such as Brazil nuts, figs, and mangosteens, as well as fibrous kapok and the drugs quinine and curare.

rain gauge Instrument used for measuring rainfall and other forms of PRECIPITATION. There are many different types, including some self-recording varieties. Most rain gauges consist of a cylinder that stands up above the ground surface to prevent splashing water bouncing in. Inside the cylinder, there is a funnel to lead the water into a collecting jar. The water collected is poured into a measuring cylinder, and the total rainfall can be measured, generally in millimeters or inches.

rainwash Process in which drops of rainwater hit rock and soil, washing away tiny particles. It contributes significantly to the EROSION of the sides of valleys, especially if they have no covering vegetation, and helps to form V-shaped valleys in areas of significant rainfall. In the absence of rainwash, as in deserts, the sides of valleys carved out by rivers remain very steep and form wadis, gorges, or canyons.

Rainwater, Leo James (1917–) US physicist who proposed a new theory of atomic structure in 1950, following the discovery that atomic nuclei are not spherical. He postulated that most nuclear PARTICLES form an inner NUCLEUS while the remaining particles form an outer nucleus, the shape of each set of particles affecting the other set. Rainwater believed that if some of the outer particles moved in similar orbits, sufficient energy could be created from their interaction to permanently deform an otherwise symmetrical nucleus. By 1953, the Danish physicists Ben MOTTELSON and Aage Niels BOHR had provided experimental evidence to support his theory. For this research, which made NUCLEAR FUSION a practical possibility, Rainwater, Mottelson, and Bohr shared the 1975 Nobel Prize for physics. *See also* NUCLEAR ENERGY; RELATIVE ATOMIC MASS (R.A.M.)

raised beach Elevated beach that is above sea level because of an uplift of the land or a retreat of the sea. A raised beach is often bounded on its inland edge by old cliffs, sometimes with caves. The uplift of land may have been the result of tectonic movements or the result of isostatic readjustment of the land after ice has melted. When the weight of ice has been removed by melting, the land gradually rises, relative to the sea. The same effect can be produced by the retreat of the sea, and this occurred when snow and ice increased during the glacial phases: as water becomes locked in ice and snow there is less water in the sea and it retreats from the beach.

RAM (Random Access Memory) INTEGRATED CIRCUITS (chips) that act as a temporary store for COMPUTER PROGRAMS and DATA (information). To run a program on a computer, the program is first transferred from a MAGNETIC DISK, or other storage device, to RAM. RAM also holds documents produced when the program is used. Another part of RAM stores the images to be displayed on the screen. The contents of RAM are lost when the computer is switched off.

Raman, Sir Chandrasekhara Venkata (1888–1970) Indian physicist who greatly influenced and contributed to the growth of science and research facilities in India. He was professor of physics (1917–33) at Calcutta University. He was director of the Indian Institute of Science, Bangalore, and in 1946 founded and directed the Raman Institute. Raman received the 1930 Nobel Prize for physics for his research on the DIFFUSION of light, and his discovery of the RAMAN EFFECT.

Raman effect Slight change in the frequency of monochromatic (single-wavelength) LIGHT that has been scattered. The effect, discovered by C.V. RAMAN, is seen as secondary spectral lines on each side of the primary spectral line. *See also* ELECTROMAGNETIC RADIATION; SCATTERING, LIGHT

ramjet engine Aircraft JET ENGINE, a reaction engine that relies on the speed of the aircraft to ram air (for COMBUSTION of a HYDROCARBON fuel) into the intake, thereby compressing it. The only moving part is the variable diameter intake, or diffuser, which varies the volume, and therefore the pressure, of the incoming air. The remainder of the engine is an elongated tube which forms the combustion chamber. The engine operates best at speeds above that of sound; it must first be accelerated by other means (such as solid-fuel ROCKETS) until sufficient pressure builds up.

Ramsay, Sir William (1852–1916) British chemist. Working with Lord RAYLEIGH he discovered ARGON in air. Later, he discovered HELIUM, NEON, and KRYPTON. Ramsay was awarded the 1904 Nobel Prize for chemistry. He was also interested in RADIOACTIVITY and showed that helium is produced by RADIOACTIVE DECAY.

Ramsden, Jesse (1735–1800) British maker of astronomical instruments. He made significant improvements to the THEODOLITE, BAROMETER, MICROMETER, and SEXTANT.

random access memory (RAM) *See* RAM

range finder Optical attachment, most commonly to guns and cameras, used for measuring distances. Optical range finders create images of the target as viewed from two positions; the amount of rotation or other movement needed to make the images merge indicates the range of the target. More advanced range finders measure the time it takes a pulse of light from a LASER to reach the target and return.

Ranger First series of US SPACE PROBES to the Moon. After several failures, in 1964 and 1965 Rangers 7, 8, and 9 transmitted thousands of photographs of the lunar surface as they approached predetermined impact points. The photographs showed details of boulders and craters as small as 1m (3ft), a much higher resolution than that attained by Earth-based TELESCOPES.

Rankine, William John Macquorn (1820–72) Scottish engineer and physicist. In 1855 he became professor of engineering at Glasgow University, where he made valuable contributions to civil and mechanical engineering and THERMODYNAMICS. Rankine wrote manuals on these subjects and also devised the absolute TEMPERATURE SCALE based on the degree FAHRENHEIT, known as the Rankine scale.

Raoult's law VAPOR PRESSURE of a SOLUTION is the sum obtained by multiplying the vapor pressure of each component by the MOLE fraction of that component and adding these products. (The mole fraction is the number of moles of the component divided by the total number of moles present.) The law holds only approximately and over a limited range of conditions, differing for different substances. Discovered (1886) by the French physical chemist François-Marie Raoult (1830–1901), the law is used to find the RELATIVE MOLECULAR MASS of a substance and has been fundamental to many theories of solution.

rapid eye movement (REM) Phenomenon observed in sleeping people. REM has been calculated to take up to 20% of a normal night's SLEEP and has been associated with the time spent dreaming. The sleeper passes in and out of REM sleep and, although it is possible to make do with very little sleep, he or she suffers if deprived of REM sleep.

rare-earth elements *See* LANTHANIDE SERIES

rare gas *See* NOBLE GAS

ratchet Mechanism that permits leverage in one direction only. It consists of a wheel or bar with teeth sloping in one direction, which is engaged by a pawl so that the wheel or bar is unable to rotate backwards. Ratchets are often parts of MACHINE TOOLS in which nonslip movement is essential. The rear sprocket of a free-wheelable bicycle incorporates a ratchet. *See also* LEVER

ratio Measure of relative size of two quantities. Ratios can be expressed using a colon. For instance, if a classroom contains twenty girls and ten boys, the ratio of girls to boys is 20:10 or 2:1 (read two to one). Ratios can also be expressed as fractions (or percentages). In this case $\frac{1}{3}$ of the pupils are boys and $\frac{2}{3}$ are girls.

rational number *See* NUMBER, RATIONAL

ratite Any of a group (Ratitae) of large, usually flightless birds with flat breastbones instead of the keel-like prominences found in most flying birds. Ratites include the ostrich, rhea, cassowary, emu, kiwi, and the unusual flying tinamou.

Ray, John (1627–1705) English naturalist whose work on plant and animal classification later influenced Carolus LINNAEUS and Georges CUVIER. Ray was the first to distinguish the two main types of flowering plants: MONOCOTYLEDONS and DICOTYLEDONS.

Rayleigh, John William Strutt, Lord (1842–1919) British physicist. His work was chiefly concerned with various forms of wave motion. Rayleigh was awarded the 1904 Nobel Prize for physics for his discovery (with the chemist William RAMSAY) of the noble gas ARGON and for work on gas densities.

rayon Fine, smooth, SYNTHETIC FIBER made from solutions of CELLULOSE from soft wood pulp or cotton linters (short fibers). The first SYNTHETIC FIBER, rayon was produced (1884) by the French chemist Hilaire de Chardonnet (1839–1924). VISCOSE rayon, the most common, is spun-dried and has a strength approaching that of NYLON. ACETATE rayon is made of filaments of cellulose acetate, while triacetate rayon has filaments of triacetate. The more costly cuprammonium rayon, known as Bemberg, resembles silk.

reactant *See* CHEMICAL REACTION

reaction propulsion Propulsion utilizing the third of Isaac NEWTON's laws of MOTION: for every action there is an equal and opposite reaction. Both ROCKETS and JET ENGINES build up internal pressure from gases or PLASMAS that are allowed to escape in one direction, creating propulsion in the opposite direction.

reaction rate In chemistry, speed at which CHEMICAL REACTIONS move to completion or to EQUILIBRIUM. The rate is accelerated by one or more of the following: the presence of a CATALYST or a promoter, heat, pressure, or REDUCTION of the concentration of the reaction products.

reaction time Interval between the onset of a STIMULUS and the conscious execution of a response.

read only memory (ROM) *See* ROM

reagent Substance that takes part in a CHEMICAL REACTION. The term is often understood to mean a common laboratory reagent, that is, a substance used in routine qualitative or quantitative analytical tests.

real number *See* NUMBER, REAL

real-time computing Description of any COMPUTER application that reacts to things as they happen. DATA is received, processed and a response given within a time set by outside events. In human terms, the reply is virtually immediate. Real-time systems are employed in air-traffic control, automatic banking, automatic pilots on aircraft, industrial robotics, and many military applications.

Réaumur, René Antoine Ferchault de (1683–1757) French scientist best known for his THERMOMETER scale, which designates zero as the freezing point of water and 80° as its boiling point. He wrote widely on natural history and conducted research in mining, metallurgy, fossils, and insects.

Reber, Grote (1911–) US radio engineer who built (1937) the first astronomical RADIO TELESCOPE, acting on the discoveries made by Karl JANSKY five years earlier. With his 9-metre steerable dish aerial, Reber found several non-optical radio sources in the Milky Way and paved the way for subsequent developments in RADIO ASTRONOMY. His later work consisted of trying to disprove the BIG BANG theory.

recapitulation Biological theory which says that the embryonic development of a species repeats the evolutionary development of that species. It was called the biogenetic law by Ernst HAECKEL.

receptacle Biological structure that serves as a container for reproductive cells or organs in plants. In ANGIOSPERMS (flowering plants), the receptacle is the enlarged end of a stalk to which the FLOWER is attached. In FERNS, it is the mass of tissue that forms the sporangium (the SPORE-bearing organ). In some SEAWEEDS, it is the part that seasonally becomes swollen and carries the reproductive organs.

receptor In higher organisms, tiny organs found on both NEURONS (nerve cells) and ENDOCRINE cells. They detect physical stimuli, such as light, sound, touch, and taste, and chemical changes occurring inside the body and on or near its surface. They then transfer the physical or chemical signals into electrical energy or the secretion of HORMONES. For example, the RODS AND CONES of the RETINA of the eye are receptors that transform light energy into the nerve impulses sent via SENSORY NEURONS to the brain. Others, such as those in muscles, the ear, and visceral organs, tell the brain how the body is positioned and how it moves.

recessive In genetics, describes a form of GENE (ALLELE) that does not express itself when paired with a DOMINANT allele. Although it is part of the GENOTYPE (genetic make-up) of a HETEROZYGOTE (an organism containing both dominant and recessive alleles for a particular gene), it does not contribute to the PHENOTYPE (physical characteristics) of the organism. It becomes expressed only when the same recessive allele form appears on both CHROMOSOMES of a HOMOZYGOTE. Recessive alleles were discovered by Gregor MENDEL, who found that a cross between purebred red- and white-flowering garden peas always produced red flowers in the offspring. The allele for red coloration is dominant; the allele for white is recessive. In humans, the alleles for blue eyes are recessive to those of brown eyes; alleles for HEMOPHILIA are also recessive.

reciprocal Name given to the quantity that results when a number is divided into 1 (unity). For example, the reciprocal of 2 is 1/2, the reciprocal of 1/2 is 2, the reciprocal of 8 is 1/8 (or 0.125) and the reciprocal of 3/7 is 7/3 and so on.

reciprocating engine Any ENGINE in which a piston moves to and fro, such as a STEAM ENGINE or an INTERNAL COMBUSTION ENGINE, commonly used to power motor vehicles and aircraft. There are two basic designs of reciprocating internal combustion engine. An **in-line** engine has either one row of CYLINDERS or two rows of cylinders in a V arrangement or horizontally opposed. A **radial** engine, used only for aircraft, has cylinders arranged in a circle.

recombinant DNA research Branch of GENETIC ENGINEERING involving the transfer of a segment of DNA from a source organism into a host organism (typically a microbe). It is also known as GENE splicing because the transferred segment is spliced into the overall DNA structure of the host, thus altering the information contained in its GENETIC CODE. When the host undergoes asexual CELL DIVISION, each product cell carries a replica of the new DNA. In this way, numerous CLONES of the new cell can be made. Gene splicing can be used to design cells to produce medicines, to digest spilled oil or waste products, and for numerous other purposes.

recombination Process that rearranges GENES to increase genetic VARIATION in sexually produced offspring. Recombination takes place during MEIOSIS, the type of CELL DIVISION that leads to the formation of sex cells (GAMETES). It is achieved by CROSSING OVER of paired CHROMATIDS. Its effect is to "shuffle" genes derived from both parents and thereby create genetic variation. Some GENETIC ENGINEERING techniques can induce recombination artificially.

recording Storing of signals that represent sound or images, on a medium such as a plastic disc or MAGNETIC TAPE. In traditional SOUND RECORDING, the sounds being recorded are converted into variations in an electric current that controls the movements of a stylus. This cuts an original lacquer disc, which is then electroplated to make a master negative for pressing out plastic copies. In magnetic recording, sound is recorded on tape with a TAPE RECORDER; a VIDEO RECORDING machine records sound and vision. Optical recording of sound is used for the soundtrack of motion pictures. The sounds cause variations in thickness or brightness of a strip of film running alongside the picture photographs. By a reverse process this is converted back into sound when the picture is run. Modern sound recording is digital. *See also* COMPACT DISC; DIGITAL AUDIO TAPE (DAT)

rectangle Four-sided geometric figure (quadrilateral), the interior angles of which are right angles and each pair of opposite sides is of equal length and is parallel. It is a special case of a PARALLELOGRAM.

rectangular coordinate system CARTESIAN COORDINATE SYSTEM in which the axes are at right angles to each other.

rectified spirit Constant-boiling mixture of ETHANOL (alcohol) and water. It is made by distilling ethanol-water mixtures and contains 95.6% ethanol. *See also* ABSOLUTE ALCOHOL

rectifier Component of an ELECTRIC CIRCUIT that converts ALTERNATING CURRENT (AC) into DIRECT CURRENT (DC). The rectifier is usually a semiconductor DIODE that presents a high RESISTANCE to a current flowing in one direction and a low resistance to current flowing in the opposite direction.

rectum In humans and many other vertebrates, last part of the large INTESTINE, where the feces accumulate prior to voiding.

recycling Natural and synthetic processes by which substances are broken down and reconstituted. In nature, elemental cycles include the CARBON CYCLE, NITROGEN CYCLE, and HYDROLOGICAL CYCLE. Natural cyclic chemical processes include the metabolic cycles, such as the KREBS CYCLE, in the bodies of living organisms. Synthetic recycling includes using bacteria to break down excreta and many kitchen and factory organic wastes to harmless, or even beneficial, substances. Large quantities of inorganic waste, such as metal scrap, glass bottles, and building refuse, are recycled. POLYMER wastes are often burned rather than recycled, with consequent atmospheric POLLUTION and loss of material resources. Controlled burning (PYROLYSIS) recovers useful substances from PLASTICS.

red algae *See* RHODOPHYTE

red blood cell Alternative common name for an ERYTHROCYTE

red dwarf STAR at the lower end of the MAIN SEQUENCE. Red dwarfs have masses of between 0.8 and 0.08 of a solar mass. They are of small diameter, relatively low surface temperature (between 2,500–5,000K), and low absolute magnitude. BARNARD'S STAR and Proxima Centauri are examples. *See also* HERTZSPRUNG-RUSSELL DIAGRAM

red giant Type of STAR in the late stage of its evolution. They have a diameter of between 10 to 100 times that of the Sun, and a luminosity 100 to 10,000 times greater. **Low-mass** red giants usually evolve into WHITE DWARFS, while those of a **high mass** go through further nuclear fusion before becoming NEUTRON STARS or BLACK HOLES. Examples include Arcturus and Aldebaran. *See also* HERTZSPRUNG-RUSSELL DIAGRAM; STELLAR EVOLUTION

redox *See* OXIDATION-REDUCTION

red shift (symbol *z*) Lengthening of the WAVELENGTH of light or other ELECTROMAGNETIC RADIATION from a source, caused either by the source moving away (the DOPPLER EFFECT) or by the expansion of the Universe (*see* EXPANDING UNIVERSE). It is defined as the change in the wavelength of a particular spectral line, divided by the unshifted, or rest, wavelength of that line. Red shifts caused by the expansion of the Universe, called **cosmological red shifts**, have nothing to do with the Doppler effect. The Doppler effect results from motion through space, whereas cosmological red shifts are caused by the expansion of space itself, which stretches the wavelengths of light traveling toward us. The longer the travel time of light, the more its wavelength is stretched, as embodied in the HUBBLE CONSTANT. GRAVITATIONAL RED SHIFT is a phenomenon predicted by EINSTEIN's general theory of RELATIVITY. Light emitted by a star has to do work to overcome the star's gravitational field. As a result, there is a loss of energy and a consequent increase in wavelength, so that all the spectral lines are shifted towards the red.

▲ **red shift** Certain red shift effects, in which light emitted from stars moves toward the longer (red) end of the spectrum, can be explained by the Doppler effect. Just as radar (A) can calculate the position of a moving object by measuring the time a transmitted signal (1) takes to return (2), in this way the movements of stars can be measured relative to Earth. The wavelength of a star that appears to be moving neither closer to Earth nor farther away (B) does not alter, and its spectrum does not change. The wavelength of a star that is moving away from Earth lengthens (C), and moves toward the red end of the spectrum (left). The wavelength of a star moving closer to Earth (D) shortens, and moves toward the blue end of the spectrum (right).

reducing agent Substance that causes REDUCTION reactions. Thus, in the reduction of oxygen to form water, $2H_2 + O_2 \rightarrow 2H_2O$, hydrogen is the reducing agent. In biological systems, some SUGARS act as reducing agents. *See also* OXIDATION-REDUCTION

reduction Chemical reaction in which an ELECTRON is added to an ATOM or an ION. It is always part of an OXIDATION-REDUCTION reaction.

reef Rocky outcrop lying in shallow water, especially one built up by CORALS or other organisms.

refinery Any industrial plant based on processes such as DISTILLATION and ELECTROLYSIS. Spirit drinks such as whisky, foods such as SUGAR, and materials such as ores, metals or PETROLEUM are produced, purified or separated into their components in refineries. *See also* OIL REFINERY

refining Process of removing impurities from metals, sugar, petroleum, and other products. For metal ORES especially, the refining process is carried out near the MINE in order to avoid unnecessary transport costs. Iron ore is between 25% to 70% pure; a low-quality ore of only 25% contains 75 metric tons of unwanted rock for every 100 metric tons extracted.

reflection Change in direction of part or all of a WAVE. This happens when a wave, such as a light or sound wave, encounters a surface separating two different media, such as air and metal, and is partly "bounced" back into the original medium. The remaining part passes into the second medium. The incident wave (striking the surface), reflected wave, and the normal (line perpendicular to the surface) all lie in the same plane; the incident wave and reflected wave make equal angles with the normal.

reflex action Rapid, involuntary response to a particular STIMULUS – for example, the "knee-jerk" reflex that occurs when the bent knee is tapped. It is controlled by a **reflex arc** found in the CENTRAL NERVOUS SYSTEM. A reflex arc consists of a **sensory receptor** (which detects the stimulus), an **afferent** (or sensory) nerve, which conveys the impulse to the grey matter of the SPINAL CORD, and an **efferent** (or motor) nerve, which carries the response impulse to the EFFECTOR, in this case the muscles in the knee, causing the leg to jerk. This type of response is only detected by the brain once the reflex has taken place. This reflex is sometimes called a **spinal** reflex. A **cranial** reflex involves the brain, whereby the initial reflex can be overridden by the brain and either stopped or altered.

reforestation Planting of trees in a region where DEFORESTATION has occurred. Throughout the world, trees have been cut down for fuel or wood pulp, or to clear land for agriculture. On sloping land with no trees, soil EROSION soon takes place as rain washes away the soil. Replanting trees prevents erosion and provides a renewable resource of timber and fuel.

reforming Conversion of straight-chain ALKANES into branched-chain alkanes by CRACKING or by catalytic reaction. Reforming is used to produce more efficient GASOLINE, since straight-chain alkanes burn unsteadily and cause engine KNOCK.

refraction Bending of a WAVE, such as a light or sound wave, when it crosses the boundary between two transparent media, such as air and glass, and undergoes a change in speed. The incident wave (striking the surface), the refracted wave and the normal (line perpendicular to the surface) all lie in the same plane. The incident wave and refracted wave make an angle of incidence, i, and an angle of refraction, r, with the normal. The index of refraction for a transparent medium is the ratio of the speed of light in a vacuum to its speed in the medium. It is also equal to sin i/sin r. Snell's law states that this ratio is constant for a given interface. *See also* REFLECTION

refractive index Measure of the ability of a transparent medium, such as glass, to refract (bend) light. It is the ratio of the speed of light in a vacuum to its speed in the medium and is calculated by dividing the SINE of the ANGLE OF REFRACTION into the sine of the ANGLE OF INCIDENCE of the incoming light.

refractometer Instrument for measuring the REFRACTIVE INDEX of transparent materials – the degree to which they bend light that passes through them. Since the refractive index of a chemical solution changes according to its composition, refractometers are frequently employed in chemical analysis.

refrigeration Process by which the temperature in a REFRIGERATOR is lowered. In a domestic refrigerator, a refrigerant gas such as AMMONIA or CHLOROFLUOROCARBON (CFC) is alternately compressed and expanded. The gas is first compressed by a pump, causing it to warm up. It is then cooled in a CONDENSER where it liquefies. It is then passed into an evaporator where it expands and boils, absorbing heat from its surroundings and thus cooling the refrigerator. It is then passed through the pump again to be compressed. *See also* AIR-CONDITIONING

refrigerator Heat-insulated container, maintained at a low temperature for preserving perishable foods and other materials. The refrigerator was invented (1859) by Ferdinand Carré (1824–1900) for preserving meat during the long sea voyage from Australia to Europe. In 1913 the domestic refrigerator appeared in the US, and was marketed in Britain in the early 1920s. In domestic refrigerators, power for the cooling process, called REFRIGERATION, is provided by electricity or gas. Some commercial refrigerators use a steam jet to power the cooling process.

regeneration Biological term for the ability of an organism to replace one of its parts if it is lost. Regeneration also refers to a form of ASEXUAL REPRODUCTION in which a new individual grows from a detached portion of a parent organism.

regolith Loose fragments of rock between the layers of soil and the BEDROCK beneath. The fragments are usually thin and have broken away from the bedrock. Eventually they break down and become mineral particles in the soil. *See also* SOIL HORIZON

regulatory gene Any GENE that controls the way in which another gene is produced. Regulatory genes activate or deactivate a group of neighboring genes called an OPERON, which functions as a unit. Usually found in BACTERIA, operons are responsible for the formation of ENZYMES that control various processes in a metabolic pathway.

Reichstein, Tadeus (1897–1996) Swiss chemist and pharmacologist, b. Poland. After teaching chemistry (1922–38) at the Institute of Technology in Zurich, he became head of the department of pharmacy at the University of Basel. Reichstein shared the 1950 Nobel Prize for physiology or medicine with Edward KENDALL and Philip S. HENCH for his work on the HORMONES of the cortex of the ADRENAL GLANDS. He was the first to synthesize (1933) ascorbic acid (vitamin C).

Reines, Frederick (1918–1998) US physicist. In the 1950s he proved that radiation from nuclear reactors contains NEUTRINOS. He developed detectors for these elusive particles and went on to discover them in COSMIC RADIATION. In recognition of this work Reines shared the 1995 Nobel Prize for physics with Martin Perl for his discovery of the TAU PARTICLE.

Reis, Johann Phillip (1834–74) German scientist who invented (*c*.1861) an early form of the TELEPHONE. Its capacity to reproduce speech, however, was limited and the first practical telephone was patented by Alexander Graham BELL in 1876.

rejection In transplantation, the IMMUNE SYSTEM's response during which a grafted tissue or organ is destroyed for being "foreign" tissue. **Acute** rejection mostly occurs within the first few weeks after surgery and can be halted with IMMUNOSUPPRESSIVE DRUGS. **Chronic** rejection, a long-term process resulting in gradual failure of the grafted part, is irreversible.

relative atomic mass (r.a.m.) (formerly atomic weight) Mass of an ATOM of the naturally occurring form of an ELEMENT divided by 1/12 of the mass of an atom of carbon-12. The naturally occurring form may consist of two or more ISOTOPES, and the r.a.m. takes this into account in averaging.

relative density (r.d.) (formerly specific gravity) Ratio of the DENSITY of a substance to the density of water. Thus, the relative density of gold is 19.3; that is it is *c*.19 times denser than water.

relative humidity *See* HUMIDITY

relative molecular mass (formerly molecular weight) Ratio of the average mass per MOLECULE of an element or compound to 1/12th of the mass of an ATOM of carbon-12. The molecular masses of REACTANTS (elements or compounds) must be known in order to make calculations about yields in a chemical reaction.

relativity Theory, proposed by Albert EINSTEIN, based on the postulate that the motion of one body can be defined only with respect to that of a second body. This led to the concept of a four-dimensional SPACE-TIME continuum in which the three space dimensions and time are treated on an equal footing. The **special theory**, put forward in 1905, is limited to the description of events as they appear to observers in a state of uniform relative motion. The more important consequences of the theory are: (1) that the velocity of light is constant for all observers; (2) that the mass of a body increases with its velocity, although appreciably only at velocities approaching that of light; (3) that mass (m) and energy (E) are equivalent, that is, $E = mc^2$, where c is the velocity of light (this shows that when mass is converted to energy, a small mass gives rise to large energy); (4) the LORENTZ-FITZGERALD CONTRACTION, that is bodies contract as their velocity increases, again only appreciably near the velocity of light; and (5) relative to a stationary observer, time passes more slowly for a moving object, "time dilation." The **general theory**, completed in 1915, is applicable to observers not in uniform relative motion (that is accelerating). This showed the relation of space and GRAVITATION. It may be put as the idea that the presence of MATTER in space causes space to "curve," forming GRAVITATIONAL FIELDS; thus gravitation becomes a property of space itself. Light is also bent by massive gravitational fields, which may explain the existence of BLACK HOLES. *See also* SPACE-TIME *See illustration on page 444*

relay Electromagnetic switch that is operated (opened or closed) by variations in an ELECTRIC CURRENT in a coil of wire. Current flowing through the wire coil sets up a MAGNETIC FIELD that moves a metal contact that can open or close the CIRCUIT.

releasing hormone HORMONE secreted by the HYPOTHALAMUS in the brain that travels to the nearby PITUITARY GLAND where it stimulates the release of another hormone. Each of the pituitary hormones has a specific releasing hormone (also called a releasing factor). For example, the production of thyrotrophin (thyroid-stimulating hormone, TSH) by the pituitary gland is triggered by thyrotrophin-releasing hormone from the hypothalamus.

▲ **relativity** Einstein's theories of relativity are based on the postulate that the motion of one body can be defined only with respect to a second body. For example (A) shows how a stationary observer sees the sports car moving at 90mph and the saloon car at 70mph. To the saloon car, however, the sports car is moving at 20mph relative to it.

Further consequences of relativity show the velocity of light is absolute. In (B), even though the supercar is travelling hundreds of mph faster, the speed of light emitted from the headlamps of both vehicles is identical, 186,000mi s^{-1}. Nuclear explosions (C) demonstrate that mass (m) and energy (E) are equivalent, that is $E = mc^2$. A small amount of

mass can produce huge amounts of energy. (D) shows how, when approaching the speed of light, a body's mass increases while at the same time the body contracts. Finally E demonstrates "time-dilation." Cosmic-ray muons are unstable, and at rest decay in two nanoseconds. They should therefore only penetrate

c.600m (2,000ft) of Earth's atmosphere before breaking down. However, muons have been detected in bubble chambers (F) 10km (6mi) below sealevel. According to relativity time runs slower for objects with speed close to that of light. To observers on Earth, therefore, it appears that a fast-moving muon lives longer than a stationary one.

relief printing Form of PRINTING whereby a print is achieved from a raised, inked surface; the ink-bearing surface stands out above the surrounding non-printing area. *See* LETTERPRESS PRINTING

REM Acronym for RAPID EYE MOVEMENT

remote sensing Any method of obtaining and recording information from a distance. The most common sensor is the CAMERA; cameras are used in aircraft, satellites, and space probes to collect information and transmit it back to Earth (often by RADIO). The resulting photographs provide a variety of information, including archaeological evidence and weather data. MICROWAVE sensors use RADAR signals that can penetrate clouds. Infrared sensors can measure temperature differences over an area. Data from sensors can be processed by computers.

renewable energy (alternative energy) ENERGY from a source that can be replenished or that replenishes itself and is generally environmentally less harmful than "traditional" energy forms such as COAL, NATURAL GAS or NUCLEAR ENERGY. SOLAR ENERGY harnesses the rays of the Sun. TIDAL POWER stations make use of the gravitational force of the Sun and Moon on the ocean. Wave power harnesses the natural movement of the sea. The power of rivers and lakes can also be tapped, usually by damming the flow and using turbines to generate HYDROELECTRICITY. WIND POWER devices have existed for centuries in the form of WIND-MILLS. A less well-known renewable source is GEOTHERMAL ENERGY produced in the Earth's crust. *See also* ENERGY SOURCES

renin Polypeptide HORMONE produced by the KIDNEY in response to lowered blood pressure. It activates ANGIOTENSIN (a powerful VASOCONSTRICTOR), which reduces venous and arterial volume with a subsequent rise in blood pressure.

Rennell, James (1742–1830) English cartographer, geographer, and oceanographer. He was an expert on the geography of W Asia and N Africa, and constructed the most accurate map of India of his time (1783). Rennell pioneered the scientific study of winds and ocean currents. *See also* OCEANOGRAPHY

reproduction Process by which living organisms create new organisms similar to themselves. Reproduction may be **sexual** or **asexual**, the first being the fusion of two special reproductive CELLS from different parents, and the second being the generation of new organisms from a single organism. ASEXUAL REPRODUCTION is the more limited, found mainly in PROTOZOA, some INVERTEBRATES, and in many plants. By contrast, almost all living organisms have the capacity for SEXUAL REPRODUCTION. In the majority

▲ **remote sensing**
Remote sensing satellites (1) view the Earth from space using various sensors and cameras. These instruments are classed as either active or passive. Passive devices, such as optical cameras and infrared scanners, pick up reflected radiation (either light or heat). Active instruments on the other hand send out radio pulses and record the returning signal. One of the strengths of active scanning is the ability to see through cloud. Powered by a solar sail (2), the satellites use orbits which take them above the whole of the Earth during a series of days (3). Images of the Earth's surfaces are beamed down to ground stations (4) in digital form (5), and are converted into pictures by computers (6).

of cases, the species has two kinds of individuals – male and female – with different sex functions. Male and female sex cells (in animals, SPERM and OVA) fuse to produce a new cell, the ZYGOTE, which contains GENETIC information from both parents, and from which a new individual develops. Alternatively, organisms may be HERMAPHRODITE (like the earthworm), each individual of the species having male and female functions, so that when two of them mate each individual fertilizes the other's eggs. Usually, animals that can reproduce asexually have alternating (not always regularly) generations of asexual and sexual individuals. Examples are the aphid, jellyfish, and some parasitic worms. Some plants have alternating life cycles. Sexually reproducing plants (or generations) are called GAMETOPHYTE; ones which reproduce asexually, SPOROPHYTE. *See also* ALTERNATION OF GENERATIONS; PARTHENOGENESIS

reproductive system Organs of a plant or animal necessary for SEXUAL REPRODUCTION. In plants, the reproductive system comprises the STAMENS and CARPALS. In mammals, the male system comprises the TESTES, the EPIDIDYMIS, and the PENIS, while the female system consists of the OVARIES, the UTERUS, and the FALLOPIAN TUBES.

reptile Any one of about 6,000 species of VERTEBRATES distributed worldwide. Reptiles are POIKILOTHERMIC (cold blooded). Most lay yolky eggs on land. Some species – particularly snakes – carry eggs in the body and bear live young. The skin is dry and covered with scales or embedded with bony plates. Their limbs are poorly developed or nonexistent. Those with limbs usually have five clawed toes on each foot. There are now four living orders: Chelonia (turtles); Rhynchocephalia (tuatara); Squamata (scaly reptiles such as snakes and lizards); and Crocodilia (alligators and crocodiles).

resin (rosin) Artificial or natural POLYMER that is generally viscous and sticky. **Artificial** resins, such as POLYESTERS and EPOXY RESINS, are used as ADHESIVES and binders. **Natural** resins are secreted by various plants. Resin is impermeable to water and when present in large amounts, such as in pine, makes wood resistant to rot and weather. Oleoresin, secreted by conifers, is distilled to produce TURPENTINE; ROSIN (a yellow, brown or black material) remains after the oil of turpentine has been distilled off. AMBER is fossilized resin.

resistance (symbol R) Property of an electric CONDUCTOR, calculated as the ratio of the VOLTAGE applied to the conductor to the current passing through it. In a conductor, this represents the opposition to the flow of ELECTRONS; the electrical energy is converted into HEAT. The SI unit of resistance is the OHM. In an ALTERNATING CURRENT circuit, resistance is the real part of the IMPEDANCE. *See also* RESISTOR

resistivity (symbol ρ) Electrical property of materials. Its value is given by $\rho = AR/l$, where A is the cross-sectional area of a CONDUCTOR, l is its length and R is its RESISTANCE in OHMS. As the temperature of the CONDUCTOR rises, the resistivity usually also rises. Resistivity is also called specific resistance and its SI UNIT is the ohm-meter.

resistor Electrical or ELECTRONIC CIRCUIT component with a specified RESISTANCE. Resistors limit the size of the current flowing in a circuit. Those for electronic circuits usually consist of finely ground CARBON particles mixed with a ceramic material and enclosed in an insulated tube. The value of the resistance is given by a coded set of colored rings on the outside of the tube. Resistors for carrying larger currents are coils of insulated wire.

resolution In chemistry, the separation of an optically inactive MIXTURE into its optically active components (*see* OPTICAL ACTIVITY). If the components have a different CRYSTAL shape, they can be separated by hand (using a microscope). More usually, the mixture is reacted with an optically active COMPOUND to produce two new nonactive compounds that can be separated. In another technique, bacteria are used that react with one component of the mixture but do not react with the other.

resolution In electronics, ability of an imaging system, such as a COMPUTER or TELEVISION screen, to distinguish between two closely spaced objects. In computing, it is expressed as the number of dots per unit length in which an image can be formed. The usual unit is **dpi** (dots per inch) and a typical value for a color screen is 80dpi (this compares with 150dpi for the resolution of the human eye at normal reading distance and up to 2,400dpi for printed illustrations in books). The total resolution of a computer screen is expressed as the number of pixels in each direction, typically 640×480 pixels.

resonance Increase in the amplitude of vibration of a mechanical or acoustic system when it is forced to vibrate by an external source. It occurs when the FREQUENCY of the applied FORCE is equal to the natural vibrational frequency of the system. Large vibrations can cause damage to the system.

resonance, particle Extremely short-lived ELEMENTARY PARTICLE produced in high-energy nuclear reactions occurring in particle ACCELERATORS. Such particles decay in about 10^{-24} seconds, which is characteristic of HADRONS. More than 150 have been detected, all since the 1960s. A resonance can be considered as an increase in the probability of interaction between two energetic particles, when the particles' combined energy reaches the "resonance energy."

respiration Series of chemical reactions by which complex MOLECULES are broken down to release ENERGY in living organisms. These reactions are controlled by ENZYMES and are an essential part of METABOLISM. There are two main types of respiration: AEROBIC and ANAEROBIC. In **aerobic** respiration, OXYGEN combines with the breakdown products and is necessary for the reactions to take place. **Anaerobic** respiration takes place in the absence of oxygen. Aerobic respiration results in a more complete breakdown of the original substrate, so it releases more energy than anaerobic respiration. If all the energy were released in a single burst, it would probably disrupt CELL reactions. Instead, energy is released in small amounts at different stages in respiration. In most living organisms, the energy released by respiration is used to convert ADENOSINE DIPHOSPHATE (ADP) into ADENOSINE TRIPHOSPHATE (ATP), a molecule that acts as a movable energy store to TRANSPORT energy around the cell. At the site where the energy is needed, ATP is converted back to ADP, with the aid of a special ENZYME, and energy is released. In plant and animal cells, the first stages of respiration take place in the CYTOPLASM and the later stages in the MITOCHONDRIA. *See also* COENZYME; GAS EXCHANGE; KREBS CYCLE; TRANSPIRATION

respiratory system System of the body concerned with GAS EXCHANGE. The respiratory tract of an air-breathing animal begins with the nose and mouth, through which air enters the body. The air then passes through the LARYNX and into the TRACHEA. At its lower end, the trachea branches into two bronchi and each BRONCHUS leads to a lung. Inside the LUNGS, the bronchi divide into many tiny bronchioles that lead in turn to bunches of tiny air sacs (ALVEOLUS), where the exchange of gases between air and blood takes place. Exhaled air leaves along the same pathway.

rest mass In RELATIVITY, the MASS of an object when it is at rest. The PHOTON is described as having a rest mass of zero, but this is misleading: the photon's speed is the speed of light in all frames of reference and it can never be brought to rest. The GRAVITON, not yet observed, would have zero rest mass, and so may the NEUTRINO.

restriction enzyme In molecular biology, an ENZYME used in GENETIC ENGINEERING to cut a molecule of DNA at specific points in order to insert or remove a piece of DNA. There are many different restriction enzymes. Each cuts the DNA at a specific sequence of BASES, allowing great precision in genetic engineering. A piece of DNA may be cut out and removed for insertion into another organism. Or the geneticist may wish to insert a different piece of DNA at that point. Pieces of DNA are joined together again with enzymes called ligases. These enzymes are also used when analysing DNA to find out its sequence of bases.

reticular formation Complex mechanism in the vertebrate CENTRAL NERVOUS SYSTEM, located in the BRAINSTEM. It consists of interconnected clusters of nerve cell bodies (gray matter) and is believed to influence many physiological functions, including sleep, blood pressure, and coordination.

reticulo-endothelial system Body system made up of reticular tissue, a specialized type of CONNECTIVE TISSUE that lines and supports the spleen, lymph nodes, and some other organs. Reticular tissue also contains CELLS that have MACROPHAGE action, destroying worn-out ERY-THROCYTES (red blood cells) in the spleen or invading organisms in the lymph nodes.

retina Inner layer of the EYE, composed mainly of different kinds of NEURONS (nerve cells), some of which are the visual receptors of the eye. Receptor cells (RODS AND CONES) are sensitive to light. **Cones** respond primarily to the spectrum of visible colors; **rods** respond mainly to shades of gray and to movement. The rods and cones connect with SENSORY NEURONS that in turn connect with the OPTIC NERVE that carries the visual stimuli to the brain.

retrorocket ROCKET engine used to slow a space vehicle, separate it from another one, place it in a desired ORBIT, or allow it to make a soft landing.

retrovirus Any of a large family of VIRUSES (Retroviridae) that, unlike other living organisms, contain the genetic material RNA (ribonucleic acid) rather than the customary DNA (deoxyribonucleic acid). In order to multiply, retroviruses make use of a special ENZYME to convert their RNA into DNA, which then becomes integrated with the DNA in the CELLS of their hosts. Diseases caused by retroviruses include ACQUIRED IMMUNE DEFICIENCY SYNDROME (AIDS) and some forms of LEUKAEMIA.

reverse osmosis Movement of a liquid (the SOLVENT) through a SEMIPERMEABLE MEMBRANE from a concentrated SOLUTION to a less concentrated solution. This movement is in the opposite direction to that in normal OSMOSIS and can be achieved only by applying external pressure to the concentrated solution. It is used in DESALINATION plants to produce drinking water from seawater. The process is generally employed only on a small scale because the pressures involved are so high (up to 25 atmospheres).

reverse transcription Method of producing a double-stranded DNA copy of RNA from a VIRUS. The technique is used often in GENETIC ENGINEERING to make DNA copies of MESSENGER RNA (mRNA). It is achieved using an ENZYME called **reverse transcriptase**, which occurs in RETROVIRUSES.

reversible process Any process that can be made to go "backwards" so that parameters defining the system change in reverse order to their original values. If there were any exchanges of ENERGY or MATTER with the surroundings during the forward process, these change in direction and order when the process is reversed. In an isolated system, all processes are irreversible, because the system's ENTROPY always increases.

reversible reaction CHEMICAL REACTION in which the products can change back into the reactants. Thus nitrogen and hydrogen can be combined to give ammonia (as in the HABER PROCESS) and ammonia may be decomposed to nitrogen and hydrogen. Such processes yield an equilibrium mixture of reactants and products. *See also* CHEMICAL EQUILIBRIUM

revolution Movement of a planet or other celestial object around its ORBIT, as distinct from ROTATION of the object on its axis. A single revolution is the planet's or satellite's "year."

Reynolds number Dimensionless quantity characterizing types of FLUID FLOW. For a body of density p and linear dimension l traveling at velocity v in a fluid of viscosity v, the Reynolds number is pvl / v. At low values of the Reynolds number, the fluid flow is laminar, or layered, and is well understood mathematically; at higher values, the flow becomes turbulent and complicated.

rhenium (symbol Re) Silver-white TRANSITION ELEMENT, first discovered in 1925. Rhenium is found in MOLYBDENITE and PLATINUM ores from which it is obtained as a by-product. It is heavy and used in thermocouples, flashlights, and filaments; it is also a useful CATALYST. Properties: at.no. 75; r.a.m. 186.2; r.d. 21.0; m.p. 3,180°C (5,756°F); b.p. 5,627°C (10,160°F); most common isotope [187]Re (62.93%).

rheology Study of the ways in which MATTER deforms and flows. It includes the investigation of such physical properties as VISCOSITY, ELASTICITY,

▲ **reversible reaction** A reversible reaction is one in which the products of one reaction can react to form the chemicals that reacted to form the products. For example, if steam (1) is passed over iron that is being heated in an enclosed tube (2), a reaction will take place producing iron oxide (rust) (3) and hydrogen (4). If hydrogen is passed over rust and heated (5), water and iron are formed (6).

and plasticity (nonelastic deformation). Although fluidity is most widely recognized as a property of liquids and gases, solids also flow to some extent and are included in rheological investigation. Rheological properties are influenced by TEMPERATURE and PRESSURE. Thixotropic liquids, such as paints, decrease in viscosity as they are stirred. These and other rheological phenomena are related to the molecular structure of materials. *See also* FLUID FLOW

rheostat Variable RESISTOR for regulating an ELECTRIC CURRENT. The resistance element may be a metal wire, carbon or a conducting liquid, depending upon the application. Rheostats are used to adjust GENERATOR characteristics, to dim lights, and to control the speeds of electric motors.

rhesus factor (Rh factor) Any of a group of ANTIGENS (substances that stimulate ANTIBODY production) found on the surface of ERYTHROCYTES (red blood cells), so called because they were first discovered in the rhesus monkey. Rh-negative blood lacks the Rh factor and Rh-positive blood contains it; Rh factor is present in *c.*85% of human beings. Rh incompatibility (an Rh-negative pregnant woman with an Rh-positive fetus) can give rise to ANEMIA in newborn babies. *See also* AGGLUTINATION

rheumatism General term for a group of disorders the symptoms of which are pain, inflammation and stiffness in the bones, joints, and surrounding tissues. Rheumatoid ARTHRITIS is an AUTOIMMUNE DISEASE featuring chronic inflammation of the joints, often with serious disability and disfigurement; there may also be damage to the eyes and other organs. It is more common in women.

rhizoid Fine, hairlike growth used for attachment to a solid surface by some simple organisms, such as certain fungi and mosses. The rhizoid lacks the conducting TISSUES of a ROOT.

rhizome Creeping, rootlike underground stem of certain plants. It usually grows horizontally, is rich in accumulated starch, and can produce new roots and stems asexually. Examples include the iris and water lily. Rhizomes differ from roots in having nodes, buds, and scale-like leaves. *See also* ASEXUAL REPRODUCTION; TUBER

rhodium (symbol Rh) Silver-white TRANSITION ELEMENT, first discovered in 1803. It occurs associated with PLATINUM and its chief source is as a by-product of nickel smelting. It resists tarnish and is used in hard platinum alloys and to plate jewelry. Properties: at.no. 45; r.a.m. 102.906; r.d. 12.4; m.p. 1,966°C (3,571°F); b.p. 3,727°C (6,741°F); most common isotope [103]Rh (100%).

rhodophyte (red algae) Any member of the ALGAE phylum Rhodophyta. They are numerous in tropical and subtropical seas. Most rhodophytes are slender, branching SEAWEEDS that form shrub-like masses. Some become encrusted with CALCIUM CARBONATE and are important in REEF formation. Rhodophytes have red and purplish pigments, which help to absorb light for PHOTOSYNTHESIS. They also have CHLOROPHYLL. They have complex life cycles with two or three distinct stages, involving ALTERNATION OF GENERATIONS.

rhodopsin (visual purple) Visual PIGMENT present in the RODS in the eye's RETINA. It absorbs light, producing a NERVE IMPULSE that is perceived as SIGHT.

rhombus Plane figure with opposite sides parallel and all of its sides equal in length. A rhombus is a particular type of PARALLELOGRAM. The diagonals of a rhombus bisect each other at right angles.

rhyolite Fine-grained, extrusive IGNEOUS ROCK with a similar composition to GRANITE. It is an uncommon, light-colored rock made up of alkali feldspars, quartz, and sometimes mica. It forms by the rapid cooling of viscous LAVA explosively ejected from a VOLCANO and as a result is made up of tiny CRYSTALS, too small to be seen with the naked eye. It may contain small cavities (vesicles) caused by gas bubbles and areas of glassy material.

rib Any of the long, curved bones that are arranged in pairs, extending sideways from the backbone, to form the framework of the chest. There are 12 pairs of ribs in humans. The first seven pairs, known as true ribs, are joined by CARTILAGE to the BREASTBONE; the next three, called false ribs, are joined to the cartilage of the seventh pair; and the last two, called floating ribs, are not attached at the front. Adjacent ribs are joined by intercostal muscles, which contract and relax to vary the capacity of the chest during BREATHING.

riboflavin VITAMIN B_2 of the B complex, lack of which impairs growth and causes skin disorders. It is a COENZYME important in transferring energy within cells. Soluble in water, riboflavin is found in milk, eggs, liver, and green vegetables.

ribonucleic acid *See* RNA

ribose ($C_5H_{10}O_5$) Pentose or five-carbon sugar occurring in the structure of RNA. It is an isomeric form of arabinose. Ribose, which can be obtained by the hydrolysis of RNA, is composed of a large string of nucleotides derived from nucleic acids.

ribosome Tiny structure in the CYTOPLASM of cells, involved in synthesizing PROTEIN molecules. Each protein is made up of a specific sequence of AMINO ACIDS. The genetic material DNA is made up of a long string of organic BASES of four different kinds. The order of these bases is like a code for all the instructions needed to make a living organism. It specifies the sequence of AMINO ACIDS for all the proteins the cell needs to make. Segments of DNA, called GENES, contain the instructions for individual proteins. The DNA molecule is too large to escape from the nucleus into the cytoplasm, but a "working copy" is made in the form of a molecule of MESSENGER RNA (mRNA), and this travels to the ribosomes in the cytoplasm. Ribosomes attach themselves to the messenger RNA, then assemble the amino acids in the correct sequence to form a particular protein. Each ribosome is itself made up of proteins and a special kind of RNA. *See also* GENETIC CODE

ribulose bisphosphate (RuBP) (ribulose disphosphate) Five-carbon sugar (ribulose) linked to two PHOSPHATE groups. It is involved in the CALVIN CYCLE in which it combines with CARBON DIOXIDE in the first stage of the dark reaction of PHOTOSYNTHESIS. The reaction is mediated by an ENZYME called ribulose biphosphate carboxylase, which, because of its presence in every green plant, is reputed to be the most abundant PROTEIN on Earth.

Richards, Theodore William (1868–1928) US chemist. From 1901 he was professor of chemistry at Harvard. Richards was awarded the 1914 Nobel Prize for chemistry for his accurate determinations of atomic weights (now called RELATIVE ATOMIC MASS). He also provided experimental evidence that supported Frederick SODDY's theory of the existence of ISOTOPES.

Richardson, Sir Owen Willans (1879–1959) British physicist who was awarded the 1928 Nobel Prize for physics for his work in THERMIONICS, his term for the phenomenon of the emission of electricity from hot bodies. In 1911 Richardson showed that ELECTRONS are emitted from hot metals, disproving the theory that they arose from the surrounding air. Also in 1911, he formulated a mathematical equation, known as **Richardson's law**, relating the absolute temperature of a metal to the rate of electron emission. It was of fundamental importance in the development of the thermionic valve or ELECTRON TUBE, and in the ELECTRON GUN of a CATHODE-RAY TUBE.

Richet, Charles Robert (1850–1935) French physiologist. He was awarded the 1913 Nobel Prize for physiology or medicine for his work on immune reactions, which led to research on ALLERGIES.

Richter, Burton (1931–) US physicist. Working with a very powerful particle ACCELERATOR, consisting of a pair of ELECTRON storage rings that could drive MATTER and antimatter particles together with extremely high energy, Richter discovered (1974) a new ELEMENTARY PARTICLE (which he named psi) with a relatively long lifetime; it is a type of MESON. Samuel TING, working independently of Richter and using a different method, simultaneously discovered the same particle, which he named the J-particle. For this work, Richter and Ting shared the 1976 Nobel Prize for physics.

Richter scale Classification of earthquake magnitude set up in 1935 by the US geologist Charles Richter (1900–85). The scale is logarithmic – each point on the scale increases by a factor of ten – and is based on the total energy released by an earthquake as opposed to a scale of intensity that measures the damage done at a particular place.

rickettsiae Group of tiny, BACTERIA-like PARASITES carried by fleas, lice, and ticks. They cause many diseases of vertebrates, including typhus, Rocky Mountain spotted fever and Q fever, all of which can be transmitted to humans.

▲ **rift valley** Rift valleys are formed by tension between two roughly parallel faults (A), causing downward earth movement resulting in the formation of a graben (trough of land between two faults). Sometimes a number of parallel faults results in land sinking in steps (B). A typical example of a step-faulted rift valley is shown (C). A series of block faults can occur on either side of a graben, sometimes tilting in the process of creating the block-faulted rift valley. The Great Rift Valley in E Africa (D), incorporating the Red Sea, is perhaps the world's best example of this geological formation.

Riemann, Georg Friedrich Bernhard (1826–66) German mathematician. With Augustin CAUCHY, Riemann formalized the theory of INTEGRATION. He also made advances in complex analysis and NUMBER THEORY. His work on DIFFERENTIAL and NON-EUCLIDEAN GEOMETRY was used by EINSTEIN to develop his general theory of RELATIVITY.

rift valley DEPRESSION formed by the subsidence of land between two parallel FAULTS. Rift valleys are believed to be formed by THERMAL currents within the Earth's MANTLE that break up the CRUST into large slabs or blocks of rock, which then become fractured. The best example on land is the Great Rift Valley in E Africa. Rift valleys are also characteristic of MID-OCEAN RIDGES. *See also* GRABEN

Rigel (Beta Orionis) Brightest star in the constellation ORION. It is of magnitude 0.12 (the seventh brightest star in the sky), and is a SUPERGIANT of spectral type B8. Its distance is not accurately known, but is thought to be about 1,000 LIGHT-YEARS. Rigel's LUMINOSITY is estimated to be over 100,000 times that of the Sun.

right ascension (RA) In astronomy, coordinate used to define the position of a celestial object. It is the angular distance measured eastwards from the vernal EQUINOX (also known as the first point in Aries) to the point where the hour circle of an object meets the CELESTIAL EQUATOR. RA is usually expressed in hours, minutes, and seconds, the circumference of the celestial equator being 24 hours. *See also* CELESTIAL SPHERE

right-hand rule *See* FLEMING'S RULES

rigor mortis Stiffening of the body after death brought about by chemical changes in MUSCLE tissue. Onset is gradual from minutes to hours, and rigor mortis disappears within about 24 hours.

rip current Narrow, swift, short-lived surface CURRENT that flows seawards at rightangles to the shoreline. It occurs at the mouth of a bay or along a coast where wind and incoming waves pile up the water until the excess rushes quickly and forcefully back into the sea.

Ritter, Johann Wilhelm (1776–1810) German physicist. In 1801 he discovered ULTRAVIOLET RADIATION beyond the violet end of the SPECTRUM. Ritter made the find while working with SILVER CHLORIDE, which decomposes in the presence of light, especially ultraviolet light. He also worked in ELECTROCHEMISTRY and in 1802 made the first DRY CELL.

river Large, natural channel containing water, which flows downhill under gravity. The water in a river is supplied by tributary streams, by direct runoff over the land, by seepage from valley side slopes, by water emerging from underground sources in springs, by water falling on the river surface, and by the melting of snow and glacier ice. A river system is a network of connecting channels. It can be divided into tributaries that collect water and SEDIMENT, the main trunk river, and the dispersing system at the river's mouth where much of the sediment is deposited. The DISCHARGE of a river is the volume of water flowing past a point in a given time. It is usually expressed as cubic meters per second (cumecs), and is calculated by multiplying the cross-sectional area of the river by the VELOCITY (speed) of the water. The velocity of a river is controlled by the slope of the river, the depth of the river and the roughness of the river bed. Rivers transport sediment as they flow, by the processes of TRACTION (rolling), saltation (jumping), SUSPENSION (carrying), and SOLUTION. A greater discharge increases the amount of sediment that can be transported. Most river sediment is transported during flood conditions, but as a river returns to normal flow it deposits sediment. A river adjusts its channel shape to be able to transport sediment most efficiently. This can result in the EROSION of a river channel or in the building up of FLOODPLAINS, sand and gravel banks. All rivers tend to flow in a twisting pattern, even if the slope is relatively steep, because water flow is naturally turbulent. The force of the water striking the river bank causes erosion and undercutting, which starts a small bend in the river channel. Over time the bend grows into a large MEANDER. The current flows faster on the outside of the bend eroding the bank, while sedimentation occurs on the inside of the bend where the current is slowest. Eventually, the river channel cuts across the loop and follows a more direct course downstream and the loop is abandoned as an OXBOW LAKE. Rivers flood when their channels cannot contain the discharge. Flood risk can be reduced by straightening the channel, dredging sediment or making the channel deeper by raising the banks. *See also* DELTA; LEVEE

▲ **river** Young rivers (A) close to their source tend to be fast-flowing, high-energy environments with rapid downward and headward erosion, despite the hardness of the rock over which they may flow. Steep-sided, "V-shaped" valleys, waterfalls and rapids are characteristic features. Mature rivers (B) are lower-energy systems. Erosion takes place on the outside of bends, creating looping meanders (1) in the soft alluvium of the river plain. Deposition occurs on the inside of bends and on the river bed. At a river's mouth (C), sediment is deposited as the velocity of the river slows. As the river becomes shallower more deposition occurs, forming islands and braiding the main channel into multiple, narrower channels. As the sediment is laid down (2), the actual mouth of the river moves away from the source into the sea or lake, forming a delta.

RNA (ribonucleic acid) Chemical (NUCLEIC ACID) that controls the synthesis of PROTEIN in a CELL and is the genetic material in some VIRUSES. The molecules of RNA in a cell are copied from DNA and consist of a single strand of NUCLEOTIDES, each containing the sugar RIBOSE, PHOSPHORIC ACID, and one of four BASES: ADENINE, GUANINE, CYTOSINE or URACIL. MESSENGER RNA (mRNA) carries the information for protein synthesis from DNA in the cell NUCLEUS to the RIBOSOMES in the CYTOPLASM. During a process known as TRANSLATION, TRANSFER RNA (tRNA) brings AMINO ACIDS to their correct position on the mRNA to form POLYPEPTIDE chains.

Roaring Forties Region between 40° and 50°S in the Southern Hemisphere where strong westerly winds blow. A series of DEPRESSIONS move continually from west to east, bringing mild temperatures but wet and windy conditions. The winds remain persistent and strong because there is no major land mass to slow them.

Robbins, Frederick Chapman (1916–) US physician. He shared the 1954 Nobel Prize for physiology or medicine with John F. ENDERS and Thomas H. WELLER for his part in the discovery of the ability of the poliomyelitis viruses to grow in cultures of various types of tissues. This discovery was vital to the development of a poliomyelitis VACCINE.

Robinson, Sir Robert (1886–1975) British chemist who was awarded the 1947 Nobel Prize for chemistry for research into ALKALOIDS and other plant chemicals. Robinson succeeded in synthesizing the alkaloid tropinone using three simple COMPOUNDS found in plants. This led to his theory of organic molecular structure based on electronic processes occurring during the formation and disruption of CHEMICAL BONDS. The theory proved important for the understanding of all biosynthetic mechanisms. His other research concerned plant PIGMENTS and the GENETICS of variations in flower colors. Robinson also assisted in the synthesis of PENICILLIN and female SEX HORMONES.

robot Automated MACHINE used to carry out various tasks. Robots are often COMPUTER-controlled devices, the most common type having a single arm that can move in any direction. Such robots are used to carry out various tasks in car manufacturing. A range of attachments enables the robots to weld metal parts, pick up and attach components, and carry out paint spraying. Scientists working in the field of NANOTECHNOLOGY hope to produce microscopic robots that will prolong lives by cleaning the arteries in human bodies.

Rochelle salt Colorless, crystalline compound, potassium sodium tartrate tetrahydrate ($KNaC_4H_4O_6.4H_2O$). It can be prepared from BAKING SODA and cream of tartar (crystallized potassium hydrogen bitartrate). Rochelle salt is used in baking powders and as a laxative. It is also used in electronics for the PIEZOELECTRIC EFFECT. It is

▲ **robot** A robotic arm can be programmed to use different tools (1) to carry out a variety of tasks. The arm, head, and base can all rotate (2) and can extend to reach different points. In a typical simple engineering setting, the robot first places the raw material into position (3) and then changes tools (4) before drilling the required holes (5).

named after La Rochelle, France, where it was first discovered. *See also* BAKING SODA

roche moutonnée (Fr. sheep rock) Rock that has been scraped smooth by ice on one side, but has been plucked into a jagged shape on the other, downhill side. They are found in glacial valleys. Roches moutonnées can be only a few meters in height, but some are much bigger, reaching 30–40m (98–130ft). It is said that roches moutonnées were so named because they resemble sheep lying down.

rock Solid material that comprises the Earth's CRUST. Although solid, it is not necessarily hard – CLAY and volcanic ASH are also considered to be rocks. Rocks are classified by origin into three major groups. IGNEOUS ROCK is any rock formed by the cooling and solidification of molten material from the Earth's interior; volcanic LAVA and GRANITE are igneous rocks. Igneous rocks can also be classified as INTRUSIVE and EXTRUSIVE ROCK. SEDIMENTARY ROCK is formed from older rock that has been transported from its original position by water, glaciers or the atmosphere, and consolidated again into rock; LIMESTONE and SANDSTONE are examples. METAMORPHIC ROCK originates from igneous or sedimentary rocks, but has been changed in texture or mineral content or both by extreme pressure and heat deep within the Earth; MARBLE, derived from limestone, is a metamorphic rock.

rocket Slender, tapering MISSILE or craft powered by a rocket ENGINE. Most of its volume contains fuel; the remainder is the payload (such as an explosive, scientific instruments or a spacecraft). **Liquid-fueled** rockets use a fuel (such as liquid hydrogen) and an oxidizer (usually liquid oxygen), which are burned together in the engine. **Solid-fueled** rockets have both fuel and oxidizer in a solid mixture. A single-stage rocket has one fuel load, or several that are used simultaneously. A multistage rocket has more than one fuel load; each is ignited singly in succession as the preceding one burns out. Because rocket engines carry their own supply of both fuel and oxidizer, they can operate in outer SPACE where there is no atmosphere. They gain THRUST from the reaction (referred to in Newton's third law of MOTION) produced by rapid, continuous output of exhaust gases. **Chemical** rocket engines are powered by solid or liquid PROPELLANTS that are burned in a COMBUSTION chamber and expelled at SUPERSONIC VELOCITIES from the exhaust nozzle. Nuclear engines heat fuel by RADIATION from NUCLEAR REACTOR cores. ION engines use thermoelectric power to expel IONS rather than gases. The walls of the combustion chamber and the exit nozzle must be cooled to protect them against the heat of the escaping gases, whose temperature may be as high as 3,000°C. Today, payloads are put into orbit by the SPACE SHUT-

TLE. In the late 19th century, Ernst MACH investigated the possibility of supersonic speed. Konstantin TSIOLKOVSKY is regarded as the "father of astronautics." He suggested that multi-stage rockets could escape Earth's GRAVITATIONAL FIELD. In 1926 Robert H. GODDARD launched the first liquid-fuel rocket. During World War II Wernher VON BRAUN and Hermann OBERTH developed the first long-range GUIDED MISSILES. After the war, von Braun helped develop the US space program. Saturn V, the largest rocket ever built, delivered 3.4 million kg (7.5 million lb) of thrust. It was used for the US Apollo missions to the Moon. *See also* SPACE EXPLORATION

rods and cones Cells in the RETINA of the EYE that are sensitive to light. Located in the pigmented layer, the rod-shaped RHODOPSIN-secreting cells are the RECEPTORS for low-intensity light, and the cone-shaped iodopsin-secreting cells are adapted to distinguish color. Rods detect only shades of black and white, but are particularly sensitive to movement.

Roebling, John Augustus (1806–69) US engineer and industrialist, b. Germany. He devised a method for the manufacture of wire CABLES and directed the construction of SUSPENSION BRIDGES across the Allegheny and Niagara rivers. As one of the earliest proponents of the construction of the Brooklyn Bridge, Roebling was appointed its chief engineer in 1869. Soon after, he died in an accident during a survey of the East River. His son, **Washington Augustus** Roebling (1837–1926), was also an engineer and industrialist. Washington Roebling assisted his father in the construction of the Allegheny Suspension Bridge. During the American Civil War, he served as a military engineer and balloonist. In 1869 he succeeded his father as chief engineer on the Brooklyn Bridge and supervised the work until its successful completion in 1883.

Roentgen, Wilhelm Konrad *See* RÖNTGEN, WILHELM KONRAD

roller bearing Component used in engineering to minimize FRICTION by limiting contact between rotating parts of a machine. It consists of two concentric metal rings separated by, and in contact with, cylindrical, barrel-shaped or tapered rollers made of polished hardened steel. *See also* BEARING

ROM (acronym for **r**ead-**o**nly **m**emory) INTEGRATED CIRCUITS (chips) that act as a permanent store for DATA (information) required by a COMPUTER. The contents of ordinary ROM chips are set by the manufacturer and cannot be altered by the user. The stored data is available to the computer's MICROPROCESSOR whenever the computer is switched on.

Röntgen, Wilhelm Konrad (1845–1923) German physicist. In 1895, while teaching at Würzburg University, he discovered X-RAYS, for which he was awarded the first Nobel Prize for physics in 1901. Röntgen also conducted important research on ELECTRICITY, the specific heats of gases, and the heat CONDUCTIVITY of crystals. The RÖNTGEN is named after him.

röntgen (symbol R) Unit once used to measure the X-RAY or GAMMA RADIATION to which a body is exposed. One röntgen causes sufficient IONIZATION to produce a total electric charge of 2.58×10^{-4} COULOMBS on all the positive (or negative) IONS in one kilogram of air. The unit, named after Wilhelm RÖNTGEN, has been replaced by the SI UNIT, the GRAY (symbol Gy).

root Underground portion of a VASCULAR PLANT that serves as an anchor and absorbs water and minerals from the soil. Some plants, such as the dandelion, have TAPROOTS with smaller lateral branches. Some of these plants, such as carrots and parsnips, have large taproots that store nutrients and are grown as food for humans and farm animals. Other plants, such as the GRASSES (including cereal crops), develop fibrous roots with lateral branches.

root In mathematics, fractional POWER of a number. The SQUARE ROOT of a number, x, is written as either $\sqrt{x}$ or $x^{\frac{1}{2}}$, and represents a number that, when multiplied by itself, gives the original number. The fourth root of x may be written in RADICAL form as $\sqrt[4]{x}$ or in power form as $x^{\frac{1}{4}}$. For example, the fourth root of 16 is 2, since $2 \times 2 \times 2 \times 2 = 16$. The term root is also used to refer to the solution of a POLYNOMIAL equation.

root-mean-square (RMS) In mathematics, the value obtained by taking the MEAN average of a set of squared values. It provides the value of a periodic quantity and was developed at the beginning of the 20th century as an accurate measure of VOLTAGE and ELECTRIC CURRENT. An RMS error of approximity is also used for equations devised to predict outcomes, such as weather forecasts.

root nodule Small swelling in the roots of various plants, such as legumes, that contain nitrogen-fixing bacteria. *See also* NITROGEN CYCLE; NITROGEN FIXATION

root pressure Means by which water travels from the soil through the roots of a plant and up its STEM. The XYLEM conducts the water, while the pressure is produced by a combination of OSMOSIS, which forces water from the soil into the CELLS of the root, and an active pumping mechanism that creates a concentration gradient of salt IONS in the xylem.

Excessive root pressure may cause "bleeding" of SAP from a wound in the plant or the collection of drops of water from undamaged leaf margins.

rosin (colophony) Yellowish, amorphous RESIN obtained as a residue from the distillation of TURPENTINE; the chief constituent is abietic acid. It is used in VARNISHES, soldering FLUXES, linoleum, and printing INKS. Properties: r.d. 1.08; m.p. 100–150°C (212–308°F).

Ross, Sir Ronald (1857–1932) British bacteriologist, b. India. In 1898 he discovered the PARASITE (*Plasmodium*) that carries MALARIA in the stomach of the *Anopheles* mosquito. Ross was awarded the 1902 Nobel Prize for physiology or medicine for his work.

rotation Turning of a celestial body about its AXIS, as distinct from its orbital REVOLUTION. In the SOLAR SYSTEM, the Sun and all the planets, with the exception of Uranus and Venus, rotate from W to E.

rotor, helicopter Engine-driven rotary wing. Each of the blades is an AIRFOIL and produces LIFT. The hub moves to permit altering of the rotors' plane of rotation in order to give sideways or forward and backward motion.

Rous, Francis Peyton (1879–1970) US pathologist who shared the 1966 Nobel Prize for physiology or medicine with Charles B. HUGGINS for the discovery of CANCER-inducing VIRUSES. In 1910 Rous began a series of experiments on SARCOMAS in hens, injecting healthy birds with a solution made from sarcomas from cancerous birds. He filtered the solution to remove all cells and bacteria, but found that the healthy hens developed cancer. This led him to postulate that cancer could be caused by viruses. The theory was rejected at the time, but later experiments proved him correct.

Royal Greenwich Observatory UK national astronomical observatory, founded at Greenwich, London, in 1675. After World War II the observatory moved to Herstmonceux in Sussex, and in 1990 it moved again to Cambridge. The RGO operates the UK TELESCOPES at the Roque de los Muchachos Observatory on La Palma in the Canary Islands.

royal jelly In honeybee colonies, the food given to female bee LARVAE destined to become queens. It is a glandular secretion of young worker bees who themselves are females, but have been fed on honey and POLLEN and so have not developed into queens.

Royal Society British society founded in 1660 and incorporated two years later. Its aim was to accumulate experimental evidence on a wide range of scientific subjects, including medicine and botany as well as the physical sciences. In the 20th century, the Royal Society became an independent organization concerned with the promotion of scientific research.

rubber Elastic solid obtained from the LATEX of the rubber tree (*Hevea brasiliensis*, Family Euphorbiaceae). **Natural** rubber consists of a POLYMER of cis-ISOPRENE and is widely used for tires and other applications, especially after VULCANIZATION. **Synthetic** rubbers are polymers tailored to excel in certain properties for specific purposes; none has the overall advantages of natural rubber. They include STYRENE-BUTADIENE rubber (SBR), NEOPRENE, NITRILE rubber, and the newer stereo-rubbers based on synthetic cis-polyisoprene.

rubidium (symbol Rb) Silver-white, metallic element of the ALKALI METALS. It was discovered (1861) by Robert BUNSEN and Gustav KIRCHHOFF. The element has few commercial uses; small amounts are used in PHOTOELECTRIC CELLS. Chemically, it resembles SODIUM but is more reactive. Properties: at.no. 37; r.a.m. 85.4678; r.d. 1.53; m.p. 38.89°C (102°F); b.p. 688°C (1,270°F); most common isotope ^{85}Rb (72.15%).

ruby Gem variety of the mineral CORUNDUM (aluminum oxide), whose characteristic red color is due to impurities of chromium and iron OXIDES. The traditional source of rubies is Burma. Today, synthetic rubies are widely used in industry, especially in LASER technology.

rudaceous Describing a coarse-grained SEDIMENTARY ROCK. Examples of rudaceous rocks include breccia, conglomerate, and gravel, as well as types of stony boulder clay. The particles in the rock range in size from large grains of sand to fragments of gravel.

Rumford, Benjamin Thompson, Count (1753–1814) British scientist and administrator, b. US. He moved to Britain during the Revolutionary War and was made a count of the Holy Roman Empire in 1791. In science, Rumford discovered that HEAT is a form of MOTION. He also introduced the standard candle, which became the unit of LUMINOSITY until 1940 (when it was replaced by the CANDELA). In 1799 Rumford founded the Royal Institution of Great Britain.

ruminant Cud-chewing, even-toed, hoofed mammals. They include the okapi, chevrotain, deer, giraffe, antelope, cattle, sheep, and goat. All except the chevrotain have four-chambered STOMACHS. They are named for the habit of rechewing food previously swallowed (cud) and stored in one of the chambers.

runner In botany, a long, thin STEM (STOLON) that extends along the surface of the soil from the axil of a plant's leaf. It serves to propagate the plant. At points (NODES) along its length, a runner has small leaves with buds that develop shoots and roots and turn into small independent plants as the runner dies. Runners are produced by such plants as strawberries and creeping buttercups. *See also* ASEXUAL REPRODUCTION

Russell, Henry Norris (1877–1957) US astronomer. He was director (1912–47) of the observatory at Harvard University. Russell devised a method for calculating the dimensions of ECLIPSING BINARY stars. He and Ejnar HERTZSPRUNG developed the HERTZSPRUNG-RUSSELL DIAGRAM.

rust In botany, group of FUNGI that live as PARASITES on many kinds of higher plants. The rust-red color of their SPORES gives them their name. Rusts are of particular concern for the damage they do to cereal crops, such as wheat and barley, and to several fruits and vegetables. They have complex life cycles that involve growth on more than one host plant.

rust Product of CORROSION of iron caused by a reaction with air and water. CARBON DIOXIDE from the air dissolves in water to form an acid solution that attacks the iron to form iron(II) (ferrous) oxide. This is then oxidized by oxygen in the air to reddish-brown iron(III) (ferric) OXIDE. Rusting may be prevented by coating the iron with ZINC, a process called GALVANIZING, when the zinc is preferentially attacked by the acid solution.

ruthenium (symbol Ru) Silver-white TRANSITION ELEMENT, first discovered in 1827. It is found in PLATINUM ores. Ruthenium is used as a CATALYST and is alloyed with other platinum metals in electrical contacts and used to color glass and ceramics. Properties: at.no. 44; r.a.m. 101.07; r.d. 12.41; m.p. 2,310°C (4,190°F); b.p. 3,900°C (7,052°F); most common isotope ^{102}Ru (31.61%).

Rutherford, Ernest, Lord (1871–1937) British physicist, b. New Zealand, who pioneered modern NUCLEAR PHYSICS. Rutherford discovered and named alpha and beta radiation, named the NUCLEUS, and proposed a theory of the radioactive transformation of ATOMS for which he received the 1908 Nobel Prize for Chemistry. He studied in Christchurch, New Zealand and in Cambridge, England. In Cambridge, under J.J. THOMSON, Rutherford discovered the URANIUM radiations. At McGill University, Canada, he formed the theory of atomic disintegration with Frederick SODDY. In 1907 Rutherford became professor at the University of Manchester. Here, he devised the nuclear theory of the atom, which was combined with QUAN-

TUM THEORY by Niels BOHR to produce the description of the atom that is still used today. In 1919 he became director of the Cavendish Laboratory, where his research team became the first to induce a change in an atom's nucleus by bombarding it with ELEMENTARY PARTICLES (HELIUM nuclei). He predicted the existence of the NEUTRON, which was later discovered by his Cambridge colleague, James CHADWICK.

rutherfordium (symbol Rf) Synthetic, radioactive, metallic element of the TRANSACTINIDE SERIES. It has a HALF-LIFE of less than a second. It is thought to have been discovered first in 1974, but its existence was only proved in 1994, the year it was officially given the name rutherfordium. The name rutherfordium had previously been used for element 104 (now known as DUBNIUM). Properties: at.no. 106

rutile (titanium dioxide, TiO_2) Black to red-brown oxide mineral, found in IGNEOUS and METAMORPHIC rocks and quartz veins. Rutile occurs as long prismatic and needle-like CRYSTALS in the tetragonal system and as granular masses. It has a metallic luster, is brittle, and is used as a gemstone. Hardness 6–6.5; r.d. 4.2.

Ružička, Leopold Stephen (1887–1976) Swiss chemist, b. Yugoslavia, who shared the 1939 Nobel Prize for chemistry with Adolf BUTENANDT for his research into ring MOLECULES and TERPENES (hydrocarbons found in the ESSENTIAL OILS of many plants). In 1916 Ružička began investigations into the active compounds in musk and civet, finding that they were ring structures consisting of 15 and 17 CARBON atoms respectively. This discovery had a profound effect on organic CHEMISTRY, as until that time rings with more than six atoms were thought to be too unstable to exist. In the 1930s he synthesized several male SEX HORMONES, including TESTOSTERONE, having previously discovered their molecular structure.

Rydberg, Johannes Robert (1854–1919) Swedish chemist who classified optical spectra and developed (1890) empirical formulae for spectral lines, using a constant ($1.097 \times 10^7 m^{-1}$) now known as the **Rydberg constant** (symbol R). He also studied hydrogen line wavelengths. The **Rydberg atom** is a highly excited hydrogen atom that exhibits chaotic behaviour in a strong MAGNETIC FIELD.

Ryle, Sir Martin (1918–84) British physicist and radio astronomer. After studying RADAR during World War II, he went to Cambridge University where he pioneered RADIO ASTRONOMY. Ryle also catalogued RADIO SOURCES, which led to his discovery of QUASARS.

Sabin, Albert Bruce (1906–93) US virologist, b. Russia. He developed a **live** VIRUS oral VACCINE against **poliomyelitis**, in contrast to Jonas SALK's inactive virus vaccine, and began testing the Sabin vaccine in 1957. In 1980 Sabin worked to make the polio vaccine available in Brazil.

saccharide Organic compound based on SUGAR molecules. Monosaccharides include GLUCOSE and FRUCTOSE, found in fruit and honey. Two sugar molecules join together to make a DISACCHARIDE, such as LACTOSE (milk sugar) or SUCROSE (cane sugar). POLYSACCHARIDES have more than two sugar molecules. *See also* CARBOHYDRATE

Sachs, Julius von (1832–97) German botanist, known for his studies of the metabolism of plants, notably the role of CHLOROPHYLL. Sachs wrote *Textbook of Botany* (1868).

sacrum (sacral bone) Triangular segment of the lower backbone formed by the fusion of five vertebrae. It articulates above with the last lumbar vertebra, below with the coccyx and, to either side, with the hip bones.

safety glass Form of GLASS that is less hazardous than ordinary glass when broken. One form of safety glass consists of two sheets of 3mm (⁄8in) plate glass bonded to a thinner central sheet of transparent plastic. If an impact breaks the glass, the plastic holds the fragments in place. **Bulletproof** glass consists of several layers of glass and plastic. **Wired** glass has an embedded wire mesh to hold the fragments in place. **Toughened** glass is glass that has been treated to make it stronger than ordinary glass. When shattered, the glass forms small blunt fragments, which are much less dangerous than the sharp splinters from ordinary glass.

safety lamp Lamp that can be used in mines. Sir Humphry DAVY invented a miner's safety lamp in the early 1800s. Its oil flame burned inside a wire mesh cylinder, which conducted away much of the heat, so that any explosive FIREDAMP (methane) in the air would not be ignited. Others improved Davy's lamp, and its use became widespread. Today, most miners use electric lamps, but a modern version of the Davy lamp is still used to check for the presence of METHANE, which turns a low flame blue.

Sagan, Carl Edward (1934–96) US astronomer and writer. He was professor of astronomy and space science (1968–96) at Cornell University. Sagan was the greatest popularizer of science of his era. He hosted the television series *Cosmos* (1980). Sagan studied many aspects of the SOLAR SYSTEM, including the physics and chemistry of planetary atmospheres and surfaces. He also investigated the origins of extraterrestrial life and maintained the possibility of extraterrestrial life-forms. Sagan's work includes *The Dragons of Eden* (1977), *Broca's Brain* (1979), and *Contact* (1985).

Sagittarius (Archer) Southern constellation on the ecliptic between Scorpio and Capricorn. Rich in STELLAR CLUSTERS, this region of the sky also contains much interstellar matter, which obscures the central region of the MILKY WAY galaxy and is only penetrable with radio telescopes. The brightest star is Epsilon Sagittarii (*Kaus Australis*), magnitude 1.8.

St. Elmo's fire (corposant) Electrical discharge illuminating the tops of tall objects. It usually occurs during a storm when the atmosphere becomes charged strongly enough to create a discharge between the air and an object. Early sailors named this phenomenon after their patron saint.

Sakharov, Andrei Dmitrievich (1921–89) Soviet physicist and social critic. His work in nuclear FUSION was instrumental in the development of the Soviet HYDROGEN BOMB. An outspoken defender of civil liberties, he created the Human Rights Committee in 1970. Sakharov received the 1975 Nobel Peace Prize. In 1989 he was elected to the Russian Congress. Sakharov's books include *Sakharov Speaks* (1974) and *My Country and My World* (1975).

Salam, Abdus (1926–96) Pakistani physicist who proposed (1967) a theory that unifies the electromagnetic and WEAK NUCLEAR FORCES within the NUCLEUS of an ATOM (*see* FUNDAMENTAL FORCES). Salam and the US physicist Steven WEINBERG worked independently on the theory (now known as the **Weinberg-Salam theory**). After the theory was proved experimentally, the pair shared the 1979 Nobel Prize for physics with US physicist Sheldon GLASHOW, who also independently came to similar conclusions. *See also* GRAND UNIFIED THEORY (GUT)

salicylic acid (1-hydroxybenzoic acid, $C_7H_6O_3$) Colorless, crystalline solid, derivatives of which are used as analgesics (including ASPIRIN, acetylsalicylic acid), antiseptics, dyes, and liniments. It occurs naturally in plants, including willow bark and oil of wintergreen.

salina (playa lake) Dried lake bed found in desert or semiarid areas. Salinas form where a stream has flowed into an area of inland drainage and then evaporated, leaving behind tiny deposits of salt. The deposits can accumulate for hundreds of years to form large salty areas.

salinity Degree of saltiness, as of the OCEANS. The average proportion of salt in the sea is about 35 parts per 1,000 (3.5%). In the Red Sea, because of high evaporation rates, the figure is over 40, but in the N parts of the Baltic Sea it is less than 10, and in the Dead Sea the figure is about 250. The salts found in the oceans include sodium chloride (common salt), which accounts for about 75%, magnesium chloride, magnesium sulfate, and calcium sulfate.

saliva Fluid released into the mouth by the SALIVARY GLANDS. In vertebrates, saliva is composed of about 99% water with dissolved traces of sodium, potassium, calcium, and the ENZYME amylase. Saliva softens and lubricates food to aid swallowing, and AMYLASE starts the digestion of starches.

salivary glands One of three pairs of GLANDS located on each side of the mouth that form and secrete SALIVA. The **parotid** gland, just below and in front of each ear, is the largest of the salivary glands and the one that becomes enlarged in mumps; the **submaxillary** gland is near the angle of the lower jaw; and the **sublingual** gland is under the side of the tongue.

Salk, Jonas Edward (1914–95) US medical researcher. In 1952 he developed the first vaccine against poliomyelitis, using an **inactive** polio VIRUS as an immunizing agent. Extensive tests of the vaccine began in 1954 and mass immunization programs followed. *See also* SABIN, ALBERT

Salmonella Genus of rod-shaped BACTERIA that cause intestinal infections in human beings and animals. *Salmonella typhi* causes typhoid, but the effects of other species may result in only mild gastroenteritis. The bacteria are transmitted by carriers, particularly flies, and in food and water.

salt Ionic COMPOUND formed, along with water, when an ACID is neutralized by a BASE. The HYDROGEN of the acid is replaced by a metal or ammonium ION (NH_4^+). The most familiar salt is SODIUM CHLORIDE (NaCl). Salts are typically high-melting crystalline compounds that tend to be soluble in water. They are formed of ions held together by electrostatic forces, and in SOLUTION they conduct electricity. *See also* ELECTROLYSIS

salt dome Body of salt that has intruded into a SEDIMENTARY ROCK overlay. The flow of the relatively plastic salt into a dome may be the result of a differ-

ence in density between the salt and the overlying rock, but the process is not completely understood.

salt lake Saline lake in a desert region. The lake originates in an area of inland drainage. The salinity increases as the water evaporates, and layers of salt may crystallize out along the lake's shores. Prolonged rain may redissolve the salt and wash most of it away. The largest salt lakes are Lake Eyre in Australia and the Great Salt Lake in Utah, USA.

saltpeter *See* NITER; POTASSIUM

samara Form of ACHENE in which the FRUIT wall dries out to form a winglike structure. Common examples include ash and elm; the sycamore samara has two wings, and is called a schizocarp.

samarium (symbol Sm) Gray-white, metallic element of the LANTHANIDE SERIES, first identified spectroscopically in 1879. Its chief ores are MONAZITE and bastnaesite. Samarium is used in carbon-arc lamps, as a moderator in NUCLEAR REACTORS and as a catalyst. Some samarium alloys are used for making powerful permanent magnets. Properties: at.no. 62; r.a.m. 150.35; r.d. 7.52; m.p. 1,072°C (1,962°F); b.p. 1,791°C (3,256°F); most common isotope ^{152}Sm (26.72%).

sand Mineral particles worn away from rocks by EROSION, individually large enough to be distinguished with the naked eye. Sand is composed mostly of QUARTZ, but black sand (containing volcanic rock) and coral sand also occur. It is usually classified according to grain size. *See also* DUNE; SANDSTONE

sand blasting Industrial method for cleaning the surfaces of materials such as metals and concrete. It is an efficient means of cleaning the outer walls of buildings, but its most frequent use is to remove the oxide scale from metals prior to painting, spraying, electroplating or dip-coating. Articles to be cleaned are subjected to a high pressure blast of SAND.

sand casting Method of molding metals and alloys in which the molten metal is poured into a MOLD made of firmly packed sand. To help the sand grains cohere, they are mixed with clay, water, and other binders. The method is used in the casting of steel, brass, bronze, and aluminum.

sandstone SEDIMENTARY ROCK composed of SAND grains cemented in such materials as SILICA or iron oxide. Sandstones are ARENACEOUS rocks, and are very variable in character as a result of differences in grain size, chemical content, and cementing materials. Most sand grains contain QUARTZ, which is hard and resistant. Other minerals in sandstones include FELDSPARS and MICAS, and iron also occurs, which tends to give sandstones a reddish or brownish color. Most sandstones are formed by the accumulation of river sediments on the seabed. They are then compressed and uplifted to form new land. There are also a few sandstones that have been formed of wind-blown materials, especially in desert regions.

Sanger, Frederick (1918–) English biochemist who became the first person to win two Nobel Prizes for chemistry. Sanger was awarded his first prize in 1958 for finding the structure of INSULIN. His second came in 1980 after work on the chemical structure of NUCLEIC ACID. Sanger and his colleagues found the entire sequence of the more than 5,400 bases in the DNA of a VIRUS and some 17,000 bases of another DNA. His work was instrumental in the breakthrough, in 1999, by an international team of scientists who successfully mapped chromosome 22.

sap In plants, fluid that circulates water and nutrients through them. Water is absorbed by the roots and carried, along with minerals, through the XYLEM to the leaves. Sap from the leaves is distributed throughout the plant; it travels upwards by OSMOSIS, ROOT PRESSURE, and pressure differences created by TRANSPIRATION.

saponin GLUCOSIDES present in some plants that produce a lather when agitated with water. Used as a SOAP substitute, it is found in horse chestnut, soapberry, soapnut, and soapwort.

sapphire Transparent to translucent gemstone variety of CORUNDUM. It has various colors produced by impurities of iron and titanium, the most valuable being deep blue. They are brilliant when cut and polished. *See also* RUBY

saprophyte Plant or FUNGI that obtains its food from dead or decaying plant or animal tissue. Generally it has no CHLOROPHYLL. Saprophytes include fungi, such as mushrooms and molds, and some flowering plants.

sarcoma Cancerous growth or TUMOR arising from muscle, fat, bone, blood or lymph vessels or connective tissue. *See also* CANCER

satellite (moon) Celestial body orbiting a PLANET or STAR. In the SOLAR SYSTEM, planets with natural satellites include the EARTH (1), MARS (2), JUPITER (16), SATURN (18), URANUS (17), NEPTUNE (8), and PLUTO (1). There are probably more small satellites of the giant planets awaiting discovery. They vary enormously in their size, orbit, surface features, and supposed origin. One, Jupiter's GANYMEDE is bigger than Mercury, and seven are bigger than Pluto, while the smallest are tiny bodies a few kilometers across. The satellites' orbits vary just as much, from Phobos,

which circles Mars' equator more than three times a day, to the outer satellites of Jupiter, which take more than two Earth years to complete their inclined, eccentric orbits. In their surface features satellites are remarkably diverse, ranging from dead, densely cratered worlds to ones that show a great variety of geological processes. Jupiter's Io is the most geologically active body known. As to origin, some, such as Jupiter's GALILEAN SATELLITES, may have condensed out of the same part of the solar NEBULA as their parent planet, while several small ones are believed to be "captured" asteroids. The MOON is currently thought to have formed out of debris flung out into space when a Mars-sized body collided with the Earth.

satellite, artificial Manmade object placed in orbit around the Earth or other celestial body. Satellites can perform many tasks and can send back data or pictures to the Earth. Hundreds of satellites of various types orbit the Earth. They may study the atmosphere, or photograph the surface for scientific or military purposes. **Communications** satellites relay radio, television, telephone, telegraph, and data signals from one part of the Earth to another. **Navigation** satellites transmit radio signals that enable navigators to determine their positions. The GLOBAL POSITIONING SYSTEM (GPS) uses satellites in this way. **Geodetic** satellites are used to make accurate measurements of the Earth's size and shape. SPUTNIK 1, launched by the Soviet Union on October 4, 1957, was the first artificial satellite. TELSTAR, launched by the USA on July 10, 1962, was the first active communications satellite. The first satellite in geosynchronous ORBIT, Early-bird, was launched in 1965. Geosynchronous satellites circle the Earth at a height of 35,900km (22,300mi), enabling them to complete one revolution every 24 hours. Satellites that orbit the Earth at heights less than about 160km (100mi) are slowed by friction with the atmosphere and eventually spiral in and burn up. In SPACE EXPLORATION, a spacecraft may be put into orbit around a planet or a moon and thus become an artificial satellite of that body. *See also* SATELLITE, WEATHER; SYNCHRONOUS ORBIT

satellite, weather Artifical SATELLITE studying weather patterns on Earth by photographic, infrared or other means. Some satellites are in GEOSTATIONARY ORBIT, which means that they remain over the same point of the Earth, travelling at the same speed as the rotation of the Earth. Other satellites circle the Earth in a polar direction, and with each ORBIT pass over the Equator about 30° W of the previous orbit. Photographic histories of hurricane development and discovery of large-scale coherence in cloud and other weather patterns have greatly helped WEATHER FORECASTING.

satellite television TELEVISION services transmitted to viewers via communications SATELLITES in orbit around the Earth. These satellites are in synchronous orbit above Earth's Equator. As a result, each one remains above a fixed point on the Equator (*see* GEOSTATIONARY ORBIT). Television companies beam their signals to the satellites from TRANSMITTERS on the ground. The satellites retransmit the signals back to viewers' dish-shaped receiving aerials. A fixed ANTENNA can receive many services from a single satellite. Some people use a movable antenna to receive services from more than one satellite.

saturated compound In organic chemistry, compounds in which the CARBON atoms are bonded to one another by single COVALENT BONDS only, never by the more reactive double or triple CHEMICAL BONDS. For this reason, they tend to be unreactive. A simple example is ETHANE (C_2H_6). In a molecule of ethane, each carbon atom is bonded to three hydrogen atoms and to the other carbon atom by single bonds. *See also* UNSATURATED COMPOUND

saturated fat Organic fatty compounds, the molecules of which contain only saturated FATTY ACIDS combined with GLYCEROL. These acids have long chains of carbon atoms that are bound together by single bonds only. *See also* SATURATED COMPOUNDS

saturated solution In chemistry, a SOLUTION containing so much of a dissolved compound (SOLUTE) that no more will dissolve at the same temperature or pressure, except in cases of SUPERSATURATION.

Saturn Sixth planet from the Sun and second largest in the SOLAR SYSTEM. It was the most remote planet known in the pre-telescope era. Viewed through a telescope, it appears as a flattened, golden-yellow disk encircled by white **rings**. Saturn's magnificent system of rings is its dominant feature. They were first observed by GALILEO in 1610, but Christiaan HUYGENS was the first to identify (1656) them as rings. As seen from Earth, the ring system consists of a few principal components, separated by a number of gaps. The rings are made up of particles ranging from dust to objects a few meters in size, all in individual orbits around Saturn. The main rings are only a kilometer or so thick. In the PIONEER PROGRAM, Pioneer 11 visited Saturn after Jupiter. The two VOYAGER PROJECT probes flew past Saturn, **Voyager 1** in 1980 and **Voyager 2** in 1981, and revealed the ring system to be made up of thousands of separate smaller rings. The gaps were not empty, but simply regions with a lower ring density. Saturn has an internal heat source, which probably drives its weather systems. Its density is much lower than for any other planet, and it also has much more of its mass concentrated at the center. It is assumed, therefore, to be composed predominantly of hydrogen, and to have an iron–silicate core about five times the Earth's mass, surrounded by an ice mantle of perhaps 20 Earth masses. Near the man-

tle, the hydrogen would be in metallic form; above this, it would exist as liquid molecular hydrogen. The upper atmosphere contains 97% hydrogen and 3% helium, with traces of other gases including methane, ethane, ammonia, and phosphine. The MAGNETIC FIELD of Saturn is 500 to 1,000 times the strength of Earth's, and it generates a MAGNETOSPHERE intermediate between those of the Earth and Jupiter in its shape, extent, and the intensity of its radiation belts of trapped particles. Saturn has 18 known SATELLITES, including TITAN, the only satellite in the Solar System to have a significant atmosphere. **Data:** Diameter (equatorial) : 120,000km; Mass: (Earth=1) 95.2; Volume: (Earth=1) 744; Density (water=1): 0.71; Orbital period: 29.46 years; Rotation period (equatorial): 10h 13m 59s; Temperature at cloud tops: 95K

▲ **satellite, artificial** Four artificial satellite orbits around the globe are illustrated. Two are equatorial: one, a geostationary satellite, orbits at 35,900km (21,500mi) (1); another has a lower orbit (2). A polar orbit, as used by remote sensing satellites, is shown running vertically around the Earth (3) and the last is an angled elliptical orbit used by communications satellites for the high latitudes of the Earth (4). The diagrams above the orbital diagram show the launching of an Intelsat communication satellite by a NASA shuttle (5) and the two main types of satellites. To prevent them being knocked off course by the fluctuations in the Earth's magnetic field, satellites remain stable in one of two ways. The first is to rotate the whole satellite, as is the case in "spinners" such as the Intelsat satellite (6). The other method is to have gyroscopes within the satellite (7).

savanna Area of tropical grassland, found between the Earth's major desert regions and the tropical forests. Savannas occur mainly between 5° and 20° N and S of the Equator, except in East Africa, where, because the land is elevated, savanna regions are situated on the Equator. The savanna climate is hot and wet in summer, when the Sun is overhead; temperatures average *c*.25°C (77°F) and the rainfall is 250mm to 1,250mm (10–50in). In winter, the tropical high-pressure system controls the climate, and so the conditions are warm but very dry. Temperatures are 15°C to 25°C (59°–77°F), and rainfall totals are less than 250mm (10in). In such a climate, grass grows very well in summer – up to 2m (6ft) in height – but it will shrivel in winter. Trees do not grow well, although baobabs and bottle trees survive by storing water in their trunks. Pastoral farming is the main activity on the savanna grasslands.

Savery, Thomas (1650–1715) English engineer. In 1698 he patented a machine for pumping mines, which employed the principle of a vacuum STEAM ENGINE. Savery developed this invention with Thomas NEWCOMEN. Savery also designed an early form of paddle-driven boat.

saw Cutting tool with a toothed blade, usually operated manually, by treadle, or by motor. Alternate teeth are bent outwards in opposite directions, so that the saw cut is slightly wider than the blade thickness. This helps to prevent the blade from sticking. The blade may be straight, circular, or formed into a band. Widely spaced teeth give a rapid, but rough cut. Closely spaced teeth give a fine cut. Applying a lubricant to the blade improves its performance.

scalar Mathematical quantity that only has a magnitude, as contrasted with a VECTOR, which also has direction. Mass, energy and speed are scalars whereas weight, force, and velocity are represented by vectors.

scale In biology, small hard plate that forms part of the external SKIN of an animal. It is usually a development of the skin layers. In most fish, scales are composed of bone in the dermal skin layer. The scales of reptiles and those on the legs of birds are horny growths of the epidermal skin layer, and are comprised mostly of the fibrous protein KERATIN.

scandium (symbol Sc) Silver-white, metallic element of group III of the periodic table, predicted (as ekaboron) by MENDELEYEV and discovered in 1879. It is found in thortveitite and, in small amounts, in other minerals. Scandium is a soft metal used as a radioactive tracer and in nickel alkaline storage batteries. Chemically, it resembles the rare-earth metals of the LANTHANIDE SERIES. Properties: at.no. 21; r.a.m. 44.956; r.d. 2.99; m.p.

1,539°C (2,802°F); b.p. 2,832°C (5,130°F); most common isotope ^{45}Sc (100%).

scanning In medicine, use of a noninvasive system to detect abnormalities of structure or function in the body. Detectable waves (X-RAYS, gamma rays, ultrasound) are passed through the part of the body to be investigated and the computer-analysed results are displayed as images on a viewing screen.

scanning tunelling microscope Type of ELECTRON MICROSCOPE that produces highly magnified images of the surface of a specimen. A tiny metal probe on a movable arm is held in near contact with the specimen, which has to be a conducting material, and scanned across it. An electrical signal, produced when electrons "tunnel" between the specimen and the probe, is kept constant by raising and lowering the arm. The necessary movements are analyzed by a computer, which generates a "contour map" of the surface displayed on the computer screen.

scapolite (wernerite) Group of SILICATE minerals related to the FELDSPARS and of variable chemical composition. They are common in metamorphosed limestones and are gray and glassy in appearance. Hardness 5–6; r.d. 2.7.

scapula (shoulder blade) In vertebrates, either of two large, roughly triangular flat bones found one on either side of the upper back, and forming part of the PECTORAL girdle. It provides for the attachment of muscles that move the forelimb.

scarp Abbreviation of ESCARPMENT

scattering Deflection of ELECTROMAGNETIC RADIATION by particles. Where the particles are very much larger than the WAVELENGTH, scattering consists of a mixture of REFLECTION and DIFFRACTION, and the amount of scattering depends very little on wavelength. Where the particles are very much smaller than the wavelength, the amount of scattering (d) is inversely proportional to the fourth power of the wavelength (λ): $d \propto 1/\lambda^4$. Thus, blue LIGHT is scattered by small particles ten times as much as red light. Scattering of sunlight by atoms and molecules in the atmosphere is what makes the sky blue (*see* SPECTRUM; TYNDALL, JOHN). It is the main cause of **atmospheric extinction**, making the Sun appear red at sunrise and sunset by preferentially scattering blue light out of the line of sight. ELEMENTARY PARTICLES themselves can be scattered by atomic nuclei or other particles. It is the means by which the structure of ATOMS was discovered. Ernest RUTHERFORD's students, Hans GEIGER and Ernest Marsden, "fired" ALPHA PARTICLES through thin metal films and noted their scattering. From the results, Rutherford deduced the existence of the atomic NUCLEUS (positively charged and centrally placed in the atom). Most knowledge of elementary particles and the discovery of new ones has been obtained by scattering experiments carried out in particle ACCELERATORS. *See also* COMPTON EFFECT; RAMAN EFFECT

Scheele, Karl Wilhelm (1742–86) Swedish chemist who discovered OXYGEN. An apothecary, his interest in chemistry led to an investigation of combustion and the discovery of oxygen in 1771, but publication of his discovery was delayed and the credit went to Joseph PRIESTLEY. Scheele made other important discoveries, including CHLORINE, GLYCEROL, and a number of organic acids.

scheelite (Ca[WO₄MoO₄]) Mineral, calcium tungstatemolybdate, an important ore of TUNGSTEN. It is found in metamorphic deposits and in pegmatites. Scheelite has a tetragonal system and bipyramidal crystals, also occurring as massive and granular aggregates. It reveals various tints with adamantine luster, fluoresces under ultraviolet light and is brittle. Hardness 4.5–5; r.d. 6.

Schiaparelli, Giovanni Virginio (1835–1910) Italian astronomer. He was an assiduous observer of the inner planets, MARS in particular. Schiaparelli prepared a map of the surface of Mars, introducing a nomenclature for the various features, that remained standard until the planet was mapped by space probes. His use of the term *canali* (It. canals) has caused much subsequent controversy. He also prepared maps of Venus and Mercury. Schiaparelli discovered the connection between meteor streams and comets, and studied double stars.

Schick, Bela (1877–1967) US pediatrician who devised (1913) the **Schick test**, which determines susceptibility to diphtheria. A small amount of diphtheria toxin is injected into the skin and swelling indicates a level of antibody production. If the subject has adequate immunity, no skin reaction occurs.

schist Large group of METAMORPHIC ROCKS that have been made cleavable, causing the rocks to split into thin plates leaving a wavy, uneven surface. Almost any type of rock will become a schist if subjected to sufficient metamorphism. Schists are named after their predominant mineral; a common example is MICA SCHIST.

Schönbein, Christian Friedrich (1799–1868) German chemist, the discoverer of OZONE. After studying at various universities, he became professor at the University of Basel (1835). Schönbein also discovered GUNCOTTON, recognizing its potential as a smokeless substitute for GUNPOWDER.

Schrieffer, John Robert (1931–) US physicist.

In 1957, working at the University of Illinois with John BARDEEN and Leon COOPER, Schrieffer helped to formulate a theory of SUPERCONDUCTIVITY, which is known as the **BCS theory** (after their initials). They all shared the 1972 Nobel Prize for physics.

Schrödinger, Erwin (1887–1961) Austrian physicist. He was professor at the Dublin Institute for Advanced Studies (1940–57). Schrödinger established the science of WAVE MECHANICS as part of the QUANTUM THEORY. **Schrödinger's equation** is a fundamental mathematical tool of QUANTUM MECHANICS. It formulates a relationship between WAVE FUNCTION, mass, and energy:

$$\nabla^2\psi + 8\pi^2 m(E - U)\psi/h^2 = 0,$$

where ∇^2 is the LAPLACE operator, ψ is the wave function, m is a particle's mass, E its total energy, U its potential energy and h PLANCK'S CONSTANT. Schrödinger shared the 1933 Nobel Prize for physics with Paul DIRAC. *See also* BROGLIE, LOUIS DE

Schultze, Max Johann Sigismund (1825–74) German biologist. He studied the nervous system of vertebrates, the electric organs of fish and the anatomy of worms and mollusks. Schultze's most

▲ **scattering** Scattering is the process by which visible light entering Earth's atmosphere is diffused by the gases and dust in the air. The shorter wavelength, higher frequency light at the blue end of the visible spectrum (1), is scattered more by particles in the atmosphere (2) than the red end of the spectrum (3), because the average size of the particles in the air corresponds with the blue light's wavelength. This diffusion is the reason why the daylight sky appears blue (4) and sunlight is perceived as having a yellow cast. When the Sun is at a low level in the sky at dawn or sunset the light has to pass through a greater thickness of atmosphere, so that only light from the red end of the spectrum reaches the Earth's surface in any quantity – hence red skies at sunset or sunrise.

important work was the identification of a CELL as an organism containing both nucleus and protoplasm and his recognition that PROTOPLASM is a basic substance of plant and animal cells.

Schwann, Theodor (1810–82) German biologist who laid the foundation of modern HISTOLOGY by establishing the CELL theory. In 1836 he became the first person to prepare an ENZYME from animal tissue when he isolated and named PEPSIN.

Schwann cells Cells that wrap around the AXONS of nerve fibers to form a fatty, protective MYELIN SHEATH. Between sections of myelin sheath are small gaps called **nodes of Ranvier**. The sheath serves to electrically insulate the nerve and allow the more rapid passage of nerve impulses. The cells were named after their discoverer, Theodor SCHWANN.

Schwarzschild radius Critical radius at which a very massive body under the influence of its own GRAVITATION becomes a BLACK HOLE. It is the radius of the EVENT HORIZON of a black hole, from which nothing can escape, not even light. The radius is given by the expression $R = 2GM/c^2$, where G is the GRAVITATIONAL CONSTANT, M is the MASS of the body and c is the speed of LIGHT. The Schwarzschild radius for the Sun is about 3km and for the Earth it is about 1cm. It is named after the German astronomer Karl Schwarzschild (1873–1916).

Schwinger, Julian Seymour (1918–94) US physicist who devised a theory of QUANTUM ELECTRODYNAMICS (QED) that explained mathematically the interaction of LIGHT and MATTER. For their work on QED, Schwinger shared the 1965 Nobel Prize for physics with Richard FEYNMANN and the Japanese physicst S. Tomonaga. He also conducted research on SYNCHROTRON radiation, a type of ELECTROMAGNETIC RADIATION emitted by charged particles in circular orbits.

scintillation (twinkling) In astronomy, slight variability in the brightness and color of the stars caused by local disturbances in the Earth's atmosphere that change its REFRACTIVE INDEX. Planets do not exhibit this phenomenon because the scintillations from different points on the surface are not in PHASE, and the fluctuations are lost in the general illumination. For astronomical purposes, the inconvenience of the effects of scintillation is reduced by siting observatories at high altitudes, and it is almost entirely avoided by using telescopes carried by balloons, rockets or artificial satellites.

scintillation counter Instrument containing a crystal that emits scintillations of light when bombarded by radiation. Each light flash, corresponding to a single particle, is converted into an electric

pulse by a PHOTOMULTIPLIER. The number of pulses, counted electronically, indicates the activity of the source of radiation.

sclerenchyma In botany, plant support-tissue, the cells of which have become rigid due to the formation of LIGNIN. Sclerenchyma occurs in veins, stems, and midribs of leaves. As sclerenchyma cells mature, they die because the lignin in the thickened cell wall does not allow the passage of gases and water, although shallow pits (**plasmodesmata**) in the wall are free from lignin and permit some exchange of substances between adjacent cells. In zoology, the hard skeletal tissue of corals is also called sclerenchyma.

sclerosis Degenerative hardening of tissue, usually due to scarring, following inflammation or as a result of aging. It can affect the brain and spinal cord, causing neurological symptoms, or the walls of the arteries. *See also* MULTIPLE SCLEROSIS (MS)

Scorpius (Scorpio, scorpion) Constellation of the s sky, situated on the ECLIPTIC between LIBRA and SAGITTARIUS. It contains several OPEN CLUSTERS and GLOBULAR CLUSTERS. The MILKY WAY also passes through the region. The brightest star is the 1st-magnitude Alpha Scorpii (Antares).

scrambler Electronic device used to render a transmitted message unintelligible to all but its sender and its receiver, used particularly on voice messages sent by radio or telephone. The message is distorted in a prearranged way by the sender's scrambler, for instance by oscillating its frequency or by rearranging its order; the receiver's scrambler reverses this process to make the message intelligible again.

scrapie Fatal disease of sheep and goats that affects the CENTRAL NERVOUS SYSTEM (CNS), causing staggering and itching that makes an infected animal scrape itself against various objects. It is caused by a slow-acting, viruslike, microscopic particle called a PRION. The animal usually dies within six months of exhibiting the first symptoms. The disease is thought to be related to BOVINE SPONGIFORM ENCEPHALOPATHY (BSE) and CREUTZFELDT-JAKOB DISEASE (CJD).

scree (talus) Loose fragments of rock that have accumulated on a hillside. Scree is made up of particles ranging in size from sand grains to boulders. It mainly results from WEATHERING processes, such as freeze-thaw activity. Scree forms more frequently on certain rock types, such as limestone.

screening In medicine, a test applied either to an individual or to groups of people who, although apparently healthy, are known to be at special risk of developing a particular disease. It is used to detect treatable diseases while they are still in the early stages. For example, middle-aged women may be offered mammography to rule out breast cancer. Screening, therefore, is an important form of preventive health care.

screw Variant of a simple MACHINE, the INCLINED PLANE. It is an inclined plane cut around a cone, usually of metal, in a helical spiral. When FORCE is exerted radially on the screw, for example by a screwdriver or the lever of a screw jack, the screw advances to an extent determined by its pitch (the distance between crests of its thread).

screw propeller Device that converts the TORQUE of a rotating shaft into propulsion or THRUST. It is based on the principle of a helical SCREW, which advances in wood when it is rotated; the PROPELLER advances in air or water when it is rotated by an engine. **Simple** propellers have a fixed pitch so that the amount of thrust increases as the speed of rotation increases; **variable-pitch** propellers are rotated at constant speed, and thrust is varied by adjusting the angle at which the blades meet the air or water.

scuba diving Diving with the use of **s**elf-contained **u**nderwater **b**reathing **a**pparatus, or scuba. The equipment usually consists of tanks of compressed air connected to a device called a **demand regulator**, which controls the air flow through the mouth. Other equipment includes a wet suit to insulate the diver from the cold, weighted belt, face mask, and flippers. Professional divers use scuba to inspect vessels below the waterline and to search underwater. Scuba diving has attracted many enthusiasts since the design of the first scuba equipment called the Aqualung by Jacques COUSTEAU and Émile Gagnan in 1943.

sea *See* OCEAN

Seaborg, Glenn Theodore (1912–) US physicist. During World War II he worked on the development of the atom bomb. In 1940 Seaborg and Edwin M. McMILLAN discovered PLUTONIUM. They went on to isolate the entire ACTINIDE SERIES, Seaborg and McMillan shared the 1951 Nobel Prize for chemistry.

seafloor Floor of the OCEANS. The major features of the seafloor are the CONTINENTAL SHELF, the CONTINENTAL RISE, the ABYSSAL floor, seamounts, oceanic TRENCHES, and MIDOCEAN RIDGES. The **abyssal** or deep ocean floor is *c*.3km (1.8mi) deep and is mostly made of basaltic rock covered with fine-grained (PELAGIC) sediment consisting of dust and the shells of marine organisms. Oceanic **trenches** are up to 11km (7mi) deep, typically 50–100km (30–60mi) wide and may be thousands of kilometers long. The slopes are usually asymmetrical with the steeper slope on the landward side and a more gentle slope on the side of the ocean basin. They are regarded as the site of plate SUBDUCTION. Oceanic **ridges** are long, linear volcanic structures that tend to occupy the middle of seafloors; they are the sites of crustal spreading. *See also* SEAFLOOR SPREADING

seafloor spreading Theory suggested (1960) by the US geologist Harry Hess to explain how CON-

Seafloor spreading in the Atlantic Ocean

Seafloor spreading in the Indian Ocean and continental plate collision

Oceanic and continental plate collision

▲ **seafloor spreading**
The vast ridges that divide the Earth's crust beneath each of the world's oceans mark the boundaries between tectonic plates that are gradually moving in opposite directions. As the plates shift apart, molten magma rises from the mantle to seal the rift and the seafloor slowly spreads toward the continental landmasses.

The rate of spreading has been calculated by magnetic analysis of the rock at *c*.3.75cm (1.5in) a year in the N Atlantic Ocean. Underwater volcanoes mark the line where the continental rise begins. As the plates meet, much of the denser oceanic crust dips beneath the continental plate and melts back into the magma.

TINENTAL DRIFT occurs. It is now a key part of PLATE TECTONICS. It proposes that the SEAFLOOR is moved laterally as new basalt rock is injected along MID-OCEAN RIDGES and so the seafloor spreads symmetrically, and becomes older with increasing distance from the ridge. In 1963 British geophysicists F.J. Vine and D.H. Matthews uncovered supporting evidence for the theory when they tested the patterns of FOSSIL magnetism in the rocks of the seafloor. They discovered a symmetrical pattern of stripes centered on midocean ridges and by matching the stripes with known dates it became possible to calculate the rate of spreading.

seal In engineering, liquidtight or gastight union, such as that between mating pipe flanges, beneath bolted or clamped hatches and between moving parts of machines. Seals are made from rubber, plastics, fibers such as cotton and flax, or combinations of these materials.

sea level Level from which topographic heights are measured. It is usually the mean sea level and is reckoned on being the average of regular sea levels taken over a long period of time. It may also be referred to as the Ordnance Datum (OD).

seaplane AIRCRAFT that can land on and take off from water. The first practical seaplane was built in the USA by Glen CURTISS and flown in 1911. There are two main types of seaplane. **Floatplanes** have large floats that support the fusilage above the water. **Flying boats** float on their boat-shaped hulls, with small floats supporting the wings. In the 1930s flying boats carried more passengers than any other type of aircraft. Some floatplanes and flying boats are fitted with wheels so that they can take off and land on ground as well as water; such aircraft are called **amphibians**.

seasonal affective disorder (SAD) Mental depression apparently linked to the changing seasons of the year. Sufferers experience depressed mood during autumn and winter, and become more cheerful in spring and summer. Some research has indicated that SAD may be caused by decreased levels of light.

seasons Four astronomical and climatic periods of the year based on differential solar heating of the Earth as it makes its annual revolution of the Sun. Due to the parallelism of the Earth's axis of rotation, pointed near the North Star throughout the year, the Northern Hemisphere receives more solar radiation when its pole is aimed towards the Sun in summer and less in winter when it is aimed away, whereas the opposite holds for the Southern Hemisphere. The seasons are conventionally initiated at the vernal (spring) and autumnal EQUINOXES and the winter and summer SOLSTICES.

seawater Solution of several minerals in the OCEAN. It has a high concentration of microscopic life and tastes salty. The dissolved salt content varies between 3.3% and 3.7%, most of which is common salt (sodium chloride, NaCl) but nearly all elements are found in seawater, including vast reserves of magnesium, potassium, and calcium. Large amounts of nitrogen, oxygen, and carbon dioxide are present as dissolved gases. The freezing point of seawater of salt content 3.5% is $-1.9°C$ ($28.6°F$). See also SALINITY

seaweed Any of many species of brown, green or red ALGAE, found in greatest profusion in shallow waters on rocky coasts. KELPS are the largest forms. Many species are important for the manufacture of fertilizers or food, or as a valuable source of chemicals such as iodine. Kingdom PROTOCTISTA.

sebaceous gland Gland in the skin producing the oily substance sebum and opening into a hair follicle. Sebum acts as a skin lubricant and also contains some antibacterial substances.

seborrhea Disorder of the SEBACEOUS GLANDS characterized by overproduction of sebum that results in red, scaly patches on the skin and dandruff. Common at puberty, it predisposes to ACNE.

secant In TRIGONOMETRY, ratio of the length of the hypotenuse to the length of the side adjacent to an acute angle in a right-angled TRIANGLE. The secant of angle A is usually abbreviated to sec$A$, and is equal to the reciprocal of its COSINE, that is $1/\cos A$.

Secchi, Pietro Angelo (1818–78) Italian astronomer who was the first to use PHOTOGRAPHY and SPECTROSCOPY in astronomy. In 1851 Secchi photographed a solar eclipse and between 1864 and 1868 he examined the spectra of more than 4,000 stars, classifying them into spectral types. His system forms the basis of the SPECTRAL CLASSIFICATION used today.

second (symbol s) SI unit of time defined as the time taken for 9,192,631,770 periods of vibration of the ELECTROMAGNETIC RADIATION emitted by a CAESIUM-133 atom. See also PHYSICAL UNITS

secondary cell See CELL

secondary consumer In a FOOD CHAIN, an organism that feeds on the PRIMARY CONSUMERS. The primary consumers, which themselves feed on PRIMARY PRODUCERS, are predominantly herbivores. Secondary consumers are, therefore, mainly CARNIVORES that feed on the herbivores. If a secondary consumer has no PREDATORS (that is, there are no **tertiary consumers**), it is known as the **top carnivore**. See also ECOSYSTEM

secondary emission In physics, emission of ELEMENTARY PARTICLES by a target, such as a gas or metal surface, that is bombarded with primary (newly generated) radiation or particles. Secondary COSMIC RAYS are emitted when primary cosmic rays bombard the atmosphere. The anode of an ELECTRON TUBE emits secondary electrons when it is bombarded by primary electrons from the cathode. This principle is employed in electron-multiplying devices, such as an image intensifier.

secondary sexual characteristics External features that identify a sexually mature animal and distinguish the sexes. Such characteristics play a part in reproductive behavior, although they are not essential for mating. Their development is brought about by the first production of SEX HORMONES in a maturing juvenile. For example, ANDROGENS in male deer stimulate the growth of antlers (except in reindeer, in which the females also bear antlers). In human beings, secondary sexual characteristics develop during PUBERTY. They include breast development in girls, and the growth of facial hair and enlargement of the penis in boys. A boy's voice also becomes deeper. *See also* SEX

secondary thickening (secondary growth) Increase in the girth of stems and roots, especially shrubs and trees, as a result of the generation of new cells in the CAMBIUM. The cambium cells differentiate to produce additional XYLEM on the inside and PHLOEM on the outside. A further cylinder of cambium near the surface of the stem or root may produce a tough outer layer of CORK cells or BARK. *See also* VASCULAR BUNDLE

secretin Polypeptide HORMONE secreted by the small intestine. It stimulates the release of pancreatic juice and also stimulates bile secretion by the GALL BLADDER.

secretion Production and discharge of a substance, usually a fluid, by a cell or a GLAND. The substance so discharged is also known as a SECRETION. Secretions include ENZYMES, HORMONES, SALIVA, and sweat.

sedative Drug used for its calming effect, to reduce anxiety and tension; at high doses it induces sleep. Sedative drugs include NARCOTICS, BARBI-TURATES, **and** BENZODIAZEPINES.

sediment In geology, a general term used to describe any material (such as gravel, sand, and clay) that is transported and deposited by water, ice, wind or gravity. Most sediment is eventually deposited in the sea. The term includes material such as lime that is transported in solution and later precipitated, and organic deposits such as coal and coral reefs.

coal
sand
sandstone-forming sediment
shale
limestone

▲ **sediment** A delta's sediments are laid down in a specific order that may be endlessly repeated if the region where deposition takes place is sinking. Limestone deposits cover the seabed when the delta is too distant to be influential (A). As the delta encroaches (B), fine-grained muds that will become shale are deposited, followed by coarser, sandstone-forming sediments as the advance continues (C). As the water shallows, current bedding (D) indicates that sand is being deposited. Once the delta builds above water level (E) it can support swamp vegetation which forms coal. When the region sinks (F) the cycle restarts.

sedimentary rock Type of rock formed by deposition of SEDIMENT derived from preexisting rocks, which may have been sedimentary, IGNEOUS or METAMORPHIC. Most sediment accumulates on the bed of the sea, having been dumped there by rivers, or having accumulated there as dead sea creatures fell to the seafloor. This accumulated sediment is consolidated and compressed. Earth movements uplift the sediments, and they may be tilted, folded, or faulted. The resulting rocks are sedimentary, and their type depends on their composition. The layers in sedimentary rocks, known as strata, may be a few centimeters or many meters in thickness. Sedimentaries consisting of land sediment are **clastic** rocks, and are GRAVELS, SANDS, SILTS, or CLAYS, according to the size of the particles. Other types of sedimentary rock include: LIMESTONE, which consists of fragments of dead sea creatures, sometimes mixed with land sediment; EVAPORITE, which comprises saline deposits; COAL, which is accumulated vegetation; coralline, which contains large quantities of CORAL; and CHALK, which is a pure form of limestone, with very little land sediment. *See also* FAULT; FOLD; STRATUM

sedimentation Any process or processes that deposit ROCK-forming materials. The materials deposited are continental in origin, the debris of landforms worn down and carried off by wind, water, or ice. Whenever the flow of the transporting medium is interrupted or diverted, the carried material is deposited on land (in desert, lake or river bed), on coasts, or as marine sediments. *See also* DEPOSITION; EROSION

Seebeck, Thomas Johann (1770–1831) German physicist, b. Russia. Seebeck is celebrated chiefly for his contribution to THERMOELECTRICITY. In 1821 he discovered that if two strips of different metals were joined in a loop and the two junctions were maintained at different temperatures, an ELECTROMOTIVE FORCE (E.M.F.) was produced that was proportional to the temperature difference. He called this thermomagnetism. Today, it is called the SEEBECK EFFECT. Seebeck falsely believed that the conductors were magnetized directly by the difference in temperature. His discovery was neglected for many years, but is the basis of the modern THERMOMETER. Seebeck also studied the polarization of stressed glass.

Seebeck effect In THERMOELECTRICITY, the generation of ELECTROMOTIVE FORCE (E.M.F.) in a circuit consisting of two different metals or semiconductors joined in a loop, when the two junctions are kept at different temperatures. It was discovered by Thomas SEEBECK (1770–1831). This type of circuit is called a **thermocouple**. *See also* PELTIER EFFECT

seed Part of a flowering plant (ANGIOSPERM) that contains the EMBRYO and food store. It is formed in the OVARY by FERTILIZATION of the female GAMETE. Food may be stored in a special tissue called the ENDOSPERM, or it may be concentrated in the swollen seed leaves (COTYLEDONS). Seeds are often capable of surviving quite harsh conditions, and provide a means of survival for many plants in extreme environments. They are also the unit of dispersal of flowering plants (angiosperms) and CONIFERS. *See also* GERMINATION

Segrè, Emilio Gino (1905–89) US physicist, b. Italy. In 1937 he discovered TECHNETIUM, the first ELEMENT to be artificially produced. In 1940 he helped to discover ASTATINE and PLUTONIUM-239. Segrè is, however, chiefly remembered for his discovery of the antiproton. In 1930 Paul DIRAC had predicted the existence of antiparticles. In 1955, using a Bevatron particle ACCELERATOR at the University of California, Segrè and Owen CHAMBERLAIN bombarded copper with high-energy PROTONS to produced the antiparticle of the proton. Chamberlain and Segrè shared the 1959 Nobel Prize for physics. *See also* ANTIMATTER

seif Elongated type of sand DUNE. Seifs may be several kilometers long and often occur in groups. The narrow ridge of the dune runs parallel to the direction of the prevailing wind, and its height and width can be increased by the action of cross-winds. Seifs are found mainly in the sand deserts of N Africa.

seismic profile Continuous record of sound waves bounced off sediments on the SEAFLOOR. The sounds become SEISMIC WAVES and as such are used to determine the thickness and structure of bottom sediments. *See also* ECHO SOUNDER

seismic survey Use of SEISMIC WAVES produced by underground explosions to detect and measure the extent of minerals (such as oil and coal) under the ground. The speed of the shock waves indicates the type of material they are passing through.

seismic wave Shock wave produced by EARTHQUAKES and man-made explosions. The velocity (speed and direction) of seismic waves varies according to the material (type of rock, molten core or oil) through which they pass. Primary (P) and secondary (S) waves are transmitted by the solid Earth. **P waves** vibrate in the direction that they are advancing; **S waves** vibrate at right angles to the direction in which they are advancing. Only P waves are transmitted through fluid (liquid or gas) zones. *See also* SEISMOLOGY

seismograph (seismometer) Instrument for measuring and recording SEISMIC WAVES caused by movement (EARTHQUAKE or explosion) in the Earth's crust. The vibrations are recorded by a pen on a revolving drum. Some seismographs are sensitive enough to pick up seismic activity thousands of kilometers away. For example, in January 1994 the earthquake in Los Angeles was picked up by seismographs in Britain. *See also* RICHTER SCALE

seismology Study of EARTHQUAKES and the SEISMIC WAVES they produce. The movement of seismic waves is detected and recorded by SEISMOGRAPHS. The detection involves the separation of events from the ever-present background of seismic noise. Pinpointing sites of events has become very accurate since the development of precise instrumentation and the establishment of the World Wide Standard Seismograph Network (WWSSN). Seismology is very important in the exploration of the Earth's internal structure.

selection In GENETICS, the probability that one ALLELE will be passed on in favor of another. *See also* DOMINANT; NATURAL SELECTION; RECESSIVE

selective breeding Process by which stockbreeders and agriculturalists improve the strains of domesticated animals and cultivated plants; it involves selection and pairing of individuals with desirable qualities in the PHENOTYPE. Today a growing knowledge of GENETICS ensures more predictable results.

selenium (symbol Se) Gray, METALLOID element of group VI of the periodic table, discovered (1817) by Jöns BERZELIUS. Its chief source is as a by-product in the electrolytic refining of copper. The element resembles sulfur. It is used extensively in PHOTOELECTRIC CELLS, SOLAR CELLS, XEROGRAPHY, and red pigments. Properties: at.no. 34; r.a.m. 78.96; r.d. 4.79; m.p. 217°C (422.6°F); b.p. 684.9°C (1,265°F); most common isotope ^{80}Se (49.82%).

selenology Study of the MOON's physical and chemical composition.

Selye, Hans (1907–82) Canadian physician, b. Austria. He is celebrated for his pioneering studies of the effects of stress on human and animal physiology. Selye outlined the "general adaptation syndrome," a series of stages the body undergoes when subjected to stress. His works include *The Stress of Life* (1956), *Physiology and Pathology of Exposure to Stress* (1960), and *Stress Without Distress* (1974).

semaphore Device or technique that communicates messages visually; a type of optical TELEGRAPH. A **railroad** semaphore signal consists of a vertical post on which a single projecting arm is mounted, whose angle indicates "all clear" or "danger." In **flag** semaphore, a signaler indicates

letters and numerals by the position of his or her outstretched arms, emphasized with flags.

semen Fluid ejaculated by the male at the climax of sexual intercourse; it contains SPERM from the TESTES and the secretions of various accessory sexual glands. In human beings, each ejaculate is normally about 3 to 6ml by volume and contains about 200 to 300 million sperm.

Semenov, Nikolai Nikolaevich (1896–1986) Russian chemist. He studied in Leningrad and later Moscow State University. Semenov shared the 1956 Nobel Prize for chemistry with Cyril HINSHELWOOD for their independent work on branched chain reactions in COMBUSTION processes.

semicircular canals Three parts of the inner EAR that function as balance organs. They are tubular ducts that project in different planes and can detect, by the movement of a liquid within them, movement in any direction in space.

semiconductor Substance with an electrical CONDUCTIVITY between that of a CONDUCTOR and an INSULATOR. The conductivity increases as temperature increases. Adding appropriate impurities also increases conductivity. A semiconductor consists of elements, such as GERMANIUM and SILICON, or compounds, such as aluminum phosphide, with a crystalline structure. At normal temperatures, some ELECTRONS break free and become carriers of the electric current. The "holes" (electron deficiencies) left by these electrons move in the opposite direction to the electrons and behave like positive charge carriers. Impurities are usually added to the semiconductor material in controlled amounts during manufacture to add more free electrons or create more holes. This process is called **doping**. There are two types of semiconductor: the *n-type,* in which the current carriers (electrons) are negative, and the *p-type,* in which the current carriers are moving, positively charged holes. A semiconductor junction is formed when there is an abrupt change from one type of impurity to the other. Such a *p-n* **junction** acts as a very efficient RECTIFIER and is the basis of the semiconductor DIODE. Semiconductors are used in COMPUTERS, TRANSISTORS, and PHOTOELECTRIC CELLS. *See also* INTEGRATED CIRCUIT (IC)

semipermeable membrane Thin, sheetlike material that permits the passage of a SOLVENT (such as water) but not larger dissolved SOLUTES (such as salt and sugar). The property of permeability depends on the molecular diameter of the dissolved substance and the nature of the membrane. Common membranes include thin palladium foil, pig's bladder, copper ferrocyanide (cyanoferrate), and the walls of plant and animal cells. *See also* OSMOSIS

semiprecious stones GEMS worn as jewelry, but not classified as precious stones. Not a strict grouping, semiprecious stones include the minerals AMETHYST, GARNET, OPAL, TOPAZ, TOURMALINE, TURQUOISE, and the nonminerals AMBER, CORAL, and JET. Precious stones are generally taken to be DIAMOND, EMERALD, SAPPHIRE, PEARL, and RUBY.

Semmelweiss, Ignaz Philipp (1818–65) Hungarian physician, probably the first to recognize the importance of ANTISEPTICS in preventing infection. In 1847, during an outbreak of puerperal fever, he ordered doctors at the Vienna General Hospital to wash their hands before touching a patient. His ideas were ridiculed until confirmed by Joseph LISTER.

sense receptor *See* RECEPTOR

senses Means by which animals gain information about their environment and physiological condition. The five senses (SIGHT, HEARING, TASTE, SMELL, and TOUCH) all rely on specialized RECEPTORS on or near the external surface of the body. SENSORY NEURONS carry information from the sense organs to the brain. Additionally, receptors within the body detect internal physical and chemical changes.

sensory neuron NERVE that carries information from a RECEPTOR organ in any part of the body to the CENTRAL NERVOUS SYSTEM (CNS). Their nerve endings are in the sense organs. *See also* MOTOR NEURON; NEURON; SENSES

sepal Modified leaf that makes up the outermost portion of a FLOWER bud. Although usually green and inconspicuous once the flower is open, in some species, such as orchids, the sepals look like the PETALS. In other species, such as anemones, the sepals are absent.

septic tank Unit for SEWAGE disposal, usually in rural locations not connected to sewers. Watertight and airtight, septic tanks are buried in the ground. ANAEROBIC bacteria in the sewage decompose solids to form a sludge, which is pumped out regularly.

sequence Numbers or terms in an organized list. Sequences may be finite (having a limited number of terms) or infinite, such as the complete sequence of counting numbers 1, 2, 3, 4, ... Infinite sequences may or may not be convergent (tend to a single limit). The sequence 0, 0.9, 0.99, 0.999, ... for example, is convergent, tending to the limit 1. Sequences are sometimes called progressions. *See also* ARITHMETIC PROGRESSION; CONVERGENCE; DIVERGENCE; GEOMETRIC PROGRESSION; HARMONIC PROGRESSION; SERIES

séracs Very irregular ice surface on a GLACIER. It may include an impenetrable system of pillars and pinnacles several meters tall. A séracs usually forms at the foot of an icefall, caused by compression of a crevassed section of the glacier. This, in turn, results from an increase in the gradient of the moving ice or because its speed is somehow reduced.

sere SUCCESSION of plant communities that eventually produces the CLIMAX COMMUNITY. Any one sere has a series of seral communities that change with time. There are various recognized types of seres, including a **hydrosere** (occurring in shallow fresh water), **lithosere** (on rocky ground), **plagiosere** (cleared ground), and **xerosere** (dry ground).

series Mathematical expression (sum) obtained by adding the terms of a SEQUENCE. Thus, the series 1 + 4 + 9 + 16 + ... is formed from the sequence 1, 4, 9, 16, ... Like a sequence, a series may involve a finite or infinite number of terms and the infinite series may or may not converge. There are many useful series for the number π, which can be used to evaluate π to several decimal places. A series formed from increasing powers of a VARIABLE is a **power series**; such power series are used to represent many functions. *See also* SUMMATION

series circuit ELECTRIC CIRCUIT in which the components are connected in sequence so that the current flows through them one after the other. The total RESISTANCE, R, of a number of resistors connected in series is given by the expression $R = R_1 + R_2 + R_3 + ...$, where R_1, R_2 and R_3 are the resistances of the individual resistors. The total CAPACITANCE, C, of capacitors in series is given by $1/C = 1/C_1 + 1/C_2 + 1/C_3 + ...$, where C_1, C_2 and C_3 are the capacitances of the component capacitors. *See also* PARALLEL CIRCUIT

serine ($HOCH_2CH(NH_2)COOH$) White, crystalline AMINO ACID found in proteins.

serotonin Chemical found in cells of the gastrointestinal tract, blood platelets, and brain tissue, concentrated in the midbrain and HYPOTHALAMUS. It is a VASOCONSTRICTOR and has an important role in the functioning of the nervous system and in the stimulation of smooth muscles.

Serpens (Serpent) Extensive equatorial constellation situated between Libra and Hercules and divided into two parts, Serpens Cauda and Serpens Caput, by Ophiuchus (the Serpent-bearer). Serpens Caput contains the bright globular cluster M5 (NGC 5904), and Serpens Cauda includes M16 (NGC 6611), an open cluster shining through surrounding nebulosity. The brightest star is Alpha Serpentii (Unukalhai), magnitude 2.7.

serpentine Group of sheet silicate minerals, hydrated magnesium silicate $(Mg_3Si_2O_5(OH)_4)$. Serpentine minerals come in various colours, usually green, although sometimes brownish, with a pattern of green mottling. They have monoclinic system crystals. Serpentines are secondary minerals, formed from minerals such as olivine and orthopyroxene, and they occur in igneous rocks containing these minerals. They are commonly used in decorative carving; fibrous varieties are used in asbestos cloth. Hardness 2.5–4; r.d. 2.5–2.6.

serum Clear fluid that separates out if blood is left to clot. It is essentially of the same composition as PLASMA, but without FIBRINOGEN and clotting factors.

servomechanism Device that provides remote control to activate a mechanism. This is done by converting an input signal, such as a radio impulse or mechanical movement, into a mechanical output like a lever movement or amplified hydraulic force. The device usually forms part of a control system. Some servomechanisms, such as the AUTOPILOT of an aircraft, incorporate a FEEDBACK mechanism that makes it independent of human control.

sessile In zoology, describing an animal that remains fixed in one place. Such sedentary animals are usually permanently attached to a surface, such as sea anemones, barnacles, limpets, and mussels. The term sessile is also used to describe the eyes of crustaceans that lack stalks and sit directly on the animal's head. In botany, sessile describes any structure that has no stalk (in cases where one might be expected) and grows directly from a stem. Examples include the acorns and leaves of some oak trees.

sets In mathematics, a defined collection of objects. The objects are called the elements or members of the set. The number of members can be finite or infinite, or even be zero (the number of members in the null set denoted $\varnothing$). Each element in a set is counted only once. Various relations can exist between two sets. Two sets, A and B, are equal $(A = B)$ if both sets contain exactly the same elements. A is a **subset** of B if all members of A are also members of B. A subset of A that is not equal to B is called a **proper subset** (written $A \subset B$). **Disjoint** sets have no members in common; **overlapping** sets have one or more members in common. The **union** of two sets A and B (written $A \cup B$) is a new set that contains all the members of both A and B. The **intersection** of A and B (written $A \cap B$) contains only those elements common to both A and B. For example, the set $A = \{3, 6, 9, 12, 15\}$ is the set of positive multiples of 3 less than 16; $B = \{2, 4, 6, 8, 10, 12, 14\}$ is the set of even numbers less than 16; $C = \{blue, green, red\}$ is the set of primary colors; $D = \{1, 2, 3, 4, ...\}$ is the infinite set of natural NUMBERS. Both A and B are proper

subsets of D ($A \subset D$ and $B \subset D$). There are no elements common to both A and C, so $A \cap C = \varnothing$. In contrast, A intersection B contains the even multiples of 3 less than 16, that is $A \cap B = \{6, 12\}$.

set theory Branch of MATHEMATICS developed by Georg CANTOR in the late 19th century. It is based on George BOOLE's work on mathematical LOGIC, but involves manipulations of SETS of abstract or real objects rather than logical propositions. It is concerned with the properties of sets, the relations between them and operations on them, and can be applied to many other branches of mathematics.

sewage Matter that flows through sewers, including drainage water, and liquid and solid waste materials (notably feces). Sewage works remove solids and treat the WATER so that it can be discharged into rivers, lakes or the sea without causing POLLUTION. The discharged liquid is called **effluent**. At a typical

▲ **servomechanism**
A servomechanism will automatically correct a deviation from a pre-set position of a machine. For example, the autopilots fitted to many aircraft utilize servomechanisms to manipulate the control surfaces of the plane. A change from the intended altitude of the plane – such as banking (1) will deflect a gyroscope (2). This will move a linkage (3) so that a sliding contact (4) moves from a central position so

that a current flows in a detector circuit (5). This current is amplified (6) and used to power a servo motor (7) – to move the ailerons (8) and correct the flight of the plane (9), returning the contact to a point where no current flows in the circuit (5). If the system was too sensitive, it could tip the plane the other way, reversing the current in the circuit and causing the plane to oscillate around the safe position – called hunting.

sewage plant, incoming sewage is screened free from stones and other hard materials. Heavy sludge is "settled off" and removed to closed digester vessels, in which bacteria break it down to a relatively inoffensive material that can be used as FERTILIZER. Liquid sewage passes on to trickle filters or activated sludge tanks, in which it is purified by the combined digestive action of bacteria and protozoa; chemical treatment may also be used. Further settling tanks remove the inactive HUMUS materials, after which the remaining water is fully purified and can be reused. *See also* CHLORINE; EUTROPHICATION

sex Classification of an organism denoting the reproductive function of the individual. There are two divisions of the classification – male and female. In mammals, the presence of sex organs (OVARIES in the female, TESTES in the male) are primary sexual characteristics. SECONDARY SEXUAL CHARACTERISTICS, such as size, coloration, and, in human beings, the development of the breasts and the growth of facial and body hair, are governed by the secretion of SEX HORMONES. In flowering plants, the female sex organs are the CARPEL (including the OVARY, STYLE, and STIGMA), and the male organs the STAMENS. Male and female organs may occur in the same flower or on separate flowers or plants. In lower organisms, such as fungi, where the differences are biochemical rather than morphological, the term "mating strains" is used instead of male and female. *See also* SEXUAL REPRODUCTION

sex determination Genetic basis for the gender of an organism. Gender depends on the combination of sex CHROMOSOMES. In mammals, which have a pair of sex chromosomes, there are two types, the X-chromosome and the Y-chromosome. All ova have one **X-chromosome**, whereas sperm have either an X- or a Y-. If the sperm that fertilizes the OVUM has an X-chromosome, the resulting ZYGOTE and the organism that develops from it will be XX, or female. If the egg is fertilized with a Y-carrying sperm, the zygote will be XY, or male.

sex hormones Chemical "messengers" secreted by the gonads (TESTES and OVARIES). They regulate sexual development and reproductive activity and influence sexual behavior. In males, they include the ANDROGEN testosterone, made by the testes; in females, the sex hormones ESTROGEN and PROGESTERONE are produced by the ovaries.

sextant Optical instrument for finding LATITUDE (angular distance N or S of the Equator). The sextant consists of a frame with a curved scale marked in degrees, a movable arm with an ordinary mirror at the pivot, a half-silvered glass, and a TELESCOPE. The instrument works on the principle that the angle of a celestial body above the HORIZON

depends on the observer's latitude. While looking through the telescope, the sextant arm is adjusted until an image of the Sun or a star just touches an image of the horizon. The angle of the Sun or star above the horizon is then read from the scale. A set of tables gives the corresponding latitude for various angles measured.

sexually transmitted disease (STD) Any disease that is transmitted by sexual activity involving the transfer of body fluids. It encompasses a range of conditions that are spread primarily by sexual contact, although they may also be transmitted in other ways. These include ACQUIRED IMMUNE DEFICIENCY SYNDROME (AIDS), pelvic inflammatory disease (PID), and viral HEPATITIS. One effect of the AIDS crisis may have been to divert attention away from some of the older STDs, such as SYPHILIS and GONORRHEA, which nonetheless remain significant public health problems.

sexual reproduction Biological process of REPRODUCTION involving the combination of genetic material from two parents. It occurs in different forms throughout the plant and animal kingdoms. This process gives rise to variations of the GENOTYPE and PHENOTYPE within a species. GAMETES, HAPLOID sex cells produced by MEIOSIS, contain only half the number of CHROMOSOMES of their parent cells (which are DIPLOID). At FERTILIZATION, the gametes, generally one from each parent, fuse to form a ZYGOTE with the DIPLOID number of chromosomes. The zygote divides repeatedly and the cells differentiate to give rise to an EMBRYO and, finally, a fully formed organism. In mammalian reproduction, the male's erect PENIS is inserted into the female's VAGINA. Rhythmic, thrusting movements of the penis cause the male to ejaculate and so introduce SEMEN (containing SPERM) into the vagina in order for fertilization of the OVUM to take place.

sexual selection Process among animals that derives from the successful finding of a mate and production of young. It is similar to NATURAL SELECTION but involves mainly the SECONDARY SEXUAL CHARACTERISTICS. For example, if a female chooses to mate with the male that has the brightest coloration and best courtship display, these features (being genetically controlled) will tend to be inherited by the male offspring. In succeeding generations the same features will be exaggerated. By the same mechanism, males may develop longer horns or louder roars to better compete with other rival males.

shadow Area screened from a light source and therefore relatively dark. It varies in size depending on the distance between the light source and the screening object. When the light source is extended, the outline of the shadow is blurred; where the source is partly visible, there is an area of mid-shadow (**penumbra**) lying outside the central, darker shadow (**umbra**).

shale Common SEDIMENTARY ROCK similar to clay. It is fine grained and consists of thin layers or sheets, with each layer probably representing a period of DEPOSITION. Shales may contain various materials such as fossils, carbonaceous matter, and oil.

Shannon, Claude Elwood (1916–) US mathematician who is best remembered for developing information theory. In the late 1940s he proposed a method of making information less sensitive to the corrupting effects of electronic noise by storing it as binary digits. He also did research on robots and seminal work on information transmission, which did not bear fruit until after the development of integrated circuits.

Shapley, Harlow (1885–1972) US astronomer who provided the first accurate model of our GALAXY. In 1914 Shapley explained how the variability of CEPHEID VARIABLES is caused by their pulsations. Later, he calibrated the period–luminosity law discovered by Henrietta LEAVITT, using Cepheid variables in GLOBULAR CLUSTERS, which enabled him to estimate of the size of our galaxy and the position in it of the SOLAR SYSTEM. Shapley also made estimates of the sizes of ECLIPSING BINARIES and worked on photometry and SPECTROSCOPY.

shearing force (shearing stress) Force tending to cause deformation of a material by slipping along a plane parallel to the imposed STRESS. In nature, the resulting shear is related to the downslope of Earth materials as well as to earthquakes.

shell In biology, hard protective case of various MOLLUSKS. The case is secreted by the epidermis of the mollusk and consists of a protein matrix strengthened by CALCIUM CARBONATE.

shellac Purified RESIN made from the secretions of the lac insect, *Laccifer lacca*. Shellac is soft and fluid (it flows) when heated but hard at room temperature. It is used in the production of adhesives, hair sprays, varnishes, and sealants.

shell, atomic Any of the groupings of orbital ELECTRONS around the nucleus of an ATOM, named K, L, M, and so on, outward, each containing a limited number of electrons of a particular energy.

Shepard, Alan Bartlett, Jr (1923–98) US astronaut. In 1961 Shepard became the first American to be launched into space. The brief, suborbital flight to a height of *c*.185km (115mi) lasted just over 15 minutes. His *Freedom 7* capsule splashed down in the Atlantic Ocean nearly 480km (300mi) from the Cape Canaveral (Florida) launchpad. Shepard commanded *Apollo 14* on its flight to the Moon in 1971, and was the fifth man to walk on the lunar surface. Among the awards he received was the Space Medal of Honor.

▲ **sexual reproduction** The male reproductive system is situated both inside and outside the pelvic region. Outside the body are the testes. They produce sperm in the seminiferous tubules. Once the sperm are matured they are stored in the epididymis. During intercourse they pass along the vas deferens to the urethra. The seminal vessicles, prostate and Cowper's glands all secrete fluids into the urethra, helping to make semen. The female reproductive organs lie solely within the pelvic girdle. The ovaries usually release just one mature egg each month, which is then transferred to the uterus via one or other of the fallopian tubes. Embedded in the ovaries of the newly born female are several hundred thousand follicles that can potentially develop into eggs. Only a few hundred of them will do so, becoming mature eggs which, when they are ripe, are caught by the fimbria at the end of the tubes, near the site where fertilization usually occurs. The woman's urinary system, unlike the male's, is separate from the genitals.

sherardizing Industrial method of coating STEEL parts with ZINC for protection from CORROSION. The parts, which may be steel castings, nuts and bolts, welded tubes or sections, are tumbled in heated drums containing zinc dust. After some hours the zinc forms a very thin intermetallic surface coating the steel. *See also* GALVANIZING

Sherrington, Sir Charles Scott (1857–1952) British physiologist, pioneer in the study of how the NERVOUS SYSTEM works. His book *The Integrative Action of the Nervous System* (1906) helped to establish physiological psychology. Sherrington shared with Edgar Adrian the 1932 Nobel Prize for physiology or medicine.

shield In geology, large, low-relief, exposed mass of PRECAMBRIAN rock, commonly having a gently convex surface and surrounded by belts of younger rock. Shields contain rock that is more than 2.5 billion years old, now changed by METAMORPHISM but originally composed of basaltic LAVA. Shields form the nucleus of continents; for example, the Canadian Shield, located between Hudson Bay and the Great Lakes, occupies two-thirds of Canada.

shield, radiation Material, such as loaded concrete (concrete containing lead, iron or barium) or lead, surrounding a radioactive source to absorb NEUTRONS and RADIATION. The core of a NUCLEAR REACTOR is surrounded by a massive concrete biological shield to protect personnel, and by an iron thermal inner shield to protect the concrete from high temperatures.

ship Vessel for conveying passengers and freight by water. Modern oceangoing ships developed from early **sailing ships**, such as the carracks (Mediterranean merchant ships) of the 13th century and the larger early ships called galleons. Fighting ships of the 17th and 18th centuries included frigates of various designs; later models had several rows of guns. Sailing freighters led to the development of the great **clippers** of the late 19th century, some of which had iron hulls. A century or so earlier, the first **steamships** had been built. They were powered by wood or coal-burning STEAM ENGINES that drove large paddle wheels, hence the term paddle wheeler. In 1819 the first Atlantic crossing was made by "steam-assisted sail" and this crossing soon became a regular service. By the mid-19th century steamships driven by PROPELLERS, or screws, were in general service and were voyaging on trade missions in all oceans. Marine steam TURBINES were developed at the turn of the 19th century and gradually replaced reciprocating (back-and-forth cranking) steam engines for large vessels, examples being ocean **liners** of the early 1900s with displacements (weight of water that a ship displaces when fully laden) up to

10,000 metric tons. Oil, rather than coal, was now the favorite fuel for large marine engines. Some of the newest military ships and icebreakers are fitted with nuclear engines in which heat from a NUCLEAR REACTOR raises steam in boilers to drive steam turbines. The larger modern ships are classified as warships, passenger vessels, bulk dry cargo ships, tankers for bulk liquids, freighters for mixed cargo and container ships. The largest of these are the supertankers carrying oil, with displacements of about half a million metric tons and lengths of several hundred meters. *See also* BOAT

shock Acute circulatory failure, possibly with collapse. Caused by disease, injury or emotional trauma, it is characterized by weakness, pallor, sweating, vomiting, and a shallow, rapid pulse. In shock, there is a drop in blood pressure to a level below that needed to oxygenate the tissues. Treatment depends on the cause, but will involve prompt blood or fluid replacement if there has been severe loss.

shock absorber Elastic mechanism for absorbing shock in order to prevent damage to a machine. The shock absorbers of motor vehicles typically have a piston, attached to the vehicle body, moving inside a cylinder attached to a wheel. The piston moves against a liquid that escapes from and returns to the cylinder through a spring-loaded valve.

Shockley, William Bradford (1910–89) US physicist. During World War II he worked on anti-submarine research for the US Navy. In 1947 Shockley and his colleagues, John BARDEEN and Walter BRATTAIN, invented a point-contact TRANSISTOR and a junction transistor at the Bell Telephone Laboratories. They shared the 1956 Nobel Prize for physics. As professor of electrical engineering (1958–75) at Stanford University, he caused controversy with his views on the link between race and intelligence.

shock wave In fluids (liquids or gases), region across which sharp discontinuities in pressure and density occur. Shock waves are brought about by objects moving at supersonic velocities. Because the surrounding fluid can propagate disturbances only at the local speed of sound, the moving object "piles up" the disturbances it is causing into a V-shaped "wake" attached to the object. The SONIC BOOM of supersonic aircraft is the passage of this shock wave past the eardrum. *See also* EARTHQUAKE; SEISMIC WAVES; SUPERSONIC FLIGHT

shooting star *See* METEOR

short circuit Unintended connection between two points in an ELECTRIC CIRCUIT, offering the easiest path for current to follow. A large current will flow and fire or damage may result if the voltage supply is not turned off quickly. FUSES or CIR-CUIT BREAKERS are used to protect against the effects of short circuits.

shoulder blade Common name for the SCAPULA.

shrub (bush) Woody, perennial plant that is smaller than a TREE. Instead of having a main STEM, it branches at or slightly above ground level into several stems. Its hard stem distinguishes it from an HERB. Rhododendrons and azaleas are popular ornamental shrubs.

sial In geology, uppermost of the two main rock classes in the Earth's CRUST. Sial rocks are so called because their main constituents are silicon and aluminum. They make up the material of the CONTINENTS and overlay the SIMA. The sial is less dense than the sima and is generally not carried beneath the surface when subduction occurs.

sickle-cell anemia Inherited blood disorder featuring an abnormality of HEMOGLOBIN. The hemoglobin is sensitive to a deficiency of oxygen and it distorts ERYTHROCYTES (red blood cells), causing them to become rigid and sickle-shaped. Sickle cells are rapidly lost from the circulation, which gives rise to anemia and jaundice. Particularly common in people of African descent, the disease gives some protection against MALARIA.

sideband Any FREQUENCY produced by MODULA-TION that is added to a CARRIER WAVE. Those frequencies above the carrier wave are called **upper sidebands**; those below, lower sidebands. Sidebands contain the information that is passed from transmitter to receiver. Such FREQUENCY MODULATION (FM) minimizes interference and provides greater fidelity than AMPLITUDE MODULATION (AM). *See also* RADIO

sidereal period Orbital period of a planet or other celestial body in relation to a background star. It is the true orbital period. **Sidereal time** is local time reckoned according to the rotation of the Earth in relation to the stars. The sidereal day is 23 hours, 56 minutes, and 4 seconds of mean solar time, nearly 4 minutes shorter than the mean solar day. The **sidereal year** is the time required for the Earth to complete one revolution around the Sun relative to the fixed stars. It is equal to 365.25636 mean solar days. *See also* ORBIT

siderite Green, brown or white mineral, iron(II) carbonate ($FeCO_3$), found in sedimentary iron ores and as vein deposits with other ores. Its crystals are rhombohedral in the hexagonal system and it occurs as massive deposits or in granular form. 3.8. IRON METEORITES are sometimes known as siderites. Hardness 4; r.d.

Siegbahn, Karl Manne Georg (1886–1978) Swedish physicist who was awarded the 1924 Nobel Prize for physics for research into X-ray SPECTROSCOPY. Siegbahn improved techniques for measuring the WAVELENGTHS of X-RAYS and used these new methods to discover groups of X-rays both less penetrating and with longer wavelengths than had been previously thought possible. This discovery supported the SHELL THEORY of atomic structure of the Danish physicist Niels BOHR. In 1924 he also proved the similarity between X-rays and visible light by showing that X-rays could be refracted by a PRISM.

Siemens German brothers associated with the electrical engineering industry. In 1849 **Ernst Werner** von Siemens (1816–92) developed an electric TELEGRAPH system. With his brother **Karl** (1829–1906), he set up subsidiaries of the family firm in London, Vienna, and Paris. **Frederich** (1826–1904) and **Karl Wilhelm** (later Sir William) (1823–83) developed a regenerative FUR-NACE that led to the OPEN-HEARTH PROCESS uesd in the steel industry. In 1843 Karl Wilhelm introduced to Britain an ELECTROPLATING process that he and Werner had developed.

siemens (symbol S) SI unit of electrical CONDUC-TANCE. It is equal to the conductance of a circuit with a RESISTANCE of 1Ω (OHM).

sievert (symbol Sv) Derived SI unit of dose equivalent, equal to one joule per kilogram ($1Jkg^{-1}$). One sievert is equal to 100 rems. *See also* RADIATION

sight SENSE by which form, color, size, movement, and distance of objects are perceived. Essentially, it is the detection of light by the EYE, enabling the formation of visual images.

significant figures (s.f.) (significant digits) Digits of a number that express it to a desired accuracy, the last figure possibly being rounded up or down. Thus 2.871828 to six significant figures is 2.87183 (6s.f.); to three significant figures it is 2.87 (3s.f.).

Sikorsky, Igor Ivanovich (1889–1972) US aeronautical engineer and designer, b. Russia. In 1913 Sikorsky built and flew the first four-engined AIRCRAFT. In 1919 he emigrated to the US. In 1923 he established the Sikorsky Aero Engineering Corporation, which was later merged with the United Aircraft Corporation. Sikorsky is celebrated for his development of the HELICOPTER. In 1939 he designed the first commercially successful helicopter. In 1941 Sikorsky set a world endurance record for sustained flight in a helicopter.

silane (silicane, SiH_4) Colorless gas that is produced by the reduction of SILICON with lithium tetrahydridoaluminate. Silane is stable in the absence of oxygen, but it is spontaneously combustible. It is insoluble in water. Silane is used for the removal of corrosion in nuclear reactor pipes. It is used often in reference to the entire class of **silicon hydride** compounds. Properties: r.d. (liquid) 0.68; m.p. −185°C (−301°F); b.p. −112°C (−170°F).

silica (SiO_2) Silicon dioxide, a compound of SILICON and oxygen. It occurs naturally as QUARTZ and CHERT (which includes flint). Silica and SILICATE minerals (silica combined with other elements) are the main constituents of 95% of all rocks and account for 59% of the Earth's crust. Silica is used in the manufacture of GLASS, CERAMICS, and SILICONE.

silica gel Amorphous chemical compound. It is produced by separating out a sol of sodium silicate

and then heating it to achieve partial dehydration. Because of its ability to absorb moisture silica GEL is used as a drying agent and as a CATALYST.

silicate Any of a large group of rock-forming minerals made up of silicon and oxygen in SiO_4 units bonded to various metals. The SiO_4 units may form single or double chains (as in PYROXENE and AMPHIBOLE), sheets (as in MICA), rings (as in BERYL) or ionic bonds to a metal (as in OLIVINE). Silicate minerals, such as FELDSPAR, GARNET, and mica, form 95% of the material of the Earth's crust. GLASS is a mixture of silicates with small amounts of other substances. Sodium silicates are used as adhesives and in the production of detergents. QUARTZ (SiO) is also usually regarded as a silicate mineral.

silicon (symbol Si) Common, gray, nonmetallic element of group IV of the periodic table, first isolated in 1824 by J.J. BERZELIUS. Silicon is found only in combinations such as SILICA and SILICATE. After FELDSPAR it is the second most abundant element in the Earth's crust (27.7% by weight), being a common constituent of minerals, such as QUARTZ. Silicon is found in many plants and animals; it is the major constituent of the cell wall (test) of DIATOMS. It is produced commercially by heating sand and coke in a furnace. It is used as an alloy and a purified form is used in the preparation of SILICONES. It is also utilized in the production of GLASS. SILICON CHIPS are used extensively in the COMPUTER industry for TRANSISTORS and other SEMICONDUCTOR devices. Properties: at.no. 14; r.a.m. 28.086; r.d. 2.33; m.p. 1,410°C (2,570°F); b.p. 2,355°C (4,271°F); most common isotope ^{28}Si (98.21%).

▲ **silicon chip** A silicon chip is manufactured by building up layers on top of a wafer of silicon (1). (A) First a layer of silicon dioxide (2), an insulator, is laid down followed by photo-sensitive photoresist (3). Photoresist hardens when hit by ultraviolet light (4). By using a mask (5) the area to be hardened can be controlled. (B) The unhardened area shielded by the mask can then be rinsed out with a solvent (6). The photoresist is then removed by hot gases. (C) The same process is used to apply a conducting polysilicon (7). Again ultraviolet light fixes the photoresist (8) in the unmasked area. (D) A solvent removes the photoresist. (E) N-type silicon, which only carries a negative charge, is then created by doping of the silicon base. (F) A third masking process creates shafts (9) to the n-type silicon. (G) An aluminum layer is then applied. (H) A fourth masking forms electrical contacts connecting the layers of silicon.Hundreds of chips are simultaneously made on a single wafer of silicon (10) before they are eventually separated (11) and mounted individually for use.

silicon chip Small piece of SILICON etched to carry many tiny ELECTRIC CIRCUITS. Silicon chips are at the heart of most electronic equipment. A personal COMPUTER will contain many different types of chip, most notably the MICROPROCESSOR that is the "brain" of the computer. Chips are etched, layer by layer, onto slivers of pure silicon. Each layer is "doped" to give it particular electrical properties, and the combination of different layers form components such as TRANSISTORS, DIODES, and their interconnections. Most chips are made by photographic etching processes, which impose a limit on the separation of components. New ways of etching chips using much finer X-rays are being investigated. *See also* CHARGE-COUPLED DEVICE (CCD); INTEGRATED CIRCUIT (IC); PRINTED CIRCUIT; SEMICONDUCTOR

silicone Odorless and colorless polymer based on SILICON. Silicones are inert and stable at high temperatures, and are used in lubricants, varnishes, adhesives, water repellents, hydraulic fluids, and artificial heart valves. The modern use of silicone gel for cosmetic breast implants was drastically reduced in 1992 following health warnings.

silk Natural FIBER produced by many creatures, notably the silkworm. Some other insects and spiders produce silk, but it is most economic to take it from the silkworm. Silk is a strong, high-quality material used to make fabric. The fiber consists mainly of fibroin (tough, elastic protein) covered with sericin (gelatinous protein). Almost all silk is obtained from silkworms reared commercially. Silkworms feed on mulberry leaves and a single COCOON can provide 600–900m (2,000–3,000ft) of filament (thread). Fiber is formed from a liquid that the caterpillar (*see* LARVA) produces from its spinning glands. It spins a silk cocoon around its body in preparation for its change into a silk moth. When the cocoons have been spun, the silk farmer heats them to kill the insects inside. The cocoons are then soaked to unstick the fibers, and the strands from several cocoons are unwound together to form a single thread of yarn. The yarn may be dyed before or after weaving into fabric. The Chinese were the first to use silk and its production was a closely guarded secret, since it could fetch its own weight in gold. Sicily was one of the first European production centers and the industry spread to Italy, Spain, and France. Silk manufacturing developed in England in the 17th century. China is still the largest producer of raw silk in the world. *See also* TEXTILES

sill Sheet-like intrusion of IGNEOUS ROCK that is parallel to the bedding or other structure of the surrounding rock. Sill rock is normally medium-grained; basic sills (DOLERITES) are the commonest.

sillimanite (fibrolite) Mineral, aluminum silicate (Al_2SiO_5), found in MICA SCHISTS and GNEISS. Its crystals are of the orthorhombic system, usually fibrous masses, and are satinlike or glossy white, brown, green or blue. A pale blue gem variety occurs in Sri Lanka. Hardness 6–7.5; r.d. 3.2.

silt Mineral particles, produced by the WEATHERING of rock. These particles, varying in size between grains of SAND and CLAY, are carried along in streams and RIVERS, to be deposited in the gently flowing lower reaches of rivers. When the river changes course or overflows its bank, the silt deposit forms very fertile land. Siltstone (flagstone) is a hard, durable stone that is formed from hardened silt. *See also* ALLUVIUM

Silurian Third oldest period of the PALEOZOIC era, lasting from 438 to 408 million years ago. Marine invertebrates resembled those of ORDOVICIAN times, and fragmentary remains show that jawless fishes (agnathans) began to evolve. The earliest land plants

(psilopsids) and first land animals (archaic mites and millipedes) developed. Mountains formed in NW Europe and Greenland. *See* GEOLOGICAL TIME

silver (symbol Ag) White, metallic element in the second series of TRANSITION ELEMENTS. It occurs in ARGENTITE (a sulfide) and horn silver (SILVER CHLORIDE), and is also obtained as a by-product in the refining of copper and lead. Silver ores are scattered throughout the world; Mexico is the major producer. It was one of the first metals to be mined and crafted by humans. Silver is an excellent conductor of heat and electricity and is used for some electrical contacts and on PRINTED CIRCUITS; other uses include jewelry, ornaments, coinage, and mirrors. Silver does not oxidize in air, but tarnishes if sulfur compounds are present. Properties: at.no. 47; r.a.m. 107.868; r.d. 10.5; m.p. 961.93°C (1,763°F), b.p. 2,212°C (4,104°F); most common isotope ^{107}Ag (51.82%).

silver chloride (AgCl) White crystalline compound, insoluble in water; it occurs naturally as the mineral horn silver. It darkens on exposure to light and for this reason is, like silver bromide, used for making photographic emulsions.

silver nitrate ($AgNo_3$) Colorless, solid compound. It is the most important salt of SILVER because it is very soluble in water. Silver nitrate is used in PHOTOGRAPHY, chemical analysis, silver-plating, mirrors, inks, and dyes. It is extremely caustic. Properties: sp.gr. 4.3; m.p. 414°F.

sima In geology, undermost of the two main rock-classes that make up the Earth's CRUST, so called because its main constituents are silicon and magnesium. It underlies the SIAL of the CONTINENTS.

simple harmonic motion (SHM) Periodic motion such as that of a PENDULUM, atomic vibrations or an oscillating electric circuit. A body has simple harmonic motion when it oscillates along a line, moving an equal distance on either side of a central point and accelerating towards that point with a speed proportional to its distance from it.

simultaneous equations Two or more equations that can be manipulated to give common solutions. In the simultaneous equations $x + 10y = 25$ and $x + y = 7$, the problem is to find values of x and y, such that these values are solutions of both the equations simultaneously. This can be done by rearranging and combining the equations to obtain each of the unknown variables in turn. In this example, one equation can be subtracted from the other to give $9y = 18$, hence $y = 2$. Substituting this value of y back into either of the equations gives $x = 5$. In general, such systems of equations may have no solutions, exactly one solution, or infinitely many solutions.

sine In TRIGONOMETRY, ratio of the length of the side opposite an acute angle to the length of the hypotenuse in a right-angled TRIANGLE. The sine of angle *A* is written sin*A*.

Singer, Isaac Merrit (1811–75) US manufacturer, inventor of a rock drill (1839) and a single-thread sewing machine (1852). Singer's machine allowed continuous and curved stitching.

▲ **skeleton** The human skeleton consists of about 206 bones divided into two broad groups, the axial and appendicular skeletons. It has three functions: it supplies support; it protects the internal organs; and, by using muscles, it gives movement. The axial skeleton, consisting of the skull, spine, and rib cage, supplies the basic structure on to which the limbs, the appendicular skeleton, are joined, via the pelvic and shoulder girdles.

sinkhole Hollow or hole in LIMESTONE formations that extends from the surface all or part of the way down to underground channels and caverns. Such holes are formed by water dissolving the limestone.

sintering Process in powder METALLURGY in which compressed particles are heated at temperatures below the melting point of the metal. This forms the particles into a coherent solid body. Glass and ceramics are nonmetals that may be sintered.

sinus Hollow space or cavity, usually in bone. Most often the term refers to the paranasal sinuses, any of the four sets of air-filled cavities in the skull near the nose.

siphon Principle by which a liquid is raised from a container at one level and delivered to a lower level; also any inverted U-shaped tube used for this purpose. The tube is filled with the liquid, and one end of it is placed below the surface of the liquid in the container. When the other end hangs below the level of the surface (outside the container), liquid flows out due to the imbalance between the weights of the portions of liquid in the two arms of the tube.

Siphonaptera Order of small, wingless insects comprising the **fleas**. They are blood-sucking, external parasites on warm-blooded animals. Fleas have hard, laterally compressed bodies with many microscopic, backward-pointing spines. Their mouthparts are adapted for piercing their host and sucking up blood. Their powerful, bristly legs enable them to perform single leaps of more than 300mm. Flea eggs are laid in dirt or in the nest of the host; the whitish larvae feed on organic matter and the feces of adult fleas. The larvae spin cocoons as part of the METAMORPHOSIS into adults. Many species can live on more than one host species. Adults can survive for several weeks without feeding. Fleas cause irritation and some carry diseases. The rat flea (*Xenopsylla cheopsis*), for instance, can transmit typhus and bubonic plague to humans. Length: to 1cm (0.4in).

Sirius (Alpha Canis Majoris, Dog Star) Brightest STAR in the sky (magnitude −1.47), situated in the N constellation of CANIS MAJOR. Its luminosity is 23 times that of the Sun. Sirius is a main-sequence star, and is the sixth-closest star system to us. It has a binary companion, Sirius B, the existence of which was deduced by Friedrich BESSEL in 1844 from its gravitational PERTURBATIONS on Sirius. Sirius B is a WHITE DWARF, the first star to be recognized as such.

SI units (*Système International d'Unités*) Internationally agreed system of units, derived from the MKS SYSTEM (meter, kilogram, and second). SI units are now used for many scientific purposes

▶ **skin** The skin consists of epidermal and dermal layers supported by a subcutaneous layer of fat cells. The epidermis is subdivided into a layer of dead flattened cells, beneath which lies a granular layer containing living, dividing cells. Pigment producing melanocytes that color the skin lie below this layer. The thicker dermis consists of connective tissues in which are embedded lymphatics, nerve fibers, sensory nerve endings, capillaries, sweat glands, and hair follicles. A sebaceous gland and an erector muscle accompany each follicle.

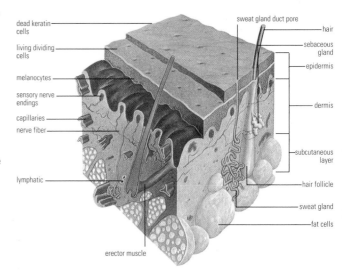

dead keratin cells
living dividing cells
melanocytes
sensory nerve endings
capillaries
nerve fiber
lymphatic
erector muscle
sweat gland duct pore
hair
sebaceous gland
epidermis
dermis
subcutaneous layer
hair follicle
sweat gland
fat cells

and have replaced the FPS SYSTEM (foot, pound, and second), and cgs system (centimeter, gram, and second). The seven basic units are: the METER (m), KILOGRAM (kg), SECOND (s), AMPERE (A), KELVIN (K), MOLE (mol), and CANDELA (cd).

skeletal muscle Alternative name for VOLUNTARY MUSCLE

skeleton Bony framework of the body of a VERTEBRATE. It supports and protects the internal organs, provides sites of attachment for MUSCLES and a system of levers to aid locomotion. In the HUMAN BODY the skeleton consists of 206 BONES and is divided into two parts. The **axial** skeleton, or main axis of the body, includes the SKULL, the SPINAL COLUMN, the STERNUM (breastbone), and the RIBS. The **appendicular** skeleton, serving for the attachment of limbs, includes the shoulder girdle and arm bones and the pelvic, or hip, girdle, and leg bones. The external skeleton of some insects is called an EXOSKELETON. *See also* ENDOSKELETON

skin Tough, elastic outer covering of the body, serving many functions. It is sometimes regarded as the largest organ of the body. The skin protects the body from injury and from the entry of some microorganisms and prevents dehydration. Nerve endings in the skin provide the sensations of TOUCH, warmth, cold, and pain, each perceived at discrete points on the surface. It helps to regulate body temperature through sweating, regulates moisture loss, and keeps itself smooth and pliable with an oily secretion from the SEBACEOUS GLANDS. Structurally, the skin consists of two main layers: an outer layer (EPIDERMIS) and an inner

layer (DERMIS). The top layer of epidermis is made of closely packed dead cells constantly shed as microscopic scales. Below this is a layer of living cells that contain pigment and nerve fibers, and which divide to replace outer, shed layers. The dermis contains dense networks of connective tissue, blood vessels, nerves, glands and hair follicles.

Skinner, Burrhus Frederic (1904–90) US psychologist. Skinner developed the concept of operant CONDITIONING (the control of behavior by its consequences or reinforcements). His many books include *Science and Human Behavior* (1953), *Beyond Freedom and Dignity* (1971) and a controversial novel about social engineering, *Walden II* (1948).

skull Skeleton of the head. In mammals, it comprises the CRANIUM casing of the BRAIN and the facial and upper jawbones (the lower jawbone, or MANDIBLE, attaches to the skull). There are 14 facial bones, mostly in symmetrical pairs. These are the zygomatic arches (cheekbones), the lacrimals around the eye socket, the nasals, nasal conchae and vomer (nasal septum) of the nose, the palatines (palate) of the mouth and the maxillae (upper jaw). The adult human cranium is formed of fused skull bones with immovable joints: the frontal, occipital, temporals, parietals, ethmoid, and splenoid. The occipital bone at the base of the skull forms a joint with the first (atlas) vetebra of the neck.

slag In the production of IRON and STEEL, liquid layer of ore impurities such as oxides, ash, and limestone. It floats on molten steel during refining, protecting the metal from oxidation and removing some unwanted substances.

slate Gray to blue, fine-grained, homogeneous METAMORPHIC ROCK, which splits into smooth, thin layers. It is formed by the metamorphosis of SHALE, and is valuable as a roofing material.

sleep Periodic state of unconsciousness from which a person or animal can be roused. During an ordinary night's rest there are intervals of deep sleep associated with RAPID EYE MOVEMENT (REM). It is during this REM sleep that dreaming occurs. Although its importance is not understood, regular sleep is essential. Studies have shown that people deprived of sleep become grossly disturbed. Sleep requirement falls sharply in old age. Difficulty in sleeping is called insomnia.

slide rule Calculating device consisting of two rules engraved with logarithmic scales of numbers, one of which slides next to the other. See also LOGARITHMS

slime mold Any of a small group of strange, basically single-celled organisms that are intermediate between the plant and animal kingdoms. During their complex life cycle they pass through several stages. These include a flagellated swimming stage, an AMOEBA-like stage, a stage consisting of a slimy mass of PROTOPLASM with many nuclei, and a flowering sporangium stage. See PROTOCTISTA

small intestine Part of the DIGESTIVE SYSTEM that extends – about 6m (20ft) coiled and looped – from the STOMACH to the large INTESTINE, or colon. Its function is the digestion and absorption of food. See also DUODENUM; ILEUM

smart card Plastic card that incorporates a computer MICROPROCESSOR and a small memory. Visually resembling an automatic teller (ATM) card, it can hold various data, such as personal identification details and the status of a bank account. It can then be used, by means of a personal identification number (PIN) as a store card, credit card, or debit card. It can have credits "stored" on it, which can then be spent through electronic tills (for example, for telephone calls or fares on public transportation) and be replenished as required.

smell (olfaction) SENSE that responds to airborne molecules. The olfactory RECEPTORS can detect even a few molecules per million parts of air. There are different receptors for different chemical groups.

smelling salts (sal volatile) Mixture of ammonium carbonate crystals, alcohol, and a perfume, often lavender. This sharp fragrance was formerly a common stimulant or restorative held to the nose of a person who had fainted or felt weak, dizzy or nauseated.

smelting Heat treatment for separating METALS

from their ORES. The ore, often with other ingredients, is heated in a FURNACE to remove nonmetallic constituents. The metal produced is later purified. When iron is smelted in a BLAST FURNACE, for example, the ore (iron oxide) is heated with COKE and limestone. The oxide combines with carbon (from the coke) to form gaseous oxides of carbon and pig iron (an alloy of carbon, iron, and other impurities).

Smith, Michael (1932–) Canadian biochemist, b. Britain. He discovered site-specific mutagenesis, a method of altering the GENETIC CODE through specific MUTATIONS instead of the previous random ones. This method has enabled the production of new proteins with a variety of functions. For this work, Smith shared the 1993 Nobel Prize for chemistry with Kary MULLIS.

Smithson, James (1765–1829) British chemist and mineralogist; the mineral SMITHSONITE was named after him. Angered at the Royal Society's rejection of a paper by him in 1826, Smithson left £105,000 to found an institution, the Smithsonian Institution in Washington, D.C., for the "increase and diffusion of knowledge among men." Today, it is one of the leading American scientific research and education centers.

smithsonite Carbonate mineral, zinc carbonate ($ZnCO_3$). It is found in ore deposits of zinc minerals, commonly associated with sphalerite, hemimorphite, galena, and calcite. Its crystals, which are rare, are rhomobohedral in the trigonal system; it usually occurs as masses. Smithsonite is generally white, but may be other colors; blue specimens from New Mexico have been used as gemstones. Hardness 4–4.5; r.d. 4.4.

smog Dense, atmospheric mixture of smoke and FOG or chemical fumes, commonly occurring in urban or industrial areas. Smog generally occurs when there is radiation fog. If the air remains calm, the fog gets worse; more industrial grime accumulates because it is unable to escape into the atmosphere. The water droplets condense around the pollutants, forming thick smog. One of the worst areas for smog is Los Angeles. This is a result of the exhaust emissions of the city's vast number of automobiles. The air pollution is exacerbated by Los Angeles' topography: it lies in a large basin, noted for inversions of temperature, which create ideal conditions for the formation of fog and smog.

smoke detector Device that gives a loud alarm when smoke particles enter it. There are various types of smoke detector. One type contains a PHOTOELECTRIC CELL. Smoke entering the detector cuts down on the amount of light entering the cell, thereby reducing the amount of current flowing through the cell. The drop in current sets off the alarm. In another type of detector, a steady current is sustained by IONS produced by a tiny radioactive source. Smoke particles become attached to the ions, reducing the current and setting off the alarm.

smooth muscle *See* INVOLUNTARY MUSCLE

snow Type of PRECIPITATION consisting of water vapor that has frozen into ICE crystals. Several ice crystals join together to form a **snowflake**, which will gradually fall to Earth. Snowflakes, which are symmetrical (usually hexagonal) crystalline structures, may melt as they fall, especially if the temperature is near freezing point. In such conditions, **sleet** occurs and CLOUDS yield snowfall on mountains, while producing rainfall on adjacent lowland areas. Snow is dry and powdery if it has come from a dryish area, and wetter if its source is over the sea. Interior regions, such as prairies, generally have dry snow, which is blown by powerful winds in blizzards. *See also* RAIN

soap Cleansing agent made of salts of FATTY ACIDS, used to remove dirt and grease. Common soaps are produced by heating fats and oils with an alkali, such as sodium hydroxide or potassium hydroxide. Soap consists of long-chain molecules, one end of which attaches to grease while the other end dissolves in the water, causing the grease to loosen and form a floating scum. *See also* DETERGENT

soapstone (steatite) Rock with a soft soapy or greasy texture. There are many types of soapstone but all contain a large proportion of magnesium silicate, often associated with various amounts of serpentine and carbonates. Food vessels and carvings made from soapstone have been found among the remains of prehistoric human cultures. *See also* TALC

Sobrero, Ascanio (1812–88) Italian chemist who discovered (1846) the explosive NITROGLYC-ERINE (glyceryl trinitrate) by adding glycerine to a mixture of nitric and sulfuric acids. Horrified at the potential of his discovery, Sobrero made no attempt to develop it. *See also* EXPLOSIVE

soda Any of several sodium compounds, especially SODIUM CARBONATE and SODIUM HYDROGEN-CARBONATE.

sodalite Glassy silicate mineral, sodium aluminum silicate with some chloride, found in alkaline IGNEOUS ROCKS. It occurs as small dodecahedral crystals in the cubic system and also as masses. It may be colorless, white, blue or pink; it is sometimes used as gem. Hardness 5–6; r.d. 2.2.

Soddy, Frederick (1877–1956) British chemist. He was awarded the 1921 Nobel Prize for chem-

istry for his theory and definition of ISOTOPES, a term he coined. Soddy conducted research in RADIOACTIVITY with Ernest RUTHERFORD, and later Sir William RAMSAY. In 1920 he produced an explanation of the RADIOACTIVE DECAY of isotopes that proved invaluable in calculating GEOLOGICAL TIME. Soddy and Ramsay found HELIUM to be a product of URANIUM decay. His classic science books include *The Interpretation of Radium* (1909), *Matter and Energy* (1912), and *The Chemistry of the Radio-Elements* (1911–14).

sodium (symbol Na) Common, silver-white metallic element, one of the ALKALI METALS, first isolated (1807) by Sir Humphry DAVY. It occurs in the sea (as salt) and in many minerals. Its chief source is SODIUM CHLORIDE (common salt), from which it is extracted by ELECTROLYSIS. Sodium is used in the manufacture of tetraethyl lead, an anti-KNOCK compound for motor engines. It also used in arc lamps for street lights, and as a heat-transfer medium in nuclear reactors. Sodium compounds, such as SODIUM CARBONATE and SODIUM CHLORIDE, have wide industrial uses such as baking and washing sodas, fertilizers, and soap. It is an essential part of the human diet. Chemically, sodium is one of the most reactive elements. Properties: at.no. 11; r.a.m. 22.9898; r.d. 0.97; m.p. 97.81°C (208.05°F); b.p. 882°C (1,620°F).

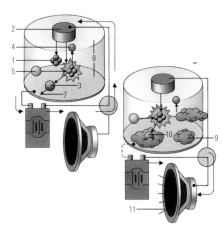

▲ **smoke detector**
A smoke detector uses radiation to detect the presence of smoke. Alpha particles (1) are emitted from a radioactive source (2). Positively charged ions (3) and electrons (4) are created when the alpha particles react with molecules (5) in the air.

The ions conduct a current (7) between two plates (8). When smoke particles (9) are present between the plates they absorb the ions (10) weakening the current. Electronic circuits register the drop in current and an alarm sounds through a loudspeaker (11).

sodium bicarbonate (sodium hydrogencarbonate, NaHCO₃, popularly known as bicarbonate of soda) White, crystalline salt that decomposes in acid or on heating to release carbon dioxide gas, which explains its use in baking powder. It has a slightly alkaline reaction, hence its other main application, as an ingredient of indigestion powders and tablets.

sodium carbonate (soda, washing soda) White, crystalline salt (Na₂CO₃), usually manufactured from SODIUM CHLORIDE (common salt, NaCl) and AMMONIA by the SOLVAY PROCESS. The anhydrous (lacking water) form (Na₂CO₃) is known as soda ash; washing soda is hydrated sodium carbonate (Na₂CO₃.10H₂O)

sodium chlorate White, crystalline compound (NaClO₃), a strong OXIDIZING AGENT. It is used as a general herbicide and soil sterilizer and as a laboratory reagent. It is made by the ELECTROLYSIS of a solution of SODIUM CHLORIDE.

sodium chloride (NaCl) Common salt. It is the major mineral component of sea water, making up 80% of its dissolved materials. Sodium chloride is also the major ELECTROLYTE of living cells, and the loss of too much salt, through evaporation from the skin or through illness, is dangerous. It is used as a seasoning, to cure and preserve foods and, in the chemical industry, to produce SODIUM, CHLORINE, and SODIUM HYDROXIDE. Sodium chloride is an essential part of an animal's diet, and is a constituent of body fluids, such as blood, sweat, and tears. Salt helps digestion by contributing to hydrochloric acid found in the stomach. Sodium chloride is combined with AMMONIA in the SOLVAY PROCESS to produce SODIUM CARBONATE.

sodium hydrogencarbonate *See* SODIUM BICARBONATE

sodium hydroxide (caustic soda, NaOH) Strong ALKALI, prepared industrially by the ELECTROLYSIS of salt (sodium chloride, NaCl). It is a white solid that burns the skin, with a slippery feel because it absorbs moisture from the air. It also absorbs atmospheric carbon dioxide, forming a crust of sodium carbonate. Sodium hydroxide is used in many industries on a large scale, such as in SOAP-making to saponify fats (turn them into soap by alkali treatments) and in bauxite-processing to manufacture ALUMINUM.

sodium thiosulfate ("hypo," Na₂S₂O₃.5H₂O) Chemical used as a photographic fixer. It is usually acidified by the addition of sodium metabisulfite.

softener, water *See* WATER SOFTENER

software COMPUTER PROGRAM and any associated data file. The term software is used to distinguish these coded instructions and data from computer HARDWARE, or equipment. In most cases, suppliers provide software on MAGNETIC DISK or CD-ROM for transfer to the built-in HARD DISK on the user's COMPUTER. Software can also be transferred via the INTERNET.

soil Surface layer of loose material resting on top of the rock that makes up the surface of the Earth. It consists of undissolved minerals produced by the WEATHERING and breakdown of surface rocks, organic matter, water, and gases. The organic remains provide the HUMUS and the inorganic particles provide vital minerals. The inorganic fraction of the soil includes CLAY, SILT, SAND, GRAVEL, and STONE. Soils are classified by structure and texture. The structure is determined by the aggregation of particles. The four main textures of soil are sand, silt, clay, and LOAM. Loam soils are best for cultivation, since they are able to retain more water and nutrients. Erosion and mismanagement are the chief causes of soil infertility. Fertility can be restored with the correct use of FERTILIZERS. A SOIL PROFILE reveals a number of distinct SOIL HORIZONS (layers).

soil horizon Layer of SOIL that shows in a SOIL PROFILE – a cross-section of the soil. Usually soil is divided into three horizons, designated A, B, and C. A is the topsoil, B the subsoil, and C the BEDROCK. Fine particles and organic HUMUS make up the **A horizon**, much more inorganic material and larger particles occur in the **B horizon**, and the **C horizon** is ROCK. There may be a layer of REGOLITH between the B and C horizons. *See also* CHERNOZEM; PODZOL

soil profile Vertical view of layers of SOIL from the surface down to the unaltered parent material. It is used in classifying soils. A layer of soil in a soil profile is known as a SOIL HORIZON.

soil testing Process of analyzing the chemical and mineral composition and porosity of SOIL. It is carried out for various reasons, including to determine the soil's ability to support plant life.

sol Colloidal SUSPENSION of particles of a solid in a liquid, such that the suspension remains liquid and does not solidify as in a GEL. *See also* COLLOID

solar cell Device that converts sunlight directly to ELECTRICITY. It normally consists of a *p*-type silicon crystal coated with an *n*-type (*see* SEMICONDUCTOR). Light radiation causes electrons to be released and creates a POTENTIAL DIFFERENCE so current can flow between electrodes connected to the two crystals. All wavelengths shorter than one micrometer can create electrical energy. Cells convert about 10% of sunlight into useful energy. Solar cells are often used to power small electronic devices such as pocket calculators. Several thousand cells may be used in panels to provide power of a few hundred watts. *See also* SOLAR ENERGY

solar constant Measure of the amount of solar energy received by a body a certain distance from the Sun. For the Earth, the solar constant is defined as the solar power received per unit area, at the top of the atmosphere, at the average Earth–Sun distance of 1AU (1 astronomical unit). The value of the solar constant is $c.1.35\text{kW m}^{-2}$. The "constant" varies from day to day with SUNSPOT activity, and also, in the longer term, with the SOLAR CYCLE.

solar cycle Periodic fluctuation in the number of SUNSPOTS and the level of other kinds of solar activity; the cycle lasts about 11 years. Over the course of a cycle, sunspots vary in both number and latitude. At solar maximum, when sunspot numbers are greatest, astronomers refer to the **active Sun**; at solar minimum, they refer to the **quiet Sun**.

solar energy Heat and light from the SUN consisting of ELECTROMAGNETIC RADIATION, including heat (infrared rays), light, and radio waves. About 35% of the energy reaching the Earth is absorbed: most is spent evaporating moisture into CLOUDS, and some is converted into organic chemical energy by PHOTOSYNTHESIS in plants. All forms of energy (except NUCLEAR ENERGY) come ultimately from the Sun. SOLAR CELLS are used to power instruments on spacecraft, and experiments are being done to store solar energy in liquids from which ELECTRICITY can be generated. The effective use of solar energy is hampered by the diurnal cycle, and by seasonal and climatic variations.

solar engine Machine that converts SOLAR ENERGY into mechanical WORK, particularly for the purpose of providing THRUST to a spacecraft. Solar engines now under development have one or more large solar collectors, which gather the energy of the Sun's rays to heat up a working fluid such as hydrogen. The heated hydrogen can be used directly to provide thrust or it can be fed to a turbogenerator.

solar flare Sudden and violent release of matter and energy from the Sun's surface, usually from the region of an active group of SUNSPOTS. In the **flash stage**, a flare builds to a maximum in a few minutes, after which it gradually fades and disappears within an hour or so. Flares emit radiation right across the ELECTROMAGNETIC SPECTRUM. Charged particles are emitted, mostly ELECTRONS and PROTONS, and smaller numbers of neutrons and atomic nuclei. A flare can cause material to be ejected in bulk, most spectacularly in the form of PROMINENCES. When energetic particles from flares reach the Earth they may cause radio interference, magnetic storms, and more intense AURORAE. Although not well understood, the origin of flares is believed to be connected with local discontinuities in the Sun's MAGNETIC FIELD.

solar heating Domestic heating system that usually employs roof-mounted heat collectors, often called solar panels. These collectors have transparent glass or plastic panels, behind which water or air circulates over black-painted metal surfaces that absorb heat energy from the Sun's rays. The warmed liquid or air passes on to a water tank or rock storage tank from which hot water or air is pumped to RADIATORS or air vents.

solar physics Study of the SUN's physical behavior. It includes study of the solar spectrum and of the nature and causes of surface phenomena. From these may be inferred the inner thermal, electrical, magnetic, nuclear, and gravitational processes.

▶ **solar cell** A solar (photovoltaic) cell (A) is made up of two silicon semiconductors between metal contacts. One of the silicon semiconductors collects positive charge (1), the other negative (2), creating a potential difference. As light photons (3) hit the *p-n* semiconductor junction between the semiconductors (4) they displace electrons which are attracted to the positive semiconductor. There is thus a potential difference (voltage) generated between the metal contacts (5).

Solar System The SUN and all the celestial bodies that revolve around it: the nine PLANETS, together with their SATELLITES and ring systems, the thousands of ASTEROIDS and COMETS, meteoroids, and other interplanetary material. The **inferior planets** are those planets that are closer than EARTH to the Sun; the rest are called **superior planets**. Astronomical distance is measured in ASTRONOMICAL UNITS (AU) – the mean distance between the Earth and the Sun. The boundaries of the Solar System lie beyond PLUTO, which orbits the Sun at a mean distance of 39 AU, to include the Kuiper Belt (100 AU) and the Oort Cloud of comets. The Solar System as a whole moves in a roughly circular orbit about the center of the GALAXY, taking 2.2×10^8 years to complete one full orbit. ARISTOTLE and PTOLEMY's notion of a geocentric UNIVERSE remained fundamentally unchallenged until the 16th century. COPERNICUS developed the first heliocentric theory of the UNIVERSE, which was championed by GALILEO. Drawing on the observations of Tycho BRAHE, Johannes KEPLER accurately described the elliptical orbit of each planet with the Sun at one of the foci. All the planets orbit the Sun in roughly the same plane (the ECLIPTIC), although Pluto's orbit is more irregular. They all move in the same direction – counterclockwise, when viewed from above. All the planets rotate about their own axes as they revolve around the Sun; the period of rotation varies (in Earth time) from under 10 hours (JUPITER) to more than 243 days (VENUS). Venus is the only planet with retrograde motion – it rotates from east to west. The equatorial plane of each planet is tilted in relation to its orbital plane: Jupiter has the smallest tilt (3°), URANUS has the largest (98°). Earth's equatorial plane is tilted at 23.5°. This tilting produces annual SEASONS. Sir Isaac NEWTON showed that all bodies in the Solar System are subject to the force of GRAVITATION. The Sun is by far the most massive component of the Solar System, accounting for 99.9% of the entire system's mass. Thus, it exerts the greatest pull. Other celestial bodies exert small variations (PERTURBATIONS) on each other's orbits. The planets are classified also by their physical properties. The inner planets (MERCURY, Venus, Earth, and MARS) are called **terrestrial planets**. They are relatively small and dense, with solid CRUSTS and molten, metallic CORES. They are composed of high-temperature condensates (mainly iron and metal silicates). The **Jovian planets** (Jupiter, SATURN, Uranus, and NEPTUNE) are massive, but with relatively low density. Jupiter is heavier than all the other planets combined. Jovian planetary atmospheres are thick and gaseous, chiefly composed of hydrogen and helium. Pluto is unique, and comparatively little is known of it. The **origin** of the Solar System is a major debate among cosmogonists. In the late 18th century, Pierre LAPLACE developed the

nebular hypothesis. According to this theory, about 5,000 million years ago a NEBULA or INTERSTELLAR CLOUD of gas and dust slowly cooled and contracted, causing gravitational collapse. The nebula began to rotate faster, throwing out rings of gaseous matter. In these rings balls of matter were formed, then shrank to form planets. At the center of the system, mass condensed to form the Sun. In the early 20th century, Forest MOULTON and Thomas Chamberlin developed an "encounter" theory called the PLANETESIMAL theory. They proposed that a star passed close to the Sun, causing huge surface tides, that ripped matter from the solar nebula and threw it into elliptical orbit. These small aggregates of matter (planetesimals) were attracted to each other by gravitation and accreted to form planets. In 1917 James JEANS proposed a similar **tidal** theory. All of these theories, however, failed to account for the distribution of angular momentum. Gerard KUIPER's **protoplanets** theory satisfied this objection. Harold UREY developed Kuiper's ideas to provide an explanation of the different physical and chemical compositions of the planets.

solar time System of time reckoning based upon the interval between successive transits of the Sun across the observer's meridian (the solar day). Because of variations in the Earth's orbital velocity and changes in the Sun's apparent position as viewed from the orbiting Earth, solar days vary in length throughout the year. A mean solar day has therefore been adopted, giving a mean solar time. As measured on the MERIDIAN at the Greenwich Observatory, mean solar time is GREENWICH MEAN TIME (GMT) or Universal time.

solar wind Steady flow of charged particles (mainly protons and electrons) accelerated by the high temperatures of the solar CORONA to velocities great enough to allow them to escape from the SUN's gravity. The solar wind deflects the tail of the Earth's MAGNETOSPHERE and the tails of COMETS away from the Sun. Some solar wind particles get trapped in planetary magnetic fields. Closer to the Earth, some are trapped in the outer VAN ALLEN RADIATION BELTS; others reach the Earth's upper ATMOSPHERE in the region of the magnetic poles and cause AURORAE. The solar wind carries away about 10^{-13} of the Sun's mass per year. Its intensity increases during periods of solar activity.

solder Metallic ALLOY used in a molten state to fuse together metal surfaces. **Soft solders** are usually alloys of LEAD and TIN and have melting points between 200°C and 300°C. **Hard solders** contain significant amounts of SILVER and have a much higher melting point. **Brazing solders** are alloys of COPPER and ZINC and melt at more than 800°C. *See also* BRAZING

solenoid ELECTROMAGNET in which a soft iron core moves so as to open or close an electric circuit, thus acting as a switch or RELAY.

solfatara Small CRATER or vent in the Earth's surface through which steam and gases escape. Solfataras occur mostly in volcanic regions and many of the gases contain sulfur. They probably indicate an area of declining volcanic activity. *See also* VOLCANO

solid State of MATTER in which a substance has a relatively fixed shape and size. The forces between atoms or molecules are strong enough to hold them in definite locations (about which they can vibrate) and to resist compression. *See also* CRYSTAL; GAS; LIQUID; PLASMA

solid angle Three-dimensional angle formed at the center of a SPHERE by the VERTEX of a CONE that has its base on the surface of the sphere. Measured in **steradians**, a solid angle is defined as the ratio of the surface area occupied by the base of the cone to the square of the radius of the sphere.

solid circuit ELECTRIC CIRCUIT that is present inside a semiconductor or other solid-state device. *See also* SEMICONDUCTIVITY

solid geometry Study of figures in three dimensions. It includes the study of planes intersecting with planes, planes intersecting with solids, solids interacting with solids and the measurement of solid figures. *See also* GEOMETRY

solid-state physics Physics of SOLID materials. From the study of the structure, binding forces, electrical, magnetic, and thermal properties of solids has come the development of the SEMICONDUCTOR, MASER, LASER, and SOLAR CELL. **Solid-state circuits** are electronic circuits in which the individual components are formed from a single, solid SILICON CHIP.

soliton Stable, solitary wave that keeps its shape and speed without widening as it propagates. Solitons behave like particles and are important in many areas of FLUID MECHANICS and SOLID-STATE PHYSICS.

solstice Either of the two days each year when the Sun is at its greatest angular distance from the CELESTIAL EQUATOR, leading to the longest day and shortest night (**summer solstice**) in one hemisphere of the Earth, and the shortest day and longest night (**winter solstice**) in the other hemisphere. In the Northern Hemisphere, the summer solstice occurs on about June 21, and the winter solstice on about December 22.

solubility Mass (grams) of a SOLUTE that will saturate 100 grams of SOLVENT under given conditions to give a SATURATED SOLUTION. Solubility generally rises with temperature, but for a few solutes, such as CALCIUM SULFATE, increasing temperature decreases solubility in water.

solute Gaseous, liquid or solid substance that is dissolved in a SOLVENT to form a SOLUTION. Ionic solids, such as common salt, and sugars and starch dissolve in water. Liquids can dissolve in liquids; for example, ethanol and water are miscible – capable of mixing – in all proportions at room temperature. Some gases, such as hydrogen chloride (HCl), are soluble in water.

solution In chemistry, a liquid (the SOLVENT) into which another substance (the SOLUTE) has dissolved. It is a liquid consisting of two or more chemically distinct compounds that are inseparable by filtering. The amount of a solute dissolved in a given volume of solvent is called the CONCENTRATION of a solution. The ability of one substance to dissolve another depends on the type of chemical bonding and the temperature. Heat can be released (an exothermic solution) or absorbed (an endothermic solution) during the formation of the solution. *See also* MIXTURE; SATURATED SOLUTION

solution In geology, a form of chemical WEATHERING. It is particularly active in limestone areas where the JOINTS can be enlarged to form GRIKES and POTHOLES. Solution can also be active in chalk areas and anywhere with rocks containing salts.

Solvay, Ernest (1838–1922) Belgian industrial chemist. He invented a technique (the SOLVAY PROCESS) for preparing SODIUM HYDROGENCARBONATE (and hence SODIUM CARBONATE) from sodium chloride (common salt) and calcium carbonate. In 1861 Solvay patented the process. By 1913 the technique was producing a large part of the world's sodium carbonate.

Solvay process (ammonia-soda process) Widely used industrial method of making SODIUM CARBONATE (soda, [Na_2CO_3]), invented by Ernest SOLVAY. Brine is saturated with AMMONIA and carbon dioxide gas (made by heating calcium carbonate) is bubbled into this solution in a Solvay tower. SODIUM HYDROGENCARBONATE is formed, which precipitates from the brine, and is then heated to make sodium carbonate.

solvent Liquid that dissolves a substance (the SOLUTE) without changing its composition to form a solution. Water is the most universal solvent, and many inorganic compounds dissolve in it. Ethanol, ether, acetone (propanone), and tetrachloromethane (carbon tetrachloride) are common solvents for organic substances. *See also* SOLUTION

somatotrophin Growth HORMONE made by cells of the anterior, or frontal, PITUITARY GLAND, lying under the brain. It stimulates growth in young animals, including humans, by increasing the amounts of PROTEIN made by the body cells and stimulating fat and carbohydrate metabolism.

Sömmering, Samuel Thomas von (1755–1828) German scientist. In 1809, using the voltaic pile invented by Alessandro VOLTA, Sömmering developed an early electric TELEGRAPH system that sent signals indicated at the receiving end by the appearance of hydrogen bubbles liberated at a gold electrode in a solution of acid.

sonar (acronym for **so**und **na**vigation and **r**anging) Underwater detection and NAVIGATION system. The system emits a high-frequency sound that is reflected by underwater objects. *See also* ECHO SOUNDER

sonic boom Sudden NOISE produced by SHOCK WAVES from an AIRCRAFT flying at a SUPERSONIC VELOCITY. The shock waves are formed by the build-up of SOUND waves at the front and back of the aircraft. These waves spread out and sweep across the ground behind the aircraft, often causing a double bang. Sometimes the shock waves are strong enough to break windows in buildings.

sound VIBRATIONS that travel through solids, liquids or gases. The human ear can respond to sounds with frequencies between about 20 and 20,000Hz. Frequencies above this audible range are called ULTRASONICS. Sound WAVES are a combination of **condensation** (close packing of molecules) and **rarefication** (spreading out of molecules). This is usually represented as a wavy line, whose crests indicate condensation and whose troughs represents rarefication. The WAVELENGTH is the distance between two crests. It is determined by dividing the velocity of sound by the FREQUENCY of vibration. A pure sound is a SINE wave of definite frequency and intensity (rate of flow of energy). Natural sounds are a mixture of sine waves characterized by PITCH and TIMBRE. The speed at which a sound wave travels through a solid depends on the ELASTICITY of the medium and its DENSITY. Solids are the best conductors of sound. If the medium is a gas, the sound wave is longitudinal and its speed depends on the gas temperature. The **speed of sound** in dry air at STANDARD TEMPERATURE AND PRESSURE (STP) is 331.4ms^{-1} (741mph). ACOUSTICS is the study of sound waves. For instance, acoustics is used in the design of a concert hall, usually to minimize the reflection of sound that can produce ECHOES and INTERFERENCE. Loudness of sound is measured in DECIBELS. *See also* DOPPLER EFFECT; HARMONICS

sound barrier Cause of an aircraft's difficulties in accelerating to a speed faster than that of SOUND (SUPERSONIC VELOCITY). When approaching the speed of sound, an aircraft experiences a sudden increase in DRAG and loss of LIFT. These effects are caused by the buildup of sound waves to form SHOCK WAVES at the front and back of the aircraft. Some aircraft designers once thought that these effects would make SUPERSONIC FLIGHT impossible. However, the problems were solved by designing aircraft with smaller surface areas, swept-back wings, and more powerful propulsion systems. *See also* MACH NUMBER; SONIC BOOM

sound card (sound board) Piece of circuitry that is required for a COMPUTER to reproduce music and other SOUNDS through LOUDSPEAKERS. A sound card is necessary for various MULTIMEDIA applications using a CD-ROM drive. Eight-bit cards give poorer reproduction than the newer 16-bit sound cards, which can provide stereo. Some cards also facilitate SOUND RECORDING using a computer.

sounding Determining the depth of water. The simplest means of sounding is by dropping a measured weighted line until it reaches the bottom. Acoustic SONAR, a sonic depth finder, is often used. *See also* ECHO SOUNDER

sound recording Conversion of SOUND waves into a form that can be stored and reproduced. In 1877 Thomas EDISON invented a phonograph that recorded sound vibrations as indentations made by a stylus on a revolving cylinder wrapped in tinfoil. Emile Berliner's gramophone improved the process by using a zinc disc instead of a cylinder. The volume was amplified by the addition of acoustical horns, which were replaced before World War I by tube AMPLIFIERS. In 1901 molded thermoplastic records were introduced, and subsequently improved plastics allowed finer grooving with reduced surface noise. In 1927 and 1928 patents were issued in the US and Germany for MAGNETIC RECORDING processes. Later innovations include HIGH-FIDELITY (HI-FI), stereophonic, and quadrophonic reproduction. Modern recordings on COMPACT DISC (CD) employ digital signals rather than the analog signals used on earlier records.

Soyuz Series of Soviet manned space missions. **Soyuz 1** was launched in April 1967. In November 1968 a Soviet cosmonaut in **Soyuz 3** completed a manual approach with the unmanned **Soyuz 2**. In January 1969 **Soyuz 4** and 5 docked in space, and two cosmonauts from **Soyuz 5** completed the first first transfer in space between two spacecraft. In July 1969 a rocket explosion destroyed the Soviet's entire launch complex. In 1970 **Soyuz 9** established the record for a manned space flight – 18 days. The

Soyuz program suffered another setback when **Soyuz 11** depressurized during reentry into Earth's atmosphere, killing the three-man crew. In July 1975 APOLLO and Soyuz spacecraft docked in space, completing the first joint US-Soviet space project.

space Boundless, three-dimensional expanse in which objects are located. RELATIVITY states that space and TIME are aspects of one entity, known as SPACE-TIME. Also, space has a non-Euclidean GEOMETRY wherever there is a gravitational field. In another sense, space – sometimes referred to as outer space – is taken to mean the rest of the Universe beyond Earth's atmosphere.

space exploration Use of spacecraft to investigate outer space and heavenly bodies. On October 4, 1957, the first artificial satellite, SPUTNIK 1, was launched into Earth orbit by the Soviet Union. The first US satellite was **Explorer I** (January 31, 1958). The "space race" began in earnest. Within the next decade the two superpowers launched more than 50 unmanned SPACE PROBES to the Moon. Soviet cosmonauts, and their American equivalents, astronauts, orbited the Earth soon after. Unmanned space probes crash-landed on the Moon, sending back television pictures to Earth during the descent. Then came soft landings, and probes made to orbit the Moon showed its "dark" side for the first time. By 1968 Soviet space scientists had developed techniques for returning a Moon orbiter safely to Earth. In 1969 the US APOLLO 11 mission became the first to place a man on the Moon. Meanwhile, the Soviet Union had sent probes to explore Mars and Venus. **Chronology**: First probe to land on the Moon: **Luna 2**, launched September 12, 1959. First manned spaceflight: Yuri GAGARIN, April 12, 1961. First close-up pictures of Mars: MARINER 4, July 14, 1965. First person to walk on the Moon: Neil ARMSTRONG, July 21, 1969. First pictures from the surface of another planet: VENERA 9, received from Venus on October 22, 1975. First probes to land on Mars: VIKING 1 and 2, July 1976. Flybys of VOYAGER 2: Jupiter (1979), Saturn (1981), Uranus (1986), Neptune (1989). First SPACE SHUTTLE: *Columbia*, launched April 12, 1981. **Giotto probe**, launched 1985, flew within 600km (375mi) of HALLEY'S COMET, sent back photographs and data. GALILEO project, launched October 1989, photographed the asteroids Gaspra (1991) and Ida (1993), dropped a subprobe (1995) into Jupiter's atmosphere and provided data on the GALILEAN SATELLITES. **Clementine probe**, launched 1994, discovered what are believed to be water-ice deposits in craters of the Moon. **Solar and Heliospheric Observatory** (SOHO) probe, launched 1995, provided data on the Sun's interior. **Mars Pathfinder** mission, launched December 1996, landed on Mars and deployed a "micro-

rover," *Sojourner*, to explore and collect rock samples. **Cassini** space probe, launched 1997, started on a seven-year journey towards Saturn. *See also* HUBBLE SPACE TELESCOPE; SPACE PROBE

space probe Unmanned craft sent to investigate the Moon, planets or interplanetary space. The first US space probe was **Explorer 1**, launched January 31, 1958; it discovered the inner VAN ALLEN RADIATION BELT. The first LUNAR PROBE was **Luna 2**, launched by the Soviet Union on September 12, 1959. The first planetary probe was MARINER 2, launched by the US on August 27, 1962, which flew past Venus. PIONEERS 10 and 11 photographed Jupiter (October 22, 1975). Pioneer 11 went on to Saturn. VIKING 1 landed on Mars and collected data to test the Martian surface for life (July 1976). VOYAGER 2 sent back images from its flybys of Jupiter (1979), Saturn (1981), Uranus (1986), and Neptune (1989). *See also* SPACE EXPLORATION

space research Scientific and technological investigations that gather knowledge through SPACE EXPLORATION. The physical makeup and behavior of stars, planets, and other cosmic objects were first determined by ground-based optical and radio TELESCOPES and later by spacecraft. Research projects have been carried into space by artificial SATELLITES, the SPACE SHUTTLE, SPACE STATIONS, and SPACE PROBES.

space shuttle Reusable, US ROCKET-powered spacecraft. The main part of the shuttle, the **orbiter** (of which six have been made, *Enterprise*, *Columbia*, *Challenger*, *Discovery*, *Atlantis*, and *Endeavour*), looks like a bulky jet aircraft with swept-back wings. It ferries people and equipment between the ground and Earth ORBIT. The orbiter has a large payload bay in which it can carry SATELLITES for release into orbit. Astronauts can leave the shuttle to repair a faulty artificial satellite, and can retrieve satellites and return them to the ground. The shuttle sometimes carries experiments and experimenters from other countries. On such missions, scientists and astronauts carry out various experiments and observations. The orbiter takes off attached to a large fuel tank, using its own three ROCKET engines, assisted by two **booster** rockets. The boosters are jettisoned about two minutes after launch and are later recovered for re-use. Six minutes later, the orbiter's main engines cut off and the external fuel tank is dumped. Maneuvering engines then put the craft into the required orbit. When it is time to return to Earth, these engines are used to provide reverse thrust. As a result, the craft slows down and descends into Earth's atmosphere. It glides down and lands on a runway. The first space shuttle, *Columbia*, was launched on 12 April 1981. On mission 25 in January 1986, the shuttle *Challenger* exploded soon

▲ space shuttle A space shuttle flight has three distinct parts. The first is reaching orbit. The orbiter (1) is propelled upward from the launch platform by two solid fuel boosters (2) and the shuttle's three engines (3), fed with liquid oxygen and liquid hydrogen fuel from the external tank (4). After the boosters have burnt out they detach (5), and float to the surface of the Earth by parachute. The external tank detaches later, also returning to Earth, usually landing in the Indian Ocean. The second phase now commences. Once in orbit, the shuttle floats upside down above the Earth with its cargo bay doors (6) open to help dispel heat. Satellites (7) are launched from the cargo bay, and can be retrieved with the use of an arm (8). While in orbit, the shuttle manoeuvers using helium-fuelled thrusters in the nose (9). Finally, the shuttle returns to land unpowered. The orbiter is shielded from the enormous heat generated on re-entering the Earth's atmosphere by ceramic tiles (10) on its outer surface. A parachute (11) slows the vehicle on the runway.

after launch, killing all seven people on board. A leak had allowed burning gases from a booster rocket to ignite the fuel in the main tank.

space station Orbiting structure in space for use by astronauts and scientists. Space stations are more spacious than most spacecraft as the occupants may live there for several months before returning to the Earth. Space laboratories, such as the American *Skylab* (launched 1972) and the Russian MIR (launched 1986), are space stations built for scientists to carry out experiments, study the Solar System and observe distant parts of the Universe, their view undistorted by Earth's atmosphere. In 1997 *Mir* suffered a series of technical problems, and the program was cancelled by the Russian government. In 1999 *Mir* was due to re-enter Earth's atmosphere, where the majority of the station was due to burn up, but in early 2000 it was suggested that *Mir* might still serve a useful scientific purpose.

spacesuit Sealed garment enabling an astronaut to function in space. The suit is multilayered, combining eight different materials. The outside is nylon treated to prevent perforations from tiny micrometeoroids. Four layers of aluminum material then provide insulation from extremes of temperature and protection from harmful solar radiation. These are backed by a fire- and tear-resistant layer. The astronaut is protected from the vacuum of space by a pressure suit of nylon, coated with polyeurathane. A network of tubes in a nylon chiffon undergarment carries water that maintains body temperature. Gas pressure is kept below Earth's normal atmospheric pressure; otherwise, with virtually no pressure on the outside, the suit would balloon out. At such low pressure, the only gas suitable for breathing is pure oxygen. A backpack propelled by gas jets may be worn to enable the astronaut to move through space.

space-time Central concept in the theory of RELATIVITY that unifies the three SPACE dimensions (length, breadth, and height) with TIME to form a **four-dimensional** frame of reference. Durations and rates of processes depend on the relative state of motion of the observer and the system observed. In 1907 Hermann MINKOWSKI, who taught Albert EINSTEIN, clarified relativity theory by describing space-time in terms of a four-dimensional geometry. An event in space-time is specified by the three coordinates of space and the time coordinate. A line drawn in this space represents a particle's path in both space and time. Einstein incorporated this viewpoint into his theory of relativity: in the **general theory**, GRAVITATION is a distortion of space-time by MATTER. A consequence of this relativistic notion of time is that two events can only occur simultaneously for one observer: another observer will not see these events as concurrent. *See also* LORENTZ CONTRACTION

spadix In some flowering plants, a spike of small flowers, generally enclosed in a sheath called a SPATHE. A familiar plant with an inflorescence of this kind is the cuckoopint (*Arum maculatum*).

spark chamber Gas-filled chamber in which ionizing radiation or charged particles can be detected by the tracks of sparks made visible after the particles cause ionization. Up to 100 parallel metal plates are stacked in the chamber, connected alternately to the positive and negative terminals of a very high-voltage supply. When a particle passes through a pair of plates, it creates a pair of IONS in its track. The gas (usually neon or helium) becomes a conductor and sparks jump between the plates. A pair of camera lenses may be used to take stereoscopic photographs of the spark tracks.

spark plug Component in an INTERNAL COMBUSTION ENGINE. It has two ELECTRODES separated by an air gap, across which electric current from the ignition discharges to form a spark. The spark plug fits into the cylinder head and the spark ignites the compressed mixture of fuel and air.

spathe Broad leaflike organ that spreads from the base of, or enfolds, the SPADIX of certain plants, such as the cuckoopint (*Arum maculatum*).

special theory of relativity Part of EINSTEIN's theory of RELATIVITY that applies only to observers who move with constant speed and direction relative to each other.

speciation Emergence of new species in EVOLUTION. It results from the separation of parts of a homogeneous population. Over many generations, NATURAL SELECTION operates within the separated groups to produce gradually increasing differences. New species can be said to have evolved when individuals of the separated groups are no longer capable of interbreeding.

species Part of the CLASSIFICATION of living organisms. Species are groups of physically and genetically similar individuals that can interbreed to produce fertile offspring under natural conditions. Each species has a unique two-part Latin name (BINOMIAL NOMENCLATURE), the first part being the GENUS name. This name is written in italics. For example, the tiger's species name is *Panthera tigris*, *Panthera* being the genus of big cats. So far, more than 1.5 million plant and animal species have been identified, but estimates of the total number on land and in the oceans run as high as 100 million.

specific gravity Former term for RELATIVE DENSITY (R.D.)

specific heat capacity (symbol *c*) Heat necessary to raise the temperature of 1kg (2.2lbs) of a substance by 1K (Kelvin). It is measured in J K^{-1} kg^{-1} (J equals JOULE). Substances with a high specific heat capacity, such as water, require more

energy to raise their temperatures than do substances of low specific heat.

spectral classification Categorization of STARS based on the characteristics of their spectra. In the 1860s Angelo SECCHI made the first spectral classification of stars by dividing them into four groups according to color and spectral lines. As SPECTROSCOPY improved, a more comprehensive system was developed at Harvard College Observatory. This was embodied in the **Henry Draper Catalogue** (1918–24) of stellar spectra. The sequence of spectral types was arranged according to the prominence or absence of certain lines in the spectra. There were seven **spectral types**, designated O, B, A, F, G, K, and M (an ordering that results from revising an earlier alphabetical sequence). This sequence is a **temperature** sequence, from the hottest stars of types O and B, which appear blue-white, to type M, the coolest stars, which appear orange-red. As a result of this ordering, O, B, and A stars are called **early-type stars**; K and M stars are called **late-type stars**. Three new types were added when it was found that some cool stars had strong absorption bands not usually seen in other stars of the same color. These were classes R and N with strong bands of molecular carbon, and class S with bands of zirconium oxide. Stellar spectra can be classified into even finer divisions within these seven types, so decimal subdivisions were introduced. G5, for example, indicates a star midway in type between G0 and K0. Further refinements include the use of additional letters as suffixes to the spectral type, giving more information about the star, for example the existence of emission lines (e), metallic lines (m), broad lines due to rotation (n and nn), or a peculiar spectrum (p). However, the Harvard system could not deal with stars of different luminosities at a given temperature (that is, DWARFS, GIANTS, and SUPERGIANTS). In 1943 William Morgan, Philip Keenan, and Edith Kellman of Yerkes Observatory redefined the spectral types and added a classification scheme for the LUMINOSITY (absolute magnitude) of stars, which they represented using Roman numerals, ranging from I (supergiants) to VI (subdwarfs). This is now known as the **Morgan–Keenan system** (MK system), and is used universally. For example, a star classified in the MK system as O9.5 IV–V has a spectral type (and therefore a temperature) midway between that of an O9 and a B0 star, and a luminosity between that of a dwarf and a subgiant. The MK system is applicable to stars of normal chemical composition, which is to say about 95% of all stars. The various types of peculiar star are given their own special classification schemes. The Harvard R and N types are now combined into one **carbon star** class, the designations for which include a temperature type and a carbon band strength, as for example in C2,4. A similar classification is used for the S stars. WHITE DWARFS are usually classified using a Harvard-type system.

spectrograph SPECTROSCOPE fitted with a camera or an electronic detector such as a CHARGE-COUPLED DEVICE (CCD) used to obtain a permanent record of the SPECTRUM of ELECTROMAGNETIC RADIATION. Astronomical spectrographs, generally designed for use with a specific telescope, record the spectrum of a celestial object. Spectographs generally operate in a band of wavelengths from the near-infrared to the near-ultraviolet. There are various designs for different purposes. **High-dispersion** instruments, such as the spectroheliograph, spread the spectral lines widely so that a narrow band of wavelengths can be studied in detail.

spectrometer Type of SPECTROSCOPE equipped to measure accurately the WAVELENGTHS and intensities of spectral lines by means of an INTERFEROMETER. Spectrometers are often loosely referred to as SPECTROGRAPHS. Since spectra are now routinely recorded on charge-coupled devices (CCDs), from which information in digital form can be analyzed by computer, the distinction between modern spectrometers and spectrographs lies in their application rather than their design.

spectrophotometer Type of SPECTROMETER that works in conjunction with a SPECTROGRAPH. It outputs a graph of intensity against a WAVELENGTH called a **line profile**.

spectroscope Instrument for producing and studying the SPECTRUM of light from ELECTROMAG-NETIC RADIATION (light or other wavelengths). SPECTROSCOPY is the use of such an instrument to probe the chemical composition and physical conditions of an object. Spectroscopes are used in astronomy to study the light from stars, and in chemistry to detect the presence of traces of various elements in samples that are too small to detect by other means. The light entering a spectroscope is collimated into a narrow beam by means of a slit and lens. The beam then passes through either a PRISM or DIFFRACTION GRATING so that it is dispersed into a spectrum. Combined with the grating or prism is a scale from which the spectral WAVELENGTHS may be read directly through the telescope that magnifies the spectrum. Astronomical spectroscopes are known as SPECTROGRAPHS or SPECTROMETERS. Strictly, a SPECTROGRAPH is a spectroscope equipped with a camera for recording a permanent record of a spectrum, whereas a SPECTROMETER incorporates devices for accurately measuring the wavelengths and intensities of the spectral lines.

spectroscopy Branch of OPTICS dealing with the measurement of the WAVELENGTH and intensity of lines in a SPECTRUM. The main tool in this study is the SPECTROSCOPE. An analysis of the spectrogram (record of the spectrum) can reveal the substances causing the spectrum by the position of emission and absorption lines and bands. *See also* SPECTROGRAPH; SPECTROMETER

spectrum (pl. spectra) Arrangement of ELECTROMAGNETIC RADIATION ordered by WAVELENGTH or FREQUENCY. The visible light spectrum is a series of colors: red, orange, yellow, green, blue, indigo, and violet. Each color corresponds to a different wavelength of light. A spectrum is seen in a rainbow or when white light passes through a PRISM. This effect, also seen when visible light passes through a DIFFRACTION GRATING, produces a **continuous** spectrum in which all wavelengths (between certain limits) are present. Spectra consisting of bright lines or bands on a dark background are called **emission** spectra. These occur when a substance is strongly heated or bombarded by electrons. An **absorption** spectrum, consisting of dark regions on a bright background, is obtained when white light passes through a semi-transparent medium that absorbs certain frequencies. A **line** spectrum is one in which only certain wavelengths or "lines" appear. **Band** spectra consist of grouped bands of lines. The emission and absorption spectra are basic characteristics of a substance and are used as identification in the science of SPECTROSCOPY. Spectra result from ELECTRONS making transitions between different energy levels in atoms or molecules of the substance, giving rise to the emission or absorption of electromagnetic radiation.

▲ **spectroscope** Light from a telescope is passed through a fine vertical line in a plate (1). A lens (2) concentrates the rays onto a prism (3), which splits them into a spectrum. A second prism (4) is positioned so that the spectrum is widened even further after passing through it. The wide spectrum is then focused by a lens (5) onto a screen (6). The screen may be replaced by a recording apparatus, this instrument is called a spectrograph.

speed Rate of movement. It is the ratio of the distance covered to the time taken by a moving object. It is a SCALAR quantity. Speed in a specified direction is VELOCITY, a VECTOR quantity.

speedometer Apparatus for recording the SPEED of a vehicle. In a traditionally designed car or truck, it is driven by a flexible cable, connected to the TRANSMISSION, which turns at a speed proportional to the road speed. Inside the speedometer the cable rotates a magnet, which partly turns a spring-loaded drum to which the indicator needle or colored strip is attached. In more modern designs, a sensor in the gearbox generates an electrical signal according to the speed of rotation of the transmission, and this signal is converted into a luminous display on the dashboard.

speleology Scientific study of CAVES and cave systems. Included also are the hydrological and geological studies concerned with the rate of formation of STALAGMITES and STALACTITES, and the influence of GROUNDWATER conditions on cave formation. A special aspect is the study of the animals that live in caves.

sperm (spermatozoon) Male sex cell (GAMETE) in sexually reproducing organisms. It corresponds to the female OVUM. The head of the sperm contains the genetic material of the male parent, while its tail or other motile structure provides the means of moving to the ovum to carry out FERTILIZATION. *See also* SEXUAL REPRODUCTION

spermatogenesis Process by which SPERM are formed. Sperm (spermatozoa) are produced by a series of cell divisions that take place in the seminiferous tubules of a male animal's TESTES. Initially, germ cells divide to produce **spermatogonia**, which further divide to form **spermatocytes**. These finally divide yet again to produce **spermatids**, which develop into spermatazoa.

spermatophyte Traditionally a member of the division of seed-bearing plants (Spermatophyta), including most trees, shrubs, and herbaceous plants. A spermatophyte has a stem, leaves, roots, and a well-developed vascular system. The dominant generation is the SPOROPHYTE. The widely accepted Five Kingdoms classification now classifies seed plants as several distinct phyla: the Angiospermophyta (ANGIOSPERMS or flowering plants), Coniferophyta (CONIFERS), Ginkgophyta (GINKGO or maidenhair tree), Cycadophyta (CYCADS), and Gnetophyta (a group of cone-bearing desert plants).

spermicide *See* CONTRACEPTIVE

Sperry, Elmer Ambrose (1860–1930) US inventor and industrialist. He is celebrated for his improvements to the GYROSCOPE. In 1910 Sperry invented the GYROCOMPASS. He also designed a number of other electrical devices, including a powerful searchlight. Sperry used gyroscopes in his development of AUTOPILOT systems for NAVIGATION. He founded the American Institute of Electrical Engineers.

sphagnum Genus of MOSS that grows on boggy soils. When decomposed under boggy conditions, sphagnum moss forms PEAT MOSS, which is used in horticulture.

sphalerite (blende) Sulfide mineral composed of zinc sulfide (ZnS); it is an important source of zinc. Sphalerite has cubic system tetrahedral crystals or granular masses. It is white when pure, but more commonly yellow, black or brown with a resinous luster. It is often found in hydrothermal veins with galena, and in limestones where it occurs by replacement. Hardness 3.5–4; r.d. 4.

Sphenophyta Phylum comprising HORSETAILS.

sphere Three-dimensional geometric figure formed by the LOCUS in space of points equidistant from a given point (the center), or equivalently a surface generated by rotation of a CIRCLE about a diameter. If the distance from the center to the surface is r, the volume is $4/3\pi r^3$ and the surface area is $4\pi r^2$. A **spheroid** is the name given to the figure enclosed by the sphere.

spherical triangle TRIANGLE formed by the intersection, on the surface of a SPHERE, of arcs of three great circles (circles that have the same RADIUS as the sphere). The sides of spherical triangles are measured in terms of the angles that these arcs subtend at the sphere's center. **Spherical trigonometry**, the branch of geometry concerned with the properties of such triangles, is used in navigation.

sphincter Ring of muscle surrounding a body orifice which can open it or seal it off. Important sphincters include the pyloric sphincter in the stomach and the anal sphincter.

sphygmomanometer Instrument used to measure BLOOD PRESSURE. The device incorporates an inflatable rubber cuff connected to a column of mercury with a graduated scale. The cuff is wrapped around the upper arm and inflated to apply pressure to a major artery. When the air is slowly released, the pressure readings can be ascertained from the scale.

spicule Spearlike column of hot gas ejected from the Sun's CHROMOSPHERE as a jet, often reaching heights of 8,000km (5,000mi) above the solar surface. Most spicules are less than 500km (300 miles) in diameter. *See also* SUN

spider Any of numerous species of terrestrial, invertebrate, ARACHNID arthropods found worldwide. Spiders have an unsegmented abdomen attached to a cephalothorax (connected head and thorax) by a slender pedicel. There are no antennae; sensory hairs are found on the appendages (four pairs of walking legs). Most species have spinnerets on the abdomen for spinning silk to make egg cases and webs.

spilite Fine-grained, extrusive IGNEOUS ROCK found near VOLCANOES. The dark-colored rock occurs as PILLOW LAVA, which is formed where LAVA from a volcanic eruption flows into the sea; it often contains cavities formed by gas bubbles. It is a basic rock containing up to 40% silica, and more sodium than basalt, the common volcanic rock.

spin (symbol s) In QUANTUM MECHANICS, intrinsic angular momentum possessed by some ELEMENTARY PARTICLES, atoms, and nuclei. This may be regarded by analogy as the spinning of the particle on its own axis. Spin is one of the quantum numbers by which a particle is specified.

spinal column See SPINE

spinal cord Tubular, central nerve cord, lying within the SPINE (vertebral column or backbone) and bathed in cerebrospinal fluid. It connects the PERIPHERAL NERVOUS SYSTEM to the BRAIN, with which it makes up the CENTRAL NERVOUS SYSTEM (CNS). It gives rise to the 31 pairs of spinal nerves, each of which has sensory and motor fibers. *See also* NERVOUS SYSTEM

spinal fluid See CEREBROSPINAL FLUID

spinal nerves Thirty-one pairs of nerves branching out from the SPINAL CORD to supply the muscles of the trunk and limbs and carry sensory information to the BRAIN. They are mixed nerves containing both sensory and motor fibers. The motor fibers arise from the anterior (frontal) side of the spinal cord and the sensory fibers from the rear.

spindle Rod-shaped structure formed from microtubules in the CYTOPLASM of cells during MITOSIS or MEIOSIS. CHROMOSOMES are attached at the bulge of the spindle (the equator). The spindle draws the chromosomes apart, causing a cell to divide. *See also* METAPHASE

spine Backbone of VERTEBRATES, extending from the base of the SKULL to the tip of the tail and enclosing the SPINAL CORD. The human spine consists of 26 small vertebrae interspersed with INTERVERTEBRAL DISKS of CARTILAGE. It comprises seven cervical, 12 thoracic, and five lumbar vertebrae. The five sacral and four vertebrae of the COCCYX

fuse together to make two solid bones. The spine articulates with the bones of the skull, ribs, and hip bones and provides points of attachment for the back muscles. *See also* SKELETON

spinel Oxide mineral, magnesium aluminum oxide ($MgAl_2O_4$), found in IGNEOUS and METAMORPHIC ROCKS. It has cubic system, frequently twinned, octahedral crystals. It is either glassy black, red, blue, brown or white in color. Ruby spinel from Sri Lanka is a valuable gemstone. Hardness 7.5–8; r.d. 3.8.

spinning Process of making thread or yarn by twisting FIBERS together. The fibers may be of animal, vegetable or synthetic origin. Machines developed for mechanizing the spinning process include the SPINNING WHEEL (a simple machine for spinning a single strand of yarn), SPINNING FRAME, SPINNING JENNY, and SPINNING MULE. Today, large machines carry out the same basic process, but at much increased speeds. Different machines are designed to suit each kind of fiber. *See also* TEXTILES

spinning frame Machine for spinning yarn. Richard ARKWRIGHT invented the spinning frame in 1768. He called his invention a water frame, as it was powered by a water wheel. The motion of the wheel turned several spindles, so that many threads could be spun at the same time.

spinning jenny Hand-operated spinning machine invented by James HARGREAVES in 1764. It operated on the same basic principles as the SPINNING WHEEL, but spun several strands of yarn simultaneously.

spinning mule (Crompton's mule) Spinning machine invented in 1779 by Samuel CROMPTON. The machine was called a mule because it was a hybrid, combining the main features of Hargreaves' SPINNING JENNY and Arkwright's SPINNING FRAME.

spinning wheel Simple machine for spinning a single strand of yarn. A hand-operated spinning wheel was used in India more than 2,500 years ago. Around 1500 a foot-operated wheel was invented in Germany. A spinning machine has a large wheel, which is linked by a belt to a spindle. The spindle turns rapidly, spinning the fibers into yarn.

spiracle External opening for RESPIRATION in various animals. In insects and spiders, it is the opening to a TRACHEA. In sharks, rays and some bony fish, water passes through a pair of spiracles during GILL respiration. In whales, the nasal opening is called a spiracle.

spiral galaxy Type of regular GALAXY as classified by Edwin HUBBLE.

spirochaete General name applied to a group of protozoa-like BACTERIA that are spiral-rod shaped and capable of flexing and wriggling their bodies as they move about. SYPHILIS is the best known disease caused by a spirochete (*Treponema pallidum*).

spit Ridge of sand or shingle on a coastline, which has been built up as a result of LONGSHORE DRIFT. A spit is attached to the land at one end, and may consist of sand, mud, shingle or any combination of these. Most spits protrude across an ESTUARY, and may eventually block the estuary to form a DELTA.

spleen Dark-red organ located on the left side of the abdomen, behind and slightly below the stomach. It is important in both the BLOOD and LYMPHATIC SYSTEMS, helping to process LYMPHOCYTES, destroying worn out or damaged ERYTHROCYTES (red blood cells) and storing iron. Removal of the spleen (splenectomy) is sometimes necessary following trauma or in the treatment of some blood disorders.

spodumene Silicate mineral, lithium aluminum silicate ($LiAlSi_2O_6$). It is the chief source of LITHIUM. Spodumene displays monoclinic system prismatic crystals or masses. It is opaque or transparent in many hues, and the transparent lilac and green varieties are used as gems. Hardness 6.5–7; r.d. 3.2.

sponge Primitive, multicellular, aquatic animal. Its extremely simple structure is supported by a skeleton of lime, silica or spongin. There is no mouth, nervous system or cellular coordination, nor are there any internal organs. Sponges reproduce sexually and by asexual budding. There are about 5,000 species, including the simple sponge genus *Leucosolenia*. Length: 1mm–2m (0.04in–6ft). Phylum: Porifera (*see* PORIFERAN).

spontaneous combustion Outbreak of FIRE without external application of HEAT. When combustible material, such as damp hay, paper or rags, is slowly oxidized by bacteria or air, the temperature may rise to the ignition point. *See also* COMBUSTION

spontaneous generation Belief, now discredited, that living organisms arise from nonliving matter. It supposedly explained the sudden appearance of maggots on decaying meat. *See* EVOLUTION

spore Small reproductive body that detaches from the parent organism to produce new offspring without having to fuse with another reproductive cell. Mostly microscopic, spores may consist of one or several cells (but do not contain an embryo) and are produced in large numbers. Some germinate rapidly, others "rest," surviving unfavorable environmental conditions. Spores are formed by ferns, horsetails, mosses, fungi, bacteria, and some protazoa. Spores are formed during the SPOROPHYTE stage in the life cycle of such organisms. *See also* GAMETE

sporophyte DIPLOID (having two sets of chromosomes) stage in the life cycle of plants or algae. In most plants and algae, the sporophyte gives rise to HAPLOID (having only one set of chromosomes) spores which germinate to produce a haploid generation (the GAMETOPHYTE stage). The haploid generation produces the GAMETES. In ferns, horsetails, conifers and flowering plants, the diploid sporophyte is the dominant phase of the life cycle, the plant body we usually see. In mosses and liverworts, the main plant body is the gametophyte. *See also* ALTERNATION OF GENERATIONS

spring Point on the Earth's surface at which underground water emerges. Springs may emerge at points on dry land or in the beds of streams or ponds. They are generally located where there is a change of rock type, and the water flows out at the top of an impervious layer, such as slate, shale or clay. They are an important part of the water cycle. The mineral content of spring water varies with the surrounding soil or rocks.

spring Mechanical device designed to be elastically compressed, extended or deflected. It may be used to store energy, absorb shock, or maintain contact between two surfaces.

spring tide *See* TIDE

Sputnik 1 World's first artificial SATELLITE, launched by the Soviet Union on October 4, 1957. Weighing 83.5kg (184lb) and equipped with a radio transmitter, Sputnik 1 circled the Earth sending back signals for 21 days. **Sputnik 2** was launched in November 1957. Weighing 508kg (1,120lb), it carried the dog Laika. *See also* SPACE EXPLORATION

Squamata Order of cold-blooded, scaly REPTILES, comprising the snakes (suborder Serpentes) and lizards (suborder Sauria). They are mostly terrestrial vertebrates that first appeared at the end of the Triassic period, c.170 million years ago. Typically, **lizards** have cylindrical bodies with four legs, a long tail, and movable eyelids. **Snakes** are legless reptiles with no external ear openings, eardrums or middle ears; sound vibrations are detected through the ground. Their eyelids are immovable. The long, forked, protractile tongue is used to detect odors. Poisonous species have hollow or grooved fangs, though which they inject venom.

square RECTANGLE with four sides of the same length. In arithmetic or algebra, a square is the result of multiplying a quantity by itself: the square of 3 is 9, and the square of x is x^2.

square root Number or quantity denoted $\sqrt{x}$ that must be multiplied by itself to give the number x. The square root of 4 is 2, i.e. $\sqrt{4} = 2$. To four decimal places, $\sqrt{2} = 1.4142$. A negative number has complex square roots. *See also* CUBE ROOT; ROOT

stabilizer In AERODYNAMICS, vertical and horizontal fins that prevent erratic rolling and pitching, respectively. They are usually placed together in the form of a tailplane. Like the wings, the horizontal fins have an AIRFOIL shape and can provide LIFT at takeoff. In flight, however, they act only as stabilizers and in some designs the air THRUST on them acts downwards. In marine engineering, fins or tanks are used to keep a submarine or ship steady in rough seas. *See also* PITCH

stack Pillar of rock that forms an islet at the end of a headland. A stack is the result of a series of events caused by EROSION. Sea waves striking a headland first carve a CAVE by opening up JOINTS in the rock. The cave gradually penetrates the headland until an arch is formed. Finally, the top of the arch collapses and leaves a stack separated from the rest of the headland. The process may be repeated several times, giving a series of stacks.

▲ **stalactite/stalagmite** Weak carbonic acid, formed when rain absorbs carbon dioxide from the atmosphere, dissolves tiny amounts of calcium carbonate from limestone rock, as it percolates down through the rock (1). If the water enters a previously eroded cave (2), it forms drips (3). Some of the dissolved carbon dioxide escapes from the water, and the calcium bicarbonate in the water is precipitated as a calcite known as travertine, consisting of calcium carbonate. They form rising columns called stalagmites (4) and descending pillars from which water drips, called stalactites (5).

stained glass Colored GLASS used for decorative effect in windows. Stained glass is most commonly found in churches. The oldest surviving examples of complete stained glass windows are in Augsburg Cathedral, Germany, and date from around 1100. In its purest form, stained glass is made by adding metal-oxide coloring agents during the manufacture of glass. Shapes cut from the resulting sheets are then arranged to form patterns or images. These shapes are joined and supported by flexible strips of lead that form dark, emphatic contours. Details are painted onto the glass surfaces in liquid ENAMEL and fused on by heat. Other techniques include the application of colored surface coatings onto clear glass. Intricate designs may be scraped into the surface. In modern windows, the pieces of glass are sometimes joined using an adhesive instead of lead strips.

stainless steel Name for a group of IRON alloys that resist CORROSION. Besides carbon, contained in all STEELS, stainless steels contain from 12% to 25% CHROMIUM. This makes the steel stainless and rust-resistant by forming a thin, protective oxide coating on the surface. Most stainless steels also contain NICKEL. Various other metals and nonmetals may be added to give the steel particular properties. Stainless steel is used to make kitchen utensils, cutlery, chemical process equipment, and fittings of various kinds, including parts for cars.

stalactite Icicle-shaped deposit of tiny CALCIUM CARBONATE crystals found hanging down from the roof of CAVES in Carboniferous limestone areas. Stalactites are formed by water slowly percolating through the rocks; as the water droplets are about to fall from the cave roof, a tiny layer of calcium precipitates out of the water and solidifies. Gradually, the deposits build up to form a column of calcium carbonate crystals known as CALCITE.

stalagmite Column of CALCIUM CARBONATE crystals that grows upwards from the floor of CAVES in a Carboniferous limestone areas. Stalagmites are formed as a result of water dripping from the ceiling of the cave and leaving tiny deposits of calcium carbonate on the floor. The deposits gradually build up into CALCITE crystals, and can sometimes extend upwards to meet the STALACTITE above, forming a column or pillar. Stalagmites tend to be fatter than stalactites, possibly because the water splashes out when it drops to the floor, causing a wider spread of calcium carbonate.

stamen POLLEN-producing element of a FLOWER. It consists of an ANTHER, in which pollen is produced, on the end of a stalklike filament. The arrangement and number of stamens is important in the classification of ANGIOSPERMS (flowering plants).

stamping Method of forming metal. A flat blank of metal is stretched over a die and struck with a movable punch into the desired shape. *See also* METALLURGY

standard deviation In STATISTICS, a measure of deviation of observed data or scores from the average or MEAN. It is written as either σ or *s*. A small standard deviation indicates that observations cluster around the MEAN, while a large one indicates that the data points are spread far from the mean. The standard deviation is equal to the square root of a quantity called the variance. The variance is the mean of the sum of the squared differences of the data points from the mean.

standard electrode (reference ELECTRODE) In ELECTROCHEMISTRY, a HALF-CELL combined with other half-cells and used as a reference. The basic standard is the hydrogen electrode – a platinum electrode immersed in a molar solution of hydrogen ions and covered with an absorbent layer of platinum black, over which is bubbled molecular hydrogen. This is taken to have a ELECTRODE POTENTIAL of zero.

standard form Scientific method of writing NUMBERS, especially very large or very small numbers. The first nonzero digit of the number is placed before the decimal point and all other digits are placed in order after the decimal point. This decimal is multiplied by an appropriate power of 10. For example, 1,721,448 is written as $1.721\ 448 \times 10^6$ in standard form and 0.003378159 is written as $3.378\ 159 \times 10^{-3}$.

standard model Model of ELEMENTARY PARTICLES and their interactions that provides the best description of current physical phenomena. Particles are divided into HADRONS (which experience the STRONG NUCLEAR FORCE and are made up of QUARKS), LEPTONS (particles such as ELECTRONS that do not experience the strong nuclear force), and particles that "carry" interactions (such as PHOTONS, W and Z BOSONS, and GLUONS; these mediate in interactions between particles in the other two categories). Physicists are striving to go beyond the standard model to achieve a GRAND UNIFIED THEORY (GUT). *See also* FUNDAMENTAL FORCES

standard temperature and pressure (STP) (formerly normal temperature and pressure, NTP) In chemistry and physics, standard conditions for measurements, especially when comparing the volumes of gases. STP is a **temperature** of 0°C (32°F) and a **pressure** of 1 standard atmosphere (760mm of mercury or 101,325 pascals).

standing wave (stationary wave) In physics, a WAVE in which the points of maximum vibration (the **antinodes**) and the points of no vibration (the **nodes**) do not move. A standing wave is formed by the interference of TRAVELING WAVES of equal frequency and intensity moving in opposite directions.

Stanley, Wendell Meredith (1904–71) US biochemist who shared the 1946 Nobel Prize for chemistry with John Northrop (1891–1987) and James SUMNER for his research into VIRUSES. In 1935 Stanley crystallized tobacco mosaic virus and showed that it consists of PROTEIN and NUCLEIC ACID molecules in a rod-shaped form. This enabled the exact molecular structure and means of proliferation of this and other viruses to be determined. Stanley also studied influenza viruses and developed a VACCINE against the disease.

Staphylococcus Genus of spherical bacteria that grow in grapelike clusters and are found on the skin and mucous membranes of human beings and other animals. Pathogenic staphylococci cause a range of local or generalized infections, including **pneumonia** and **septicemia**. They are the most common cause of food poisoning. They may be destroyed by ANTIBIOTICS, although some strains have become resistant. They require oxygen for growth.

star Hot, self-luminous ball of gas, held together by its own GRAVITATION and giving off ELECTROMAGNETIC RADIATION produced by NUCLEAR ENERGY, mainly the conversion of hydrogen into helium. The stars visible to the naked eye are all in the MILKY WAY. They are classified according to their MAGNITUDE; from the brightest (first magnitude) to the dimmest (sixth magnitude). Their brightness varies because of their relative distance from Earth and their different LUMINOSITIES. Frequently, stars are grouped together in STELLAR CLUSTERS. The velocity of the movement of a star through space is measured by its PROPER MOTION and RADIAL VELOCITY. The temperatures and luminosities of stars are prescribed by their masses. The most massive stars are about 100 solar masses (a hundred times heavier than the Sun). Above this mass, a star is not stable and is likely to break up. Stars with the greatest mass are very luminous and hot (*c*.20,000°C). The hotter the star, the shorter the wavelength at which light appears most intense. Thus, the hottest stars appear blue. RED GIANTS are extremely large stars, with a diameter between 10 and 100 times that of the Sun, but are very cool. Stars with an average mass, such as the Sun, are yellow, while small stars are a dull red (*c*.3,000°C). The smallest stars (WHITE DWARFS and NEUTRON STARS) contain less than one-twentieth of a solar mass; below this the temperature does not become high enough for nuclear reactions to take place, and the object becomes a BROWN DWARF. The colors, and hence the temperatures, of stars give rise to the scheme of SPECTRAL CLASSIFI-

CATION. Most stars are born not singly but in pairs (BINARY STARS) or groups. VARIABLE STARS (such as SUPERNOVAS) have fluctuating brightness. *See also* BLACK BODY; GALAXY; HERTZSPRUNG-RUSSELL DIAGRAM; SPECTRUM; STELLAR EVOLUTION

starch CARBOHYDRATE, stored in many plants, that provides about 70% of humankind's food in such forms as rice, potatoes, and cereals. Animals and plants convert it to GLUCOSE for energy (RESPIRATION). Consisting of linked glucose units, starch exists in two forms: **amylose**, in which the glucose chains are unbranched, and **amylopectin**, in which they are branched. It is made commercially from cereals, corn, potatoes, and other plants. Starch is used in the manufacture of adhesives and foods (as a thickening agent) and to stiffen laundered clothes and other fabrics.

star cluster *See* STELLAR CLUSTER

Stark, Johannes (1874–1957) German physicist. He received the 1919 Nobel Prize for physics for his discovery (1913) of the STARK EFFECT and for his work demonstrating the DOPPLER EFFECT. In the 1920s Stark rejected the theory of RELATIVITY and QUANTUM THEORY.

Stark effect In physics, splitting of spectral lines of an ATOM or MOLECULE when it is placed in a strong electric field. It was discovered (1913) by Johannes STARK. It follows from the ELECTRON theory of Hendrik LORENTZ. *See also* ZEEMAN EFFECT

Starley, James (1830–81) British engineer and inventor of BICYCLES. Working in Coventry from 1857, he produced his first bicycle (*The Coventry*) in 1868. His "ordinary" or "penny farthing" bicycle of 1871 had center pivot steering. Starley introduced the *Coventry* tricycle in 1876, adding differential gears in 1877.

Starling, Ernest Henry (1866–1927) British physiologist who contributed greatly to the knowledge of body functions. Starling made major discoveries about body TISSUES, the mechanical controls of the HEART and the ENDOCRINE SYSTEM.

static electricity Study of ELECTRIC CHARGES at rest, not flowing as an ELECTRIC CURRENT. Ordinary uncharged ATOMS have equal numbers of positive and negative ELECTRONS. On the everyday (macroscopic) scale, electrically charged objects are those that have either too many or too few electrons. COULOMB'S LAW relates the forces that charged bodies have on each other and to their charge and the distance between them. Static ELECTRICITY can be produced by friction, such as by combing hair (which acquires a negative charge) or rubbing

nylon material. Electrons may then jump off as a spark, shocking anyone touching the object. Lightning is the result of a large-scale buildup of static electricity. This form of electricity is studied in the branch of physics known as ELECTROSTATICS.

statistical mechanics Branch of physics that studies large-scale properties of MATTER based on the statistical laws of large numbers. The large number of molecules in such a system allows the use of statistics to predict the probability of finding the system in any state. The ENTROPY (disorder or randomness) of the system is related to the number of its possible states; a system left to itself will tend to approach the most probable distribution of energy states. *See also* THERMODYNAMICS

statistics Science of collecting and classifying numerical data. Statistics can be **descriptive** (summarizing the data obtained) or **inferential** (leading to conclusions or inferences about larger numbers of which the data obtained are a sample). Inferential statistics are used to give a greater degree of confidence to conclusions, since statistics make it possible to calculate the PROBABILITY that a conclusion is in error.

Staudinger, Hermann (1881–1965) German chemist who was awarded the 1953 Nobel Prize for chemistry for his work on the structure of PLASTICS. He showed that plastics, rather than being random aggregations of MOLECULES, are POLYMERS, similar to CELLULOSE, in which the molecules are bonded together to form a long chain. This discovery contributed greatly to the growth of the plastics industry.

staurolite Silicate mineral, iron magnesium aluminum silicate, found in SCHISTS and GNEISSES. It has orthorhombic system, prismatic and flattened crystals, and is glassy brown in color. Hardness 7–7.5; r.d. 3.6.

steady-state theory Cosmological theory put forward (1948) by the Austrian astronomers Hermann Bondi (1919–) and Thomas Gold (1920–), and further developed by Fred HOYLE and others. According to this theory, the UNIVERSE has always existed; it had no beginning and will continue forever. Although the Universe is expanding, it maintains its average density – its steady state – through the continuous creation of new MATTER. Most cosmologists now reject this theory because it cannot explain naturally the COSMIC MICROWAVE BACKGROUND or the observation that the appearance of the Universe has changed with TIME. The BIG BANG theory is the currently accepted theory. *See also* COSMOLOGY; EXPANDING UNIVERSE

stealth technology Methods used to render an aircraft nearly invisible to RADAR, heat, and visual detection. To achieve "invisibility," a stealth aircraft must have sympathetic airframe design (all radar-reflecting "hard" edges smoothed away), engine exhaust dampers that mask and disperse jet efflux, and radar absorbent material that "holds" electronic emissions rather than reflecting them. This coating has to be regularly applied, since it is sensitive to water and wear. In 1983 the American Lockheed F-117A became the first "stealth" aircraft to enter frontline service; it was followed a decade later by the Northrop B-2. The F-19 stealth fighter was used in the Gulf War (1991). Few aerodynamic advances have had such a profound effect on military aviation as stealth technology.

steam Gaseous form of WATER. As commonly seen in the home, steam contains minute drops of water and is technically called **wet steam**. The pure gas, **dry steam**, is obtained by heating steam at temperatures well above the boiling point of water, 100°C (212°F). It is used widely in industry for heating, transporting heat (in heat exchangers) and to drive TURBINES.

steam engine ENGINE powered by steam. Steam, generated by heating water, is used to produce movement. In some engines, the steam forces pistons to move along cylinders. This results in a reciprocating (back-and-forth) motion. A mechanism usually changes this into rotary motion. Steam LOCOMOTIVES use RECIPROCATING ENGINES. STEAM TURBINES are engines that produce rotary motion directly by using the steam to turn sets of fan-like wheels. Steam turbines drive power station GENERATORS and ships' propellers. In any steam engine, some of the HEAT used to turn water into steam in a boiler is converted into energy of motion. The heat may be produced by burning fuel in a FURNACE, or may come from a NUCLEAR REACTOR. The first steam engine was a form of pump, used to remove water from mines. It was invented (1689) by Thomas SAVERY. In his simple machine, steam condensed to form a very small volume of water, thus producing a partial vacuum. This was used to suck water up a mine shaft. In 1712 Thomas NEWCOMEN invented a steam-operated pump with pistons. From the 1760s, James WATT improved on Newcomen's ideas and produced more efficient steam engines. This led to the use of steam engines to power machinery in factories. In 1884 British engineer Charles Parsons (1854–1931) invented the first practical steam TURBINE. His machines were so efficient that turbines soon started to replace reciprocating steam engines in power stations. The most astonishing development in steam power has been the manufacture of a completely enclosed working steam engine of microscopic size. Japanese scientists made the device using techniques developed for manufacturing INTEGRATED CIRCUITS. A small

current passed through an electric heating element changes a small droplet of water into steam to move a piston. Practical applications for this device are still awaiting discovery.

steam turbine STEAM ENGINE with rotating blades that are turned by steam. *See also* TURBINE

stearic acid (octadecanoic acid) Common, saturated, FATTY ACID ($CH_3(CH_2)_{16}CO_2H$) commonly found as a glyceride in animal and vegetable fats. It is used in ointments, lubricants, creams, candles, and soap. Properties: r.d. 0.85; m.p. 70°C (158°F).

stearin (glyceryl tristearate, $C_{57}H_{110}O_6$) Triglyceride of STEARIC ACID ($CH_3(CH_2)_{16}CO_2H$). It is present in many animal and vegetable fats, particularly the hard ones such as tallow and cocoa butter. Stearin is prepared from stearic acid and GLYCEROL, using aluminum oxide as a catalyst.

steatite *See* SOAPSTONE

steel Group of IRON alloys containing a little CARBON. The great strength of steel makes it an extremely important material in construction and manufacturing. The most common type is called **plain carbon** steel, since carbon is the main alloying material. This kind of steel usually contains less than 1% of carbon by weight, with smaller amounts of other alloying materials, such as manganese, silicon, sulfur, and phosphorus. ALLOY steels contain some carbon, but owe their special properties to the presence of some combination of the alloying metals manganese, nickel, chromium, vanadium, and molybdenum. **Low-alloy** steels, with less than 5% of alloying metals, are exceptionally strong and are used in buildings, bridges, and machine parts. **High-alloy** steels contain more than 5% of alloying metals. This group includes various forms of STAINLESS STEEL. Steel has been produced for more than 2,000 years, but early methods were slow and laborious, and produced only small quantities. Large-scale steelmaking became possible in the mid-1800s, when the BESSEMER PROCESS and OPEN-HEARTH PROCESS were introduced. Today, most steel is produced by the BASIC OXYGEN PROCESS. An ELECTRIC FURNACE is used for some special steels containing materials that would become oxidized in other steelmaking processes.

Stefan–Boltzmann law (Stefan's law) In physics, principle that the total radiated energy from a BLACK BODY per unit time per unit area is directly proportional to the fourth power of its THERMODYNAMIC temperature (T^4). The constant of proportionality (the **Stefan–Boltmann constant**) is equal to 5.6997×10^{-8} Js^{-1} m^{-2} K^{-4}. It is named after the Austrian physicist Josef Stefan (1835–93) and was derived for thermodynamic purposes by Ludwig BOLTZMANN.

Stein, William Howard (1911–80) US biochemist. He shared the 1972 Nobel Prize for chemistry with Christian ANFINSEN and Stanford MOORE for research into the molecular structure of PROTEINS. Stein developed techniques for the analysis of PEPTIDES and AMINO ACIDS obtained from proteins and applied these methods to determining the molecular structure of the ENZYME ribonuclease. This was of great importance to the understanding of processes involved in protein synthesis within CELLS.

stellar cluster (star cluster) Any of innumerable collections of gravitationally associated STARS occurring within GALAXIES. Stellar clusters are of two main types: the OPEN CLUSTER (or galactic cluster) and GLOBULAR CLUSTER. **Open clusters**, usually found in the spiral arms of galaxies, consist of up to several thousand young stars belonging to Population I; **globular clusters**, which are much more concentrated, are found in the halo surrounding the centers of galaxies and consist of old Population II stars.

stellar evolution Various phases of a STAR over its lifetime. A star "lives" for between a few thousand years to billions of years depending on its MASS. Stars form when an INTERSTELLAR CLOUD of gas and dust collapses under its own GRAVITATION. As the cloud collapses, atoms collide and generate heat. This process continues until the heat generated is sufficient to cause nuclear FUSION reactions, converting hydrogen to helium. The reactions from the core throw out considerable amounts of radiation, which stops further collapse. This stage (the MAIN SEQUENCE phase) is the longest in a star's lifetime. Eventually the mainly hydrogen core of the star is depleted, and fusion can no longer occur. With the central energy source removed, the core collapses under gravitation and heats itself further until hydrogen fusion is able to occur in a spherical shell surrounding the core. As this change takes place, the outer layers of the star expand considerably and the star becomes either a RED GIANT, or in the most massive stars, a SUPERGIANT. During this phase the core can reach temperatures of 100 million K, hot enough for fusion of helium to carbon. When this second process of fusion is complete, the core collapses once again and heats up. In **low-mass** stars the temperature will not rise sufficiently for carbon fusion to take place, and the red giant loses its outer layers leaving a WHITE DWARF. In **high-mass** stars, carbon fusion is initiated, converting the carbon into elements with atomic weights close to that of iron. At this stage no further fusion is possible, and the core collapses explosively, throwing off outer layers in a SUPERNOVA explosion. The resulting super dense core forms either a NEUTRON STAR or a BLACK HOLE.

stem Main, upward-growing part of a plant that bears leaves, buds, and flowers or other reproductive structures. In VASCULAR PLANTS, the stem contains conducting tissues (XYLEM and PHLOEM). In flowering plants, this vascular tissue is arranged in a ring (in DICOTYLEDONS) or scattered (in MONOCOTYLEDONS). Stems are usually erect, but may be climbing (vine) or prostrate (stolon). They may also be SUCCULENT (cactus) or modified into underground structures (RHIZOMES, TUBERS, CORMS, BULBS). Stems vary in size from the threadlike stalks of aquatic plants to tree trunks. Woody stems have an outer layer of corky cells, called BARK. They increase in girth every year as specialized layers of CAMBIUM cells produce extra tissues in the cortex, and sometimes also just below the EPIDERMIS.

Stephenson, George (1781–1848) British engineer and inventor, regarded as the father of the LOCOMOTIVE. In 1814 he built his first locomotive, *Blucher*. This locomotive, the first to have flanged wheels, ran on a tramway. Stephenson's most famous locomotive, *Rocket*, was built in 1829. It reached a top speed of 47km/h (29mph). It later ran on the Liverpool to Manchester line, one of the many railroad lines that he engineered.

Stephenson, Robert (1803–59) British engineer, son of George STEPHENSON. From 1827 Robert Stephenson managed his father's locomotive works. He built several railroad lines and tubular bridges, including the Britannia Bridge over the Menai Strait, Wales.

steppe Area of temperate grassland found in Kazakhstan, Russia, Ukraine, and neigboring parts of southern Europe. The landscape is often very flat and open and the climate is too dry to support much vegetation. When it is plowed, the steppe can become very rich farmland, especially for growing cereals. Steppes are very similar to **prairies**, and have a similar continental climate. Winters are very cold, but summers are quite warm. Rainfall is quite light, and falls mainly in the summer months.

Steptoe, Patrick Christopher (1913–88) British gynaecologist. In 1968 he began working with Robert Edwards on the problems of infertility in women. The final outcome of their researches was the development of IN VITRO FERTILIZATION, in which eggs collected from a woman's ovary are fertilized in the laboratory and then implanted in her womb. In 1978 Louise Brown became the first "test-tube baby" to be born using this technique. Within 10 years, more than a thousand others had been born.

stereochemistry Study of the chemical and physical properties of COMPOUNDS as affected by the ways in which the atoms of their molecules are arranged in space. Such arrangements can result in two or more compounds having the same numbers and kinds of atoms, but differently shaped molecules; these are known as **stereoisomers**. Stereochemistry also deals with optical isomerism, in which the configuration of one molecule is the mirror image of another. *See also* ISOMERS

stereophonic sound Method of SOUND REPRODUCTION that gives the illusion of both location and direction; at least two separate channels are required. In stereophonic LONG-PLAYING RECORDS, the groove is modulated in two planes – the lateral and the vertical (groove depth). Depth variations correspond to the difference between the left and right channels of the stereo-recording and lateral modulations correspond to their sum. Two transducers in the pick-up cartridge feed two separate AMPLIFIERS and two separated LOUDSPEAKERS. *See also* COMPACT DISC (CD); MAGNETIC TAPE

stereoscope Optical device that produces an apparently three-dimensional image by presenting two slightly different plane images, usually photographs, one to each eye. Early stereoscopes of the 1830s used mirrors and prisms, and some modern ones use POLARIZED LIGHT to project images that are viewed through polarizing filters.

stereotype In PRINTING, duplicate printing plate made from the original plate (which is reserved for making further duplicates as necessary). A traditional method of making stereotypes is to press papier mâché into the original and produce a negative mold from which a duplicate plate is electroplated or cast.

sterility Inability to reproduce. It may be due to INFERTILITY or to surgical intervention.

sterilization Surgical intervention that terminates a person's ability to reproduce. In women, the usual procedure is **tubal ligation**: sealing or tying off the FALLOPIAN TUBES so that fertilization can no longer take place. In men, a VASECTOMY is performed to block the release of sperm. The term is also applied to the practice of destroying microorganisms in order to prevent the spread of infection. Techniques include heat treatment, irradiation, and the use of disinfecting agents. Sterile conditions are necessary in surgery, in the food and pharmaceutical industries, and in some laboratory experiments.

Stern, Otto (1888–1969) US physicist, b. Germany. In 1933 he resigned from the University of Hamburg and emigrated to the United States. Stern became professor of physics at the Carnegie Institute of Technology. A leading experimental physicist, Stern was awarded the 1943 Nobel Prize for physics for developing the molecular beam as a tool for studying the characteristics of the MOLECULE and for his measurement of the magnetic moment of the PROTON.

sternum (breastbone) Flat, narrow bone extending from the base of the neck to just below the diaphragm in the center of the chest. It consists of three sections: the **manubrium**, or upper part; the **body** or gladiolus; and the **xiphoid process**, the lower and more flexible cartilaginous part. The top of the manubrium is attached by ligaments to the collarbones and the body is joined to the ribs by seven pairs of costal cartilages.

steroid Class of organic compounds characterized by a basic molecular structure of 17 carbon atoms arranged in four rings and bounded by up to 28 hydrogen atoms. Steroids are widely distributed in animals and plants, the most abundant being the sterols (steroid alcohols) such as CHOLESTEROL. Another important group are the steroid HORMONES, including the CORTICOSTEROIDS, secreted by the adrenal cortex, and the sex hormones (ESTROGEN, PROGESTERONE, and TESTOSTERONE). Synthetic steroids are widely used in medicine. Athletes sometimes abuse steroids to increase their muscle mass, strength, and stamina; but there are harmful side effects. Unauthorized use of steroids is illegal in sports.

stethoscope Instrument that enables an examiner to listen to the action of various parts of the body, principally the heart and lungs, to help evaluate their condition. It consists of two earpieces attached to flexible rubber tubes, which lead to either a disk or a cone.

Stevenson, Robert (1772–1850) Scottish civil engineer who designed LIGHTHOUSES. He designed and built 23 lighthouses, and systems for making the light flash an identifying signal. Stevenson also produced designs for canals and harbors.

stibnite Sulfide mineral, ANTIMONY sulfide (Sb_2S_3), found in low-temperature veins and rock impregnations. Stibnite occurs as orthorhombic system aggregates of either prismatic crystals or granular masses. It is opaque, sometimes iridescent, and gray or black in color. Stibnite is an important ore of antimony. Hardness 2; r.d. 4.6.

stigma In botany, the free upper part of the STYLE of the female organs of a FLOWER, to which a pollen grain adheres before FERTILIZATION.

Still, Andrew Taylor (1828–1917) US founder of OSTEOPATHY. After three of his children died in an epidemic of spinal MENINGITIS, he devised a treatment based on the theory that most physical disorders were caused by "structural derangement" of the body, which could be remedied by manipulations of bones and muscles without the use of drugs.

stillbirth Delivery of a dead FETUS of more than 24 weeks gestation. Death of a fetus before this point is termed MISCARRIAGE.

stimulant Substance that increases mental alertness and activity. There are a number of stimulants that act on the CENTRAL NERVOUS SYSTEM (CNS), notably drugs in the AMPHETAMINE group (now rarely prescribed). Many common beverages, including tea and coffee, contain small quantities of the stimulant CAFFEINE.

stimulus In psychology, any change in energy, either external or internal, that influences the RECEPTORS of the NERVOUS SYSTEM.

stoichiometric compound Chemical COMPOUND in which the atoms combine in simple whole-number ratios – that is, any compound with an exact chemical formula. *See also* NONSTOICHIOMETRIC COMPOUND

Stokes, Sir George Gabriel (1819–1903) British mathematician and physicist, b. Ireland. He was professor of mathematics (1849–1903) at Cambridge and served as president (1885–92) of the Royal Society. Stokes is celebrated for his law

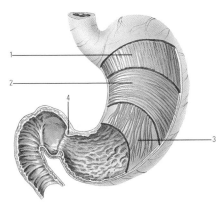

▲ **stomach** The stomach, like most of the digestive tract, is walled with involuntary muscle. The fibers of the stomach wall are built up in three layers: longitudinal (1), circular (2), and oblique (3). These muscular layers work in collaboration, producing a wavelike movement (peristalsis) and forcing food through the stomach, past the circular muscle valve sphincter (4) at the base and into the duodenum (5).

of VISCOSITY (*see* STOKES'S LAW), which describes the movement of a solid sphere in a fluid. It forms the basis of the modern theory of viscous fluids. He also developed an important theorem of VECTOR ANALYSIS, accurately described and named FLUORESCENCE, and proposed the concept of an ETHER. Stokes' book, *Dynamic Theory of Diffraction* (1849), was an important work in the wave theory of light. He was also a pioneer in GEODESY.

Stokes' law In fluid mechanics, a formula that gives the terminal (final) velocity at which particles fall in a fluid (liquid or gas). For particles of radius r (less than 0.1mm) the terminal velocity is $2(d_1 - d_2)gr^2/9\eta$, where d_1 and d_2 are the densities of the particle and the fluid, g is the acceleration due to gravitation, and η is the viscosity of the fluid. It is named after Sir George STOKES.

STOL aircraft (short takeoff and landing aircraft) AIRCRAFT with the ability to operate out of extremely small airfields. Present STOL aircraft are capable of taking off and climbing over a 15-m (50-ft) obstacle in less than 150m (500ft). Wings of great lifting ability and powerful engines allow for unusual performance without sacrificing load-carrying capability. *See also* VTOL AIRCRAFT

stolon Modified, horizontal, underground or aerial stem, growing from the basal node of a plant. Aerial stolons, or RUNNERS, may be slender, such as strawberry plants, or stiff and arching, such as brambles. The stolon produces a new plant at its tip, which puts out ADVENTITIOUS ROOTS (roots that arise from nodes on the stem) to anchor itself. Underground stolons are characteristic of white potato plants. *See also* TUBER; VEGETATIVE REPRODUCTION

stoma (pl. stomata) In botany, pores found mostly on the undersides of leaves through which atmospheric gases pass in and out, for RESPIRATION and PHOTOSYNTHESIS. Surrounding each stoma are two guard cells, which can swell and close the stoma to prevent excessive loss of water vapor. *See also* GAS EXCHANGE; TRANSPIRATION

stomach J-shaped organ, lying to the left and slightly below the DIAPHRAGM in human beings; one of the organs of the DIGESTIVE SYSTEM. It is connected at its upper end to the gullet (ESOPHAGUS) and at the lower end to the SMALL INTESTINE. The stomach itself is lined by three layers of muscle (longitudinal, circular, and oblique) and a folded mucous layer that contains gastric GLANDS. These glands secrete HYDROCHLORIC ACID that destroys some food bacteria, and makes possible the action of PEPSIN, the ENZYME secreted by the glands, which digests PROTEINS. Gastric gland secretion is controlled by the sight, smell, and taste of food, and by hormonal stimuli, chiefly the HORMONE gastrin. As the food is digested, it is churned by muscular action into a thick liquid state called chyme; then it passes into the small intestine.

stoneware Opaque, nonporous, high-fired clay ware, intermediate between PORCELAIN and EARTHENWARE. Stoneware is often salt-glazed for utility pieces and is used for vessels and tiles.

Stoney, George Johnstone (1826–1911) British physicist who coined the term "ELECTRON" for the fundamental unit of ELECTRICITY (the term was later applied to an ELEMENTARY PARTICLE). Stoney worked on the KINETIC THEORY of gases and estimated the number of molecules in a volume of gas at STANDARD TEMPERATURE AND PRESSURE (STP). His estimate of an electron's charge was incorrect because he used an incorrect value for the mass of the hydrogen atom.

stony-iron meteorite METEORITE consisting of approximately equal proportions of SILICATES (stony material) and metals, mostly nickel-iron. There are two main subtypes. **Pallasites**, which consist of olivine (a magnesium–iron silicate) mixed with nickel–iron, may have originated near the core/mantle interface of a planetary body. **Mesosiderites** are a much coarser combination of chunks of various silicates and nickel–iron, and could have been produced by impacts on a planetary surface.

stony meteorite (aerolite) METEORITE that consists mostly of SILICATES (stony material, mainly olivine, pyroxene, and feldspar), with only a small amount of chemically uncombined metals (typically 5% nickel–iron). There are two subtypes: chondrites and achondrites. In **chondrites**, the rocky minerals take the form of small spherical grains (chondrules), and there is some nickel and iron. **Achondrites** are coarser-grained and have no chondrules or metal. Both types are about 4.6 billion years old, dating from the formation of the Solar System.

STP Abbreviation of STANDARD TEMPERATURE AND PRESSURE

strain Change in dimensions of an object subjected to STRESS. Linear strain is the ratio of the change in length of a bar to its original length while being stretched or compressed. Shearing strain describes the change in shape of an object whose opposite faces are pushed in different directions. HOOKE'S LAW for elastic materials states that strain is proportional to stress up to the material's elastic limit.

strain gauge Device used to measure mechanical STRAIN such as that caused by the movement of a bridge or glacier. **Electrical** strain gauges use the phenomenon (first noted by Lord KELVIN in 1856) that if a wire is strained, its electrical resistance is changed. An electrical strain gauge usually consists of a grid of wire filaments printed onto a metal foil. The gauge is cemented onto the structural surface whose strain is to be measured. As the strain in the structure changes, the strain in the gauge changes too, thus producing changes in resistance that can be measured.

strangeness (symbol s) Property of HADRONS (ELEMENTARY PARTICLES) that experience the STRONG NUCLEAR FORCE). Particles are assigned a strangeness of $+1$, -1 or 0. In a strong interaction, strangeness may be conserved (the total strangeness must be the same before and after the process). If strangeness is not conserved, the process can proceed only by the WEAK NUCLEAR FORCES, and hence only very slowly.

strange particles ELEMENTARY PARTICLES that have nonzero STRANGENESS. They have an anomalously long lifetime (10^{-10} to 10^{-7} s) compared to that characteristic of most other HADRONS (10^{-23} s). *See also* CONSERVATION LAW, NUCLEAR

Strassmann, Fritz (1902–80) German chemist. In 1938 he ushered in the nuclear age when he discovered, with Otto HAHN, the process of nuclear FISSION by bombarding URANIUM atoms with NEUTRONS. Strassmann shared the 1966 Fermi Award of the US Atomic Energy Commission with Otto Hahn and Lise MEITNER.

stratification Layering in ROCKS. It occurs in SEDIMENTARY ROCKS and in IGNEOUS ROCKS that formed from LAVA flows and volcanic fragmental deposits. Separations between individual layers are called stratification planes. They parallel the strata they bound, being horizontal near flat layers and exhibiting inclination on a sloping surface. *See also* STRATIGRAPHY; STRATUM

stratigraphy Branch of geology concerned with stratified or layered ROCKS. It deals with the correlation of rocks from different localities using FOSSILS and distinct rock types. *See* STRATUM

stratosphere Section of the Earth's ATMOSPHERE between the TROPOSPHERE and the higher MESOSPHERE. It is about 40km (25mi) thick and for half this distance the temperature remains fairly constant. The stratosphere contains most of the atmosphere's OZONE LAYER.

stratum In biology, a layer of tissue, such as the *stratum basale* of the uterus.

stratum (pl. strata) In geology, a single layer or bed of SEDIMENTARY ROCK, usually one of a series of layers that have been deposited one on top of another. Each stratum indicates the conditions that prevailed at the time of its deposition. The study of the strata is called STRATIGRAPHY. The stratigraphical or geological column contains all the periods of GEOLOGICAL TIME. Rock strata may be very thin, of just a few centimeters, or they can be several meters in extent. Initially, they are usually horizontal, but they are likely to be affected by folding and faulting.

streptococcus Genus of spherical or oval BACTERIA that grow in pairs or beadlike chains. They live mainly as parasites in the mouth, respiratory tract, and intestine. Some are harmless but others are pathogenic, causing strep throat and other infections. Treatment is with ANTIBIOTICS.

streptomycin Antibiotic drug used to treat certain bacterial diseases resistant to PENICILLIN. It may produce side effects such as deafness and giddiness. Its discovery (1944) by Selman WAKSMAN led to the development of other related ANTIBIOTICS. Streptomycin is used, along with other drugs, to treat tuberculosis.

stress Force per unit area applied to an object. **Tensile** stress stretches an object, **compressive** stress squeezes it and **shearing** stress deforms it sideways. In a fluid, no shearing stress is possible because the fluid slips sideways, so all fluid stresses are PRESSURES (positive or negative).

string theory Recent branch of PARTICLE PHYSICS, largely developed by Ed Witten, in which ELEMENTARY PARTICLES can be represented by the vibrations of a one-dimensional string. The concept of a string is completely theoretical since no string has ever been detected experimentally: the stringlike nature of these objects is believed to be billions of times smaller than the scale observable in current experiments. *See also* SUPERSTRING THEORY

stroboscope (strobe) Device that emits regular flashes of light. Stroboscopes usually have a calibrated scale from which the number of flashes per minute can be read. They are used in photography to make multiple exposures of moving subjects and in engineering to seemingly "slow down" moving objects for observation. A strobe is also used to find the speed of a rotating wheel by adjusting the rate of flashing until it equals the rotation speed (when the wheel seems to be stationary).

strong nuclear force One of the four FUNDAMENTAL FORCES in nature. The strongest of the four forces, it binds together protons and neutrons within the NUCLEUS of an atom. Like the WEAK NUCLEAR FORCE, it operates at very short distances (a mil-

lionth of a millionth of a centimeter) and therefore occurs only within the nucleus. It is mediated by GLUONS. *See also* GRAND UNIFIED THEORY (GUT)

strontianite Carbonate mineral, STRONTIUM carbonate ($SrCO_3$). It has orthorhombic system, hexagonal twinned crystals, and it also occurs as massive or columnar aggregates. It can be pale green, white, gray, yellow or brown. Strontianite is found in low-temperature hydrothermal veins, often in limestone. Hardness 3.5–4; r.d. 3.7.

strontium (symbol Sr) Silver-white, metallic element of the ALKALINE-EARTH METALS in group II of the periodic table. Resembling CALCIUM physically and chemically, strontium occurs naturally in STRONTIANITE and celestite and is extracted by electrolysis. Strontium salts are used to give a red color to flares and fireworks. The isotope ^{90}Sr (half-life 28 years) is a harmful radioactive element present in fallout, from which it is absorbed into milk and bones; it is used in nuclear reactors. Properties: at.no. 38, r.a.m. 87.62, r.d. 2.554, m.p. 769°C (1,416°F), b.p. 1,384°C (2,523°F).

Strutt, John William, Lord Rayleigh *See* RAYLEIGH, JOHN WILLIAM STRUTT, LORD

Struve, Friedrich Georg Wilhelm von (1793–1864) German astronomer who catalogued more than 3,000 stars, including many previously undiscovered DOUBLE STARS. In 1839 Struve became one of the first astronomers to measure the PARALLAX of a star, Vega. His son, **Otto Wilhelm** von Sturve (1819–1905), succeeded him as director (1862–99) of the Pulkovo Observatory, Russia. Otto discovered a satellite of Uranus. Friedrich's grandson, Otto Struve (1897–1963) was director (1932–47) of the Yerkes Observatory and professor of astrophysics at the University of California (1950–59). He made important discoveries in the fields of stellar evolution and radial velocities.

strychnine Poisonous ALKALOID obtained from the plant *Strychnos nux-vomica*. In the past, it was believed to have therapeutic value in small doses as a tonic. Strychnine poisoning causes symptoms similar to those of tetanus, with death occurring due to spasm of the breathing muscles.

Sturgeon, William (1783–1850) British electrical engineer who demonstrated (1825) the first ELECTROMAGNET capable of useful work. It consisted of a soft, iron core insulated with varnish and wound with 18 turns of bare copper wire; it was capable of supporting 4kg (9lb) and was powered by a single cell. Sturgeon also built an electric motor, invented the COMMUTATOR, and the suspended coil GALVANOMETER, and studied charges in clouds by flying kites.

style In botany, the slender tube of a flower that connects the pollen-receiving STIGMA at its tip to the OVARY at its base.

styrene ($C_6H_5CH=CH_2$) Colorless, liquid HYDROCARBON, important as a MONOMER of POLYSTYRENE. It is also a monomer of styrene-butadiene rubber (SBR), the artificial RUBBER most commonly used in pneumatic tires. Styrene is a derivative of BENZENE. Properties: m.p. 145.2°C (293.4°F), b.p. −30.6°C (−23.1°F).

subatomic particle *See* ELEMENTARY PARTICLE

subduction In PLATE TECTONICS, descent of rocky material from the edge of one of the Earth's tectonic plates into the semimolten ASTHENOSPHERE below. It occurs where two plates collide, and one rides over the other (**subduction zone**); the lower one is subducted. Often the upper plate is a continental plate and the lower one is an oceanic plate, and subduction leads to the formation of an ocean TRENCH. This type of subduction is associated with volcanic eruptions, earthquakes, and other seismic activity. *See also* CONTINENT

subdwarf STAR that is less luminous by one to two MAGNITUDES than a normal DWARF STAR of the same spectral type. Subdwarfs are mainly of types F, G, and K, and lie below the MAIN SEQUENCE on the HERTZSPRUNG–RUSSELL DIAGRAM. Most are Population II stars. They are placed in LUMINOSITY class VI, although an alternative designation is to prefix their spectral type with the letters "sd." *See also* SPECTRAL CLASSIFICATION

subgiant Star of smaller radius and lower LUMINOSITY than a normal GIANT STAR of the same spectral type. They are mainly of types G and K, and lie between the MAIN SEQUENCE and the giants on the HERTZSPRUNG–RUSSELL DIAGRAM. Subgiants are placed in luminosity class IV. *See also* SPECTRAL CLASSIFICATION

sublimation Direct change from SOLID to GAS, without an intervening LIQUID phase. **Dry ice** (solid carbon dioxide) sublimes from the solid phase directly to the gaseous phase. Most substances can sublimate at certain pressures, but usually not at atmospheric pressure. *See also* CONDENSATION; EVAPORATION

submarine Seagoing warship capable of travelling both on and under the water, armed principally with torpedoes. Submarines were planned in the 16th century, but they were not built until the Dutch inventor Cornelis Drebbel (1572–1633) took a greased, leather-covered boat underwater in 1620. Experimental submarines were used in warfare in the late 18th and 19th centuries, to little effect.

Technical advances of the late 19th century led to the general spread of underwater craft in the world's navies. Early submarines were essentially surface ships with a limited ability to remain submerged. Once underwater, they depended on battery-powered electric motors for propulsion; with a limited air supply, they were soon forced to surface. Submerging is accomplished by letting air out of internal ballast tanks; trimming underwater is done by regulating the amount of water in the ballast tanks with pumps; and surfacing is accomplished by pumping ("blowing") the water out of the tanks. In 1954 the first nuclear-powered submarine, the Nautilus, was launched. Today's nuclear-powered submarines are propelled by steam turbines and are capable of speeds of up to 30 knots. The British *Trident* submarines have multiple-head nuclear missiles with a range of *c*.11,000km (6,750mi).

submersible Small craft for underwater exploration, research or engineering. Modern submersibles have evolved from simple devices. Diving bells were open-bottomed devices in which people were lowered into the water. A device called the BATHYSPHERE, invented in the 1930s by William BEEBE, was a spherical observation chamber. The bathyscaphe, invented in the 1940s by Auguste PICCARD, also had a spherical chamber, but this was attached to a much larger hull, which was used as a buoyancy control device. The craft also had propellers powered by electric motors. A new generation of submersibles has evolved since the late 1950s for engineering and research work. A typical craft has a spherical passenger capsule capable of withstanding water pressure down to *c*.3,600m (12,000ft). Attached to this is a structure containing batteries, an electric motor with propeller, lighting, a mechanical arm for gathering samples, and other technical equipment. A specially designed support ship launches and retrieves the submersible. Some submersibles are unmanned and operated by remote control from the surface.

subscript Small letter, number or other symbol written or printed below and to the left or right of another symbol, as in a chemical formula (such as C_2H_5) or to indicate different values, x_1, x_2, x_3, of a mathematical variable. *See also* SUPERSCRIPT

substitution reaction CHEMICAL REACTION in which one atom or group of atoms replaces (usually in the same structural position) another atom or group in a molecule or ion.

substrate In biochemistry, a reactant that is acted on by an ENZYME or other CATALYST. In biology, a substrate is a nutrient medium used to grow microorganisms, or the surface on which a sessile organism (such as a limpet) lives and grows. In electronics, a substrate is a single crystal or SEMICONDUCTOR used as a basis for making INTEGRATED CIRCUITS or TRANSISTORS.

subtraction Arithmetical operation signified by $-$, interpreted as the inverse process of ADDITION. The difference of two numbers, $a - b$, is the number that has to be added to b to give a. a is called the **minuend** and b the **subtrahend**. *See also* ARITHMETIC

subway The underground part of the transportation system, including RAILROADS and elevated trains, or **els**, that carries people through congested urban areas. The first subway opened in London, England, in 1863, and was steam-powered. Subways now operate on electricity provided through a third rail that runs between the two regular ones. The US's biggest subway system, in New York City, has 238 miles of track and 469 stations.

succession In ecology, orderly change in plant and animal life in a biotic COMMUNITY over a long time period. It is the result of modifications in the community ENVIRONMENT. The process ends in establishment of a stable ECOSYSTEM, known as a CLIMAX COMMUNITY.

succulent Plant that stores water in its tissues in order to resist periods of drought. Usually PERENNIAL and evergreen, most of the succulent plant body is made up of water-storage cells, which give it a fleshy appearance. A well-developed CUTICLE and low rate of daytime TRANSPIRATION also conserve water. Swollen leaves and stems have a smaller surface area to volume ratio than thin ones, thus further reducing water loss. Succulent plants include cactus and stonecrop.

sucker In botany, an upward-growing shoot that forms at the base of the plant stem or from an ADVENTITIOUS ROOT. In zoology, a disklike structure found on the tentacles of cephalopods used for grasping objects.

sucrase ENZYME secreted by the small intestine, which breaks down SUCROSE (ordinary sugar) into GLUCOSE and FRUCTOSE so that they can be absorbed during DIGESTION.

sucrose Common, white, crystalline SUGAR, a DISACCHARIDE ($C_{12}H_{22}O_{11}$) consisting of linked GLUCOSE and FRUCTOSE molecules. It occurs in many plants, but its principal commercial sources are sugar cane and sugar beet. It is also obtained from maple trees, date palms, and sorghum, and is widely used for food sweetening and in the manufacture of confectionery and preserves.

sudden infant death syndrome (SIDS) (Crib death) Unexpected death of an apparently healthy baby, usually during sleep. The cause is unknown,

but a number of risk factors have been identified, including prematurity and respiratory infection. Evidence suggests that laying a baby on its back when putting him/her to sleep can reduce the risk of SIDS.

sugar Sweet-tasting, soluble, crystalline, mono-SACCHARIDE or disaccharide CARBOHYDRATE. The common sugar used in food and beverages is the disaccharide SUCROSE. This is also the main sugar transported in plant tissues. The main sugar transported around the bodies of animals to provide energy in RESPIRATION is the monosaccharide GLUCOSE. *See also* SACCHARIDE

sulfa drug *See* SULFONAMIDE DRUG

sulfate Salt of SULFURIC ACID (H_2SO_4). Common sulfates include copper(II) sulfate ($CuSO_4$) and iron(II) sulfate ($FeSO_4$).

sulfide Chemical compound derived from HYDROGEN SULFIDE (H_2S) in which another element or group replaces the hydrogen. An inorganic example is lead sulfide (PbS), which occurs as the mineral GALENA; an organic example is ethane thiol (ethyl mercaptan, C_2H_5SH).

sulfite Salt or ester of sulfurous acid (H_2SO_3). Sulfurous acid is known systematically as trioxosulfuric(IV) acid and its salts, the sulfites, contain the trioxosulfate(IV) ion (SO_3^{2-}). Many sulfites are reducing agents, used as disinfectants and bleaches. Hydrogensulfates(IV) (bisulfites, HSO_3^-) also exist.

sulfonamide drug (sulfa drug) Any of a group of DRUGS derived from sulfanilamide, a red textile dye, that prevent the growth of bacteria. Discovered by Gerhard DOMAGK in the 1930s, the sulfonamides were the first commercially available antibacterials, prescribed to treat a range of infections. However, their use was associated with a number of unpleasant side effects, and with the arrival of the less toxic and more effective ANTIBIOTICS, they were gradually replaced. Today, their use is limited chiefly to the treatment of urinary tract infections.

sulfonation Introduction of a sulfonic acid group, $-SO_3H$, into an organic molecule, widely used in the manufacture of dyes, SULFONAMIDE DRUGS, and so on. Many detergents are manufactured by sulfonation of a HYDROCARBON.

sulfonic acids Organic chemical compounds having the general formula RSO_3H, where R is a HYDROCARBON group. Examples are ethane sulfonic acid, $C_2H_5.SO_3H$, and benzene-sulfonic acid, $C_6H_5.SO_3H$. Sulfonic acids are used as chemical intermediates and their salts, the sulfonates, are widely used in DETERGENTS.

sulfur (symbol S) Nonmetallic element in group VI of the PERIODIC TABLE, known from earliest time (as brimstone). It occurs naturally as a free element and in sulfide minerals, such as GALENA and PYRITE, and sulfate minerals, such as GYPSUM. The main commercial source is native (free) sulfur, extracted by the FRASCH PROCESS. It is used in the VULCANIZATION of rubber and in the manufacture of SULFURIC ACID, drugs, matches, dyes, fungicides, insecticides, and fertilizers. Properties: at.no. 16; r.a.m. 32.064; r.d. 2.07; m.p. 112.8°C (235.0°F); b.p. 444.7°C (832.5°F). Most common isotope ^{32}S (95.1%).

sulfur cycle Continual movement of SULFUR between living and nonliving things. Most sulfur in the nonliving (**abiotic**) environment occurs as SULFUR DIOXIDE (SO_2) in the atmosphere, derived from the burning of FOSSIL FUELS, or as deposits underground and in rocks. Photochemical oxidation of sulfur dioxide in the atmosphere produces SULFATES (SO_4^{2-}), which are carried by rain into the ground; oxidation and WEATHERING of underground deposits also produces sulfates. In the ground, sulfates are absorbed by plants in the living (**biotic**) environment to become part of sulfur-containing proteins. Animals that eat the plants assimilate the sulfur. Dead plants and animals (and their feces) are broken down by special bacteria that return sulfur to the soil (abiotic) as HYDROGEN SULFIDE (H_2S). Finally, other bacteria convert the hydrogen sulfide back to sulfate or native sulfur.

sulfur dioxide Colorless, poisonous gas (SO_2) with a pungent odour. It is used industrially in the manufacture of SULFURIC ACID and as a refrigerant, bleaching agent, and preservative. Properties: m.p. −75.5°C (−103.9°F); b.p. −10.0°C (14.8°F); density 2.2 (air = 1).

sulfuric acid (formerly oil of vitriol) Colorless, odorless, poisonous liquid (H_2SO_4), one of the most corrosive acids known. It is produced by the OXIDATION of SULFUR DIOXIDE (SO_2) in the contact process. Sulfuric acid is a major industrial chemical, the most produced chemical by tonnage, and is used in the manufacture of many acids, fertilizers, detergents, drugs, and a wide range of chemicals. Properties: r.d. 1.84; m.p. 10.3°C (50.5°F); b.p. 330°C (626°F).

sulfur trioxide (SO_3) Inorganic compound, the ANHYDRIDE of SULFURIC ACID, that is, it reacts with water to make the acid. At room temperatures sulfur trioxide can exist either as a liquid or as a solid. It dissolves in sulfuric acid to make fuming sulfuric acid, or oleum ($H_2S_2O_7$).

summation (symbol Σ) In mathematics, finding the total of a sequence or array of numbers, or of an infinite SERIES of terms.

Sumner, James Batcheller (1887–1955) US biochemist who shared the 1946 Nobel Prize for chemistry with John Northrop (1891–1987) and Wendell STANLEY for his discovery that ENZYMES could be crystallized. In 1926 Sumner isolated and crystallized the enzyme urease; it was the first enzyme to be crystallized and this achievement finally proved the PROTEIN nature of enzymes.

Sun Star at the center of the SOLAR SYSTEM, around which all other Solar System bodies

▲ **sulfur** Naturally occuring sulfur is extracted from rock formations by a method known as the Frasch process. Boreholes (1) are made at the sulfur-rich areas (2). A specialized head (3) is dropped down the borehole. First, pressurized water at 155°C (310°F) is pumped (4) into the rock melting the sulfur. The water is followed by compressed air (5) ,which forces the liquid containing the sulfur to the surface (6).

revolve in their orbits. The apparent daily motion of the Sun across the sky and its annual motion along the ECLIPTIC are caused by the Earth rotating on its axis and revolving in its orbit. The Sun does, however, have its own motion in the galaxy of about 20km s^{-1} (12.5mi s^{-1}) towards the point in the sky known as the APEX. The amount of sunlight normally reaching the Earth is quantified as the SOLAR CONSTANT. The Sun is a DWARF STAR, on the MAIN SEQUENCE of the HERTZSPRUNG–RUSSELL DIAGRAM, and is thus a typical, average star. It consists of about 70% hydrogen (by weight) and 28% helium, with the remainder mostly oxygen and carbon. Its temperature, pressure and density increase towards the center, where the values are about 15 × 10^6 K, 10^{11} bar and 150g cm^{-3}, respectively. Like all stars, the Sun's energy is generated by nuclear FUSION reactions taking place under the extreme conditions in the core. The core is c.400,000km (250,000mi) across. Energy released from the core passes up through the radiative layer, which is about 300,000km (nearly 200,000mi) thick, then passes through the 200,000km (125,000mi) thick convective layer to the surface, the PHOTOSPHERE (meaning "sphere of light"), from where it is radiated into space. Most of the Sun's visible activity takes place in the 500km (300mi) thick photosphere. SUNSPOTS are darker, cooler regions of the photosphere where the local magnetic field is enhanced. Associated with sunspots are SOLAR FLARES – sudden, violent releases of energy and material. Above the photosphere lies the CHROMOSPHERE (meaning "sphere of color," so-called because of its rosy tint when seen during a total solar ECLIPSE), which consists of hot gases and extends for thousands of kilometers. It is generally in a state of turbulence, the temperature rising extremely rapidly with height. This is the realm of SPICULES and PROMINENCES, huge eruptions of material from the Sun. Extending outward from the chromosphere for millions of kilometers is the CORONA. A continuous emission from the corona of particles and radiation is known as the SOLAR WIND. The solar wind and the Sun's magnetic field dominate a region of space called the HELIOSPHERE, which extends to the boundaries of the Solar System. The Sun must never be viewed directly through any optical instrument. Temporary or permanent blindness can result from direct viewing or from the use of unsuitable filters.

sundial Instrument first used about 5,000 years ago in the Near East to measure the time of day. Traditionally a sundial is a short, flat-topped pillar on which is mounted the **gnomon**, or time pointer. This slants upwards away from the dial at an angle determined by the latitude; it points due North in N latitudes and due South in S latitudes. The position of the shadow of the gnomon indicates the time on a scale marked around the top of the sundial.

sun dog (parhelion or mock sun) Either of two bright coloured spots observed at 22° on each side of the Sun. They are caused by REFRACTION of sunlight by ice crystals in the Earth's ATMOSPHERE.

sunspot Region in the Sun's PHOTOSPHERE that is cooler than its surroundings and therefore appears darker. Sunspots consists of a dark central region, the **umbra**, and a gray outer region, the **penumbra**. They vary in size from 1,000 to 50,000km (600 to 30,000mi), and are occasionally up to 200,000km (125,000mi). Their duration varies from a few hours to a few weeks, or months for the very biggest. Sunspots are seen to move across the face of the Sun as it rotates. The number of spots visible depends on the stage of the 11-year SOLAR CYCLE. Sunspots occur where there is a local strengthening of the Sun's magnetic field.

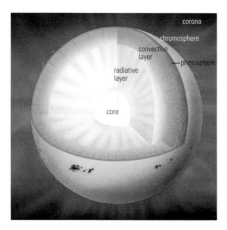

corona
chromosphere
convective layer
photosphere
radiative layer
core

▲ **Sun** Cross-section of the Sun (left), showing the core, radiative layer, convective layer, photosphere, chromosphere, and corona. The sunspot observations of 1947 show that, on 11 February (A), the identities of the leading and following sunspots were still in doubt. But as the lines indicate, the magnetic polarities were clear. From 9 March (B) to 7 April (C), the leader (1) and follower (2) are distinct. By 5 May (D), activity has ceased.

superalloys Name given to ALLOYS of cobalt and nickel used as surface materials for spacecraft to withstand the rapid heating and high temperatures caused by passage through the Earth's atmosphere, particularly on re-entry from space. These alloys oxidize in air above 700°C (1,300°F) but the oxide coating initially formed prevents further oxidation.

supercluster *See* GALAXY CLUSTER

superconductivity Electrical behavior in metals and alloys that are cooled to very low temperatures. In a superconducting circuit, an electric current flows indefinitely because there is no electrical RESISTANCE. Superconductivity was first observed in materials near ABSOLUTE ZERO. Research continues to develop superconductors that function at higher temperatures.

supercooling Lowering of the temperature of a liquid below its freezing point without the liquid solidifying. Adding particles of dust, scratching the inside surface of the container or seeding the super-cooled liquid with a small crystal makes it solidify immediately and return to its freezing temperature. The term supercooling is also applied to the lowering of the temperature of a SATURATED SOLUTION without crystallization occurring. Instead, a SUPER-SATURATED SOLUTION is formed. Dust particles, scratching and seeding with a crystal will all make the supersaturated solution crystallize. Finally a third meaning of supercooling is the lowering of the temperature of a vapor until it is supersaturated without CONDENSATION occurring. The introduction of a particle or a disturbance leads to condensation. This is the principle of the CLOUD CHAMBER.

superfluidity Property of a liquid that has no VIS-COSITY and therefore no resistance to flow. HELIUM II – liquid helium at temperatures less than 2K, or $-271°C$ ($-456°F$) – was the first known superfluid. Its unusual properties were discovered independently by Peter KAPITZA in 1937 and by the British physicist John Allen in 1938. They found that helium II apparently defies gravity by flowing up a slope, and if placed in a container, it climbs up the inside walls and overflows. It also appears to contravene the laws of THERMODYNAMICS, by flowing from a cool region to a warmer one. See also CRYOGENICS

supergiant Extremely luminous STAR of large diameter and low density. They can be of any spectral type, from O to M. RIGEL (type B) and BETELGEUSE (type M) are examples. The LUMINOSITIES of supergiants are several magnitudes greater than those of giant stars of the same SPECTRAL CLASSIFICATION, and so they lie at the top of the HERTZSPRUNG–RUSSELL DIAGRAM. Red (M-type) supergiants, such as Betelgeuse, have the largest diameters, around 1,000 times that of the Sun. Supergiants are assigned to luminosity class I. The brightest are often given the separate class Ia, and the others placed in class Ib.

superheterodyne receiver Most common type of RADIO receiver. An oscillator produces signals of varying frequency such that the difference between them and the frequencies of the incoming modulated CARRIER WAVES – the broadcast signal – is a constant, called the intermediate-frequency (i-f). The

"i-f" can then be amplified and demodulated to produce the output signal for the loudspeaker.

supernova Stellar explosion in which virtually an entire STAR is disrupted. For a week or so, a supernova may outshine all the other stars in its galaxy. The luminosity is some 23 magnitudes (1,000 million times) brighter than the Sun, and the energy released in the explosion is the same as is released over the star's entire previous life. After a couple of years the supernova has expanded so much that it becomes thin and transparent. For hundreds or thousands of years the ejected material remains visible as a **supernova remnant**. A supernova is about 1,000 times brighter than a NOVA. There is probably about one supernova every 30 years in a galaxy like our own, but most of them are concealed by dust. Supernovas are of two main types, classified by their light curves and spectra.

superphosphate (superphosphate of lime) Mixture of dihydrogen calcium phosphate (obtained by treating calcium phosphate with sulfuric acid) and calcium sulfate. It is used as a fertilizer.

superposition, law of In geology, law that states that in undisturbed layers of sedimentary deposits, younger beds overlie older ones. See also STRATUM

supersaturation Term that describes the condition of a SOLUTION that contains more SOLUTE (dissolved substance) than a normal SATURATED SOLUTION. See also SUPERCOOLING

superscript Small letter, number or other symbol written or printed above and to the left or right of a number, letter or symbol, as indicating a mathematical exponent, such as 3^2 or x^n, or an element's mass number, such as ^{238}U. See also SUBSCRIPT

supersonic flight Flight at a speed greater than the local speed of sound – SUPERSONIC VELOCITY. As an aircraft approaches the speed of sound, sound waves that normally move away from the aircraft build up to form SHOCK WAVES. Engineers once thought that these waves formed a SOUND BARRIER that would make supersonic flight impossible. Flight at supersonic speed is often expressed in terms of a MACH NUMBER. This indicates that an aircraft is flying at a certain number of times the local speed of sound.

supersonic velocity Velocity greater than that of the local speed of sound. In dry air at 0°C (32°F), this speed is about $330 m\ s^{-1}$ ($1,080 ft\ s^{-1}$) or $1,188 km/h$ ($736 mph$). Its magnitude is usually expressed as a MACH NUMBER. This is the ratio of the speed of a body to the speed of sound in a medium such as air. A velocity in excess of Mach 5 is said to be **hypersonic**. Any object traveling at supersonic velocity leaves behind a shock wave that a ground observer hears as a SONIC BOOM. *See also* SUPERSONIC FLIGHT

supersonic wave Wave set up in the air by any object moving at or above SUPERSONIC VELOCITY. At these speeds, sound waves created by the object's speed pile up ahead of it and are released violently behind it as a SHOCK WAVE. Observers on the ground see a supersonic aircraft pass overhead before they experience the shock wave that spreads out and reaches them as a SONIC BOOM.

superstring theory Physical theory that attempts to explain the properties of ELEMENTARY PARTICLES and the interactions between them. It combines QUANTUM THEORY and RELATIVITY, especially to explain nuclear forces and the force of gravity (*see* FUNDAMENTAL FORCES). Superstrings are hypothetical one-dimensional objects $c.10^{-35}$m long, which require a 10-dimensional Universe to accommodate them. Their use in quantum calculations predicts the existence of GRAVITONS, the particles believed to be involved in gravitational interaction. *See also* UNIFIED FIELD THEORY

supersymmetry Physical theory that interrelates the basic matter particles (QUARKS and LEPTONS), both of which are classified as FERMIONS, with force-carrying BOSONS such as PHOTONS and GLUONS. Every boson is assigned a corresponding fermion, and each fermion has a corresponding boson. The bosons associated with the leptons and quarks are designated the superparticles **sleptons** and **squarks** (by adding an initial *s*). The fermions that partner the bosons, gluons, and photons are called **gluinos** and **photinos**. None of these particles has yet been found, but the theory predicts an underlying superforce that would combine the electromagnetic, gravitational, and strong and weak nuclear forces. *See also* FUNDAMENTAL FORCES; GRAND UNIFIED THEORY (GUT)

supplementary angles Two angles the sum of which is 180°. Each is the supplement of the other.

suprarenal glands Alternative name for the ADRENAL GLANDS.

surface tension Molecular force associated with the boundary layer of a liquid. It makes a liquid behave as if there were a "skin" on the surface. Cohesive forces in the skin tend to resist disruption, so that a needle or razor blade placed carefully on the surface "floats" even though its density is many times that of the liquid. *See also* MENISCUS

surfactant Substance that lowers the SURFACE TENSION of a SOLVENT in which it is dissolved. In water, SOAPS and DETERGENTS act as surfactants. Their molecules concentrate at the surface of oil and water molecules and act as emulsifying or foaming agents, to produce lathering and frothing.

surgery Branch of medical practice concerned with treatment by operation. Traditionally it has mainly involved open surgery: gaining access to the operative site by way of an incision. However, the practice of using ENDOSCOPES has enabled the development of MINIMAL ACCESS SURGERY. Surgery is carried out under sterile conditions, using local or general ANESTHESIA. There is a growing number of surgical specialties, including general, gynecological, orthopedic, cardiothoracic, neuro, ophthalmologic, plastic, reconstructive, and transplantation surgery.

surveying Accurate measurement of the Earth's surface. It is used in establishing land boundaries, the TOPOGRAPHY of landforms, and for major construction and civil engineering work, such as dams, bridges, and highways. Measurements are linear or angular, applying principles of GEOMETRY and TRIGONOMETRY. For smaller areas, the land is treated as a horizontal plane. Large areas involve considerations of the Earth's curved shape, a technique referred to as GEODESIC SURVEYING.

suspension In chemistry, liquid (or gas) medium in which small solid (or liquid) particles are uniformly dispersed. Examples are dust in air, and fine particles of sand in water. The particles are larger than those found in a COLLOID and will settle if the suspension stands undisturbed.

suspension In geology, manner in which small sedimentary particles are carried along by rivers. Turbulence and upward eddies keep particles suspended and make the water look muddy. Worldwide, the amount of material transported in this way is tremendous, estimated at 80 metric tons of solids per square kilometer of the Earth's surface every year.

suspension bridge BRIDGE that has its roadway (deck) suspended from two or more CABLES, which usually pass over towers and are anchored at the ends. The cables are of wire twisted spirally, and the deck is usually made rigid by stiffening trusses. The Verrazano-Narrows Bridge in New York City has a span of 1,297m (4,260ft) and is the US's longest suspension bridge.

Svedberg, Theodor (1884–1971) Swedish chemist who was awarded the 1926 Nobel Prize for chemistry for his development (1923) of the ultracentrifuge. Svedberg used it to study COLLOIDS and large MOLECULES and it enabled the RELATIVE MOLECULAR MASSES (molecular weights) of large PROTEINS to be determined for the first time. He also helped to develop ELECTROPHORESIS techniques for separating proteins on the basis of electrical charge.

swallow hole Large POTHOLE containing a flowing stream or river. A pothole is an enlarged JOINT, usually in limestone, formed by running water and possibly chemical WEATHERING. It may be up to 100m (330ft) deep and is termed a swallow hole if water continues to pour down it. *See also* SINKHOLE

Swammerdam, Jan (1637–80) Dutch naturalist. His research and his book *General History of Insects* provided a system for classifying insects and laid the foundations of ENTOMOLOGY, the scientific study of insects. Swammerdam also discovered (1658) the existence of red blood corpuscles.

Swan, Sir Joseph Wilson (1828–1914) British scientist who designed the first electric LIGHTBULB. Starting in 1848, he used a carbon filament inside an evacuated glass envelope and by 1860, some 20 years earlier than Thomas EDISON, he had produced a workable bulb, though with a short lifetime. His bulbs were used to light the House of Commons in 1881 and the British Museum in 1882.

swarm Gathering of many honeybees around a queen bee, which flies off from a hive to found another colony. This happens when a young queen supplants the old queen, and it is the latter who leads off her retinue in a swarm.

sweat gland One of many small GLANDS that open on the SKIN surface through pores and release sweat, composed mainly of water and some salt, to regulate body temperature. In humans, they are found all over the body, especially in places such as the armpits. In some mammals they are found only on the soles of the feet. The glands extract water, salts, and urea from the blood capillaries that supply them and excrete them as sweat. The process is under the control of the AUTONOMIC NERVOUS SYSTEM, and forms an important part of the body's temperature control mechanism. *See also* HOMEOSTASIS

sweating Loss of water, salts, and urea from the body surface of many mammals as a result of the action of SWEAT GLANDS. The evaporation of sweat cools both the skin and the blood passing through capillaries close to the skin surface. Sweat evaporates most rapidly when the outside air is warm and dry, and when it is windy. Excessive sweating must be compensated for by increased intake of water and salt. Sweating may also increase in response to stress or anxiety.

swim bladder (air bladder) Air-filled sac that controls the buoyancy of bony FISH. The swim bladder is located above the gut. A connection from the bladder to the gut allows air to enter and leave the bladder. Air may also travel to the bladder along capillary blood vessels. The amount of air in the bladder can be changed so that the fish's specific gravity matches that of the water surrounding it. In lungfish the swim bladder also acts as a rudimentary lung.

swing-wing aircraft Aircraft with wings the geometry of which can be changed to suit a wide range of flying speeds. A swing wing, also known as a variable sweep or variable geometry wing, can be moved forwards or backwards in flight. Moved forwards it creates a high aspect ratio (ratio of span to width) with good LIFT at low speeds for takeoff and landing. Moved backwards it forms a highly swept back wing with low DRAG, enabling high-speed flight.

Sydenham, Thomas (1624–89) English physician, often called the English Hippocrates. He initiated the cooling method of treating SMALLPOX, made a thorough study of all aspects of epidemics, and wrote descriptions of many diseases, including malaria, smallpox, and GOUT.

syenite Uncommon coarse-grained, intrusive IGNEOUS ROCK. Composed mainly of alkali FELDSPAR or PLAGIOCLASE with up to 65% silica, syenite is a pink or gray rock also containing some AMPHIBOLE, which forms dark patches. A blue-gray form with large feldspar crystals, called larvikite, is used as an ornamental stone.

symbiosis Relationship between two or more different organisms living in close contact with each other. Such relationships may be MUTUALISTIC, PARASITIC, or COMMENSALISTIC. They include that of the cleaner fish that pick food from the teeth of larger fish, thus obtaining a meal while providing a service; or the presence of nitrogen-fixing bacteria in the roots of many plants, where the plant gains nitrogen compounds and the bacteria are supplied with food materials such as carbohydrates. *See illustration on page 508*

symmetry In biology, anatomical description of body form or geometrical pattern of a plant or animal. It can be used in the classification of living things (TAXONOMY). In mathematics, symmetry is a similarity or correspondence between parts of an object. A symmetrical figure is one that has a line or point about which certain operations produce the same figure. These operations include reflection, rotation, and translation. For example, a circle has rotational symmetry about its center point: it can be rotated any amount around this point without its figure changing. It also has reflectional symmetry along its diameter: its form is unchanged by reflection along this line.

sympathetic nervous system One of the two parts of the AUTONOMIC NERVOUS SYSTEM, the other being the PARASYMPATHETIC NERVOUS SYSTEM. Both are involved with the action of INVOLUNTARY MUSCLES. The sympathetic nervous system controls muscles that prepare the body for action, by, for example, increasing heart rate and slowing digestion. The parasympathetic nervous system does the reverse.

synapse Connection between the nerve ending of one NEURON and the next or between a nerve cell and a muscle. It is the site at which NERVE IMPULSES are transmitted, using chemicals (NEUROTRANSMITTERS), such as acetylcholine, secreted by structures in the nerve endings.

synchrocyclotron Type of particle ACCELERATOR, a modification of the CYCLOTRON, in which the frequency of the electric field is slowly changed as the particles spiral around the device. This counteracts the increase in particle mass at relativistic velocities and prevents the particles getting out of phase with the field, the radius of the path being proportional to velocity and mass. Energies can reach 700 MeV.

synchronous orbit Path of an artificial SATELLITE with an ORBIT taking the same time as one rotation of the planet about which it moves. If a synchronous satellite is placed in a circular orbit above the equator of a planet, and moves in the same direction as

the planet spins, then the satellite will keep up with the planet's movement and will always be above the same point on the equator. In the case of the Earth, this condition occurs when a satellite is orbiting at a height of about 35,900km (22,300mi) above the Equator. Such a satellite is said to be in GEOSTATIONARY ORBIT. Synchronous satellites are useful for telecommunications purposes as they can be arranged to cover specific areas. Geostationary satellites are most commonly used. Because these satellites remain in fixed positions relative to the Earth, their aerials, and those used on the ground to communicate with them, point in fixed directions. Geostationary satellites provide numerous telephone, telegraph, television, radio, and data communication services around the world.

synchrotron Type of particle ACCELERATOR in which a beam of ELECTRONS or PROTONS is focused and guided around a fixed circular path by changing magnetic fields. Millions of revolutions are made. A high-frequency electric field at one point in the path accelerates the particles. Proton energies can reach hundreds of GeV, and electron energies reach tens of GeV.

synchrotron radiation In physics, stream of ELECTROMAGNETIC RADIATION generated by high-energy ELECTRONS constantly accelerating around a MAGNETIC FIELD. It can take the form of X-rays, radio waves or visible light. Such radiation has uses in medicine and the electronics industry. In

▶ **symbiosis** *Iridomyrmex* ants and the *Myrmecodia* (ant plant) benefit mutually from a symbiotic relationship. The ants feed on the sugary nectar of the plant. This is produced in nectaries (1) which develop at the base of the flower (2) after the petals and sepals have fallen off. The plant benefits from the vital minerals in the ants' defecation and waste materials (3), absorbed through the warty inner surface of the chambers (4). The ant plant is epiphytic, growing suspended from trees in upland rain forests, where the soils are often lacking in nutrients. The mineral nutrients provided by the ants supplement the plant's poor diet. As the plant grows, its stem

enlarges and develops cavities that are invaded by the ants (5). These

chambers do not interconnect but have separate passages to the

outside (6). A complete ant colony soon becomes established in the plant.

astronomy, the term is used to describe cosmic RADIO WAVES of a similar SPECTRUM.

syncline Downward FOLD in rocks. When rock layers fold down into a troughlike form, it is known as a syncline. (An upward arch-shaped fold is called an ANTICLINE.) The sides of the syncline are called **limbs**, and the median line between the limbs along the trough is known as the axis of the fold.

synecology Branch of ecology that studies the ABIOTIC FACTORS as well as the BIOTIC FACTORS of a COMMUNITY, and the ways in which they interact with each other and the environment. *See also* AUTECOLOGY

Synge, Richard Laurence Millington (1914–94) British biochemist. He shared the 1952 Nobel Prize for chemistry with Archer MARTIN for identifying the individual AMINO ACIDS in INSULIN.

synodic period Period of apparent revolution of one heavenly body about another as observed from the Earth, for example from one opposition or conjunction to the next. A **synodic month** is the period between two identical phases of the Moon. The same as the duration of one lunation, its length is 29.53059 mean solar days.

synovial fluid Viscous, colorless fluid that lubricates the movable joints between bones. It is secreted by the synovial membrane, which links the bones at a freely movable joint. Synovial fluid is also found in the bursae, membranous sacs that help to reduce friction in major joints such as the shoulder, hip, and knee.

synthesizer In music, an electronic instrument capable of producing a wide variety of different sounds, pitches, and timbres. Although a form of synthesizer had been conceived as early as 1929, the modern instrument was invented in 1964 by the US electronic designer Robert Moog (1934–). Computer technology is now used to control the instrument's different functions, enabling more sophisticated synthesizers to replicate the sounds of other (often non-electronic) instruments. Because they are electronic instruments, synthesizers can easily be controlled by a computer, using a program called a **sequencer**, to play rhythms and musical pieces.

synthetic fiber Materials intended to substitute for, supplement or improve on the quality of natural fibers such as cotton, wool, and silk. The manufacture of synthetic fibers began early in this century with RAYON, still a familiar textile, which (like cotton) is made of the natural POLYMER cellulose. The first truly synthetic, or artificial-polymer, fiber was NYLON – a revolutionary material that, weight for weight, is stronger than steel. POLYESTERS, synthetics like nylon

widely used for textiles, are also melt-spun. The equally popular ACRYLICS decompose before they melt, so they are wet-spun like rayon. POLYETHENE (polythene), POLYPROPENE and VINYL PLASTICS such as POLYVINYL CHLORIDE (PVC) can also be made as fibers although they are more familiar as film or sheet. Inorganic synthetic fibers include FIBERGLASS, spun from molten glass; CARBON FIBER, used to reinforce ceramics and metals; and metallic fibers, used mainly for decorative purposes in clothing.

syrinx Vocal organ of a bird, consisting of thin, vibrating muscles at the base of the windpipe.

systems analysis Method of analyzing a process or operation in order to improve efficiency, particularly with the aid of a COMPUTER, which has the capacity to handle the large number of interdependent activities involved in a complex operation.

systems engineering Application of such methods of analysis as operations research, CYBERNETICS, and INFORMATION THEORY to the solution of problems associated with large, complex, artificial systems in which there is considerable interaction between the component parts of the system. These solutions often involve the use of COMPUTERS, feedback CONTROL SYSTEMS, and SERVOMECHANISMS. Originally developed during World War II for military applications, it has broadened into a variety of fields, including telephony, traffic control, and social administration. *See also* FEEDBACK

systole Time in the cardiac cycle when the HEART muscle contracts, propelling blood from the heart into the circulatory system.

Szent-Györgyi, Albert von (1893–1986) US biochemist, b. Hungary. He was awarded the 1937 Nobel Prize for physiology or medicine for his work on biological oxidation processes and the isolation of vitamin C. He also studied the biochemistry of MUSCLE, discovering the muscle protein ACTIN, which is responsible for muscular contraction when combined with the muscle protein MYOSIN.

Szilard, Leo (1898–1964) US physicist, b. Hungary. His early work established the relation between information transfer and ENTROPY (disorder or randomness of a system). He devised a means of separating isotopes of artificial radioactive elements, developed the **chemostat** (a type of culture vessel in which organisms are grown), and proposed theories of aging, recall, and memory. He was also involved at the University of Chicago, along with Enrico FERMI, in the creation of the first sustained nuclear CHAIN REACTION based on uranium FISSION. Szilard urged the US government to develop the atom bomb but later campaigned against nuclear weapons.

tachometer Device that measures the rate of rotation of a wheel or shaft. One type of tachometer is a simple revolution counter, with which average speeds of rotation can be measured with a stopwatch. More sophisticated instruments show instantaneous speeds of rotation. These instruments are used widely in high-performance cars and in industry to check machine efficiency. They can be made on mechanical, electrical or electronic principles.

tachycardia Increase in HEART rate beyond the normal. It may occur after exertion or due to excitement or illness, particularly during fever; or it may result from a heart condition.

tachyon Hypothetical ELEMENTARY PARTICLE that travels faster than the speed of light. It might be detectable by the CERENKOV RADIATION it emits. Such a particle would be consistent with the special theory of RELATIVITY. It would require energy to slow it down, so it could never be brought down to the speed of light.

taiga Coniferous forest extending over thousands of square kilometers in Siberia, Russia, Finland, Sweden, Norway, and Canada. Climatic conditions here are unsuitable for deciduous woodland, but coniferous trees can survive. Summer temperatures are around 15°C (59°F), and January averages −10°C to −20°C (14°F to −4°F). There are several months with temperatures below freezing point, and the subsoil is frozen for much of the year. The roots of the coniferous trees spread out horizontally because they cannot penetrate vertically into the ground. The main types of tree are spruce, larch, fir, and pine. In the S parts of the taiga, the trees grow to 15m (50ft) in height, but further N the trees become smaller until they are little more than bushes, and TUNDRA vegetation can be seen. Wherever coniferous forests occur, the needles fall to the ground to produce the type of soil known as PODZOL, which is acid and ashy gray in color. *See also* CONIFER; EVERGREEN

Talbot, William Henry Fox (1800–77) British scientist. Talbot improved on the work of Nicéphore Niepce (1765–1833) and Louis Daguerre (1789–1851) by inventing the first photographic process capable of producing an infinite number of positive prints from a single original negative. *See* PHOTOGRAPHY

talc Sheet silicate mineral, hydrous magnesium silicate ($Mg_3Si_4O_{10}(OH)_2$). Talc occurs as rare tabulate crystals in a monoclinic system and as masses. It is either white, green, blue or brown in color. It is used as base for talcum powder and in ceramics. Massive, fine-grained talc is known as SOAPSTONE. Hardness 1; r.d. 2.6.

talus *See* SCREE

Tamm, Igor Yevgenyevich (1895–1971) Russian theoretical physicist. Tamm shared the 1958 Nobel Prize for physics with Pavel CERENKOV and I.M. Frank for the discovery and interpretation (1937) of CERENKOV RADIATION, showing that light is emitted by charged particles traveling at very high speeds in a transparent medium.

tangent In TRIGONOMETRY, the RATIO between the length of the sides opposite and adjacent to an acute angle within a right-angle TRIANGLE. (The third side is the hypotenuse.) The expression denoting the tangent of angle A is commonly abbreviated to $\tan(A)$; $\tan(A) = \sin(A)/\cos(A)$. Also, the tangent at a point on a curve is the straight line whose slope matches that of the curve at that point.

tank Tracked armored vehicle mounting a single primary weapon, usually an artillery piece, and one or more machine guns. Modern tanks have an enclosed fully revolving turret and are heavily armored; main battle tanks weigh from 35 to 50 tons and usually have a crew of four. Developed in great secrecy by the British during World War I, tanks were first employed at the Battle of the Somme (1916).

tanker Ship designed to carry liquid cargo. The largest tankers can carry up to 500,000 tons of cargo. This usually consists of PETROLEUM (crude oil) products. Spillage during loading and unloading causes considerable WATER POLLUTION. Other tanker cargoes include natural gas (liquefied by cooling) and wine. Combination carriers are tankers that can carry solid cargoes, such as grain and ore, as well as liquids.

tannin (tannic acid) Any of a group of complex organic compounds derived from tree bark, roots, and galls (such as the oak), unripe fruit, tea, and coffee. Yellow and astringent, tannin has the property of converting the protein GELATIN into an insoluble compound that will not putrefy, and for this reason is used in TANNING to cure hides and make leather. It is also used in inks and dyes, and as an astringent in medicine.

tanning Process of converting skins and hides into leather. Traditionally, tanning liquids are based on TANNIN and, although the process takes weeks, it is still used for heavy leathers. Light leathers are now tanned in a few hours using chromium salts.

tantalum (symbol Ta) Rare, lustrous, blue-gray metallic element, first discovered in 1802. Its chief ore is columbite-tantalite. Hard but malleable, tantalum is used as a wire and in electrical components, chemical equipment, and medical instruments. Properties: at.no. 73; r.a.m. 180.948; r.d. 16.6; m.p. 2,996°C (5,425°F); b.p. 5,425°C (9,797°F); most common isotope ^{181}Ta (99.988%).

tape, magnetic *See* MAGNETIC TAPE

tape recorder Device that records and plays back sound on MAGNETIC TAPE. Sound is transformed into an electric signal by a MICROPHONE and fed to a TRANSDUCER, which converts it into the magnetic variations that magnetize the particles on the treated tape. *See also* SOUND RECORDING

tapetum Reflecting structure in the EYES of some nocturnal vertebrates, which serves to improve night vision. Occurring on the retinal side of the choroid, the tapetum contains guanine crystals that reflect light back onto the RETINA. This process makes eyes shine in the dark. The eyes of some night-flying insects also contain a tapetum.

tapeworm Parasite of the genus *Taenia* that colonizes the intestines of vertebrates, including human beings. Caught from eating raw or undercooked meat, it may cause serious disease. Class Sestoda.

taproot First ROOT of a plant that develops from the RADICLE. The taproot grows directly downwards and remains the main root of the plant, sending off lateral side roots to extend the root system. In BIENNIAL plants, whose leaves usually die down in the first winter, the root is the part of the plant that remains alive underground ready to grow new leaves the following year. In some plants (such as beets, carrots, and parsnips) the taproot develops into a fleshy organ for storing STARCH. In root vegetables, it forms the edible part of the plant.

tar Black or dark brown complex liquid mixture of HYDROCARBON compounds, derived from wood, coal, and other organic materials by heating them with little or no air, then condensing the tar from the distilled vapors. Tar, from petroleum, is a major source of HYDROCARBONS for the synthesis of pharmaceuticals, pesticides, and plastics; cruder tar compounds such as pitch are used for road surfacing and protecting timber against rot and pests. Wood tar yields creosote and PARAFFIN. *See also* BITUMEN

Tarantula Nebula (NGC 2070 or 30 Doradus) Collection of gas and dust that is 1,000 light-years across and 500,000 times the mass of the Sun. It is bigger and brighter than any NEBULA in our GALAXY or any other nearby galaxy. Faintly visible to the naked eye on the SE edge of the Large MAGELLANIC CLOUD, it has a complex filamentary structure with a cluster of stars, R136, at its center.

tarmac Mixture of GRAVEL and TAR mainly used to cover roads, drives, and paths. It is hard-wearing, impervious to water and provides an ideal surface for the rubber tires of motor vehicles. *See also* BITUMEN; MCADAM, JOHN LOUDON

tarsals In terrestrial vertebrates, the seven bones that make up the ankle and adjoining part of the foot. Strong, compact bones, they are arranged so that the foot can be rotated (to a limited extent) in any direction. The tarsals articulate with the lower leg bones above and with the METATARSALS of the foot below.

tar sands Porous, often black, rocks such as limestones, sands, and sandstones found on the surface. They contain deposits of BITUMEN (asphalt) in the spaces between the grains and commonly have a distinct odour of TAR. Extensive deposits are found in North America, primarily in Texas and Alberta, Canada. Estimates of the amount of bitumen contained in these deposits are extremely high, but its high viscosity makes extraction difficult.

Tarski, Alfred (1902–83) US mathematician and philosopher, b. Poland. He is celebrated for his development of a semantic method for metamathematics, a branch of mathematical LOGIC. His publications include *Introduction to Logic and the Methodology of Deductive Science* (1936) and *Logic, Semantics, Metamathematics* (1956).

Tartaglia, Niccolò Fontana (1499–1557) Italian mathematician, the first to obtain a general solution to the cubic equation of the form $x^3 + ax^2 + bx + c = 0$. His method was announced by Girolamo CARDANO. From 1534 he taught in Venice. In *Nova Scientia* (1537), Tartaglia applied mathematics to ballistics. Other publications include *Trattato di Numeri et Misure* (1556–60).

tartaric acid (dihydroxybutanedioic acid) Colorless, optically active organic compound, $(CHOH)_2 (COOH)_2$, occurring naturally in fruits. It can be made from maleic anhydride and hydrogen peroxide or by ENZYME action on a succinic acid. Tartaric acid is widely used in baking powder, cream of tartar, soft drinks, and in the textile industry. Properties: r.d. 1.76; m.p. 171°C (340°F).

tartrate Salt or ester of TARTARIC ACID. Tartrates, such as potassium bitartrate (cream of tartar) and sodium bitartrate, are used in baking powder, in the food industry and in tanning.

taste One of the five SENSES, responding to the chemical constituents of food. In human beings, the taste buds of the tongue differentiate four qualities: sweetness, saltiness, bitterness, and sourness. The sense of taste is supplemented by the sense of smell.

Tatum, Edward Lawrie (1900–75) US geneticist and biochemist. He shared the 1958 Nobel Prize for physiology or medicine with George BEADLE and Joshua LEDERBERG for his part in the discovery that GENES act by regulating specific chemical processes, a basic principle explaining how genes determine hereditary characteristics. *See also* HEREDITY

tau particle ELEMENTARY PARTICLE that has a mass approximately 1.9 times that of the PROTON. It is a LEPTON, meaning that it does not experience the STRONG NUCLEAR FORCE. It decays via the WEAK NUCLEAR FORCE with a lifetime of $c.3 \times 10^{-23}$ s. It has a unit negative charge.

Taurus (Bull) Constellation of the N sky, on the ecliptic between Aries and Gemini. It contains the Pleiades and Hyades stellar clusters and the Crab Nebula. The brightest star is the 1st-magnitude Alpha Tauri (Aldebaran).

tautomerism Property exhibited by a COMPOUND that exists as a mixture of two ISOMERS in EQUILIBRIUM. If either isomer is removed from the mixture, the equilibrium is momentarily disturbed but is restored by the spontaneous conversion (due to the migration of an ION or FREE RADICAL) of some of the remaining isomer into the removed type.

taxis Movement of a single cell, such as a GAMETE or bacterium, in response to an external stimulus. Movement towards the stimulus is **positive** taxis and away from the stimulus is **negative** taxis. Notable stimuli include light (**phototaxis**) and chemicals (**chemotaxis**). Various types of algae, for example, exhibit positive phototaxis – they swim towards a light source in order to increase photosynthesis.

taxonomy Organization of organisms into categories based on similarities of either MORPHOLOGY and ANATOMY (**classical** taxonomy), protein and nucleic acid structure (**biochemical** taxonomy), the behavior and morphology of chromosomes (**cytotaxonomy**), or the analysis of numerical data (**numerical** taxonomy). Carolus LINNAEUS developed the first taxonomic system during the 1750s. *See also* BINOMIAL NOMENCLATURE; CLASSIFICATION

Taylor, Sir Geoffrey Ingram (1886–1975) British physicist and meteorologist. Specializing in FLUID MECHANICS, Taylor made significant contributions to the understanding of turbulence and diffusion in fluids.

T-cell (Thymus-cell or T-lymphocyte) Type of LYMPHOCYTE that is the key to the defense mechanism of the IMMUNE SYSTEM. There are three main kinds of T-cells designated **Th** (helper T-cells), **Ts** (suppressor T-cells), and **Tc** (cytotoxic T-cells). **Helper** T-cells recognize foreign ANTIGENS and help other immune cells to act; **suppressor** T-cells prevent specific immune reactions; and **cytotoxic** T-cells, also called effector T-cells, kill cancerous cells or cells infected with a virus. T-cells are produced in the bone marrow and then move to the THYMUS GLAND. *See also* ACQUIRED IMMUNE DEFICIENCY SYNDROME (AIDS)

tear fault (lateral fault, wrench fault) Geological FAULT in which the movement is along the horizontal. Tear faults occur frequently in areas of TECTONIC activity. The amount of movement may only be a few meters, but can be several kilometers in some cases. The San Andreas Fault in California is a well-known example of a tear fault.

tears Salty fluid, secreted by the LACHRYMAL GLANDS, that moistens the surface of the eye. Having antibacterial properties, they cleanse and disinfect the surface of the eye and also bring nutrients to the CORNEA.

technetium (symbol Tc) Silver-gray TRANSITION ELEMENT. It was first made in 1937 by bombarding MOLYBDENUM with deuterons (nuclei of atoms of DEUTERIUM), and it was the first element to be synthesized in a CYCLOTRON. Technetium is found in the fission products of uranium and is present in some stars. It is used in radioactive tracer studies. There are 16 known isotopes. Properties: at. no. 43; r.a.m. 98.9062; r.d. 11.5; m.p. 2,172°C (3,942°F); b.p. 4,877°C (8,811°F); most stable isotope ^{99}Tc (half-life 2.6×10^6 yr).

tectonics Deformation within the Earth's CRUST and the geological structures produced by deformation, including FOLDS, FAULTS, and the development of mountain chains. PLATE TECTONICS originates from the study of the main structural features of the Earth's crust, such as MIDOCEAN RIDGES, major tear faults, ocean TRENCHES, continental blocks, earthquake belts, and so on, but is now used to describe a theory that can explain the distribution and evolution of these features.

teeth Hard, bonelike structures embedded in the jaws of vertebrates, used for chewing food, defense, or other purposes. The teeth of all vertebrates have a similar structure, consisting of three layers. Mammalian teeth have an outer layer of hard ENAMEL. The middle layer consists of dentine, a bonelike substance that is capable of regeneration. The core of a tooth contains pulp that is softer and has a blood supply and nerves; it provides nutrients for the dentine. *See also* DENTITION

tektite Generally dark, glassy objects, ranging in diameter from 40 micrometers to 2mm (microtektites) and larger (to 10cm), believed to be of either lunar origin or formed from splashes of liquefied rock during meteorite impact on Earth. They occur in limited areas, called strewn-fields, on continents and on ocean floors.

telecommunications Technology involved in the sending of information over a distance. The information comes in a variety of forms, such as digital signals, sounds, printed words, or images. The sending is achieved through TELEGRAPH, TELEPHONE, and RADIO and the medium may be wires or RADIO WAVES, or a combination of the two. Telegraphy was developed in the mid-19th century and radio arrived at the end of the century. TELEVISION manufacture was perfected in the 1930s. There are two basic types of message: DIGITAL SIGNALS, in which the message to be sent is converted in simple coded pulses and then sent (as in MORSE code); and ANALOG SIGNALS, in which the message – for example, a voice speaking – is converted into a series of electrical pulses that are similar in waveform to the MODULATIONS of the original message. *See also* FAX; SATELLITE, ARTIFICIAL

telegraph Any communications system that transmits and receives visible or audible coded signals over a distance. The first, optical, telegraphs were forms of SEMAPHORE. Although most modern devices are electrically operated and connected by wires or cables, telegraph messages can be sent via RADIO WAVES, microwaves, and artificial SATELLITES. Pioneering work was performed by many, but credit for the electric telegraph and its code is credited usually to Samuel MORSE, who inaugurated (1844) the first public line between Washington and Baltimore. In 1866 the first permanent telegraph cable was laid across the Atlantic. In 1874 Thomas EDISON invented the "quadruplex" method of simultaneously transmitting four messages over the same wire.

telemetry System of transmitting data, usually measurements, over a distance, such as to Earth from a spacecraft via RADIO WAVES. In this example, SOLAR CELLS or chemical batteries provide power for operating special ANTENNAE on the spacecraft, which beam high-frequency (but low-power) radio waves to large receiving stations on the ground.

telephone Instrument that communicates speech sounds by means of wires or MICROWAVES. In 1876 Alexander Graham BELL invented the prototype, which employed a diaphragm of soft iron that vibrated to sound waves. These VIBRATIONS caused disturbances in the MAGNETIC FIELD of a nearby bar magnet, causing an ELECTRIC CURRENT of fluctuating intensity in the thin copper wire wrapped around the magnet. This current could be transmitted along wires to a distant identical device, which reversed the process to reproduce audible sound. Later improvements separated the TRANSMITTER from the RECEIVER and replaced the bar magnet with BATTERIES. With the invention of the MOBILE TELEPHONE, telephones have become an integral part of social and professional life, as well as a lifestyle accessory.

telephone exchange Central switching system where incoming telephone calls are routed to their destinations, either by telephone operators or, increasingly, by automatic mechanisms that are called into operation by direct dialling codes.

telephoto lens CAMERA lens with a long FOCAL LENGTH. A true telephoto LENS has a focal length longer than the physical length of the lens, as opposed to a **long-focus** lens, in which the focal length is equal to the physical length. There is no theoretical upper limit to the focal length possible, but sheer size tends to limit the actual focal lengths practicable. For a 35mm camera, any lens with a focal length of more than about 80mm may be regarded as a telephoto lens. For larger format cameras, the focal length may be as much as 1,000mm. *See also* PHOTOGRAPHY

Key
1 enamel
2 capillaries, nerves, lymphatics
3 pulp
4 gum
5 dentine
6 jaw
7 cementum
8 root canal

crown
neck
root

▲ **teeth** A human tooth consists of the crown, neck and root. The crown is made up of a dense mineral, enamel, surrounding the hard dentine, which has a soft center – the pulp. The pulp is filled with blood vessels, lymphatics, and the nerve, which reach the tooth through the root canal. The neck adheres to the gum, and the root penetrates the bone, where it is held in place by a ligament and cementum.

teleprinter Machine having a keyboard similar to that of a TYPEWRITER, on which messages can be typed and telegraphed to distant RECEIVERS, and that also acts as a receiver, printing out incoming messages.

telescope Instrument for producing an enlarged image of a distant object or studying electromagnetic radiation from a distant source. Telescopes are categorized according to the area of the electromagnetic spectrum they cover. **Optical** telescopes can use lenses (**refracting** telescopes) or mirrors (**reflecting** telescopes); **catadioptric** telescopes use both in combination. The LENS or mirror is the telescope's main light-gathering part (**objective**), and its diameter is known as the APERTURE of the telescope and determines its light-gathering ability. The point at which the objective concentrates the light from the source is its FOCUS, and the distance from the focus to the objective is its FOCAL LENGTH. The light-gathering power of a telescope is proportional to the square of the aperture. Its ability to resolve (separate) two closely spaced objects into distinct images also depends on the size of the aperture: the larger the aperture, the better the resolution. Refracting telescopes were extensively used after versions were invented by Hans LIPPERSHEY (1608) and GALILEO (1609). The main disadvantage with refracting telescopes was chromatic ABERRATION, in which light of different colors is brought to a focus at different points, causing a single lens to produce a partly distorted image. This problem was solved by combining lenses so their aberrations cancel each other out. Reflecting telescopes are free of chromatic aberration – though, like lenses, they suffer from spherical aberration. In 1668 Sir Isaac NEWTON built an early astronomical reflector. Refracting telescopes also have a limit to the size of lenses: the world's largest refractor has a diameter of 1m (39in), much less than one of the larger reflectors; the Hale reflector in Mount Palomar, California, has an aperture of 508cm (200in). Earth-bound telescopes have limitations because the incoming radiation has to pass through the Earth's atmosphere. Advanced telescopes combat turbulence in the atmosphere (which makes stars twinkle) with adaptive OPTICS. Their mirrors consist of segments that are jiggled at high speed to stabilize the image. Major modern OBSERVATORIES are built on mountain peaks – for example, in Chile, the Canaries, and Hawaii – in order to improve "seeing" and to observe INFRARED RADIATION from celestial bodies. Most of this infrared light is absorbed in the denser, lower part of the atmosphere. This ceases to be a problem with telescopes in Earth orbit, such as the HUBBLE SPACE TELESCOPE. Orbiting telescopes can also detect other types of electromagnetic radiation more easily, such as ULTRAVIOLET RADIATION, X RAYS, and GAMMA RADIATION. RADIO TELESCOPES are complex electronic systems that detect and analyze radio waves from beyond the Earth. Incoming RADIO WAVES are collected from a small area of the sky and focused at the center of a large parabolic, steerable dish aerial. The signals are then amplified and analyzed. The first radio telescope was built in 1937 by the US radio engineer Grote Reber.

Telescopium (the Telescope) Faint constellation of the S sky. Its brightest star is Alpha, magnitude 3.5.

television System that transmits and receives visual images by RADIO WAVES or CABLE. A television CAMERA converts the images from light rays into electrical signals. The basis of most television cameras is an **image orthicon tube**. Light rays are focused onto thin sheets of photoelectric material that emit ELECTRONS in proportion to the amount of light striking them. Behind these sheets is a positively charged target that emits additional electrons. The target has a positive charge in proportion to the amount of light in the original scene. Behind the image orthicon tube is an **electron gun** emitting a beam that scans the target. The beam is deflected back to the gun and collected by a multiplier that amplifies its intensity. The strength of this beam depends on the positive charges on the target and thus the original scene is converted into electrical signals. These are amplified and transmitted as VERY HIGH FREQUENCY (VHF) or ULTRA HIGH FREQUENCY (UHF) radio waves. Typically, a television channel has a BANDWIDTH of 5MHz (5 million cycles per second). The RECEIVER (TV set) operates in reverse to the camera. On reception, the signals are amplified and converted to light again in a CATHODE-RAY TUBE. Color television has three synchronized image orthicon tubes in the camera, one for each of the three primary colors – red, blue, and green. The tube of the receiver has three electron guns and the face of the tube is covered with a mosaic of fine phosphors in groups of three, each emitting only red, blue or green light when struck by a beam. These primary colors merge on the face of the screen to reconstitute the originally transmitted image. *See also* RADIO

telex (acronym for **tele**type-writer **ex**change service) Telegraphic system using telephone lines through which direct current pulses representing characters typed on a typewriter keyboard are sent. It has been superseded by the FAX machine.

Telford, Thomas (1757–1834) Scottish civil engineer who built roads, bridges, canals, docks, and harbors. His most notable achievements were the Caledonian Canal in Scotland, and the 177m (580ft) Menai Strait suspension bridge connecting Anglesey with mainland Wales. Telford was a founder and the first president of the Institution of Civil Engineers.

Teller, Edward (1908–) US physicist, b. Hungary, who has been called "the father of the HYDROGEN BOMB." In 1926 he received the Enrico Fermi Award. In 1935 he left Europe and settled in the USA, where he conducted research on solar energy. During World War II, Teller contributed to atom bomb research with Enrico FERMI and was then involved in the MANHATTAN PROJECT at Los Alamos to produce the bomb. He was a central figure in developing and testing (1952) the hydrogen bomb. Teller was a major supporter of Ronald Reagan's abortive Strategic Defense Initiative ("star wars" project).

tellurium (symbol Te) Silver-white, metalloid element, discovered in 1782. It occurs naturally combined with gold in sylvanite, and its chief source is as a by-product of the electrolytic refining of copper. The lustrous, brittle element is used in semiconductor devices, as a catalyst in petroleum cracking and as an additive to increase the ductility of steel. Properties: at.no. 52; r.a.m. 127.60; r.d. 6.24; m.p. 449.5°C (841.1°F); b.p. 989.8°C (1,814°F).

telophase Stage in CELL DIVISION, following ANAPHASE. It is the final phase of MITOSIS and MEIOSIS. In **mitosis**, telophase involves the division of the CYTOPLASM to form two daughter cells with the same number of CHROMOSOMES as the original cell nucleus. The separated CHROMATIDS accumulate at the poles of the SPINDLE and a nuclear membrane forms to separate the two groups. In **meiosis**, there are two stages of telophase. In the first stage, two daughter cells are produced when a membrane forms between the separated chromatids. In the second telophase, these daughter cells divide to produce four cells, which are HAPLOID.

Telstar First active communications SATELLITE, launched by the US on July 10, 1962. It contained a microwave radio receiver, amplifier, and transmitter for relaying telephone and television signals. It operated for about 18 weeks, failed for five weeks, and then worked again for a further seven weeks before failing for good.

Temin, Howard Martin (1934–) US biologist. He shared the 1975 Nobel Prize for physiology or medicine with David BALTIMORE and Renato DULBECCO for his studies of the interaction between tumor viruses and the genetic material of the cell.

temperate zone Either of two regions of the Earth. The N temperate zone lies between the Arctic Circle and the Tropic of Cancer, and the S zone lies between the Tropic of Capricorn and the Antarctic Circle. A **temperate climate** is a moderate climate, as occurs in most temperate zones. The summers tend to be warm, but the winters are cool. Temperate climates have neither very high nor very low temperatures.

temperature In biology, intensity of heat. In warm-blooded (HOMEOTHERMIC) animals such as birds and mammals, body temperature is maintained within narrow limits regardless of the temperature of their surroundings. This is accomplished by muscular activity, the operation of cooling mechanisms such as vasodilation and vasoconstriction (changes in the diameter of superficial blood vessels) and sweating and metabolic activity. In humans, the normal body temperature is 36.9°C (98.4°F), but this may vary with degree of activity, reaching 40°C (104°F) during exercise, and falling slightly below normal during sleep. In so-called cold-blooded (POIKILOTHERMAL) animals, such as reptiles, body temperature varies between wider limits, depending on the temperature of the surroundings (air or water).

temperature In physics, measure of the hotness or coldness of an object. Strictly, it describes the number of energy states available to a substance or system. Two objects placed in thermal contact exchange heat energy initially, but eventually arrive at thermal EQUILIBRIUM where both are said to have the same temperature – each is losing and gaining heat at equal rates so that neither has a net gain or loss of heat. At equilibrium, the most probable distribution of energy states of the atoms and molecules composing the objects has been attained. At high temperatures, the number of energy states available to the atoms and molecules of a system is large; at lower temperatures, fewer states are available (molecules become locked into position and liquids change to solids). At a sufficiently low temperature, all parts of the system are at their lowest energy levels, the ABSOLUTE ZERO of temperature. *See also* TEMPERATURE MEASUREMENT; TEMPERATURE SCALE

temperature coefficient In physics, change in any physical quantity per degree rise in temperature.

temperature inversion Anomalous increase in TEMPERATURE with height. Normally the temperature of the air decreases from ground level upwards. The average rate of decrease is 1°C (1.8°F) for every 160m (525ft). In certain meteorological conditions, this situation is reversed. On a clear, calm anticyclonic night, the cool air may roll downhill and accumulate in valleys, and the air temperature will be lower near the valley bottom than it is 100m to 200m (328–656ft) higher. Above the cold layer there will be warmer air, which is likely to form cloud or haze. Evidence of a temperature inversion can be seen if there is smoke rising from a bonfire. The smoke will rise vertically and then bend horizontally when it reaches the "inversion layer." If this situation develops on a larger scale, dust and dirt rising into the atmosphere are trapped and unable to escape, causing serious pollution.

temperature measurement Method by which TEMPERATURE is specified. Several TEMPERATURE SCALES have been established, each of which uses a property (such as electrical resistance or volume expansion) of a thermometric substance (such as a metal wire or a liquid) to define a unit of temperature. The constant-volume gas THERMOMETER is the standard thermometer. Its temperatures are expressed in KELVIN degrees, but it is inconvenient for practical purposes. The International Practical Temperature Scale (IPTS), first adopted in 1968 and regularly revised since, provides scientists and technologists with a more convenient scale. The IPTS defines 0°C (32°F) as 273.15K and defines a number of reference temperatures including the boiling point of oxygen, 90.20K (−182.96°C, −297.33°F), the melting and boiling points of water and the melting points of zinc, silver, and gold. It also specifies the thermometers that are used to span the range. At higher temperatures, such as in furnaces, optical pyrometers that detect variation in color of the emitted light are used.

temperature scale Graduated scale of degrees for measuring temperature. The establishment of any temperature scale requires a thermometric parameter that varies linearly with temperature (such as the volume of a gas at constant pressure, or the expansion of a liquid in a tube); two or more fixed points (readily reproducible reference points such as the boiling and the freezing points of water); and the assignment of arbitrary divisions (called degrees)

between the fixed points. Gas, alcohol, mercury, electrical resistance, and wavelength of light have been used as thermometric parameters. Common temperature scales include the FAHRENHEIT, CELSIUS (formerly centigrade), and KELVIN (or absolute) scales; these are abbreviated to °F, °C, and K. The Fahrenheit scale originally used as fixed points the freezing point of water (taken to be 32°F) and the human body temperature (96°F, although later found to be 98.6°F). The interval between these was divided into 64 degrees; by extrapolation, the boiling point of water is 212°F. The Celsius scale uses 0°C and 100°C as the freezing and the boiling point of water, respectively; the interval is divided into 100 degrees. Zero on the Kelvin (or thermodynamic) scale (−273.15°C, −459.67°F) coincides with ABSOLUTE ZERO, the lower limit of temperature; the kelvin represents the same temperature difference as the degree Celsius. To convert Fahrenheit to Celsius: $C = 5(F − 32)/9$; to convert Celsius to Fahrenheit, $F = (9C/5) + 32$.

tempering Heat treatment to alter the hardness of an ALLOY. The treatment alters the crystal structure of the alloy. Tempering may make the alloy harder or softer. The effect produced depends on the composition of the alloy, the temperature to which it is heated, and the rate at which it is cooled. Usually, the metal is heated slowly to a specific temperature, then cooled rapidly. Tool steel is hardened in this way.

template (templet) Mold or pattern for making repeated copies of an object. It is usually cut from thin sheets of metal, wood or paper. In architecture, a template is a supporting structure such as a beam placed over a doorway.

temporal lobes Prominent lobes of the CEREBRAL CORTEX that, in humans, lie one on either side, directly behind the temples. The temporal lobes are involved in the interpretation and generation of language. *See also* BRAIN

tendon Strong, flexible band of CONNECTIVE TISSUE that joins MUSCLE to BONE.

tendril Coiling part of stem or leaf, a slender, threadlike structure used by climbing plants for support.

Tennant, Charles (1768–1838) British industrial chemist, who invented (1799) a process for manufacturing bleaching powder from chlorine and slaked lime.

tensile strength Resistance that a material offers to tensile STRESS. It is defined as the smallest tensile stress (force divided by unit cross-sectional area) required to break the body.

▲ **tendon** The tendon jerk is the simplest reflex action, involving only a sensory receptor neuron (1) and a motor neuron (2). Impulses, such as those created by a hammer tapping a knee, run to and from the muscles (3) and traverse only one segment of the spinal cord (4). This reflex is independent of the brain.

tentacle Any slender, flexible organ of an animal, most notably those of the octopus and some other cephalopods, capable of feeling and grasping.

teratogen Agent that increases the likelihood of deformities developing in the fetus. Teratogens taken in by the mother during pregnancy cause malformations of the fetus. They include certain chemicals and drugs (such as ALCOHOL), some pathogenic microorganisms (such as the virus that causes rubella, or German measles), and various kinds of ionizing radiation (such as X rays in high doses).

terbium (symbol Tb) Silver-gray, metallic element of the LANTHANIDE SERIES. Discovered in 1843, it is found in such minerals as monazite, gadolinite, and apatite. The soft element is used in semiconductors; sodium terbium borate is used in lasers. Properties: at.no. 65; r.a.m. 158.9254; r.d. 8.234; m.p. 1,360°C (2,480°F); b.p. 3,041°C (5,506°F). Single natural isotope ^{159}Tb.

Tereshkova, Valentina Vladimirovna (1937–) Soviet cosmonaut and the first woman in space. In June 1963 she made 45 orbits of the Earth in Vostok 6, spending nearly 71 hours in space.

terminal velocity Maximum velocity attainable by a falling body or powered aircraft. It is dependent on the shape of the body, the resistance of the air through which it is moving, and (in the case of aircraft) the THRUST of the engines.

terminator Boundary between the sunlit and dark sides of a planet or satellite. With a body lacking an atmosphere, such as the Moon, the terminator is distinct, although often broken up because of reflections from craters or mountains. Bodies with atmospheres have less well-defined terminators because atmospheric scattering causes twilight.

terpenes Group of unsaturated HYDROCARBONS related to ISOPRENE. They occur in most ESSENTIAL OILS and are colorless liquids. Examples are pinene, which is the chief ingredient of TURPENTINE, and limonene, which is found in the oils of citrus fruits.

terracing In geology, process of DEPOSITION or EROSION that produces steplike formations on a slope. Alluvial terracing is usually caused by periodic reductions in the area of the floodplain of a river.

terrigenous deposits Accumulations of sand, silt or mud that form in the sea near land as a result of EROSION.

territory In zoology, the restricted life space of an organism. It is an area selected for mating, nesting, roosting, hunting, or feeding and may be occupied by one or more organisms and defended against others of the same or a different species. The area may be defended or indicated by noise-making, chemical scent, physical displays or aggression. Many invertebrates and most vertebrates display this behavior.

Tertiary Early period of the CENOZOIC era, lasting from 65 million to about 2 million years ago. It is divided into five epochs, starting with the PALEOCENE, followed by the EOCENE, OLIGOCENE, MIOCENE, and PLIOCENE. Early Tertiary times were marked by great mountain-building activity (Rockies, Andes, Alps, and Himalayas). Both marsupial and placental mammals diversified greatly. Archaic forms of carnivores and herbivores flourished, along with early primates, bats, rodents, and whales. *See also* GEOLOGICAL TIME

tertiary consumer In a FOOD CHAIN, a CARNIVORE that preys on other carnivores (SECONDARY CONSUMERS). Secondary consumers are also carnivores, but they prey on PRIMARY CONSUMERS.

tesla (symbol T) SI unit of MAGNETIC FLUX density equal to a density of one WEBER of magnetic flux per square metre.

Tesla, Nikola (1856–1943) US electrical engineer and inventor, b. Croatia, who pioneered the applications of high-voltage ELECTRICITY. Tesla developed arc lighting, the first generator of alternating current (AC) and a system of transmitting electric power without wires. He also devised the **Tesla coil**, a device for producing high-frequency currents.

testis (pl. testes) Male sex GLAND, found as a pair located in a pouch, the SCROTUM, external to the body. The testes are made up of seminiferous tubules in which SPERM are formed and mature, after which they drain into ducts and are stored in the epididymis prior to being discharged.

testosterone Steroid HORMONE secreted mainly by the mammalian TESTIS. It is responsible for the growth and development of male sex organs and male SECONDARY SEXUAL CHARACTERISTICS, such as voice change and growth of facial hair. Synthetic testosterone has been used illegally by athletes to increase their muscular development.

tetrachloromethane (carbon tetrachloride, CCl_4) Colourless, nonflammable liquid with a characteristic odour, prepared by the chlorination of methane or the catalytic reaction of carbon disulphide and chlorine. It is used as a refrigerant, insecticide, degreaser and dry-cleaning fluid. Properties: r.d. 1.59; m.p. −23°C (−9.4°F); b.p. 76.8°C (170.2°F).

tetracyclines Group of broad-spectrum antibiotics effective against a wide range of bacterial infections.

tetraethyllead (Pb(C$_2$H$_5$)$_4$) Colorless, oily poisonous liquid. It is manufactured either by treating an alloy of lead and sodium with ethyl chloride or by the action of lead chloride on a GRIGNARD REAGENT. It is used in leaded gasoline as an anti-KNOCK agent. Properties: r.d. 1.65; b.p. 200°C (400°F).

tetrahedrite Sulfide mineral, composed of varying amounts of copper, iron, zinc, silver, antimony, and arsenic sulfides. It is found in medium- to low-temperature ore veins. Tetrahedrite displays cubic system, well-formed, tetrahedral crystals, and also appears as masses. In color it is metallic gray to black. It is an important ore of copper. Hardness 3–5–4; r.d. 4.9.

tetrapod Any animal that has four limbs. Tetrapods include most mammals, amphibians, and reptiles (except snakes). Some authorities include also bats and birds (because their forelimbs have become adapted into wings) as well as whales and snakes (because they evolved from ancestors that once had four limbs). **Quadrupeds** use all four limbs for walking.

tripod
telescope
laser
spirit levels
upper horizontal plate
lower horizontal plate
horizontal adjusters

▲ **theodolite** Theodolite is a surveying instrument used to measure angles between different points accurately, enabling them to be plotted in three dimensions. The telescope is used to locate measuring points on surveying poles; to align the telescope it is rotated about a horizontal or vertical axis. The precise angles of rotation in these planes are measured by using a micrometric microscope to read a scale on glass protractors. Modern computerized theodolites take these measurements electronically, and use lasers to measure distance.

tetrode ELECTRON TUBE that has a fourth electrode, a screen grid, between the anode and the control grid; the cathode is the other electrode. Tetrodes are used in power amplification and oscillator circuits.

textiles Fabrics, especially those produced by weaving yarn. The yarn is made by SPINNING natural or artificial FIBERS. Fabrics made by other processes, such as lacemaking or knitting, are now also regarded as textiles. Textiles are used to make clothing, curtains, carpets, and many other products. Powered machines for spinning and weaving were introduced in the 18th century. Most textiles are now produced in factories. Twentieth-century advances in textile manufacturing have included the development of flame-resistant fabrics and "easy-care" materials that resist creasing.

thalamus One of two ovoid masses of gray matter located deep on each side of the forebrain. Sometimes called the sensory-motor receiving areas, they fulfill important relay and integration functions for sensory messages that reach the BRAIN.

Thales (*c.*636–*c.*546 BC) First Greek scientist and philosopher of whom we have any knowledge. His discoveries in geometry included that the angles at the base of an isosceles TRIANGLE are equal. He predicted the ECLIPSE of the Sun that occurred in 585 BC.

thallium (symbol Tl) Shiny, metallic element of group III of the periodic table, discovered in 1861. Soft and malleable, it is obtained as a by-product of processing zinc or lead ores. It is used in electronic components, infrared detectors, and optical and infrared glasses. Thallium is an extremely toxic compound, and thallium sulfide is used as a rodent and ant poison. Properties: at.no. 81; r.a.m. 204.37; r.d. 11.85; m.p. 303.5°C (578.3°F); b.p. 1,457°C (2,655°F); most common isotope ^{205}Tl (70.5%).

thallophyte Obsolete term for a subkingdom of plants that lack clearly differentiated roots, stems or leaves, and range in size from one-celled plants to 61m (200ft) SEAWEEDS. Asexual reproduction is by spores and SEXUAL REPRODUCTION is by FUSION of GAMETES. Chlorophyll-containing thallophytes are ALGAE, Euglenoids, Dinoflagellates, and LICHENS. Thallophytes also included organisms formerly considered plants, lacking in chlorphyll such as BACTERIA, FUNGI, and SLIME MOULDS. *See also* THALLUS

thallus Nonvascular plant body of a SEAWEED. Usually flat or ribbon-shaped, it is not differentiated into root, stem or leaves.

Thenard, Louis Jacques (1777–1857) French chemist who discovered hydrogen peroxide and the porcelain pigment now known as Thenard's blue. Thenard worked with Joseph GAY-LUSSAC and wrote the standard chemical textbook of his time, *Traité de chimie élémentaire* (4 vols, 1813–16).

theodolite Surveying instrument dating back to the 16th century, used to measure horizontal and vertical angles. Its modern form consists of a telescope (with crosshairs in the lens for accurate alignment) mounted to swivel in both directions and levelled with a spirit level.

theorem Statement or proposition that is to be proved by logical reasoning from given facts and AXIOMS. *See also* FERMAT'S LAST THEOREM

therm Unit of heat used until recently, particularly for expressing the calorific value of town gas; it is now obsolete. It is equal to 100,000 British Thermal Units (BTU, also obsolete) or 25,200 kilocalories.

thermal Small-scale, rising current of air produced by local heating of the Earth's surface. Thermals are often used by gliding birds and human-built GLIDERS.

thermal capacity In physics, capacity of an object to absorb heat. It is the product of the object's mass and its SPECIFIC HEAT CAPACITY.

thermal diffusion (Soret effect) DIFFUSION caused by a temperature difference setting up a concentration gradient in a fluid.

thermal power Literally, heat power. Any heat engine, for example a STEAM ENGINE, employs thermal power. The heat generated by burning a fuel is converted into useful energy; in this case the energy drives the locomotive's pistons.

thermic lance Long, metal tube tipped with a nozzle, through which oxygen is fed at pressure. Its most important application is in the BASIC OXYGEN PROCESS of steelmaking, in which the lance, surrounded by a water jacket for cooling, jets oxygen on to a bath of molten steel in a FURNACE.

thermionic emission Emission of ELECTRONS from a substance's surface as a result of heating it.

thermionics Study of the emission of ELECTRONS or IONS from a heated CONDUCTOR. This is the principle on which ELECTRON tubes work. The heated conductor is the CATHODE and the emitted electrons are attracted to the ANODE. Such tubes, known as thermionic valves, have now largely been replaced by TRANSISTORS, although electron tubes are still employed as CATHODE-RAY TUBES. A more modern aim for thermionics is the design and construction of thermionic power generators, which will convert heat directly into electricity.

thermistor Type of SEMICONDUCTOR whose resistance sharply decreases with increasing TEMPERATURE. At 20°C the resistance may be of the order of a thousand OHMS and at 100°C it may be only 10 ohms. Thermistors are used to measure temperature and to compensate for temperature changes in other parts of a circuit.

thermite process SMELTING process in which a metallic oxide ore is reduced to the metal by heating with finely divided aluminum powder. It is the basis of the **Goldschmidt process** for extracting such high melting-point metals as chromium, manganese, molybdenum, and vanadium. It can also be used for welding cracks in metal, such as large castings and rails.

thermochemistry Branch of physical chemistry that deals with the HEAT effects that accompany chemical changes. Examples include the heat given off or absorbed during a chemical reaction, the dissolving of a substance, or changes of state, such as from a liquid to a gas.

thermocline Middle layers of OCEAN water, between surface and deep waters, which are defined by differing densities and temperatures. The thermocline is up to 1,000m (3,300ft) thick with a temperature only a few degrees above freezing. It is important as a stable boundary that tends to prevent interchange between layers.

thermocouple THERMOMETER made from two wires of different metals joined at one end, with the other two ends maintained at constant temperature. The junction between the two wires is placed in the substance whose temperature is to be measured. An ELECTROMOTIVE FORCE (voltage) is generated that can be measured by a millivoltmeter and that is, in turn, a measure of temperature.

thermodynamics Branch of physics that studies HEAT and how it is transformed to and from other forms of ENERGY. In general, a thermodynamic system is defined by its temperature, volume, pressure and chemical composition. The original laws of thermodynamics were conceived by observing large-scale properties of systems and with no understanding of the underlying molecular structure. The KINETIC THEORY of gases was developed in the mid-19th century. In general, the TEMPERATURE of a body is a measure of its internal energy. A body's internal energy, however, can increase if mechanical work is performed on it, and decrease if it is converted into mechanical energy. The **first law** of thermodynamics, basically a restatement of the CONSERVATION law of energy, states that the change in a system's internal energy is equal to the heat that flows into the system plus the work done on the system. Environmental sources of heat, such as oceans, are referred to as **heat reservoirs**. Any change in external conditions effects a shift in the thermodynamic system. The two main forms of change are ADIABATIC and isothermal processes. Adiabatic processes involve no change of heat, since the system is thermally insulated from the environment. The **second law** says that in an isolated system its ENTROPY (disorder or randomness) tends to increase. The **third law**, formulated by Walter NERNST, states that a system at ABSOLUTE ZERO would effectively have an entropy of zero. *See also* CARNOT CYCLE; CLAUSIUS, RUDOLF; KELVIN, WILLIAM THOMSON, 1ST BARON

thermoelectricity Phenomena involving the conversion of HEAT into ELECTRICITY or vice versa. James JOULE described the irreversible conversion of heat into electricity. There are, however, three reversible effects. In the SEEBECK EFFECT, a current flows if different temperatures are maintained at the junctions of a circuit containing two different metals; this is the basis of the THERMOCOUPLE (thermometer). The PELTIER EFFECT acts in the reverse manner, converting electrical energy to heat energy. Lord KELVIN discovered the third thermoelectric effect, called the **Thomson effect**: if the ends of a wire conductor have different temperatures, a potential difference is created along the wire. If an electric current flows from the cooler to the hotter part of the wire, the effect is called the positive Thomson effect; the reverse is the negative Thomson effect.

thermoelectric propulsion Any of several ROCKET propulsion systems combining heat and electrical means to accelerate particles to high velocities. An arc JET ENGINE uses an electric arc to heat liquid hydrogen to 50,000°C; the resulting PLASMA of ionized hydrogen is accelerated through a conventional nozzle or, in the plasma engine, through a magnetic field for greater force. *See also* ION

thermogram Photographic or other record of the heat radiated from an object, usually obtained with an infrared camera. For example, a thermogram can show the different temperature zones of the surface of the human body, which may be useful in the diagnosis of various disorders. Or it may show isothermal contours of the Earth's surface, as recorded by an orbiting spacecraft.

thermometer Instrument for measuring TEMPERATURE. Any substance with physical properties that change with varying temperature may be used as a measure, provided the change is correctly monitored and scaled. A mercury thermometer, for example, depends on the expansion of the mercury

metal that is held in a glass bulb connected to a narrow, graduated tube. Temperatures are also measured by gas thermometers and resistance thermometers, which measure the RESISTANCE of a CONDUCTOR. Common scales are CELSIUS (formerly centrigrade), FAHRENHEIT, and KELVIN.

thermonuclear reaction *See* FUSION

thermopile Device used to measure radiant heat, consisting of several THERMOCOUPLES connected together in series. Alternate junctions are blackened for absorbing radiant heat, the other junctions are shielded from the RADIATION. The ELECTROMOTIVE FORCE (EMF) generated by the temperature difference between the junctions can be measured. From this the temperature of the blackened junctions can be calculated, and thus the intensity of the radiation measured.

thermoplastic resin Type of POLYMER that softens on being heated and can be repeatedly melted or softened by heat without change of properties. Examples include NYLON, POLYVINYL CHLORIDE (PVC), POLYETHENE, FLUOROCARBON PLASTIC, and POLYSTYRENE. *See also* PLASTIC; THERMOSETTING RESIN

thermoreceptors In anatomy, sensory nerve endings, sited in the skin and in the deep body tissues, that detect changes in temperature and signal these to the brain. There are two types: receptors that are stimulated by cold and receptors responsive to heat.

thermosetting resin Type of POLYMER that sets hard, losing its plasticity, under heat and pressure. Examples include POLYESTER, EPOXY RESIN, and SILICONE. *See also* PLASTIC; THERMOPLASTIC RESIN

thermosphere Shell of light gases between the MESOSPHERE and the EXOSPHERE, between 100km (60mi) and 400km (250mi) above the Earth's surface. The temperature steadily rises with height in the thermosphere.

thermostat Device for maintaining a constant TEMPERATURE. A common type contains a strip of two metals, one of which expands and contracts more than the other. Thus the strip bends and breaks the circuit at a set temperature. As it cools, the strip straightens, makes contact, and the heating begins again once the circuit is complete.

thiamine VITAMIN B_1 of the B complex, required for carbohydrate METABOLISM. Its deficiency causes the disease beriberi. Thiamine is found in grains and seeds, nuts, liver, yeast, and legumes.

thin-layer chromatography Type of CHROMATOGRAPHY in which the components of a mixture of liquids are separated by differential absorption onto a thin layer of material supported on a glass plate. The usual absorbing material is ALUMINA, which is made into a paste with water, spread on the plate and allowed to dry. A spot of the liquid mixture is placed near one edge of the plate, which is stood vertically in a trough of SOLVENT. As the solvent rises up the plate by CAPILLARITY, the components travel with it, but at different rates depending on how readily they are absorbed by the alumina. This results in a series of spots that can be identified by a reagent or by the distance they have moved in a particular time. The technique is very similar to PAPER CHROMATOGRAPHY.

thiol (thio alcohols, mercaptans) Organic compound having a thiol (–SH) group bound to a carbon atom. Thiols are the sulfur analogs of ALCOHOLS. They have unpleasant odors resembling rotten cabbage or garlic. Ethane thiol is a typical thiol, having the formula C_2H_5SH.

Thomas, Sidney Gilchrist (1850–85) British metallurgist. In 1875 he invented a method, known as the basic OPEN HEARTH PROCESS, for removing phosphorus impurities from iron ore during conversion in the BESSEMER PROCESS.

Thompson, Benjamin, Count Rumford See RUMFORD, BENJAMIN THOMPSON, COUNT

Thomson, Sir Charles Wyville (1830–82) Scottish naturalist. He was appointed the director of the CHALLENGER voyage (1872–76). Thomson published *The Depths of the Sea* (1873) and an account of his expedition, *The Voyage of the Challenger* (1877).

Thomson, Elihu (1853–1937) US electrical engineer and inventor, b. Britain. With Edwin J. Houston he developed an arc lighting system that led to the founding of the American Electric Company, which became the Thomson-Houston Electric Company, and later merged with the Edison Company to form the General Electric Company (GE). Thomson invented the first high-frequency DYNAMOS and TRANSFORMERS.

Thomson, Sir George Paget (1892–1975) British physicist, son of Sir Joseph John THOMSON. He shared the 1937 Nobel Prize for physics with Clinton DAVISSON for their independent work on diffracting ELECTRONS (1927). This confirmed the wave nature of particles first predicted (1923) by Louis de BROGLIE. Thomson fired fast electrons through thin gold leaf to obtain a DIFFRACTION pattern.

Thomson, Sir Joseph John (1856–1940) British physicist, father of George THOMSON. He succeeded James Clerk MAXWELL as professor of experimental physics (1884–1919) at Cambridge. Thomson's dis-

covery (1897) of the ELECTRON is regarded as the birth of PARTICLE PHYSICS. He established that cathode rays consisted of a stream of particles. Thomson went on to prove that the electron was negatively charged and that its mass was about 2,000 times smaller than the smallest atom (hydrogen). He was awarded the 1906 Nobel Prize for physics for his investigations into the electrical conductivity of gases. Thomson and Francis ASTON produced evidence of ISOTOPES of neon. He transformed the Cavendish Laboratory into a major center for atomic research, attracting scientists of the calibre of Ernest RUTHERFORD.

Thomson, William, Lord Kelvin See KELVIN, WILLIAM THOMSON, 1ST BARON

Thomson effect Potential difference that develops between two points on a metal conductor if the two points are maintained at different temperatures. It is named after William Thomson (Lord KELVIN). *See also* THERMOELECTRICITY

thomsonite Mineral, hydrated aluminum sodium silicate, of the ZEOLITE group. It is found in cavities of basaltic rocks. Its crystals are of the orthorhombic system and it is snow-white when pure. Hardness 5–5.5; r.d. 2.2.

thorax In anatomy, part between the neck and the abdomen. In mammals it is formed by the ribcage and contains the lungs, heart, and esophagus. It is separated from the abdomen by the DIAPHRAGM. In arthropods it consists of several segments to which legs and other appendages are attached.

thorium (symbol Th) Radioactive, metallic element of the ACTINIDE SERIES, first discovered in 1828. The chief ore is monazite (a phosphate). The metal is used in photoelectric and thermionic emitters. One decay product is RADON-220. Thorium is sometimes used in radiotherapy, and it is increasingly used for conversion into uranium for nuclear FISSION. Chemically reactive, it burns in air but reacts slowly in water. Properties: at.no. 90; r.a.m. 232.0381; r.d. 11.72; m.p. 1,750°C (3,182°F); b.p. 4,790°C (8,654°F); most stable isotope ^{232}Th; (half-life 1.41 × 10^{10} yr).

three-body problem Fundamental problem in CELESTIAL MECHANICS: to determine the motions of three bodies under the influence only of their mutual gravitational attractions. While there is no exact general solution – only solutions to special cases – highly accurate approximations can be achieved with modern computers. The three-body problem was examined first by Isaac NEWTON and has been tackled by many astronomers and mathematicians. Work on the problem was stimulated by the need to understand the orbit of the Moon and, more recently, the orbits of artificial satellites.

threonine ($CH_3CH(OH)CH(NH_2)COOH$) Soluble, crystalline, essential AMINO ACID found in proteins.

throat *See* PHARYNX

thrombin Blood ENZYME that converts FIBRINOGEN to fibrin during the formation of blood clots.

thrombocyte *See* PLATELET

throttle Another name for the accelerator on any vehicle powered by an INTERNAL COMBUSTION ENGINE, so-called because such engines have a throttle valve, worked by the accelerator control, that governs the amount of fuel-air mixture entering

▲ **tidal power** The power of the sea can be harnessed to generate electricity. **Tidal power** uses a barrage (1) across an estuary or bay. The barrage contains turbines that can spin with a flow of water in either direction. As the tide comes in, gates on the barrage remain closed until a head of water has built up on the sea side of the structure (2). The gates are then opened (3), and the incoming tide flows through the barrage driving the turbines (4). As the tide falls, the process is

reversed with the gates closed until the sea has fallen below the level of water retained in the estuary (5). The second method of utilizing the power of the sea harnesses **wave power** (6). The key difference is that the turbine (7) is air driven, not turned by water. As a wave hits the shore, the force of the water (8) drives air (9) through the turbine blades (10). When the water level drops, air is sucked back down through the turbine spinning it again.

the CARBURETOR. Opening the throttle valve admits more fuel and makes the engine run more quickly.

thrust Driving force resulting from operation of a PROPELLER, JET ENGINE or ROCKET engine. An aircraft propeller forces air backwards, and jet and rocket engines expel gases backwards. Thrust is produced in a forward direction in accordance with the third of Newton's laws of MOTION. This states that for every action, there is an equal and opposite reaction.

thulium (symbol Tm) Lustrous, silver-white, metallic element of the LANTHANIDE SERIES. First discovered in 1879, its chief ore is monazite. Soft, malleable and ductile, it combines with oxygen and the halogens. It is used in arc lighting; thulium-170, which emits X rays, is used in portable X-ray units. Properties: at.no. 69; r.a.m. 168.9342; r.d. 9.31 (25°C); 1,545°C (2,813°F); b.p. 1,947deg;C (3,537°F); most stable isotope ^{169}Tm (100%).

thunderstorm Electrical storm, commonly experienced as LIGHTNING and thunder. Thunderstorms are caused by the separation of electrical charges in CLOUDS. Water drops are carried by updrafts to the top of a cloud, where they become ionized and accumulate into positive charges – the base of the cloud being negatively charged. An electrical discharge (a spark) between clouds, or a cloud and the ground, is accompanied by light (seen as a lightning stroke) and heat, which expands the air explosively and causes it to reverberate and produce sounds and echoes called thunder. Thunderstorms are usually accompanied by heavy rain; ozone and the oxides of nitrogen are produced in the air.

thymine In molecular biology, one of the four nitrogen bases in the nucleic acid DNA (the others are ADENINE, GUANINE, and CYTOSINE). In RNA thymine is replaced by uracil in the base sequence.

thymus gland One of the endocrine (ductless) GLANDS, located in the upper chest in mammals; it is large in infancy and shrinks after puberty. In childhood, it controls the development of lymphoid tissue and the immune response to infection. It secretes a HORMONE, thymosin, that stimulates the activity of defensive T-CELLS. Disorder of the thymus may be associated with AUTOIMMUNE DISEASES (those caused by the body's own antibodies). *See also* ENDOCRINE SYSTEM

thyroid gland H-shaped gland of the ENDOCRINE SYSTEM lying in the base of the neck, straddling the TRACHEA (windpipe) below the Adam's apple. It secretes thyroid hormones, principally THYROXINE which is essential for growth and development and for the regulation of metabolism.

thyroid-stimulating hormone (TSH) Hormone secreted by the frontal (anterior) lobe of the PITUITARY GLAND. It stimulates the THYROID GLAND in the neck to produce the hormone thyroxine, which helps to control metabolism.

thyroxine Hormone secreted by the THYROID GLAND. It contains iodine and helps regulate the rate of METABOLISM; it is essential for normal growth and development.

tibia (shinbone) Inner and larger of the two lower leg bones. It articulates with the FEMUR, or upper leg bone, at the knee and extends to the ankle, where its lower end forms the projecting ankle bone on the inside of the leg. *See also* FIBULA

tidal bore Flow of tidal water from the ocean into a funnel-shaped river mouth or estuary which, opposing out-flowing river water, builds into a surface "wall" that accelerates upstream. Three bores have been observed in North America – at the heads of the Bay of Fundy in New Brunswick, Canada; the Gulf of California; and Cook Inlet, Alaska. The world's largest known tidal bore is in Tsientang Kiang, China. *See also* TIDE

tidal flat Extensive, nearly flat, barren land area that is alternately covered and uncovered by the action of the tide. It consists of mud and sand. A tidal marsh has a covering of salt-tolerant plants and grasses.

tidal power Energy harnessed and used by man from tidal movement of the Earth's OCEANS. This form of power is cost-effective only where the tidal range is greater than 4.6m (15ft). Modern projects involve the use of turbogenerators driven by the passage of water through a tidal barrage. The La Rance power plant in N France has been working successfully since 1966, and in 1969 a small tidal power plant was completed by the Soviet Union on the White Sea. Special TURBINES have been developed that can be driven by water flowing in either of two directions. However, all such plants can have very serious effects on the ecology of the ESTUARY that they enclose.

tide Periodic rise and fall of the sea caused by the pull exerted on Earth by the Moon, and, to a lesser extent, by the Sun. In most parts of the world there are two high tides and two low tides every day. The time of each high tide is 12 hours 20–25 minutes later than the preceding tide, because the position of the Moon relative to the Earth will have changed by a small amount after 12 hours has elapsed.

till (boulder clay) In geology, SEDIMENT consisting of an unsorted mixture of clay, sand, gravel, and boulders deposited directly by the ice of GLACIERS.

▲ **tide** The daily rise and fall of the ocean's tides are the result of the gravitational pull of the Moon as it orbits the Earth and that of the Sun, though the effect of the latter is less than half as strong as that of the Moon. The effect is greatest on the hemisphere facing the Moon and causes a tidal "bulge". When the Sun, Earth, and Moon are in line, tide-raising forces are at a maximum, and spring tides occur; high tide reaches the highest values, and low tide falls to low levels. When lunar and solar forces are least coincidental, with the Sun and Moon at an angle (near the Moon's first and third quarters), neap tides occur, which have a small tidal range.

It is not deposited in layers. **Tillite** is till that has become solid rock.

time Non-spatial order in which all events occur; also the interval between two events. Time falls within the disciplines of physics, psychology, philosophy, and biology. Until the theory of RELATIVITY was devised by Albert EINSTEIN, time was perceived as absolute: a constant one-direction (past to future) flow. Since then the concept of time linked with distance in space (SPACE-TIME) has connected time with the relative velocities of those perceiving it. For clocks at velocities approaching that of light, time expands from the point of view of a stationary observer: that is, the moving clock records a smaller interval of time between two events than does the clock of the stationary observer. The passage of time (or equivalently, the rate of an accurate clock) is also slowed in a gravitational field.

timescale *See* GEOLOGICAL TIME

time-space distortion In the theory of RELATIVITY, the bending of the SPACE-TIME continuum so that its geometry is no longer Euclidean. This effect is caused by very strong gravitational fields such as that found near a BLACK HOLE, where the distortion is so great that light leaving the black hole is bent round itself and cannot escape. *See also* EUCLIDEAN GEOMETRY

time zone One of 24 divisions of the Earth's surface, each of which is 15° of LONGITUDE wide and within which the time of day is reckoned to be the same. In Washington DC in 1884 the meridian of Greenwich, England, was adopted as the zero of longitude, and zones of longitude were established.

Standard time in each successive zone westward is one hour behind that in the preceding zone. Large countries such as the US and Russia span several time zones. The adoption of time zones results in a discrepancy at longitude 180° (the International Date Line), which is resolved by omitting one day from the calendar when crossing from west to east, or repeating one day if the crossing is from east to west. *See also* GREENWICH MEAN TIME (GMT)

tin (symbol Sn) Metalloid element of group IV of the periodic table, known from ancient times. Its chief ore is CASSITERITE (an oxide). Soft, malleable and resistant to corrosion, tin is used as a protective coating for iron, steel, copper, and other metals, and in such alloys as solder, pewter, and bronze. It forms two series of salt, termed tin(II), or stannous, and tin(IV), or stannic. Tin compounds are used as fungicides, in glass coatings and as a tooth decay preventive (tin(II) fluoride) in toothpaste. Three allotropes exist: the common lustrous metallic form (white tin) changes slowly below 13.2°C (55.8°F) to a powder (gray tin); above 161°C, (321.8°F) a brittle form exists. Tin has been used by humans since the Bronze Age. Until the 20th century, the tin mines of Cornwall, England, were the world's leading source. Properties: at.no. 50; r.a.m. 118.69; r.d. 7.29; m.p. 232°C (449.6°F); b.p. 2,270°C (4,118°F); most common isotope ^{118}Sn (24.03%).

Tinbergen, Nikolaas (1907–88) Dutch ethologist. He shared with Konrad LORENZ and Karl von FRISCH the 1973 Nobel Prize for physiology or medicine for his pioneering work in ETHOLOGY. Tinbergen studied how in animals certain stimuli evoke specific responses, and emphasized the importance of observing animals under natural conditions.

tin chloride Either of two chlorides: tin(II) or stannous chloride ($SnCl_2$) or tin(IV) or stannic chloride ($SnCl_4$). **Tin(II) chloride** is a white, soluble solid, which can be produced by dissolving TIN in hydrochloric acid. It is used as a reducing agent and also as a constituent of tin-plating electrolytes. **Tin(IV) chloride** is a colorless, fuming liquid. It is used in the preparation of other inorganic tin compounds and organic tin (organotin) compounds. When sprayed onto glass and fired, it provides a conductive coating.

Ting, Samuel Chao Chung (1936–) US physicist who researched ELEMENTARY PARTICLES. He used a SYNCHROTRON to fire PROTONS at a BERYLLIUM target and then observed the resultant decaying particles. From the results Ting found evidence for the existence of a heavy ELEMENTARY PARTICLE, which he named the **J particle**. At the same time but working independently, Burton RICHTER discovered the same particle, which he named **psi**. For this discovery Ting and Richter shared the 1976 Nobel Prize for physics.

tin oxide Either of two oxides: tin(II) or stannous oxide, SnO, a black powder used as a reducing agent; tin(IV) or stannic oxide, SnO_2, a white powder occuring naturally as the mineral CASSITERITE. It is used in ceramics, glass, and cosmetics.

tire Air-filled rubber and fabric cushion that fits over the wheels of vehicles to grip the road and absorb shock. The pneumatic tire was invented in 1845 but was not much used until the end of the century – solid tires were more popular. The pneumatic tire consists of a layer of fabric – rayon, nylon or polyester – surrounded by a layer of rubber treated with chemicals to harden it and decrease wear and tear.

Tiselius, Arne Wilhelm Kaurin (1902–71) Swedish biochemist. In 1937, while working with Theodor SVEDBERG on ELECTROPHORESIS, he

▶ **tire** In cross-ply tires (A) the fabric bracing the cords of the tire run across the tire in overlapping layers, or plies (1). Additional strips and fillers stiffen the casing and metal bands at the tire's edge (2). These tires have largely been replaced by radial tires (B). A radial tire has cords reinforcing the sidewalls at right angles to the tire (1), with reinforcing fiber or steel beneath the tread (2), which stop the tire moving about on the road.

observed that a current passing through a colloidal solution causes different proteins to move at different speeds and therefore separate. Tiselius built a machine that would take advantage of the effect and for this work received the 1948 Nobel Prize for chemistry. *See also* COLLOID

tissue Material of a living body consisting of a group of similar and often interconnected CELLS, usually supporting a similar function. Tissues vary greatly in structure and complexity. In animals, they may be loosely classified according to function into epithelial, connective, skeletal, muscular, nervous, and glandular tissues, although each of these categories contains more than one different type of cell.

tissue culture In biology, artificial cultivation of living TISSUE in sterile conditions. Tissue culture in laboratories is used for biological research or to help in the diagnosis of diseases. Tissue culture is also used a means of propagating plant CLONES; it avoids the mingling of genetic material that occurs with POLLINATION and normal SEXUAL REPRODUCTION, and is faster than waiting for seed set and germination. It is used in the production of genetically engineered crop plants. *See also* GENETIC ENGINEERING

Titan Largest SATELLITE of SATURN, and the second-largest satellite in the Solar System, discovered (1655) by Christiaan HUYGENS. Titan is unique among planetary satellites in having a substantial atmosphere. It is composed of rock and water-ice in roughly equal proportions. The space probe, VOYAGER 1, found no gaps in an opaque, reddish cloud layer 200km (125mi) above the surface. The atmosphere consists mostly of nitrogen, with some methane and other hydrocarbon compounds. The surface temperature is 95K, at which methane can exist as solid, liquid or gas, so methane may play the role that water does on Earth, forming clouds, rain, lakes, and even snow.

Titania Largest of the five satellites of the planet URANUS, discovered (1787) by William HERSCHEL. It orbits at a distance of 438,370km (272,270mi) from Uranus and has a diameter of about 1,800km (1,120mi). Comprising rock and water-ice, it is almost OBERON's twin in size but its surface resembles Ariel's.

titanium (symbol Ti) Lustrous, silver-gray TRANSITION ELEMENT, first discovered in 1791. It is a common element found in many minerals, chief sources being ILMENITE and RUTILE. Resistant to corrosion and heat, titanium is used in steels and other alloys, especially in aircraft, spacecraft, and guided missiles, where strength must be combined with lightness. Titanium(IV) oxide is used as a white pigment in paints and as the gemstone tita-

nia. Properties: at.no. 22; r.a.m. 47.90; r.d. 4.54; m.p. 1,660°C (3,020°F); b.p. 3,287°C (5,949°F); most common isotope ^{48} Ti (73.94%).

titanium oxide Either of two oxides. Titanium(II) oxide (titanium monoxide, TiO) is a white powder formed by reducing the dioxide at 1,500°C (2,732°F). It has no industrial uses. Titanium(IV) oxide (titanium dioxide, TiO_2) is a white or gray powder, according to its purity. It occurs naturally as the mineral RUTILE, from which it is extracted by chlorination. It can be extracted from the mineral ILMENITE by treating it with sulfuric acid. It is used as a white pigment in ceramics, glass, cosmetics, paper, and paints.

titration Method used in volumetric analysis to determine the concentration of a COMPOUND in a SOLUTION by measuring the amount needed to complete a reaction with another compound. A solution of known concentration is added in measured amounts from a burette (a graduated glass tube) to a measured volume of liquid of unknown concentration until the reaction is complete (as shown by an INDICATOR or electrochemical device). The volume added enables the unknown concentration to be calculated.

TNT (2,4,6–trinitrotoluene, $C_7H_5N_3O_6$) EXPLOSIVE organic compound made from TOLUENE by using sulfuric and nitric acids. It was discovered in 1863; its resistance to shock – it requires a detonator to set it off – makes it one of the safest high explosives.

tobacco mosaic virus (TMV) Simple VIRUS used in experiments concerning the transference of the GENETIC CODE. It consists of a single helix of RNA containing some 6,400 NUCLEOTIDES. This is coated with some 2,100 molecules of a single PROTEIN, each molecule of which comprises a polypeptide chain of 158 AMINO ACIDS. The sequence of these acids has been determined.

Tokamak Soviet design of experimental nuclear FUSION reactor consisting of a TORUS-shaped vessel, in which a PLASMA (an extremely hot ionized gas that can reach temperatures that facilitate nuclear fusion) is contained by a magnetic field produced by coils wound over the reaction vessel. Several versions of these devices now exist outside the former Soviet Union.

toluene (methylbenzene) Aromatic HYDROCARBON ($C_6H_5CH_3$) derived from coal tar and petroleum. It is a colorless, flammable liquid widely used as an industrial solvent and in aircraft and motor fuels. Toluene is also used in the manufacture of TNT and chemicals such as PHENOL and BENZENE. Properties: r.d. 0.87; m.p. −94.5°C (−138.1°F); b.p. 110.7°C (231.3°F).

Tombaugh, Clyde William (1906–97) US astronomer. In 1930, nearly a year into a search based on predictions by PERCIVAL LOWELL, he discovered the planet PLUTO. He continued to search for other planets for more than ten years, discovering in the process star clusters, clusters of galaxies, a comet, and hundreds of asteroids. After World War II he developed telescopic cameras for tracking rockets after launch.

tombolo Bar connecting an island with the mainland. It is usually formed by a sand spit that grows until it reaches the island.

tomography Technique of X-RAY photography in which details of only a single slice or plane of body tissue are shown. *See also* COMPUTERIZED AXIAL TOMOGRAPHY (CAT)

tongue Muscular organ usually rooted to the floor of the mouth. The tongue contains the taste buds, groups of cells that distinguish the four basic tastes: bitter, tasted on the back of the tongue; sweet and salty, tasted on the tip and front of the tongue; and sour, tasted mainly on the sides of the tongue. The tongue helps to move food around the mouth for chewing and swallowing; animals also

▲**topology** Topology explores the properties of geometric shapes that have been deformed but not torn. If A¹ and B¹ are folded so that the arrows point in the same direction, a collar A² and a twisted band B² result. A² has an inner and outer surface and two edges, but B² has only one of each – precisely because of its twist. B² is called a Möbius strip and, unlike A², a path beginning at S and traced round the band will, after one revolution, end on the other "side" of the band at T. A Möbius strip stretched as in C¹ is distorted further into one half of a Klein bottle, (C²) which again has only one edge and one surface – compare a normal vase (C³).

use it for lapping fluids and for grooming. In human beings, the tongue is vital for the production of speech. *See also* SENSES

tonsils Two masses of lymphoid tissue located at the back of the PHARYNX. They have a pitted surface that frequently becomes infected (tonsillitis). They act as a filter against disease microorganisms. *See also* ADENOIDS

tooth *See* TEETH

topaz Transparent, glassy mineral, aluminum fluosilicate, $Al_2SiO_4(F,OH)_2$, found in pegmatites. Its crystals are orthorhombic system, columnar prisms; it occurs as granular masses. Topaz is colorless, white, blue or yellow, and some large crystals are of gem quality. Hardness 8; r.d. 3.5.

topocentric Relating to observations made from a point on the surface of the Earth. Topocentric coordinates contrast with geocentric coordinates, which are measured from the center of the Earth.

topographic mapping Representation of the surface of the Earth in relief, using contour lines drawn through points of equal elevation. Intervals are arbitrary. Most topographic maps are made from aerial photographs. Special plotting instruments allow the cartographer to follow a "floating" dot along an elevation, producing contour lines in sequence.

topography Study of surface features such as hills, valleys, rivers, roads, and lakes. It is also the representation of such features on a relief map or a plan for construction. The terrain of a region is explored using surveyors' instruments or aerial PHOTOGRAMMETRY (plotting elevations from photographs). *See also* SURVEYING

topology Branch of mathematics concerned with those properties of geometric figures that remain unchanged after a continuous deformation process such as squeezing, stretching, or twisting (but not joining, tearing or breaking). A cup with a handle is topologically equivalent to a doughnut with a hole in it; a cube, a solid cone and a solid cylinder are topologically equivalent to a sphere. These figures can be considered sets of points, each point of one set being continuously transformable into one point in another set. *See also* GEOMETRY

torbernite Complex, radioactive, phosphate mineral that contains copper and uranium. It occurs as aggregates that easily cleave to give green, brittle, mica-like flakes, which have a vitreous luster. The platy crystals (monoclinic) are translucent or transparent. Hardness: 2.0–2.5; s.g. 3.3.

tornado (twister) Funnel-shaped, violently rotating storm extending downwards from the cumulonimbus cloud in which it forms. At the ground its diameter may be only about 100m (300ft). Rotational wind speeds range from 150 to 500km/h (100–300mph). The center of a tornado is an area of extreme low pressure, which sucks up dust to form a blackish funnel rising to the sky. Around this funnel of rising air are very strong winds that destroy crops and sometimes buildings. Tornadoes occur in deep low-pressure areas, associated with FRONTS or other instabilities. They are associated with intense heating in continental areas in late summer, when land masses are at their hottest. Tornadoes are most common in the interior of the USA, but also occur in India and many other countries. When tornadoes cross over water they become WATERSPOUTS. The average life of a tornado or a waterspout is 4 minutes. The violent storms the occur at the start of the rainy season in West Africa are also called tornadoes. They bring strong winds and torrential rain, which is caused by mild air coming in from the sea and meeting the dry NE air from the Sahara Desert.

torpedo Self-propelled, underwater MISSILE used by submarines, small surface warships, and aircraft to destroy enemy vessels. Modern torpedoes may be launched by rocket boosters and often have internal electronic equipment for guiding the missile to the target.

torque Turning effect of a FORCE. An example is a turbine that produces a torque on its rotating shaft to turn a generator. The output of a rotary engine, such as the familiar FOUR-STROKE ENGINE or an electric motor, is rated by the torque it can develop. The unit of measurement is Nm (newton-meter).

Torrey, John (1796–1873) US botanist and chemist. He conducted major studies of North American plant life and amassed one of the most valuable botanical libraries and herbariums of his time. His works include *Flora of the Northern and Middle Sections of the United States* (1824) and *Flora of the State of New York* (1843), which he wrote after being appointed state botanist (1836).

Torricelli, Evangelista (1608–47) Italian physicist. Assistant and secretary to GALILEO during the latter's last three months of life, he is credited with designing the first artificial VACUUM (the Torricellian vacuum). In 1643 Torricelli invented the mercury BAROMETER. He also constructed a primitive MICROSCOPE and improved the TELESCOPE. His work in geometry eventually contributed to the development of INTEGRAL CALCULUS.

torsion In mechanics, strain in material that is subjected to a twisting FORCE. In a rod or shaft, such as an engine drive shaft, the torsion angle of **twist** is inversely proportional to the fourth power of the rod diameter multiplied by the shear modulus (a constant) of the material. Torsion bars are used in the spring mechanism of some automobile suspensions.

torsion balance Sensitive device for measuring small forces, whether gravitational, magnetic or electrical. A TORSION balance has a horizontal arm, suspended at its centre by a fine wire or fiber, which is twisted about a vertical axis by the force being measured.

torus (toroid, anchor ring) Solid figure generated by rotating a circle about an external line in its plane. It resembles a ring doughnut or the inner tube of a tire. The volume of a torus is $2\pi^2Rr^2$ and the surface area is $4\pi^2Rr$, where R is the distance between the center of the circle and the line, and r is the radius of the circle.

total internal reflection REFLECTION without REFRACTION of light at a boundary. When light passes from a dense medium, such as glass, to a less dense medium, such as water or air, a range of angles of incidence exists in which no light passes through the boundary; all the light is reflected within the denser medium.

touch One of the five SENSES, functioning by means of specialized nerve receptors in the SKIN.

touch screen COMPUTER INTERFACE device that allows the user to give instructions by touching the screen of the computer display or monitor. The user can simply point to an icon, menu option or piece of data in order to select it. Usually the screen is covered by a pressure-sensitive, double-skinned membrane. When it is touched, fine conducting lines on the inside surfaces of the membrane come into contact and pass an electrical signal. Other types of touch screen use capacitive surface devices or a grid of intersecting light beams in front of the screen.

tourmaline Silicate mineral, sodium or calcium aluminum borosilicate, found in IGNEOUS and METAMORPHIC rocks. Its crystals are hexagonal system and glassy, either opaque or transparent. Tourmaline may be black, red, green, brown or blue. Some crystals are prized as gems. Hardness 7.5; r.d. 3.1.

Townes, Charles Hard (1915–95) US physicist. In 1953 he invented the first operational MASER, for which he shared the 1964 Nobel Prize for physics with Alexsandr PROKHOROV and Nikolai BASOV. The MICHELSON-MORLEY EXPERIMENT was accurately confirmed with the aid of masers.

toxin Poisonous substance produced by a living organism. The unpleasant symptoms of many bacterial diseases are due to the release of toxins into the body by the BACTERIA. The toxins may be secreted into the bloodstream or they may be released when the bacteria die. Many molds, some larger FUNGI (such as the death cap), and seeds of some higher plants (such as LABURNUM and castor oil) produce toxins. The VENOMS of many snakes contain powerful toxins. *See also* BLOOM

trace element Chemical element that is essential to life but only in small quantities normally obtainable from the diet. They include boron, cobalt, copper, iodine, magnesium, manganese, molybdenum, and zinc. They are essential to the reactions of ENZYMES and HORMONES.

tracer, radioactive Radioactive substance that is introduced into the body so that its progress can be tracked by special diagnostic equipment. Its course is followed in the body by the detection and identification of metabolic compounds "labeled" with the isotope. This technique may help in the diagnosis of conditions such as thyroid disease. Radioactive tracers are used in a similar way to follow chemical reactions in complex organic industrial processes. In 1904 Franz Knoop used synthetic derivatives of fatty acids to trace the metabolic pathway. In 1923 George HEVESY became the first person to use a radioactive ISOTOPE (thorium) to investigate the metabolism.

trachea (windpipe) Airway that extends from the LARYNX to about the middle of the STERNUM (breastbone). Reinforced with rings of CARTILAGE, it is lined with hairlike CILIA that prevent dirt and other substances from entering the LUNGS. At its lower end, the trachea splits into two branches, the bronchi, which lead to the lungs.

tracheophyte In certain classification systems, any VASCULAR PLANT of the phylum Tracheophyta, defined by having some kind of vascular tissue. Within this phylum are: psilopsids (leafless, rootless primitive forms, such as whisk fern), sphenopsids (such as HORSETAIL), lycopsids (such as CLUB MOSS), pteropsids (such as FERN), GYMNOSPERMS, and flowering plants. In the Five KINGDOMS classification, these groups now constitute separate phyla in their own right, and the Tracheophyta is no longer considered to be a separate entity.

trachoma Chronic eye infection, similar to conjunctivitis, caused by the microorganism *Chlamydia trachomatis*, characterized by inflammation of the cornea with the formation of pus. A disease of dry tropical regions, it is the major cause of blindness in the developing world.

trachyte Uncommon, fine-grained, extrusive IGNEOUS ROCK of volcanic origin. Usually light-colored, it is rich in alkali feldspar and contains small amounts of quartz, nepheline, and other minerals; its silica content is up to 60%. Trachyte is formed, along with basalt, in lava flows from island volcanoes, and it also occurs in small sills and dikes.

tracking In aerospace technology, following a moving object, usually with RADAR or radio using aerials that "lock" onto the object. The correct sweep speed of the ANTENNAE is calculated by computer from information on the object's position and velocity. Satellite tracking is of two kinds: **active** tracking uses radar to locate an object; **passive** tracking uses signals from the object itself.

traction In medicine, gradual exertion of force on a limb to overcome muscular tension and so ensure proper alignment of a fractured bone. The force is exerted gradually over a period of time, generally using weights on cords over pulleys.

traction engine Large, heavy vehicle powered by steam, once used for drawing farm wagons and other heavy loads, but now obsolete. Traction engines were also used at carnivals and circuses to drive dynamos for electricity. *See also* STEAM ENGINE

tractor Four-wheeled or tracked vehicle for moving and operating heavy implements. Tractors are used mostly in farming and construction. The first tractors were steam-powered TRACTION ENGINES built in the 1870s. Modern tractors have gasoline or diesel engines and can haul and power a wide range of implements, including hay balers, crop sprayers, and mowing machines. Some tractors operate without drivers; they have a computer to control their work and use a form of GLOBAL POSITIONING SYSTEM (GPS) to track their positions.

Tradescant, John Name of two British botanists and travelers, father and son, who became successive gardeners to Charles I. John (*c.*1570–1638) travelled to Russia and Algeria gathering plants, among them the "Algiers Apricot," and established a garden of exotic plants at his Lambeth home. His son, also named John (1608–62), travelled to Virginia for plants for the Lambeth garden. He wrote a book about the Lambeth collection, *Musaeum Tradescantium* (1656). Their collection, transferred to Oxford after their deaths, formed the nucleus of the Ashmolean Museum.

trade winds Steady winds that blow from the tropical high-pressure zones (HORSE LATITUDES) to the equatorial low-pressure zones (DOLDRUMS). In the Northern Hemisphere, the air moving from the

Tropic of CANCER towards the equator is deflected to the right, making it a NE wind. In the Southern Hemisphere, the air moving from the Tropic of CAPRICORN to the equator is turned to the left to become a SE wind. The trades are persistent winds – all are easterly and tropical – blowing from the deserts towards the tropical forest zone.

train *See* RAILROAD

trajectory Path of a projectile. On Earth, and in the absence of air resistance, all trajectories would be portions of an ELLIPSE with one focus at the center of the Earth. Since this is 6,400km (4,000mi) deep, which is usually much greater than the height of the trajectory, the mathematically simpler PARABOLA provides an adequate approximation.

tram Passenger vehicle that runs on rails that are usually sunk into the road. Horse-drawn trams first appeared in New York in 1832. Some later trams were powered by steam locomotives. Trams with electric motors, supplied by overhead cables, became popular in the early 1900s. In some places, trams became unpopular in the 1920s. Because they run on fixed rails, they tended to cause traffic jams. Buses, which not are confined to rails and so offer more flexibility, steadily became more popular, although many European cities retained tramway systems. A new generation of trams is appearing in some cities. With increasing concern over air pollution and public health, electric trams are attracting renewed interest. Unlike gasoline and diesel engines, the electric motors in trams do not emit polluting exhaust gases.

tranquilizer Drug prescribed to reduce anxiety or tension and generally for its calming effects. Tranquilizers are used to control the symptoms of severe mental disturbance, such as schizophrenia or manic depression. They are also prescribed to relieve depression. Prolonged use of tranquilizers can produce a range of unwanted side effects, including dependence.

transactinide element Element beyond LAWRENCIUM in the PERIODIC TABLE – that is, with an ATOMIC NUMBER of more than 103. Elements 104 (DUBNIUM), 105 (HAHNIUM), and 106 (RUTHERFORDIUM) have been produced in particle accelerators, and heavier elements (ELEMENTS 107 to 112) have been claimed. All are unstable and radioactive, with a very short half-life.

transatlantic cable Any submarine cable laid under the Atlantic Ocean and used to carry electrical signals. The first successful one was the telegraph cable laid in 1866.

transcendental number *See* NUMBER, TRANSCENDENTAL

transcription (DNA transcription) Stage in the synthesis of PROTEINS within a EUKARYOTE cell in which the GENETIC CODE, represented by NUCLEOTIDES on DNA, is copied into single-stranded MESSENGER RNA (mRNA). It takes place in the cell nucleus (because the DNA-containing chromosomes are located there). RNA polymerase enzymes make an RNA copy from the original unwound strand of DNA that forms the template. The section to be copied is indicated on the DNA template strand by start and termination "signals." The mRNA then migrates to the RIBOSOMES, where actual protein synthesis takes place. Different RNA polymerase ENZYMES synthesize different types of RNA. *See also* TRANSLATION

transducer Device for converting any nonelectrical signal, such as sound or light, into an electrical signal, and vice versa. Examples are microphones, loudspeakers, and various measuring instruments used in ACOUSTICS.

transduction In microbiology, natural process by which genetic material is transferred between one host CELL and another by a VIRUS.

transfer orbit (transfer ellipse) Path followed by a spacecraft while changing from one orbital path to another. The tangential ELLIPSE is a tangent to the arrival and departure ORBITS and uses the least energy. Transfer ellipses intersecting the arrival and departure at higher angles use more energy but may be advantageous in other ways.

transfer RNA *See* RNA

transfinite number (symbol $\aleph$) Number denoting the size of an infinite SET, with a subscript. $\aleph_0$ represents the set of all INTEGERS and $\aleph_1$ represents the size of the set of all real NUMBERS. Although both sets are infinite, the set of real numbers is, in a sense, bigger than the set of integers (it can be shown that for every real number there is not an integer).

transformer Device for converting alternating current (AC) at one VOLTAGE to another voltage at the same frequency. It consists of two coils of wire coupled together magnetically. The input current is fed to one coil (the primary), the output being taken from the other coil (the secondary). If core losses are ignored, the ratio of the input voltage to the output voltage is equal to the ratio of the number of turns in the primary coil to the number of turns in the secondary coil. *See also* ELECTRIC CURRENT

transform fault Special class of strike-slip FAULT characteristic of MIDOCEAN RIDGES. Because of the transform faults, which are at right angles to the ridge itself, the MID-ATLANTIC RIDGE does not run in a straight line but in offset steps. Some geologists think that the four major structures of the Earth's crust are mountains, deep-sea trenches, mid-ocean ridges, and strike-slip faults, and that they form continuous networks.

transgenic Describing an organism whose GENOME contains GENES artificially transferred from another species. Using GENETIC ENGINEERING techniques, transgenic organisms are created by inserting isolated DNA from another species into the fertilized egg or early embryo. Transgenic animals and plants are usually created for commercial purposes. For example, it is possible to insert the genes for extra growth hormone into the genomes of livestock in order to increase meat production.

transistor Electronic device made of SEMICONDUCTOR material that can amplify electrical signals. The material, such as SILICON or GERMANIUM, is "doped" with minute amounts of PHOSPHOROUS, ARSENIC or ANTIMONY to produce *n*-type material, in which current is carried by negative charges, ELECTRONS; or with ALUMINUM, GALLIUM or INDIUM to give a *p*-type material. Joining together a piece of each produces a DIODE. Sandwiching one type between two of the other produces a transistor. These can thus be of two kinds: a *p-n-p* or *n-p-n* transistor. The middle region is known as the base, one of the side regions as the emitter and the other as the collector. In an *n-p-n* transistor, the signal to be amplified is fed across the collector, maintained at a constant voltage with respect to the base. The amplified signal comes out across the base and the emitter. Transistors were first developed in 1948 by John BARDEEN, Walter BRATTAIN, and William SHOCKLEY, making possible many advances in technology, especially in computers, portable radios and televisions, satellites, industrial control systems, and navigation.

transit In astronomy, passage of either of the inner planets, Mercury and Venus, across the face of the Sun. Transits of Mercury occur every 13 years, those of Venus every 100 years. The term also signifies the motion of any celestial object across the observer's celestial meridian.

transition In atomic physics, movement from one ENERGY LEVEL, or quantum state, to another. A transition from a lower to a higher energy level usually involves the absorption of a PHOTON. To move from a higher to a lower energy level, a photon is usually released.

transition elements Metallic elements that have incomplete inner electron shells. Transition elements are characterized by variable VALENCIES (combining power) and the formation of colored ions. They include elements with atomic numbers from 21 to 112. The ACTINIDE SERIES of elements and the rare-earth metals of the LANTHANIDE SERIES are sometimes called the inner transition elements. *See also* PERIODIC TABLE

translation Part of the process of protein synthesis that takes place in living cells. MESSENGER RNA (mRNA) carries a genetic code, in the form of a series of **codons** (triplets of NUCLEOTIDES), which ultimately determines the correct sequence for AMINO ACIDS to form protein. RIBOSOMES in the cell cytoplasm scan the mRNA for its codons, relaying the information to amino acid-bearing TRANSFER RNA (tRNA). The strands of tRNA have their own **anti-codons**, which form pairs with the codons of the mRNA. In effect, the tRNA (along with its amino acid), by pairing and attaching to the mRNA, "translates" the mRNA code to form the correct sequence of amino acids that form peptides, polypeptides, and eventually proteins. As the polypeptide chain grows, the tRNA falls away.

translocation In botany, movement of food materials in solution through the tissues of VASCULAR PLANTS from one part to another. It includes the passage upwards of minerals and other inorganic salts from the soil and roots in the XYLEM tissue, and the movement of sugars and other organic compounds from the leaves to other parts of the plant in the PHLOEM tissue.

translocation In genetics, process that alters the genetic makeup of a CHROMOSOME. A section of a chromosome breaks off, leaving it with less genetic information. The fragment joins onto another chromosome, also altering its genetic makeup. Rearrangement of genes within a single chromosome is also called translocation.

translucent Describing a material that allows the passage of light (or other electromagnetic radiation) but scatters it so that no object is visible through the material. *See also* TRANSPARENT

transmission, automobile Device (a gear or gears and shaft or driving chain) for transmitting power from the engine of a vehicle to the wheels. Speed can usually be varied in discrete steps, with gears or chains providing fixed speed ratios.

transmitter Electrical apparatus that generates RADIO WAVES. It has three main parts: an **oscillator**, which generates high-frequency alternating electric current (a carrier wave); a **modulator**, which encodes the carrier wave with the pattern or sound to be broadcast; and an **amplifier**, which increases the power of the modulated carrier (signal). *See also* MODULATION

transmutation In physics, formation of one element or ISOTOPE from another by RADIOACTIVE DECAY or by bombardment with energetic particles. The term was formerly applied to alchemists' attempts to convert base metals into gold.

transparent Describing a material that allows the passage of light (or other electromagnetic radiation) with little or no scattering so that objects are clearly visible through it. Different materials may be transparent to different wavelengths of radiation. *See also* TRANSLUCENT

transpiration In plants, loss of moisture as water vapor from leaf surfaces or other plant parts. Most of the water entering plant roots is lost by transpiration. The process is speeded up in light, warm, and dry conditions. The flow of water from the roots to the leaf pores (stomata) is called the **transpiration stream**. Several forces drive this flow. As moisture evaporates from the stomata in the leaves, water moves in by OSMOSIS from adjacent cells, setting up a flow of water by osmosis across the leaf from the XYLEM tubes to the stomata, pulling water out of the xylem. The force of attraction between the water molecules as they pass up the xylem tubes from the root to the leaves makes it difficult to break the water column. These tubes are also very narrow, so water rises in them by capillary action. In the roots, water being drawn into the xylem creates a lower concentration of water in adjacent root cells; thus water is drawn across the root from the root hairs by osmosis.

transplant Surgical operation to introduce organ or tissue from one person (the donor) to another (the recipient); it may also refer to the transfer of tissues from one part of the body to another, as in grafting of skin or bone. Major transplants are performed to save the lives of patients facing death from end-stage organ disease. Organs routinely transplanted include the kidneys, heart, lungs, liver, and pancreas. Experimental work continues on some other procedures, including small bowel grafting. Many other tissues are commonly grafted, including heart valves, bone, and bone marrow. The oldest transplant procedure (and still the most successful) is corneal grafting, undertaken to restore the sight of one or both eyes. Most transplant material is acquired from dead people, although kidneys, part of the liver, bone marrow, and corneas may be taken from living donors. The two main problems in transplantation

are rejection and infection. Rejection refers to the body's tendency to destroy "foreign" tissue and this cannot always be controlled by drugs given to suppress the immune system.

transport In chemistry and physics, movement of measurable entities such as molecules, ions, isotopes, electrical charges, mass, momentum or energy through or across a medium, because of the natural or applied nonuniform conditions existing within it. Transport properties include VISCOSITY, diffusivity (*see* DIFFUSION), and thermal CONDUCTIVITY. ACTIVE TRANSPORT in biochemistry is the movement of a substance against a concentration gradient.

transposon (transposable element) Section of DNA that can insert itself at many different sites on a CHROMOSOME. The presence of a transposon alters the genetic composition of the chromosome and can affect the surrounding GENES, causing some to be omitted or altered.

transuranic element (transuranium element) Any of those elements with atomic numbers higher than that of URANIUM (92), the best known of which are members of the ACTINIDE SERIES (atomic numbers 89 to 103). All transuranic elements are radioactive. Only NEPTUNIUM and PLUTONIUM occur naturally (in minute amounts) and have been synthesized in great quantity. Neptunium was the first of the transuranic elements to be produced, in 1940 by the US physicist Edwin MCMILLAN and US physical chemist Philip Abelson, who bombarded uranium with neutrons in a synchrocyclotron. The other elements have since been produced in a similar manner. The only commercially important element in the group is plutonium, which is used in NUCLEAR WEAPONS and as a fuel for NUCLEAR REACTORS.

trapezoid Four-sided plane figure with one pair of opposite sides parallel. The area of a trapezoid is half the sum of its parallel sides multiplied by the perpendicular distance between them.

traveling wave WAVE that continuously carries electrical energy, light energy, and so on, away from a source.

Travers, Morris William (1872–1961) British chemist who, in collaboration with Sir William RAMSAY, discovered the NOBLE GASES of the atmosphere. Those found included argon, helium, neon, krypton, and xenon. These researches are well described in his book *Discovery of the Rare Gases* (1928).

travertine SEDIMENTARY ROCK consisting almost entirely of crystals of CALCITE (calcium carbonate,

CaCO₃). It has a very light color, unless stained yellowish by iron compounds, and is often porous. It usually forms in hot springs from water rich in dissolved calcium carbonate. *See also* TUFA

tree Woody, PERENNIAL plant with one principal stem or trunk and smaller, offshooting branches. The trunk increases in diameter each year, and the leaves may be EVERGREEN or DECIDUOUS. The largest trees, redwoods, can grow more than 110m (420ft) tall; the bristlecone pine can live for over 5,000 years.

trematode Any member of the class of invertebrates Trematoda, which includes nearly 6,000 species of parasitic FLATWORMS and FLUKES. Parasitic trematodes have a thick outer cuticle and one or more suckers for attaching to the tissues of the host. They are bilaterally symmetrical, and are commonly found as endoparasites or ectoparasites in all classes of vertebrates, including human beings. Length: to 10cm (4in).

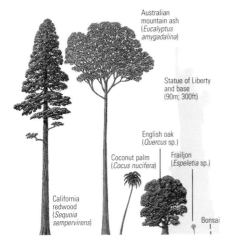

Australian mountain ash (*Eucalyptus amygadalina*)

Statue of Liberty and base (90m; 300ft)

English oak (*Quercus* sp.)

Coconut palm (*Cocus nucifera*)

Frailjon (*Espeletia* sp.)

California redwood (*Sequoia sempervirens*)

Bonsai

▲ **tree** Trees grow taller than any other living thing, but can still survive in miniature form. If the roots are restricted either artificially, as in bonsai perfected in Japan, or by natural means, as when a seed germinates in very thin soil on a mountain, a fully formed tree only a few centimetres high will result. The California redwood, the world's tallest tree, is closely rivalled by a eucalyptus, such as the mountain ash of Australia. The coconut palm reaches its height of 27m (90ft) in a few years. The English oak is one of 450 species of oak that grow as trees, bushes and shrubs. It enlarges slowly – c.4.5m (15ft) – in ten years, but produces wood of prodigious strength. Espeletia grows on snowy ledges more than 400m (1,300ft) up in the Sierra Nevada of California.

tremor Involuntary shaking of part of the body. It may be caused by fatigue, anxiety, shock, debility or any one of a range of neurological disorders, especially PARKINSON'S DISEASE.

trench Deep, V-shaped depression of the SEAFLOOR. In PLATE TECTONIC theory trenches are places where one plate is being pushed under another. They are the deepest – to 11km (7mi) – formations on Earth and are found primarily along the borders of the Pacific Ocean.

Trevithick, Richard (1771–1833) British engineer and designer of steam engines. Trevithick worked in the Cornish tin mines, where the use of steam-operated pumps to remove water led to his interest in pumps and steam engines. In 1801 Trevithick built a steam-powered road vehicle. The following year, he patented a high-pressure steam engine, his most important invention. In 1803 Trevithick built the first steam railroad LOCOMOTIVE. In 1816 he went to Peru to install his steam engines in mines. On returning to England in 1827, Trevithick found that others were profiting from his engine, and he died in poverty.

triangle Plane figure bounded by three straight lines. The sum of the interior angles totals 180°. The area is measured by either (1) half the product of one of the sides and the perpendicular onto it from the opposite vertex ($\frac{1}{2}$ × base × height), or (2) half the product of two of the sides and the sine of the angle between them.

triangle of vectors Triangle, the sides of which represent the magnitude and direction of three VECTORS about a point that are in the same plane and are in equilibrium. A triangle of vectors is commonly used to represent forces or velocities. If the magnitude and direction of two forces (or velocities) are known, then two sides of the triangle can be drawn. Using scale drawing or trigonometry, the magnitude and direction of the third force (or velocity) can be calculated. *See also* PARALLELOGRAM OF VECTORS

triangulation In navigation and land surveying, a method of determining distance. The area under survey is divided into triangles. The baseline of one triangle is measured and the angles that it makes to its vertex are measured with a THEODOLITE. The distances from each end of the baseline to the vertex can be calculated using trigonometry.

Triangulum (Triangle) Inconspicuous N constellation situated between Andromeda and Aries. Its brightest star, Beta (magnitude 3.0), forms a well-marked triangle with Alpha (3.4) and Gamma (4.0).

The main feature of interest is the spiral galaxy M33, 2.7 million light years away in our Local Group.

Triangulum Australe (Southern Triangle) Southern circumpolar constellation situated s of Norma. The three brightest stars, Alpha (magnitude 1.9), Beta and Gamma (both 2.9), form a conspicuous triangular figure. The open cluster NGC 6025 is visible through binoculars.

Triassic First period of the MESOZOIC era, lasting from 248 to 213 million years ago. Following a wave of extinctions at the close of the PERMIAN period, many new kinds of animals developed. On land lived the first DINOSAURS. MAMMAL-like reptiles were common and by the end of the period the first true mammals existed. In the seas lived the first ichthyosaurs, placodonts and nothosaurs. The first frogs, turtles, crocodilians, and lizards also appeared. Plant life consisted mainly of primitive GYMNOSPERMS, with ferns and conifers predominating.

tricarboxylic acid cycle (TCA cycle) *See* KREBS CYCLE

trichloromethane (chloroform, $CHCl_3$) Colorless, volatile, sweet-smelling liquid prepared by the chlorination of methane. Formerly a major anesthetic and then always called chloroform, it is used in the manufacture of fluorocarbons, in cough medicines, and as a solvent. Properties: r.d. 1.48; m.p. $-63.5°C$ ($-82.3°F$); b.p. $61.2°C$ ($142.2°F$).

tricyclic compound Organic chemical that has three rings as its primary structure. Examples are ANTHRACENE and phenanthrene.

triglyceride *See* LIPID

trigonometric function Six ratios of the sides of a right-angled TRIANGLE containing a given acute ANGLE – they are the SINE, COSINE, TANGENT, COTANGENT, SECANT, and COSECANT of the angle. These functions can be extended to cover angles of any size. Thus the sine of an angle A increases from 0 to 1 as A increases from 0° to 90°, decreases to 0 as A increases from 90° to 180°, decreases further to -1 as A increases to 270°, and increases again to 0 as A increases to 360°. The trigonometric functions can also be defined when the argument (the angle) is measured in RADIANS. Many relationships exist between different trigonometric functions, including the identities $\sin^2 A + \cos^2 A = 1$; $1 + \tan^2 A = \sec^2 A$; and $\tan A = \sin A / \cos A$.

trigonometric parallax Means of determining the distance of a star using the principle of TRIANGULATION. As the Earth revolves around the Sun, nearby stars show a small change in posi-

tion with respect to more distant stars. This change in position can be measured from photographs taken six months apart, that is when the Earth is on opposite sides of the Sun. The angular displacement is a direct measure of the star's PARALLAX (π). Since the baseline (the distance from the Sun to the Earth) is known, the distance d of the star can be determined by TRIGONOMETRY. The first star to be measured by this method was 61 Cygni, by Friedrich BESSEL in 1838. Parallax is measured in arc seconds. The inverse of the parallax is the distance in PARSECS. Alpha Centauri is the star with the largest known parallax, $0.752''$, which corresponds to a distance of 1.3 parsecs. Typical errors in measuring parallax photographically are about $0.01''$, but by using many plates this error can be reduced to about $0.004''$. The parallax of a star at 25 parsecs is $0.04''$, so the uncertainty is around 10%. The most accurate parallax determinations have been made by the Hipparcos satellite.

trigonometry Use of ratios of the sides of a right-angled TRIANGLE to calculate lengths and angles in geometrical figures. If three sides, or two sides and the included angle, or one side and two angles of a triangle are known, then all the other sides and angles may be found.

triiodomethane (iodoform, CHI_3) Yellow, crystalline chemical compound with a characteristic odor. It is used as an antiseptic.

trilobite Any of an extinct group of ARTHROPODS found as fossils in marine deposits, ranging in age from Cambrian through Permian times. The body is mostly oval, tapering towards the rear; it was covered by a skeleton made of chitin. The name refers to the division of the body into three – a central axis and two lateral lobes. Transverse divisions show segmentation, with each segment bearing a pair of jointed limbs. Most species were bottom-crawling, shallow-water forms that ranged in size from 6mm (0.25in) to 75cm (30in).

trinitrotoluene *See* TNT

triode *See* ELECTRON TUBE

triple-alpha process Nuclear reaction in which three alpha particles (helium nuclei) are transformed into carbon with the release of energy; it is thought to be the dominant energy-producing process in RED GIANTS. It takes place after all the hydrogen of a star's core has been exhausted. The core contracts and its temperature rises to over 100 million K, when the triple-alpha reaction begins. In each reaction two helium nuclei combine to form a beryllium nucleus, which, in turn, captures a further helium nucleus to form a carbon nucleus. Additional helium captures can produce oxygen, neon, and a number of heavier elements. The process is also known as the Salpeter process after the US physicist Edwin Salpeter (1924–), who described it.

triple point Temperature and pressure at which all three PHASES (solid, liquid, and gas) of a substance can coexist. For water, the triple point occurs at 273.16K and 610Pa.

trisomic Describing a DIPLOID organism that has three of one type of chromosome instead of two. Trisomy is the cause of several human genetic disorders, notably DOWN'S SYNDROME.

tritium (symbol T) Radioactive isotope of hydrogen, the nucleus of which consists of one proton and two neutrons. Only one atom in 10^{17} of natural hydrogen is tritium. Tritium compounds are used in radioactive tracing. Properties: half-life 12.3 yr.

triton Nucleus of a TRITIUM atom, consisting of one proton and two neutrons.

Triton Largest of Neptune's eight satellites, discovered in 1846. It has an inclined orbit in the opposite direction to Neptune's own spin (retrograde), and was almost certainly a separate planet "captured" by Neptune. Its diameter is 2,705km (1,680mi), and it probably comprises three parts rock to one part water ice. Triton is the coldest world so far visited by probes: Voyager 2's instruments measured a surface temperature of 35K (–238°C). There is evidence of recent volcanic activity and in the s polar region eruptive plumes send fine dark material rising to altitudes of 8km (5mi).

trophic level Position that an organism occupies in a FOOD CHAIN. It is usually defined in terms of the food supply. The first trophic level is occupied by PRIMARY PRODUCERS, which are green plants that use photosynthesis to obtain energy from sunlight. Herbivores (PRIMARY CONSUMERS), which eat the plants, occupy the second level. SECONDARY CONSUMERS, on the third trophic level, are carnivores, which eat the herbivores. There may be a fourth level (TERTIARY CONSUMERS) occupied by other carnivores that eat those on the third level.

tropics See CANCER, TROPIC OF; CAPRICORN, TROPIC OF

tropism (tropic response) Response in growth and orientation of a plant or a part of it in relation to a directional, external stimulus. Such stimuli include light (PHOTOTROPISM), gravity (GEOTROPISM) or water (HYDROTROPISM).

tropomyosin Filamentous protein associated with ACTIN in MUSCLE fibers. When the muscle is at rest, tropomyosin separates actin from MYOSIN. When the muscle contracts, TROPONIN displaces tropomyosin, and the actin and myosin come into contact.

troponin Protein involved in MUSCLE contraction. It consists of three polypeptide chains. When a muscle contracts, troponin binds with calcium to displace the TROPOMYOSIN, thus allowing ACTIN and MYOSIN to interact.

troposphere Lowest part of the Earth's ATMOSPHERE. In the troposphere, temperature decreases with height, and it is within this region that clouds and other weather phenomena occur. The troposphere extends to about 16km (10mi) above the Earth's surface.

trough In meteorology, area of low atmospheric pressure, usually an extension to a DEPRESSION. The opposite are ridges of high pressure.

truss In engineering and building, structural member made up of straight pieces of metal or timber formed into a series of triangles that lie in a vertical plane. The triangles resist distortion through stress, making the truss capable of sustaining great loads over long spans. The system was probably first used almost 5,000 years ago in lake dwellings.

trypsin Digestive enzyme secreted by the PANCREAS. It is secreted in an inactive form (so it does not damage tissues en route to the intestine), which is converted into active trypsin by an enzyme in the small intestine. It breaks down peptide bonds on the amino acids lysine and arginine. *See also* ALIMENTARY CANAL; DIGESTION; DIGESTIVE SYSTEM

tryptophan AMINO ACID ($C_{13}H_{10}O_2N_2$), first isolated in 1902, which is necessary for the synthesis of the antipellagra vitamin NICOTINIC ACID. Tryptophan itself cannot be synthesized by animals and is therefore an essential amino acid that must be supplied in the diet. It is found in most natural proteins, but some cereals, notably corn and sorghum, contain particularly low levels.

TSH Abbreviation of THYROID STIMULATING HORMONE

Tsiolkovsky, Konstantin Eduardovich (1857–1935) Russian scientist who provided the theoretical basis for space travel. In 1898 he became the first person to stress the importance of liquid propellants in ROCKETS. He also proposed the idea of using multistage rockets to overcome the effect of GRAVITATION, and he described the principles of SPACE STATIONS.

tsunami WAVE caused by a submarine EARTHQUAKE, subsidence, or volcanic eruption. Tsunamis spread radially from their source in ever-widening circles. In midocean they are shallow, 30–60cm (1–2ft) high, and are rarely detected. In shallower water they build up in force and height, occasionally reaching 30m (100ft), crashing on shore and causing enormous damage. Tsunami is a Japanese word; such waves are also called seismic sea waves, but tidal wave is an erroneous term for them, as they are never the result of the activity of the TIDES.

tuber In plants, short, swollen, sometimes edible underground stem (modified for the storage of food as in the potato) or swollen root (such as the dahlia). In stem tubers new plants develop from the buds, or eyes, growing in the axils of the scale leaves. Tubers are propagated by sections containing at least one eye. In the wild, they enable the plant to survive an adverse season (winter or dry season), providing food for the development of new shoots and roots later.

Tubulidentata Order of insectivorous mammals, found in central and S Africa. The aardvark is the only member of the order. A nocturnal, bristly-haired animal, the aardvark feeds on termites and ants, which it scoops up with its sticky, 30cm- (12in-) long tongue. Length: up to 1.5m (5ft); weight: up to 70kg (155lb).

Tucana (Toucan) Constellation of the S sky. Notable for the presence of the small MAGELLANIC CLOUD and the superb globular cluster 47 Tucanae (NGC 104). It lies 15,000 light years away.

tufa SEDIMENTARY ROCK composed mainly of CALCITE (calcium carbonate, $CaCO_3$). It is a porous, nonbedded rock, similar to but less dense than TRAVERTINE. It forms when dissolved calcium carbonate precipitates from lime-rich water in limestone caves. It may rapidly encrust small growing plants or it may contain embedded sediment grains and pebbles. The presence of iron oxides gives it a yellow or red color.

tuff SEDIMENTARY ROCK made up of particles of IGNEOUS ROCK from volcanic eruptions. The particles vary in size from fine to coarse but are generally smaller than 4mm (0.16in). They may be either stratified or heterogeneous in their arrangement. *See also* PYROCLASTIC

Tull, Jethro (1674–1741) British agriculturalist. He influenced agricultural methods through his innovations and writings. He invented a mechanical drill for sowing in 1701 and advocated the use of manure and thorough tilling during the growing period. He is considered a pioneer of the AGRICULTURAL REVOLUTION.

tumor Any uncontrolled, abnormal proliferation of cells, often leading to the formation of a lump. Tumors are classified as either benign (noncancerous) or malignant (*see* CANCER).

tundra Treeless, level or gently undulating plain characteristic of Arctic and sub-Arctic regions. The main swath of tundra occurs in the Northern Hemisphere, running across N North America and Eurasia. Summer days are quite long and often sunny, but average temperatures for July do not rise above 10°C (50°F). During winter, temperatures can fall to −30°C (−86°F). For six to nine months the average temperature is below freezing point, and there is a thick layer of PERMAFROST, of which only the top few centimeters thaw out in the summer. Where melting occurs, the land is wet and marshy, an ideal breeding ground for mosquitoes. The wet tundra soils are **gleys**, and waterlogging is characteristic. Generally, trees cannot grow in the tundra because of the unfavorable conditions. Mosses and lichens are common, and there are many small flowering plants in the summer.

tungsten (wolfram, symbol W) Silver-gray TRANSITION ELEMENTS, first isolated in 1783. Its chief ores are WOLFRAMITE and SCHEELITE. Tungsten has the highest melting point of all metals and is used for lamp filaments and in special alloys. TUNGSTEN CARBIDE is used in high-speed cutting tools. The sulfide is used as a lubricant. Chemically tungsten is fairly unreactive; it oxidizes only at high temperatures. Properties: at.no. 74; r.a.m. 183.85; r.d. 19.3; m.p. 3,410°C (6,170°F); b.p. 5,660°C (10,220°F); most common isotope ^{184}W (30.64%).

tungsten carbide Extremely hard, inert, gray powder used for making ABRASIVES, dyes, drill tips, and armor-piercing shells. Tungsten carbide (WC) is made by heating tungsten and lampblack (powdered carbon) in an electric furnace. Properties: r.d. 15.6; m.p. 2,780°C (5,036°F).

tunnel effect Effect in which electrons cross a potential barrier that, according to the rules of classical physics, is impossible to cross. Using QUANTUM MECHANICS, however, it can be shown that there is a probability that the electrons will cross the barrier. The effect is utilized in the **tunnel diode** (*see* ESAKI, LEO).

tunnelling Construction of passages under the ground or underwater. Tunnels in hard rock are usually made by blasting, while boring machines cut tunnels through softer material. Tunnels in soft earth have to be shored up to prevent collapse. During construction, a cylindrical steel shield is pushed through the earth, immediately followed by a permanent cast iron or concrete lining. Flooding is a danger when constructing river and sea tunnels. Usually, the air pressure inside is kept higher than the water pressure outside. This means that if the tunnel is not quite sealed, air will escape, but no water will enter. Some underwater tunnels are formed by sinking sealed, prefabricated steel or concrete sections into the riverbed or seabed. The sections are joined together and then the structure is covered with earth.

turbellarian Any member of the class Turbellaria, comprising the PLANARIANS.

turbine Rotary device turned by a moving fluid (liquid or gas). Turbines enable the energy of moving wind, water, steam, or other fluids to do useful work. The WATER WHEEL is a simple form of turbine. In early designs, a stream of water flowed over or under a vertical wheel with blades or buckets on its circumference, making it turn. The movement was used to operate simple machinery. For example, the WATER MILL could turn a stone to grind grain to make flour. The modern form of water turbine is like a multibladed propeller and is used to generate HYDROELECTRICITY. In power stations that burn fuels to produce electricity, the energy released by the burning is harnessed by the blades of jet engine-like steam turbines. As they spin, the turbines turn GENERATORS that produce electricity. WINDMILLS are simple turbines that use wind energy to grind grain. Modern wind generators produce electricity when the wind turns their rotors. In gas turbines, hot gases from burning fuel turn turbines that can operate generators or other machinery. Jet engines use gas turbines to drive compressors, fans or propellers.

turbocharger Device that boosts the performance of an INTERNAL COMBUSTION ENGINE. A TURBINE driven by exhaust gases compresses the fuel/air mixture before it passes through the inlet valve.

turbofan engine TURBINE-engine that developed from, and is more efficient than, the TURBOJET ENGINE. It has a compressor, which forces air from the intake into the combustion chamber, and a turbine, which is driven by the exhaust gases and drives the compressor. Extra thrust is provided by large blades at the front, driven by the turbine, which push air along a bypass round the combustion chamber and into the tailpipe.

turbojet engine Aircraft engine (a type of gas TURBINE) that produces power through the reactive force of expanding gases. Air is taken in at the front through a compressor, forced into a combustion chamber, mixed with fuel, and burned, producing a rush of expanding gas that propels the aircraft in a direction opposite to that of the rapid outflow. To maintain the cycle, the expanding gas also turns a turbine which drives the air compressor. See also TURBOFAN ENGINE

turboprop engine Aircraft engine that has a PROPELLER (airscrew) driven by a gas TURBINE through gears. The turbine compresses air which is mixed with fuel and burned in a combustion chamber, producing exhaust gases that drive the turbine, compressor, and propeller. See also TURBOFAN ENGINE; TURBOJET ENGINE

turbulent flow In physics, type of fluid flow in which the movement of the particles is irregular. It is characteristic of a fluid with a high REYNOLDS NUMBER. See also LAMINAR FLOW

high-pressure turbine

intermediate-pressure turbine

low-pressure turbine

electrical generator

▲ **turbine** Turbines and generators are the components that convert the rotary motion produced by steam or water power into electricity. To obtain the maximum energy from the water or steam, there are several stages of turbines – up to five in a large power station – each using water or steam at a slightly lower pressure than the previous one. Super-heated steam at up to 600°C (1,112°F) gives up much of its energy in a high-pressure turbine. The exhaust steam from this stage is reheated and passed to an intermediate-pressure turbine and then to the low-pressure turbine. The output shaft drives a generator.

turgor pressure Hydrostatic pressure generated in cells of plants and bacteria as a result of the uptake of water by OSMOSIS. Water diffuses through the semipermeable membrane of the cell, causing the cell to swell. The increase in volume is resisted by the limited elasticity of the cell wall, which generates the pressure, just as pumping air into a bicycle tire causes the tire to become stiff and resist squeezing. It is turgor pressure that supports nonwoody plants. When water is lost, a plant's cells collapse, and it wilts.

Turing, Alan (1912–54) British mathematician and logician who formulated basic theories about COMPUTERS. In 1937 he invented the **Turing machine**, a hypothetical machine that could modify a set of input instructions. It was the forerunner of the modern computer. Turing also used the concept of a computer to give an alternative and simpler proof of GÖDEL's incompleteness theorem. Turing played a major role in deciphering the Enigma code – a complex method of encryption used by Germany during World War II. In 1948 he helped to construct one of the world's first computers. In 1950 he devised the **Turing test**: a proposed test of a computer's ability to think. The test is essentially that a human would not be able to distinguish between the dialogue of the computer and that of another human. This work paved the way for the foundation of ARTIFICIAL INTELLIGENCE (AI). Turing also worked in the area of theoretical biology. In *The Chemical Basis of Morphogenesis* (1952) he proposed a model that described the origin of pattern formation in biology. Such models have since been shown to explain many patterns seen in nature. Turing committed suicide after being prosecuted for homosexuality.

Turner's syndrome Hereditary condition in females, in which there is only one X chromosome instead of two. It results in infertility (due to lack of ovaries), short stature, and various developmental defects, including the lack of secondary sexual characteristics.

turpentine (gum turpentine) Sticky liquid obtained from coniferous trees; it contains ROSIN and a volatile oil. Oil of turpentine, often called merely turpentine, is obtained by DISTILLATION of the gum and is used as a paint thinner, solvent, and in varnishes and lacquers.

turquoise Blue mineral, hydrated basic copper aluminum phosphate, found in aluminum-rich rocks in deserts. Its crystal system is triclinic and it occurs as tiny crystals and dense masses. Its color ranges from sky-blue and blue-green to a greenish gray, and it is a popular gemstone. Turquoise

▲ **turbocharger** The diesel turbocharger uses the energy of the exhaust gases to force air into the cylinder via linked impellers (1). First air is sucked into the engine (A). The air, as it is compressed into a space 22 times smaller (B), is heated. Diesel fuel is then injected in a swirling manner to maximize mixing, accentuated by the shape of the piston head (C). Combustion occurs spontaneously, without the need of a spark plug, pushing the piston down (D). The rotation of the crankshaft pushes the piston up, expelling the waste gases (E) that turn one of the linked impellors.

occurs as a secondary mineral in veins in association with aluminum IGNEOUS or SEDIMENTARY rocks that have undergone considerable alteration. Hardness 6; r.d. 2.7.

twins Two offspring that are born within a short time (usually minutes) of each other. In human beings, it is the most usual form of MULTIPLE BIRTH. **Identical twins** are those that develop from the same OVUM; **fraternal twins** are those that develop from two separately fertilized ova.

two-stroke engine Engine in which the operation of each piston is in two stages. This kind of operation is called the **two-stroke cycle**. Many smaller gasoline engines use the two-stroke principle. In the two-stroke cycle, a piston moves up a cylinder to compress a fuel-air mixture in the top. At the same time, more of the mixture is sucked in below the piston through ports in the cylinder. A spark ignites the compressed mixture, causing an explosion. This sends the piston back down the cylinder. The piston forces the fresh fuel-air mixture out from beneath it and along a transfer port leading to the top part of the cylinder. The mixture forces the exhaust gases out from the top of the cylinder. The process then repeats. *See also* FOUR-STROKE ENGINE

tympanic membrane (tympanum) Medical term for the eardrum. It is a thin membrane that in mammals closes the inner end of the EAR canal (EUSTACHIAN TUBE), separating the outer ear from the middle ear. When sound waves vibrate the tympanic membrane, the vibrations are picked up in the middle ear by three small bones called OSSICLES, which pass them on to the COCHLEA in the inner ear. In frogs and other amphibians, the tympanic membrane is on the surface of the skin at the side of the skull.

Tyndall, John (1820–1893) Irish physicist who correctly suggested that the blue color of the sky is due to the SCATTERING of light by particles of dust and other colloidal particles. He also studied the magnetic properties of crystals, and the scattering of light by colloidal particles in solution (now known as the **Tyndall effect**). By 1881 he had helped to disprove the theory of spontaneous generation by showing that food does not decay in germ-free air.

type metal Alloy of lead, containing from 2.5–12% of tin and 2.5–25% of antimony, used for printing type. Both tin and antimony lower the melting temperature, which is usually between 240° and 250°C (460–475°F). Antimony also hardens the alloy and enhances its molding properties, reducing shrinkage during solidification.

typesetting Part of the PRINTING process in which metal or other type is set, or composed, into lines, paragraphs, and pages. In early printing, wooden or metal type was set by hand. In hot-metal type processes, the LINOTYPE machine casts

a complete line of molten type metal into a mould, and the MONOTYPE machine casts individual letters and spaces and arranges these in lines. Most type is now set using photographic or computerized processes. The text is put on computer and printed on sheet film. The film is used in making printing plates.

typhoid fever Acute, sometimes epidemic, communicable disease of the digestive system. Caused by *Salmonella typhi*, which is transmitted in contaminated water or food, typhoid fever is characterized by bleeding from the bowel and enlargment of the spleen. Treatment is with antibiotics. The TAB vaccine protects against typhoid fever and the related but milder condition known as paratyphoid fever.

typhoon Name given in the NW Pacific Ocean to a HURRICANE, a violent tropical cyclonic storm.

typhus Any of a group of infectious diseases caused by RICKETTSIAE and spread by PARASITES of the human body such as lice, fleas, ticks, and mites. Epidemic typhus, the result of infection by *Rickettsia prowazekii*, is the most serious manifestation, affecting the brain, heart, lungs, and kidneys. Associated with dirty, overcrowded conditions, it is mainly seen during times of war or famine.

Tyrian purple Natural dye obtained from the shell of a small, purple, marine mollusk, *Murex brandaris*, which must be crushed and exposed to light. Originally made from Mediterranean rock whelks by the Phoenicians, it takes its name from the ancient city of Tyre, where it was reputedly discovered.

tyrosine ($OHC_6H_4CH_2CH(NH_2)COOH$) Crystalline AMINO ACID found in proteins. It was isolated in 1849 from CASEIN.

► **two-stroke engine**
In a two-stroke engine cycle the upstroke (A) opens the inlet port (1) and compresses the mixture already in the cylinder. A spark plug ignites the fuel (B). After combustion the descending piston (C) first opens the exhaust port (2), and then the transfer port (3) letting in fresh fuel-air mixture. In this way the piston

also acts as the engine's valves, and oil is added to

the fuel to lubricate it. In many two-stroke engines

the cylinder is air-cooled via fins attached to it.

UFO Abbreviation of UNIDENTIFIED FLYING OBJECT

UHF Abbreviation of ULTRA HIGH FREQUENCY

ulna Long bone of the inner side of the forearm. At its upper end it articulates with the HUMERUS and with the RADIUS, the slightly shorter bone of the outer side of the forearm. At its other end it articulates only with the radius, not directly with the carpal bones of the wrist.

ultrabasic Igneous rock containing very little QUARTZ or FELDSPAR. Ultrabasic rocks are usually PLUTONIC in origin.

ultrafiltration Separation of small particles from a suspension or colloidal solution using filtration under pressure. Small molecules, ions, and water are forced across a semipermeable membrane or colloidal filter against a concentration gradient. The larger molecules cannot cross the membrane.

ultra high frequency (UHF) Radio waves in the frequency band 300–3,000MHz, with wavelengths of about 0.1–1m (0.3–3ft). UHF waves have a range of about 80km (50mi) and are used for TELEVISION broadcasting and tracking spacecraft.

ultrasonics Study of sound waves with frequencies beyond the upper limit of human hearing (above 20,000Hz). In medicine, ultrasonics are used to locate a tumor, to scan a pregnant woman's abdomen in order to produce a picture of the fetus, and to treat certain neurological disorders. Other applications of ultrasonics include the agitation of liquids to form emulsions, detection of flaws in metals (the ultrasonic wave passed through a metal is reflected by a hairline crack), cleaning small objects by vibrating them ultrasonically in a solvent, echo sounding in deep water (*see* ECHO SOUNDER) and soldering aluminum.

ultraviolet astronomy *See* ASTRONOMY

ultraviolet lamp Lamp that emits radiation of wavelength 220–400nm (nanometers). It is used as a germicidal lamp, for fluorescent effects, and for producing an artificial suntan. It uses a mercury vapor discharge lamp, the walls of which are made of quartz, which is much more transparent to ultraviolet light than is glass.

ultraviolet radiation Type of ELECTROMAGNETIC RADIATION of shorter wavelength and higher frequency than visible light. Typical wavelengths range from roughly 4–400nm (nanometers), and those of visible light range from 400nm (violet) to 700nm (red). Ultraviolet radiation affects living matter in a number of important ways: it kills bacteria and many other parasites (and so is used medically to sterilize equipment); it tans the skin by stimulating the formation of the pigment MELANIN; and it helps to make VITAMIN D in the body, which plays a part in the prevention of rickets. Ultraviolet light also causes certain materials to fluoresce (emit visible light),

▲ **ultrasonics** Ultrasonic, or ultrasound, scanners send out beams of sound (1) and read the returning echoes to build up a picture of structures below the surface (2). They are commonly used to view fetuses in the womb with the picture displayed on a real-time monitor (3). The sound waves used are above the range of human hearing and are created by piezoelectric crystals in the handset (4). These crystals change shape when the voltage is switched off. An ultrasonic scanner uses an oscillating voltage to make the crystals vibrate, producing sound of the right frequency. The process is reversed when an echo hits the transducer and the sound wave is converted back into a voltage, which the monitor reads to build up an image. Each handset has dozens of the piezoelectric crystals at its face and by altering the timing of the oscillations of the voltage, the beam can be steered. When the voltage hits all the crystals at the same time a flat beam is issued (5). By sending the voltage to one end of the row of crystals before the other the beam can be steered (6). The beam of sound can also be focused by hitting the end crystals before the ones on the center (7).

notably the brighteners added to some detergents. Sunlight contains ultraviolet rays, most of which are filtered by the OZONE LAYER. If the ozone layer is weakened, enough ultraviolet can reach the ground to harm living things. Excessive exposure to sunlight can cause sunburn and skin cancer.

Ulysses European Space Agency probe launched in October 1990 to study the polar regions of the Sun. It was thrown out of the ecliptic by a flyby of Jupiter in February 1992 and looped over the south pole of the Sun in September 1994, moving over its north pole a year later.

umbilical cord Long, thick cord that connects a developing FETUS with the PLACENTA. The umbilical cord contains two large arteries and one vein. At birth, the cord is clamped and cut from the placenta; the part of the cord remaining on the baby's abdomen dries and falls off, leaving the scar known as the navel.

uncertainty principle In physics, law that states that an object does not have a measurable position and momentum at the same time, because the act of measuring disturbs the system. The product of these two uncertainties is always greater than PLANCK'S CONSTANT. This principle was first stated by Werner HEISENBERG.

uncommitted logic array (ULA) In computing, a CIRCUIT consisting of logic gates connected in all possible ways. It consists of a single chip (INTEGRATED CIRCUIT) in which all the elements are identical. To make a circuit for a particular function, the unwanted connections are burned away. Alternatively, the chip may merely have all the elements in place and the required connections are then added to complete the required circuit.

unconformity In geology, break in the time sequence of rocks layered one above the other. The time gap may be caused by interruptions in the deposition of sediment, ancient erosion, earth movements, or other activity. An angular unconformity occurs when successive strata (*see* STRATUM) dip at different angles.

ungulate MAMMAL with hoofed feet. Most ungulates, including cattle, sheep, pigs, and deer, are members of the order ARTIODACTYLA (with an even number of toes). The order PERISSODACTYLA (ungulates with an odd number of toes) consists of horses, tapirs, and rhinoceroses. The orders PROBOSCIDEA and Hyracoidea, collectively known as subungulates, contain elephants and hyraxes.

unidentified flying object (UFO) Any flying object that cannot readily be explained as either a human-made craft or a natural phenomenon. Reports of UFOs have been documented since ancient times. With the development of aeronautics and astronautics the number of sightings has increased enormously. The vast majority of supposed UFO sightings have various rational explanations, including clouds, spy planes, optical floaters in the observer's own eye, weather balloons, and artificial satellites.

unified field theory Attempt to extend the general theory of RELATIVITY to give a simultaneous representation of both gravitational and electromagnetic fields. A more comprehensive theory would also include the strong and weak nuclear forces. Although success has been achieved in unifying the electromagnetic and weak nuclear forces, the general problem is still unsolved. *See also* GRAND UNIFIED THEORY (GUT)

unit of measurement One of many kinds of weights or measurements agreed upon as a standard. Commonly used units include pounds, feet, seconds, meters, miles, and grams. Today most scientists use SI units.

units *See* WEIGHTS AND MEASURES

universal joint Connection between two shafts that allows them to rotate together even when they are at an angle to each other.

universal set In mathematics, a SET containing all the elements with a certain property. Also the name given to a hypothetical set that could include everything. Such an all-encompassing set contradicts the basic notion of a set.

universal time System of time reckoning based on the mean solar day, the average interval between two successive transits of the Sun across the Greenwich (0°) meridian.

Universe Aggregate of all matter, energy, and space, consisting of vast cold, empty regions with a distribution of high-temperature stars and other objects grouped in GALAXIES. On a large scale the universe is considered uniform: it is identical in every part. It is believed to be expanding at a uniform rate, the galaxies all receding from each other. The origin, evolution and future characteristics of the universe are considered in several cosmological theories. Recent developments in astronomy imply a finite universe, as postulated in the BIG BANG theory. *See also* COSMOLOGY; STEADY-STATE THEORY

unknown Variable with values that are to be found by solving a given EQUATION or set of equations.

unnilquadium *See* DUBNIUM

unsaturated compound In organic chemistry, compound in the molecule of which two or more carbon atoms are linked, or bonded together, with double or triple bonds. Simple examples are ethene (ethylene) and acetylene (ethyne). By breakage of their double or triple bonds, unsaturated compounds can add on more atoms to their molecules – and for this reason are chemically reactive. *See also* SATURATED COMPOUND

upper atmosphere Region of the atmosphere, extending upwards from about 50km (30mi), which is free of disturbances caused by the WEATHER. It includes the MESOSPHERE, THERMOSPHERE, and IONOSPHERE. At this altitude the air is rarefied, with temperatures ranging from −110°C (−170°F) in the lower regions to 250–1,500°C (500–2,700°F) higher up. The behavior of the upper atmosphere is greatly affected by such extraterrestrial phenomena as solar and COSMIC RADIATION, which cause the gas molecules to produce the ionosphere, and atmospheric tides, which cause turbulence.

upwelling Process that brings water of greater density and lower temperature up to the surface of the ocean. It is especially characteristic of the w side of continents where winds blow parallel to the coast and the water carried away by the surface current is replaced by bottom water.

uracil One of the four bases in the NUCLEOTIDES that make up RNA (ribonucleic acid). A PYRIMIDINE derivative, uracil accounts for the major difference between RNA and DNA: in DNA uracil is replaced by thymine (the other three bases, ADENINE, CYTOSINE, and GUANINE, are common to both RNA and DNA). In forming the double-stranded molecule of RNA, uracil pairs with adenine.

uraninite (pitchblende) Dense radioactive mineral form of uranium oxide, UO_2, the chief ore of uranium and the most important source for uranium and radium. The blackish and lustrous ore occurs in several varieties: crystallized varieties are called uraninite, massive varieties are called pitchblende. They usually contain small amounts of radium, lead, thorium, polonium, and sometimes the gases helium and argon. Both varieties are found around the world in veins (often quartz and silver), especially in the US, Canada, Britain, Europe, Australia, Congo, and South Africa. Hardness 5–6; r.d. 6.5–8.5 (pitchblende), 8–10 (uraninite)

uranium (symbol U) Radioactive metallic element, one of the ACTINIDE SERIES. It was discovered in 1789 by Martin KLAPROTH, and it is now

important because of its use in nuclear reactors and bombs. Soft and silvery-white, uranium is the most abundant radioactive element in the Earth's crust, the chief ores being URANINITE, CARNOTITE, autunite, and TORBERNITE. The ISOTOPE ^{238}U makes up more than 99% of natural uranium. Chemically uranium is a reactive metal: it oxidizes in air and reacts with cold water. It is the heaviest of the naturally occurring elements and is malleable and ductile. ^{235}U is fissionable and will sustain a neutron chain reaction as a fuel for reactors. ^{238}U is converted in certain nuclear reactors into the plutonium isotope ^{239}Pu, a nuclear fuel. ^{235}U and ^{239}Pu are needed for atom bombs, while ^{238}U and ^{235}U are also used to date rocks and determine the age of the Earth. Uranium is used to synthesize the TRANSURANIC ELEMENTS (atomic numbers greater than uranium). Properties: at.no. 92; r.a.m. 238.029; r.d. 19.05; m.p. 1,132°C (2,070°F); b.p. 3,818°C (6,904°F); most stable isotope ^{238}U (half-life 4.51×10^9 yr).

uranium oxide One of a series of compounds of which UO_2, U_4O_9, U_3O_8, and UO_3 are the most common, and of which U_3O_8 is the most stable. The latter is green, brown or black with an orthorhombic crystalline structure. URANINITE and PITCHBLENDE (containing UO_2 and UO_3) occur naturally and are used as a source of uranium.

uranium-thorium-lead dating Method of assessing geological age up to many millions of years. The proportions of URANIUM and its radioactive decay products, THORIUM and LEAD, show how long it is since the rock was laid down. *See also* DATING, RADIOACTIVE

Uranus Seventh planet from the Sun, discovered (1781) by Sir William HERSCHEL. Uranus is visible to the naked eye under good conditions. Through a telescope it appears as a small, featureless, greenish blue disk. In size, mass, atmosphere and color, Uranus resembles NEPTUNE. Like all the giant planets, it possesses a ring system and a retinue of SATELLITES. Uranus was the first planet to be discovered since ancient times. Although it had been observed on several occasions (and had once been catalogued as a star), its nonstellar nature was first recognized by Herschel. Like Pluto, Uranus's axis of rotation is steeply inclined, and its poles spend 42 years in sunlight followed by 42 years in darkness. Highly exaggerated seasonal variations are therefore experienced by both the planet and its satellites. The flyby of the Voyager 2 probe in 1986 provided most of our current knowledge of the planet. The upper atmosphere is about 83% molecular hydrogen and 15% helium, with the other 2% mostly methane, whose strong absorption of red light gives Uranus its predominant hue. Other images obtained from the ground and from the Hubble Space Telescope have since revealed dark spots and bright clouds similar to those visible when Voyager flew past Neptune. The five largest satellites were known before the Voyager encounter, which led to the discovery of ten more. All 15 are regular satellites orbiting in or close to Uranus's equatorial plane. They are all darkish bodies composed of ice and rock. Two further moons were discovered by powerful telescopes in 1997. The main components of Uranus's ring system were discovered in 1977 and others were imaged by Voyager. The brightest and outermost is the Epsilon Ring; the innermost is a very diffuse sheet of material. Between the two are nine narrow, darkish rings. **Data:** Diameter (equatorial) : 51,118km; Mass: (Earth=1) 14.53; Volume: (Earth=1) 62.18; Density (water=1): 1.27; Orbital period: 84.01 years; Rotation period (equatorial): 17h 14m; Temperature at cloud tops: 55K

urea Organic compound $(CO(NH_2)_2)$, a white crystalline solid excreted in URINE. Most vertebrates excrete most of their nitrogen wastes as urea; human urine contains about 25 grams of urea to a litre. Because it is so high in nitrogen, urea is a good fertilizer. It is also used to make urea-formaldehyde resins and barbiturates.

ureter In vertebrates, the long, narrow duct that connects the KIDNEY to the urinary BLADDER. The ureters transport URINE from the kidneys to the bladder, where it is stored until it is discharged by way of the URETHRA.

urethra Duct through which URINE is discharged from the bladder in mammals. Urine is produced in the KIDNEY, stored in the bladder until pressure in the bladder triggers specific neural responses that cause urine, under voluntary control, to be released through the urethra. In males the urethra is also the tube through which SEMEN is ejaculated.

Urey, Harold Clayton (1893–1981) US chemist. He became professor at Columbia University and in the same year (1934) was awarded the Nobel Prize for chemistry for his isolation of deuterium. Thereafter he worked on isotope separation at the universities of Chicago and California. Concerned that his work on isotopes had aided the construction of nuclear weapons, he later turned to geophysics.

uric acid Insoluble end product of PROTEIN metabolism. It is the main excretory substance in birds and reptiles, but only small quantities of it are normally produced by humans (who, like most mammals, mainly excrete soluble urea). An excessive buildup of uric acid in human blood causes gout, the deposition of urate salts around the joints.

urine Fluid filtered out from the bloodstream by the KIDNEY. It consists mainly of water, salts, and waste products such as UREA. From the kidneys it passes through the URETERS to the BLADDER for discharging by way of the URETHRA. In DIABETES, urine contains glucose, and to diagnose the disease urine is tested for glucose.

urogenital system Organs comprising the body's urinary and reproductive systems. The urinary system, which provides an important route of excretion, consists of the KIDNEYS, URETERS, the BLADDER, and the URETHRA. In males, the reproductive system consists of paired TESTES located in an external pouch known as the scrotum; accessory glands; and the PENIS, the organ through which both urine and sperm are discharged to the exterior. In females, the reproductive system consists of paired OVARIES, one on each side of the pelvic cavity; FALLOPIAN TUBES, which provide a passage from the ovaries to the UTERUS, located in the pelvic cavity between the bladder and rectum; the CERVIX (lower part of the uterus); and the VAGINA. In females, the urethral and vaginal openings are separate but close to each other.

Ursa Major (Great Bear) Famous N constellation, the main pattern of which is known as the Big Dipper or the Drinking Gourd. This feature consists of seven stars. Five of the stars make up a moving CLUSTER, a group of stars that share the same motion through the galaxy.

Ursa Minor (Little Bear) Constellation that contains the north celestial pole. Its brightest star is the pole star, called Polaris or the North Star, magnitude 2.0. The constellation's seven main stars, including Polaris, form a pattern that is called the Little Dipper.

uterus (womb) Hollow muscular organ located in the pelvis of female mammals. It protects and nourishes the growing FETUS until birth. The upper part is broad and branches out on each side into the FALLOPIAN TUBES. The lower uterus narrows into the CERVIX, which leads to the VAGINA. Its muscular walls are lined with mucous membrane (ENDOMETRIUM), to which the egg attaches itself after fertilization. *See also* MENSTRUAL CYCLE

utricle Fluid-filled chamber in the mammalian inner EAR. The three SEMICIRCULAR CANALS, the organs of balance, loop out from the utricle. Sensory cells within the utricle detect changes in the direction of movement of the ear (and hence the head). They also sense how fast the fluid in the canals is moving. All the information passes along part of the auditory nerve to the brain, which processes it to provide the sense of balance. *See also* EQUILIBRIUM SENSE

vaccination Injection of a VACCINE in order to produce IMMUNITY against a disease. In many countries today children are vaccinated routinely against infectious diseases such as polio, whooping cough, measles, mumps, and German measles. *See also* IMMUNIZATION; JENNER, EDWARD

vaccine Agent used to give immunity against various diseases without producing symptoms. A vaccine consists of modified disease organisms, such as a live attenuated VIRUS (low in virulence), or a dead one (still able to induce the production of specific ANTIBODIES within the blood). The first vaccine to come into general use was for SMALLPOX, developed in the late 18th century. *See also* IMMUNE SYSTEM

vacuole Membrane-bounded cavity within the CYTOPLASM of a CELL. Depending on the organism, vacuoles contain either liquid, gas or food particles. Vacuoles perform various functions. In single-celled organisms, such as an amoeba, **contractile** vacuoles fill with water before rapidly contracting and expelling excess water or wastes. In this way vacuoles perform OSMOREGULATION and excretion. In plants, vacuoles allow individual cells to increase in size without amassing bulk that would hinder cell METABOLISM.

vacuum Area of extremely low pressure. Interstellar space is a high vacuum, with an average density of less than 1 molecule per cubic centimeter; the best artificial vacuums contain less than 100,000 molecules per cubic centimeter. Evangelista TORRICELLI is credited with developing the first man-made vacuum in a mercury BAROMETER. In 1650 the German physicist Otto von Guericke (1602–86) invented the first vacuum pump. Vacuums are used widely in scientific research and industry, such as in the vacuum packing of foodstuffs.

vacuum deposition Method of producing a thin adherent coating, generally of a metal, on plastics, ceramics, and other materials. The object to be plated is placed in a vacuum chamber that also contains a length of metal wire inside a heavy-duty coil. When a high-voltage electric current is passed through the coil, the metal vaporizes and coats the object. *See also* ELECTROPLATING

vacuum flask Container for keeping things (usually liquids) hot or cold. A vacuum flask is made with double, silvered glass walls separated by a near vacuum. This vessel, which has a close-fitting stopper, is held in an insulated metal or plastic case. It reduces the transfer of heat between the contents and the surroundings. The vacuum prevents heat transfer by CONDUCTION and CONVECTION, and the silvering on the glass minimizes heat transfer by radiation. The vacuum flask was invented (1892) by James DEWAR, and it is sometimes called a Dewar flask. It is also known as a Thermos, after the trade name of the first commercial brand.

vacuum pump Device that exhausts air from an enclosed space, which becomes a partial VACUUM. In the simplest vacuum pump, water or steam enters the chamber and it forces air out, thus creating a partial vacuum. The first vacuum pump was invented by the German physicist Otto von Guericke (1602–86).

vagina Portion of the female reproductive tract, running from the CERVIX of the UTERUS to the exterior of the body. Tubelike in shape, it receives the erect PENIS during sexual intercourse. The muscular walls of the vagina enable it to dilate massively to allow the passage of the baby during childbirth. *See also* SEXUAL REPRODUCTION

vagus nerve Tenth CRANIAL NERVE, running from the BRAIN to the abdomen. It contains motor, secretory and sensory fibers and supplies branches to the lungs, heart, stomach, and other abdominal organs.

valence (valency) Measure of the "combining power" of a particular element, equal to the number of (individual) CHEMICAL BONDS one ATOM can form. The valence of an atom is determined by the number of ELECTRONS in the outermost (valence) shell. The valence of many elements is determined by their ability to combine with hydrogen or displace it in compunds (hydrogen has a valence of 1). For example, one carbon atom combines with four hydrogen atoms to make methane (CH_4) so the valence of carbon is given as 4. *See also* ATOMIC NUMBER; COVALENT BOND; IONIC BOND; OXIDATION-REDUCTION

valine ($[CH_3]_2CH(NH_2)COOH$) White, crystalline essential AMINO ACID found in proteins.

Valium Proprietary name for diazepam, a sedative drug in the BENZODIAZEPINE group. It is used in the treatment of anxiety, muscle spasms, and EPILEPSY and may also feature as part of the premedication given to patients before surgery.

valley Elongated, gently sloping depression of the Earth's surface, commonly situated between mountains or hills. It often contains a stream or river that receives the drainage from the surrounding heights. A **U-shaped** valley was probably formed by a glacier, a **V-shaped** one by a stream. The term may also be applied to a broad, generally flat area that is drained by a river, such as the Mississippi Valley.

valve, electronic *See* ELECTRON TUBE

valve In a gasoline or diesel ENGINE, device that cyclically opens and closes the intake and exhaust parts of an engine's combustion chamber or cylinder. It consists of a disk attached to a shank and held against a seat by a spring. A rotating cam pushes on the bottom of the shank, raising the valve from its seat and permitting the fuel to flow past it.

valves In anatomy, the structures that prevent the backflow of blood in the HEART and VEINS. Heart valves separate and connect the two atria and ventricles, the right ventricle and the pulmonary artery, and the left ventricle and the aorta.

▲ **Van de graaf generator** Extremely high voltages can be produced with the Van de Graaff generator (B). If a body having an excess of positive ions is placed inside a container, the inside acquires electrons (A) and the outside an equal number of positive ions. If the charged body touches the inside, all the free electrons flow into it, thus making it neutral. The outside of the container still retains its positive ions. In the Van de Graaff generator, positive ions are sprayed from a suitable source (1) onto an endless conveyor belt which carries them inside a metal sphere. The belt connects to the inside wall through a conductor in the form of a comb (2), thus permitting an electron flow to the belt. This causes positive ions to form on the sphere's outside wall (3). The effect may be enhanced by using two generators connected as shown (C).

vanadinite (lead chlorovanadate, $Pb_5Cl(VO_4)_3$) found in the upper zone of lead ore deposits. It occurs in hexagonal system prisms and needlelike masses, resinous orange to brown. It is a commercial source of VANADIUM; hardness 2.7–3; r.d. 7.

vanadium (symbol V) Silver-white, TRANSITION ELEMENT, first discovered in 1801. A malleable and ductile metal, vanadium is found in IRON, LEAD, and URANIUM ores and in coal and petroleum. It is used in steel alloys to add strength and heat resistance. Vanadium pentoxide is an important oxidation CATALYST in the chemical industry and is used in glass, ceramics, and dyes. Properties: at.no. 23; r.a.m. 50.9414; r.d. 6.1 at 18.7°C; m.p. 1,890°C (3,434°F); b.p. 3,380 °C (6,116°F); most common isotope ^{51}V (99.76%).

Van Allen, James Alfred (1914–) US physicist. During World War II, he helped to develop the radio proximity fuse, which used radio waves to detonate an explosive projectile when it neared a target. After the war, Van Allen supervised experiments with captured German V-2 ROCKETS, leading to research with artificial SATELLITES and his discovery of the two zones of radiation encircling the Earth, known as the VAN ALLEN RADIATION BELTS. They were discovered by a detector in the first US satellite, Explorer I, which he helped launch in 1958.

Van Allen radiation belts Two doughnut-shaped regions of RADIATION trapped by Earth's MAGNETIC FIELD in the upper ATMOSPHERE, named after James VAN ALLEN, who discovered them in 1958. The belts contain particles carrying energies of 10,000 to several million electron volts. Artificial satellites are insulated against this radiation. The inner belt (of ELECTRONS and PROTONS) extends from 1,000km to 4,000km (600–2,500mi) above the Equator. The outer belt (of ELECTRONS) extends from 15,000 to 25,000km (9,000–15,000mi) above the Equator. It is thought that the particles come from SOLAR FLARES and are carried by the SOLAR WIND.

Van de Graaff generator Machine that generates high voltages by concentrating electrical charges on the outside of a hollow CONDUCTOR. The Cockcoft-Walton ACCELERATOR, built by John COCKCROFT and Ernest WALTON, produced high voltages through a bank of charged CAPACITORS placed in series. US physicist Robert Van de Graaff (1901–67) improved on this design by spraying positive or negative charges onto a continuously moving belt that carried them up to a large, hollow metal sphere where the voltage built up. In this manner, an applied voltage of 50,000 volts generated up to 1 million electron volts. Today, the Van de Graaff generator is used primarily to "inject" particles into more powerful, linear accelerators. *See also* LAWRENCE, ERNEST

Van der Waals, Johannes Diderik
(1837–1923) Dutch physicist who was awarded
the 1910 Nobel Prize for physics for his work on
gases and the gas equation that he derived. The
Van der Waals equation takes into account intermolecular attraction and repulsion, which were
ignored by the KINETIC THEORY of gases. *See also*
IDEAL GAS LAW; VAN DER WAALS FORCES

Van der Waals forces Weak forces of mutual
attraction that contribute towards cohesion between
neighboring ATOMS or MOLECULES. They are named
after Johannes VAN DER WAALS who investigated
such phenomena during the 19th century. Van der
Waals forces are caused by the distribution of electrons in neighboring atoms or molecules.

van't Hoff, Jacobus Henricus *See* HOFF,
JACOBUS HENRICUS, VAN'T

Van Vleck, John Hasbrouck (1899–1980) US
mathematician and physicist. A professor at Harvard University, he studied how electrons behave
in noncrystalline, magnetic materials. Van Vleck
was the first scientist to use QUANTUM MECHANICS
to explain the phenomenon of MAGNETISM. He later
developed the LIGAND field theory, which explains
chemical bonding in molecules, and studied free
molecules by means of their spectra. For his work
on magnetism, Van Vleck shared the 1977 Nobel
Prize for physics with the US physicist Philip
Anderson and the British physicist Nevill MOTT.

vapor Gas that is below its CRITICAL POINT and so
can be liquefied by an increase in PRESSURE. It
does not obey the IDEAL GAS LAW.

vapor pressure Pressure exerted by a vapor
when it evaporates from a liquid or solid. In the
instance where as many molecules leave to form
vapor as return (in an enclosed space), this EQUI-
LIBRIUM is termed a **saturated** vapor pressure.
When a solid is dissolved in a liquid, the vapor
pressure of the liquid is reduced by an amount proportional to the solid's RELATIVE MOLECULAR MASS.
This is important in a chemical analysis.

vaporization (volatilization) Conversion of a
liquid or solid into its vapor, such as water into
steam. When heated, some solids (such as ammonium chloride) pass directly into the vapor state, a
process known as SUBLIMATION. Vaporization of a
solid or liquid can also be achieved by lowering
the surrounding pressure or by EVAPORATION.

Varenius, Bernhardus (1622–50) German geographer. He is celebrated for his *Geographia gen-
eralis* (1650), in which he attempted to structure
geography on a scientific basis and coordinate its

various branches. Varenius developed the concept
of what is now known as regional geography.

variable In mathematics, symbol used to represent a quantity that may take any of several values
in a range. For example, in the expression $y = x^2 +
x + 1$, the quantity x may be assigned the value of
any real number; here x is said to be an **independent** variable, and y is a **dependent** variable
because its value depends on the value of x. Types
of variables include SCALAR, VECTOR, and MATRIX.

variable In statistics, item that may change or
vary – unlike a constant, that maintains a given
value. The mark given to each student in a class is
a variable: it changes from student to student.

variable star Star whose brightness varies with
time. The approximately 30,000 known variable
stars fall into three broad classes according to their
light curve: pulsating variables, eruptive variables,
and eclipsing variables. Most variable stars are
pulsating variables, which have slight instabilities
that cause the star to alternately contract and
expand. The majority of pulsating variables are
long-term variables, such as the RED GIANTS.
CEPHEID VARIABLES are unusual short-term pulsating variables. Most short-term variables are found
in STELLAR CLUSTERS. **Eruptive** variables, such as
nova and supernova, are extremely unstable stars
whose fluctuations in brightness are unpredictable.
Both pulsating and eruptive variables are **intrinsic**
variables – their variations are due to inherent
structural features. **Eclipsing** variables are extrin-
sic variables: their apparent fluctuation in brightness is due to the fact that they are BINARY STARS.

variation In biology, differences between members of the same SPECIES. Variation naturally
occurs due to HEREDITY and to differences in the
environment during development. The difference
may be in physical appearance, behavior or fertility. *See also* ADAPTATION; EVOLUTION

variegation Term used to describe parts of plants,
especially leaves, that display different colors. It
occurs naturally, mostly as the result of a lack of
the green pigment CHLOROPHYLL, the most common
effect being yellow, cream or white patches, streaks
or spots on green leaves. Variegation is exploited
for ornamental purposes, but such decorative plants
are usually weaker than the uniformly colored
plants. The variegated petals of some flowers can
often be the result of mutation or viral infection.

variscite Glassy white to green phosphate mineral, hydrous aluminum phosphate with iron impurities, found in aluminum-rich rocks near the Earth's
surface. It occurs as crystals in the orthorhombic

system. It resembles TURQUOISE, but is greener, and is used in jewelry. Hardness 3.5–4.5; r.d. 2.5.

varnish Solution of a RESIN or a PLASTIC that dries to form a hard, transparent, protective and often decorative coating. Varnishes may have a matt or glossy finish. Pigments are often added to color the varnish. Drying takes place by evaporation of the solvent or by chemical action. French polish is a varnish made by dissolving SHELLAC in alcohol (ethanol). It dries when the alcohol evaporates into the air.

varve Term applied to a layer of sediment deposited in a single year in a body of still water. Specifically a varve consists of two layers of sediment, a coarse layer deposited in the summer and a fine layer deposited in the winter in glacial meltwater lakes. Varves have been used to date the age of Pleistocene glacial deposits.

vascular bundle Strand of conductive tissue that transports water and dissolved mineral salts and nutrients throughout a vascular plant. Vascular bundles are the equivalent of the blood and lymph vessels found in animals. They extend from the roots, through the stem, and out to the leaves. They consist of two types of tissue: XYLEM, which conducts water from the roots to the shoot and is located towards the center of the bundle, and PHLOEM, which conducts salts and nutrients and forms the outer regions of the bundle. In plants that exhibit SECONDARY THICKENING, the xylem and phloem are separated by a thin layer of CAMBIUM, from which new vascular tissue is generated.

vascular plant Plant with vessels or tubes to carry water and nutrients within it. All higher plants – FERNS, CONIFERS, and FLOWERING PLANTS – have a vascular system (XYLEM and PHLOEM). This system exists in roots, stems, and leaves.

vascular system Network of vessels through which fluids circulate in the body of an animal or plant. They include BLOOD vessels (arteries, veins, and capillaries) in the blood vascular system and lymph vessels in the LYMPHATIC SYSTEM. The network of tubes that carry fluid and nutrients in the PHLOEM and XYLEM tissues of plants is also called a vascular system or VASCULAR BUNDLE.

vas deferens Long, narrow tube that carries SPERM from the TESTES, around the bladder, and opens into the seminal vesicles, where the sperm are stored until ejaculation.

vasoconstrictor Any substance that causes constriction of BLOOD VESSELS and, therefore, decreased blood flow. Examples include NORADRENALINE, ANGIOTENSIN and the hormone VASO-

PRESSIN (also known as antidiuretic hormone). Vasoconstricting drugs are used for a number of reasons, such as to raise the BLOOD PRESSURE in circulatory disorders or for treating SHOCK.

vasodilator Any substance that causes widening of the blood vessels, permitting freer flow of blood. Vasodilator drugs are mostly used to treat HYPERTENSION and ANGINA.

vasopressin HORMONE secreted by the posterior lobe of the PITUITARY GLAND situated under the brain. It is also called antidiuretic hormone or ADH, its action being to prevent excessive water loss by reabsorption of water in the KIDNEY tubules during the filtration of urine.

vector In mathematics, a quantity that has both magnitude (size) and direction, as contrasted with a SCALAR, which has the property of magnitude only. For example, the VELOCITY of an object is specified by its speed and the direction in which it is moving at any instant. Similarly, a force has both magnitude and direction, hence weight (the force of gravity on a body) is a vector. In contrast, mass and temperature are scalar quantities. A vector at a point in three-dimensional space is represented by a list of three numbers that give the direction and magnitude (the value of which depends also on the position of the point in question). Geometrically, the three values specify an arrow with direction and magnitude. A vector must have two or more entries. *See also* PARALLELOGRAM OF VECTORS

vector analysis Study of the mathematical properties of VECTORS. The addition of two vectors is the diagonal of a parallelogram formed with the two vectors as adjacent sides. Multiplication of a vector by a SCALAR a gives a vector with the same direction but its magnitude multiplied by a. Two vectors can also be multiplied, and this may occur in two ways: the scalar, or dot product of vectors a and b (written $a . b$) is a scalar given by $ab \cos\theta$, θ being the angle between the vectors and a and b being their magnitudes; the vector, or cross product (written ab or $a \times b$), is a vector given by $ab \sin \theta$, directed at right-angles to the plane of a and b.

vector In disease, any agent that carries an infectious agent from one host to another. It may be an insect such as a mosquito or flea, or an inanimate object such as a cup. It also may serve as an intermediate host for the infectious agent.

vegetative reproduction (vegetative propagation) Form of ASEXUAL REPRODUCTION in higher plants. It involves an offshoot or a piece of the original plant (from leaf, stem or root) separating and giving rise to an entire new plant. It may occur naturally, as in strawberries reproducing by runners, or artificially, as in a houseplant cutting yielding a new plant. The reproductive structure may include a food store, as in a BULB, CORM or TUBER.

vein In mammals, vessel that carries deoxygenated blood to the heart. An exception is the pulmonary vein, which carries oxygenated blood from the lungs to the left upper chamber of the heart. Veins have valves on their inner walls to prevent any backflow of blood. *See also* ARTERY; VENA CAVA

velocity Rate of motion of a body in a certain direction. Its symbol is *v*. Velocity is a VECTOR (having magnitude and direction), whereas speed, which does not specify direction, is called a SCALAR.

velocity ratio *See* DISTANCE RATIO

vena cava Main VEIN of vertebrates that supplies the HEART with deoxygenated blood, emptying into its right atrium.

veneer Extremely thin sheet of wood, or a thin sheet of a precious material such as ivory or tortoiseshell, which gives furniture or other objects the appearance of being more valuable than they are. Plain, cheaper woods or substitutes (such as plywood, chipboard or blockboard) may be completely covered in a richly colored or grained veneer to give it the appearance of a decorative hardwood. Veneers may also be used as decorative shapes inlaid into a surface.

Venera space missions Series of SPACE PROBES sent to Venus by the Soviet Union after 1965. **Venera 3**, launched in 1965, was the first spacecraft to make an impact on another planet (Venus, 1966). Veneras 9 and 10 made soft landings on Venus in 1975 and transmitted pictures and atmospheric data for an hour. *See also* SPACE EXPLORATION

Vening Meinesz, Felix Andries (1887–1966) Dutch geophysicist and geodesist who developed (1923) a method of measuring gravity in ocean basins. He is also noted for research on the effect of solar movements on the Earth's surface and his studies of convection currents within the Earth.

Venn diagram Diagrammatical representation of the relations between mathematical SETS or logical statements, named after the British logician John Venn (1834–1923). The sets are drawn as geometrical figures, usually circles, which overlap whenever different sets share some elements.

venom, snake Toxic substance produced in the poison glands of snakes and injected into the victim through ducts in or along their fangs. Many venoms

are dangerous to people, and some can be lethal unless counteracted by antiserums. The effects of snake venom vary according to the species and the constituents of the poison. Blood coagulation, respiratory effects, and hemorrhage are the commonest.

ventifact In geology, any stone that has been shaped and sculptured by windblown sand.

ventilation In biology, the process by which air or water is taken into and expelled from the body of an animal and passed over a surface across which GAS EXCHANGE takes place. Ventilation mechanisms include BREATHING, by which air is drawn into the LUNGS for gas exchange across the wall of the alveoli; the movements of the floor of a FISH's mouth, coupled with those of its GILL covers, which draw water across the gills; and the pumping movements of the ABDOMEN of some insects, which draw air through SPIRACLES that carry it to the tissues.

ventilator In medicine, a machine that helps an ill or injured person to breathe. A **life-support machine** is a positive-pressure ventilator. It works by forcing air along a tube into the patient's lungs; when the pressure is regularly released, the patient breathes out. An **iron lung** is a negative-pressure ventilator. A chamber encloses the patient's chest; as the pressure in the chamber is decreased, the chest wall expands and "sucks" air into the lungs.

▲ **vein** Veins carry blood to the heart. The returning venous blood moves slowly due to low pressure, and the veins can collapse or expand to accommodate variations in blood flow. Movement relies on the surrounding muscles, which contract (1) and compress the vein. Pulsation of adjacent arteries (2) has a regular pumping effect. Semilunar valves (3) are found at regular intervals throughout the larger veins, and these allow the blood to move only in one direction.

When the pressure is reapplied, the chest wall moves back and the patient breathes out.

ventral Describing that part of an organism that normally faces the ground. In plants, the ventral surface of a leaf is the under surface (which usually faces away from the light). In fish, the ventral fin is the fin pointing downwards. In four-legged animals, it is the lower or under surface. In humans and other bipedal animals, it is the forward-facing surface, more usually described as anterior. *See also* DORSAL.

ventricle Either of the two lower chambers of the HEART. The term is also used for the four fluid-filled cavities within the BRAIN.

Venturi tube Mechanism for mixing a fine spray of liquid with a gas or measuring fluid flow. In an automobile CARBURETOR, gasoline is passed through a jet into two narrowing sections of tube joined by a slim throat. In the throat, the gasoline mixes with air from the INTERNAL COMBUSTION ENGINE. The pressure differential produced by liquid flowing through the constriction gives a measure of its rate of flow. It is named after its inventor, Italian physicist G.B. Venturi (1746–1822).

Venus Second planet from the Sun and almost as large as the Earth. Visible around dawn or dusk as the so-called **morning star** or **evening star**, it is the most conspicuous celestial object, after the Sun and Moon. At its brightest it is even visible to the naked eye in the daytime, when the Sun is high in a clear sky. It has no known satellite. A telescope shows the planet's dazzling yellowish-white cloud cover, with faint markings. Venus' very high surface temperature was indicated by measurements at radio wavelengths in 1958, but SPACE PROBES were needed to reveal more about the surface. A gently undulating plain covers two-thirds of Venus. Highlands account for a further quarter, and depressions and chasms the remainder. Although there is a mountain range and some craters, most of the surface features are volcanic in origin and Venus is almost certainly still volcanically active. The atmosphere consists of 96% carbon dioxide and 3.5% nitrogen, with traces of helium, argon, neon, and krypton. Data: Diameter: 12,104km; Density: 5.25g/cm^3; Mass (Earth=1): 0.8149; Volume (Earth=1): 0.8568; Orbital period: 224.7 days; Rotation period: 243d 0h 30m (retrograde); Surface temperature: 750K

verdigris Green, basic COPPER carbonate that forms on objects made of copper or bronze that have been exposed to atmospheric corrosion. The word also refers to the blue-green powder (copper ACETATE or ethanoate) formed from metallic copper and acetic (ethanoic) acid. *See also* PATINA

vermiculite Clay mineral. Its flakes are light and are used in plaster and insulation, and as a packing material. The mineral is also used widely for conditioning soil and as a starting medium for seeds.

vernal equinox Spring EQUINOX in the N Hemisphere (September 23 in S Hemisphere).

Verneuil process Method for making gemstones artificially, particularly rubies and sapphires. Both of these gems are essentially forms of aluminum oxide (Al_2O_3) colored with small amounts of other metal oxides. These are fused together in an oxy-hydrogen flame to form molten droplets of the colored substance, which builds up gradually into a crystalline boule that is cut to make the gemstone.

Vernier scale On instruments for measuring length, small scale that moves alongside the main scale. Named after its French inventor Pierre Vernier (1580–1637), its function is to allow more precise measurement. On MICROMETERS the Vernier scale is on the stem, at right-angles to the spiral main scale.

vertebra One of the bones making up the SPINE (backbone) or vertebral column. Each vertebra is composed of a large, solid body from the top of which winglike projections jut to either side. It has a hollow center through which the SPINAL CORD passes. The human backbone is composed of 26 vertebrae (the five sacral and four vertebrae of the coccyx fuse together to form two solid bones), which are held together by ligaments and INTERVERTEBRAL DISCS.

vertebral column *See* SPINE

vertebrate Animal with individual disks of bone or cartilage called VERTEBRA surrounding or replacing the embryonic NOTOCHORD to form a jointed backbone that encloses the spinal column. The principal division is between the aquatic, fishlike forms and partly land-adapted forms (AMPHIBIANS), and the wholly land-adapted forms (REPTILES, BIRDS, and MAMMALS), although some mammals, such as whales, have adapted for a totally aquatic existence. Birds and mammals are the only warm-blooded vertebrates with circulatory, respiratory, and excretory systems that allow for constant high body temperatures. Phylum CHORDATA; subphylum Vertebrata.

vertex (pl. vertices) Point at which two sides of a triangle or other polygon intersect, or at which three or more sides of a pyramid or other polyhedron intersect. It is also the point of a cone.

very high frequency (VHF) Range or band of RADIO WAVES of frequencies between 30 and 300MHz and of wavelength between one and 10m (3–33ft). This band is used for TELEVISION and FRE-

QUENCY MODULATION (FM) radio broadcasts to provide high-quality reception.

Vesalius, Andreas (1514–64) Flemish anatomist. His *De Corporis Humani Fabrica*, stating the results of his study and dissection of the human body, marked the beginning of modern anatomy.

vesicle Small sac within the CYTOPLASM of a CELL, usually containing containing serous fluid. *See also* GOLGI BODY

vestigial structure In biology, an organ or limb that is deformed or degenerate in appearance and no longer has any recognizable function. Human vestigial structures include the tonsils and the appendix. *See also* DEGENERATION

vesuvianite *See* IDOCRASE

VHF Abbreviation of VERY HIGH FREQUENCY

viaduct BRIDGE (usually resting on several narrow, concrete or masonry arches) with high supporting piers, which carries roads or railways. Steel bridges made up of short spans carried on high towers are also referred to as viaducts. *See also* AQUEDUCT

video Used in TELEVISION and computing to refer to electronic vision signals, and to equipment and software associated with visual displays. The picture component of a television signal is often referred to as the video. Video (or visual) display unit (VDU) is another name for a COMPUTER screen and its associated equipment. *See also* VIDEO RECORDING

video disc Vinyl disc coated with a reflective metallic surface, on one side of which a spiral of microscopic pits is etched corresponding to digital information that can be picked up by a laser scanner and converted electronically to video pictures and sound. Commercially marketed since 1978, most video discs are 30cm (12in) in diameter and consist of recordings of feature films or music presentations. Since the late 1980s, however, they have been almost entirely superseded by the smaller, more comprehensive type of COMPACT DISC (CD) called a CD-ROM. *See also* RECORDING

videotape Magnetic tape that records TELEVISION signals. The signals are passed through the recording head (an ELECTROMAGNET) of a videotape recorder, and by INDUCTION, a magnetic pattern is produced on the tape. These patterns generate the original signals when the tape passes the replay head (also an electromagnet).

videotext General term for the different methods by which information can be brought up onto a TELE-VISION screen. Information that is transmitted by the television station in parallel with the ordinary TV signals, and that may be screened simultaneously with or independently from other channels, is known as **teletext**. The system that brings information to the screen from a computer databank via a telephone landline is called **videotex** (formerly viewdata); the telephone link provides an interactive facility that allows armchair shopping and banking.

Vieta, Franciscus (1540–1603) French mathematician. He devised methods for solving algebraic equations to the fourth degree and discovered several trigonometric results. He introduced the terms COEFFICIENT and "negative."

Vigneaud, Vincent du (1901–78) US biochemist who was awarded the 1955 Nobel Prize for chemistry for determining the structure and synthesis of VASOPRESSIN and oxytocin, two polypeptide hormones. In the 1930s he discovered the metabolic

▲ **videotape recording** A camcorder converts an image into an electrical signal, which can then be stored on magnetic tape (1). Light from an image (2) is focused by a series of lenses (3) and then split into its component colors by a prism (4). The red, green, and blue light strike separate light-sensitive chips (5) that reproduce them in electronic form. The magnetic tape is housed in a protective case (6) that opens when inserted into the camcorder. The recording head (7) is angled and records information onto the tape in diagonal bands of magnetic particles. The helical scanning allows more information to be stored on a length of tape. A microphone (8) picks up sound, which is laid down in parallel to the visual information. Camcorders have a small television screen in the eyepiece (9) that allows the operator to play back and review the pictures taken. The camcorder can also be plugged directly into the television and the images played back.

processes by which a methyl group ($-CH_3$) is moved from one compound to another, called transmethylation. In 1943 Vigneaud deduced the two-ring structure of biotin. He also studied INSULIN, PENICILLIN synthesis, and AMINO ACID metabolism.

Viking space mission US space project to investigate conditions on Mars, launched in 1975. Two spacecraft, Viking 1 and Viking 2, each attached to separate vehicles that orbited the planet, made the first successful landings on Mars in July and September 1976. They transmitted much information to Earth, including weather conditions and dramatic photographs of the surface. Samples of soil from the landers were analyzed on board, but no organic compounds or other signs of life were detected.

Villard, Paul (1860–1934) French physicist. In 1900, while studying radiation emitted from uranium, he found that some of it was more penetrating than either alpha or beta rays and, unlike these two, was not deflected by a magnetic field. This new form of radiation was similar to X rays and later came to be called GAMMA RADIATION.

villi In anatomy, small, fingerlike projections of a mucous membrane such as that which lines the inner walls of the SMALL INTESTINE or the CHORION. They serve to increase the absorptive surface area of a particular organ. In digestion, intestinal villi absorb most of the products of food broken down in the STOMACH, DUODENUM, and ILEUM. Each villus contains blood capillaries, which receive PROTEIN and CARBOHYDRATE, and a lymph vessel, which receives FAT.

▲ **virtual reality** A data glove measures the movements of the wearer's hand and allows the user to manipulate objects in virtual reality. Fiber optic cables (1) on the glove detect the flexing of the hand. Light travels up and down the cables (2). When the cables are bent (3) they no longer reflect light back to the interface board (4). A position sensor (5) detects the movement of the glove in three dimensions. Fingertip padding (6) convinces the wearer they are touching an actual object.

vinyl chloride *See* CHLOROETHENE

vinyl plastics Any of several substances composed of large molecules linked by the POLYMERIZATION of vinyl compounds. They are used in plastic wrap, upholstery, floor tiles, toys, buttons, and in fibers. The most common is POLYVINYL CHLORIDE (PVC).

viral infection Disease caused by a VIRUS. Viral infections are among the greatest health problems facing humans, since ANTIBIOTICS have achieved some measure of control over bacterial infections. Many viral infections can be treated only with specific VACCINES and antisera. Notable among viral infections affecting human beings are YELLOW FEVER, RABIES, influenza (FLU), MEASLES, and POLIOMYELITIS. Viruses are also responsible for many diseases in plants, and these are particularly difficult to control. The role of viruses in causing CANCER has been the subject of extensive research since this effect was discovered in 1908. Living cells gain some protection by their own production of INTERFERON, which leads to the formation of another protein virus inhibitor.

Virchow, Rudolf (1821–1902) German pathologist. His discovery that all CELLS arise from other cells completed the formulation of the cell theory and repudiated the theories of SPONTANEOUS GENERATION. Virchow also studied the nature of disease at a cellular level and established the science of cellular pathology.

Virgo (Virgin) Equatorial constellation on the ecliptic between Leo and Libra. It lies in a region of the sky that has many GALAXIES and GALAXY CLUSTERS. The brightest star in Virgo is the 1st-magnitude Alpha Virginis, or Spica.

virology Study of VIRUSES. The existence of viruses was established (1892) by the Russian botanist D. Ivanovski, who found that the causative agent of tobacco mosaic disease could pass through a porcelain filter impermeable to BACTERIA. Bacteriophages (viruses that infect bacteria) were discovered in 1915. The introduction of the electron microscope in the 1940s made it possible to view viruses.

Virtanen, Artturi Ilmari (1895–1973) Finnish biochemist who was awarded the 1945 Nobel Prize for chemistry for his work on fodder preservation. During the 1920s Virtanen discovered that fermentation in green fodder, which ruins silage stores, could be prevented by the addition of hydrochloric and sulfuric acids. He also found that this technique, known as the AIV method, had no detrimental effects on the nutritional qualities or edibility of the fodder.

virtual particle Particle that is not directly observable because it exists for an extremely short time, during an interaction between observable particles. Because of the Heisenberg UNCERTAINTY PRINCIPLE, it is not subject to the law of CONSERVATION of energy. Virtual PHOTONS, for example, are exchanged between charged particles when they interact electromagnetically.

virtual reality Use of COMPUTER GRAPHICS to simulate a three-dimensional environment that users can explore as if it were real. A virtual reality system can allow an architect to see what the inside of a building will look like before construction begins. The computer images are produced using the architect's drawings of the building. Some entertainment machines use virtual reality to simulate space-flight adventures and various games. In many cases, sensors in a special helmet and glove detect the player's movements and automatically control the screen display. Twin screens inside a special helmet show slightly different images, calculated to give a convincing three-dimensional effect.

virus Submicroscopic infectious organism. Viruses vary in size from 10 to 300 nanometres and contain only genetic material in the form of DNA or RNA. This is enclosed in a protein coat known as the capsid. Viruses are incapable of independent existence. They can grow and reproduce only when they enter another cell, such as a bacterium or animal cell, because they lack energy-producing and protein-synthesizing functions. When they enter a cell, the host's metabolism is subverted so that viral reproduction is favored. Pathogenesis is the result of cell death or altered metabolism as the virus multiplies. Control of viruses is difficult because harsh measures are required to kill them. The animal body has, however, evolved some protective measures, such as production of INTERFERON and of ANTIBODIES directed against specific viruses. In most cases, one attack of a viral disease confers lifelong IMMUNITY. Where the specific agent can be isolated VACCINES can be developed, but some viruses (such as those responsible for causing influenza) change so rapidly that vaccines become ineffective. There are few antiviral drugs. *See also* PATHOGEN;

virus In computing, a program that, unknown to the user, reproduces itself and can be transferred to another COMPUTER. The virus may be transferred on a floppy disk, via a modem or communicated to any other computer on a network. Viruses can corrupt data or modify programs and may even erase a memory or cause a computer to crash. Software is available to detect and counteract known viruses, but new ones are being introduced all the time. To knowingly spread a virus is a crime in some countries.

viscose *See* RAYON

viscosity Resistance of a fluid to flow because of internal friction. The more viscous the fluid, the slower it flows. Viscosity is large for liquids and extremely small for gases. In most liquids, it increases with decreasing temperature.

vision *See* SIGHT

vitamin Organic compound that is essential in small amounts to the maintenance and healthy growth of all animals. Vitamins are classified as either water-soluble (B and C) or fat-soluble (A, D, E, and K). They are usually found in a healthy DIET, but today most can be made synthetically. Some are synthesized in the body. Many vitamins act as coenzymes, helping ENZYMES in RESPIRATION and other metabolic processes such as the synthesis of cell components and the detoxification of body wastes. Lack of a particular vitamin can lead to a deficiency disease. **Vitamin E** is important in reproduction and many other biological processes. **Vitamin D** helps the body absorb phosphorus and calcium. It is essential for the normal growth of bone and teeth. Existing in human skin (activated by sunlight), vitamin D is also found in fish-liver oil, yeast, and egg yolk. Deficiency causes such bone diseases as rickets, while an excess could cause kidney damage. **Vitamin C** or ascorbic acid is commonly found in many fruits and vegetables. It helps the body resist infection and stress, and is essential to normal metabolism. A deficiency causes scurvy. Vitamin C is destroyed by heating, and a lot is lost during cooking. **Vitamin B** is actually a group of 12 vitamins, known collectively as the **vitamin B complex**, important in assisting the process by which energy is produced in the body (RESPIRATION). Vitamin B1 or THIAMINE occurs in yeast and cereals. Lack of it causes a disease called beriberi. Another B vitamin is **niacin** (nicotinic acid) found in milk, meat, and green vegetables. Lack of this vitamin causes pellagra. **Vitamin B_{12}** is needed for the formation of blood cells. It is found especially in meat, liver, and eggs. Deficiency causes anemia. **Vitamin A** (retinol) is important for healthy eyes, and shortage of the vitamin caused night-blindness. It is found in fish-liver oil, but the body can convert carotene, the red pigment in carrots, mangoes, and red peppers, into Vitamin A. *See also* RIBOFLAVIN

viticulture Worldwide cultivation of GRAPES. Grapes are grown throughout the warmer temperate regions. The main centers of viticulture are southern Europe, California, South Africa, Australia, New Zealand, and South America. Grapes are grown mostly for wine, though in some places they are also used in various dried forms, such as currants and raisins.

vitreous humor Transparent, jellylike medium that fills the eyeball between the lens and the retina. It constitutes the vitreous body, which serves to hold the retina in position and combines with the lens to ensure the clear passage of light to the receptor cells of the retina. *See also* AQUEOUS HUMOR

vitrification In ceramics, process of producing a glassy look or close crystallization by firing at a high temperature. Vitrified clay becomes non-porous and loses its plasticity.

viviparity Process or trait among animals of giving birth to live young. Placental mammals show the highest development of viviparity, in which the off-spring develops inside the body within the mother's UTERUS. The young receive nutrition via the mother's PLACENTA. Viviparity is also exhibited by certain arthropods, amphibians, fish, and reptiles. Viviparous plants are those whose seeds germinate before separating from the mother plant or those that produce plantlets rather than flowers and seeds.

vivisection Dissection of living bodies. Work with laboratory animals in testing drugs, vaccines, and pharmaceuticals frequently involves such dissections. Smaller animals such as rats, mice, and hamsters are usually killed painlessly after the survey of the test results. The ethical issue of experimenting on living animals is still a matter of controversy, because alternative methods may be available, if not as cheap or convenient to use.

voice In human beings, sound produced by air from the lungs passing through the vocal cords, causing them to vibrate. The cords are two wedge-shaped bands of fibrous elastic tissue attached to the cartilage

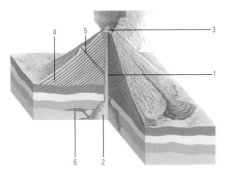

▲ **volcano** Volcanoes are formed when molten lava (1) from a magma chamber (2) in the Earth's crust forces its way to the surface (3). The classic, cone-shaped volcano is formed of alternating layers of cooled lava and cinders (4) thrown out during an eruption. Side vents (5) can occur and when offshoots of lava are trapped below the surface, laccoliths (6) are formed.

of the LARYNX (voice box), which bulges in the front of the neck as the Adam's apple. The respiratory tract above the larynx acts as a resonating chamber, lending the voice its particular timbre. The voice of a bird is produced differently by a vocal apparatus called the SYRINX, located at the base of the windpipe.

volcanic plug Hard mass of solidified LAVA in the vent of a VOLCANO. Over a period of millions of years the volcano will erode, but the hard plug will survive as a small isolated hill in the crater.

volcanism Volcanic activity. The term is a general one and includes all aspects of the process: the eruption of molten and gaseous matter, the building up of cones and mountains, and the formation of lava flows, geysers, and hot springs.

volcano Vent from which molten rock or LAVA, solid rock debris, and gases issue. The term is also applied to the pile of rock around the vent. Volcanoes may be of the **central vent type**, where the material erupts from a single pipe, or of the **fissure type**, where material is extruded along an extensive fracture, building plains, and plateaus. Volcanoes are usually classed as active, dormant, or extinct.

volt (symbol V) SI unit of electric potential and ELECTROMOTIVE FORCE (EMF), or voltage. It is the difference of potential between two points on a conducting wire carrying a current of 1 ampere when the power dissipated is 1 watt.

Volta, Count Alessandro Giuseppe Antonio Anastasio (1745–1827) Italian physicist who invented (1800) the first electric BATTERY, the voltaic pile. Volta also invented a charge-accumulating machine, which incorporated the principle of some modern electrical CONDENSERS. His investigations into ELECTRICITY led him to interpret correctly Luigi GALVANI's experiments with muscles, showing that the metal electrodes and not the tissue generated the current. The VOLT was named in his honour.

voltage Measure of the electrical POTENTIAL DIFFERENCE between two points in a circuit. Two points are at a potential difference of one volt if one coulomb of electricity (ELECTRIC CHARGE) does one joule of work in flowing between them. Voltage is also given by the RESISTANCE multiplied by the CURRENT (using OHM'S LAW).

voltameter (coulometer) Electrolytic CELL for measuring the magnitude of an ELECTRIC CHARGE by determining the total amount of decomposition on the CATHODE arising from the passage of the charge through the cell.

voltmeter Instrument for measuring the voltage

(POTENTIAL DIFFERENCE) between two points in an electric circuit. Voltmeters are always connected across (in parallel with) the components whose voltages are being measured. A voltmeter has a high internal RESISTANCE (compared with the resistance across which it is connected). Digital voltmeters are now common. *See also* AMMETER

volume Measure of the amount of space taken up by a body. Volume is measured in cubic units, for example cm^3 or cubic meters.

volumetric analysis In chemistry, method of finding the concentration of solutions. *See also* QUANTITATIVE ANALYSIS; TITRATION

voluntary muscle In human beings and other mammals, the most plentiful of the three types of MUSCLE, comprising the bulk of the body. It is also known as **skeletal** muscle or **striated** muscle because of its characteristic striped appearance under the microscope. Mostly attached to BONE, either directly or through TENDONS, it includes FLEXORS to bend joints and EXTENSORS to straighten them. *See also* INVOLUNTARY MUSCLE; SKELETAL MUSCLE

vomiting Act of bringing up the contents of the stomach by way of the mouth. Vomiting is a reflex mechanism that may be activated by any of a number of stimuli, including dizziness, pain, gastric irritation, or shock.

von Braun, Wernher (1912–77) German-born engineer, known for his expertise in the development of ROCKETS for SPACE EXPLORATION. In World War II, von Braun worked on the development of the V2 ROCKET. In 1945 he went to the US to work on MISSILE development, becoming a US citizen in 1955. Von Braun developed the Jupiter rocket that took the first US SATELLITE, Explorer 1, into space (1958). In 1960 he joined the NATIONAL AERONAUTICS AND SPACE ADMINISTRATION (NASA) and developed the Saturn rocket that carried astronauts to the Moon.

Von Frisch, Karl *See* FRISCH, KARL VON

Von Neumann, John (1903–57) US mathematician, b. Hungary. In 1930 he left Hungary to teach at Princeton University (1930–33) and then at the Institute for Advanced Studies. In 1937 he became a US citizen. In mathematics, Von Neumann helped found GAME THEORY. His early contribution to QUANTUM THEORY was followed by work on the atomic bomb at Los Alamos. Von Neumann is celebrated chiefly, however, for his development of high-speed electronic COMPUTERS. One of his first designs, the maniac (**m**athematical

analyzer **n**umerical **i**ntegrator **a**nd **c**omputer), was used to test (1952) the first hydrogen bomb.

vortex Eddy or whirlpool observed in fluid motion. Vortices cannot occur in ideal (nonviscous) fluid motion, but they are important in the study of real fluids. In particular, the vortices occurring behind airfoils are of great interest in aerodynamic design. The study of turbulence shows that energy is dissipated most efficiently in smaller vortices.

Vostok I First of a series of manned spacecraft placed into Earth orbit by the Soviet Union on April 12, 1961. It was launched by the two-stage Vostok rocket and carried cosmonaut Yuri GAGARIN.

Voyager project SPACE EXPLORATION program using two unmanned craft to study Jupiter, Saturn, Uranus, and Neptune, launched in 1977. The two Voyager probes were launched 16 days apart from Kennedy Space Center at Cape Canaveral, Florida, and beamed back stunning close-up pictures of Jupiter in 1979. The probes then passed Saturn and showed the structure of the planet's rings. Although that was Voyager 1's last planetary encounter, Voyager 2 went on to study Uranus in 1986 and Neptune in 1989. Too far from the Sun to generate power from solar cells, it carried small nuclear generators. Both probes have now left the Solar System.

Vries, Hugo de *See* DE VRIES, HUGO

VTOL aircraft (**v**ertical **t**ake-**o**ff and **l**anding aircraft) Experimental AIRCRAFT designed to perform vertical take-offs and landings but to maintain flight speed and payload capabilities superior to those of a HELICOPTER. There are designs that permit rotation of the wing and engines from the vertical for take-off, to the horizontal for high-speed cruising. Another design allows for the diversion downward of the exhaust of fixed jet engines. A third style has two separate systems of thrust for upward and forward movement. *See also* STOL AIRCRAFT

vulcanization Chemical process, developed (1839) by Charles Goodyear, of heating sulfur or its compounds with natural or synthetic rubber to improve its durability and resilience. The process was used in the development (1845) of the first pneumatic TIRE.

vulva In females, the external genitalia. Extending downwards from the clitoris (a sensitive, erectile organ), a pair of fleshy lips (*labia majora*) surround the vulvar orifice. Within them two smaller folds of skin (*labia minora*) surround a small depression called the vestibule, within which are the urethral and vaginal openings.

Waals, Johannes Diderik van der *See* VAN
DER WAALS, JOHANNES DIDERIK

wadi (arroyo) Narrow, steep-sided valley found in
deserts and semiarid regions. As rainfall is infre-
quent in desert areas, wadis are dry for much of the
year. When it does rain, however, it falls in heavy
showers. Because much of the land has no vegetation, surface runoff is rapid. The water carries
away sand and small rock fragments, which causes
EROSION. Vertical CORRASION forms the narrow,
steep-sided wadis, which can be several hundred
meters in depth and extend for long distances.

waggle dance (bee dance) Method of communication performed by worker honeybees to indicate
to other worker bees the exact position and rich-
ness of sources of pollen and nectar. The dances
rely on a knowledge of the position of the Sun.

Waksman, Selman Abraham (1888–1973) US
microbiologist, b. Russia. He was awarded the 1952
Nobel Prize for physiology or medicine for his discovery of the antibiotic STREPTOMYCIN in 1943.
Waksman developed techniques for extracting
ANTIBIOTICS from various microorganisms, and he
discovered several new ones, including neomycin.

Wald, George (1906–) US biologist who shared
the 1967 Nobel Prize for physiology or medicine
with the US physiologists Haldan Hartline
(1903–83) and Ragnar Granit (1900–91) for
research into the chemical processes of vision. He
discovered that vitamin A is an essential component
of visual pigments and later determined the chemical reactions that occur in the rods of the retina.

Wallace, Alfred Russel (1823–1913) British
naturalist and evolutionist. Wallace developed a
theory of NATURAL SELECTION independently of but
at the same time as Charles DARWIN. He wrote
Contributions to the Theory of Natural Selection
(1870), which, with Darwin's *Origin of Species*,
comprised the fundamental explanation and understanding of the theory of EVOLUTION.

Wallace's line Imaginary line dividing the
islands of Borneo and Bali from Sulawesi and

Lombok. It was drawn by Alfred Russel WALLACE
to demarcate animal populations that live close to
each other and yet are extraordinarily different.
Earlier this observation had influenced Wallace to
state his theory of natural selection, which was
coincident with that of Charles DARWIN. He argued
that homogeneous animal populations could, for
various reasons, divide and diversify gradually
until they had evolved into separate species.

Wallach, Otto (1847–1931) German chemist
who was awarded the 1910 Nobel Prize for chemistry for his research into TERPENES. In 1884 he
began investigating the ESSENTIAL OILS in plants
and, by a method of repeated distillation, he isolated the components of these complex compounds.

Wallis, Sir Barnes Neville (1887–1979) British
aeronautical engineer and inventor, best known for
his invention of the bouncing bomb during World
War II. In the 1920s he designed the AIRSHIP R100
and after the war the first SWING-WING AIRCRAFT.

Wallis, John (1616–1703) English mathematician,
who together with Robert BOYLE formed a society
where scientists congregated to discuss current scientific ideas. It led to the foundation of the Royal
Society in 1660. In mathematics Wallis developed
various series expressions for PI and worked on
problems in number theory, mechanics, and algebra.

Walton, Ernest Thomas Sinton (1903–95)
British physicist who shared the 1951 Nobel Prize
for physics with John COCKCROFT for the development of the first nuclear particle ACCELERATOR. After
two unsuccessful attempts they succeeded in 1929,
using a voltage multiplier to accelerate protons. In
1931 they produced the first artificial nuclear reaction, using high-energy protons to bombard lithium
nuclei. This process was of great importance for
studying nuclear structure; its modern counterparts
are among the most useful tools of NUCLEAR PHYSICS.

Wang, An (1920–90) US computer engineer, b.
China. He emigrated to the US in 1945. In 1948 he
devised the core memory, which came into universal use in mainframe computers before integrated
circuit chips were introduced. In 1951 he established Wang Laboratories, which grew to be one of
the world's major computer companies.

Wankel rotary engine Gasoline engine with
rotors instead of pistons. German engineer Felix
Wankel (1902–88) invented this engine in the
1950s. Each triangular rotor turns inside a close-fitting casing. Gaps between the casing and rotor form
three crescent-shaped combustion chambers. Each
chamber goes through a sequence of events similar
to those in a FOUR-STROKE ENGINE with pistons. The

movement of the rotor draws in the gasoline-air mixture through an inlet port and then compresses it. A spark ignites the mixture, producing expanding gases that force the rotor around. The gases then pass out of an exhaust port, and the process repeats.

Warburg, Otto Heinrich (1883–1970) German biochemist who was awarded the 1931 Nobel Prize for physiology or medicine for his discovery of respiratory enzymes. He made significant contributions to the understanding of the mechanisms of cellular respiration and energy transfer, and of photosynthesis.

warm-blooded *See* HOMEOTHERMIC

warning coloration Markings of an animal that warn potential predators that it is dangerous. It may indicate that the animal tastes bad or is poisonous. Red, yellow and black are common warning colors, as in some venomous insects and snakes. Some nonpoisonous animal species mimic the warning colors of other animals in order to trick potential predators into thinking they are not suitable prey. *See also* MIMICRY

Wassermann, August von (1866–1925) German bacteriologist. He made important contributions to immunology, the best known of which was his development of a diagnostic test for syphilis, the Wassermann test.

waste disposal, nuclear FISSION reactors and their fuel preparation plants produce waste residues containing highly radioactive substances. After URANIUM, PLUTONIUM and any other useful fission products have been removed some long-lived radioactive products remain, such as CAESIUM-137 and STRONTIUM-90. These wastes have to be disposed of so that they will not contaminate the soil, seas, or atmosphere. International regulations exist for disposing of the wastes, with separate regulations applying to liquids, gases, and solids. In general, disposal consists of burying the materials under controlled conditions in underground tanks, in disused mines, and in the ocean.

water Odorless, colorless liquid (H_2O) that covers over 70% of the Earth's surface and is the most widely used solvent. Essential to life, it makes up from 60–70% of the human body. It is a compound of hydrogen and oxygen with the two H-O links of the molecule forming an angle of 105°. This asymmetry results in polar properties and a force of attraction (hydrogen bond) between opposite ends of neighboring water molecules. These forces maintain the substance as a liquid, in spite of its low molecular weight, and account for its unusual property of having its maximum density at 4°C (39.2°F). Properties: r.d. 1.000; m.p. 0.0°C (32°F); b.p. 100°C (212°F).

water cycle *See* HYDROLOGICAL CYCLE

waterfall Steep or vertical descent of the water of a stream or river. Where there is a change of rock type on the bed of a stream, there are different rates of EROSION. The softer rock is eroded more quickly, causing an abrupt change in the gradient. If there is a vertical or nearly vertical STRATUM of hard rock, only a small waterfall will be created. If the harder rock is horizontal, a much larger waterfall can form.

watermark Translucent impression made in otherwise opaque PAPER, either by machine or by hand. A watermark is formed during the making of paper by a raised design on wires over which the paper is passed while still in a pulpy state. Watermarks have been used since the latter part of the 13th century as a guarantee of authenticity, as in banknotes, and to identify the makers of high-quality papers.

▲ **Wankel rotary engine**
The Wankel rotary engine (A) compresses and ignites a gasoline/air mixture with a spark plug like a normal combustion engine, but does so with a three-sided rotor (1), rather than an in-line piston action. The explosion of the fuel/air mixture drives a crankshaft (2) that passes through the center of the cylinder. (3) The movement of the rotor sucks air into the cylinder and, as it continues to rotate, seals the inlet (so valves are not needed) and then compresses the mixture as it continues to turn. The spark plugs (4) then ignite the mixture, which expands, rotating the piston and driving the crankshaft. The burned fuel is expelled through an outlet (5) as the piston turns, pulling in more air to repeat the process (6). The seals (7) at the edge of the piston's faces are very important in creating the vacuum needed to pull in the fuel/air mixture and in compressing the mixture. A complete cycle is shown (B).

water mill Machinery powered by the flow of water past a WATERWHEEL. Water mills were invented about 2,000 years ago to grind grain into flour. Water also powered looms at textile mills beginning in the early 1700s, but was generally replaced by steam engines in the 1800s.

water of crystallization Definite molecular proportion of water that is chemically combined with certain substances in the crystalline state. Much of this water will be lost on heating to 100°C (212°F) but some, the water of constitution, will be retained to a much higher temperature. Copper(II) sulfate ($CuSO_4.5H_2O$) loses four molecules of water at 100°C (212°F), becoming $CuSO_4H_2O$.

water pollution Contamination of water by harmful wastes. The chief source of water pollution is **industrial waste**. Toxic chemicals, such as POLYCHLORINATED BIPHENYL (PCB), are discharged as effluent and cannot be disinfected with CHLORINE. The burning of fossil fuels causes ACID RAIN. Untreated or partially treated SEWAGE is another source of water pollution. Sewage treatment is unable to prevent the spread of viruses and of some phosphorus-based detergents that cause EUTROPHICATION. Agricultural chemicals and wastes, such as PESTICIDES and FERTILIZERS, are another major cause of pollution. Once pollution has affected GROUNDWATER it spreads more rapidly.

▲ **water table** Artesian springs and wells are found where groundwater is under pressure. The water-table (1) in the confined aquifer (2) lies near the top of the dipping layers. A well (4) drilled through the top impervious layer (3) is not an artesian well because the head of the hydrostatic pressure (6) is not sufficient to force water to the surface. In such wells, the water must be pumped or drawn to the surface. The top of an artesian well (5) lies below the level of the head of hydrostatic pressure and so water gushes to the surface. Artesian springs (8) may occur along joints or faults (7), where the head of hydrostatic pressure is sufficient to force the water up along the fault. Areas with artesian wells are called artesian basins. In the London and Paris artesian basins, the water has been so heavily tapped that the water level has dropped below the level of the well heads.

Oil spills and ocean dumping are major causes of marine pollution. *See also* DDT

water potential Force that brings about OSMOSIS in a biological system such as a cell. It makes water tend to move between solutions of different concentrations. It is the difference between the chemical potential of pure water (which is zero) and the chemical potential of water in a solution (which is negative). During osmosis, water moves down the water potential gradient, which is measured in pressure units (kilopascals or bars).

waterproofing Rendering materials such as fabrics, leather, and wood impervious to water by the application of substances such as rubber, synthetic resins, waxes, and metallic compounds.

water power *See* HYDROELECTRICITY

watershed (water divide, water parting) High ground that forms the boundary or dividing line between two river basins. On one side of the boundary line the water will drain in one direction; on the other side of the boundary line it will drain in the opposite direction. The watershed is normally a ridge or a piece of ground that is higher than the surrounding areas. It is often a very irregular line. The **Continental Divide** is the US watershed that separates westward- and eastward-flowing rivers; it largely follows the spine of the Rocky Mountains.

water softener Substance added to water to reduce its hardness – that is, its inability to form a lather with soap as a result of the presence of dissolved calcium and magnesium compounds. Some hardness, mostly due to the bicarbonates of these metals, can be removed by boiling. The remaining "permanent" hardness is mostly due to sulfates of the metals and is reduced by adding such water-softening compounds as sodium carbonate, sodium phosphate or zeolites (hydrated silicates of calcium and aluminum) to remove or sequester the metallic ions. *See also* HARDNESS OF WATER

waterspout Funnel-shaped mass of water, which is similar in formation to a TORNADO. It occurs over water when a heated patch of air rises, and whirls around in a circular or corkscrew fashion. Waterspouts are mostly seen in tropical areas, especially in late summer. When waterspouts go on to land they become tornadoes, and when tornadoes go over lakes or the ocean, they become waterspouts. Waterspouts are generally less than 30m (98ft) in height, and may last for half an hour or so. They can cause great damage to shipping and coastal installations

water table In geology, level below which the rock is saturated. After rain has fallen, the water trickles downward through the pores in the rocks. The level at which all pores are full is the water table. The height of the water table gradually changes, moving up or down depending on the recent rainfall. Water located below the water table is called GROUNDWATER.

water treatment Removal of undesirable elements from a water supply. Unpleasant tastes and odors are removed by aeration. Bacteria are destroyed by the addition of chlorine (the taste of chlorine is then removed with sodium sulfite). Hardness is reduced with lime or by using zeolite as a WATER SOFTENER. Other organic and mineral matter is removed with alum. FLUORIDE is added to the water supply of many communities to reduce dental caries. *See also* SEWAGE

water vapor Water in its gaseous state. It occurs in the atmosphere and determines humidity and the formation of clouds as it condenses. Most of the atmosphere's water vapor is found in the TROPOSPHERE, mainly below an altitude of 8km (5mi). Water vapor absorbs infrared radiation and holds it in the atmosphere, because of the GREENHOUSE EFFECT, and thus plays a vital role in the transfer of energy and the Earth's heat balance.

waterwheel Mechanical device used since early Roman times to supply power, mainly for grinding cereals. The first waterwheels were laid horizontally in flowing water, their shafts pointing upwards and surmounted by a millstone. Later waterwheels were vertical, either undershot (in which the water passed under the wheel) or overshot (in which the water flowed through a funnel to the top of the wheel). The latter had the advantage of being operable almost independently of the level of water in the stream or river that supplied it.

Watson, James Dewey (1928–) US geneticist and biophysicist. He is known for his role in the discovery of the double-helix molecular structure of deoxyribonucleic acid (DNA) and shared the 1962 Nobel Prize for physiology or medicine with Francis CRICK and Maurice WILKINS. Watson later helped to break the GENETIC CODE of the DNA base sequences and found the ribonucleic acid messenger (MESSENGER RNA) that carries the DNA code to the cell's protein-forming structures.

Watson-Watt, Sir Robert Alexander (1892–1973) British physicist. As scientific adviser to the Air Ministry in 1940, he was a major influence in the rapid development of RADAR, which helped to detect German warplanes during World War II. In 1941 he helped to establish the US radar system.

Watt, James (1736–1819) Scottish engineer. In 1765 Watt invented the condensing STEAM ENGINE,

but he did not build a full-scale working model until some ten years later. The new engine was much more efficient than earlier designs and reduced the cost of running steam-powered machines by 75%. In 1782 he invented the double-acting engine, in which steam pressure acted alternately on each side of a piston. With Matthew BOULTON, Watt coined the term HORSEPOWER. The unit of POWER is called the WATT in his honor.

watt (symbol W) SI unit of POWER. A machine consuming one JOULE of energy per second has a power output of one watt. One horsepower corresponds to 746 watts. For example, an appliance that uses 12 amps at 250 volts expends a power of 3,000 watts (or 3 kilowatts).

wattle and daub Ancient method of building, in which mud or clay (daub) is plastered on to wooden laths or branches (wattle) to form walls.

wave In oceanography, moving disturbance traveling on or through water that does not move the water itself. Wind causes waves by frictional drag. The strength and size of the waves depends on the wind speed and the distance of open water across which the wind blows. Waves not under pressure from strong winds are called swells. Waves begin to break on shore or "feel bottom" when they reach a depth shallower than half the wave's length. When the water depth is about 1.3 times the wave height, the wave front is so steep that the top falls over and the wave breaks.

wave In physics, an oscillating disturbance in a material medium or in an electromagnetic field. In a **traveling** wave, the oscillations spread out (propagate) from the source. The velocity depends on the type of wave and on the medium. Electromagnetic waves, such as light, consist of varying magnetic and electric fields at right angles to each other and to the direction of motion; they are **transverse** waves. Sound waves are transmitted by the vibrations of the particles of the medium itself; because the vibrations are in the direction of wave motion they are called **longitudinal** waves. Sound waves, unlike electromagnetic waves, cannot travel through a vacuum and cannot undergo POLARIZATION. Waves can undergo REFRACTION and REFLECTION, and give rise to INTERFERENCE phenomena. A wave is characterized by its WAVELENGTH and FREQUENCY, the VELOCITY of wave motion being the product of wavelength and frequency. *See also* ELECTROMAGNETIC RADIATION; STANDING WAVE

wave amplitude Peak value of a periodically varying quantity. This peak value may be either positive or negative, as the quantity varies either above or below its zero value (maximum or minimum).

wave dispersion Separation of the WAVELENGTHS in a wave because the REFRACTIVE INDEX of the medium is different for different wavelengths. It occurs with all electromagnetic radiation, but is most obvious at visible wavelengths, causing light to be separated into its component colors. Dispersion is seen when a beam of light passes through a refracting medium, such as a glass PRISM, and forms a SPECTRUM. Each color has a different wavelength, and so the prism bends each color in the light a different amount. Red (long wavelength) is bent less than violet (short wavelength). Dispersion can cause chromatic ABERRATION in lenses. *See also* REFRACTION

wave frequency Number of complete oscillations or wave cycles produced in 1 second, measured in HERTZ. It can be calculated from the wave velocity divided by wavelength. According to QUANTUM THEORY, the frequency of any ELECTROMAGNETIC RADIATION (such as light, radio waves, and X-rays) is proportional to the energy of the component photons. Many of the characteristics of electromagnetic radiation depend on frequency.

wave front Contour on which at every point the medium or electromagnetic field is vibrating in step (that is, in phase). The wave front is usually at right angles to the direction of wave motion, and may be plane, spherical, or otherwise. Each point is a source of secondary wavelets, which give rise to a new wave front position a short time later.

wave function In QUANTUM MECHANICS, function that represents the probability of a quantum system being in a particular state s at time t, usually written $\psi(s)$ or $\Psi(s, t)$. The wave function is used in Erwin SCHRÖDINGER's equation.

wave interference Phenomenon in which two WAVES of the same WAVELENGTH and generally from a single source that have traveled along different paths interact at a point (superpose). If the two waves have a constant PHASE relationship – that is, if they are "in step" or "out of step" by a constant amount – they will interfere with each other. If maximum coincides with maximum, and minimum with minimum, the two waves will reinforce each other and the amplitude will be doubled; if maximum and minimum coincide, the two waves will cancel each other out, producing zero amplitude. With two light waves, an interference pattern of light and dark bands or rings can be seen on a screen in the overlap region.

wavelength (symbol λ) Distance between successive points of equal PHASE of a WAVE. For example, the wavelength of water waves could be measured as the distance from crest to crest. The wavelength of a visible light wave, which can range from 390 to 780nm (nanometers), deter-

mines its color. Wavelength is equal to the wave VELOCITY divided by the WAVE FREQUENCY.

wave mechanics Version of QUANTUM MECHANICS developed in 1926 by Erwin SCHRÖDINGER. It explains the behaviour of electrons in terms of their wave properties. Although quickly superseded by a more complex formulation by physicist Paul DIRAC, it is still widely used in calculations.

wave-particle duality Concept that combines the corpuscular theory, first outlined by Sir Isaac NEWTON, with the WAVE theory described by Christiaan HUYGENS. ELECTROMAGNETIC RADIATION, such as light, behaves like a wave when being transmitted or moved, and like particles when interacting with matter. INTERFERENCE, DIFFRACTION and POLARIZATION can be described best in terms of waves, whereas the PHOTOELECTRIC EFFECT and the COMPTON EFFECT can be described in terms of PHOTONS. *See also* PARTICLE PHYSICS; PLANCK'S CONSTANT; QUANTUM THEORY

wax Solid, insoluble substance of low melting point that is moldable and water-repellent. There are three forms of wax: animal, vegetable, and mineral. The first two are simple LIPIDS consisting of esters of fatty acids; animal waxes include beeswax, wool wax (LANOLIN), AND spermaceti wax from the sperm whale; vegetable waxes include carnauba (wax palm) from the leaves of the Brazilian carnauba palm tree, and candelilla from a Mexican rush. Mineral waxes include PARAFFIN wax made from petroleum. Synthetic waxes are of diverse origins and include POLYETHENES. Waxes are used in the manufacture of lubricants, polishes, cosmetics, and candles, to waterproof leather, and to coat paper.

W boson (symbol W or W^-) ELEMENTARY PARTICLE that is a carrier of the WEAK NUCLEAR FORCE. It carries a unit electric charge and has a mass approximately 80 times that of the PROTON. *See also* BOSON

weak nuclear force (weak interaction) One of the four FUNDAMENTAL FORCES in physics. The weak nuclear force causes radioactive decay. It can be observed only in the subatomic realm, being of very short range. It is weaker than the ELECTROMAGNETIC FORCE and the STRONG NUCLEAR FORCE (the strongest of the forces) but stronger than GRAVITATION.

weather State of the ATMOSPHERE at a given locality or over a broad area, particularly as it affects human activity. Weather refers to short-term states (days or weeks) as opposed to long-term CLIMATE conditions. Weather involves such elements as atmospheric temperature, pressure, humidity, precipitation, cloudiness, brightness, visibility, and wind. *See also* CLIMATOLOGY; METEOROLOGY

weather forecasting Technique of anticipating and describing future WEATHER for a given time and place. There is a threefold classification depending on the length of the period covered. **Short-term forecasts** cover the next 24 hours and include wind velocity, cloud cover, and temperature. **Medium-range forecasts** look two to five days ahead and contain information on the dominant circulatory conditions – the paths of HURRICANES and TORNADOES. **Long-range forecasts** cover periods of five days or more and are more extensive than medium-range forecasts. Pressures and circulatory patterns are usually given as averages. Artificial SATELLITES have increased the accuracy of weather forecasting, especially when dealing with severe turbulences such as hurricanes.

weathering In geology and physical geography, the breakdown and chemical disintegration of rocks and minerals at the Earth's surface by physical, chemical, and organic processes. It is important in the formation of soil and plays a major part in shaping landscapes. In **mechanical** (physical) weathering in cold, wet climates, water seeping into cracks in the rock expands on freezing, so causing the rock to crack further and to crumble. Extreme temperature changes in drier regions, such as deserts, also cause rocks to fragment. Cracks opened by the weather can be exploited by plants, the roots of which place further stress on the rock. **Chemical** weathering can lead to a weakening of the rock structure by altering the minerals of a rock and changing their size, volume, and ability to hold shape. The best-known example is the dissolution of limestone in acid rainwater. Chemical weathering processes include oxidation, hydration, silication, desilication, and carbonation. **Organic** weathering describes the breakdown of rock and soil by plants, such as by root action, and animals, such as worms, which are important for the breaking up of soil. Unlike EROSION, weathering does not involve the transportation of broken-down material.

weather observations Furnishing the data required for WEATHER description, analysis, and WEATHER FORECASTING from thousands of weather stations and instruments around the globe, such as ships, aircraft, balloons, and weather SATELLITES. Weather stations measure and record weather conditions in their vicinity and survey with radar up to 161 to 320km (100 to 200mi). Information about upper-air conditions is obtained from RADIOSONDE balloons, from high-flying aircraft, and from stratospheric balloons operating for long periods at heights up to 32km (20mi). Weather satellites obtain day and night observations in visible and infrared wavelengths. They provide pictures of cloud cover and storm disturbances, data on vertical temperature, and moisture profiles of the atmosphere, making accurate long-term weather forecasts possible.

weather vane Device to indicate wind direction. It comprises an upright rod, crossmembers bearing the cardinal points of the compass, and a revolving arm. The arm is arrow-shaped at one end and has a vane on the other, the large surface area of which is designed to react to wind pressure and point the arrow in the direction from which the wind is blowing. One of the oldest meteorological instruments, vanes are still widely used.

weber (symbol W) SI unit of MAGNETIC FLUX. It is defined so that an ELECTROMOTIVE FORCE (EMF) of 1 volt is induced in a circuit of a single loop when the flux changes through it at a steady rate of 1 W s^{-1}.

Weber, Ernst Heinrich (1795–1878) German physiologist who pioneered the study of sensation and perception. He laid the foundations for the branch of psychology called "psychophysics" and encouraged psychologists to be more scientific and methodical.

Weber, Wilhelm Eduard (1804–91) German physicist who, in 1846, standardized the units used in ELECTRICITY, relating them to the fundamental dimensions of mass, length, charge, and time. He was the first physicist to regard electricity as composed of elementary units, each carrying a charge, now known as ELECTRONS. The SI UNIT of magnetic flux is named after him.

◀ **weathering** Weathering is the break down of rock in place. It occurs in two ways: physical (A and C) and chemical (B), and usually in combination. At the surface, plant roots and animals break down rock turning it into soil (A). In chemical weathering (B), soluble rocks such as limestone (1) are dissolved by ground water. Acid rain caused by sulfate pollution (2) also attacks the rock. Both heat and cold can cause physical weathering (C). When temperatures drop below freezing, freeze-thaw weathering can split rocks such as granite (4). Water which settles in cracks and joints during the day expands as it freezes at night (5). This expansion cleaves the rock along naturally occurring joints (6).

web offset Method of PRINTING in which the press utilizes continuous rolls (webs) of paper and lithographic printing plates on rotary printing machines. Ink is transferred to the paper by offset rubber rollers. Web offset is widely used in the magazine and newspaper industry, where speed is vital. *See also* LITHOGRAPHY; OFFSET PRINTING

weed Uncultivated or unwanted plant (a pest). Weeds are a threat to commercial crops because they compete for water and sunlight and harbor pests and diseases that can spread to the crop plants. HERBICIDES are used to kill weeds.

Wegener, Alfred Lothar (1880–1930) German geologist, meteorologist, and explorer. In *The Origin of Continents and Oceans* (1915) he was the first to use well-presented evidence and scientific argument in support of a theory of CONTINENTAL DRIFT.

weight Force of attraction on a body due to GRAVITY. A body's weight is the product of its MASS and the gravitational field strength at that point. Mass remains constant, but weight depends on the object's position on the Earth's surface, decreasing with increasing altitude.

weightlessness Condition experienced by an object for which the effects of WEIGHT are not apparent. Weightlessness can be experienced in space and during a free fall, even though gravitational attraction on the "weightless" body is still present. Astronauts are trained in the weightless conditions that exist briefly in an aircraft flying in an arc and in water tanks. The adverse effects on the human body of prolonged weightlessness include decreased circulation of blood, less water retention in tissues and the bloodstream, loss of bone calcium and loss of muscle tone.

weights and measures Accepted units for expressing the amount of some quantity, such as capacity, length or weight. Early measurements were based on body measurements, plant grains, and so on. The French introduced the METRIC SYSTEM in 1799 in which the unit of length, the METER, was taken as one ten-millionth of the distance from the Equator to the North Pole. The unit of mass was the GRAM, defined as the mass of a cubic centimeter of water. The metric system has developed into the system of SI UNITS.

Weinberg, Steven (1933–) US physicist who in 1967, independently of Pakistani physicist Abdus SALAM, proposed a theory, known as the ELECTROWEAK THEORY, that unifies the electromagnetic and weak interactions between subatomic particles. Later experiments proved the Salam-Weinberg hypothesis to be true. In 1979 they shared the Nobel Prize for physics with Sheldon GLASHOW, who had earlier proposed a similar theory.

Weismann, August (1834–1914) German biologist. His essay discussing germ plasm theory, *The Continuity of the Germ Plasm* (1885), proposed the immortality of the germ line cells as opposed to body cells. It was influential in the development of modern genetic study.

welding Technique for joining metal parts, usually by controlled melting. Vehicles, domestic appliances, bridges and electronic components are just some of the many items that have parts joined by welding. Several welding processes can be used. In **fusion welding**, the parts to be joined are heated together until the metal starts to melt. On cooling, the molten metal solidifies to form a permanent bond between the parts. Such welds are usually strengthened with filler metal from a welding rod or wire. In **arc welding**, an electric arc produced by an arc welder heats the work and filler metal. In **oxyacetylene welding**, the heat is provided by burning acetylene gas in oxygen; in **resistance** or **spot welding**, the heat is generated by passing an electric current through the joint. Spot welds are made at regular intervals to join sheet metal. In brazing and soldering, the temperature used is sufficient to melt the filler metal, but not the parts that it joins. Special techniques include the use of friction, electron beams, and ultrasonics to cause bonding. Some plastics can also be welded.

well Shaft sunk vertically in the Earth's CRUST through which water, oil, natural gas, brine, sulfur or other mineral substances can be extracted. Artesian wells are sunk into water-bearing rock strata, the so-called AQUIFERS, from which water rises under pressure in the wells to the surface. *See also* OIL WELL

Weller, Thomas Huckle (1915–) US bacteriologist and virologist. He shared the 1954 Nobel Prize for physiology or medicine with John F. ENDERS and Frederick C. ROBBINS for the discovery that the poliomyelitis virus can grow in cultures of various types of tissue, so making it possible to develop a VACCINE for polio.

Welsbach, Carl Auer, Freiherr von (1858–1929) Austrian chemist, inventor of the Welsbach mantle – a shaped gauze that when heated by gas produces a bright light. In 1885 he discovered NEODYMIUM and PRASEODYMIUM.

Werner, Abraham Gottlob (1750–1817) German geologist who was the first to classify minerals systematically. His theory of the Earth's origins, called **neptunism**, proposed that the Earth was originally a vast ocean from which solid rocks were precipitated to form land.

Werner, Alfred (1866–1919) Swiss chemist who was awarded the 1913 Nobel Prize for chemistry. After working with Marcelin BERTHELOT in Paris, he returned to a professorship at Zurich University. Werner was awarded the Nobel Prize for his coordination theory of VALENCE in which he correctly suggested that some metals can form secondary, or coordinate, bonds. *See also* CO-ORDINATION COMPOUND

wernerite *See* SCAPOLITE

Westinghouse, George (1846–1914) US engineer and inventor. In 1865 he received his first patent, for a rotary engine, but the best known of his hundreds of inventions was the air brake, which made high-speed rail travel safe. He formed the Westinghouse Electric Company in 1886.

Weston standard cell (CADMIUM cell) Primary voltaic CELL, used as a standard of ELECTROMOTIVE FORCE (EMF). It produces a constant 1.018636 volts at 20°C (68°F). The Weston cell comprises a mercury anode and a cadmium amalgam cathode. The electrolyte is a saturated solution of cadmium sulfate.

wetland Marshy ground in an intertidal zone that has prolific vegetation; coastal wetlands are said to contain a greater concentration of living matter, both flora and fauna, than any other kind of terrain.

Weyl, Herman (1885–1955) US mathematician, b. Germany. After holding professorships at Zurich and Göttingen universities he moved to Princeton University, US, in 1933. He worked in GROUP THEORY and QUANTUM THEORY, and also on NUMBER THEORY and on GEOMETRY particularly relevant to general RELATIVITY. His findings play an important role in modern PARTICLE PHYSICS, particularly in STRING THEORY. His *Group Theory and Quantum Mechanics* (1928) is a classic text of quantum theory.

Wheatstone, Sir Charles (1802–75) British physicist and inventor. In 1843, with William COOKE, he improved a device that accurately measures electrical RESISTANCE. Widely used in laboratories, it became known as the WHEATSTONE BRIDGE. By 1837 they had patented an electric TELEGRAPH. Wheatstone also coined the term "microphone" for a sound magnifier.

Wheatstone bridge Electric circuit used for measuring RESISTANCE, named after Sir Charles WHEATSTONE. It consists of four resistances arranged to form a square with an ELECTROMOTIVE FORCE (EMF) (voltage from a battery) across one diagonal, and a GALVANOMETER across the other. When the galvanometer shows no deflection, the ratio of the values of one adjacent pair of resis-

tances equals the ratio of the values of the other two. By adjusting the ratio of one pair (with a sliding contact along a wire which forms one of the adjacent pairs of resistances) the unknown resistance of one of the other two can be calculated.

wheel Circular structure that revolves around a central axis. Before the wheel was invented, heavy loads were sometimes moved by rolling them on logs or on rounded stones. More than 5,000 years ago sections of tree trunks were cut to form the first wheels for carts. Spoked wheels were introduced several hundred years later. Eventually, the wheel was used in simple machines, such as the waterwheel and potter's wheel.

wheel and axle Machine based on the principle that a small force applied to the rim of a wheel will exert a larger force on an object attached to the axle. The FORCE RATIO, or mechanical advantage, is the ratio of the radius of the wheel to that of the axle.

Whinfield, John Rex (1901–66) British chemist who, with fellow British scientist John Dickson, invented the synthetic fiber TERYLENE in 1941. His interest was stimulated by the discovery of nylon by Wallace CAROTHERS in 1935. In 1946 terylene was made in the US and sold as Fiber V and then as Dacron, and in 1947 Imperial Chemical Industries (ICI) began to make and market it to the rest of the world.

Whipple, Fred Lawrence (1906–) US astronomer best known for his "dirty snowball" theory of comets – proposed in 1949 but not proven until 1986, when space probes were sent to HALLEY'S COMET. Whipple discovered several comets and worked on cometary orbits. He also studied planetary nebulae, stellar evolution, and the Earth's upper atmosphere.

whirlpool Circular motion of a fluid. Whirlpools in rivers occur in regions where waterfalls or sharp breaks in topographic continuity make steady flow impossible. *See also* VORTEX

White, Gilbert (1720–93) British naturalist. His famous book *The Natural History and Antiquities of Selborne* (1789) consists of letters to his fellow naturalists.

white blood cell Alternative name for a LEUCOCYTE

white dwarf Type of star about the size of the Earth, but with a mass about that of the Sun. As a result, its density is enormously greater than that of any terrestrial material. This is because the normal atomic structure is broken down completely, with electrons and nuclei packed tightly together. A white dwarf cannot have a mass of more than about 1.4 solar masses (the CHANDRASEKHAR limit). For larger masses, gravity would always overwhelm the pressure of the electrons and the star would collapse under its own weight, forming a NEUTRON STAR or BLACK HOLE. White dwarfs are of low luminosity and gradually cool down to become cold, dark objects. They represent the final stage in the evolution of low-mass stars after they have lost their outer layers. The first white dwarf to be discovered was the companion of SIRIUS. *See also* STELLAR EVOLUTION

white hole Hypothetical localized point in space at which matter is suddenly created with explosive violence. It can be considered as being similar to a BLACK HOLE but with time reversed. The cosmological BIG BANG is an analogous phenomenon that embraced the whole Universe rather than being localized.

Whitney, Eli (1765–1825) US inventor and manufacturer. Whitney invented the cotton gin (1793), which revolutionized cotton picking in the South and turned cotton into a profitable export. After 1798 he manufactured muskets at a factory in New Haven, Connecticut, which was one of the first to use MASS PRODUCTION methods with interchangeable parts.

Whittle, Sir Frank (1907–96) British aeronautical engineer who pioneered the development of the TURBOJET ENGINE. Whittle entered the Royal Air Force (RAF) as an apprentice and later studied at the RAF engineering school and at Cambridge University. In 1930 he patented his first turbojet engine. In 1936 he founded a company called Power Jets and was later given special permission to carry out development work on jet propulsion. A plane powered by the Whittle engine had its first flight in May 1941 and it entered service with the RAF in 1944.

Whitworth, Sir Joseph (1803–87) British engineer. After studying engineering in Manchester he went to London, where he invented a technique for making perfectly flat metal surfaces. He returned to Manchester, where he produced extremely accurate machine tools. A system of threads for nuts and bolts is named after him.

whole number *See* NUMBER, NATURAL

Widmanstätten patterns Lines surrounded by dark bands found in the class of METEORITES called IRON METEORITES (composed entirely of metal). They are formed when the liquefied outer surface of a meteorite cools on entering the Earth's atmosphere, and they can be used to determine the meteorite's motions in flight. They appear when the meteorite is treated with nitric acid. Widmanstätten patterns are named after Count Aloys Joseph von Widmanstätten (1754–1849), who observed them in 1808.

Wieland, Heinrich Otto (1877–1957) German chemist who was awarded the 1927 Nobel Prize for chemistry for his research into BILE acids. In 1912 he discovered that the three bile acids then known were closely related in structure. He showed them to have a STEROID skeleton and thus found that they were also structurally related to CHOLESTEROL. Wieland also did research into oxidation reactions (*see* OXIDATION-REDUCTION) occurring in living tissues and discovered that such oxidation consisted of dehydrogenation (removal of hydrogen atoms), not the addition of oxygen.

Wien, Wilhelm (1864–1928) German physicist who was awarded the 1911 Nobel Prize for physics for his work on RADIATION. He investigated the radiation from BLACK BODIES and in 1893 calculated that the peak frequency of radiation increases with temperature. This is called **Wien's displacement law**.

Wiener, Norbert (1894–1964) US mathematician and originator of CYBERNETICS. He contributed much to the study and development of the COMPUTER and to the understanding of FEEDBACK systems that control the behavior of humans and machines. Wiener developed a mathematical model of BROWNIAN MOVEMENT and also explained ENTROPY, which is a measure of disorder.

Wigner, Eugene Paul (1902–95) US physicist, b. Hungary, who worked on the MANHATTAN PROJECT during World War II to develop the atom bomb. Wigner was the first physicist to apply GROUP THEORY to QUANTUM MECHANICS. With this technique he discovered the law of conservation of PARITY. For his work on the structure of the atomic nucleus, Wigner shared the 1963 Nobel Prize for physics with the German physicist Hans Jensen and the German-born US physicist Maria Goeppert-Mayer. He also received the 1960 Atoms for Peace Award.

Wigner effect In nuclear engineering, the change in the physical properties of a MODERATOR (such as graphite) as a consequence of irradiation. In a NUCLEAR REACTOR, high-energy NEUTRONS displace atoms in the CRYSTAL LATTICE resulting in the accumulation of stored energy (**Wigner energy**). This accumulation changes the lattice and the overall size of the moderator. It was described first by Eugene WIGNER.

Wiles, Andrew (1953–) British mathematician, the number theorist who proved FERMAT'S LAST THEOREM. As a child, Wiles had been fascinated with NUMBER THEORY and, in particular, the conjectured last theorem of Fermat. After many years, working largely alone, he announced in 1993 that he had proved the theorem. Though subsequent checks revealed a gap in the proof, Wiles soon submitted a completed version, which was widely accepted in 1995. His achievement is one of the greatest in 20th-century mathematics.

Wilkins, Maurice Hugh Frederick (1916–) British biophysicist, b. New Zealand. During World War II he worked on the separation of uranium isotopes for the Manhattan Project. Wilkins is celebrated for his work in molecular biology, in particular the structure of NUCLEIC ACID. He extracted some fibers from a gel of DNA while working at the Biophysics Research Unit at King's College, London. Using X-ray diffraction, he noted the helical structure of the strands. Francis CRICK and James WATSON used this research to build a model of the structure of DNA and to describe its action. For this work the three men shared the 1962 Nobel Prize for Physiology or Medicine.

Wilkinson, Sir Geoffrey (1921–96) British chemist. He shared the 1973 Nobel Prize for chemistry with Ernst FISCHER for their independent work on ORGANOMETALLIC sandwich compounds. Wilkinson demonstrated the new chemical structure in which an iron atom is "sandwiched" between two rings of carbon and hydrogen (organometallic compounds have metal or metalloid atoms bound to carbon atoms or organic groups). He was appointed assistant professor at Harvard University in 1951 and became professor of inorganic chemistry at Imperial College, London, in 1955.

will-o'-the-wisp (jack-o'-lantern) Mysterious light sometimes seen at night in marshy areas. It is not well understood but is thought to be due to the spontaneous combustion of marsh gas (METHANE). Its elusive nature makes it the basis for various superstitious beliefs.

Willstätter, Richard (1872–1942) German chemist who revived the technique of CHROMATOGRAPHY, first discovered by the Russian chemist Mikhail Tsvett (1872–1920). Willstätter used it to partially deduce the structure of CHLOROPHYLL. He showed that it contained a magnesium atom surrounded by pyrrole rings and demonstrated its relationship to HEMOGLOBIN. For this work Willstätter was awarded the 1915 Nobel Prize for chemistry.

Wilson, Charles Thomson Rees (1869–1959) British physicist who invented the Wilson CLOUD CHAMBER used to study radioactivity, X-rays, and cosmic rays. It uses water droplets to track ions left by passing radiation. For this work, Wilson shared the 1927 Nobel Prize for physics with Arthur

COMPTON. Wilson also devised a way of protecting World War II barrage balloons from lightning.

Wilson, Edmund Beecher (1856–1939) US biologist whose research mainly concerned embryology and cytology. He traced the formation of different kinds of tissues from individual precursor cells. He also studied the relationship of CHROMOSOMES to sex determination and the hereditary function of chromosomes.

Wilson, John Tuzo (1908–) Canadian geophysicist and geologist who determined global patterns of faulting and the structure of the continents. His investigations have influenced theories of CONTINENTAL DRIFT, SEAFLOOR SPREADING, and convection currents within the Earth.

Wilson, Robert Woodrow (1936–) US physicist. He shared the 1978 Nobel Prize for physics with Arno PENZIAS for their discovery, in 1964, of the residual background radiation known as COSMIC MICROWAVE BACKGROUND. This phenomena provides strong evidence for the BIG BANG theory.

wilting Drooping of plants due to shortage of water. Insufficient water causes a decline in the TURGOR PRESSURE of plant cells.

winch Drum that turns to pull or release a rope, cable or chain. A winch is used to raise, lower or pull heavy loads. It may be motor-driven or operated by hand. A winch normally has a pawl and RATCHET wheel to prevent accidental release of a load.

wind Air current that moves rapidly parallel to the Earth's surface. (Air currents in vertical motion are called updrafts or downdrafts.) Wind direction is indicated by wind or WEATHER VANES, wind speed by ANEMOMETERS and wind force by the BEAUFORT WIND SCALE. Steady winds in the tropics, such as

▶ **wind** Onshore winds (A) generally occur during the day. The land is heated by the Sun, causing the air over it to rise. As the warm air rises, it is replaced by cooler air overlying the sea. At night, because the land loses heat quicker than the sea, air flows down hillsides out to sea, where the air is relatively warmer, generating offshore breezes (B).

those in the DOLDRUMS, are called TRADE WINDS. MONSOONS are seasonal winds that bring predictable rains in Asia. **Foehns** (föhns) are warm, dry winds produced by adiabatic compression (compression accompanied by temperature rise) as air descends the lee of mountainous areas in the Alps; they are called **chinooks** in the Rockies. **Siroccos** are hot, dry Mediterranean winds. The MISTRAL is a cold, northerly wind that sweeps from the Massif Central of France down the Rhône valley during winter.

Windaus, Adolf (1876–1959) German chemist. In 1907 he synthesized histamine and in 1928 he was awarded the Nobel Prize for chemistry for his work on the structure of STEROIDS and the photochemistry of VITAMIN D.

windchill factor Perceived cooling effect of WIND in low-temperature conditions. The air temperature feels much colder if there is a wind blowing, so the windchill factor is obtained by the relationship between the temperature and the windspeed. If the wind reaches speeds around 40km/h (25mph) temperatures of 5°C can feel very cold, and in sub-zero weather even moderate winds of 10km/h (6mph) can significantly reduce effective temperatures. With a wind speed of 80km/h (50mph), a zero temperature would feel like –20°C (–4°F).

windlass Hauling device operated by hand, most familiar as that used to raise buckets of water from a well. It consists of a cylinder around which

▲ **wind power** A wind generator converts the energy of the wind into electricity. Three-bladed, variable pitch designs are the most efficient. Variable pitch means the attitude of the blades (1) can be changed. By altering the pitch (2) of the blades, they can generate at maximum efficiency in varying wind conditions. The whole rotor assembly rotates (3) into the wind. The blades turn a prop shaft (4), which links to a generator (5) through gearing (6). The largest wind farms have hundreds of linked turbines and can produce the same power as a fossil fuel power station.

the haulage rope is wound and that is turned by a hand lever or crank, using the principle of the WHEEL AND AXLE.

windmill Machine powered by the wind acting on sails or vanes. The earliest known windmills were built in the Middle East in the 7th century. The idea spread to Europe in the Middle Ages. Their use was widespread during the early years of the Industrial Revolution, but declined with the development of the steam engine in the 19th century. Today modern windmills are used to generate electricity on wind farms. *See also* RENEWABLE ENERGY; WATER MILL

window In computing, a rectangle displayed on the screen of a computer that displays what is stored in a particular part of the machine's memory or in some other storage device. A window may show text, graphics or other work in progress. The contents of a window are shown in text, often accompanied by symbols called icons.

windpipe *See* TRACHEA

wind power Technology of harnessing the WIND to provide ENERGY to drive machines and to generate electricity. Since the 1970s, advanced aerodynamic designs have been used to build wind turbines that generate electricity. The largest of these is on the island of Hawaii. It has two blades, each 50m (160ft) long, attached to a 20-story high tower. Individual turbines are often grouped in strategic locations (wind farms) to maximize the generating potential. Wind power is a cheap form of RENEWABLE ENERGY, but cannot as yet produce sufficiently large amounts of electricity to provide a realistic alternative to fossil fuel and nuclear power stations. *See also* WINDMILL

wind sock Flexible, cone-shaped device that indicates wind direction and force. The sock is designed to swivel so that the small end of the cone points downwind. An idea of wind speed can be judged from the sock's extension. *See also* ANEMOMETER; WEATHER VANE

wind tunnel Chamber in which scale models and even full-size aircraft and road vehicles are tested in a controlled airflow. Some wind tunnels can reproduce extreme conditions of wind speed, temperature, and pressure. Models of bridges and other large structures are also tested in wind tunnels to check that winds cannot set up destructive vibrations.

wings In biology, specialized organs for flight that are possessed by most birds, many insects, and certain mammals and reptiles. The forelimbs of a bird have developed into such structures. Bats have membranous tissue supported by the digits ("fingers") of the forelimbs. Insects may have one or two pairs of veined or membranous wings.

wire Strand of metal, made by drawing a rod through progressively smaller holes in metal dies. The drawing process toughens steel, so that a rope or cable made from steel wire is much stronger than an undrawn steel rod of the same diameter. Copper and aluminum wires are used to make electric CABLES. If flexibility is important, each conductor is made of several fine strands of wire instead of one thick one.

Withering, William (1741–99) British doctor who introduced the use of digitalis (a drug obtained from the leaves of the foxglove) to treat cardiac disorders, giving details of his work in *An Account of the Foxglove* (1785). He was the first to establish a connection between dropsy and heart disease. Also a keen botanist, he published *Botanical Arrangement* (1776).

witherite Glassy white, yellow or gray mineral, barium carbonate (BaCO), found in low-temperature lead and barite ore veins. It occurs as twinned, near-hexagonal crystals in the orthorhombic system. Hardness 3–3.5; r.d. 4.5.

Wöhler, Friedrich (1800–82) German chemist who first isolated ALUMINUM and BERYLLIUM and discovered calcium carbide. His synthesis of UREA (from the inorganic substance ammonium cyanate) in 1828 was the first synthesis of an organic chemical compound from an inorganic one, and contributed to the foundation of modern organic chemistry.

Wolf, Max (Maximilian Franz Joseph Cornelius) (1863–1932) German astronomer. He was a pioneer of photographic methods in astronomy. In 1891 he made the first photographic discovery of an asteroid (Brucia, no. 323); in all he made 232 asteroid discoveries. He also detected several new nebulae.

wolfram *See* TUNGSTEN

wolframite Black to brown mineral, iron-manganese tungstate, (Fe,Mn)WO. It is the chief ore of the metal TUNGSTEN. It occurs as crystals in the monoclinic system or as granular masses. It is found in quartz veins and pegmatites associated with granitic rocks and also in high-temperature hydrothermal veins in association with other minerals. It is found extensively in Australia, China, and India. Hardness 5–5.5; r.d. 7–7.5

Wolf–Rayet star (WR star, W star) Type of star, the spectrum of which contains bright emission lines rather than dark absorption ones. WR stars are divided into two kinds. In the WN type, emission lines from nitrogen dominate the spectrum. In the WC type, emission lines of carbon and oxygen predominate. Both types have strong lines of helium and a few have moderate or weak lines of hydrogen as well. WR stars are very hot, with surface temperatures between 25,000 and 50,000K, luminosities between 100,000 and 1 million times the Sun's, and masses from 10 to 50 solar masses. They have strong stellar winds that carry away 3 solar masses per million years. Many central stars of planetary nebulae are WR stars. Wolf–Rayet stars are named after their French discoverers, Charles Joseph Etienne Wolf (1827–1918) and Georges Antoine Pons Rayet (1839–1906).

Wollaston, William Hyde (1766–1828) British scientist. He developed a method of making PLATINUM malleable, discovered PALLADIUM and RHODIUM, and invented the reflecting GONIOMETER, the CAMERA LUCIDA, and a lens (named after him) for correcting spherical ABERRATION. He also did research in the fields of pharmacology, pathology, physiology, and botany.

womb *See* UTERUS

wood Hard substance that forms the trunks of TREES; it is the XYLEM (the vascular tissue of a woody plant) that comprises the bulk of the stems and roots, supporting the plant. It consists of fine cellular tubes arranged vertically within the trunk, which accounts for the grain found in all wood. The relatively soft, light-colored wood is called **sapwood**. The nonconducting, older, darker wood is called **heartwood**, and is generally filled with RESIN, gums, mineral salts, and TANNIN (tannic acid). Easily worked softwood, generally from a CONIFER such as pine, is composed of simple tracheids that provide support and conduct water and food. More durable hardwood, generally from a DECIDUOUS species such as oak, derives support from woody fibers; water and food are conducted through separate vessels. Wood is still commonly used as a building material, fuel, to make some types of PAPER, and as a source of CHARCOAL, CELLULOSE, ESSENTIAL OIL, LIGNIN, tannins, DYES, AND SUGAR.

Woodward, Sir Arthur Smith (1864–1944) British geologist. He was a curator in the department of Geology at the British Museum, but is popularly remembered for his part in the Piltdown Man controversy; his belief that the remains, which appeared to be the "missing link" in the evolution of humans from the apes, were human contributed to the success of the hoax. In 1953 it was proved that the jaw belonged to a modern ape, stained to appear like a prehistoric bone.

Woodward, Robert Burns (1917–79) US chemist. He became professor at Harvard in 1950 and was awarded the 1965 Nobel Prize for chemistry in recognition of his synthesis of a number of complex organic substances, including quinine, cholesterol, cortisone, strychnine, lysergic acid, reserpine, chlorophyll, and tetracycline.

wool Soft, generally white, brown or black animal fiber that forms the fleece (coat) of sheep. Wool is also the name of the yarns and textiles made from the fibers after spinning, dying, and weaving. In most sheep, selective breeding has reduced or eliminated a coarse outer coat of long hairs (still present in the moufflon) leaving a woolly undercoat. Composed chiefly of KERATIN, the fibers are treated to remove a fat called LANOLIN, which is used in some ointments.

word processor COMPUTER system used for writing and printing text. The system may be designed just for this purpose, in which case it is called a **dedicated word processor**. More common is a general-purpose personal computer running a **word processing program**. Text typed on the computer keyboard is displayed on the screen. Any errors are easily corrected before a printout or hard copy is produced on a printer connected to the computer. If required, the text can be stored in code on a magnetic disk for future use.

work In physics, ENERGY transferred in moving the point of application of a FORCE. It equals the magnitude of the force multiplied by the distance moved in the direction of the force. If the force opposing movement is the object's weight mg (where m is the object's mass and g is the acceleration due to gravity), the work done in raising it a height h is mgh. This work has been transferred to the object in the form of POTENTIAL ENERGY; if the object falls a distance x, the KINETIC ENERGY that it gains equals the work done in raising it through the height x.

work function Energy required to liberate an ELECTRON from a material. It is important in the PHOTOELECTRIC EFFECT and in THERMIONICS.

World Health Organization (WHO) Intergovernmental, specialized agency of the United Nations. Founded in 1948, it collects and shares medical and scientific information and promotes the establishment of international standards for drugs and vaccines. WHO has made major contributions to the prevention of diseases such as malaria, polio, leprosy, and tuberculosis. It helps to fund research into other diseases such as AIDS. Its headquarters is in Geneva, Switzerland.

World Meteorological Organization (WMO) Intergovernmental organization, founded as the International Meteorological Organization in 1873. It has been a specialized agency of the United Nations since 1950. WMO promotes international cooperation in meteorology through the establishment of a network of meteorological stations throughout the world, and by the mutual exchange of weather information. Its headquarters is in Geneva, Switzerland.

World Wide Web (WWW) Name given to a series of COMPUTER NETWORKS that can be accessed via local servers, which in turn are serviced by telephone lines. It consists of a network of sites that users can access via the INTERNET to retrieve or post data. Web documents may include text, graphics, and sound, and have HYPERTEXT the user can click on to access further related information from other Web documents at the same site or another site anywhere in the world.

Worthington, Henry Rossiter (1817–80) US engineer and inventor. He invented a direct steam pump and set up a factory for its manufacture in New York in 1859. Worthington also developed the duplex steam feed pump, still much used in oil pipelines and in waterworks.

Wright brothers Wilbur (1867–1912) and Orville (1871–1948), US aviation pioneers. The Wright brothers assembled their first aircraft in their bicycle factory. They experimented with model gliders and built a WIND TUNNEL to study the effects of using various wing shapes. In 1903 Orville made the first recorded piloted flight in a power-driven plane at Kitty Hawk, North Carolina. This flight lasted just 12 seconds, and attracted little attention. But the brothers eventually convinced manufacturers to invest in powered aircraft.

wrought iron Commercial form of smelted IRON (the other is cast iron), containing less than 0.3% carbon with 1 or 2% SLAG mixed with it. Originally it was made from ore in a forge, and later in a puddling furnace, where it never becomes molten. Wrought iron replaced bronze in Asia Minor (c.2000 BC) at the beginning of the IRON AGE. In the 19th century wrought iron began to be used in building construction, but was replaced by steel after the invention of the BESSEMER and OPEN-HEARTH PROCESSES. Today wrought iron is used principally for decoration, as in ornamental gates and railings.

xanthine Yellow substance found in plants, in most body tissues and fluids, and in urinary tract stones, and that can be oxidized to form URIC ACID. It is a muscle stimulant, particularly of cardiac muscle, and its synthetic derivatives are used as DIURETICS and to dilate blood vessels and bronchi.

xanthophyll Plant pigment responsible for the yellow and brown colors of leaves in autumn. Chemically it is a CAROTENOID, a group of compounds that includes CAROTENE, another pigment that gives the red color to carrots and tomatoes. All carotenoids, including xanthophyll, also play a part in PHOTOSYNTHESIS.

X chromosome One of the two kinds of sex CHROMOSOME in human CELL nuclei, the other being the Y CHROMOSOME. A normal human female has two X chromosomes in her body cells, so that each of her OVA (eggs) is HAPLOID, having only one X chromosome. When she passes this on to one of her offspring, it influences the development of female characteristics. Her male offspring will have one X and one Y chromosome in his body cells. Nonsexual characteristics are also carried on the X chromosome. Examples are the genes for colorblindness and hemophilia. X chromosomes occur in other organisms as well, such as fruitflies. Where one sex has two identical chromosomes and the other has one of each type of sex chromosome, the former chromosomes are called the X chromosomes. Sometimes sex is determined by the presence or absence of one particular chromosome, as in grasshoppers; this is also called the X chromosome.

xenomorphism Formation of a mineral crystal in a restricted space so that it cannot develop crystal faces. It occurs in the late-forming minerals of IGNEOUS ROCKS.

xenon (symbol Xe) Gaseous, nonmetallic element, one of the NOBLE GASES. Discovered in 1898, xenon is present in the Earth's atmosphere (about one part in 20 million) and is obtained by FRACTIONATION of liquid air. Colorless and odorless, it is used in lightbulbs, LASERS, and movie projectors. The element, which has 9 stable isotopes, forms some compounds, mostly with FLUORINE. Properties: at.no. 54;

r.a.m. 131.30; r.d. 5.88; m.p. $-111.9°C$ $(-169.42°F)$; b.p. $-107.1°C$ $(-160.8°F)$; most common isotope ^{132}Xe (26.89%).

xerography Most common process used for PHOTOCOPYING.

xerophyte Any plant that is adapted to survive in dry conditions, in areas subject to drought or in physiologically dry areas such as saltmarshes and acid bogs, where saline or acid conditions make the uptake of water difficult. SUCCULENTS, such as cacti, have thick fleshy leaves and stems for storing water. Other adaptations include the ability to reduce water loss by shedding leaves during drought, having waxy or hairy leaf coatings or reduced leaf area.

xi particle *See* HYPERON

X ray In medicine, X RAYS are directed toward the body to provide images of internal structures or to destroy diseased or unwanted tissue. Passed through the body onto a photographic plate, X rays reveal such abnormalities as fractures, tumors, foreign bodies, and enlargement of organs. The capacity of X rays to break up cells makes them useful also in the treatment of some forms of cancer.

X ray In physics, form of ELECTROMAGNETIC RADIATION of much shorter wavelength, or higher frequency, than visible light, produced when a beam of electrons hits a solid target. X rays were discovered in 1895 by the German physicist Wilhelm RÖNTGEN. They are normally produced for scientific use in X-RAY TUBES. Because they are able to penetrate matter that is opaque to light, X rays are used to investigate inaccessible areas, especially of the body. *See also* RADIOGRAPHY; X-RAY ASTRONOMY

X-ray astronomy *See* ASTRONOMY

X-ray crystallography Use of X RAYS to discover the molecular structure of CRYSTALS. It uses the phenomenon of **X-ray diffraction**, the scattering of an X-ray beam by the atomic structure of a crystal.

X-ray diffractometer Instrument used in the analysis of the atomic arrangement that determines the crystal structure of minerals. X rays, when passing through the symmetrically arranged atoms of a crystal, are deflected in a regular pattern, with the atoms acting as a DIFFRACTION GRATING. Photos of these patterns permit deductions of interatomic dimensions, spacing, and bonding arrangements within crystals. *See also* X-RAY CRYSTALLOGRAPHY

X-ray telescope *See* TELESCOPE

X-ray tube Evacuated tube used to provide a source of X RAYS for medical or other purposes. It consists of an electron gun producing a stream of ELECTRONS that strike an ANODE, part of which is made of a heavy metal such as TUNGSTEN. The tungsten emits X rays when it is bombarded by the stream of high-energy electrons.

▲ X-ray crystallography Each substance has a different crystalline structure. X-ray crystallography allows the details of the structure to be recorded by bombarding a crystal sample (1) with a beam of X-rays (2). The X-ray is created by bombarding a tungsten anode (3) with electrons in a vacuum (4). A slit (5) focusses the X-rays on the crystal. The distance between the atomic planes in the crystal either reinforces (6) or cancels out (7) the X-rays. When the ray is reinforced the emerging X-ray (8) creates a spot (9) on photographic film. The pattern of the spots can be used by scientists to deduce the exact structure of molecules.

xylem Transport TISSUE of a plant, which conducts water and minerals from the roots to the rest of the plant and provides support for it. The most important cells are long, thin tapering cells called xylem vessels. These cells are dead and have no cross-walls; they are arranged in columns to form long tubes, up which water is drawn. As water evaporates from the leaves (TRANSPIRATION), water is drawn across the leaf by OSMOSIS to replace it, drawing water out of the xylem. This process creates TURGOR PRESSURE in the xylem vessels, which, along with rings of LIGNIN that reinforce side walls and prevent them from collapsing, provides support to the plant. Tiny holes in the walls of the xylem vessels, called **pits**, allow water to cross from one tube to another. Ferns and conifers do not have xylem vessels. Instead they have similar cells called **tracheids**, which do not lose their end walls, so water has to travel through the pits, which slows the flow. The lignin makes the walls of vessels and tracheids strong and rigid, an important support as the plant grows bigger. Xylem tissue also contains non-conducting fibers, dead cells thickened with lignin for extra support. In trees, the xylem becomes blocked with age, and new xylem forms towards the outside of the trunk to replace it. The core of dead, non-functioning xylem remains an essential part of the support system. *See also* PHLOEM; VASCULAR BUNDLE

xylene ($C_6H_4(CH_3)_2$) Organic chemical compound obtained from the distillates of coal tar and petroleum, and important as a solvent. Chemically it is dimethyl benzene, which exists in three isomeric forms: ortho-, meta-, and para-xylene. The ISOMERS have different physical properties.

Yagi antenna Basic form of ANTENNA used in simple RADIO TELESCOPES. It consists of several parallel elements mounted on a straight member, and often forms the basis of cheap arrays used in **aperture synthesis** (in which arrays of radio dishes are used to synthesize the resolution of a much larger telescope); the Yagi antenna is also a familiar form of television aerial. It was developed by the Japanese engineer Hidetsugu Yagi (1886–1976).

Yale, Linus (1821–68) US inventor of various types of locks, including the modern cylinder LOCK that bears his name.

Yalow, Rosalyn (1921–) US biochemist who shared the 1977 Nobel Prize for physiology or medicine for her role in the development of a method of detecting peptide HORMONES in the blood. In the 1950s Yalow found that some people who received regular INSULIN injections developed ANTIBODIES against the hormone. She discovered that insulin, labeled with radioactive iodine, combined with the antibodies; from this she developed radioimmunological tests to detect and measure the amount of insulin present.

Yang, Chen Ning (1922–) US physicist, b. China, who shared the 1957 Nobel Prize for physics with Tsung-Dao LEE for their discovery of violations of the conservation of PARITY. They studied the decay of K MESONS, which seemed to break down in two different ways, and in 1956 they concluded that in these WEAK NUCLEAR FORCE interactions parity need not be conserved.

yard Unit of length equal to 3 feet or 36 inches. 1 yard (yd) also equals 0.9144m.

yardang Elongated ridge formed by wind erosion in arid areas. Rock is eroded into a series of ridges and furrows lying parallel to the prevailing wind direction.

Y chromosome One of the two kinds of sex CHROMOSOME in the nuclei of DIPLOID human CELLS, the other being the X CHROMOSOME. A normal male mammal has one X and one Y chromosome in his body cells, and his SPERM cells are HAPLOID, containing either an X or a Y chromosome. Since female OVA (eggs) always contain an X chromosome, the resulting offspring is either XY (male) or XX (female). The Y chromosome is smaller than the X chromosome and contains fewer GENES.

year Length of time taken by the Earth to circle once round the Sun in its orbit. Various years, defined according to the choice of reference point, are given in the table. The civil or calendar year averages 365.2425 mean solar days.

yeast Any of a group of single-celled, microscopic FUNGI found in all parts of the world in the soil and in organic matter. Yeasts reproduce asexually by budding or fission. Yeasts are also produced commercially for use in baking, brewing, and wine-making. They occur naturally as a bloom (white covering) on grapes and other fruit. See also ASEXUAL REPRODUCTION

yellow fever Acute infectious disease marked by sudden onset of headaches, fever, muscle and joint pain, jaundice, and vomiting; the kidneys and heart may also be affected. It is caused by a VIRUS transmitted by mosquitoes in tropical and subtropical regions. It may be prevented by vaccination.

yellow-green algae Variety of flagellate ALGAE (division Xanthophyta). They differ from GREEN ALGAE by their characteristic, motile bodies and unequal FLAGELLA.

Yerkes, Robert Mearns (1876–1956) US biologist and psychologist. He was a pioneer in the comparative study of apes and the development of methods to test the abilities of lower animals and humans. His publications include *The Mind of a Gorilla* (1927) and *Chimpanzees: A Laboratory Colony (1943)*.

Yersin, Alexandre Émile John (1863–1943) French bacteriologist who studied with Louis PASTEUR. In Hong Kong Yersin discovered (independently of Shibasaburo KITASATO) the plague bacillus and developed a serum against it in 1895.

yolk Rich substance found in the EGGS or OVA of most animals except those of placental MAMMALS. It consists of fats and proteins and serves as a store of food for the developing embryo.

yolk sac Membranous, saclike structure in the EGGS of most animals. It is attached directly to the ventral surface or gut of the developing embryo in the eggs of birds, reptiles, and some fish, and contains YOLK. The term also refers to an analogous, saclike membrane that develops below the mammalian embryo. It contains no yolk but is connected to the umbilical cord.

Young, Charles Augustus (1834–1908) US astronomer. He made many solar observations, including the earliest spectroscopic studies of the corona of the SUN. He also used spectral line measurements to determine the Sun's rotation period. *See also* SPECTROSCOPY

Young, Thomas (1773–1829) British physicist and physician. He revived the wave theory of LIGHT first put forward in the 17th century by Christiaan HUYGENS. He helped present the Young-Helmholtz theory of COLOR vision and detailed the cause of ASTIGMATISM. He studied ELASTICITY, giving his name to the tensile elastic modulus (*see* YOUNG'S MODULUS). Young was also an Egyptologist who helped decipher the Rosetta stone.

Young's modulus Ratio of the STRESS exerted on a body to the longitudinal STRAIN produced.

ytterbium (symbol Yb) Silver-white, metallic element of the LANTHANIDE SERIES. First isolated in 1878, ytterbium's chief ore is monazite. The shiny, soft element is malleable and ductile is used to produce steel and other alloys. Properties: at.no. 70; r.a.m. 173.04; r.d. 6.97; m.p. 824°C; (1,515°F), b.p. 1,193°C (2,179°F); most common isotope ^{174}Yb (31.84%).

yttrium (symbol Y) Silver-gray, metallic element of group III of the PERIODIC TABLE. First isolated in 1828, it is found associated with lanthanide elements (*see* LANTHANIDE SERIES) in monazite sand, bastnaesite, and gadolinite, and resembles the lanthanides in its chemistry. Yttrium was found in lunar rock samples collected by the Apollo 11 space mission. Its compounds are used in phosphors and communications devices, such as color TELEVISON picture tubes and superconducting ceramics. Properties: at.no. 39; r.a.m. 88.9059; r.d. 4.47; m.p. 1,523°C (2,773°F); b.p. 3,337°C (6,039°F); most common isotope ^{89}Y (100%).

Yukawa, Hideki (1907–81) Japanese physicist who was awarded the 1949 Nobel Prize for physics for his prediction of the existence of the MESON. In the 1930s he proposed that there was a nuclear force of very short range (less than 10^{-15} m), strong enough to overcome the repulsive force of protons, which diminishes rapidly with distance. He predicted that this force manifests itself by the transfer of particles between neutrons and protons. In 1947 Yukawa's theory was confirmed by the discovery of the PION (pi meson) by Cecil POWELL..

ful in investigating the charge/mass ratio and MAGNETIC MOMENT of an ELECTRON.

Zeiss, Carl (1816–88) German manufacturer of optical instruments. In 1846 he established a factory at Jena, Germany, and made various optical components, including lenses, binoculars, and microscopes.

zenith In astronomy, point on the CELESTIAL SPHERE that is directly overhead. The zenith distance of a heavenly body is the angle it makes with the zenith. In common use it means the highest point.

zeolite Group of alumino-silicates containing loosely held water that can be continuously expelled on heating. They also contain sodium, calcium or barium. Some are fibrous aggregates while others form robust, nonfibrous crystals. There are many zeolites varying in hardness from 3 to 5 and in specific gravity from 2 to 2.4. Zeolites include analcime ($NaAlSi_2O_6.H_2O$), stilbite ($NaCa_2(Al_5Si_{13}36.14H_2O$), and natrolite ($Na_2Al_2Si_3O_{10}.2H_2O$).

Z boson (symbol Z^0) ELEMENTARY PARTICLE that is a carrier of the WEAK NUCLEAR FORCE. The superscript zero (^{0}) indicates that it has no electric charge. Its mass is about 96 times that of the PROTON.

Zeeman, Pieter (1865–1943) Dutch physicist. He shared the 1902 Nobel Prize for physics with his teacher Hendrik LORENTZ for their 1896 discovery of the ZEEMAN EFFECT. This is the splitting of spectral lines when a light source is placed in a strong MAGNETIC FIELD. He also detected the magnetic fields at the surface of the SUN.

Zeeman effect In physics, effect produced by a strong MAGNETIC FIELD on the light emitted by a radiant body, observed as a splitting of its spectral lines. It was first observed in 1896 by the Dutch physicist Pieter ZEEMAN. The effect has been use-

Zeppelin, Ferdinand, Count von (1838–1917) German army officer, inventor, and aeronaut. While an observer with the Union army during the Civil War (1861–65) made his first balloon ascent. In 1900 he invented the first rigid AIRSHIP, named Zeppelin after him.

Zermelo, Ernst (1871–1956) German mathematician who was one of the founders of axiomatic SET THEORY. The Zermelo-Fraenkel AXIOMS stipulate which collections of objects constitute sets.

Zernike, Frits (1888–1966) Dutch physicist who in 1935 developed the PHASE-CONTRAST MICROSCOPE, in which objects being viewed (often living-cell biological specimens) take on a different color from their surroundings. For this work, he received the 1953 Nobel Prize for physics.

zero In mathematics, the number that, when added to any number x leaves the number unchanged: $x + 0 = x$. Multiplication of a number by zero gives zero. Division by zero is undefined. The concept of zero was unknown to the ancient Greeks; it was introduced to the West by the Arabs, who had learned about it in India. The ancient Babylonians and the Mayans also discovered the concept of zero.

zero gravity (zero g) Condition of experiencing no apparent gravitational force. It may be attained by airplane pilots in aircraft briefly during free fall, or by astronauts in a spacecraft, when the engines have been turned off and the craft is "falling" freely. Gravitational force is still present, however. See also WEIGHTLESSNESS

Zeta Cancri Group of four stars in the constellation of Cancer consisting of two BINARY STAR systems. Zeta-one has a period of 59.6 years, Zeta-two is a close binary with a period of 17.6 years.

Ziegler, Karl (1898–1973) German chemist who shared the 1963 Nobel Prize for chemistry with the Italian chemist Giulio NATTA (his colleague) for research into POLYMERS. Ziegler discovered a technique of using a RESIN with metal ions attached as a CATALYST in the production of POLYETHYLENE. He also did research into AROMATIC COMPOUNDS and developed ORGANOMETALLIC COMPOUNDS (in which metal or metalloid atoms are bound to carbon atoms or organic groups) that are more reactive than GRIGNARD REAGENTS.

zinc (symbol Zn) Blue-white, metallic element of group II of the periodic table, known from early times. Chief ores are sphalerite (zinc blende), smithsonite, and calamine. Zinc is a vital TRACE ELEMENT, found in erythrocytes (red blood cells). The shiny element is used in many alloys, including brass, bronze, nickel, silver, and soft solder. It is corrosion-resistant and used in galvanizing iron. Zinc oxide is used in cosmetics, pharmaceuticals, paints, inks, pigments, and plastics. Zinc chloride is used in dentistry and to manufacture batteries and fungicides. Properties: at.no. 30; r.a.m. 65.37; r.d. 7.133; m.p. 419.6°C (787.3°F); b.p. 907°C (1,665°F); most common isotope ^{64}Zn (48.89%).

zinc oxide (ZnO) White powder used in ointments as a mild antiseptic, astringent, and protection against ECZEMA and other skin conditions.

zircon Orthosilicate mineral, zirconium silicate ($ZrSiO_4$) found in IGNEOUS and METAMORPHIC ROCKS and in sand and gravel. It displays tetragonal system prismatic crystals. Its color is variable: it is usually light or reddish brown but can be colorless, gray, yellow or green. It is used widely as a gemstone because of its hardness and high refractive index. Hardness 7.5; r.d. 4.6.

zirconium (symbol Zr) Gray-white TRANSITION ELEMENT, first discovered in 1789. Its chief source is the gemstone ZIRCON (zirconium silicate). Lunar rocks collected during the Apollo space missions show a higher content of zirconium than Earth ones, and zirconium exists in meteorites and stars, including the Sun. Chemically similar to titanium, it is used in ceramics and absorption of NEUTRONS in nuclear REACTORS. Properties: at.no. 40; r.a.m. 91.22; r.d. 6.51; m.p. 1,852°C (3,366°F); b.p. 4,377 °C (7,911°F); most common isotope Zr^{90} (51.46%).

zodiac Belt on the celestial sphere that forms the background for the motions of the Sun, Moon, and planets (except Pluto). The zodiac is divided into 12 **signs**, which are named after the constellations they contained at the time of the ancient Greeks: Aries, Taurus, Gemini, Cancer, Leo, Virgo, Libra, Scorpio, Sagittarius, Capricorn, Aquarius, and Pisces. The constellations now inside the zodiac do not correspond to those named by the ancients, because PRECESSION of the Earth's axis has meanwhile tilted the Earth in a somewhat different direction. Most of the zodiacal signs represent animals and the name zodiac derives from a Greek phrase meaning "circle of animals." To modern astronomers the zodiac has only historical significance. See also ASTROLOGY; ASTRONOMY

zodiacal light Cone of faint light, usually fainter than the Milky Way, visible at all seasons in the tropics in the absence of moonlight. It can be seen on the w horizon after dusk, or on the E horizon before sunrise. The spectrum of the zodiacal light resembles the Sun's, indicating that the phenomenon results from the SCATTERING of sunlight by particles. These particles constitute the zodiacal dust cloud, and probably originate partly from matter ejected by the Sun, and partly from the decay of comets and asteroids.

zoisite Orthosilicate mineral, hydrous calcium aluminum silicate, found in METAMORPHIC ROCKS. Orthorhombic system prismatic crystals and masses. Glassy, transparent gray, white, brown, green, or pink; hardness 6–6.5; r.d. 3.2. A vivid blue variety from Tanzania, called tanzanite, is a gemstone.

zonation In ecology, occurrence of bands of characteristic FLORA and FAUNA in an area. Examples include the bands of different types of seaweed that occur progressively up a shore from the low-tide to high-tide levels. Species from the low-tide zone would dry out and die at the high-tide level, whereas high-tide species could not withstand prolonged immersion in seawater in the low-tide zone.

zone refining Method of purifying crystals, especially for use in SEMICONDUCTOR devices. The material is placed in a long tube and passed through a furnace in which hot and cold zones alternate. As the rod moves through the furnace, impurities remain in the molten state while melting and recrystallization of the material takes place; the impurities are thus transferred to one end of the tube.

zoogeography (animal geography) Study of the geographic distribution of animals. Formerly, the approach of this science was descriptive; today it utilizes various data, including isotope DATING and ocean-bottom core sampling.

zoology Study of animals; combined with BOTANY, it comprises the science of BIOLOGY. It is concerned with the structure of the animal and the way in which animals behave, reproduce and function, their evolution and their role in interactions with humankind and their environment. There are various subdivisions of the discipline, including: TAXONOMY; ECOLOGY; PALEONTOLOGY; ANATOMY, and ZOOGEOGRAPHY, the distribution of animals. ANTHROPOLOGY is an extension of zoology. *See also* EMBRYOLOGY; GENETICS; MORPHOLOGY

zoonosis Any infection or infestation of other vertebrates that under normal conditions is transmissible to human beings. Sources of zoonoses include pets, livestock, and wild animals; examples are: cat scratch disease (cats), brucellosis (cows, goats or pigs), and psitticosis (birds).

zooplankton Animal portion of the PLANKTON. It consists of a wide variety of microorganisms, such as COPEPOD and larval forms of higher animals. It is a central constituent of the oceans' FOOD CHAINS. There are few levels or areas of the ocean that have no zooplankton. *See also* PHYTOPLANKTON

Zsigmondy, Richard Adolf (1865–1929) Austrian chemist who won the 1925 Nobel Prize for chemistry for his work on COLLOIDS. While employed at a glass manufacturing company (1897–1900), he discovered a water suspension of gold and proposed that the shape and size of colloids could be deduced from the way in which the particles scatter light. To aid such studies he helped to develop the **ultramicroscope**, in which the particles are suspended in a gas or liquid, through powerful light is shone. The particles scatter the light ensuring they appear to the viewer due to the **Tyndall effect**. *See also* SCATTERING; TYNDALL, JOHN

Zwicky, Fritz (1898–1974) Swiss-American astronomer, b. Bulgaria, who in 1934, with Walter BAADE suggested that what is left after a SUPERNOVA explosion is a NEUTRON STAR. This was confirmed in 1968 with the discovery of the PULSAR in the Crab Nebula. In the mid-1960s, he discovered many supernovas and began a study of CLUSTER GALAXIES.

Zworykin, Vladimir Kosma (1889–1982) US physicist and inventor, b. Russia, who was a pioneer of TELEVISION. In 1929, he joined the Radio Corporation of America (RCA), becoming its director of development and in 1947 a vice president. Zworykin developed the **iconoscope**, the forerunner of the modern television camera tube, and the **kinescope**, a CATHODE-RAY TUBE for TV sets. In 1928, he patented a color television system. He also invented the ELECTRON MICROSCOPE. In 1967, Zworykin received the National Medal of Science for his inventions and contributions to research.

zygote In SEXUAL REPRODUCTION, a CELL formed by fusion of a male and a female GAMETE. It contains a DIPLOID number of CHROMOSOMES, half contributed by SPERM, half by the OVUM. Through successive CELL DIVISIONS, a zygote will develop into an EMBRYO.